Lecture Notes in Computer Science 4666

Mike E. Davies Christopher J. James
Samer A. Abdallah Mark D. Plumbley (Eds.)

Independent Component Analysis and Signal Separation

7th International Conference, ICA 2007
London, UK, September 9-12, 2007
Proceedings

Springer

Volume Editors

Mike E. Davies
Edinburgh University
IDCOM & Joint Research Institute for Signal and Image Processing
King's Buildings, Mayfield Road, Edinburgh EH9 3JL, UK
E-mail: mike.davies@ed.ac.uk

Christopher J. James
University of Southampton
Signal Processing and Control Group, ISVR
Southampton, SO17 1BJ, UK
E-mail: C.James@soton.ac.uk

Samer A. Abdallah
Mark D. Plumbley
Queen Mary University of London
Department of Electronic Engineering
Centre for Digital Music
Mile End Road, London, E1 4NS, UK
E-mail: {samer.abdallah,mark.plumbley}@elec.qmul.ac.uk

Library of Congress Control Number: 2007933693

CR Subject Classification (1998): C.3, F.1.1, E.4, F.2.1, G.3, H.1.1, H.2.8, H.5.1, I.2.7

LNCS Sublibrary: SL 1 – Theoretical Computer Science and General Issues

ISSN 0302-9743
ISBN-10 3-540-74493-2 Springer Berlin Heidelberg New York
ISBN-13 978-3-540-74493-1 Springer Berlin Heidelberg New York

Springer is a part of Springer Science+Business Media

springer.com

Printed in Germany

Typesetting: Camera-ready by author, data conversion by Scientific Publishing Services, Chennai, India
Printed on acid-free paper SPIN: 12113033 06/3180 5 4 3 2 1 0

Preface

This volume contains the papers presented at the 7th International Conference on Independent Component Analysis (ICA) and Source Separation held in London, 9–12 September 2007, at Queen Mary, University of London.

Independent Component Analysis and Signal Separation is one of the most exciting current areas of research in statistical signal processing and unsupervised machine learning. The area has received attention from several research communities including machine learning, neural networks, statistical signal processing and Bayesian modeling. Independent Component Analysis and Signal Separation has applications at the intersection of many science and engineering disciplines concerned with understanding and extracting useful information from data as diverse as neuronal activity and brain images, bioinformatics, communications, the World Wide Web, audio, video, sensor signals, or time series.

This year's event was organized by the EPSRC-funded UK ICA Research Network (`www.icarn.org`). There was also a minor change to the conference title this year with the exclusion of the word 'blind'. The motivation for this was the increasing number of interesting submissions using non-blind or semi-blind techniques that did not really warrant this label. Evidence of the continued interest in the field was demonstrated by the healthy number of submissions received, and of the 149 papers submitted just over two thirds were accepted.

These proceedings have been organized into 6 sections: theory, algorithms, sparse methods, biomedical applications, speech and audio applications, and miscellaneous. Within each section, papers have been organized alphabetically by the first author's last name. However the strong interaction between theory, method and application inevitably means that many papers could have equally been placed in alternative categories.

In this year's papers there was a significant growth in development of sparsity as a tool for source separation, while the application areas were once again dominated by submissions focusing on speech and audio, and on biomedical processing. The organizing committee decided to reflect this in their choice of keynote speakers and were pleased to secure keynote talks from two leading researchers in these fields: Shoji Makino from NTT Communication Science Laboratories, Kyoto, Japan; and Scott Makeig from the Swartz Center for Computational Neuroscience, Institute for Neural Computation, UCSD, USA.

Following the successful additions made to the 2006 event, the 2007 conference continued to offer a "Student Best Paper Award" and included two tutorial sessions on the day preceding the main conference. This year's tutorials covered two topics closely connected with recent research progress in ICA and source separation: "Information Filtering," lectured by José Principe of the University of Florida; and "Sparse Representations" lectured by Rémi Gribonval of INRIA at IRISA, Rennes. A further innovative aspect of the 2007 event was the introduction of the "Stereo Audio Source Separation Evaluation Campaign", which aimed to compare the performance of source separation algorithms from different researchers when applied to stereo under-determined mixtures. The different contributions to the challenge were presented in a special poster session on the last day of the conference, which was then followed by a panel discussion. The overall results of the evaluation campaign are also summarized in a paper in this volume.

There are many people that should be thanked for their hard work, which helped to produce the high quality scientific program. First and foremost we would like to thank all the authors who have contributed to this volume. Without them there would be no proceedings. In addition, we thank the members of the organizing committee and the reviewers for their efforts in commissioning the reviews, and for their help in selecting the very best papers for inclusion in this volume. We are also grateful to the organizers of the "Stereo Audio to Source Separation Evaluation Campaign" for managing both the submissions to and the evaluation of this work.

Thanks also go to the members of the ICA international steering committee for their continued advice and ongoing support for the ICA conference series. All these contributions went towards making the conference a great success. Last but not least, we would like to thank Springer for their rapid transformation of a collection of papers into this volume in time for the conference.

June 2007

Mike Davies
Christopher James
Samer Abdallah
Mark Plumbley

ICA 2007 Committee Listings

Organizing Committee

Mark Plumbley, Queen Mary, University of London, UK
Mike Davies, IDCOM, University of Edinburgh, UK
Christopher James, ISVR, University of Southampton, UK
Samer Abdallah, Queen Mary, University of London, UK
Betty Woessner, Queen Mary, University of London, UK

UK Advisory Committee

Paul Baxter, Qinetiq, UK
Richard Everson, University of Exeter, UK
Colin Fyfe, University of Paisley, UK
Mark Girolami, University of Glasgow, UK
Asoke Nandi, University of Liverpool, UK
Saeid Sanei, Cardiff University, UK

International ICA Steering Committee

Luis Almeida, INESC-ID, Lisbon, Portugal
Shun-ichi Amari, Riken BSI, Japan
Jean-François Cardoso, ENST, Paris, France
Andrzej Cichocki, Riken BSI, Japan
Scott Douglas, Southern Methodist University, USA
Simon Haykin, McMaster University, Canada
Christen Jutten, INPG, France
Te-Won Lee, UCSD, USA
Shoji Makino, NTT CS Labs, Japan
Klaus-Robert Müller, Fraunhofer FIRST, Germany
Noboru Murata, Waseda University, Japan
Erkki Oja, Helsinki University of Technology, Finland
Liqing Zhang, Shangai Jiaotong University, China

Referees

Samer Abdallah
Karim Abed-Meraim
Pierre-Antoine Absil
Tulay Adali
Luis Almeida
Jörn Anemüller
Shoko Araki
Simon Arberet
Francis Bach
Radu Balan
Paul Baxter
Adel Belouchrani

Raphael Blouet
Thomas Blumensath
Markus Borschbach
Jean-François Cardoso
María Carrión
Marc Castella
Taylan Cemgil
Jonathon Chambers
Chong-Yung Chi
Seungjin Choi
Heeyoul Choi
Andrzej Cichocki
Pierre Comon
Sergio Cruces-Alvarez
Adriana Dapena
Mike Davies
Lieven De Lathauwer
Yannick Deville
Konstantinos Diamantaras
Scott Douglas
Ivan Duran-Diaz
Deniz Erdogmus
Jan Eriksson
Da-Zheng Feng
Cédric Févotte
Colin Fyfe
Pando Georgiev
Mark Girolami
Pedro Gomez Vilda
Juan Manuel Gorriz Saez
Rémi Gribonval
Anne Guérin-Dugué
Zhaoshui He
Kenneth Hild
Antti Honkela
Tetsuya Hoya
Patrik Hoyer
Aapo Hyvarinen
Shiro Ikeda
Alexander Ilin
Mika Inki
Yujiro Inouye
Maria Jafari
Christopher James
Tzyy-Ping Jung
Christian Jutten
Juha Karhunen
Wlodzimierz Kasprzak
Mitsuru Kawamoto
Taesu Kim
Kevin Knuth
Visa Koivunen
Kostas Kokkinakis
Ercan Kuruoglu
Elmar Lang
Soo-Young Lee
Intae Lee
Yuanqung Li
Ju Liu
Hui Luo
Ali Mansour
Rubén Martín Clemente
Kiyotoshi Matsuoka
Oleg Michailovich
Nikolaos Mitianoudis
Dmitri Model
Eric Moreau
Ryo Mukai
Juan José Murillo-Fuentes
Klaus-Robert Müller
Wakako Nakamura
Noboru Nakasako
Asoke Nandi
Klaus Obermayer
Erkki Oja
Umut Ozertem
Sunho Park
Barak Pearlmutter
Dinh-Tuan Pham
Ronald Phlypo
Mark Plumbley
Carlos Puntonet
Martin Pyka
Stephen Roberts
Justinian Rosca
Diego Ruiz Padillo
Tomasz Rutkowski
Jaakko Särelä
Moussaoui Saïd
Fathi Salem
Reza Sameni
Ignacio Santamaria
Hiroshi Saruwatari
Hiroshi Sawada
Mikkel Schmidt
Christine Serviere
Hosseini Sharam
Shohei Shimizu
Paris Smaragdis
Jordi Solé-Casals
Toshihisa Tanaka
Fabian Theis
Nadege Thirion-Moreau
Alle-Jan van der Veen
Jean-Marc Vesin
Ricardo Vigário
Emmanuel Vincent
Tuomas Virtanen
Frédéric Vrins
DeLiang Wang
Wenwu Wang
Yoshikazu Washizawa
Lei Xu
Kehu Yang
Arie Yeredor
Vicente Zarzoso
Liqing Zhang
Yimin Zhang
Yi Zhang
Xu Zhu
Michael Zibulevsky
Andreas Ziehe

Sponsoring Institutions

UK ICA Research Network
UK Engineering and Physical Sciences Research Council
European Association for Signal Processing (EURASIP)
IEEE Region 8 UK & Republic of Ireland Section Engineering in Medicine and Biology Society Chapter
Queen Mary, University of London, UK

Table of Contents

Theory

Algorithms

Sparse Methods

Speech and Audio Applications

Biomedical Applications

Miscellaneous

Keynote Talk

A Flexible Component Model for Precision ICA

Jean-François Cardoso[1,2] and Maude Martin[2,*]

[1] CNRS / LTCI , France
[2] Univ. Paris 7 / APC, France

Abstract. We describe an ICA method based on second order statistics which was originally developed for the separation of components in astrophysical images but is appropriate in contexts where accuracy and versatility are of primary importance. It combines several basic ideas of ICA in a new flexible framework designed to deal with complex data scenarios. This paper describes our approach and discusses its implementation in terms of a *library of components*.

1 Introduction

Objectives. This paper describes a framework for component separation which has been designed with the following objectives in mind:

- the ability to model components with much more flexibility than in most basic ICA techniques. This includes, in particular, the case of correlated or multidimensional components,
- the ability to take noise into account. The noise is not necessarily well characterized: its correlation structure may have to be estimated,
- the ability to deal with signals/images of varying resolution.
- the ability to incorporate easily prior information about the components,
- 'Relative' speed (enabling error estimates via Monte-Carlo simulations).

Motivation. The original motivation for the work presented here is the processing of spherical maps resulting from all-sky surveys of the microwave emission. The object of interest is the Cosmic Microwave Background (CMB). The space-based Planck mission of the European Space Agency will provide observations of the CMB in 9 frequency channels which will be used as inputs to a component separation method. This is needed because the cosmological background is observed together with several other astrophysical emissions, dubbed 'foregrounds', both of galactic and extra-galactic origins. These foregrounds, together with the CMB, are the components which are to be separated. The CMB map itself is very well modeled as a realization of a (spherical) Gaussian stationary random field but this is not the case of the other components.

Our method, however, is not specific to this particular problem and may be considered for application to any situation where 'expensive' data deserve special care and have to be fitted by a complex component model.

Section 2 describes the statistical framework while section 3 discusses implementation in terms of a library of components.

* Maude Martin is partly supported by the Cosmostat project funded by CNRS.

M.E. Davies et al. (Eds.): ICA 2007, LNCS 4666, pp. 1–8, 2007.

2 The Additive Component Model

2.1 A Component Based Model

We introduce a special notation for the ICA model which is more convenient to our framework, in particular to deal with correlated sources. Traditionally, the noisy ICA model is denoted as

$$X = \mathbf{A}S + N$$

where matrix $\mathbf{A}$ is $m \times n$ for m channels (sensors) and n sources. This is a multiplicative model in the sense that the mixing matrix $\mathbf{A}$ multiplies an hypothetical vector S of 'sources'. The ith source S_i contributes $\mathbf{a}_i S_i$ to the observation vector X where $\mathbf{a}_i$ is the ith column of $\mathbf{A}$. Hence, the model can be rewritten 'additively' as a superposition of $C = n + 1$ random components:

$$X = \sum_{c=1}^{C} X^c \qquad (1)$$

where $X^i = \mathbf{a}_i S_i$ for $1 \leq i \leq n$ and $X^{n+1} = N$. Such a reformulation is worthless if all sources are modeled as independent. Assume now that two sources, say i and j, are modeled as statistically *dependent*. They contribute $\mathbf{a}_i S_i + \mathbf{a}_j S_j$ to the observed X. We decide to lump them into a single component denoted X^c for some index c: $X^c = \mathbf{a}_i S_i + \mathbf{a}_j S_j$. Then X can still be written as in eq. (1) and all the *components* X^c are independent, *again*. However, the new component X^c can no longer be written as one-dimensional, *i.e.* as the product of a single fixed column vector multiplied by a random variable. Instead, it can be written as $[S_i S_j]^\dagger$ left multiplied by the $m\times 2$ matrix $[\mathbf{a}_i \mathbf{a}_j]$. Such a component is termed 'bi-dimensional' and we could obviously define multidimensional components of any dimension.

In eq. (1), we have included the noise term as one of the components. Note that if the noise is uncorrelated from channel to channel, as is often the case, then the noise component is m-dimensional (the largest possible dimension).

More generally, our model does not require any component to be low dimensional. Rather, our model is a plain superposition of C components as in eq. (1). None of these components is required to have any special structure, one-dimensional or otherwise. *We only require that they are mutually uncorrelated.* In other words, we rely on the property

$$\mathbf{R} = \sum_{c=1}^{C} \mathbf{R}^c \qquad (2)$$

where $\mathbf{R}$ (*resp.* $\mathbf{R}^c$) is the covariance matrix of the data vector X (*resp.* of cth component X^c).

2.2 Component Restoration by Wiener Filtering

The best (in the mean-square sense) linear estimate $\widehat{X}^c$ of X^c based on X is well known to be $\mathrm{Cov}(X^c, X)\mathrm{Cov}(X, X)^{-1}X$ which reduces here to

$$\widehat{X}^c = \mathbf{R}^c \mathbf{R}^{-1} X \qquad (3)$$

Hence optimal linear recovery of X^c from X requires only the determination of matrices $\mathbf{R}$ and $\mathbf{R}^c$. The total covariance matrix $\mathbf{R}$ can often be estimated directly from the data so that, in order to restore component c from the mixture, one "only" has to estimate its covariance matrix $\mathbf{R}^c$.

2.3 Localized Statistics, Localized Separation

In practice, we do not consider a *single* covariance matrix. Rather, in order to capture better the correlation structure, we compress the data into a set $\widehat{\mathcal{R}} = \{\widehat{\mathbf{R}}_q\}_{q=1}^{Q}$ of Q covariance matrices of size $m \times m$. For instance, one would estimate the covariance of X over several domains (time intervals for time series, spatial domains for images) or in Fourier space over several frequency bands. More generally, one could localize the covariance matrices in both domains using a wavelet basis or just plain windowed transforms. Index q can be thought of as a time interval (or a spatial zone), a Fourier band, a wavelet domain, etc... In our application (see introduction), we would consider a hundred of angular frequency bands localized over a few zones on the sky and Q would be the product of these two numbers.

This can be formalized by denoting $X(i)$ the ith coefficient of the data in some localized basis of p elements

$$X(i) \qquad i \in [1, 2, \ldots, p] = \cup_{q=1}^{Q} \mathcal{D}_q \tag{4}$$

where the set $[1, \ldots, p]$ of all coefficient indices is partitioned into Q domains $\mathcal{D}_1, \ldots, \mathcal{D}_Q$. For instance, $X(i)$ is an $m \times 1$ vector of Fourier coefficients and $\mathcal{D}_q$ is a set of discrete Fourier frequencies in a particular band. The sample covariance matrix for the qth domain and its expected value are defined/denoted as

$$\widehat{\mathbf{R}}_q = \frac{1}{p_q} \sum_{i \in \mathcal{D}_q} X(i) X(i)^\dagger, \quad \mathbf{R}_q = E\widehat{\mathbf{R}}_q \tag{5}$$

where p_q is the number of coefficients in $\mathcal{D}_q$. The same notation is also used for each component so that, these being mutually uncorrelated by assumption, one has the decomposition

$$\mathbf{R}_q = \sum_{c=1}^{C} \mathbf{R}_q^c \tag{6}$$

There are two strong reasons for localizing the statistics.

First, if the strength of the various components and the SNR vary with time, space, frequency,..., reconstruction is improved by localizing the filter in time, space frequency,... More specifically, the cth component is reconstructed from its coefficients $\widehat{X}^c(i)$ estimated by

$$\widehat{X}^c(i) = \mathbf{R}_q^c \mathbf{R}_q^{-1} X(i) \quad \text{if} \quad i \in \mathcal{D}_q. \tag{7}$$

i.e. the reconstruction filter also is localized, taking advantage of the 'local SNR conditions' on domain $\mathcal{D}_q$.

Second, the diversity of the statistics of the components over several domains is precisely what may make this model blindly identifiable. For instance, if all components are one-dimensional and there is no noise, we are back to the standard ICA model. Then, if $X(i)$ are Fourier coefficients and $\mathcal{D}_q$ are spectral bands, it is known that spectral diversity (no two components have identical spectrum) is a sufficient condition for blind identifiability.

2.4 Model Identification

So far, the separation of components has been discussed without any blindness ingredient. However, we saw that computing the MSE-optimal separating filter for component c in domain q requires only, by eq. (7), the determination of $\mathbf{R}_q^c$. A generic procedure for identifying these matrices is to assume some parametric model for each component: the set $\mathcal{R}^c$ of localized covariance matrices for the cth component is parametrized by a vector θ^c of parameters and the component model is some well thought of function $\theta^c \rightarrow \mathcal{R}^c(\theta^c) = \{\mathbf{R}_q^c(\theta^c)\}_{q=1}^Q$. Some examples are given at section 3.1.

A parametric model $\mathcal{R}(\theta)$ follows by assuming component decorrelation (6) and taking the global parameter θ as the concatenation of the parameters of each component: $\theta = (\theta^1, \ldots, \theta^c)$, so that $\mathcal{R}(\theta) = \{\mathbf{R}_q(\theta)\}_{q=1}^Q = \{\sum_c \mathbf{R}_q^c(\theta^c)\}_{q=1}^Q$. The unknown parameters are found by matching model to data, that is, by minimizing some measure of discrepancy between $\widehat{\mathcal{R}}$ and $\mathcal{R}(\theta)$. More specifically:

$$\widehat{\theta} = \arg\min \phi(\theta) \quad \text{where} \quad \phi(\theta) = \sum_{q=1}^{Q} w_q K(\widehat{\mathbf{R}}_q, \mathbf{R}_q(\theta)). \tag{8}$$

Here, $K(\cdot,\cdot)$ is measure of mismatch between two positive matrices and w_q are positive weights (example below).

2.5 Summary. Blind... or Not?

At this stage, (most of) the statistical framework is in place but our method is not well defined yet because many options are available:

1. choice of a basis to obtain coefficients $X(i)$ and of domains $\{\mathcal{D}_q\}_{q=1}^Q$ to define their second-order statistics $\widehat{\mathcal{R}}$,
2. choice of a model $\theta^c \rightarrow \mathcal{R}^c(\theta^c)$, for each component contribution,
3. choice of weights w_q and matrix mismatch $K(\cdot,\cdot)$ in criterion $\phi(\theta)$.

Regarding point 3, our most common choice is to use $w_q = p_q$ and $K(\mathbf{R}_1, \mathbf{R}_2) = \frac{1}{2}[\mathrm{trace}(\mathbf{R}_1^{-1}\mathbf{R}_2) - \log\det(\mathbf{R}_1^{-1}\mathbf{R}_2) - m]$. Then, $\phi(\theta)$ is the negative log-likelihood of the model where $X(i) \sim \mathcal{N}(0, \mathbf{R}_q)$ for $i \in \mathcal{D}_q$ and is independent from $X(i')$ for $i \neq i'$.

Another design choice is to implement the recovery (7) of individual components either as $\widehat{X}^c(i) = \mathbf{R}_q^c(\widehat{\theta}^c)\mathbf{R}_q(\widehat{\theta})^{-1}X(i)$ or as $\widehat{X}^c(i) = \mathbf{R}_q^c(\widehat{\theta}^c)\widehat{\mathbf{R}}_q^{-1}X(i)$.

Is this a *blind* component separation method? It all depends on the component model. If all components are modeled as 'classic' ICA components (see 3.1), then the method is as blind as regular ICA. Our approach, however, leaves open the possibility of tuning the blindness level at will by specifying more or less stringent models $\theta^c \to \mathcal{R}^c$ for some or all of the components.

3 Implementation

We are 'only' left with the numerical issue of actually minimizing $\phi(\theta)$ using an arbitrary library of components. This is the topic of next section.

We call a collection of models $\theta^c \to \mathcal{R}^c(\theta^c)$ a *library* of components. In practice, each member of the library must not only specify a function $\theta^c \to \mathcal{R}^c(\theta^c)$ but also its gradient and other related quantities, as we shall see next.

3.1 A Library of Components

Typical examples of component models are now listed.

1. The 'classic' ICA component is one dimensional $X^c(i) = \mathbf{a}_c S_c(i)$. Denoting σ^2_{qc} the average variance of $S_c(i)$ over the qth domain, the contribution $\mathbf{R}^c_q$ of this component to $\mathbf{R}_q$ is the rank-one matrix

$$\mathbf{R}^c_q = \mathbf{a}_c \mathbf{a}_c^\dagger \sigma^2_{qc}$$

   This component can be described by an $(m+Q) \times 1$ vector θ^c of parameters containing the m entries of $\mathbf{a}_c$ and the Q variance values σ^2_{qc}. Such a parametrization is redundant, but we leave this issue aside for the moment.
2. A d-dimensional component can be modeled as

$$\mathbf{R}^c_q = A_c P_{qc} A_c^\dagger$$

   where A_c is an $m \times d$ matrix and P_{qc} is an $d \times d$ positive matrix varying freely over all domains. This can be parametrized by a vector θ^c of $m \times d + Q \times d(d+1)/2$ scalar parameters (the entries of A_c and of P_{qc}). Again, this is redundant, but we ignore this issue for the time being.
3. Noise component. A simple noise model is given by

$$\mathbf{R}^c_q = \mathrm{diag}(\sigma^2_1, \ldots, \sigma^2_m)$$

   that is, uncorrelated noise from channel to channel, with the same level in all domains but not in all channels. This component is described by a vector θ^c of only m parameters. In our application, we also use $\mathbf{R}^c_q = \mathrm{diag}(\sigma^2_{1q}, \ldots, \sigma^2_{mq})$ meaning that the noise changes from domain to domain. We then need a parameter vector θ^c of length $mQ \times 1$.
4. As a final example, for modeling 'point sources', we also use $\mathbf{R}^c_q = \mathbf{R}^c_\star$. This component contributes identically in all channels. If, for instance, we assume that this contribution $\mathbf{R}^c_\star$ is known, then the parameter vector θ^c is void. If $\mathbf{R}^c_\star$ is known up to a scale factor, then θ^c is just a scalar, etc...

3.2 Optimization

For a noise-free model containing only 'classic ICA' components, criterion $\phi(\theta)$ is a joint diagonalization criterion for which a very efficient algorithm exists [3]. In the noisy case, this is no longer true but it is possible, for simple component models, to use the EM algorithm. The EM algorithm, however, is not convenient for general component models and, in addition, EM appears too slow for our purposes. Specialized optimization algorithms are thus required.

The Conjugate Gradient (CG) algorithm has been found well suited for minimizing $\phi(\theta)$. Its implementation requires that $\partial\phi/\partial\theta$ be computed. Also, CG only works well when properly pre-conditioned by (some approximation of) the inverse of $\partial^2\phi/\partial\theta^2$. Since $\phi(\theta)$ actually is a negative log-likelihood in disguise, its Hessian can be approximated by $\mathbf{F}(\theta)$, the Fisher information matrix (FIM). The FIM is also useful for computing (approximate) error bars on $\widehat{\theta}$.

Hence we need to compute $\partial\phi/\partial\theta$ and (possibly an approximation of) $\partial^2\phi/\partial\theta^2$. This computation offers no particular difficulty in theory but our aim is to implement it in the framework of a library of components. It means that we seek to organize the computations in such a way that each component model works as a 'plug-in'.

Computing the gradient. Slightly abusing the notation, the derivative with respect to θ^c takes the form

$$\frac{\partial\phi(\theta)}{\partial\theta^c} = \sum_{q=1}^{Q} \text{trace}\left(\mathbf{G}_q(\theta)\frac{\partial\mathbf{R}_q^c(\theta^c)}{\partial\theta^c}\right) \tag{9}$$

where matrix $\mathbf{G}_q(\theta)$ is defined as

$$\mathbf{G}_q(\theta) = \frac{1}{2}w_q\mathbf{R}_q^{-1}(\theta)\left(\mathbf{R}_q(\theta) - \widehat{\mathbf{R}}_q\right)\mathbf{R}_q^{-1}(\theta) \tag{10}$$

Hence the computation of $\partial\phi/\partial\theta$ at a given point $\theta = (\theta^1, \ldots, \theta^C)$ can be organized as follows. A first loop through all components computes $\mathcal{R}(\theta)$ by adding up the contribution $\mathcal{R}^c(\theta^c)$ of each component. Then, a second loop over all Q domains computes matrices $\{\mathbf{G}_q(\theta)\}_{q=1}^Q$ which are stored in a common work space. Finally, a third loop over all components concatenates all partial gradients $\partial\phi/\partial\theta^c$, each component implementing the computation of the right hand side of (9) in the best possible way, using the available matrices $\{\mathbf{G}_q(\theta)\}_{q=1}^Q$.

Computing an (approximate) Hessian. The Fisher information matrix can be partitioned component-wise with the off-diagonal block $[\mathbf{F}(\theta)]_{cc'}$ depending on components c and c'. This seems to be a problem for a plug-in architecture because its principle requires that new component models can be introduced (plugged in) independently of each other. Nonetheless, this requirement can be full-filled because, the (c, c') block of the FIM is

$$[\mathbf{F}(\theta)]_{cc'} = \frac{1}{2}\sum_q w_q \text{trace}\left(\frac{\partial\mathbf{R}_q^c(\theta^c)}{\partial\theta^c}\mathbf{R}_q^{-1}(\theta)\frac{\partial\mathbf{R}_q^{c'}(\theta^{c'})}{\partial\theta^{c'}}\mathbf{R}_q^{-1}(\theta)\right) \tag{11}$$

Hence, the FIM can be computed by a double loop over c and c' since it is only necessary that the code for each component be able to return $\{\frac{\partial \mathbf{R}_q^c(\theta^c)}{\partial \theta^c}\}_{q=1}^Q$.

A straightforward implementation of this idea may be impractical, though, because $\{\frac{\partial \mathbf{R}_q^c(\theta^c)}{\partial \theta^c}\}_{q=1}^Q$ is a set of $|\theta^c| \times Q$ matrices, possibly very large. This problem can be largely alleviated in the frequent case where components have 'local' variables, that is whenever θ^c can be partitioned as $\theta^c = (\theta_0^c, \theta_1^c, \ldots, \theta_Q^c)$ where, for $q > 0$, vector θ_q^c influences only $\mathbf{R}_q^c$ and where θ_0^c collects all the remaining parameters, *i.e.* those which affect the covariance matrix over two or more domains (the simplest example is the 'classic' ICA component: $\mathbf{R}_q^c = \mathbf{a}_c\mathbf{a}_c^\dagger\sigma_{qc}^2$, for which $\theta_0^c = \mathbf{a}$ and $\theta_q^c = \sigma_{qc}^2$ for $q = 1, \ldots, Q$). In that case, vector θ can be partitioned into a 'global part' $\theta_0 = (\theta_0^1, \ldots, \theta_0^C)$ and Q local parts $\theta_q = (\theta_q^1, \ldots, \theta_q^C)$. With such a partitioning, the FIM has many zero blocks since then $[\mathbf{F}(\theta)]_{qq'} = \mathbf{0}$ for $1 \leq q \neq q' \leq Q$ and the computations can be organized much more efficiently. Space is lacking for giving more details here.

Indeterminations and penalization. We saw at section 3.1 that 'natural' component parametrization often are redundant. From a statistical point of view, this is irrelevant: we seek ultimately to identify $\mathcal{R}^c = \{\mathbf{R}_q^c\}_{q=1}^Q$ as a member of a family described by a mapping $\theta^c \rightarrow \mathcal{R}^c(\theta^c)$ but this mapping does not need to be one-to-one. The simplest example again is for $\mathbf{R}_q^c = \mathbf{a}_c\mathbf{a}_c^\dagger\sigma_{qc}^2$ which is invariant if one changes $\mathbf{a}_c$ to $\alpha\mathbf{a}_c$ and σ_{qc} to $\alpha^{-1}\sigma_{qc}$. This is the familiar ICA scale indetermination but $\mathbf{R}_q^c$ itself is perfectly well defined [2].

The only serious concern about over-parametrization is from the optimization point of view. Redundancy makes the $\phi(\theta)$ criterion *flat* in the redundant directions and it makes the FIM a singular matrix. Finding non redundant reparametrizations is a possibility, but it is often simpler to add a penalty function to $\phi(\theta)$ for any redundantly parametrized component. For instance, the scale indetermination of the classic ICA component $\mathbf{R}_q^c = \mathbf{a}_c\mathbf{a}_c^\dagger\sigma_{qc}^2$ when parametrized $\theta_0^c = \mathbf{a}_c$ and $\theta_q^c = \sigma_{qc}^2$ $(q > 1)$ is fixed by adding $\phi^c(\theta^c) = g(\|\mathbf{a}_c\|^2)$ to $\phi(\theta)$, where $g(u)$ is any reasonable function which has a single minimum at, say, $u = 1$.

4 Conclusion

Our technique for component separation gains a lot of its flexibility from realizing that one can start with *covariance matrix separation* —*i.e.* the identification of individual component terms in the domain-wise decomposition (6)— followed by *data separation* according to (3). It is thus sufficient to identify matrices $\mathbf{R}_q^c$. Whether or not minimizing the covariance matching criterion $\phi(\theta)$ leads to *uniquely* identified components depends on the particular component models picked from a 'library of components'. Uniqueness (identifiability) can only be decided on a case-by-case basis, either from analytical considerations or by inspection of the Fisher information matrix which can be numerically computed using the component library. By using more or less constrained components, the method ranges from totally blind to semi-blind, to non-blind.

Some strong points of the approach are the following. **Speed**: the method is potentially fast because large data sets are compressed into $\widehat{\mathcal{R}}$, a possibly much smaller object. **Accuracy**: the method is potentially accurate because it can model complex components and then recover separated data via local Wiener filtering. **Flexibility**: the method is flexible because it can be implemented via a library of components with arbitrary structure. **Noise**: the method can take noise into account without increased complexity since noise is not processed differently from any other component. **Prior**: the implementation also allows for easy inclusion of prior information about a component c if it can be cast in the form of a prior probability distribution $p_c(\theta^c)$ in which case one only need to subtracting $\log p_c(\theta^c)$ from $\phi(\theta)$ and the related changes can be delegated to the component code. **Varying resolution**: in our application, and possibly others, the input channels are acquired by sensors with channel-dependent resolution. Accurate component separation can only be achieved if this effect is taken into account. This can be achieved with relative simplicity if the data coefficients entering in $\widehat{\mathbf{R}}_q$ are Fourier coefficients.

This paper combines several ideas already known in the ICA literature: lumping together correlated components into a single multidimensional component is in [2]; minimization of a covariance-matching contrast $\phi(\theta)$ derived from the log-likelihood of a simple Gaussian model is found for instance in [3]; the extension to noisy models is already explained in [4]. The current paper goes one step further by showing how arbitrarily structured components can be separated and how the related complexity can be managed at the software level by a library of components.

References

1. Pham, D.T., Cardoso, J.F.: Blind separation of instantaneous mixtures of non stationary sources. IEEE Trans. on Sig. Proc. 49(9), 1837–1848 (2001)
2. Cardoso, J.F.: Multidimensional independent component analysis. In: Proc. ICASSP '98. Seattle (1998)
3. Pham, D.: Blind separation of instantaneous mixture of sources via the Gaussian mutual information criterion. Signal Processing 4, 855–870 (2001)
4. Delabrouille, J., Cardoso, J.F., Patanchon, G.: Multi–detector multi–component spectral matching and applications for CMB data analysis. MNRAS 346(4), 1089–1102 (2003), http://arXiv.org/abs/astro-ph/0211504

Blind Separation of Instantaneous Mixtures of Dependent Sources

Marc Castella[1] and Pierre Comon[2]

[1] GET/INT, UMR-CNRS 5157, 9 rue Charles Fourier, 91011 Évry Cedex, France
marc.castella@int-evry.fr
[2] CNRS, I3S, UMR 6070, BP.121, Sophia-Antipolis cedex, France
pcomon@i3s.unice.fr

Abstract. This paper deals with the problem of Blind Source Separation. Contrary to the vast majority of works, we do not assume the statistical independence between the sources and explicitly consider that they are dependent. We introduce three particular models of dependent sources and show that their cumulants have interesting properties. Based on these properties, we investigate the behaviour of classical Blind Source Separation algorithms when applied to these sources: depending on the source vector, the separation may be sucessful or some additionnal indeterminacies can be identified.

1 Introduction

Independent Component Analysis (ICA) is now a well recognized concept, which has fruitfully spread out to a wide panel of scientific areas and applications. Contrary to other frameworks where techniques take advantage of a strong information on the diversity, for instance through the knowledge of the array manifold in antenna array processing, the core assumption in ICA is much milder and reduces to the statistical mutual independence between the inputs.

However, this assumption is not mandatory in Blind Source Separation (BSS). For instance, in the case of static mixtures, sources can be separated if they are only decorrelated when their nonstationarity or their color can be exploited. Other properties such as the fact that sources belong to a finite alphabet can alternatively be utilized [1,2] and do not require statistical independence.

Inspired from [3,4], we investigate the case of dependent sources, without assuming nonstationarity nor color. To our knowledge, only few references have tackled this issue [5,6].

2 Mixture Model and Notations

We consider a set of N source signals $(s_i(n))_{n\in\mathbb{Z}}, i = 1, \ldots, N$. The dependence on time of the signals will not be made explicit in the paper. The sources are mixed, yielding a P-dimensional observation vector $\mathbf{x} = (\mathbf{x}(n))_{n\in\mathbb{Z}}$ according to the model:

$$\mathbf{x} = \mathbf{A}\mathbf{s} \tag{1}$$

M.E. Davies et al. (Eds.): ICA 2007, LNCS 4666, pp. 9–16, 2007.

where $\mathbf{s} = (s_1, \ldots, s_N)^\intercal$, $\mathbf{x} = (x_1, \ldots, x_P)^\intercal$ and $\mathbf{A}$ is a $P \times N$ matrix called the mixing matrix. We assume that $\mathbf{A}$ is left-invertible.

Source separation consists in finding a $N \times P$ separating matrix $\mathbf{B}$ such that its output $\mathbf{y} = \mathbf{Bx}$ corresponds to the original sources. When only the observations are used for this, the problem is referred to as the BSS problem. Introducing the $N \times N$ global matrix $\mathbf{G} \triangleq \mathbf{BA}$, the BSS is problem is solved if $\mathbf{G}$ is a so-called trivial matrix, i.e. the product of a diagonal matrix with a permutation: these are well known ambiguities of BSS.

In this paper, we will study separation criteria as functions of $\mathbf{G}$. Source separation sometimes proceeds iteratively, extracting one source at a time (e.g. deflation approach). In this case, we will write $y = \mathbf{bx} = \mathbf{gs}$ where $\mathbf{b}$ and $\mathbf{g} = \mathbf{bA}$ respectively correspond to a row of $\mathbf{B}$ and $\mathbf{G}$ and y denotes the only output of the separating algorithm. In this case, the separation criteria are considered as functions of $\mathbf{g}$. Finally, we denote by $\mathrm{E}\{.\}$ the expectation operator and by $\mathrm{Cum}\{.\}$ the cumulant of a set of random variables. $\mathrm{Cum}_4\{y\}$ is equivalent to $\mathrm{Cum}\{y, y, y, y\}$ and, for complex variables, $\mathrm{Cum}_{2,2}\{y\}$ stands for $\mathrm{Cum}\{y, y, y^*, y^*\}$.

3 Examples and Properties of Dependent Sources

We introduce in this section different cases of vector sources that are dependent and that will be considered in this paper.

3.1 Three Dependent Sources

Binary phase shift keying (BPSK) signals have specificities that will allow us to obtain source vectors with desired properties. In this paper, we will consider BPSK sources that take values $s = +1$ or $s = -1$ with equal probability1/2. We define the following source vector:

A1. $\mathbf{s} \triangleq (s_1, s_2, s_3)^\intercal$ where s_1 is BPSK; s_2 is real-valued non Gaussian, independent of s_1 and satisfies $\mathrm{E}\{s_2\} = \mathrm{E}\{s_2^3\} = 0$; and $s_3 = s_1 s_2$.

Interestingly, the following lemma holds true:

Lemma 1. *The sources s_1, s_2, s_3 defined by A1 are obviously mutually dependent. Nevertheless they are decorrelated and their fourth-order cross-cumulants vanish, that is:*

$$\mathrm{Cum}\{s_i, s_j\} = 0 \text{ except if } i = j, \tag{2}$$

$$\mathrm{Cum}\{s_i, s_j, s_k, s_l\} = 0 \text{ except if } i = j = k = l. \tag{3}$$

Proof. Using the definition of s_1, s_2 and their independence, one can easily check that $\mathrm{E}\{s_1\} = \mathrm{E}\{s_2\} = \mathrm{E}\{s_3\} = 0$. For these centered random variables, it is known that cumulants can be expressed in terms of moments:

$$\mathrm{Cum}\{s_i, s_j\} = \mathrm{E}\{s_i s_j\} \tag{4}$$

$$\mathrm{Cum}\{s_i, s_j, s_k, s_l\} = \mathrm{E}\{s_i s_j s_k s_l\} - \mathrm{E}\{s_i s_j\}\mathrm{E}\{s_k s_l\} - \mathrm{E}\{s_i s_k\}\mathrm{E}\{s_j s_l\} - \mathrm{E}\{s_i s_l\}\mathrm{E}\{s_j s_k\} \tag{5}$$

Using again the definition of s_1, s_2 and their independence, it is then easy to check all cases of equations (4) and (5) and to verify that these fourth order cross-cumulants are indeed null. On the other hand, the third order cross-cumulant reads:

$$\mathrm{Cum}\{s_1, s_2, s_3\} = \mathrm{E}\{s_1 s_2 s_3\} = \mathrm{E}\{s_1^2 s_2^2\} = \mathrm{E}\{s_1^2\}\,\mathrm{E}\{s_2^2\} > 0 \qquad (6)$$

and this proves that s_1, s_2, s_3 are mutually dependent. □

Depending on s_2, more can be proved about the source vector defined by A1. For example, if the probability density function of s_2 is symmetric, then s_1 and s_3 are independent. On the contrary s_2 and s_3 are generally not independent.

An even more specific case is obtained when s_2 is itself BPSK. In this case, one can check that the sources (s_1, s_2, s_3) are pairwise independent, although not mutually independent.

3.2 Pairwise Independent Sources

We now investigate further the case of pairwise independent sources and introduce the following source vector:

A2. $\mathbf{s} = (s_1, s_2, s_3, s_4)^\mathsf{T}$ where s_1, s_2 and s_3 are independent BPSK and $s_4 = s_1 s_2 s_3$.

This case has been considered in [3], where it has been shown that

$$\forall i \in \{1, \ldots, 4\}, \mathrm{Cum}\{s_i, s_i, s_i, s_i\} = -2\,, \quad \mathrm{Cum}\{s_1, s_2, s_3, s_4\} = 1 \qquad (7)$$

and all other cross-cumulants vanish. The latter cumulant value shows that the sources are mutually dependent; although it can be shown that they are pairwise independent. It should be clear that pairwise independence is not equivalent to mutual independence but in an ICA context, it is relevant to recall the following proposition, which is a direct consequence of Darmois' theorem [7, p.294]:

Proposition 1. *Let* $\mathbf{s}$ *be a random vector with mutually independent components, and* $\mathbf{x} = \mathbf{G}\mathbf{s}$*. Then the mutual independence of the entries of* $\mathbf{x}$ *is equivalent to their pairwise independence.*

Based on this proposition, the ICA algorithm in [7] searches for an output vector with pairwise independent component. Let us stress that this holds only if the source vector has *mutually* independent components: pairwise independence is indeed not sufficient to ensure identifiability as we will see in Section 4.2.

3.3 Complex Valued Sources

We consider quaternary phase shift keying (QPSK) sources which take their values in $\{e^{\imath\frac{\pi}{4}}, e^{-\imath\frac{\pi}{4}}, e^{\imath\frac{5\pi}{4}}, e^{-\imath\frac{5\pi}{4}}\}$ with equal probability $1/4$. We then define the following source vector:

A3. $\mathbf{s} = (s_1, s_2, s_3, s_4)^\mathsf{T}$ where s_1, s_2 and s_3 are mutually independent QPSK and $s_4 = s_1 s_2 s_3$.

Based on the Equations (4) and (5) which hold for the above centered sources, one can check the following proposition:

Lemma 2. *The sources in A3 are dependent and* $\text{Cum}\{s_1, s_2, s_3, s_4^*\} = 1$. *However, they are second-order decorrelated and all their fourth order circular cross-cumulants (i.e. with as many conjugates as non-conjugates) vanish, that is:*

$$\text{Cum}\{s_i, s_j^*\} = 0 \text{ and } \text{Cum}\{s_i, s_j\} = 0 \text{ except if } i = j, \tag{8}$$

$$\text{Cum}\{s_i, s_j, s_k^*, s_l^*\} = 0 \text{ except if } i = j = k = l. \tag{9}$$

Actually, we can prove that the above Lemma, as well as Lemma 4 and Proposition 6 still hold in the case when s_1, s_2 and s_3 are second order circular and have unit modulus: this is not detailed for reasons of space and clarity.

4 ICA Algorithms and Dependent Sources

4.1 Independence Is Not Necessarily Required

The sources given by A1 provide us with a specific example of dependent sources that are sucessfully separated by several ICA methods:

Proposition 2. *Let* $y = \mathbf{g}\mathbf{s}$ *where the vector of sources is defined by A1. Then, the function*

$$\mathbf{g} \mapsto |\text{Cum}_4\{y\}|^\alpha, \quad \alpha \geq 1 \tag{10}$$

defines a MISO contrast function, that is, its maximization over the set of unit norm vectors ($\|\mathbf{g}\|^2 = 1$) *leads to a vector* $\mathbf{g}$ *with only one non-zero component.*

Proof. The above proposition follows straightforwardly from Lemma 1 since the proof of the validity of the above contrast functions only relies on the property in Equation (3). □

Considering again the argument in the proof, one should easily notice that the above proposition can be generalized to the case of sources which satisfy:

A4. $\mathbf{s} = (s_1, \ldots, s_{3K})$ where: $\forall i \in \{0, \ldots, K-1\}$, s_{3i+1} is BPSK; s_{3i+2} is non Gaussian and satisfies $\text{E}\{s_{3i+2}\} = \text{E}\{s_{3i+2}^3\} = 0$; $s_{3i+3} = s_{3i+1}s_{3i+2}$; and the random variables $\{s_{3i+1}, s_{3i+2}\ ;\ i = 0, \ldots, K-1\}$ are mutually independent.

In addition, the above result can be generalized convolutive systems and to MIMO (multiple input/multiple output) contrast functions as defined in [7,2]:

Proposition 3. *Let* $\mathbf{y} = \mathbf{G}\mathbf{s}$ *where the vector of sources is defined by A1. Then the function:*

$$\mathbf{G} \mapsto \sum_{i=1}^{N} |\text{Cum}_4\{y_i\}|^\alpha, \quad \alpha \geq 1 \tag{11}$$

is a MIMO contrast, that is, its maximization over the group of orthogonal matrices leads to a solution $\mathbf{G}$ *which is a trivial matrix (permutation, scaling).*

Many classical algorithms for BSS or ICA first whiten the data: it is known that in so doing, they constrain matrix $\mathbf{G}$ to be orthogonal. In particular so does the algorithm proposed in [7], which relies on the contrast function in (11). It justifies that this algorithm successfully separates the sources A1. Actually, any algorithm relying on a prewhitening and associated with a contrast function based on the vanishing of the fourth-order cross cumulants (e.g. JADE) is able to separate sources such as A1.

4.2 Pairwise Independence Is Not Sufficient

We now consider the pairwise independent sources given by A2 and show that pairwise independence is not sufficient to ensure identifiability of the ICA model. We first have the following preliminary result:

Lemma 3. *Let $y = \mathbf{g}\mathbf{s}$ where the vector of sources is defined by A2. Assume that the vector (s_1, s_2, s_3) takes all 2^3 possible values. If the signal y has values in $\{-1, +1\}$, then $\mathbf{g} = (g_1, g_2, g_3, g_4)$ is either one of the solutions below:*

$$\begin{cases} \exists i \in \{1, \dots, 4\} \quad g_i = \pm 1, & \text{and: } \forall j \neq i, g_j = 0 \\ \exists i \in \{1, \dots, 4\} \quad g_i = \pm 1/2, & \text{and: } \forall j \neq i, g_j = -g_i \end{cases} \tag{12}$$

Proof. If $y = \mathbf{g}\mathbf{s}$, using the fact that $s_i^2 = 1$ for $i = 1, \dots, 4$, we have with the particular sources given by A2:

$$y^2 = g_1^2 + g_2^2 + g_3^2 + g_4^2 + 2\Big[(g_1g_2 + g_3g_4)s_1s_2 + (g_1g_3 + g_2g_4)s_1s_3 + (g_2g_3 + g_1g_4)s_2s_3\Big]$$

Since (s_1, s_2, s_3) take all possible values in $\{-1, 1\}^3$, we deduce from $y^2 = 1$ that the following equations necessarily hold:

$$\begin{cases} g_1^2 + g_2^2 + g_3^2 + g_4^2 = 1 \\ g_1g_2 + g_3g_4 = g_1g_3 + g_2g_4 = g_2g_3 + g_1g_4 = 0 \end{cases} \tag{13}$$

First observe that values given in (12) indeed satisfy (13). Yet, if a polynomial system of N equations of degree d in N variables admits a finite number of solutions[1], then there can be at most d^N distinct solutions. Hence we have found them all in (12), since (12) provides us with 16 solutions for (g_1, g_2, g_3, g_4). □

Using the above result, we are now able to specify the output of classical ICA algorithms when applied to a mixture of sources which satisfy A2.

Constant modulus and contrasts based on fourth order cumulants. The constant modulus (CM) criterion is one of the most known criteria for BSS. In the real valued case, it simplifies to:

$$J_{\mathrm{CM}}(\mathbf{g}) \triangleq \mathrm{E}\left\{\left(y^2 - 1\right)^2\right\} \qquad \text{with: } y = \mathbf{g}\mathbf{s} \tag{14}$$

[1] One can show that the number of solutions of (13) is indeed finite.

Proposition 4. *For the sources given by A2, the minimization of the constant modulus criterion with respect to* $\mathbf{g}$ *leads to either one of the solutions given by Equation* (12).

Proof. We know that the minimum value of the constant modulus criterion is zero and that this value can be reached (for $\mathbf{g}$ having one entry being ± 1 and other entries zero). Moreover, the vanishing of the constant modulus criterion implies that $y^2 - 1 = 0$ almost surely and one can then apply Lemma 3. □

A connection can now be established with the fourth-order autocumulant if we impose the following constraint:

$$\mathrm{E}\left\{y^2\right\} = 1 \quad \text{(or equivalently } \|g\| = 1 \text{ since } y = \mathbf{gs}) \tag{15}$$

Because of the scaling ambiguity of BSS, the above normalization can be freely imposed. Under (15), we have $\mathrm{Cum}_4\{y\} = \mathrm{E}\left\{\left(y^2 - 1\right)^2\right\} - 2$ and minimizing $J_{\mathrm{CM}}(\mathbf{g})$ thus amounts to maximizing $-\mathrm{Cum}_4\{y\}$. Unfortunately, since $\mathrm{Cum}_4\{y\}$ may be positive or negative, no simple relation between $|\mathrm{Cum}_4\{y\}|$ and $J_{\mathrm{CM}}(\mathbf{g})$ can be deduced from the above equation. However, we can state:

Proposition 5. *Let* $y = \mathbf{gs}$ *where the vector of sources is defined by A2. Then, under the constraint* (15) *(*$\|\mathbf{g}\| = 1$*), we have:*

(i) The maximization of $\mathbf{g} \mapsto -\mathrm{Cum}_4\{y\}$ *leads to either one of the solutions given by Equation* (12).
(ii) $|\mathrm{Cum}_4\{y\}| \leq 2$ *and the equality* $|\mathrm{Cum}_4\{y\}| = 2$ *holds true if and only if* $\mathbf{g}$ *is one of the solutions given in Equation* (12).

Proof. Part (i) follows from the arguments given above. In addition, using multilinearity of the cumulants and (7), we have for $y = \mathbf{gs}$:

$$\mathrm{Cum}_4\{y\} = -2\left(g_1^4 + g_2^4 + g_3^4 + g_4^4\right) + 24\left(g_1 g_2 g_3 g_4\right) \tag{16}$$

The result then follows straightfowardly from the study of the polynomial function in Equation (16). Indeed, optimizing (16) leads to the following Lagrangian:

$$\mathcal{L} = -2\sum_{i=1}^{4} g_i^4 + 24\prod_{i=1}^{4} g_i - \lambda\left(\sum_{i=1}^{4} g_i^2 - 1\right) \tag{17}$$

After solving the polynomial system which cancels the Jacobian of the above expresssion, one can check that all solutions are such that $|\mathrm{Cum}_4\{y\}| \leq 2$. Details are omitted for reasons of space. Part (ii) of the proposition easily follows. □

Similarly to the previous section, the above proposition can be generalized to MIMO contrast functions. In particular, this explains why, for a particular set of mixing matrices such as that studied in [3], the pairwise maximization algorithm of [7] still succeeded: a separation has luckily been obtained for the considered mixing matrices and initialization point of the algorithm, but it actually would not succeed in separating BPSK dependent sources for general mixing matrices.

Let us stress also that the results in this section are specific to the contrast functions given by (10) or (11). In particular, these results do no apply to algorithms based on other contrast functions such as JADE, contrary to the results in Sections 4.1 and 4.3.

4.3 Complex Case

The output given by separation algorithms in case of complex valued signals may differ from the previous results which have been proved for real valued signals only. Indeed, complex valued BSS does not always sum up to an obvious generalization of the real valued case [8]. We illustrate it in our context and show that, quite surprisingly, blind separation of the sources given by A3 can be achieved up to classical inderterminations of ICA. This is in contrast with the result in Equation (12) where additionnal indeterminacies appeared. First, we have:

Lemma 4. *Let $y = \mathbf{g}\mathbf{s}$ where the vector of sources is defined by A3. Assume that the vector (s_1, s_2, s_3) takes all 4^3 possible values. If the signal y is such that its values satisfy $|y|^2 = 1$, then $\mathbf{g} = (g_1, g_2, g_3, g_4)$ satisfies:*

$$\exists i \in \{1, \ldots, 4\} \quad |g_i| = 1, \textit{ and: } \forall j \neq i, g_j = 0 \tag{18}$$

Proof. If $y = \mathbf{g}\mathbf{s}$, using the fact that $|s_i|^2 = 1$ for $i = 1, \ldots, 4$, we have with the particular sources given by A3:

$$|y|^2 = \sum_{i=1}^{4} |g_i|^2 + \sum_{i \neq j} g_i g_j^* s_i s_j^* \tag{19}$$

Since (s_1, s_2, s_3) take all possible values in $\{1, \imath, -1, -\imath\}^3$, we deduce from $|y|^2 = 1$ that the following equations necessarily hold:

$$\begin{cases} |g_1|^2 + |g_2|^2 + |g_3|^2 + |g_4|^2 = 1 \\ g_1 g_2^* = g_1 g_3^* = g_1 g_4^* = g_2 g_3^* = g_2 g_4 * = g_3 g_4^* = 0 \end{cases} \tag{20}$$

Solving for the polynomial system in the variables $|g_1|, |g_2|, |g_3|$ and $|g_4|$, we obtain that the solutions are the ones given in Equation (18). □

Constant modulus and fourth-order cumulant based contrasts. In contrast with Propositions 4 and 5 we have the following result:

Proposition 6. *Let $y = \mathbf{g}\mathbf{s}$ where the sources satisfy A3. Then, the functions:*

$$\mathbf{g} \mapsto -\mathrm{E}\left\{ \left| |y|^2 - 1 \right|^2 \right\} \qquad \textit{and:} \tag{21}$$

$$\mathbf{g} \mapsto |\mathrm{Cum}_{2,2}\{y\}| \textit{ under constraint } \mathrm{E}\left\{|y|^2\right\} = 1 \tag{22}$$

are contrast functions, that is, their maximization leads to $\mathbf{g}$ *satisfying* (18).

Proof. The validity of the first contrast function is obtained with the same arguments as in the proof of Proposition 4: we have $|y|^2 \overset{m.s.}{=} 0$, which yields (20) via

(19). In the case of independent sources, the proof of the validity of the second contrast involves only cumulants with equal number of conjugate and non conjugate variables: invoking Lemma 2, one can see that the same proof still holds here. □

Note that the same arguments can be applied to ICA methods such as the pairwise algorithm in [2] or JADE [9]. Figure 1 illustrates our result.

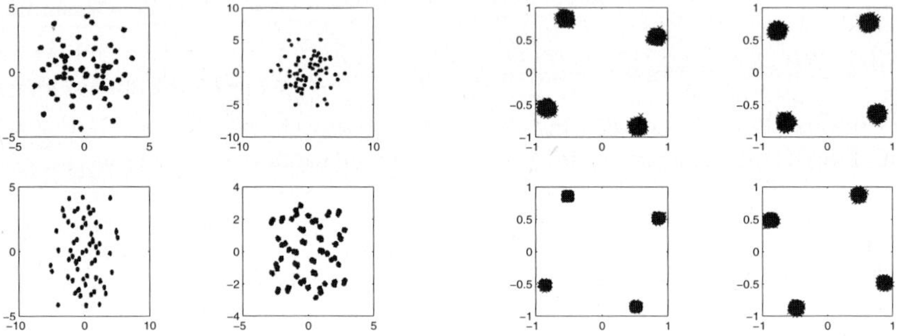

Fig. 1. Typical observed separation result of the sources A3 with the algorithm JADE (left: sensors, right: separation result)

References

1. Li, T.H.: Finite-alphabet information and multivariate blind deconvolution and identification of linear systems. IEEE Trans. on Information Theory 49(1), 330–337 (2003)
2. Comon, P.: Contrasts, independent component analysis, and blind deconvolution. Int. Journal Adapt. Control Sig. Proc. 18(3) 225–243 (April 2004) special issue on Signal Separation: Preprint: I3S Research Report RR-2003-06 (2004), `http://www3.interscience.wiley.com/cgi-bin/jhome/4508`
3. Comon, P.: Blind identification and source separation in 2 × 3 under-determined mixtures. IEEE Trans. Signal Processing 52(1), 11–22 (2004)
4. Comon, P., Grellier, O.: Non-linear inversion of underdetermined mixtures. In: Proc. of ICA'99, Aussois, France, pp. 461–465 (January 1999)
5. Hyvärinen, A., Shimizu, S.: A quasi-stochastic gradient algorithm for variance-dependent component analysis. In: Kollias, S., Stafylopatis, A., Duch, W., Oja, E. (eds.) ICANN 2006. LNCS, vol. 4131, pp. 211–220. Springer, Heidelberg (2006)
6. Cardoso, J.F.: Multidimensional independent component analysis. In: Proc. ICASSP '98. Seattle (1998)
7. Comon, P.: Independent component analysis, a new concept. Signal Processing 36(3), 287–314 (1994)
8. Eriksson, J., Koivunen, V.: Complex random vectors and ICA models: Identifiability, uniqueness and separability. IEEE Trans. on Information Theory 52(3), 1017–1029 (2006)
9. Cardoso, J.F., Souloumiac, A.: Blind beamforming for non gaussian signals. In: IEEE- Proceedings-F. vol. 140, pp. 362–370 (1993)

The Complex Version of the Minimum Support Criterion

Sergio Cruces, Auxiliadora Sarmiento, and Iván Durán

Dpto. de Teoría de la Señal y Comunicaciones*, Escuela de Ingenieros,
Universidad de Sevilla, Camino de los descubrimientos. s/n,
41092-Sevilla, Spain
sergio@us.es

Abstract. This paper addresses the problem of the blind signal extraction of sources by means of an information theoretic and geometric criterion. Our main result is the extension of the minimum support criterion to the case of mixtures of complex signals. This broadens the scope of its possible applications in several fields, such as communications.

1 Introduction

The paradigm of linear ICA consists in the decomposition of the observations into a linear combination of independent components (or sources), plus some added noise. The problem is named blind signal separation (BSS) when one tries to recover all the involved sources, whereas, it is named blind signal extraction (BSE) when one is interested in one or a subset of sources.

In the late 1970s, a powerful contrast function was proposed to solve the problem of blind deconvolution [1]. This contrast function, which minimizes the Shannon entropy of the output under a variance constraint on its signal component, was a direct consequence of the entropy power inequality [2]. A similar principle was much latter rediscovered in the field of ICA, where the minimization of the mutual information of the outputs, under a covariance constraint, was seen as a natural contrast function to solve the BSS problem [3]. Indeed, provided that the inverse system exists, there is a continuum of contrast functions based on marginal entropies which allows the simultaneous extraction of an arbitrary number of source signals [4].

Since them, the ICA literature explored the properties of other generalized entropy measures, like Renyi's entropies, to obtain novel information theoretic contrast functions [5,6]. A criterion, which involved the minimization of the sum of ranges of the outputs, was proposed in [7] for solving the BSS problem with order statistics. Some time latter, we independently proposed a similar criterion (the minimum support criterion) which minimizes zero order Renyi's entropy of the output for solving the problem of the blind extraction of one of the sources [8]. In [9] the minimum range criterion for extraction was rediscovered and proved

* Part of this research was supported by the MCYT Spanish project TEC2004-06451-C05-03.

M.E. Davies et al. (Eds.): ICA 2007, LNCS 4666, pp. 17–24, 2007.

to be free of erroneous minima, a very desirable property. The minimum support and the minimum range criteria coincide only when all the involved signals have convex support, otherwise they differ [10].

In this paper, we retake the minimum support criterion and extend its role as contrast function for mixtures of complex source signals.

The paper is organized as follows. In section 2 we present the signal model. Section 3 and section 4 detail some useful results and geometrical object definitions. Section 5 presents the complex version of the minimum support criterion and other extensions. Section 6 presents the simulations, and finally, section 7 discusses the conclusions.

2 Signal Model and Notation

We consider the standard linear mixing model of complex stationary processes in a noiseless situation. The observations random vector obeys the following equation

$$\boldsymbol{X} = \mathbf{A}\boldsymbol{S}, \tag{1}$$

where $\boldsymbol{S} = [S_1, \cdots, S_n]^T \in \mathbb{C}^{n\times 1}$ is a random vector with independent components, and $\mathbf{A} \in \mathbb{C}^{n\times n}$ is a mixing matrix of complex elements.

In order to extract one non-Gaussian source from the mixture, one can compute the inner product of the observations with the vector $\mathbf{u}$, to obtain the output random variable or estimated source

$$Y = \mathbf{u}^H \boldsymbol{X} = \mathbf{g}^H \boldsymbol{S}, \tag{2}$$

where $\mathbf{g}^H = \mathbf{u}^H \mathbf{A}$ denotes the vector with the coefficients of the mixture of the sources at the output.

The Darmois-Skitovitch theorem [3] guarantees the identifiability of non-Gaussian complex sources, up to a permutation, scaling and phase term. Let $\mathbf{e_i}$, $i = 1, \ldots, n$, denote the coordinate vectors; one source is extracted when

$$\mathbf{g} = \|\mathbf{g}\| e^{j\theta} \mathbf{e_i}, \quad i \in \{1, \ldots, n\}. \tag{3}$$

3 Support Sets and Geometric Inequalities

Consider two m-dimensional vectors of random variables $\boldsymbol{A}$ and $\boldsymbol{B}$, whose respective densities are $f_{\boldsymbol{A}}(\boldsymbol{a})$ and $f_{\boldsymbol{B}}(\boldsymbol{b})$.

Definition 1. *The support set of a random vector $\boldsymbol{A}$, which we denote by $\mathcal{S}_{\boldsymbol{A}} = supp\{\boldsymbol{A}\}$, is the set of points for which its probability density function is nonzero, i.e., $\mathcal{S}_{\boldsymbol{A}} = \{\boldsymbol{a} \in \mathbb{R}^m : f_{\boldsymbol{A}}(\boldsymbol{a}) > 0\}$.*

Definition 2. *The convex hull of the set $\mathcal{S}_{\boldsymbol{A}}$, which we denote by $\mathcal{S}_{\breve{A}} = conv\, \mathcal{S}_{\boldsymbol{A}}$, is the intersection of all convex sets in $\mathbb{R}^m$ which contain $\mathcal{S}_{\boldsymbol{A}}$.*

In this paper, we will consider that all the support sets of our interest are compact (bounded and closed), thus we will make no distinction between *convex hull* and the *convex closure.*

Definition 3. *The Minkowski sum of two given sets $\mathcal{S}_{\boldsymbol{A}}$ and $\mathcal{S}_{\boldsymbol{B}}$ is defined as the set $\mathcal{S}_{\boldsymbol{A}} \oplus \mathcal{S}_{\boldsymbol{B}} = \{\boldsymbol{a} + \boldsymbol{b} : \boldsymbol{a} \in \boldsymbol{A}, \boldsymbol{b} \in \boldsymbol{B}\}$ which contains all the possible sums of the elements of $\mathcal{S}_{\boldsymbol{A}}$ with the elements of $\mathcal{S}_{\boldsymbol{B}}$.*

In the case of two independent random vectors $\boldsymbol{A}$ and $\boldsymbol{B}$, it is easy to observe that the support of their sum $\mathcal{S}_{\boldsymbol{A}+\boldsymbol{B}}$ is equal to the Minkowski sum of the original support sets $\mathcal{S}_{\boldsymbol{A}} \oplus \mathcal{S}_{\boldsymbol{B}}$.

The following famous theorem in geometry establishes the superadditivity of the n-th root of the volume of a Minkowsky sum of two sets.

Theorem 1 (Brunn-Minkowski inequality in $\mathbb{R}^m$). *Let $\mathcal{S}_{\boldsymbol{A}}$ and $\mathcal{S}_{\boldsymbol{B}}$ be non-empty bounded Lebesgue measurable sets in $\mathbb{R}^m$ such that $\mathcal{S}_{\boldsymbol{A}} \oplus \mathcal{S}_{\boldsymbol{B}}$ is also measurable. Then*

$$\mu_m(\mathcal{S}_{\boldsymbol{A}} \oplus \mathcal{S}_{\boldsymbol{B}})^{1/m} \geq \mu_m(\mathcal{S}_{\boldsymbol{A}})^{1/m} + \mu_m(\mathcal{S}_{\boldsymbol{B}})^{1/m} \tag{4}$$

The Brunn-Minkowski inequality is formulated for nonempty bounded measurable sets in $\mathbb{R}^m$. However, we want to apply it to obtain a criterion that works for complex data. The next section will help us in this task.

4 Isomorphisms Between Real and Complex Sets

The following bijective mapping

$$c = \Re\{c\} + j\Im\{c\} \mapsto T_1(c) = \begin{pmatrix} \Re\{c\} \\ \Im\{c\} \end{pmatrix}. \tag{5}$$

defines a well-known isomorphism between the space of complex scalar numbers $\mathbb{C}$ and the vector space $\mathbb{R}^2$ with the operation of addition and multiplication by a real number. However, the multiplication of two complex numbers is not naturally carried in $\mathbb{R}^2$. Hopefully, there is another isomorphism between the space of complex scalar numbers $c \in \mathbb{C}$ and the subfield of the M^2 vector space of real 2×2 which carries the operation of multiplication. It is defined by the following bijective mapping

$$c = \Re\{c\} + j\Im\{c\} \mapsto T_2(c) = \begin{pmatrix} \Re\{c\} & -\Im\{c\} \\ \Im\{c\} & \Re\{c\} \end{pmatrix}. \tag{6}$$

The two previously presented isomorphisms allow one to express the following operation of complex random variables

$$Y = \sum_{i=1}^{n} g_i^* S_i \tag{7}$$

as the equivalent real operation between real vectors of random variables

$$\begin{pmatrix} \Re\{Y\} \\ \Im\{Y\} \end{pmatrix} = \sum_{i=1}^{n} \begin{pmatrix} \Re\{g_i\} & \Im\{g_i\} \\ -\Im\{g_i\} & \Re\{g_i\} \end{pmatrix} \begin{pmatrix} \Re\{S_i\} \\ \Im\{S_i\} \end{pmatrix}. \tag{8}$$

Moreover, to any given set of complex numbers $S_{\boldsymbol{A}}$ we can associate an area $\mu_2(A)$ which represents the area of the equivalent set $T_1(S_{\boldsymbol{A}}) = \{T_1(a) : a \in S_{\boldsymbol{A}}\}$ of $\mathbb{R}^2$ defined by the real and imaginary pairs of coordinates. Thus, the measure of the support of a complex scalar random variable is defined as the measure of support of the random vector formed by its real and imaginary parts

$$\mu_1^c(\mathcal{S}_C) \equiv \mu_2 \left(\text{supp} \left\{ \begin{pmatrix} \Re\{C\} \\ \Im\{C\} \end{pmatrix} \right\} \right). \tag{9}$$

Note that the measure of the support of the complex scalar multiplication $g_i^* S_i$ is invariant to the phase of the complex scalar g_i^*, because the phase term only implies a rotation of the space. This can be better seen from the fact that

$$\mu_1^c \left(\mathcal{S}_{(g_i^* S_i)} \right) = \begin{vmatrix} \Re\{g_i\} & \Im\{g_i\} \\ -\Im\{g_i\} & \Re\{g_i\} \end{vmatrix} \mu_2 \left(\text{supp} \left\{ \begin{pmatrix} \Re\{S_i\} \\ \Im\{S_i\} \end{pmatrix} \right\} \right) = |g_i|^2 \, \mu_1^c(\mathcal{S}_{S_i})$$

5 The Complex Version of the Minimum Support Criterion

Now we are ready to apply the Brunn-Minkowski theorem. We will implicitly assume complex sources whose densities have bounded Lebesgue measurable and non-empty supports. Under these conditions, we can exploit the previously defined isomorphisms, between real and complex sets, to rewrite the Brunn-Minkowski inequality in $\mathbb{R}^2$ (see equation (4)) as an inequality for the measure of the support of complex random variables

$$(\mu_1^c(\mathcal{S}_Y))^{\frac{1}{2}} \geq \sum_{i=1}^{n} \left(\mu_1^c(\mathcal{S}_{g_i^* S_i})\right)^{\frac{1}{2}} = \sum_{i=1}^{n} |g_i| \left(\mu_1^c(\mathcal{S}_{S_i})\right)^{\frac{1}{2}}. \tag{10}$$

A theorem, originally formulated by Lusternik and whose proof was later corrected by Henstock and Macbeath [13], establishes the general conditions for the equality to hold in the Brunn-Minkowski theorem.

Theorem 2 (Conditions for equality). *Let $\mathcal{S}_{\boldsymbol{A}}$ and $\mathcal{S}_{\boldsymbol{B}}$ be nonempty bounded Lebesgue m-dimensional measurable sets, let $\mathcal{S}'_{\boldsymbol{A}}$ and $\breve{\mathcal{S}}_{\boldsymbol{A}}$ denote, respectively, the complement and the convex closure of $\mathcal{S}_{\boldsymbol{A}}$.*

a) *If $\mu_m(\mathcal{S}_{\boldsymbol{A}}) = 0$ and $0 < \mu_m(\mathcal{S}_{\boldsymbol{B}}) < \infty$, then the necessary and sufficient condition for the equality in Brunn-Minkowski theorem is that $\mathcal{S}_{\boldsymbol{A}}$ should consist of one point only.*

b) *If* $0 < \mu_m(\mathcal{S}_{\boldsymbol{A}})\mu_m(\mathcal{S}_{\boldsymbol{B}}) < \infty$ *the equality in Brunn-Minkowski theorem holds if and only if*

$$\mu_m(\breve{\mathcal{S}}_{\boldsymbol{A}} \cap \mathcal{S}'_{\boldsymbol{A}}) = \mu_m(\breve{\mathcal{S}}_{\boldsymbol{B}} \cap \mathcal{S}'_{\boldsymbol{B}}) = 0,$$

and the convex closures $\breve{\mathcal{S}}_{\boldsymbol{A}}$ *and* $\breve{\mathcal{S}}_{\boldsymbol{B}}$ *are homothetic*[1]*.*

By the application of theorem 2, the equality in (10) is only obtained when one of the following conditions is true:

Case a) The mixture at the output is trivial, i.e.,

$$Y = g_i^* S_i, \quad i \in \{1, \ldots, n\}, \tag{11}$$

which happens when the output is an arbitrary scaled and rotated version of only one the sources.

Case b) When the sources whose contribution to the output does not vanish have support sets which are all convex and homothetic.

The connection between the zero order Rényi's entropy of a random vector in $\mathbb{R}^2$ and the volume of its support set (see [11]) leads us to identify the zero order entropy of a complex random variable with the joint zero order entropy of its real and imaginary parts,

$$h_0^c(Y) = \log \mu_1^c(\mathcal{S}_Y) \equiv h_0(\Re\{Y\}, \Im\{Y\}) \,. \tag{12}$$

Then, we can use equation (10) to obtain a different inequality which relates the zero order entropy of the output with those of the sources and which, at the same time, prevents the equality to hold true for the situations described in the case b). This new inequality is at the heart of the following result.

Theorem 3. *If the measure of the support set of the complex sources if finite and does not vanish for at least* $n-1$ *of them,*

$$\mu_1^c(\mathcal{S}_{S_{\pi_i}}) \neq 0, \quad i = 1, \ldots, n-1, \quad \pi \textit{ perm. of } \{1, \ldots, n\}, \tag{13}$$

the zero order entropy of the normalized output

$$\Psi(\boldsymbol{X}, \mathbf{u}) = h_0^c \left(\frac{\mathbf{u}^H}{\|\mathbf{u}\|_2} \boldsymbol{X} \right) = h_0^c \left(\frac{\boldsymbol{Y}}{\|\mathbf{u}\|_2} \right) \tag{14}$$

is a contrast function for the extraction of one of the sources. The global minimum of this contrast function is obtained for the source (or sources) with smallest scaled measure of support, i.e.,

$$\min_{\mathbf{u}} \Psi(\boldsymbol{X}, \mathbf{u}) = \min_i h_0^c \left(S_i / \|\mathbf{a}_i^-\|_2 \right), \tag{15}$$

where $\mathbf{a}_i^-$ *denotes the* i*th column of* $\mathbf{A}^{-H}$*, the inverse hermitian transpose of the mixing matrix.*

[1] They are equal sets up to translation and dilation.

Due to the lack of space, its proof is omitted. The result tells us that we can extract one of the sources by minimizing the area of the support set of the output.

Note that the theorem does not require the typical ICA assumption of the circularity of the complex sources nor the mutual independence between their real and imaginary parts.

The minimum support contrast function does not work for discrete sources (drawn from alphabets of finite cardinality) because they are of zero measure, a case not covered by the conditions of the theorem. Nevertheless, after replacing the support sets of the original random variables by its convex hull, we return to the conditions of the theorem, obtaining the well-behaved contrast function

$$\Psi(\breve{\boldsymbol{X}}, \mathbf{u}) = \log \mu_1^c(\breve{\mathcal{S}}_{Y/\|\mathbf{u}\|_2}) \equiv h_0^c\left(\frac{\breve{Y}}{\|\mathbf{u}\|_2}\right). \tag{16}$$

Indeed, in all of our experiments, and in similarity with the minimum range contrast for the case of real mixtures [9], this contrast function was apparently free of deceptive minima. Although we still don't know whether this property is true in general, we succeeded in proving the following result.

Theorem 4. *For a mixture of n complex sources with bounded circular convex hull, the minima of the contrast function $\Psi(\breve{\boldsymbol{X}}, \mathbf{u})$ can only be attained at the solutions of the extraction problem, i.e., there are no local deceptive minima.*

6 Simulations

In order to optimize the contrast function we first parametrized a complex unit norm vector $\mathbf{u}$ in terms of $2n - 2$ angles (ignoring a common phase term). Let $\mathbf{R}(1, k+1, \alpha_k, \beta_k)$, for $k = 1, \ldots, n-1$, denote a class of planar rotation matrices, then

$$\mathbf{u} = \mathbf{e_1}^T \mathbf{R}(1, n, \alpha_{n-1}, \beta_{n-1}) \cdots \mathbf{R}(1, 2, \alpha_1, \beta_1).$$

Since the extraction solutions are non-differentiable points of the contrast function, we used the downhill simplex method of Nelder and Mead to optimize it in low dimensions [14]. In high dimensions, an improved convergence is obtained when combining the previous optimization technique with numerical gradient and line-search methods. Each function evaluation requires the computation of the planar convex hull of a set of T outputs. The optimal algorithms for this task, have, in the worst case, a computational complexity of $O(T \log V)$ where V is the number of vertices of the convex hull [15].

Consider the sample experiment of 200 observations of a complex mixture of two 16QAM sources (a typical constellation used in communications). The illustration of figure 1 presents the graph of the contrast function $\Psi(\breve{\boldsymbol{X}}, \mathbf{u})$ which periodically tessellates the (α_1, β_1)-plane. The figure shows a contrast function with no local deceptive minima, which is non-differentiable at those points where

Fig. 1. Graph of the contrast function, with respect the parameters (α_1, β_1), for a mixture of two 16QAM sources. The solutions to the extraction problem are at the minima of the function.

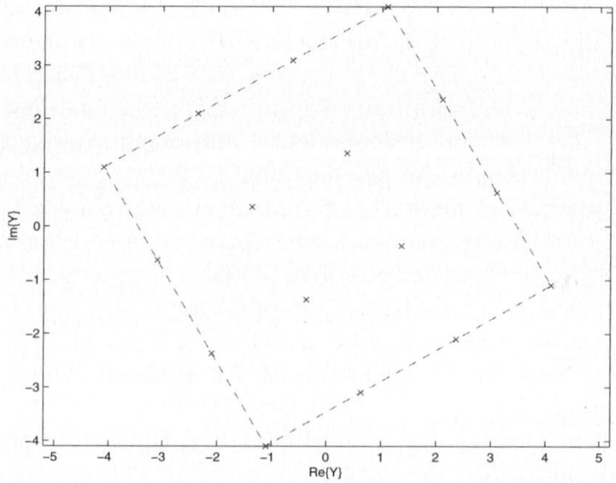

Fig. 2. The 16QAM source recovered by the extraction algorithm and the frontier of the convex hull of its support (dashed line)

the Brunn-Minkowski equality holds true. The illustration of figure 2 presents the 16QAM source extracted by the previously described algorithm and the frontier of the convex hull of its support.

7 Conclusions

We have presented a geometric criterion for the extraction of one independent component from of a linear mixture of complex and mutually independent signals. The criterion favors the extraction of the source signals with minimum scaled support and does not require the mutual independence between their real and imaginary parts. Under certain given conditions, the criterion is proved to be free of defective local minima, although, a general proof is still elusive.

References

1. Donoho, D.: On minimum entropy deconvolution. In: Findley, D.F. (ed.) Applied Time Series Analysis II, pp. 565–608. Academic Press, New York (1981)
2. Blachman, N.M.: The convolution inequality for entropy powers. IEEE Trans. on Information Theory IT-11, 267–271 (1965)
3. Comon, P.: Independent component analysis, a new concept? Signal Processing 3(36), 287–314 (1994)
4. Cruces, S., Cichocki, A., Amari, S-i.: From blind signal extraction to blind instantaneous signal separation: criteria, algorithms and stability. IEEE Trans. on Neural Networks 15(4), 859–873 (2004)
5. Bercher, J.-F., Vignat, C.: A Renyi entropy convolution inequality with application. In: Proc. of EUSIPCO, Toulouse, France (2002)
6. Erdogmus, D., Principe, J.C., Vielva, L.: Blind deconvolution with minimum Renyi's entropy. In: Proc. of EUSIPCO, Toulouse, France, vol. 2, pp. 71–74 (2002)
7. Pham, D.T.: Blind separation of instantaneous mixture of sources based on order statistics. IEEE Trans. on Signal Processing 48(2), 363–375 (2000)
8. Cruces, S., Durán, I.: The minimum support criterion for blind signal extraction. In: proc. of the int. conf. on Independent Component Analysis and Blind Signal Separation, Granada, Spain, pp. 57–64 (2004)
9. Vrins, F., Verleysen, M., Jutten, C.: SWM: A class of convex contrasts for source separation. In: proc. of the Int. Conf. on Acoustics, Speech and Signal Processing, Philadelphia (USA), vol. V, pp. 161–164 (2005)
10. Cruces, S., Sarmiento, A.: Criterion for blind simultaneous extraction of signals with clear boundaries. Electronics Letters 41(21), 1195–1196 (2005)
11. Cover, T.M., Thomas, J.A.: Elements of Information Theory. Wiley series in telecommunications. John Wiley, Chichester (1991)
12. Gardner, R.J.: The Brunn-Minkowski Inequality. Bulletin of the American Mathematical Society 39(3), 355–405 (2002)
13. Henstock, R., Macbeath, A.M.: On the measure of sum-sets. I. The theorems of Brunn, Minkowski, and Lusternik. In: Proceedings of the London Mathematical Society, third series, vol. 3, pp. 182–194 (1953)
14. Nelder, J.A., Mead, R.: A simplex method for function minimization. Computer Journal 7, 308–313 (1965)
15. Chan, T.M.: Optimal output-sensitive convex hull algorithms in two and three dimensions. Discrete and Computational Geometry 16, 361–368 (1996)

Optimal Joint Diagonalization of Complex Symmetric Third-Order Tensors. Application to Separation of Non Circular Signals

Christophe De Luigi and Eric Moreau

STD, ISITV, av. G.Pompidou, BP56, F-83162 La Valette du Var Cedex, France
deluigi@univ-tln.fr, moreau@univ-tln.fr

Abstract. In this paper, we address the problem of blind source separation of non circular digital communication signals. A new Jacobi-like algorithm that achieves the joint diagonalization of a set of symmetric third-order tensors is proposed. The application to the separation of non-gaussian sources using fourth order cumulants is particularly investigated. Finally, computer simulations on synthetic signals show that this new algorithm improves the STOTD algorithm.

1 Introduction

In the classical blind source separation problem, see e.g. [1] [2] [3] and [4], statistics based matrices or tensors often have an identical decomposition. This known decomposition is then used through a Jacobi-like algorithm to estimate the so-called mixing matrix. Perhaps one of the most popular algorithms of that kind is given in [1]. It is called JADE and its goal is to joint-diagonalize a set of hermitian matrices. The algorithm in [5] is intended to "joint-diagonalize" a set of complex symmetric matrices. The ICA algorithm in [2] is intended to diagonalize a fixed order (cumulant) tensor. The STOTD algorithm in [3] is intended to "joint-diagonalize" a particular set of (cumulant) third order tensor.

Actually, principally in wireless telecommunication applications, non circular signals are of importance, see e.g. [5][6][7][8]. The main goal of this paper is to propose a novel approach that can combine "non-circular" statistics to circular one easily for separation. It is based on a particular decomposition of symmetric third order tensors. Notice that the circular part corresponds to the STOTD algorithm [3] while the non-circular one is original.

We apply the proposed algorithm and compare it with STOTD using computer simulations. They illustrate the usefulness to consider both kind of statistics.

2 The Proposed Algorithm

2.1 The "Non Circular" Algorithm

We consider N_1 symmetric complex third-order tensors $\mathbf{T}_l$, $l = 1, \cdots, N_1$, of dimension $N \times N \times N$ decomposed linearly as:

$$\mathbf{D}_l = \mathbf{T}_l \times_1 \mathbf{U} \times_2 \mathbf{U} \times_3 \mathbf{U} \tag{1}$$

M.E. Davies et al. (Eds.): ICA 2007, LNCS 4666, pp. 25–32, 2007.

where $\mathbf{D}_l$ are also symmetric complex third-order tensors and $\mathbf{U}$ is a complex unitary matrix. The notation in (1) is defined component-wise as

$$(\mathbf{D}_l)_{j_1 j_2 j_3} = \sum_{k_1,k_2,k_3} (\mathbf{T}_l)_{k_1 k_2 k_3} (\mathbf{U})_{j_1 k_1} (\mathbf{U})_{j_2 k_2} (\mathbf{U})_{j_3 k_3} . \tag{2}$$

It is important to notice that our decomposition is different from the one in [3]. Indeed, there, the considered complex third-order tensors satisfy the following decomposition

$$\mathbf{D}_l = \mathbf{T}_l \times_1 \mathbf{U} \times_2 \mathbf{U}^* \times_3 \mathbf{U}^* \tag{3}$$

where $\mathbf{D}_l$ are also symmetric complex third-order tensors.

The goal is to estimate a unitary matrix $\mathbf{U}$ in such a way that tensors $\mathbf{D}_l$ are (approximately) diagonal. For that task, it is rather classical to consider the following quadratic criterion

$$\mathcal{C}(\mathbf{U}, \{\mathbf{T}\}) = \sum_{l=1}^{N_1} \sum_{i=1}^{N} |(\mathbf{D}_l)_{iii}|^2 \tag{4}$$

to be maximized. It corresponds to the maximization of the sum of the squared norm of all diagonals of the set of tensors $\{\mathbf{D}\}$ hence to the minimization of the squared norm of all off diagonal components.

As we consider a Jacobi-like algorithm, we study the case $N = 2$. In that case the unitary matrix $\mathbf{U}$ can be parameterized as

$$\mathbf{U} = \begin{pmatrix} \cos(\alpha) & -\sin(\alpha)\exp(j\phi) \\ \sin(\alpha)\exp(-j\phi) & \cos(\alpha) \end{pmatrix} \tag{5}$$

where α and ϕ are two angles.

We can now propose the following result.

Proposition 1. *The criterion in (4) can be written as*

$$\mathcal{C}(\mathbf{U}, \{\mathbf{T}\}) = \mathbf{u}^T \mathbf{B}_1 \mathbf{u} \tag{6}$$

where

$$\mathbf{u} = (\cos(2\alpha) \quad \sin(2\alpha)\sin(\phi) \quad \sin(2\alpha)\cos(\phi))^{\mathrm{T}} \tag{7}$$

and $\mathbf{B}_1$ is a real symmetric matrix whose expression is given in the proof.

Proof
With the property of symmetry of the tensor $\mathbf{T}_l$ that is to say

$$(\mathbf{T}_l)_{ppk} = (\mathbf{T}_l)_{pkp} = (\mathbf{T}_l)_{kpp} , \tag{8}$$

we can write the two elements of the diagonals of one single $(2 \times 2 \times 2)$ tensor $\mathbf{D}_l$

$$\begin{aligned} (\mathbf{D}_l)_{111} &= (\mathbf{T}_l)_{111} \cos^3(\alpha) - 3(\mathbf{T}_l)_{112} \sin(\alpha)\cos^2(\alpha)\mathbf{e}^{j\phi} \\ &\quad +3(\mathbf{T}_l)_{122} \sin^2(\alpha)\cos(\alpha)\mathbf{e}^{j2\phi} - (\mathbf{T}_l)_{222} \sin^3(\alpha)\mathbf{e}^{j3\phi} \\ (\mathbf{D}_l)_{222} &= (\mathbf{T}_l)_{111} \sin^3(\alpha)\mathbf{e}^{-j3\phi} + 3(\mathbf{T}_l)_{112} \cos(\alpha)\sin^2(\alpha)\mathbf{e}^{-2j\phi} \\ &\quad +3(\mathbf{T}_l)_{122} \cos^2(\alpha)\sin(\alpha)\mathbf{e}^{-j\phi} + (\mathbf{T}_l)_{222} \cos^3(\alpha). \end{aligned} \tag{9}$$

Then, we obtain the squared norms of the diagonals of this single $(2 \times 2 \times 2)$ tensor

$$\begin{aligned}|(\mathbf{D}_l)_{111}|^2 + |(\mathbf{D}_l)_{222}|^2 &= \left(|(\mathbf{T}_l)_{111}|^2 + |(\mathbf{T}_l)_{222}|^2\right)\left(\cos^6(\alpha) + \sin^6(\alpha)\right) \\ &+ 2.25\left(|(\mathbf{T}_l)_{112}|^2 + |(\mathbf{T}_l)_{222}|^2\right)\sin^2(\alpha) \\ &+ 3\ \mathrm{Re}\left\{(\mathbf{T}_l)_{222}(\mathbf{T}_l)^*_{122}\mathbf{e}^{j\phi} - (\mathbf{T}_l)_{111}(\mathbf{T}_l)^*_{112}\mathbf{e}^{-j\phi}\right\}\sin(2\alpha)\cos(2\alpha) \\ &+ 1.5\ \mathrm{Re}\left\{(\mathbf{T}_l)_{111}(\mathbf{T}_l)^*_{122}\mathbf{e}^{-j2\phi} + (\mathbf{T}_l)_{222}\ (\mathbf{T}_l)^*_{112}\mathbf{e}^{j\phi}\right\}\sin^2(2\alpha)\end{aligned} \tag{10}$$

which can be written

$$|(\mathbf{D}_l)_{111}|^2 + |(\mathbf{D}_l)_{222}|^2 = u^{\mathrm{T}}\ \mathbf{B}_l\ u \tag{11}$$

where u is a real (3×1) vector such that $u^{\mathrm{T}} u = 1$ and defined by (7), and $\mathbf{B}_l$ is a real symmetric matrix (3×3) defined by

$$\begin{aligned}(\mathbf{B}_l)_{11} &= |(\mathbf{T}_l)_{111}|^2 + |(\mathbf{T}_l)_{222}|^2 \\ (\mathbf{B}_l)_{12} &= -1.5\ \mathrm{Im}\left\{(\mathbf{T}_l)_{222}(\mathbf{T}_l)^*_{122} + (\mathbf{T}_l)_{111}(\mathbf{T}_l)^*_{112}\right\} \\ (\mathbf{B}_l)_{13} &= 1.5\ \mathrm{Re}\left\{(\mathbf{T}_l)_{222}(\mathbf{T}_l)^*_{122} - (\mathbf{T}_l)_{111}(\mathbf{T}_l)^*_{112}\right\} \\ (\mathbf{B}_l)_{22} &= 0.25\left(|(\mathbf{T}_l)_{111}|^2 + |(\mathbf{T}_l)_{222}|^2\right) \\ &\quad +2.25\left(|(\mathbf{T}_l)_{112}|^2 + |(\mathbf{T}_l)_{122}|^2\right) \\ &\quad -1.5\ \mathrm{Re}\left\{(\mathbf{T}_l)_{111}(\mathbf{T}_l)^*_{122} + (\mathbf{T}_l)_{222}(\mathbf{T}_l)^*_{112}\right\} \\ (\mathbf{B}_l)_{23} &= 1.5\ \mathrm{Im}\left\{(\mathbf{T}_l)_{111}(\mathbf{T}_l)^*_{122} - (\mathbf{T}_l)_{222}(\mathbf{T}_l)^*_{112}\right\} \\ (\mathbf{B}_l)_{33} &= 0.25\left(|(\mathbf{T}_l)_{111}|^2 + |(\mathbf{T}_l)_{222}|^2\right) \\ &\quad +2.25\left(|(\mathbf{T}_l)_{112}|^2 + |(\mathbf{T}_l)_{122}|^2\right) \\ &\quad +1.5\ \mathrm{Re}\left\{(\mathbf{T}_l)_{111}(\mathbf{T}_l)^*_{122} + (\mathbf{T}_l)_{222}(\mathbf{T}_l)^*_{112}\right\}\end{aligned} \tag{12}$$

So, the maximization of the sum of the squared norms of the diagonals of the set of tensors $\{\mathbf{D}_l\}$ is obtained by

$$\sum_{l=1}^{N_1} |(\mathbf{D}_l)_{111}|^2 + |(\mathbf{D}_l)_{222}|^2 = u^{\mathrm{T}}\ \mathbf{B_1}\ u \tag{13}$$

where the real matrix $\mathbf{B_1}$ is defined by

$$\mathbf{B_1} = \sum_{l=1}^{N_1} \mathbf{B}_l \tag{14}$$

which completes the proof of proposition.

The maximization of the criterion in (6) can be easily find by computing the eigenvector associated with the largest eigenvalue and then the angles are obtained using

$$\begin{aligned} \cos(\alpha) &= \sqrt{1+\frac{1}{2}u(1)} \\ \sin(\alpha)\exp(j\phi) &= \frac{1}{2\cos(\alpha)}\,(u(3)+j\;u(2)) \end{aligned} \tag{15}$$

with $\alpha \in [-\pi/4, \pi/4]$.

Finally, the unknown unitary matrix $\mathbf{U}$ is obtained from the accumulation of the successives Jacobi matrix which are taken transposed and conjugated.

2.2 The General Algorithm

In general one has to use all available useful statistics to solve a problem. Hence we propose to combine by an optimal way our hereabove developments with them of the STOTD algorithm [3]. As now seen, this can be done very easily. For third order tensors that can be decomposed as in (3), it was shown in [3], that in the case of $N = 2$, the criterion in (4) is written as

$$\mathcal{C}(\mathbf{U}, \{\mathbf{T}\}) = \mathbf{u}^T \mathbf{B}_2 \mathbf{u} \tag{16}$$

where $\mathbf{B}_2$ is a real symmetric matrix.

Hence an optimal combination of the two kind of tensors can be considered altogether by simply searching the eigenvector of $(1-\lambda)\mathbf{B}_1 + \lambda\, \mathbf{B}_2$ associated with the largest eigenvalue, where λ is a real parameter with $\lambda \in [0\ 1]$.

We can see that $\lambda = 0$ corresponds to the "non circular" algorithm called NC-STOTD and $\lambda = 1$ corresponds to the STOTD algorithm.

In this paper, the optimal coefficient λ will be found by simulations (in fact it would be possible to propose to find it by the minimization of a norm of the covariance matrix from the parameters α and ϕ).

3 Link with Source Separation

In the source separation problem, an observed signal vector $\mathbf{x}[n]$ is assumed to follow the linear model

$$\mathbf{x}[n] = \mathbf{A}\mathbf{s}[n] \tag{17}$$

where $n \in \mathbb{Z}$ is the discrete time, $\mathbf{s}[n]$ the $(N, 1)$ vector of $N \neq 2$ unobservable *complex* input signals $s_i[n]$, $i \in \{1, \ldots, N\}$, called sources, $\mathbf{x}[n]$ the $(N, 1)$ vector of observed signals $x_i[n]$, $i \in \{1, \ldots, N\}$ and $\mathbf{A}$ the (N, N) square mixing matrix assumed *invertible.*

It is classical to consider that the sources $s_i[n]$, with $i \in \{1, \ldots, N\}$, are zero-mean, unit power, stationary and statistically mutually independent.

We also assume that the sources possess non zero high order cumulant (of order under consideration) *i.e.* $\forall i \in \{1, \ldots, N\}$, the R-th cumulant

$$\mathsf{Cum}\{\underbrace{s_i[n], \ldots, s_i[n]}_{R \text{ terms}}\} = \mathsf{C}_R\{s_i\} \tag{18}$$

is non zero for all i and for a fixed $R \geq 3$.

We also assume that the matrix $\mathbf{A}$ is unitary. This can always be done assuming that a first whitening stage is applied onto the observations.

The blind source separation problem consists now in estimating a *unitary* matrix $\mathbf{H}$ in such a way that the vector

$$\mathbf{y}[n] = \mathbf{H}\mathbf{x}[n] \tag{19}$$

restores one of the different sources on each of its different components.

Perhaps one of the most useful way to solve the separation problem consists in the use of a contrast functions. They correspond to objective functions which depend on the outputs of the separating system and they have to be maximized to get a separating solution. Let us now propose the following result.

Proposition 2. *Let R be an integers such that $R \geq 3$, using the notation*

$$\mathsf{C}_R\{\mathbf{y}, i, j\} = \mathsf{Cum}\{y_i, y_i, y_i, \underbrace{y_{j_1}, \ldots, y_{j_{R-3}}}_{R-3 \text{ terms}}\} \tag{20}$$

the function

$$\mathcal{J}_R(\mathbf{y}) = \sum_{i, j_1, \ldots, j_{R-3}=1}^{N} |\mathsf{C}_R\{\mathbf{y}, i, j\}|^2 \tag{21}$$

is a contrast for white vectors $\mathbf{y}$.

The proof is reported in a forthcoming paper. Now we show that contrast $\mathcal{J}_R(\mathbf{y})$ is linked to a joint-diagonalization criterion of a set of symmetric third order tensor. Such a joint-diagonalization criterion is defined as in (4). This equivalence is given according to the following result.

Proposition 3. *With $R \geq 3$, let $\mathcal{T}_R$ be the set of $M = N^{R-3}$ third order tensors*

$$\mathbf{T}(j_1, \ldots, j_{R-3}) = (T_{i,j,k}(j_1, \ldots, j_{R-3}))$$

defined as

$$T_{i,j,k}(j_1, \ldots, j_{R-3}) = \mathsf{Cum}\{x_i, x_j, x_k, \underbrace{x_{j_1}, \ldots, x_{j_{R-3}}}_{R-3 \text{ terms}}\} \; . \tag{22}$$

Then, if $\mathbf{H}$ *is a unitary matrix, we have*

$$\mathcal{C}(\mathbf{H}, \mathcal{T}_R) = \mathcal{J}_R(\mathbf{H}\mathbf{x}) \; . \tag{23}$$

Hence the joint-diagonalization of third order symmetric tensors is a sufficient condition for separation. Moreover different order of cumulant can be considered onto the same framework.

4 Simulations

We illustrate the performances of the proposed algorithm in comparison with the STOTD algorithm (case where $\lambda = 1$) and the NC-STOTD one (case where $\lambda = 0$) by Monte Carlo simulations in which we average over 500 iterations. In our experiment, we consider two independent complex source signals which are non circular and two noisy mixtures. We have taken the mixing matrix unitary to avoid the whitening step which may degrade the performances.

The objective is to emphasize the existence of an optimal parameter λ which allows an optimal performance of the general algorithm.

We get into two situations: one with the 5 states source distribution S_1 defined as:

$\{-1; -j; 0; \beta; j\beta\}$ with the probabilities $\left\{\frac{1}{2(1+\beta)}; \frac{1}{2(1+\beta)}; \frac{\beta-1}{\beta}; \frac{1}{2\beta(1+\beta)}; \frac{1}{2\beta(1+\beta)}\right\}$,

and the other one with the 4 states source distribution S_2 defined as:

$\{-1; -j; 0; \beta; j * \beta\}$ with uniform probabilities.

These two sources are non-circular and in S_1 the parameter β may be chosen such that the cumulant $C_4^0\{\cdot\}$ is more weighty than the cumulant $C_2^2\{\cdot\}$ while in S_2 whatever the parameter β, the cumulant $C_2^2\{\cdot\}$ is more weighty than the cumulant $C_4^0\{\cdot\}$.

At each process, we take 10000 samples for each of the chosen source and we take the same unitary mixing 2-by-2 matrix. The noise distribution is a zero-mean Gaussian distribution. The signal to noise ratio (SNR) goes to obtain a power of noise equal to: 0, 1/8, 1/4, 1/2 and 3/4 of the power of the source signal. In order to find the optimal coefficient we vary λ from 0 to 1 with a 1/40 step.

We consider the following index of performance [4] which evaluates the proximity of the estimated matrix $\hat{\mathbf{A}}$, which is the separating matrix to the mixing matrix A:

$$I(\hat{\mathbf{A}}\mathbf{A}) = \frac{1}{N(N-1)}\left(\sum_{i=1}^{N}\left(\sum_{j=1}^{N}\frac{|(\hat{\mathbf{A}}\mathbf{A})_{i,j}|^2}{\max_\ell |(\hat{\mathbf{A}}\mathbf{A})_{i,\ell}|^2} - 1\right) + \sum_{j=1}^{N}\left(\sum_{i=1}^{N}\frac{|(\hat{\mathbf{A}}\mathbf{A})_{i,j}|^2}{\max_\ell |(\hat{\mathbf{A}}\mathbf{A})_{\ell,j}|^2} - 1\right)\right), \tag{24}$$

with N the dimension of the considered matrix $\hat{B}A$.

In Fig.1 and Fig.2, we plot for different power of noise this index of performance for the general algorithm called G-STOTD versus λ.

First, we can see in the two cases S_1 and S_2 that it exists an optimal coefficient λ which gives a better performance that the STOTD and the NC-STOTD algorithms. So, we can tell that combining in an optimal way the statistics of symmetry allows to improve the results of the ICA algorithm in the case of non-circular sources.

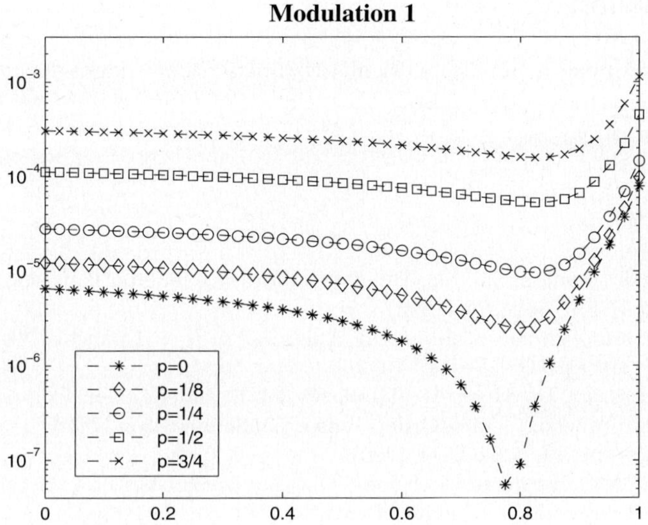

Fig. 1. Performance of the G-STOTD algorithm versus λ at different power of noise for S_1

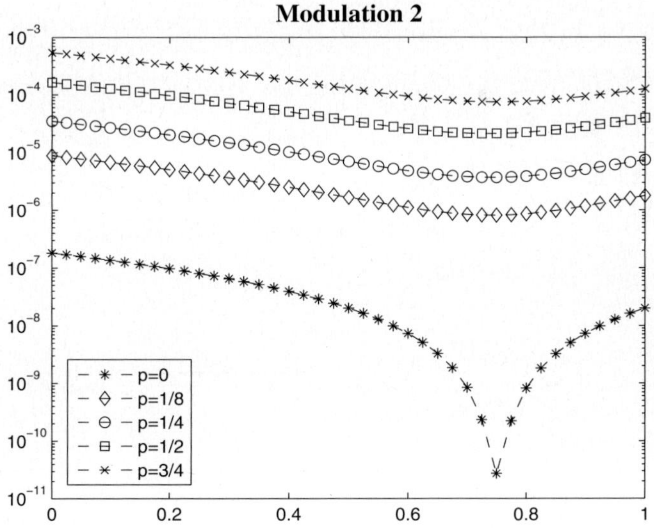

Fig. 2. Performance of the G-STOTD algorithm versus λ at different power of noise for S_2

5 Conclusion

This paper propose a new general algorithm of joint diagonalization of complex symmetric third-order tensors that allow not only to improve the STOTD algorithm but opens new perspectives for non-circular sources.

References

1. Cardoso, J.-F., Souloumiac, A.: Blind beamforming for non Gaussian signals. IEE Proceedings-F 140, 362–370 (1993)
2. Comon, P.: Independent Component Analysis, A New Concept? Signal Processing 36, 287–314 (1994)
3. De Lathauwer, L., De Moor, B., Vanderwalle, J.: Independent Component Analysis and (Simultaneous) Third-Order Tensor Diagonalization. IEEE Transactions on Signal Processing 49, 2262–2271 (2001)
4. Moreau, E.: A Generalization of Joint-Diagonalization Criteria for Source Separation. IEEE Transactions on Signal Processing 49, 530–541 (2001)
5. De Lathauwer, L., De Moor, B., Vanderwalle, J.: ICA techniques for more sources than sensors. In: Proceeding of the IEEE Signal Processing Workshop on Higher-Order Statistics (HOS'99), June 1999, Caesarea, Israel, pp. 121–124 (1999)
6. De Lathauwer, L., De Moor, B.: On the Blind Separation of Non-circular Sources. In: Proceeding of EUSIPCO-02, (September 2002), Toulouse, France, vol. II, pp. 99–102 (2002)
7. Chevalier, P.: Optimal Ttime invariant and widely linear spatial filtering for radiocommunications. In: Proc. EUSIPCO'96, Trieste, Italy, September 1996, pp. 559–562 (1996)
8. Chevalier, P.: Optimal array processing for non stationary signals. In Proc. ICASSP'96, Atlanta, pp. 2868–2871 (May 1996)

Imposing Independence Constraints in the CP Model*

Maarten De Vos[1], Lieven De Lathauwer[2], and Sabine Van Huffel[1]

[1] ESAT, Katholieke Universiteit Leuven, Leuven, Belgium
[2] CNRS - ETIS, Cergy-Pontoise, France
maarten.devos@esat.kuleuven.be

Abstract. We propose a new algorithm to impose independence constraints in one mode of the CP model, and show with simulations that it outperforms the existing algorithm.

1 Introduction

One of the most fruitful tools in linear algebra-based signal processing is the Singular Value Decomposition (SVD) (4). Most other important algebraic concepts use the SVD as building block, generalise or refine this concept for analysing quantities that are characterised by only two variables. When the data has an intrinsically higher dimensionality, higher-order generalizations of the SVD can be used. An example of a multi-way decomposition method is the **CP** model (also known as **C**anonical Decomposition (CANDECOMP) (3) or **P**arallel Factor Model (PARAFAC) (5)). Recently, a new interesting concept arose in the biomedical field. In (1), the idea of combining Independent Component Analysis (ICA) and the CP model was introduced. However, the multi-way structure was imposed after the computation of the independent components. In this paper, we propose an algorithm to impose the CP structure during the ICA computation. We also performed some numerical experiments to compare our algorithms to the algorithm proposed in (1).

* This research is funded by a PhD grant of the Institute for the Promotion of Innovation through Science and Technology in Flanders (IWT-Vlaanderen). Research supported by **Research Council KUL**: GOA-AMBioRICS, CoE EF/05/006 Optimization in Engineering, IDO 05/010 EEG-fMRI; **Flemish Government**: FWO: projects, G.0407.02 (support vector machines), G.0360.05 (EEG, Epileptic), G.0519.06 (Noninvasive brain oxygenation), FWO-G.0321.06 (Tensors/Spectral Analysis), G.0341.07 (Data fusion), research communities (ICCoS, ANMMM); IWT: PhD Grants; **Belgian Federal Science Policy Office** IUAP P6/04 ('Dynamical systems, control and optimization', 2007-2011); **EU**: BIOPATTERN (FP6-2002-IST 508803), ETUMOUR (FP6-2002-LIFESCIHEALTH 503094), Healthagents (IST-2004-27214), FAST (FP6-MC-RTN-035801); **ESA**: Cardiovascular Control (Prodex-8 C90242).

M.E. Davies et al. (Eds.): ICA 2007, LNCS 4666, pp. 33–40, 2007.

1.1 Basic Definitions

Definition 1. A 3rd-order tensor $\mathcal{T}$ has rank 1 if it equals the outer product of 3 vectors A_1, B_1, C_1: $t_{ijk} = a_i b_j c_k$ for all values of the indices. The outerproduct of A_1, B_1 and C_1 is denoted by $A_1 \circ B_1 \circ C_1$.

Definition 2. The rank of a tensor is defined as the minimal number of rank-1 terms in which the tensor can be decomposed.

Definition 3. The *Kruskal rank* or *k-rank* of a matrix is the maximal number r such that any set of r columns of the matrix is linearly independent.

Definition 4. The Frobenius norm of a tensor $\mathcal{T} \in \mathbb{R}^{I \times J \times K}$ is defined as

$$||\mathcal{T}||_F = (\sum_{i=1}^{I} \sum_{j=1}^{J} \sum_{k=1}^{K} t_{ijk}^2)^{\frac{1}{2}} \tag{1}$$

Notation Scalars are denoted by lower-case letters (a, b, ...), vectors are written as capitals (A, B, ...) (italic shaped), matrices correspond to bold-face capitals ($\mathbf{A}$, $\mathbf{B}$, ...) and tensors are written as calligraphic letters ($\mathcal{A}$, $\mathcal{B}$, ...). This notation is consistently used for lower-order parts of a given structure. For instance, the entry with row index i and column index j in a matrix $\mathbf{A}$, i.e. $(\mathbf{A})_{ij}$, is symbolized by a_{ij} (also $(A)_i = a_i$ and $(\mathcal{A})_{i_1 i_2 \ldots i_N} = a_{i_1 i_2 \ldots i_N}$). The ith column vector of a matrix $\mathbf{A}$ is denoted as A_i, i.e. $\mathbf{A} = [A_1 A_2 \ldots]$. Italic capitals are also used to denote index upper bounds (e.g. $i = 1, 2, \ldots, I$).

$\odot$ is the Khatri-Rao or column-wise Kronecker product.

1.2 Independent Component Analysis

Assume the basic linear statistical model

$$Y = \mathbf{M} \cdot X + N \tag{2}$$

where $Y \in \mathbb{R}^I$ is called the observation vector, $X \in \mathbb{R}^J$ the source vector and $N \in \mathbb{R}^I$ additive noise. $\mathbf{M} \in \mathbb{R}^{I \times J}$ is the mixing matrix.

The goal of Independent Component Analysis is to estimate the mixing matrix $\mathbf{M}$, and/or the source vector X, given only realizations of Y. In this study, we assume that $I \geqslant J$.

Blind identification of $\mathbf{M}$ in (2) is only possible when some assumptions about the sources are made. One assumption is that the sources are mutually statistically independent, as well as independent from the noise components and that at most one source is gaussian (2).

For more details, we refer to (9; 6).

1.3 The CP Model

The model. The CP model (5; 3; 15) of a three-way tensor $\mathcal{T} \in \mathbb{R}^{I \times J \times K}$ is a decomposition of $\mathcal{T}$ as a linear combination of a minimal number R of rank-1 terms:

$$\mathcal{T} = \sum_{r=1}^{R} \lambda_r \, A_r \circ B_r \circ C_r \; (+\mathcal{E}) \tag{3}$$

A pictorial representation of the CP model for third-order tensors is given in figure 1.

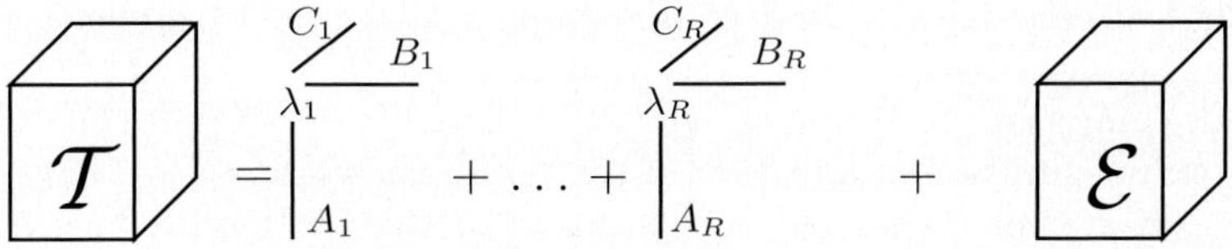

Fig. 1. Pictorial representation of the CP model

Consider a third-order $(I \times J \times K)$ tensor $\mathcal{T}$ of which the CP model can be expressed as

$$t_{ijk} = \sum_{r=1}^{R} a_{ir} b_{jr} c_{kr}, \qquad \forall i, j, k \tag{4}$$

in which $\mathbf{A} \in \mathbb{R}^{I \times R}$, $\mathbf{B} \in \mathbb{R}^{J \times R}$ and $\mathbf{C} \in \mathbb{R}^{K \times R}$. Another equivalent and useful expression of the same CP model is given with the Khatri-Rao product. We assume that $\min(IJ, K) \geqslant R$.

Associate with $\mathcal{T}$ a matrix $\mathbf{T} \in \mathbb{R}^{IJ \times K}$ as follows:

$$(\mathbf{T})_{(i-1)J+j,k} = \mathcal{T}_{ijk}. \tag{5}$$

This matrix has following structure:

$$\mathbf{T} = (\mathbf{A} \odot \mathbf{B}) \cdot \mathbf{C}^T. \tag{6}$$

Comparing the number of free parameters of a generic tensor and the CP model, it can be seen that this model is very restricted. The advantage of this model is its uniqueness under mild conditions (7; 14):

$$\text{rank}_k(\mathbf{A}) + \text{rank}_k(\mathbf{B}) + \text{rank}_k(\mathbf{C}) \geqslant 2R + 2 \tag{7}$$

with $\text{rank}_k(\mathbf{A})$ the k-rank of matrix $\mathbf{A}$ and R the rank of the tensor.

Computation of the CP decomposition. Originally, an alternating least-squares (ALS) algorithm was proposed in order to minimize the least squares cost function for the computation of the CP decomposition:

$$\min_{\mathbf{A},\mathbf{B},\mathbf{C}} ||\mathbf{T} - \mathbf{A}(\mathbf{B} \odot \mathbf{C})^T||^2. \tag{8}$$

Due to the symmetry of the model in the different modes, the updates for all modes are essentially identical with the role of the different modes shifted. Assume that $\mathbf{B}$ and $\mathbf{C}$ are fixed, the estimate of the other can be optimized with a classical linear least squares problem:

$$\min_{\mathbf{A}} ||\mathbf{X} - \mathbf{A}\mathbf{Z}^T||^2 \tag{9}$$

where $\mathbf{Z}$ equals $\mathbf{B} \odot \mathbf{C}$. This has to be repeated until convergence while matrices in other modes are kept fixed in order to compute all factors of the decomposition.

Afterwards, it was also shown that the CP decomposition can be in theory computed from an eigen value decomposition (EVD) (12; 13) under certain assumptions among which the most restricting is that $R \leqslant \min\{I, J\}$. This results in a faster computation. However, when the model is only approximately valid, this will only form the initialization of the ALS-procedure.

In (11), it is shown that the computation of the CP model, based on a simultaneous EVD is actually more robust than a single EVD. This again implied the rank condition $R \leqslant \min\{I, J\}$. As we will need this algorithm in the further developments, we review the computational scheme here. Substitution of (5) in (4) shows that any vector in the range of $\mathbf{T}$, can be represented by an $I \times J$ matrix that can be decomposed as:

$$\mathbf{V} = \mathbf{A} \cdot \mathbf{D} \cdot \mathbf{B}^T \tag{10}$$

with $\mathbf{D}$ diagonal. If the range is spanned by K matrices $\mathbf{V}_1, \mathbf{V}_2, \ldots, \mathbf{V}_K$, the computation of the canonical decomposition can be obtained by the simultaneous decomposition of the set $\{\mathbf{V}_k\}_{(1 \leqslant k \leqslant K)}$.

$$\mathbf{V}_1 = \mathbf{A} \cdot \mathbf{D}_1 \cdot \mathbf{B}^T \tag{11}$$
$$\mathbf{V}_2 = \mathbf{A} \cdot \mathbf{D}_2 \cdot \mathbf{B}^T \tag{12}$$
$$\vdots$$
$$\mathbf{V}_K = \mathbf{A} \cdot \mathbf{D}_K \cdot \mathbf{B}^T \tag{13}$$

The best choice for these matrices in order to span the full range of this mapping consists of the K dominant left singular matrices of the mapping in (5) (10). In order to deal with these equations in a numerical proper way, the problem can be formulated in terms of orthogonal unknowns (17; 11). Introducing a QR-factorization $\mathbf{A} = \mathbf{Q}^T\mathbf{R}$ and an RQ-factorization $\mathbf{B}^T = \tilde{\mathbf{R}}\mathbf{Z}^T$, leads to a simultaneous generalized Schur decomposition:

$$\mathbf{Q}\mathbf{V}_1\mathbf{Z} = \mathbf{R} \cdot \mathbf{D}_1 \cdot \tilde{\mathbf{R}} \tag{14}$$
$$\mathbf{Q}\mathbf{V}_2\mathbf{Z} = \mathbf{R} \cdot \mathbf{D}_2 \cdot \tilde{\mathbf{R}} \tag{15}$$
$$\vdots$$
$$\mathbf{Q}\mathbf{V}_K\mathbf{Z} = \mathbf{R} \cdot \mathbf{D}_K \cdot \tilde{\mathbf{R}}. \tag{16}$$

This simultaneous generalized Schur decomposition can be computed by an extended QZ-iteration (17).

Recently, in (8) it is shown that the canonical components can be obtained from a simultaneous matrix diagonalization with a much less severe restriction on R.

2 Combination of ICA and CP Model

In this section, we review the tensorial extension of ICA (§2.1), called 'tensor pICA', as it was introduced by Beckmann (1). Then we present a new algorithm to compute the CP decomposition of a tensor $\mathcal{T}$ where independence is imposed to the factors in one mode (§2.2). For notational convenience, we will restrict us to the three-way real case, but generalization to higher dimensions or complex tensors is straightforward. In the following, we always assume that the components of the third mode are independent. Due to the symmetric structure of the PARAFAC model, equivalent equations can be derived for the other two modes.

In formulas, we consider the matricized version of the real tensor $\mathcal{T}$, given by equation (6) where matrix $\mathbf{C}$ contains the independent source values, and the mixing matrix $\mathbf{M}$ equals $(\mathbf{A} \odot \mathbf{B})$.

2.1 Tensor pICA

In (1), a generalization of the standard bilinear (two-way) exploratory analysis to higher dimensions was derived as follows.

1. Perform an iteration step for the decomposition of the full data using the twodimensional probabilistic ICA approach for the decomposition into a compound mixing matrix $\mathbf{M}^{IJ\times R}$ and the associated source signals $\mathbf{C}^{K\times R}$: $\mathbf{X}^{IJ\times K} = \mathbf{M}\mathbf{C}^T + \tilde{\mathbf{E}}_1$.
2. Decompose the estimated mixing matrix $\mathbf{M}$ such that $\mathbf{M} = (\mathbf{A} \odot \mathbf{B}) + \tilde{\mathbf{E}}_2$ via a column-wise rank-1 eigenvalue decomposition: each column in $(\mathbf{A} \odot \mathbf{B})$ is formed by K scaled repetitions of a single column from $\mathbf{A}$. In order to obtain $\mathbf{A}$ and $\mathbf{B}$, the matrices $\mathbf{G_1}, \ldots, \mathbf{G_R} \in \mathbb{R}^{I\times J}$ can be introduced as
$$(\mathbf{G_r})_{ij} = m_{(i-1)J+j,r} \qquad \forall i,j,r \tag{17}$$
$\mathbf{A}_r$ and $\mathbf{B}_r$ can then be computed as the dominant left and right singular vector of $\mathbf{G_r}, 1 \leqslant r \leqslant R$.
3. iterate decomposition of $\mathbf{X}^{IJ\times K}$ and $\mathbf{M}$ untill convergence, i.e. when $||\mathbf{A}^{\text{new}} - \mathbf{A}^{\text{old}}||_F + ||\mathbf{B}^{\text{new}} - \mathbf{B}^{\text{old}}||_F + ||\mathbf{C}^{\text{new}} - \mathbf{C}^{\text{old}}||_F < \epsilon$.

2.2 ICA-CP

The ordinary ICA problem is solved by diagonalising the fourth-order cumulant (9). This cumulant can be written as following CP decomposition:

$$\mathcal{C}_y^{(4)} = \sum_{r=1}^{R} \kappa_{x_r} M_r \circ M_r \circ M_r \circ M_r \tag{18}$$

With a mixing matrix $\mathbf{M} = \mathbf{A} \odot \mathbf{B}$, this fourth-order cumulant can be expressed as an eighth-order tensor with CP structure:

$$\mathcal{C}_y^{(8)} = \sum_{r=1}^{R} \kappa_{x_r} A_r \circ B_r \circ A_r \circ B_r \circ A_r \circ B_r \circ A_r \circ B_r \tag{19}$$

This can be seen as follows.

Define matrices $\mathbf{E_1}, \ldots, \mathbf{E_R} \in \mathbb{R}^{I \times J}$ as

$$(\mathbf{E_r})_{ij} = m_{(i-1)J+j,r} \qquad \forall i, j, r \tag{20}$$

When the model in (6) is exactly satisfied, $\mathbf{E}_r$ can be decomposed as

$$(\mathbf{E_r}) = A_r B_r^T \qquad r = 1, \ldots, R \tag{21}$$

which explains the CP decomposition in equation (19).

This CP decomposition can be computed in different ways, depending on the rank R of the tensor. It is even not necessary to compute the full decomposition. Once $\mathbf{A}$ and $\mathbf{B}$ are known, the mixing matrix (matrix $\mathbf{A} \odot \mathbf{B}$) can be computed and the independent sources can be estimated from equation (6).

Rank R restricted by $R \leqslant \min\{I, J\}$. In order to compute the mixing matrix ($\mathbf{A} \odot \mathbf{B}$) from (19), associate a matrix $\mathbf{H} \in \mathbb{R}^{IJ \times I^3 J^3}$ with $\mathcal{C}_y^{(8)}$ as follows:

$$\mathbf{H}_{(i-1)J+j,(k-1)I^2J^3+(l-1)I^2J^2+(m-1)IJ^2+(n-1)IJ+(o-1)J+p} = (\mathcal{C}_y^{(8)})_{ijklmnop} \tag{22}$$

This mapping can be represented by a matrix $\mathbf{H} \in \mathbb{R}^{IJ \times I^3 J^3}$:

$$\mathbf{H} = (\mathbf{A} \odot \mathbf{B}) \cdot \mathbf{\Lambda} \cdot (\mathbf{A} \odot \mathbf{B} \odot \mathbf{A} \odot \mathbf{B} \odot \mathbf{A} \odot \mathbf{B})^T. \tag{23}$$

with $\mathbf{\Lambda} = diag\{\kappa_1, \ldots, \kappa_R\}$. Substituting (22) in (19) shows that any vector in the range of $\mathcal{C}_y^{(8)}$ can be represented by an $(I \times J)$ matrix that can be decomposed as:

$$\mathbf{V} = \mathbf{A} \cdot \mathbf{D} \cdot \mathbf{B}^T \tag{24}$$

with $\mathbf{D}$ diagonal. Any matrix in this range can be diagonalized by congruence with the same loading matrices $\mathbf{A}$ and $\mathbf{B}$. A possible choice of $\{\mathbf{V}_k\}_{(1 \leqslant k \leqslant K)}$ consist of 'matrix slices' obtained by fixing the 3rd to 8th index. An optimal approach would be to estimate the R dominant left singular values of (23). The joint decomposition of the matrices $\{\mathbf{V}_k\}_{(1 \leqslant k \leqslant K)}$ will give a set of equations similar to equations (11) - (13). We have explained in §1.3 how to solve these equations simultaneously.

3 Numerical Experiments

In this section we illustrate the performance of our algorithm by means of numerical experiments and compare it to 'tensor pICA'.

Rank-R tensors $\tilde{\mathcal{T}} \in \mathbb{R}^{5 \times 3 \times 100}$, of which the components in the different modes will be estimated afterwards, are generated in the following way:

$$\tilde{\mathcal{T}} = \frac{\mathcal{T}}{||\mathcal{T}||_F} + \sigma_N \frac{\mathcal{N}}{||\mathcal{N}||_F}, \tag{25}$$

in which $\mathcal{T}$ exactly satisfies the CP model with $R = 3$ independent sources in the third mode ($\mathbf{C}$) and $\mathcal{N}$ represents gaussian noise. All the source distributions are binary (1 or -1), with an equal probability of both values. The sources are zero mean and have unit variance. The entries of the two other modes (**A** and **B**) are drawn from a zero-mean unit-variance gaussian distribution.

We conduct Monte Carlo simulations consisting of 500 runs. We evaluate the performance of the different algorithms by means of the normalized Frobenius norm of the difference between the estimated and the real sources:

$$error_C = \frac{||\mathbf{C} - \hat{\mathbf{C}}||_F}{||\mathbf{C}||_F} \tag{26}$$

In figure 2, we plot the mean value of the 500 simulations. The previously proposed method tensor pICA is clearly outperformed by the new algorithm.

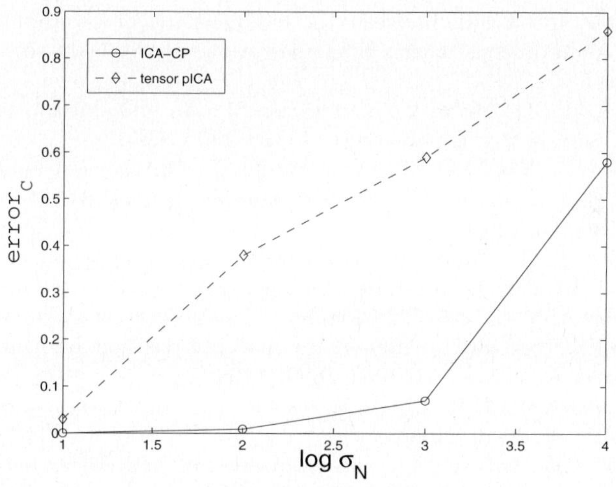

Fig. 2. The mean value of $error_C$ as a function of the noise level σ_N for the algorithms ICA-CP (solid) and tensor pICA (dash-dash)

4 Conclusion

We proposed a new algorithm to impose the CP structure already during the ICA computation for the case that the rank R was restricted by $R \leqslant \min\{I, J\}$. We showed with simulations that by taking this structure into account, the algorithm outperformed tensor pICA. A follow-up paper will discuss an algorithm for the case the rank $R \geqslant \min\{I, J\}$. For a detailed comparison between CP and the combination of ICA-CP, we refer to (16).

References

[1] Beckmann, C.F., Smith, S.M.: Tensorial extensions of independent component analysis for multisubject fmri analysis. Neuroimage 25, 294–311 (2005)

[2] Cardoso, J.-F., Souloumiac, A.: Blind beamforming for non-gaussian signals. IEE Proc. F 140, 362–370 (1994)

[3] Carroll, J.D., Chang, J.: Analysis of individual differences in multidimensional scaling via an n-way generalization of 'eckart-young' decomposition. Psychometrika 35, 283–319 (1970)
[4] Golub, G.H., Van Loan, C.F.: Matrix computations, 3rd edn. The Johns Hopkins University Press, Baltimore (1996)
[5] Harshman, R.A.: Foundations of the parafac procedure: models and conditions for an 'explanation' multi-modal factor analysis. UCLA Working Papers in Phonetics, 16 (1970)
[6] Hyvärinen, A., Karhunen, J., Oja, E.: Independent Component Analysis. John Wiley & Sons, Chichester (2001)
[7] Kruskal, J.B.: Three-way arrays: rank and uniqueness of trilinear decompositions, with applications to arithmetic complexity and statistics. Psychometrika 18, 95–138 (1977)
[8] De Lathauwer, L.: A link between the canonical decomposition in multilinear algebra and simultaneous matrix diagonalization. SIAM J Matrix Anal Appl 28, 642–666 (2006)
[9] De Lathauwer, L., De Moor, B., Vandewalle, J.: An introduction to independent component analysis. J Chemometrics 14, 123–149 (2000)
[10] De Lathauwer, L., De Moor, B., Vandewalle, J.: On the best rank-1 and rank-$\{R_1, R_2, \ldots, R_N\}$ approximation of higher-order tensors. SIAM J Matrix Anal Appl 21, 1324–1342 (2000)
[11] De Lathauwer, L., De Moor, B., Vandewalle, J.: Computation of the canonical decomposition by means of a simultaneous generalized schur decomposition. SIAM J Matrix Anal Appl 26, 295–327 (2004)
[12] Leurgans, S.E., Ross, R.T., Abel, R.B.: A decomposition for three-way arrays. SIAM J Matrix Anal Appl 14, 1064–1083 (1993)
[13] Sanchez, E., Kowalski, B.R.: Tensorial resolution: a direct trilinear decomposition. J Chemometrics 4, 29–45 (1990)
[14] Siridopoulos, N.D., Bro, R.: On the uniqueness of multilinear decomposition of n-way arrays. J Chemometrics 14, 229–239 (2000)
[15] Smilde, A., Bro, R., Geladi, P.: Multi-way Analysis with applications in the Chemical Sciences. John Wiley & Sons, Chichester (2004)
[16] Stegeman, A.: Comparing independent component analysis and the parafac model for artificial multi-subject fmri data. Technical Report, Heymans Institute, University of Groningen, the Netherlands (2007)
[17] Van Der Veen, A.-J., Paulraj, A.: An analytical constant modulus algorithm. IEEE Trans Signal Proc 44, 1136–1155 (1996)

Blind Source Separation of a Class of Nonlinear Mixtures

Leonardo Tomazeli Duarte* and Christian Jutten

GIPSA-lab, INPG-CNRS, Grenoble, France
Leonardo.duarte@lis.inpg.fr, Christian.Jutten@inpg.fr

Abstract. In this work, we deal with blind source separation of a class of nonlinear mixtures. The proposed method can be regarded as an adaptation of the solutions developed in [1,2] to the considered mixing system. Also, we provide a local stability analysis of the employed learning rule, which permits us to establish necessary conditions for an appropriate convergence. The validity of our approach is supported by simulations.

1 Introduction

The problem of blind source separation (BSS) concerns the retrieval of an unknown set of source signals by using only samples that are mixtures of these original signals. A great number of methods has been proposed for the case wherein the mixture process is of linear nature. The cornerstone of the majority of these techniques is the independent component analysis (ICA) [3]. In contrast to the linear case, the recovery of the independence, which is the very essence of ICA, does not guarantee, as a rule, the separation of the sources when the mixture model is nonlinear. In view of this limitation, a more reasonable approach is to consider constrained mixing systems as, for example, post-nonlinear (PNL) mixtures [4] and linear-quadratic mixtures [2].

In this work, we investigate the problem of BSS in a particular class of nonlinear systems which is related to a chemical sensing application. More specifically, the contributions of this paper are the adaptation of the ideas presented in [1,2] to the considered mixing system, as well as a study on some necessary conditions for a proper operation of the obtained separating method. Concerning the organization of the document, we begin, in Section 2, with a brief description of the application that has motivated us. After that, in Section 3, we expose the separation method and also a stability analysis of the learning rule. In Section 4, simulations are carried out in order to verify the viability of the proposal. Finally, in Section 5, we state our conclusions and remarks.

2 Motivation and Problem Statement

The classical methods for chemical sensing applications are generally based on the use of an unique high-selective sensor. As a rule, these techniques demand

* Leonardo Tomazeli Duarte would like to thank CNPq (Brazil) for the financial support.

M.E. Davies et al. (Eds.): ICA 2007, LNCS 4666, pp. 41–48, 2007.

sophisticated laboratory analysis, which makes them expensive and time consuming. An attractive alternative to these methods relies on the use of an array of less-selective sensors combined with a post-processing stage whose purpose is exactly to extract the relevant information from the acquired data.

In [5,6], post-processing stages based on BSS methods were considered in the problem of estimating the concentrations of several ions in a solution. In this sort of application, a device called ion-sensitive field-effect transistor (ISFET) [5] may be employed as sensor. In short, the ISFET is built on a MOSFET by replacing the metallic gate with a membrane sensitive to the ion of interest, thus permitting the conversion of chemical information into electrical one.

The Nikolsky-Eisenman (NE) model [5] provides a very simple and yet adequate description of the ISFET operation. According to this model, the response of the i-th ISFET sensor is given by:

$$x_i = c_{i1} + c_{i2} \log\left(s_i + \sum_{j,j \neq i} a_{ij} s_j^{\frac{z_i}{z_j}}\right), \tag{1}$$

where s_i and s_j are the concentration of the ion of interest and of the concentration of the j-th interfering ion, respectively, and where z_i and z_j denote the valence of the ions i and j, respectively. The selective coefficients a_{ij} model the interference process; c_{i1} and c_{i2} are constants that depends on some physical parameters. Note that when the ions have the same valence, then the model (1) can be seen as a particular case of the class of PNL systems, as described in [6].

In the present work, we envisage the situation in which $z_i \neq z_j$. According to the NE model, one obtains a tough nonlinear mixing model in this case. For the sake of simplicity, we assume, in this paper, that the coefficients c_{i1} and c_{i2} are known (even if their estimations are not so simple). Considering a mixture of two ions, such simplification leads to the following nonlinear mixing system that will be considered in this work

$$\begin{aligned} x_1 &= s_1 + a_{12} s_2^k \\ x_2 &= s_2 + a_{21} s_1^{\frac{1}{k}} \end{aligned}, \tag{2}$$

where $k = z_1/z_2$ and is known. We consider that k takes only positive integer values. Indeed, in many actual applications, typical target ions are H_3O^+, NH_4^+, Ca^{2+}, K^+, etc. Consequently, many cases correspond to $k \in \mathbb{N}$ and, in this paper, we will focus on this case. Also, the sources are supposed positives, since they represent concentrations. Finally, it is assumed that s_i are mutually independent, which is equivalent to assume that there is no interaction between the ions.

3 Separation Method

For separating sources s_i from mixtures (2), we propose a parametric recursive model (see (3) below), whose parameters w_{ij} will be adjusted by a simple ICA algorithm. Consequently, equilibrium points and their stability are depending both on a structural condition (due to the recursive nature of (3)) and on the learning algorithm, as explained in subsection 3.3.

3.1 Separating Structure

In this work, we adopted the following recurrent network as separating system:

$$\begin{aligned} y_1(m+1) &= x_1 - w_{12} y_2(m)^k \\ y_2(m+1) &= x_2 - w_{21} y_1(m)^{\frac{1}{k}} \end{aligned}, \tag{3}$$

where $[w_{12}\ w_{21}]^T$ are the parameters to be adjusted. In order to understand how this structure works, let $\mathbf{s} = [s_1\ s_2]^T$ denote a sample of the sources. By considering (2), one can easily check that when $[w_{12}\ w_{21}]^T = [a_{12}\ a_{21}]^T$, then $\mathbf{s}$ corresponds to an equilibrium point of (3). This wise approach to counterbalance the action of the mixing system without relying on its direct inversion was firstly developed in [1] regarding linear BSS. Its extension to the nonlinear case was proposed in [2], in the context of source separation of linear-quadratic mixtures.

Naturally, an ideal operation of (3) as a separating system requires that $\mathbf{s} = [s_1\ s_2]^T$ be the only equilibrium point when $[w_{12}\ w_{21}]^T = [a_{12}\ a_{21}]^T$. Unfortunately, this is not the case as can be checked by setting $y_1(m+1) = y_1(m) = y_1$ and $y_2(m+1) = y_2(m) = y_2$ in (3). From this, one observes that the determination of the equilibrium points of (3) leads to the following equation:

$$y_1 = x_1 - a_{12}\left(x_2 - a_{21} y_1^{(1/k)}\right)^k. \tag{4}$$

After straightforward calculation, including a binomial expansion, (4) becomes

$$(1 + a_{12} b_0) y_1 + a_{12} \sum_{i=1}^{k-1} b_i y_1^{1-\frac{i}{k}} + (a_{12} b_k - x_1) = 0, \tag{5}$$

where $b_i = \binom{k}{i} x_2^i (-a_{21})^{(k-i)}$.

By considering the transformation $u = y_1^{\frac{1}{k}}$ in (5), one can verify that the solution of this expression is equivalent to the determination of the roots of a polynomial of order k and, as a consequence, the number of equilibrium points grows linearly as k increases. Thus, it becomes evident that the use of (3) is appropriate only for small values of k. For instance, when $k = 2$ there are just two equilibrium points: one corresponds to the sources themselves and the other one corresponds to a mixture of these sources. In the next step of our investigation, we shall verify the conditions to be satisfied so that the equilibrium point associated with the sources be stable.

In view of the difficulty embedded in a global analyze of stability, we consider the study of the local stability in the neighborhood of the equilibrium point $\mathbf{s} = [s_1\ s_2]^T$ based on the first-order approximation of the nonlinear system (3). This linearization can be expressed by using a vectorial notation as follows:

$$\mathbf{y}(m+1) \approx \mathbf{c} + \mathbf{J}\mathbf{y}(m), \tag{6}$$

where $\mathbf{y}(m) = [y_1(m)\ y_2(m)]^T$, $\mathbf{c}$ is a constant vector and $\mathbf{J}$ is the Jacobian matrix of (3) evaluated at $[s_1\ s_2]^T$, which is given by:

$$\mathbf{J} = \begin{bmatrix} 0 & -a_{12} k s_2^{(k-1)} \\ -\frac{1}{k} a_{21} s_1^{(\frac{1}{k}-1)} & 0 \end{bmatrix}. \tag{7}$$

It can be proved that a necessary and sufficient condition for local stability of a discrete system is that the absolute values of the eigenvalues of the Jacobian matrix evaluated at the equilibrium point of interest be smaller than one [7]. Applying this result on (7), the following condition of local stability is obtained:

$$|a_{12}a_{21}s_1^{(\frac{1}{k}-1)}s_2^{k-1}| < 1. \tag{8}$$

This is a first constraint of our strategy, given that this condition must be satisfied for each sample $[s_1\ s_2]^T$. In order to illustrate this limitation, the stability boundaries in the (a_{12}, a_{21}) plane for several cases are depicted in Figure 1.

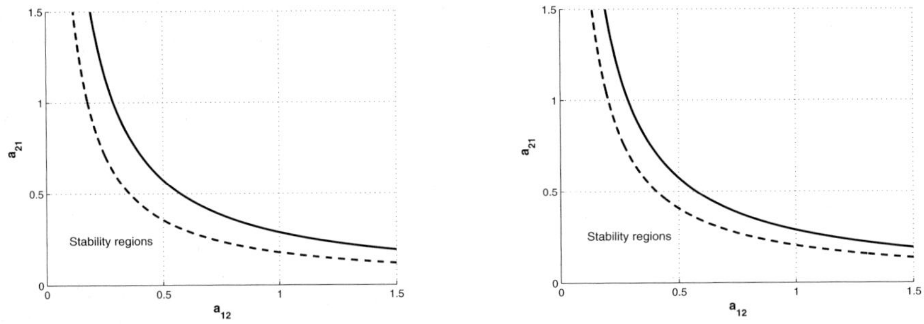

(a) Influence of k: sources distributed between $(0.1, 1.1)$ with $k = 2$ (solid) and $k = 3$ (dash)

(b) For $k = 2$: sources distributed between $(0.1, 1.1)$ (solid) and between $(0.2, 2.2)$ (dash)

Fig. 1. Stability boundaries in the (a_{12}, a_{21}) plane

3.2 Learning Algorithm

We consider a learning rule founded on the cancellation of nonlinear correlations, given by $E\{f(y_i)g(y_j)\}$, between the retrieved sources [1]. The following nonlinear functions were chosen: $f(\cdot) = (\cdot)^3$ and $g(\cdot) = (\cdot)$. Therefore, at each time n, the iteration of the separating method consists of: 1) the computation of y_i, for each sample of the mixtures, according to the dynamics (3) and 2) the update of the parameters w_{ij} according to:

$$\begin{aligned} w_{12}(n+1) &= w_{12}(n) + \mu E\{y_1^3\bar{y}_2\} \\ w_{21}(n+1) &= w_{21}(n) + \mu E\{y_2^3\bar{y}_1\} \end{aligned}, \tag{9}$$

where μ corresponds to the learning rate, $[y_1\ y_2]^T$ denotes the equilibrium point of (3) and $\bar{y}_i$ is a centering version of y_i[1,2]. One can check[3] that (9) converges

[1] More specifically, we adopt the following notation $\bar{y}_i^r = y_i^r - E\{y_i^r\}$.

[2] Given that the signals are not supposed zero-mean, the centering of one the variables in (9) is necessary, so that it converges when y_1 and y_2 are mutually independent.

[3] Note that (9) converges when $E\{y_i^3\bar{y}_j\} = E\{y_i^3 y_j\} - E\{y_i^3\}\ E\{y_j\} = 0$.

when $E\{y_1^3 y_2\} = E\{y_1^3\}E\{y_2\}$ and $E\{y_2^3 y_1\} = E\{y_2^3\}E\{y_1\}$. Obviously, these conditions are only necessary ones for the statistical independence between the sources and, as a consequence, there may be particular sources for which such strategy fails. On the other hand, this strategy provides a less complex algorithm than those that deal directly with a measure of statistical independence.

In the last section, a stability condition concerning the separation structure was provided. Likewise, as it will be seen in the sequel, it is possible to analyze the stability of the learning rule (9). This study will permit us to determine whether the separating equilibrium point, i.e., $[w_{12}\ w_{21}]^T = [a_{12}\ a_{21}]^T$, corresponds to a stable one and, as a consequence, whether it is attainable for the learning rule.

3.3 Stability Analysis of the Learning Rule

According to the ordinary differential equation theory, it is possible, by assuming that μ is sufficiently small, to rewrite (9) as:

$$\begin{aligned} \frac{dw_{12}}{dt} &= E\{y_1^3 \bar{y}_2\} \\ \frac{dw_{21}}{dt} &= E\{y_2^3 \bar{y}_1\}. \end{aligned} \tag{10}$$

A first point to be stressed is that the determination of all equilibrium points of (10) is a rather difficult task. Even when $k = 1$ in (2), which corresponds to the linear BSS problem, this calculation demands a great deal of effort [8].

Secondly, we are interested in the stability of the point $[w_{12}\ w_{21}]^T = [a_{12}\ a_{21}]^T$, but one must keep in mind that there are structural conditions to be assured so that it corresponds to an equilibrium point of (10). For example, when $k = 2$, we observed through simulations that this ideal adjustment of the separating system usually guarantees the separation of the sources when the local condition (8) is satisfied. Thus, in this situation and under the hypothesis of independent sources, it is assured that $E\{y_i^3 \bar{y}_j\} = E\{s_i^3 \bar{s}_j\} = 0$.

As in Section 3.1, the local stability analysis is based on a first-order approximation of (10). However, since we are dealing with a continuous dynamics in this case, a given equilibrium point of the learning rule is locally stable when the real parts of all eigenvalues of the Jacobian matrix are negatives [7]. After straightforward calculations, one obtains the Jacobian matrix evaluated at the equilibrium point $[a_{12}\ a_{21}]^T$

$$\mathbf{J} = \begin{bmatrix} (3E\{y_1^2 \bar{y}_2 \frac{\partial y_1}{\partial a_{12}}\} + E\{\bar{y}_1^3 \frac{\partial y_2}{\partial a_{12}}\}) & (3E\{y_1^2 \bar{y}_2 \frac{\partial y_1}{\partial a_{21}}\} + E\{\bar{y}_1^3 \frac{\partial y_2}{\partial a_{21}}\}) \\ (3E\{y_2^2 \bar{y}_1 \frac{\partial y_2}{\partial a_{12}}\} + E\{\bar{y}_2^3 \frac{\partial y_1}{\partial a_{12}}\}) & (3E\{y_2^2 \bar{y}_1 \frac{\partial y_2}{\partial a_{21}}\} + E\{\bar{y}_2^3 \frac{\partial y_1}{\partial a_{21}}\}) \end{bmatrix}. \tag{11}$$

Note that, assuming an ideal operation of the separating system, $[y_1\ y_2]^T$ could be replaced by $[s_1\ s_2]^T$, which permits us to express the stability conditions of (9) in terms of some statistics of the sources.

The entries of the Jacobian matrix can be calculated by applying the chain rule property on (3). For instance, it is not difficult to verify from that:

$$\frac{\partial y_1}{\partial a_{12}} = -(y_2^k + a_{12} k y_2^{k-1} \frac{\partial y_2}{\partial a_{12}}). \tag{12}$$

Given that

$$\frac{\partial y_2}{\partial a_{12}} = -\frac{1}{k} a_{21} y_1^{\frac{1}{k}-1} \frac{\partial y_1}{\partial a_{12}}, \tag{13}$$

and substituting this expression in (12), one obtains:

$$\frac{\partial y_1}{\partial a_{12}} = \frac{-y_2^k}{1 - a_{12}a_{21}y_1^{\frac{1}{k}-1} y_2^{k-1}}. \tag{14}$$

By conducting similar calculations, one obtains the other derivatives:

$$\frac{\partial y_2}{\partial a_{12}} = \frac{a_{21}y_1^{\frac{1}{k}-1} y_2^k}{k(1 - a_{12}a_{21}y_1^{\frac{1}{k}-1} y_2^{k-1})} \tag{15}$$

$$\frac{\partial y_1}{\partial a_{21}} = \frac{k a_{12} y_1^{\frac{1}{k}} y_2^{k-1}}{1 - a_{12}a_{21}y_1^{\frac{1}{k}-1} y_2^{k-1}} \tag{16}$$

$$\frac{\partial y_2}{\partial a_{21}} = \frac{-y_1^{\frac{1}{k}}}{1 - a_{12}a_{21}y_1^{\frac{1}{k}-1} y_2^{k-1}} \tag{17}$$

As it would be expected, when $k = 1$, one obtains from the derived expressions the same conditions developed in [8] and [9] for the stability of the Hérault-Jutten algorithm for linear source separation.

4 Experimental Results

Aiming to assess the performance of the proposed solution, experiments were conducted for the cases $k = 2$ and $k = 3$. In both situations, the efficacy of the obtained solutions was quantified according to the following index:

$$SNR_i = 10 \log \left(\frac{E\{s_i^2\}}{E\{(s_i - y_i)^2\}} \right). \tag{18}$$

From this, a global index can be defined as $SNR = 0.5(SNR_1 + SNR_2)$.

k = 2. In a first scenario, we consider the separation of two sources uniformly distributed between $[0.1, 1.1]$. The mixing parameters are given by $a_{12} = 0.5$ and $a_{21} = 0.5$; a set of 3000 samples of the mixtures was considered and the number of iterations regarding the learning algorithm (9) was defined to 3500 with $\mu = 0.05$. The initial conditions of the dynamics (3) were chosen as $[y_1(1)\ y_2(1)]^T = [0\ 0]^T$. The results of this first case are expressed in the first row of Table 1. In Figure 2, the joint distributions of the mixtures and of the retrieved signals are depicted for a typical case (SNR = 35dB). Note that the outputs of the separating system are almost uniformly distributed, which indicates that the separation task was fulfilled. Also, we performed experiments by considering on each sensor an additive white Gaussian noise with a signal-to-noise ratio of $17dB$. The results for this second scenario are depicted in the second row of Table 1.

Table 1. Average SNR results over 100 experiments and standard deviation (STD)

	SNR_1	SNR_2	SNR	$STD(SNR)$
$k = 2$ (Scenario 1)	37.18	33.05	35.12	8.90
$k = 2$ (Scenario 2)	17.98	15.35	16.67	1.72
$k = 2$ (Scenario 3)	36.49	31.84	34.17	4.92
$k = 3$ (Scenario 1)	22.46	21.04	21.75	5.02

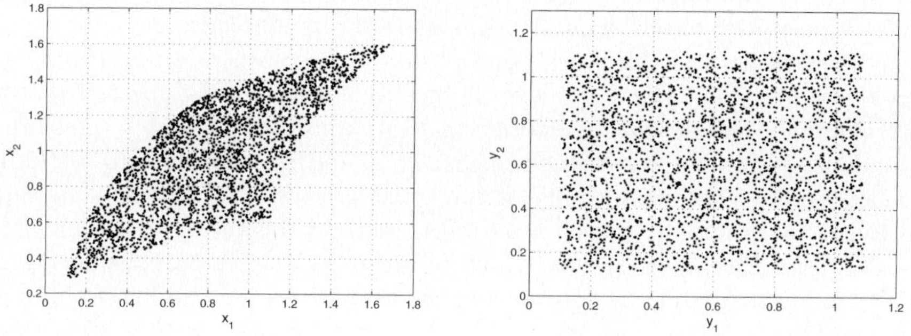

(a) Mixed signals (b) Retrieved signals

Fig. 2. First scenario - $k = 2$

A third scenario was composed by a uniformly distributed source between $[0.3, 1.3]$ and a sinusoidal source varying in the range $[0.2, 1.2]$. In this case, the mixing parameters are given by $a_{12} = 0.6$ and $a_{21} = 0.6$ and the constants related to the separating system were adjusted as in the first experiment. Again, the separation method was able to separate the original sources, as can be seen in the third row of Table 1.

k = 3. The problem becomes more tricky when $k = 3$. Firstly, we observed through simulations that, even for a separating point $[w_{12}\ w_{21}]^T = [a_{12}\ a_{21}]^T$ that satisfies the equilibrium condition (8), the structure (3) does not guarantee source separation, since there can be another stable equilibrium solution that has no relation with the sources. In this particular case, we observed, after performing some simulations, that the adopted network may be attracted by a stable limit cycle and, also, that it is possible to overcome this problem by changing the initial conditions of (3) when a periodic equilibrium solution occurs.

A second problem in this case is related to the convergence of the learning rule. Some simulations suggested the existence of spurious minima in this case. These two problems result in a performance degradation of the method when compared to the case $k = 2$, as can be seen in the last row of Table 1. In this case, we considered a scenario with two sources uniformly distributed between $[0.3, 1.3]$ and mixing parameters given by $a_{12} = 0.5$ and $a_{21} = 0.5$. The initial conditions of (3) were defined as $[0.5\ 0.5]^T$. Also, we considered 3000 samples of the mixtures and 10000 iterations of the learning algorithm with $\mu = 0.01$.

5 Conclusions

The aim of this work was to design a source separation strategy for a class of nonlinear systems that is related to a chemical sensing application. Our approach was based on the ideas presented in [1,2] in such a way that it may be viewed as an extension of these works to the particular model considered herein. Concerning the proposed technique, we investigated the stability of the separation structure as well as the stability of the learning algorithm. This study permitted us to obtain necessary conditions for a proper operation of the separation method. Finally, the viability of our approach was attested by simulations.

A first perspective of this work concerns its application in a real problem of chemical sensing. Also, there are several questions that deserve a detailed study as, for example, the design of algorithms that minimizes a better measure of independence between the retrieved sources (e.g. mutual information), including an investigation of the separability of the considered model. Another envisaged extension is to provide a source separation method for the most general case of the Nikolsky-Eisenman model, which is given by (1): 1) by considering the logarithmic terms; and 2) by considering the cases $k \in \mathbb{Q}$. Actually, preliminary simulations show that our proposal works for simple cases in $k \in \mathbb{Q}$, such as $k = 1/3$ and $k = 2/3$. However, there are tricky points in the theoretical analysis conducted in this paper that are not appropriate to this new situation.

References

1. Jutten, C., Hérault, J.: Blind separation of sources, part I: An adaptive algorithm based on neuromimetic architecture. Signal Processing 24, 1–10 (1991)
2. Hosseini, S., Deville, Y.: Blind separation of linear-quadratic mixtures of real sources using a recurrent structure. In: Mira, J.M., Álvarez, J.R. (eds.) IWANN 2003. LNCS, vol. 2687, pp. 241–248. Springer, Heidelberg (2003)
3. Comon, P.: Independent component analysis: a new concept? Signal Processing 36, 287–314 (1994)
4. Taleb, A., Jutten, C.: Source separation in postnonlinear mixtures. IEEE Trans. Signal Processing 47, 2807–2820 (1999)
5. Bermejo, S., Jutten, C., Cabestany, J.: ISFET source separation: Foundations and techniques. Sensors and Actuators B Chemical 113, 222–233 (2006)
6. Bedoya, G., Jutten, C., Bermejo, S., Cabestany, J.: Improving semiconductor-based chemical sensor arrays using advanced algorithms for blind source separation. In: Proc. of the ISA/IEEE Sensors for Industry Conference (SiCon04), pp. 149–154 (2004)
7. Hilborn, R.C.: Chaos and Nonlinear Dynamics: An Introduction for Scientists and Engineers, 2nd edn. Oxford University Press, New York (2000)
8. Sorouchyari, E.: Blind separation of sources, part III: Stability Analysis. Signal Processing 24, 21–29 (1991)
9. Fort, J.C.: Stabilité de l'algorithme de séparation de sources de Jutten et Hérault (in French). Traitement du Signal 8, 35–42 (1991)

Independent Subspace Analysis Is Unique, Given Irreducibility

Harold W. Gutch and Fabian J. Theis

Max Planck Institute for Dynamics and Self-Organization, 37073 Göttingen, Germany
harold.gutch@ds.mpg.de, fabian@theis.name

Abstract. Independent Subspace Analysis (ISA) is a generalization of ICA. It tries to find a basis in which a given random vector can be decomposed into groups of mutually independent random vectors. Since the first introduction of ISA, various algorithms to solve this problem have been introduced, however a general proof of the uniqueness of ISA decompositions remained an open question. In this contribution we address this question and sketch a proof for the separability of ISA. The key condition for separability is to require the subspaces to be not further decomposable (irreducible). Based on a decomposition into irreducible components, we formulate a general model for ISA without restrictions on the group sizes. The validity of the uniqueness result is illustrated on a toy example. Moreover, an extension of ISA to subspace extraction is introduced and its indeterminacies are discussed.

With the increasing popularity of Independent Component Analysis, people started to get interested in extensions. Cardoso [2] was the first to formulate an extension denoted here as Independent Subspace Analysis. The general idea is that for a given observation $\mathbf{X}$ we try to find an invertible matrix $\mathbf{W}$ such that $\mathbf{WX} = (\mathbf{S}_1^T, \ldots, \mathbf{S}_k^T)^T$ with mutually independent random vectors $\mathbf{S}_i$. If all $\mathbf{S}_i$ are one-dimensional, this is ICA, and we have the well-known separability results of ICA [3]. However without dimensionality restrictions, if mutual independence of the vectors $\mathbf{S}_i$ is the only restriction imposed on $\mathbf{W}$, ISA cannot produce meaningful results: if $\mathbf{W}$ simply is the identity and $k = 1$, then $\mathbf{S}_1 = \mathbf{X}$, which is independent of the (non-existing) rest. So, further restrictions are required for a meaningful model. A common approach is to fix the group size in advance, see [5] for a short review of ISA models. Here, we propose a more general concept based on [5], namely irreducibility of the recovered sources $\mathbf{S}_i$ that is the requirement that any $\mathbf{S}_i$ cannot be further decomposed. Our main contribution is a sound proof for the separability of this model together with a confirming simulation, thereby giving the details for the proposed ISA model from [5].

The manuscript is organized as follows. In the next section, we motivate the existence of such a separability result by studying a toy example. Then we give the sketch of the proof, and finally extend it to blind subspace extraction.

M.E. Davies et al. (Eds.): ICA 2007, LNCS 4666, pp. 49–56, 2007.

1 Motivation

Usually ISA is seen as a byproduct of ICA algorithms, which are assumed to decompose signals into components 'as independent as possible'; the components are then simply sorted to give a decomposition into higher-dimensional subspaces. However this approach is not as straight-forward as it might seem, as, strictly speaking, if ICA is performed on a data set that cannot be completely decomposed into one-dimensional independent components, we are applying ICA to a data set that does not follow the ICA model and have no theoretical results predicting the behavior of ICA algorithms. Here we present some simulations, which give a hint that indeed ISA might not be so unproblematic.

We generated a toy data set consisting of two independent sources, each of which were not further decomposable. The first data set consisted of a wireframe model of a 3-dimensional cube, the second data set was created from a solid 2-dimensional circle, see figure 1. We uniformly picked $N = 10.000$ samples and mixed them in batch runs by applying $M = 200.000$ uniformly sampled orthogonal matrices. The 200.000 matrices were sampled, by choosing random matrices $\mathbf{B}$ with entries normally sampled with mean 0 and variance 1, which then were symmetrically orthogonalized by $\mathbf{A} = (\mathbf{B}\mathbf{B}^T)^{-0.5}\mathbf{B}$. A mixture with an orthogonal matrix deviating from the block-structure should also deviate from independence within the blocks. As an ad-hoc measure for dependence within the blocks, we used the forth-order cumulant tensor:

$$\delta_{\mathrm{D}}(\mathbf{X}) = \sum_{i=1}^{3}\sum_{j=4}^{5}\sum_{k=1}^{5}\sum_{l=1}^{5} \operatorname{cum}^2(X_i, X_j, X_k, X_l) \ .$$

This is motivated by the well-known and in ICA often used fact that the crosscumulant tensor is zero, i.e. $\operatorname{cum}^2(\mathbf{Y}_1, \mathbf{Y}_2, \mathbf{Y}_*, \mathbf{Y}_*) = 0$, if $\mathbf{Y}_1$ and $\mathbf{Y}_2$ are independent. We measured the deviation of our mixing matrices from block-structure by simply taking the Frobenius-norm of the off-block-diagonal blocks:

$$\operatorname{off}(\mathbf{A}) := \sum_{i=1}^{3}\sum_{j=4}^{5}(a_{ij}^2 + a_{ji}^2) \ .$$

If ISA actually guarantees a unique block-structure in fourth order, we should get a dependence of 0 only if the mixing matrix itself is block-diagonal that is if $\operatorname{off}\mathbf{A} = 0$. However, due to sampling errors, this is of course never reached, so we estimate the minima of δ_{D}. Figure 2 shows the relation of $\operatorname{off}\mathbf{A}$ and $\delta_{\mathrm{D}}(\mathbf{AS})$, and here we observe not only the expected minimum at $\operatorname{off}\mathbf{A} = 0$, but two additional minima at $\operatorname{off}\mathbf{A} = 2$ and $\operatorname{off}\mathbf{A} = 4$. In order to take a closer look at these three points, we chose three matrices $\mathbf{A}_0$, $\mathbf{A}_2$ and $\mathbf{A}_4$, corresponding to the three local minima of the plot in Fig. 2. Starting with these matrices, we performed in their neighborhood a search for matrices with a lower model deviation. Again we sampled random orthogonal matrices, but this time biased them to be close to the identity matrix, as we wanted to search locally. We therefore again orthogonalized matrices as above, however chose the matrices $\mathbf{B}$

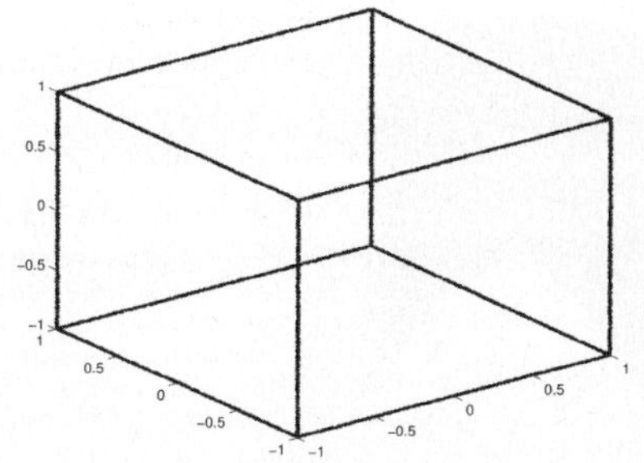

(a) 3-dimensional sources $\mathbf{S}_1$

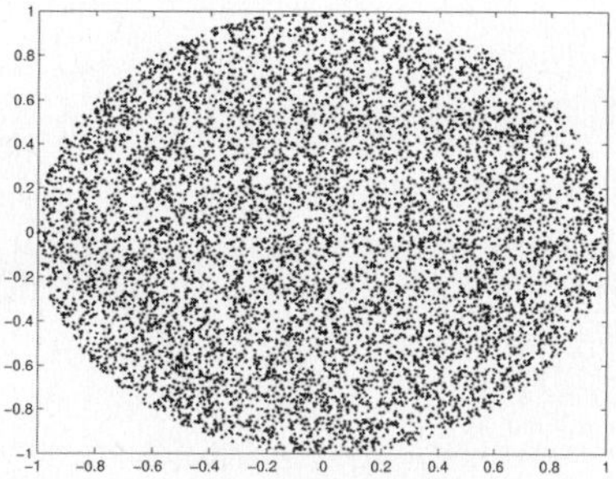

(b) 2-dimensional sources $\mathbf{S}_2$

Fig. 1. Toy data set

to be not arbitrarily normally sampled, but took matrices whose entries were normally sampled with mean 0 and variance v, which we then added to the identity matrix, modifying $\mathbf{A}_0$, $\mathbf{A}_2$ and $\mathbf{A}_4$ in every step only if it would perform a better block-independence. We evaluated this for $v = 0.1$, $v = 0.01$ and $v = 0.001$, each time running for 20.000 steps. The result of this is plotted in Fig. 3, and we indeed observe considerably better block-independence in the order of 1.5 magnitudes in the neighborhood of $\mathbf{A}_0$ than in the neighborhoods of $\mathbf{A}_2$ and $\mathbf{A}_4$. While the three local minima found by random sampling show only small difference ($\delta_{\mathrm{D}}(\mathbf{A}_0) = 0.0135$, $\delta_{\mathrm{D}}(\mathbf{A}_2) = 0.0107$, $\delta_{\mathrm{D}}(\mathbf{A}_4) = 0.0285$), local searches show up better minima for all three areas ($\delta_{\mathrm{D}}(\mathbf{A}_0) = 0.0002$, $\delta_{\mathrm{D}}(\mathbf{A}_2) = 0.0055$, $\delta_{\mathrm{D}}(\mathbf{A}_4) = 0.0053$), especially the area around off $\mathbf{A} = 0$. As a side note, the final matrices $\mathbf{A}_2$ and $\mathbf{A}_4$ correspond to the product of a block-diagonal matrix and a permutation matrices where one, respectively two indices in each of the two off-diagonal blocks are non-zero.

This shows us that while we observe local minima of our block-dependency measure on our data set, a closer inspection reveals that these minima are of different quality and we actually have only a single global minimum. We conclude that separability of ISA indeed should hold.

2 Uniqueness of ISA

In this section we present the proof of uniqueness of ISA. After explaining the notion of *irreducibility* of a random vector, we show why this idea is essential for the separability of ISA.

2.1 The ICA Model

Let us quickly repeat a few facts about ICA. The linear, noiseless ICA model can be described by the equation $\mathbf{X} = \mathbf{AS}$, where $\mathbf{S} = (S_1, \ldots, S_n)^T$ denotes a random vector with mutually independent components S_i (sources) and an

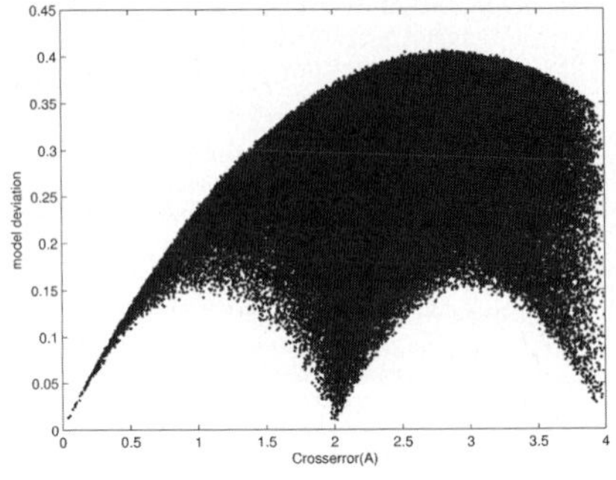

Fig. 2. Relation between block-crosserror and block-independence. Note the two additional minima at off $\mathbf{A} = 2$ and off $\mathbf{A} = 4$.

invertible mixing matrix $\mathbf{A}$. The task of ICA is the recovery of $\mathbf{S}$, given only the observations $\mathbf{X}$. This is obviously only possible up to the indeterminacies scaling and permutation, and it is well-known that recovery is possible up to exactly these permutations if $\mathbf{S}$ is square-integrable and contains at most one Gaussian component [3, 4].

2.2 The ISA Model

Loosening the requirement of mutual independence of the sources naturally brings up the idea of describing ISA through the same equation $\mathbf{X} = \mathbf{AS}$, where now $\mathbf{S} = (\mathbf{S}_1^T, \mathbf{S}_2^T, \ldots, \mathbf{S}_n^T)^T$ with mutually independent random vectors $\mathbf{S}_i$, however this time dependencies within the multidimensional $\mathbf{S}_i$ are allowed. Obvious indeterminacies of such a model are invertible linear transforms within the subspaces $\mathbf{S}_i$ (which can be seen as a generalization of scaling to higher dimensions) and permutations of subspaces of the same size (which, again, is the higher dimensional generalization of the regular permutation seen in ICA). However this model is not complete, since for any observation $\mathbf{X}$ a decomposition into mutually independent subspaces where dependencies within the subspaces are allowed is given simply by $\mathbf{X}$ itself. Realizing this naturally brings up the requirement of $\mathbf{S}$ to be 'as independent as possible'. This is formally described by the following definition.

Definition 1. *A random vector* $\mathbf{S}$ *is said to be* irreducible *if it contains no lower-dimensional independent component. An invertible matrix* $\mathbf{W}$ *is called a (general)* independent subspace analysis *of* $\mathbf{X}$ *if* $\mathbf{WX} = (\mathbf{S}_1^T, \ldots, \mathbf{S}_k^T)^T$ *with mutually independent, irreducible random vectors* $\mathbf{S}_i$. *Then* $(\mathbf{S}_1^T, \ldots, \mathbf{S}_k^T)$ *is called an irreducible decomposition of* $\mathbf{X}$.

Irreducibility is a key property in uniqueness of ISA and indeed, if we additionally assume irreducibility, we can show that this essentially allows for separability of ISA up to the above mentioned indeterminacies of higher dimensional scaling and permutation of subspaces of the same size.

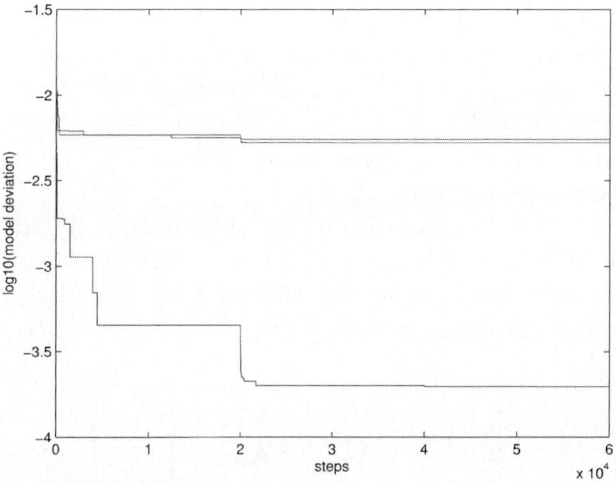

Fig. 3. Search for local minima around off $\mathbf{A} = 0$ (lower graph) and off $\mathbf{A} = 2$ respectively off $\mathbf{A} = 4$ (upper two graphs)

2.3 Uniqueness of ISA

We will now prove uniqueness of Independent Subspace Analysis under the additional assumption of no independent Gaussian components. Indeed, any orthogonal transformation of two decorrelated (and hence independent) Gaussians is again independent, so for such random vectors clearly such a strong identification result would not be possible.

Theorem 1. *Given a random vector* $\mathbf{X}$ *with existing covariance and no Gaussian independent component, then an ISA of* $\mathbf{X}$ *exists and is unique except for scaling and permutation.*

Existence holds trivially, but uniqueness is not obvious. Defining the equivalence relation $\sim$ on random vectors as $\mathbf{X} \sim \mathbf{Y} :\Leftrightarrow \mathbf{X} = \mathbf{A}\mathbf{Y}$ for some $\mathbf{A} \in Gl(n)$, we are easily able to show uniqueness given the following lemma:

Lemma 1. *Let* $\mathbf{S} = (\mathbf{S}_1^T, \ldots, \mathbf{S}_N^T)^T$ *be a square-integrable decomposition of* $\mathbf{S}$ *into irreducible, mutually independent components* $\mathbf{S}_i$ *where no* $\mathbf{S}_i$ *is a one-dimensional Gaussian. If* $(\mathbf{X}_1^T, \mathbf{X}_2^T)^T$ *is an independent decomposition of* $\mathbf{S}$*, then there is some permutation* π *of* $\{1, \ldots, N\}$ *such that* $\mathbf{X}_1 \sim (\mathbf{S}_{\pi(1)}^T, \ldots, \mathbf{S}_{\pi(l)}^T)^T$ *and* $\mathbf{X}_2 \sim (\mathbf{S}_{\pi(l+1)}^T, \ldots, \mathbf{S}_{\pi(N)}^T)^T$ *for some* l.

So, given an irreducible decomposition of a random variable $\mathbf{S}$ with no independent Gaussian components, *any* decomposition of it into independent (not necessarily irreducible) components 'splits along the irreducible components'.

Using this lemma, Theorem 1 is easy to show: Given two irreducible decompositions $(\mathbf{X}_1^T, \ldots, \mathbf{X}_N^T)^T$ and $(\mathbf{S}_1^T, \ldots, \mathbf{S}_M^T)^T$, we search for the smallest irreducible component appearing, which we may assume to be $\mathbf{X}_1$. We then group

$(\mathbf{X}_2^T, \dots, \mathbf{X}_N^T)^T$ into a (larger) random vector. As this *independent* decomposition splits along the irreducible components $\mathbf{S}_i$ and for all j, $\dim(\mathbf{X}_1) \leq \dim(\mathbf{S}_j)$, $\mathbf{X}_1$ is identical to one of the $\mathbf{S}_j$. We may remove both of these and go on iteratively, thus proving the theorem.

The more complicated part is the proof of Lemma 1, and due to space restrictions we can only sketch the proof.

Before starting, we note that due to the assumption of existing covariance, we may whiten both $\mathbf{X}$ and $\mathbf{S}$, in which case it is easy to observe that $\mathbf{A}$ is orthogonal. For notational reasons, we will split up the mixing matrix $\mathbf{A}$ into submatrices, the sizes of which are according to the sizes of $\mathbf{S}_i$ and $\mathbf{X}_j$:

$$\begin{pmatrix} \mathbf{X}_1 \\ \mathbf{X}_2 \end{pmatrix} = \begin{pmatrix} \mathbf{A}_{11} & \dots & \mathbf{A}_{1N} \\ \mathbf{A}_{21} & \dots & \mathbf{A}_{2N} \end{pmatrix} \begin{pmatrix} \mathbf{S}_1 \\ \vdots \\ \mathbf{S}_N \end{pmatrix} \tag{1}$$

so $\mathbf{X}_i = \sum_{k=1}^{N} \mathbf{A}_{ik}\mathbf{S}_k$. We now claim that in every pair $\{\mathbf{A}_{1j}, \mathbf{A}_{2j}\}$ one of the two matrices is zero.

We fix $k = k_0$ and show this claim for k_0. Let us assume the converse, that is that both $\operatorname{rank}(\mathbf{A}_{1k_0}) \neq 0$ and $\operatorname{rank}(\mathbf{A}_{2k_0}) \neq 0$. As $\mathbf{A}$ has full rank, $\operatorname{rank}(\mathbf{A}_{1k_0}) + \operatorname{rank}(\mathbf{A}_{2k_0}) \geq \dim(\mathbf{S}_{k_0}) =: D$. This leaves us with two cases to handle, $\operatorname{rank}(\mathbf{A}_{1k_0}) + \operatorname{rank}(\mathbf{A}_{2k_0}) = D$ and $\operatorname{rank}(\mathbf{A}_{1k_0}) + \operatorname{rank}(\mathbf{A}_{2k_0}) > D$. Let us first address the first case and show that this contradicts the irreducibility of $\mathbf{S}_{k_0}$.

Lemma 2. *Assume*

$$\mathbf{S} = (\mathbf{A}_1 | \mathbf{A}_2) \begin{pmatrix} \mathbf{X}_1 \\ \mathbf{X}_2 \end{pmatrix}$$

with independent random vectors $\mathbf{X}_1$ *and* $\mathbf{X}_2$ *and* $\mathbf{A}_1$, $\mathbf{A}_2$ *such that* $\operatorname{rank}(\mathbf{A}_1^T) + \operatorname{rank}(\mathbf{A}_2^T) = \dim(\mathbf{S})$ *and* $\operatorname{rank}(\mathbf{A}_1 | \mathbf{A}_2) = \dim(\mathbf{S})$. *Then* $\mathbf{S}$ *is reducible.*

Proof. Let $D := \dim(\mathbf{S})$ and $d := \dim\left(\ker(\mathbf{A}_1^T)\right)$. Then $\dim\left(\ker(\mathbf{A}_2^T)\right) = D - d$, and we can find a linearly independent set $\{\mathbf{v}_1, \dots, \mathbf{v}_d\}$ such that $\mathbf{v}_i^T\mathbf{A}_1 = 0$ for any $1 \leq i \leq d$, and similarly a linearly independent set $\{\mathbf{v}_{d+1}, \dots, \mathbf{v}_D\}$ such that $\mathbf{v}_j^T\mathbf{A}_2 = 0$ for any $d+1 \leq j \leq D$. These two sets are guaranteed to be disjoint, as $\operatorname{rank}(\mathbf{A}_1|\mathbf{A}_2) = \dim(\mathbf{S})$. Using these vectors, we define

$$\mathbf{T} := \begin{pmatrix} \mathbf{v}_1^T \\ \vdots \\ \mathbf{v}_D^T \end{pmatrix} .$$

Then

$$\mathbf{TS} = (\mathbf{TA}_1 | \mathbf{TA}_2) \begin{pmatrix} \mathbf{X}_1 \\ \mathbf{X}_2 \end{pmatrix} = \left(\begin{array}{c|c} \mathbf{T}_1 & 0 \\ \hline 0 & \mathbf{T}_2 \end{array}\right) \begin{pmatrix} \mathbf{X}_1 \\ \mathbf{X}_2 \end{pmatrix} = \begin{pmatrix} \mathbf{T}_1\mathbf{X}_1 \\ \mathbf{T}_2\mathbf{X}_2 \end{pmatrix}$$

with some full rank matrices $\mathbf{T}_1$ and $\mathbf{T}_2$. It follows that $\mathbf{S}$ is reducible, as $\mathbf{X}_1$ and $\mathbf{X}_2$ are independent and $\mathbf{T}$ is invertible. □

The other case, $\text{rank}(\mathbf{A}_{1k_0}) + \text{rank}(\mathbf{A}_{2k_0}) > D$ is harder to prove and follows some of the ideas presented in [4].

Lemma 3. *Given (1), if there is some* $1 \leq k_0 \leq N$ *such that* $\text{rank}(\mathbf{A}_{1k_0}) + \text{rank}(\mathbf{A}_{2k_0}) > \dim(\mathbf{S}_{k_0})$, *then* $\mathbf{S}_{k_0}$ *contains an irreducible Gaussian component.*

This concludes the proof of Theorem 1.

2.4 Dealing with Gaussians

The section above explicitly excluded independent Gaussian components in order to avoid additional indeterminacies. Recently, a general decomposition model dealing with Gaussians was proposed in the form of the so-called *non-Gaussian component analysis (NGCA)* [1]. It tries to detect a whole non-Gaussian subspace within the data, and no assumption of independence within the subspace is made. More precisely, given a random vector $\mathbf{X}$, a factorization $\mathbf{X} = \mathbf{AS}$ with an invertible matrix $\mathbf{A}$, $\mathbf{S} = (\mathbf{S}_N, \mathbf{S}_G)$ and $\mathbf{S}_N$ a square-integrable m-dimensional random vector is called an m-decomposition of $\mathbf{X}$ if $\mathbf{S}_N$ and $\mathbf{S}_G$ are stochastically independent and $\mathbf{S}_G$ is Gaussian. In this case, $\mathbf{X}$ is said to be *m-decomposable* and $\mathbf{X}$ is denoted to be *minimally n-decomposable* if $\mathbf{X}$ is not $(n-1)$-decomposable. According to our previous notation, $\mathbf{S}_N$ and $\mathbf{S}_G$ are independent components of $\mathbf{X}$. It has been shown that the subspaces of such decompositions are unique [6]:

Theorem 2. *The mixing matrix* $\mathbf{A}$ *of a minimal decomposition is unique except for transformations in each of the two subspaces.*

Moreover, explicit algorithms can be constructed for identifying the subspaces [6]. This result enables us to generalize Theorem 1 and to get a general decomposition theorem, which characterizes solutions of ISA.

Theorem 3. *Given a random vector* $\mathbf{X}$ *with existing covariance, an ISA of* $\mathbf{X}$ *exists and is unique except for permutation of components of the same dimension and invertible transformations within each independent component and within the Gaussian part.*

Proof. Existence is obvious. Uniqueness follows after first applying Theorem 2 to $\mathbf{X}$ and then Theorem 1 to the non-Gaussian part. □

3 Independent Subspace Extraction

Having shown uniqueness of the decomposition, we are able to introduce Independent (Irreducible) Subspace Extraction, which separates independent (irreducible) subspaces out of the random vector.

Definition 2. *A pseudo-invertible* $(n \times m)$ *matrix* $\mathbf{W}$ *is said to be an Independent Subspace Extraction of an m-dimensional random vector* $\mathbf{X}$, *if* $\mathbf{WX}$ *is an independent component of* $\mathbf{X}$. *If* $\mathbf{WX}$ *even is irreducible, then* $\mathbf{W}$ *is called an Irreducible Subspace Extraction of* $\mathbf{X}$.

This could lead to a wider variety of algorithms like deflationary approaches which are already common in standard ICA. The interesting aspect here is that we only strive to extract a single component, so Independent (Irreducible) Subspace Extraction could prove to be simpler to handle algorithmically than a complete Independent Subspace Analysis, and thus play an important role in applications (such as dimension reduction) that need to extract only a single component or subspacespace.

4 Conclusion

Although Independent Subspace Analysis has become a common practice in the last few years, separability of it has not been fully shown. We presented examples that showed that ISA is not as unproblematic as it seems. Additionally we proved uniqueness – up to higher-dimensional generalizations of the indeterminacies of ICA – of ISA, given no independent Gaussians and showed how to combine this together with existing theoretical results on NGCA to a full ISA uniqueness result. Using these results, it is now possible to speak of *the* ISA of any given random vector. Moreover, theorem 3 now gives an complete characterization of decompositions of distributions into independent factors, which might prove to be a useful result in general statistics.

Now that uniqueness of ISA has been shown for the theoretical limit of perfect knowledge of the recordings, the next obvious step is the conversion to the real-world case, where only a finite number of samples of the observations are known. Here, a decomposition of the mixtures $\mathbf{X}$ such that $\mathbf{X} = \mathbf{AS}$ where $\mathbf{S} = (\mathbf{S}_1^T, \ldots, \mathbf{S}_N^T)^T$ with irreducible (or merely independent) $\mathbf{S}_i$ cannot be expected, as in this case we expect to always see some dependency due to sampling errors. Due to uniqueness of ISA in the asymptotic case, identification of the underlying sources should hold here too, given enough samples, but additional work is required to show this in the future.

References

1. Blanchard, G., Kawanabe, M., Sugiyama, M., Spokoiny, V., Müller, K.R.: In search of non-gaussian components of a high-dimensional distribution. Journal of Machine Learning Research 7, 247–282 (2006)
2. Cardoso, J.F.: Multidimensional independent component analysis. In: Proc. of ICASSP '98, Seattle (1998)
3. Comon, P.: Independent component analysis - a new concept? Signal Processing 36, 287–314 (1994)
4. Theis, F.J.: A new concept for separability problems in blind source separation. Neural Computation 16, 1827–1850 (2004)
5. Theis, F.J.: Towards a general independent subspace analysis. In: Proc. NIPS 2006 (2007)
6. Theis, F.J., Kawanabe, M.: Uniqueness of non-gaussian subspace analysis. In: Rosca, J., Erdogmus, D., Príncipe, J.C., Haykin, S. (eds.) ICA 2006. LNCS, vol. 3889, pp. 917–925. Springer, Heidelberg (2006)

Optimization on the Orthogonal Group for Independent Component Analysis

Michel Journée[1], Pierre-Antoine Absil[2], and Rodolphe Sepulchre[1]

[1] Dept. of Electrical Engineering and Computer Science, University of Liège, Belgium
[2] Dept. of Mathematical Engineering, Université catholique de Louvain, Belgium

Abstract. This paper derives a new algorithm that performs independent component analysis (ICA) by optimizing the contrast function of the RADICAL algorithm. The core idea of the proposed optimization method is to combine the global search of a good initial condition with a gradient-descent algorithm. This new ICA algorithm performs faster than the RADICAL algorithm (based on Jacobi rotations) while still preserving, and even enhancing, the strong robustness properties that result from its contrast.

Keywords: Independent Component Analysis, RADICAL algorithm, optimization on matrix manifolds, line-search on the orthogonal group.

1 Introduction

Independent Component Analysis (ICA) was originally developed for the blind source separation problem. It aims at recovering independent source signals from linear mixtures of these. As in the seminal paper of Comon [1], a linear instantaneous mixture model will be considered in this paper,

$$X = AS, \tag{1}$$

where X, A and S are matrices in $\mathbb{R}^{n\times N}$, $\mathbb{R}^{n\times p}$ and $\mathbb{R}^{p\times N}$ respectively, with p less or equal to n. The rows of S are assumed to be samples of independent random variables. Thus, ICA provides a linear representation of the data X in terms of components S that are statistically independent.

ICA algorithms are based on the inverse of the mixing model (1),

$$Z = W^T X,$$

where Z and W are matrices in $\mathbb{R}^{p\times N}$ and $\mathbb{R}^{n\times p}$, respectively. The aim of ICA algorithms is to optimize over W the statistical independence of the p random variables, whose samples are given in the p rows of Z. The statistical independence is measured by a cost function

$$\gamma : \mathbb{R}^{n\times p} \rightarrow \mathbb{R} : W \mapsto \gamma(W),$$

termed the *contrast function*.

M.E. Davies et al. (Eds.): ICA 2007, LNCS 4666, pp. 57–64, 2007.

In the remainder of this paper, we assume that the data matrix X has been preprocessed by means of prewhitening and its dimensions have been reduced by retaining the dominant p-dimensional subspace. Consequently, the contrast function γ is defined on a set of *square* matrices, i.e,

$$\gamma : \mathbb{R}^{p \times p} \to \mathbb{R} : W \mapsto \gamma(W).$$

Several contrast functions for ICA can be found in the literature. In this paper, we consider the RADICAL contrast function proposed in [2]. Advantages of this contrast are a strong robustness to outliers as well as to the lack of samples.

A good contrast for γ is not enough to make an efficient ICA algorithm. The other ingredient is a suitable numerical method to compute an optimizer of γ. This is the topic of the present paper. The authors of [2] optimize their contrast by means of Jacobi rotations combined with an exhaustive search. This yields the complete *Robust Accurate Direct ICA aLgorithm* (RADICAL). We propose a new steepest-descent-based optimization method that reduces the computational load of RADICAL.

The paper is organized as follows. The contrast function of RADICAL is detailed in Section 2. Section 3 describes a gradient-descent optimization algorithm. In Section 4, this local optimization is integrated within a global optimization framework. The performance of this new ICA algorithm is briefly illustrated in Section 5.

2 A Robust Contrast Function

Like many other measures of statistical independence, the contrast of RADICAL [2] is derived from the mutual information [3]. The mutual information $I(Z)$ of a multivariate random variable $Z = (z_1 \ldots, z_p)$ is defined as the Kullback-Leibler divergence between the joint distribution and the product of the marginal distributions,

$$I(Z) = \int p(z_1, \ldots, z_p) \log \frac{p(z_1, \ldots, z_p)}{p(z_1) \ldots p(z_p)} dz_1 \ldots dz_p. \tag{2}$$

This quantity presents all the required properties for a contrast function: it is nonnegative and equals zero if and only if the variables Z are statistically independent. Hence, its global minimum corresponds to the solution of the ICA problem.

The challenge is to get a good estimator of $I(Z)$. A possible approach is to express the mutual information in terms of the differential entropy of a univariate random variable z,

$$S(z) = \int p(z) \log(p(z)) dz, \tag{3}$$

for which efficient statistical estimators are available.

According to definitions (2) and (3), the following holds,

$$I(Z) = \sum_{i=1}^{p} S(z_i) - S(z_1, \ldots, z_p). \tag{4}$$

The introduction of the demixing model $Z = W^T X$ within (4) results in

$$\gamma(W) = \sum_{i=1}^{p} S^{(i)}(W) - \log(|W|) - S(x_1, \ldots, x_p), \tag{5}$$

where $S^{(i)}(W) = S(e_i^T W^T X)$ and e_i is the ith basis vector. The last term of (5) is constant and its evaluation can be skipped by the ICA algorithm. An estimator for the differential entropy of univariate variables was derived in [2] by considering order statistics. Given a univariate random variable z defined by its samples, the order statistics of z is the set of samples $\{z^1, \ldots, z^N\}$ rearranged in non-decreasing order, i.e., $z^1 \leq \ldots \leq z^N$. The differential entropy of z can be estimated by the simple formula

$$S(z) = \frac{1}{N-m} \sum_{j=1}^{N-m} \log\left(\frac{N+1}{m}(z^{(j+m)} - z^{(j)})\right), \tag{6}$$

where m is typically set to $\sqrt{N}$. Function (5) with the differential entropies being estimated by (6) is the contrast of the RADICAL algorithm [2].

This contrast presents several assets in terms of robustness. Its robustness to outliers was underlined in the original paper [2]. Robustness to outliers means that the presence of some corrupted entries in the observations data set X has little influence on the position of the global minimizer of that contrast. This is a key feature in many applications, especially for the analysis of gene expression data [4], where each entry in the observation matrix results from individual experiments that are likely to sometimes fail. The RADICAL contrast brings also advances in terms of robustness to the lack of samples. This will be illustrated in Section 5.

3 A Line-Search Optimization Algorithm

In accordance with the fact that the independence between random variables is not altered by scaling, the contrast function (5) presents the scale invariance property

$$\gamma(W) = \gamma(W\Lambda),$$

for all invertible diagonal matrices Λ. Optimizing a function with such an invariance property is a degenerate problem, which entails difficulties of theoretical (convergence analysis) and practical nature unless some constraints are introduced. In the case of prewhitening-based ICA, it is common practice to restrict the matrix W to be orthonormal [1], i.e., $W^T W = I$. Classical constrained optimization methods could be used. We favor the alternative to incorporate the constraints directly into the search space and to perform unconstrained optimization over the orthogonal group, i.e.,

$$\min_{W \in \mathcal{O}_p} \gamma(W) \quad \text{with } \mathcal{O}_p = \{W \in \mathbb{R}^{p \times p} | W^T W = I\}. \tag{7}$$

Most classical unconstrained optimization methods — such as gradient-descent, Newton, trust-region and conjugate gradient methods — have been generalized to the optimization over matrix manifolds (see [5] and references therein).

The remainder of this section deals with the derivation of a line-search optimization method on the orthogonal group for the RADICAL contrast function (5). Line-search on a nonlinear manifold is based on the update formula

$$W_+ = R_W(t\eta), \tag{8}$$

which consists in moving from the current iterate $W \in \mathcal{O}_p$ in the search direction η with a certain step size t to identify the next iterate $W_+ \in \mathcal{O}_p$. t is a scalar and η belongs to $T_W\mathcal{O}_p = \{W\Omega | \Omega \in \mathbb{R}^{p \times p}, \Omega^T = -\Omega\}$, the tangent space to $\mathcal{O}_p$ at W. The retraction R_W is a mapping from the tangent space to the manifold. More details about this notion can be found in [5]. Our algorithm selects the Armijo point t^A as step size and the opposite of the gradient of the cost function γ at the current iterate as search direction.

The Armijo step size is defined by $t^A = \beta^m \alpha$, with the scalars $\alpha > 0$, $\beta \in (0,1)$ and m being the first nonnegative integer such that

$$\gamma(W) - \gamma(R_W(\beta^m \alpha)) \geq -\sigma \langle \mathrm{grad}\gamma(W), \beta^m \alpha \eta \rangle_W,$$

where W is the current iterate on $\mathcal{O}_p$ and $\sigma \in (0,1)$. This step size ensures a sufficient decrease of the cost function at each iteration. The resulting line-search algorithm converges to the set of points where the gradient of γ vanishes [5].

An analytical expression of the gradient of the RADICAL contrast (5) has been derived in [6]. Let us just sketch the main points of this computation. First, because of the orthonormality condition, the second term of (5) vanishes. Furthermore, since the last term is constant, we have

$$\mathrm{grad}\gamma(W) = \sum_{i=1}^{p} \mathrm{grad}S^{(i)}(W).$$

The gradient of $S^{(i)}$ is given by

$$\mathrm{grad}S^{(i)}(W) = P_{T_W}\left(\mathrm{grad}\tilde{S}^{(i)}(W)\right),$$

where $\tilde{S}^{(i)}$ is the extension of $S^{(i)}$ over $\mathbb{R}^{p \times p}$, i.e., $\tilde{S}^{(i)} = S^{(i)}|_{\mathcal{O}_p}$, and $P_{T_W}(Z)$ is the projection operator, namely, in case of the orthogonal group, $P_{T_W}(Z) = \frac{1}{2}W(W^T Z - Z^T W)$. The evaluation of the gradient in the embedding manifold is performed by means of the identity

$$\mathrm{D}\tilde{S}^{(i)}(W)[Z] = \langle \mathrm{grad}\tilde{S}^{(i)}(W), Z \rangle,$$

with the metric $\langle Z_1, Z_2 \rangle = \mathrm{tr}(Z_1^T Z_2)$ and where

$$\mathrm{D}\tilde{S}^{(i)}(W)[Z] = \lim_{t \to 0} \frac{\tilde{S}^{(i)}(W + tZ) - \tilde{S}^{(i)}(W)}{t}$$

is the standard directional derivative of $\tilde{S}^{(i)}$ at W in the direction Z. Since one wants to compute the gradient on the orthogonal group, the direction Z can be restricted to the tangent plane at the current iterate, i.e., $Z \in T_W \mathcal{O}_p$.

As we have shown in [6], the gradient of the differential entropy estimator on the orthogonal group $\mathcal{O}_p$ is finally given by

$$\mathrm{grad}S^{(i)}(W) = P_{T_W} \left(\frac{1}{N-m} \sum_{j=1}^{N-m} \frac{(x^{(k_{j+m})} - x^{(k_j)}) e_i^T}{e_i^T W (x^{(k_{j+m})} - x^{(k_j)})} \right),$$

where $x^{(k)}$ denotes the kth column of the data matrix X. The indices k_{j+m} and k_j point to the samples of the estimated source z_i, which are respectively at positions $j+m$ and j in the order statistics of z_i. The computational cost for the gradient is of the same order as for the contrast, namely $\mathcal{O}(pN \log N)$.

More details about the Armijo point, the computation of gradients and, more generally, about line-search algorithms on manifolds can be found in [5].

4 Towards a Global Optimization Scheme

The algorithm described in the previous section inherits all the local convergence properties of line-search optimization methods [5]. Nevertheless, the contrast of RADICAL presents many spurious local minima that do not properly separate the observations X into independent sources. The line-search algorithm may thus fail in the context of ICA. Nevertheless, it leads to an efficient ICA algorithm when it is initialized within the basin of attraction of the global minimizer W_*. It is therefore essential to find good initial candidates for the line-search algorithm. The procedure proposed in this paper rests on empirical observations about the shape of the contrast function $\gamma(W)$. Figure 1 represents the evolution of this function as well as of the norm of its gradient along geodesic curves on the orthogonal group $\mathcal{O}_p$ for a particular benchmark setup (p=6, N=1000).

Figure 1 and extensive simulations not included in the present paper incite us to view the contrast function of RADICAL as possessing a very deep global minimum surrounded by many small local minima. Furthermore, the norm of the gradient tends to be much larger within the basin of attraction of the global minimizer. The norm of the gradient thus provides a criterion to discriminate between points that are inside this basin of attraction and those that are outside.

Our algorithm precedes the gradient optimization with the global search of a point where the gradient has a large magnitude. The search is performed along particular geodesics of the orthogonal group, exploiting the low numerical cost of Jacobi rotations. All geodesics on the orthogonal group $\mathcal{O}_p$ have the form $\Gamma(t) = We^{tB}$, where $W \in \mathcal{O}_p$ and B is a skew-symmetric matrix of the same

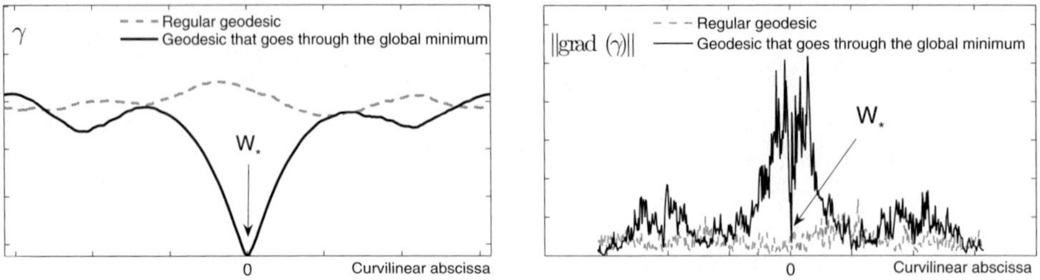

Fig. 1. Evolution of the contrast and the norm of its gradient along geodesics of $\mathcal{O}_p$

size as W. Jacobi rotations correspond to B having only zero elements except one element in the upper triangle and its symmetric counterpart, i.e., $B(i,j) = 1$ and $B(j,i) = -1$ with $i < j$. The contrast function γ evaluated along such geodesics has a periodicity of $\frac{\pi}{2}$, i.e.,

$$\gamma(We^{tB}) = \gamma(We^{(t+\frac{\pi}{2})B})$$

Such a geodesic is in fact a Jacobi rotation on the two-dimensional subspace spanned by the directions i and j. This periodicity is an interesting feature for an exhaustive search over the curvilinear abscissa t since it allows to define upper and lower bounds for t.

Our algorithm evaluates the gradient at a fixed number of points that are uniformly distributed on randomly selected geodesics of periodicity $\frac{\pi}{2}$. This process is pursued until a point with sufficient steepness is found. The steepness is simply evaluated by the Frobenius norm of the gradient of γ. Such a point is expected to belong to the basin of attraction of the global minimum and serves as initialization for the line-search algorithm of the previous section.

5 Some Benchmark Simulations

This section evaluates the performance of the new algorithm against the performance of the RADICAL algorithm. All results are obtained on benchmark setups that artificially generate observations X by linear transformation of known statistically independent sources S.

Figure 2 illustrates that the new algorithm reaches the global minimum of the contrast with less than half the computational effort required by the RADICAL algorithm. These results are based on a benchmark with $N = 1000$ samples while the dimension p of the problem varies from 2 to 8. For each p, five different data matrices X are obtained by randomly mixing p sources chosen as sinusoids of random frequencies and random phases. The indicated computational time is an average over these five ICA runs.

Figure 3 highlights the robustness properties of the contrast discussed in Section 2. The left graph results from a benchmark with $p = 6$ sources and

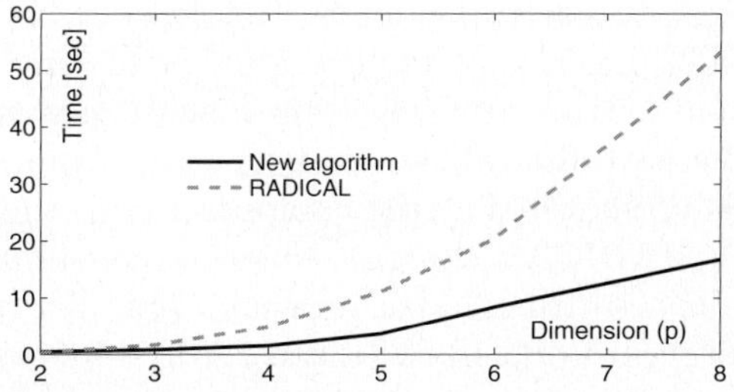

Fig. 2. Reduced computational time of the new ICA algorithm

$N = 1000$ samples. A given percent of the entries of the data set have been artificially corrupted to simulate outliers. The right graph considers a benchmark with $p = 6$ sources, no outliers and a varying number of samples. The quality of the ICA separation is measured by an index α[1], which stands for a good performance once it is close to zero. The left graph indicates that both the new algorithm and the RADICAL algorithm are robust to these outliers while classical ICA algorithms such as JADE [7] or FastICA [8] collapse immediately. It should be noted that the new algorithm supports up to 3% of outliers on the present benchmark and is thus more robust than RADICAL. Similarly, the right graph of Figure 3 suggests that the new algorithm is more robust to the lack of samples than RADICAL.

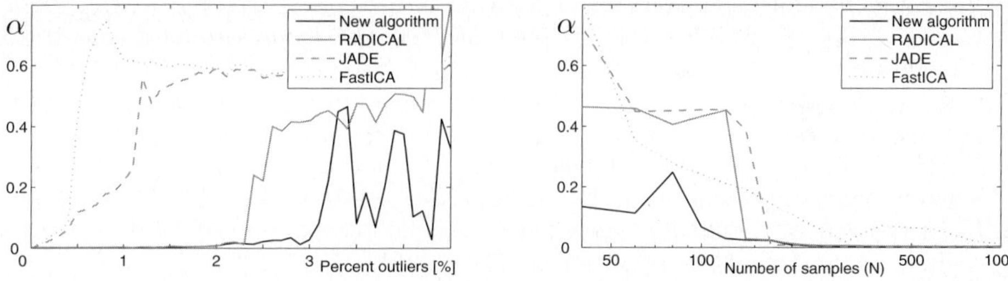

Fig. 3. Robustness properties of the new ICA algorithm

6 Conclusions

The RADICAL algorithm [2] presents very desirable robustness properties: robustness to outliers and robustness to the lack of samples. These are essential

[1] Given the demixing matrix W^* and the matrix W identified by the ICA algorithm,

$$\alpha(W, W^*) = \min_{\Lambda, P} \frac{\|W \Lambda P - W^*\|_F}{\|W^*\|_F},$$

where Λ is a non-singular diagonal matrix and P a permutation matrix.

for some applications, in particular for the analysis of biological data that are usually of poor quality because of the few number of samples available and the presence of corrupted entries resulting from failed experiments [4]. The RADICAL algorithm inherits these robustness properties from its contrast function. In this paper, we have shown that the computation of the demixing matrix by optimization of the RADICAL contrast function can be performed in a more efficient manner than with the Jacobi rotation approach considered in [2]. Our new optimization process works in two stages. It first identifies a point that supposedly belongs to the basin of attraction of the global minimum and performs afterwards the local optimization of the contrast by gradient-descent from this point. This new ICA algorithm requires less computational effort and seems to enhance the robustness margins.

Acknowledgments

M. Journée is a research fellow of the Belgian National Fund for Scientific Research (FNRS). This paper presents research results of the Belgian Network DYSCO (Dynamical Systems, Control, and Optimization), funded by the Interuniversity Attraction Poles Programme, initiated by the Belgian State, Science Policy Office. The scientific responsibility rests with its authors.

References

1. Comon, P.: Independent Component Analysis, a new concept. In: Signal Processing, vol. 36(3), pp. 287–314. Elsevier, Amsterdam (1994) (Special issue on Higher-Order Statistics)
2. Learned-Miller, E.G., Fisher III, J.W.: ICA using spacings estimates of entropy. Journal of Machine Learning Research 4, 1271–1295 (2003)
3. Cover, T.M., Thomas, J.A.: Elements of Information Theory (Wiley Series in Telecommunications and Signal Processing). Wiley-Interscience, Chichester (2006)
4. Liebermeister, W.: Linear modes of gene expression determined by independent component analysis. Bioinformatics 18, 51–60 (2002)
5. Absil, P.-A., Mahony, R., Sepulchre, R.: Optimization Algorithms on Matrix Manifolds. Princeton University Press (to appear)
6. Journée, M., Teschendorff, A.E., Absil, P.-A., Sepulchre, R.: Geometric optim. methods for ICA applied on gene expression data. In: Proc. of ICASSP (2007)
7. Cardoso, J.-F.: High-order contrasts for independent component analysis. Neural Computation 11(1), 157–192 (1999)
8. Hyvärinen, A., Karhunen, J., Oja, E.: Independent Component Analysis. John Wiley & Sons, Chichester (2001)

Using State Space Differential Geometry for Nonlinear Blind Source Separation

David N. Levin

University of Chicago
d-levin@uchicago.edu

Abstract. Given a time series of multicomponent measurements of an evolving stimulus, nonlinear blind source separation (BSS) usually seeks to find a "source" time series, comprised of statistically independent combinations of the measured components. In this paper, we seek a source time series that has a *phase-space* density function equal to the product of density functions of individual components. In an earlier paper, it was shown that the phase space density function induces a Riemannian geometry on the system's state space, with the metric equal to the local velocity correlation matrix of the data. From this geometric perspective, the vanishing of the curvature tensor is a necessary condition for BSS. Therefore, if this data-derived quantity is non-vanishing, the observations are not separable. However, if the curvature tensor is zero, there is only one possible set of source variables (up to transformations that do not affect separability), and it is possible to compute these explicitly and determine if they do separate the phase space density function. A longer version of this paper describes a more general method that performs nonlinear multidimensional BSS or independent subspace separation.

1 Introduction

Consider a set of data consisting of $\tilde{x}(t)$, a time-dependent multiplet of n measurements ($\tilde{x}_k$ for $k = 1, 2, \ldots, n$). The usual objectives of nonlinear BSS are: 1) to determine if these observations are instantaneous mixtures of n statistically independent source components $x(t)$

$$\tilde{x}(t) = f[x(t)] \qquad (1)$$

where f is an unknown, possibly nonlinear, n-component mixing function, and, if so, 2) to compute the mixing function. In most approaches to this problem [1,2], the desired source components are required to be statistically independent in the sense that their state space density function $\rho(x)$ is the product of the density functions of the individual components. However, it is well known that this problem always has many solutions (see [3] and references therein). Specifically, any observed density function can be integrated in order to construct an entire family of functions f^{-1} that transform it into a separable (i.e., factorizable) form.

M.E. Davies et al. (Eds.): ICA 2007, LNCS 4666, pp. 65–72, 2007.

The observed trajectories of many classical physical systems [4] can be characterized by density functions in *phase space* (i.e., $(\tilde{x}, \dot{\tilde{x}})$-space). Furthermore, if such a system is composed of non-interacting subsystems, the state space variables can be chosen so that the system's phase space density function is separable (i.e., is the product of the phase space density functions of the subsystems). This fact motivates the approach to BSS described in this paper [5]: we search for a function of the state space variable $\tilde{x}$ that transforms the observed phase space density function $\tilde{\rho}(\tilde{x}, \dot{\tilde{x}})$ into a separable form. Unlike conventional BSS, this "phase space BSS problem" has a unique solution in the following sense: either the data are inseparable, or they can be separated by a mixing function that is unique, up to transformations that do not affect separability (translations, permutations, and possibly nonlinear rescaling of individual source components). This form of the BSS problem has a unique solution because separability in phase space is a stronger requirement than separability in state space. In other words, if a choice of variables x leads to a separable phase space density function, it also produces a separable state space density function; however, the converse is not true. In particular, the above-mentioned procedure of using integrals of the state space density function to transform it into separable form [3] cannot be used to separate the phase space density function.

It was previously demonstrated [6] that the phase space density function of a time series induces a Riemannian metric on the system's state space and that this metric can be directly computed from the local velocity correlation matrix of the data. In the following Section, we show how this differential geometry can be used to determine if there is a source coordinate system in which the phase space density function is separable and, if so, to find the transformation between the coordinate systems of the observed variables and the source variables. In a technical sense, the method is straight-forward. The data-derived metric is differentiated to compute the affine connection and curvature tensor on state space. If the curvature tensor does not vanish, the observed data are not separable. On the other hand, if the curvature tensor does vanish, there is only one possible set of source variables (up to translations, permutations, and transformations of individual components), and it is possible to compute these explicitly and determine if they do separate the phase space density function. A longer version of this paper [5] describes the solution of a more general BSS problem (sometimes called multidimensional independent component analysis [MICA] or independent subspace analysis) in which the source components can be partitioned into groups, so that components from different groups are statistically independent but components belonging to the same group may be dependent [7,8,9].

As mentioned above, this paper exploits a stronger criterion of statistical independence than conventional approaches (i.e., separability of the phase space density function instead of separability of the state space density function). Furthermore, the new method differs from earlier approaches on the technical level. For example, the proposed method exploits statistical constraints on source time derivatives that are *locally* defined in the state space, in contrast to the usual criteria for statistical independence that are *global* conditions on the source time series or its time derivatives [10]. Furthermore, the nonlinearities of the mixing function are unraveled by imposition of local second-order statistical constraints, unlike many conventional approaches that rely on higher-order statistics [1,2]. In addition, the constraints of statistical independence are used to construct the mixing function in a "deterministic" manner, without the need for parameterizing it (with a neural network architecture or other means) and without using probabilistic learning methods [11,12]. And, the new method is quite general, unlike some other techniques that are limited to the separation of post-nonlinear mixtures [13] or other special cases. Finally, the use of differential geometry in this paper should not be confused with existing applications of differential geometry to BSS. In our case, the observed measurement trajectory is used to derive *a metric on the system's state space*, and the vanishing of the curvature tensor is shown to be a necessary condition for separability of the data. In contrast, other authors [14] define *a metric on a completely different space, the search space of possible mixing functions,* so that "natural" (i.e., covariant) differentiation can be used to expedite the search for the function that optimizes the fit to the observed data.

2 Method

This Section describes how the phase space density function of the observed data induces a Riemannian geometry on the state space and shows how to compute the metric and curvature tensor of this space from the observed time series. Next, we show that, if curvature tensor is non-vanishing, the observed data are not separable. However, if the curvature tensor vanishes, we show how to determine whether the data are separable, and, if they are, we show how to find the mixing function, which is essentially unique.

Let $x = x(t)$ (x_k for $k = 1, 2, \ldots, n$) denote the trajectory of a time series. Suppose that there is a phase space density function $\rho(x, \dot{x})$, which measures the fraction of total time that the trajectory spends in each small neighborhood $dx d\dot{x}$ of $(x, \dot{x})$-space (i.e., phase space). As discussed in [6], most classical physical systems in thermal equilibrium with a "bath" have such a phase space density function: namely, the Maxwell-Boltzmann distribution [4]. Next, define $g^{kl}(x)$ to be the local second-order velocity correlation matrix [6]

$$g^{kl}(x) =< (\dot{x}_k - \bar{\dot{x}}_k)\,(\dot{x}_l - \bar{\dot{x}}_l) >_x \tag{2}$$

where the bracket denotes the time average over the trajectory's segments in a small neighborhood of x and where $\bar{\dot{x}} =< \dot{x} >_x$, the local time average of $\dot{x}$. In other words, g^{kl} is a combination of first and second moments of the local velocity distribution. Because this correlation matrix transforms as a symmetric contravariant tensor, it can be taken to be a contravariant metric on the system's state space. Furthermore, as long as the local velocity distribution is not confined to a hyperplane in velocity space, this tensor is positive definite and can be inverted to form the corresponding covariant metric g_{kl}. Thus, under these conditions, the time series induces a non-singular metric on state space. This metric can then be used to compute the affine connection Γ^k_{lm} and Riemann-Christoffel curvature tensor $R^k{}_{lmn}$ of state space by means of the standard formulas of differential geometry [15]

$$\Gamma^k_{lm}(x) = \frac{1}{2} g^{kn} (\frac{\partial g_{nl}}{\partial x_m} + \frac{\partial g_{nm}}{\partial x_l} - \frac{\partial g_{lm}}{\partial x_n}) \tag{3}$$

and

$$R^k{}_{lmn}(x) = -\frac{\partial \Gamma^k{}_{lm}}{\partial x_n} + \frac{\partial \Gamma^k{}_{ln}}{\partial x_m} + \Gamma^k{}_{im} \Gamma^i{}_{ln} - \Gamma^k{}_{in} \Gamma^i{}_{lm} \tag{4}$$

where we have used the Einstein convention of summing over repeated indices.

Now, assume that the data are separable and that x represents a set of source variables; i.e., assume that the phase space density function ρ is equal to the product of density functions of each component of x. It follows from definition (2) that the metric $g^{kl}(x)$ is diagonal and has positive diagonal elements, each of which is a function of the corresponding coordinate component. Therefore, the individual components of x can be transformed in order to create a new state space coordinate system in which the metric is the identity matrix and the curvature tensor (4) vanishes. It follows that the curvature tensor must vanish in every coordinate system, including the coordinate system $\tilde{x}$ defined by the observed data

$$\tilde{R}^k{}_{lmn}(\tilde{x}) = 0 \tag{5}$$

In other words, the vanishing of the curvature tensor is a necessary consequence of separability. Therefore, if this data-derived quantity does not vanish, the data cannot be transformed so that their phase space density function is separable.

On the other hand, if the data do satisfy (5), there is only one possible separable coordinate system (up to transformations that do not affect separability), and it can be explicitly constructed from the observed data $\tilde{x}(t)$. To see this, first note that, on a flat manifold (e.g., (5)) with a positive definite metric, it is always possible to explicitly construct a "Euclidean" coordinate system for which the metric is the identity matrix. Furthermore, if a coordinate system has a diagonal metric with positive diagonal elements that are functions of the corresponding coordinate components, it can be derived from this Euclidean one by means of an n-dimensional rotation, followed by transformations that do not affect separability (i.e., translations, permutations, and transformations of individual components). Therefore, because every separable coordinate system must have a diagonal metric with the aforementioned properties, all possible separable coordinate systems can be found by constructing a Euclidean coordinate system and then finding all rotations of it that are separable. The first step is to construct a Euclidean coordinate system in the following manner: 1) at some arbitrarily-chosen point $\tilde{x}_0$, select n small vectors $\delta\tilde{x}_{(i)}$ $(i = 1, 2, \ldots, n)$ that are orthonormal with respect to the metric at that point (i.e., $\tilde{g}_{kl}(\tilde{x}_0)\delta\tilde{x}_{(i)k}\delta\tilde{x}_{(j)l} = \delta_{ij}$, where δ_{ij} is the Kronecker delta); 2) starting at $\tilde{x}_0$, use the affine connection to repeatedly parallel transfer all $\delta\tilde{x}$ along $\delta\tilde{x}_{(1)}$; 3) starting at each point along the resulting geodesic path, repeatedly parallel transfer these vectors along $\delta\tilde{x}_{(2)}$; ... continue the parallel transfer process along other directions ... n+1) starting at each point along the most recently produced geodesic path, parallel transfer these vectors along $\delta\tilde{x}_{(n)}$. Finally, each point is assigned the geodesic coordinate s ($s_k, k = 1, 2, \ldots, n$), where s_k represents the number of parallel transfers of the vector $\delta\tilde{x}_{(k)}$ that was required to reach it. Differential geometry [15] guarantees that the metric of a flat, positive definite manifold will be the identity matrix in a geodesic coordinate system constructed in this way. We can now transform the data into this Euclidean coordinate system and examine the separability of all possible rotations of it. The easiest way to do this is to compute the second-order correlation matrix

$$\sigma_{kl} = < (s_k - \bar{s}_k)(s_l - \bar{s}_l) > \tag{6}$$

where the brackets denote the time average over the entire trajectory and $\bar{s} = < s >$. If this data-derived matrix is not degenerate, there is a unique rotation that diagonalizes it, and the corresponding rotation of the s coordinate system is the only candidate for a separable coordinate system (up to transformations that do not affect separability). Its separability can be determined by explicitly computing the data's phase space density function in order to see if it factorizes in this rotated coordinate system. Alternatively, we can use higher-order statistical criteria to see if the rotated s components are truly independent.

In summary, the BSS problem can be solved by the following procedure:

1. Use the data $\tilde{x}(t)$ to compute the metric, affine connection, and curvature of the state space [(2-4)].
2. If the curvature does not vanish at each point, the data are not separable.
3. If the state space curvature does vanish:
   (a) Compute the transformation to a Euclidean coordinate system s and transform the data into it.
   (b) Find the rotation that diagonalizes the second-order correlation matrix σ and transform to the corresponding rotation of the s coordinate system.
   (c) Compute the phase space density function of the data in the rotated s coordinate system.
   (d) If the density function factorizes, the data are separable, and the rotated s coordinates are the unique source variables (up to translations, permutations, and transformations of individual components). If the density function does not factorize, the data are not separable.

3 Discussion

This paper outlines a new approach to nonlinear BSS that is based on a notion of statistical independence, which is characteristic of a wide variety of classical non-interacting physical systems. Specifically, the new method seeks to determine if the observed data are mixtures of source variables that have a *phase-space* density function equal to the product of density functions of individual components. This criterion of statistical independence is stronger than that of conventional approaches to BSS, in which only the *state-space* density function is required to be separable. Because of the relative strength of this requirement, the new approach to BSS produces a unique solution in each case (i.e., data are either inseparable or are separable by a unique mixing function), unlike the conventional approach that always finds an infinite number of mixing functions. Given a time series of observations in a measurement-defined coordinate system ($\tilde{x}$) on the system's state space, the basic problem is to determine if there is another coordinate system (a source coordinate system x) in which the density function is factorizable. The existence (or non-existence) of such a source coordinate system is a coordinate-system-independent property of the time series of data (i.e., an intrinsic or "inner" property). This is because, in *all* coordinate systems, there either is or is not a transformation to such a source coordinate system. In general, differential geometry provides mathematical machinery for determining whether a manifold has a coordinate-system-independent property like this. In the case at hand, we can induce a geometric structure on the state space by identifying its metric with the local second-order correlation matrix of the data's velocity [6]. Then, a necessary condition for BSS is that the curvature tensor vanishes in all coordinate systems (including the measurement coordinate system). Therefore, if this data-derived quantity is non-vanishing, the observations are not separable. However, if the curvature tensor is zero, the data are separable if and only if the density function is seen to factorize in a coordinate system that can be explicitly constructed from the data-derived affine connection. In that case, these coordinates are the unique source variables (up to transformations that do not affect separability).

A longer version of this paper [5] describes the solution of a more general BSS problem (sometimes called multidimensional ICA or independent subspace analysis) in which the source components are only required to be partitioned into groups that are statistically independent of one another but contain statistically interdependent variables [7,8,9]. The possible separable coordinate systems are a subset of all coordinate systems in which the metric is *block*-diagonal (instead of fully diagonal as in this paper). All of these "block-diagonal coordinate systems" can be derived from geodesic coordinate systems constructed from geodesics along a finite number of special directions in state space, and these special directions can be computed from algebraic equations involving the curvature tensor. Thus, it is possible to construct every block-diagonal coordinate system and then explicitly determine if the density function is separable in it. An exceptional situation arises if the metric can be transformed into a block-diagonal form with two or more one-dimensional blocks. In this case, there is an unknown rotation on this two-dimensional (or higher dimensional) subspace that is not determined by the requirement of metric block-diagonality. However, much as in Sect. 2, this rotation can be determined by applying other statistical requirements of separability, such as block diagonality of the second-order state variable correlation matrix or block-diagonality of higher-order local velocity correlation functions. In reference [5], this procedure for performing multidimensional ICA is described in detail, and it is illustrated with analytic examples, as well as with a detailed numerical simulation of an experiment.

What are the limitations of the applicability of this method? It is certainly critical that there be a well-defined metric on state space. However, this will be the case if the measurement time series is described by a phase space density function, a requirement that is satisfied by the trajectories of a wide variety of physical systems [6]. In practical applications, the measurements must cover state space densely enough to be able to compute the metric, as well as its first and second derivatives (required to calculate the affine connection and curvature tensor). In the numerical simulation in [5], approximately 8.3 million short trajectory segments (containing a total of 56 million points) were used to compute the metric and curvature tensor on a three-dimensional state space. Of course, if the dimensionality of the state space is higher, even more data will be needed. So, a relatively large amount of data may be required in order to be able to determine their separability. There are few other limitations on the applicability of the technique. For example, computational expense is not prohibitive. The computation of the metric is the most CPU-intensive part of the method. However, it can be distributed over multiple processors by dividing the observed data into "chunks" corresponding to different time intervals, each of which is sent to a different processor where its contribution to the metric (2) is computed. As additional data are accumulated, they can be processed separately and then added into the time average of the data that were used to compute the earlier estimate of the metric. Thus, the earlier data need not be processed again, and only the latest observations need to be kept in memory.

Acknowledgments

The author is indebted to Michael A. Levin for insightful comments and for a critical reading of the manuscript. Aapo Hyvärinen and Shun-ichi Amari also read the manuscript, and their comments on the uniqueness issue are appreciated. This work was performed with the partial support of the National Institute of Deafness and Communicative Disorders.

References

1. Hyvärinen, A., Karhunen, J., Oja, E.: Independent Component Analysis. Wiley, New York (2001)
2. Jutten, C., Karhunen, J.: Advances in nonlinear blind source separation. In: Proceedings of the 4th International Symposium on Independent Component Analysis and Blind Signal Separation (ICA 2003), Nara, Japan (April 2003)
3. Hyvärinen, A., Pajunen, P.: Nonlinear independent component analysis: existence and uniqueness results. Neural Networks 12, 429–439 (1999)
4. Sears, F.W.: Thermodynamics, the Kinetic Theory of Gases, and Statistical Mechanics, 2nd edn. Addison-Wesley, Reading, MA (1959)
5. Levin, D.N.: Using state space differential geometry for nonlinear blind source separation (2006), `http://arxiv.org/abs/cs/0612096`
6. Levin, D.N.: Channel-independent and sensor-independent stimulus representations. J. Applied Physics 98, 104701 (2005), Also, see the papers posted at `http://www.geocities.com/dlevin2001/`

7. Cardoso, J-F.: Multidimensional independent component analysis. In: Proc. 1998 IEEE Int. Conf. on Acoustics, Speech and Signal Processing (ICASSP'98), Seattle, pp. 1941–1944. IEEE Computer Society Press, Los Alamitos (1998)
8. Bingham, E., Hyvärinen, A.: A fast fixed-point algorithm for independent component analysis of complex-valued signals. Int. J. of Neural Systems 10, 1–8 (2000)
9. Nishimori, Y., Akaho, S., Plumbley, M.D.: Riemannian optimization method on the flag manifold for independent subspace analysis. In: Rosca, J., Erdogmus, D., Príncipe, J.C., Haykin, S. (eds.) ICA 2006. LNCS, vol. 3889, pp. 295–302. Springer, Heidelberg (2006)
10. Lagrange, S., Jaulin, L., Vigneron, V., Jutten, C.: Analytic solution of the blind source separation problem using derivatives. In: Puntonet, C.G., Prieto, A.G. (eds.) ICA 2004. LNCS, vol. 3195, pp. 81–88. Springer, Heidelberg (2004)
11. Haykin, S.: Neural Networks - A Comprehensive Foundation, 2nd edn. Prentice Hall, New York (1998)
12. Yang, H., Amari, S.I., Cichocki, A.: Information-theoretic approach to blind separation of sources in non-linear mixture. Signal Processing 64, 291–300 (1998)
13. Taleb, A., Jutten, C.: Source separation in post nonlinear mixtures. IEEE Trans. on Signal Processing 47, 2807–2820 (1999)
14. Amari, S.: Natural gradient works efficiently in learning. Neural Computation 10, 251–276 (1998)
15. Weinberg, S.: Gravitation and Cosmology - Principles and Applications of the General Theory of Relativity. Wiley, New York (1972)

Copula Component Analysis

Jian Ma and Zengqi Sun

Department of Computer Science,
Tsinghua University, Beijing 100084, China
majian03@mails.tsinghua.edu.cn

Abstract. A framework named copula component analysis (CCA) for blind source separation is proposed as a generalization of independent component analysis (ICA). It differs from ICA which assumes independence of sources that the underlying components may be dependent by certain structure which is represented by Copula. By incorporating dependency structure, much accurate estimation can be made in principle in the case that the assumption of independence is invalidated. A two phrase inference method is introduced for CCA which is based on the notion of multi-dimensional ICA. Simulation experiments preliminarily show that CCA can recover dependency structure within components while ICA does not.

1 Introduction

Blind source separation (BSS) is to recover the underlying components from their mixtures, where the mixing matrix and distribution of the components are unknown. To solve this problem, independent component analysis (ICA) is the most popular method to extract those components under the assumption of statistically independence[1,2,3,4,5]. However, in practice, the independence assumption of ICA cannot always be fully satisfied and thus strongly confines its applications. Many works have been contributed to generalize the ICA model,[6] such as Tree-ICA[7], Topology ICA[8]. A central problem of those works is how to relax the independent assumption and to incorporate different kinds of dependency structure into the model.

Copula [9] is a recently developed mathematical theory for multivariate probability analysis. It separates joint probability distribution function into the product of marginal distributions and a Copula function which represents the dependency structure of random variables. According to *Sklar theorem*, given a joint distribution with margins, there exists a copula uniquely determined. Through Copula, we can clearly represent the dependent relation of variables and analysis multivariate distribution of the underlying components.

The aim of this paper is to use Copula to model the dependent relations between elements of random vectors. By doing this, we transform BSS into a parametric or semi-parametric estimation problem which mainly concentrate on the estimation of dependency structure besides identification of the underlying components as ICA do.

M.E. Davies et al. (Eds.): ICA 2007, LNCS 4666, pp. 73–80, 2007.

This paper is organized as follows: we briefly review ICA and its extensions in section 2. The main conclusions of copula theory are briefly introduced in section 3. In section 4, we propose a new model for BSS, named copula component analysis (CCA) which takes dependency among components into consideration. Inference method for CCA is presented in section 5. Simulation experiments are presented in section 6. Finally, we conclude the paper and give some further research directions.

2 ICA and Its Extensions

Given a random vector $\mathbf{x}$, ICA is modeled as

$$\mathbf{x} = \mathbf{A}\mathbf{s}, \tag{1}$$

where the source signals $\mathbf{s} = \{s_1, \ldots, s_n\}$ assume to be mutually independent, $\mathbf{A}$ and $\mathbf{W} = \mathbf{A}^{-}$ is the invertible mixing and demixing matrix to be solved so that the recovered underlying components $\{s_1, \ldots, s_n\}$ is estimated as statistically independent as possible.

Statistical independence of sources means that the joint probability density of $\mathbf{x}$ and $\mathbf{s}$ can be factorized as

$$\begin{aligned} p(\mathbf{x}) = p(\mathbf{A}\mathbf{s}) =\mid det(\mathbf{W}) \mid p(\mathbf{s}) \\ p(\mathbf{s}) = \textstyle\prod_{i=1}^{n} p_i(s_i) \end{aligned} \tag{2}$$

The community has presented many extensions of ICA with different types of dependency structures. For example, Bach and Jordan [7] assumed that dependency can be modeled as a tree (or a forest). After the contrast function is extended with T-mutual information, Tree-ICA tries to find both a mixing matrix A and a tree structure T by embedding a Chow-Liu algorithm into algorithm. Hyvärinen etc. [8] introduced the variance into ICA model so as to model dependency structure. Cardoso generalized the notion of ICA into multidimensional ICA using geometrical structure.[6]

3 A Brief Introduction on Copula

Copula is a recently developed theory which separates the margin law and the joint law and therefore gives dependency structure as a function. According to Nelson [9], it is defined as follows:

Definition 1 (Copula). *A bidimensional copula is a function $C(x,y) : I^2 \mapsto I$ with following properties:*

1. $(x,y) \subset I^2$
2. $C(x_2,y_2) - C(x_1,y_2) - C(x_2,y_1) + C(x_1,y_1) \geq 0$, *for* $x_1 \leq x_2$ *and* $y_1 \leq y_2$;
3. $C(x,1) = x$ *and* $C(1,y) = y$.

It's not hard to know that such defined $C(x,y)$ is a cdf on I^2. Multidimensional version can be generalized in a same manner which presents in [9].

Theorem 1 (Sklar Theorem). *Given a multidimensional random vector* $\mathbf{x} = (x_1, \ldots, x_n) \in R^n$ *with its corresponding distribution function and density function* $u_i = F_i(x_i)$ *and* $p_i(x_i), i = 1, \ldots, n$. *Let* $F(x) : R^n \mapsto I$ *denotes the joint distribution, then there exists a copula* $C(\cdot) : I^n \mapsto I$ *so that*

$$F(\mathbf{x}) = C(\mathbf{u}). \tag{3}$$

where $\mathbf{u} = (u_1, \ldots, u_n)$.

If the copula is differentiable, the joint density function of $F(\mathbf{x})$ *is*

$$P_{1,\ldots,n}(\mathbf{x}) = \prod_{i=1}^{n} p_i(x_i) C'(\mathbf{u}). \tag{4}$$

where $C'(\mathbf{u}) = \frac{\partial C(\mathbf{u})}{\partial u_1, \ldots, \partial u_n}$.

Given a random vector $\mathbf{x} = (x_1, \ldots, x_n)$ with mutually independent variables, and their cdf $F(\mathbf{x}) = \prod_i F_i(x_i)$. It is easy to obtain that the corresponding copula function called *Product Copula* is $C(\mathbf{u}) = \prod_i u_i$ and $C'(\mathbf{u}) = 1$.

4 Copula Component Analysis

4.1 Geometry of CCA

As previously stated, ICA assumes that the underlying components are mutually independent, which can be represented as (1). CCA also use the same representation (1) as ICA, but without independence assumption. Here, Let the joint density function represents by Copula:

$$p_c(\mathbf{x}) = \prod_{i=1}^{N} p_i(x_i) C'(\mathbf{u}) \tag{5}$$

where the dependency structure is modeled by function $C(\mathbf{u})$.

The goal of estimation is to minimize the distance between the 'real' pdf of random vector $\mathbf{x}$ and its counterpart of the proposed model. Given a random vector $\mathbf{x}$ with pdf $p(\mathbf{x})$, the distance between $p(\mathbf{x})$ and $p_c(\mathbf{x})$ in a sense of *K-L* divergence can be represented as

$$\begin{aligned} D(p \| p_c) &= E_{p(\mathbf{x})} \log \frac{p(\mathbf{x})}{p_c(\mathbf{x})} \\ &= E_{p(\mathbf{x})} \log \frac{p(\mathbf{x})}{\prod_i p_i(x_i)} - E_{p(\mathbf{x})} \log C'(\mathbf{u}) \end{aligned} \tag{6}$$

The first term on the right of (6) is corresponding to the K-L divergence between $p(x)$ and ICA model and the second term is corresponding to entropy of copula $C(x)$.

Theorem 2. *Given a random vector* $\mathbf{x} = (x_1, \dots, x_n) \in R^n$ *with pdf* $p(\mathbf{x})$ *and its joint pdf* $p_c(\mathbf{x}) = \prod_{i=1}^{n} p_i(x_i) C'(\mathbf{u})$*, where* $u_i = F_i(x_i)$ *is the cdf of* x_i *and dependency structure is presented by copula function* $C(\mathbf{u}) : I^n \mapsto I, \mathbf{u} \in R^n$ *and* $C'(\mathbf{u}) = \frac{\partial^n C(\mathbf{u})}{\partial u_1, \dots, \partial u_n}$ *is the derivative of* $C(\mathbf{u})$*. The K-L divergence* $D(p \| p_c)$ *is as*

$$D(p \| p_c) = I(x_1, \dots, x_n) + H(C'(\mathbf{u})). \tag{7}$$

where $H(\cdot)$ *is the Shannon differential entropy.*

That is, the K-L divergence between $p(\mathbf{x})$ and $p_c(\mathbf{x})$ equal to the sum of the mutual information $I(\mathbf{x})$ and copula entropy H for function $\mathbf{u} \sim C'$.

Using the invariant of K-L divergence, we now have the following corollary to *Theorem 2* for BSS problem $\mathbf{s} = \mathbf{W}\mathbf{x}$.

Corollary 1. *With the same denotation of Theorem 2, the K-L divergence for BSS problem is*

$$D(p \| p_c) = I(s_1, \dots, s_n) + H(C'(\mathbf{u}_s)), \tag{8}$$

where $\mathbf{u}_s$ *denotes the marginal variable for sources* $\mathbf{s}$*. Assume that the number of sources equals to that of observations.*

In other words, the distance between ICA model and the true model is presented by dependency structure and its value equals to entropy of the underlying copula function. It can be easily learned from (7) that if dependency structure was incorporated into model, the distance between data and model can be further closer than that of ICA model.

ICA is a special case when it assumes mutual independence of underlying components. Actually, ICA only minimizes the first part of (7) under the assumption of independence. This also explains why sometime ICA model is not applicable when dependency relations between source components exist.

4.2 Multidimensional ICA

From the notion of multidimensional ICA generalized from ICA by Cardoso [6], it can be derived that

$$\begin{aligned} p(\mathbf{x}) &= \prod_{k=1}^{m} p_k(\mathbf{x}_k) = \prod_{k=1}^{m} p_k(x_{i_k}, \dots, x_{i_{k+1}-1}) \\ &= \prod_{k=1}^{m} \prod_{l=i_k}^{i_{k+1}-1} p_k(x_l) C'_k(\mathbf{u}_k) = \prod_{i=1}^{n} p_i(x_i) \prod_{k=1}^{m} C'_k(\mathbf{u}_k) \end{aligned} \tag{9}$$

where $C_k(\cdot)$ is the copula with respect to $p_k(\cdot)$. On the other side, the definition of copula gives

$$p(\mathbf{x}) = \prod_{i=1}^{n} p_i(x_i) C'(\mathbf{u}) \tag{10}$$

According to Sklar theorem, if all $p_i(\cdot)$ exist, then $C(\cdot)$ is unique. Therefore, we can derive the following result.

Theorem 3. *The copula corresponding to multidimensional ICA is factorial if all the marginal pdf of component exist, that is*

$$C'(\mathbf{u}) = \prod_{k=1}^{m} C'_k(\mathbf{u}_k) \tag{11}$$

Proof. Because of the unique of C, the above (11) can be easily derived by comparing (9) and (10).

The theorem can guide hypothesis selection of copula. That is, Copula should be factorized as a product of sub-function with different type for dependency structure of different sub-space.

Combining (7) and (11), we can derived the following:

$$D(p\|p_c) = I(u_1, \ldots, u_n) + \sum_{k=1}^{m} H(C'_k) \tag{12}$$

It means that the distance between the true model and ICA model composes of entropy of Copulas which corresponds to every un-factorial ICs spaces. Therefore, if we want to derive a model much closer to the 'true' one than ICA, we should find dependency structure of each space, that is, approach the goal step by step. This is one of the guide principles for designing algorithm of copula component analysis.

5 Inference of CCA

5.1 General Framework

In this section, we study inference method for CCA based on the notion of multidimensional ICA. Suppose the underlying copula function parameterized by $\theta \in \Theta$, thus the estimation of CCA should infer the demixing matrix $\mathbf{W}$ and θ. According to theorem 2, estimation of the underlying sources through our model requires the minimization of the K-L divergence of (7) or (12). Thus the objective function is

$$\min D(p\|p_c; \mathbf{W}, \theta) \tag{13}$$

which composes of two sub-objective: $\min I(x_1, \ldots, x_n; \mathbf{W})$ and $\min H(C'(\mathbf{u}); \mathbf{W}, \theta)$. Because $\mathbf{u}$ in the latter objective depends on the structure of IC spaces derived from the former objective, we should handle the optimal problem min $I(x_1, \ldots, x_n; \mathbf{W})$ at first. The first objective can be achieved by ICA algorithm. For the second one we proposed the Infomax like principle given a parametric family of copula.

We propose that the framework of CCA composes of two phrases:

1. Solve $\mathbf{W}$ through minimization of mutual information .
2. Determine $\mathbf{W}$ and θ so that the objective function (13) is minimized.

5.2 Maximum Likelihood Estimation

Given the parametric model of Copula, maximum likelihood estimation can be deployed under the constraint of ICA. Consider a group of independent observations $x_1, \ldots, x_T$ of $n \times 1$ random vector $\mathbf{X}$ with a common distribution $\mathcal{P} = C'_\theta(\mathbf{x}) \prod_{i=1}^{T} p_i(x_i); \theta \in \Theta$ where $p_i(x_i)$ is marginal distribution associated with x_i, and the log-likelihood is

$$
\begin{aligned}
\mathcal{L}(\mathbf{W}, \theta) &= \frac{1}{T} \log C'_\theta(\mathbf{x}) \prod_{i=1}^{T} p_i(x_i) \\
&= \frac{1}{T} \sum_{i=1}^{T} \log p_i(x_i) + \frac{1}{T} \log C'_\theta(\mathbf{x})
\end{aligned}
\tag{14}
$$

The representation is consist with two-phrase CCA framework in that the first term on the right of equation (14) implies mutual information of $\mathbf{x}$ and that the second term is empirical estimation of entropy of $\mathbf{x}$. It is not hard to proof that

$$
\min D(p \| p_c) \Leftrightarrow \max \mathcal{L}(\mathbf{W}, \theta) \tag{15}
$$

5.3 Estimation of Copula

Suppose the IC subspaces have been correctly determined by ICA and then we can identify the copula by minimizing the second term on the right of (7). Given a class of Copula $C(\mathbf{u}; \theta)$ with parameter vector $\theta \in \Theta$, and a set of sources $\mathbf{s} = (s_1, \ldots, s_n)$ identified from data set $\mathbf{X}$, the problem is such a optimization one

$$
\max_{\mathbf{W}, \theta} E_{p(\mathbf{s})}(C'(\mathbf{u_s}; W, \theta)) \tag{16}
$$

By using *Sklar* theorem, the copula to be identified has been separate with marginal distributions which are known except non-Gaussianity in ICA model. Therefore, the problem here is a semi-parametric one and only need identifying the copula.

Parametric method is adopted. First, we should select a hypothesis for copula among many types of copula available. The selection depends on many factors, such as priori knowledge, computational ease, and individual preference. Due to space limitations, only few of them are introduced here. For more detail please refer to [9].

When a set of sources $\mathbf{s}$ and a parametric copula $C(\cdot; \theta)$ is prepared, the optimization of (16) becomes an optimization problem which can be solved as follows:

$$
\sum_{s_i=1}^{n} \frac{\partial C'}{\partial \theta}(\mathbf{u}; \theta) = 0 \tag{17}
$$

where many readily methods can be utilized.

6 Simulation Experiments

In this section, simulation experiments are designed to compare CCA and ICA on two typical cases to investigate whether CCA can perform better than ICA as previous stated. One cases is with independent components and the other is where there are dependent components.

We first apply both methods on independent components recovery from their mixtures and then on recovery of components with dependency structure. In both experiments, the basic case of BSS with two components are considered. Two components are generated by bi-variate distribution associated with Gumbel copula:

$$C(u, v) = \exp\left(\left((-\ln u)^\theta + (-\ln v)^\theta\right)^{-\theta}\right) \tag{18}$$

where $\theta = 1, 5$ respectively. Note that two components such generated are independent when $\theta = 1$ and thus compose of sources of ICA problem. The marginal density of components are uniform. Sources are mixed by randomly generated and invertible 2×2 matrix $\mathbf{A}$. In our experiments, $\mathbf{A}$ is

$$\mathbf{A} = \begin{pmatrix} 0.4936 \; 0.9803 \\ 0.4126 \; 0.5470 \end{pmatrix}$$

Both ICA and CCA are used to recover the components from their mixtures. Without the attention to study model selection, Gumbel copula family is adopted in CCA method.

The results are illustrated in Figure 1. Due to space limitations, we only present copula density structure of sources and their recoveries by both methods

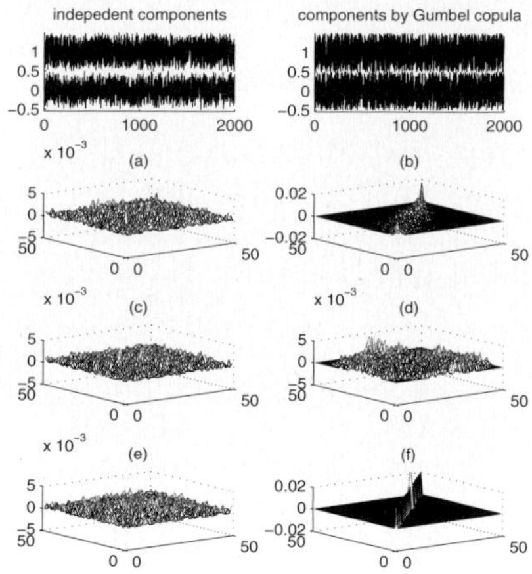

Fig. 1. Simulation experiments. The left column is for independent component experiments and the right column is for the experiment of components by Gumbel copula. The top two sub-figure is sources and (a) and (b) is their corresponding copula density. (c) and (d) is for ICA and (e) and (f) is for CCA.

in Figure 1. Note that copula density structure should be a plane if two components are independent, that is, $C(u, v) = 1$. It can be learned from figure 1 that both methods works well when components are mutually independent and more importantly that ICA always try to extracts components mutually independent while CCA can recover the dependency between components successfully.

7 Conclusions and Further Directions

In this paper, a framework named Copula Component Analysis for blind source separation is proposed as a generalization of ICA. It differs from ICA which assumes independence of sources that the underlying components may be dependent with certain structure which is represented by Copula. By incorporating dependency structure, much accurate estimation can be made, especially in the case where the assumption of independence is invalidated. A two phrase inference method is introduced for CCA which is based on the notion of multidimensional ICA. A preliminary simulated experiment demonstrates the advantage of CCA over ICA on dependency structure discovery. Many problems remain to be studied in the future, such as Identifiability of the method, selection of copula model and applications.

References

1. Comon, P.: Independent component analysis - A new concept? Signal Processing 36, 287–314 (1994)
2. Bell, A., Sejnowski, T.: An information-maximization approach to blind separation and blind deconvolution. Neural Comp. 7, 1129–1159 (1995)
3. Amari, S., Cichocki, A., Yang, H.H.: A new learning algorithm for blind source separation. In: Advances in Neural Information Processing, pp. 757–763. MIT Press, Cambridge (1996)
4. Cardoso, J.-F., Laheld, B.H.: Equivariant adaptive source separation. IEEE Transactions on Signal Processing 44, 3017–3030 (1996)
5. Pham, D.T., Garat, P.: Blind separation of mixture of independent sources through a quasi-maximum likelihood approach. IEEE Transactions on Signal Processing 45, 1712–1725 (1997)
6. Cardoso, J.-F.: Multidimensional independent component analysis, Acoustics, Speech, and Signal Processing. In: ICASSP'98. Proceedings of the 1998 IEEE International Conference on, vol. 4, pp. 1941–1944 (1998)
7. Bach, F.R., Jordan, M.I.: Beyond independent components: Trees and clusters. Journal of Machine Learning Research 4, 1205–1233 (2003)
8. Hyvärinen, A., Hoyer, P.O., Inki, M.: Topographic Independent Component Analysis. Neural Computation 13, 1527–1558 (2001)
9. Nelsen, R.B.: An Introduction to Copulas. Lecture Notes in Statistics. Springer, New York (1999)

On Separation of Signal Sources Using Kernel Estimates of Probability Densities

Oleg Michailovich[1] and Douglas Wiens[2]

[1] University of Waterloo, Waterloo, Ontario N2L 3G1, Canada
[2] University of Alberta, Edmonton, Alberta T6G 2G1, Canada

Abstract. The discussion in this paper revolves around the notion of *separation problems*. The latter can be thought of as a unifying concept which includes a variety of important problems in applied mathematics. Thus, for example, the problems of classification, clustering, image segmentation, and discriminant analysis can all be regarded as separation problems in which one is looking for a decision boundary to be used in order to separate a set of data points into a number of (homogeneous) subsets described by different conditional densities. Since, in this case, the decision boundary can be defined as a hyperplane, the related separation problems can be regarded as *geometric*. On the other hand, the problems of source separation, deconvolution, and independent component analysis represent another subgroup of separation problems which address the task of separating *algebraically* mixed signals. The main idea behind the present development is to show conceptually and experimentally that both geometric and algebraic separation problems are very intimately related, since there exists a general variational approach based on which one can recover either geometrically or algebraically mixed sources, while only little needs to be modified to go from one setting to another.

1 Introduction

Let $X = \{x_i \in \mathbb{R}^d, i = 1, \ldots, N\}$ be a set of N observations of a random variable $\mathcal{X}$ which is described by M conditional densities $\{p_k(x) \stackrel{def}{=} p(x \mid \mathcal{X} \in C_k)\}_{k=1}^M$, with C_k denoting a *class* to which a specific realization of $\mathcal{X}$ may belong. In other words, the set X can be viewed as a *mixture* of realizations of M random variables associated with different classes described by corresponding probability densities. In this case, the problem of classification (or, equivalently, *separation*) refers to the task of ascribing each observation x_i to the class C_k which it has *most likely* come from. The most challenging version of the above problem occurs in the case when the decision has to be made given the observed set X alone.

The setting considered above is standard for a variety of important problems in applied mathematics. Probably, the most famous examples here are unsupervised machine learning and data clustering [1,2]. Signal detection and image segmentation are among other important examples of the problems which could be embedded into the same separation framework [3,4]. It should be noted that, although a multitude of different approaches have been proposed previously to

M.E. Davies et al. (Eds.): ICA 2007, LNCS 4666, pp. 81–88, 2007.

address the above problems, most of them are similar at the conceptual level. Specifically, viewing the observations $\{x_i\}$ as points on either a linear or a non-linear manifold Ω, the methods search for such a partition of the latter so that the points falling at different subsets of Ω are most likely associated with different classes C_k. Moreover, the boundaries of the partition, which are commonly referred to as *decision boundaries*, are usually defined by means of geometric descriptors. The latter, for example, can be hyperplanes in machine learning [1] or active contours [4] in image segmentation. For this reason, we refer to the problems of this type as the problems of *geometric source separation* (GSS), in which case the data set X is considered to be a *geometric mixture* of unknown sources.

In parallel to the case of GSS, there exists an important family of problems concerned with separating sources that are mixed *algebraically* [5]. In a canonical setting, the problem of *algebraic source separation* (ASS) can be formulated as follows. Let $\mathbf{S}$ be a vector of M signals (sources) $[s_1(t), s_2(t), \ldots, s_M(t)]^T$, with $t = 1, \ldots, T$ being either a temporal or a spatial variable. Also, let $\mathbf{A} \in \mathbb{R}^{M \times M}$ be an unknown *mixing* matrix of full rank. Subsequently, the problem of blind source separation consists in recovering the sources given an observation of their *mixtures* $\mathbf{X} = [x_1(t), x_2(t), \ldots, x_M(t)]^T$ acquired according to[1]:

$$\mathbf{X} = \mathbf{A}\,\mathbf{S}. \tag{1}$$

Note that, in (1), neither the sources $\mathbf{S}$ nor the matrix $\mathbf{A}$ are known, and hence the above estimation problem is conventionally referred to as *blind.* Note that the problem of (algebraic) blind source separation constitutes a specific instance of *Independent Component Analysis*, which is a major theory encompassing a great number of applications [5]. Moreover, when $M = 1$ and $\mathbf{A}$ is defined to be a convolution operator, the resulting problem becomes the problem of blind deconvolution [6], which can also be inscribed in our framework of separation problems.

The main purpose of this paper is to show conceptually and experimentally that both GSS and ASS problems are intimately interrelated, since they can be solved using the same tool based on variational analysis [7]. To define this tool, let us first introduce an abstract, geometric separation operator $\varphi : X \mapsto \{S_k\}_{k=1}^M$ that "sorts" the points of X into M complementary and mutually exclusive subsets $\{S_k\}_{k=1}^M$ which represent estimates of the geometrically mixed sources. On the other hand, in the case of ASS, the separation operator is defined algebraically as a de-mixing matrix $\mathbf{W} \in \mathbb{R}^{M \times M}$ such that:

$$\mathbf{S} \simeq \mathbf{W}\,\mathbf{X}, \tag{2}$$

with $\mathbf{S}$ and $\mathbf{X}$ defined to be $\mathbf{S} = [s_1(t), \ldots, s_M(t)]^T$ and $\mathbf{X} = [x_1(t), \ldots, x_M(t)]^T$, respectively.

Additionally, let y_k be an estimate of either a geometric or an algebraic k-th source signal, computed via applying either φ or $\mathbf{W}$ to the data. This estimate

[1] Here and hereafter, the matrix $\mathbf{A}$ is assumed to be square which is merely a technical assumption which can be dropped; this is discussed in the sequel.

can be characterized by its *empirical* probability density function (*pdf*) which can be computed as given by:

$$\tilde{p}_k(z) = \frac{1}{N_k} \sum_{t=1}^{N_k} K(z - y_k(t)), \quad z \in \mathbb{R}^d, \tag{3}$$

where N_k is the size of the estimate (that is independent of k in the case of ASS). Note that (3) defines a *kernel based estimate* of the *pdf* of y_k when the *kernel* function $K(z)$ is normalized to have unit integral [8]. There exist a number of possibilities for choosing $K(z)$, among which the most frequent one is to define the kernel in the form of a Gaussian density function. Accordingly, this choice of $K(z)$ is used throughout the rest of this paper.

The core idea of the preset approach is quite intuitive and it is based on the assumption that the "overlap" between the informational contents of the estimated sources has to be minimal. To minimize this "overlap", we propose to find the optimal separation operator (viz. either φ or $\mathbf{W}$) as a minimizer of the cumulative Bhattacharyya coefficient between the empirical *pdfs* of the estimated sources, which is defined to be [9]:

$$B_M = \frac{2}{M(M-1)} \sum_{i<j} \int_{\mathbb{R}^d} \sqrt{\tilde{p}_i(z)\,\tilde{p}_j(z)}dz, \quad i,j = 1,\ldots,M. \tag{4}$$

It should be noted that, apart from the Bhattacharyya coefficient, a number of alternative metrics are available to assess the distance between the probability densities. Thus, for example, the Kullback-Leibler divergence was employed in [10] and [5] to solve the problems of image segmentation and blind source separation, respectively. However, for the reasons discussed below, we prefer using (4), since it results in comparatively more stable and reliable separation. To demonstrate how B_M can be used to unify the concept of separation, as it appears in both geometric and algebraic settings, we turn to some specific examples, among which the problem of image segmentation is chosen to be first.

2 Geometric Source Separation: Image Segmentation

In order to facilitate the discussion, we confine the derivations below to the case of two segmentation classes. In this case, the values of a vector-valued image $I(u) : \Omega \subseteq \mathbb{R}^2 \to \mathbb{R}^d$ are viewed as a geometric mixture of two sources, viz. the object of interest and its background. Consequently, the segmentation problem can be reformulated as the problem of partitioning the domain of definition Ω of $I(u)$ (with $u \in \Omega$) into two mutually exclusive and complementary subsets Ω_- and Ω_+. These subsets can be represented by their respective characteristic functions χ_- and χ_+, which can, in turn, be defined as $\chi_-(u) = \mathcal{H}(-\varphi(u))$ and $\chi_+(u) = \mathcal{H}(\varphi(u))$, with $\mathcal{H}$ standing for the Heaviside function.

Given a level-set function $\varphi(u)$, its *zero level set* $\{u \mid \varphi(u) \equiv 0, u \in \Omega\}$ is used to *implicitly* represent a curve – *active contour* – embedded into Ω. For the sake of concreteness, we associate the subset Ω_- with the support of the object of interest, while Ω_+ is associated with the support of corresponding background. In this case, the objective of active-contour-based image segmentation is, given an initialization $\varphi_0(u)$, to construct a convergent sequence of level-set functions $\{\varphi_t(u)\}_{t>0}$ (with $\varphi_t(u)_{t=0} = \varphi_0(u)$) such that the zero level-set of $\varphi_\infty(u)$ coincides with the boundary of the object of interest.

The above sequence of level-set functions can be conveniently constructed using the variational framework. Specifically, the sequence can be defined by means of a *gradient flow* that minimizes the value of the cost functional (4). In the case of two segmentation classes, the optimal level set $\varphi^\star(u)$ is defined as:

$$\varphi^\star(u) = \arg \inf_{\varphi(u)} \{B_2(\varphi(u))\}, \tag{5}$$

where

$$B_2(\varphi(u)) = \int_{z \in \mathbb{R}^N} \sqrt{p_-(z \mid \varphi(u))\, p_+(z \mid \varphi(u))}\, dz. \tag{6}$$

with $p_-(z \mid \varphi(u))$ and $p_+(z \mid \varphi(u))$ being the kernel-based estimates of the *pdf*s of the class and background sources.

In order to contrive a numerical scheme for minimizing (5), its first variation should be computed first. The first variation of $B_2(\varphi(u))$ (with respect to $\varphi(u)$) can be shown to be given by:

$$\frac{\delta B_2(\varphi(u))}{\delta \varphi(u)} = \delta(\varphi(u))\, V(u), \tag{7}$$

where

$$V(u) = \frac{1}{2} B_2(\varphi(u))(A_-^{-1} - A_+^{-1}) + \frac{1}{2} \int_{z \in \mathbb{R}^d} K(z - I(u))\, L(z \mid \varphi(u))\, dz, \tag{8}$$

with

$$L(z \mid \varphi(u)) = \frac{1}{A_+} \sqrt{\frac{p_-(z \mid \varphi(u))}{p_+(z \mid \varphi(u))}} - \frac{1}{A_-} \sqrt{\frac{p_+(z \mid \varphi(u))}{p_-(z \mid \varphi(u))}}. \tag{9}$$

Note that, in the equations above, $\delta(\cdot)$ stands for the delta function, and A_- and A_+ are the areas of Ω_- and Ω_+ given by $\int_\Omega \chi_-(u)\, du$ and $\int_\Omega \chi_+(u)\, du$, respectively.

Finally, introducing an artificial time parameter t, the gradient flow of $\varphi(u)$ that minimizes (5) is given by:

$$\varphi_t(u) = -\frac{\delta B_2(\varphi(u))}{\delta \varphi(u)} = -\delta(\varphi(u))\, V(u), \tag{10}$$

where the subscript t denotes the corresponding partial derivative, and $V(u)$ is defined as given by (8).

It should be added that, in order to regularize the shape of the active contour, it is common to constrain its length and to replace the theoretical delta function

$\delta(\cdot)$ by its smoothed version $\bar{\delta}(\cdot)$. In this case, the final equation for the evolution of the active contour becomes:

$$\varphi_t(u) = \bar{\delta}(\varphi(u))\left(\alpha\,\kappa(u) - V(u)\right), \tag{11}$$

where $\kappa(u)$ is the curvature of the active contour given by $\kappa(u) = -\mathrm{div}\left\{\frac{\nabla\varphi(u)}{\|\nabla\varphi(u)\|}\right\}$ and $\alpha > 0$ is a user-defined regularization parameter. Note that, in the segmentation results reported in this paper, α was set to be equal to 1.

3 Blind Separation of Algebraically Mixed Sources

It is surprising how little has to be done to modify the separation approach of the previous section to suit the ASS setting. Indeed, let $\mathbf{Y} = [y_1(t), y_2(t), \ldots, y_M(t)]^T$ be the matrix of estimated sources computed as $\mathbf{Y} = \mathbf{W}\,\mathbf{X}$. Additionally, let $\{p(z;\mathbf{w}_i) \stackrel{def}{=} \tilde{p}_i(z \mid \mathbf{W})\}_{i=1}^M$ (where $\mathbf{w}_i^T$ is the ith row of $\mathbf{W}$) be the set of empirical densities computed as at (3) and that correspond to the source estimates in $\mathbf{Y}$. Consequently, the optimal separation matrix $\mathbf{W}^*$ can be found as:

$$\mathbf{W}^\star = \arg\inf_{\mathbf{W}}\{B_M(\mathbf{W})\}, \tag{12}$$

where

$$B_M(\mathbf{W}) = \frac{2}{M(M-1)} \int_{\mathbb{R}^d} \sum_{i<j} \sqrt{p(z;\mathbf{w}_i)\,p(z;\mathbf{w}_j)},\ i,j = 1,\ldots M. \tag{13}$$

It should be noted that intrinsic in blind (algebraic) source separation is the problem of permutation and normalization, as, using (2), the sources can only be recovered in an arbitrary order and up to arbitrary multiplication factors. While the order of the sources is rarely of importance, the normalization could become an issue, especially from the viewpoint of numerical minimization. To overcome this difficulty, it is common to *prewhiten* the mixtures X before they are passed into the computations. In this case, it can be easily shown that the optimal solution $\mathbf{W}^\star$ becomes a member of the orthogonal group $\mathbf{O}(M) = \{\mathbf{W} \in \mathbb{R}^{M\times M} \mid \mathbf{W}\mathbf{W}^T = \mathbf{I}\}$.

We solve this constrained minimization problem with the aid of Lagrange multipliers $\{\lambda_{\alpha\beta}\}_{\alpha\leq\beta}$. Consider the problem of minimizing

$$F\left(\mathbf{w}_1, ..., \mathbf{w}_M, \lambda\right) = B_M(\mathbf{W}) + \sum_{\alpha\leq\beta} \lambda_{\alpha\beta}\left(\mathbf{w}_\alpha^T\mathbf{w}_\beta - \delta_{\alpha\beta}\right), \tag{14}$$

where $\delta_{\alpha\beta}$ is Kronecker's delta. Solving the equations

$$\frac{\partial}{\partial\mathbf{w}_i} F\left(\mathbf{w}_1, ..., \mathbf{w}_M, \lambda\right) = \mathbf{0}^T\ \ (\in \mathbb{R}^{1\times M}), \tag{15}$$

together with $\mathbf{W}\mathbf{W}^T = \mathbf{I}$ (details available from authors) leads to the characterization of $\mathbf{W}^*$ as a fixed point of the function

$$G(\mathbf{W}) = (\mathbf{P}\mathbf{P}^T)^{-1/2}\mathbf{P}, \tag{16}$$

where $(\mathbf{P}\mathbf{P}^T)^{1/2}$ is a *symmetric* square root and $\mathbf{P} = \mathbf{P}(\mathbf{W})$ is defined as follows. Let $\dot{\mathbf{K}}(z)$ be the $N \times M$ matrix with $(i, j)^{th}$ element $K'(z - \mathbf{y}_j(i))$ and let $\mathbf{D}(z)$ be the diagonal matrix with diagonal elements

$$d_i(z) = \frac{\sum_{j=1,\ldots,M;j\neq i} \sqrt{p(z;\mathbf{w}_j)}}{\sqrt{p(z;\mathbf{w}_i)}}. \tag{17}$$

Then

$$\mathbf{P}^T = \frac{1}{NM\,(M-1)}\mathbf{X} \int_{\mathbb{R}^d} \dot{\mathbf{K}}(z)\, \mathbf{D}(z)\, dz. \tag{18}$$

We solve (16) by iteration:

1. Initialize $\mathbf{W}$, say $\mathbf{W}_{(0)} = \mathbf{I}_M$.
2. For $l = 0, 1, \ldots$ to convergence, compute $\mathbf{P}_{(l)}$ from (18), and update $\mathbf{W}_{(l)}$ to $\mathbf{W}_{(l+1)} = G(\mathbf{W}_{(l)}) = (\mathbf{P}_{(l)}\mathbf{P}_{(l)}^T)^{-1/2}\mathbf{P}_{(l)})$.

4 Results

4.1 Image Segmentation

The image of Lizard shown in Subplot A of Fig. 1 is considered to be relatively hard to segment due to the multimodality of the *pdf* related to the object class. Moreover, the intensity distributions of the object and background classes of the image are very similar, which makes it impossible to segment the image based on gray-level information alone. To overcome this difficulty, the input image $I(u)$ was defined to be the bivariate image of the partial derivatives of Lizard, which are shown in Subplots B and C of the figure.

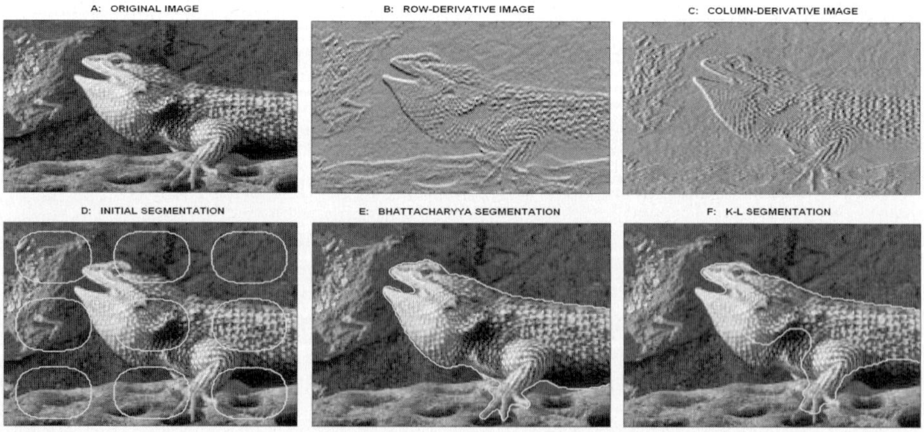

Fig. 1. (*Subplot A*) Original image of Lizard; (*Subplot B*) Row-derivative of the image; (*Subplot C*) Column-derivative of the image; (*Subplot D*) Initial segmentation; (*Subplot E*) Separation by the Bhattacharyya flow; (*Subplot F*) Separation by the K-L flow

The initial segmentation of Lizard and its segmentation obtained using the proposed method are shown in Subplots D and E of Fig. 1, respectively. For the sake of comparison, we have also segmented the image of Lizard using the active contour that maximized the Kullback-Leibler (K-L) divergence between the empirical *pdf*s of the object and background classes. The resulting segmentation is shown in Subplot F of Fig.1. It is obvious that the proposed approach (i.e., the one that exploits the Bhattacharyya metric) is the best performer here.

It is worthwhile noting that the relatively worse performance of the image segmentation using the K-L divergence seems to be stemming from the properties of the functions involved in its definition, viz. of the logarithm. In particular, the latter is known to be very sensitive to variations of its argument in vicinity of relatively small values of the latter. Moreover, the logarithm is undefined at zero, which makes computing the K-L gradient flow prone to the errors caused by inaccuracies in estimating the *tails* of probability densities. On the other hand, the square root is a well-defined function in vicinity of zero. Moreover, for relatively small values of its argument, the variability of the square root is considerably smaller than that of the logarithm. As a result, the Bhattacharyya flow is much less susceptible to the influence of the inaccuracies mentioned above.

Fig. 2. (*Subplots A1-A3*) Original image sources; (*Subplot B1-B3*) Corresponding mixtures; (*Subplot C1-C3*) Estimated sources

4.2 Blind Source Separation

Subplots A1-A3 of Fig. 2 show the original source images which have been used to test the performance of the proposed separation methodology. The corresponding mixtures obtained using a random mixing matrix **A** are shown in Subplots B1-B3 of the same figure, whereas Subplots C1-C3 of Fig.2 show the source images estimated by applying 50 iterations of the fixed point algorithm described in Section 3. One can see that the algorithm results in virtually perfect reconstruction of the image sources. For this case, the average interference-to-signal ratio (ISR) was found to be equal to 0.0024, while minimizing the mutual information between the estimated sources resulted in ISR equal to 0.036.

5 Conclusions

The present study has demonstrated the applicability and practicability of the method for separating different components of a data signal based on the notion of a distance between probability distributions. The latter was defined by means of the Bhattacharyya coefficient which was shown to be advantageous over the K-L divergence (and, hence, over the related criterion of mutual information) in practical settings, in which class-conditional densities have to be estimated in a non-paramentric manner. Additionally, the versatility of the proposed criterion was demonstrated via its application to the problems of blind separation of both geometrically and algebraically mixed sources. Thus, from a certain perspective, the proposed method can be seen as unifying for the problems of both classes.

References

1. Duda, R., Hart, R., Stork, D.: Pattern Recognition. Willey, New York (2001)
2. Han, J., Kamber, M.: Data mining: Concepts and techniques. Morgan Kaufmann, San Francisco (2001)
3. Tuzlukov, V.: Signal detection theory. Springer, Heidelberg (2001)
4. Sapiro, G.: Geometric partial differential equations and image analysis. Cambridge University Press, Cambridge (2001)
5. Hyvarinen, A., Karhunen, J., Oja, E.: Independent component analysis. John Wiley and Sons, Chichester (2001)
6. Haykin, S.: Blind deconvolution. Prentice Hall, Englewood Cliffs (1994)
7. Gelfand, I., Fomin, S., Silverman, R.: Calculus of variations. Prentice-Hall, Englewood Cliffs (1975)
8. Silverman, B.: Density estimation for statistics and data analysis. CRC Press, Boca Raton (1986)
9. Bhattacharyya, A.: On a measure of divergence between two statistical populations defined by their probability distributions. Bull. Calcutta Math. Soc. 78, 99–109 (1943)
10. Kim, L., Fisher, J., Yezzi, A., Cetin, M., Willsky, A.: A nonparametric statistical method for image segmentation using information theory and curve evolution IEEE Proc. Image Processing 4(10), 1486–1502 (2005)

Shifted Independent Component Analysis

Morten Mørup, Kristoffer H. Madsen, and Lars K. Hansen

Technical University of Denmark
Informatics and Mathematical Modelling
Richard Petersens Plads, Building 321
DK-2800 Kgs. Lyngby, Denmark
{mm,khm,lkh}@imm.dtu.dk

Abstract. Delayed mixing is a problem of theoretical interest and practical importance, e.g., in speech processing, bio-medical signal analysis and financial data modelling. Most previous analyses have been based on models with integer shifts, i.e., shifts by a number of samples, and have often been carried out using time-domain representation. Here, we explore the fact that a shift τ in the time domain corresponds to a multiplication of $e^{-i\omega\tau}$ in the frequency domain. Using this property an algorithm in the case of sources≤sensors allowing arbitrary mixing and delays is developed. The algorithm is based on the following steps: 1) Find a subspace of shifted sources. 2) Resolve shift and rotation ambiguity by information maximization in the complex domain. The algorithm is proven to correctly identify the components of synthetic data. However, the problem is prone to local minima and difficulties arise especially in the presence of large delays and high frequency sources. A Matlab implementation can be downloaded from [1].

1 Introduction

Factor analysis is widely used to reconstruct latent effects from mixtures of multiple effects based on the model

$$\mathbf{X}_{n,m} = \sum_{d} \mathbf{A}_{n,d}\mathbf{S}_{d,m} + \mathbf{E}_{n,m}, \tag{1}$$

where $\mathbf{E}_{n,m}$ is additive noise. However, this decomposition is not unique since $\widetilde{\mathbf{A}} = \mathbf{AQ}$ and $\widetilde{\mathbf{S}} = \mathbf{Q}^{-1}\mathbf{S}$ yields same approximation as $\mathbf{A}, \mathbf{S}$. Consequently, constraints have been imposed such as Varimax rotation for Principal Component Analysis (PCA) [2], statistical independence of the sources $\mathbf{S}$ as in Independent Component Analysis (ICA)[3,4]. A related strategy is sparse coding where the objective of minimizing the error is combined with a term penalizing the non-sparsity of $\mathbf{S}$ [5].

Factor analysis in the setting of ICA is often illustrated by the so-called cocktail party problem. Here mixtures of several speakers are recorded in several microphones forming the measured signal $\mathbf{X}$. The task is to identify the sources

M.E. Davies et al. (Eds.): ICA 2007, LNCS 4666, pp. 89–96, 2007.

$\mathbf{S}$ of each original speaker. However, even in an anechoic environment the mixing model is typically not accurate because of different delays in the microphones. Consider two microphones placed at distance L and $L+h$ from a given speaker. Under normal atmospheric conditions, the speed of sound is approximately $c = 344$ m/s while a typical sampling rate is $f_s = 22$ kHz. Then the delay in samples between the two microphones is given by: #samples$=\frac{f_s h}{c}$ such that the delay increases linearly with the difference in distance. Consequently, a distance of 1 cm gives a delay of 0.6395 samples while $h = 1$m leads to a delay of 63.95 samples. Harshman and Hong [6] proposed a generalization of the factor models in which the underlying sources have specific delays when they reach the sensors. The model is called shifted factor analysis (SFA), and reads

$$\mathbf{X}_{n,m} = \sum_d \mathbf{A}_{n,d}\mathbf{S}_{d,m-\widetilde{\boldsymbol{\tau}}_{n,d}} + \mathbf{E}_{n,m}. \tag{2}$$

In real acoustic environments we expect echoes due to paths that are created by reflection off surfaces. To account for general delay mixing effects, the ICA model has been generalized to convolutive mixtures, see e.g., [7,8,9]

$$\mathbf{X}_{n,m} = \sum_{\tau,d} \mathbf{A}^{\tau}_{n,d}\mathbf{S}_{d,m-\tau} + \mathbf{E}_{n,m}. \tag{3}$$

Here $\mathbf{A}^{\tau}$ is a filter that accounts for the presence of each source in the sensors at time delay τ. The shifted factor model, thus is a special case of the convolutive model where the filter coefficients $\mathbf{A}^{\tau}_{n,d} = \mathbf{A}_{n,d}$ if $\widetilde{\boldsymbol{\tau}}_{n,d} = \tau$ else $\mathbf{A}^{\tau}_{n,d} = 0$.

In fact shifted mixtures are also seen in many other contexts. For instance, astronomy where star motion Doppler effects induce frequency red shifts that can be modelled using SFA. Here we will focus on the delayed source model. In [6] strong support was found for the conjecture that the incorporation of shifts can strengthen the model enough to make the parameters identifiable up to scaling and permutation (essential uniqueness). We will demonstrate that this conjecture is not correct when allowing for arbitrary shifts. Indeed, the model is, as for regular factor analysis, ambiguous. In [10] an algorithm was proposed to estimate the model. However, the algorithm has the following drawbacks.

1. All potential shifts have to be specified in the model.
2. Exhaustive integer search for the delays is expensive.
3. The model only accounts for shifts by whole samples.
4. The model is in general not unique.

Prior to the work of [6,10] Bell and Sejnowski [4] sketched how to handle time delays in networks based on a model similar to equation 2. This was further explored in [11]. Although their algorithms derive gradients to search for the delays (alleviating the first two drawbacks above) the models are still based on pure integer delays. In [12] a different model based on equally mixed sources, i.e. $\mathbf{A} = \mathbf{1}$, formed by moving averages incorporated non-integer delays by signal interpolation. Yeredor [13] solved the SFA model by joint diagonalization of

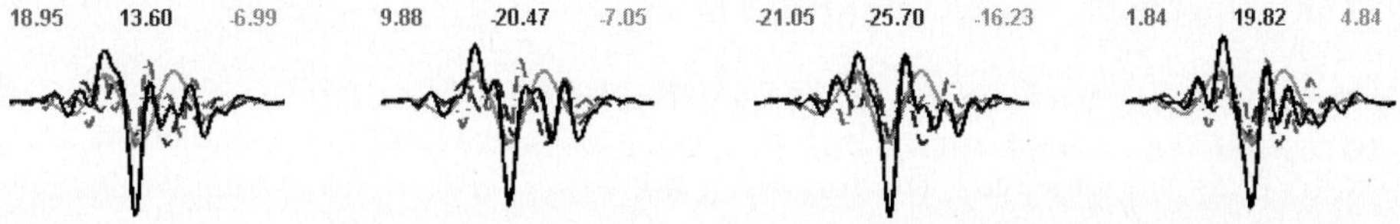

Fig. 1. Example of activities obtained (black graph) when summing three components (gray, blue dashed and red dash-dotted graphs) each shifted to various degrees (given in samples by the colored numbers). Clearly, the resulting activities are heavily impacted by the shifts such that a regular instantaneous ICA analysis would be inadequate.

the source cross spectra based on the AC-DC algorithm with non-integer shifts for the 2×2 system. This approach was extended to complex signals in [14]. The algorithm is least squares optimal for equal number of sensors and sources. More sensors than sources is not a problem for conventional ICA; we simply reduce dimension by variance decomposition, this procedure is exact for noiseless mixing. Due to the delays projection based dimensional reduction will not reproduce the simple single delay structure, but rather lead to a more general convolutive mixture. We will therefore aim at an algorithm for finding a shift invariant subspace. Hence, solve equation 2 by use of the fact that a shift τ in the time domain can be approximated by multiplication by the complex coefficients $e^{-i\omega\tau}$ in the frequency domain. This alleviates the first three drawbacks of the SFA algorithm. We will denote this algorithm a Shift Invariant Subspace Analysis (SISA). To further deal with shift and rotation ambiguities, we impose independence in the complex domain based on information-maximization (IM) [4]. Hence, we form an algorithm for ICA with shifted sources (SICA). Notice, that algorithms for ICA in the complex domain without shifts have previously been derived, see for instance [9,15] and references therein.

2 Method and Results

In the following $\mathbf{U}$ will denote a matrix in the time domain, while $\widetilde{\mathbf{U}}$ denotes the corresponding matrix in the frequency domain. $\mathcal{U}$ and $\widetilde{\mathcal{U}}$ denotes 3-way arrays in the time and frequency domains respectively. Furthermore, $\mathbf{U} \bullet \mathbf{V}$ denotes the direct product, i.e. element-wise multiplication. Also, $\omega = 2\pi\frac{f-1}{M}$ such that $\widetilde{\mathbf{U}}^{(f)} = \mathbf{U} \bullet e^{-i2\pi\frac{f-1}{M}\boldsymbol{\tau}}$. Finally, the ith row of a matrix will be denoted $\mathbf{U}_{i,:}$.

2.1 Shift Invariant Subspace Analysis (SISA)

In the following we will device an algorithm to find a shift invariant subspace based on the SFA model. Consider the SFA model and its frequency transformed

$$\mathbf{X}_{n,m} = \sum_d \mathbf{A}_{n,d}\mathbf{S}_{d,m-\boldsymbol{\tau}_{n,d}} + \mathbf{E}_{n,m}, \quad \widetilde{\mathbf{X}}_{n,f} = \sum_d \mathbf{A}_{n,d}\widetilde{\mathbf{S}}_{d,f}e^{-i2\pi\frac{f-1}{M}\boldsymbol{\tau}_{n,d}} + \widetilde{\mathbf{E}}_{n,f}. \quad (4)$$

In matrix notation this can be stated as

$$\widetilde{\mathbf{X}}_f = \widetilde{\mathbf{A}}^{(f)}\widetilde{\mathbf{S}}_f + \widetilde{\mathbf{E}}_f. \tag{5}$$

Due to Parseval's identity the following holds

$$C_{ls} = \sum_{n,m} \|\mathbf{E}_{n,m}\|_F^2 = \frac{1}{M}\sum_{n,f} \|\widetilde{\mathbf{E}}_{n,f}\|_F^2. \tag{6}$$

Thus, minimizing the least square error in the time and frequency domain is equivalent. The algorithm will be based on alternatingly solving for $\mathbf{A}$, $\mathbf{S}$ and $\boldsymbol{\tau}$.

S update: According to equation 5, $\mathbf{S}_f$ can be estimated as

$$\widetilde{\mathbf{S}}_f = \widetilde{\mathbf{A}}^{(f)^\dagger}\widetilde{\mathbf{X}}_f. \tag{7}$$

Although, $\mathbf{S}$ is updated in the frequency domain the updated version has to remain real when taking the inverse FFT. For $\mathbf{S}$ to be real valued the following has to hold

$$\widetilde{\mathbf{S}}_{M-f+1} = \widetilde{\mathbf{S}}_f^*, \tag{8}$$

where * denotes complex conjugate. This constraint is enforced by updating the first $\lfloor M/2 \rfloor + 1$ elements, i.e. up to the Nyquist frequency, while setting the remaining elements according to equation 8.

A update: Let $\widetilde{\mathbf{S}}_{d,f}^{(n)}$ denote the delayed version of the source signal $\widetilde{\mathbf{S}}_{d,f}$ to the nth channel, i.e. $\widetilde{\mathbf{S}}_{d,f}^{(n)} = \widetilde{\mathbf{S}}_{d,f} e^{-i2\pi\frac{f-1}{M}\boldsymbol{\tau}_{n,d}}$. Then equation 2 can be restated as

$$\mathbf{X}_{n,:} = \mathbf{A}_{n,:}\mathbf{S}^{(n)} + \mathbf{E}_{n,:}, \tag{9}$$

This is the regular factor analysis problem giving the update

$$\mathbf{A}_{n,:} = \mathbf{X}_{n,:}\mathbf{S}^{(n)^\dagger}. \tag{10}$$

$\boldsymbol{\tau}$ update: The least square error for the model stated in equation 5, is given by

$$C_{ls} = \frac{1}{M}\sum_f (\widetilde{\mathbf{X}}_f - \widetilde{\mathbf{A}}^{(f)}\widetilde{\mathbf{S}}_f)^H(\widetilde{\mathbf{X}}_f - \widetilde{\mathbf{A}}^{(f)}\widetilde{\mathbf{S}}_f), \tag{11}$$

where ^{H} denotes the conjugate transpose. Define $\mathbf{T}^{ND\times 1} = vec(\boldsymbol{\tau})$, i.e. the vectorized version of the matrix $\boldsymbol{\tau}$ such that $\mathbf{T}_{n+(d-1)N} = \boldsymbol{\tau}_{n,d}$. Let further

$$\widetilde{\mathcal{Q}}_{n,d,f} = \widetilde{\mathbf{A}}_{n,d}^{(f)}\widetilde{\mathbf{S}}_{d,f}, \quad \widetilde{\mathbf{E}}_f = \widetilde{\mathbf{X}}_f - \widetilde{\mathbf{A}}^{(f)}\widetilde{\mathbf{S}}_f. \tag{12}$$

Then the gradient of C_{ls} with respect to $\boldsymbol{\tau}_{n,d}$ is given as

$$\mathbf{g}_{n+(d-1)N} = \frac{\partial C_{ls}}{\partial \mathbf{T}_{n+(d-1)N}} = \frac{\partial C_{ls}}{\partial \boldsymbol{\tau}_{n,d}} = \frac{-1}{M}\sum_f 2\omega\Im[\widetilde{\mathcal{Q}}_{n,d,f}\widetilde{\mathbf{E}}_{n,f}^*] \tag{13}$$

The Hessian has the following structure

$$\mathbf{H}_{n+(d-1)N,n'+(d'-1)N} = \begin{cases} \frac{-2}{M}\sum_f \omega^2 \Re[\widetilde{\mathcal{Q}}_{n,d,f}\widetilde{\mathcal{Q}}^*_{n',d',f}] \text{ if } \mathrm{n} \neq \mathrm{n}' \wedge \mathrm{d} \neq \mathrm{d}' \\ \frac{-2}{M}\sum_f \omega^2 \Re[\widetilde{\mathcal{Q}}_{n,d,f}(\widetilde{\mathcal{Q}}^*_{n',d',f} + \widetilde{\mathbf{E}}^*_{n',f})] \text{ if } \mathrm{n}{=}\mathrm{n}' \wedge \mathrm{d}{=}\mathrm{d}' \end{cases} \tag{14}$$

As a result, $\boldsymbol{\tau}$ can be estimated using the Newton-Raphson method

$$\mathbf{T} \leftarrow \mathbf{T} - \eta \mathbf{H}^{-1}\mathbf{g}, \tag{15}$$

where η is a step size parameter that is tuned to keep decreasing the cost function. The above iterative update for $\boldsymbol{\tau}$ is sensitive to local minima. Thus, to improve the algorithm from being stuck in suboptimal solutions $\boldsymbol{\tau}$ was re-estimated by the following cross-correlation procedure every 10th iteration. Let

$$\widetilde{\mathbf{R}}_{n,f} = \widetilde{\mathbf{X}}_{n,f} - \sum_{d \neq d'} \widetilde{\mathbf{A}}^{(f)}_{n,d}\widetilde{\mathbf{S}}_{d,f}, \tag{16}$$

i.e. the signal at the nth sensor at frequency f when projecting all but the d' source out of $\widetilde{\mathbf{X}}$. The cross-correlation between the d' source and n^{th} sensor is given as $\widetilde{\mathbf{c}}_f = \widetilde{\mathbf{R}}^*_{n,f}\widetilde{\mathbf{S}}_{d',f}$, such that $\boldsymbol{\tau}_{n,d'}$ can be estimated as

$$t = \arg\max_m |\mathbf{c}_m|, \quad \boldsymbol{\tau}_{n,d'} = t - (M+1), \quad \mathbf{A}_{n,d'} = \frac{\mathbf{c}_t}{\mathbf{S}_{d',:}\mathbf{S}^T_{d',:}}. \tag{17}$$

I.e. as the delay corresponding to maximum cross-correlation between the sensor and source. The value of $\mathbf{A}_{n,d'}$ corresponding to this delay is also given above.

2.2 SISA Is Not Unique

According to equation 5, the reconstructed signal in the complex domain is given as $\widetilde{\mathbf{X}}_f \approx \widetilde{\mathbf{A}}^{(f)}\widetilde{\mathbf{S}}_f = \widetilde{\mathbf{A}}^{(f)}\widetilde{\mathbf{W}}^{(f)}\widetilde{\mathbf{W}}^{(f)^{-1}}\widetilde{\mathbf{S}}_f$.Such that $\widetilde{\mathbf{W}}^{(f)} = \mathbf{W} \bullet e^{-i2\pi\frac{f-1}{M}\hat{\boldsymbol{\tau}}}$ is a rotation, scaling and shift matrix. Assume the inverse of $\widetilde{\mathbf{W}}^{(f)}$ is also a rotation, scaling and shift matrix, i.e. $\widetilde{\mathbf{W}}^{(f)^{-1}} = \mathbf{V} \bullet e^{-i2\pi\frac{f-1}{M}\check{\boldsymbol{\tau}}}$. Since $\widetilde{\mathbf{W}}^{(f)}\widetilde{\mathbf{W}}^{(f)^{-1}} = \mathbf{I}$, we find

$$\sum_{d''} \mathbf{W}_{d,d''}\mathbf{V}_{d',d''}e^{-i2\pi\frac{f-1}{M}(\hat{\boldsymbol{\tau}}_{d,d''}+\check{\boldsymbol{\tau}}_{d',d''})} = \begin{cases} 0 \text{ for } d \neq d' \forall\, f \\ 1 \text{ for } d = d' \forall\, f \end{cases} \tag{18}$$

From $f = 1$ we obtain the relation $\mathbf{V} = \mathbf{W}^{-1}$. For the remaining frequencies this expression can only be valid if $\hat{\boldsymbol{\tau}}_{dd''} + \check{\boldsymbol{\tau}}_{d''d} = 0$ (diagonal elements) and $\hat{\boldsymbol{\tau}}_{dd''} + \check{\boldsymbol{\tau}}_{d''d'} = k_{dd'}$ (off diagonal elements) where $k_{dd'}$ denotes an arbitrary constant. The first relation gives the constraint that $\hat{\boldsymbol{\tau}} = -\check{\boldsymbol{\tau}}^T$. The second relation further constraints all the elements of the columns of $\hat{\boldsymbol{\tau}}$ to be equal. Thus the ambiguity is given by $\widetilde{\mathbf{W}}^{(f)} = [\mathbf{W}\,\mathrm{diag}(e^{-i2\pi\frac{f-1}{M}\widehat{\boldsymbol{\tau}}})]$. Where $\widehat{\boldsymbol{\tau}}$ is a vector describing the shift ambiguity.

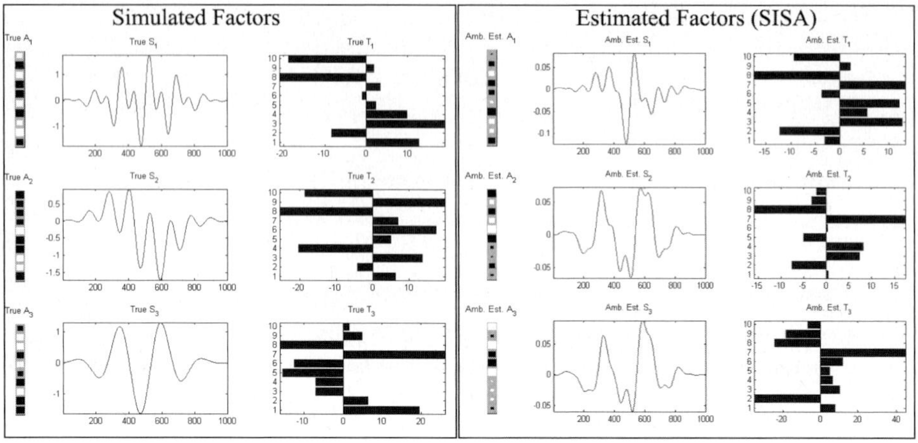

Fig. 2. Results obtained by a shift invariant subspace analysis (SISA). Left panel: the true factors forming a synthetic data set. To the left, the strength of the mixing **A** of each source is indicated in gray color scale. In the middle, the three sources are shown and to the right is given the time delays of each source to each channel. Right panel: The estimated factors from the SISA analysis. Although, all the variance is explained the decomposition has not identified the true underlying components but an ambiguous mix. Clearly, as for regular factor analysis the SISA is not unique.

2.3 Shifted Independent Component Analysis (SICA)

A common approach to ICA is the maximum likelihood (ML) method [16] which corresponds to the approach of maximizing information proposed in [4]. In the framework of ML a non-gaussian distribution on the sources is assumed such that ambiguity can be resolved up to the trivial ambiguities of scale, permutation and source shifting relative to the time delays.

Define, $\widetilde{\mathbf{U}}_f = \widetilde{\mathbf{W}}^{(f)}\widetilde{\mathbf{S}}_f$, i.e. the sources at frequency f when transformed according to the rotation and shift ambiguity described in the previous section. The ambiguity can be resolved by maximizing the log-likelihood assuming the (non-gaussian) Laplace distribution $p(\widetilde{\mathbf{U}}_f) \propto e^{-|\widetilde{\mathbf{U}}_{d,f}|}$, i.e.

$$p(\widetilde{\mathbf{S}}_f|\mathbf{W},\widehat{\boldsymbol{\tau}}) = \prod_f p(\widetilde{\mathbf{S}}_f|\mathbf{W},\widehat{\boldsymbol{\tau}}) = \prod_f |det(\widetilde{\mathbf{W}}^{(f)})| p(\widetilde{\mathbf{W}}^{(f)}\widetilde{\mathbf{S}}_f) \tag{19}$$

Such that the log-likelihood as a function of $\mathbf{W}$ and $\widehat{\boldsymbol{\tau}}$ becomes

$$\mathcal{L}(\mathbf{W},\widehat{\boldsymbol{\tau}}) = \sum_f \ln|\det(\widetilde{\mathbf{W}}^{(f)})| - \sum_d |\widetilde{\mathbf{W}}^{(f)}\widetilde{\mathbf{S}}_f|_d \tag{20}$$

By maximizing $\mathcal{L}(\mathbf{W},\widehat{\boldsymbol{\tau}})$ $\mathbf{W}$ and $\widehat{\boldsymbol{\tau}}$ is estimated and a new unambiguous $\mathbf{S}$ solution found by $\widetilde{\mathbf{S}}_f = \widetilde{\mathbf{W}}^{(f)}\widetilde{\mathbf{S}}_f$. The corresponding mixing and delays can be estimated alternating between the $\mathbf{A}$ and $\boldsymbol{\tau}$ update. We initialized $\mathbf{A}$ as $\mathbf{A} = \mathbf{A}\mathbf{W}^{-1}$ and $\boldsymbol{\tau}_{i,d}$ by the cross-correlation procedure.

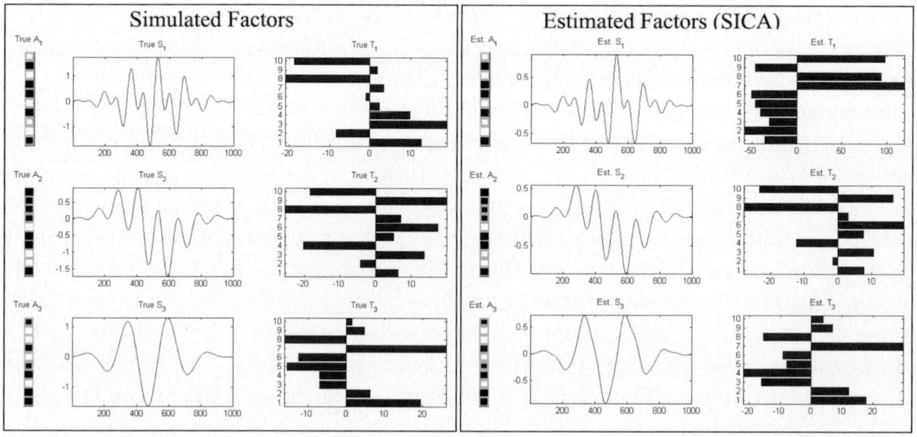

Fig. 3. Result obtained using the SICA on the decomposition found using SISA. By imposing independence, e.g., requiring the amplitudes in the frequency domain to be sparse, the rotation and shift ambiguity inherited in the model is resolved. Clearly the true underlying components and their respective mixing are correctly identified. However, a local minimum has been found, resulting in errors in the estimation of the delays particularly for the first component.

3 Discussion

Traditionally, ICA analysis is based on subspace analysis often using singular value decomposition (SVD). The sources are then found by rotating the vectors spanning the subspace according to a measure of independence. Similarly, we derived the SISA algorithm to find a shift invariant subspace by alternating least squares. Shift and rotation ambiguities were solved by imposing independence on the amplitudes of the frequency transform of the sources. While SVD has a closed form solution the SISA algorithm is non-convex. Estimating both $\mathbf{A}$, $\mathbf{S}$ and each delay in $\boldsymbol{\tau}$ using the cross-correlation procedure has a closed form solution for fixed values of $\boldsymbol{\tau}$, $\mathbf{S}$ and $\mathbf{A}$. While the cross correlation procedure only finds integer delays the Newton-Rhapson procedure can estimate the non-integer delays. The cross-correlation procedure greatly reduces the algorithm's vulnerability to local minima, however due to the alternating least squares estimation the problem cannot be circumvented completely. Furthermore, the problem becomes increasingly difficult for high frequency sources and large shifts due to additional local minima. In an example we saw this happen: The SICA algorithm failed in correctly identifying the delays of the first component; the component with the highest frequencies. A multistart strategy was invoked, we choose the best of ten random initializations to obtain a good initial solution for the estimation of the shift invariant subspace. While our algorithm was based on likelihood maximization, Yeredor [13] developed an algorithm based on joint diagonalization. The present SISA is potentially useful as a preprocessing step for this latter algorithm when estimating less sources than sensors.

Previous work based on integer shifts conjectured the decomposition to be unique [6]. When using integer shifts some shifts might perform better than others due to a better integer rounding error. Hence, this might be why the integer shifts formed seemingly unique solutions. However, as demonstrated in figure 2 the shifted factor analysis model is not in general unique. But, by imposing independence unique solutions can be obtained up to trivial permutation, scaling and specific onset relative to the delays of the sources as demonstrated in figure 3. The shift/delay model may prove useful for a wide range of data where ICA already has been employed. Furthermore, the extra information of delays can be useful for spatial source localization when combined with information of position of the sensors. Future work will focus on implementing additional constraints such as non-negativity and attempt to further improve the identifiability in the presence of many local minima. The current algorithm can be downloaded from [1].

References

1. Mørup, M., Madsen, K.H.: Algorithm for sica (2007), `www2.imm.dtu.dk/pubdb/views/publication_details.php?id=5206`
2. Kaiser, H.F.: The varimax criterion for analytic rotation in factor analysis. Psychometrica 23, 187–200 (1958)
3. Comon, P.: Independent component analysis, a new concept? Signal Processing 36, 287–314 (1994)
4. Bell, A.J., Sejnowski, T.J.: An information maximization approach to blind source separation and blind deconvolution. Neural Computation 7, 1129–1159 (1995)
5. Olshausen, B.A., Field, D.: Emergence of simple-cell receptive field propertises by learning a sparse code for natural images. Nature 381, 607–609 (1996)
6. Harshman, R., Hong, S., Lundy, M.: Shifted factor analysis—part i: Models and properties. Journal of Chemometrics 17, 363–378 (2003)
7. Attias, H., Schreiner, C.: Blind source separation and deconvolution: the dynamic component analysis algorithm. Neural Computation 10(6), 1373–1424 (1998)
8. Parra, L., Spence, C., Vries, B.: Convolutive blind source separation based on multiple decorrelation. In: IEEE Workshop on Neural Networks and Signal Processing, pp. 23–32 (1998)
9. Anemuller, J., Sejnowski, T.J., Makeig, S.: Complex independent component analysis of frequency-domain electroencephalographic data. Neural Networks 16(9), 1311–1323 (2003)
10. Harshman, R., Hong, S., Lundy, M.: Shifted factor analysis—part ii: Algorithms. Journal of Chemometrics 17, 379–388 (2003)
11. Torkkola, K.: Blind separation of delayed sources based on information maximization. Acoustics, Speech, and Signal Processing. ICASSP-96 6, 3509–3512 (1996)
12. Emile, B., Comon, P.: Estimation of time delays between unknown colored signals. Signal Processing 68(1), 93–100 (1998)
13. Yeredor, A.: Time-delay estimation in mixtures. ICASSP 5, 237–240 (2003)
14. Yeredor, A.: Blind source separation in the presence of doppler frequency shifts. ICASSP 5, 277–280 (2005)
15. Cardoso, J.F., Tulay, A.: The maximum likelihood approach to complex ica. ICASSP, pp. 673–676 (2006)
16. Hyvarinen, A., Karhunen, J., Oja, E.: Independent Component Analysis. John Wiley and Sons, Chichester (2001)

Modeling and Estimation of Dependent Subspaces with Non-radially Symmetric and Skewed Densities

Jason A. Palmer[1], Ken Kreutz-Delgado[2], Bhaskar D. Rao[2], and Scott Makeig[1]

[1] Swartz Center for Computational Neuroscience
University of California San Diego, La Jolla, CA 92093
{jason,scott}@sccn.ucsd.edu
[2] Department of Electrical and Computer Engineering
University of California San Diego, La Jolla, CA 92093
{kreutz,brao}@ece.ucsd.edu

Abstract. We extend the Gaussian scale mixture model of dependent subspace source densities to include non-radially symmetric densities using Generalized Gaussian random variables linked by a common variance. We also introduce the modeling of skew in source densities and subspaces using a generalization of the Normal Variance-Mean mixture model. We give closed form expressions for subspace likelihoods and parameter updates in the EM algorithm.

1 Introduction

The Gaussian scale mixture representation can be extended to vector subspaces to yield a model of non-affine dependency, sometimes referred to as "variance dependency" [8]. Hyvärinen [8,9] has recently proposed such a model for Independent Subspace Analysis of images. A similar approach is developed by Eltoft, Kim et al. [11,6], which is referred to as Independent Vector Analysis (IVA). In the IVA model, the EM algorithm is used with a particular case of the multivariate Gaussian scale mixture involving a Gamma mixing density.

In [10] a method is proposed for convolutive blind source separation in reverberative environments using a frequency domain approach with sources having variance (scale) dependency across frequencies. Typically in the frequency domain approach to blind deconvolution, the "permutation problem" arises when the signals are unmixed separately at each frequency. Due to the permutation ambiguity inherent in ICA [1], the frequency components of each source are output in arbitrary order at each frequency, and some matching heuristic must be employed to reconstruct the complete spectra of the sources. The IVA model allows the modeling of dependency of the frequency components of sources, while maintaining the mutual independence of the sources.

Variance dependency also arises in EEG/MEG analysis. In particular, the electromagnetic signals generated by muscles in the scalp, face, and ears will commonly activate together in various facial expressions. In this case, the individual muscle signals are not related or dependent in phase, but their variance increases and decreases together as the components are activated and deactivated together. Variance dependency may also exist among cortex regions that are simultaneously active in certain contexts.

M.E. Davies et al. (Eds.): ICA 2007, LNCS 4666, pp. 97–104, 2007.

The densities employed in models proposed previously for speech use only a particular dependent subspace density model, which may limit the flexibility of the model in application to more general domains such as communications and biological signal processing. We propose a general method for constructing multivariate Gaussian scale mixtures, giving an example of a multivariate dependent logistic density.

We also propose a scale mixture of Generalized Gaussians model, in which a generalized Gaussian random vector with independent components, is multiplied by a common scalar variance parameter, which is distributed Generalized Inverse Gaussian. This yields a generalization of the *generalized hyperbolic density* of Barndorff-Nielsen [3].

Finally we show how to use the Normal variance-mean mixtures to model skew in dependent subspaces. The location and "drift" parameters can be updated in closed form using the EM algorithm and exploiting the conditional Gaussianity and closed form formula for the posterior moment in terms of derivatives of the multivariate density function.

2 General Dependent Gaussian Scale Mixtures

In this section we show how general dependent multivariate densities can be derived using scalar Gaussian scale mixtures,

$$x = \xi^{1/2} z$$

where z is a standard Normal random variable, and ξ is a non-negative random variable.

2.1 Example Densities

Examples of Gaussian scale mixtures include the *generalized Gaussian* density, which has the form,

$$\mathcal{GG}(x;\rho) = \frac{1}{2\Gamma(1+\frac{1}{\rho})} e^{-|x|^\rho}$$

It is a Gaussian scale mixture for $0 < \rho \leq 2$. The scale mixing density is related to a positive alpha stable density of order $\rho/2$.

The *generalized Cauchy* has the form,

$$\mathcal{GC}(x;\alpha,\nu) = \frac{\alpha\Gamma(\nu+1/\alpha)}{2\Gamma(\nu)\Gamma(1/\alpha)} \frac{1}{(1+|x|^\alpha)^{\nu+1/\alpha}}$$

The Generalized Cauchy is a Gaussian scale mixture for $\nu > 0$ and $0 < \alpha < 2$. The scale mixing density is related to the Gamma density.

The *generalized Logistic*, also called the symmetric Fisher's z distribution [3], has the form,

$$\mathcal{GL}(x;\alpha) = \frac{\Gamma(2\alpha)}{\Gamma(\alpha)^2} \frac{e^{-\alpha x}}{(1+e^{-x})^{2\alpha}}$$

The Generalized Logistic is a Gaussian scale mixture for all $\alpha > 0$. The scale mixing density is related to the Kolmogorov-Smirnov distance statistic [2,3,7].

2.2 Multidimensional Analogues

If x is distributed according to the Gaussian scale mixture density $p(x)$, then,

$$p(\sqrt{x}) = \frac{1}{(2\pi)^{1/2}} \int_0^\infty \xi^{-1/2} e^{-\frac{1}{2}\xi^{-1}x} p(\xi) d\xi \tag{1}$$

We can construct a random vector by multiplying the same scalar random variable $\xi^{1/2}$ by a Gaussian random vector,

$$\mathbf{x} = \xi^{1/2}\mathbf{z}$$

where $\mathbf{z} \sim \mathcal{N}(\mathbf{0}, \mathbf{I})$. For the density of $\mathbf{x}$ we then have,

$$p(\mathbf{x}) = \frac{1}{(2\pi)^{d/2}} \int_0^\infty \xi^{-d/2} e^{-\frac{1}{2}\xi^{-1}\|\mathbf{x}\|^2} p(\xi) d\xi$$

If ξ^{-1} is a Gamma random variable, then the density of $\mathbf{x}$ can be written in terms of the modified Bessel function of the second kind [6].

In general, taking the kth derivative of both sides of (1), we find,

$$\frac{d^k}{dx^k} p(\sqrt{x}) = \frac{(-2)^{-k}}{(2\pi)^{1/2}} \int_0^\infty \xi^{-k-1/2} e^{-\frac{1}{2}\xi^{-1}x} p(\xi) d\xi$$

Thus, if d is odd, then with $k = (d-1)/2$,

$$\pi^{-(d-1)/2}(-D)^{(d-1)/2} p(\sqrt{x}) = \frac{1}{(2\pi)^{d/2}} \int_0^\infty \xi^{-d/2} e^{-\frac{1}{2}\xi^{-1}x} p(\xi) d\xi$$

and we can write the density of $p(\mathbf{x})$

$$d\ \textbf{odd}: \qquad p(\mathbf{x}) = \pi^{-(d-1)/2}(-D)^{(d-1)/2} p(\sqrt{x})\big|_{x=\|\mathbf{x}\|^2} \tag{2}$$

For even d, the density of $p(\mathbf{x})$ can be written formally in terms of the Weyl fractional derivative. However as the fractional derivative is is not generally obtainable in closed form, we consider a modification of the original univariate scale density $p(\xi)$,

$$\tilde{p}(\xi) = \frac{\xi^{-1/2} p(\xi)}{\int_0^\infty \xi^{-1/2} p(\xi) d\xi}$$

With this modified scale density, the density of x evaluated at $\sqrt{x}$ becomes,

$$p(\sqrt{x}) = \frac{\mathcal{Z}}{(2\pi)^{1/2}} \int_0^\infty e^{-\frac{1}{2}\xi^{-1}x} \tilde{p}(\xi) d\xi \tag{3}$$

where,

$$\mathcal{Z} = \int_0^\infty \xi^{-1/2} p(\xi) d\xi$$

Proceeding as we did for odd d, taking the kth derivative of both sides of (3), with $k = d/2$, we get,

$$d\ \textbf{even}: \quad p(\mathbf{x}) = \mathcal{Z}^{-1}\sqrt{2}\pi^{-(d-1)/2}(-D)^{d/2} p(\sqrt{x})\big|_{x=\|\mathbf{x}\|^2} \tag{4}$$

2.3 Posterior Moments of Gaussian Scale Mixtures

To use scale mixtures in the EM context, it is necessary to calculate posterior moments of the scaling random variable. This section indicates how this is accomplished [5]. Differentiating under the (absolutely convergent) integral we get,

$$p'(x) = \frac{d}{dx}\int_0^\infty p(x|\xi)p(\xi)d\xi \;=\; -\int_0^\infty \xi^{-1}x\,p(x,\xi)\,d\xi$$
$$= -x\,p(x)\int_0^\infty \xi^{-1}p(\xi|x)\,d\xi$$

Thus, with $p(x) = \exp(-f(x))$, we see that,

$$E(\xi_i^{-1}|x_i) \;=\; \int_0^\infty \xi_i^{-1}p(\xi_i|x_i)\,d\xi_i \;=\; -\frac{p'(x_i)}{x_i p(x_i)} \;=\; \frac{f'(x_i)}{x_i} \tag{5}$$

Similar formulae can be derived for higher order posterior moments, and moments of multivariate scale parameters. These results are used in deriving EM algorithms for fitting univariate and multivariate Gaussian scale mixtures.

2.4 Example: 3D Dependent Logistic

Suppose we wish to formulate a dependent Logistic type density on $\mathbb{R}^3$. The scale mixing density in the Gaussian scale mixture representation for the Logistic density is related to the Kolmogorov-Smirnov distance statistic [2,3,7], which is only expressible in series form. However, we may determine the multivariate density produced from the product,

$$\mathbf{x} = \xi^{1/2}\mathbf{z}$$

where $\mathbf{x}, \mathbf{z} \in \mathbb{R}^3$, and $\mathbf{z} \sim \mathcal{N}(\mathbf{0}, \mathbf{I})$. Using the formula (2) with $d = 3$, we get,

$$p(\mathbf{x}) = \frac{1}{8\pi}\,\frac{\sinh\left(\frac{1}{2}\|\mathbf{x}\|\right)}{\|\mathbf{x}\|\cosh^3\left(\frac{1}{2}\|\mathbf{x}\|\right)}$$

3 Non-radially Symmetric Dependency Models

A possible limitation of the Gaussian scale mixture dependent subspace model is the implied radial symmetry of vectors in the subspace, which leads to non-identifiability of features within the subspace—only the subspace itself can be identified. However, a similar approach using multivariate Generalized Gaussian scale mixtures can be developed, in which the multivariate density becomes a function of the p-norm of the subspace vector rather than the radially symmetric 2-norm, maintaining the directionality and identifiability of the within-subspace features, while preserving their (non-affine) dependence.

The mixing density of the generalized hyperbolic distribution is the generalized inverse Gaussian, which has the form,

$$\mathcal{N}^{\dagger}(\delta,\kappa,\lambda) = \frac{(\kappa/\delta)^{\lambda}}{2K_{\lambda}(\delta\kappa)}\,\xi^{\lambda-1}\exp\left(-\tfrac{1}{2}\left(\delta^2\xi^{-1}+\kappa^2\xi\right)\right) \tag{6}$$

where K_λ is the Bessel K function, or modified Bessel function of the second kind. The moments of the generalized inverse Gaussian [6] are given by,

$$E(\xi^r) = \left(\frac{\delta}{\kappa}\right)^r \frac{K_{\lambda+r}(\delta\kappa)}{K_\lambda(\delta\kappa)} \tag{7}$$

The isotropic generalized hyperbolic distribution [3] in dimension d,

$$\mathcal{GH}(\delta,\kappa,\lambda) = \frac{1}{(2\pi)^{d/2}} \frac{\kappa^{d/2}}{\delta^\lambda K_\lambda(\delta\kappa)} \frac{K_{\lambda-d/2}\left(\kappa\sqrt{\delta^2 + \|\mathbf{x}\|^2}\right)}{\left(\delta^2 + \|\mathbf{x}\|^2\right)^{d/4-\lambda/2}} \tag{8}$$

is derived as a Gaussian scale mixture with $\mathcal{N}^\dagger(\delta,\kappa,\lambda)$ mixing density. Now, for a generalized Gaussian scale mixture,

$$p(\mathbf{x}) = \frac{1}{\mathcal{Z}(\mathbf{p})} \int_0^\infty \xi^{-\sum_i p_i^{-1}} \exp\left(-\xi^{-1} \textstyle\sum_i |x_i|^{p_i}\right) p(\xi)\, d\xi \tag{9}$$

where,

$$\mathcal{Z}(\mathbf{p}) = 2^d \prod_{i=1}^d \Gamma(1 + 1/p_i)$$

with $\mathcal{N}^\dagger$ mixing density $p(\xi)$, the posterior density of ξ given $\mathbf{x}$ is also $\mathcal{N}^\dagger$,

$$p(\xi|\mathbf{x}) = \mathcal{N}^\dagger\left(\sqrt{\delta^2 + 2\,\|\mathbf{x}\|_{\mathbf{p}}^{\bar{p}}}\,, \kappa, \lambda - d/\bar{p}\right) \tag{10}$$

where $\bar{p}$ is the harmonic mean $d/\sum_i p_i^{-1}$, and

$$\|\mathbf{x}\|_{\mathbf{p}} \triangleq \left(\sum_{i=1}^d |x_i|^{p_i}\right)^{1/\bar{p}}$$

For $\mathbf{x}$ we then get the anisotropic distribution,

$$p(\mathbf{x};\delta,\kappa,\lambda,\mathbf{p}) = \frac{1}{\mathcal{Z}(\mathbf{p})} \frac{\kappa^{d/\bar{p}}}{\delta^\lambda K_\lambda(\delta\kappa)} \frac{K_{\lambda-d/\bar{p}}\left(\kappa\sqrt{\delta^2 + 2\,\|\mathbf{x}\|_{\mathbf{p}}^{\bar{p}}}\right)}{\left(\delta^2 + 2\,\|\mathbf{x}\|_{\mathbf{p}}^{\bar{p}}\right)^{(d/\bar{p}-\lambda)/2}} \tag{11}$$

Using (7) and (10), we have,

$$E\left(\xi^{-1}|\mathbf{x}\right) = \frac{\kappa}{\sqrt{\delta^2 + 2\,\|\mathbf{x}\|_{\mathbf{p}}^{\bar{p}}}} \frac{K_{\lambda-d/\bar{p}-1}\left(\kappa\sqrt{\delta^2 + 2\,\|\mathbf{x}\|_{\mathbf{p}}^{\bar{p}}}\right)}{K_{\lambda-d/\bar{p}}\left(\kappa\sqrt{\delta^2 + 2\,\|\mathbf{x}\|_{\mathbf{p}}^{\bar{p}}}\right)} \tag{12}$$

The EM algorithm does not require that the complete log likelihood be maximized at each step, but only that it be increased, yielding the generalized EM (GEM) algorithm [4,13]. We employ this method here to increase the complete likelihood in (9) (see [13,12]).

4 Skew Models

4.1 Construction of Multivariate Skew Densities from Gaussian Scale Mixtures

Given a Gaussian scale mixture $\mathbf{x} = \xi^{1/2}\mathbf{z}$,

$$p(\mathbf{x}) = \frac{1}{(2\pi)^{d/2}|\Sigma|^{1/2}} \int_0^\infty \xi^{-d/2} \exp\left(-\tfrac{1}{2}\xi^{-1}\mathbf{x}^T\Sigma^{-1}\mathbf{x}\right)p(\xi)\,d\xi$$

we have, trivially, for arbitrary β,

$$\frac{p(\mathbf{x})\exp(\beta^T\Sigma^{-1}\mathbf{x})}{\varphi\left(\frac{1}{2}\beta^T\Sigma^{-1}\beta\right)} = \frac{1}{(2\pi)^{d/2}|\Sigma|^{1/2}}\times$$

$$\int_0^\infty \xi^{-d/2}\exp\left(-\tfrac{1}{2}\xi^{-1}\mathbf{x}^T\Sigma^{-1}\mathbf{x} + \beta^T\Sigma^{-1}\mathbf{x} - \tfrac{1}{2}\xi\beta^T\Sigma^{-1}\beta\right)\frac{p(\xi)\exp\left(\frac{1}{2}\xi\beta^T\Sigma^{-1}\beta\right)}{\varphi\left(\frac{1}{2}\beta^T\Sigma^{-1}\beta\right)}\,d\xi \tag{13}$$

where $\varphi(t) = E\exp t\xi$ is the moment generating function of ξ. Now (13) can be written,

$$\tilde{p}(\mathbf{x}) = \frac{1}{(2\pi)^{d/2}|\Sigma|^{1/2}} \int_0^\infty \xi^{-d/2}\exp\left(-\tfrac{1}{2}\xi^{-1}\|\mathbf{x} - \xi\beta\|^2_{\Sigma^{-1}}\right)\tilde{p}(\xi;\beta)\,d\xi \tag{14}$$

where,

$$\tilde{p}(\mathbf{x}) = \frac{p(\mathbf{x})\exp(\beta^T\Sigma^{-1}\mathbf{x})}{\varphi\left(\frac{1}{2}\|\beta\|^2_{\Sigma^{-1}}\right)}, \quad \tilde{p}(\xi;\beta) = \frac{p(\xi)\exp\left(\frac{1}{2}\xi\|\beta\|^2_{\Sigma^{-1}}\right)}{\varphi\left(\frac{1}{2}\|\beta\|^2_{\Sigma^{-1}}\right)}$$

We have thus constructed a skewed density $\tilde{p}(\mathbf{x})$ in terms of the isotropic density $p(\mathbf{x})$ and the moment generating function φ of the scale mixing density $p(\xi)$. The skewed density is a that of a location-scale mixture [3] of the Gaussian $\mathbf{z} \sim \mathcal{N}(\mathbf{0}, \Sigma)$,

$$\mathbf{x} = \xi^{1/2}\mathbf{z} + \xi\beta.$$

4.2 EM Algorithm Posterior Updates

We now assume arbitrary location parameter μ, along with drift β, and structure matrix Σ. To use the EM algorithm with the Gaussian complete log likelihood in (14), we need to calculate posterior expectation of ξ^{-1}.

We do this using the method of §2.3. If we take the derivative of $-\log p(\mathbf{x} - \mu)$ with respect to $\frac{1}{2}\|\mathbf{x} - \mu\|^2_{\Sigma^{-1}}$, then we get,

$$\frac{\partial}{\partial\frac{1}{2}\|\mathbf{x}-\mu\|^2_{\Sigma^{-1}}}\left(-\log\int p(\mathbf{x}-\mu,\xi)d\xi\right)$$

$$= \frac{\int \xi^{-1}p(\mathbf{x}-\mu,\xi)\,d\xi}{\int p(\mathbf{x}-\mu,\xi)\,d\xi} = \frac{\int \xi^{-1}\tilde{p}(\mathbf{x}-\mu,\xi)\,d\xi}{\int \tilde{p}(\mathbf{x}-\mu,\xi)\,d\xi} = E(\xi^{-1}|\mathbf{x})$$

Thus, from (2) and (4), with $k \triangleq \lfloor d/2 \rfloor$ (the greatest integer less than $d/2$) we have,

$$E(\xi^{-1}|\mathbf{x}) = \frac{-1}{\|\mathbf{x}-\mu\|_{\Sigma^{-1}}}\frac{p^{(k+1)}\left(\|\mathbf{x}-\mu\|_{\Sigma^{-1}}\right)}{p^{(k)}\left(\|\mathbf{x}-\mu\|_{\Sigma^{-1}}\right)}$$

where $p^{(k)}$ is kth derivative of the univariate scale mixture $p(x)$.

4.3 Closed Form Parameter Updates

Given N observations $\{\mathbf{x}_k\}_{k=1}^N$, the μ that maximizes the complete log likelihood is found to be,

$$\mu = \frac{\frac{1}{N}\sum_k \gamma_k \mathbf{x}_k - \beta}{\frac{1}{N}\sum_k \gamma_k} \tag{15}$$

where $\gamma_k = E(\xi^{-1}|\mathbf{x}_k)$.

The estimation equation to be solved for β, which does not involve the posterior estimate of ξ_k, is,

$$\frac{\varphi'\left(\frac{1}{2}\|\beta\|^2_{\Sigma^{-1}}\right)}{\varphi\left(\frac{1}{2}\|\beta\|^2_{\Sigma^{-1}}\right)}\,\beta \;=\; \mathbf{c} - \mu \tag{16}$$

where $\mathbf{c} = \frac{1}{N}\sum_k \mathbf{x}_k$. This gives β in terms of μ up to a scale factor. Given μ, the optimal β, denoted β^*, may be found by first determining $\zeta \triangleq \frac{1}{2}\|\beta^*\|^2_{\Sigma^{-1}}$ from,

$$h(\zeta) \;\triangleq\; \left(\frac{\varphi'(\zeta)}{\varphi(\zeta)}\right)^2 \zeta \;=\; \tfrac{1}{2}\,\|\mathbf{c} - \mu\|^2_{\Sigma^{-1}}$$

assuming that the univariate function h is invertible. Then β^* is given as,

$$\beta^* = \frac{\varphi(\zeta)}{\varphi'(\zeta)}(\mathbf{c} - \mu)$$

Given β^*, we may determine the optimal μ^* by substituting β^* into (15). Repeated iteration constitutes a coordinate ascent EM algorithm for μ and β.

An alternative method suggests itself: if we fix the norm of β in the mixing density, then we can solve for new estimates of μ and β simultaneously. Let,

$$a \triangleq \tfrac{1}{N}\textstyle\sum_k \gamma_k\,, \quad \mathbf{b} \triangleq \tfrac{1}{N}\textstyle\sum_k \gamma_k \mathbf{x}_k\,, \quad \tau \triangleq \frac{\varphi'\left(\frac{1}{2}\|\beta\|^2_{\Sigma^{-1}}\right)}{\varphi\left(\frac{1}{2}\|\beta\|^2_{\Sigma^{-1}}\right)}$$

Then from (15) and (16), we have,

$$\begin{aligned} a\,\mu^* + \beta^* &= \mathbf{b} \\ \mu^* + \tau\beta^* &= \mathbf{c} \end{aligned}$$

Solving for the components μ_i, β_i, $i = 1, \ldots, d$, we get,

$$\begin{bmatrix} a & 1 \\ 1 & \tau \end{bmatrix}\begin{bmatrix} \mu_i^* \\ \beta_i^* \end{bmatrix} = \begin{bmatrix} b_i \\ c_i \end{bmatrix} \quad\Rightarrow\quad \mu_i^* = \frac{\tau b_i - c_i}{a\tau - 1}\,, \quad \beta_i^* = \frac{a c_i - b_i}{a\tau - 1}$$

For the structure matrix, Σ, setting the complete log likelihood gradient to zero, we get,

$$\begin{aligned} \Sigma &= \tfrac{1}{N}\textstyle\sum_k \gamma_k(\mathbf{x}_k - \mu)(\mathbf{x}_k - \mu)^T - \tfrac{2}{N}\textstyle\sum_k(\mathbf{x}_k - \mu)\beta^T \\ &= \tfrac{1}{N}\textstyle\sum_k \gamma_k(\mathbf{x}_k - \mu - \gamma_k^{-1}\beta)(\mathbf{x}_k - \mu - \gamma_k^{-1}\beta)^T - \left(\textstyle\sum_k \gamma_k^{-1}\right)\beta\beta^T. \end{aligned}$$

5 Conclusion

We have shown how to derive general multivariate Gaussian scale mixtures in terms of scalar Gaussian scale mixtures, and how to optimize them using an EM algorithm. We generalized the spherically (or ellipsoidally) symmetric Gaussian scale mixture by introducing a generalization of Barndorff-Nielsen's generalized hyperbolic density using Generalized Gaussian scale mixtures, yielding a multivariate dependent anisotropic model. We also introduced the modeling of skew in ICA sources, deriving a general form of skewed multivariate Gaussian scale mixture, and an EM algorithm to update the location, drift, and structure parameters.

References

1. Amari, S.-I., Cichocki, A.: Adaptive blind signal processing—neural network approaches. Proceedings of the IEEE 86(10), 2026–2047 (1998)
2. Andrews, D.F., Mallows, C.L.: Scale mixtures of normal distributions. J. Roy. Statist. Soc. Ser. B 36, 99–102 (1974)
3. Barndorff-Nielsen, O., Kent, J., Sørensen, M.: Normal variance-mean mixtures and z distributions. International Statistical Review 50, 145–159 (1982)
4. Dempster, A.P., Laird, N.M., Rubin, D.B.: Maximum likelihood from incomplete data via the EM algorithm. Journal of the Royal Statistical Society, Series B 39, 1–38 (1977)
5. Dempster, A.P., Laird, N.M., Rubin, D.B.: Iteratively reweighted least squares for linear regression when errors are Normal/Independent distributed. In: Krishnaiah, P.R. (ed.) Multivariate Analysis V, pp. 35–57. North Holland Publishing Company, Amsterdam (1980)
6. Eltoft, T., Kim, T., Lee, T.-W.: Multivariate scale mixture of Gaussians modeling. In: Rosca, J., Erdogmus, D., Príncipe, J.C., Haykin, S. (eds.) ICA 2006. LNCS, vol. 3889, pp. 799–806. Springer, Heidelberg (2006)
7. Gneiting, T.: Normal scale mixtures and dual probability densities. J. Statist. Comput. Simul. 59, 375–384 (1997)
8. Hyvärinen, A., Hoyer, P.O.: Emergence of phase- and shift-invariant features by decomposition of natural images into independent feature subspaces. Neural Computation 12, 1705–1720 (2000)
9. Hyvärinen, A., Hoyer, P.O., Inki, M.: Topographic independent component analysis. Neural Computation 13(7), 1527–1558 (2001)
10. Kim, T., Attias, H., Lee, S.-Y., Lee, T.-W.: Blind source separation exploiting higher-order frequency dependencies. IEEE Transactions on Speech and Audio Processing, 15(1) (2007)
11. Kim, T., Eltoft, T., Lee, T.-W.: Independent vector analysis: An extension of ICA to multivariate components. In: Rosca, J., Erdogmus, D., Príncipe, J.C., Haykin, S. (eds.) ICA 2006. LNCS, vol. 3889, pp. 165–172. Springer, Heidelberg (2006)
12. Palmer, J.A., Kreutz-Delgado, K., Makeig, S.: Super-Gaussian mixture source model for ICA. In: Rosca, J., Erdogmus, D., Príncipe, J.C., Haykin, S. (eds.) ICA 2006. LNCS, vol. 3889, pp. 854–861. Springer, Heidelberg (2006)
13. Palmer, J.A., Kreutz-Delgado, K., Wipf, D.P., Rao, B.D.: Variational EM algorithms for non-gaussian latent variable models. In: Advances in Neural Information Processing Systems, NIPS 2005, MIT Press, Cambridge (2006)

On the Relationships Between Power Iteration, Inverse Iteration and FastICA

Hao Shen[1,2] and Knut Hüper[1,2]

[1] Department of Information Engineering
Research School of Information Sciences and Engineering,
The Australian National University, Canberra ACT 0200, Australia
[2] Canberra Research Laboratory, National ICT Australia
Locked Bag 8001, Canberra ACT 2612, Australia
Hao.Shen@rsise.anu.edu.au, Knut.Hueper@nicta.com.au

Abstract. In recent years, there has been an increasing interest in developing new algorithms for digital signal processing by applying and generalising existing numerical linear algebra tools. A recent result shows that the FastICA algorithm, a popular state-of-the-art method for linear Independent Component Analysis (ICA), shares a nice interpretation as a Newton type method with the Rayleigh Quotient Iteration (RQI), the latter method wellknown to the numerical linear algebra community. In this work, we develop an analogous theory of single vector iteration ICA methods. Two classes of methods are proposed for the one-unit linear ICA problem, namely, power ICA methods and inverse iteration ICA methods. By means of a *scalar shift*, scalar shifted versions of both power ICA method and inverse iteration ICA method are proposed and proven to be locally quadratically convergent to a correct demixing vector.

1 Introduction

Independent Component Analysis (ICA) is a standard statistical tool for solving the Blind Source Separation (BSS) problem. Recently, there has been an increasing interest in developing new algorithms for digital signal processing by applying and generalising existing numerical linear algebra tools. In this work, we develop a theory of one-unit linear ICA algorithms in the framework of single vector iteration methods, which are efficient numerical linear algebra tools for computing one eigenvalue-eigenvector pair of a real symmetric matrix.

The FastICA algorithm, developed by the Finnish school, is one of the most popular algorithms for the linear ICA problem. Recent work in [1,2] suggests a strong connection between FastICA and the Rayleigh Quotient Iteration (RQI) method, which is wellknown to the numerical linear algebra community. A deeper result further shows that FastICA shares a nice interpretation as a Newton type method similar to RQI [3]. Other than being interpreted as a Newton method, RQI was originally developed as a single vector iteration method, specifically, a scalar shifted inverse iteration method [4]. In this work, we propose two classes of single vector iteration method for the one-unit linear ICA problem, namely, power ICA methods and inverse iteration ICA methods.

M.E. Davies et al. (Eds.): ICA 2007, LNCS 4666, pp. 105–112, 2007.

This paper is organised as follows. Section 2 briefly introduces the one-unit linear ICA model with the motivation of developing single vector iteration ICA methods. By means of a *scalar shift*, scalar shifted versions of both power ICA method and inverse iteration ICA method are proposed in Section 3 and Section 4, respectively. Both scalar shifted ICA methods are proven to be locally quadratically convergent to a correct demixing vector. As an aside, the standard FastICA can be considered as a special case of the scalar shifted power ICA method. Finally in Section 5, several numerical experiments are provided to verify the theoretical results regarding the local convergence properties of the proposed algorithms.

2 The One-Unit Linear ICA Model

We consider the standard noiseless linear instantaneous ICA model, $Z = AS$, where $S \in \mathbb{R}^{m\times n}$ represents n samples of m sources with $m \ll n$, the full rank matrix $A \in \mathbb{R}^{m\times m}$ is the mixing matrix, and $Z \in \mathbb{R}^{m\times n}$ is the observed mixtures, see [5]. The source signals S are assumed to be unknown, having zero mean and unit variance, being mutually statistically independent, and at most one being Gaussian.

The task of linear ICA is to recover the sources S by estimating the mixing matrix A given only the mixtures Z. By finding a matrix $V \in \mathbb{R}^{m\times m}$ such that $W = VZ = VAS$ with $\mathbb{E}[ww^\top] = I$, the whitened demixing linear ICA model can be formulated as $Y = X^\top W$, where $W \in \mathbb{R}^{m\times n}$ is the whitened mixture, $X \in \mathbb{R}^{m\times m}$ is the demixing matrix with $X^\top X = I$, and $Y \in \mathbb{R}^{m\times n}$ is the recovered signal.

Let us denote by $S^{m-1} := \{x \in \mathbb{R}^m | \, \|x\| = 1\}$ the $(m-1)$-dimensional unit sphere and by $X = [x_1, \ldots, x_m]$ the orthogonal demixing matrix. In this work, we study the so-called one-unit linear ICA problem, which estimates only one source at one time. It is equivalent to seeking an $x \in S^{m-1}$ which gives a correct estimation of one single source. A generic contrast function of the one-unit linear ICA, which was proposed for developing the FastICA algorithm [5], can be given as follows

$$f\colon S^{m-1} \to \mathbb{R}, \qquad f(x) := \mathbb{E}[G(x^\top w)], \tag{1}$$

where $G\colon \mathbb{R} \to \mathbb{R}$ is usually assumed to be even and differentiable. Under certain weak assumptions, it has been shown that a correct demixing vector $x^* \in S^{m-1}$ is a critical point of f, refer to [2] for details.

Recall the critical point condition of the contrast function f as follows

$$\mathbb{E}\left[G'(x^\top w)w\right] = \gamma x, \tag{2}$$

with G' being the first derivative of G and $\gamma \in \mathbb{R}$. One might consider the expression $\mathbb{E}[G'(x^\top w)w]$ as a nonlinear operator acting on $x \in \mathbb{R}^m$. Thus, a solution (γ, x) of the critical point equation as in (2) can be treated as an eigenvalue-eigenvector pair of this nonlinear operator. We can rewrite (2) as

$$\mathbb{E}\left[\frac{G'(x^\top w)w^\top x}{x^\top w} w\right] = \gamma x \qquad \Longleftrightarrow \qquad \mathbb{E}\left[\frac{G'(x^\top w)}{x^\top w} w\, w^\top\right] x = \gamma x. \tag{3}$$

It is known that the operator $\mathbb{E}\left[\frac{G'(x^\top w)}{x^\top w} w\, w^\top\right] \in \mathbb{R}^{m\times m}$ can be made positive definite by choosing the function G carefully [2], e.g., two functions widely used for FastICA, $G(a) = \log\cosh(a)$ and $G(a) = a^4$ both do the job. In this way, the expression $\mathbb{E}[G'(x^\top w)w]$ can be decomposed as a product of a positive definite matrix with a vector.

Let $H\colon \mathbb{R} \to \mathbb{R}$ be smooth with $H(a) \geq 0$ for all $a \in \mathbb{R}$, we define

$$B\colon S^{m-1} \to \mathbb{R}^{m\times m}, \qquad B(x) := \mathbb{E}\left[H(x^\top w) w\, w^\top\right]. \tag{4}$$

Similar to (3) we then define

$$F\colon S^{m-1} \to \mathbb{R}^{m}, \qquad F(x) := B(x)x. \tag{5}$$

Notice that such a $B(x)$ is a real symmetric matrix. Therefore the key work of this paper is to develop a theory of single vector iteration methods for solving the one-unit linear ICA problem, which is completely analogous to the numerical linear algebra tools for the real symmetric eigenvalue problem.

3 Power ICA Methods

According to the multiplicative decomposition of the nonlinear operator F suggested in (5), a simple power method applied to the matrix part $B(x)$ can be formulated as follows

$$\eta\colon S^{m-1} \to S^{m-1}, \qquad x \mapsto \frac{B(x)x}{\|B(x)x\|}. \tag{6}$$

Let $x^* \in S^{m-1}$ be a correct demixing vector. By the assumption of zero mean and unit variance of the sources, the expression $B(x^*)$ gives an invertible matrix with two positive eigenvalues, namely, $\lambda_1 = \mathbb{E}\left[H(x^{*\top} w)\right] > 0$ occurring with multiplicity $m-1$ and single $\lambda_2 = \mathbb{E}\left[H(x^{*\top} w)(x^{*\top} w)^2\right] > 0$. The corresponding eigenvector of the eigenvalue λ_2 is x^*, i.e. $B(x^*)x^* = \lambda_2 x^*$ holds. We therefore have proven.

Lemma 1. *Let $x^* \in S^{m-1}$ be a correct demixing vector. Then x^* is a fixed point of the power ICA method η.* □

By taking the first derivative of η at x^* in any direction $\xi \in T_{x^*}S^{m-1}$, one gets $\mathrm{D}\,\eta(x^*)\xi \neq 0$. That is, by a Taylor-type argument, the algorithmic map η does not correspond to a locally quadratically fast algorithm. Here, $T_x S^{m-1} = \{\xi \in \mathbb{R}^m | x^\top \xi = 0\}$ denotes the tangent space of the unit sphere S^{m-1} at a point $x \in S^{m-1}$.

In the rest of this section, we will modify the power ICA method (6) to obtain second order convergence in the framework of a *scalar shift strategy*, which has been successfully used in developing the RQI [4] and generalising a simple one-unit ICA method proposed by Regalia and Kofidis [6,3]. Let us define a smooth function $\rho\colon S^{m-1} \to \mathbb{R}$. We construct a scalar shifted nonlinear operator acting on S^{m-1} as follows

$$F_{\mathrm{s}}\colon S^{m-1} \to \mathbb{R}^m, \qquad F_{\mathrm{s}}(x) := (B(x) - \rho(x)I)\,x. \tag{7}$$

Let $x = x^*$, one gets $F_{\mathrm{s}}(x^*) = \lambda_{\mathrm{s}} x^*$ with $\lambda_{\mathrm{s}} = \mathbb{E}\left[H({x^*}^\top w)({x^*}^\top w)^2\right] - \rho(x^*)$. To formulate a well-defined power method based on the operator as in (7), it is necessary to have $\lambda_{\mathrm{s}} \neq 0$, i.e., $\rho(x^*) \neq \mathbb{E}\left[H({x^*}^\top w)({x^*}^\top w)^2\right]$. Moreover, if $\lambda_{\mathrm{s}} < 0$, the corresponding power method is then not differentiable at the point x^* following a similar argument as for the standard FastICA in [3]. Therefore, by introducing a sign correction term, see [3], we formulate the scalar shifted power ICA method as follows

$$\eta_{\mathrm{s}}\colon S^{m-1} \to S^{m-1}, \qquad x \mapsto \frac{\frac{1}{\tau(x)}\left(B(x)x - \rho(x)x\right)}{\|\frac{1}{\tau(x)}\left(B(x)x - \rho(x)x\right)\|}, \tag{8}$$

where $\tau(x) = x^\top B(x)x - \rho(x)$. The following lemma is then immediate.

Lemma 2. *Let $x^* \in S^{m-1}$ be a correct demixing vector and $\rho : S^{m-1} \to \mathbb{R}$ a smooth map with $\rho(x^*) \neq \mathbb{E}\left[H({x^*}^\top w)({x^*}^\top w)^2\right]$. Then x^* is a fixed point of the scalar shifted power ICA method η_s.* □

Now we will study the additional conditions on the scalar shift ρ, which fulfils already the condition stated in Lemma 2, such that the algorithmic map η_{s} is locally quadratically convergent to a correct demixing vector x^*. Define

$$\widetilde{F}_{\mathrm{s}}(x) = \frac{F_{\mathrm{s}}(x)}{\tau(x)}. \tag{9}$$

By a straightforward computation, the first derivative of η_{s} at x^* in direction $\xi \in T_{x^*}S^{m-1}$ can be computed as

$$\mathrm{D}\,\eta_{\mathrm{s}}(x)\xi|_{x=x^*} = \frac{1}{\|\widetilde{F}_{\mathrm{s}}(x^*)\|}\underbrace{\left(I - \frac{\widetilde{F}_{\mathrm{s}}(x^*)}{\|\widetilde{F}_{\mathrm{s}}(x^*)\|}\frac{\widetilde{F}_{\mathrm{s}}(x^*)^\top}{\|\widetilde{F}_{\mathrm{s}}(x^*)\|}\right)}_{=:P(x^*)}\mathrm{D}\,\widetilde{F}_{\mathrm{s}}(x)\xi|_{x=x^*}, \tag{10}$$

where $P(x^*)$ is an orthogonal projection operator onto the orthogonal complement of the span of x^*. Thus one has

$$\mathrm{D}\,\eta_{\mathrm{s}}(x)\xi|_{x=x^*} = 0 \qquad \Longleftrightarrow \qquad \mathrm{D}\,\widetilde{F}_{\mathrm{s}}(x)\xi|_{x=x^*} = \gamma x^*. \tag{11}$$

Now by the chain rule, we compute

$$\mathrm{D}\,\widetilde{F}_{\mathrm{s}}(x)\xi|_{x=x^*} = \mathrm{D}\,\tfrac{1}{\tau(x)}\xi|_{x=x^*}F_{\mathrm{s}}(x^*) + \tfrac{1}{\tau(x^*)}\,\mathrm{D}\,F_{\mathrm{s}}(x)\xi|_{x=x^*} \tag{12}$$

with

$$\begin{aligned}\mathrm{D}\,F_{\mathrm{s}}(x)\xi|_{x=x^*} =& (\mathbb{E}[H({x^*}^\top w)] + \mathbb{E}[H'({x^*}^\top w)({x^*}^\top w)])\xi \\ & - \mathrm{D}\,\rho(x)\xi|_{x=x^*}x^* - \rho(x^*)\xi.\end{aligned} \tag{13}$$

According to the fact that the first summand in (12) and the second summand in (13) give already a scalar multiple of x^*, the expression in (10) vanishes if and only if

$$\rho(x^*) = \mathbb{E}[H({x^*}^\top w)] + \mathbb{E}[H'({x^*}^\top w)({x^*}^\top w)]. \tag{14}$$

Therefore, following a Taylor-type argument, we conclude

Theorem 1. *Let $x^* \in S^{m-1}$ be a correct demixing vector and $\rho : S^{m-1} \to \mathbb{R}$ a smooth map such that*

$$\rho(x^*) \neq \mathbb{E}\left[H({x^*}^\top w)({x^*}^\top w)^2\right], \text{ and}$$
$$\rho(x^*) = \mathbb{E}[H({x^*}^\top w)] + \mathbb{E}[H'({x^*}^\top w)({x^*}^\top w)].$$

Then the scalar shifted power ICA method η_s is locally quadratically convergent to x^*. □

Naturally, a simple choice of the scalar shift ρ to make η_{s} locally quadratically convergent can be constructed by

$$\rho_{\mathrm{p}} \colon S^{m-1} \to \mathbb{R}, \qquad \rho_{\mathrm{p}}(x) := \mathbb{E}[H(x^\top w)] + \mathbb{E}[H'(x^\top w)(x^\top w)]. \tag{15}$$

We denote the corresponding algorithmic map by $\widehat{\eta}_{\mathrm{s}}$ using ρ_{p} as the scalar shift.

Remark 1. If one defines $H(a) = \frac{G'(a)}{a}$ as suggested in (3), the resulting algorithm is essentially the same as the FastICA/ANPICA algorithm in [3].

4 Inverse Iteration ICA Methods

Recall the result in Section 3 that the expression $B(x^*)$ is indeed an invertible matrix. In this section, we propose power-type methods applied on the inverse of $B(x)$ and its scalar shifted generalisations. We call them inverse iteration ICA methods. Firstly, we define a nonlinear operator acting on S^{m-1} as

$$K \colon S^{m-1} \to \mathbb{R}^m, \qquad K(x) := B(x)^{-1}x. \tag{16}$$

Note that the above operator K is locally well defined at least in an open neighborhood $U_\varepsilon(x^*) \subset S^{m-1}$ around x^*. It is also worthwhile to notice that $K(x)$ can be computed efficiently by solving the following linear system for $u \in \mathbb{R}^m$,

$$B(x)u = x. \tag{17}$$

Thus a simple inverse iteration ICA method based on K can be formulated as

$$\zeta \colon S^{m-1} \supset U_\varepsilon(x^*) \to S^{m-1}, \qquad x \mapsto \frac{B(x)^{-1}x}{\|B(x)^{-1}x\|}. \tag{18}$$

By the fact that $B(x^*)^{-1}x^* = 1/\lambda_1 x^*$, we just have proven

Lemma 3. *Let $x^* \in S^{m-1}$ be a correct demixing vector. Then x^* is a fixed point of the inverse iteration ICA method ζ.* □

Again, by showing that the first derivative of ζ at x^* in any $\xi \in T_{x^*}S^{m-1}$ does not vanish in general, i.e., $\mathrm{D}\,\zeta(x^*)\xi \neq 0$, we conclude that the algorithmic map ζ is not locally quadratically convergent to x^*. Once more, we will modify the inverse iteration ICA method as in (18) in the framework of *scalar shift strategy* to obtain second order convergence.

Let $\rho\colon S^{m-1} \to \mathbb{R}$ be smooth. We define a scalar shifted nonlinear operator as follows

$$K_{\mathrm{s}}\colon S^{m-1} \to \mathbb{R}^m, \qquad K_{\mathrm{s}}(x) := \left(B(x) - \rho(x)I\right)^{-1} x. \tag{19}$$

Such an operator is well defined if and only if, for any $x \in S^{m-1}$, $B(x) - \rho(x)I$ is nonsingular. Now let $x = x^*$. As discovered in Section 3, the matrix $B(x^*)$ has only two positive eigenvalues $\lambda_1 = \mathbb{E}\left[H(x^{*\top} w)\right]$ and $\lambda_2 = \mathbb{E}\left[H(x^{*\top} w)(x^{*\top} w)^2\right]$. Further analysis shows

(i) If $\rho(x^*) = \lambda_1$, the resulting operator $B(x^*) - \rho(x^*)I$ is of rank one, i.e. K_{s} is not defined at x^*;
(ii) If $\rho(x^*) = \lambda_2$, the matrix $B(x^*) - \rho(x^*)I$ is of rank $m - 1$. Although the operator K_{s} is still not defined, one can rescue the situation, i.e. remove the pole, by replacing the inversion by the classical *adjoint*, which has been used to handle a similar situation when analysing RQI in [7].

We therefore define

$$\widetilde{K}_{\mathrm{s}}(x) := \operatorname{adj}\left(B(x) - \rho(x)I\right) x. \tag{20}$$

Note that $\widetilde{K}_{\mathrm{s}}$ is locally well defined in $U_\varepsilon(x^*) \subset S^{m-1}$ and $\widetilde{K}_{\mathrm{s}}(x^*) = (\lambda_1 - \lambda_2)^{m-1} x^*$. We now propose the following iteration method, still called the scalar shifted inverse iteration ICA method,

$$\zeta_{\mathrm{s}}\colon S^{m-1} \supset U_\varepsilon(x^*) \to S^{m-1}, \qquad x \mapsto \frac{\operatorname{adj}\left(B(x) - \rho(x)I\right) x}{\|\operatorname{adj}\left(B(x) - \rho(x)I\right) x\|}. \tag{21}$$

We can now state

Lemma 4. *Let $x^* \in S^{m-1}$ be a correct demixing vector and $\rho : S^{m-1} \to \mathbb{R}$ a smooth map with $\rho(x^*) \neq \mathbb{E}\left[H(x^{*\top} w)\right]$. Then x^* is a fixed point of the scalar shifted inverse iteration ICA method ζ_s.* □

By similar arguments as in Section 3, to make the first derivation of ζ_{s} at x^* in direction $\xi \in T_{x^*} S^{m-1}$ vanish, i.e. $\mathrm{D}\,\zeta_{\mathrm{s}}(x)\xi|_{x=x^*} = 0$, is equivalent to requiring

$$\mathrm{D}\,\widetilde{K}_{\mathrm{s}}(x)\xi|_{x=x^*} = \gamma x^*. \tag{22}$$

A tedious computation shows that the equation in (22) holds true if and only if

$$\rho(x^*) = \mathbb{E}[H(x^{*\top} w)(x^{*\top} w)^2] - \mathbb{E}[H'(x^{*\top} w)(x^{*\top} w)]. \tag{23}$$

Therefore we just proved

Theorem 2. *Let $x^* \in S^{m-1}$ be a correct demixing vector and $\rho : S^{m-1} \to \mathbb{R}$ a smooth map such that*

$$\rho(x^*) \neq \mathbb{E}\left[H(x^{*\top} w)\right], \quad \text{and}$$
$$\rho(x^*) = \mathbb{E}[H(x^{*\top} w)(x^{*\top} w)^2] - \mathbb{E}[H'(x^{*\top} w)(x^{*\top} w)].$$

Then the scalar shifted inverse iteration ICA method ζ_s is locally quadratically convergent to x^*. □

A natural and simple choice of the scalar shift ρ to make ζ_s locally quadratically convergent to x^* can be constructed by choosing

$$\rho_i\colon S^{m-1} \to \mathbb{R}, \qquad \rho_i(x) := \mathbb{E}[H(x^\top w)(x^\top w)^2] + \mathbb{E}[H'(x^\top w)(x^\top w)]. \qquad (24)$$

It is clear that, in general, $\rho_i(x^*) \neq \mathbb{E}[H(x^{*\top} w)(x^{*\top} w)^2]$. Therefore one can construct the following algorithmic map locally, using ρ_i as the scalar shift,

$$\widehat{\zeta}_s\colon S^{m-1} \supset U_\varepsilon(x^*) \to S^{m-1}, \qquad x \mapsto \frac{\frac{1}{\kappa(x)}\left((B(x) - \rho_i(x)I)^{-1}x\right)}{\|\frac{1}{\kappa(x)}\left((B(x) - \rho_i(x)I)^{-1}x\right)\|}, \qquad (25)$$

with $\kappa(x) = x^\top(B(x) - \rho_i(x)I)^{-1}x$. The convergence properties as in Theorem 2 apply to $\widehat{\zeta}_s$ as well.

5 Numerical Experiments

In this section, we will verify the results in Theorem 1 and 2 by several experiments. Local convergence properties of two scalar shifted single vector iteration ICA methods, namely $\widehat{\eta}_s$ and $\widehat{\zeta}_s$, are investigated and compared with the classical FastICA/ANPICA. We specify the function H for both single vector iteration ICA methods by choosing $H(a) = \log\cosh(a)$ and the function G for FastICA/ANPICA by $G(a) = \log\cosh(a)$, as well. A toy set of sources is illustrated in Fig. 1(a). It is well known that the mutual statistical independence can only be ensured if the sample size n tends to infinity. In this experiment, therefore, we set $n = 10^7$ artificially high.

All three methods are initialised by the same demixing vector. However it is worthwhile to notice that these algorithms can converge to different correct demixing vectors. We only show a case where it happens that they all converge

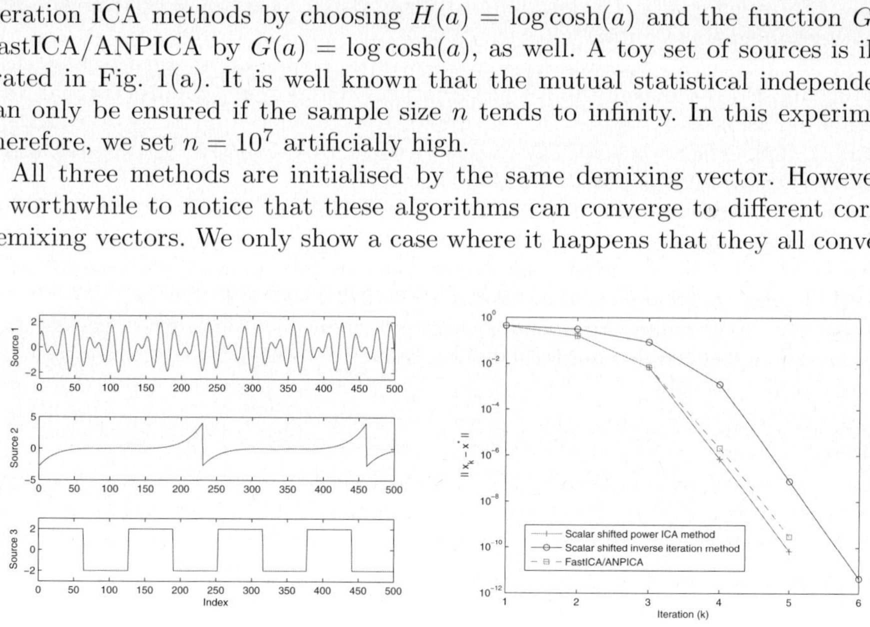

(a) Source signals (b) Convergence properties

Fig. 1. Local convergence properties of scalar shifted single vector iteration ICA methods (scalar shifted power ICA method vs. scalar shifted inverse iteration ICA method) and FastICA/ANPICA to the same correct demixing vector x^*, see Fig. 1. The error is measured by the distance of the accumulation point x^* to the current iterate x_k, i.e., by the norm $\|x_k - x^*\|$. The numerical results in Fig. 1(b) show that both scalar shifted single vector iteration ICA methods, namely $\widehat{\eta}_s$ and $\widehat{\zeta}_s$, share the the same local quadratic convergence properties with the classical FastICA/ANPICA.

Acknowledgment

National ICT Australia is funded by the Australian Government's Department of Communications, Information Technology and the Arts and the Australian Research Council through *Backing Australia's Ability* and the ICT Research Centre of Excellence programs.

References

1. Douglas, S.: Relationships between the FastICA algorithm and the Rayleigh Quotient Iteration. In: Rosca, J., Erdogmus, D., Príncipe, J.C., Haykin, S. (eds.) ICA 2006. LNCS, vol. 3889, pp. 781–789. Springer, Heidelberg (2006)
2. Shen, H., Hüper, K.: Local convergence analysis of FastICA. In: Rosca, J., Erdogmus, D., Príncipe, J.C., Haykin, S. (eds.) ICA 2006. LNCS, vol. 3889, pp. 893–900. Springer, Heidelberg (2006)
3. Shen, H., Hüper, K., Seghouane, A.K.: Geometric optimisation and FastICA algorithms. In: Proceedings of the 17th International Symposium of Mathematical Theory of Networks and Systems (MTNS 2006), Kyoto, Japan, pp. 1412–1418 (2006)
4. Parlett, B.N.: The Symmetric Eigenvalue Problem. SIAM Classic. In: Applied Mathematics Series, Philadelphia, vol. 20 (1998)
5. Hyvärinen, A., Karhunen, J., Oja, E.: Independent Component Analysis. Wiley, New York (2001)
6. Regalia, P., Kofidis, E.: Monotonic convergence of fixed-point algorithms for ICA. IEEE Transactions on Neural Networks 14(4), 943–949 (2003)
7. Hüper, K.: A Calculus Approach to Matrix Eigenvalue Algorithms. Habilitation Dissertation, Department of Mathematics, University of Würzburg, Germany (2002)

A Sufficient Condition for the Unique Solution of Non-Negative Tensor Factorization

Toshio Sumi and Toshio Sakata*

Faculty of Design, Kyushu University, Japan
sumi@design.kyushu-u.ac.jp, sakata@design.kyushu-u.ac.jp

Abstract. The applications of Non-Negative Tensor Factorization (NNTF) is an important tool for brain wave (EEG) analysis. For it to work efficiently, it is essential for NNTF to have a unique solution. In this paper we give a sufficient condition for NNTF to have a unique global optimal solution. For a third-order tensor T we define a matrix by some rearrangement of T and it is shown that the rank of the matrix is less than or equal to the rank of T. It is also shown that if both ranks are equal to r, the decomposition into a sum of r tensors of rank 1 is unique under some assumption.

1 Introduction

In the past few years, Non-Negative Tensor Factorization (NNTF) is becoming an important tool for brain wave (EEG) analysis through Morlet wavelet analysis (for example, see Miwakeichi [MMV] and Morup [MHH]). The NNTF algorithm is based on Non-Negative Matrix Factorization (NNMF) algorithms, amongst the most well-known algorithms contributed by Lee-Seung [LS]. Recently, Chichoki et al. [CZA] deals with a new NNTF algorithm using Csiszar's divergence. Furthermore, Wang et al. [WZZ] also worked on NNMF algorithms and its interesting application in preserving privacy in datamining fields. These algorithms converged to some stationary points and do not converge to a global minimization point. In fact, it is easily shown that the problem has no unique minimization points in general (see [CSS]). In applications of NNTF for EEG analysis, it is important for NNTF to have a unique solution. However this uniqueness problem has not been addressed sufficiently as far as the authors are aware of. Similarly as in Non-Negative Matrix Factorization (NNMF), it seems that the uniqueness problem has not been solved. However we managed to obtain the uniqueness and proved it. (see Proposition 1). In this paper we give a sufficient condition for NNTF to have a unique solution and for the usual NNTF algorithm to find its minimization point in the case when NNTF exists strictly, not approximation (see Theorem 3).

* The research is supported partially by User Science Institute in Kyushu University (Special Coordination Funds for Promoting Science and Technology).

M.E. Davies et al. (Eds.): ICA 2007, LNCS 4666, pp. 113–120, 2007.

2 Quadratic Form

As the NNMF problem is a minimization of a quadratic function, we shall first review quadratic functions generally. Let us consider the quadratic form defined by $f(x) = x^T Ax - 2b^T x$ where A is a $n \times n$ symmetric matrix and b is a n vector. The symmetric matrix A is a diagonalized by an orthogonal matrix P as

$$PAP^T = \mathrm{diag}(e_1, \ldots, e_n).$$

Then by assigning $y = (y_1, \ldots, y_n)^T = Px$ and $c = (c_1, \ldots, c_n)^T = Pb$, we obtain the equality

$$f(x) = y^T(PAP^T)y - 2c^T y = \sum_i (e_i y_i^2 - 2c_i y_i) = \sum_i \left(e_i (y_i - \frac{c_i}{e_i})^2 - \frac{c_i^2}{e_i} \right).$$

We assume that the matrix A is positive definite. Then,when $f(x)$ reaches its minimum at $y = (PAP^T)^{-1}c = (PAP^T)^{-1}Pb = PA^{-1}b$ in $\mathbb{R}^n$, with the value $f(A^{-1}b) = -b^T A^{-1}b$ at $x = P^T y = A^{-1}b \in \mathbb{R}^n$. The minimal value is under the condition $x \geq 0$. Here, some basic facts will be explained. Let $a \in \mathbb{R}^n$ and let $h \colon \mathbb{R}^n \to \mathbb{R}$ be a function defined as $h(x) = \|x - a\|^2$, where $\|\cdot\|$ stands for the common Euclidean norm.

Lemma 1. *On the arbitrary closed set S of $\mathbb{R}^n$, $h(t)$, $t \in S$ takes a global minimal value in S.*

Proof. Choose an arbitrary $t_0 \in S$, and set $s = h(t_0)$ and $U = h^{-1}([0, s]) \cap S$. The set U is a closed subset of $\mathbb{R}^n$. By triangular inequality, we know that $h(t) \geq \|t\| - \|a\|$. Since $s \geq h(t)$ for $t \in U$, it holds that $\|t\| \leq s + \|a\|$ which shows that U is bounded. Hence, since U is bounded and closed, it is compact. Thus $h(t)$, $t \in U$ becomes a closed map, and $h(U)$ is also compact. That is, $h(t)$, $t \in U$ takes a global minimum value, say s_0. Thus, it holds that for $t \in S$, $h(t) > s$ if $t \notin U$, and $h(t) \geq s_0$ if t in U. This means that s_0 is the global minimum of h on S. □

Lemma 2. *Let S be a closed convex subset of $\mathbb{R}^n$. Then $h(t)$, $t \in S$ reaches a global minimal value at a unique point in S.*

Proof. The existence of a global minimal value follows from Lemma 1. Let x and y be points in $\mathbb{R}^n$ which attain a global minimal value $r := \min_{z \in S} f(z)$. Note that $x, y \in S \cap \partial B_r(z_0)$, where $B_r(a) := \{x \mid \|x - a\| \leq r\}$ and $\partial B_r(a) := \{x \mid \|x - a\| = r\}$. Since $S \cap B_r(a)$ is also convex, $tx + (1-t)y \in S \cap B_r(a)$ for each $0 \leq t \leq 1$. If $x \neq y$, then $\|a - (x+y)/2\| < r$, which is contradiction. Therefore $x = y$. □

Let $D = \mathrm{diag}(\sqrt{e_1}, \ldots, \sqrt{e_n})$, $z = DPx$ and $S = \{ z \in \mathbb{R}^n \mid x \geq 0 \}$. Note that S is a convex set of $\mathbb{R}^n$ and $f(x) = h(z)$ for $a = D^{-1}Pb$. Therefore $f(x)$ reaches a global minimal value at a unique point under the condition $x \geq 0$.

The following are some basic facts about matrix decompositions. Let A, W and H be $m \times n$, $m \times r$ and $r \times n$ matrix respectively. Then $\| A - WH \|$ for any H with $H \geq O$ and any W with $W \geq O$ reaches a global minimal value at a unique $m \times n$ matrix WH but W and H are not unique. We state this precisely below.

Proposition 1. *The following properties hold:*

1. *If* $r = \text{rank}(A)$, *there exist* W *and* H *such that* $A = WH$.
2. *If* $r > \text{rank}(A)$, *there exists an infinite number of pairs of* W *and* H *such that* $A = WH$.
3. *Let* $r = \text{rank}(A)$. *Then if* $A = WH = W'H'$ *there exists a non-singular matrix* X *such that* $W' = WX$, $H' = X^{-1}H$ *([CSS, Full-Rank Decomposition Theorem]).*
4. *If* $r < \text{rank}(A)$, *there exists no pair of* W *and* H *such that* $A = WH$.

Proof. (1) In this case, let W be a matrix whose columns are linearly independent vectors of length m. From the assumption it is clear that the columns of A are expressed as linear combination of columns of W hence $A = WH$.

(2) In this case, put $s = \text{rank}(A)$, $(r > s)$. By property (1), we know there exists a $m \times s$ matrix W_1 and a $s \times m$ matrix H_1 such that $A = W_1 H_1$. Place $W = \begin{pmatrix} W_1 & W_2 \end{pmatrix}$ and $H = \begin{pmatrix} H_1 \\ H_2 \end{pmatrix}$ where W_2 and H_2 are $m \times (r - s)$ matrix and $(r - s) \times n$ matrix respectively and satisfy $W_2 H_2 = 0$. There are infinitely many such pairs of (W_2, H_2), and for all of those it clearly holds that $A = WH$.

(3) From $r = \text{rank}(A)$, in the expression of $A = WH = W'H'$, the columns of W and W' are linearly independent respectively. Hence we have $W' = WX$ for some regular $r \times r$ matrix X. From this the rest of (3) is derived trivially.

(4) Since $\text{rank}(WH) \leq r$, it is impossible to have $A = WH$. □

3 Non-Negative Matrix Factorization

It is well known NNMF (Non-Negative Matrix Factorization) is not unique ([CSS]). Let V, W and H be a $m \times n$, $m \times r$ and $r \times n$ matrix respectively. For a matrix A, we denote by A_{ij} the (i, j)-component of A and its Frobenius norm is defined by

$$\| A \|_F := \sqrt{tr(A^T A)} = \sqrt{\sum_{i,j} A_{ij}^2},$$

where tr takes the sum of all diagonal entries.

Lemma 3. *Fixing* H, $f(W) = \| V - WH \|_F$ *attains the minimum at the solution* W *of the equation* $W(HH^T) = VH^T$. *Especially, if* HH^T *is non-singular, the minimum is attained at the unique point* $W = VH^T(HH^T)^{-1}$.

Proof. It holds that

$$\begin{aligned} f(W) &= \sum_{i,j}(V_{ij} - \sum_p W_{ip}H_{pj})^2 \\ &= \sum_{i,j}\left(\sum_{p,q} W_{ip}H_{pj}W_{iq}H_{qj} - 2\sum_p V_{ij}W_{ip}H_{pj} + V_{ij}^2\right) \\ &= \sum_{i,p,q}(HH^T)_{pq}W_{ip}W_{iq} - 2\sum_{i,p}(VH^T)_{ip}W_{ip} + \sum_{i,j} V_{ij}^2. \end{aligned}$$

Therefore $f(W)$ is a quadratic function of W_{ij} $(i = 1, 2, \cdots, m,\ j = 1, 2, \cdots, r)$. Put

$$\begin{aligned} x &= (W_{11}, \ldots, W_{1r}, \ldots, W_{m1}, \ldots, W_{mr})^T \in \mathbb{R}^{mr}, \\ a &= ((VH^T)_{11}, \ldots, (VH^T)_{1r}, \ldots, (VH^T)_{m1}, \ldots, (VH^T)_{mr})^T \in \mathbb{R}^{mr} \end{aligned}$$

and define a $mr \times mr$ matrix M by $\mathrm{diag}(HH^T, \ldots, HH^T)$. Then, M is positive semidefinite and $f(W)$ is expressed as

$$f(W) = x^T Mx - 2a^T x + \sum_{i,j} V_{ij}^2.$$

Assume that HH^T is non-singular. Then M is positive definite and thus the minimum of $f(W)$ is attained at the unique point $x = M^{-1}a$, that is, $W^T = (HH^T)^{-1}(VH^T)^T$, equivalent to, $W = VH^T(HH^T)^{-1}$. The minimum value is

$$f(W) = \| V \|_F^2 - \| WH \|_F^2 \tag{1}$$

and we also have $\| WH \|_F^2 = tr(W^T VH^T) = tr((HH^T)^{-1}(VH^T)^T(VH^T))$. □

Since $\| V - WH \|_F = \| V^T - H^T W^T \|_F$, fixing W, $\| V - WH \|_F$ attains the minimum at the unique point $V = (W^T W)^{-1} W^T V$ if $W^T W$ is non-singular.

We recall the Lee-Seung NNMF Algorithm for the Frobenius norm property.

Theorem 1 ([LS]). *The Frobenius norm* $\| V - WH \|_F$ *is non-increasing under the update rules:*

$$H_{ij} \leftarrow H_{ij}\frac{(W^T V)_{ij}}{(W^T WH)_{ij}} \quad W_{ij} \leftarrow W_{ij}\frac{(VH^T)_{ij}}{(WHH^T)_{ij}}$$

Now we propose the following improvement of the Lee-Seung NNMF Algorithm for the Frobenius norm property. For matrices X with $X \geq 0$ and Y, let $t_{max}(X, Y) = \max\{t \mid (1 - t)X + tY \geq O,\ 0 \leq t \leq 1\}$.

Theorem 2. *The Frobenius norm* $\| V - WH \|_F$ *is non-increasing under the update rules:*

$$H \leftarrow \begin{cases} (1-h_0)(W^TW)^{-1}W^TV + h_0H, & \text{if } W^TW \text{ is non-singular and } h_0 > 0 \\ H_{ij}\dfrac{(W^TV)_{ij}}{(W^TWH)_{ij}}, & \text{otherwise} \end{cases}$$

$$W \leftarrow \begin{cases} (1-w_0)VH^T(HH^T)^{-1} + w_0W, & \text{if } HH^T \text{ is non-singular and } w_0 > 0 \\ W_{ij}\dfrac{(VH^T)_{ij}}{(WHH^T)_{ij}}, & \text{otherwise} \end{cases}$$

where $h_0 = t_{max}((W^TW)^{-1}W^TV, H)$ *and* $w_0 = t_{max}(VH^T(HH^T)^{-1}, W)$.

Proof. If either W^TW is singular or $h_0 = 0$, the claim follows from Theorem 1. Suppose both W^TW is non-singular and $h_0 > 0$. By Lemma 3, fixing W, $\| V - WH \|_F$ takes minimum at $(W^TW)^{-1}W^TV$ without the assumption $x \geq 0$. Let us denote $H' = (1-h_0)(W^TW)^{-1}W^TV + h_0H$ for clarity. On the line from H to H', the Frobenius norm decreases and thus $\| V - WH \|_F \geq \| V - WH' \|_F$. Clearly $H' \geq 0$ which follows from the definition of h_0. □

4 Non-Negative Tensor Factorization

4.1 Existence of a Global Optimal Solution

Let $\mathbb{R}_{\geq 0}$ be the set of all non-negative real numbers. Let T be a third-order tensor in $\mathbb{R}_{\geq 0}^{a \times b \times c}$. Let $X = \begin{pmatrix} x_1 \dots x_r \end{pmatrix}$, $Y = \begin{pmatrix} y_1 \dots y_r \end{pmatrix}$ and $Z = \begin{pmatrix} z_1 \dots z_r \end{pmatrix}$ be $a \times r$, $b \times r$ and $c \times r$ matrices, respectively. We define a function f over $\mathbb{R}_{\geq 0}^{(a+b+c)r}$ as

$$f(X,Y,Z) = \sum_{ijk} \left(t_{ijk} - \sum_{\ell} X_{i\ell} Y_{j\ell} Z_{k\ell} \right)^2 .$$

Let $S_b = \{x \in \mathbb{R}_{\geq 0}^b \mid \| x \| = 1\}$ be an intersection of an unit sphere in $\mathbb{R}^b$ with $\mathbb{R}_{\geq 0}^b$. Put $S = (\mathbb{R}^a)^{\times r} \times (S_b)^{\times r} \times (S_c)^{\times r}$ for short, where $M^{\times r} = M \times \cdots \times M$ (r times). Then S is a closed subspace of $\mathbb{R}_{\geq 0}^{(a+b+c)r}$ and the image $f(S)$ coincides with the full image $f(\mathbb{R}_{\geq 0}^{(a+b+c)r})$. Let $(X,Y,Z) \in S$. Then $X \geq O$, $Y \geq O$, $Z \geq O$ and $\| y_j \| = \| z_j \| = 1$ for all j. Noting that

$$\sum_{i,j,k} \left(\sum_{\ell} X_{i\ell} Y_{j\ell} Z_{k\ell} \right)^2 \geq \sum_{i,j,k} (X_{i\ell} Y_{j\ell} Z_{k\ell})^2 = \| x_\ell \|^2$$

if $\sum_{i,j,k} \left(\sum_\ell X_{i\ell} Y_{j\ell} Z_{k\ell} \right)^2$ is bounded, $\| X \|_F$ is also bounded and thus so is S. Hence, we can apply the proof of Lemma 1 for the function f on S instead of h and we obtain an existence of a global minimal value.

4.2 Uniqueness

We show the uniqueness under some assumption. First,several facts are presented. For convenience, we define

$$X_1 \circ \cdots \circ X_k = (x_1^{(1)} \otimes \cdots \otimes x_1^{(k)}, \ldots, x_r^{(1)} \otimes \cdots \otimes x_r^{(k)})$$

for matrices $X_1 = (x_1^{(1)}, \ldots, x_r^{(1)})$, ..., $X_k = (x_1^{(k)}, \ldots, x_r^{(k)})$ with r-columns. For $u = (1, \ldots, 1)^T \in \mathbb{R}^r$, we have $f(X,Y,Z) = \| T - (X \circ Y \circ Z)u \|_F^2$. For a transformation $M_\sigma = (m_{ij})$ among $\{1, \ldots, r\}$ σ, a permutation matrix M_σ is defined by $m_{ij} = \delta_{i\sigma(j)}$. For a permutation matrix M_σ it does hold that

$$M_\sigma^T = M_{\sigma^{-1}} = M_\sigma^{-1}.$$

Proposition 2. *In a general P, the following equation does not hold*

$$(X_1 P) \circ \cdots \circ (X_r P) = (X_1 \circ \cdots \circ X_r)P.$$

However, if P is a permutation matrix, and $P_1, \ldots, P_r$ are diagonal matrices,

$$(X_1 P) \circ \cdots \circ (X_r P) = (X_1 \circ \cdots \circ X_r)P$$
$$(X_1 P_1) \circ \cdots \circ (X_r P_r) = (X_1 \circ \cdots \circ X_r)P_1 \cdots P_r$$

does hold.

Lemma 4. *Let A and C be $m \times r$ matrices and B and D be $n \times r$ matrices, and Q be $r \times r$ non-singular matrix. Assume that $A \circ B = (C \circ D)Q$ and* $\text{rank}(C) = \text{rank}(C \circ D) = r$ *Then there exists a permutation matrix $P = M_\sigma$ such that both of PQ and QP^{-1} become diagonal matrices and $A = CQX$ and $B = DP^{-1}X^{-1}$ hold for some diagonal matrix X. Further suppose that $A, B, C, D \geq 0$. Let $Q_{1/2}$ be a $r \times r$ matrix whose (i,j)-component is the square root of the (i,j)-component of Q. Then $Q_{1/2}$ is a real matrix, and both $A = CQ_{1/2}X$ and $B = DQ_{1/2}X^{-1}$ hold for some diagonal matrix X.*

Proof. We use the notations $A = (a_1, \ldots a_r)$, $B = (b_1, \ldots b_r) = (b_{ij})$, $C = (c_1, \ldots c_r)$, $D = (d_1, \ldots d_r) = (d_{ij})$, $Q = (q_{ij})$. Since $\otimes$ is a bilinear operation and $\text{rank}(C \circ D) = r$, it holds that $d_k \neq 0$ $({}^\forall k)$ and that $d_k \,//\, d_\ell$ implies $k = \ell$. Since $A \circ B = (C \circ D)Q$, we have

$$a_k \otimes b_k = \sum_\ell q_{\ell k} c_\ell \otimes d_\ell, \quad {}^\forall k, \quad \text{and} \quad b_{ik} a_k = \sum_\ell q_{\ell k} d_{i\ell} c_\ell, \quad {}^\forall i, k.$$

Since Q is non-singular, for each k there exists a permutation $\sigma(k)$ such that $q_{\sigma(k)k} \neq 0$. Now we will show that for each ℓ there exists an i such that $b_{i\ell} \neq 0$. Assume that $b_s = 0$ for some s. Then, it holds that $\sum_\ell q_{\ell s} c_\ell \otimes d_\ell = 0$, and since $\text{rank}(C \circ D) = r$, it holds that $q_{\ell s} = 0$ $({}^\forall \ell)$. This contradicts to the fact that Q is non-singular. Therefore, for each ℓ, there exists a $\tau(\ell)$ such that $b_{\tau(\ell)\ell} \neq 0$. Then, it follows

$$q_{\ell k} b_{ik} d_{\tau(k)\ell} = q_{\ell k} b_{\tau(k)k} d_{i\ell}, \quad {}^\forall i, k, \ell$$

from the equality

$$b_{\tau(k)k} b_{ik} a_k = \sum_{\ell} b_{ik} q_{\ell k} d_{\tau(k)\ell} c_\ell = \sum_{\ell} b_{\tau(k)k} q_{\ell k} d_{i\ell} c_\ell, \quad {}^{\forall} i, k$$

and $\mathrm{rank}(C) = r$. On the assumption of $q_{\ell k} \neq 0$, since $d_{i\ell} = \dfrac{d_{\tau(k)\ell}}{b_{\tau(k)k}} \cdot b_{ik}$ for all i it holds that $d_\ell = \dfrac{d_{\tau(k)\ell}}{b_{\tau(k)k}} b_k$. Especially it holds that $d_{\tau(k)\ell} \neq 0$. That is, it holds that $d_\ell \,//\, b_k$. Hence, by $\mathrm{rank}(C \circ D) = r$, if $q_{\ell k} \neq 0$, then $\ell = \sigma(k)$. This implies that there exists a permutation matrix $P = M_\sigma$ such that both of PQ and QP^{-1} are diagonal. Then, if we choose $X = \mathrm{diag}\left(\dfrac{d_{\tau(k)\sigma(k)}}{b_{\tau(k)k}}\right)$, it holds that

$$a_k = q_{\sigma(k)k} \cdot \frac{d_{\tau(k)\sigma(k)}}{b_{\tau(k)k}} c_{\sigma(k)}, \quad b_k = \frac{b_{\tau(k)k}}{d_{\tau(k)\sigma(k)}} d_{\sigma(k)}, \qquad {}^{\forall} k$$

that is, it holds that $A = CQX$, $B = DP^{-1}X^{-1}$. Further, on the assumption of $Q \geq 0$, if we choose $Y = \mathrm{diag}\left(\dfrac{\sqrt{q_{\sigma(k)k}}\, d_{\tau(k)\sigma(k)}}{b_{\tau(k)k}}\right)$, it holds that $A = CQ_{1/2}Y$, $B = DQ_{1/2}Y^{-1}$. These completes the proof of Lemma 4. □

We should note that the factorization $(X \circ Y \circ Z)$ has the scalar uncertainty such that for scalars a, b, c, it holds

$$(a'X) \circ (b'Y) \circ (c'Z) = (abc)(X \circ Y \circ Z).$$

where (a', b', c') denotes any permutation of (a, b, c). Now we give a sufficient condition that NNTF has the unique global solution. From now set $u = (1, \ldots, 1)^T \in \mathbb{R}^r$ and let $fl_1(T)$ be a $a \times bc$ matrix whose $(i, j + b(k-1))$-component is t_{ijk}. Then the following theorem holds.

Theorem 3. *For* $f(X, Y, Z) = \| T - (X \circ Y \circ Z)u \|_F^2$, *we assume* $\mathrm{rank}(fl_1(T)) = r$ *and* $\min f(X, Y, Z) = 0$. *Then, under the condition* $\mathrm{rank}(Y) = \mathrm{rank}(Y \circ Z) = r$, *the optimal global point is unique up to permutations and scalar uncertainty.*

Proof. By triangular inequality we have

$$\| (X_1 \circ Y_1 \circ Z_1)u - (X_0 \circ Y_0 \circ Z_0)u \|_F \leq f(X_0, Y_0, Z_0) + f(X_1, Y_1, Z_1) = 0,$$

and thus $(X_0 \circ Y_0 \circ Z_0)u = (X_1 \circ Y_1 \circ Z_1)u$ which is equivalent to the following equation $X_0(Y_0 \circ Z_0)^T = X_1(Y_1 \circ Z_1)^T$. By Proposition 1 (3), there exists a non-singular matrix Q such that $X = X_0(Q^T)^{-1}$, $Y \circ Z = (Y_0 \circ Z_0)Q$. From Lemma 4, for some permutation matrix P and diagonal matrix D_1, it holds that $D_2 := PQ$ is a diagonal matrix and $Y = Y_0 Q D_1$ and $Z = Z_0 P^{-1} D_1^{-1}$. Hence, noting $P^{-1} = P^T$, it holds that $X = X_0 P^{-1} D_2^{-1}$, $Y = Y_0 P^{-1} D_2 D_1$, $Z = Z_0 P^{-1} D_1^{-1}$. Up to scalar uncertainty, (X, Y, Z) is equal to $(X_0 P^{-1}, Y_0 P^{-1}, Z_0 P^{-1}) = (X_0, Y_0, Z_0)P^{-1}$, and also it is, up to permutation, equal to (X_0, Y_0, Z_0). □

In general, it does not hold $(X_0 \circ Y_0 \circ Z_0)u = (X_1 \circ Y_1 \circ Z_1)u$, but we can show the following property.

Proposition 3. *For the function* $f(X, Y, Z) = \| T - (X \circ Y \circ Z)u \|_F^2$*, assume that* (X_0, Y_0, Z_0)*,* (X_1, Y_1, Z_1) *are two stationary points which attain the minimal value such that* $f(X_0, Y_0, Z_0) = f(X_1, Y_1, Z_1)$*. Then it holds that*

$$\| (X_0 \circ Y_0 \circ Z_0)u \|_F = \| (X_1 \circ Y_1 \circ Z_1)u \|_F .$$

Proof. Since $f(X, Y, Z) = \| fl_1(T) - X(Y \circ Z)^T \|_F^2$, from the equation (1), we have $\| fl_1(T) \|_F^2 - \| X_0(Y_0 \circ Z_0)^T \|_F^2 = \| fl_1(T) \|_F^2 - \| X_1(Y_1 \circ Z_1)^T \|_F^2$. That is, it holds that $\| X_0(Y_0 \circ Z_0)^T \|_F = \| X_1(Y_1 \circ Z_1)^T \|_F$. □

Finally we remark that the equality

$$\| X_0(Y_0 \circ Z_0)^T \|_F = \| Y_0(Z_0 \circ X_0)^T \|_F = \| Z_0(X_0 \circ Y_0)^T \|_F .$$

5 Conclusion

For a third-order tensor T and each r, there exists a sum of r tensors of rank 1 which is the closest to T in the sense of Frobenius norm (Existence property). Generally, a global optimal solution is not unique for NNTF. For this problem we proved that if T is of rank r the rank of the matrix made by an arrangement of T is less than or eaual to r, and that if the equality of both ranks holds the decomposition of T into a sum of r tensors of rank 1 is unique under some condition (Uniqueness property).

References

[CSS] Cao, B., Shen, D., Sun, J.-T., Wang, X., Yang, Q., Chen, Z.: Detect and Track Latent Factors with Online Nonnegative Matrix Factorization, IJCAI 2007, pp. 2689–2694 (2007)

[CZA] Chichoki, A., Zdunek, R., Amari, S.-i.: Non-Negative Tensor Factorization Using Csiszar's Divergence. In: Rosca, J., Erdogmus, D., Príncipe, J.C., Haykin, S. (eds.) ICA 2006. LNCS, vol. 3889, pp. 32–39. Springer, Heidelberg (2006)

[LS] Lee, D.D., Seung, H.S.: Algorithms for Non-negative Matrix Factorization. In: Leen, T.K., Dietterich, T.G., Tresp, V. (eds.) Advances in Neural Information Processing Systems, vol. 13, pp. 556–562. MIT Press, Cambridge (2001)

[MMV] Miwakeichi, F., Martínez-Montes, E., Valdés-Sosa, P.A., Nishiyama, N., Mizuhara, H., Yamaguchi, Y.: Decomposing EEG Data into Space–Time–Frequency Components Using Parallel Factor Analysis. NeuroImage 22, 1035–1045 (2004)

[MHH] Morup, M., Hansen, L.K., Herman, C.S., Parnas, J., Arnfred, S.M.: Parallel Factor Analysis as an Exploratory Tool for Wavelet Transformed Event-related EEG. NeuroImage 29, 938–947 (2006)

[WZZ] Wang, J., Zhong, W., Zhang, J.: NNMF-Based Factorization techniques for High-Accuracy Privacy Protection on Non-negative-valued Datasets. In: Perner, P. (ed.) ICDM 2006. LNCS (LNAI), vol. 4065, pp. 513–517. Springer, Heidelberg (2006)

Colored Subspace Analysis

Fabian J. Theis[1] and M. Kawanabe[2]

[1] Bernstein Center for Computational Neuroscience
MPI for Dynamics and Self-Organisation, Göttingen, Germany
[2] FraunhoferFIRST.IDA, Germany
fabian@theis.name
http://fabian.theis.name

Abstract. With the advent of high-throughput data recording methods in biology and medicine, the efficient identification of meaningful subspaces within these data sets becomes an increasingly important challenge. Classical dimension reduction techniques such as principal component analysis often do not take the large statistics of the data set into account, and thereby fail if the signal space is for example of low power but meaningful in terms of some other statistics. With 'colored subspace analysis', we propose a method for linear dimension reduction that evaluates the time structure of the multivariate observations. We differentiate the signal subspace from noise by searching for a subspace of non-trivially autocorrelated data; algorithmically we perform this search by joint low-rank approximation. In contrast to blind source separation approaches we however do not require the existence of sources, so the model is applicable to any wide-sense stationary time series without restrictions. Moreover, since the method is based on second-order time structure, it can be efficiently implemented even for large dimensions. We conclude with an application to dimension reduction of functional MRI recordings.

1 Introduction

Dimension reduction considers the question of removing a noise subspace from a larger multivariate signal. Classically, a signal is differentiated from noise by having a higher variance, and algorithms such as principal component analysis (PCA) in the linear case remove the low-variance components. This can be extended to nonlinear settings, which results in methods including nonlinear PCA [1], kernel PCA [2] and ISOMAP [3], to name but a few. These techniques are well-developed and powerful if the noise is comparatively low (in terms of power i.e. variance) when compared to the signal; in other words a signal manifold has to be 'visible' in the local covariance matrix. However the methods fail to capture signals that are deteriorated by noise of similar or stronger power.

Broadly speaking, there are two solutions to extract signals from higher-variance noise: (a) use higher-order statistics of the data to differentiate signal from noise, or (b) use additional information of the data such as temporal structure to define a signal manifold. (a) leads to the recently proposed *non-Gaussian*

M.E. Davies et al. (Eds.): ICA 2007, LNCS 4666, pp. 121–128, 2007.

component analysis (NGCA) [4,5,6], which is a semiparametric statistical framework for searching non-Gaussian subspaces—there are a few algorithmic implementations such as the multi-index projection pursuit. The noise subspace is characterized simply by being Gaussian. NGCA tries to detect the non-Gaussian signal subspace within the data, and in contrast to independent component analysis no assumption of independence within the subspace is made.

More precisely, given a random vector $\mathbf{x}$, a factorization $\mathbf{x} = \mathbf{As}$ with an invertible matrix $\mathbf{A}$, $\mathbf{s} = (\mathbf{s}_N, \mathbf{s}_G)$ and $\mathbf{s}_N$ a square-integrable n-dimensional random vector is called an *n-decomposition* of $\mathbf{x}$ if $\mathbf{s}_N$ and $\mathbf{s}_G$ are stochastically independent and $\mathbf{s}_G$ is Gaussian. In this case, $\mathbf{x}$ is said to be *n-decomposable.* $\mathbf{x}$ is denoted to be *minimally n-decomposable* if $\mathbf{x}$ is not $(n-1)$-decomposable. It has been shown that the minimal NGCA signal subspaces of a minimally n-decomposable decomposition are unique [5]. This method is clearly the only available alternative to second-order approaches if i.i.d. signals are given.

However, if the observations possess additional structure such as temporal dependencies, approach (b) provides an often simpler dimension reduction framework. Frequently, it is implicitly taken by methods that preprocess the data by transforming them into for example a Fourier or a Wavelet basis, which uses the time structure only in the preprocessing step. The assumption is that in the transformed domain, variance-based methods then suffice.

Here, we take approach (b) in a more direct fashion, and propose a novel method that takes the idea of NGCA and its underlying algorithms [4,6], namely the decomposition into a maximally white and 'another' signal, to the temporal domain, and apply it to the extraction of the signal subspace of fMRI data sets.

2 Colored Subspace Analysis (CSA)

The goal of CSA is to determine a basis of a random process such that in this basis as many components as possible are white (i.i.d.). The remaining components then span the 'colored subspace', onto which we project for dimension reduction.

Let $\mathbf{x}(t)$ be an (observed) d-dimensional real stochastic process and $\mathbf{A}$ an invertible real matrix such that $\mathbf{x}(t) = \mathbf{As}(t)$. As in NGCA, an *$n$-temporal-decomposition* of $\mathbf{s}(t)$ is defined by $\mathbf{s}(t) = (\mathbf{s}_C(t), \mathbf{s}_W(t))$. Here $\mathbf{s}_C(t)$ is an n-dimensional square-integrable wide-sense stationary random process and $\mathbf{s}_W(t)$ is i.i.d., such that the auto-crosscorrelation of $\mathbf{s}_W(t)$ and $\mathbf{s}_C(t)$ vanishes. Splitting up $\mathbf{A} = (\mathbf{A}_C, \mathbf{A}_W)$ accordingly yields the generative model $\mathbf{x}(t) = \mathbf{A}_C\mathbf{s}_C(t) + \mathbf{A}_W\mathbf{s}_W(t)$. With $\mathbf{W} := \mathbf{A}^{-1} =: (\mathbf{W}_C^\top, \mathbf{W}_W^\top)^\top$, the dimension reduction consists of projecting $\mathbf{x}(t)$ onto the lower-dimensional signal $\mathbf{s}_C(t) = \mathbf{W}_C\mathbf{x}(t)$. Note that the more traditional model $\mathbf{x}(t) = \mathbf{A}_G\mathbf{s}_G(t) + \mathbf{n}(t)$ using full-rank noise $\mathbf{n}(t)$ is included in the above model, simply by adding the n-dimensional part of $\mathbf{n}(t)$ lying in the image of $\mathbf{A}_G$ to $\mathbf{s}_G(t)$. Out claim then is that we cannot distinguish between signal and noise in the signal subspace without additional assumptions.

2.1 Indeterminacies

The subspace given by the range of $\mathbf{W}_C$ is denoted as the *colored subspace* of $\mathbf{x}(t)$. Clearly, the coefficients of $\mathbf{A}$ or $\mathbf{W}$ cannot be unique. However, from similar arguments as below, it can be shown that the colored subspacc itself is unique if the n-temporal decomposition is minimal in the sense that no $(n-1)$-temporal-decomposition of $\mathbf{x}(t)$ exists; we have to assume that the noise subspace is maximal as we do not make any assumptions on $\mathbf{s}_C(t)$.

2.2 Algorithm

The key assumption of the model is that the $\mathbf{s}_C(t)$ and $\mathbf{s}_W(t)$ have no common autocorrelations, i.e.—after centering—that $\mathbf{R}_\mathbf{s}(\tau) := \mathbf{E}(\mathbf{s}(t+\tau)\mathbf{s}(t)^\top)$ is block diagonal of the form

$$\mathbf{R}_\mathbf{s}(\tau) = \begin{array}{|c|c|}\hline \mathbf{R}_{\mathbf{s}_C}(\tau) & 0 \\ \hline 0 & \mathbf{R}_{\mathbf{s}_W}(\tau) \\ \hline \end{array} \tag{1}$$

for all τ. Moreover, the noise component $\mathbf{s}_W(t)$ is characterized by being i.i.d., hence $\mathbf{R}_{\mathbf{s}_W}(\tau) = 0$ for $\tau \neq 0$. It can be shown that minimality of the colored subspace is guaranteed if n is chosen maximal such that there still exists a $\tau \neq 0$ with full-rank $\mathbf{R}_{\mathbf{s}_C}(\tau)$. The factorization model now provides that the observed autocorrelations $\mathbf{R}_\mathbf{x}(\tau)$ can be factorized into

$$\mathbf{R}_\mathbf{x}(\tau) = \mathbf{A}\mathbf{R}_\mathbf{s}(\tau)\mathbf{A}^\top. \tag{2}$$

As preprocessing, we first remove correlations by PCA, which guarantees that $\mathbf{R}_\mathbf{x}(0) = \mathbf{I}$. Since the basis in the signal and noise subspaces are non-unique, we may choose coordinates as normalization such that without loss of generality $\mathbf{R}_{\mathbf{s}_C}(0) = \mathbf{I}$ and $\mathbf{R}_{\mathbf{s}_W}(0) = \mathbf{I}$, hence $\mathbf{R}_\mathbf{s}(0) = \mathbf{I}$ according to (1). Then $\mathbf{A}$ is orthogonal, because $\mathbf{A}\mathbf{A}^\top = \mathbf{A}\mathbf{R}_\mathbf{s}(0)\mathbf{A}^\top = \mathbf{R}_\mathbf{x}(0) = \mathbf{I}$.

So the model factorization (2) together with the block structure (1) implies that $\mathbf{A}$ and hence the colored subspace can be algorithmically detected by block-diagonalizing one symmetrized $\bar{\mathbf{R}}_\mathbf{x}(\tau) = 1/2(\mathbf{R}_\mathbf{x}(\tau) + \mathbf{R}_\mathbf{x}(\tau)^\top)$. Robustness can be increased by performing orthogonal *joint* block-diagonalization [7,8] of multiple or all $\bar{\mathbf{R}}_\mathbf{x}(\tau)$ for $\tau \neq 0$.

The dimension n of the signal subspace can be determined as

$$n := \max_{\tau \neq 0} \operatorname{rank} \bar{\mathbf{R}}_\mathbf{x}(\tau),$$

which in practice has to be replaced by a thresholded or adaptive rank calculation to allow for noise and finite-sample effects. Using the fact that $\mathbf{R}_{\mathbf{s}_W}(\tau) = 0, \tau \neq 0$ more explicitly, we get

$$\mathbf{R}_\mathbf{x}(\tau) = (\mathbf{A}_C, \mathbf{A}_W)\mathbf{R}_\mathbf{s}(\tau)(\mathbf{A}_C, \mathbf{A}_W)^\top = \mathbf{A}_C\mathbf{R}_{\mathbf{s}_C}(\tau)\mathbf{A}_C^\top.$$

Hence after joint block-diagonalization, the colored subspace is given by the non-zero eigenvalues—which in the finite-sample case has to be approximated.

This model is closely related to the BSS-algorithms AMUSE [9], SFA [10] for one and SOBI [11], TDSEP [12] for multiple autocovariance matrices. The difference is that no generative data model is necessary—CSA is applicable to any wide-sense stationary random process; the signal subspace is automatically and uniquely determined, and additional assumptions *within* the data subspace (such as autodecorrelated sources) are not necessary. This is analogous to the step from ICA to NGCA as discussed in the introduction.

Interestingly, the two models of ICA and autodecorrelation can also be combined, see e.g. JADE$_{TD}$ [13], where JADE and TDSEP are combined with $\mathbf{R}(\tau), \tau \neq 0$ for ICA in the presence of i.i.d. Gaussian noise. Similar combinations are possible for corresponding dimension reduction frameworks. A review of related cost functions is given in [14].

2.3 Block-Diagonalization by Joint Low-Rank Approximation

Recently, the authors have presented a method for extracting a single non-zero block from a set of matrices distributed by unitary transformations [14]. There we focused on the NGCA problem and proposed a procedure called joint low-rank approximation (JLA) with a set $\{\mathbf{M}_k\}_{k=1}^K$ of transformed block matrices as $\mathbf{R}_{\mathbf{x}}(\tau)$ in Eq.(2) for $\tau \neq 0$. The reduction matrix $\mathbf{W}_0$, which extracts the non-Gaussian part of the data $\mathbf{x}$ can be determined by maximizing $\mathcal{L}(\mathbf{W}_0) = \sum_{k=1}^K \|\mathbf{W}_0\mathbf{M}_k\mathbf{W}_0^\top\|_{\mathrm{Fro}}^2$ over Stiefel manifold $\mathbf{W}_0 \in V_n(\mathbb{R}^d)$, where $\|\mathbf{C}\|_{\mathrm{Fro}}^2 = \mathrm{tr}(\mathbf{C}\mathbf{C}^*)$. It can be shown that the true reduction matrix is the unique maximizer of $\mathcal{L}$ up to equivalence in the Stiefel manifold. By taking derivative of $\mathcal{L}$, we get the equation

$$\mathbf{W}_0 \sum_{k=1}^K \mathcal{M}_k(\mathbf{W}_0) = \Lambda \mathbf{W}_0,$$

which can be solved by iterative eigenvalue decomposition, where $\mathcal{M}_k(\mathbf{W}_0) := \mathbf{M}_k\mathbf{W}_0^\top\mathbf{W}_0\mathbf{M}_k^* + \mathbf{M}_k^*\mathbf{W}_0^\top\mathbf{W}_0\mathbf{M}_k$. Examples of such matrix sets for NGCA case are:

(a) fourth-order cumulant tensor, i.e. $\mathbf{Q}^{(kl)} := (\mathrm{cum}(x_i, x_j, x_k, x_l))$ for all (k, l),
(b) Hessian of log characteristic function, i.e. $\mathbf{M}_k := \frac{\partial^2}{\partial\boldsymbol{\zeta}\partial\boldsymbol{\zeta}^\top} \log \mathrm{E}[\exp(i\boldsymbol{\zeta}^\top\mathbf{x})] + \mathbf{I}_d$.

For the second case, we developed somewhat sophisticated choices and updates of the frequency vectors $\boldsymbol{\zeta}_k$ which is necessary to improve the performance of JLA. In the case of CSA, we commonly fix the autocovariance matrices in advance, but informative lags τ can be chosen by a similar idea. Algorithm 1 shortly summarizes how JLA is applied to our autocovariance data set.

3 Simulations

As a simple toy example, we consider $n = 3$-dimensional colored signals in $d = 10$-dimensional data. The colored signals are three sinusoids of equal frequency and varying phase, which have been instantaneously gaussianized, see figure 1(a), so methods based on higher-order statistics such as NGCA cannot work. They have been embedded in white Gaussian noise of approximately equal power. The resulting 10-dimensional data set is then mixed by a matrix $\mathbf{A}$ with coefficients

Algorithm 1: Joint low-rank approximation for CSA

Input: $d \times T$ sample matrix $\mathbf{X}$ of a multivariate time series, number of autocovariances K, source dimension n
Output: CSA projection $\mathbf{W}$

prewhiten data
calculate eigenvalue decomposition (EVD) of covariance $\mathbf{E}_0\boldsymbol{\Lambda}_0\mathbf{E}_0^\top = \mathrm{Cov}(\mathbf{X})$
$\mathbf{V} \leftarrow \boldsymbol{\Lambda}_0^{-1/2}\mathbf{E}_0^\top$
$\mathbf{Y} \leftarrow \mathbf{V}\mathbf{X}$

estimate autocovariance matrices
for $\tau \leftarrow 1 \dots K$ **do**
  $\mathbf{M}_\tau \leftarrow (T-\tau)^{-1}\mathbf{Y}(:, 1 : T-\tau+1)\mathbf{Y}(:, \tau : T)^\top$

initialize JLA algorithm
calculate EVD $\mathbf{E}\boldsymbol{\Lambda}\mathbf{E}^\top = \sum_\tau \mathbf{M}_\tau + \mathbf{M}_\tau^\top$
$\mathbf{W} \leftarrow \mathbf{E}(:, 1 : n)^\top$
$I \leftarrow \{1, \dots, K\}$

iterate JLA, possibly quit loop earlier
for $i \leftarrow 1 \dots K$ **do**
  $\mathcal{M} \leftarrow \sum_{\tau \in I} \mathbf{M}_\tau\mathbf{W}^\top\mathbf{W}\mathbf{M}_\tau^\top + \mathbf{M}_\tau^\top\mathbf{W}^\top\mathbf{W}\mathbf{M}_\tau$
  calculate EVD $\mathcal{M} = \mathbf{E}_i\boldsymbol{\Lambda}_i\mathbf{E}_i^\top$
  $\mathbf{W} \leftarrow \mathbf{E}_i(:, 1 : n)^\top$
  determine τ_0 with minimal $\|\mathbf{W}\mathbf{M}_\tau\mathbf{W}^\top\|_F^4 / \|\mathbf{M}_\tau\|_F^2$
  remove τ_0 from I

$\mathbf{W} \leftarrow \mathbf{W}\mathbf{V}$

chosen from an standard normal distribution; the mixtures $\mathbf{x}(t)$ are shown in figure 1(b). The resulting SNR is -5dB, so distinction of signal from the noise subspace by power ($\rightarrow$ PCA) cannot work either, as will also be shown later.

We first apply CSA with $K = 10$ autocovariance matrices and known signal subspace dimension $n = 3$. If we multiply the recovered projection $\mathbf{W}_C$ with the mixing matrix $\mathbf{A}$, we expect $\mathbf{W}_C\mathbf{A}$ to have strong contributions in the first $n \times n$-block and close to zero entries everywhere else. This is the case indicating that CSA works fine, see Hinton-diagram in figure 1(c). Indeed a possible error-index $e(\mathbf{W}_{\mathrm{CSA}}) := \|(\mathbf{W}_{\mathrm{CSA}}\mathbf{A})(:, n+1 : d)\|_F$ is low (0.0139): If we perform similar joint block diagonalization-based search for the projection, extending the SOBI algorithm, we also achieve an approximate signal projection, however with an increased error of 0.0363. If however only PCA is applied, the resulting error is high with $e(\mathbf{W}_{\mathrm{PCA}}) = 5.12$, see figure 1(d).

A more systematical comparison of the three methods is achieved when we perform the above experiment for a batch run of length 100, with randomly chosen $\mathbf{A}$ in each run. The resulting statistics, figures 1(e-f), confirm the superior performance of CSA in terms of recovery error, as well as computational time (with respect to the extension of SOBI).

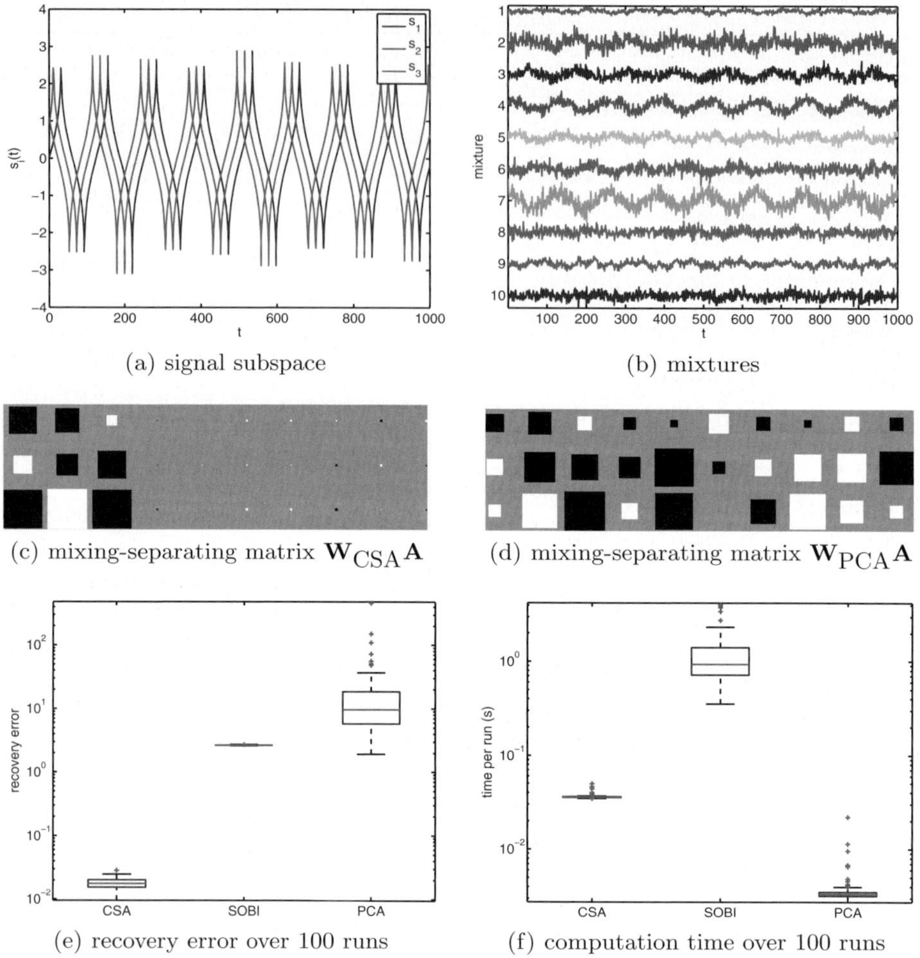

(a) signal subspace

(b) mixtures

(c) mixing-separating matrix $\mathbf{W}_{\mathrm{CSA}}\mathbf{A}$

(d) mixing-separating matrix $\mathbf{W}_{\mathrm{PCA}}\mathbf{A}$

(e) recovery error over 100 runs

(f) computation time over 100 runs

Fig. 1. Toy example of an $n = 3$-dimensional signal (a) in $d = 10$ dimensions (b). CSA outperforms the other methods (c-f). See text for details.

4 Signal-Subspaces in fMRI Data

Functional magnetic-resonance imaging (fMRI) can be used to measure brain activity. Multiple MRI scans are taken in various functional conditions; the extracted task-related component reveals information about the task-activated brain regions. Classical power-based methods fail to blindly recover the task-related component as it is very small with respect to the total signal, usually around one percent in terms of variance. Hence we propose to use the autocovariance structure (in this case spatial autocovariances) in combination with CSA to properly reduce the data dimension.

fMRI data with 98 images (TR/TE = 3000/60 msec) were acquired with five periods of rest and five photic simulation periods with rest. Simulation and rest periods comprised 10 repetitions each, i.e. 30s. Resolution was $3 \times 3 \times 4$ mm. The slices were oriented parallel to the calcarine fissure. Photic stimulation was performed using an 8 Hz alternating checkerboard stimulus with a central fixation point and a dark background with a central fixation point during the control periods [15]. The first scans were discarded for remaining saturation effects. Motion artifacts were compensated by automatic image alignment.

In order to compare the performance of CSA versus standard PCA-based dimension reduction in varying source data dimension, we reduce the total data to $p \in \{2, 5, 10, 20, 50, 98\}$ dimensions by PCA. Then we either apply CSA or PCA and order the components in decreasing order of the eigenvalues (of total autocovariance or covariance respectively). We analyze how well the task-related component with the known task vector $\mathbf{v} \in \{0,1\}^{98}$ is contained in a component by $f(i) := (\mathbf{W}_0(i,:)\mathbf{v})^2$, where $\mathbf{W}_0$ is the separating matrix. In order to allow for finite-sample effects, we compare the recovered subspace for all varying reduced dimensions n by comparing it to the total power by plotting $c(n) = \sum_{i=1}^{n} f(i) / \sum_{i=1}^{p} f(i)$ versus n, see figure 2.

For strongly reduced data $p \leq 5$, both methods capture the task component for low n, PCA more so than CSA. But in more realistic data settings $p \geq 10$, necessary for full data evaluation, CSA consistently needs $n = 5$ components to guarantee that the task-related component is contained in the signal subspace (with cumulative contribution ratio .8 for $p \leq 20$), whereas PCA needs already $n = 18$ components to guarantee the same for $p = 20$, and more so for larger p.

This illustrates that CSA can be used as preprocessing tool for fMRI data much more efficiently than PCA, albeit at a somewhat higher computational cost.

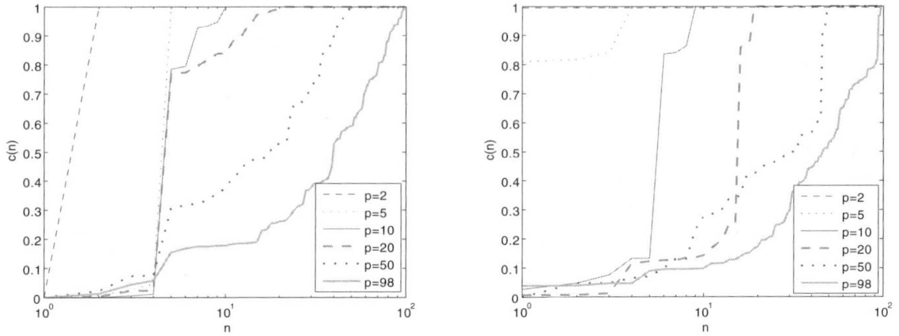

(a) CSA (b) PCA

Fig. 2. Comparison of CSA (left) and PCA (right) for dimension reduction

Conclusions. We have presented a generally applicable, efficient method for linear dimension reduction that separates a subspace with nontrivial autocorrelations (color) from the white remainder. Results on toy and real data are promising. Presently, we are working on a statistically sound estimation of the subspace dimension as well as on a generalization without prewhitening. Moreover, we are planning to study the performance of CSA on other medical imaging applications. We believe that the method may provide a useful tool for preprocessing to allow for more efficient analysis in a lower-dimensional signal subspace still capturing fine-grained and low-power statistics of the observations.

References

1. Oja, E.: Nonlinear pca criterion and maximum likelihood in independent component analysis. In: Proc. ICA 1999, Aussois, France, pp. 143–148 (1999)
2. Mika, S., Schölkopf, B., Smola, A.J., Müller, K., Scholz, M., Rätsch, G.: Kernel PCA and de-noising in feature spaces. In: Kearns, M.S., Solla, S.A., Cohn, D.A (eds.) Advances in Neural Information Processing Systems, vol. 11, MIT Press, Cambridge (1999)
3. Tenenbaum, J., de Silva, V., Langford, J.: A global geometric framework for nonlinear dimensionality reduction. Science 290(5500), 2319–2323 (2000)
4. Blanchard, G., Kawanabe, M., Sugiyama, M., Spokoiny, V., Müller, K.: In search of non-gaussian components of a high-dimensional distribution. Journal of Machine Learning Research 7, 247–282 (2006)
5. Theis, F., Kawanabe, M.: Uniqueness of non-gaussian subspace analysis. In: Rosca, J., Erdogmus, D., Príncipe, J.C., Haykin, S. (eds.) ICA 2006. LNCS, vol. 3889, pp. 917–925. Springer, Heidelberg (2006)
6. Kawanabe, M., Theis, F.: Estimating non-gaussian subspaces by characteristic functions. In: Rosca, J., Erdogmus, D., Príncipe, J.C., Haykin, S. (eds.) ICA 2006. LNCS, vol. 3889, pp. 157–164. Springer, Heidelberg (2006)
7. Abed-Meraim, K., Belouchrani, A.: Algorithms for joint block diagonalization. In: Proc. EUSIPCO 2004, Vienna, Austria, pp. 209–212 (2004)
8. Févotte, C., Theis, F.: Orthonormal approximate joint block-diagonalization. Technical report, GET/Télécom Paris (2007)
9. Tong, L., Liu, R.W., Soon, V., Huang, Y.F.: Indeterminacy and identifiability of blind identification. IEEETransactions on Circuits and Systems 38, 499–509 (1991)
10. Wiskott, L., Sejnowski, T.: Slow feature analysis: Unsupervised learning of invariances. Neural Computation 14, 715–770 (2002)
11. Belouchrani, A., Meraim, K.A., Cardoso, J.F., Moulines, E.: A blind source separation technique based on second order statistics. IEEE Transactions on Signal Processing 45(2), 434–444 (1997)
12. Ziehe, A., Mueller, K.R.: TDSEP – an efficient algorithm for blind separation using time structure. In: Niklasson, L., Bodén, M., Ziemke, T. (eds.) Proc. of ICANN'98, Skövde, Sweden, pp. 675–680. Springer, Berlin (1998)
13. Müller, K.R., Philips, P., Ziehe, A.: Jadetd: Combining higher-order statistics and temporal information for blind source separation (with noise). In: Proc. of ICA 1999, Aussois, France, pp. 87–92 (1999)
14. Theis, F., Inouye, Y.: On the use of joint diagonalization in blind signal processing. In: Proc. ISCAS 2006, Kos, Greece (2006)
15. Wismüller, A., Lange, O., Dersch, D., Leinsinger, G., Hahn, K., Pütz, B., Auer, D.: Cluster analysis of biomedical image time-series. International Journal on Computer Vision 46, 102–128 (2002)

Is the General Form of Renyi's Entropy a Contrast for Source Separation?

Frédéric Vrins[1], Dinh-Tuan Pham[2], and Michel Verleysen[1]

[1] Machine Learning Group
Université catholique de Louvain
Louvain-la-Neuve, Belgium
{vrins,verleysen}@dice.ucl.ac.be
[2] Laboratoire de Modélisation et Calcul
Centre National de la Recherche Scientifique
Grenoble, France
dinh-tuan.pham@imag.fr

Abstract. Renyi's entropy-based criterion has been proposed as an objective function for independent component analysis because of its relationship with Shannon's entropy and its computational advantages in specific cases. These criteria were suggested based on "convincing" experiments. However, there is no theoretical proof that globally maximizing those functions would lead to separate the sources; actually, this was implicitly conjectured. In this paper, the problem is tackled in a theoretical way; it is shown that globally maximizing the Renyi's entropy-based criterion, in its general form, does not necessarily provide the expected independent signals. The contrast function property of the corresponding criteria simultaneously depend on the value of the Renyi parameter, and on the (unknown) source densities.

1 Introduction

Blind source separation (BSS) aims at recovering underlying source signals from mixture of them. Under mild assumptions, including the mutual independence between those sources, it is known from Comon [1] that finding the linear transformation that minimizes a dependence measure between outputs can solve the problem, up to acceptable indeterminacies on the sources. This procedure is known as Independent Component Analysis (ICA).

This problem can be mathematically expressed in a very simple way. Consider the square, noiseless BSS mixture model: a K-dimensional vector of independent unknown sources $\mathbf{S} = [S_1, \ldots, S_K]^{\mathrm{T}}$ is observed via an instantaneous linear mixture of them $\mathbf{X} = \mathbf{AS}$, $\mathbf{X} = [X_1, \ldots, X_K]^{\mathrm{T}}$, where $\mathbf{A}$ is the full-rank square mixing matrix. Many separation methods are based on the maximization (or minimization) of a criterion. A specific class of separation criteria is called "contrast functions" [1]. The contrast property ensures that a given criterion is suitable to achieve BSS. Such a function i) is scale invariant, ii) only depends on the demixing matrix B and of the mixture densities iii) reaches its global maximum if and only if the transfer matrix $\mathbf{W} = \mathbf{BA}$ is non-mixing [1]. A matrix $\mathbf{W}$

M.E. Davies et al. (Eds.): ICA 2007, LNCS 4666, pp. 129–136, 2007.

is said non-mixing if it belongs to the subgroup $\mathcal{W}$ of the general linear group $\mathcal{GL}(K)$ of degree K, and is defined as:

$$\mathcal{W} \doteq \{\mathbf{W} \in \mathcal{GL}(K) : \exists \mathbf{P} \in \mathcal{P}^{\mathbf{K}},\ \boldsymbol{\Lambda} \in \mathcal{D}^{\mathbf{K}}, \mathbf{W} = \mathbf{P}\boldsymbol{\Lambda}\} \tag{1}$$

In the above definition $\mathcal{P}^K$ and $\mathcal{D}^K$ respectively denote the groups of permutation matrices and of regular diagonal matrices of degree K.

Many contrast functions have been proposed in the literature. One of the most known contrast function is the opposite of mutual information $I(\mathbf{Y})$ [2] where $\mathbf{Y} = \mathbf{BX}$, which can be equivalently written as a sum of differential entropies $h(.)$:

$$I(\mathbf{Y}) \doteq \sum_{i=1}^{K} h(Y_i) - h(\mathbf{Y}) = \sum_{i=1}^{K} h(Y_i) - \log|\det \mathbf{B}| - h(\mathbf{X}). \tag{2}$$

The differential (Shannon) entropy of $X \sim p_X$ is defined by

$$h(X) \doteq -\mathrm{E}[\log p_X]. \tag{3}$$

Since $I(\mathbf{Y})$ has to be minimized with respect to $\mathbf{B}$, its minimization is equivalent to the following optimization problem under a prewhitening step:

$$\max_{\mathbf{B} \in \mathcal{SO}(K)} C(\mathbf{B}),\quad C(\mathbf{B}) \doteq -\sum_{i=1}^{K} h(\mathbf{b}_i\mathbf{X}), \qquad \text{problem 1} \tag{4}$$

where $\mathbf{b}_i$ denotes the i-th row of $\mathbf{B}$ and $\mathcal{SO}(K)$ is the special orthogonal group

$$\mathcal{SO}(K) \doteq \{\mathbf{W} \in \mathcal{GL}(K) : \mathbf{W}\mathbf{W}^T = \mathbf{I}_K,\ \det \mathbf{W} = +1\}$$

with $\mathbf{I}_K$ the identity matrix of degree K. The $\mathbf{B} \in \mathcal{SO}(K)$ restriction, yielding $\log|\det \mathbf{B}| = 0$, results from the fact that, without loss of generality, the source can be assumed to be centered and unit-variance ($\mathrm{E}[\mathbf{S}\mathbf{S}^{\mathrm{T}}] = \mathbf{I}_K$ and $\mathbf{A} \in \mathcal{SO}(K)$ if the mixtures are whitened [8]). Clearly, if $\mathbf{W} = \mathbf{BA}$, problem 1 is equivalent to problem 2:

$$\max_{\mathbf{W} \in \mathcal{SO}(K)} \widetilde{C}(\mathbf{W}),\quad \widetilde{C}(\mathbf{W}) \doteq -\sum_{i=1}^{K} h(\mathbf{w}_i\mathbf{S}). \qquad \text{problem 2}$$

Few years ago, it has been suggested to replace Shannon's entropy by Renyi's entropy [4,5]. More recent works still focus on that topic (see e.g. [7]). Renyi's entropy is a generalization of Shannon's one in the sense that they share the same key properties of information measures [10]. The Renyi entropy of index $r \geq 0$ is defined as:

$$h_r(X) \doteq \frac{1}{1-r} \log \int_{\Omega(X)} p_X^r(x)dx, \tag{5}$$

where $r \geq 0$ and $\lim_{r\to 1} h_r(X) = h_1(X) = h(X)$ and $\Omega(X) \doteq \{x : p_X(x) > 0\}$. Based on simulation results, some researchers have proposed to modify the above BSS contrast $C(\mathbf{B})$ defined in problem 1 by the following modified criterion

$$C_r(\mathbf{B}) \doteq -\sum_{i=1}^{K} h_r(Y_i), \tag{6}$$

assuming implicitly that the contrast property of $C_r(\mathbf{B})$ is preserved even for $r \neq 1$. This is clearly the case for the specific values $r = 1$ (because obviously $C_1(\mathbf{B}) = C(\mathbf{B})$) and $r = 0$ (under mild conditions); this can be easily shown using the Entropy Power and the Brunn-Minkowski inequalities [3], respectively. However, there is no formal proof that the contrast property of $C_r(\mathbf{B})$ still holds for other values of r.

In order to check if this property may be lost in some cases, we restrict ourselves to see if a necessary condition ensuring that $C_r(\mathbf{B})$ is a contrast function is met. More specifically, the criterion $\widetilde{C}_r(\mathbf{W})$ should admit a local maximum when $\mathbf{W} \in \mathcal{W}$. To see if this condition is fulfilled, a second order Taylor development of $\widetilde{C}_r(\mathbf{W})$ is provided around a non-mixing point $\mathbf{W}^\star \in \mathcal{W}$ in the next section. For the sake of simplicity, we further assume $K = 2$ and that a prewhitening is performed so that we shall constraint $\mathbf{W} \in \mathcal{SO}(2)$ since it is sufficient for our purposes, as shown in the example of Section 3 (the extension to $K \geq 3$ is easy).

2 Taylor Development of Renyi's Entropy

Setting $K = 2$, we shall study the variation of the criterion $\widetilde{C}_r(\mathbf{W})$ due to a slight deviation of $\mathbf{W}$ from any $\mathbf{W}^\star \in \mathcal{W} \cap \mathcal{SO}(2)$ of the form $\mathbf{W} \leftarrow \mathcal{E}\mathbf{W}^\star$ where

$$\mathcal{E} \doteq \begin{bmatrix} \cos\theta & \sin\theta \\ -\sin\theta & \cos\theta \end{bmatrix} \tag{7}$$

and $\theta \simeq 0$ is a small angle. This kind of updates covers the neighborhood of $\mathbf{W}^\star \in \mathcal{SO}(K)$: if $\mathbf{W}, \mathbf{W}^\star \in \mathcal{SO}(2)$, there always exists $\boldsymbol{\Phi} \in \mathcal{SO}(2)$ such that $\mathbf{W} = \boldsymbol{\Phi}\mathbf{W}^\star$; $\boldsymbol{\Phi}$ can be written as $\mathcal{E}$ and if $\mathbf{W}$ is further restricted to be in the neighborhood of $\mathbf{W}^\star$, θ must be small enough. In order to achieve that aim, let us first focus on a first order expansion of the criterion, to analyse if non-mixing matrices are stationary points of the criterion. This is a obviously a necessary condition for $C_r(\mathbf{B})$ to be a contrast function.

2.1 First Order Expansion: Stationarity of Non-mixing Points

Let Z be a random variable independent from Y. From the definition of Renyi's entropy given in eq. (5), it comes that Renyi's entropy of $Y + \epsilon Z$ is

$$h_r(Y + \epsilon Z) = \frac{1}{1-r} \log \int p^r_{Y+\epsilon Z}(x)dx, \tag{8}$$

where the density $p_{Y+\epsilon Z}$ reduces to, up to first order in ϵ [9]:

$$p_{Y+\epsilon Z}(y) = p_Y(y) - \epsilon \left. \frac{\partial \mathrm{E}[(Z|Y = x)p_Y(x)]}{\partial x} \right|_{x=y} + o(\epsilon). \tag{9}$$

Therefore, we have:

$$p^r_{Y+\epsilon Z}(y) = p^r_Y(y) - r\epsilon p^{r-1}_Y(y) \left.\frac{\partial[\mathrm{E}(Z|Y=x)p_Y(x)]}{\partial x}\right|_{x=y} + \phi(\epsilon, y), \quad (10)$$

where $\phi(\epsilon, y)$ is $o(\epsilon)$. Hence, noting that $\log(1+a) = a + o(a)$ as $a \to 0$, equations (8) and (9) yield[1]

$$h_r(Y+\epsilon Z) = h_r(Y) - \epsilon\frac{r}{1-r}\frac{\int p^{r-1}_Y(y)[\mathrm{E}(Z|Y)p_Y]'(y)dy}{\int p^r_Y(y)dy} + o(\epsilon). \quad (11)$$

But, by integration by parts, one gets

$$\frac{1}{r-1}\int p^{r-1}_Y(y)[\mathrm{E}(Z|Y)p_Y]'(y)dy = -\int p^{r-1}_Y(y)\mathrm{E}(Z|Y=y)p'_Y(y)dy, \quad (12)$$

yielding

$$-\epsilon\frac{r}{1-r}\frac{\int p^{r-1}_Y(y)[\mathrm{E}(Z|Y)p_Y]'(y)dy}{\int p^r_Y(y)dy} = -\epsilon r\frac{\int p^{r-1}_Y(y)\mathrm{E}(Z|Y=y)p'_Y(y)dy}{\int p^r_Y(y)dy}. \quad (13)$$

From the general iterated expectation lemma (p. 208 of [6]), the right-hand side of the above equality equals

$$-\epsilon r\frac{\mathrm{E}[p^{r-2}_Y(Y)p'_Y(Y)Z]}{\int p^r_Y(y)dy} = \epsilon\mathrm{E}[\psi_r(Y)Z], \quad (14)$$

if we define the r-score function $\psi_r(Y)$ of Y as

$$\psi_r(Y) \doteq -\frac{rp^{r-2}_Y(Y)p'_Y(Y)}{\int p^r_Y(y)dy} = -\frac{1}{p_Y(Y)}\frac{(p^r_Y)'(Y)}{\int p^r_Y(y)dy}. \quad (15)$$

Observe that the 1-score reduces to the score function of Y, defined as $-(\log p_Y)'$.

Then, using eq. (11), noting $\mathbf{Y} = \mathcal{E}\mathbf{W}^\star\mathbf{S}$, $\cos\theta = 1 + o(\theta)$ and $\sin\theta = \theta + o(\theta)$, the criterion $\widetilde{C}_r(\mathcal{E}\mathbf{W}^\star)$ becomes up to first order in θ:

$$\begin{aligned}\widetilde{C}_r(\mathcal{E}\mathbf{W}^\star) &= -h_r(Y_1) - h_r(Y_2) \\ &\approx \widetilde{C}_r(\mathbf{W}^\star) \pm \theta\Big\{\mathrm{E}[\psi_r(S_1)S_2] - \mathrm{E}[\psi_r(S_2)S_1]\Big\}. \end{aligned} \quad (16)$$

The sign of θ in the last equation depends on matrix $\mathbf{W}^\star$; for example, if $\mathbf{W}^\star = \mathbf{I}_2$, it is negative, and if the rows of $\mathbf{I}_2$ are permuted in the last definition of $\mathbf{W}^\star$, it is positive.

Remind that the criterion is not sensitive to a left multiplication of its argument by a scale and/or permutation matrix. For instance, $\widetilde{C}_r(\mathbf{W}^\star) = \widetilde{C}_r(\mathbf{I}_2) = -h_r(S_1) - h_r(S_2)$. It results that since independence implies non-linear decorrelation, both expectations vanish in eq. (16) and $\widetilde{C}_r(\mathbf{W})$ admits a stationary point whatever is $\mathbf{W}^\star \in \mathcal{W}$.

[1] Provided that there exist $\epsilon^\star > 0$ and an integrable function $\Phi(y) > 0$ such that for all $y \in \mathbb{R}$ and all $|\epsilon| < \epsilon^\star$, $\phi(\epsilon, y)/\epsilon < \Phi(y)$. It can be shown that this is indeed the case under mild regularity assumptions.

2.2 Second Order Expansion: Characterization of Non-mixing Points

Let us now characterize these stationary points. To this end, consider the second order expansion of $p_{Y+\epsilon Z}$ provided in [9] (Z is assumed to be zero-mean to simplify the algebra):

$$p_{Y+\epsilon Z} = p_Y + \frac{1}{2}\epsilon^2 \mathrm{E}(Z^2) p''_Y + o(\epsilon^2). \tag{17}$$

Therefore, since Renyi's entropy is not sensitive to translation we have, for $r > 0$:

$$h_r(Y + \epsilon Z) = h_r(Y) + \frac{\epsilon^2}{2} \underbrace{\frac{r}{1-r} \frac{\int p_Y^{r-1}(y) p''_Y(y) dy}{\int p_Y^r(y) dy}}_{\doteq J_r(Y)} \mathrm{var(Z)} + \mathrm{o}(\epsilon^2), \tag{18}$$

where $J_r(Y)$ is called the r-th order information of Y. Observe that the first order information reduces to $J_1(Y) = \mathrm{E}[\psi^2_{Y,r}]$, i.e. to Fisher's information [2].

In order to study the "nature" of the stationary point reached at $\mathbf{W}^\star$ (minimum, maximum, saddle), we shall check the variation of $\widetilde{C}_r$ resulting from the update $\mathbf{W} \leftarrow \mathcal{E}\mathbf{W}^\star$ up to second order in θ. Clearly, $\cos\theta = 1 - \theta^2/2 + o(\theta^2)$ and $\tan\theta = \theta + o(\theta^2)$, the criterion then becomes:

$$\begin{aligned}
\widetilde{C}_r(\mathcal{E}\mathbf{W}^\star) &= -h_r(Y_1) - h_r(Y_2) \\
&= -h_r(S_1 + \tan\theta S_2) - h_r(S_2 - \tan\theta S_1) - 2\log|\cos\theta| \\
&= \widetilde{C}_r(\mathbf{W}^\star) - \frac{\theta^2}{2}\left[J_r(S_1)\mathrm{var(S_2)} + \mathrm{J_r(S_2)var(S_1)}\right] - 2\log|1 - \frac{\theta^2}{2}| + o(\theta^2) \\
&= \widetilde{C}_r(\mathbf{W}^\star) - \frac{\theta^2}{2}\left[J_r(S_1)\mathrm{var(S_2)} + \mathrm{J_r(S_2)var(S_1)} - 2\right] + o(\theta^2)
\end{aligned} \tag{19}$$

where we have used $H_r(\alpha Y) = H_r(Y) + \log|\alpha|$, for any real number $\alpha > 0$. This clearly shows that if the sources share a same density with variance var(S) and r-th order information $J_r(S)$, the sign of $\widetilde{C}_r(\mathcal{E}\mathbf{W}^\star) - C_r(\mathbf{W}^\star)$ equals, up to second order in θ to $\mathrm{sign}(1 - J_r(S)\mathrm{var(S)})$. In other words, the criterion reaches a local minimum at any $\mathbf{W} \in \mathcal{W}$ if $J_r(S_i)\mathrm{var(S_i)} < 1$, instead of an expected global maximum. In this specific case, maximizing the criterion does not yield the seeked sources.

3 Example

A necessary and sufficient condition for a scale invariant criterion $f(\mathbf{W})$, $\mathbf{W} \in \mathcal{SO}(K)$ to be an orthogonal contrast function is that the set of its global maximum points matches the set of the orthogonal non-mixing matrices, i.e. $\mathrm{argmax}_{\mathbf{W}\in\mathcal{SO}(K)} f(\mathbf{W}) = \mathcal{W}$. Hence, in the specific case where the two sources share the same density p_S, it is necessary that the criterion admits (at least) a local maximum at non-mixing matrices. Consequently, according to the results derived in the previous section, the $J_r(S)\mathrm{var(S)} < 1$ inequality implies that the sources cannot be recovered through the maximization of $C_r(\mathbf{B})$.

3.1 Theoretical Characterization of Non-mixing Stationary Points

The above analysis would be useless if the $J_r(S)\text{var}(S) < 1$ inequality is never satisfied for non-Gaussian sources. Actually, it can be shown that simple and common non-Gaussian densities satisfies this inequality. This is e.g. the case of the triangular density. We assume that both sources S_1, S_2 share the same triangular density p_T[2]:

$$p_T(s) \doteq \begin{cases} \frac{1-|s/\sqrt{6}|}{\sqrt{6}} & \text{if } |s| \leq \sqrt{6} \\ 0 & \text{otherwise} . \end{cases} \tag{20}$$

Observe that $E[S_i] = 0$, $\text{var}(S_i) = 1$, $i \in \{1, 2\}$, and note that using integration by parts, the r-th order information can be rewritten as

$$J_r(Y) = r \frac{\int p_Y^{r-2}(y)[p'_Y(y)]^2 dy}{\int p_Y^r(y) dy} .$$

Then, for $S \in \{S_1, S_2\}$ and noting $u \doteq 1 - s/\sqrt{6}$:

$$J_r(S) = \frac{r}{6} \frac{\int_0^{\sqrt{6}} (1 - s/\sqrt{6})^{r-2} ds}{\int_0^{\sqrt{6}} (1 - s/\sqrt{6})^r ds} = \frac{r}{6} \frac{\int_0^1 u^{r-2} du}{\int_0^1 u^r du} = \begin{cases} r(r+1)/[6(r-1)] & \text{if } r > 1 \\ \infty & \text{if } r \leq 1 \end{cases}$$

Thus $J_r(S)\text{var}(S) < 1$ if and only if $r(r+1)/[6(r-1)] < 1$. But for $r \geq 1$, the last inequality is equivalent to $0 > r(r+1) - 6(r-1) = (r-2)(r-3)$. Therefore $J_r(S)\text{var}(S) < 1$ if and only if $2 < r < 3$, as shown in Figure 1(a). We conclude that for a pair of triangular sources, the criterion $C_r(\mathbf{B})$ is not a contrast for $2 < r < 3$.

3.2 Simulation

Let us note $\mathbf{Y} = [Y_1, Y_2]^T$, $\mathbf{Y} = \mathbf{W}_\theta \mathbf{S}$, where $\mathbf{W}_\theta$ is a 2D rotation matrix of angle θ of the same form of $\mathcal{E}$ but where θ can take arbitrary values $[0, \pi]$. The criterion $-(h_r(Y_1) + h_r(Y_2))$ is plotted with respect to the transfer angle θ. Obviously, the set of non-mixing points reduces to $\mathcal{W} = \{\mathbf{W}_\theta : \theta \in \{k\pi/2 | k \in \mathbb{Z}\}\}$. Drawing this figure requires some approximations, and we are aware about the fact that it does not constitute a proof of the violation of the contrast property by itself; this proof is provided in the above theoretical development where it is shown that out of any problem of e.g. density estimation or exact integration approximation, Renyi's entropy is not always a contrast for BSS. The purpose of this plot is, complementary to Section 2, to show that *in practice, too*, the use of Renyi's entropy with arbitrary value of r might be dangerous.

Figure 1(b) has been drawn as follows. For each angle $\theta \in [0, \pi]$, the exact triangular probability density function p_T is used to compute the pdf of $\sin\theta S$

[2] This density is piecewise differentiable and continuous. Therefore, even if the density expansions are not valid everywhere, eq. (19) is still of use.

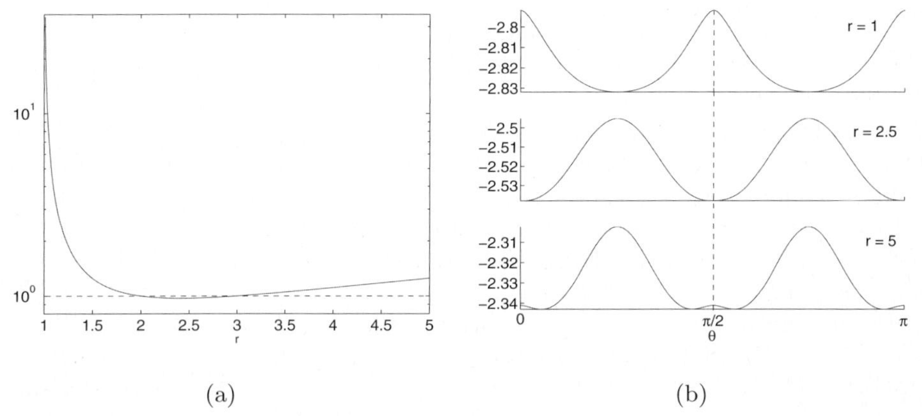

Fig. 1. Triangular sources. (a): $\log(J_r(S)\text{var(S)})$ vs r. (b): Estimated Renyi's criterion $\widetilde{C}_r(\mathcal{E})$ vs θ. The criterion is not a contrast function for $r = 2.5$ and $r = 5$.

and $\cos\theta S$, $S \sim p_T$, by using the well-known formula of the pdf of a transformation of random variables. Then, the output pdfs are obtained by convoluting the independent sources scaled by $\sin\theta$ and $\cos\theta$. Finally, Renyi's entropy is computed by replacing exact integration by Riemannian summation restricted on points were the output density is larger than $\tau = 10^{-4}$ to avoid numerical problems resulting from the log operator. At each step, it is checked that the pdfs of $\sin\theta S$, $\cos\theta S$, Y_1 and Y_2 integrate to one with an error smaller than τ and that the variance of the outputs deviates from unity with an error smaller than τ. Note that at non-mixing points, the exact density p_T is used as the output pdf to avoid numerical problems.

The two last plots of Figure 1(b) clearly indicate that the problem could be emphasized even when dealing with an approximated form of Renyi's entropy. On the top of the figure ($r = 1$), the criterion $\widetilde{C}_r(\mathbf{W}) = \widetilde{C}(\mathbf{W})$ (or more precisely, $C_r(\mathbf{B}) = C(\mathbf{B})$) is a contrast function, as expected. On the middle plot ($r = 2.5$), $\widetilde{C}_r(\mathbf{W}_\theta)$ admits a local minimum point when $\mathbf{W}_\theta \in \mathcal{W}$ (this results from $J_r(S)\text{var(S)} < 1$), and thus violates a necessary requirement of a contrast function. Finally, on the last plot (r=5), the criterion is not a contrast even though $J_r(S)\text{var(S)} > 1$ since the set of *global* maximum points of the criterion does not correspond to the set $\mathcal{W}$.

4 Conclusion

In this paper, the contrast property of a well-known Renyi's entropy based criterion for blind source separation is analyzed. It is proved that at least in one realistic case, *globally* maximizing the related criterion does not provide the expected sources, whatever is the value of Renyi's exponent; the transfer matrix $\mathbf{W}$ globally maximizing the criterion might be a mixing matrix, with possibly more than one non-zero element per row. Even worst, it is not guaranteed that the criterion reaches a *local* maximum at non-mixing solutions ! Actually, the only thing we are sure is that the criterion is stationary for non-mixing matrices. This is a mere information since if the criterion has a local maximum (resp. minimum) point at mixing matrices, then a stationary point might also exist at mixing solution, i.e. at $\mathbf{W}$ such that the components of $\mathbf{WS}$ are not proportional to distinct sources. Consequently, the value of Renyi's exponent has to be chosen with respect to the source densities in order to satisfy $\sum_{i=1}^{K} J_r(S_i)\mathrm{var}(\mathrm{S_i}) > \mathrm{K}$ (again, this is not a sufficient condition: it does not ensure that the local maximum is global). Unfortunately, the problem is that the sources are unknown. Hence, nowadays, the only way to guarantee that $C_r(\mathbf{B})$ is a contrast function is to set $r = 1$ (mutual information criterion) or $r = 0$ (log-measure of the supports criterion, this requires that the sources are bounded); it can be shown that counter-examples exist *for any other value of* r*, including* $r = 2$. To conclude, we would like to point out that contrarily to the kurtosis criterion case, it seems that it does not exist a simple mapping $\phi[.]$ (such as e.g. the absolute value or even powers) that would match the set $\mathrm{argmax}_{\mathbf{B}}\, \phi[C_r(\mathbf{B})]$ to the set $\{\mathbf{B} : \mathbf{BA} \in \mathcal{W}\}$ where $\mathcal{W}$ is the set of non-mixing matrices, because there is no information about the sign of the relevant local optima.

References

1. Comon, P.: Independent component analysis, a new concept? Signal Processing 36(3), 287–314 (1994)
2. Cover, T.M., Thomas, J.A.: Elements of Information Theory. Wiley Series in Telecommunications. Wiley and Sons, Inc, Chichester (1991)
3. Dembo, A., Cover, T.M., Thomas, J.A.: Information theoretic inequalities. IEEE Transactions on Information Theory 37(6), 1501–1518 (1991)
4. Erdogmus, D., Hild, K.E., Principe, J.: Blind source separation using renyi's mutual information. IEEE Signal Processing Letters 8(6), 174–176 (2001)
5. Erdogmus, D., Hild, K.E., Principe, J.: Blind source separation using renyi's α-marginal entropies. Neurocomputing 49(49), 25–38 (2002)
6. Gray, R., Davisson, L.: An Introduction to Statistical Signal Processing. Cambridge University Press, Cambridge (2004)
7. Hild, K.E., Erdogmus, D., Principe, J.: An analysis of entropy estimators for blind source separation. Signal Processing, 86, 174–176, 182–194 (2006)
8. Hyvärinen, A., Karhunen, J., Oja, E.: Independent component analysis. John Willey and Sons, Inc. New York (2001)
9. Pham, D.-T.: Entropy of a variable slightly contaminated with another. IEEE Signal Processing Letters 12(7), 536–539 (2005)
10. Renyi, A.: On measures of entropy and information. Selected papers of Alfred Renyi 2, 565–580 (1976)

A Variational Bayesian Algorithm for BSS Problem with Hidden Gauss-Markov Models for the Sources

Nadia Bali and Ali Mohammad-Djafari

Laboratoire des Signaux et Systèmes,
Unité mixte de recherche 8506 (CNRS-Supélec-UPS)
Supélec, Plateau de Moulon, 91192 Gif-sur-Yvette, France

Abstract. In this paper we propose a Variational Bayesian (VB) estimation approach for Blind Sources Separation (BSS) problem, as an alternative method to MCMC. The data are M images and the sources are N images which are assumed piecewise homogeneous. To insure these properties, we propose a piecewise Gauss-Markov model for the sources with a hidden classification variable which is modeled by a Potts-Markov field. A few simulation results are given to illustrate the performances of the proposed method and some comparison with other methods (MCMC and VBICA) used for BSS, are presented.

Introduction

We consider the problem of sources separation in the case of instantaneous mixture with noisy images. We propose to use the Bayesian inference which gives the possibility to take into account uncertainties and all prior knowledge on the model of sources and observations. We assign priors to the noise, to the sources, to the mixing matrix and to all the hyperparameters of the model and obtain the expression of the posterior of all the unknowns. However, using this posterior to compute its mode, its mean or its exploration needs approximation. Classical methods of approximations are *i*) numerical methods such as MCMC which requires in practice a great computational effort, *ii*) analytical methods such as asymptotic approximation of Laplace which is more easily practicable but makes typically a rough approximation.

In this paper we propose an alternative approximation method which is based on a variational approach which offers a practical framework in term of efficiency and precision[1],[2]. Indeed the goal is to give an approximation to marginal likelihood or the model evidence which is the distribution of the data knowing the model. The model evidence is obtained by integrating over the hidden variables of the joined posterior law. Using variational approximation of the posterior law is not new in image processing and in particular in image segmentation [3]. However, in our knowledge its use in blind image separation with a particular mixture of Gaussians with a Potts Markov model is new. The hidden Potts Markov model is and many works exist on different approximations such Mean

M.E. Davies et al. (Eds.): ICA 2007, LNCS 4666, pp. 137–144, 2007.

Field Approximation (MFA) [4],[5] to resolve the problem. Here we propose a global variational approach that insures us to maximize free energy with a more complex hierarchical model since our problem is extended to sources separation problem.

1 Sources Separation Model

An instantaneous sources separation problem can be written as:

$$\boldsymbol{x}(\boldsymbol{r}) = \boldsymbol{A}\boldsymbol{s}(\boldsymbol{r}) + \boldsymbol{\epsilon}(\boldsymbol{r}), \tag{1}$$

Where:

- $\boldsymbol{x}(\boldsymbol{r}) = \{x_i(\boldsymbol{r}), i = 1, \cdots, M\}$ is a set of M images (observations) and $\boldsymbol{r} \in \mathcal{R} = \{1, \cdots, R\}$ is a pixel position with R the total number of pixels.
- $\boldsymbol{A}$ is an unknown mixture matrix with dimension (M, N),
- $\boldsymbol{s}(\boldsymbol{r}) = \{s_{j,r}, j = 1, \cdots, n\}$ is a set of N unknown components (sources images);

In the following, we assume that the errors $\boldsymbol{\epsilon}(\boldsymbol{r})$ are centered, white, Gaussian with inverse covariance matrix $\boldsymbol{\Sigma}_\epsilon = \text{diag}\left[\frac{1}{\sigma^2_{\epsilon_1}}, \cdots, \frac{1}{\sigma^2_{\epsilon_M}}\right]$.

Now, if we note by $\underline{\boldsymbol{x}} = \{\boldsymbol{x}(\boldsymbol{r}), \boldsymbol{r} \in \mathcal{R}\}$, $\underline{\boldsymbol{s}} = \{\boldsymbol{s}(\boldsymbol{r}), \boldsymbol{r} \in \mathcal{R}\}$ and $\underline{\boldsymbol{\epsilon}} = \{\boldsymbol{\epsilon}(\boldsymbol{r}), \boldsymbol{r} \in \mathcal{R}\}$, then we can write

$$\underline{\boldsymbol{x}} = \boldsymbol{A}\underline{\boldsymbol{s}} + \boldsymbol{\epsilon}. \tag{2}$$

and

$$p(\underline{\boldsymbol{x}}|\underline{\boldsymbol{s}}, \boldsymbol{\Sigma}_\epsilon) = \prod_{\boldsymbol{r}} \mathcal{N}(\boldsymbol{A}\boldsymbol{s}(\boldsymbol{r}), \boldsymbol{\Sigma}_\epsilon^{-1}) \tag{3}$$

We note that $\boldsymbol{\Sigma}_\epsilon$ is an inverse covariance matrix.

1.1 Sources Modelization

We propose to model the prior marginal law of each source $s_j(\boldsymbol{r})$ by a mixture of Gaussians model:

$$p(s_j(\boldsymbol{r})) = \sum_{k=1}^{K} p(z_j(\boldsymbol{r}) = k)\, \mathcal{N}(m_{j\,k}, \sigma^2_{j\,k}) \tag{4}$$

which implies that $p(s_j(\boldsymbol{r})|z_j(\boldsymbol{r}) = k) = \mathcal{N}(m_{j\,k}, \sigma^2_{j\,k})$ where $p(z_j(\boldsymbol{r}) = k) = \alpha_{j,k}$ and $\sum_k \alpha_{j,k} = 1$. This model is appropriate for the image sources which we consider, where the discrete valued hidden variables $z_j(\boldsymbol{r}) \in \{1, \cdots, K_j\}$ represent the classification labels of the source images pixels $s_j(\boldsymbol{r})$. To insure some spatial regularity to these labels, they are modelized by a Potts-Markov random field:

$$p(z_j(\boldsymbol{r})|z_j(\boldsymbol{r}'), \boldsymbol{r}' \neq \boldsymbol{r}, \boldsymbol{r} \in \mathcal{R}) \propto \exp\left[\beta_j \sum_{\boldsymbol{r}' \in \mathcal{V}(\boldsymbol{r})} \delta(z_j(\boldsymbol{r}) - z_j(\boldsymbol{r}'))\right].$$

The parameters β_j controls the mean size of regions.

1.2 Prior Models for the Mixing Matrix and the Hyperparameters

Mixing matrix model: We consider a Gaussian distribution law for mixture matrix, so the prior distribution of $\boldsymbol{A}$ is given by:

$$\pi(\boldsymbol{A}|\boldsymbol{A}_0, \boldsymbol{\Gamma}_0) = \mathcal{N}(\boldsymbol{A}_0, \boldsymbol{\Gamma}_0). \tag{5}$$

Inverse covariance noise model: We assign a Wishart distribution to the covariance of noise $\boldsymbol{\Sigma}_{\boldsymbol{\epsilon}}$

$$\pi(\boldsymbol{\Sigma}_{\boldsymbol{\epsilon}}|\nu_{\epsilon_0}, \boldsymbol{\Sigma}_{\epsilon_0}) \propto |\boldsymbol{\Sigma}_{\boldsymbol{\epsilon}}|^{\frac{(\nu_{\epsilon_0}-M-1)}{2}} \exp -\frac{1}{2}\mathrm{Tr}\left\{\boldsymbol{\Sigma}_{\boldsymbol{\epsilon}}\boldsymbol{\Sigma}_{\epsilon_0}^{-1}\right\} \tag{6}$$

Where ν_{ϵ_0} is the number of degrees of freedom and $\boldsymbol{\Sigma}_{\epsilon_0}$ is the prior covariance matrix.

Means and variances of different classes: We assign Gaussian laws to the means:

$$\pi(\boldsymbol{m}_z|\boldsymbol{\mu}_0, \boldsymbol{T}_0) = \mathcal{N}(\boldsymbol{\mu}_0, \boldsymbol{T}_0), \tag{7}$$

and Wishart law to the inverse covariances

$$\pi(\boldsymbol{\Sigma}_z|\nu_0, \boldsymbol{V}_0) = \mathcal{W}(\nu_0, \boldsymbol{V}_0). \tag{8}$$

2 Variational Bayesian Algorithm

Our goal is to obtain a separable approximation $q(\underline{\boldsymbol{s}}, \underline{\boldsymbol{z}}, \boldsymbol{\theta}) = q(\underline{\boldsymbol{s}}, \underline{\boldsymbol{z}})\, q(\boldsymbol{\theta})$ for the joint posterior law $p(\underline{\boldsymbol{s}}, \underline{\boldsymbol{z}}, \boldsymbol{\theta}|\underline{\boldsymbol{x}}, \mathcal{M})$ of $(\underline{\boldsymbol{s}}, \underline{\boldsymbol{z}})$ and $\boldsymbol{\theta} = (\boldsymbol{A}, \boldsymbol{\Sigma}_{\boldsymbol{\epsilon}}, \boldsymbol{m}_z, \boldsymbol{\Sigma}_z)$. The idea is thus to minimize the Kullback Leibler divergence between the approximate distribution law $q(\underline{\boldsymbol{s}}, \underline{\boldsymbol{z}})q(\boldsymbol{\theta})$ and the joint posterior law on the hidden variables and the parameters $p(\underline{\boldsymbol{s}}, \underline{\boldsymbol{z}}, \boldsymbol{\theta}|\underline{\boldsymbol{x}})$:

$$KL[q(\underline{\boldsymbol{s}}, \underline{\boldsymbol{z}}|\underline{\boldsymbol{x}}, \mathcal{M})q(\boldsymbol{\theta}|\underline{\boldsymbol{x}}, \mathcal{M})||p(\underline{\boldsymbol{s}}, \underline{\boldsymbol{z}}, \boldsymbol{\theta}|\underline{\boldsymbol{x}}))] = \int d\boldsymbol{\theta} \int d\boldsymbol{s} \sum_{\boldsymbol{z}} q(\underline{\boldsymbol{s}}, \underline{\boldsymbol{z}}|\underline{\boldsymbol{x}}, \mathcal{M})q(\boldsymbol{\theta}|\underline{\boldsymbol{x}}, \mathcal{M}) \ln \frac{q(\underline{\boldsymbol{s}}, \underline{\boldsymbol{z}}|\underline{\boldsymbol{x}}, \mathcal{M})q(\boldsymbol{\theta}|\underline{\boldsymbol{x}}, \mathcal{M})}{p(\underline{\boldsymbol{s}}, \underline{\boldsymbol{z}}, \boldsymbol{\theta}|\underline{\boldsymbol{x}})}$$

where $\mathcal{M}$ is the model. In case of sources separation $\mathcal{M}$ represents the number of sources N. Developing this expression at one side and looking to the expression of the evidence $p(\underline{\boldsymbol{x}}|\mathcal{M})$ at the other-side, it is easy to show that:

$$KL[q(\underline{\boldsymbol{s}}, \underline{\boldsymbol{z}}|\underline{\boldsymbol{x}}, \mathcal{M})q(\boldsymbol{\theta}|\underline{\boldsymbol{x}}, \mathcal{M})||p(\underline{\boldsymbol{s}}, \underline{\boldsymbol{z}}, \boldsymbol{\theta}|\underline{\boldsymbol{x}}))] = \ln p(\underline{\boldsymbol{x}}|\mathcal{M}) - \mathcal{F}(q(\underline{\boldsymbol{s}}, \underline{\boldsymbol{z}}), q(\boldsymbol{\theta})) \tag{9}$$

where $\mathcal{F}$ is given by:

$$\mathcal{F}(q(\underline{\boldsymbol{s}}, \underline{\boldsymbol{z}}), q(\boldsymbol{\theta})) = \int d\boldsymbol{\theta} q(\boldsymbol{\theta}) \left[\int d\underline{\boldsymbol{s}} \sum_{\underline{\boldsymbol{z}}} q(\underline{\boldsymbol{s}}, \underline{\boldsymbol{z}}) \ln \frac{p(\underline{\boldsymbol{x}}, \underline{\boldsymbol{s}}, \underline{\boldsymbol{z}}|\boldsymbol{\theta}, \mathcal{M})}{q(\underline{\boldsymbol{s}}, \underline{\boldsymbol{z}})} + \ln \frac{p(\boldsymbol{\theta}|\mathcal{M})}{q(\boldsymbol{\theta})}\right] \tag{10}$$

which is called free energy of the model. From these relations we see that minimizing KL is equivalent to maximizing $\mathcal{F}$. The distribution of the variational approximation $q(\underline{\boldsymbol{s}}, \underline{\boldsymbol{z}})$ and $q(\boldsymbol{\theta})$ must belong to a family of distributions simpler than that of the posterior distribution $p(\underline{\boldsymbol{s}}, \underline{\boldsymbol{z}}, \boldsymbol{\theta}|\boldsymbol{x})$. Obtaining expressions for $q(\underline{\boldsymbol{s}}, \underline{\boldsymbol{z}})$ and $q(\boldsymbol{\theta})$ is done iteratively. The family of distributions is selected such that q be in the same family than the true posterior distributions. [2] [6] noted that important simplifications are made when updating the variational equations if the choice of the distributions of the variables conditionally to their parameters is done from conjugated exponential families model. In this case, the posterior distribution has analytically stables and intuitive forms.

To optimize $\mathcal{F}(q(\underline{\boldsymbol{s}}, \underline{\boldsymbol{z}}), q(\boldsymbol{\theta}))$ we simply equate to zero the functional derivatives with respect to each distribution q. In summary, the two main hypothesis of the proposed method are: $i)$ posterior independence between $(\underline{\boldsymbol{s}}, \underline{\boldsymbol{z}})$ and $\boldsymbol{\theta}$, $ii)$ separable conjugate priors for all the hyperparameters.

This last hypothesis associated with $\frac{d\mathcal{F}}{dq} = 0$ results to:

$$q(\theta_i) \propto \exp(< \log p(\underline{\boldsymbol{x}}|\boldsymbol{\theta}, \mathcal{M}) >_{q(\theta_{|i})})\pi(\theta_i), \tag{11}$$

where $< f(x) >_q = \mathrm{E}_q\{f(x)\} = \int q(x) f(x) dx$.

2.1 Approximate Posterior Laws for Mixing Matrix and Hyperparameters

Approximation posterior law for mixing matrix: We note by $\boldsymbol{A}_v$ the vector wise representation of a matrix defined by :$\boldsymbol{A}_v = [\boldsymbol{A}_{(1,.)}, \cdots, \boldsymbol{A}_{(m,.)}]^t$. By taking the functional derivative of Eq.(10) and equating to zero $\frac{d\mathcal{F}}{dq(\boldsymbol{A})} = 0$, we get the update:

$$q(\boldsymbol{A}) \propto \pi(\boldsymbol{A}) \exp(< \ln p(\underline{\boldsymbol{x}}|\underline{\boldsymbol{s}}, \boldsymbol{\Sigma}_{\boldsymbol{\epsilon}}) >_{q(\boldsymbol{\Sigma}_{\boldsymbol{\epsilon}})q(\underline{\boldsymbol{s}})}) \tag{12}$$

With the appropriate conjugate prior $\pi(\boldsymbol{A})$ that we chosen in (5), it is easy to see that $q(\boldsymbol{A}|\widetilde{\boldsymbol{A}}, \widetilde{\boldsymbol{\Sigma}}_{\boldsymbol{A}}) = \mathcal{N}(\widetilde{\boldsymbol{A}}, \widetilde{\boldsymbol{\Sigma}}_{\boldsymbol{A}})$, with:

$$\widetilde{\boldsymbol{\Sigma}}_{\boldsymbol{A}} = R\widetilde{\boldsymbol{\Sigma}}_{\epsilon} \otimes (\sum_{\boldsymbol{z}(\boldsymbol{r})} q(\boldsymbol{z}(\boldsymbol{r}))\widetilde{\boldsymbol{\Sigma}}_{s|z}^{-1}) + \sum_{r} \widetilde{\boldsymbol{\Sigma}}_{\epsilon} \otimes (\sum_{\boldsymbol{z}(\boldsymbol{r})} q(\boldsymbol{z}(\boldsymbol{r}))\widetilde{\boldsymbol{s}}_z(\boldsymbol{r})\widetilde{\boldsymbol{s}}_z^t(\boldsymbol{r})) + \boldsymbol{\Sigma}_{\boldsymbol{A}} \tag{13}$$

$$\text{and} \quad \widetilde{\boldsymbol{A}}_v = \widetilde{\boldsymbol{\Sigma}}_{\boldsymbol{A}}^{-1} \left[\sum_{r} \widetilde{\boldsymbol{\Sigma}}_{\epsilon}(\boldsymbol{x}^t(\boldsymbol{r}) \otimes (\sum_{\boldsymbol{z}(\boldsymbol{r})} q(\boldsymbol{z}(\boldsymbol{r}))\widetilde{\boldsymbol{s}}_z^t(\boldsymbol{r}))) + \boldsymbol{\Gamma}_p \boldsymbol{A}_p \right] \tag{14}$$

Approximate posterior law for noise inverse covariance: $q(\boldsymbol{\Sigma}_{\boldsymbol{\epsilon}})$ is obtained by equating to zero $\frac{d\mathcal{F}}{dq(\boldsymbol{\Sigma}_{\boldsymbol{\epsilon}})} = 0$ which results to:

$$\ln(q(\boldsymbol{\Sigma}_{\boldsymbol{\epsilon}})) = \frac{(R + \nu_{\epsilon_0} - m - 1)}{2} \ln|\boldsymbol{\Sigma}_{\boldsymbol{\epsilon}}| - \frac{1}{2}(\boldsymbol{\Sigma}_{\boldsymbol{\epsilon}}\boldsymbol{Q}_{\epsilon}) \tag{15}$$

where $\boldsymbol{Q}_{\epsilon} = \boldsymbol{\Sigma}_{\epsilon_0}^{-1} + \boldsymbol{Q} + R\boldsymbol{D}_{\boldsymbol{A}}(\sum_{\boldsymbol{z}} q(\boldsymbol{z})\widetilde{\boldsymbol{\Sigma}}_{s|z}^{-1}, \widetilde{\boldsymbol{\Sigma}}_{\boldsymbol{A}}^{-1}, \widetilde{\boldsymbol{A}})$. For the definition of $\boldsymbol{D}_{\boldsymbol{A}}$ and for more details see [7], defining a posterior wishart distribution with a mean matrix $\widetilde{\boldsymbol{\Sigma}}_{\epsilon} = (R + \nu_{\epsilon_0})\boldsymbol{Q}_{\epsilon}^{-1}$.

Approximate posterior laws for $\boldsymbol{m}_z$ and $\boldsymbol{\Sigma}_z$: By writing $\mathcal{F}$ as a function of $\boldsymbol{m}_z$ and $\boldsymbol{\Sigma}_z$ only we can differentiate with respect to these hyperparameters to yield the following update equations : $q(\boldsymbol{m}_z|\widetilde{\boldsymbol{m}}_z, \widetilde{\boldsymbol{T}}_z) = \mathcal{N}(\widetilde{\boldsymbol{m}}_z, \widetilde{\boldsymbol{T}}_z)$, $q(\boldsymbol{\Sigma}_z|\widetilde{\nu}_z, \widetilde{\boldsymbol{V}}_z) = \mathcal{W}(\widetilde{\nu}_z, \widetilde{\boldsymbol{V}}_z)$. The details of deriving the update equations are omitted due to the space constraints. They can be obtained in [8].

2.2 Approximate Posterior Laws for Hidden Variables

Approximate posterior distribution $q(\underline{s}|\underline{z})$: The expression of $q(\underline{s}|\underline{z})$ is obtained by $\frac{d\mathcal{F}}{dq(\boldsymbol{s}(\boldsymbol{r})|\boldsymbol{z}(\boldsymbol{r}))} = 0$ which results to :

$$\ln q(\boldsymbol{s}(\boldsymbol{r})|\boldsymbol{z}(\boldsymbol{r})) = -\frac{1}{2}\boldsymbol{s}(\boldsymbol{r})^t \mathcal{Q} \boldsymbol{s}(\boldsymbol{r}) + (\widetilde{\boldsymbol{A}}^t \widetilde{\boldsymbol{\Sigma}}_\epsilon \boldsymbol{x}(\boldsymbol{r}))^t \boldsymbol{s}(\boldsymbol{r}) - \frac{1}{2}(\boldsymbol{s}(\boldsymbol{r}) - \boldsymbol{m_z})^t \boldsymbol{\Sigma_z}(\boldsymbol{s}(\boldsymbol{r}) - \boldsymbol{m_z})$$

with $\mathcal{Q} = \widetilde{\boldsymbol{A}}' \widetilde{\boldsymbol{\Sigma}}_\epsilon \widetilde{\boldsymbol{A}} + F(\widetilde{\boldsymbol{\Sigma}}_\epsilon, \widetilde{\boldsymbol{\Sigma}}_{\boldsymbol{A}}^{-1})$ which is quadratic in $\boldsymbol{s}(\boldsymbol{r})$. For the definition of $F(\widetilde{\boldsymbol{\Sigma}}_\epsilon, \widetilde{\boldsymbol{\Sigma}}_{\boldsymbol{A}}^{-1})$ and for more details see [7]. In summary, we obtain $q(\boldsymbol{s}(\boldsymbol{r})|\boldsymbol{z}(\boldsymbol{r})) = \mathcal{N}(\widetilde{\boldsymbol{s}}_z(\boldsymbol{r}), \widetilde{\boldsymbol{\Sigma}}_{s|z})$ with

$$\widetilde{\boldsymbol{\Sigma}}_{s|z} = \mathcal{Q} + \boldsymbol{\Sigma_z} \tag{16}$$

$$\widetilde{\boldsymbol{s}}_z(\boldsymbol{r}) = \widetilde{\boldsymbol{\Sigma}}_{s|z}^{-1}[\widetilde{\boldsymbol{A}}^t \widetilde{\boldsymbol{\Sigma}}_\epsilon \boldsymbol{x}(\boldsymbol{r}) + \boldsymbol{\Sigma_z m_z}] \tag{17}$$

Approximate posterior law for labels variables: Deriving an expression for $q(\underline{z}|\underline{\boldsymbol{x}}, \mathcal{M})$ is the most difficult task in this paper due to its Markovian model. Hopefully, using the four nearest neighbors neighborhood system often used in image processing, it is easy to divide $\underline{z}$ into two subsets $\boldsymbol{z}_N$ and $\boldsymbol{z}_B$ in the manner of a chess-board. From this, we only need to work with the distributions $q_{z_B}(\underline{\boldsymbol{z}}_B)$ and $q_{z_N}(\underline{\boldsymbol{z}}_N)$. Thus, each white pixel (respectively black) has its four black neighbors (respectively white). All the white pixels (respectively black), knowing the black pixels (respectively white), are thus independent. We can thus write:

$q_{\boldsymbol{z}}(\underline{z}|\underline{\boldsymbol{x}}, \mathcal{M}) = q_{z_N}(\underline{\boldsymbol{z}}_N|\underline{\boldsymbol{x}}, \mathcal{M}) q_{z_B}(\underline{\boldsymbol{z}}_B|\underline{\boldsymbol{x}}, \mathcal{M})$.

The expression of q_{z_B} is obtained by $\frac{d\mathcal{F}}{dq_{z_B}} = 0$:

$$q_{z_B}(\underline{\boldsymbol{z}}_B|\underline{\boldsymbol{x}}, m) \propto \exp\{< \ln p(\underline{\boldsymbol{z}}_B|\underline{\boldsymbol{z}}_N, \beta) >_{q(\underline{\boldsymbol{z}}_N)} + \mathcal{H}_B(\boldsymbol{r})\} \tag{18}$$

where:

$$\mathcal{H}_B(\boldsymbol{r}) = < \ln p(\underline{\boldsymbol{s}}|\underline{z}, \boldsymbol{m_z}, \boldsymbol{\Sigma_z}) + \ln p(\underline{\boldsymbol{x}}|\underline{\boldsymbol{s}}, \underline{z}, \boldsymbol{A}, \boldsymbol{\Sigma_\epsilon}) >_{q(\boldsymbol{\theta}), q(\underline{\boldsymbol{s}}|\underline{z}), q(\underline{\boldsymbol{z}}_N)}$$

Expanding these expressions, we obtain:

$$q_{z_B}(\underline{\boldsymbol{z}}_B|\underline{\boldsymbol{x}}, m) \propto \prod_{r \in \mathcal{R}_B} \exp\left\{\beta \sum_{r'} \sum_{\boldsymbol{k}} \delta(\boldsymbol{z}_r - \boldsymbol{z}_{r'}) q(\boldsymbol{z}_{r'} = \boldsymbol{k}) + \mathcal{H}_B(\boldsymbol{r})\right\}$$

with:

$$\begin{aligned}
\mathcal{H}_B(\boldsymbol{r}) &= -\frac{1}{2}\widetilde{\boldsymbol{s}}_z^t(\boldsymbol{r})\widetilde{\boldsymbol{\Sigma}}_{\boldsymbol{z}}\widetilde{\boldsymbol{s}}_z(\boldsymbol{r}) - \frac{1}{2}\mathrm{Tr}\left\{\widetilde{\boldsymbol{\Sigma}}_{\boldsymbol{z}}\widetilde{\boldsymbol{\Sigma}}_{s|z}^{-1}\right\} + \widetilde{\boldsymbol{m}}_{\boldsymbol{z}}^t\widetilde{\boldsymbol{\Sigma}}_{\boldsymbol{z}}\widetilde{\boldsymbol{s}}_z(\boldsymbol{r}) \\
&- \frac{1}{2}\widetilde{\boldsymbol{m}}_{\boldsymbol{z}}^t\widetilde{\boldsymbol{\Sigma}}_{\boldsymbol{z}}\widetilde{\boldsymbol{m}}_{\boldsymbol{z}} - \frac{1}{2}\mathrm{Tr}\left\{\widetilde{\boldsymbol{\Sigma}}_{\boldsymbol{z}}\boldsymbol{T}_z^{-1}\right\} \\
&+ \frac{R}{2}\sum_{i=1}^{M}\Psi\left(\frac{\nu+R+1-i}{2}\right) + M\ln 2 + \ln|\widetilde{\boldsymbol{\Sigma}}_\epsilon| \\
&- \frac{R}{2}\mathrm{Tr}\left\{\widetilde{\boldsymbol{\Sigma}}_\epsilon\boldsymbol{D}_{\boldsymbol{A}}\Big(\sum_{\boldsymbol{z}(\boldsymbol{r})} q(\boldsymbol{z}(\boldsymbol{r}))\widetilde{\boldsymbol{\Sigma}}_{s|z}^{-1}, \widetilde{\boldsymbol{\Sigma}}_{\boldsymbol{A}}^{-1}, \widetilde{\boldsymbol{A}}\Big)\right\} - \frac{1}{2}\mathrm{Tr}\left\{\widetilde{\boldsymbol{\Sigma}}_\epsilon\boldsymbol{Q}\right\}
\end{aligned}$$

In the same manner, we obtain an expression for $q_{z_N}(\underline{\boldsymbol{z}}_N|\underline{\boldsymbol{x}},\mathcal{M})$

$$q_{z_N}(\underline{\boldsymbol{z}}_N|\underline{\boldsymbol{x}},\mathcal{M}) \propto \prod_{r\in\mathcal{R}_N}\exp\left\{\beta\sum_{r'}\sum_{\boldsymbol{k}}\delta(\boldsymbol{z}(\boldsymbol{r})-\boldsymbol{z}(\boldsymbol{r}'))q(\boldsymbol{z}(\boldsymbol{r}')=\boldsymbol{k}) + \mathcal{H}_N(\boldsymbol{r})\right\} \tag{19}$$

where:

$$\mathcal{H}_N(\boldsymbol{r}) = <\ln p(\underline{\boldsymbol{s}}|\underline{\boldsymbol{z}},\boldsymbol{m}_{\boldsymbol{z}},\boldsymbol{\Sigma}_{\boldsymbol{z}}) + \ln p(\underline{\boldsymbol{x}}|\underline{\boldsymbol{s}},\underline{\boldsymbol{z}},\boldsymbol{A},\boldsymbol{\Sigma}_\epsilon) >_{q(\boldsymbol{\theta}),q(\underline{\boldsymbol{s}}|\underline{\boldsymbol{z}}),q(\underline{\boldsymbol{z}}_B)}$$

Given all these expressions, the general iterative algorithm obtained and proposed consists in updating successively

$$\begin{cases}
q(\underline{\boldsymbol{s}}|\underline{\boldsymbol{z}},\underline{\boldsymbol{x}},\boldsymbol{\theta},\mathcal{M}), \text{using } (16,\ 17) \\
q_{z_B}(\underline{\boldsymbol{z}}_B|\underline{\boldsymbol{z}}_N,\underline{\boldsymbol{x}},\mathcal{M}), \text{using } (18) \\
q_{z_N}(\underline{\boldsymbol{z}}_N|\underline{\boldsymbol{z}}_B,\underline{\boldsymbol{x}},\mathcal{M}), \text{using } (19) \\
q(\boldsymbol{\theta}_i|\underline{\boldsymbol{s}},\underline{\boldsymbol{z}},\underline{\boldsymbol{x}},\boldsymbol{\theta}_{|i},\mathcal{M}), \text{using } (13,\ 14,\ 15)
\end{cases}$$

Once these iterations converged, one can use these laws to provide estimation for $\{\underline{\boldsymbol{s}}, \underline{\boldsymbol{z}}$ and $\boldsymbol{\theta}\}$.

3 Free Energy Estimation

The estimate of $\mathcal{F}$ enables us to have a criterion of convergence. The maximization of the free energy is made by an iterative procedure (Figure: 2, (*a*)), following a total iteration which contains an update of all the parameters. We use a threshold on $\frac{\Delta\mathcal{F}}{\mathcal{F}}$ to stop the iterations. Since calculations of the parameters for all the q functional are already made, it is easy to calculate $\mathcal{F}$:

$$\begin{aligned}
\mathcal{F} &= <\ln p(\underline{\boldsymbol{x}}|\underline{\boldsymbol{s}},\boldsymbol{z},\boldsymbol{A},\boldsymbol{\Sigma}_\epsilon) >_{q(\underline{\boldsymbol{s}}|\boldsymbol{z}),q(\boldsymbol{A}),q(\boldsymbol{\Sigma}_\epsilon)} - < KL(q(\underline{\boldsymbol{s}}|\boldsymbol{z})||p(\underline{\boldsymbol{s}}|\boldsymbol{z})) >_{q(z)} \\
&- KL(q(\boldsymbol{A})||p(\boldsymbol{A})) - KL(q(\boldsymbol{\Sigma}_\epsilon)||p(\boldsymbol{\Sigma}_\epsilon)) - KL(q(\boldsymbol{m}_z)||p(\boldsymbol{m}_z)) \\
&- KL(q(\boldsymbol{\Sigma}_z)||p(\boldsymbol{\Sigma}_z))
\end{aligned}$$

We also use the final values of $\mathcal{F}$ for model selection.

4 Results

To evaluate the performance of our method we generate synthetic data with:

$$\boldsymbol{A} = \begin{bmatrix} 0.86 & 0.44 \\ 0.50 & 0.89 \end{bmatrix}, \boldsymbol{\Sigma}_\epsilon = \begin{bmatrix} 10 & 0 \\ 0 & 10 \end{bmatrix}, \boldsymbol{m}_z = \begin{bmatrix} -1 & 1 \\ 1 & -1 \end{bmatrix}, \boldsymbol{\Sigma}_z = \begin{bmatrix} 0.01 & 0.01 \\ 0.01 & 0.01 \end{bmatrix} \text{ and}$$

$\beta = 1.5$. We choose to compare with MCMC method that had been proposed with the same modelization (Figure 1: (*d*), (*e*)) This enables us to see a gain in computation time.

We also compared with the method (VBICA [8]) (Figure 1: (*c*), (*E*)). This method uses a mixture of Gaussian with an independent model on the coefficients

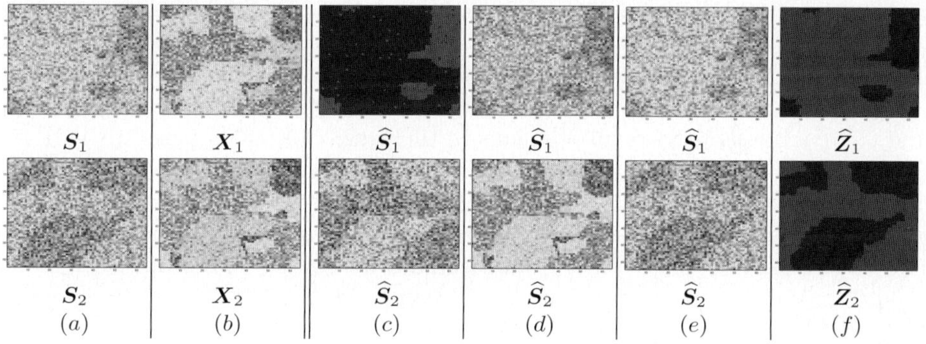

Fig. 1. Results of separation for two images $\boldsymbol{X}_1$ et $\boldsymbol{X}_2$ (*b*) obtained by an instantaneous mixture of $\boldsymbol{S}_1$ et $\boldsymbol{S}_2$ (*a*) every images is modelized by Markov model. In (*c*), (*d*) and (*e*) are represented sources obtained by three different methods VB ICA, MCMC and the proposed approach with their associated segmentation.

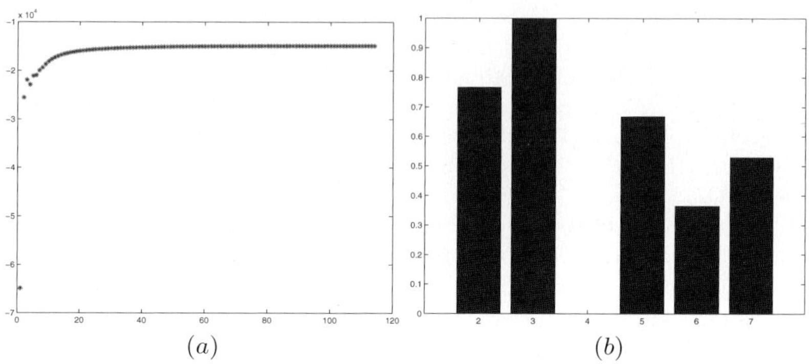

Fig. 2. (*a*) evolution of $\mathcal{F}$ during the iterations (*b*) Final values of $\mathcal{F}$ for simulated data composed with 4 mixed images obtained from 3 sources. $\mathcal{F}$ has been normalized with respect to its minimum value for $N = 4$. The maximum is obtained for the right sources number 3.

of mixture contrary to the method we propose. This type of model is often used in a variational estimation[1],[2] considering the *a priori* law are always selected in separable families. Our method converges at the end of 120 iterations (Figure 2: (a)) whereas method (VBICA) reached 500 iterations without convergence for a degree of tolerance on $\mathcal{F}$ of $1e^{-4}$. $\mathcal{F}$ enables us to make the suitable choice of the model. For a simulated example with three sources we compute the converged values of $\mathcal{F}$ for different number of sources from 2 to 7. As it is shown in the (figure 2: (b)), maximum of $\mathcal{F}$ is 3 which is the good result.

5 Conclusion

In this paper we propose a new approach for sources separation problem based on Bayesian estimation and variational approximation. Variational approximation is applied to the posterior law obtained with a more complex source model. The proposed source model is marginally a mixture of Gaussians but also we assigned a Potts model to its hidden variable which enforces the homogeneity and compactness of the pixel positions in the same class. Compared to VBICA, our source model is reacher but also due to the Potts model, it is non-separable. This makes the expressions of the posterior law more complex. However, the main benefice of the complex modelization of the sources is that the hidden variables now can be used as a non-supervised segmentation result. The proposed method then does simultaneously separation and segmentation.

References

1. Miskin, J.W.: Ensemble Learning for Independent Component Analysis, Phd thesis, Cambridge (2000), `http://www.inference.phy.cam.ac.uk/jwm1003/`
2. Attias, H.: A variational Bayesian framework for graphical models. In: Solla, S., Leen, T.K., Muller, K.L. (eds.) Advances in Neural Information Processing Systems, vol. 12, pp. 209–215. MIT Press, Cambridge (2000)
3. Cheng, L., Jiao, F., Schuurmans, D., Wang, S.: Variatonal bayesian image modelling. In: Proceedings of the 22nd international conference on Machine learning, Bonn Germany, August 2005, vol. 119, pp. 129–136. ACM Press, New York (2005)
4. Peyrard, N.: Approximations de type champ moyen des problèmes de champ de Markov pour la segmentation des données spatiales, Ph.D. thesis, Université Joseph Fourrier (2001)
5. Bali, N., Mohammad-Djafari, A.: Mean Field Approximation for BSS of images with compound hierarchical Gauss-Markov-Potts model. In: Proceedings of MaxEnt05, august 2005, pp. 249–256 (2005)
6. Ghahramani, Z., Beal, M.J.: Propagation algorithms for variational Bayesian learning. In: Leen, T.K., Dietterich, T., Tresp, V. (eds.) Advances in Neural Information Processing Systems, vol. 13, pp. 507–513. MIT Press, Cambridge (2001)
7. Ichir, M.M., Mohammad-Djafari, A.: A mean field approximation approach to blind source separation with l_p priors. In: Eusipco, Antalya, Turkey (September 2005)
8. Choudrey, R.A., Roberts, S.J.: Bayesian ICA with Hidden Markov Model Sources. In: ICA, NARA, JAPAN (April 2003)

A New Source Extraction Algorithm for Cyclostationary Sources

C. Capdessus[1], A.K. Nandi[2], and N. Bouguerriou[1]

[1] Laboratoire d'Electronique, Signaux et Images
21 rue Loigny la Bataille
28000 Chartres – France
Cecile.Capdessus@univ-orleans.fr
[2] Department of Electrical Engineering and Electronics
The University Liverpool, Brownlow Hill
Liverpool, L69 3GJ, UK
A.Nandi@liverpool.ac.uk

Abstract. Cyclostationary signals can be met in various domains, such as telecomunications and vibration analysis. Cyclostationarity allows to model repetitive signals and hidden periodicities such as those induced by modulation for communications and by rotating machines for vibrations. In some cases, the fundamental frequency of these repetitive phenomena can be known. The algorithm that we propose aims at extracting one cyclostationary source, whose cyclic frequency is *a priori* known, from a set of observations. We propose a new criterion based on second order statistics of the measures which is easy to estimate and leads to extraction with very good accuracy.

Keywords: Source extraction, cyclostationarity, second order statistics, vibration analysis.

1 Introduction

The method we propose here has been developed within the frame of vibration analysis applied to rotating machinery monitoring. Rotating machines produce vibrations whose characteristics depend on the machine state. These vibrations can thus be used to monitor systems such as engines, roller bearings, toothed gearings [1]. One of the problems that arises for complex systems is that each vibration sensor receives a mixture of the vibrations produced by the different parts of the system. These vibrations are usually wideband and spread over much of the spectrum.

Though these different contributions can be separated neither in the time domain nor in the frequency domain, some hope still lays in their cyclostationary features. Indeed, due to their symmetric geometry and repetitive movements, each part of the system produces random but repetitive vibrations which are cyclostationary at specific frequencies [2], [3], [4]. In order to avoid early wearing, the different parts are usually designed with different characteristics, ensuring that the system will seldom come back to its initial position. This precaution also ensures that these all

M.E. Davies et al. (Eds.): ICA 2007, LNCS 4666, pp. 145–151, 2007.

produce vibrations cyclostationary at different cyclic frequencies. Furthermore, the rotation speed is ususally known or easy to measure, so that these frequencies can be computed.

We are interested here in the case when one peculiar part of the system, whose cyclic frequency is a priori known, is to be monitored, and no other assumption about the other parts is made except that, if cyclostationary, their cyclic frequencies are different from the one we are interested in. The methods previously developped [5], [6], which suppose that all the cyclic frequencies of the sources are known, are not fitted to this case. We thus developped a new source extraction algorithm based on the only cyclostationary properties of the source to be extracted. In order to derive an algorithm as simple and robust as possible, we chose to base the extraction criterion on second order statistics. We first present the principle of the method, with thorough calculations in the two sources two mixtures case and application to three different cases.

2 Principle of the Method

2.1 Criterion

The method will be described and proofs will be given in the two sources and two mixtures case. We assume that the two mixtures are additive and we call the source vector $S = \lfloor s_i \rfloor$, the mixture vector $X = \lfloor x_i \rfloor$ and the mixing matrix $A = \lfloor a_{ij} \rfloor$ (all real valued), where $(i, j) \in \{1,2\}^2$ and represent respectively lines and columns. They are related by:

$$X = AS \tag{1}$$

The source s_1 is supposed to be cyclostationary at a known frequency α_0. The source s_2 can be either stationary or cyclostationary at any cyclic frequency except to α_0. The sources are supposed to be uncorrelated and their respective powers are denoted σ_1^2 and σ_2^2. Estimating both sources would mean estimating a B matrix such that $Z = BX$ is an estimate of the source vector S. Our goal is only to retrieve the first source on the first estimate vector component, i.e. to find coefficients b_{11} and b_{12} such that

$$z_1 = b_{11}x_1 + b_{12}x_2 \tag{2}$$

is an estimate of s_1. The method that we propose consists in maximizing cyclostationarity at frequency α_0 on z_1 and minimizing the power of z_1, in order to ensure that only the cyclostationary source is kept on that estimate.

The criterion that we chose to minimize is given by:

$$C(b_{11}, b_{12}) = \frac{\left|R_{z_1}(0)\right|}{\left|R_{z_1}^{\alpha_0}(0)\right|} \tag{3}$$

where $R_{z_1}(0)$ and $R_{z_1}^{\alpha_0}(0)$ are coefficients of the Fourier series decomposition of the autocorrelation function of z_1 respectively for cyclic frequencies 0 and α_0 and zero time lag.

2.2 Theoretical Validation of the Criterion

In the 2*2 case, the estimate z_1 can be written as a function of the mixing matrix A, the demixing coefficients b_{11} and b_{12}, and the sources.

$$z_1 = (b_{11}a_{11} + b_{12}a_{21})s_1 + (b_{11}a_{12} + b_{12}a_{22})s_2 \tag{4}$$

There exist an infinity of coefficients pairs (b_{11}, b_{12}) that can lead to the extraction of s_1 on z_1, which are all the pairs satisfying

$$\frac{b_{11}}{b_{12}} = -\frac{a_{22}}{a_{12}} \tag{5}$$

Let us show that minimizing the chosen criterion leads to these solutions. From eq. (3) and (4), we derive:

$$C(b_{11}, b_{12}) = \frac{\sigma_1^2}{\left|R_{s_1}^{\alpha_0}(0)\right|} + \frac{(b_{11}a_{12} + b_{12}a_{22})^2}{(b_{11}a_{11} + b_{12}a_{21})^2} \frac{\sigma_2^2}{\left|R_{s_1}^{\alpha_0}(0)\right|} \tag{6}$$

It is easy to see that this criterion reaches an absolute minimum for any coefficients pair (b_{11}, b_{12}) that satisfies equation (5). Therefore, for a fixed value of one of the coefficients, the criterion exhibits an absolute minimum versus the other one, and this minimum corresponds to one of the solutions.

2.3 Criterion Evaluation

The criterion is estimated from the statistics of the measures. The different statistics to be estimated are:

$$R_x^{\alpha}(0) = \frac{1}{T_0} \int_0^{T_0} E\left[x_i(t) x_j^*(t)\right] e^{-2\pi j \alpha t} dt \tag{7}$$

with

$$T_0 = \frac{1}{\alpha_0} \tag{8}$$

where the cyclic frequency $\alpha \in \{0, \alpha_0\}$ and i and j can take the values $\{1,2\}$. Assuming that the cyclostationary signal s_1 is cycloergodic, the ensemble averaging $E[...]$ can be replaced by temporal synchronized averaging, i.e. averaging over cyclic periods of the signal, which leads to the following estimator:

$$\hat{R}_x^{\alpha}(0) = \frac{1}{NT_0} \int_0^{NT_0} x_i(t) x_j^*(t) e^{-2\pi j \alpha t} dt \tag{9}$$

The criterion can then be computed for any value of (b_{11}, b_{12}) by :

$$\hat{C}(b_{11}, b_{12}) = \frac{b_{11}^2 \hat{R}_{x_1}^0(0) + b_{12}^2 \hat{R}_{x_2}^0(0) + b_{11} b_{12} \hat{R}_{x_1 x_2}^0(0)}{b_{11}^2 \hat{R}_{x_1}^{\alpha_0}(0) + b_{12}^2 \hat{R}_{x_2}^{\alpha_0}(0) + b_{11} b_{12} \hat{R}_{x_1 x_2}^{\alpha_0}(0)} \tag{10}$$

The coefficient b_{11} is arbitrarily set to 1 and the criterion is computed for different values of b_{12} (with a step 0.01) until it reaches a minimum. If b_{12} increases until it reaches a given threshold, the strategy is reversed: b_{12} is set to one and b_{11} made to vary.

3 Simulations

In the following sections, the method is applied to three different cases, each with two sources and two mixtures. The first case includes one artificial cyclostationary source and one stationary source. The second one includes two cyclostationary sources at different cyclic frequencies. In the last case, one of the sources is a real, vibrations signal and the other one is an artificial stationary source.

In each case, we apply the algorithm to a set of different mixing matrices randomly generated. Their four coefficients a_{ij} are random real numbers equally distributed over the interval $[-1, 1]$. In order to evaluate the performance of the method, we compute the mean square error between the cyclostationary source to be extracted and its estimate. Both are normalised before computing the error, in order to take into account the indeterminacy over the estimate amplitude. This error is computed for each of the random matrices and given in dB relatively to the power of the source. The parameters were estimated over 100 realisations for all the simulations.

3.1 One Cyclostationary Source and One Stationary Source

The first simulation was performed over a simple cyclostationary signal

$$s_1(t) = a(t)\cos(2\pi f_0 t) \tag{11}$$

with $a(t)$ a random white noise. This signal is second order cyclostationary at frequency $\alpha_0 = 2f_0$. The second source is chosen to be a random white noise. Both sources are uncorrelated and have power equal to one.

Note that the two sources are both wideband and cannot be separated directly in the frequency domain, so that classical second order source separation methods such as SOBI fail to separate these two peculiar sources.

The algorithm was applied to these sources with 100 different mixing matrices generated as previously described. Fig. 1 shows the estimation error versus the determinant of the mixing matrix. It shows that the source was estimated with a –20 dB accuracy for 98% of the matrices, and with a –30 dB accuracy for 95% of the matrices. For most of the mixing matrices, the source was estimated with very good accuracy, close to –40 dB. The only matrices leading to a poor estimation are ill conditioned ones, whose determinant is close to zero.

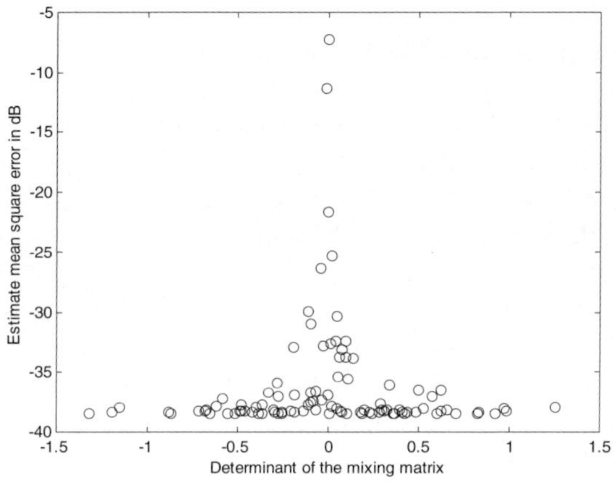

Fig. 1. Mean square error (in dB) between source s_1 and its estimate represented versus the determinant of the mixing matrix for one cyclostationary source and one stationary source

3.2 Two Cyclostationary Sources

Source s_1 is the same cyclostationary source as in section 3.1, while source s_2 is a cyclostationary source built the same way with cyclic frequency $\alpha_2 \neq \alpha_0$. Both sources have power equal to one.

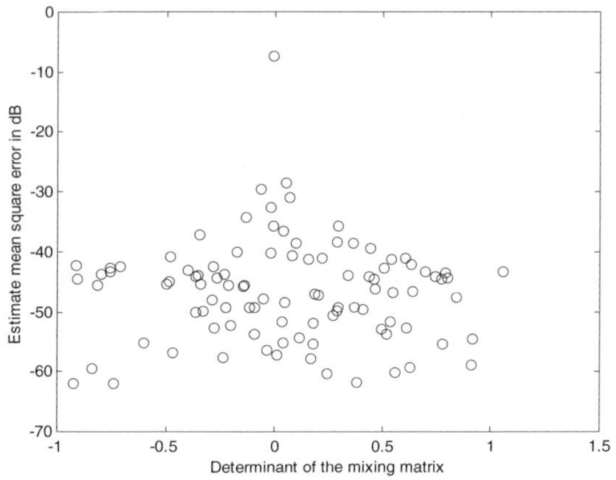

Fig. 2. Mean square error (in dB) between source s_1 and its estimate represented versus the determinant of the mixing matrix for two cyclostationary sources

Fig. 2 shows the results obtained over 100 randomly generated mixing matrices. This figure shows that the results are really good, since source s_1 was estimated with –20 dB accuracy for 99% of the mixing matrices, and –30 dB accuracy for 97% of the mixing matrices. The estimation accuracy is between –30 dB and –60 dB for well conditioned matrices.

3.3 Real Vibration Source and Stationary Source

For this last simulation, we mixed the same stationary source s_2 as in section 3.1 with a real damaged roller bearing vibration that was the source to be estimated. The vibration was recorded at 100kHz sampling frequency and the vibration spectrum spreads over the whole recorded frequency range.

Damaged roller bearing vibrations were shown to be wide sense cyclostationary and one of their cyclic frequencies, which corresponds to the frequency of the shocks over the damaged part, can be computed from the rotation speed and the location of the damage. For the vibration that we used, this frequency is $\alpha_0 = 195Hz$. We computed the criterion at this very frequency in order to achieve extraction. Both sources were normalised so as to have the same power and separation was performed with 100 randomly generated mixing matrices.

As in the artificial signal cases, the source is estimated with very good accuracy. Though the roller bearing vibration is a complex signal, that exhibits several cyclostationary frequencies, the knowledge of one of them is enough to extract the vibration from a set of additive mixtures.

Table 1. sumarises the results obtained in the three studied cases and shows that the proposed algorithm achieves extraction with very good accuracy whatever the nature of the second source can be, provided that it is not cyclostationary at the same frequency as the one to be extracted.

Table 1. Percentage of good estimates for a given accuracy depending on the nature of the sources

	Cyclostationary / Stationary	Both cyclostationary	Vibration / Stationary
- 20 dB	98 %	99 %	98 %
- 30 dB	95 %	97 %	92 %

4 Conclusion

We presented here a new source extraction method for cyclostationary source. The method is based on second order statistics of the sources and only two hypotheses are made about the sources : they are uncorrelated at order two and the source to be extracted exhibits at least one cyclic frequency that it does not share with the other sources and that is *a priori* known. These hypotheses are realistic when coping with vibration signals. The method has been presented here in the two sources and two additive mixtures cases. The criterion to be minimised in order to achieve the extraction was shown to exhibit a unique minimum leading to perfect extraction. The method was applied to different mixtures of artificial or real sources and shown to achieve proper estimation for most of the mixing matrices. It can be easily extended to the N*N case. The set of solutions is then given by a set of (N-1) equations with N unknown variables. This will be presented in further publications.

References

1. Tandon, N., Choudhury, A.: A review of vibration and acoustic measurement methods for the detection of defects in rolling element bearings. Tribology International 32, 469–480 (1999)
2. Capdessus, C., Sidahmed, M., Lacoume, J.L.: Cyclostationary processes : application in gear faults early diagnosis. Mechanical Systems and Signal Processing 14(3), 371–385 (2000)
3. McCormick, C., Nandi, A.K: Cyclostationarity in rotating machine vibrations. Mechanical Systems and Signal Processing 12(2), 225–242 (1998)
4. Antoni, J., Bonnardot, F., Raad, A., El Badaoui, M.: Cyclostationary modelling of rotating machine vibration signals. Mechanical Systems and Signal Processing 18, 1285–1314 (2004)
5. Abed-Meraim, K., Xiang, Y., Manton, J.H., Hua, Y.: Blind Source separation Using Second-Order Cyclostationary Statistics. IEEE Trans on Signal Processing, 49(4) (2001)
6. Jafari, M.G., Chambers, J.A.: Normalised natural gradient algorithm for the separation of cyclostationary sources. In: IEEE International Conference on Acoustics, Speech, and Signal Processing, Ap, vol. 5, pp. 301–304 (2003)

A Robust Complex FastICA Algorithm Using the Huber M-Estimator Cost Function

Jih-Cheng Chao[1] and Scott C. Douglas[2]

[1] Semiconductor Group, Texas Instruments, Dallas, Texas 75243, USA
[2] Department of Electrical Engineering, Southern Methodist University, Dallas, Texas 75275, USA

Abstract. In this paper, we propose to use the Huber M-estimator cost function as a contrast function within the complex FastICA algorithm of Bingham and Hyvarinen for the blind separation of mixtures of independent, non-Gaussian, and proper complex-valued signals. Sufficient and necessary conditions for the local stability of the complex-circular FastICA algorithm for an arbitrary cost are provided. A local stability analysis shows that the algorithm based on the Huber M-estimator cost has behavior that is largely independent of the cost function's threshold parameter for mixtures of non-Gaussian signals. Simulations demonstrate the ability of the proposed algorithm to separate mixtures of various complex-valued sources with performance that meets or exceeds that obtained by the FastICA algorithm using kurtosis-based and other contrast functions.

1 Introduction

In complex-valued blind source separation (BSS), one possesses a set of measured signal vectors

$$\mathbf{x}(k) = \mathbf{A}\mathbf{s}(k) + \boldsymbol{\nu}(k), \tag{1}$$

where $\mathbf{A}$ is an arbitrary complex-valued $(m \times m)$ mixing matrix, such that $\mathbf{A} = \mathbf{A}_R + j\mathbf{A}_I$, $\mathbf{s}(k) = [s_1(k) \ \cdots \ s_m(k)]^T$ is a complex-valued signal of sources, and $s_i(k) = s_{R,i}(k) + js_{I,i}(k)$, where $j = \sqrt{-1}$, and $\boldsymbol{\nu}(k)$ contains circular Gaussian uncorrelated noise. In most treatments of the complex-valued BSS task, the $\{s_i(k)\}$ are assumed to be statistically-independent, and $\mathbf{A}$ is full rank. The goal is to obtain a separating matrix $\mathbf{B}$ such that

$$\mathbf{y}(k) = \mathbf{B}\mathbf{x}(k) \tag{2}$$

contains estimates of the source signals. In independent component analysis (ICA), the linear model in (1) may not hold, yet the goal is to produce signal features in $\mathbf{y}(k)$ that are as independent as possible.

One of the most-popular procedures for complex-valued BSS is the complex circular FastICA algorithm in [1]. This algorithm first prewhitens the mixtures $\mathbf{x}(k)$ to obtain $\mathbf{v}(k) = \mathbf{P}\mathbf{x}(k)$ such that $E\{\mathbf{v}(k)\mathbf{v}^H(k)\} = \mathbf{I}$, after which the

M.E. Davies et al. (Eds.): ICA 2007, LNCS 4666, pp. 152–160, 2007.

rows of a unitary separation matrix $\mathbf{W}$ are adapted sequentially such that $\mathbf{y}(k) = \mathbf{W}\mathbf{v}(k)$ contains the separated sources. For mixtures of sources that are proper, such that $E\{s_i^2(k)\} = 0$ for all i, this algorithm appears to separate such complex mixtures given enough snapshots N for an appropriate choice of algorithm nonlinearity. Several algorithm nonlinearities are suggested as possible candidates, although little work has been performed to determine the suitability of these choices for general complex-valued source signals. More recently, several researchers have explored the structure of the complex-valued BSS task for mixtures of non-circular sources, such that $E\{s_i^2(k)\} \neq 0$ [2]–[4]. In what follows, we limit our discussion to the complex-circular source distribution case, as several practical applications involve mixtures of complex-circular sources.

In this paper, we extend our recent work on employing the Huber M-estimator cost function from robust statistics as a FastICA algorithm contrast [5] to the complex-valued BSS task for mixtures of proper sources ($E\{s_i^2(k)\} = 0$). We provide the complete form of the local stability condition for the complex-circular FastICA algorithm omitted in [1]. We then propose a single-parameter nonlinearity for the algorithm and show through both theory and simulations that the algorithm's performance is largely independent of the cost function's threshold parameter for many source distributions, making it a robust choice for separating complex-valued mixtures with unknown circularly-symmetric source p.d.f.'s. Simulations comparing various contrast choices for the complex circular FastICA algorithm show that ours based on the Huber M-estimator cost often works better than others based on kurtosis maximization or heuristic choice.

2 Complex Circular FastICA Algorithm

We first give the general form of the single-unit FastICA algorithm for extracting one non-Gaussian-distributed proper source from an m-dimensional complex linear mixture [1] and study its local stability properties. The algorithm assumes that the source mixtures have been prewhitened by a linear transformation $\mathbf{P}$ where $\mathbf{v}(k) = \mathbf{P}\mathbf{x}(k)$ contains uncorrelated entries, such that the sample covariance of $\mathbf{v}(k)$ is the identity matrix. For the vector $\mathbf{w}_t = [w_{1t} \ \cdots \ w_{mt}]^T$, the complex circular FastICA update is

$$y_t(k) = \mathbf{w}_t^T \mathbf{v}(k) \tag{3}$$

$$\widetilde{\mathbf{w}}_t = E\{y_t(k) g(|y_t(k)|^2) \mathbf{v}^*(k)\} - E\{g(|y_t(k)|^2) + |y_t(k)|^2 g'(|y_t(k)|^2)\}\mathbf{w}_t \tag{4}$$

$$\mathbf{w}_{t+1} = \frac{\widetilde{\mathbf{w}}_t}{\sqrt{\widetilde{\mathbf{w}}_t^H \widetilde{\mathbf{w}}_t}}, \tag{5}$$

where $y_t(k)$ is the estimated source at time k and algorithm iteration t, $g(u)$ is a real-valued nonlinearity, $g'(u) = dg(u)/du$, and the expectations in (4) are computed using N-sample averages. This algorithm is formulated in [1] as the solution to the following optimization problem:

$$\text{maximize} \ \left|E\{G(|y_t(k)|^2)\} - E\{G(|n|^2)\}\right|^2 \tag{6}$$

$$\text{such that} \ E\{|y_t(k)|^2\} = 1, \tag{7}$$

where n has a circularly-symmetric unit-variance Gaussian distribution and $G(u)$ is a real-valued even-symmetric but otherwise "arbitrary non-linear contrast function" [1] producing $g(u) = dG(u)/du$. The criterion in (6) is described as the square of a simple estimate of the negentropy of $y_t(k)$. Several cost functions are suggested as possible choices for $G(u)$, including $G(u) = \sqrt{a_1 + u}$ for $a_1 \approx 0.1$, $G(u) = \log(a_2 + u)$ for $a_2 \approx 0.1$, and the kurtosis-based $G(u) = 0.5u^2$, although no verification of (9) for the first two choices of $G(u)$ and any well-known non-Gaussian distributions has been given.

In [1], the authors give the following necessary condition for the above algorithm to be locally-stable at a separating solution, where s_i possesses the distribution of the source extracted in $y_t(k)$:

$$(E\{g(|s_i|^2) + |s_i|^2 g'(|s_i|^2) - |s_i|^2 g(|s_i|^2)\}) \neq 0. \quad (8)$$

This condition is not sufficient, however, for local stability of the algorithm, as the curvature of the cost function has not been considered in [1]. Although omitted for brevity, we can show that the necessary and sufficient local stability conditions for the algorithm about a separating solution are

$$\begin{aligned} &[E\{g(|s_i|^2) + |s_i|^2 g'(|s_i|^2) - |s_i|^2 g(|s_i|^2)\}] \\ &\times [E\{G(|s_i|^2)\} - E\{G(|n|^2)\}] < 0. \end{aligned} \quad (9)$$

This result can be compared to that for the real-valued FastICA algorithm in [6], which shows a somewhat-different relationship. Thus, it is necessary and sufficient for the two real-valued quantities on the left-hand-side of the inequality in (9) to be non-zero and have different signs for the complex circular FastICA algorithm to be locally-stable.

3 A Huber M-Estimator Cost Function for the Complex Circular FastICA Algorithm

In [5], a novel single-parameter cost function based on the Huber M-estimator cost in robust statistics [7] was proposed for the real-valued FastICA algorithm. Unlike most other cost functions, the one chosen in [5] has certain nice practical and analytical properties. In particular, it is possible to show that there always exists a nonlinearity parameter for the cost function such that two sufficient conditions for local stability of the algorithm are met. We now extend this work to design a novel cost function for the complex-circular FastICA algorithm.

As the algorithm in [1] implicitly assumes mixtures of proper source signals, we propose to choose $G(|y_t(k)|^2)$ such that the amplitude of $y_t(k)$ is maximized according to the Huber M-estimator cost. Thus, we have

$$G(u) = \begin{cases} \dfrac{u}{2} & u < \theta^2 \\ \theta u^{1/2} - \dfrac{\theta^2}{2} & u \geq \theta^2 \end{cases} \quad (10)$$

where $\theta > 0$ is a threshold parameter designed to trade off the parameter estimation quality with the estimate's robustness to outliers and lack of prior distributional knowledge. The corresponding algorithm nonlinearities are

$$g(u) \equiv \frac{\partial G(u)}{\partial u} = \begin{cases} \frac{1}{2} & u < \theta^2 \\ \frac{\theta}{2} u^{-1/2} & u \geq \theta^2 \end{cases} \tag{11}$$

$$g'(u) \equiv \frac{dg(u)}{du} = \begin{cases} 0 & u < \theta^2 \\ -\frac{\theta}{4} u^{-3/2} & u \geq \theta^2. \end{cases} \tag{12}$$

After some simplification, we can implement the circular complex FastICA update using the above nonlinearities as

$$\widetilde{\mathbf{w}}_t = 2E\{y_t(k) h_\theta(|y_t(k)|) \mathbf{v}^*(k)\} - E\{t_\theta(|y_t(k)|) + h_\theta(|y_t(k)|)\} \mathbf{w}_t \tag{13}$$

$$h_\theta(u) = \begin{cases} 1 & u < \theta \\ \frac{\theta}{u} & u \geq \theta \end{cases}, \quad t_\theta(u) = \begin{cases} 1 & u < \theta \\ 0 & u \geq \theta. \end{cases} \tag{14}$$

The functions $h_\theta(u)$ and $t_\theta(u)$ depend on the threshold parameter θ, and the choice of this nonlinearity will be considered in the next section. Table 1 lists a short MATLAB script for implementing the multiple-unit version of this algorithm, in which the QR decomposition is used for signal deflation.

Table 1. Complex circular FastICA algorithm with Huber M-estimator cost

```
%----------------------------------------------------------------------
[N,m]=size(x); R = (1/N)*(x'*x); v = x/chol(0.5*(R+R')); W = eye(m);
for i=1:iter
    y = v*W;
    absy = abs(y);
    t = (absy<theta);
    h = t + theta*(1-t)./absy;
    W = 2*(v'*(y.*h)) - W*diag(sum(t+h));
    [W,T] = qr(W);
end
%----------------------------------------------------------------------
```

4 On the Local Stability of the Huber M-Estimator Cost for FastICA

Given the new stability condition in (9), what can be said about the circularly-symmetric Huber M-estimator cost function when it is used in the complex FastICA algorithm? The following two theorems, proven in the Appendix, illustrate two properties about this cost. These theorems make statements about the p.d.f. of $u = |s_i|^2$, the squared amplitude of the extracted source. The theorems are non-trivial extensions of the theorems presented in [5].

Theorem 1: *Let $g(u)$ and $g'(u)$ have the forms in (11) and (12), respectively. Then, so long as the random variable u is not exponentially-distributed, there always exists a value of θ such that*

$$E\{g(u)\} + E\{ug'(u)\} - E\{ug(u)\} \neq 0. \tag{15}$$

Theorem 2: *Let $G(u)$ have the form in (10). Then, so long as the random variable u is not exponentially-distributed, there always exists a value of θ such that*

$$E\{G(u)\} - E\{G(|n|^2)\} \neq 0. \tag{16}$$

Note that if s_i is unit-variance circular Gaussian, the p.d.f. of $u = |s_i|^2$ is exponential ($p(u) = e^{-u}$ for $u \geq 0$). Taken together, these two theorems *do not* ensure (9) for all non-Gaussian proper source distributions. They suggest, however, that the design range for θ could be significant for many distributions. We substantiate this claim through the analysis below and by simulations in the next section. These results are significant because, to our knowledge, few if any statements about the stability of a specific non-kurtosis-based cost function within the complex FastICA algorithm have been given in the scientific literature. Moreover, it is unlikely that such results could be easily found given the complexity of the integrals for other $g(y)$ choices (*e.g.* $g(y) = 0.5(a_1 + y)^{-1/2}$ for $a_1 \approx 1$).

We have evaluated the range of θ values for which (9) is satisfied for five well-known zero-mean, unit-power, non-Gaussian distributions: 4-QAM-$\{\pm 1\} + j\{\pm 1\}$, 16-QAM-$\{\pm\frac{1}{\sqrt{10}} \pm \frac{3}{\sqrt{10}}\} + j\{\pm\frac{1}{\sqrt{10}} \pm \frac{3}{\sqrt{10}}\}$, 64-QAM-$\{\pm\frac{1}{\sqrt{42}} \pm \frac{3}{\sqrt{42}} \pm \frac{5}{\sqrt{42}} \pm \frac{7}{\sqrt{42}}\} + j\{\pm\frac{1}{\sqrt{42}} \pm \frac{3}{\sqrt{42}} \pm \frac{5}{\sqrt{42}} \pm \frac{7}{\sqrt{42}}\}$, the uniform amplitude circular distribution such that $|s_i|$ is equally probable for $0 \leq |s_i| \leq \sqrt{2}$ and is zero otherwise, and the exponential amplitude distribution in which $|s_i|$ is exponentially-distribution with $E\{|s_i|^2\} = 1$. For all of these five distributions, the Huber M-estimator cost produces an algorithm that is locally-stable for θ in the range $[0, |s_{max}|)$, where s_{max} is the maximum possible value of $s_i(k)$ admitted by the source p.d.f. Thus, any positive value of θ that places part of the nonlinear portion of $g(u)$ within the range of $|s_i(k)|^2$ often results in a locally-convergent algorithm. Again, this evaluation does not guarantee that the chosen cost function will always work, but it suggests that one does not need to design specific values of θ to achieve separation.

In practice, one may not know what θ value to choose to obtain separation of a particular source mixture. As was suggested in [5] in the real-valued case, we recommend that one *randomize* the value of θ over a range of positive values during coefficient adaptation. The main observed effect using such randomization is a slight slowdown in convergence speed.

5 Simulations

We now explore the performance of the FastICA algorithm with various cost function via simulations. In these simulations, $m = 15$-source mixtures were

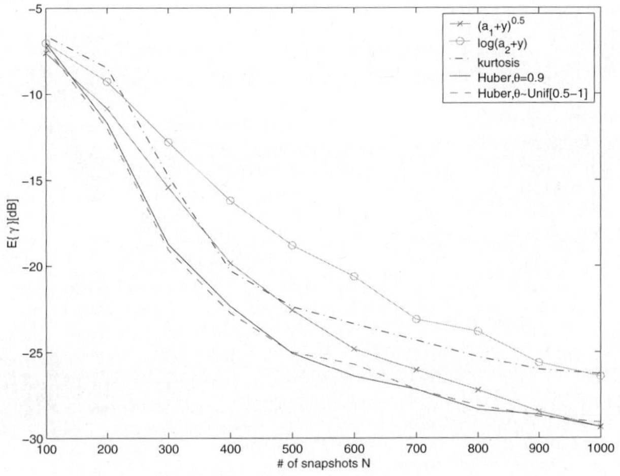

Fig. 1. $E\{\gamma\}$ vs. number of snapshots N for the various algorithms in the simulation example

generated consisting of three 4-QAM, three 16-QAM, three 64-QAM, three uniform and three exponential amplitude circular-distributed independent sources, and a random mixing matrix. The multi-unit FastICA procedure was applied to this data for numbers of snapshots ranging from $N = 100$ to $N = 5000$ and for different θ values. The performance factor computed is the separation cost

$$\gamma = \frac{1}{2m}\left(\sum_{i=1}^{m}\sum_{l=1}^{m}\frac{|c_{il}|^2}{\max\limits_{1\le i\le m}|c_{il}|^2} + \frac{|c_{il}|^2}{\max\limits_{1\le l\le m}|c_{li}|^2}\right) - 1 \tag{17}$$

with $\mathbf{C} = \mathbf{WPA}$ as obtained at convergence of the algorithm. One hundred iterations were averaged to obtain each data point shown.

Fig. 1 compares the performance of FastICA with the Huber cost function and $\theta = 0.9$ and with the Huber cost function and a uniformly-randomized θ in the range $0.5 \le \theta \le 1$ at each iteration with three other versions of FastICA – using $G(y) = \sqrt{a_1 + y}$ or $g(y) = \frac{1}{2\sqrt{a_1+y}}$ for $a_1 \approx 0.1$, $G(y) = \log(a_2 + y)$ or $g(y) = \frac{1}{a_2+y}$ for $a_2 \approx 0.1$, and the kurtosis-based choice $G(y) = 0.5y^2$ or $g(y) = y$. As can be seen, the Huber cost function-based versions outperform the algorithms based on previously-proposed contrast functions. More significantly, our algorithm version with a randomized threshold parameter θ provides good separation performance across all sample sizes; performance deviations were less than $\pm$1dB from the algorithm with a fixed $\theta = 0.9$ value.

Fig. 2 illustrates the performance sensitivity of the FastICA algorithm with Huber M-estimator cost to the value of θ for these signal mixtures. As can be seen, the algorithm performs well for values of θ satisfying $0.1 \le \theta \le 1$, and its performance degrades gracefully for higher θ values.

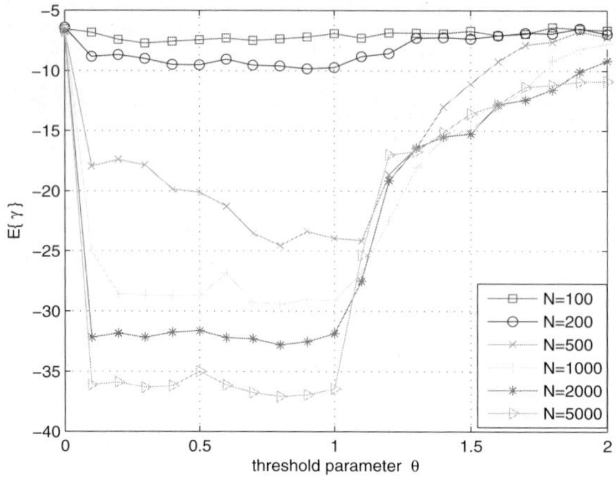

Fig. 2. $E\{\gamma\}$ vs. θ for the FastICA algorithm with Huber M-estimator cost in the simulation example

6 Conclusions

In many blind source separation and independent component analysis algorithms, the cost function used to measure signal independence is a design parameter. In this paper, we have considered Huber's single-parameter M-estimator cost function for use within the complex-valued FastICA algorithm for proper source mixtures. The algorithm obtained is computationally-simple, and the procedure works well for a wide range of threshold parameters θ. The reasons for the algorithm's robust behavior for a wide range of the threshold parameter is indicated through a stability analysis.

References

1. Bingham, E., Hyvarinen, A.: A fast fixed-point algorithm for independent component analysis of complex valued signals. Int J. Neural Systems 10(1), 1–8 (2000)
2. De Lathauwer, L., De Moor, B.: On the blind separation of non-circular sources. In: Proc. EUSIPCO-02, Toulouse, France (September 2002)
3. Novey, M., Adali, T.: ICA by maximization of nongaussianity using complex functions. In: Proc. IEEE Workshop Machine Learning for Signal Processing, Mystic, CT, September 2005, pp. 21–26. IEEE Computer Society Press, Los Alamitos (2005)
4. Eriksson, J., Koivunen, V.: Complex random vectors and ICA models: Identifiability, uniqueness and separability. IEEE Trans. Inform. Theory 52, 1017–1029 (2006)
5. Chao, J., Douglas, S.C.: A simple and robust fastICA algorithm using the Huber M-estimator cost function. In: Proc. IEEE Int. Conf. Acoust. Speech, Signal Processing, Toulouse, France, vol. 5, pp. 685–688 (May 2006)
6. Hyvärinen, A.: Fast and robust fixed-point algorithms for independent component analysis. IEEE Trans. Neural Networks 10, 626–634 (1999)
7. Huber, P.: Robust Statistics. Wiley, New York (1981)

7 Appendix

Proof of Theorem 1: Assume without loss of generality that u is unit variance. Consider the terms on the left-hand-side of (15) for the nonlinearities in (11) and (12), and define $f_1(\theta) = 2(E\{g(u)\} + E\{ug'(u)\} - E\{ug(u)\})$. Then, we obtain

$$f_1(\theta) = \int_{\theta^2}^{\infty} u^{-1/2}(u^{3/2} - \theta u - u^{1/2} + \frac{\theta}{2})p(u)du. \tag{18}$$

For Eq. (15) not to hold, $f_1(\theta) = 0$ for all possible values of θ. Suppose that the slightly-more-general condition

$$f_1(\theta) = c_1\theta + c_2 \tag{19}$$

is true, where c_1 and c_2 are unknown constants. Such a condition justified when $p(u)$ is smooth, as $f_1(\theta)$ an then be modeled by a polynomial approximation - see the comment below. Then,

$$\frac{\partial f_1(\theta)}{\partial\theta} = p(\theta^2)\theta + \int_{\theta^2}^{\infty} (\frac{1}{2}u^{-1/2} - u^{1/2})p(u)du = c_1 \tag{20}$$

$$\frac{\partial^2 f_1(\theta)}{\partial\theta^2} = p'(\theta^2) + p(\theta^2) = 0, \tag{21}$$

which yields the relationship

$$p'(u) = -p(u). \tag{22}$$

The only distribution $p(u)$ satisfying (22) is the exponential distribution, i.e. $p(u) = e^{-u}$ for $u \geq 0$. Thus, the theorem follows. Note that if s_i is circular Gaussian-distributed, $|s_i|^2$ has an exponential distribution, although other distributions for s_i could lead to an exponential distribution for $|s_i|^2$.

Proof of Theorem 2: Substituting (10) into the left-hand-side of (16), defining $f_2(\theta) = 2E\{G(u) - G(|n|^2)\}$, and simplifying yields the expression

$$f_2(\theta) = -\int_{\theta^2}^{\infty} (u^{1/2} - \theta)^2[p(u) - p_n(u)]du, \tag{23}$$

where $p_n(u) = e^{-u}$ for $u \geq 0$. For Eq. (16) not to hold, $f_2(\theta) = 0$ for all possible values of θ. Suppose that the slightly-more-general condition

$$f_2(\theta) = c_1\theta + c_2 \tag{24}$$

is true, where c_1 and c_2 are unknown constants. Then,

$$\frac{\partial f_2(\theta)}{\partial\theta} = 2\int_{\theta^2}^{\infty} (u^{1/2} - \theta)[p(u) - p_n(u)]du \;=\; c_1 \tag{25}$$

$$\frac{\partial^2 f_2(\theta)}{\partial\theta^2} = -2\int_{\theta^2}^{\infty} [p(u) - p_n(u)]du \;=\; 0. \tag{26}$$

For (24) to hold for all $\theta > 0$, we must have

$$p(u) = p_n(u), \tag{27}$$

which results in $c_1 = 0$, $c_2 = 0$, and finally $f_2(\theta) = 0$. Thus, the theorem follows.

Comment: In both of the above proofs, $f_i(\theta)$ is a continuous function of θ given a continuous smooth amplitude-squared distribution $p(u)$. Thus, we can express $f_i(\theta)$ as a polynomial function of θ with coefficients c_i. Now, for the condition $f_i(\theta) = 0$, we must have all $c_i = 0$. Clearly, it is impossible that $c_0 = 0$ and all c_i not equal to 0 for $i > 0$ and the condition $f_i(\theta) = 0$, because any change in θ would make $f_i(\theta)$ not equal to zero. Hence, $f_i(\theta)$ defines only one function and therefore only one distribution $p(u)$ has $f_i(\theta) = 0$. In both of the above proofs, the exponential distribution yields $f_i(\theta) = 0$.

Stable Higher-Order Recurrent Neural Network Structures for Nonlinear Blind Source Separation

Yannick Deville and Shahram Hosseini

Université Paul Sabatier Toulouse 3 - Observatoire Midi-Pyrénées - CNRS, Laboratoire d'Astrophysique de Toulouse-Tarbes (UMR 5572), 14 Av. Edouard Belin, 31400 Toulouse, France
ydeville@ast.obs-mip.fr, shahram.hosseini@ast.obs-mip.fr

Abstract. This paper concerns our general recurrent neural network structures for nonlinear blind source separation, especially suited to polynomial mixtures. We here focus on linear-quadratic mixtures. We introduce an extended structure, with additional free parameters as compared to the structure that we previously proposed. We derive the equilibrium points of our new structure, thus showing that it has no spurious fixed points. We analyze its stability in detail and propose a practical procedure for selecting its free parameters, so as to guarantee the stability of a separating point. We thus solve the stability issue of our previous structure. Numerical results illustrate the effectiveness of this approach.

1 Introduction

Blind source separation (BSS) methods aim at restoring a set of N unknown source signals $s_j(n)$ from a set of P observed signals $x_i(n)$ which are mixtures of these source signals [1], with $P = N$ in the standard configuration considered hereafter. In the simplest case, the observed signals are Linear Instantaneous (LI) mixtures of the source signals. Denoting $A = [a_{ij}]$ the matrix composed of these unknown mixture coefficients a_{ij} and $s(n) = [s_1(n) \dots s_N(n)]^T$ and $x(n) = [x_1(n) \dots x_P(n)]^T$ the source and observation vectors respectively, the mixing model reads in matrix form

$$x(n) = As(n). \tag{1}$$

One of the very first solutions to this LI-BSS problem reported in the literature is the Hérault-Jutten artificial neural network [2]. This network has a recurrent (or feedback) structure, i.e. each of its outputs $y_i(n)$ consists of an LI combination of input $x_i(n)$ and of all *other* outputs $y_j(n)$ with $j \neq i$, using adequate combination coefficients estimated from the outputs by means of an unsupervised algorithm. Such a recurrent structure is not mandatory however, i.e. the same class of LI mappings from the signals $x_i(n)$ to the signals $y_i(n)$ may also be achieved by a feedforward structure, where each output $y_i(n)$ is derived only as an LI combination of all inputs $x_i(n)$. The latter structure is simpler than the recurrent one

M.E. Davies et al. (Eds.): ICA 2007, LNCS 4666, pp. 161–168, 2007.

and is therefore the one mainly used today for LI mixtures. The same evolution occurred for convolutive mixtures, i.e. an approach based on a recurrent structure was first proposed by Nguyen and Jutten [3] and extended by Charkani and Deville [4],[5] but other approaches, based on feedforward structures, were then preferred, since they also provide the considered linear (convolutive) mappings.

The situation is quite different for nonlinear mixtures, which are now receiving increasing attention. We especially showed in [6]-[8] that recurrent structures are much more attractive than feedforward ones for polynomial mixtures. Building upon our above-mentioned experience on recurrent structures, we proposed in [6]-[8] a general approach for polynomial mixtures and we investigated it in more detail in the case of linear-quadratic mixtures. The higher-order recurrent neural network that we proposed for such mixtures was shown to be promising, but its applicability is limited by stability constraints.

This paper also focuses on linear-quadratic mixtures (references of previously published BSS approaches for such mixtures are provided in [7]-[8]). Our main contributions then consist in : (i) introducing a more general higher-order recurrent neural network structure than in [6]-[8], (ii) providing a detailed theoretical analysis of its fixed points and their stability, depending on its parameter values, and (iii) deriving a practical method for selecting these values so as to guarantee stability at a separating point. As a spin-off, we also obtain stability conditions for our previous network. This confirms that it cannot operate with any signals, while its extended version can.

2 Mixing and Separating Models

We here consider an instantaneous mixing model, where two observed signals $x_1(n)$ and $x_2(n)$ consist of linear combinations of two source signals $s_1(n)$ and $s_2(n)$, added to quadratic terms, i.e. terms proportional to $s_1(n)s_2(n)$. Taking into account BSS scale indeterminacies, the rescaled mixing model reads [7]-[8]

$$x_1(n) = s_1(n) - L_{12}s_2(n) - Q_1 s_1(n)s_2(n) \tag{2}$$

$$x_2(n) = -L_{21}s_1(n) + s_2(n) - Q_2 s_1(n)s_2(n). \tag{3}$$

For each time n, a recurrence is performed to compute the values of the outputs y_i of the feedback network that we now introduce in this paper. We denote as m the index associated to this recurrence and $y_i(m)$ the successive values of each output in this recurrence at time n[1]. This recurrence reads

$$y_1(m+1) = x_1(n) + l_{11}y_1(m) + l_{12}y_2(m) + q_1 y_1(m)y_2(m) \tag{4}$$

$$y_2(m+1) = x_2(n) + l_{21}y_1(m) + l_{22}y_2(m) + q_2 y_1(m)y_2(m) \tag{5}$$

where l_{ij} and q_i are the adaptive weights of the proposed neural network.

[1] These successive output values therefore also depend on n. This index n is omitted in the notations $y_i(m)$ however, in order to improve readability and to focus on the recurrence on output values for given input values $x_1(n)$ and $x_2(n)$.

As compared to our previous papers [6]-[8], we here consider the same mixing model (2)-(3), but we propose an extended version (4)-(5) of our previous recurrent neural network, where we introduce the linear feedback terms l_{11} and l_{22} from each output $y_i(m)$ to the input with the *same* index. In other words, our previous network is a specific case of the new one, obtained by forcing

$$l_{11} = 0 \quad \text{and} \quad l_{22} = 0. \tag{6}$$

By analyzing the stability of our previous and extended networks, we will show below that the constraint (6) does not make it possible to handle all signal values and we will introduce adequate values of l_{11} and l_{22} for solving this problem.

3 Fixed and Separating Points

The stability of the recurrence (4)-(5) is analyzed for fixed points (i.e. equilibrium points) of this recurrence. The first step of this investigation therefore consists in determining these fixed points, i.e. the points (y_1^E, y_2^E) which are such that

$$y_1(m+1) = y_1(m) = y_1^E \quad \text{and} \quad y_2(m+1) = y_2(m) = y_2^E. \tag{7}$$

As may be seen by combining (4)-(5) and (7), the fixed points of the recurrence (4)-(5) depend: (i) on the inputs $x_i(n)$, and therefore on the source values and mixing parameters through the mixing equations (2)-(3), and (ii) also on the network weights l_{ij} and q_i. Our eventual goal will be to adapt the network weights l_{ij} and q_i so as to achieve BSS, i.e. so as to make the network outputs $y_i(m)$ equal to the source signals, up to BSS indeterminacies. Therefore, before considering that adaptation, we here determine all network weights which are such that one of the associated fixed points corresponds to BSS without permutation, i.e. to

$$y_1^E = k_1 s_1(n) \quad \text{and} \quad y_2^E = k_2 s_2(n) \tag{8}$$

where k_1 and k_2 are two arbitrary scale factors. In other words, for given mixture parameters in (2)-(3), we look for all network weights l_{ij} and q_i and scale factors k_i such that Eq. (2)-(5) and (7)-(8) are met whatever the source values $s_i(n)$. Our calculations, which are skipped here due to space limitations, yield

$$l_{11} = -\frac{1}{k_1} + 1 \tag{9}$$

$$l_{12} = \frac{L_{12}}{k_2} = L_{12} l'_{22} \tag{10}$$

$$q_1 = \frac{Q_1}{k_1 k_2} = Q_1 l'_{11} l'_{22} \tag{11}$$

$$l_{21} = \frac{L_{21}}{k_1} = L_{21} l'_{11} \tag{12}$$

$$l_{22} = -\frac{1}{k_2} + 1 \tag{13}$$

$$q_2 = \frac{Q_2}{k_1 k_2} = Q_2 l'_{11} l'_{22} \tag{14}$$

with

$$l'_{11} = 1 - l_{11} \quad \text{and} \quad l'_{22} = 1 - l_{22}. \tag{15}$$

The first expressions in (10)-(12) and (14) show that we thus obtain an infinite number of solutions, due to the two arbitrary scale factors k_i with which the sources appear in the network outputs. There is a one-to-one correspondence between these factors and the two network weights l_{11} and l_{22} as shown by (9) and (13). One may therefore consider the weights l_{11} and l_{22} as the primary parameters, select them (freely at this stage), and then assign accordingly the other network weights, using the second expressions in (10)-(12) and (14).

For these weight values, we know by construction that the network has at least one fixed point, i.e. the point defined by (8) with (9) and (13). We must then determine *all* fixed points for these weight values, because the network may converge to any of these points (depending on their stability) and we should especially determine whether each of them yields separated sources. This topic is addressed by looking for all solutions of Eq. (2)-(5), (7), (10)-(12) and (14)-(15). Long calculations then yield two solutions, more easily expressed by defining

$$\alpha = Q_2 + Q_1 L_{21} \tag{16}$$

$$\beta = -(Q_2 L_{12} + Q_1) \tag{17}$$

$$\gamma = 1 - L_{12} L_{21} \tag{18}$$

$$\epsilon_{y_1} = \pm 1 \tag{19}$$

$$\epsilon_{T1} = \epsilon_{y_1} \, sgn(l'_{11} l'_{22}) \; sgn(-\alpha s_1(n) + \beta s_2(n) + \gamma). \tag{20}$$

These solutions then read

$$y_1^E = \frac{1}{2 l'_{11} \alpha} \{ [\alpha s_1(n) + \beta s_2(n) + \gamma] - \epsilon_{T1} [-\alpha s_1(n) + \beta s_2(n) + \gamma] \} \tag{21}$$

$$y_2^E = \frac{1}{\beta l'_{22}} \{ -\alpha l'_{11} y_1^E + [\alpha s_1(n) + \beta s_2(n)] \}. \tag{22}$$

These two solutions correspond to the two possible values of ϵ_{y_1} in (19) and are more easily expressed with respect to ϵ_{T1}. The first solution, corresponding to $\epsilon_{T1} = 1$, reads

$$y_1^E = \frac{1}{l'_{11}} s_1(n) \quad \text{and} \quad y_2^E = \frac{1}{l'_{22}} s_2(n). \tag{23}$$

This solution yields BSS without permutation (and with scale factors). It is nothing but the above solution (8), as shown by (9), (13) and (15). The second solution, corresponding to $\epsilon_{T1} = -1$, reads

$$y_1^E = \frac{1}{l'_{11}} \left[\frac{\beta}{\alpha} s_2(n) + \frac{\gamma}{\alpha} \right] \quad \text{and} \quad y_2^E = \frac{1}{l'_{22}} \left[\frac{\alpha}{\beta} s_1(n) - \frac{\gamma}{\beta} \right]. \tag{24}$$

This additional solution yields BSS with a permutation (and with scale factors and additive constants).

We thus only obtain the classical BSS solutions, with indeterminacies associated to nonlinear mixtures. In other words, for these weight values, this network does not yield spurious fixed points, i.e. fixed points such that the outputs are still mixtures of the sources. We now analyze the stability of these fixed points.

4 Stability Condition

The considered network is a two-dimensional nonlinear dynamic system, since the evolution of its state vector $[y_1(m), y_2(m)]^T$ is defined by the nonlinear equations (4)-(5). The linear stability of any such system at a given fixed point may be analyzed by considering a first-order approximation of its evolution equations at that point. The new value of a small disturbance added to the state vector is thus expressed as the product of a matrix $\mathbf{H}$, which defines the first-order approximation of the system, by the former value of that disturbance (see e.g. details in [9]). The (asymptotic) stability of the system at the considered point is guaranteed by constraining the moduli of both eigenvalues of the corresponding matrix $\mathbf{H}$ to be lower than 1. Our calculations show that this condition may be expressed as

$$\begin{cases} T > -D - 1 \\ T < D + 1 \\ D < 1 \end{cases} \tag{25}$$

where D and T are resp. the determinant and trace of $\mathbf{H}$. The stability region thus obtained in the (D, T) plane is bounded by a triangle which includes the origin (see the figure in [9], which also indirectly confirms (25)).

By computing the matrix $\mathbf{H}$ of the system defined by (4)-(5) and applying condition (25) to each of the above two fixed points (here derived with respect to the observations $x_1(n)$ and $x_2(n)$, i.e. without using (2)-(3)), long calculations yield the following results. The fixed point corresponding to $\epsilon_{y_1} = -1$ never meets the stability condition (25). For the fixed point corresponding to $\epsilon_{y_1} = 1$, this condition reads

$$|l'_{11}l'_{22}|\sqrt{\delta_{y_1}} - 2Al'_{11} - 2Bl'_{22} + 4 > 0 \tag{26}$$

$$|l'_{11}l'_{22}|\sqrt{\delta_{y_1}} - Al'_{11} - Bl'_{22} < 0 \tag{27}$$

with

$$\delta_{y_1} = [Q_2x_1(n) - Q_1x_2(n) + \gamma]^2 - 4\alpha[x_1(n) + x_2(n)L_{12}] \tag{28}$$

$$A = \frac{Q_1}{2\beta}\left[-Q_2x_1(n) + Q_1x_2(n) + \gamma - sgn(l'_{11}l'_{22})\sqrt{\delta_{y_1}}\right] + 1 \tag{29}$$

$$B = \frac{Q_2}{2\alpha}\left[-Q_2x_1(n) + Q_1x_2(n) - \gamma + sgn(l'_{11}l'_{22})\sqrt{\delta_{y_1}}\right] + 1. \tag{30}$$

5 Limitations of Our Previous Network

The stability condition (26)-(27) especially applies to the more specific network that we proposed in [6]-[8], which corresponds to (6) as explained above, and therefore to $l'_{11} = 1$ and $l'_{22} = 1$. Inserting these values in (26)-(27) shows that our previous network yields a stable fixed point only for some values of the mixing coefficients and observed signals, and therefore of the source signals. To clearly illustrate this phenomenon with an example, one may consider the simple case

$$L_{12} = 0, \quad L_{21} = 0, \quad Q_1 = Q_2 = Q > 0. \tag{31}$$

It may be shown that our previous network then only yields stability for a limited domain of source values, which consists of a strip in the $(s_1(n), s_2(n))$ plane, defined by

$$-s_1(n) - \frac{1}{Q} < s_2(n) < -s_1(n) + \frac{3}{Q}. \tag{32}$$

6 A Method for Stabilizing Our New Network

6.1 Analysis of Stability Condition

Condition (26)-(27) is of high theoretical interest, because it completely defines the stability of the considered fixed point. However, it does not show easily if and how l'_{11} and l'_{22} may be selected in order to ensure stability at the considered fixed point for any given observed signal values. To address that topic, we consider l'_{11} as the primary variable and l'_{22} as the secondary variable, and we express it as $l'_{22} = \lambda l'_{11}$, where λ is a parameter. For any fixed λ, we first investigate whether there exist values of l'_{11} such that (26)-(27) is met, in order to eventually determine if there exist values of l'_{11} and λ (and therefore l'_{22}) such that (26)-(27) is met. In other words, we first determine the intersection of the part of the (l'_{11}, l'_{22}) plane where (26)-(27) is met and of a given line in that plane, defined by $l'_{22} = \lambda l'_{11}$ (with $\lambda \neq 0$). Using the latter expression of l'_{22}, Eq. (26)-(27) become

$$|\lambda|\sqrt{\delta_{y_1}}(l'_{11})^2 - 2(A + B\lambda)l'_{11} + 4 > 0 \tag{33}$$

$$|\lambda|\sqrt{\delta_{y_1}}(l'_{11})^2 - (A + B\lambda)l'_{11} < 0. \tag{34}$$

For a given λ, (28)-(30) show that A and B do not depend on l'_{11} and l'_{22}. (33)-(34) then yield two inequalities with respect to l'_{11}, which are solved as follows. For (34), any (non-zero) λ is suitable and the solutions of (34) for a given λ are

$$l'_{11} = \mu \frac{A + B\lambda}{|\lambda|\sqrt{\delta_{y_1}}} \quad \text{with} \quad 0 < \mu < \mu_{max} \tag{35}$$

where $\mu_{max} = 1$. It may then be shown that, taking (33) into account in addition still yields (35), but now with μ_{max} defined as follows. Denoting

$$C(\lambda) = \frac{(A + B\lambda)^2}{|\lambda|\sqrt{\delta_{y_1}}} \tag{36}$$

we have

$$\mu_{max} = \begin{cases} 1 & \text{if} \qquad C(\lambda) \leq 4 \\ 1 - \sqrt{1 - \frac{4}{C(\lambda)}} & \text{otherwise} . \end{cases} \tag{37}$$

The above analysis shows that any value of λ yields a non-empty interval of solutions for l'_{11}. For a given λ, a simple and safe approach therefore consists in selecting for l'_{11} the value situated in the middle of the allowed range (35), i.e.

$$l'_{11} = \frac{\mu_{max}}{2} \frac{A + B\lambda}{|\lambda|\sqrt{\delta_{y_1}}} \tag{38}$$

with μ_{max} defined by (37). We should eventually propose a method for selecting λ. As stated above, any (non-zero) value of λ is acceptable. A simple solution is $\lambda = \pm 1$, which gives the same "weight" to l'_{11} and $l'_{22} = \lambda l'_{11}$. Moreover, the sign of λ may be chosen as follows. As explained above, the considered fixed point correspond to $\epsilon_{y_1} = 1$. Since $l'_{22} = \lambda l'_{11}$ in addition, (20) here reduces to

$$\epsilon_{T1} = sgn(\lambda)\ sgn(-\alpha s_1(n) + \beta s_2(n) + \gamma). \tag{39}$$

Therefore, if λ has the same sign as $(-\alpha s_1(n) + \beta s_2(n) + \gamma)$, then $\epsilon_{T1} = 1$, so that the considered stable fixed point yields the non-permuted sources defined in (23). Otherwise, the permuted sources of (24) are obtained. We thus guarantee that both solutions are obtained by successively applying our approach with two opposite values of λ. Moreover, if the source and mixture coefficients are such that the sign of $(-\alpha s_1(n) + \beta s_2(n) + \gamma)$ is known, then selecting λ also with that sign guarantees (local) convergence to the non-permuting point. Otherwise, this version of our approach yields a permutation issue, to be further investigated.

6.2 Summary of Proposed Method

Based on the above analysis, the following procedure is guaranteed to yield local convergence towards a separating point:

1. Select λ as explained above.
2. Set l'_{11} according to (38), taking into account $sgn(l'_{11} l'_{22}) = sgn(\lambda)$.
3. Set $l'_{22} = \lambda l'_{11}$.
4. Set all other network parameters according to (10)-(12) and (14)-(15).

7 Numerical Results

To illustrate the performance of our approach, we consider observations defined by (2)-(3) at a single time n (the same test could then be repeated for different

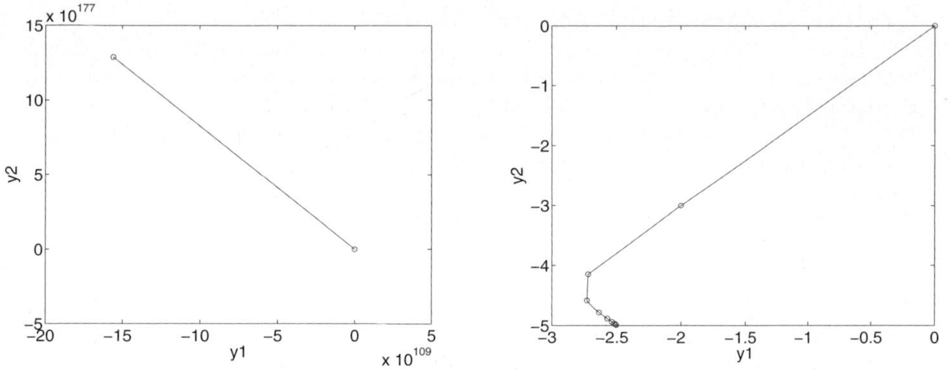

Fig. 1. Divergence of previous network (left), convergence of new network (right)

times). We use $s_1(n) = -1$, $s_2(n) = -2$ and mixing coefficients defined by (31), with $Q = 0.5$. We first implement our previous network [6]-[8] by running 10 steps of the recurrence (4)-(5) with (6), (10)-(12) and (14), starting from $y_1(1) = 0$, $y_2(1) = 0$. The resulting trajectory of $(y_1(m), y_2(m))$ is provided in Fig. 1. This shows that the network diverges very rapidly. This is in agreement with the fact that condition (32) is not met here. We then implement our new network by running the recurrence (4)-(5) as above but with parameters selected as explained in Section 6.2 (with $\lambda = 1$). The network then converges to the solution (23), as shown in Fig. 1 (we here have $l'_{11} = l'_{22} = 0.4$). Moreover, it converges very rapidly. This shows the effectiveness of the proposed approach.

8 Conclusion

In this paper, we introduced a new BSS neural network structure which solves the stability issue of our previous network. We plan to investigate whether other aspects of its operation may be further improved by taking advantage of its free parameters, especially λ. We will also develop algorithms to adapt the network weights so that they converge towards separating points.

Acknowledgment. Y. Deville would like to thank Y. Naudet for drawing his attention to linear-quadratic mixtures, which first resulted in the approach presented in [6], which was then developed in [7]-[8].

References

1. Hyvarinen, A., Karhunen, J., Oja, E.: Independent Component Analysis. Wiley, New York (2001)
2. Jutten, C., Hérault, J.: Blind separation of sources, Part I: An adaptive algorithm based on neuromimetic architecture. Signal Processing 24, 1–10 (1991)
3. Thi, H.L.N., Jutten, C.: Blind source separation for convolutive mixtures. Signal Processing 45(2), 209–229 (1995)
4. Charkani, N., Deville, Y.: Self-adaptive separation of convolutively mixed signals with a recursive structure. Part I: stability analysis and optimization of asymptotic behaviour. Signal Processing 73(3), 225–254 (1999)
5. Charkani, N., Deville, Y.: Self-adaptive separation of convolutively mixed signals with a recursive structure. Part II: Theoretical extensions and application to synthetic and real signals. Signal Processing 75(2), 117–140 (1999)
6. Deville, Y.: Méthode de séparation de sources pour mélanges linéaires-quadratiques, private communication (September 6, 2000)
7. Hosseini, S., Deville, Y.: Blind separation of linear-quadratic mixtures of real sources using a recurrent structure. In: Mira, J., Alvarez, J.R. (eds.) IWANN 2003. LNCS, vol. 2687, pp. 241–248. Springer, Heidelberg (2003)
8. Hosseini, S., Deville, Y.: Blind maximum likelihood separation of a linear-quadratic mixture. In: Puntonet, C.G., Prieto, A.G. (eds.) ICA 2004. LNCS, vol. 3195, pp. 22–24. Springer, Heidelberg (2004)
9. Thompson, J.M.T., Stewart, H.B.: Nonlinear dynamics and chaos. Wiley, Chichester, England (2002)

Hierarchical ALS Algorithms for Nonnegative Matrix and 3D Tensor Factorization

Andrzej Cichocki[1], Rafal Zdunek[2], and Shun-ichi Amari[3]

[1] Dept. of EE, Warsaw University of Technology, and IBS PAN Warsaw, Poland
[2] Institute of Telecommunications, Teleinformatics and Acoustics, Wroclaw University of Technology, Poland
[3] RIKEN Brain Science Institute, Wako-shi, Saitama, Japan
{cia,zdunek,amari}@brain.riken.jp

Abstract. In the paper we present new Alternating Least Squares (ALS) algorithms for Nonnegative Matrix Factorization (NMF) and their extensions to 3D Nonnegative Tensor Factorization (NTF) that are robust in the presence of noise and have many potential applications, including multi-way Blind Source Separation (BSS), multi-sensory or multi-dimensional data analysis, and nonnegative neural sparse coding. We propose to use local cost functions whose simultaneous or sequential (one by one) minimization leads to a very simple ALS algorithm which works under some sparsity constraints both for an under-determined (a system which has less sensors than sources) and over-determined model. The extensive experimental results confirm the validity and high performance of the developed algorithms, especially with usage of the multi-layer hierarchical NMF. Extension of the proposed algorithm to multidimensional Sparse Component Analysis and Smooth Component Analysis is also proposed.

1 Introduction - Problem Formulation

Nonnegative Matrix Factorization (NMF) and its multi-way extensions: Nonnegative Tensor Factorization (NTF) and Parallel Factor analysis (PARAFAC) models with sparsity and/or non-negativity constraints have been recently proposed as promising and quite efficient tools for processing sparse signals, images, or general data [1,2,3,4,5,6,7,8]. From a viewpoint of data analysis, NMF/NTF provides nonnegative and usually sparse common factors or hidden (latent) components with physiological meaning and interpretation [6,9]. NMF, NTF, and Sparse Component Analysis (SCA) are used in a variety of applications, ranging from neuroscience and psychometrics to chemometrics [10,1,6,7,9,11,12].

In this paper, we propose new Hierarchical Alternating Least Squares (HALS) algorithms for NMF/NTF. By incorporating the regularization and penalty terms into the local squared Euclidean norms, we are able to achieve sparse and local representations of the desired solution, and to alleviate the problem of getting stuck in local minima.

M.E. Davies et al. (Eds.): ICA 2007, LNCS 4666, pp. 169–176, 2007.

We impose nonnegativity and sparsity constraints to the following NTF (i.e., standard PARAFAC with nonnegativity constraints) model [3]:

$$\boldsymbol{X}_q = \boldsymbol{A}\boldsymbol{D}_q\tilde{\boldsymbol{S}} + \boldsymbol{E}_q, \qquad (q = 1, 2, \ldots, Q) \tag{1}$$

where $\boldsymbol{X}_q \in \mathbb{R}_+^{I\times T}$ are frontal slices (matrices) of the observed 3D tensor data or signals $\underline{\boldsymbol{X}} \in \mathbb{R}^{I\times T\times Q}$, $\boldsymbol{D}_q \in \mathbb{R}_+^{J\times J}$ are diagonal scaling matrices, $\boldsymbol{A} = [\boldsymbol{a}_1, \boldsymbol{a}_2, \ldots, \boldsymbol{a}_J] \in \mathbb{R}_+^{I\times J}$ is a mixing or basis matrix, $\tilde{\boldsymbol{S}} \in \mathbb{R}_+^{J\times T}$ represents unknown sources or hidden (nonnegative and sparse) components, and $\boldsymbol{E}_q \in \mathbb{R}^{I\times T}$ represents the q-th frontal slice of the tensor $\underline{\boldsymbol{E}} \in \mathbb{R}^{I\times T\times Q}$ representing a noise or error. In the special case for $Q = 1$, the model simplifies to the standard NMF model. The objective is to estimate the set of all nonnegative matrices: $\boldsymbol{A}$, $\{\boldsymbol{D}_q\}$, $\tilde{\boldsymbol{S}}$[1]. The problem can be converted to a tri-NMF model by applying averaging of frontal slices: In this section, we develop the alternative algorithm which converts the problem to a simple tri-NMF model (under condition that all frontal slices $\boldsymbol{X}_q$ have the same dimension):

$$\boldsymbol{X} = \boldsymbol{A}\boldsymbol{D}\tilde{\boldsymbol{S}} + \boldsymbol{E} = \boldsymbol{A}\boldsymbol{S} + \boldsymbol{E}, \tag{2}$$

where $\boldsymbol{X} = \sum_{q=1}^{Q} \boldsymbol{X}_q$, $\boldsymbol{D} = \sum_{q=1}^{Q} \boldsymbol{D}_q = \mathrm{diag}\{d_{q1}, d_{q2}, \ldots, d_{qJ}\}$, $\boldsymbol{E} = \sum_{q=1}^{Q} \boldsymbol{E}_q$, and $\boldsymbol{S} = \boldsymbol{D}\tilde{\boldsymbol{S}}$ is a scaled matrix of sources. The above system of linear algebraic equations can be represented in an equivalent scalar form as follows $x_{it} = \sum_j a_{ij} s_{jt} + e_{it}$, or equivalently in the vector form: $\boldsymbol{X} = \sum_j \boldsymbol{a}_j \; \underline{\boldsymbol{s}}_j + \boldsymbol{E}$ where $\underline{\boldsymbol{s}}_j$ are rows of $\boldsymbol{S}$, and $\boldsymbol{a}_j$ are columns of $\boldsymbol{A}$ $(j = 1, 2, \ldots, J)$. Such a simple model provides improved performance if the noise (in the frontal slices) is not correlated.

The majority of NMF/NTF algorithms for BSS applications works only if the following assumption $T >> I \geq J$ is held, where J is known or can be estimated using SVD. In the paper, we propose the NMF algorithm that can work also for an under-determined case, i.e. $T >> J > I$, if signal representations are enough sparse. Our objective is to estimate the mixing (basis) matrix $\boldsymbol{A}$ and the sources $\boldsymbol{S}$, subject to nonnegativity and sparsity constraints.

2 Locally Regularized ALS Algorithm

The most of known and used adaptive algorithms for NMF are based on alternating minimization of the squared Euclidean distance expressed by the Frobenius norm:

$$D_F(\boldsymbol{X}||\boldsymbol{A}\boldsymbol{S}) = \frac{1}{2}||\boldsymbol{X} - \boldsymbol{A}\boldsymbol{S}||_F^2 + \alpha_A||\boldsymbol{A}||_1 + \alpha_S||\boldsymbol{S}||_1, \tag{3}$$

subject to nonnegativity constraints of all the elements in $\boldsymbol{A}$ and $\boldsymbol{S}$, where $||\boldsymbol{A}||_1 = \sum_{ir} a_{ir}$, $||\boldsymbol{S}||_1 = \sum_{jt} s_{jt}$, and α_A and α_S are nonnegative regularization coefficients controlling sparsity of the matrices [9].

[1] Usually, the common factors, i.e., matrices $\boldsymbol{A}$ and $\tilde{\boldsymbol{S}}$ are normalized to unit length column vectors and rows, respectively, and are forced to be as sparse as possible.

The basic approach to NMF is alternating minimization or alternating projection: the specified cost function is alternately minimized with respect to two sets of the parameters $\{s_{jt}\}$ and $\{a_{ij}\}$, each time optimizing one set of arguments while keeping the other one fixed [9,1].

In this paper we consider minimization of the set of local squared Euclidean cost functions:

$$D_F^{(j)}(\boldsymbol{X}^{(j)}||\boldsymbol{a}_j\underline{\boldsymbol{s}}_j) = \frac{1}{2}\|(\boldsymbol{X}^{(j)} - \boldsymbol{a}_j\underline{\boldsymbol{s}}_j)\|_F^2 + \alpha_A^{(j)}\ J_A(\boldsymbol{a}_j) + \alpha_S^{(j)}\ J_S(\underline{\boldsymbol{s}}_j), \tag{4}$$

for $j = 1, 2, \ldots, J$, subject to nonnegativity constraints for all elements: $a_{ij} \geq 0$ and $s_{jt} \geq 0$, where

$$\boldsymbol{X}^{(j)} = \boldsymbol{X} - \sum_{p \neq j} \boldsymbol{a}_p \underline{\boldsymbol{s}}_p, \tag{5}$$

$\boldsymbol{a}_j \in \mathbb{R}^{I\times 1}$ are columns of the basis mixing matrix $\boldsymbol{A}$, $\underline{\boldsymbol{s}}_j \in \mathbb{R}^{1\times T}$ are rows of $\boldsymbol{S}$, $\alpha_A^{(j)} \geq 0$ and $\alpha_S^{(j)} \geq 0$ are local parameters controlling a sparsity level of the individual vectors, and the penalty terms $J_A(\boldsymbol{a}_j) = \sum_i a_{ij}$ and $J_S(\underline{\boldsymbol{s}}_j) = \sum_t s_{jt}$ enforce sparsification of the columns in $\boldsymbol{A}$ and the rows in $\boldsymbol{S}$, respectively. The construction of such a set of local cost functions follows from the simple observation that the observed data can be decomposed approximately as follows $\boldsymbol{X} = \sum_{p=1}^{J} \boldsymbol{a}_p \underline{\boldsymbol{s}}_p + \boldsymbol{E}$ or more generally $\boldsymbol{X} = \sum_{p=1}^{J} \lambda_p \boldsymbol{a}_p \underline{\boldsymbol{s}}_p + \boldsymbol{E}$ with $\lambda_1 \geq \lambda_2 \geq \ldots \geq \lambda_J > 0$.

The gradients of the cost function (4) with respect to the unknown vectors $\boldsymbol{a}_j$ and $\underline{\boldsymbol{s}}_j$ are expressed by

$$\frac{\partial D_F^{(j)}(\boldsymbol{X}^{(j)}||\boldsymbol{a}_j\underline{\boldsymbol{s}}_j)}{\partial \underline{\boldsymbol{s}}_j} = \boldsymbol{a}_j^T \boldsymbol{a}_j \underline{\boldsymbol{s}}_j - \boldsymbol{a}_j^T \boldsymbol{X}^{(j)} + \alpha_S^{(j)}, \tag{6}$$

$$\frac{\partial D_F^{(j)}(\boldsymbol{X}^{(j)}||\boldsymbol{a}_j\underline{\boldsymbol{s}}_j)}{\partial \boldsymbol{a}_j} = \boldsymbol{a}_j \underline{\boldsymbol{s}}_j \underline{\boldsymbol{s}}_j^T - \boldsymbol{X}^{(j)} \underline{\boldsymbol{s}}_j^T + \alpha_A^{(j)}, \tag{7}$$

where the scalars $\alpha_S^{(j)}$ and $\alpha_A^{(j)}$ are added/substracted component-wise. By equating the gradient components to zero and assuming that we enforce the nonnegativity constraints with a simple "half-rectifying" nonlinear projection, we obtain a new set of sequential learning rules:

$$\underline{\boldsymbol{s}}_j \leftarrow \left[\frac{1}{\boldsymbol{a}_j^T \boldsymbol{a}_j}(\boldsymbol{a}_j^T \boldsymbol{X}^{(j)} - \alpha_S^{(j)})\right]_+ \qquad \boldsymbol{a}_j \leftarrow \left[\frac{1}{\underline{\boldsymbol{s}}_j \underline{\boldsymbol{s}}_j^T}(\boldsymbol{X}^{(j)} \underline{\boldsymbol{s}}_j^T - \alpha_A^{(j)})\right]_+, \tag{8}$$

for $j = 1, 2, \ldots, J$, where $[\xi]_+ = \max\{\epsilon, \xi\}$, and ϵ is a small constant to avoid numerical instabilities (usually $\epsilon = 10^{-16}$).

Remark 1. In practice, it is necessary to normalize in each iteration step the column vectors $\boldsymbol{a}_j$ and the row vectors $\underline{\boldsymbol{s}}_j$ to unit length vectors (in the sense of norm l_p norm ($p = 1, 2, ..., \infty$)). In the special case of l_2 norms the above algorithms can be further simplified by neglecting the denominator in (8). After estimating the normalized matrices $\boldsymbol{A}$ and $\widetilde{\boldsymbol{S}}$, we estimate the diagonal matrices as follows:

$$\boldsymbol{D}_q = \left[\text{diag}\{\boldsymbol{A}^+ \, \boldsymbol{X}_q \, \tilde{\boldsymbol{S}}^+\}\right]_+, \qquad (q = 1, 2, \ldots, Q) \tag{9}$$

Remark 2. In this paper we have applied a simple nonlinear half-wave rectifying projection $[s_{jt}]_+ = \max\{\epsilon, s_{jt}\}$, $\forall t$ (element-wise) in order to impose nonnegativity constraints. However, other nonlinear projections or filtering can be applied to extract sources (not necessary nonnegative) with specific properties. First of all, the proposed method can be easily extended for semi-NMF and semi-NTF, where nonnegativity constraints are imposed only for some preselected sources, i.e, rows of the matrix $\boldsymbol{S}$ and/or some selected columns of the matrix $\boldsymbol{A}$ if some *a priori* information is available. Furthermore, instead of using the simple nonlinear half-rectifying projection, we can apply more complex nonlinear projections and filtering to estimate bipolar sources which have some specific properties, for example, sources can be bounded, sparse or smooth. In order to estimate the sparse bipolar sources, we can apply well-known adaptive (soft or hard) shrinking nonlinear transformations (e.g, the nonlinear projection can be defined as: $P_{sr}(s_{jt}) = s_{jt}$ for $|s_{jt}| > \delta$ and $P_{sr}(s_{jt}) = 0$ otherwise, with the adaptive threshold $\delta > 0$). Alternatively, we may apply a power nonlinear element-wise transformation: $P_{sp}(s_{jt}) = \text{sign}(s_{jt})|s_{jt}|^{1+\gamma_s}$, $\forall t$, where γ_s is a small coefficient which controls a sparsity/density level of individual sources [11]. In order to achieve smoothness of the estimated sources, we may apply a local averaging operator (such as MA or ARMA models) or low pass filtering which gradually enforces some level of smoothness during an iterative process.

3 Possible Extensions and Improvements

To deal with the factorization problem (1) efficiently, we adopt several approaches from constrained optimization and multi-criteria optimization, where we minimize simultaneously several cost functions using alternating switching between the sets of parameters: $\{\boldsymbol{A}\}$, $\{\boldsymbol{S}\}$.

The above simple algorithm can be further extended or improved (in respect of convergence rate and performance). First of all, different cost functions can be used for estimation of the rows in the matrix $\boldsymbol{S}$ and the columns in the matrix $\boldsymbol{A}$. Furthermore, the columns of $\boldsymbol{A}$ can be estimated simultaneously, instead one by one. For example, by minimizing the set of cost functions in (4) with respect to $\underline{\boldsymbol{s}}_j$, and simultaneously the cost function (3) with normalization of the columns $\boldsymbol{a}_j$ to unit l_2-norm, we obtain the new ALS learning algorithm in which the rows of $\boldsymbol{S}$ are updated locally (row by row) and the matrix $\boldsymbol{A}$ is updated globally (all columns $\boldsymbol{a}_j$ simultaneously):

$$\underline{\boldsymbol{s}}_j \leftarrow \left[\boldsymbol{a}_j^T \boldsymbol{X}^{(j)} - \alpha_S^{(j)}\right]_+, \;(j = 1, \ldots, J), \quad \boldsymbol{A} \leftarrow \left[(\boldsymbol{X}\boldsymbol{S}^T - \alpha_A)(\boldsymbol{S}\boldsymbol{S}^T)^{-1}\right]_+ \tag{10}$$

with normalization (scaling) of the columns in $\boldsymbol{A}$ to the unit length.

Secondly, instead of the standard gradient descent approach we can apply the Quasi-Newton method [13,14] for estimation of matrix $\boldsymbol{A}$. Since the Hessian $\nabla^2_A(D_F) = \boldsymbol{I}_I \otimes \boldsymbol{S}\boldsymbol{S}^T \in \mathbb{R}^{IJ \times IJ}$ of $D_F(\boldsymbol{X}||\boldsymbol{A}\boldsymbol{S})$ has the diagonal block structure with the same blocks, we can simplify the update of $\boldsymbol{A}$ with the Newton method to the very simple form:

$$\boldsymbol{A} \leftarrow \left[\boldsymbol{A} - \nabla_A(D_F(\boldsymbol{X}||\boldsymbol{A}\boldsymbol{S}))\boldsymbol{H}_A^{-1}\right]_+, \tag{11}$$

where $\nabla_A D_F(\boldsymbol{X}||\boldsymbol{A}\boldsymbol{S}) = (\boldsymbol{A}\boldsymbol{S} - \boldsymbol{X})\boldsymbol{S}^T \in \mathbb{R}^{I\times J}$, and $\boldsymbol{H}_A = \boldsymbol{S}\boldsymbol{S}^T \in \mathbb{R}^{J\times J}$. The matrix $\boldsymbol{H}_A$ may be ill-conditioned, especially if $\boldsymbol{S}$ is sparse, and due to this the Levenberg-Marquardt approach is used to control ill-conditioning of the Hessian. Thus we have developed the following NMF/NTF algorithm:

$$\underline{\boldsymbol{s}}_j \leftarrow \left[\boldsymbol{a}_j^T \boldsymbol{X}^{(j)} - \alpha_S^{(j)}\right]_+, \quad \boldsymbol{A} \leftarrow \left[\boldsymbol{A} - (\boldsymbol{A}\boldsymbol{S} - \boldsymbol{X})\boldsymbol{S}^T(\boldsymbol{S}\boldsymbol{S}^T + \lambda \boldsymbol{I}_J)^{-1}\right]_+, \tag{12}$$

for $j = 1, \ldots, J$, where $\lambda \leftarrow \lambda_0 \exp\{-\tau k\}$, k is an index of a current alternating step, and $\boldsymbol{I}_J \in \mathbb{R}^{J\times J}$ is an identity matrix.

Since the alternating minimization technique in NMF is not convex, the selection of initial conditions is very important. Our algorithms are initialized with random uniform matrices. Thus, to minimize the risk of getting trapped in local minima of the cost functions, we use some steering technique that comes from a simulated annealing approach. The solution is triggered with the exponential rule. For our problems, we set heuristically $\lambda_0 = 100$ and $\tau = 0.02$.

3.1 Multi-layer NMF/NTF

In order to improve the performance of the NTF algorithms proposed in this paper, especially for ill-conditioned and badly scaled data and also to reduce risk of getting stuck in local minima in non-convex alternating minimization computations, we have developed a simple hierarchical multi-stage procedure [15] combined together with multi-start initializations, in which we perform a sequential decomposition of nonnegative matrices as follows. In the first step, we perform the basic decomposition (factorization) $\boldsymbol{X}_q \approx \boldsymbol{A}^{(1)}\boldsymbol{D}_q^{(1)}\boldsymbol{S}^{(1)}$ using any available NTF algorithm. In the second stage, the results obtained from the first stage are used to build up a new tensor $\widehat{\underline{\boldsymbol{S}_1}}$ from the estimated frontal slices defined as $\widehat{\boldsymbol{X}}_q^{(1)} = \boldsymbol{S}_q^{(1)} = \boldsymbol{D}_q^{(1)}\boldsymbol{S}^{(1)}$, $(q = 1, 2, \ldots, Q)$ and in the next step we perform the similar decomposition for the new available frontal slices: $\widehat{\boldsymbol{X}}_q^{(1)} = \boldsymbol{S}_q^{(1)} \approx \boldsymbol{A}^{(2)}\boldsymbol{D}_q^{(2)}\boldsymbol{S}^{(2)}$, using the same or different update rules. We continue our decomposition taking into account only the last achieved components. The process can be repeated arbitrarily many times until some stopping criteria are satisfied. In each step, we usually obtain gradual improvements of the performance. Thus, our NTF model has the form:

$$\boldsymbol{X}_q \approx \boldsymbol{A}^{(1)}\boldsymbol{A}^{(2)}\cdots\boldsymbol{A}^{(L)}\boldsymbol{D}_q^{(L)}\boldsymbol{S}^{(L)}, \qquad (q = 1, 2, \ldots, Q), \tag{13}$$

with final results $\boldsymbol{A} = \boldsymbol{A}^{(1)}\boldsymbol{A}^{(2)}\cdots\boldsymbol{A}^{(L)}$, $\boldsymbol{S} = \boldsymbol{S}^{(L)}$ and $\boldsymbol{D}_q = \boldsymbol{D}_q^{(L)}$.

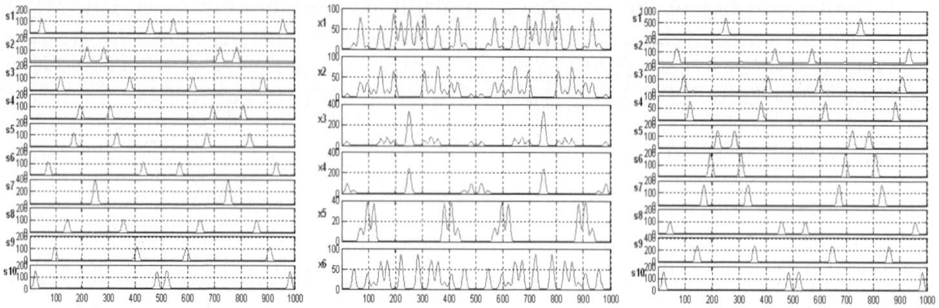

Fig. 1. (left) Original 10 sparse source signals ; (middle) observed 6 mixed signals with randomly generated mixing matrix $\boldsymbol{A} \in \mathbb{R}^{6\times 10}$ (under-determined case); (right) estimated 10 source signals using our new algorithm (12); For 10 layers we achieved the following performance: SIRs for $\boldsymbol{A}$ and $\boldsymbol{S}$ are as follows: $SIR_A = 38.1, 37.0, 35.9, 32.4, 28.2, 33.1, 34.5, 41.2, 25.1, 25.1$[dB] and $SIR_S = 23.1, 32.7, 35.2, 26.2, 29.8, 22.5, 41.8, 29.3, 30.2, 32.5$[dB], respectively

Physically, this means that we build up a system that has many layers or cascade connections of L mixing subsystems. The key point in our approach is that the learning (update) process to find parameters of matrices $\boldsymbol{S}^{(l)}$ and $\boldsymbol{A}^{(l)}$ is performed sequentially, i.e. layer by layer. In fact, we found that the hierarchical multi-layer approach is necessary to apply in order to achieve high performance for all the proposed algorithms.

4 Simulation Results

All the NTF algorithms presented in this paper have been extensively tested for many difficult benchmarks for signals and images with various statistical distributions and additive noise, and also for preliminary tests with real EEG data. Due to space limitations we present here only the selected simulations results in Figs.1–2. The synthetic benchmark illustrated in Fig.1(left) contains sparse nonnegative and weakly statistically dependent 10 source components. The sources have been mixed by the randomly generated full rank matrix $\boldsymbol{A} \in \mathbb{R}_+^{6\times 10}$. The typical mixed signals are shown in Fig.1(middle). The results obtained with the new algorithm (12) with $\alpha_S^{(j)} = 0.05$ are illustrated in Fig.1(right) with average Signal-to-Interference (SIR) level greater than 25 [dB].

Since the proposed algorithms (alternating techniques) perform a non-convex optimization, the estimated components are initial condition dependent. To estimate the performance in a statistical sense, we present the histograms of 100 mean-SIR samples for estimation of $\boldsymbol{S}$ (Fig.2). We tested the two different algorithms (combination of the algorithms) – algorithm (10): ALS for $\boldsymbol{A}$ and HALS for $\boldsymbol{X}$ ($\alpha_A = 0,\ \alpha_S^{(j)} = 0.05$), and algorithm (12): quasi-Newton for $\boldsymbol{A}$ and HALS for $\boldsymbol{S}$.

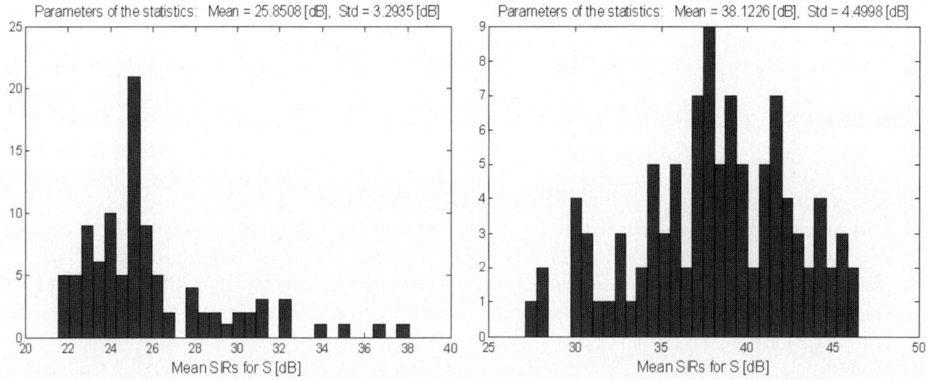

Fig. 2. Histograms of 100 mean-SIR samples for estimating $\boldsymbol{S}$ from Monte Carlo analysis performed using the following algorithms with 10 layers: (left) ALS for $\boldsymbol{A}$ and HALS for $\boldsymbol{S}$ (10); (right) quasi-Newton for $\boldsymbol{A}$ and HALS for $\boldsymbol{S}$ (12)

5 Conclusions and Discussion

The main objective and motivation of this paper is to derive simple algorithms which are suitable both for under-determined and over-determined cases. We have proposed the generalized and flexible cost function (controlled by sparsity penalty) that allows us to derive a family of robust and efficient alternating least squares algorithms for NMF and NTF. Exploiting gradient and Hessian properties, we have derived a family of efficient algorithms for estimating nonnegative sources even if the number of sensors is smaller than the number of hidden nonnegative sources under assumption that the sources are sufficiently sparse and not strongly overlapped. This is the unique modification of the standard ALS algorithm, and to the authors' best knowledge, the first time such a cost function and algorithms have been applied to NMF and NTF. The proposed algorithm gives also better performance (SIRs ans speed) than the ordinary ALS algorithm for NMF, and also some applications of the FOCUSS algorithm [16,17]. We implemented the discussed algorithms in our NMFLAB/NTFLAB MATLAB Toolboxes [18]. The algorithms may be also promising for other applications, such as Sparse Component Analysis, Smooth Component Analysis and EM Factor Analysis because they relax the problems of getting stuck to in local minima much better than the standard ALS algorithm.

We have motivated the use of the proposed models in three areas of data analysis (especially, EEG and fMRI) and signal/image processing: (i) multi-way blind source separation, (ii) model reduction and selection, and (iii) sparse image coding. Our preliminary experiments are promising. The models can be further extended by imposing additional, natural constraints such as smoothness, continuity, closure, unimodality, local rank - selectivity, and/or by taking into account a prior knowledge about specific 3D, or more generally, multi-way data.

Obviously, there are many challenging open issues remaining, such as global convergence, an optimal choice of the associated parameters.

References

1. Cichocki, A., Amari, S.: Adaptive Blind Signal And Image Processing (New revised and improved edition). John Wiley, New York (2003)
2. Dhillon, I., Sra, S.: Generalized nonnegative matrix approximations with Bregman divergences. In: Neural Information Proc. Systems, Vancouver, Canada (2005)
3. Hazan, T., Polak, S., Shashua, A.: Sparse image coding using a 3D non-negative tensor factorization. In: International Conference of Computer Vision (ICCV), pp. 50–57 (2005)
4. Heiler, M., Schnoerr, C.: Controlling sparseness in non-negative tensor factorization. In: Leonardis, A., Bischof, H., Pinz, A. (eds.) ECCV 2006. LNCS, vol. 3951, pp. 56–67. Springer, Heidelberg (2006)
5. Hoyer, P.: Non-negative matrix factorization with sparseness constraints. Journal of Machine Learning Research 5, 1457–1469 (2004)
6. Morup, M., Hansen, L.K., Herrmann, C.S., Parnas, J., Arnfred, S.M.: Parallel factor analysis as an exploratory tool for wavelet transformed event-related EEG. NeuroImage 29, 938–947 (2006)
7. Smilde, A., Bro, R., Geladi, P.: Multi-way Analysis: Applications in the Chemical Sciences. John Wiley and Sons, New York (2004)
8. Oja, E., Plumbley, M.D.: Blind separation of positive sources by globally convergent gradient search. Neural Computation 16, 1811–1825 (2004)
9. Lee, D.D., Seung, H.S.: Learning the parts of objects by nonnegative matrix factorization. Nature 401, 788–791 (1999)
10. Berry, M., Browne, M., Langville, A., Pauca, P., Plemmons, R.: Algorithms and applications for approximate nonnegative matrix factorization. Computational Statistics and Data Analysis (in press, 2006)
11. Cichocki, A., Amari, S., Zdunek, R., Kompass, R., Hori, G., He, Z.: Extended SMART algorithms for non-negative matrix factorization. In: Rutkowski, L., Tadeusiewicz, R., Zadeh, L.A., Zurada, J.M. (eds.) ICAISC 2006. LNCS (LNAI), vol. 4029, pp. 548–562. Springer, Heidelberg (2006)
12. Kim, M., Choi, S.: Monaural music source separation: Nonnegativity, sparseness, and shift-invariance. In: Rosca, J., Erdogmus, D., Príncipe, J.C., Haykin, S. (eds.) ICA 2006. LNCS, vol. 3889, pp. 617–624. Springer, Heidelberg (2006)
13. Zdunek, R., Cichocki, A.: Non-negative matrix factorization with quasi-Newton optimization. In: Rutkowski, L., Tadeusiewicz, R., Zadeh, L.A., Zurada, J.M. (eds.) ICAISC 2006. LNCS (LNAI), vol. 4029, pp. 870–879. Springer, Heidelberg (2006)
14. Zdunek, R., Cichocki, A.: Nonnegative matrix factorization with constrained second-order optimization. Signal Processing 87, 1904–1916 (2007)
15. Cichocki, A., Zdunek, R.: Multilayer nonnegative matrix factorization. Electronics Letters 42, 947–948 (2006)
16. Murray, J.F., Kreutz-Delgado, K.: Learning sparse overcomplete codes for images. Journal of VLSI Signal Processing 45, 97–110 (2006)
17. Kreutz-Delgado, K., Murray, J.F., Rao, B.D., Engan, K., Lee, T.W., Sejnowski, T.J.: Dictionary learning algorithms for sparse representation. Neural Computation 15, 349–396 (2003)
18. Cichocki, A., Zdunek, R.: NTFLAB for Signal Processing. Technical report, Laboratory for Advanced Brain Signal Processing, BSI, RIKEN, Saitama, Japan (2006)

Pivot Selection Strategies in Jacobi Joint Block-Diagonalization

Cédric Févotte[1,⋆] and Fabian J. Theis[2]

[1] GET/Télécom Paris (ENST), 37-39 rue Dareau, 75014 Paris, France
fevotte@tsi.enst.fr
http://www.tsi.enst.fr/~fevotte/
[2] Bernstein Center for Computational Neuroscience
MPI for Dynamics and Self-Organisation, Göttingen, Germany
fabian@theis.name
http://fabian.theis.name

Abstract. A common problem in independent component analysis after prewhitening is to optimize some contrast on the orthogonal or unitary group. A popular approach is to optimize the contrast only with respect to a single angle (Givens rotation) and to iterate this procedure. In this paper we discuss the choice of the sequence of rotations for such so-called Jacobi-based techniques, in the context of joint block-diagonalization (JBD). Indeed, extensive simulations with synthetic data, reported in the paper, illustrates the sensitiveness of this choice, as standard cyclic sweeps appear to often lead to non-optimal solutions. While not being able to guarantee convergence to an optimal solution, we propose a new schedule which, from empirical testing, considerably increases the chances to achieve global minimization of the criterion. We also point out the interest of initializing JBD with the output of joint diagonalization (JD), corroborating the idea that JD could in fact perform JBD up to permutations, as conjectured in previous works.

1 Introduction

Joint diagonalization techniques have received much attention in the last fifteen years within the field of signal processing, and more specifically within the fields of independent component analysis (ICA) and blind source separation (BSS). JADE, one of the most popular ICA algorithms developed by Cardoso and Souloumiac [1], is based on orthonormal joint diagonalization (JD) of a set of cumulant matrices. To this purpose the authors designed a Jacobi algorithm for approximate joint diagonalization of a set of matrices [2]. In a BSS parlance, JADE allows for separation of determined linear instantaneous mixtures of mutually independent sources, exploiting fourth-order statistics. Other standard BSS techniques involving joint diagonalization include the SOBI algorithm [3], TDSEP [4], stBSS [5] and TFBSS [6], which all rely on second-order statistics

⋆ Cédric Févotte acknowledges support from European Commission under contract FP6-027026-K-SPACE.

M.E. Davies et al. (Eds.): ICA 2007, LNCS 4666, pp. 177–184, 2007.

of the sources, namely covariance matrices in the first through third case and spatial Wigner-Ville spectra in the fourth case; see [7] for a review.

Joint block-diagonalization (JBD) came into play in BSS when Abed-Meraim, Belouchrani and co-authors extended the SOBI algorithm to overdetermined convolutive mixtures [8]. Their idea was to turn the convolutive mixture into an overdetermined linear instantaneous mixture of block-dependent sources, the second-order statistics matrices of the source vector thus becoming block-diagonal instead of diagonal. Hence, the joint diagonalization step in SOBI needed to be replaced by a JBD step. Another area of application can be found in the context of multidimensional ICA or independent subspace analysis [9,10]. Its goal is to linearly transform an observed multivariate random vector such that its image is decomposed into groups of stochastically independent vectors. It has been shown that by using fourth-order cumulants to measure the independence, JADE now translates into a JBD problem [11]; similarly also SOBI and other JD-based criteria can be extended to this group ICA setting [12,13].

Abed-Meraim et al. have sketched several Jacobi strategies in [14, 15, 16]: the JBD problem is turned into a minimization problem, where the matrix parameter (the joint block-diagonalizer) is constrained to be unitary (because of spatial prewhitening). The minimizer is searched for iteratively, as a product of Givens rotations, each rotation minimizing a block-diagonality criterion around a fixed axis, which we refer to as 'pivot'. Convergence of the algorithm is easily shown, but convergence to an optimal solution (which minimizes the chosen JBD criterion) is not guaranteed. In fact, we observed that results vary widely according to the choice of the successive pivots (which we refer to as 'schedule') and the initialization of the algorithm, which is not discussed in previous works [14,15,16]. The main contributions of this paper are 1) to point out that the choice of the rotation schedule is a sensitive issue which greatly influences the convergence properties of the Jacobi algorithm, as illustrated on extensive simulations with synthetic data, 2) to propose a new schedule, which, from empirical testing, offers better chances to converge to the optimal solution (while still not guaranteeing it), as compared to the standard cyclic Jacobi technique. We also point out the interest of initializing JBD with the output of JD, corroborating the idea that JD could in fact perform JBD up to permutations, as suggested by Cardoso in [10], more recently conjectured by Abed-Meraim and Belouchrani in [16] and partially proved in [11,17].

The paper is organized as follows. Section 2 briefly describes the Jacobi approach to approximate JBD, with fixed equal block sizes. Section 3 compares the convergence results obtained with three choices of initialization/schedule on generated sets of matrices exactly joint block-diagonalizable, with various size, block size and set dimension. Section 4 reports conclusions.

2 Jacobi Approximate Joint Block-Diagonalization

2.1 Approximate Joint Block-Diagonalization

Let $\mathcal{A} = \{\mathbf{A}_1, \ldots, \mathbf{A}_K\}$ be a set of K complex matrices of size $n \times n$. The problem of approximate JBD consists of finding a unitary matrix $\mathbf{U} \in \mathbb{C}^{n\times n}$ such that

$\forall k \in [\![1, K]\!] := \{1, \ldots, K\}$, the matrices

$$\mathbf{U}\,\mathbf{A}_k\,\mathbf{U}^H = \mathbf{B}_k$$

are as block-diagonal as possible. More precisely, let us denote L the (fixed) length of the diagonal blocks and $m = n/L$ the number of blocks. Writing for $k \in [\![1, K]\!]$

$$\mathbf{A}_k = \begin{bmatrix} \mathbf{A}_{k11} & \ldots & \mathbf{A}_{k1m} \\ \vdots & & \vdots \\ \mathbf{A}_{km1} & \ldots & \mathbf{A}_{kmm} \end{bmatrix}$$

where $\mathbf{A}_{kij}$ is a subblock of dimensions $L \times L$, $\forall (i,j) \in [\![1, m]\!]^2$, our block-diagonality criterion is chosen as

$$\text{boff}\,(\mathbf{A}_k) := \sum_{1 \le i \ne j \le m} \|\mathbf{A}_{kij}\|_F^2. \tag{1}$$

Here $\|\mathbf{B}\|_F^2 = \sum_{ij} |b_{ij}|^2$ denotes the Frobenius norm. We look for $\mathbf{U}$ by minimizing the cost function

$$C_{\text{jbd}}(\mathbf{V}; \mathcal{A}) := \sum_{i=1}^{K} \text{boff}\left(\mathbf{V}\,\mathbf{A}_i\,\mathbf{V}^H\right)$$

with respect to $\mathbf{V} \in U(n)$, where $U(n)$ is the set of unitary $n \times n$-matrices.

2.2 The Jacobi Approach

Jacobi approaches rely on the fact that any unitary matrix $\mathbf{V} \in U(n)$ can be written as a product of complex Givens matrices $\mathbf{G}(p, q, c, s) \in U(n)$, $1 \le p < q \le n$, defined as everywhere equal to the identity $\mathbf{I}_n$ except for $[\mathbf{G}(p, q, c, s)]_{pp} = [\mathbf{G}(p, q, c, s)]_{qq} = c$, $[\mathbf{G}(p, q, c, s)]_{pq} = \bar{s}$, $[\mathbf{G}(p, q, c, s)]_{qp} = -s$, with $(c, s) \in \mathbb{R} \times \mathbb{C}$ such that $c^2 + |s|^2 = 1$. The Jacobi approach consists of iteratively applying the same Givens rotation to all the matrices in set $\mathcal{A}$, with (p, q) chosen as to minimize criterion C_{jbd}. In other words, for fixed p and q, one iteration of the method consists of the following two steps:

- compute $(c^\star, s^\star) = \text{argmin}_{c,s}\, C_{\text{jbd}}(\mathbf{G}(p, q, c, s); \mathcal{A})$
- $\forall k \in [\![1, K]\!]$, $\mathbf{A}_k \leftarrow \mathbf{G}(p, q, c^\star, s^\star)\,\mathbf{A}_k\,\mathbf{G}(p, q, c^\star, s^\star)^H$

Let $I_1, \ldots, I_m$ be the partition of $[\![1, n]\!]$ defined by $I_i = [\![(i-1)\,L + 1, i\,L]\!]$, and let $i(k) = \lceil i/L \rceil$ give the index i of the interval I_i to which k belongs. Let $\mathbf{B}_k = \mathbf{G}(p, q, c, s)\,\mathbf{A}_k\,\mathbf{G}(p, q, c, s)^H$, $k \in [\![1, K]\!]$. $\mathbf{B}_k$ is everywhere equal to $\mathbf{A}_k$, except for its pth and q^{th} rows and columns, which depend on c and s, such that [18, 17]

$$
\begin{aligned}
b_{kpp} &= c^2\, a_{kpp} + |s|^2\, a_{kqq} + c\, s\, a_{kpq} + c\, \bar{s}\, a_{kqp} \\
b_{kqq} &= c^2\, a_{kqq} + |s|^2\, a_{kpp} - c\, s\, a_{kpq} - c\, \bar{s}\, a_{kqp} \\
b_{kpj} &= c\, a_{kpj} + \bar{s}\, a_{kqj} \quad (j \in I_{i(p)}, j \neq p) \\
b_{kjp} &= c\, a_{kjp} + s\, a_{kjq} \quad (j \in I_{i(p)}, j \neq p) \\
b_{kqj} &= -s\, a_{kpj} + c\, a_{kqj} \quad (j \in I_{i(q)}, j \neq q) \\
b_{kjq} &= -\bar{s}\, a_{kjp} + c\, a_{kjq} \quad (j \in I_{i(q)}, j \neq q)
\end{aligned}
$$

Using the fact that the Frobenius norm is invariant to rotations, minimization of criterion $C_{\mathrm{jbd}}(\mathbf{G}(p,q,c,s);\mathcal{A})$ with respect to (c,s) can be shown to be equivalent to the maximization of

$$
C'_{\mathrm{jbd}}(c,s) := \sum_{k=1}^{K} \left\{ |b_{kpp}|^2 + |b_{kqq}|^2 + \sum_{j \in I_{i(p)}, j \neq p} |b_{kpj}|^2 + |b_{kjp}|^2 + \sum_{j \in I_{i(q)}, j \neq q} |b_{kqj}|^2 + |b_{kjq}|^2 \right\}
$$

However, the latter criterion is constant if p and q belong to the same interval $I_i(p)$ (i.e, $i(p) = i(q)$). Details of above derivations can be found in [18, 17].

It may be shown [15, 16] that the maximization of $C'_{\mathrm{jbd}}(c,s)$ boils down to the constrained maximization of a linear quadratic form. This optimization can be achieved using Lagrange multipliers. The computation of the latter requires solving a polynomial of degree 6 in the complex case (i.e, $\mathbf{U} \in \mathbb{C}^{n \times n}$), and of degree 4 in the real case (i.e, $\mathbf{U} \in \mathbb{R}^{n \times n}$). First order approximations of the criterion are also considered in [15, 16] to simplify its maximization. A tensorial rank-1 approximation is also found in [19]. For real matrices, when both $\mathbf{A}_k$ and $\mathbf{U}$ belong to $\mathbb{R}^{n \times n}$, maximization of $C'_{\mathrm{jbd}}(c,s)$ directly amounts to rooting a polynomial of order 4 (without requiring a Lagragian parametrization), as sketched in [19] and developed in [18, 17].

So far, the indices p and q have been fixed. However, the important issue appears not to be how to maximize $C'_{\mathrm{jbd}}(c,s)$, which can be done exactly in a way or another, but how to choose these pivots (p,q). Similarly to JD, the convergence of the proposed (joint) block-diagonalization scheme is guaranteed by construction, whatever the chosen rotation schedule [18, 17]. If convergence to the *global* minimum was in practice usually observed with joint diagonalization schemes, this is certainly not the case for joint block-diagonalization, where we found convergence to be very sensitive to initialization and rotation schedule, as illustrated in the next section.

3 Simulations

The employed algorithms as well as some of the following examples are freely available for download at [20]. The programs have been realized in MATLAB, and sufficient documentation is given to reproduce the results and extend the algorithms. We propose to test the following initialization/schedule strategies.

(M1) The first method is inspired from the standard cyclic Jacobi approach [2, 21], which consists of systematically sweeping the pivots one after the other, except for the fact that the couples (p, q) are chosen not to include the diagonal blocks. The algorithm is initialized with the identity matrix, i.e $\mathbf{U} = \mathbf{I}_n$. The algorithm is stopped when all the values of $s^\star$ are lower than 10^{-4} within a sweep.

(M2) The second method is identical to (M1) except for the fact that the algorithm is initialized with the matrix $\mathbf{U}_{\text{jdr}}$ provided by joint diagonalization of $\mathcal{A}$, as obtained from [2].

(M3) The third method is inspired from the classical Jacobi method for the diagonalization of a normal matrix [21] and consists, after initialization as in (M2), of choosing at each iteration the pivot (p, q) ensuring a maximum decrease of criterion C_{jbd}. This requires computing all the differences $|\sum_{k=1}^{K} \text{boff}(\mathbf{B}_k) - \text{boff}(\mathbf{A}_k)|$ for all couples (p, q) and to pick up the couple which yields the largest difference value. The algorithm stops when 20 successive values of $s^\star$ are all lower than 10^{-4}.

For simplicity, the three methods are tested in the real case. The three methods are applied to 100 random draws of K real matrices *exactly* joint block-diagonalizable in a real common orthogonal basis (optimal rotation angles are thus computed by rooting a polynomial of order 4 like in [18, 17]). Various values of L (size of the blocks), m (number of blocks) and K (number of matrices) are considered. The number of failures over the 100 realizations (i.e, the number of times the methods do not converge to a solution such that $C_{\text{jbd}} = 0$) is reported in Table 1.

Table 1. Number of failures of methods M1, M2 and M3 over 100 random realizations of K matrices exactly block-diagonalizable in a common orthonormal basis

m	2														
L	2					4					6				
K	1	3	6	12	24	1	3	6	12	24	1	3	6	12	24
M1	1	4	4	1	2	32	33	25	10	11	55	33	21	24	16
M2	0	0	0	0	0	11	1	0	0	0	43	2	0	0	0
M3	0	0	0	0	0	5	0	0	0	0	14	0	0	0	0

m	3														
L	2					4					6				
K	1	3	6	12	24	1	3	6	12	24	1	3	6	12	24
M1	3	14	11	18	8	68	54	38	33	32	84	60	48	51	52
M2	0	0	0	0	0	29	5	1	2	0	53	10	8	7	8
M3	0	0	0	0	0	15	1	0	3	1	44	0	0	2	8

m	4														
L	2					4					6				
K	1	3	6	12	24	1	3	6	12	24	1	3	6	12	24
M1	5	30	21	19	16	87	75	68	60	59	99	83	77	77	75
M2	0	0	0	0	0	47	7	6	4	2	88	15	8	4	10
M3	0	0	0	0	0	21	5	4	2	3	65	8	2	0	5

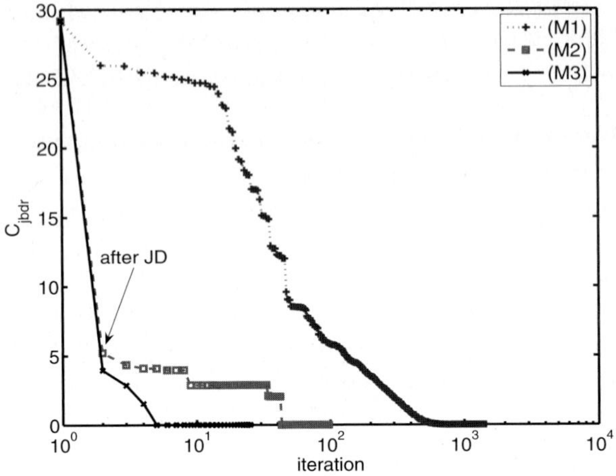

Fig. 1. Evolution of criterion C_{jbd} for a random set $\mathcal{A}$ such that $m = 3$, $L = 4$, $K = 3$. Using a 2.60 GHz Pentium 4 with 1 Go RAM, the computation times for this particular dataset are: (M1 - 1.2 s), (M2 - 0.3 s), (M3 - 1.2 s). The three methods succeed in minimizing the criterion.

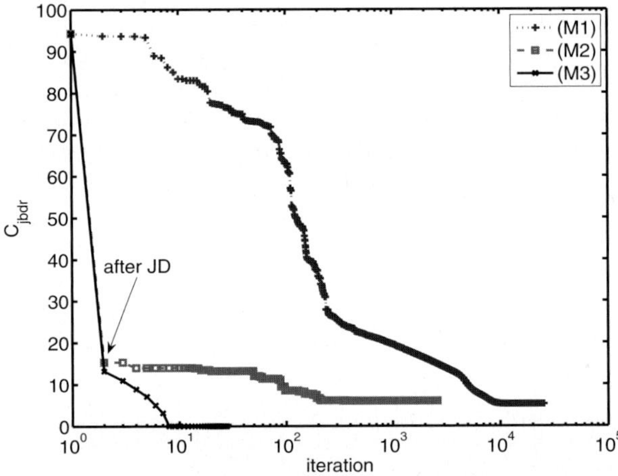

Fig. 2. Evolution of criterion C_{jbd} for a random set $\mathcal{A}$ such that $m = 4$, $L = 6$, $K = 3$. Using a 2.60 GHz Pentium 4 with 1 Go RAM, the computation times for this particular dataset are: (M1 - 28.4 s), (M2 - 4.1 s), (M3 - 6.9 s). Only (M3) succeeds in minimizing the criterion.

The results emphasize the importance of the initialization and the choice of the schedule. Failure rates of (M1) are very high, in particular when m and L increase. (M2) and (M3), which are both initialized by joint diagonalization, give much better results, with (M3) being in nearly every case more reliable than

(M2). However, none of the two methods systematically converge to a global minimum of C_{jbd} when $m \geq 3$, and, interestingly, the methods do not usually fail on the same data sets. Also, Fig. 1 and Fig. 2 show that (M3) only need a few iterations after JD to minimize C_{jbd}. This indicates the validity of the claim from [16], that JD minimizes the joint block-diagonality C_{jbd}, however only up to a permutation. In the above simulation, the permutation is then discovered by application of the JBD algorithm — this also explains why in Figures 1 and 2, when (M2) is used, the cost function after JD only decreases in discrete steps, corresponding to identified permutations.

Audio results of the separation of a convolutive mixture with 3 observations and 2 sources, obtained with the generalization of SOBI using our pivot selection scheme and followed by a SIMO identification step removing filtering ambiguities are found at [22], following the approach described in [23].

4 Conclusions

The main algorithmic conclusion of this paper is: Jacobi algorithms for joint block-diagonalization bring up convergence problems that do not occur in joint diagonalization and that still need to be properly addressed. However we proposed a strategy (method (M3)) which considerably reduces the failure rates of the straightforward approach (M1). The fact that lower failure rates are obtained with (M2) and (M3), which are initialized with joint diagonalization, tend to corroborate the conjecture that JBD diagonalization could be achieved up to an arbitrary permutation of columns via JD [10, 16], but it still does not explain why this permutation cannot be solved by minimization of C_{jbd}. This is a question we are currently working on, and for which partial results exist already [11, 17]. Moreover, extensions to the case of varying, possibly unknown block sizes are interesting [11], with respect to both the optimization and the application in the field of ICA.

References

1. Cardoso, J.F., Souloumiac, A.: Blind beamforming for non Gaussian signals. IEE Proceedings-F 140, 362–370 (1993)
2. Cardoso, J.F., Souloumiac, A.: Jacobi angles for simultaneous diagonalization. SIAM J. Mat. Anal. Appl. 17, 161–164 (1996)
3. Belouchrani, A., Abed-Meraim, K., Cardoso, J.F., Moulines, E.: A blind source separation technique based on second order statistics. IEEE Trans. Signal Processing 45, 434–444 (1997)
4. Ziehe, A., Mueller, K.R.: TDSEP – an efficient algorithm for blind separation using time structure. In: Niklasson, L., Bodén, M., Ziemke, T. (eds.) Proc. of ICANN'98, Skövde, Sweden, pp. 675–680. Springer, Berlin (1998)
5. Theis, F., Gruber, P., Keck, I., Meyer-Bäse, A., Lang, E.: Spatiotemporal blind source separation using double-sided approximate joint diagonalization. In: Proc. EUSIPCO 2005, Antalya, Turkey (2005)

6. Févotte, C., Doncarli, C.: Two contributions to blind source separation using time-frequency distributions. IEEE Signal Processing Letters 11(3) (2004)
7. Theis, F., Inouye, Y.: On the use of joint diagonalization in blind signal processing. In: Proc. ISCAS 2006, Kos, Greece (2006)
8. Bousbiah-Salah, H., Belouchrani, A., Abed-Meraim, K.: Jacobi-like algorithm for blind signal separation of convolutive mixtures. Electronics Letters 37(16), 1049–1050 (2001)
9. De Lathauwer, L., Callaerts, D., De Moor, B., Vandewalle, J.: Fetal electrocardiogram extraction by source subspace separation. In: Proc. IEEE Signal Processing / ATHOS Workshop on Higher-Order Statistics, pp. 134–138. IEEE Computer Society Press, Los Alamitos (1995)
10. Cardoso, J.F.: Multidimensional independent component analysis. In: Proc. ICASSP (1998)
11. Theis, F.: Towards a general independent subspace analysis. In: Proc. NIPS 2006 (2007)
12. Theis, F.: Blind signal separation into groups of dependent signals using joint block diagonalization. In: Proc. ISCAS 2005, Kobe, Japan, pp. 5878–5881 (2005)
13. Févotte, C., Doncarli, C.: A unified presentation of blind source separation methods for convolutive mixtures using block-diagonalization. In: Proc. 4th Symposium on Independent Component Analysis and Blind Source Separation (ICA'03), Nara, Japan (April 2003)
14. Belouchrani, A., Amin, M.G., Abed-Meraim, K.: Direction finding in correlated noise fields based on joint block-diagonalization of spatio-temporal correlation matrices. IEEE Signal Processing Letters 4(9), 266–268 (1997)
15. Belouchrani, A., Abed-Meraim, K., Hua, Y.: Jacobi-like algorithms for joint block diagonalization: Application to source localization. In: Proc. International Symposium on Intelligent Signal Processing and Communication Systems (1998)
16. Abed-Meraim, K., Belouchrani, A.: Algorithms for joint block diagonalization. In: Proc. EUSIPCO'04, Vienna, Austria, pp. 209–212 (2004)
17. Févotte, C., Theis, F.J.: Orthonormal approximate joint block-diagonalization. Technical Report GET/Télécom Paris 2007D007 (2007), `http://service.tsi.enst.fr/cgi-bin/valipub_download.cgi?dId=34`
18. Févotte, C.: Approche temps-fréquence pour la séparation aveugle de sources non-stationnaires. Thèse de doctorat de l'École Centrale de Nantes (2003)
19. De Lathauwer, L., Févotte, C., De Moor, B., Vandewalle, J.: Jacobi algorithm for joint block diagonalization in blind identification. In: Proc. 23th Symposium on Information Theory in the Benelux, Louvain-la-Neuve, Belgium, (Mai 2002), pp. 155–162 (2002)
20. `http://www.biologie.uni-regensburg.de/Biophysik/Theis/researchjbd.html`
21. Golub, G.H., Loan, C.F.V.: Matrix Computations, 3rd edn. The Johns Hopkins University Press, Baltimore (1996)
22. `http://www.tsi.enst.fr/~fevotte/bass_demo.html`
23. Févotte, C., Debiolles, A., Doncarli, C.: Blind separation of FIR convolutive mixtures: application to speech signals. In: Proc. 1st ISCA Workshop on Non-Linear Speech Processing, Le Croisic, France (May 20-23, 2003)

Speeding Up FastICA by Mixture Random Pruning

Sabrina Gaito and Giuliano Grossi

Dipartimento di Scienze dell'Informazione
Università degli Studi di Milano
Via Comelico 39, I-20135 Milano, Italy
grossi@dsi.unimi.it

Abstract. We study and derive a method to speed up kurtosis-based FastICA in presence of information redundancy, i.e., for large samples. It consists in randomly decimating the data set as more as possible while preserving the quality of the reconstructed signals. By performing an analysis of the kurtosis estimator, we find the maximum reduction rate which guarantees a narrow confidence interval of such estimator with high confidence level. Such a rate depends on a parameter β easily computed a priori combining together the fourth and the eighth norms of the observations.

Extensive simulations have been done on different sets of real world signals. They show that actually the sample size reduction is very high, preserves the quality of the decomposition and impressively speeds up FastICA. On the other hand, the simulations also show that, decimating data more than the rate fixed by β, the decomposition ability of FastICA is compromised, thus validating the reliability of the parameter β. We are confident that our method will follow to better approach real time applications.

1 Introduction

Independent Component Analysis (ICA) ([1,2,3,4]) is a method to identify a set of unknown and generally non-Gaussian source signals whose mixtures are observed, under the only assumption that they are mutually independent. ICA has become more and more popular and, thanks to the few assumptions needed and its feasibility, it is applied in many areas such as blind source separation (BSS) which we are interested in [5].

More in general, ICA aim is to describe a very large set of data in terms of variables better capturing the essential structure of the problem. In many cases, due to the huge amount of data, it is crucial to make ICA analysis as fast as possible. From this point of view, one of the most popular algorithm is the well-known FastICA [6], which is based on the optimization of some nonlinear contrast functions [7] characterizing the non-Gaussianity of the components. Because of its widespread uses, in this paper we refer only to the kurtosis-based FastICA [6].

M.E. Davies et al. (Eds.): ICA 2007, LNCS 4666, pp. 185–192, 2007.

Our aim is to speed up FastICA by a suitable pruning of the linear mixtures that preserves the output quality. Essentially, the method proposed consists in randomly select a subset of data of size d' less than the original size d whose sample kurtosis is not too far from the right one. More in details, we perform an analysis of the kurtosis estimator on the sub-sample with the purpose to find the minimum reduction ratio $\rho = \frac{d'}{d}$ which guarantees a narrow confidence interval with high confidence level.

In particular, we identify a data-dependent parameter, called β, which combines both fourth and eighth norms of the observations, from which the reduction rate depends on.

The main step in our method is to compute β on the mixed signals and obtain the actual reduction ratio $\rho = \frac{\beta}{\delta\epsilon^2}$, where ϵ and δ are the fixed confidence interval parameters of the sub-sample kurtosis. Then we randomly decimate the sample and we apply FastICA to the reduced dataset.

To assess the reliability of β many simulations have been done on different sets of both real world and artificial signals. The experiments show that, accordingly to the β, a consistent ratio of reduction can be normally applied when the sample size is considerable, achieving a great benefit in terms of computation time. Furthermore, since β (and consequently ρ) decreases also with respect to the number of signals n, the simulations show that the computation time is weakly affected by n. Moreover, the experiments give also prominence that when forcing the reduction ratio over the bounds derived by our analysis, the reconstruction error of FastICA grows noticeably.

Section 2 describes the pruning methodology. The effect of the data reduction will be analyzed in term of analysis of the kurtosis estimator in Section 3. In the same section the statistical meaning of the parameter β is explained. In Section 4 we apply the method on a large set of real signals extracted from audio signals showing the performance of the proposed method.

2 Random Pruning

The model we assume for ICA is instantaneous and the mixture is linear and noiseless:

$$\mathbf{X} = \mathbf{AS},$$

where the $n \times d$ matrices $\mathbf{X}$ and $\mathbf{S}$ are respectively the observed mixtures and the mutually independent unknown signals, while $\mathbf{A}$ is a full rank $n \times n$ mixing matrix. Thus, n is the number of mixed non-Gaussian signals and d is their length. Therefore, for each $i \in [1..n]$ the i-th row $\boldsymbol{x}_i$ of $\mathbf{X}$ represents a i.i.d. sample of size d of the random variable x_i representing the i-th mixture.

The goal of ICA is to estimate the demixing matrix $\hat{\mathbf{W}} \approx \mathbf{A}^{-1}$ in order to reconstruct the original sources signals

$$\hat{\mathbf{S}} = \hat{\mathbf{W}}\mathbf{X}.$$

Kurtosis-based FastICA is a very simple fixed-point algorithm with satisfactory performance, but it is time consuming for large scale real signals because its computational complexity is $\mathcal{O}(nd^3)$ [6].

In order to spare running time, before running FastICA we operate a random pruning on the mixtures procedure reducing the data by decimating the sample up to the minimum size allowed by β.

Denoting with $\|\boldsymbol{x}_i\|_p$ the usual p-norm, the overall procedure, with the preprocessing pruning preliminary phase, can be summarized in the following steps:

`Pruning preprocessing`

1. $\beta(\boldsymbol{x}_i) = \dfrac{\|\boldsymbol{x}_i\|_8^8}{\|\boldsymbol{x}_i\|_4^8} \quad \forall i \in [1\,..\,n]$
2. $\beta = \max_{\boldsymbol{x}_i} \beta(\boldsymbol{x}_i)$
3. $d' = \dfrac{1}{\delta\varepsilon^2}(d\beta - 1) \approx \dfrac{d\beta}{\delta\varepsilon^2}$
4. random draw $I_{d'} \subseteq [1\,..\,d]$ of size d'
5. $\forall i \in [1\,..\,n]\ \forall j \in I_{d'}\ \ y_{ij} = x_{ij}$ so that $\boldsymbol{y}_i = (y_{ij_1}, \dots, y_{ij_{d'}})$

`FastICA`

1. Perform FastICA on the matrix $\mathbf{Y}$ (whose i-th row is $\boldsymbol{y}_i$) instead of $\mathbf{X}$, obtaining $\hat{\mathbf{W}}$ by maximizing the sequence kurt $\left[\boldsymbol{w}_i^T \mathbf{Y}\right]$, where $\boldsymbol{w}_i^T$ is the i-th row of $\hat{\mathbf{W}}$
2. Reconstruct the signals $\hat{\mathbf{S}} = \hat{\mathbf{W}}\mathbf{X}$.

Note that the decimation process throws away the same set of intermediate data points in all mixtures.

3 Theoretical Motivation

In this section we look for a lower bound for the reduction ratio ρ. The main step in FastICA where the sample size is relevant is when the kurtosis is being estimated on the data set.

Assuming, as usual, that each mixture $\boldsymbol{x}_i$ has zero mean and unitary variance, the kurtosis of each random variable x_i reduces to its fourth moment $\mathsf{M}_4[x_i]$. Thus we analyze the effects coming from the use of a reduced data set in terms of confidence interval of the sample fourth moment.

The fourth moment estimate is generally performed on the whole sample $\boldsymbol{x}_i$ of size d via the sample fourth moment $\hat{\mathsf{M}}_4^d[\boldsymbol{x}_i]$:

$$\hat{\mathsf{M}}_4^d[\boldsymbol{x}_i] = \frac{1}{d}\sum_{t=1}^{d} \boldsymbol{x}_{it}^4,$$

having the following mean and variance:

$$\mathsf{E}\left[\hat{\mathsf{M}}_4^d[\boldsymbol{x}_i]\right] = \mathsf{M}_4[x_i], \qquad \mathsf{var}\left[\hat{\mathsf{M}}_4^d[\boldsymbol{x}_i]\right] = \frac{1}{d}(\mathsf{M}_8[x_i] - (\mathsf{M}_4[x_i])^2).$$

Let us now estimate $\mathsf{M}_4[x_i]$ on the basis of the sub-sample $\boldsymbol{y}_i$.

Using the Chebyschev inequality we obtain the probability bounds:

$$\mathsf{Pr}\left\{\mathsf{M}_4[x_i](1-\varepsilon) \le \hat{\mathsf{M}}_4^{d'_i}[\boldsymbol{y}_i] \le \mathsf{M}_4[x_i](1+\varepsilon)\right\} \ge 1 - \frac{\mathsf{var}\left[\hat{\mathsf{M}}_4^{d'_i}\right]}{\varepsilon^2(\mathsf{M}_4[x_i])^2}$$
$$= 1 - \frac{\mathsf{M}_8[x_i] - (\mathsf{M}_4[x_i])^2}{d'\varepsilon^2(\mathsf{M}_4[x_i])^2}.$$

Setting the previous term equal to the confidence $1-\delta$, fixing the margin of error ε and introducing the sample moments, we derive the minimum sample size d'_i which respects the probability bound above:

$$d'_i = \frac{\hat{\mathsf{M}}_8^d[\boldsymbol{x}_i] - (\hat{\mathsf{M}}_4^d[\boldsymbol{x}_i])^2}{\delta\varepsilon^2(\hat{\mathsf{M}}_4[\boldsymbol{x}_i])^2}.$$

Expressing the sample moments in terms of norms:

$$\hat{\mathsf{M}}_4^{d'}[\boldsymbol{x}_i] = \frac{1}{d'}\|\boldsymbol{x}_i\|_4^4 \quad \text{and} \quad \hat{\mathsf{M}}_8^{d'}[\boldsymbol{x}_i] = \frac{1}{d'}\|\boldsymbol{x}_i\|_8^8,$$

we obtain:

$$d'_i = \frac{1}{\delta\varepsilon^2}\left(\frac{d\|\boldsymbol{x}_i\|_8^8}{\|\boldsymbol{x}_i\|_4^8} - 1\right).$$

It is evident that the minimum allowed sample size depends on the ratio of the two norms $\|\boldsymbol{x}_i\|_8^8$ and $\|\boldsymbol{x}_i\|_4^8$. Their statistical meaning is related to the variance of the estimator of the fourth moments estimated on the whole sample as:

$$\mathsf{var}\left[\hat{M}_4^d[\boldsymbol{x}_i]\right] = \frac{1}{d^2}(\|\boldsymbol{x}_i\|_8^8 - \frac{1}{d}\|\boldsymbol{x}_i\|_4^8)$$

Of course a low variance implies a good estimate and the possibility of highly reduce the sample size d'_i.

Since it holds that:

$$\frac{1}{d} \le \frac{\|\boldsymbol{x}_i\|_8^8}{\|\boldsymbol{x}_i\|_4^8} \le 1,$$

we note that the better ratio for the variance is $\|\boldsymbol{x}_i\|_8^8 = \frac{1}{d}\|\boldsymbol{x}_i\|_4^8$. On the other side, the variance of the estimator is highest when $\|\boldsymbol{x}_i\|_8^8 = \|\boldsymbol{x}_i\|_4^8$.

Introducing the parameter

$$\beta = \max_{\boldsymbol{x}_i} \frac{\|\boldsymbol{x}_i\|_8^8}{\|\boldsymbol{x}_i\|_4^8}$$

the minimum allowed sample size is:

$$d' = \frac{1}{\delta\varepsilon^2}(d\beta - 1) \approx \frac{d\beta}{\delta\varepsilon^2}$$

and the reduction ratio is:

$$\rho = \frac{d'}{d} = \frac{\beta}{\delta\varepsilon^2}.$$

4 Numerical Experiments

In this section we report the summary of extensive computer simulations obtained from the executions of FastICA on different set of sampled source signals: speech, musical and environmental sounds of various nature, mixed with randomly generated matrix. All the experiments have been carried out on Pentium P4 (2GHz, 1GB RAM) through software environment MATLAB 7.0.1.

The main purpose of the simulations is to apply the preprocessing pruning technique in order to appreciate the performance of FastICA both in terms of computation complexity and of quality of the reconstructed signals. Specifically, we are interested in validating the reliability of the parameter β observing the performance decay. This attitude may find application in real time scenarios where high sampling rate can make prohibitive the use of the ICA technique.

All signals considered in the experiments are very big (order of magnitude 10^5 and 10^6) because for short sample size FastICA sometimes fails to converge or gets stuck at saddle points [8].

To measure the accuracy of the demixing matrix we use the performance index reported in [9], which represents a plausible measure of discrepancy between the product matrix $\mathbf{P} = (p_{ij})_{n\times n} = \mathbf{A}\hat{\mathbf{W}}$ and the identity matrix, defined as:

$$\mathrm{Err} = \sum_{i=1}^{n}\left(\sum_{j=1}^{n}\frac{|p_{ij}|}{\max_k |p_{ik}|} - 1\right) + \sum_{j=1}^{n}\left(\sum_{i=1}^{n}\frac{|p_{ij}|}{\max_k |p_{kj}|} - 1\right).$$

Due to the limit of space we present here only the most illustrative example, which examines signals of size $d = 10^6$. Table 1 shows the results on different groups of n signals (with $2 \le n \le 10$).

Table 1. Average performance index and average computation time of FastICA on various groups of signals (from 2 to 10 with $d = 10^6$). Second column reports the reduction ratio $\rho < 1$, third and fourth columns report the performance index both with full and reduced sample size respectively. The last two columns report the computation times in both the cases. The numbers between brackets are the standard deviations calculated on the 30 trials.

n	$\rho < 1$	Err ($\rho = 1$)	Err ($\rho < 1$)	Time ($\rho = 1$)	Time ($\rho < 1$)
2	0.03 (0.01)	0.02 (0.05)	0.03 (0.02)	2.5 (0.9)	0.1 (0.0)
3	0.27 (0.01)	0.04 (0.02)	0.05 (0.02)	4.5 (0.8)	1.3 (0.6)
4	0.25 (0.07)	0.11 (0.11)	0.11 (0.05)	6.7 (0.8)	1.7 (0.6)
5	0.22 (0.07)	0.18 (0.07)	0.33 (0.63)	9.4 (1.3)	2.1 (0.7)
6	0.19 (0.07)	0.37 (0.15)	0.46 (0.14)	12.0 (1.7)	2.4 (0.9)
7	0.16 (0.06)	0.62 (0.70)	0.97 (0.97)	14.7 (1.1)	2.4 (1.0)
8	0.16 (0.06)	1.08 (0.75)	1.44 (1.12)	18.5 (2.1)	2.9 (1.1)
9	0.12 (0.04)	1.23 (1.30)	1.75 (2.70)	26.5 (3.8)	2.8 (0.9)
10	0.11 (0.04)	1.43 (0.29)	1.91 (2.23)	33.5 (3.4)	2.8 (1.0)

For each group we randomly generated 30 mixtures in order to observe, on average, both the time of convergence and the performance index of FastICA for the whole and the reduced samples respectively. All the experiments are obtained at confidence level 0.9 and margin of error 0.1.

Based on the simulations we can draw the following conclusions.

1. Sample size is highly reduced (up to one hundred times) while the quality of the decomposition is preserved, as highlighted by the performance index. Here, in particular, $\beta = \rho * 10^{-3}$ is sufficiently small, lying in the range between 10^{-5} and 10^{-4}.
2. The discrepancy between the error given by the whole sample and that given by the pruned sample increases very slowly with n (number of signals) as shown graphically in Fig. 1 (the lowest two errors corresponding to the third and fourth column of Table 1).
3. To assess the reliability of β, in the same figure we report the data obtained with a reduction ratio of one order of magnitude under that provided by analysis, i.e., with $\rho_{sub} = 10^{-1}\rho$ (highest error in the graphic). This experiment shows that the error grows noticeably.
4. As far as computation time is concerned, Fig. 2 (average times corresponding to the fifth and sixth column of Table 1) highlights the impressive gain of the computational cost. This gain depends on the fact that the computational cost is cubic with respect to sample size. Moreover, it can be noticed that in our pruning FastICA the computation time depends weakly on the number of signals because β decreases with respect to n.

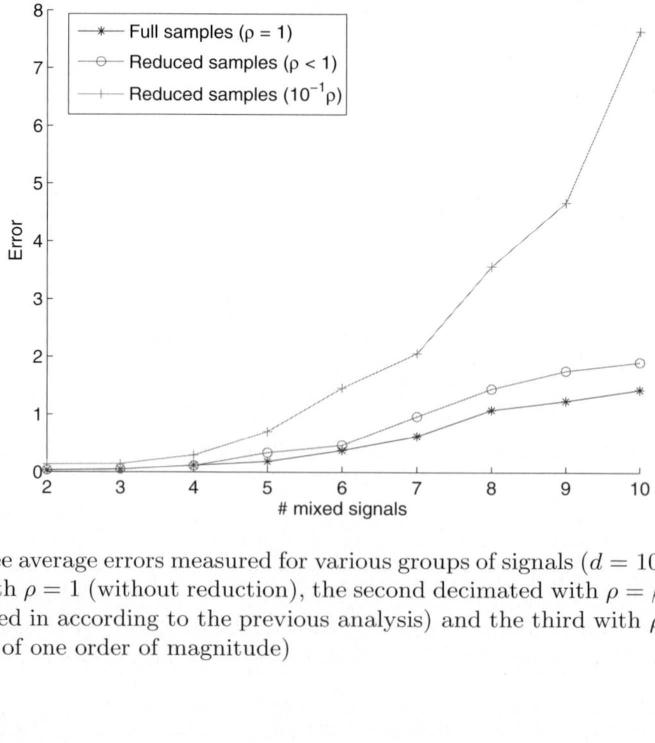

Fig. 1. Three average errors measured for various groups of signals ($d = 10^6$): the first is obtained with $\rho = 1$ (without reduction), the second decimated with $\rho = \beta * 10^3$ (where β is computed in according to the previous analysis) and the third with $\rho_{sub} = \beta * 10^2$ (reducing β of one order of magnitude)

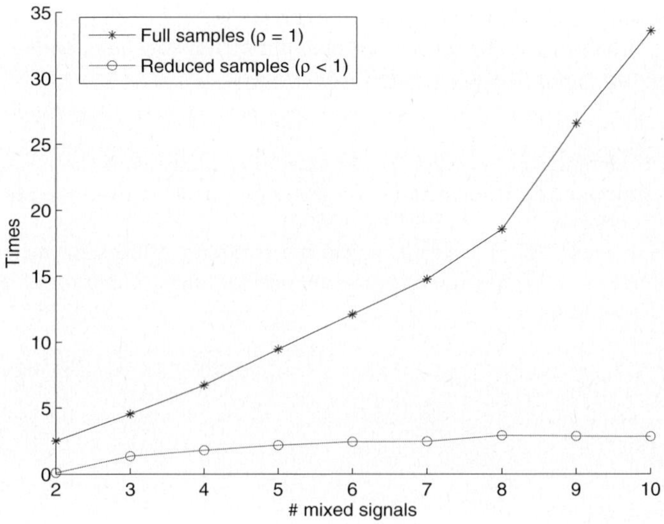

Fig. 2. Average times of FastICA on different groups of signals of full and reduced size: the first is obtained with $\rho = 1$ (without reduction), the second by decimation with $\rho = \beta * 10^3$

5 Conclusions

The contribution of this paper is the derivation of a signal-dependent parameter useful to randomly decimate high-dimensional mixtures in order to reduce the time in kurtosis-based FastICA executions. Such a parameter has been validated both in terms of rigorous high-order moments analysis and by means of computer simulations on real word signals. The results encourage to study the pruning technique deeper by exploring different sub-sampling methodologies and different contrast functions used in ICA. Finally, we are confident that our method can be used in real-time applications dealing with high sampling rate, where the online decimation permits to reasonably reduce the mixture size enabling FastICA to operate tightly.

References

1. Comon, P.: Independent component analysis - a new concept? Signal Processing 36, 287–314 (1994)
2. Jutten, C., Herault, J.: Blind separation of sources, part i: An adaptive algorithm based on neuromimetic architecture. Signal Processing 24, 1–10 (1991)
3. Hyvrinen, A., Karhunen, J., Oja, E.: Independent Component Analysis. John Wiley & Sons, Chichester (2001)
4. Cichocki, A., Amari, S.: Adaptive Blind Signal and Image Processing: Learning Algorithms and Applications. John Wiley & Sons, Chichester (2002)

5. Cardoso, J.: Eigen-structure of the fourth-order cumulant tensor with application to the blind source separation problem (1990)
6. Hyvärinen, A., Oja, E.: A fast fixed-point algorithm for independent component analysis. Neural Computation 9, 1483–1492 (1997)
7. Hyvärinen, A.: Fast and robust fixed-point algorithms for independent component analysis. IEEE Transactions on Neural Networks 10(3), 626–634 (1999)
8. Tichavsky, P., Koldovsky, Z., Oja, E.: Performance analysis of the fastica algorithm and cramr-rao bounds for linear independent component analysis. IEEE Transaction on Signal Processing 54(4), 1189–1202 (2006)
9. Amari, S., Cichocki, A.: Recurrent neural networks for blind separation of sources. In: Proceedings of International Symposium on Nonlinear Theory and Applications. vol. I, pp. 37–42 (1995)

An Algebraic Non Orthogonal Joint Block Diagonalization Algorithm for Blind Separation of Convolutive Mixtures of Sources

Hicham Ghennioui[2,3], El Mostafa Fadaili[1],
Nadège Thirion-Moreau[2], Abdellah Adib[3,4], and Eric Moreau[2]

[1] IBISC, CNRS FRE 2873 40 rue du Pelvoux, F-91020 Evry-Courcouronnes, France
[2] STD, ISITV, av. G. Pompidou, BP56, F-83162 La Valette du Var Cedex, France
ghennioui@gmail.com, {fadaili,thirion,moreau}@univ-tln.fr
[3] GSCM-LRIT, FSR, av. Ibn Battouta, BP1014, Rabat, Maroc
[4] DPG, IS, av. Ibn Battouta, BP703, Rabat, Maroc
adib@israbat.ac.ma

Abstract. This paper deals with the problem of the blind separation of convolutive mixtures of sources. We present a novel method based on a new non orthogonal joint block diagonalization algorithm ($\mathsf{NO}-\mathsf{JBD}$) of a given set of matrices. The main advantages of the proposed method are that it is more general and a preliminary whitening stage is no more compulsorily required. The proposed joint block diagonalization algorithm is based on the algebraic optimization of a least mean squares criterion. Computer simulations are provided in order to illustrate the effectiveness of the proposed approach in three cases: when exact block-diagonal matrices are considered, then when they are progressively perturbed by an additive Gaussian noise and finally when estimated correlation matrices are used. A comparison with a classical orthogonal joint block-diagonalization algorithm is also performed, emphasizing the good performances of the method.

1 Introduction

In the signal processing community, many works have been recently dedicated to the study of the problem of joint decomposition of matrices or tensors because of their numerous applications especially in blind source separation and array processing [1]-[14].

Here, we are interested in the problem of the blind separation of convolutive mixtures of sources. That is why this communication is dedicated to the so-called joint block-diagonalization of matrices problem. In such a decomposition, the wanted matrices are block diagonal ones[1]. Such a problem has been already considered in [1][4][7] but under the constraint that the joint-block diagonalizer is an orthogonal (unitary in the complex case) matrix. Our purpose, here, is to

[1] A block diagonal matrix is a block matrix in which the off-diagonal block terms are zero matrices and the diagonal matrices are square.

M.E. Davies et al. (Eds.): ICA 2007, LNCS 4666, pp. 193–200, 2007.

discard this unitary constraint. To that aim, we show how the (non necessarily orthogonal) joint-block diagonalizer can be algebraically estimated by minimizing a least mean squares criterion, leading to a new non-orthogonal joint block-digonalization algorithm. Some computer simulations are provided in order to illustrate the good behaviour of the proposed algorithm. Then, it is shown how this algorithm finds application in blind source separation where it is applied, here, to a set of observations correlation matrices at different time delays.

The rest of this communication is organized as follows. The problem statement and the proposed joint block-diagonalization algorithm are both introduced in the Section 2. In the Section 3, we show how this algorithm can be applied to solve the problem of blind separation of convolutive mixtures of sources. Computer simulations are provided in both sections to illustrate the effectiveness of the proposed algorithm and to compare it with another one based on an orthogonal joint block-diagonalization.

2 Non-orthogonal Joint Block-Diagonalization Problem

2.1 Problem Statement

The non-orthogonal joint block-diagonalization problem is stated in the following way: let us consider a set $\mathcal{M}$ of N_m, $N_m \in \mathbb{N}^*$ square invertible matrices $\mathbf{M}_i \in \mathbb{R}^{M\times M}$, $i \in \{1,\ldots,N_m\}$ which all admit the following decomposition:

$$\mathbf{M}_i = \mathbf{A}\mathbf{D}_i\mathbf{A}^T \;, \quad \text{or} \quad \mathbf{D}_i = \mathbf{B}\mathbf{M}_i\mathbf{B}^T \;, \quad \forall i \in \{1,\ldots,N_m\} \tag{1}$$

where $\mathbf{D}_i = \begin{pmatrix} \mathbf{D}_{i1} & \ldots & \mathbf{0} \\ & \ddots & \\ \mathbf{0} & \ldots & \mathbf{D}_{ir} \end{pmatrix}$, $\forall i \in \{1,\ldots,N_m\}$, are $N\times N$ block diagonal matrices with $\mathbf{D}_{ij}, i \in \{1,\ldots,N_m\}$, $j \in \{1,\ldots,r\}$ are $n_j \times n_j$ square matrices so that $n_1 + \ldots + n_r = N$ (in our case, we will assume that all the matrices have the same size *i.e* $N = r \times n_j, \forall j \in \{1,\ldots,r\}$) and where $\mathbf{0}$ denotes the $n_j \times n_j$ null matrix. $\mathbf{A}$ is the $M \times N$ $(M \geq N)$ full rank matrix and $\mathbf{B}$ is its pseudo-inverse (or generalized Moore-Penrose inverse).

The non-orthogonal joint block-diagonalization problem consists in estimating the matrix $\mathbf{A}$ and the matrices $\mathbf{D}_{ij}$, $i \in \{1,\ldots,N_m\}, j \in \{1,\ldots,r\}$ (or more simply the matrix $\mathbf{B}$ only) from the matrices set $\mathcal{M}$. The case of an orthogonal matrix $\mathbf{A}$ has been already considered in [7] where a first solution is proposed.

2.2 Joint Block-Diagonalization Algorithm

In this communication, we propose to consider the following cost function

$$\mathcal{C}_{BD}(\mathbf{C}) = \sum_{k=1}^{N_m} \|\mathsf{OffBdiag}\{\mathbf{C}^T\mathbf{M}_k\mathbf{C}\}\|^2, \tag{2}$$

where the operator $\mathsf{OffBdiag}\{\cdot\}$ denotes the zero-block-diagonal matrix and $\mathbf{C} = \mathbf{B}^T$. Thus:

$$\mathbf{M} = \begin{pmatrix} \mathbf{M}_{11} & \mathbf{M}_{12} & \dots & \mathbf{M}_{1r} \\ \mathbf{M}_{21} & \dots & \dots & \vdots \\ \vdots & \dots & \dots & \vdots \\ \mathbf{M}_{r1} & \mathbf{M}_{r2} & \dots & \mathbf{M}_{rr} \end{pmatrix} \Rightarrow \mathsf{OffBdiag}\{\mathbf{M}\} = \begin{pmatrix} \mathbf{0} & \mathbf{M}_{12} & \dots & \mathbf{M}_{1r} \\ \mathbf{M}_{21} & \ddots & & \vdots \\ \vdots & & \ddots & \vdots \\ \mathbf{M}_{r1} & \mathbf{M}_{r2} & \dots & \mathbf{0}, \end{pmatrix}. \quad (3)$$

Let $\mathbf{C} = [\mathbf{C}_1, \cdots, \mathbf{C}_r]$, where $\mathbf{C}_j, j \in \{1, \cdots, r\}$, are r block matrices of dimension $M \times n_j$. The cost function (2) can be rewritten as:

$$\mathcal{C}_{BD}(\mathbf{C}) = \sum_{k=1}^{N_m} \sum_{i,j=1(i \neq j)}^{r} \|\mathbf{C}_i^T \mathbf{M}_k \mathbf{C}_j\|^2 = \sum_{k=1}^{N_m} \sum_{m=1}^{n_i} \sum_{n=1}^{n_j} \sum_{i,j=1(i \neq j)}^{r} |(\mathbf{c}_i^m)^T \mathbf{M}_k \mathbf{c}_j^n|^2 \quad (4)$$

where $\mathbf{c}_j^n$, $\forall n \in \{1, \dots, n_j\}$ stand for the n_j column vectors of matrices $\mathbf{C}_j$, $\forall j \in \{1, \dots, r\}$. Then:

$$\begin{aligned} \mathcal{C}_{BD}(\mathbf{C}) &= \sum_{k=1}^{N_m} \sum_{m,n=1}^{n_i,n_j} \sum_{i,j=1(i \neq j)}^{r} ((\mathbf{c}_i^m)^T \mathbf{M}_k \mathbf{c}_j^n)((\mathbf{c}_i^m)^T \mathbf{M}_k \mathbf{c}_j^n)^T \\ &= \sum_{k=1}^{N_m} \sum_{m,n=1}^{n_i,n_j} \sum_{i,j=1(i \neq j)}^{r} (\mathbf{c}_i^m)^T (\mathbf{M}_k \mathbf{c}_j^n (\mathbf{c}_j^n)^T \mathbf{M}_k^T) \mathbf{c}_i^m \\ &= \sum_{m=1}^{n_i} \sum_{i=1}^{r} (\mathbf{c}_i^m)^T \left[\sum_{j=1(j \neq i)}^{r} \sum_{n=1}^{n_j} \sum_{k=1}^{N_m} \mathbf{M}_k \mathbf{c}_j^n (\mathbf{c}_j^n)^T \mathbf{M}_k^T \right] \mathbf{c}_i^m \\ &= \sum_{m=1}^{n_i} \sum_{i=1}^{r} (\mathbf{c}_i^m)^T \mathbf{Q}_i(\mathbf{C}_{\bar{i}}) \mathbf{c}_i^m \end{aligned} \quad (5)$$

where $\mathbf{Q}_i(\mathbf{C}_{\bar{i}}) = \sum_{j=1(j \neq i)}^{r} \sum_{n=1}^{n_j} \sum_{k=1}^{N_m} \mathbf{M}_k \mathbf{c}_j^n (\mathbf{c}_j^n)^T \mathbf{M}_k^T$ is a symmetric matrix.

As $\mathbf{c}_j^n (\mathbf{c}_j^n)^T$ is rank one, $\forall j = 1, \dots, r$, and $\forall n = 1, \dots, n_j$, the matrix $\mathbf{Q}_i(\mathbf{C}_{\bar{i}})$ possesses $N - (r-1)n_j = n_j$ eigenvectors associated with null eigenvalues. Then, the minimization of this quadratic form under the unit norm constraint can be achieved by taking the n_j unit eigenvectors associated with the n_j smallest eigenvalues of $\mathbf{Q}_i(\mathbf{C}_{\bar{i}})$. However since matrix $\mathbf{Q}_i$ for a given i also depends on column vectors of matrix $\mathbf{C}$, we propose to use an iterative procedure. The proposed non-orthogonal joint block-diagonalization (denoted by $\mathsf{NO-JBD}$) writes:

$\forall i \in \{1, \dots, r\}$ with $l \in \mathbb{N}^*$ and given $\mathbf{C}_{\bar{i}}^{(0)}$ an initial matrix, do (a) and (b)

(a) Calculate $\mathbf{Q}_i(\mathbf{C}_{\bar{i}}^{(l)})$
(b) Find the n_i lowest eigenvalues $\lambda_i^{m\,(l)}, m \in \{1, \dots, n_i\}$ and the associated eigenvectors $\mathbf{c}_i^{m\,(l)}, m \in \{1, \dots, n_i\}$ of matrix $\mathbf{Q}_i(\mathbf{C}_{\bar{i}}^{(l)})$

Stop after a given number of iterations or when $|\lambda_i^{m\,(l)} - \lambda_i^{m\,(l-1)}| \leq \varepsilon$ where ε is a given small positive threshold.

2.3 Computer Simulations

We present simulations to illustrate the effectiveness of the proposed algorithm. We consider a set $\mathbf{D}$ of $N_m = 11$ (resp. 31, 56, 96) matrices, randomly chosen (according to a Gaussian law) of mean 0 and variance 1. Initially these matrices are exactly block-diagonal, then random noise matrices of mean 0 and variance σ_b^2 are added. A signal to noise ratio can be defined as $\mathsf{SNR} = 10\log(\frac{1}{\sigma_b^2})$. To measure the quality of the separation, the following performance index (which is an extension of the one introduced in [10]) is used:

$$I(\mathbf{G}) = \frac{1}{r(r-1)}\left[\sum_{i=1}^{r}\left(\sum_{j=1}^{r}\frac{\|(\mathbf{G})_{i,j}\|^2}{\max_{\ell}\|(\mathbf{G})_{i,\ell}\|^2}-1\right)+\sum_{j=1}^{r}\left(\sum_{i=1}^{r}\frac{\|(\mathbf{G})_{i,j}\|^2}{\max_{\ell}\|(\mathbf{G})_{\ell,j}\|^2}-1\right)\right]$$

where $(\mathbf{G})_{i,j}\forall i,j \in \{1,\ldots,r\}$ is the (i,j)-th (square) block matrix of $\mathbf{G} = \hat{\mathbf{C}}^T\mathbf{A}$. All the displayed results have been averaged over 30 Monte-Carlo trials. On the Fig. 1, the performance index of algorithm $\mathsf{NO-JBD}$ is displayed versus the number of used matrices (left) and versus the SNR (right). These curves illustrate the good behaviour of the algorithm since $I \approx -110$ dB at high SNR.

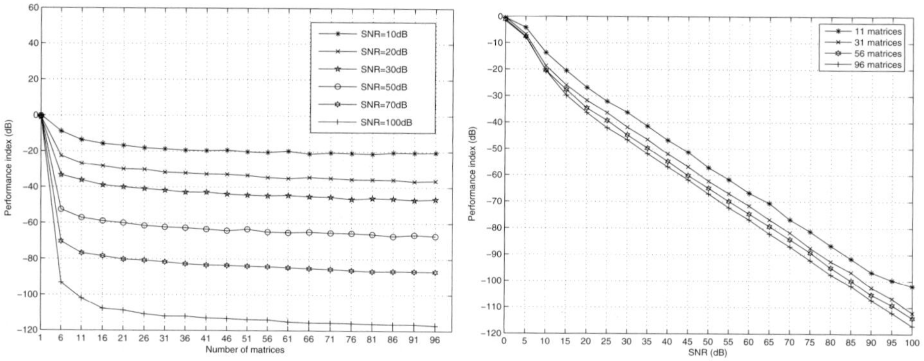

Fig. 1. Left: performance index versus number of matrices, right: performance index versus SNR

3 Separation of Convolutive Mixtures of Sources

3.1 Model and Assumptions

We consider a convolutive finite-duration impulse response (FIR) model given by

$$x_i(t) = \sum_{j=1}^{n}\sum_{\ell=0}^{L} h_{ij}(\ell)s_j(t-\ell) + n_j(t),\ \forall i = 1,\ldots,m \tag{6}$$

where $s_j(t)$, $\forall j = 1,\ldots,n$ are the n sources, $x_i(t)$, $i = 1,\ldots,m$, are the $m > n$ observed signals, $h_{ij}(t)$ is the real transfer function between the j-th source and i-th sensor with an overall extent of $(L+1)$ taps. $n_i(t)$, $\forall i = 1,\ldots,m$ are additive noises. Our developments are based on the two following assumptions:

Assumption A: Each source signal is a real temporally coherent signal. Moreover they are uncorrelated two by two, *i.e.*, for all pairs of sources $(s_i(t),s_j(t))$ with $i \neq j$, for all time delay τ_{ij}, we have $R_{ij}(t,\tau_{ij}) = 0$, where $R_{ij}(t,\tau)$ denotes the cross-correlation function between the sources $s_i(t)$ and $s_j(t)$. It is defined as follows: $R_{ij}(t,\tau) = \mathsf{E}\{s_i(t)s_j(t+\tau)\}$, where $\mathsf{E}\{.\}$ stands for the mathematical expectation.

Assumption B: The noises $n_i(t), i = 1,\ldots,m$, are assumed real stationary white random signals, mutually uncorrelated, independent from the sources, with the same variance σ_n^2. The noises correlation matrix can be written as:

$$\mathbf{R}_n(\tau) = \mathsf{E}\{\mathbf{n}(t)\mathbf{n}^T(t+\tau)\} = \sigma_n^2\delta(\tau)\mathbf{I}_m \tag{7}$$

where $\delta(\tau)$ stands for the Delta impulse, $\mathbf{I}_m$ for the $m \times m$ identity matrix and $(.)^T$ for the transpose operator.

Let us now recall how the convolutive mixing model can be reformulated into an instantaneous one [4][7].

Considering the vectors $\mathbf{S}(t)$, $\mathbf{X}(t)$ and $\mathbf{N}(t)$ respectively defined as:

$$\begin{aligned}
\mathbf{S}(t) &= [s_1(t),\ldots,s_1(t-(L+L')+1),\ldots,s_n(t-(L+L')+1)]^T \\
\mathbf{X}(t) &= [x_1(t),\ldots,x_1(t-L'+1),\ldots,x_m(t-L'+1)]^T \\
\mathbf{N}(t) &= [n_1(t),\ldots,n_1(t-L'+1),\ldots,n_m(t-L'+1)]^T
\end{aligned}$$

and the $(M \times N)$ matrix $\mathbf{A}$, where $M = mL'$ and $N = n(L+L')$:

$$\mathbf{A} = \begin{pmatrix} \mathbf{A}_{11} & \ldots & \mathbf{A}_{1n} \\ \vdots & \ddots & \vdots \\ \mathbf{A}_{m1} & \ldots & \mathbf{A}_{mn} \end{pmatrix}$$

where

$$\mathbf{A}_{ij} = \begin{pmatrix} h_{ij}(0) & \ldots\ldots & h_{ij}(L) & 0 & \ldots & 0 \\ 0 & \ddots\ddots & \ddots & \ddots\ddots & & \vdots \\ \vdots & \ddots\ddots & \ddots & \ddots\ddots & & 0 \\ 0 & \ldots & 0 & h_{ij}(0) & \ldots\ldots & h_{ij}(L) \end{pmatrix} \tag{8}$$

are $(L' \times (L+L'))$ matrices, the model described by Eq. (6) can be written in matrix form as:

$$\mathbf{X}(t) = \mathbf{A}\mathbf{S}(t) + \mathbf{N}(t) \tag{9}$$

In order to have an over-determined model, L' must be chosen such that $mL' \geq n(L+L')$. We assume, here, that the matrix $\mathbf{A}$ is full rank. Because of the Assumption A, all the components of $\mathbf{S}(t)$ are temporally coherent signals. Moreover, two different components of this vector are correlated at least in one non null time delay. With regard to the noise vector $\mathbf{N}(t)$, the Assumption B holds for each of its components involving that its correlation matrix $\mathbf{R_N}(\tau)$ reads:

$$\mathbf{R}_N(\tau) = \mathsf{E}\{\mathbf{N}(t)\mathbf{N}^T(t+\tau)\}$$
$$= \begin{pmatrix} \sigma_n^2\tilde{\mathbf{I}}_{L'}(\tau) & \mathbf{0}_{L'} & \dots & \mathbf{0}_{L'} \\ \mathbf{0}_{L'} & \ddots & \ddots & \vdots \\ \vdots & \ddots & \ddots & \mathbf{0}_{L'} \\ \mathbf{0}_{L'} & \dots & \mathbf{0}_{L'} & \sigma_n^2\tilde{\mathbf{I}}_{L'}(\tau) \end{pmatrix} \tag{10}$$

where $\tilde{\mathbf{I}}_{L'}(\tau)$ is the $L' \times L'$ matrix which contains ones on the τth superdiagonal if $0 \leq \tau < L'$ or on the $|\tau|^{th}$ subdiagonal if $-L' \leq \tau \leq 0$ and zeros elsewhere. Then, we have:

$$\mathbf{R}_X(t,\tau) - \mathbf{R}_N(\tau) = \mathbf{A}\mathbf{R}_S(t,\tau)\mathbf{A}^T = \mathbf{R}_Y(t,\tau) \tag{11}$$

Because sources signals are spatially uncorrelated and temporally coherent, the matrices $\mathbf{R}_S(t,\tau)$, $\forall\tau$ are block diagonal matrices. To recover the mixing matrix $\mathbf{A}$, the matrices $\mathbf{R}_Y(t,\tau)$, $\forall\tau$ and $\forall t$ can be joint block diagonalized without any unitarity constraint about the wanted matrix $\mathbf{A}$.

Notice that in this case, the recovered sources after inversion of the system are obtained up to a permutation and up to a filter but we will not discuss about these indeterminations in this communication.

3.2 Computer Simulations

We present simulations to illustrate the effectiveness of the proposed algorithm in the blind source separation context and to establish a comparison with another algorithm ($\mathsf{O-JBD}$) for the orthogonal joint block diagonalization of matrices. While our algorithm is directly applied on the correlation matrices of the observations, the second algorithm is applied after a pre-whitening stage on the correlation matrices of the pre-whitened observations. We consider $m = 4$ mixtures of $n = 2$ speech source signals sampled at 8 kHz, $L = 2$ and $L' = 4$. These signal sources are mixed according to the following transfer function matrix whose components are randomly generated:

$$\mathbf{A}[z] = \begin{pmatrix} 0.9772 + 0.2079z^{-1} - 0.0439z^{-2} & -0.6179 + 0.7715z^{-1} + 0.1517z^{-2} \\ -0.2517 - 0.3204z^{-1} + 0.9132z^{-2} & -0.1861 + 0.4359z^{-1} - 0.8805z^{-2} \\ 0.0803 - 0.7989z^{-1} - 0.5961z^{-2} & 0.5677 + 0.6769z^{-1} + 0.4685z^{-2} \\ -0.7952 + 0.3522z^{-1} + 0.4936z^{-2} & -0.2459 + 0.8138z^{-1} - 0.5266z^{-2} \end{pmatrix}$$

where $\mathbf{A}[z]$ stands for the z transform of $\mathbf{A}(t)$. On the Fig. 2, we have displayed the performance index versus the number of matrices (left) and versus the SNR. One can check that the obtained performance are better with the $\mathsf{NO-JBD}$ algorithm than with the $\mathsf{O-JBD}$ algorithm. One can also evaluate the block-diagonalization error defined as:

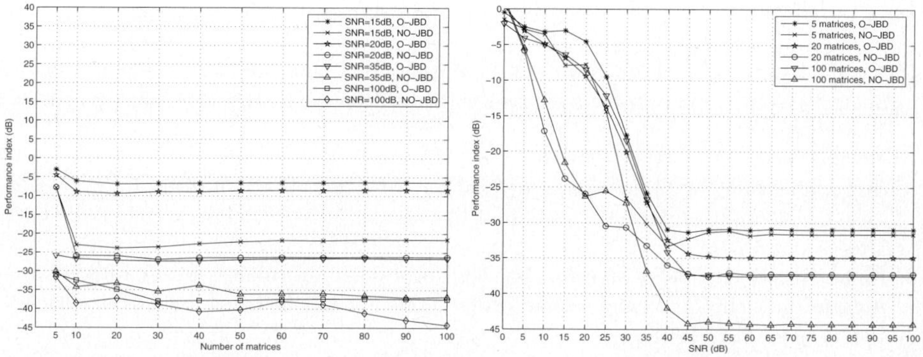

Fig. 2. Left: performance index versus number of matrices, right: performance index versus SNR

$\mathcal{E} = 10\log_{10}\{\frac{1}{N_m}\sum_{k=1}^{N_m} \|\mathsf{OffBdiag}\{\mathbf{B}\mathbf{R}_Y(t,\tau_k)\mathbf{B}^T\|_F^2\}$ where $\mathbf{B}$ is the pseudo-inverse of the mixing matrix $\mathbf{A}$ and $\|.\|_F$ denotes the Frobenius norm. Finally, a comparaison of the block-diagonalization error with the $\mathsf{NO}-\mathsf{JBD}$ and $\mathsf{O}-\mathsf{JBD}$ algorithms versus the number of matrices (resp. SNR) is given in the left of Fig. 3 (resp. its right).

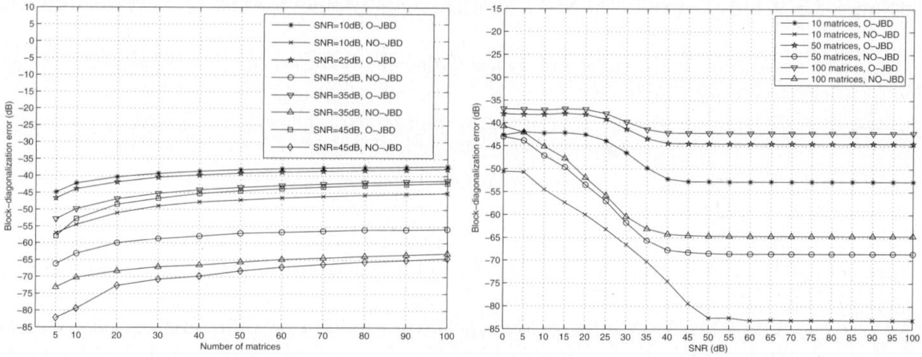

Fig. 3. Left: block-diagonalization error versus number of matrices, right: block-diagonalization error versus SNR

4 Discussion and Conclusion

In this paper, we have proposed a new joint block diagonalization algorithm for the separation of convolutive mixtures of sources that does not rely upon a unitary constraint. We have illustrated the usefulness of the proposed approach thanks to computer simulations: the considered algorithm has been applied to source separation using the correlation matrices of speech sources evaluated over different time delays.

References

1. Abed-Meraïm, K., Belouchrani, A., Leyman, R.: Time-frequency signal analysis and processing: a comprehensive reference. In: Boashash, B. (ed.) chapter in Blind source separation using time-frequency distributions, Prentice-Hall, Oxford (2003)
2. Belouchrani, A., Abed-Meraïm, K., Amin, M., Zoubir, A.: Joint anti-diagonalization for blind source separation. In: Proc. IEEE International Conference on Acoustics, Speech and Signal Processing (ICASSP'2001), Salt Lake City, Utah (May 2001)
3. Belouchrani, A., Abed-Meraïm, K., Cardoso, J.-F., Moulines, E.: A blind source separation technique using second order statistics. IEEE Trans. on Signal Processing 45, 434–444 (1997)
4. Bousbiah-Salah, H., Belouchrani, A., Abed-Meraim, K.: Blind separation of non stationary sources using joint block diagonalization. In: Proc. IEEE Workshop on Statistical Signal Processing, pp. 448–451 (2001)
5. Cardoso, J.-F., Souloumiac, A.: Blind Beamforming for non Gaussian signals. IEEE Proceedings-F 40, 362–370 (1993)
6. Comon, P.: Independant component analysis, a new concept? Signal Processing 36, 287–314 (1994)
7. DeLathauwer, L., Févotte, C., De Moor, B., Vandewalle, J.: Jacobi algorithm for joint block diagonalization in blind identification. In: 23rd Symposium on Information Theory in the Benelux, Louvain-la-Neuve, Belgium (May 2002)
8. Fadaili, E.-M., Thirion-Moreau, N., Moreau, E.: Algorithme de zéro-diagonalisation conjointe pour la séparation de sources déterministes, dans les Proc. du 20ème colloque GRETSI, Louvain-La-Neuve, Belgique, Septembre 2005, pp. 981–984 (2005)
9. Fadaili, E.-M., Thirion-Moreau, N., Moreau, E.: Non orthogonal joint diagonalization/zero-diagonalization for source separation based on time-frequency distributions. To appear in IEEE Trans. on Signal Processing 55(5) (2007)
10. Moreau, E.: A generalization of joint-diagonalization criteria for source separation. IEEE Trans. Signal Processing 49(3), 530–541 (2001)
11. Pham, D.-T.: Joint approximate diagonalization of positive definite matrices. SIAM Journal on Matrix Analysis and Appli 22(4), 1136–1152 (2001)
12. Yeredor, A.: Non-orthogonal joint diagonalization in the least square sense with application in blind source separation. IEEE Transactions on signal processing 50(7), 1545–1553 (2002)
13. Yeredor, A., Ziehe, A., Müller, K.R.: Approximate joint diagonalization using a natural gradient approach. In: Puntonet, C.G., Prieto, A.G. (eds.) ICA 2004. LNCS, vol. 3195, pp. 89–96. Springer, Heidelberg (2004)
14. Ziehe, A., Laskov, P., Nolte, G.G., Müller, K.-R.: A fast algorithm for joint diagonalization with non-orthogonal transformations and its application to blind source separation Journal of Machine Learning Research, No. 5, pp. 801–818 (July 2004)

Non Unitary Joint Block Diagonalization of Complex Matrices Using a Gradient Approach

Hicham Ghennioui[1,2], Nadège Thirion-Moreau[1], Eric Moreau[1], Abdellah Adib[2,3], and Driss Aboutajdine[2]

[1] STD, ISITV, av. G. Pompidou, BP56, F-83162 La Valette du Var Cedex, France
ghennioui@gmail.com, {thirion,moreau}@univ-tln.fr
[2] GSCM-LRIT, FSR, av. Ibn Battouta, BP1014, Rabat, Maroc
[3] DPG, IS, av. Ibn Battouta, BP703, Rabat, Maroc
adib@israbat.ac.ma, aboutaj@ieee.org

Abstract. This paper addresses the problem of the non-unitary approximate joint block diagonalization ($\mathsf{NU}-\mathsf{JBD}$) of matrices. Such a problem occurs in various fields of applications among which blind separation of convolutive mixtures of sources and wide-band signals array processing. We present a new algorithm for the non-unitary joint block-diagonalization of complex matrices based on a gradient-descent algorithm whereby the optimal step size is computed algebraically at each iteration as the rooting of a 3rd-degree polynomial. Computer simulations are provided in order to illustrate the effectiveness of the proposed algorithm.

1 Introduction

In the recent years, the problem of the joint decomposition of matrices or tensors sets have found interesting solutions through signal processing applications in blind source separation and array processing.

One of the first considered problem was the joint-diagonalization of matrices under the unitary constraint, leading to the nowadays well-known JADE [4] and SOBI [2] algorithms. The following works have addressed either the problem of the joint-diagonalization of tensors [5][7][12] or the problem of the joint-diagonalization of matrices but discarding the unitarity constraint [6][10][14] [15][16][17].

A second type of matrices decomposition has proven to be useful in blind source separation, telecommunications and cryptography. It consists in joint zero-diago-nalizing several matrices either under the unitary constraint [1] or not [9][10]. Most of the proposed (unitary) joint-diagonalization and/or zero-diagonalization algorithms have been applied to the problem of the blind separation of instantaneous mixtures of sources.

Finally, a third particular type of matrices decomposition arises in both the wide-band sources localization in correlated noise fields and the blind separation

M.E. Davies et al. (Eds.): ICA 2007, LNCS 4666, pp. 201–208, 2007.

of convolutive mixtures of sources problems. It is called joint block-diagonalization since the wanted matrices are block diagonal matrices[1] in such a decomposition. Such a problem has been considered in [3][8] where the block-diagonal matrices under consideration have to be positive definite and hermitian matrices and the required joint-block diagonalizer is a unitary matrix.

In this paper, our purpose is to discard this unitary constraint. To that aim, we generalize the non unitary joint-diagonalization approach proposed in [16] to the non-unitary joint block-diagonalization of several complex hermitian matrices. The resulting algorithm is based on a gradient-descent approach whereby the optimal step size is computed algebraically at each iteration as the rooting of a 3rd-degree polynomial. The main advantage of the proposed algorithm is that it is relatively general since the only needed assumption about the complex matrices under consideration is their hermitian symmetry. Finally, the use of the optimal step size speeds up the convergence.

The paper is organized as follows. We state the considered problem in the Section 2. In the Section 3, we present the algebraical derivations leading to the proposed non-unitary joint block-diagonalization algorithm. Computer simulations are provided in the Section 4 in order to illustrate the behaviour of the proposed approach.

2 Problem Statement

The non-unitary joint block-diagonalization problem is stated in the following way: let us consider a set $\mathcal{M}$ of N_m, $N_m \in \mathbb{N}^*$ square matrices $\mathbf{M}_i \in \mathbb{C}^{M\times M}$, $i \in \{1,\ldots,N_m\}$ which all admit the following decomposition:

$$\mathbf{M}_i = \mathbf{A}\mathbf{D}_i\mathbf{A}^H \quad \text{or} \quad \mathbf{D}_i = \mathbf{B}\mathbf{M}_i\mathbf{B}^H\ , \quad \forall i \in \{1,\ldots,N_m\} \tag{1}$$

where $\mathbf{D}_i = \begin{pmatrix} \mathbf{D}_{i1} & \ldots & \mathbf{0} \\ & \ddots & \\ \mathbf{0} & \ldots & \mathbf{D}_{ir} \end{pmatrix}$, $\forall i \in \{1,\ldots,N_m\}$, are $N \times N$ block diagonal matrices with $\mathbf{D}_{ij}, i \in \{1,\ldots,N_m\}$, $j \in \{1,\ldots,r\}$ are $n_j \times n_j$ square matrices so that $n_1 + \ldots + n_r = N$ (in our case, we assume that all the matrices have the same size *i.e.* $N = r \times n_j, j \in \{1,\ldots,r\}$) and where $\mathbf{0}$ denotes the $n_j \times n_j$ null matrix. $\mathbf{A}$ is the $M \times N$ ($M \geq N$) full rank matrix and $\mathbf{B}$ is its pseudo-inverse (or generalized Moore-Penrose inverse).

The non-unitary joint bloc-diagonalization problem consists in estimating the matrix $\mathbf{A}$ and the matrices $\mathbf{D}_{ij}$, $i \in \{1,\ldots,N_m\}, j \in \{1,\ldots,r\}$ from only the matrices set $\mathcal{M}$. The case of a unitary matrix $\mathbf{A}$ has been considered in [8] where a first solution is proposed.

---

[1] A block diagonal matrix is a square diagonal matrix in which the diagonal elements are square matrices of any size (possibly even), and the off-diagonal elements are 0. A block diagonal matrix is therefore a block matrix in which the blocks off the diagonal are the zero matrices and the diagonal matrices are square.

3 Non-Unitary Joint Block-Diagonalization Using a Gradient Approach

In this section, we present a new algorithm to solve the problem of the non-unitary joint block-diagonalization. We propose to consider the following cost function

$$\mathcal{C}_{BD}(\mathbf{B}) = \sum_{i=1}^{N_m} \|\mathsf{OffBdiag}\{\mathbf{B}\mathbf{M}_i\mathbf{B}^H\}\|_F^2, \tag{2}$$

where $\|.\|_F$ stands for the Frobenius norm and the operator $\mathsf{OffBdiag}\{\cdot\}$ denotes the zero block-diagonal matrix. Thus:

$$\mathbf{M} = \begin{pmatrix} \mathbf{M}_{11} & \mathbf{M}_{12} \dots & \mathbf{M}_{1r} \\ \mathbf{M}_{21} & \dots \; \dots & \vdots \\ \vdots & \dots \; \dots & \vdots \\ \mathbf{M}_{r1} & \mathbf{M}_{r2} \dots & \mathbf{M}_{rr} \end{pmatrix} \Rightarrow \mathsf{OffBdiag}\{\mathbf{M}\} = \begin{pmatrix} \mathbf{0} & \mathbf{M}_{12} \dots & \mathbf{M}_{1r} \\ \mathbf{M}_{21} & \ddots & \vdots \\ \vdots & \ddots & \vdots \\ \mathbf{M}_{r1} & \mathbf{M}_{r2} \dots & \mathbf{0} \end{pmatrix} \triangleq \mathbf{E} \tag{3}$$

Our aim is to minimize the cost function (2).

To make sure that the found matrix $\mathbf{B}$ keeps on being invertible, it is updated according to the following scheme (see [17]):

$$\mathbf{B}^{(m)} = (\mathbf{I} + \mathbf{W}^{(m-1)})\mathbf{B}^{(m-1)} \quad \forall m = 1, 2, \dots, \tag{4}$$

where $\mathbf{B}^{(0)}$ is some initial guess, $\mathbf{B}^{(m)}$ denotes the estimated matrix $\mathbf{B}$ at the m-th iteration, $\mathbf{W}^{(m-1)}$ is a sufficiently small (in terms of Frobenius norm) zero-block diagonal matrix and $\mathbf{I}$ is the identity matrix.

Denoting $\mathbf{M}_i^{(m)} = \mathbf{B}^{(m-1)}\mathbf{M}_i\mathbf{B}^{(m-1)H}$ $\forall i = 1, \dots, N_m$ and $\forall m = 1, 2, \dots$, where $(\cdot)^H$ stands for the transpose conjugate operator, then at the m-th iteration, the cost function can be expressed versus $\mathbf{W}^{(m-1)}$ rather than $\mathbf{B}^{(m)}$. We now have:

$$\mathcal{C}_{BD}(\mathbf{W}^{(m-1)}) = \sum_{i=1}^{N_m} \|\mathsf{OffBdiag}\{(\mathbf{I} + \mathbf{W}^{(m-1)})\mathbf{M}_i^{(m)}(\mathbf{I} + \mathbf{W}^{(m-1)})^H\}\|_F^2 \tag{5}$$

or more simply $\mathcal{C}_{BD}^{(m)}(\mathbf{W}) \triangleq \sum_{i=1}^{N_m} \|\mathsf{OffBdiag}\{(\mathbf{I} + \mathbf{W})\mathbf{M}_i^{(m)}(\mathbf{I} + \mathbf{W})^H\}\|_F^2$.

At each iteration, the wanted matrix $\mathbf{W}$ is then updated according to the following adaptation rule:

$$\mathbf{W}^{(m)} = -\mu\nabla\mathcal{C}_{BD}(\mathbf{W}^{(m-1)}) \quad \forall m = 1, 2, \dots \tag{6}$$

where μ is the step size or adaptation coefficient and where $\nabla\mathcal{C}_{BD}(\mathbf{W}^{(m-1)})$ stands for the complex gradient matrix defined, like in [13], by:

$$\nabla\mathcal{C}_{BD}(\mathbf{W}^{(m-1)}) = 2\frac{\partial\mathcal{C}_{BD}(\mathbf{W}^{(m-1)})}{\partial\mathbf{W}^{(m-1)*}} \quad \forall m = 1, 2 \dots \tag{7}$$

where $(\cdot)^*$ is the complex conjugate operator. We now have to calculate the complex gradient matrix $\nabla\mathcal{C}_{BD}^{(m)}(\mathbf{W}) = 2\frac{\partial\mathcal{C}_{BD}^{(m)}(\mathbf{W})}{\partial\mathbf{W}^*}$.

3.1 Gradient of the Cost Function $\mathcal{C}_{BD}^{(m)}(\mathbf{W})$

Let $\mathbf{D}_i^{(m)}$ and $\mathbf{E}_i^{(m)}$ respectively denote the block-diagonal and zero block-diagonal matrices extracted from the matrix $\mathbf{M}_i^{(m)}$ ($\mathbf{M}_i^{(m)} = \mathbf{E}_i^{(m)} + \mathbf{D}_i^{(m)}$). As $\mathbf{W}$ is a zero-block diagonal matrix too, the cost function $\mathcal{C}_{BD}^{(m)}(\mathbf{W})$ can be expressed as:

$$
\begin{aligned}
\mathcal{C}_{BD}^{(m)}(\mathbf{W}) &= \sum_{i=1}^{N_m} \|\mathsf{OffBdiag}\{\mathbf{M}_i^{(m)}\} + \mathsf{OffBdiag}\{\mathbf{M}_i^{(m)}\mathbf{W}^H\} + \mathsf{OffBdiag}\{\mathbf{W}\mathbf{M}_i^{(m)}\} \\
&+ \mathsf{OffBdiag}\{\mathbf{W}\mathbf{M}_i^{(m)}\mathbf{W}^H\}\|_F^2 \\
&= \sum_{i=1}^{N_m} \|\mathbf{E}_i^{(m)} + \mathbf{D}_i^{(m)}\mathbf{W}^H + \mathbf{W}\mathbf{D}_i^{(m)} + \mathbf{W}\mathbf{E}_i^{(m)}\mathbf{W}^H\|_F^2 \\
&= \sum_{i=1}^{N_m} \mathsf{tr}\{(\mathbf{E}_i^{(m)} + \mathbf{D}_i^{(m)}\mathbf{W}^H + \mathbf{W}\mathbf{D}_i^{(m)} + \mathbf{W}\mathbf{E}_i^{(m)}\mathbf{W}^H)^H(\mathbf{E}_i^{(m)} + \mathbf{D}_i^{(m)}\mathbf{W}^H \\
&+ \mathbf{W}\mathbf{D}_i^{(m)} + \mathbf{W}\mathbf{E}_i^{(m)}\mathbf{W}^H)\}
\end{aligned} \tag{8}
$$

where $\mathsf{tr}\{.\}$ stands for the trace operator. Then, using the linearity property of the trace and assuming to simplify the derivations that the considered matrices are hermitian, we finally find that:

$$
\begin{aligned}
\mathcal{C}_{BD}^{(m)}(\mathbf{W}) = \sum_{i=1}^{N_m} & \mathsf{tr}\{\mathbf{E}_i^{(m)H}\mathbf{E}_i^{(m)}\} + 2\mathsf{tr}\{\mathbf{E}_i^{(m)H}(\mathbf{D}_i^{(m)}\mathbf{W}^H + \mathbf{W}\mathbf{D}_i^{(m)})\} \\
&+ \mathsf{tr}\{\mathbf{W}\mathbf{D}_i^{(m)H}\mathbf{D}_i^{(m)}\mathbf{W}^H + \mathbf{D}_i^{(m)H}\mathbf{W}^H\mathbf{W}\mathbf{D}_i^{(m)}\} \\
&+ 2\mathsf{tr}\{\mathbf{E}_i^{(m)H}\mathbf{W}\mathbf{E}_i^{(m)}\mathbf{W}^H\} \\
&+ \mathsf{tr}\{\mathbf{W}\mathbf{D}_i^{(m)H}\mathbf{W}\mathbf{D}_i^{(m)} + \mathbf{D}_i^{(m)H}\mathbf{W}^H\mathbf{D}_i^{(m)}\mathbf{W}^H\} \\
&+ 2\mathsf{tr}\{\mathbf{W}\mathbf{E}_i^{(m)H}\mathbf{W}^H(\mathbf{D}_i^{(m)}\mathbf{W}^H + \mathbf{W}\mathbf{D}_i^{(m)})\} \\
&+ \mathsf{tr}\{\mathbf{W}\mathbf{E}_i^{(m)H}\mathbf{W}^H\mathbf{W}\mathbf{E}_i^{(m)}\mathbf{W}^H\}
\end{aligned} \tag{9}
$$

Using now the following properties [11]

$$\mathsf{tr}\{\mathbf{PQR}\} = \mathsf{tr}\{\mathbf{RPQ}\} = \mathsf{tr}\{\mathbf{QRP}\} \tag{10}$$

$$\frac{\partial \mathsf{tr}\{\mathbf{P}\mathbf{X}^H\}}{\partial \mathbf{X}^*} = \mathbf{P} \tag{11}$$

$$\frac{\partial \mathsf{tr}\{\mathbf{PX}\}}{\partial \mathbf{X}^*} = \mathbf{0} \tag{12}$$

$$d\mathsf{tr}\{\mathbf{P}\} = \mathsf{tr}\{d\mathbf{P}\} \tag{13}$$

$$d\mathsf{tr}\{\mathbf{P}\mathbf{X}^H\mathbf{QX}\} = \mathsf{tr}\{\mathbf{P}d\mathbf{X}^H\mathbf{QX} + \mathbf{P}\mathbf{X}^H\mathbf{Q}d\mathbf{X}\} \tag{14}$$

$$\frac{\partial \mathsf{tr}\{\mathbf{P}\mathbf{X}^H\mathbf{QX}\}}{\partial \mathbf{X}^*} = \mathbf{QXP} \tag{15}$$

It finally leads to the following result:

$$\nabla\mathcal{C}_{BD}^{(m)}(\mathbf{W}) = 4\sum_{i=1}^{N_m}\Big(\mathbf{E}_i^{(m)H}\mathbf{D}_i^{(m)} + \mathbf{W}\mathbf{D}_i^{(m)H}\mathbf{D}_i^{(m)} + \mathbf{E}_i^{(m)H}\mathbf{W}\mathbf{E}_i^{(m)}$$
$$+ \mathbf{W}\mathbf{E}_i^{(m)H}\mathbf{W}^H\mathbf{D}_i^{(m)} + \mathbf{W}\mathbf{E}_i^{(m)}\mathbf{W}^H\mathbf{W}\mathbf{E}_i^{(m)H} + \mathbf{D}_i^{(m)}\mathbf{W}^H\mathbf{D}_i^{(m)H}$$
$$+ \mathbf{D}_i^{(m)}\mathbf{W}^H\mathbf{W}\mathbf{E}_i^{(m)H} + \mathbf{W}\mathbf{D}_i^{(m)}\mathbf{W}\mathbf{E}_i^{(m)H}\Big). \quad (16)$$

3.2 Seek of the Optimal Step Size

The expression (16) is then used in the gradient descent algorithm (6). To accelerate its convergence, the optimal step size μ is computed algebraically at each iteration. To that aim, one has to calculate $\mathcal{C}_{BD}^{(m)}(\mathbf{W} \leftarrow -\mu\nabla\mathcal{C}_{BD}^{(m)}(\mathbf{W}))$, but here we use $\mathcal{C}_{BD}^{(m)}(\mathbf{W} \leftarrow \mu\mathbf{F}^{(m)} = -\mu\mathsf{OffBdiag}\{\nabla\mathcal{C}_{BD}^{(m)}(\mathbf{W})\})$. $\mathbf{F}^{(m)}$ is the anti-gradient matrix. We use $\mathsf{OffBdiag}\{\nabla\mathcal{C}_{BD}^{(m)}(\mathbf{W})\}$ instead of $\nabla\mathcal{C}_{BD}^{(m)}(\mathbf{W})$ because $\mathbf{W}$ is a sufficiently small (in terms of norm) zero block-diagonal matrix and thus only the off block-diagonal terms are involved in the descent of the criterion. We now have to seek for the optimal step μ ensuring the minimization of the cost function $\mathcal{C}_{BD}^{(m)}(\mu\mathbf{F}^{(m)})$. This step is determined by the rooting of the 3rd-degree polynomial (18) which is obtained as the derivative of the 4rd-degree polynomial $\mathcal{C}_{BD}^{(m)}(\mu\mathbf{F}^{(m)})$ with respect to μ:

$$\mathcal{C}_{BD}^{(m)}(\mu\mathbf{F}^{(m)}) = a_0^{(m)} + a_1^{(m)}\mu + a_2^{(m)}\mu^2 + a_3^{(m)}\mu^3 + a_4^{(m)}\mu^4, \quad (17)$$

$$\frac{\partial\mathcal{C}_{BD}^{(m)}(\mu\mathbf{F}^{(m)})}{\partial\mu} = 4a_4^{(m)}\mu^3 + 3a_3^{(m)}\mu^2 + 2a_2^{(m)}\mu + a_1^{(m)}, \quad (18)$$

where the coefficients have been found to be equal to:

$$a_0^{(m)} = \sum_{i=1}^{N_m}\mathsf{tr}\{\mathbf{E}_i^{(m)H}\mathbf{E}_i^{(m)}\} \quad (19)$$

$$a_1^{(m)} = \sum_{i=1}^{N_m}\mathsf{tr}\{\mathbf{E}_i^{(m)H}(\mathbf{D}_i^{(m)}\mathbf{F}^H + \mathbf{F}\mathbf{D}_i^{(m)}) + (\mathbf{D}_i^{(m)}\mathbf{F}^H + \mathbf{F}\mathbf{D}_i^{(m)})^H\mathbf{E}_i^{(m)}\} \quad (20)$$

$$a_2^{(m)} = \sum_{i=1}^{N_m}\mathsf{tr}\Big\{\mathbf{E}_i^{(m)H}\mathbf{F}\mathbf{E}_i^{(m)}\mathbf{F}^H + \mathbf{F}\mathbf{E}_i^{(m)H}\mathbf{F}^H\mathbf{E}_i^{(m)}$$
$$+ (\mathbf{D}_i^{(m)}\mathbf{F}^H + \mathbf{F}\mathbf{D}_i^{(m)})^H(\mathbf{D}_i^{(m)}\mathbf{F}^H + \mathbf{F}\mathbf{D}_i^{(m)})\Big\} \quad (21)$$

$$a_3^{(m)} = \sum_{i=1}^{N_m}\mathsf{tr}\Big\{(\mathbf{D}_i^{(m)}\mathbf{F}^H + \mathbf{F}\mathbf{D}_i^{(m)})^H\mathbf{F}\mathbf{E}_i^{(m)}\mathbf{F}^H$$
$$+ \mathbf{F}\mathbf{E}_i^{(m)H}\mathbf{F}^H(\mathbf{D}_i^{(m)}\mathbf{F}^H + \mathbf{F}\mathbf{D}_i^{(m)})\Big\} \quad (22)$$

$$a_4^{(m)} = \sum_{i=1}^{N_m}\mathsf{tr}\{\mathbf{F}^{(m)}\mathbf{E}_i^{(m)H}\mathbf{F}^{(m)H}\mathbf{F}^{(m)}\mathbf{E}_i^{(m)}\mathbf{F}^{(m)H}\}. \quad (23)$$

The optimal step μ corresponds to the root of the polynomial (18) attaining the absolute minimum in the polynomial (17).

3.3 Summary of the Proposed Algorithm

The proposed non-unitary joint block-diagonalization based on a gradient algorithm denoted by $\mathsf{JBD}_{\mathsf{NU,G}}$ is now presented below:

Denote the N_m square matrices as $\mathbf{M}_1^{(0)}, \mathbf{M}_2^{(0)}, \ldots, \mathbf{M}_{N_m}^{(0)}$
Given initial estimates $\mathbf{W}^{(0)} = \mathbf{0}$ and $\mathbf{B}^{(0)} = \mathbf{I}$
For $m = 1, 2, \ldots$
  For $i = 1, \ldots, N_m$
    Compute $\mathbf{M}_i^{(m)}$ as

$$\mathbf{M}_i^{(m)} = \mathbf{B}^{(m-1)}\mathbf{M}_i^{(m-1)}\mathbf{B}^{(m-1)H}$$

    Compute $\nabla\mathcal{C}_{BD}^{(m)}(\mathbf{W})$ whose expression is given by equation (16)
  EndFor
Set $\mathbf{F}^{(m)} = -\mathsf{OffBdiag}\{\nabla\mathcal{C}_{BD}^{(m)}(\mathbf{W})\}$
Compute the coefficients $a_0^{(m)}, \ldots, a_4^{(m)}$ thanks to (19), (20), (21), (22) and (23)
Set the optimal step μ by the research of the root of the polynomial (18) attaining the absolute minimum in the polynomial (17)
Set $\mathbf{W}^{(m)} = \mu\mathbf{F}^{(m)}$ and $\mathbf{B}^{(m)} = (\mathbf{I} + \mathbf{W}^{(m-1)})\mathbf{B}^{(m-1)}$
EndFor

4 Computer Simulations

In this section, we perform simulations to illustrate the behaviour of the proposed algorithm. We consider a set $\mathbf{D}$ of $N_m = 11$ (resp. 31, 101) matrices, randomly chosen (according to a Gaussian law of mean 0 and variance 1). Initially these matrices are exactly block-diagonal, then matrices with random entries chosen from a Gaussian law of mean 0 and variance σ_b^2 are added. The signal to noise ratio (SNR) is then defined by $\mathsf{SNR} = 10\log(\frac{1}{\sigma_b^2})$. We use the following performance index which is an extension of that introduced in [12]:

$$I(\mathbf{G}) = \frac{1}{r(r-1)}\left[\sum_{i=1}^{r}\left(\sum_{j=1}^{r}\frac{\|(\mathbf{G})_{i,j}\|^2}{\max_{\ell}\|(\mathbf{G})_{i,\ell}\|^2} - 1\right) + \sum_{j=1}^{r}\left(\sum_{i=1}^{r}\frac{\|(\mathbf{G})_{i,j}\|^2}{\max_{\ell}\|(\mathbf{G})_{\ell,j}\|^2} - 1\right)\right]$$

where $(\mathbf{G})_{i,j}\forall i, j \in \{1, \ldots, r\}$ is the (i, j)-th matrix block (square) of $\mathbf{G} = \hat{\mathbf{B}}\mathbf{A}$. The displayed results are averaged over 30 Monte-Carlo trials. In this example, they were obtained considering $M = N = 12$, $r = 3$ and real and symmetric matrices. On the left of Fig. 1 we display the performance index obtained with the proposed algorithm versus the number of used matrices for different values of the SNR. On its right we have plotted the evolution of the performance index versus the SNR.

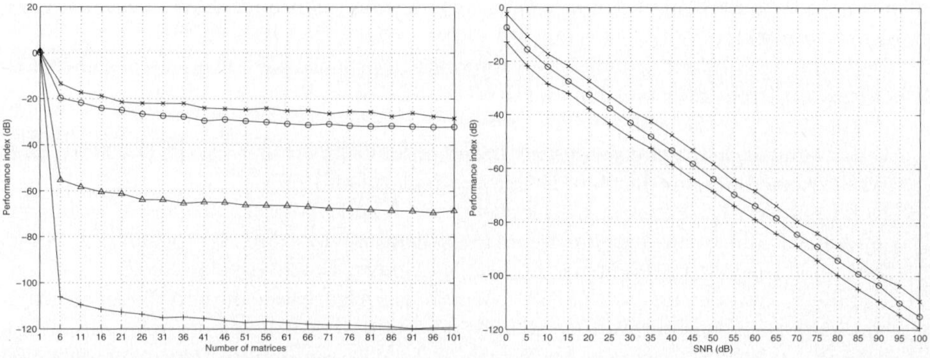

Fig. 1. Left: performance index versus number N_m of used matrices for different values of the SNR (SNR=10 dB ($\times$), 20 dB ($\circ$), 50 dB ($\triangle$) and 100 dB ($+$)). Right: performance index versus SNR for different size of the matrices set to be joint block-diagonalized (N_m=11 ($\times$), 31 ($\circ$), 101 ($+$)).

5 Discussion and Conclusion

In this paper, we have proposed a new algorithm (named $\mathsf{JBD}_{\mathsf{NU,G}}$) based on a gradient approach to perform the non-unitary joint block-diagonalization of a given set of complex matrices. One of the main advantages of this algorithm is that it applies to complex hermitian matrices. This algorithm finds application in blind separation of convolutive mixtures of sources and in array processing. In the context of blind sources separation, it should enable to achieve better performances by discarding the unitary constraint. In fact, starting with a pre-whitening stage is a possible way to amount to a unitary square mixture of sources to be able to use unitary joint-decomposition algorithms. But such a pre-whitening stage imposes a limit on the attainable performances that can be overcome thanks to non-unitary algorithms.

References

1. Belouchrani, A., Abed-Meraïm, K., Amin, M., Zoubir, A.: Joint anti-diagonalization for blind source separation. In: Proc. IEEE International Conference on Acoustics, Speech and Signal Processing (ICASSP'2001), Salt Lake City, Utah (May 2001)
2. Belouchrani, A., Abed-Meraïm, K., Cardoso, J.-F., Moulines, E.: A blind source separation technique using second order statistics. IEEE Transactions on Signal Processing 45, 434–444 (1997)
3. Bousbiah-Salah, H., Belouchrani, A., Abed-Meraïm, K.: Blind separation of non stationary sources using joint block diagonalization. In: Proc. IEEE Workshop on Statistical Signal Processing, pp. 448–451 (2001)
4. Cardoso, J.-F., Souloumiac, A.: Blind beamforming for non Gaussian signals. IEEE Proceedings-F 40, 362–370 (1993)

5. Comon, P.: Independant component analysis, a new concept? Signal Processing 36, 287–314 (1994)
6. Dégerine, S.: Sur la diagonalisation conjointe approchée par un critère des moindres carrés. In: Proc. 18ème Colloque GRETSI, Toulouse, Septembre 2001, pp. 311–314 (2001)
7. DeLathauwer, L.: Signal processing based on multilinear algebra. PhD Thesis, Université Catholique de Leuven, Belgique (September 1997)
8. DeLathauwer, L., Févotte, C., De Moor, B., Vandewalle, J.: Jacobi algorithm for joint block diagonalization in blind identification. In: 23rd Symposium on Information Theory in the Benelux, Louvain-la-Neuve, Belgium (May 2002)
9. Fadaili, E.-M., Thirion-Moreau, N., Moreau, E.: Algorithme de zéro-diagonalisation conjointe pour la séparation de sources déterministes. In: dans les Proc. du 20ème colloque GRETSI, Louvain-La-Neuve, Belgique, Septembre 2005, pp. 981–984 (2005)
10. Fadaili, E.-M., Thirion-Moreau, N., Moreau, E.: Non orthogonal joint diagonalization/zero-diagonalization for source separation based on time-frequency distributions. To appear in IEEE Transactions on Signal Processing, 55(4) (April 2007)
11. Joho, M.: A systematic approach to adaptive algorithms for multichannel system identification, inverse modeling and blind identification. PHD Thesis, Swiss Federal Institute of Technology, Zürich (December 2000)
12. Moreau, E.: A generalization of joint-diagonalization criteria for source separation. IEEE Trans. Signal Processing 49(3), 530–541 (2001)
13. Petersen, K.B., Pedersen, M.S.: The matrix cookbook (January 5, 2005)
14. Pham, D.-T.: Joint approximate diagonalization of positive definite matrices. SIAM Journal on Matrix Analysis and Applications 22(4), 1136–1152 (2001)
15. Yeredor, A.: Non-orthogonal joint diagonalization in the least square sense with application in blind source separation. IEEE Transactions on Signal Processing 50(7), 1545–1553 (2002)
16. Yeredor, A., Ziehe, A., Muller, K.R.: Approximate joint diagonalization using a natural gradient approach. In: Puntonet, C.G., Prieto, A.G. (eds.) ICA 2004. LNCS, vol. 3195, pp. 89–96. Springer, Heidelberg (2004)
17. Ziehe, A., Laskov, P., Nolte, G.G., Müller, K.-R.: A fast algorithm for joint diagonalization with non-orthogonal transformations and its application to blind source separation. Journal of Machine Learning Research, No. 5, 801–818 (July 2004)

A Toolbox for Model-Free Analysis of fMRI Data

P. Gruber[1], C. Kohler[1], and F.J. Theis[2]

[1] Institute of Biophysics, University of Regensburg, D-93040 Regensburg
[2] Max Planck Institute for Dynamics and Self-Organization, D-37073 Göttingen

Abstract. We introduce Model-free Toolbox (MFBOX), a Matlab toolbox for analyzing multivariate data sets in an explorative fashion. Its main focus lies on the analysis of functional Nuclear Magnetic Resonance Imaging (fMRI) data sets with various model-free or data-driven techniques. In this context, it can also be used as plugin for SPM5, a popular tool in regression-based fMRI analysis. The toolbox includes BSS algorithms based on various source models including ICA, spatiotemporal ICA, autodecorrelation and NMF. They can all be easily combined with higher-level analysis methods such as reliability analysis using projective clustering of the components, sliding time window analysis or hierarchical decomposition. As an example, we use MFBOX for the analysis of an fMRI experiment and present short comparisons with the SPM results. The MFBOX is freely available for download at `http://mfbox.sf.net`.

1 Introduction

With the increasing complexity and dimensionality of large-scale biomedical data sets, classical analysis techniques yield way to more powerful methods that can take into account higher-order relationships in the data. Such often explorative methods have been popular in the field of engineering and statistics for quite some time, however they perpetrate into the application areas such as psychology, biology or medicine in a much slower fashion. This is partially due to the fact that visual and simple tools for more complex analyses are rarely available. In this contribution, we present a toolbox, MFBOX, for performing blind source separation of complex tasks with appropriate pre- and postprocessing methods. One of our main goals is to provide a simple toolbox that allows for the model-free analysis of fMRI data sets [14], although MFBOX may just as well be applied to other recordings. The graphical user interface of MFBOX enables users to easily try out various model-free algorithms, together with additional pre- and postprocessing and reliability analysis. The design of the toolbox is modular, so it can be easily extended to include your algorithm of choice. It can integrate into SPM5 [1] and can be used to perform model-free analysis of biomedical image time series such as fMRI or PET data sets. The toolbox is realized in MATLAB, and has been tested on Linux, Windows and Mac OS. The paper itself is organized as follows: In the next section, we present the core features of MFBOX,

M.E. Davies et al. (Eds.): ICA 2007, LNCS 4666, pp. 209–217, 2007.

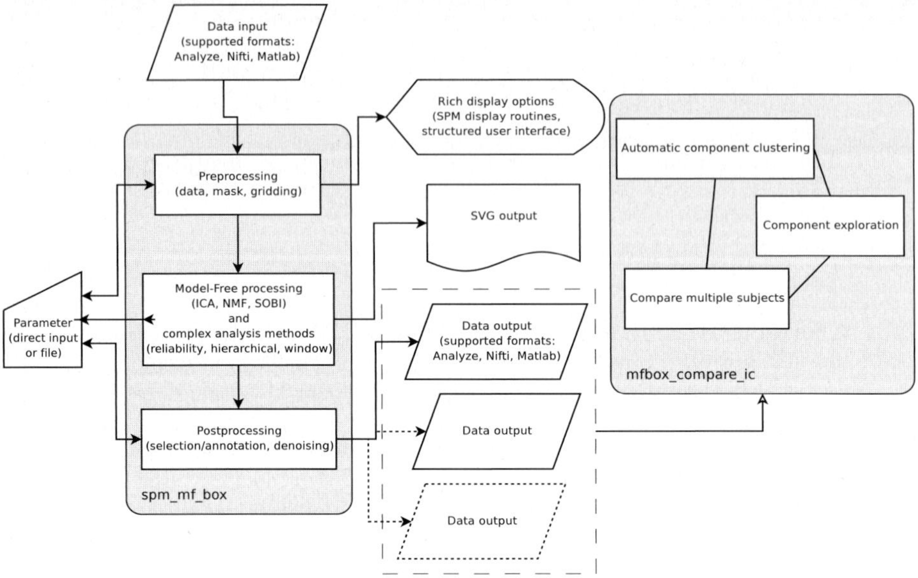

Fig. 1. The data flow of the MFBOX application. The user interface is divided into the `spm_mf_box` part and the `mfbox_compare_ic` part.

from preprocessing to Blind Source Separation (BSS) and higher-order analysis itself to postprocessing. We then illustrate the usage of MFBOX on a complex fMRI experiment, and conclude with an outlook.

2 Features of the MFBOX

The MFBOX application includes two main graphical interfaces, `spm_mf_box` which gives access to all processing possibilities and `mfbox_compare_ic` which allows for easy comparison of different results. Any of the algorithms can also be used separately from the main toolbox interface and additionally a batch run command `mfbox_runbatch` is provided to ease the analysis of multiple data sets. The workflow can be divided into modular stages, also see figure 1:

- Preprocessing
- Model-free processing and higher-level analysis
- Postprocessing

2.1 Preprocessing

The preprocessing stage can include one or more preprocessing steps in a row where the precedence can be easily controlled. The purpose of this stage is to select regions of interest to apply the model-free algorithms on or enhance the effectiveness of the main processing algorithms by denoising or transforming the original sequence data.

datavalue mask selection by bounds on voxel values
denoise high quality Local ICA (lICA) based 3d denoising [4]
infomap gridding using a Self-Organizing Maps (SOM) [12] based on the information content of the voxels
remmean different mean removal options
roiselect mask selection by loadable masks
selectslices rectangular mask selection in voxel space
varthreshold mask selection by bounds on voxel variances

Recommended standard options are to mask out the parts of the signal which are outside the brain and regions uninteresting for the Blood Oxygenation Level Dependent Contrast (BOLD) effect like blood support system. Moreover the analysis can be enhanced by only using the white matter voxels. How this mask selection is achieved depends on the available data. If structural data is available the preprocessing option roiselect can use the data from a segmentation step. Otherwise the mask selection can be accomplished by datavalue or varthreshold selection.

2.2 Model-Free Analysis and Higher-Level Analysis

Given a multivariate data set $\mathbf{x}(t)$, our goal is to find a new basis $\mathbf{W}$ such that $\mathbf{W}\mathbf{x}(t)$ fulfills some statistical properties. If we for example require the transformed random vector to be statistically independent, this is denoted as Independent Component Analysis [8]. Independent Component Analysis (ICA) can be applied to solve the BSS problem, where $\mathbf{x}(t)$ is known to be the mixture of some underlying hidden independent sources. Depending on the data set also other models are of interest, as for e.g. depending on the data set. The MFBOX currently includes three different paradigms for tackling the BSS problem and for each of these fundamental types it contains different algorithms to perform the separation.

1. ICA algorithms
   PearsonICA spatial ICA algorithm which employs a fast fixed point algorithm as extension of FastICA [8] where the nonlinearity is estimated from the data and automatically adjusted at each step [9]
   JADE spatial ICA algorithm which is based on the approximate diagonalization of the fourth cumulant tensor [3]
   stJADE spatiotemporal version of the JADE algorithm [15], for an impression on how the spatiotemporal weighting can enhance the separation of a real data set see figure 2
   TemporalICA temporal ICA [2] optionally using Projection Approximation Subspace Tracking (PAST) [17] for the temporal Principal Component Analysis (PCA) reduction
   hrfICA semi-blind spatial ICA algorithm using the Haemoglobin Response Function (HRF) function to incorporate prior knowledge about the assumed sources

2. second order algorithms
   mdSOBI Multidimensional version of the Second Order Blind Identification (SOBI) [16] algorithm, which is based on second-order autodecorrelation
   stSOBI Spatio-temporal version of the SOBI algorithm [15]
3. other decomposition algorithms
   hNMF a Nonnegative Matrix Factorization (NMF) decomposition algorithm using hyperplane clustering [5]

These base BSS algorithms can be combined with different types of higher-level analysis methods. These three methods share the fact that they apply a previously selected BSS algorithm multiple times in order to extract additional statistics e.g. for reliability. The methods will be explained in the following.

Reliability analysis with projective k-means clustering. A common problem in explorative data analysis is how to access the reliability of the obtained results. A proven method of reliability analysis in statistics is bootstrapping i.e. to randomly subsample the same methods as before and compare the results from multiple different random sub sample runs, see e.g. [6, 7]. For most BSS algorithms this leads to a permutation problem (namely identifying corresponding Independent Component (IC)) since there is an inherent permutation and scaling, especially sign, invariance of BSS. Here we use projective k-means clustering [5] for the assignment and to evaluate the quality of a component using the resulting cluster size. It is essentially a k-means-type algorithm, which acts

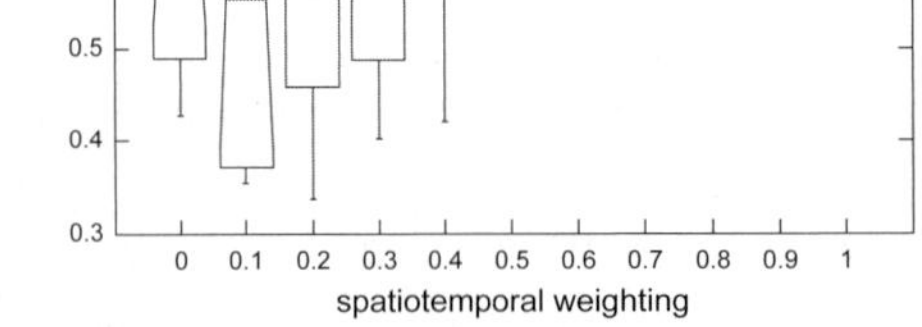

(a) Comparison of the maximal temporal reliability measure with varying weighting α

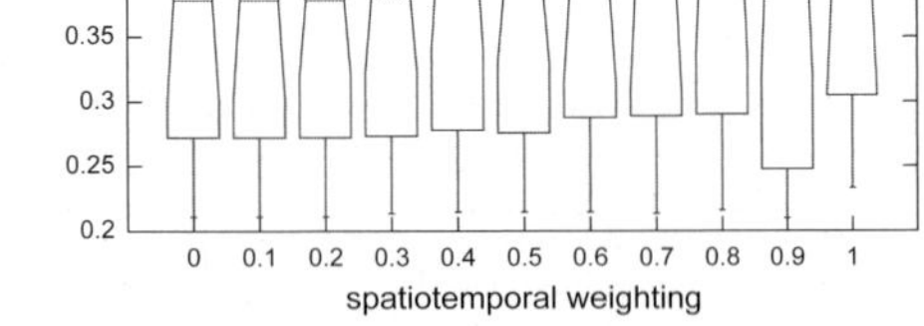

(b) Comparison of the maximal correlation with the external stimulus

Fig. 2. Evaluation of some of the higher level algorithms present in the MFBOX on a real data set using the stJADE algorithm with 10 components. In both graphs $\alpha = 1$ is the situation where only temporal ICA is performed whereas at $\alpha = 0$ only spatial ICA is performed.

on samples in the projective space $\mathbb{RP}^n$ of all lines of $\mathbb{R}^n$. This models the scaling indeterminacy of ICA more precisely than projection onto its hypersphere since it also deals with the sign invariance. The result (figure 2) is explained by the fact that the spatial dimension of the data is strongly smoothed and reveals a some structure. Hence it does not follow the usual linear ICA model of independently drawn samples from an identically distributed random variable. The temporal dimension does not expose such additional structure and is less smoothed by the data acquisition and so the temporal ICA should be more stable.

Hierarchical analysis of component maps. Another common issue in data-driven BSS methods is how to choose the number of components and how to evaluate the large amount of possibly interesting components the process might yield without means to identify the ones which are interesting in the current problem. The hierarchical analysis tries to overcome this problem by extracting different numbers of components and extracting a hierarchical structure form the relations between the timecourses and the component maps of the extracted sources. This yields a tree structure of the components which can be used to easily navigate to the components which are of interest in an experiment. For more detailed implementation details see also [11].

Sliding time-window analysis. Usualy ICA cannot deal with non-stationary data, so most commonly approximations or windowing techniques are used. In ordinary ICA, the whole data set is transformed, and as many component maps as time steps are reconstructed. Window ICA [10] groups the observations in windows of a fixed size, and extracts components out of each of them. Correlation analysis then gives corresponding components over the whole time interval, thus yielding additional structure in a single full-scale component map.

2.3 Import, Export and Postprocessing

The MFBOX has rich im- and export possibilities to load and save data, masks, designs, reference brain masks, and parameters from different formats as Analyze, Nifti, plain MATLAB, and its own file format.

selectcomp semi-manual selection, labeling and grouping of extracted components

denoise lICA based 3d denoising [4] of the extracted components

3 Using the MFBOX on fMRI Recording of an Wisconsin Card Sorting Test (WCST)

In this part we will demonstrate the results from using the MFBOX on a real world data set. After a short introduction into the nature of the data we will present the result and compare it to the standard SPM result. The WCST is

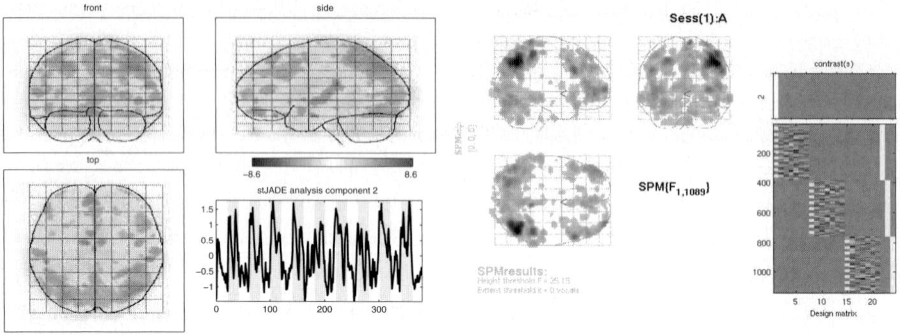

(a) The stimulus component as exported by MFBOX

(b) SPM result of the stimulus component

Fig. 3. Result of a stJADE run with α=1.0, selecting the whole brain area as mask, and after applying spatial and temporal mean removal. The component network in the prefrontal cortex and a more compact network at the parietal/occipital lobe is clearly visible.

a traditional neuropsychological test that is sensitive to frontal lobe dysfunction and allows to assess the integrity of the subjects' executive functions. The WCST-task primarily activates the dorsolateral prefrontal cortex that is responsible for executive working memory operations and cognitive control functions. The given fMRI data set originates from a modified version of the WCST [13]. Its aim is to segregate those network components participating in the above mentioned process. At first, a number of stimulus cards are presented to the participants. Afterwards the subject is asked to match additional cards to one of the stimulus cards with respect to a sorting criterion unknown to the subject. In fact, the subject has to discover it by trial and error (Sorting dimensions include: the color of symbols, their shape or the number of displayed symbols on the card). Sorting dimension changes if a previously defined number of correct answers have been given consecutively. Three different test variants of the WCST were applied

Task A. No instructions of dimensions (very close to the original WCST)
Task B. Instruction of dimensional change as sorting criterion changes.
Task C. Reminder of dimension prior to each trial; subject knows in advance the attribute that was searched for in the test.

Our results largely verify the findings of [13], as can be seen in figure 3(a). In addition to the stimulation of the prefrontal cortex, an increased activity in the parietal lobe as well as in the occipital lobe was revealed by our stJADE algorithm. The increased activity in the rear section of the brain (see figure 3(b)) was also discovered by the model-free algorithm. The complete summary of one stJADE analysis of the data set is shown in figure 4. The classification into

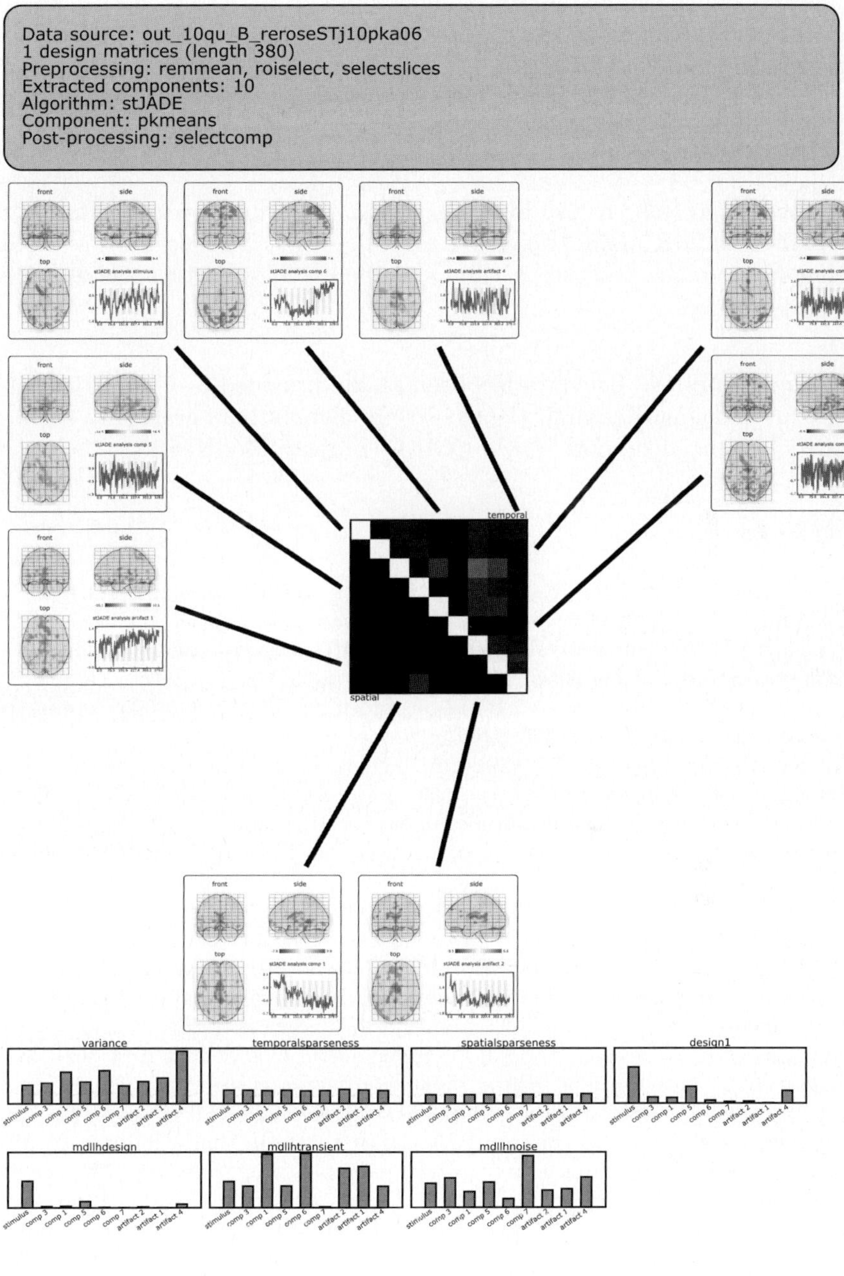

Fig. 4. A complete BSS analysis output as provided by the SVG export function of the MFBOX. The component labeled stimulus indicates the one which has the highest correlation with the stimulus. The components labeled artifact are most likely artifacts, two of them are also grouped together such that the number of displayed components is 9 although 10 components were extracted by stJADE.

stimulus component and artifacts was done after the analysis using the reliability analysis and a Minimum Description Length (MDL)-likelihood based noise estimator included in the MFBOX.

4 Conclusion

The MFBOX is an easy to use but nonetheless powerful instrument for explorative analysis of fMRI data. It includes several modern BSS algorithms and due to its highly modular structure it can easily be extended with novel as well as classical approaches.

Acknowledgements. The data has been kindly provided by Chuh Hyoun Lie, Institute of Medicine, Research Centre Jülich. The authors gratefully acknowledge partial financial support by the BMBF (project 'ModKog').

References

[1] SPM5. edn. 5 (July 2005), `http://www.fil.ion.ulc.ac.uk/spm/spm5.html`
[2] Calhoun, V.D., Adali, T., Pearlson, G.D., Pekar, J.J.: Spatial and temporal independent component analysis of functional MRI data containing a pair of task-related waveforms. Hum.Brain Map. 13, 43–53 (2001)
[3] Cardoso, J.-F., Souloumiac, A.: Blind beamforming for non gaussian signals. IEE Proceedings-F 140(6), 362–370 (1993)
[4] Gruber, P., Stadlthanner, K., Böhm, M., Theis, F.J., Lang, E.W., Tom é, A.M., Teixeira, A.R., Puntonet, C.G., Górriz, J.M.: Denoising using local projective subspace methods. Neurocomputing 69, 1485–1501 (2006)
[5] Gruber, P., Theis, F.J.: Grassmann clustering. In: Proc. of EUSIPCO, Florence, Italy (2006)
[6] Harmeling, S., Meinecke, F., Müller, K.R.: Injecting noise for analysing the stability of ICA components. Signal Processing 84, 255–266 (2004)
[7] Himberg, J., Hyvärinen, A., Esposito, F.: Validating the independent components of neuroimaging time-series via clustering and visualization. NeuroImage 22, 1214–1222 (2004)
[8] Hyvärinen, A., Karhunen, J., Oja, E.: Independent Component Analysis (2001)
[9] Karvanen, J., Koivunen, V.: Blind separation methods based on pearson system and its extensions. Signal Processing 82(4), 663–673 (2002)
[10] Karvanen, J., Theis, F.J.: Spatial ICA of fMRI data in time windows. In: Proc. MaxEnt 2004 AIP conference proceedings, Garching, Germany, vol. 735, pp. 312–319 (2004)
[11] Keck, I.R., Lang, E.W., Nassabay, S., Puntonet, C.G.: Clustering of signals using incomplete independent component analysis. In: Cabestany, J., Prieto, A.G., Sandoval, F. (eds.) IWANN 2005. LNCS, vol. 3512, pp. 1067–1074. Springer, Heidelberg (2005)
[12] Kohonen, T.: Self-Organizing Maps. Springer, New York, Inc. Secaucus, NJ (2001)
[13] Lie, C.-H., Specht, K., Marshall, J.C., Fink, G.R.: Using fMRI to decompose the neural processes underlying the wisconsin card sorting test. NeuroImage 30, 1038–1049 (2006)

[14] McKeown, M.J., Sejnowski, T.J.: Independent component analysis of FMRI data: Examining the assumptions. Human Brain Mapping 6, 368–372 (1998)
[15] Theis, F.J., Gruber, P., Keck, I.R., Meyer-Bäse, A., Lang, E.W.: Spatiotemporal blind source separation using double-sided approximate joint diagonalization. In: Proc. of EUSIPCO, Antalya, Turkey (2005)
[16] Theis, F.J., Meyer-Bäse, A., Lang, E.W.: Second-order blind source separation based on multi-dimensional autocovariances. In: Puntonet, C.G., Prieto, A.G. (eds.) ICA 2004. LNCS, vol. 3195, pp. 726–733. Springer, Heidelberg (2004)
[17] Yang, B.: Projection approximation subspace tracking. IEEE Trans. on Signal Processing 43(1), 95–107 (1995)

An Eigenvector Algorithm with Reference Signals Using a Deflation Approach for Blind Deconvolution

Mitsuru Kawamoto[1], Yujiro Inouye[2], Kiyotaka Kohno[2], and Takehide Maeda[2]

[1] National Institute of Advanced Industrial Science and Technology (AIST), Central 2, 1-1-1 Umezono, Tsukuba, Ibaraki, 305-8568, Japan
m.kawamoto@aist.go.jp
http://staff.aist.go.jp/m.kawamoto/

[2] Department of Electronic and Control Systems Engineering, Shimane University, 1060 Nishikawatsu, Matsue, 690-8504, Japan
inouye@riko.shimane-u.ac.jp,
kohno@yonago-k.ac.jp, maeda@helen.ocn.ne.jp

Abstract. We propose an eigenvector algorithm (EVA) with reference signals for blind deconvolution (BD) of multiple-input multiple-output infinite impulse response (MIMO-IIR) channels. Differently from the conventional EVAs, each output of a deconvolver is used as a reference signal, and moreover the BD can be achieved without using whitening techniques. The validity of the proposed EVA is shown comparing with our conventional EVA.

1 Introduction

This paper deals with a blind deconvolution (BD) problem for a multiple-input and multiple-output (MIMO) infinite impulse response (IIR) channels. To solve this problem, we use eigenvector algorithms (EVAs) [6,7,12]. The first proposal of the EVA was done by Jelonnek et al. [6]. They have proposed the EVA for solving blind equalization (BE) problems of single-input single-output (SISO) channels or single-input multiple-output (SIMO) channels. In [12], several procedures for the blind source separation (BSS) of instantaneous mixtures, using the generalized eigenvalue decomposition (GEVD), have been introduced. Recently, the authors have proposed an EVA which can solve BSS problems in the case of MIMO static systems (instantaneous mixtures) [8]. Moreover, based on the idea in [8], an EVA was derived for MIMO-IIR channels (convolutive mixtures) [9].

In the EVAs in [8,9], an idea of using reference signals was adopted. Researches applying this idea to solving blind signal processing (BSP) problems, such as the BD, the BE, the BSS, and so on, have been made by Jelonnek et al. (e.g., [6]), Adib et al. (e.g., [2]), Rhioui et al. [13], and Castella, et al. [3]. In [8,9], differently from the conventional methods, only one reference signal was utilized for recovering all the source signals simultaneously.

However, the EVA in [9] has difference performances for a different choice of the reference signal (see section 4), and in order to recover all source signals, it

M.E. Davies et al. (Eds.): ICA 2007, LNCS 4666, pp. 218–226, 2007.

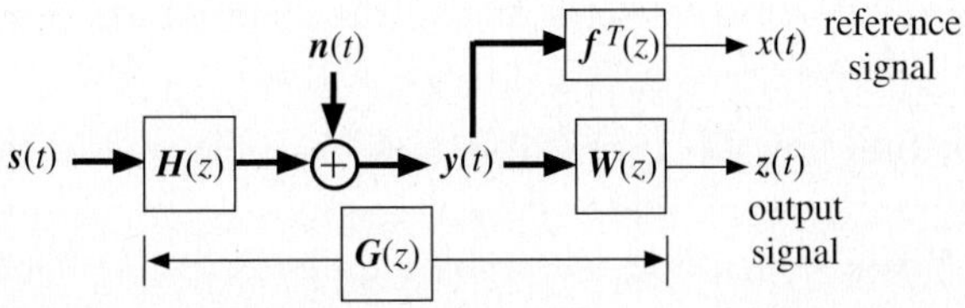

Fig. 1. The composite system of an unknown system and a deconvolver, and a reference system

must be taken into account how to select appropriate eigenvectors from the set of eigenvectors calculated by the EVA. In this paper, in order to circumvent such a tedious (or nasty) task, the output of a deconvolver which is used to recover source signals is used as a reference signal. Accordingly, deflation techniques are needed to recover all source signals. The method proposed in [3] is almost same as the proposed EVA. However, the proposed EVA can achieve the BD without using whitening techniques. Moreover, the proposed EVA provides good performances compared with our conventional EVA [9] (see section 4).

The present paper uses the following notation: Let Z denote the set of all integers. Let C denote the set of all complex numbers. Let $\boldsymbol{C}^n$ denote the set of all n-column vectors with complex components. Let $\boldsymbol{C}^{m\times n}$ denote the set of all $m \times n$ matrices with complex components. The superscripts T, $*$, and H denote, respectively, the transpose, the complex conjugate, and the complex conjugate transpose (Hermitian) of a matrix. The symbols block-diag$\{\cdots\}$ and diag$\{\cdots\}$ denote respectively a block diagonal and a diagonal matrices with the block diagonal and the diagonal elements $\{\cdots\}$. The symbol cum$\{x_1,x_2,x_3,x_4\}$ denotes a fourth-order cumulant of x_i's. Let $i = \overline{1,n}$ stand for $i = 1, 2, \cdots, n$.

2 Problem Formulation and Assumptions

We consider a MIMO channel with n inputs and m outputs as described by

$$\boldsymbol{y}(t) = \textstyle\sum_{k=-\infty}^{\infty} \boldsymbol{H}^{(k)} \boldsymbol{s}(t-k) + \boldsymbol{n}(t), \quad t \in Z, \tag{1}$$

where $\boldsymbol{s}(t)$ is an n-column vector of input (or source) signals, $\boldsymbol{y}(t)$ is an m-column vector of channel outputs, $\boldsymbol{n}(t)$ is an m-column vector of Gaussian noises, and $\{\boldsymbol{H}^{(k)}\}$ is an $m \times n$ impulse response matrix sequence. The transfer function of the channel is defined by $\boldsymbol{H}(z) = \sum_{k=-\infty}^{\infty} \boldsymbol{H}^{(k)} z^k$, $z \in C$.

To recover the source signals, we process the output signals by an $n \times m$ deconvolver (or equalizer) $\boldsymbol{W}(z)$ described by

$$\begin{aligned} \boldsymbol{z}(t) &= \textstyle\sum_{k=-\infty}^{\infty} \boldsymbol{W}^{(k)} \boldsymbol{y}(t-k) \\ &= \textstyle\sum_{k=-\infty}^{\infty} \boldsymbol{G}^{(k)} \boldsymbol{s}(t-k) + \sum_{k=-\infty}^{\infty} \boldsymbol{W}^{(k)} \boldsymbol{n}(t-k), \end{aligned} \tag{2}$$

where $\{\boldsymbol{G}^{(k)}\}$ is the impulse response matrix sequence of $\boldsymbol{G}(z) := \boldsymbol{W}(z)\boldsymbol{H}(z)$, which is defined by $\boldsymbol{G}(z) = \sum_{k=-\infty}^{\infty} \boldsymbol{G}^{(k)} z^k$, $z \in C$. The cascade connection of the unknown system and the deconvolver is illustrated in Fig. 1.

Here, we put the following assumptions on the channel, the source signals, the deconvolver, and the noises.

A1) The transfer function $\boldsymbol{H}(z)$ is stable and has full column rank on the unit circle $|z| = 1$, where the assumption **A1)** implies that the unknown system has less inputs than outputs, i.e., $n \leq m$, and there exists a left stable inverse of the unknown system.
A2) The input sequence $\{\boldsymbol{s}(t)\}$ is a complex, zero-mean and non-Gaussian random vector process with element processes $\{s_i(t)\}$, $i = \overline{1,n}$ being mutually independent. Each element process $\{s_i(t)\}$ is an i.i.d. process with a variance $\sigma_{s_i}^2 \neq 0$ and a nonzero fourth-order cumulant $\gamma_i \neq 0$ defined as

$$\gamma_i = \text{cum}\{s_i(t), s_i(t), s_i^*(t), s_i^*(t)\} \neq 0. \tag{3}$$

A3) The deconvolver $\boldsymbol{W}(z)$ is an FIR channel of sufficient length L so that the truncation effect can be ignored.
A4) The noise sequence $\{\boldsymbol{n}(t)\}$ is a zero-mean, Gaussian vector stationary process whose component processes $\{n_j(t)\}$, $j = \overline{1,m}$ have nonzero variances $\sigma_{n_j}^2$, $j = \overline{1,m}$.
A5) The two vector sequences $\{\boldsymbol{n}(t)\}$ and $\{\boldsymbol{s}(t)\}$ are mutually statistically independent.

Under **A3)**, the impulse response $\{\boldsymbol{G}^{(k)}\}$ of the cascade system is given by

$$\boldsymbol{G}^{(k)} := \textstyle\sum_{\tau=L_1}^{L_2} \boldsymbol{W}^{(\tau)} \boldsymbol{H}^{(k-\tau)}, \quad k \in Z, \tag{4}$$

where the length $L := L_2 - L_1 + 1$ is taken to be sufficiently large. In a vector form, (4) can be written as

$$\tilde{\boldsymbol{g}}_i = \tilde{\boldsymbol{H}} \tilde{\boldsymbol{w}}_i, \quad i = \overline{1,n}, \tag{5}$$

where $\tilde{\boldsymbol{g}}_i$ is the column vector consisting of the ith output impulse response of the cascade system defined by $\tilde{\boldsymbol{g}}_i := [\boldsymbol{g}_{i1}^T, \boldsymbol{g}_{i2}^T, \cdots, \boldsymbol{g}_{in}^T]^T$,

$$\boldsymbol{g}_{ij} := [\cdots, g_{ij}(-1), g_{ij}(0), g_{ij}(1), \cdots]^T, \quad j = \overline{1,n} \tag{6}$$

where $g_{ij}(k)$ is the (i,j)th element of matrix $\boldsymbol{G}^{(k)}$, and $\tilde{\boldsymbol{w}}_i$ is the mL-column vector consisting of the tap coefficients (corresponding to the ith output) of the deconvolver defined by $\tilde{\boldsymbol{w}}_i := \left[\boldsymbol{w}_{i1}^T, \boldsymbol{w}_{i2}^T, \cdots, \boldsymbol{w}_{im}^T\right]^T \in \boldsymbol{C}^{mL}$,

$$\boldsymbol{w}_{ij} := [w_{ij}(L_1), w_{ij}(L_1+1), \cdots, w_{ij}(L_2)]^T \in \boldsymbol{C}^L, \tag{7}$$

$j = \overline{1,m}$, where $w_{ij}(k)$ is the (i,j)th element of matrix $\boldsymbol{W}^{(k)}$, and $\tilde{\boldsymbol{H}}$ is the $n \times m$ block matrix whose (i,j)th block element $\boldsymbol{H}_{ij}$ is the matrix (of L columns and possibly infinite number of rows) with the (l,r)th element $[\boldsymbol{H}_{ij}]_{lr}$ defined by $[\boldsymbol{H}_{ij}]_{lr} := h_{ji}(l-r)$, $l = 0, \pm 1, \pm 2, \cdots$, $r=\overline{L_1, L_2}$, where $h_{ij}(k)$ is the (i,j)th element of the matrix $\boldsymbol{H}^{(k)}$.

In the multichannel blind deconvolution problem, we want to adjust $\tilde{\boldsymbol{w}}_i$'s (i $= \overline{1,n}$) so that

$$[\tilde{\boldsymbol{g}}_1, \cdots, \tilde{\boldsymbol{g}}_n] = \tilde{\boldsymbol{H}}[\tilde{\boldsymbol{w}}_1, \cdots, \tilde{\boldsymbol{w}}_n] = [\tilde{\boldsymbol{\delta}}_1, \cdots, \tilde{\boldsymbol{\delta}}_n]\boldsymbol{P}, \tag{8}$$

where $\boldsymbol{P}$ is an $n \times n$ permutation matrix, and $\tilde{\boldsymbol{\delta}}_i$ is the n-block column vector defined by

$$\tilde{\boldsymbol{\delta}}_i := [\boldsymbol{\delta}_{i1}^T, \boldsymbol{\delta}_{i2}^T, \ldots, \boldsymbol{\delta}_{in}^T]^T, \qquad i = \overline{1, n} \tag{9}$$

$$\boldsymbol{\delta}_{ij} := \begin{cases} \hat{\boldsymbol{\delta}}_i, & \text{if} \quad i = j, \\ (\cdots, 0, 0, 0, \cdots)^T, & \text{otherwise.} \end{cases} \tag{10}$$

Here, $\hat{\boldsymbol{\delta}}_i$ is the column vector (of infinite elements) whose rth element $\hat{\delta}_i(r)$ is given by $\hat{\delta}_i(r) = d_i \delta(r - k_i)$, where $\delta(t)$ is the Kronecker delta function, d_i is a complex number standing for a scale change and a phase shift, and k_i is an integer standing for a time shift.

3 Eigenvector Algorithms (EVAs)

3.1 Analysis of Eigenvector Algorithms with Reference Signals for MIMO-IIR Channels

In order to solve the BD problem, the following cross-cumulant between $z_i(t)$ and a reference signal $x(t)$ (see Fig. 1) is defined;

$$D_{z_i x} = \text{cum}\{z_i(t), z_i^*(t), x(t), x^*(t)\}, \tag{11}$$

where $z_i(t)$ is the ith element of $\boldsymbol{z}(t)$ in (2) and the reference signal $x(t)$ is given by $\boldsymbol{f}^T(z)\boldsymbol{y}(t)$, using an appropriate filter $\boldsymbol{f}(z)$. The filter $\boldsymbol{f}(z)$ is called a *reference system.* Let $\boldsymbol{a}(z) := \boldsymbol{H}^T(z)\boldsymbol{f}(z) = [a_1(z), a_2(z), \cdots, a_n(z)]^T$, then $x(t) = \boldsymbol{f}^T(z)\boldsymbol{H}(z)\boldsymbol{s}(t) = \boldsymbol{a}^T(z)\boldsymbol{s}(t)$. The element $a_i(z)$ of the filter $\boldsymbol{a}(z)$ is defined as $a_i(z) = \sum_{k=-\infty}^{\infty} a_i(k) z^k$ and the reference system $\boldsymbol{f}(z)$ is an m-column vector whose elements are $f_j(z) = \sum_{k=L_1}^{L_2} f_j(k) z^k$, $j = \overline{1, m}$.

Jelonnek et al. [6] have shown in the single-input case that by the Lagrangian method, the maximization of $|D_{z_i x}|$ under $\sigma_{z_i}^2 = \sigma_{s_{\rho_i}}^2$ leads to a closed-form solution expressed as a generalized eigenvector problem, where $\sigma_{z_i}^2$ and $\sigma_{s_{\rho_i}}^2$ denote the variances of the output $z_i(t)$ and a source signal $s_{\rho_i}(t)$, respectively, and ρ_i is one of integers $\{1, 2, \cdots, n\}$ such that the set $\{\rho_1, \rho_2, \cdots, \rho_n\}$ is a permutation of the set $\{1, 2, \cdots, n\}$. In our case, $D_{z_i x}$ and $\sigma_{z_i}^2$ can be expressed in terms of the vector $\tilde{\boldsymbol{w}}_i$ as, respectively,

$$D_{z_i x} = \tilde{\boldsymbol{w}}_i^H \tilde{\boldsymbol{B}} \tilde{\boldsymbol{w}}_i, \quad \sigma_{z_i}^2 = \tilde{\boldsymbol{w}}_i^H \tilde{\boldsymbol{R}} \tilde{\boldsymbol{w}}_i, \tag{12}$$

where $\tilde{\boldsymbol{B}}$ is the $m \times m$ block matrix whose (i, j)th block element $\boldsymbol{B}_{ij}$ is the matrix with the (l, r)th element $[\boldsymbol{B}_{ij}]_{lr}$ calculated by $\text{cum}\{y_i^*(t - L_1 - l + 1), y_j(t - L_1 - r + 1), x^*(t), x(t)\}$ $(l, r = \overline{1, L})$ and $\tilde{\boldsymbol{R}} = E[\tilde{\boldsymbol{y}}^*(t)\tilde{\boldsymbol{y}}^T(t)]$ is the covariance matrix of m-block column vector $\tilde{\boldsymbol{y}}(t)$ defined by

$$\tilde{\boldsymbol{y}}(t) := \left[\boldsymbol{y}_1^T(t), \boldsymbol{y}_2^T(t), \cdots, \boldsymbol{y}_m^T(t)\right]^T \in \boldsymbol{C}^{mL}, \tag{13}$$

$$\boldsymbol{y}_j(t) := [y_j(t\text{-}L_1), y_j(t\text{-}L_1\text{-}1), \cdots, y_j(t\text{-}L_2)]^T \in \boldsymbol{C}^L, \tag{14}$$

$j = \overline{1,m}$. Therefore, by the similar way to as in [6], the maximization of $|D_{z_i x}|$ under $\sigma^2_{z_i} = \sigma^2_{s_{\rho_i}}$ leads to the following generalized eigenvector problem;

$$\tilde{\boldsymbol{B}}\tilde{\boldsymbol{w}}_i = \lambda_i \tilde{\boldsymbol{R}}\tilde{\boldsymbol{w}}_i. \tag{15}$$

Moreover, Jelonnek et al. have shown that the eigenvector corresponding to the maximum magnitude eigenvalue of $\tilde{\boldsymbol{R}}^\dagger \tilde{\boldsymbol{B}}$ becomes the solution of the blind equalization problem in [6], which is referred to as an *eigenvector algorithm* (EVA). Note that since Jelonnek et al. have dealt with SISO-IIR channels or SIMO-IIR channels, the constructions of $\tilde{\boldsymbol{B}}$, $\tilde{\boldsymbol{w}}_i$, and $\tilde{\boldsymbol{R}}$ in (15) are different from those proposed in [6,7]. In this paper, we want to show how the eigenvector algorithm (15) works for the BD of the MIMO-IIR channel (1).

To this end, we use the following equalities;

$$\tilde{\boldsymbol{R}} = \tilde{\boldsymbol{H}}^H \tilde{\boldsymbol{\Sigma}} \tilde{\boldsymbol{H}}, \;\; \tilde{\boldsymbol{B}} = \tilde{\boldsymbol{H}}^H \tilde{\boldsymbol{\Lambda}} \tilde{\boldsymbol{H}}, \tag{16}$$

where $\tilde{\boldsymbol{\Sigma}}$ is the block diagonal matrix defined by

$$\tilde{\boldsymbol{\Sigma}} := \text{block-diag}\{\boldsymbol{\Sigma}_1, \boldsymbol{\Sigma}_2, \cdots, \boldsymbol{\Sigma}_n\}, \tag{17}$$

$$\boldsymbol{\Sigma}_i := \text{diag}\{\cdots, \sigma^2_{s_i}, \sigma^2_{s_i}, \sigma^2_{s_i}, \cdots\}, \quad i = \overline{1,n}, \tag{18}$$

and $\tilde{\boldsymbol{\Lambda}}$ is the block diagonal matrix defined by

$$\tilde{\boldsymbol{\Lambda}} := \text{block-diag}\{\boldsymbol{\Lambda}_1, \boldsymbol{\Lambda}_2, \cdots, \boldsymbol{\Lambda}_n\}, \tag{19}$$

$$\boldsymbol{\Lambda}_i := \text{diag}\{\cdots, |a_i(-1)|^2\gamma_r, |a_i(0)|^2\gamma_i, |a_i(1)|^2\gamma_i, \cdots\}, \tag{20}$$

$i = \overline{1,n}$. Since both $\tilde{\boldsymbol{\Sigma}}$ and $\tilde{\boldsymbol{\Lambda}}$ become diagonal, (16) shows that the two matrices $\tilde{\boldsymbol{R}}$ and $\tilde{\boldsymbol{B}}$ are simultaneously diagonalizable.

Here, let the eigenvalues of the diagonal matrix $\tilde{\boldsymbol{\Sigma}}^{-1}\tilde{\boldsymbol{\Lambda}}$ is denoted by

$$\lambda_i(k) := |a_i(k)|^2 \gamma_i / \sigma^2_{s_i}, \qquad i = \overline{1,n}, \quad k \in Z. \tag{21}$$

We put the following assumption on the eigenvalues $\lambda_i(k)'s$.
A6) All the eigenvalues $\lambda_i(k)'s$ are distinct for $i = \overline{1,n}$ and $k \in Z$.

Theorem 1. *Suppose the noise term $\boldsymbol{n}(t)$ is absent and the length L of the deconvolver is infinite (that is, $L_1 = -\infty$ and $L_2 = \infty$). Then, under the assumptions* **A1)** *through* **A6)**, *the n eigenvector $\tilde{\boldsymbol{w}}_i$'s corresponding to the n nonzero eigenvalues $\lambda_i(k)'s$ of matrix $\tilde{\boldsymbol{R}}^\dagger \tilde{\boldsymbol{B}}$ for $i = \overline{1,n}$ and an arbitrary $k \in Z$ become the vectors $\tilde{\boldsymbol{w}}_i$'s satisfying (8).*

Outline of the proof: Based on (15), we consider the following eigenvector problem;

$$\tilde{\boldsymbol{R}}^\dagger \tilde{\boldsymbol{B}}\tilde{\boldsymbol{w}}_i = \lambda_i \tilde{\boldsymbol{w}}_i. \tag{22}$$

Then, from (16), (22) becomes

$$(\tilde{\boldsymbol{H}}^H \tilde{\boldsymbol{\Sigma}} \tilde{\boldsymbol{H}})^\dagger \tilde{\boldsymbol{H}}^H \tilde{\boldsymbol{\Lambda}} \tilde{\boldsymbol{H}} \tilde{\boldsymbol{w}}_i = \lambda_i \tilde{\boldsymbol{w}}_i. \tag{23}$$

Under $L_1 = -\infty$ and $L_2 = \infty$, we have the following equations;

$$(\tilde{\boldsymbol{H}}^H \tilde{\boldsymbol{\Sigma}} \tilde{\boldsymbol{H}})^\dagger = \tilde{\boldsymbol{H}}^\dagger \tilde{\boldsymbol{\Sigma}}^\dagger \tilde{\boldsymbol{H}}^{H\dagger}, \quad \tilde{\boldsymbol{H}}^{H\dagger} \tilde{\boldsymbol{H}}^H = \boldsymbol{I}, \tag{24}$$

which are shown in [11] along with their proofs. Then it follows from (23) and (24);

$$\tilde{\boldsymbol{H}}^\dagger \tilde{\boldsymbol{\Sigma}}^{-1} \tilde{\boldsymbol{\Lambda}} \tilde{\boldsymbol{H}} \tilde{\boldsymbol{w}}_i = \lambda_i \tilde{\boldsymbol{w}}_i. \tag{25}$$

Multiplying (25) by $\tilde{\boldsymbol{H}}$ from the left side and using (24), (25) becomes

$$\tilde{\boldsymbol{\Sigma}}^{-1} \tilde{\boldsymbol{\Lambda}} \tilde{\boldsymbol{H}} \tilde{\boldsymbol{w}}_i = \lambda_i \tilde{\boldsymbol{H}} \tilde{\boldsymbol{w}}_i. \tag{26}$$

By (22), $\tilde{\boldsymbol{\Sigma}}^{-1} \tilde{\boldsymbol{\Lambda}}$ is a diagonal matrix with diagonal elements $\lambda_i(k)$, $i = \overline{1,n}$ and $k \in Z$, and thus (22) and (26) show that its diagonal elements $\lambda_i(k)'s$ are eigenvalues of matrix $\tilde{\boldsymbol{R}}^\dagger \tilde{\boldsymbol{B}}$. Here we use the following fact;

$$\lim_{L\to\infty} (\text{rank } \tilde{\boldsymbol{R}})/L = n, \tag{27}$$

which is shown in [10] and its proof is found in [4]. Using this fact, the other remaining eigenvalues of $\tilde{\boldsymbol{R}}^\dagger \tilde{\boldsymbol{B}}$ are all zero. From the assumption **A6)**, the n nonzero eigenvalues $\lambda_i(k) \neq 0$, $i = \overline{1,n}$, obtained by (26), that is, the n nonzero eigenvectors $\tilde{\boldsymbol{w}}_i$, $i = \overline{1,n}$, corresponding to n nonzero eigenvalues $\lambda_i(k) \neq 0$, $i = \overline{1,n}$, obtained by (22) become n solutions of the vectors $\tilde{\boldsymbol{w}}_i$ satisfying (8).

3.2 How to Choose a Reference Signal

In [9], a reference system $\boldsymbol{f}(z)$ is appropriately chosen, and then all source signals can be recovered simultaneously from the observed signals. However, the performances obtained by the EVA in [9] change with the way of choosing a reference system (see section 4) and moreover, the EVA has such a complicated task that the way of selecting appropriate eigenvectors from the set of eigenvectors calculated from the EVA must be taken into account.

In this paper, by adopting $x_i(t) = \tilde{\boldsymbol{w}}_i^T \tilde{\boldsymbol{y}}_i(t)$ as a reference signal, we want to circumvent such a tedious (or nasty) task. To this end, (11) can be reformulated as

$$D_{z_i x_i} = \text{cum}\{z_i(t), z_i^*(t), x_i(t), x_i^*(t)\}, \ i = \overline{1,n}, \tag{28}$$

The vector $\tilde{\boldsymbol{w}}_i$ in $x_i(t)$ is given by an eigenvector obtained from the EVA at the previous time, that is, $x_i(t) = \tilde{\boldsymbol{w}}_i^T(t-1)\tilde{\boldsymbol{y}}_i(t)$, where the valuc of $\tilde{\boldsymbol{w}}_i^T(t-1)$ is assumed to be fixed. By using $x_i(t)$, the matrix $\tilde{\boldsymbol{B}}$ is calculated, which is denoted by $\tilde{\boldsymbol{B}}_i(t)$, and then the eigenvector $\tilde{\boldsymbol{w}}_i^T(t)$ at time t can be obtained from the EVA using $\tilde{\boldsymbol{B}}_i(t)$. By repeating this operation, the BD can be achieved. Then it can be seen that as the EVA works successfully, $x_i(t)$ gradually becomes a source signals $s_{\rho_i}(t-k_i)$. Namely, the diagonal elements of $\tilde{\boldsymbol{\Lambda}}$ in (19) gradually become zeros except for one element corresponding to $s_{\rho_i}(t-k_i)$. This means that when the eigenvectors of $\tilde{\boldsymbol{R}}^\dagger \tilde{\boldsymbol{B}}_i(t)$ are calculated for achieving the BD, it is only enough that we select the eigenvector corresponding to the absolute maximum eigenvalue of $\tilde{\boldsymbol{R}}^{\dagger}\tilde{\boldsymbol{B}}_i(t)$. This is the reason why we can circumvent the tedious task by using the reference signal. After all, the EVA is implemented as follows:

Set initial values: $\tilde{\boldsymbol{w}}_i(0)$, $\tilde{\boldsymbol{R}}(0)$, $\tilde{\boldsymbol{B}}_i(0)$
for $t_l = 1 : t_{l_{all}}$
  for $t = t_d(t_l - 1)+1{:}t_d t_l$
    $x_i(\mathrm{t}) = \tilde{\boldsymbol{w}}_i^T(t_l - 1)\tilde{\boldsymbol{y}}_i(t)$
    Calculate $\tilde{\boldsymbol{R}}(t)$ and $\tilde{\boldsymbol{B}}_i(t)$ by a moving average.
  end
  Calculate the eigenvector $\tilde{\boldsymbol{w}}_i(t_l)$ associated with the absolute maximum eigenvalue $|\lambda_i|$ from (22).
end

where $t_{l_{all}}$ denotes the total number of iterations and t_d denotes the number of data samples for estimating the matrices $\tilde{\boldsymbol{R}}(t)$ and $\tilde{\boldsymbol{B}}_i(t)$. Note that $\tilde{\boldsymbol{R}}$ is not needed to estimate iteratively, but for the sake of our convenience, this way is adopted.

Here it is worth noting that when the above algorithm is implemented, it may happen that each output of a deconvolver provides the same source signal. Therefore, in order to avoid such a situation, we apply a deflation approach, that is, the Gram-Schmidt decorrelation [1] to the eigenvectors $\tilde{\boldsymbol{w}}_i(t_l)$ for $i = \overline{1, n}$.

4 Simulation Results

To demonstrate the validity of the proposed method, many computer simulations were conducted. Some results are shown in this section. The unknown system $\boldsymbol{H}(z)$ was set to be the same channel with two inputs and three outputs as in [9]. Also, other setup conditions, that is, the source signals $s_i(t)$'s, the noises $n_i(t)$'s,

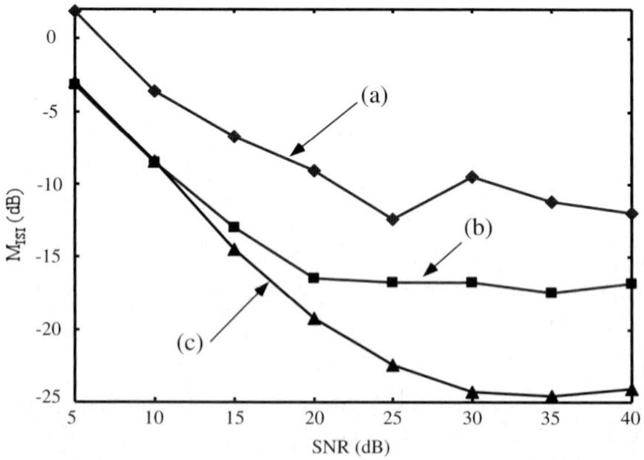

Fig. 2. The performances of the proposed EVA and our conventional EVA with varying SNR levels, in the cases of 5,000 data samples

and their SNR levels were the same as in [9]. As a measure of performances, we used the *multichannel intersymbol interference* (M_{ISI}) [5], which was the average of 30 Monte Carlo runs. In each Monte Carlo run, the number of iterations $t_{l_{all}}$ was set to be 10, and the number of data samples t_d was set to be 5,000. For comparison, our conventional EVA in [9] was used, where the conventional EVA does not need deflation approaches.

Fig. 2 shows the results of performances of the EVAs when the SNR levels were respectively taken to be 5 through 40 dB for every 5 dB, where there are three kinds of reference signals, (a) $x(t) = \sum_{i=1}^{3} f_i(5)y_i(t-5)$, where each parameter $f_i(5)$ was randomly chosen from a Gaussian distribution with zero mean and unit variance, (b) $x(t) = f_2(2)y_2(t-2)$, where $f_2(2)$ also was randomly chosen from the Gaussian distribution, (c) $x_i(t) = \tilde{\boldsymbol{w}}_i^T(t-1)\tilde{\boldsymbol{y}}_i(t)$, $i = \overline{1,3}$. The last reference signal (c) corresponds to the proposed EVA, while the other two (a) and (b) correspond to our conventional EVA.

From Fig. 2, it can be seen that the proposed EVA provides better performances than our conventional EVA [9].

5 Conclusions

We have proposed an EVA for solving the BD problem. Using the output of a deconvolver as a reference signal, the tedious task of our conventional EVA can be circumvented. The simulation results have demonstrated the effectiveness of the proposed EVA. However, from the simulation results, one can see that all our EVAs have such a drawback that it is sensitive to Gaussian noise. Therefore, as a further work, we will propose an EVA having such a property that the BD can be achieved as little insensitive to Gaussian noise as possible.

Acknowledgments. This work is supported by the Research Projects, Grant-IN AIDs, No. 18500146[1] and No. 1850054[2] for Scientific Research of the JSPS.

References

1. Hyvärinen, A.: Fast and robust fixed-point algorithms for independent component analysis. IEEE Trans. Neural Networks 10(3), 62–634 (1999)
2. Adib, A., et al.: Source separation contrasts using a reference signal. IEEE Signal Processing Letters 11(3), 312–315 (2004)
3. Castella, M., et al.: Quadratic Higher-order criteria for iterative blind separation of a MIMO convolutive mixture of sources. IEEE Trans. Signal Processing 55(1), 218–232 (2007)
4. Inouye, Y.: Autoregressive model fitting for multichannel time series of degenerate rank: Limit properties. IEEE Trans. Circuits and Systems 32(3), 252–259 (1985)
5. Inouye, Y., Tanebe, K.: Super-exponential algorithms for multichannel blind deconvolution. IEEE Trans. Sig. Proc. 48(3), 881–888 (2000)
6. Jelonnek, B., Kammeyer, K.D.: A closed-form solution to blind equalization. Signal Processing 36(3), 251–259 (1994)

7. Jelonnek, B., Boss, D., Kammeyer, K.D.: Generalized eigenvector algorithm for blind equalization. Signal Processing 61(3), 237–264 (1997)
8. Kawamoto, M., et al.: Eigenvector algorithms using reference signals. In: Proc. ICASSP 2006, vol. V, pp. 841–844 (May 2006)
9. Kawamoto, M., et al.: Eigenvector algorithms for blind deconvolution of MIMO-IIR systems. In: Proc. ISCAS 2007, pp. 3490–3493, (May 2007), This manuscript is downloadable at `http://staff.aist.go.jp/m.kawamoto/manuscripts/ISCAS2007.pdf`
10. Kohno, K., et al.: Adaptive super-exponential algorithms for blind deconvolution of MIMO systems. In: Proc. ISCAS 2004, vol. V, pp. 680–683 (May 2004)
11. Kohno, K., et al.: Robust super-exponential methods for blind equalization of MIMO-IIR systems. In: Proc. ICASSP 2006, vol. V, pp. 661–664 (2006)
12. Parra, L., Sajda, P.: Blind source separation via generalized eigenvalue decomposition. Journal of Machine Learning, No. 4, 1261–1269 (2003)
13. Rhioui, S., et al.: Quadratic MIMO contrast functions for blind source separation in a convolutive contest. In: Rosca, J., Erdogmus, D., Príncipe, J.C., Haykin, S. (eds.) ICA 2006. LNCS, vol. 3889, pp. 230–237. Springer, Heidelberg (2006)

Robust Independent Component Analysis Using Quadratic Negentropy

Jaehyung Lee, Taesu Kim, and Soo-Young Lee

Department of Bio & Brain Engineering, KAIST, Republic of Korea
{jaehyung.lee, taesu.kim, sylee}@kaist.ac.kr

Abstract. We present a robust algorithm for independent component analysis that uses the sum of marginal quadratic negentropies as a dependence measure. It can handle arbitrary source density functions by using kernel density estimation, but is robust for a small number of samples by avoiding empirical expectation and directly calculating the integration of quadratic densities. In addition, our algorithm is scalable because the gradient of our contrast function can be calculated in O(LN) using the fast Gauss transform, where L is the number of sources and N is the number of samples. In our experiments, we evaluated the performance of our algorithm for various source distributions and compared it with other, well-known algorithms. The results show that the proposed algorithm consistently outperforms the others. Moreover, it is extremely robust to outliers and is particularly more effective when the number of observed samples is small and the number of mixed sources is large.

1 Introduction

In the last decade, Independent Component Analysis (ICA) has shown to be a great success in many applications, including sound separation, EEG signal analysis, and feature extraction. ICA shows quite a good performance for simple source distributions, if given assumptions hold well, but its performance is degraded for sources with skewed or complex density functions [1]. Several ICA methods are currently available for arbitrary distributions, but these methods have not yet shown practical performance when the number of sources is large and the number of observed samples is small, thus preventing their application to more challenging real-world applications, such as blind source separation for non-stationary mixing environments and frequency-domain BSS for convolutive mixtures [2].

The problem of ICA for arbitrary distributions mainly arises from the difficulty of estimating marginal entropies that usually appear in the contrast function derived from mutual information. Direct estimation of marginal entripies without parametric assumptions involves excessive computation, including numerical integration, and is sensitive to outliers because of the log terms. Several approximations are available, but these still rely on higher order statistical terms that are also sensitive to outliers. Different estimators of entropy [3] or dependence measure based on canonical correlations [1] have been suggested to

M.E. Davies et al. (Eds.): ICA 2007, LNCS 4666, pp. 227–235, 2007.

overcome this problem and have shown promising performance. In addition, there have been approaches using nonparametric mutual information via Renyi's entropy [4] for ICA [5]. However, this method requires sign correction by kurtosis because Renyi's entropy does not have a maximum at a Gaussian distribution [6].

In this paper, we define the concept of quadratic negentropy, replace the original negentropy with quadratic negentropy in the original definition of mutual information, and obtain a new contrast function for ICA. Using kernel density estimation along with quadratic negentropy can reduce the integration terms into sums of pairwise interactions between samples. The final contrast function can be calculated efficiently using the fast Gauss transform, guaranteeing scalability. The performance of our algorithm consistently outperforms the best existing algorithms for various source distributions and the existence of outliers, especially when the number of observed samples is small and the number of mixed sources is large.

This paper is organized as follows. In Section 2, we review the basic problem of ICA and the contrast function using negentropy. In Section 3, we define a new contrast function for ICA using quadratic negentropy along with kernel density estimation. We also apply the fast Gauss transform to reduce computation. In Section 4, we evaluate the performance of the derived algorithm on various source distributions, varying the number of sources and the number of samples, to compare the proposed algorithm with other, well-known algorithms, such as FastICA and KernelICA.

2 Background on ICA

In this section, we briefly review the basic problem of ICA and the contrast function using original negentropy.

2.1 The Basic Problem of ICA

Let $s_1, s_2, ..., s_L$ be L statistically independent source random variables that are linearly mixed by some unknown but fixed mixing coefficients to form m observed random variables $x_1, x_2, ..., x_L$. For example, source variables can be the voices of different people at a location and observation variables represent the recordings from several microphones at the location. This can be written in matrix form as

$$\mathbf{x} = \mathbf{A}\mathbf{s} \tag{1}$$

where $\mathbf{x} = (x_1, x_2, ..., x_L)^T$, $\mathbf{s} = (s_1, s_2, ..., s_L)^T$, and $\mathbf{A}$ is an L $\times$ L matrix. The basic problem of ICA is to determine $\mathbf{W}$, the inverse of mixing matrix $\mathbf{A}$, to recover the original sources from observations, by using N samples of observation $\mathbf{x}$ under the assumption that sources are independent of each other.

2.2 Contrast Function Using Negentropy

Mutual information between components of estimated source vectors is known to be a natural contrast function for ICA because it has a zero value when the components are independent and a positive value otherwise. In addition, it is well known that mutual information can be represented using joint and marginal negentropies [7], as follows:

$$I(\mathbf{x}) = J(\mathbf{x}) - \sum_{i=1}^{N} J(x_i) + \frac{1}{2} \log \frac{\prod V_{ii}}{\det V} \tag{2}$$

where $\mathbf{x}$ is a vector random variable of dimension N, x_i is the i-th component of $\mathbf{x}$, V is the covariance matrix of $\mathbf{x}$, and $J(\mathbf{x})$ is the negentropy of a random variable $\mathbf{x}$, which can be represented using Kullback-Leibler divergence, as shown below. The proof is based on the fact that only the first and second order moment of Gaussian density are nonzero and that $\log p_\phi(\boldsymbol{\xi})$ is a polynomial of degree 2 [7].

$$J(\mathbf{x}) = D_{KL}(p_x \| p_\phi) = \int p_x(\boldsymbol{\xi}) \log \frac{p_x(\boldsymbol{\xi})}{p_\phi(\boldsymbol{\xi})} d\boldsymbol{\xi} \tag{3}$$

where $\boldsymbol{\phi}$ is a Gaussian random variable that has the same mean and variance with $\mathbf{x}$, and p_ϕ is the pdf of $\boldsymbol{\phi}$. As a result, it is nonnegative, invariant to invertible transforms and zero if $p_x \equiv p_\phi$.

If we assume $\mathbf{x}$ be whitened, then the last term of Eq. (2) becomes zero and only negentropy terms remain. Now, we define the contrast function of ICA using mutual information, as

$$C(\hat{\mathbf{W}}) = -I(\hat{\mathbf{s}}) = \sum_{i=1}^{L} J(\hat{s}_i) - J(\hat{\mathbf{s}}). \tag{4}$$

In Eq. (4), $\hat{\mathbf{s}} = \hat{\mathbf{W}}\mathbf{x}$ is the estimated sources using the current estimate of the unmixing matrix $\hat{\mathbf{W}}$, and $\hat{s}_i$ is the i-th component of $\hat{\mathbf{s}}$. We assume that the observation is whitened and thus can restrict the unmixing matrix to rotations only, thus making the first term constant and the third term zero in Eq. (2). The final contrast function of ICA using negentropy can be interpreted as the total nongaussianity of the estimated source components.

3 ICA Using Quadratic Negentropy

3.1 Contrast Function Using Quadratic Negentropy

We replace the KL divergence with the L_2 distance in Eq. (3) and obtain quadratic negentropy defined as

$$J_q(\mathbf{x}) = \int \left(p_x(\boldsymbol{\xi}) - p_\phi(\boldsymbol{\xi})\right)^2 d\boldsymbol{\xi}. \tag{5}$$

We can easily show that it is nonnegative, invariant under rotational transform, and zero if $p_x \equiv p_\phi$. Assuming $\mathbf{x}$ is whitened and using quadratic negentropy instead of the original negentropy in Eq. (4), we obtain

$$C_q(\hat{\mathbf{W}}) = -I_q(\hat{\mathbf{s}}) = \sum_{i=1}^{L} J_q(\hat{s}_i) - J_q(\hat{\mathbf{s}}). \tag{6}$$

In addition, $J_q(\hat{\mathbf{s}})$ is constant because the quadratic negentropy is invariant under a rotational transform. Ignoring the constant gives us

$$C_q(\hat{\mathbf{W}}) = \sum_{i=1}^{L} J_q(\hat{s}_i) = \sum_{i=1}^{L} \int (\hat{p}_{\hat{s}_i}(\xi) - \frac{1}{\sqrt{2\pi}} e^{-\xi^2/2})^2 d\xi \tag{7}$$

where $\hat{p}_{\hat{s}_i}$ is the estimated marginal pdf of $\hat{s}_i$. Here $\hat{s}_i$ has zero mean and unit variance because $\hat{\mathbf{W}}$ is rotation and $\mathbf{x}$ is whitened. Thus p_ϕ in (5) becomes a standard Gaussian pdf.

To be a contrast function, Eq. (6) and (7) should have a global maximum when components are independent. We hope this can be proved for general source distributions, but currently we have proof only for Laplacian distributions and further work is needed.

3.2 Kernel Density Estimation

Using kernel density estimation, $\hat{p}_{\hat{s}_i}$ can be estimated as

$$\hat{p}_{\hat{s}_i}(y) = \frac{1}{N} \sum_{n=1}^{N} G(y - \hat{s}_i(n), \sigma^2) \tag{8}$$

where N is the number of observed samples, $\hat{s}_i(n)$ is the n-th observed sample of i-th estimated source, and $G(y, \sigma^2)$ is a Gaussian kernel defined as

$$G(y, \sigma^2) = \frac{1}{\sqrt{2\pi}\sigma} e^{-y^2/2\sigma^2}. \tag{9}$$

Interestingly, the calculation of integration involving quadratic terms of $\hat{p}_{\hat{s}_i}$ estimated as (8) can be simplified as pairwise interactions between samples [8]. Simplifying Eq. (7) using this yields

$$C_q(\hat{\mathbf{W}}) = \sum_{i=1}^{L} \left(\frac{1}{2\sqrt{\pi}} + \frac{1}{N^2} \sum_{n=1}^{N} \sum_{m=1}^{N} G(\hat{s}_i(n) - \hat{s}_i(m), 2\sigma^2) - \frac{2}{N} \sum_{n=1}^{N} G(\hat{s}_i(n), 1+\sigma^2) \right), \tag{10}$$

which is our final contrast function to maximize. Obtaining the partial derivative of $C_q(\hat{\mathbf{W}})$ with respect to w_{ij} yields

$$\begin{aligned} \frac{\partial C_q}{\partial w_{ij}} = \sum_{n=1}^{N} \Bigg(& \frac{2 \cdot G(\hat{s}_i(n), 1+\sigma^2) \hat{s}_i(n)}{N \cdot (1+\sigma^2)} \\ & - \sum_{m=1}^{N} \frac{G(\hat{s}_i(n) - \hat{s}_i(m), 2\sigma^2)(\hat{s}_i(n) - \hat{s}_i(m))}{N^2 \cdot \sigma^2} \Bigg) x_j(n) \end{aligned} \tag{11}$$

where symmetry with respect to m and n is utilized to simplify equation. Also note that $\hat{s}_i(n) = \sum_{j=1}^{L} w_{ij} x_j(n)$.

3.3 Efficient Computation Using Fast Gauss Transform

It takes $O(LN^2)$ time to directly compute the gradient given in Eq. (11). To reduce the computation we use the fast Gauss transform [9] that evaluates the following in $O(N + N')$ time, given 'source' points $\mathbf{x} = \{x_1, ..., x_N\}$ and 'target' points $\mathbf{y} = \{y_1, ..., y_{N'}\}$.

$$FGT(y_j, \mathbf{x}, \mathbf{q}, h) = \sum_{i=1}^{N} q_i e^{-(y_j - x_i)^2/h^2}, j = 1, ..., N' \tag{12}$$

where $\mathbf{q} = \{q_1, ..., q_N\}$ are weight coefficients and h is the bandwidth parameter.

Using Eq. (12), Eq. (10) can be rewritten as

$$C_q(\hat{\mathbf{W}}) = \sum_{i=1}^{L} \left(\frac{1}{2\sqrt{\pi}} + \frac{1}{N^2} \sum_{n=1}^{N} \frac{FGT(\hat{s}_i(n), \hat{\mathbf{s}}_\mathbf{i}, \mathbf{1}, \sqrt{2}\sigma)}{2\sqrt{\pi}\sigma} - \frac{2}{N} \sum_{n=1}^{N} G(\hat{s}_i(n), 1+\sigma^2) \right), \tag{13}$$

and the partial derivative in Eq. (10) can be rewritten as

$$\begin{aligned} \frac{\partial C_q}{\partial w_{ij}} = \sum_{n=1}^{N} \Bigg(& \frac{2 \cdot G(\hat{s}_i(n), 1 + \sigma^2)\hat{s}_i(n)}{N \cdot (1 + \sigma^2)} \\ & - \frac{FGT(\hat{s}_i(n), \hat{\mathbf{s}}_\mathbf{i}, \hat{\mathbf{s}}_\mathbf{i}, \sqrt{2}\sigma) - FGT(\hat{s}_i(n), \hat{\mathbf{s}}_\mathbf{i}, \mathbf{1}, \sqrt{2}\sigma)}{2\sqrt{\pi} \cdot N^2 \cdot \sigma^3} \Bigg) x_j(n) \end{aligned} \tag{14}$$

where $\hat{\mathbf{s}}_\mathbf{i} = \{\hat{s}_i(1), ..., \hat{s}_i(N)\}$ and $\mathbf{1}$ is an N-dimensional one vector.

Now, Eq. (13) and Eq. (14) can be computed in $O(LN)$ by performing the fast Gauss transform $2L$ times.

3.4 Steepest Descent on Stiefel Manifold

The set of orthogonal matrices is a special case of the Stiefel manifold and a gradient of a function can be computed based on the canonical metric of the Stiefel manifold [10]. Unconstrained optimization on the Stiefel manifold is more efficient than orthogonalizing the weight matrix per each iteration. In this paper, we used the steepest descent with a bracketed backtracking line search along geodesics.

3.5 Parameter Selection and Convergence Criterion

Our learning rule has one parameter: the bandwidth parameter σ of the kernel density estimation. We used $\sigma = 1.06 \times N^{-1/5}$ [11].

We calculated the value of the contrast function per each iteration to check convergence. If the difference between iterations becomes less than a given ratio $\tau = 10^{-8}$ of the contrast function, then it is regarded as convergence.

In general, ICA contrast functions have multiple local maxima. This is also true for our contrast function, and we needed a fixed number of restarts to find a good local optimum. We restarted our algorithm four times with a convergence criterion $\tau = 10^{-6}$ and picked the best one as an initial estimate for final optimization.

4 Experimental Results

We conducted an extensive set of simulation experiments using a variety of source distributions, sample numbers, and components. The 18 source distributions used in our experiment were adopted from the KernelICA paper [1]. They include subgaussian, supergaussian and nearly Gaussian source distributions and

Table 1. LEFT: The normalized Amari errors (×100) for mixtures of identical source distributions (top left) and random source distributions (bottom left). L: number of mixed components, N: number of samples, Fast: FastICA, Np: NpICA, Kgv: KernelICA-KGV, Imax: extended infomax ICA, QICA: our method. For identical sources, simulation is repeated 100 times for each of the 18 source distributions for $L = \{2, 4\}$, 50 times for $L = 8$, and 20 times for $L = 16$. For random sources, simulation is repeated 2000 times for $L = \{2, 4\}$, 1000 times for $L = 8$, and 400 times for $L = 16$. **RIGHT:** Amari errors for each source distributions for $L = 2$ and $N = 1000$.

L	N	Fast	Np	Kgv	Imax	QICA
	100	20.6	20.3	16.3	21.3	**15.7**
2	250	13.0	12.9	8.6	14.4	**7.7**
	1000	6.5	9.8	3.0	8.5	**2.9**
	100	28.6	23.0	28.4	23.1	**18.9**
4	250	16.8	13.9	19.2	14.5	**9.8**
	1000	6.9	6.5	7.2	8.7	**3.6**
	250	30.2	20.9	31.3	18.1	**15.9**
8	1000	10.6	7.8	20.6	8.2	**4.7**
	2000	6.4	4.7	14.4	6.2	**2.8**
	1000	26.2	17.3	30.4	**11.1**	12.4
16	2000	11.8	12.5	26.1	**6.6**	6.9
	4000	7.1	6.9	21.3	4.8	**4.3**

L	N	Fast	Np	Kgv	Imax	QICA
	100	18.0	13.6	13.4	19.0	**12.0**
2	250	11.3	7.2	6.3	13.1	**6.1**
	1000	5.6	2.8	**2.4**	6.7	2.5
	100	24.5	18.1	26.3	21.2	**14.9**
4	250	13.7	8.5	14.1	13.1	**6.9**
	1000	5.7	2.6	3.4	5.9	**2.5**
	250	25.4	14.8	30.0	16.0	**10.1**
8	1000	6.3	2.9	13.4	6.0	**2.7**
	2000	4.0	**1.7**	5.6	4.1	1.8
	1000	12.5	8.5	27.9	7.9	**4.1**
16	2000	4.3	2.6	27.0	4.3	**2.3**
	4000	2.9	**1.2**	20.3	2.9	2.0

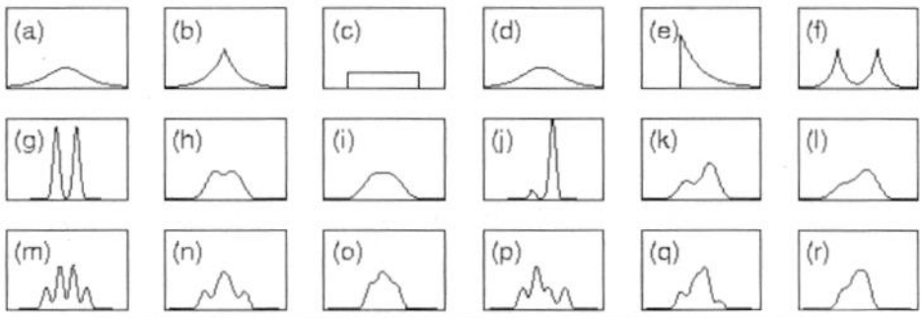

pdfs	Fast	Np	Kgv	Imax	QICA
a	4.7	5.6	3.0	**2.1**	2.7
b	5.5	4.1	3.0	2.7	**2.4**
c	2.3	3.1	**1.6**	3.0	2.1
d	7.2	8.8	**5.7**	6.4	6.4
e	5.7	**0.9**	1.3	3.3	1.6
f	4.7	26.9	**1.5**	1.6	**1.5**
g	1.7	30.0	1.3	**1.1**	1.3
h	5.8	5.7	4.5	**3.4**	3.6
i	9.4	14.9	9.5	**6.9**	7.3
j	7.0	29.7	**1.4**	11.4	**1.4**
k	5.8	3.3	2.8	4.9	**2.7**
l	12.1	**4.8**	5.5	8.2	**4.8**
m	3.5	14.9	**1.4**	4.3	**1.4**
n	5.7	10.7	**1.8**	22.3	1.9
o	4.4	**3.1**	3.6	4.2	3.9
p	3.8	**1.1**	1.5	8.0	1.6
q	21.8	4.3	**2.1**	53.2	2.5
r	6.0	3.5	**2.9**	5.1	3.5
mean	6.5	9.8	3.0	8.5	**2.9**
std	4.5	9.7	2.1	12.2	**1.7**

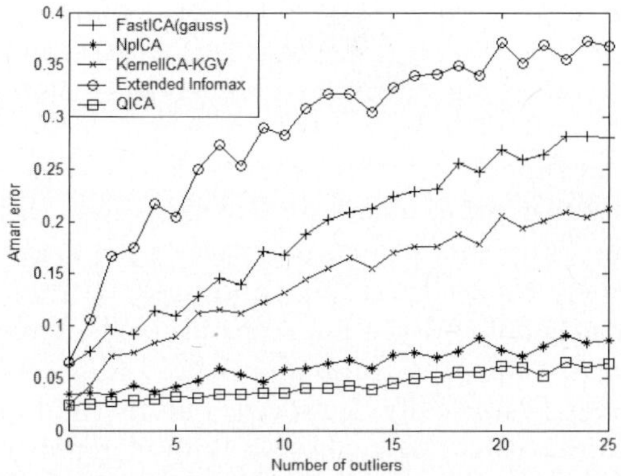

Fig. 1. Robustness to outliers for $L = 2$, $N = 1000$. Up to 25 observations are corrupted by adding +5 or -5. The experiment is repeated 1000 times with random source distributions.

unimodal, multimodal, symmetric, and skewed sources. We varied the number of samples from 100 to 4000 and the number of components from 2 to 16.

Comparisons were made with four existing ICA algorithms: the FastICA algorithm [12], the KernelICA-KGV algorithm [1], the extended infomax algorithm [13] using tanh nonlinearity, and the NpICA algorithm [14]. Software programs were downloaded from corresponding authors' websites and were used with default parameters, except for the extended infomax algorithm, which is our own implementation. Note that KernelICA-KGV also has four restarts as a default to obtain initial estimates. The performance was measured using the Amari error [15], which is invariant to permutation and scaling, lies between 0 and L−1 and is zero for perfect demixing. We normalized the Amari error by dividing it by L−1, where L is the number of independent components.

We summarized our results in Table 1. Consistent performance improvement over existing algorithms was observed. The improvement was significant if the number of components was large and the number of observations was small. However, the performance gain became smaller as the number of observations increased. Amari errors for each source pdf are also shown separately for two-components and 1000 observations. The proposed method showed the smallest standard deviation among the five methods. All of the methods, except for KernelICA-KGV and the proposed method had problems with specific pdfs.

Another interesting result was the high performance of the extended infomax algorithm for a large number of components. For $L = 16$, it showed the best performance among the five methods. But further experiments with outliers discouraged its practical use.

Fig. 1 shows the result of the outlier experiment. We randomly chose up to 25 observations and added the value +5 or -5 to a single component in the observation, which was the same as the one in the KernelICA paper. The results show that our method is extremely robust to outliers.

5 Conclusions

We have proposed a robust algorithm for independent component analysis that uses the sum of marginal quadratic negentropies as a dependence measure. The proposed algorithm can handle arbitrary source distributions and is scalable with respect to the number of components and observations. Experimental results have shown that the proposed algorithm consistently outperforms others. In addition, it is extremely robust to outliers and more effective when the number of observed samples is small and the number of mixed sources is large.

The proposed contrast function is not guaranteed to have the same maximum with the original one. Empirically, however, our method shows good performance and can be applied to cases where a limited number of observations is available.

Acknowledgment

This research was supported as a Brain Neuroinformatics Research Program by Korean Ministry of Commerce, Industries, and Energy.

References

1. Bach, F.R., Jordan, M.I: Kernel independent component analysis. Journal of Machine Learning Research 3, 1–48 (2002)
2. Araki, S., Makino, S., Nishikawa, T., Saruwatari, H.: Fundamental limitation of frequency domain blind source separation for convolutive mixture of speech. In: Proc. ICASSP. vol. 5, pp. 2737–2740 (2001)
3. Learned-Miller, E.G.: Ica using spacings estimates of entropy. Journal of Machine Learning Research 4, 1271–1295 (2003)
4. Torkkola, K.: Feature extraction by non-parametric mutual information maximization. Journal of Machine Learning Research 3, 1415–1438 (2003)
5. Hild II, K.E., Erdogmus, D., Principe, J.C.: Blind source separation using renyi's mutual information. IEEE Signal Processing Letters 8(6), 174–176 (2001)
6. Hild II, K.E., Erdogmus, D., Principe, J.C.: An analysis of entropy estimators for blind source separation. Signal Processing 86, 182–194 (2006)
7. Comon, P.: Independent component analysis, a new concept? Signal Processing 36, 287–314 (1994)
8. Principe, J.C., Fisher III, J.W., Xu, D.: Information theoretic learning. In: Haykin, S. (ed.) Unsupervised Adaptive Filtering, Wiley, New York (2000)
9. Greengard, L., Strain, J.: The fast gauss transform. SIAM Journal on Scientific and Statistical Computing 12(1), 79–94 (1991)
10. Edelman, A., Arias, T.A., Smith, S.T.: The geometry of algorithms with orthogonality constraints. SIAM Journal on Matrix Analysis and Applications 20(2), 303–353 (1998)
11. Silverman, B.W.: Density Estimation for Statistics and Data Analysis. Chapman and Hall, Sydney (1986)
12. Hyvarinen, A., Oja, E.: A fast fixed-point algorithm for independent component analysis. Neural Computation 9, 1483–1492 (1997)

13. Lee, T.-W., Girolami, M., Sejnowski, T.J.: Independent component analysis using an extended infomax algorithm for mixed sub-gaussian and super-gaussian sources. Neural Computation 11, 417–441 (1999)
14. Boscolo, R., Pan, H., Roychowdhury, V.P.: Independent component analysis based on nonparametric density estimation. IEEE Transactions on Neural Networks 15(1), 55–65 (2004)
15. Amari, S., Cichocki, A., Yang, H.: A new learning algorithm for blind source separation. In: Advances in Neural Information Processing 8 (Proc. NIPS'95), pp. 757–763 (1996)

Underdetermined Source Separation Using Mixtures of Warped Laplacians

Nikolaos Mitianoudis and Tania Stathaki

Communications and Signal Processing Group,
Imperial College London,
Exhibition Road, London SW7 2AZ, UK
n.mitianoudis@imperial.ac.uk

Abstract. In a previous work, the authors have introduced a Mixture of Laplacians model in order to cluster the observed data into the sound sources that exist in an underdetermined two-sensor setup. Since the assumed linear support of the ordinary Laplacian distribution is not valid to model angular quantities, such as the Direction of Arrival to the set of sensors, the authors investigate the performance of a Mixture of Warped Laplacians to perform efficient source separation with promising results.

1 Introduction

Assume that a set of M microphones $\boldsymbol{x}(n) = [x_1(n), \ldots, x_M(n)]^T$ observes a set of N sound sources $\boldsymbol{s}(n) = [s_1(n), \ldots, s_N(n)]^T$. The case of instantaneous mixing, i.e. each sensor captures a scaled version of each signal with no delay in transmission, will be considered with negligible additive noise. The instantaneous mixing model can thus be expressed in mathematical terms, as follows:

$$\boldsymbol{x}(n) = \boldsymbol{A}\boldsymbol{s}(n) \tag{1}$$

where $\boldsymbol{A}$ represents the *mixing matrix* and n the sample index. The blind source separation problem provides an estimate of the source signals $\boldsymbol{s}$, based on the observed microphone signals and some general source statistical profile.

The underdetermined source separation problem ($M < N$) is a challenging problem. In this case, the estimation of the mixing matrix A is not sufficient for the estimation of nonGaussian source signals $\boldsymbol{s}$, as the pseudo-inverse of $\boldsymbol{A}$ can not provide a valid solution. Hence, this blind estimation problem can be divided into two sub-problems: i) estimating the mixing matrix $\boldsymbol{A}$ and ii) estimating the source signals $\boldsymbol{s}$.

In this study, we will assume a two sensor instantaneous mixing approach. The combination of several instruments into a stereo mixture in a recording studio follows the instantaneous mixing model of (1). The proposed approach can thus be used to decompose a studio recording into the separate instruments that exist in the mixture for many possible applications, such as music transcription, object-based audio coding and audio remixing.

M.E. Davies et al. (Eds.): ICA 2007, LNCS 4666, pp. 236–243, 2007.

The solution of the two above problems is facilitated by moving to a sparser representation of the data, such as the *Modified Discrete Cosine Transform* (MDCT). In the case of sparse sources, the density of the data in the mixture space shows a tendency to cluster along the directions of the mixing matrix columns. It has been demonstrated [6] that the phase difference θ_n between the two sensors can be used to identify and separate the sources in the mixture.

$$\theta_n = \mathrm{atan}\frac{x_2(n)}{x_1(n)} \tag{2}$$

Using the phase difference information between the two sensors is equivalent to mapping all the observed data points on the unit-circle. The strong super-Gaussian characteristics of the individual components in the MDCT domain are preserved in the angle representation θ_n. We can also define the amplitude r_n of each point $\boldsymbol{x}(n)$, as follows:

$$r_n = \sqrt{x_1(n)^2 + x_2(n)^2} \tag{3}$$

In a previous work [6], we proposed a clustering approach on the observed θ_n to perform source separation. In order to model the sparse characteristics of the source distributions, we introduced the following Mixture of Laplacians (MoL) that was trained using an Expectation-Maximisation (EM) algorithm on the observed angles θ_n of the input data.

$$p(\theta_n) = \sum_{i=1}^{N} \alpha_i \mathcal{L}(\theta, c_i, m_i) = \sum_{i=1}^{N} \alpha_i c_i e^{-2c_i|\theta_n - m_i|} \tag{4}$$

where N is the number of the Laplacians in the mixture, m_i defines the mean and $c_i \in \mathbf{R}^+$ controls the "width" of the distribution. Once the model is trained each of the Laplacians of the MoL should be centred on the Direction of Arrival (DOA) of the sources in the majority of the cases, i.e. the angles denoted by the columns of the mixing matrix. One can perform separation using optimal detection approaches for the individual trained Laplacians.

There is a shortcoming in the previous assumed model. The model in (4) assumes a linear support for θ_n, which is not valid as the actual support for θ_n wraps around $\pm 90^o$. The linear support is not a problem if the sources are well contained within $[-90^o, 90^o]$. To overcome this problem, we proposed a strategy in [6], where in each update we check whether any of the centres are closer to any of the boundaries ($\pm 90^o$). In this case, all the data points and the estimated centres m_i are rotated, so that the affected boundary (-90° or 90°) is mapped to the middle of the centres m_i that feature the greatest distance. This seemed to alleviate the problem in the majority of cases, however, it still serves as a heuristic solution.

To address this problem in a more eloquent manner, one can introduce wrapped *distributions* to provide a more complete solution. In the literature, there exist several "circular" distributions, such as the von Mises distribution (also known as the circular normal distribution). However, this definition is rather difficult to optimise in an EM perspective. In this study, we examine the use of an approximate warped-Laplacian distribution to model the periodicity of 180° that exists in atan$(\cdot)$ with encouraging results.

2 A Mixture of Warped Laplacians

The observed angles θ_n of the input data can be modelled, as a Laplacian wrapped around the interval $[-90^o, 90^o]$ using the following additive model:

$$\begin{aligned}\mathcal{L}_w(\theta, c, m) &= \frac{1}{2T-1}\sum_{t=-T}^{T} c e^{-2c|\theta - m - 180t|} \\ &= \frac{1}{2T-1}\sum_{t=-T}^{T} \mathcal{L}(\theta - 180t, c, m) \qquad \forall\ \theta_n \in [-90^o, 90^o] \end{aligned} \tag{5}$$

where $T \in \mathbf{Z}^+$ denotes the number of ordinary Laplacians participating in the wrapped version. The above expression models the wrapped Laplacian by an ordinary Laplacian and its periodic repetitions by 180°. This is an extension of the wrapped Gaussian distribution proposed by Smaragdis and Boufounos [7] for the Laplacian case. The addition of the wrapping of the distribution aims at mirroring the wrapping of the observed angles at $\pm 90^o$, due to the atan$(\cdot)$ function. In general, the model should have $T \rightarrow \infty$ components, however, it seems that in practice a small range of values for T can successfully approximate the full warped probability density function.

In a similar fashion to Gaussian Mixture Models (GMM), one can introduce *Mixture of Warped Laplacians* (MoWL) in order to model a mixture of angular or circular sparse signals. A *Mixture of Warped Laplacians* can thus be defined, as follows:

$$p(\theta) = \sum_{i=1}^{N} \alpha_i \mathcal{L}_w(\theta, c_i, m_i) = \sum_{i=1}^{N} \alpha_i \frac{1}{2T-1} \sum_{t=-T}^{T} c_i e^{-2c_i|\theta - m_i - 180t|} \tag{6}$$

where α_i, m_i, c_i represent the weight, mean and width of each Laplacian respectively and all weights should sum up to one, i.e. $\sum_{i=1}^{N} \alpha_i = 1$. The *Expectation-Maximization* (EM) algorithm has been proposed as a valid method to train a mixture model [1]. Consequently, the EM can be employed to train a MoWL over a training set. We derive the EM algorithm, based on Bilmes's analysis [1] for the estimation of a GMM. Bilmes estimates Maximum Likelihood mixture density parameters using the EM [1]. Assuming K training samples for θ_n and Mixture of Warped Laplacians densities (6), the log-likelihood of these training samples θ_n takes the following form:

$$I(\alpha_i, c_i, m_i) = \sum_{n=1}^{K} \log \sum_{i=1}^{N} \alpha_i \mathcal{L}_w(\theta_n, c_i, m_i) \tag{7}$$

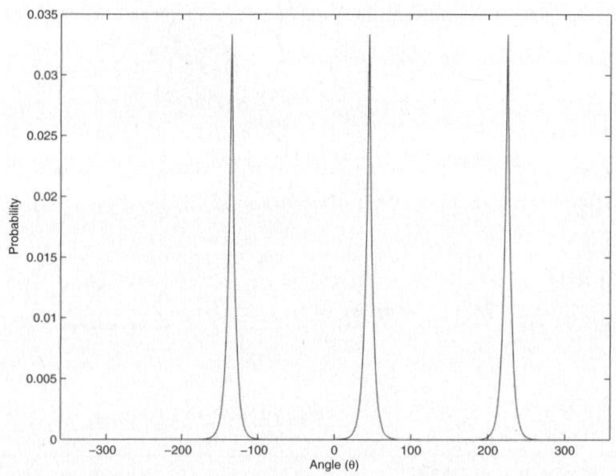

Fig. 1. An example of the Wrapped Laplacian for $T = [-1, 0, 1]$ $c = 0.01$ and $m = 45^o$

Introducing unobserved data items that can identify the components that "generated" each data item, we can simplify the log-likelihood of (7) for Warped Laplacian Mixtures, as follows:

$$J(\alpha_i, c_i, m_i) = \sum_{n=1}^{K} \sum_{i=1}^{N} \left(\log \alpha_i - \log(2T + 1) + \log \sum_{t=-T}^{T} \mathcal{L}(\theta - 180t, c_i, m_i) \right) p(i|\theta_n) \tag{8}$$

where $p(i|\theta_n)$ represents the probability of sample θ_n belonging to the ith Laplacian of the MoWL. In a similar manner, we can also introduce unobserved data items to identify the individual Laplacian of the ith Warped Laplacian that depends on θ_n.

$$H(\alpha_i, c_i, m_i) = \sum_{n=1}^{K} \sum_{i=1}^{N} (\log \alpha_i - \log(2T + 1) + \log c_i \tag{9}$$

$$- \sum_{t=-T}^{T} 2c_i |\theta - 180t - m_i| p(t|i, \theta_n)) p(i|\theta_n) \tag{10}$$

where $p(t|i, \theta_n)$ represents the probability of sample θ_n belonging to the ith Warped Laplacian and the tth individual Laplacian. The updates for $p(t|i, \theta_n)$, $p(i|\theta_n)$ and α_i can be given by the following equations:

$$p(t|i, \theta_n) = \frac{\mathcal{L}(\theta_n - t\pi, m_i, c_i)}{\sum_{t=-T}^{T} \mathcal{L}(\theta_n - 180t, m_i, c_i)} \tag{11}$$

$$p(i|\theta_n) = \frac{\alpha_i \mathcal{L}_w(\theta_n, m_i, c_i)}{\sum_{i=1}^{N} \alpha_i \mathcal{L}_w(\theta_n, m_i, c_i)} \tag{12}$$

$$\alpha_i \leftarrow \frac{1}{K} \sum_{n=1}^{K} p(i|\theta_n) \tag{13}$$

In a similar manner to [6], one can set $\partial H(\alpha_i, c_i, m_i)/\partial m_i = 0$ and solve for m_i for the recursive update for m_i, as follows:

$$\frac{\partial H}{\partial m_i} = \sum_{n=1}^{K} \sum_{t=-T}^{T} 2c_i \mathrm{sgn}(\theta_n - 180t - m_i) p(t|i, \theta_n) p(i|\theta_n) = 0 \Rightarrow \tag{14}$$

$$\sum_{n=1}^{K} \sum_{t=-T}^{T} \frac{\theta_n - 180t}{|\theta_n - 180t - m_i|} p(t|i, \theta_n) p(i|\theta_n) = m_i \sum_{n=1}^{K} \sum_{t=-T}^{T} \frac{p(t|i, \theta_n) p(i|\theta_n)}{|\theta_n - 180t - m_i|} \Rightarrow \tag{15}$$

$$m_i \leftarrow \frac{\sum_{n=1}^{K} \sum_{t=-T}^{T} \frac{\theta_n - 180t}{|\theta_n - 180t - m_i|} p(t|i, \theta_n) p(i|\theta_n)}{\sum_{n=1}^{K} \sum_{t=-T}^{T} \frac{1}{|\theta_n - 180t - m_i|} p(t|i, \theta_n) p(i|\theta_n)} \tag{16}$$

Similarly, one can set $\partial H(\alpha_i, c_i, m_i)/\partial c_i = 0$, to solve for the estimate of c_i:

$$\frac{\partial H}{\partial c_i} = \sum_{n=1}^{K} (c_i^{-1} - 2 \sum_{t=-T}^{T} |\theta_n - 180t - m_i| p(t|i, \theta_n)) p(i|\theta_n) = 0 \Rightarrow \tag{17}$$

$$c_i \leftarrow \frac{\sum_{n=1}^{K} p(i|\theta_n)}{2 \sum_{n=1}^{K} \sum_{t=-T}^{T} |\theta_n - 180t - m_i| p(t|i, \theta_n) p(i|\theta_n)} \tag{18}$$

Once the MoWL is trained, optimal detection theory and the estimated individual Laplacians can be employed to provide estimates of the sources. The centre of each warped Laplacian m_i should represent a column of the mixing matrix A in the form of $[\cos(m_i)\ \sin(m_i)]^T$. Each warped Laplacian should model the statistics of each source in the transform domain and can be used to perform underdetermined source separation.

A "Winner takes all" strategy attributes each point (r_n, θ_n) to only one of the sources. This is performed by setting a hard threshold at the intersections between the trained Warped Laplacians. Consequently, the source separation problem becomes an *optimal decision* problem. The decision thresholds θ_{ij}^{opt} between the i-th and the j-th neighbouring Laplacians are given by the following equation:

$$\theta_{ij}^{opt} = \frac{\ln \frac{\alpha_i c_i}{\alpha_j c_j} + 2(c_i m_i + c_j m_j)}{2(c_i + c_j)} \tag{19}$$

Using these thresholds, the algorithm can attribute the points with $\theta_{ij}^{opt} < \theta_n < \theta_{jk}^{opt}$ to source j, where i, j, k are neighbouring Laplacians (sources). Having attributed the points $\boldsymbol{x}(n)$ to the N sources, using the proposed thresholding technique, the next step is to reconstruct the sources. Let $S_i \cap K$ represent the point indices that have been attributed to the ith source. We initialise $u_i(n) = 0, \forall$ $n = 1, \ldots, K$ and $i = 1, \ldots, N$. The source reconstruction is performed by substituting:

$$u_i(S_i) = [\cos(m_i)\ \sin(m_i)] \boldsymbol{x}(S_i) \qquad \forall\ i = 1, \ldots, N \tag{20}$$

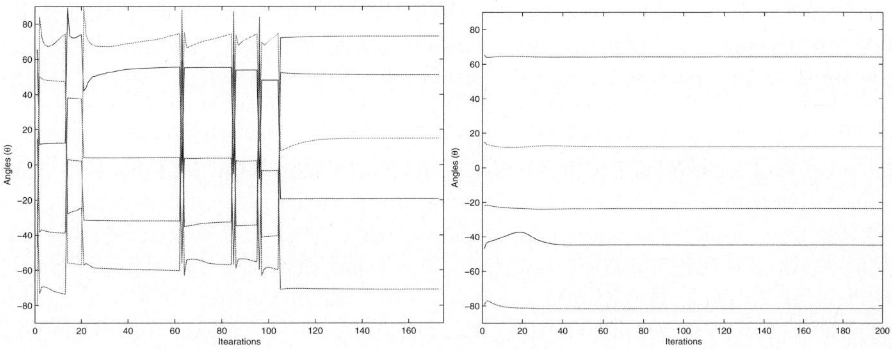

Fig. 2. Estimation of the mean using MoL with the shifting strategy (left) and the warped MoL (right)

3 Experiments

In this section, we evaluate the algorithm proposed in the previous section. We will use Hyvärinen's clustering approach [4], O'Grady and Pearlmutter's [5] Soft LOST algorithm's and the MoL-EM_Hard as proposed in a previous work [6], to demonstrate several trends using artificial mixtures or publicly available datasets[1]. In order to quantify the performance of the algorithms, we are estimating the *Signal-to-Distortion Ratio* (SDR) from the BSS_EVAL Toolbox [2]. The frame length for the MDCT analysis is set to 64 msec for the test signals sampled at 16 KHz and to 46.4 msec for those at 44.1 KHz. We initialise the parameters of the MoL and MoWL, as follows: $\alpha_i = 1/N$ and $c_i = 0.001$ and $T = [-1, 0, 1]$ (for MoWL only). The centres m_i were initialised in both cases using a *K-means* step. The initialisation of m_i is important, as if we choose two initial values for m_i that are really close, then it is very probable that the individual Laplacians may not converge to different clusters. To provide a more accurate estimation of m_i, training is initially performed using a "reduced" dataset, containing all points that satisfy $r_n > 0.2$, provided that the input signals are scaled to $[-1, 1]$. The second phase is to use the "complete" dataset to update the values for α_i and c_i.

3.1 Artificial Experiment

In this experiment, we use 5 solo audio uncorrelated recordings (a saxophone, an accordion, an acoustic guitar, a violin and a female voice) of sampling frequency 16 KHz and duration 8.19 msec. The mixing matrix is constructed as in (21), choosing the angles in Table 1. Two of the sources are placed close to the wrapping edges $(-80^o, 60^o)$ and three of them are placed rather closely at

[1] All the experimental audio results are available online at: `http://www.commsp.ee.ic.ac.uk/~nikolao/lmm.htm`

$-40^o, -20^o, 10^o$, in order to test the algorithm's resilience to the wrapping at $\pm 90^o$. In Table 1, we can see the estimated angles of the original MoL_Hard with the shifting solution and the MoWL. In both cases, the algorithms estimate approximately the same means m_i, which are very close to the original ones. In Fig. 2, the convergence of the means m_i in the two cases is depicted. The proposed warped solution seems to converge smoothly and faster without the perturbations caused by the shifting solution in the previous algorithm. Note that Fig. 2(a) depicts the angles after the rotating steps to demonstrate the shifting of ψ_i in the original MoL solution. Their performance in terms of SDR is depicted in Table 2. Hyvärinen's approach is very prone to initialisation, however, the results are acquired using the best run of the algorithm. This could be equally avoided by using a K-means initialisation step. The Soft_Lost algorithm managed to separate the sources in most cases, however, there were some audible artifacts and clicks that reduced the calculated quality measure. To appreciate the results of this rather difficult problem, we can spot the improvement performed by the methods compared to the input signals. It seems that the proposed algorithm performs similarly to MoL_Hard and the Hyvärinen's approach, which implies that the proposed solution to approximate the wrapping of the pdf is valid.

$$A = \begin{bmatrix} \cos(\psi_1) \cos(\psi_2) \ldots \cos(\psi_N) \\ \sin(\psi_1) \sin(\psi_2) \ldots \sin(\psi_N) \end{bmatrix} \tag{21}$$

Table 1. The five angles used in the artificial experiment and their estimates using the MoL and MoWL approaches

	ψ_1	ψ_2	ψ_3	ψ_4	ψ_5
`Original`	-80°	-40°	-20°	10°	60°
`Estimated MoL`	-81.52^o	-45.45^o	-23.45^o	12.59^o	64.18^o
`Estimated MoWL`	-81.59^o	-44.98^o	-23.59^o	12.18^o	64.19^o

3.2 Real Recording

In this section, we tested the algorithms with the *Groove* dataset, available by (BASS-dB) [3], sampled at 44.1 KHz. The "Groove" dataset features four widely spaced sources: bass (far left), distortion guitar (center left), clean guitar (center right) and drums (far right). In Table 2, we can see the results for the four methods in terms of SDR. The proposed MoWL approach managed to perform similarly to the previous MoL_EM, despite the small spacing of the sources and the source being placed at the edges of the solution space, which implies that the warped Laplacian model manages to model the warping of θ_n without any additional steps. The proposed MoL approaches managed to outperform Hyvärinen and Soft_LOST approach for the majority of the sources. Again, the LOST approach still introduces several audio artifacts and clicks.

Table 2. The proposed MoWL approach is compared in terms of SDR (dB) with MoL-EM_hard, Hyvärinen's, soft_LOST and the average SDR of the mixtures

	Artificial experiment					Groove Dataset			
	s_1	s_2	s_3	s_4	s_5	s_1	s_2	s_3	s_4
Mixed Signals	-6.00	-13.37	-26.26	-6.67	-6.81	-30.02	-10.25	-6.14	-21.24
MoWL-EM_hard	6.07	-2.11	5.62	4.09	6.15	4.32	-4.35	-1.16	3.27
MoL-EM_hard	6.69	0.32	7.66	3.65	6.03	2.85	-4.47	-0.86	3.28
Hyvärinen	6.53	-1.16	7.60	4.14	5.79	3.79	-3.72	-1.13	1.49
soft_LOST	4.58	-4.01	5.09	1.67	3.93	4.54	-5.77	-1.74	3.62

4 Conclusions

The problem of underdetermined source separation is examined in this study. In a previous work, we proposed to address the two-sensor problem by clustering using a Mixture of Laplacian approach on the source Direction of Arrival (DOA) θ_n to the sensors. In this study, we address the problem of wrapping of θ_n using a Warped Mixture of Laplacians approach. The new proposed approach features similar performance and faster convergence to MoL_hard and seems to be able to separate sources that are close to the boundaries ($\pm 90^0$) without any extra trick and therefore serves as a valid solution to the problem.

References

1. Bilmes, J.A.: A gentle tutorial of the EM algorithm and its application to parameter estimation for Gaussian Mixture and Hidden Mixture Models, Tech. Rep. Dep. of Electrical Eng. and Computer Science, U.C. Berkeley, California (1998)
2. Févotte, C., Gribonval, R., Vincent, E.: BSS EVAL Toolbox User Guide, Tech. Rep. IRISA Technical Report 1706, Rennes, France (April 2005), `http://www.irisa.fr/metiss/bsseval/`
3. Vincent, E., Gribonval, R., Fevotte, C., Nesbit, A., Plumbley, M.D., Davies, M.E., Daudet, L.: BASS-dB: the blind audio source separation evaluation database, Available at `http://bass-db.gforge.inria.fr/BASS-dB/`
4. Hyvärinen, A.: Independent component analysis in the presence of Gaussian noise by maximizing joint likelihood. Neurocomputing 22, 49–67 (1998)
5. O'Grady, P.D., Pearlmutter, B.A.: Soft-LOST: EM on a mixture of oriented lines. In: Proc. Int. Conf. on Independent Component Analysis and Blind Source Separation, Granada, Spain, pp. 428–435 (2004)
6. Mitianoudis, N., Stathaki, T.: Underdetermined Source Separation using Laplacian Mixture Models, IEEE Trans. on Audio, Speech and Language (to appear)
7. Smaragdis, P., Boufounos, P.: Position and Trajectory Learning for Microphone Arrays. IEEE Trans. Audio, Speech and Language Processing 15(1), 358–368 (2007)

Blind Separation of Cyclostationary Sources Using Joint Block Approximate Diagonalization

D.T. Pham

Laboratoire Jean Kuntzmann - CNRS/INPG/UJF, BP 53, 38041 Grenoble Cedex, France
Dinh-Tuan.Pham@imag.fr

Abstract. This paper introduces an extension of an earlier method of the author for separating stationary sources, based on the joint approximated diagonalization of interspectral matrices, to the case of cyclostationary sources, to take advantage of their cyclostationarity. the proposed method is based on the joint block approximate diagonlization of cyclic interspectral density. An algorithm for this diagonalization is described. Some simulation experiments are provided, showing the good performance of the method.

1 Introduction

Blind source separation aims at recovering sources from their unknown mixtures [1]. All separation methods are based on some "non properties" of the source signals. Early methods which do not exploit the time structure of the signals would require non Gaussianity of the sources. However, by exploiting the time structure, one can separate mixtures of Gaussian sources provided that the sources are *not* independent identically distributed (iid) in time., that is one (or both) of the two "i" in "iid" is not met. If only the first "i" is not met, one has stationary correlated sources and separation can be achieved by considering the lagged covariances or inter-spectra between mixtures signals. This is the basis of most second order separation methods [2, 3, 4]. If the second "i" in "iid" is not fulfilled, one has nonstationary sources and separation methods can again be developed using only second order statistics [5, 6]. However, "nonstationarity" is a too general non property to be practical, the above works actually focus only on a particular aspect of it: They assume temporal independence (or more accurately ignore possible temporal dependency) and focus only on the variation of variance of the signal in time and assume that this variation is slow enough to be adequately estimated nonparametrically. In this paper, we consider another aspect of non stationarity: the cyclostationarity. The variance of the source is also variable in time but in an (almost) periodic manner. Further, the autocovariance between the source at different time points does not depend only on the delay as in the stationary case, but also on time as well and again in a (almost) periodic manner. Thus the "nonstationary" method in [6] may not work as this source variance can vary rapidly since the period (frequency) can be short (high). Moreover, such method ignores the lagged autocovariance of the sources, which provide important useful information for the separation. The "stationary" methods [2, 3, 4] still work in general if one takes as lagged covariances the average lagged covariances

M.E. Davies et al. (Eds.): ICA 2007, LNCS 4666, pp. 244–251, 2007.

over time. In fact the usual lagged covariance estimator when applied to cyclostationary signal actually estimates the average lagged covariance. However, such methods ignore the specificity of cyclostationary signals and thus don't benefice from it and further would fail if the sources are noncorrelated (but has variance varying periodically with time). Our method is specially designed to exploit this specificity. There have been several works on blind separation of cyclostationary sources [7, 8, 9, 10]. Our work is different in that we work with cyclic inter-spectral densities while the above works are mainly based on cyclic cross-covariances. Our work may be viewed as an extension of our earlier work for blind separation of stationary sources [4] based on the joint approximate diagonalization of a set of inter-spectral matrices. As said earlier, this method still works for cyclostationary sources, provided that their average spectra are different up to a constant factor. The present method exploits the extra information of cyclostationarity and thus yields better performance and also can be dispensed with the above restriction.

2 Cyclostationary Signals

A discrete time (possibly complex value) process $\{X(t)\}$ is said to be cyclostationary (or almost periodically correlated) if its mean function $t \mapsto \mathrm{E}[X(t)]$ and its covariance functions $t \mapsto \mathrm{cov}\{X(t+\tau), X(t)\}$ are almost periodic [11]. The definition of almost periodicity is rather technical, but here we consider only zero mean cyclostationary process with a finite "number of cycles", for which an equivalent definition is that there exists a finite subset $\mathcal{A}$ of $(-1/2, 1/2]$ such that

$$\mathrm{E}[X(t+\tau)X^*(t)] = \sum_{\alpha\in\mathcal{A}} R(\alpha;\tau)e^{i2\pi\alpha t}, \quad \forall t,\ \forall\tau. \tag{1}$$

where * denotes the complex conjugate. The function $\tau \mapsto R(\alpha;\tau)$ is called the cyclic autocovariance function of cycle α. From (1), it can be computed as

$$R(\alpha;\tau) = \lim_{T\to\infty} \frac{1}{T}\sum_{t=1}^{T} \mathrm{E}[X(t+\tau)X^*(t)]e^{-i2\pi\alpha t} \tag{2}$$

Note that for $\alpha \notin \mathcal{A}$, the last right hand side yields zero by (1). Thus we may define $R(\alpha,\tau)$ for all α,τ by the above right hand side, and $\mathcal{A}$ as the set $\{\alpha : R(\alpha;\cdot) \neq 0\}$.

We shall assume that the function $R(\alpha;\cdot)$ admits a Fourier transform $f(\alpha;\cdot)$, called the cyclic spectral density of cycle α:

$$f(\alpha;\nu) = \sum_{\tau=-\infty}^{\infty} R(\alpha;\tau)e^{-i2\pi\nu\tau} \quad \Leftrightarrow \quad R(\alpha;\tau) = \int_0^1 f(\alpha;\nu)e^{i2\pi\nu\tau}d\nu.$$

Note It can be seen from (2) that $R(-\alpha;\tau) = R^*(\alpha;-\tau)e^{-i2\pi\alpha\tau}$. This means that if $\mathcal{A}$ contains α, it must contain $-\alpha$.

Let $\alpha_1,\dots,\alpha_q$ be in $\mathcal{A}$, the matrix of general j,k element $R(\alpha_k - \alpha_j;\tau)e^{i2\pi\alpha_j\tau}$ can be seen to be the average autocovariance of lag τ of the vector process $\{[X(t)e^{i2\pi\alpha_1 t} \ \cdots \ X(t)e^{i2\pi\alpha_q t}]^T\}$, since

$$R(\alpha_k - \alpha_j; \tau)e^{i2\pi\alpha_j\tau} = \lim_{T\to\infty} \frac{1}{T}\sum_{t=1}^{T} \mathrm{E}[X(t+\tau)X^*(t)]e^{i2\pi\alpha_j(t+\tau)}e^{-i2\pi\alpha_k t}$$

Therefore this matrix as a function of τ is of type positive and it follows that its Fourier transform is a non negative (matrix) function. In other words:

$$\begin{bmatrix} f(0;\nu-\alpha_1) & \cdots & f(\alpha_q-\alpha_1;\nu-\alpha_1) \\ \vdots & \ddots & \vdots \\ f(\alpha_1-\alpha_q;\nu-\alpha_q) & \cdots & f(0;\nu-\alpha_q) \end{bmatrix} \geq \mathbf{0} \tag{3}$$

In particular, $f(0;\cdot) \geq 0$. The functions $R(0;\cdot)$ and $f(0;\cdot)$ may be viewed as the average covariance function and spectral density of the process $\{X(t)\}$. Since $R(0;0) = \lim_{T\to\infty} T^{-1}\sum_{t=1}^{T} \mathrm{E}[|X(t)|^2] > 0$, $0 \in \mathcal{A}$. By taking $\alpha_1 = 0$, one see that the matrix in (3) can contain all the cyclic spectral densities of cycle in $\mathcal{A}$ and possibly some other vanishing cyclic spectral densities (since its cycle is not in $\mathcal{A}$) as well.

The natural estimator of $R(\alpha;\tau)$ based on an observed sample $X(1),\ldots,X(T)$ is

$$\hat{R}(\alpha;\tau) = \frac{1}{T}\sum_{t=\max(1,1-\tau)}^{\min(T,T-\tau)} X(t+\tau)X^*(t)e^{-i2\pi\alpha t}. \tag{4}$$

From this estimator, one may construct an estimator for $f(\alpha;\nu)$ as

$$\hat{f}(\alpha;\nu) = \sum_{\tau=1-T}^{T-1} k_M(\tau)\hat{R}(\alpha;\tau)e^{-i2\pi\nu\tau} \tag{5}$$

where $k_M(\cdot)$ is a given lag windows, often of the form $k(\cdot/M)$ with k being some given even function taking the value 1 at 0, and M is a window width parameter.

3 The Mixture Model and Separation Method

We consider the simplest mixture model in which the mixing is instantaneous without noise and there is a same numbers of mixtures as the sources: $\mathbf{X}(t) = \mathbf{AS}(t)$ where $\mathbf{X}(t)$ and $\mathbf{S}(t)$ denote the vectors of mixtures and of sources at time t, and $\mathbf{A}$ is a square matrix. The sources are assumed to be independent cyclostationary processes. It is easily seen that the observed mixtures are also cyclostationary, with the set of cycle frequencies contained in the union of the sets of cycle frequencies of the sources, which we denote by $\mathcal{A}$. The goal is to recover the sources from their mixtures. For simplicity, we shall assume that $\mathcal{A}$ is known. In practice, such set can be estimated. Further, it is not important that $\mathcal{A}$ be accurately known.

We define the cyclic autocovariance function $\mathbf{R_X}(\alpha;\cdot)$ of cycle α of the vector process $\{\mathbf{X}(t)\}$ similar to (2) except that $X(t)$ is replaced by $\mathbf{X}(t)$ and * is understood as the transpose conjugate. Clearly $\mathbf{R_X}(\alpha;\tau) = \mathbf{AR_S}(\alpha;\tau)\mathbf{A}^*$ where $\mathbf{R_S}(\alpha;\cdot)$ is the cyclic autocovariance function of cycle α of the vector source process $\{\mathbf{S}(t)\}$. The independence of the sources implies that the matrices $\mathbf{R_S}(\alpha;\tau)$ are diagonal for all α,τ

(of course if $\alpha \notin \mathcal{A}$ this matrix vanishes and is of no interest). Similarly, we define the cyclic spectral density of cycle α of the vector process $\{\mathbf{X}(t)\}$ as the Fourier transform $\mathbf{f_X}(\alpha;\cdot)$ of $\mathbf{R_X}(\alpha;\cdot)$. Again, we have $\mathbf{f_X}(\alpha;\nu) = \mathbf{A}\mathbf{f_S}(\alpha;\nu)\mathbf{A}^*$ where $\mathbf{f_S}(\alpha;\cdot)$ is the cyclic spectral density of cycle α of the vector process $\{\mathbf{S}(t)\}$, which is diagonal for all frequencies and all α.

The analogue of the matrix in (3) is the block matrix

$$\mathbf{C}(\nu) = \begin{bmatrix} \mathbf{C}_{11}(\nu) & \cdots & \mathbf{C}_{1K}(\nu) \\ \vdots & \ddots & \vdots \\ \mathbf{C}_{K1}(\nu) & \cdots & \mathbf{C}_{KK}(\nu) \end{bmatrix} \tag{6}$$

where

$$\mathbf{C}_{jk}(\nu) = \begin{bmatrix} f_{X_jX_k}(0;\nu-\alpha_1) & \cdots & f_{X_jX_k}(\alpha_q-\alpha_1;\nu-\alpha_1) \\ \vdots & \ddots & \vdots \\ f_{X_jX_k}(\alpha_1-\alpha_q;\nu-\alpha_q) & \cdots & f_{X_jX_k}(0;\nu-\alpha_q) \end{bmatrix} \tag{7}$$

$f_{X_jX_k}$ denoting the jk element of $\mathbf{f}_X$. The relation $\mathbf{f_X}(\alpha;\nu) = \mathbf{A}\mathbf{f_S}(\alpha;\nu)\mathbf{A}^*$ implies that $\mathbf{C}(\nu) = (\mathbf{A}\otimes\mathbf{I}_q)\mathbf{D}(\nu)(\mathbf{A}^*\otimes\mathbf{I}_q)$ where $\mathbf{D}$ is defined similar to $\mathbf{C}$ but with $f_{S_jS_k}$ (the jk element of $\mathbf{f}_S$) in place of $\mathbf{f}_X$, $\mathbf{I}_q$ is the identity matrix of order q and $\otimes$ denotes the Kronecker product:

$$\mathbf{A}\otimes\mathbf{M} = \begin{bmatrix} A_{11}\mathbf{M} & A_{12}\mathbf{M} & \cdots \\ A_{21}\mathbf{M} & A_{22}\mathbf{M} & \cdots \\ \vdots & \vdots & \ddots \end{bmatrix},$$

A_{ij} being the elements of $\mathbf{A}$. The independence of the sources implies that the matrix $\mathbf{D}$ is block diagonal ($\mathbf{D}_{jk} = \mathbf{0}$ except when $j = k$). Thus our idea is to find a separation matrix $\mathbf{B}$ such that $\mathbf{B}\otimes\mathbf{I}_q$ block diagonalizes all the matrices $\mathbf{C}(\nu)$ in the sense that the matrices $(\mathbf{B}\otimes\mathbf{I}_q)\mathbf{C}(\nu)(\mathbf{B}^*\otimes\mathbf{I}_q)$ are block diagonal (of block size q) for all ν.

In practice, the matrices $\mathbf{C}(\nu)$ have to be replaced by their estimators $\hat{\mathbf{C}}(\nu)$. This estimator is naturally built from the estimators $\hat{f}_{\mathbf{X}}(\alpha;\nu)$ of $f_{\mathbf{X}}(\alpha;\nu)$, defined similarly as in (5) with $\hat{R}(\alpha;\tau)$ replaced by $\hat{R}_{\mathbf{X}}(\alpha;\tau)$, the estimator of $R_{\mathbf{X}}(\alpha;\tau)$. The last estimator is defined similarly as in (4) with $X(t)$ replaced by $\mathbf{X}(t)$. As the lag window k_M in (5) has the effect of a smoothing, the (cyclic) spectral density estimator at a given frequency actually does not estimate the spectral density at this frequency but the average density over a frequency band centered around it. Therefore, we shall limit ourselves to the matrices $\hat{\mathbf{C}}(\nu)$ for ν on some finite grid, so that we have only a finite set of matrices to be block diagonalized. The spacing of the grid would be directly related to the resolution of the spectral estimator. Of course, since the $\hat{\mathbf{C}}(\nu)$ are not exactly equal to $\mathbf{C}(\nu)$, one cannot block diagonalize them exactly but only approximately, according to some block diagonality measure, which will be introduced below.

It is important that the estimator $\hat{\mathbf{C}}(\nu)$ be non negative as $\mathbf{C}(\nu)$ is. One can ensure that this is the case regardless of the data, by chosing the (real) window k_M in (5) such that $\sum_\tau k_M(\tau)e^{-2\pi\nu\tau}$ is real and non negative for all ν. Indeed, there then exists a real window $k_M^{1/2}$ (not unique) such that $\sum_\tau k_M(\tau)e^{-2\pi\nu\tau} = |\sum_\tau k_M^{1/2}(\tau)e^{-2\pi\nu\tau}|^2$ or $k_M(\tau) = \sum_u k_M^{1/2}(u-\tau)k_M^{1/2}(u)$. Therefore

$$\hat{f}_{\mathbf{X}}(\alpha;\nu) = \frac{1}{T}\sum_{\tau}\Big[\sum_{u} k_M^{1/2}(u-\tau)k_M^{1/2}(u)\Big]\Big[\sum_{v}\tilde{\mathbf{X}}(v+\tau)\tilde{\mathbf{X}}^*(v)e^{-i2\pi\alpha v}\Big]e^{-i2\pi\nu\tau}$$

where $\tilde{\mathbf{X}}(t) = \mathbf{X}(t)$ for $1 \le t \le T$, $= \mathbf{0}$ otherwise. The last right hand side equals, after summing up with respect to τ: $T^{-1}\sum_u\sum_v(k_M \star \tilde{\mathbf{X}}_\nu)(v+u)\, k_M^{1/2}(u)\mathbf{X}^*(v)e^{i2\pi(\nu-\alpha)v}$ where $\tilde{\mathbf{X}}_\nu(t) = \tilde{\mathbf{X}}(t)e^{-i2\pi\nu t}$ and $\star$ denotes the convolution. Let $t = u+v$ and summing up again first with respect to u, one gets

$$\hat{f}_{\mathbf{X}}(\alpha;\nu) = \frac{1}{T}\sum_{t}(k_M \star \tilde{\mathbf{X}}_\nu)(t)\ (k_M^{1/2} \star \mathbf{X}^*_{\nu-\alpha})(t).$$

This formula shows that $\hat{\mathbf{C}}(\nu)$ is the sample covariance of certain vector sequence, hence is non negative, and can be used for the calculation of $\hat{\mathbf{C}}(\nu)$.

4 Joint Block Approximate Diagonalization

The separation method in previous section leads to the problem of joint approximate block diagonalizing a set of positive definite block matrices $\hat{\mathbf{C}}(\nu_m), m = 1,\dots,M$, of block size q, by a matrix of the form $\mathbf{B}\otimes\mathbf{I}_q$. Following [4] we take as the measure of block diagonality of a Hermitian non negative block matrix $\mathbf{M}$: $(1/2)[\log\det\mathrm{Diag}(\mathbf{M}) - \log\det(\mathbf{M})]$ where Diag denotes the operator which builds a bloc diagonal matrix from its argument. This measure is always positive and can be zero if and only if the matrix $\mathbf{M}$ is block diagonal. Indeed, each diagonal block $\mathbf{M}_{ii}$ of $\mathbf{M}$, being non negative, can be diagonalized by a unitary matrix $\mathbf{U}_i$. Thus the matrices $\mathbf{U}_i\mathbf{M}_{ii}\mathbf{U}_i^*$ are diagonal with diagonal elements being also those of $\mathbf{UMU}^*$ where $\mathbf{U}$ is the block diagonal matrix with diagonal block $\mathbf{U}_i$. Hence by the Hadamard inequality [12], $\prod_i\det(\mathbf{U}_i\mathbf{M}_{ii}\mathbf{U}_i^*) \ge \det\mathbf{UMU}^*$ with equality if and only if $\mathbf{UMU}$ is diagonal. This yields the announced result, since the right and left hand sides of the above inequality are no other than $\det\mathrm{Diag}(\mathbf{M})$ and $\det(\mathbf{M})$, and $\mathbf{UMU}^*$ diagonal is the same as $\mathbf{M}$ is block diagonal.

Therefore we consider the joint block diagonality criterion

$$\frac{1}{2}\sum_{m=1}^{M}\{\log\det\mathrm{Diag}[(\mathbf{B}\otimes\mathbf{I}_q)\hat{\mathbf{C}}(\nu_m)(\mathbf{B}\otimes\mathbf{I}_q)] - \log\det[(\mathbf{B}\otimes\mathbf{I}_q)\hat{\mathbf{C}}(\nu_m)(\mathbf{B}^*\otimes\mathbf{I}_q)]\}. \tag{8}$$

Note that the last term in the above curly bracket $\{\ \}$ may be replaced by $2q\log\det|\mathbf{B}|$ since these two terms differ by $\log\det[\mathbf{C}(\nu_m)]$ which does not depend on $\mathbf{B}$.

The algorithm in [13] can be adapted to solve the above problem. For lack of space, we here only describe how it works. Starting from a current value of $\mathbf{B}$, it consists in performing successive transformations, each time on a pair of rows of $\mathbf{B}$, the i-th row $\mathbf{B}_{i\cdot}$ and the j-th row $\mathbf{B}_{j\cdot}$ say, according to

$$\begin{bmatrix}\mathbf{B}_{i\cdot}\\ \mathbf{B}_{j\cdot}\end{bmatrix} \leftarrow \mathbf{T}_{ij}\begin{bmatrix}\mathbf{B}_{i\cdot}\\ \mathbf{B}_{j\cdot}\end{bmatrix},$$

where $\mathbf{T}_{ij}$ is a 2×2 non singular matrix, chosen such that the criterion is decreased and whose expression is given later. Once this is done, the procedure is repeated with another pair of rows. The processing of all the $K(K-1)/2$ is called a sweep. The algorithm consists of repeated sweeps until convergence is achieved. Put

$$g_{ij} = \sum_{m=1}^{M} \frac{1}{Mq} \operatorname{tr}[\mathbf{C}_{ii}^{-1}(m;\mathbf{B})\mathbf{C}_{ij}(m;\mathbf{B})], \qquad 1 \le i \ne j \le K, \tag{9}$$

$$\omega_{ij} = \sum_{m=1}^{M} \frac{1}{Mq} \operatorname{tr}[\mathbf{C}_{ii}^{-1}(m;\mathbf{B})\mathbf{C}_{jj}(m;\mathbf{B})], \qquad 1 \le i \ne j \le K.$$

where $\mathbf{C}_{ij}(m;\mathbf{B})$ stands for the ij block of $(\mathbf{B} \otimes \mathbf{I}_q)\mathbf{C}(\nu_m)(\mathbf{B}^* \otimes \mathbf{I}_q)$ for short. The matrix is $\mathbf{T}_{ij}$ given by

$$\begin{bmatrix} 1 & 0 \\ 0 & 1 \end{bmatrix} - \frac{2}{1 + h_{ij}h_{ji} - h_{ij}^* h_{ji}^* + \sqrt{(1 + h_{ij}h_{ji} - h_{ij}^* h_{ji}^*)^2 - 4h_{ij}h_{ji}}} \begin{bmatrix} 0 & h_{ij} \\ h_{ji} & 0 \end{bmatrix}$$

where h_{ij} and h_{ji} are the solution of

$$\begin{bmatrix} \omega_{ij} & 1 \\ 1 & \omega_{ji} \end{bmatrix} \begin{bmatrix} h_{ij} \\ h_{ji}^* \end{bmatrix} = \begin{bmatrix} g_{ij} \\ g_{ji}^* \end{bmatrix}.$$

Note 1. In the case where the signal $\mathbf{X}(t)$ is real, $\hat{\mathbf{R}}_{\mathbf{X}}(\alpha;\tau) = \hat{\mathbf{R}}_{\mathbf{X}}^T(-\tau)e^{i2\pi\alpha\tau}$, ^{T} denoting the transpose, hence $\mathbf{f}_{\mathbf{X}}(\alpha;-\nu) = \mathbf{f}_{\mathbf{X}}^T(\alpha;\alpha+\nu)$. It follows that

$$f_{X_j X_k}(\alpha_m - \alpha_l; -\nu - \alpha_l) = f_{X_k X_j}(\alpha_m - \alpha_l; \nu + \alpha_m).$$

We already know that if $\mathcal{A}$ contain α it must contain $-\alpha$. Thus it is of interest to choose $\alpha_1 = 0$ and $\alpha_j = -\alpha_{q+2-j}$, $2 \le j \le q$ (which implies that q is odd, unless $1/2 \in \mathcal{A}$, in this case q may be even with $\alpha_{q/2+1} = 1/2$)[1]. Then the above right hand side can be written as $f_{X_k X_j}(\alpha_{q+2-l} - \alpha_{q+2-m}; \nu - \alpha_{q+2-m})$, with $\alpha_{q+1} = 0$ by convention. Therefore by (7): $\mathbf{C}_{jk}(-\nu) = \mathbf{\Pi}\mathbf{C}_{kj}^T(\nu)\mathbf{\Pi}^T$ for some permutation matrix $\mathbf{\Pi}$, hence $\mathbf{C}(-\nu) = (\mathbf{I}_K \otimes \mathbf{\Pi})\mathbf{C}^T(\nu)(\mathbf{I}_K \otimes \mathbf{\Pi}^T)$. It follows that for a *real* matrix $\mathbf{B}$

$$(\mathbf{B} \otimes \mathbf{I}_q)\mathbf{C}(-\nu)(\mathbf{B}^* \otimes \mathbf{I}_q) = (\mathbf{I}_K \otimes \mathbf{\Pi})[(\mathbf{B} \otimes \mathbf{I}_q)\mathbf{C}(\nu)(\mathbf{B}^* \otimes \mathbf{I}_q)]^T(\mathbf{I}_K \otimes \mathbf{\Pi}^T),$$

and thus the measure of block diagonality of the matrix in the above left hand side is the same as that of $(\mathbf{B} \otimes \mathbf{I}_q)\mathbf{C}(\nu)(\mathbf{B}^* \otimes \mathbf{I}_q)$. It is then of interest to consider a grid of frequencies $\nu_1, \ldots, \nu_M$ with M even and $\nu_m = -\nu_{M+1-m} \bmod 1$, so as to reduce the number of matrices to be block diagonalized by half, since the term corresponding to ν_m in (8) can be grouped with the one corresponding to ν_{M+1-m}. One may take $\nu_m = (m - 1/2)/M$ which yield a regular grid of spacing $1/M$.

Note 2. In the case where the signals are real, the matrix $\mathbf{B}$ must be constrained to be real, that is the minimization of (8) must be done over the set of real matrices. It can be shown that the algorithm is the same as before but the g_{ij} are now defined as the real part of the right hand side of (9).

[1] $\{\alpha_1, \ldots, \alpha_q\}$ need not be equal to $\mathcal{A}$ but can be a subset of $\mathcal{A}$.

5 Some Simulation Examples

We consider two cyclostationary sources constructed as Gaussian stationary autoregressive (AR) processes of second order, modulated with sine waves $\cos(\alpha_2 \pi t)$ and $\cos(\alpha_3 \pi t)$ respectively. Thus they have cycle frequencies $0, \pm\alpha_2$ and $0, \pm\alpha_3$ respectively. We take $\alpha_2 = 0.3/\pi = 0.0955$ and $\alpha_3 = 0.7/\pi = 0.2228$ (the same as in [9]). The AR coefficients are $1.9\cos(0.16\pi), -0.95^2$ and $\cos(0.24\pi), -0.5^2$ for the first and second sources, respectively. This corresponds to the AR polynomials with roots $0.95e^{\pm i0.16\pi}$ and $0.5e^{\pm i0.24\pi}$ respectively.

Four hundred series of length 256 are generated for each source. The 2 sources are mixed according to the mixing matrix $\mathbf{A} = \begin{bmatrix} 1 & 1 \\ -1 & 1 \end{bmatrix}$ and our method is applied to obtain the separation matrix $\mathbf{B}$. The number of positive frequency bins is set to 4. To quantify the quality of the separation, we introduce two contamination coefficients c_{12} and c_{21} defined as follows. First the global matrix $\mathbf{G} = \mathbf{BA}$ is formed, then its rows is eventually permuted such that $|G_{11}G_{22}| \geq |G_{12}G_{21}|$, G_{ij} denoting the elements of $\mathbf{G}$. Finally $c_{12} = G_{12}/G_{11}$ and $c_{21} = G_{21}/G_{22}$.

Table 1 shows the mean square of the contamination coefficients and of their products, all multiplied by 256 which is the length of the observed series (since the variance of the estimator should be asymptotically inversely proportional to this length). The mean number of iterations is also listed. For comparison, the values for the stationary method in [4] is also given. This method amounts to running our algorithm with no cycle frequency: $q = 1$ and $\alpha_1 = 0$, which means that one just ignore the cyclostationarity of the sources and considers them as stationary (with spectrum being the average spectrum over time). It can be seen that cyclostationary method yields better results than the stationary method. However, the algorithm converges a little more slowly and each iteration is also more costly computationally.

Table 1. Mean square of the contamination coefficients and of their products and mean number of iterations, obtained from the cyclostationary and stationary methods. The sources are modulated AR processes.

	256(mean c_{12}^2)	256(mean c_{21}^2)	256(mean $c_{12}c_{21}$)	mean # iterations
cyclostationary method	0.3707	0.0310	0.0010	5.86
stationary method	0.5513	0.1250	−0.0628	3.97

In a second test, we consider two cyclostationary sources constructed as (temporally) independent Gaussian processes of unit variance, modulated in the same way as before. Thus the sources are uncorrelated but have variance varying periodically. Therefore, the stationary methods, which amount to considers the sources as stationary with spectrum being the average spectrum over time, would fail since the average sources spectra are constant. Table 2 compares the results of the cyclostationary and stationary methods. It can be seen the stationary method fails completely, as expected. The cyclostationary still works reasonably well, although less well than in the case where the sources are modulated AR processes. The "nonstationarity" method in [6] is also not suitable since the variance function vary too fast. Indeed, the variance function of the sources have frequencies α_1 and α_2 respectively, which corresponds to the periods $1/\alpha_1 = \pi/0.3 = 10.472$ and $1/\alpha_1 = \pi/0.7 = 4.4880$. Thus in order to "see" the variation of the source variances one has to estimate them in a moving window of size less than 4 which is to short.

Table 2. Mean square of the contamination coefficients and of their products and mean number of iterations, obtained from the cyclostationary and stationary methods. The sources are modulated independent Gaussian processes of unit variance.

	256(mean c_{12}^2)	256(mean c_{21}^2)	256(mean $c_{12}c_{21}$)	mean # iterations
cyclostationary method	0.6638	0.6131	-0.2586	8.27
stationary method	74.4639	72.8648	-72.5062	4.43

References

1. Cardoso, J.F.: Blind signal separation: statistical principles. Proceedings of the IEEE 9, 2009–2025 (1998)
2. Tong, L., Soon, V.C., Huang, Y.F., Liu, R.: Amuse: A new blind identification algorithm. In: Proc. IEEE ISCAS, New Orleans, LA, USA, pp. 1784–1787. IEEE Computer Society Press, Los Alamitos (1990)
3. Belouchrani, A., Meraim, K.A., Cardoso, J.F., Moulines, E.: A blind source separation technique based on second order statistics. IEEE Trans. on Signal Processing 45, 434–444 (1997)
4. Pham, D.T.: Blind separation of instantaneous mixture of sources via the Gaussian mutual information criterion. Signal Processing 81, 850–870 (2001)
5. Matsuoka, K., Ohya, M., Kawamoto, M.: A neural net for blind separation of nonstationary signals. Neural networks 8, 411–419 (1995)
6. Pham, D.T., Cardoso, J.F.: Blind separation of instantaneous mixtures of non stationary sources. IEEE Trans. Signal Processing 49, 1837–1848 (2001)
7. Liang, Y.C., Leyman, A.R., Soong, B.H.: Blind source separation using second-order cyclic-statistics. In: First IEEE Signal Processing Workshop on Advances in Wireless Communications, Paris, France, pp. 57–60. IEEE Computer Society Press, Los Alamitos (1997)
8. Ferreol, A., Chevalier, P.: On the behavior of current second and higher order blind source separation methods for cyclostationary sources. IEEE Trans. on Signal Processing 48, 1712–1725 (2000)
9. Abed-Meraim, K., Xiang, Y., Manton, J.H., Hua, Y.: Blind source-separation using second-order cyclostationary statistics. IEEE Trans. on Signal Processing 49, 694–701 (2001)
10. Ferreol, A., Chevalier, P., Albera, L.: Second-order blind separation of first- and second-order cyclostationary sources-application to AM, FSK, CPFSK, and deterministic sources. IEEE Trans. on Signal Processing 52, 845–861 (2004)
11. Dehay, D., Hurd, H.L.: Representation and estimation for periodically and almost periodically correlated random processes. In: Gardner, W.A. (ed.) Cyclostationarity in Communications and Signal Processing, IEEE Press, Los Alamitos (1993)
12. Cover, T., Thomas, J.: Elements of Information Theory. Wiley, New-York (1991)
13. Pham, D.T.: Joint approximate diagonalization of positive definite matrices. SIAM J. on Matrix Anal. and Appl. 22, 1136–1152 (2001)

Independent Process Analysis Without a Priori Dimensional Information

Barnabás Póczos, Zoltán Szabó, Melinda Kiszlinger, and András Lőrincz*

Department of Information Systems
Eötvös Loránd University, Budapest, Hungary
{pbarn,szzoli}@cs.elte.hu, {kmelinda,andras.lorincz}@elte.hu

Abstract. Recently, several algorithms have been proposed for independent subspace analysis where hidden variables are i.i.d. processes. We show that these methods can be extended to certain AR, MA, ARMA and ARIMA tasks. Central to our paper is that we introduce a cascade of algorithms, which aims to solve these tasks without previous knowledge about the number and the dimensions of the hidden processes. Our claim is supported by numerical simulations. As an illustrative application where the dimensions of the hidden variables are unknown, we search for subspaces of facial components.

1 Introduction

Independent Subspace Analysis (ISA) [1] is a generalization of Independent Component Analysis (ICA). ISA assumes that certain sources depend on each other, but the dependent groups of sources are still independent of each other, i.e., the independent groups are multidimensional. The ISA task has been subject of extensive research [1,2,3,4,5,6,7,8]. In this case, one assumes that the hidden sources are independent and identically distributed (i.i.d.) in time. Temporal independence is, however, a gross oversimplification of real sources including acoustic or biomedical data. One may try to overcome this problem, by assuming that hidden processes are, e.g., autoregressive (AR) processes. Then we arrive to the AR Independent Process Analysis (AR-IPA) task [9,10]. Another method to weaken the i.i.d. assumption is to assume moving averaging (MA). This direction is called Blind Source Deconvolution (BSD) [11], in this case the observation is a temporal mixture of the i.i.d. components.

The AR and MA models can be generalized and one may assume ARMA sources instead of i.i.d. ones. As an additional step, these models can be extended to non-stationary integrated ARMA (ARIMA) processes, which are important, e.g., for modelling economic processes [12].

In this paper, we formulate the AR-, MA-, ARMA-, ARIMA-IPA generalizations of the ISA task, when (i) one allows for multidimensional hidden components and (ii) the dimensions of the hidden processes are not known. We show that in the undercomplete case, when the number of 'sensors' is larger than the number of 'sources', these tasks can be reduced to the ISA task.

* Corresponding author.

M.E. Davies et al. (Eds.): ICA 2007, LNCS 4666, pp. 252–259, 2007.

2 Independent Subspace Analysis

The ISA task can be formalized as follows:

$$\mathbf{x}(t) = \mathbf{A}\mathbf{e}(t), \text{ where } \mathbf{e}(t) = \left[\mathbf{e}^1(t); \ldots; \mathbf{e}^M(t)\right] \in \mathbb{R}^{D_e} \tag{1}$$

and $\mathbf{e}(t)$ is a vector concatenated of components $\mathbf{e}^m(t) \in \mathbb{R}^{d_e^m}$. The total dimension of the components is $D_e = \sum_{m=1}^M d_e^m$. We assume that for a given m, $\mathbf{e}^m(t)$ is i.i.d. in time t, and sources $\mathbf{e}^m$ are jointly independent, i.e., $I(\mathbf{e}^1, \ldots, \mathbf{e}^M) = 0$, where $I(.)$ denotes the mutual information (MI) of the arguments. The dimension of the observation $\mathbf{x}$ is D_x. Assume that $D_x > D_e$, and $\mathbf{A} \in \mathbb{R}^{D_x \times D_e}$ has rank D_e. Then, one may assume without any loss of generality that both the observed ($\mathbf{x}$) and the hidden ($\mathbf{e}$) signals are white. For example, one may apply Principal Component Analysis (PCA) as a preprocessing stage. Then the ambiguities of the ISA task are as follows [13]: Sources can be determined up to permutation and up to orthogonal transformations within the subspaces.

2.1 The ISA Separation Theorem

We are to uncover the independent subspaces. Our task is to find orthogonal matrix $\mathbf{W} \in \mathbb{R}^{D_e \times D_x}$ such that $\mathbf{y}(t) = \mathbf{W}\mathbf{x}(t)$, $\mathbf{y}(t) = \left[\mathbf{y}^1(t); \ldots; \mathbf{y}^M(t)\right]$, $\mathbf{y}^m = [y_1^m; \ldots; y_{d_e^m}^m] \in \mathbb{R}^{d_e^m}$, $(m = 1, \ldots, M)$ with the condition that components $\mathbf{y}^m$ are independent. Here, y_i^m denotes the ith coordinate of the mth estimated subspace. This task can be solved by means of a cost function that aims to minimize the mutual information between components:

$$J_1(\mathbf{W}) \doteq I(\mathbf{y}^1, \ldots, \mathbf{y}^M). \tag{2}$$

One can rewrite $J_1(\mathbf{W})$ as follows:

$$J_2(\mathbf{W}) \doteq I(y_1^1, \ldots, y_{d_e^M}^M) - \sum_{m=1}^M I(y_1^m, \ldots, y_{d_e^m}^m). \tag{3}$$

The first term of the r.h.s. is an ICA cost function; it aims to minimize mutual information for all coordinates. The other term is a kind of *anti-ICA* term; it aims to maximize mutual information within the subspaces. One may try to apply a heuristics and to optimize (3) in order: (1) Start by any 'infomax' ICA algorithm and minimize the first term of the r.h.s. in (3). (2) Apply only permutations to the coordinates such that they optimize the second term. Surprisingly, this heuristics leads to the global minimum of (2) in many cases. In other words, in many cases, ICA that minimizes the first term of the r.h.s. of (3) solves the ISA task apart from the grouping of the coordinates into subspaces. This feature was observed by Cardoso, first [1]. The extent of this feature is still an open issue. Nonetheless, we call it '*Separation Theorem*', because for elliptically symmetric sources and for some other distribution types one can prove that it is rigorously true [14]. (See also, the result concerning local minimum points [15]). Although there is no proof for general sources as of yet, a number of algorithms apply this heuristics with success [1,3,15,16,17,18].

2.2 ISA with Unknown Components

Another issue concerns the computation of the second term of (3). If the d_e^m dimensions of subspaces $\mathbf{e}^m$ are known then one might rely on multi-dimensional entropy estimations [8], but these are computationally expensive. Other methods deal with implicit or explicit pair-wise dependency estimations [16,15]. Interestingly, if the observations are indeed from an ICA generative model, then the minimization of the pair-wise dependencies is sufficient to get the solution of the ICA task according to the Darmois-Skitovich theorem [19]. This is not the case for the ISA task, however. There are ISA tasks, where the estimation of pair-wise dependencies is insufficient for recovering the hidden subspaces [8]. Nonetheless, such algorithms seem to work nicely in many practical cases.

A further complication arises if the d_e^m dimensions of subspaces $\mathbf{e}^m$ are not known. Then the dimension of the entropy estimation becomes uncertain. Methods that try to apply pair-wise dependencies were proposed to this task. One can find a block-diagonalization method in [15], whereas [16] makes use of kernel estimations of the mutual information.

Here we shall assume that the separation theorem is satisfied. We shall apply ICA preprocessing. This step will be followed by the estimation of the pair-wise mutual information of the ICA coordinates. These quantities will be considered as the weights of a weighted graph, the vertices of the graph being the ICA coordinates. We shall search for clusters of this graph. In our numerical studies, we make use of Kernel Canonical Correlation Analysis [4] for the MI estimation. A variant of the Ncut algorithm [20] is applied for clustering. As a result, the mutual information within (between) cluster(s) becomes large (small).

The problem is that this ISA method requires i.i.d. hidden sources. Below, we show how to generalize the ISA task to more realistic sources. Finally, we solve this more general problem when the dimensions of the subspaces are not known.

3 ISA Generalizations

We need the following notations: Let z stand for the time-shift operation, that is $(z\mathbf{v})(t) := \mathbf{v}(t-1)$. The N order polynomials of $D_1 \times D_2$ matrices are denoted as $\mathbb{R}[z]_N^{D_1\times D_2} := \{\mathbf{F}[z] = \sum_{n=0}^N \mathbf{F}_n z^n, \mathbf{F}_n \in \mathbb{R}^{D_1\times D_2}\}$. Let $\nabla^r[z] := (\mathbf{I}-\mathbf{I}z)^r$ denote the rth order difference operator, where $\mathbf{I}$ is the identity matrix, $r \geq 0$, $r \in \mathbb{Z}$.

Now, we are to estimate unknown components $\mathbf{e}^m$ from observed signals $\mathbf{x}$. We always assume that $\mathbf{e}$ takes the form like in (1) and that $\mathbf{A} \in \mathbb{R}^{D_x\times D_s}$ is of full column rank.

1. AR-IPA: The AR generalization of the ISA task is defined by the following equations: $\mathbf{x} = \mathbf{A}\mathbf{s}$, where $\mathbf{s}$ is a multivariate AR(p) process i.e, $\mathbf{P}[z]\mathbf{s} = \mathbf{Q}\mathbf{e}$, $\mathbf{Q} \in \mathbb{R}^{D_s\times D_e}$, and $\mathbf{P}[z] := \mathbf{I}_{D_s} - \sum_{i=1}^p \mathbf{P}_i z^i \in \mathbb{R}[z]_p^{D_s\times D_s}$. We assume that $\mathbf{P}[z]$ is stable, that is $\det(\mathbf{P}[z] \neq 0)$, for all $z \in \mathbb{C}$, $|z| \leq 1$. For $d_e^m = 1$ this task was investigated in [9]. Case $d_e^m > 1$ is treated in [10]. The special case of $p = 0$ is the ISA task.

2. MA-IPA or Blind Subspace Deconvolution (BSSD) task: The ISA task is generalized to blind deconvolution task (moving average task, MA(q)) as follows: $\mathbf{x} = \mathbf{Q}[z]\mathbf{e}$, where $\mathbf{Q}[z] = \sum_{j=0}^{q} \mathbf{Q}_j z^j \in \mathbb{R}[z]_q^{D_x \times D_e}$.
3. ARMA-IPA task: The two tasks above can be merged into a model, where the hidden $\mathbf{s}$ is multivariate ARMA(p,q): $\mathbf{x} = \mathbf{A}\mathbf{s}$, $\mathbf{P}[z]\mathbf{s} = \mathbf{Q}[z]\mathbf{e}$. Here $\mathbf{P}[z] \in \mathbb{R}[z]_p^{D_s \times D_s}$, $\mathbf{Q}[z] \in \mathbb{R}[z]_q^{D_s \times D_e}$. We assume that $\mathbf{P}[z]$ is stable. Thus the ARMA process is stationary.
4. ARIMA-IPA task: In practice, hidden processes $\mathbf{s}$ may be non-stationary. ARMA processes can be generalized to the non-stationary case. This generalization is called integrated ARMA, or ARIMA(p,r,q). The assumption here is that the rth difference of the process is an ARMA process. The corresponding IPA task is then

$$\mathbf{x} = \mathbf{A}\mathbf{s}, \text{ where } \mathbf{P}[z]\nabla^r[z]\mathbf{s} = \mathbf{Q}[z]\mathbf{e}. \tag{4}$$

4 Reduction of ARIMA-IPA to ISA

We show how to solve the above tasks by means of ISA algorithms. We treat the ARIMA task. Others are special cases of this one. In what follows, we assume that: (i) $\mathbf{P}[z]$ is stable, (ii) the mixing matrix $\mathbf{A}$ is of full column rank, and (iii) $\mathbf{Q}[z]$ has left inverse. In other words, there exists a polynomial matrix $\mathbf{W}[z] \in \mathbb{R}[z]^{D_e \times D_s}$ such that $\mathbf{W}[z]\mathbf{Q}[z] = \mathbf{I}_{D_e}$.[1]

The route of the solution is elaborated here. Let us note that differentiating the observation $\mathbf{x}$ of the ARIMA-IPA task in Eq. (4) in r^{th} order, and making use of the relation $z\mathbf{x} = \mathbf{A}(z\mathbf{s})$, the following holds:

$$\nabla^r[z]\mathbf{x} = \mathbf{A}\left(\nabla^r[z]\mathbf{s}\right), \text{ and } \mathbf{P}[z]\left(\nabla^r[z]\mathbf{s}\right) = \mathbf{Q}[z]\mathbf{e}. \tag{5}$$

That is taking $\nabla^r[z]\mathbf{x}$ as observations, one ends up with an ARMA-IPA task. Assume that $D_x > D_e$ (undercomplete case). We call this task uARMA-IPA. Now we show how to transform the uARMA-IPA task to ISA. The method is similar to that of [22] where it was applied for BSD.

Theorem. *If the above assumptions are fulfilled then in the uARMA-IPA task, observation process* $\mathbf{x}(t)$ *is autoregressive and its innovation* $\tilde{\mathbf{x}}(t) := \mathbf{x}(t) - E[\mathbf{x}(t)|\mathbf{x}(t-1), \mathbf{x}(t-2), \ldots] = \mathbf{A}\mathbf{Q}_0\mathbf{e}(t)$*, where* $E[\cdot|\cdot]$ *denotes the conditional expectation value. Consequently, there is a polynomial matrix* $\mathbf{W}_{\text{AR}}[z] \in \mathbb{R}[z]^{D_x \times D_x}$ *such that* $\mathbf{W}_{\text{AR}}[z]\mathbf{x} = \mathbf{A}\mathbf{Q}_0\mathbf{e}$.

Due to lack of space the proof is omitted here. Thus, AR fit of $\mathbf{x}(t)$ can be used for the estimation of $\mathbf{A}\mathbf{Q}_0\mathbf{e}(t)$. This innovation corresponds to the observation of an undercomplete ISA model $(D_x > D_e)$[2], which can be reduced to a complete

[1] One can show for $D_s > D_e$ that under mild conditions $\mathbf{Q}[z]$ has an inverse with probability 1 [21]; e.g., when the matrix $[\mathbf{Q}_0, \ldots, \mathbf{Q}_q]$ is drawn from a continuous distribution.

[2] Assumptions made for $\mathbf{Q}[z]$ and $\mathbf{A}$ in the uARMA-IPA task implies that $\mathbf{A}\mathbf{Q}_0$ is of full column rank and thus the resulting ISA task is well defined.

ISA ($D_x = D_e$) using PCA. Finally, the solution can be finished by any ISA procedure. The reduction procedure implies that hidden components $\mathbf{e}^m$ can be recovered only up to the ambiguities of the ISA task: components of (identical dimensions) can be recovered only up to permutations. Within each subspaces, unambiguity is warranted only up to orthogonal transformations.

The steps of our algorithm are summarized in Table 1.

Table 1. Pseudocode of the undercomplete ARIMA-IPA algorithm

Input of the algorithm
Observation: $\{\mathbf{x}(t)\}_{t=1,\ldots,T}$
Optimization
Differentiating: for observation $\mathbf{x}$ calculate $\mathbf{x}^* = \nabla^r[z]\mathbf{x}$
AR fit: for $\mathbf{x}^*$ estimate $\hat{\mathbf{W}}_{\text{AR}}[z]$
Estimate innovation: $\tilde{\mathbf{x}} = \hat{\mathbf{W}}_{\text{AR}}[z]\mathbf{x}^*$
Reduce uISA to ISA and whiten: $\tilde{\mathbf{x}}' = \hat{\mathbf{W}}_{\text{PCA}}\tilde{\mathbf{x}}$
Apply ICA for $\tilde{\mathbf{x}}'$: $\mathbf{e}^* = \hat{\mathbf{W}}_{\text{ICA}}\tilde{\mathbf{x}}'$
Estimate pairwise dependency e.g., as in [16] on $\mathbf{e}^*$
Cluster $\mathbf{e}^*$ **by Ncut**: the permutation matrix is $\mathcal{P}$
Estimation
$\hat{\mathbf{W}}_{\text{ARIMA-IPA}}[z] = \mathcal{P}\hat{\mathbf{W}}_{\text{ICA}}\hat{\mathbf{W}}_{\text{PCA}}\hat{\mathbf{W}}_{\text{AR}}[z]\nabla^r[z]$
$\hat{\mathbf{e}} = \hat{\mathbf{W}}_{\text{ARIMA-IPA}}[z]\mathbf{x}$

5 Results

In this section we demonstrate the theoretical results by numerical simulations.

5.1 ARIMA Processes

We created a database for the demonstration: Hidden sources $\mathbf{e}^m$ are 4 pieces of 2D, 3 pieces of 3D, 2 pieces of 4D and 1 piece of 5D stochastic variables, i.e., $M = 10$. These stochastic variables are independent, but the coordinates of each stochastic variable $\mathbf{e}^m$ depend on each other. They form a 30 dimensional space together ($D_e = 30$). For the sake of illustration, the 3D (2D) sources emit random samples of uniform distributions defined on different 3D geometrical forms (letters of the alphabet). The distributions are depicted in Fig. 1a (Fig. 2b). 30,000 samples were drawn from the sources and they were used to drive an ARIMA(2,1,6) process defined by (4). Matrix $\mathbf{A} \in \mathbb{R}^{60\times 60}$ was randomly generated and orthogonal. We also generated the polynomial $\mathbf{Q}[z] \in \mathbb{R}[z]_5^{60\times 30}$ and the stable polynomial $\mathbf{P}[z] \in \mathbb{R}[z]_1^{60\times 60}$ randomly. The visualization of the 60 dimensional process is hard: a typical 3D projection is shown in Fig. 1c. The task is to estimate original sources $\mathbf{e}^m$ using these non-stationary observations. r^{th}-order differencing of the observed ARIMA process gives rise to an ARMA process. Typical 3D projection of this ARMA process is shown Fig. 1d. Now, one can execute the other steps of Table 1 and these steps provide the estimations of the hidden components $\hat{\mathbf{e}}^m$. Here, we estimated the AR process and its order by the methods detailed in [23]. Estimations of the 3D (2D) components are provided in Fig. 1e (Fig. 1f). In the ideal case, the product of matrix $\mathbf{AQ}_0$ and the matrices provided by PCA and ISA, i.e., $\mathbf{G} := (\mathcal{P}\hat{\mathbf{W}}_{\mathrm{ICA}}\hat{\mathbf{W}}_{\mathrm{PCA}})\mathbf{AQ}_0 \in \mathbb{R}^{D_e \times D_e}$ is a block permutation matrix made of $d_e^m \times d_e^m$ blocks. This is shown in Fig. 1g.

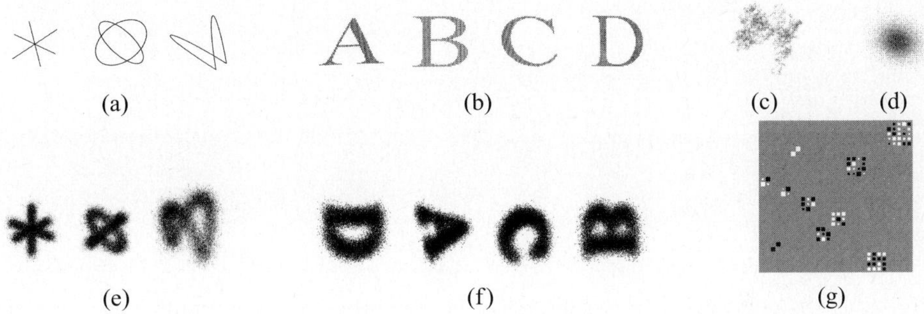

Fig. 1. (a-b) components of the database. (a): 3 pieces of 3D geometrical forms, (b): 4 pieces of 2D letters. Hidden sources are uniformly distributed variables on these objects. (c): typical 3D projection of the observation. (d): typical 3D projection of the rth-order difference of the observation, (e): estimated 3D components, (f): estimated 2D components, (g): Hinton diagram of **G**, which – in case of perfect estimation – becomes a block permutation matrix.

5.2 Facial Components

We were interested in the components that our algorithm finds when independence *is a crude approximation* at best. We have generated another database using the FaceGen[3] animation software. In our database we had 800 different front view faces with the 6 basic facial expressions. We had thus 4,800 images in total. All images were sized to 40×40 pixel. Figure 2a shows samples of the database. A large $\mathbf{X} \in \mathbb{R}^{4800 \times 1600}$ matrix was compiled; rows of this matrix were 1600 dimensional vectors formed by the pixel values of the individual images. The *columns* of this matrix were considered as mixed signals. This treatment replicates the experiments in [24]: Bartlett et al., have shown that in such cases, undercomplete ICA finds components resembling to what humans consider facial components. We were interested in seeing the components grouped by undercomplete ISA algorithm. The observed 4800 dimensional signals were compressed by PCA to 60 dimensions and we searched for 4 pieces of ISA subspaces using the algorithm detailed in Table 1.

The 4 subspaces that our algorithm found are shown in Fig. 2b. As it can be seen, the 4 subspaces embrace facial components which correspond mostly to mouth, eye brushes, facial profiles, and eyes, respectively. Thus, ICA finds interesting components and MI based ISA groups them sensibly. The generalization up to ARIMA-IPA processes is straightforward.

[3] `http://www.facegen.com/modeller.htm`

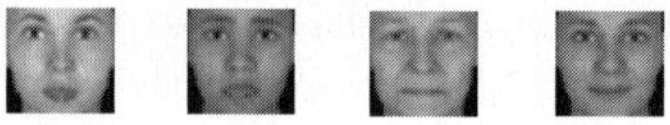

(a)

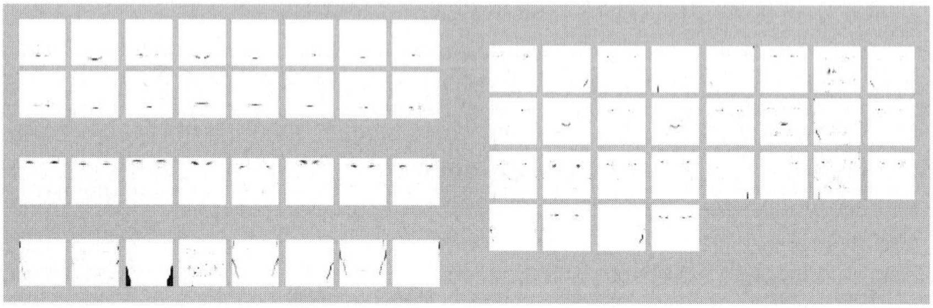

(b)

Fig. 2. (a) Samples from the database. (b) Four subspaces of the components. Distinct groups correspond mostly to mouth, eye brushes, facial profiles, and eyes, respectively.

6 Conclusions

We have extended the ISA task in two ways. (1) We solved problems where the hidden components are AR, MA, ARMA, or ARIMA processes. (2) We suggested partitioning of the graph defined by pairwise mutual information to identify the hidden ISA subspaces under certain conditions. The algorithm does not require previous knowledge about the dimensions of the subspaces. An artificially generated ARIMA process was used for demonstration. The algorithm provided sensible grouping of the estimated components for facial expressions.

Acknowledgment

This material is based upon work supported by the EC FET and NEST grants, 'New Ties' (No. 003752) and 'PERCEPT' (No. 043261). Any opinions, findings and conclusions expressed in this material are those of the authors and do not necessarily reflect the views of the EC, or other members of the EC projects.

References

1. Cardoso, J.F.: Multidimensional independent component analysis. In: Proc. of ICASSP, vol. 4, pp. 1941–1944 (1998)
2. Vollgraf, R., Obermayer, K.: Multi-dimensional ICA to separate correlated sources. In: Proc. of NIPS, vol. 14, pp. 993–1000. MIT Press, Cambridge (2001)
3. Stögbauer, H., Kraskov, A., Astakhov, S.A., Grassberger, P.: Least dependent component analysis based on mutual information. Phys. Rev. E, 70 (2004)
4. Bach, F.R., Jordan, M.I.: Beyond independent components: Trees and clusters. Journal of Machine Learning Research 4, 1205–1233 (2003)

5. Theis, F.J.: Blind signal separation into groups of dependent signals using joint block diagonalization. In: Proc. of ISCAS, pp. 5878–5881 (2005)
6. Hyvärinen, A., Köster, U.: FastISA: A fast fixed-point algorithm for independent subspace analysis. In: Proc. of ESANN, Evere, Belgium (2006)
7. Nolte, G., Meinecke, F.C., Ziehe, A., Müller, K.R.: Identifying interactions in mixed and noisy complex systems. Physical Review E, 73 (2006)
8. Póczos, B., Lőrincz, A.: Independent subspace analysis using geodesic spanning trees. In: Proc. of ICML, pp. 673–680. ACM Press, New York (2005)
9. Hyvärinen, A.: Independent component analysis for time-dependent stochastic processes. In: Proc. of ICANN, pp. 541–546. Springer, Berlin (1998)
10. Póczos, B., Takács, B., Lőrincz, A.: Independent subspace analysis on innovations. In: Gama, J., Camacho, R., Brazdil, P.B., Jorge, A.M., Torgo, L. (eds.) ECML 2005. LNCS (LNAI), vol. 3720, pp. 698–706. Springer, Heidelberg (2005)
11. Choi, S., Cichocki, A., Park, H.-M., Lee, S.-Y.: Blind source separation and independent component analysis. Neural Inf. Proc. - Lett. Rev. 6, 1–57 (2005)
12. Mills, T.C.: Time Series Techniques for Economists. Cambridge University Press, Cambridge (1990)
13. Theis, F.J.: Uniqueness of complex and multidimensional independent component analysis. Signal Processing 84(5), 951–956 (2004)
14. Szabó, Z., Póczos, B., Lőrincz, A.: Separation theorem for $\mathbb{K}$-independent subspace analysis with sufficient conditions. Technical report (2006), `http://arxiv.org/abs/math.ST/0608100`
15. Theis, F.J.: Towards a general independent subspace analysis. In: Proc. of NIPS, vol. 19 (2007)
16. Bach, F.R., Jordan, M.I.: Finding clusters in Independent Component Analysis. In: Proc. of ICA2003, pp. 891–896 (2003)
17. Szabó, Z., Póczos, B., Lőrincz, A.: Cross-entropy optimization for independent process analysis. In: Rosca, J., Erdogmus, D., Príncipe, J.C., Haykin, S. (eds.) ICA 2006. LNCS, vol. 3889, pp. 909–916. Springer, Heidelberg (2006)
18. Abed-Meraim, K., Belouchrani, A.: Algorithms for joint block diagonalization. In: Proc. of EUSIPCO, pp. 209–212 (2004)
19. Comon, P.: Independent Component Analysis, a new concept. Signal Processing, Elsevier 36(3), 287–314 (1994) (Special issue on Higher-Order Statistics)
20. Yu, S., Shi, J.: Multiclass spectral clustering. In: Proc. of ICCV (2003)
21. Rajagopal, R., Potter, L.C.: Multivariate MIMO FIR inverses. IEEE Transactions on Image Processing 12, 458–465 (2003)
22. Gorokhov, A., Loubaton, P.: Blind identification of MIMO-FIR systems: A generalized linear prediction approach. Signal Processing 73, 105–124 (1999)
23. Neumaier, A., Schneider, T.: Estimation of parameters and eigenmodes of multivariate autoregressive models. ACM Trans. on Math. Soft. 27(1), 27–57 (2001)
24. Bartlett, M., Movellan, J., Sejnowski, T.: Face recognition by independent component analysis. IEEE Trans. on Neural Networks 13(6), 1450–1464 (2002)

An Evolutionary Approach for Blind Inversion of Wiener Systems

Fernando Rojas[1], Jordi Solé-Casals[2], and Carlos G. Puntonet[1]

[1] Computer Architecture and Technology Department, University of Granada, 18071, Spain
{frojas, carlos}@atc.ugr.es
http://atc.ugr.es/
[2] Signal Processing Group, University of Vic (Spain)
jordi.sole@uvic.es
http://www.uvic.es/

Abstract. The problem of blind inversion of Wiener systems can be considered as a special case of blind separation of post-nonlinear instantaneous mixtures. In this paper, we present an approach for nonlinear deconvolution of one signal using a genetic algorithm. The recovering of the original signal is achieved by trying to maximize an estimation of mutual information based on higher order statistics. Analyzing the experimental results, the use of genetic algorithms is appropriate when the number of samples of the convolved signal is low, where other gradient-like methods may fail because of poor estimation of statistics.

Keywords: Independent component analysis, signal deconvolution, blind source separation, Wiener systems, genetic algorithms, mutual information.

1 Introduction

This paper deal with a particular class of nonlinear systems, composed by a linear subsystem followed by a memoryless nonlinear distortion. This class of nonlinear systems, also known as Wiener systems (Figure 1), can model a considerable range of actual systems in nature, such as the activity of individual primary neurons in response to prolonged stimuli [1], the dynamic relation between muscle length and tension [2], and other situations in biology, industry and psychology.

Fig. 1. A Wiener system (linear filter + nonlinear distortion)

The inverse configuration for a Wiener system is known as a Hammerstein system and consists of a nonlinear distortion followed by a linear filter.

M.E. Davies et al. (Eds.): ICA 2007, LNCS 4666, pp. 260–267, 2007.

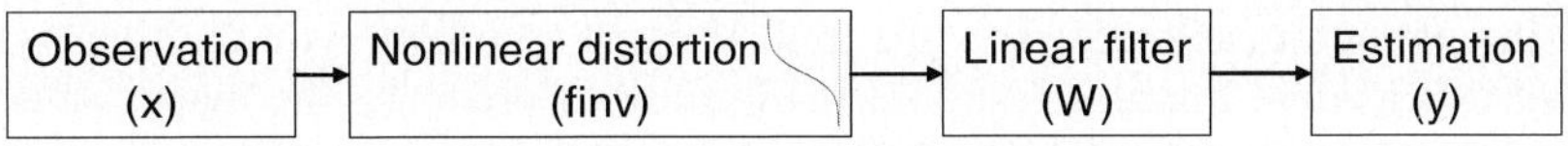

Fig. 2. A Hammerstein system (nonlinear distortion + linear filter)

We propose in this contribution the use of genetic algorithms (GA) for the nonlinear blind inversion of the (nonlinear) convolved mixed signal. The aim of the genetic algorithm is to give more aptitude to those individuals who represent a solution giving a minimal result of a mutual information estimation measure between a known number of pieces of the observed signal (**x**). In this way, the best solution given by the genetic algorithm must represent a valid estimation signal (**y**) of the original source (**s**).

Genetic algorithms have been previously applied in linear deconvolution [3] and linear and nonlinear blind source separation [4,5,6,7]. The theoretic framework for using source separation techniques in the case of blind deconvolution is presented in [8]. There, a quasi-nonparametric gradient approach is used, minimizing the mutual information of the output as a cost function to deal with the problem. This work was extended in [9].

The paper has been organized as follows: Section 2 describes the preliminary issues and the inversion main guidelines. Section 3 explains the genetic algorithm which will accomplish blind inversion, specially concerning chromosome representation and fitness function expression. Section 4 presents the experimental results showing the performance of the method. As a final point, section 5 presents the conclusions of this contribution.

2 Blind Inversion of Nonlinear Channels

2.1 Nonlinear Convolution

We suppose that the original source (**s**) is an unknown, non-Gaussian, time independent and identically distributed (i.i.d.) process. The filter H is linear, unknown and invertible. Finally, the nonlinear distortion f is invertible and differentiable. Following this notation, the observation **x** can be modeled as:

$$\mathbf{x} = f(\mathbf{H} \cdot \mathbf{s}) \tag{1}$$

where f is the nonlinear distortion and:

$$\mathbf{H} = \begin{pmatrix} \cdots & \cdots & \cdots & \cdots & \cdots \\ \cdots & h(t+1) & h(t) & h(t-1) & \cdots \\ \cdots & h(t+2) & h(t+1) & h(t) & \cdots \\ \cdots & \cdots & \cdots & \cdots & \cdots \end{pmatrix} \tag{2}$$

is an infinite dimension Toeplitz matrix which represents the action of the filter h to the signal *s(t)*. The matrix **H** is non-singular provided that the filter h is invertible, i.e. satisfies $h^{-1}(t) * h(t) = h(t) * h^{-1}(t) = \delta(t)$, where $\delta(t)$ is the Dirac impulse. The infinite dimension of vectors and matrix is due to the lack of assumption on the filter order. If the filter h is a finite impulse response (FIR) filter of order N_h, the matrix dimension can be reduced to the size N_h. In practice, because infinite-dimension equations are not tractable, we have to choose a pertinent (finite) value for N_h.

2.1 Nonlinear Deconvolution

The process of nonlinear blind inversion of the observed signal (x) assumes solving a Hammerstein system (Figure 1). Therefore, the algorithm should calculate the unknown inverse nonlinear distortion (finv), followed by the unknown inverse filter (**W**):

$$\mathbf{y} = \mathbf{W}(g \cdot \mathbf{x}) \tag{3}$$

The goal of the proposed algorithm will be to minimize mutual information (Eqn. 4), as it is assumed that mutual information vanishes when the samples in the measured signal are mutually statistically independent:

$$I(Y) = \lim_{T \to \infty} \frac{1}{2T+1} \left\{ \sum_{t=-T}^{T} H(y(t)) - H(y_{-T}, \ldots, y_T) \right\} = H(y(\tau)) - H(Y) \tag{4}$$

where τ is arbitrary due to the stationary assumption.

3 Genetic Algorithm for Nonlinear Blind Deconvolution

Nonlinear blind deconvolution can be handled by a genetic algorithm which evolves individuals corresponding to different inverse filters and nonlinear distortions. Each individual represent a potential solution and it is evaluated according to a measure of statistical independence. This is a problem of global optimization: minimizing or maximizing a real valued function $f(\mathbf{x})$ in the parameter space $\mathbf{x} \in P$. This particular type of problems is suitable to be solved by a genetic algorithm. GAs are designed to move the population away from local minima that a traditional hill climbing algorithm might get stuck in. They are also easily parallelizable and their evaluation function can be any that assigns to each individual a real value into a partially ordered set.

3.1 GA Characterization

The proposed canonical genetic algorithm can be generally characterized by the following features:

- Encoding Scheme. The genes will represent the coefficients of the unknown deconvolution filter **W** (real coding) and the unknown nonlinear distortion f. This nonlinear function may be approximated by n-th order odd polynomials. An initial decision must therefore be taken about the length of the inverse filter (s) and the order of the polynomial (n).
- Initialization Procedure. Genes within the chromosome are randomly initialized.

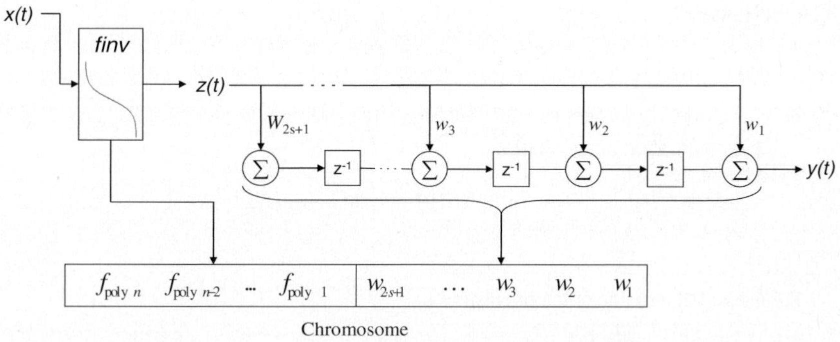

Fig. 3. Encoding scheme in the genetic algorithm for the odd coefficients of the polynomial ($f_{\text{poly}\ j}$) approximating the inverse nonlinearity and the linear filter coefficients (w_i). The values of the variables stored in the chromosome are real numbers.

– Fitness Function. The chosen evaluation function must give higher scores for those chromosomes representing estimations which maximize statistical independence. According to the central limit theorem, a simple approach as maximizing a higher order statistic like kurtosis absolute value demonstrates to be a good estimator. However, as dependence is supposed to exist between the samples due to time delays, a better estimator is achieved by dividing the observed signal in several "sub-signals" (in the simulations 10 partitions were made) and then computing kurtosis in each of them. Finally, the expression for the chosen fitness function is:

$$\text{eval}_{\text{Kurt}}(w) = \sum_{i=1}^{n} \left| kurt(y_i) \right| \tag{5}$$

where $\left| kurt(x) \right| = \left| \dfrac{\text{E}(x^4)}{\text{E}(x^2)^2} - 3 \right|$ and the y_i are each of the partitions of the estimated signal obtained after applying the nonlinear inverse function (*finv*) and the inverse filter *w* (both encoded in the chromosome w) to the observation x.

– Genetic Operators. Typical crossover and mutation operators will be used for the manipulation of the current population in each iteration of the GA. The crossover operator is "Simple One-point Crossover". The mutation operator is "Non-Uniform Mutation" [10]. This operator makes the exploration-exploitation trade-off be more favorable to exploration in the early stages of the algorithm, while exploitation takes more importance when the solution given by the GA is closer to the optimal.

– Parameter Set. Population size, number of generations, probability of mutation and crossover and other parameters relative to the genetic algorithm operation were chosen depending on the characteristics of the mixing problem. Generally a population of 50-80 individuals was used, stopping criteria was set between 100-150 iterations, crossover probability is 0.8 per chromosome and mutation probability is typically set between 0.05 and 0.08 per gene.

3.2 GA Scheme

The operation of the algorithm can be illustrated by the following figure:

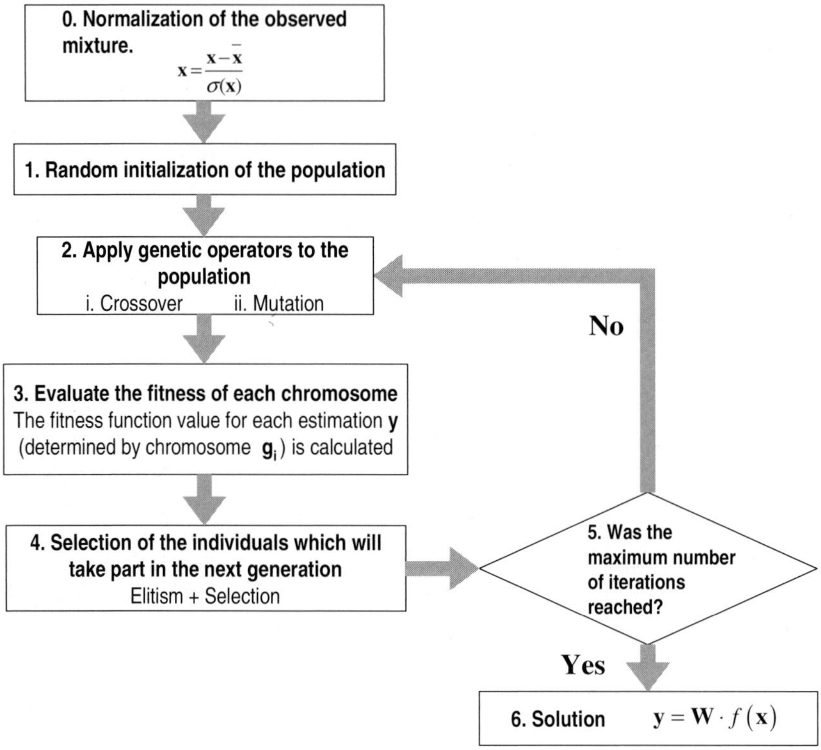

Fig. 4. Genetic algorithm operation for nonlinear blind deconvolution

4 Experimental Results

After the description of the proposed algorithm, some experimental results using uniform random sources are presented. In all the experiments, the source signal is an uniform random source with zero mean and unit variance. As the performance criterion, we have used the crosstalk between the estimations (**y**) and the sources (**s**), measured in decibels. Also, the unknown filter is the low-pass filter $H(Z)=1+0.5z^{-1}$ and we applied a strong non-linearity such as $f(x)=\operatorname{atanh}(10x)$.

Genetic algorithm was configured with a crossover probability of 0.8, mutation probability 0.08 per gene, population size is 50, and the stopping criterion is 100 generations.

In the first experiment, we applied the former configuration to a random generated signal of t=1900 samples. As the algorithm is non-deterministic, it was run with the same configuration 10 times. The average crosstalk is shown in the equation below:

$$CTalk(\mathbf{s}(t), \mathbf{y}(t)) = -13.6dB \tag{6}$$

Figure 5 shows the evolution of the fitness of the best individual (left) and the average fitness (right) of each generation along each iteration, showing the smooth convergence of the algorithm.

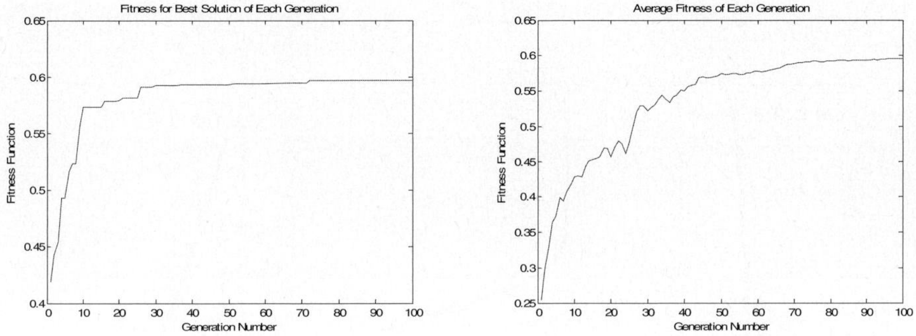

Fig. 5. Genetic algorithm operation for nonlinear blind deconvolution

Figure 6 shows how the algorithm cancels the nonlinear part of the Wiener system (which it is the most difficult component).

Secondly, the number of samples was decreased to t=500 sample, maintaining the rest of parameters. Thus, we can determine whether the algorithm still works when the number of samples is low.

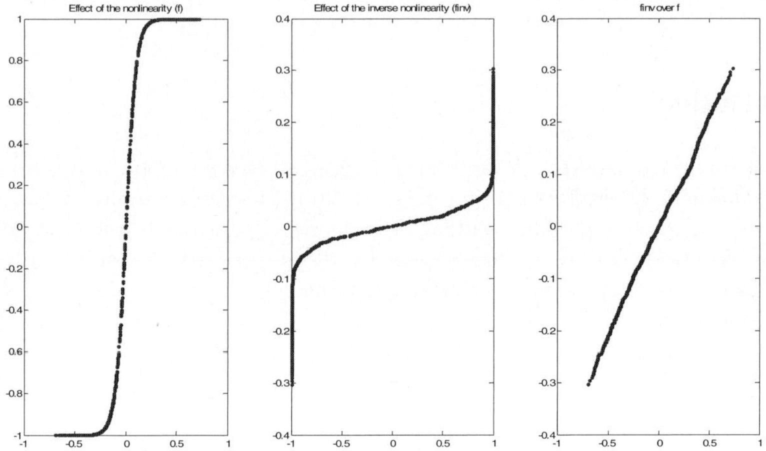

Fig. 6. First experiment, when t=1900 samples. Left: effect of the original distortion (f) over the filtered signal. Center: effect of the inverse nonlinearity (finv) over the observed signal (x). Right: composition of finv over f. In the ideal situation, should be linear.

The algorithm was again run 10 times, obtaining an average crosstalk of:

$$CTalk(\mathbf{s}(t), \mathbf{y}(t)) = -10.7dB \tag{7}$$

Although, this results would not be satisfactory for a linear convolution situation, the nature of this case where exists a strong nonlinearity and the number of samples is not sufficient makes the operation of the algorithm quite acceptable.

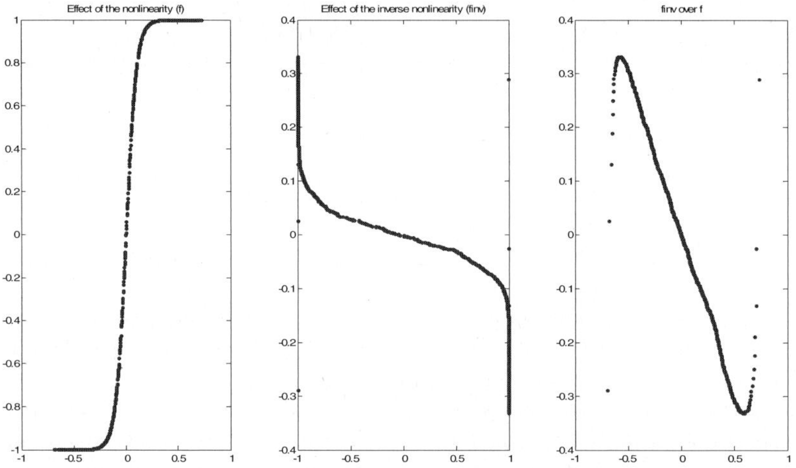

Fig. 7. Second experiment, when t=500 samples. Left: effect of the original distortion (f) over the filtered signal. Center: effect of the inverse nonlinearity (finv) over the observed signal (x). Right: composition of finv over f. In the ideal situation, should be linear.

5 Conclusion

This contribution discusses an appropriate application of genetic algorithms to the complex problem of nonlinear blind deconvolution. Using a simple approach based on kurtosis calculation and mutual information approximation, the proposed algorithm demonstrates, as it was shown by the experimental results, an effective operation even when the number of samples is low.

References

1. Segal, B.N., Outerbridge, J.S.: Vestibular (semicircular canal) primary neurons in bullfog: nonlinearity of individual and population response to rotation. J Neurophys 47, 545–562 (1982)
2. Hunter, I.W.: Experimental comparison of Wiener and Hammerstein cascade models of frog muscle fiber mechanics. Biophys J, 49(81)a (1986)

3. Rojas, F., Solé-Casals, J., Monte-Moreno, E., Puntonet, C.G., Prieto, A.: A Canonical Genetic Algorithm for Blind Inversion of Linear Channels. In: Rosca, J., Erdogmus, D., Príncipe, J.C., Haykin, S. (eds.) ICA 2006. LNCS, vol. 3889, pp. 238–245. Springer, Heidelberg (2006)
4. Rojas, F., Puntonet, C.G., Rodríguez-Álvarez, M., Rojas, I.: Evolutionary Algorithm Using Mutual Information for Independent Component Analysis. In: Mira, J.M., Álvarez, J.R. (eds.) IWANN 2003. LNCS, vol. 2687, pp. 302–9743. Springer, Heidelberg (2003)
5. Martín, R., Rojas, F., Puntonet, C.G.: Post-nonlinear Blind Source Separation using methaheruristics. Electronics Letters 39, 1765–1766 (2003)
6. Rojas, F., Puntonet, C.G., Rodríguez, M., Rojas, I., Martín-Clemente, R.: Blind Source Separation in Post-Nonlinear Mixtures using Competitive Learning, Simulated Annealing and Genetic Algorithms. IEEE Transactions on Systems, Man and Cybernetics (Part C) 34(4), 407–416 (2004)
7. Rojas, F., Puntonet, C.G., Gorriz, J.M., Valenzuela, O.: Assessing the Performance of Several Fitness Functions in a Genetic Algorithm for Nonlinear Separation of Sources. In: Wang, L., Chen, K., Ong, Y.S. (eds.) ICNC 2005. LNCS, vol. 3612, pp. 863–872. Springer, Heidelberg (2005)
8. Taleb, A., Solé-Casals, J., Jutten, C.: Quasi-Nonparametric Blind Inversion of Wiener Systems. IEEE Transactions on Signal Processing 49(5), 917–924 (2001)
9. Solé-Casals, J., Jutten, C., Pham, D.T: Fast approximation of nonlinearities for improving inversion algorithms of PNL mixtures and Wiener systems. Signal Processing 85(9), 1780–1786 (2005)
10. Michalewicz, Z.: Genetic Algorithms + Data Structures = Evolution Programs, 3rd edn. Springer, Heidelberg (1996)

A Complexity Constrained Nonnegative Matrix Factorization for Hyperspectral Unmixing

Sen Jia and Yuntao Qian

College of Computer Science, Zhejiang University
Hangzhou 310027, P.R. China
zjujiasen@hotmail.com, ytqian@cs.zju.edu.cn

Abstract. Hyperspectral unmixing, as a blind source separation (BSS) problem, has been intensively studied from independence aspect in the last few years. However, independent component analysis (ICA) can not totally unmix all the materials out because the sources (abundance fractions) are not statistically independent. In this paper a complexity constrained nonnegative matrix factorization (CCNMF) for simultaneously recovering both constituent spectra and correspondent abundances is proposed. Three important facts are exploited: First, the spectral data are nonnegative; second, the variation of the material spectra and abundance images is smooth in time and space respectively; third, in most cases, both of the material spectra and abundances are localized. Experimentations on real data are provided to illustrate the algorithm's performance.

1 Introduction

Hyperspectral sensors collect imagery simultaneously in hundreds of narrow and contiguously spaced spectral bands, with wavelengths ranging from 0.4 to 2.5μm. Owing to the low spatial resolution of the sensor, disparate substances may contribute to the spectrum measured from a single pixel, causing it to be a "mixed" pixel. Consequently, the extraction of constituent spectra from a mixed spectrum, as well as their proportions, is important to both civilian and military applications in which subpixel detail is valuable. So hyperspectral unmixing is a necessary preprocessing step for hyperspectral applications [1,2].

Linear spectral unmixing [1] is a commonly accepted approach to analyze the massive volume of data. In addition, due to the difficulty of obtaining a priori knowledge about the endmembers, unsupervised hyperspectral unmixing tries to identify the endmembers and abundances directly from the observed data without any user interaction [3]. Unsupervised linear hyperspectral unmixing falls into the class of blind source separation (BSS) problems [4]. Therefore, independent component analysis (ICA) is naturally selected to unmix hyperspectral data [5,6]. However, the source (abundance fraction) independence assumption of ICA is violated because the sum of abundance fractions is constant, implying statistical dependence among them, which compromises the applicability of ICA to hyperspectral data [7].

M.E. Davies et al. (Eds.): ICA 2007, LNCS 4666, pp. 268–276, 2007.

Recently, Stone [8] has proposed a complexity based BSS algorithm, called complexity pursuit, which studies the complexity of sources instead of the independence. We introduces the algorithm to unmix hyperspectral data [9], and proves that the undercomplete case, i.e., the number of bands (mixtures number) is much larger than that of endmembers (sources number), is an advantage rather than a limitation. However, it neglects the nonnegativity of both spectra and abundances, which could lead to unmixing results with negative values that are obviously meaningless. Lately, nonnegative matrix factorization (NMF) [10,11], which is a useful technique in representing high dimensional data into nonnegative components, has been extended to unmix hyperspectral data by enforcing the smoothness constraints in spectra and abundances, called SCNMF [12]. For each estimated endmember, it selects the closest signature from a spectral library as the endmember spectrum. However, the best match is not reliable due to the strong atmospheric and environmental variations [13]. The results of the method are given for comparative analysis in the experimental section.

In this paper, we develop a complexity constrained NMF (CCNMF) that incorporates the complexity and locality constraints into NMF. Three important facts are exploited: First, the spectral data are nonnegative; second, the variation of endmember spectra and abundances is smooth due to the high spectral resolution and low spatial resolution of hyperspectral data [14]; third, in many cases, both of the endmember spectra and abundances are localized: part of the reflectance spectra are small due to the atmospheric and environmental effects; and the fractions of each endmember are often localized in the scene. The experimental results validate the efficiency of the approach.

The rest of the paper is organized as follows. Section 2 reviews the complexity based hyperspectral unmixing. Section 3 presents the CCNMF algorithm. Section 4 evaluates the performance of our proposed algorithm on two real data. Section 5 sets out our conclusion.

2 Complexity Based BSS Algorithm for Hyperspectral Unmixing

In this section, we first present the linear superposition model for the reflectances, then give a brief introduction of the complexity based BSS algorithm for hyperspectral unmixing.

2.1 Linear Spectral Mixture Model

Hyperspectral data is a three dimensional array with the width and length corresponding to spatial dimensions and the spectral bands as the third dimension, which are denoted by I, J and L in sequence. Let $\mathbf{R}$ be the image cube with each spectrum $\mathbf{R}_{ij}$ being an $L \times 1$ pixel vector. Let $\mathbf{M}$ be an $L \times P$ spectral signature matrix that each column vector $\mathbf{M}_p$ corresponds to an endmember spectrum and P is the number of endmembers in the image. Let $\mathbf{S}$ be the abundance cube (the length of each dimension is I, J and P respectively) and every column $\mathbf{S}_{ij}$ be a $P \times 1$ abundance vector associated with $\mathbf{R}_{ij}$, with each element denoting the abundance fraction of relevant endmember present in $\mathbf{R}_{ij}$. The simplified linear spectral mixture model for the pixel with coordinate (i, j) can be written as

$$\mathbf{R}_{ij} = \mathbf{M}\mathbf{S}_{ij} + \mathbf{n} \tag{1}$$

where $\mathbf{n}$ is noise that can be interpreted as receiver electronic noise. Meanwhile, endmember spectra and fractional abundances are subject to

$$\mathbf{M}_{lp} \geq 0 \; (1 \leq l \leq L), \quad \mathbf{S}_{ijp} \geq 0, \; \sum_{p=1}^{P} \mathbf{S}_{ijp} = 1 \tag{2}$$

which are called nonnegativity and full additivity respectively [2].

2.2 Complexity Based BSS Algorithm

Instead of the assumption of independence in ICA, complexity based BSS algorithm makes the complexity of the extracted signal to be as low as possible. One simple measure of complexity can be formulated in terms of predictability. The predictability of the hyperspectral data is composed of two parts

$$F(\mathbf{R}) = F(\mathbf{M}) + F(\mathbf{S}) \tag{3}$$

$F(\mathbf{M})$ is the predictability of the spectral signatures, which is defined as

$$F(\mathbf{M}) = \sum_{p=1}^{P} \ln \frac{\sum_{l=1}^{L} (\overline{\mathbf{M}}_p - \mathbf{M}_{lp})^2}{\sum_{l=1}^{L} (\widetilde{\mathbf{M}}_{lp} - \mathbf{M}_{lp})^2} = \sum_{p=1}^{P} \ln \frac{V_p}{U_p} \tag{4}$$

$$\widetilde{\mathbf{M}}_{lp} = \lambda_S \widetilde{\mathbf{M}}_{(l-1)p} + (1 - \lambda_S)\mathbf{M}_{(l-1)p} \quad 0 \leq \lambda_S \leq 1 \tag{5}$$

V_p is the overall variance of the spectral signature $\mathbf{M}_p$ in which $\overline{\mathbf{M}}_p$ is the mean value. U_p is a measure of the temporal "roughness" of the signature. $\widetilde{\mathbf{M}}_{lp}$ is the short-term moving average to predict $\mathbf{M}_{lp}$, with λ_S being the predictive rate. Maximizing the ratio V_p/U_p means: (i) $\mathbf{M}_p$ has a nonzero range, (ii) the values in $\mathbf{M}_p$ change slowly. Consequently, $F(\mathbf{M})$ characterizes the smoothness of the spectral signature.

$F(\mathbf{S})$ is the predictability of the abundance cube, which is defined as

$$F(\mathbf{S}) = \sum_{p=1}^{P} \ln \frac{\sum_{i,j=1}^{I,J} (\overline{\mathbf{S}}_p - \mathbf{S}_{ijp})^2}{\sum_{i,j=1}^{I,J} \chi(\mathbf{S}_{ijp})} \tag{6}$$

where $\overline{\mathbf{S}}_p$ is the mean value of the pth abundance image except $\mathbf{S}_{ijp}$. The energy function of Gibbs distribution is used to formulate $\chi(\mathbf{S}_{ijp})$, which measures the local correlation of $\mathbf{S}_{ijp}$. That is,

$$\chi(\mathbf{S}_{ijp}) = \sum_{i'j' \in \mathcal{N}_{ijp}} \omega_{i'j'} \phi(\mathbf{S}_{ijp} - \mathbf{S}_{i'j'p}, \delta) \tag{7}$$

where $\mathcal{N}_{ijp}$ is the nearest neighborhood of $\mathbf{S}_{ijp}$, $\omega_{i'j'}$ is a weighting factor, δ is a scaling factor, and $\phi(\xi, \delta)$ is the potential function, which takes the form [15]

$$\phi(\xi, \delta) = \delta \ln[\cosh(\xi/\delta)] \tag{8}$$

We assume $\delta = 0.1$, $\omega_{i'j'} = 1$ and $\mathcal{N}_{ijp} = \{(i-1)j, (i+1)j, i(j-1), i(j+1)\}$. Similarly, $F(\mathbf{S})$ characterizes the spatial correlation of each abundance image.

3 Complexity Constrained NMF (CCNMF)

NMF uses alternating minimization of a cost function subject to nonnegativity constraints. The most widely used cost function is the euclidean distance function (two matrices $\mathbf{R}'$ and $\mathbf{S}'$ are introduced that each row is obtained by converting the band and abundance images of $\mathbf{R}$ and $\mathbf{S}$ into vectors respectively. The dimensions of them are $L \times K$ and $P \times K$, where K is the number of pixels)

$$E(\mathbf{M}, \mathbf{S}') = \frac{1}{2}\|\mathbf{R}' - \mathbf{MS}'\|^2 = \frac{1}{2}\sum_{l,k}(\mathbf{R}'_{lk} - (\mathbf{MS}')_{lk})^2 \tag{9}$$

To represent the local constraints of the spectra and abundances, nonsmooth NMF [16], which explicitly controls the degree of locality, is used.

$$nsE(\mathbf{M}, \mathbf{S}') = \frac{1}{2}\|\mathbf{R}' - \mathbf{MCS}'\|^2, \quad \mathbf{C} = (1-\alpha)\mathbf{I} + \frac{\alpha}{P}\mathbf{1}\mathbf{1}^T \tag{10}$$

where $\mathbf{C} \in \mathbb{R}^{P \times P}$ is a "nonsmoothing" matrix, $\mathbf{I}$ is the identity matrix, $\mathbf{1}$ is a vector of ones, the notation $(\cdot)^T$ is matrix transposition, and the parameter α $(0 \le \alpha \le 1)$ explicitly controls the extent of nonsmoothness of $\mathbf{C}$, which is set to 0.5 in the experiments.

Taking into consideration complexity constraints, the cost function of CCNMF can be formulated as

$$D(\mathbf{M}, \mathbf{S}') = nsE(\mathbf{M}, \mathbf{S}') + \theta_M J_M(\mathbf{M}) + \theta_{S'} J_{S'}(\mathbf{S}') \tag{11}$$

$$J_M(\mathbf{M}) = \frac{1}{2}\|\widetilde{\mathbf{M}} - \mathbf{M}\|^2, \quad J_{S'}(\mathbf{S}') = \frac{|\chi(\mathbf{S}')|}{\|\overline{\mathbf{S}'} - \mathbf{S}'\|^2} \tag{12}$$

θ_M and $\theta_{S'}$ are regularization parameters, $\widetilde{\mathbf{M}}$ and $\chi(\mathbf{S}')$ are $L \times P$ and $P \times K$ matrices, with the element $\widetilde{\mathbf{M}}_{lp}$ and $\chi(\mathbf{S}'_{pk})$ at the (l, p) and (p, k) position respectively, $|\cdot|$ is the sum of matrix elements, and $\overline{\mathbf{S}'}$ is a $P \times K$ matrix with the entries in the pth row equaling to $\overline{\mathbf{S}_p}$. (12) are the approximations of the reciprocals of (4) and (6) respectively. The numerator of (4) is neglected because function $nsE(\mathbf{M}, \mathbf{S}')$ ensures the extracted spectrum is not constant, while that of (6) is reserved to avoid the abundances of adjacent pixels being equal.

The general additive update rules can be constructed as

$$\mathbf{M} \leftarrow \mathbf{M} - \boldsymbol{\mu}.*\frac{\partial D(\mathbf{M},\mathbf{S}')}{\partial \mathbf{M}}, \quad \mathbf{S}' \leftarrow \mathbf{S}' - \boldsymbol{\nu}.*\frac{\partial D(\mathbf{M},\mathbf{S}')}{\partial \mathbf{S}'} \tag{13}$$

where ".*" denotes element-wise multiplication. Taking the derivatives of $D(\mathbf{M},\mathbf{S}')$ with respect to $\mathbf{M}$ and $\mathbf{S}'$ and after some algebraic manipulations, the gradients about $\mathbf{M}$ and $\mathbf{S}'$ are

$$\frac{\partial D(\mathbf{M},\mathbf{S}')}{\partial \mathbf{M}} = -(\mathbf{R}' - \mathbf{MCS}')(\mathbf{CS}')^T - \theta_M(\widetilde{\mathbf{M}} - \mathbf{M}) \tag{14}$$

$$\frac{\partial D(\mathbf{M},\mathbf{S}')}{\partial \mathbf{S}'} = -(\mathbf{MC})^T(\mathbf{R}' - \mathbf{MCS}') + \theta_{\mathbf{S}'}\frac{\partial J_{S'}(\mathbf{S}')}{\partial \mathbf{S}'} \tag{15}$$

where

$$\frac{\partial J_{S'}(\mathbf{S}')}{\partial \mathbf{S}'} = \left(\frac{\frac{\partial|\chi(\mathbf{S}')|}{\partial \mathbf{S}'}}{||\overline{\mathbf{S}'} - \mathbf{S}'||^2} + \frac{2|\chi(\mathbf{S}')|(\overline{\mathbf{S}'} - \mathbf{S}')}{\left(||\overline{\mathbf{S}'} - \mathbf{S}'||^2\right)^2} \right) \tag{16}$$

Choosing the step sizes from [11], the multiplicative rules are given below

$$\mathbf{M} \leftarrow \mathbf{M}.*(\mathbf{R}'(\mathbf{CS}')^T + \theta_M(\widetilde{\mathbf{M}} - \mathbf{M}))./(\mathbf{MCS}'(\mathbf{CS}')^T) \tag{17}$$

$$\mathbf{S}' \leftarrow \mathbf{S}'.*((\mathbf{MC})^T\mathbf{R}' - \theta_{S'}\frac{\partial J_{S'}(\mathbf{S}')}{\partial \mathbf{S}'})./((\mathbf{MC})^T\mathbf{MCS}') \tag{18}$$

where "./" denotes element-wise division. In addition, to ensure the full additivity of abundance cube, $\mathbf{S}'$ should be normalized

$$\mathbf{S}'_{pk} \leftarrow \frac{S'_{pk}}{\sum_{p=1}^{P} \mathbf{S}'_{pk}}, \quad 1 \le p \le P, 1 \le k \le K \tag{19}$$

One hurdle of the NMF problem is the existence of local minima due to the nonconvexity of the objective function. But through adding complexity and local constraints, the feasible solution set is confined and the nonuniqueness of solution is alleviated. At last, we summarize the CCNMF algorithm.

1. Use virtual dimensionality (VD) method [17] to find the number of endmembers P involved in the mixture data.
2. Initialize $\mathbf{M}$ and $\mathbf{S}'$ with non-negative values.
3. For $t = 1, 2, \ldots,$ until $D(\mathbf{M},\mathbf{S}') \le tol$, for a tolerance value $tol \in \mathbb{R}$, update $\mathbf{M}$ and $\mathbf{S}'$ by (17) and (18), and then normalize $\mathbf{S}'$ by (19).

4 Experimental Results

In this section, the data being analyzed were collected by the Hyperspectral Digital Imagery Collection Experiment (HYDICE) system. It is composed of 210 channels with spectral resolution 10 nm acquired in the 0.4-2.5 micron region. To evaluate the performance of SCNMF and CCNMF, spectral angle distance (SAD) and Root-Mean-Square Error (RMSE) are employed (The definitions are omitted here due to the length of the paper. Readers are referred to [18] for details). The ground truth of the data are computed according to [9].

4.1 Washington D.C. Data

Figure 1 shows a subscene of size 30 × 30 extracted from the Washington D.C. data set. After low signal-to-noise ratio (SNR) bands are removed, only 191 bands remain (i.e., L=191). Using VD method to estimate P, it is equal to 4.

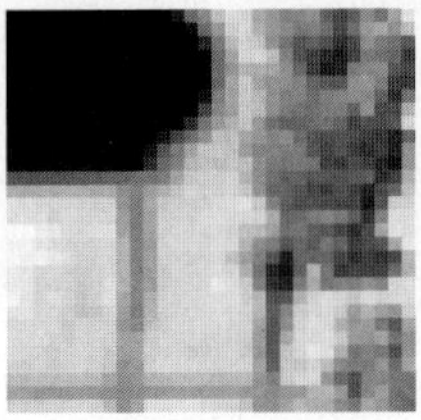

Fig. 1. The subscene (30 × 30) extracted from Washington D.C. data set

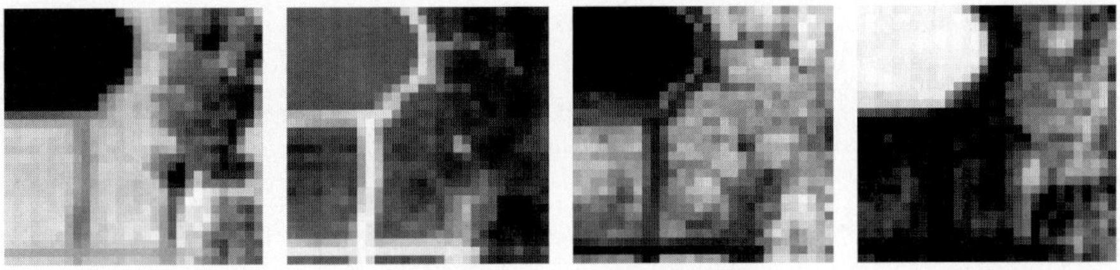

(a) grass (b) trail (c) tree (d) water

Fig. 2. Abundance maps estimated using SCNMF

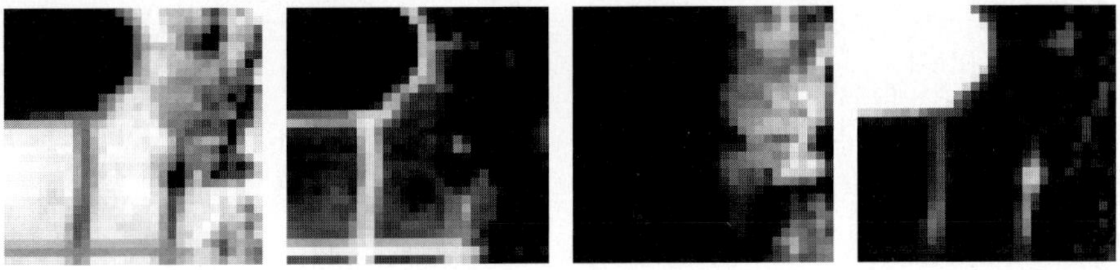

(a) grass (b) trail (c) tree (d) water

Fig. 3. Abundance maps estimated using CCNMF

Firstly, SCNMF is applied to the data set. Figure 2 presents the estimated abundance maps. Except that the grass and trail are detected in Figure 2(a)

Fig. 4. Urban scene (307 × 307) extracted from HYDICE data set

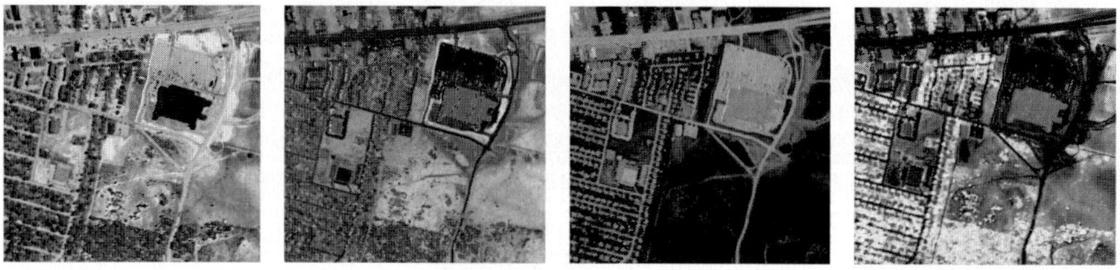

(a) asphalt (b) grass (c) roof (d) tree

Fig. 5. Abundance maps estimated using SCNMF

and 2(b), the other two maps are still mixtures. Then CCNMF is utilized, and the results are displayed in Figure 3. Different from Figure 2, all the four abundances: grass, trail, tree and water are successfully extracted. Table 1 quantifies the unmixing results using the two performance metrics.

Table 1. SAD-based similarity and RMSE-based error scores between the unmixing results of Washington D.C. data by SCNMF and CCNMF and the ground truth (The numbers in bold represent the best performance)

Method	Endmember				
	grass	trail	tree	water	
SCNMF	0.1349	**0.1075**	0.24	0.2926	(SAD)
	0.2435	**0.1628**	0.2299	0.2152	(RMSE)
CCNMF	**0.1031**	0.1299	**0.1886**	**0.1673**	
	0.2272	0.2071	**0.1403**	**0.0975**	

4.2 Urban Data

The data to be used were obtained from a HYDICE scene of 307 × 307 pixels shown in Figure 4. After low signal-to-noise ratio (SNR) bands are removed, a total of 162 bands are used in the experiment. The estimated number of endmembers using the VD method is 4.

(a) asphalt (b) grass (c) roof (d) tree

Fig. 6. Abundance maps estimated using CCNMF

The estimated abundances using SCNMF are illustrated in Figure 5. Only grass and tree are detected in Figure 5(b) and 5(d), the other two are still unmixed. Contrarily, all the four endmembers are separated out by CCNMF, as displayed in Figure 6. Likewise, Table 2 quantifies the unmixing results.

Table 2. SAD-based similarity and RMSE-based error scores between the unmixing results of urban data by SCNMF and CCNMF and the ground truth (The numbers in bold represent the best performance)

Method	Endmember				
	asphalt	grass	roof	tree	
SCNMF	0.2704	0.2608	0.32	0.1887	(SAD)
	0.2468	0.2093	0.2301	0.1792	(RMSE)
CCNMF	**0.189**	**0.111**	**0.2942**	**0.1005**	
	0.1559	**0.1247**	**0.2228**	**0.1017**	

5 Conclusion

We have presented a complexity constrained NMF (CCNMF) for hyperspectral unmixing. The algorithm extends the original NMF by incorporating the complexity and locality constraints, which accord with the three characteristics of hyperspectral data. Its effectiveness has been tested by comparison to SCNMF with data from HYDICE data sets. The experimental results show that CCNMF has the potential of providing more accurate estimates of both endmember spectra and abundance maps.

References

1. Keshava, N., Mustard, J.F.: Spectral unmixing. IEEE Signal Processing Mag. 19(3), 44–57 (2002)
2. Keshava, N.: A survey of spectral unmixing algorithms. Lincoln Lab. J. 14(1), 55–73 (2003)

3. Parra, L., Spence, C., Sajda, P., Ziehe, A., Müller, K.R.: Unmixing hyperspectral data. In: Adv. Neural Inform. Process. Syst. 12, Denver, Colorado, USA, pp. 942–948. MIT Press, Cambridge (1999)
4. Cichocki, A., Amari, S.: Adaptive Blind Signal and Image Processing: Learning Algorithms and Applications. John Wiley & Sons, Chichester (2002)
5. Chiang, S.-S., Chang, C.-I., Smith, J.A., Ginsberg, I.W.: Linear spectral random mixture analysis for hyperspectral imagery. IEEE Trans. Geosci. Remote Sensing 40(2), 375–392 (2002)
6. Chang, C.-I: Hyperspectral Imaging: Techniques for Spectral Detection and Classification. Kluwer Academic/Plenum Publishers, New York (2003)
7. Nascimento, J.M.P., Dias, J.M.B.: Does independent component analysis play a role in unmixing hyperspectral data? IEEE Trans. Geosci. Remote Sensing 43(1), 175–187 (2005)
8. Stone, J.V.: Blind source separation using temporal predictability. Neural Comput. 13(7), 1559–1574 (2001)
9. Jia, S., Qian, Y.T.: Spectral and spatial complexity based hyperspectral unmixing. IEEE Trans. Geosci. Remote Sensing (to appear)
10. Lee, D.D., Seung, H.S.: Learning the parts of objects by non-negative matrix factorization. Nature 401, 788–791 (1999)
11. Lee, D.D., Seung, H.S.: Algorithms for non-negative matrix factorization. In: Adv. Neural Inform. Process. Syst. vol. 13, pp. 556–562 (2000)
12. Paura, V.P., Piper, J., Plemmons, R.J.: Nonnegative matrix factorization for spectral data analysis. Linear Algebra and Applications 416(1), 29–47 (2006)
13. Miao, L.D., Qi, H.R.: Endmember extraction from highly mixed data using minimum volume constrained nonnegative matrix factorization. IEEE Trans. Geosci. Remote Sensing 45(3), 765–777 (2007)
14. Ferrara, C.F.: Adaptive spatial/spectral detection of subpixel targets with unknown spectral characteristics. In: Proc. SPIE, vol. 2235, pp. 82–93 (1994)
15. Green, P.J.: Bayesian reconstructions from emission tomography data using a modified em algorithm. IEEE Trans. Med. Imag. 9(1), 84–93 (1990)
16. Montano, A.P., Carazo, J.M., Kochi, K., Lehmann, D., P.-Marqui, R.D.: Nonsmooth nonnegative matrix factorization (nsNMF). IEEE Trans. Pattern Anal. Machine Intell. 28(3), 403–415 (2006)
17. Chang, C.-I, Du, Q.: Estimation of number of spectrally distinct signal sources in hyperspectral imagery. IEEE Trans. Geosci. Remote Sensing 42(3), 608–619 (2004)
18. Plaza, A., Martinez, P., Perez, R., Plaza, J.: A quantitative and comparative analysis of endmember extraction algorithms from hyperspectral data. IEEE Trans. Geosci. Remote Sensing 42(3), 650–663 (2004)

Smooth Component Analysis as Ensemble Method for Prediction Improvement

Ryszard Szupiluk[1,2], Piotr Wojewnik[1,2], and Tomasz Ząbkowski[1,3]

[1] Polska Telefonia Cyfrowa Ltd., Al. Jerozolimskie 181, 02-222 Warsaw, Poland
{rszupiluk,pwojewnik,tzabkowski}@era.pl
[2] Warsaw School of Economics, Al. Niepodleglosci 162, 02-554 Warsaw, Poland
[3] Warsaw Agricultural University, Nowoursynowska 159, 02-787 Warsaw, Poland

Abstract. In this paper we apply a novel smooth component analysis algorithm as ensemble method for prediction improvement. When many prediction models are tested we can treat their results as multivariate variable with the latent components having constructive or destructive impact on prediction results. We show that elimination of those destructive components and proper mixing of those constructive can improve the final prediction results. The validity and high performance of our concept is presented on the problem of energy load prediction.

1 Introduction

The blind signal separation methods have applications in telecommunication, medicine, economics and engineering. Starting from separation problems, BSS methods are used in filtration, segmentation and data decomposition tasks [5,11]. In this paper we apply the BSS method for prediction improvement in case when many models are tested.

The prediction problem as other regression tasks aims at finding dependency between input data and target. This dependency is represented by a specific model e.g. neural networks [7,13]. In fact, in many problems we can find different acceptable models where the ensemble methods can be used to improve final results [7]. Usually solutions propose the combination of a few models by mixing their results or parameters [1,8,18]. In this paper we propose an alternative concept based on the assumption that prediction results contain the latent destructive and constructive components common to all the model results [16]. The elimination of the destructive ones should improve the final results. To find the latent components we apply blind signal separation methods with a new algorithm for smooth component analysis (SmCA) which is addressed for signals with temporal structure [4]. The full methodology will be tested in load prediction task [11].

2 Prediction Results Improvement

We assume that after the learning process each prediction result includes two types of latent components: constructive, associated with the target, and destructive, associated

M.E. Davies et al. (Eds.): ICA 2007, LNCS 4666, pp. 277–284, 2007.

with the inaccurate learning data, individual properties of models, missing data, not precise parameter estimation, distribution assumptions etc. Let us assume there is m models. We collect the results of particular model in column vector $\mathbf{x}_i$, i=1,…, m, and treat such vectors as multivariate variable $\mathbf{X} = [\mathbf{x}_1, \mathbf{x}_2 \ldots \mathbf{x}_m]^T$, $\mathbf{X} \in R^{m \times N}$, where N means the number of observations. We describe the set of latent components as $\mathbf{S} = [\hat{\mathbf{s}}_1, \hat{\mathbf{s}}_2, ..., \hat{\mathbf{s}}_k, \mathbf{s}_{k+1}, \mathbf{s}_n]^T$, $\mathbf{S} \in R^{n \times N}$, where $\hat{\mathbf{s}}_j$ denotes constructive component and $\mathbf{s}_i$ is destructive one [3]. For simplicity of further considerations we assume $m = n$. Next we assume the relation between observed prediction results and latent components as linear transformation

$$\mathbf{X} = \mathbf{A}\mathbf{S}, \tag{1}$$

where matrix $\mathbf{A} \in R^{n \times n}$ represents the mixing system. The (1) means matrix $\mathbf{X}$ decomposition by latent components matrix $\mathbf{S}$ and mixing matrix $\mathbf{A}$.

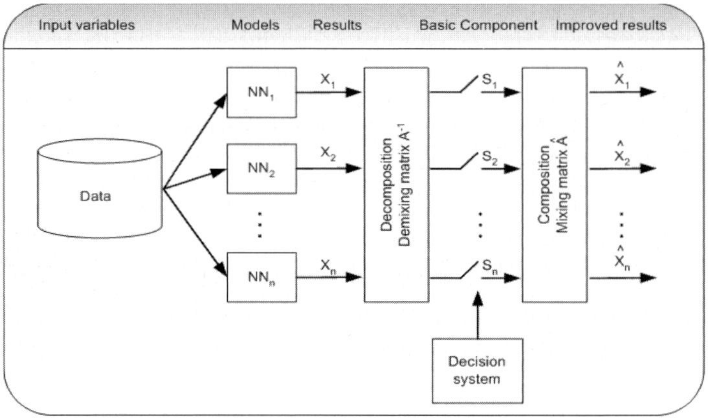

Fig. 1. The scheme of modelling improvement method by multivariate decomposition

Our aim is to find the latent components and reject the destructive ones (replace them with zero). Next we mix the constructive components back to obtain improved prediction results as

$$\hat{\mathbf{X}} = \mathbf{A}\hat{\mathbf{S}} = \mathbf{A}[\hat{\mathbf{s}}_1, \hat{\mathbf{s}}_2, ..., \hat{\mathbf{s}}_k, \mathbf{0}_{k+1}, ..., \mathbf{0}_n]^T. \tag{2}$$

The replacement of destructive signal by zero is equivalent to putting zero in the corresponding column of $\mathbf{A}$. If we express the mixing matrix as $\mathbf{A} = [\mathbf{a}_1, \mathbf{a}_2 \ldots \mathbf{a}_n]$ the purified results can be described as

$$\hat{\mathbf{X}} = \hat{\mathbf{A}}\mathbf{S} = [\mathbf{a}_1, \mathbf{a}_2, ..., \mathbf{a}_k, \mathbf{0}_{k+1}, ..., \mathbf{0}_n]\mathbf{S}, \tag{3}$$

Where $\hat{\mathbf{A}} = [\mathbf{a}_1, \mathbf{a}_2 \ldots \mathbf{a}_p, \mathbf{0}_{p+1}, \mathbf{0}_{p+2} \ldots \mathbf{0}_n]$. The crucial point of the above concept is proper $\mathbf{A}$ and $\mathbf{S}$ estimation. It is difficult task because we have not information which decomposition is most adequate. Therefore we must test various transformations giving us components of different properties. The most adequate methods to solve the first problem seem to be the blind signal separation (BSS) techniques.

3 Blind Signal Separation and Decomposition Algorithms

Blind signals separation (BSS) methods aim at identification of the unknown signals mixed in the unknown system [2,4,10,15]. There are many different methods and algorithms used in BSS task. They explore different properties of data like: independence [2,10], decorrelation [3,4], sparsity [5,19], smoothness [4], non-negativity [12] etc. In our case, we are not looking for specific real signals but rather for interesting analytical data representation of the form (1). To find the latent variables **A** and **S** we can use a transformation defined by separation matrix $\mathbf{W} \in R^{n \times n}$, such that

$$\mathbf{Y} = \mathbf{WX}. \tag{4}$$

where **Y** is related to **S**. We also assume that **Y** satisfies the following relation

$$\mathbf{Y} = \mathbf{PDS}, \tag{5}$$

where **P** is a permutation matrix and **D** is a diagonal matrix [4,10]. The relation (5) means that estimated signals can be rescaled and reordered in comparison to the original sources. These properties are not crucial in our case, therefore **Y** can be treated directly as estimated version of sources **S**. There are some additional assumptions depending on particular BSS method. We focus on methods based on decorrelation, independent component analysis and smooth component analysis.

Decorrelation is one of the most popular statistical procedures for the elimination of the linear statistical dependencies in the data. It can be performed by diagonalization of the correlation matrix $\mathbf{R}_{xx}=E\{\mathbf{XX}^T\}$. It means that matrix **W** should satisfy the following relation

$$\mathbf{R}_{yy} = \mathbf{WR}_{xx}\mathbf{W}^T = \mathbf{E}, \tag{6}$$

where **E** is any diagonal matrix. There are many methods utilizing different matrix factorisation leading to the decorrelation matrix **W**, Table 1 [6,17]. The decorrelation is not effective separation method and it is used typically as preprocessing, in general. However, we find it very useful for our analytical representation.

Table 1. Methods of decorrelation possible for models decomposition

Method	Form correlation	Cholesky	EIG (PCA)
Factorisation	$\mathbf{R}_{xx}= \mathbf{R}_{xx}^{1/2}\mathbf{R}_{xx}^{1/2}$	$\mathbf{R}_{xx}=\mathbf{G}^T\mathbf{G}$	$\mathbf{R}_{xx}=\mathbf{U\Sigma U}^T$
Decorrelation	$\mathbf{W}= \mathbf{R}_{xx}^{-1/2}$	$\mathbf{W}=\mathbf{G}^{-T}$	$\mathbf{W}=\mathbf{U}^T$

Independent component analysis, ICA, is a statistical tool, which allows decomposition of observed variable **X** into independent components $\mathbf{Y} = [\mathbf{y}_1,\mathbf{y}_2\ldots\mathbf{y}_n]^T$ [2,4,10]. Typical algorithms for ICA explore higher order statistical dependencies in a dataset, so after ICA decomposition we have got signals (variables) without any linear and non-linear statistical dependencies. To obtain independent components we explore the fact that the joint probability of independent variables can be factorized by the product of the marginal probabilities

$$\overbrace{p_1(\mathbf{y}_1)p_2(\mathbf{y}_2)\ldots p_n(\mathbf{y}_n)}^{q_{\mathbf{y}}(\mathbf{Y})} = \overbrace{p_{1\ldots n}(\mathbf{y}_1,\mathbf{y}_2,\ldots,\mathbf{y}_n)}^{p_{\mathbf{y}}(\mathbf{Y})}. \tag{7}$$

One of the most popular method to obtain (8) is to find such **W** that minimizes the Kullback-Leibler divergence between $p_{\mathbf{y}}(\mathbf{Y})$ and $q_{\mathbf{y}}(\mathbf{Y})$ [5]

$$\mathbf{W}_{opt} = \min_{\mathbf{W}} D_{KL}(p_{\mathbf{y}}(\mathbf{WX}) \,\|\, q_{\mathbf{y}}(\mathbf{WX})) = \min_{\mathbf{W}} \int_{-\infty}^{+\infty} p_{\mathbf{y}}(\mathbf{Y}) \log \frac{p_{\mathbf{y}}(\mathbf{Y})}{q_{\mathbf{y}}(\mathbf{Y})} d\mathbf{Y}. \tag{8}$$

There are many numerical algorithms estimating independent components like Natural Gradient, FOBI, JADE or FASTICA [2,4,10].

Smooth Component Analysis, SmCA, is a method of the smooth components finding in a multivariate variable [4]. The analysis of signal smoothness is strongly associated with the definitions and assumptions about such characteristics [9,17]. For signals with temporal structure we propose a new smoothness measure

$$P(\mathbf{y}) = \frac{\frac{1}{N}\sum_{k=2}^{N} |\,\mathbf{y}(k) - \mathbf{y}(k-1)\,|}{\max(\mathbf{y}) - \min(\mathbf{y}) + \delta(\max(\mathbf{y}) - \min(\mathbf{y}))}, \tag{9}$$

where symbol $\delta(.)$ means Kronecker delta, and $P(\mathbf{y}) \in [0,1]$. Measure (9) has simple interpretation: it is maximal when the changes in each step are equal to range (maximal change), and is minimal when data are constant. The Kronecker delta term is introduced to avoid dividing by zero. The range calculated in denominator is sensitive to local data, what can be avoided using extremal values distributions.

The components are taken as linear combination of signals $\mathbf{x}_i$ and should be as smooth as possible. Our aim is to find such $\mathbf{W} = [\mathbf{w}_1, \mathbf{w}_2 \ldots \mathbf{w}_n]$ that for $\mathbf{Y} = \mathbf{WX}$ we obtain $\mathbf{Y} = [\mathbf{y}_1, \mathbf{y}_2 \ldots \mathbf{y}_n]^T$ where $\mathbf{y}_1$ maximizes $P(\mathbf{y}_1)$ so we can write

$$\mathbf{w}_1 = \arg\max_{\|\mathbf{w}\|=1} (P(\mathbf{w}^T \mathbf{x})). \tag{10}$$

Having estimated the first $n-1$ smooth components the next one is calculated as most smooth component of the residual obtained in Gram-Schmidt orthogonalization:

$$\mathbf{w}_n = \arg\max_{\|\mathbf{w}\|=1} (P(\mathbf{w}^T (\mathbf{x} - \sum_{i=1}^{n-1} \mathbf{y}_i \mathbf{y}_i^T \mathbf{x}))), \tag{11}$$

where $\mathbf{y}_i = \mathbf{w}_i^T \mathbf{x}, i = 1 \ldots n$. As the numerical algorithm for finding $\mathbf{w}_n$ we can employ the conjugate gradient method with golden section as a line search routine. The algorithm outline for initial $\mathbf{w}_i(0) = rand$, $\mathbf{p}_i(0) = -\mathbf{g}_i(0)$ is as follows:

1. Identify the indexes l for extreme signal values:

$$l^{\max} = \arg\max_{l \in 1\ldots N} \mathbf{w}_i^T(k)\mathbf{x}(l), \tag{12}$$

$$l^{\min} = \arg\min_{l \in 1\ldots N} \mathbf{w}_i^T(k)\mathbf{x}(l), \tag{13}$$

2. Calculate gradient of $P(\mathbf{w}_i^T\mathbf{x})$:

$$\mathbf{g}_i = \frac{\partial P(\mathbf{w}_i^T\mathbf{x})}{\partial \mathbf{w}_i} = \frac{\sum_{l=2}^{N}\Delta\mathbf{x}(l)\cdot sign(\mathbf{w}_i^T\Delta\mathbf{x}(l)) - P(\mathbf{w}_i^T\mathbf{x})\cdot(\mathbf{x}(l^{\max}) - \mathbf{x}(l^{\min}))}{\max(\mathbf{w}_i^T\mathbf{x}) - \min(\mathbf{w}_i^T\mathbf{x}) + \delta(\max(\mathbf{w}_i^T\mathbf{x}) - \min(\mathbf{w}_i^T\mathbf{x}))}, \tag{14}$$

where $\Delta\mathbf{x}(l) = \mathbf{x}(l) - \mathbf{x}(l-1)$,

3. Identify the search direction (Polak-Ribiere formula[19])

$$\mathbf{p}_i(k) = -\mathbf{g}_i(k) + \frac{\mathbf{g}_i^T(k)(\mathbf{g}_i(k)-\mathbf{g}_i(k-1))}{\mathbf{g}_i^T(k-1)\mathbf{g}_i(k-1)}\mathbf{p}_i(k-1), \tag{15}$$

4. Calculate the new weights:

$$\mathbf{w}_i(k+1) = \mathbf{w}_i(k) + \alpha(k)\cdot\mathbf{p}_i(k), \tag{16}$$

where $\alpha(k)$ is found in golden search.

The above optimization algorithm should be applied as a multistart technique with random initialization [14].

4 Component Classification

After latent component are estimated by e.g. SmCA we need to label them as destructive or constructive. The problem with proper signal classification can be difficult task because obtained components might be not pure constructive or destructive due to many reasons like improper linear transformation assumption or other statistic characteristics than explored by chosen BSS method [21]. Consequently, it is possible that some component has constructive impact on one model and destructive on the other. There may also exist components destructive as a single but constructive in a group. Therefore, it is advisable to analyze each subset of the components separately. In particular, we eliminate each subset (use the matrix $\hat{\mathbf{A}}$) and check the impact on the final results. Such process of component classification as destructive or constructive is simple and works well but for many components it is computationally extensive.

5 Generalized Mixing

As was mentioned above, the latent components can be not pure so their impact should have weight other than 0. It means that we can try to find the better mixing system than described by $\hat{\mathbf{A}}$. The new mixing system can be formulated more general than linear, e.g. we can employ MLP neural network:

$$\hat{\mathbf{X}} = \mathbf{g}^{(2)}(\mathbf{B}^{(2)}[\mathbf{g}^{(1)}(\mathbf{B}^{(1)}\mathbf{S} + \mathbf{b}^{(1)})] + \mathbf{b}^{(2)}), \tag{17}$$

where $\mathbf{g}^{(i)}(.)$ is a vector of nonlinearities, $\mathbf{B}^{(i)}(.)$ is a weight matrix and $\mathbf{b}^{(i)}(.)$ is a bias vector respectively for *i*-th layer, *i*=1,2. The first weight layer will produce results related to (4) if we take $\mathbf{B}^{(1)} = \hat{\mathbf{A}}$. But we employ also some nonlinearities and the second layer, so in comparison to the linear form the mixing system gains some flexibility. If we learn the whole structure starting from system with initial weights of $\mathbf{B}^{(1)}(0) = \hat{\mathbf{A}}$, we can expect the results will be better, see Fig. 2.

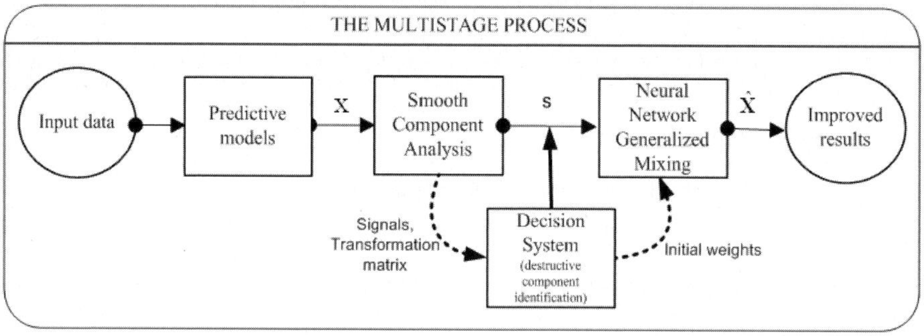

Fig. 2. The concept of filtration stage

6 Electricity Consumption Forecasting

The tests of proposed concept were performed on the problem of energy load prediction [11]. Our task was to forecast the hourly energy consumption in Poland in 24 hours basing on the energy demand from last 24 hours and calendar variables: month, day of the month, day of the week, and holiday indicator. We learned six MLP neural networks using 1851 instances in training, 1313 – in validation, and 1313 – in testing phase. The networks have the structure: M1=MLP(5,12,1) M2=MLP(5,18,1), M3=MLP(5,24,1), M4=MLP(5,27,1), M5=MLP(5,30,1), M6=MLP(5,33,1), where in parenthesis you can find the number of neurons in each layer. The quality of the results is measured with Mean Absolute Percentage Error:

$$MAPE = \frac{1}{N} \cdot \sum_{i=1}^{N} \left| \frac{y_i - \hat{y}_i}{y_i} \right|, \tag{18}$$

where i is the index of observation, N- number of instances, y_i - real load value, and $\hat{y}_i$ - predicted value.

In Table 2 we can observe the MAPE values for primary models, effects of improving the modelling results with particular decomposition, and with decomposition supported by neural networks remixing. The last column in Table 2 shows percentage improvement of the best results from each method versus the best primary result.

Table 2. Values of MAPE for primary models and after

Methods	Models						Best result	%
	M1	M2	M3	M4	M5	M6		
Primary results	2.392	2.365	2.374	2.402	2.409	2.361	2.361	-
Decorr.	2.304	2.256	2.283	2.274	2.255	2.234	2.234	5.4
Smooth	2.301	2.252	2.357	2.232	2.328	2.317	2.232	5.5
ICA	2.410	2.248	2.395	2.401	2.423	2.384	2.248	4.8
Decorr&NN	2.264	2.241	2.252	2.247	2.245	2.226	2.226	5.7
Smooth&NN	2.224	2.227	2.223	2.219	2.232	2.231	2.219	6.0
ICA&NN	2.327	2.338	2.377	2.294	2.299	2.237	2.237	5.3

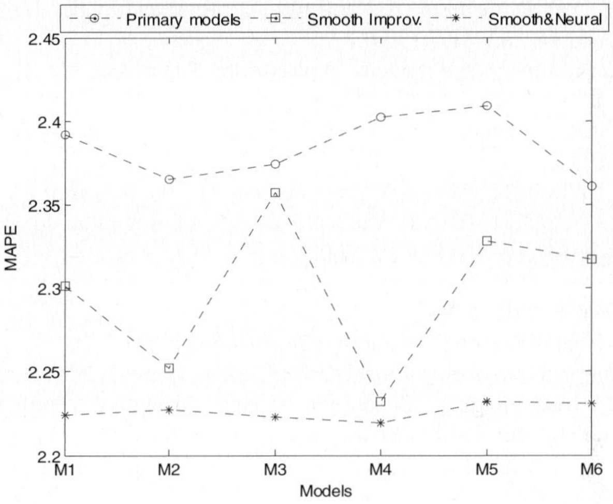

Fig. 3. The MAPE for primary models, improvement with SmCA, and improvement by SmCA&NN

To compare the obtained results with other ensemble methods we applied also bagging and boosting techniques for the presented problem of energy load prediction. They produced predictions with MAPE of 2.349 and 2.226, respectively, what means results slightly worse than SmCA with neural generalisation.

7 Conclusions

The Smooth Component Analysis as well as the other Blind Signal Separation methods can be successfully used as a novel methodology for prediction improvement. The practical experiment with the energy load prediction confirmed the validity of our method. Due to lack of space we compare SmCA approach only with basis BSS methods like decorrelation and ICA. For the same reason extended comparison with other ensemble methods was left as the further work.

References

1. Breiman, L.: Bagging predictors. Machine Learning 24, 123–140 (1996)
2. Cardoso, J.F.: High-order contrasts for independent component analysis. Neural Computation 11, 157–192 (1999)
3. Choi, S., Cichocki, A.: Blind separation of nonstationary sources in noisy mixtures. Electronics Letters 36(9), 848–849 (2000)
4. Cichocki, A., Amari, S.: Adaptive Blind Signal and Image Processing. John Wiley, Chichester (2002)
5. Donoho, D.L., Elad, M.: Maximal Sparsity Representation via l1 Minimization. The Proc. Nat. Acad. Sci. 100, 2197–2202 (2003)
6. Golub, G.H., Van-Loan, C.F.: Matrix Computations. Johns Hopkins, Baltimore (1996)
7. Haykin, S.: Neural nets: a comprehensive foundation. Macmillan, NY (1994)
8. Hoeting, J., Mdigan, D., Raftery, A., Volinsky, C.: Bayesian model averaging: a tutorial. Statistical Science 14, 382–417 (1999)
9. Hurst, H.E.: Long term storage capacity of reservoirs. Trans. Am. Soc. Civil Engineers 116 (1951)
10. Hyvärinen, A., Karhunen, J., Oja, E.: Independent Component Analysis. John Wiley, Chichester (2001)
11. Lendasse, A., Cottrell, M., Wertz, V., Verdleysen, M.: Prediction of Electric Load using Kohonen Maps – Application to the Polish Electricity Consumption. In: Proc. Am. Control Conf. Anchorage AK, pp. 3684–3689 (2002)
12. Lee, D.D., Seung, H.S.: Learning of the parts of objects by non-negative matrix factorization. Nature, 401 (1999)
13. Mitchell, T.: Machine Learning. McGraw-Hill, Boston (1997)
14. Scales, L.E.: Introduction to Non-Linear Optimization. Springer, NY (1985)
15. Stone, J.V.: Blind Source Separation Using Temporal Predictability. Neural Computation 13(7), 1559–1574 (2001)
16. Szupiluk, R., Wojewnik, P., Zabkowski, T.: Model Improvement by the Statistical Decomposition. In: Rutkowski, L., Siekmann, J.H., Tadeusiewicz, R., Zadeh, L.A. (eds.) ICAISC 2004. LNCS (LNAI), vol. 3070, pp. 1199–1204. Springer, Heidelberg (2004)
17. Therrien, C.W.: Discrete Random Signals and Statistical Signal Processing. Prentice Hall, New Jersey (1992)
18. Yang, Y.: Adaptive regression by mixing. Journal of American Statistical Association, 96 (2001)
19. Zibulevsky, M., Kisilev, P., Zeevi, Y.Y., Pearlmutter, B.A.: BSS via multinode sparse representation. Adv. in Neural Information Proc. Sys. 14, 185–191 (2002)

Speed and Accuracy Enhancement of Linear ICA Techniques Using Rational Nonlinear Functions

Petr Tichavský[1], Zbyněk Koldovský[1,2], and Erkki Oja[3]

[1] Institute of Information Theory and Automation, Pod vodárenskou věží 4, P.O. Box 18, 182 08 Praha 8, Czech Republic
p.tichavsky@ieee.org
http://si.utia.cas.cz/Tichavsky.html

[2] Faculty of Mechatronic and Interdisciplinary Studies Technical University of Liberec, Hálkova 6, 461 17 Liberec, Czech Republic

[3] Adaptive Informatics Research Centre, Helsinki University of Technology, P.O. Box 5400, 02015 HUT, Finland
Erkki.Oja@hut.fi

Abstract. Many linear ICA techniques are based on minimizing a nonlinear contrast function and many of them use a *hyperbolic tangent (tanh)* as their built-in nonlinearity. In this paper we propose two *rational* functions to replace the *tanh* and other popular functions that are tailored for separating supergaussian (long-tailed) sources. The advantage of the rational function is two-fold. First, the rational function requires a significantly lower computational complexity than *tanh*, e.g. *nine* times lower. As a result, algorithms using the rational functions are typically *twice* faster than algorithms with *tanh*. Second, it can be shown that a suitable selection of the rational function allows to achieve a better performance of the separation in certain scenarios. This improvement might be systematic, if the rational nonlinearities are selected adaptively to data.

1 Introduction

In this paper, a square linear ICA is treated (see e.g. [4,3]),

$$\mathbf{X} = \mathbf{AS}, \tag{1}$$

where $\mathbf{X}$ is a $d \times N$ data matrix. The rows of $\mathbf{X}$ are the observed mixed signals, thus d is the number of mixed signals and N is their length or the number of samples in each signal. Similarly, the unknown $d \times N$ matrix $\mathbf{S}$ includes samples of the original source signals. $\mathbf{A}$ is an unknown regular $d \times d$ mixing matrix.

As usual in linear ICA, it is assumed that the elements of $\mathbf{S}$, denoted s_{ij}, are mutually independent i.i.d. random variables with probability density functions (pdf) $f_i(s_{ij})$ $i = 1, \ldots, d$. The row variables s_{ij} for all $j = 1, \ldots, N$, having the same density, are thus an i.i.d. sample of one of the independent sources denoted by s_i. It is assumed that at most one of the densities $f_i(\cdot)$ is Gaussian, and the unknown matrix $\mathbf{A}$ has full rank. In the following, let $\mathbf{W}$ denote the demixing matrix, $\mathbf{W} = \mathbf{A}^{-1}$.

M.E. Davies et al. (Eds.): ICA 2007, LNCS 4666, pp. 285–292, 2007.

Many popular ICA methods use a nonlinear contrast function to blindly separate the signals. Examples include FastICA [5], an enhanced version of the algorithm named EFICA [7], and recursive algorithm EASI [2], Extended Infomax [1], and many others. Adaptive choices of the contrast functions were proposed in [6,8].

In practical large-scale problems, the computational speed of an algorithm is a factor that limits its applications. The main goal of this paper is to propose suitable *rational* functions that can be quickly evaluated when used instead of *tanh* and other nonlinearities, and yet achieve the same or better performance. We design such suitable rational nonlinearities for algorithms FastICA and EFICA, based on our recent analytical results on their asymptotic performances, see [5,7]. It is believed that the nonlinearities proposed here will work well when applied to other methods as well.

The structure of the paper is as follows. In section II we present a brief description of algorithms FastICA and EFICA, and the analytic expressions that characterize the asymptotic performance of the methods. In section III we propose A) two general-purpose rational nonlinearities that have similar performance as *tanh*, and B) nonlinearities that are tailored for separation of supergaussian (heavy tailed) sources.

2 FastICA, EFICA, and Their Performance

In general, the FastICA algorithm is based on minimization/maximization of the criterion $c(\mathbf{w}) = \mathrm{E}[G(\mathbf{w}^T\mathbf{Z})]$, where $G(\cdot)$ is a suitable nonlinearity, called a contrast function, applied elementwise to the row vector $\mathbf{w}^T\mathbf{Z}$; see [4]. Next, $\mathbf{w}$ is the unitary vector of coefficients to be found that separates one of the independent components from a mixture $\mathbf{Z}$. Here $\mathbf{Z}$ denotes a mean-removed and decorrelated data, $\mathbf{Z} = \mathbf{C}^{-1/2}(\mathbf{X} - \overline{\mathbf{X}})$ where $\widehat{\mathbf{C}}$ is the sample covariance matrix, $\widehat{\mathbf{C}} = (\mathbf{X} - \overline{\mathbf{X}})(\mathbf{X} - \overline{\mathbf{X}})^T/N$ and $\overline{\mathbf{X}}$ is the sample mean of the mixture data.

In the following, in accordance with the standard notation [4], $g(\cdot)$ and $g'(\cdot)$ denote the first and the second derivative of the function $G(\cdot)$. The application of $g(\cdot)$ and $g'(\cdot)$ to the vector $\mathbf{w}^T\mathbf{Z}$ is elementwise. Classical widely used functions $g(\cdot)$ include "pow3", i.e. $g(x) = x^3$ (then the algorithm performs kurtosis minimization), "tanh", i.e. $g(x) = \tanh(x)$, and "gauss", $g(x) = x\exp(-x^2/2)$.

The algorithm FastICA can be considered either in one unit form, where only one row $\mathbf{w}$ of the estimated demixing matrix $\widehat{\mathbf{W}}$ is computed, or in symmetric form, which estimates the whole matrix $\widehat{\mathbf{W}}$. The outcome of the symmetric FastICA obeys the orthogonality condition meaning that the sample correlations of the separated signals are exactly zeros.

Recently, it was proposed to complete the symmetric FastICA by a test of saddle points that eliminates convergence to side minima of the cost function, which may occur for most nonlinearities $g(\cdot)$ [9]. The test consists in checking if possible saddle points exist for each pair of the signal components exactly halfway between them in the angular sense. This test requires multiple evaluations of the primitive (integral) function of $g(\cdot)$, i.e. $G(\cdot)$. If the intention is to perform the test of saddle points in a fast manner, then it is desired that G is a rational function as well.

We introduced recently a novel algorithm called EFICA [7], which is essentially an elaborate modification of the FastICA algorithm employing a data-adaptive choice of the associated nonlinearities used in FastICA, and thus reaching a very small asymptotic error. The algorithm is initialized by performing a symmetric FastICA with a fixed nonlinearity. After that, the algorithm uses an idea of a generalized symmetric FastICA, and an adaptive choice of nonlinearities, which may be different for each signal component separately. The final demixing matrix does not obey the orthogonality condition. See [7] for details. For the purpose of this paper we shall assume that the adaptive selection of the nonlinearity in the EFICA algorithm is turned off and a fixed nonlinearity $g(\cdot)$ is used instead.

Assume now, for simplicity, that all signal components have the the same probability distribution with the density $f(\cdot)$. It was shown in [9] and in [7] that the asymptotic interference-to-signal ratio of the separated signals (one off-diagonal element of the ISR matrix) for the one-unit FastICA, for the symmetric FastICA and for EFICA is, respectively,

$$\mathrm{ISR}_{1\mathrm{U}} = \frac{1}{N}\frac{\gamma}{\tau^2}, \qquad \mathrm{ISR}_{\mathrm{SYM}} = \frac{1}{2N}\left[\frac{1}{2} + \frac{\gamma}{\tau^2}\right] \tag{2}$$

$$\mathrm{ISR}_{\mathrm{EF}} = \frac{1}{N}\frac{\gamma(\gamma+\tau^2)}{\tau^2\gamma + \tau^2(\gamma+\tau^2)} \tag{3}$$

where

$$\begin{array}{ll} \gamma = \beta - \mu^2 & \mu = \int s\, g(s)\, f(s)\, ds \\ \tau = |\mu - \rho| & \rho = \int g'(s)\, f(s)\, ds \\ & \beta = \int g^2(s)\, f(s)\, ds \end{array}$$

and the integration proceeds over the real line[1].

It can be easily seen that

$$\mathrm{ISR}_{\mathrm{EF}} = \mathrm{ISR}_{1\mathrm{U}} \frac{1/N + \mathrm{ISR}_{1\mathrm{U}}}{1/N + 2\,\mathrm{ISR}_{1\mathrm{U}}} \tag{4}$$

and

$$\mathrm{ISR}_{\mathrm{EF}} \leq \min\left\{\mathrm{ISR}_{1\mathrm{U}}, \mathrm{ISR}_{\mathrm{SYM}}\right\}. \tag{5}$$

It is well known that all three ISR's are simultaneously minimized, when the nonlinearity $g(\cdot)$ is proportional to the score function of the distribution of the sources, $g(x) = \psi(x) = -f'(x)/f(x)$. To be more accurate the optimum nonlinearity may have the form $g(x) = c_1\psi(x) + c_2 x$, where c_1 and c_2 are arbitrary

[1] Note that it is the orthogonality constraint that makes the ISR of the symmetric FastICA lower bounded by $1/(4N)$.

constants, $c_1 \neq 0$. The choice of the constants c_1, c_2 does not make any influence on the algorithm performance. For this case it was shown EFICA is maximally efficient: the ISR in (3) in fact equals the respective Cramér-Rao-induced lower bound [9,7].

3 Optimality Issues

From the end of the previous section it is clear that it is not possible to suggest a nonlinearity that would be optimum for all possible probability distributions of the sources. The opposite is true, however: for each nonlinearity g there exists a source distribution f_g such that all other nonlinearities, that are not linear combinations of g and x, perform worse in separating the data having this distribution (in the sense of mean ISR). The density f_g can be found by solving the equation

$$g(x) = -c_1 \frac{f_g'(x)}{f_g(x)} + c_2\, x = -c_1 \frac{d}{dx}[\log f_g(x)] + c_2\, x \tag{6}$$

and has the solution

$$f_g(x) = \exp\left\{ -\frac{1}{c_1} \int g(x)dx + \frac{c_2}{2c_1}\, x^2 + c_3 \right\}. \tag{7}$$

The constants c_1, c_2, and c_3 should be selected in the way that f is a valid pdf, i.e. is nonnegative, its integral over the real line is one and have zero mean and the variance one.

For example, the nonlinearity *tanh* is optimum for the source distributions of the form

$$f_{\tanh} = C_0 \exp(-C_1 x^2)(\cosh x)^{C_2} \tag{8}$$

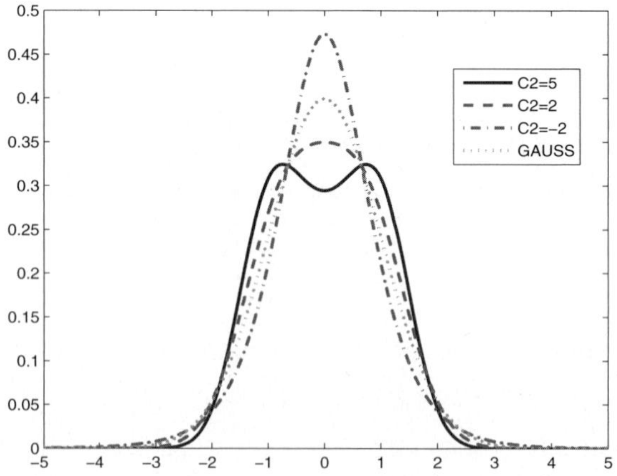

Fig. 1. Probability density functions (8) for which *tanh* is the optimum nonlinearity

where C_0, C_1, and C_2 are suitable constants. It can be shown that for any C_2 it is possible to find C_0 and C_1 such that $f_{\tanh}$ is a valid density function.

Examples of probability densities for which the *tanh* is the optimum nonlinearity are shown in Figure 1. The pdf's are compared with the standard Gaussian pdf, which would be obtained for $C_2 = 0$. The figure explains why *tanh* works very well for so many different pdf's: it includes supergaussian distributions for $C_2 < 0$ and subgaussian, even double modal distributions for $C_2 > 0$.

4 All-Purpose Nonlinearities

In this subsection we propose two rational functions that can replace *tanh* in FastICA and in other ICA algorithms,

$$g_1(x) = \frac{x}{1 + x^2/4}, \qquad g_2(x) = \frac{x(2 + |x|)}{(1 + |x|)^2} \,. \tag{9}$$

The former one has very similar behavior as *tanh* in a neighborhood of zero, and the latter one has a global behavior that is more similar to *tanh*, see Figure 2. For example, if $x \to \pm\infty$, then $g_2(x)$ approaches ± 1. These rational functions will be called RAT1 and RAT2, for easy reference.

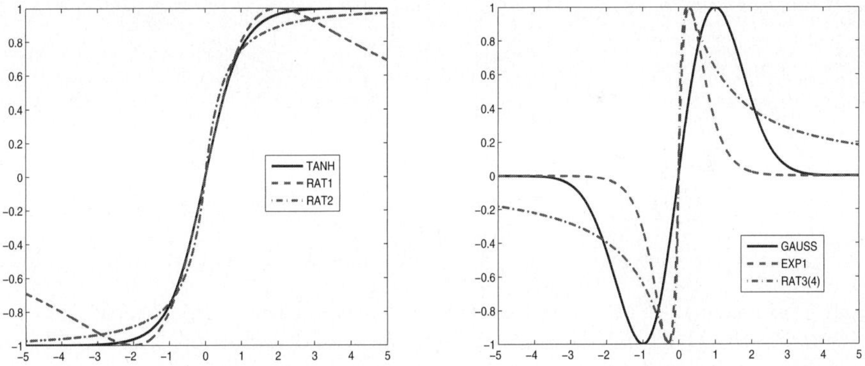

(a) (b)

Fig. 2. Comparison of nonlinearities (a) TANH, RAT1 and RAT2 and (b) GAUSS, EXP3 and RAT3(4), discussed in Section 5. In diagram (b), the functions were scaled to have the same maximum value, 1.

The speed of evaluation of tanh and the rational functions can be compared as follows. In the matlab notation, put $x = randn(1, 1000000)$. It was found that the evaluation of the command $y = tanh(x)$; takes 0.54 s, evaluation of RAT1 via command $y = x./(1 + x.\hat{}2/4)$; requires 0.07 s and evaluation of RAT2 via the pair of commands $h = x. * sign(x) + 1$; and $y = x. * (h + 1)./h.\hat{}2$; requires 0.11 s.

The computations were performed on HP Unix workstation, using a matlab profiler. We can conclude that evaluation of RAT1 is nine times faster than *tanh*, and evaluation of RAT2 is 5 times faster than *tanh*. As a result, FastICA using nonlinearity RAT1 is about twice faster that FastICA using *tanh*.

Performance of the algorithms using nonlinearities RAT1 and RAT2 appears to be very similar to that of the same algorithms using *tanh* for many probability distributions of the sources.

Assume, for example, that the source distribution belongs to the class of generalized Gaussian distribution with parameter α, which will be denoted GG(α) for easy reference. The pdf of the distribution is proportional to $\exp(-\beta_\alpha|x|)^\alpha$ where β_α is a suitable function of α such that the distribution has unit variance.

The asymptotic variance of one-unit FastICA (2) with the three nonlinearities is plotted as a function of α in Figure 3 (a). The variance is computed for $N = 1000$. We can see that performance of the algorithm with nonlinearity RAT1 is very similar to that of nonlinearity TANH. Performance of RAT2 is slightly better than the previous two ones, if the sources are supergaussian (spiky), i.e. for $\alpha < 2$, and slightly worse when the distribution is subgaussian ($\alpha > 2$).

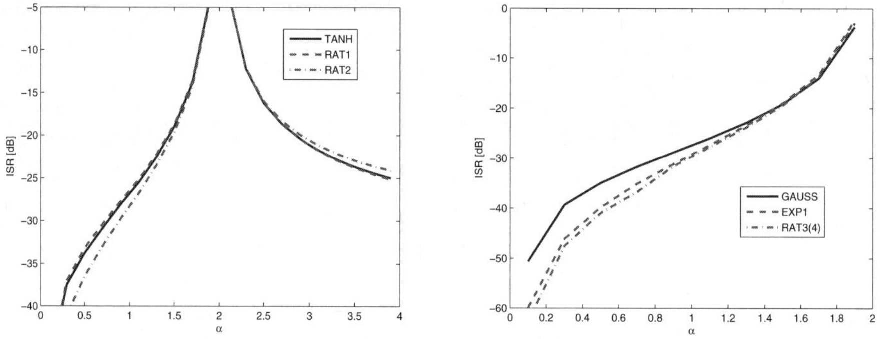

(a) (b)

Fig. 3. Performance of one unit FastICA with nonlinearities (a) TANH, RAT1 and RAT2 and (b) GAUSS, EXP1 and RAT3(4), discussed in Section 5, for sources with distribution GG(α) as a function of the parameter α

The advantage of RAT2 compared to RAT1 is that while the primitive function of $g_1(x)$ is $G_1(x) = 2\log(1 + x^2/4)$ and it is relatively complex to evaluate, the primitive function of $g_2(x)$ is rational, $G_2(x) = x^2/(1+|x|)$ and can be evaluated faster. This might be important for the test of saddle points. It is, however, possible to combine both approaches and use RAT1 in the main iteration, and the primitive function of RAT2 in the test of saddle points.

It can be shown that the asymptotic variance $\mathrm{ISR}_{1\mathrm{U}}$ goes to infinity for any nonlinearity in rare cases, when the source pdf is a linear combination of a supergaussian and a subgaussian distributions (τ in (2) is zero). An example is shown in Figure 4, where $\mathrm{ISR}_{1\mathrm{U}}$ is plotted for sources $\mathbf{s} = \beta\mathbf{b} + \sqrt{1-\beta^2}\mathbf{l}$ as a function of parameter β, where **b** and **l** stand for binary (BPSK) and Laplacean random variables, respectively. Performances of nonlinearities TANH and RAT1 appear to be very similar, while a performance of RAT2 is slightly different.

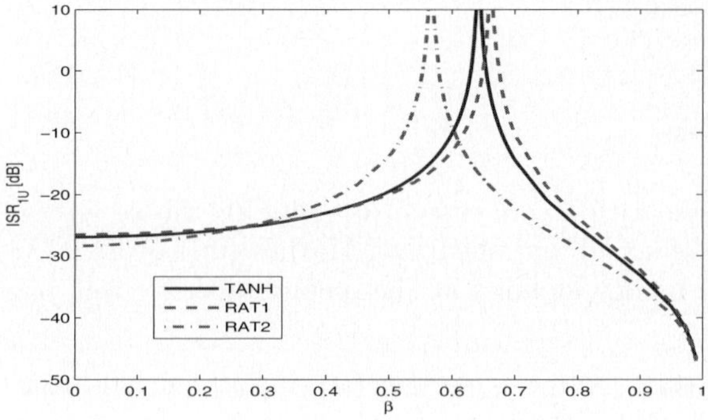

Fig. 4. Performance of one unit FastICA with nonlinearities TANH, RAT1 and RAT2 for sources of the type $s = \beta\mathbf{b} + \sqrt{1-\beta^2}\mathbf{l}$ and $N = 1000$

5 Nonlinearities for Supergaussian Sources

In [5] the following nonlinearity was proposed for separation of supergaussian (long-tailed) sources,

$$g(x) = x\exp(-x^2/2). \tag{10}$$

For a long time, this nonlinearity was considered the best known one for the separation of supergaussian sources. In [7] it was suggested to use the nonlinearity

$$g(x) = x\ \exp(-\eta|x|) \tag{11}$$

where $\eta = 3.348$ was selected. This nonlinearity will be referred as EXP1. This constant is the optimum constant for the nonlinearity of the form (11) provided that the distribution of the sources is Laplacean, i.e. GG(1). It was shown that the latter nonlinearity outperforms the former one for most of distributions GG(α) where $0 < \alpha < 2$. It was also shown in [7] that for the sources with the distribution GG(α) with $\alpha \in (0, 1/2]$ the asymptotic performance of the algorithm monotonically grows with increasing η.

In this paper we suggest to use the following nonlinearity, denoted as RAT3(b), for easy reference,

$$g_{3b}(x) = \frac{x}{(1+b|x|)^2}. \tag{12}$$

We note that like in the case of the nonlinearity EXP, the slope of the function at $x = 0$ increases with growing parameter b. This phenomenon improves the asymptotic performance of the algorithm in separation of highly supergaussian (long-tailed) sources, but makes the convergence of the algorithm more difficult.

We found that the choice $b = 4$ is quite good a trade-off between the performance and the ability to converge.

Evaluation of the nonlinearity RAT3(b) was found to be about five times faster than evaluation of EXP1. Performance of the algorithm using the 3 nonlinearities in separating sources with the distribution GG(α), $\alpha < 2$, is shown in Figure 3(b).

6 Conclusions

The rational nonlinearities were shown to be highly viable alternatives to classical ones in terms of speed and accuracy. Matlab code of EFICA, utilizing these nonlinearities can be downloaded at the second author's web page.

Acknowledgements. This work was supported by Ministry of Education, Youth and Sports of the Czech Republic through the project 1M0572 and by Grant Agency of the Czech Republic through the project 102/07/P384.

References

1. Bell, A.J., Sejnowski, T.J.: An information-maximization approach to blind separation and blind deconvolution. Neural Computation 7, 1129–1159 (1995)
2. Cardoso, J.-F., Laheld, B.: Equivariant adaptive source separation. IEEE Tr. Signal Processing 45, 434–444 (1996)
3. Cichocki, A., Amari, S.-I.: Adaptive signal and image processing: learning algorithms and applications. Wiley, New York (2002)
4. Hyvärinen, A., Karhunen, J., Oja, E.: Independent Component Analysis. Wiley-Interscience, New York (2001)
5. Hyvärinen, A., Oja, E.: A Fast Fixed-Point Algorithm for Independent Component Analysis. Neural Computation 9, 1483–1492 (1997)
6. Karvanen, J., Eriksson, J., Koivunen, V.: Maximum likelihood estimation of ICA model for wide class of source distributions. Neural Networks for Signal Processing X 1, 445–454 (2000)
7. Koldovský, Z., Tichavský, P., Oja, E.: Efficient Variant of Algorithm FastICA for Independent Component Analysis Attaining the Cramér-Rao Lower Bound. IEEE Tr. Neural Networks 17, 1265–1277 (2006)
8. Pham, D.T., Garat, P.: Blind separation of mixture of independent sources through a quasi-maximum likelihood approach. IEEE Trans. Signal Processing 45, 1712–1725 (1997)
9. Tichavský, P., Koldovský, Z., Oja, E.: Performance Analysis of the FastICA Algorithm and Cramér-Rao Bounds for Linear Independent Component Analysis. IEEE Tr. on Signal Processing 54, 1189–1203 (2006)
10. Vrins, F.: Contrast properties of entropic criteria for blind source separation, Ph.D. Thesis, Université catholique de Louvain (March 2007)

Comparative Speed Analysis of FastICA

Vicente Zarzoso and Pierre Comon

Laboratoire I3S, CNRS/UNSA
Les Algorithmes – Euclide-B, BP 121
06903 Sophia Antipolis Cedex, France
{zarzoso,pcomon}@i3s.unice.fr

Abstract. FastICA is arguably one of the most widespread methods for independent component analysis. We focus on its deflation-based implementation, where the independent components are extracted one after another. The present contribution evaluates the method's speed in terms of the overall computational complexity required to reach a given source extraction performance. FastICA is compared with a simple modification referred to as RobustICA, which merely consists of performing exact line search optimization of the kurtosis-based contrast function. Numerical results illustrate the speed limitations of FastICA.

1 Introduction

Independent component analysis (ICA) aims at decomposing an observed random vector into statistically independent variables [1]. Among its numerous applications, ICA is the most natural tool for blind source separation (BSS) in instantaneous linear mixtures when the source signals are assumed to be independent. Under certain identifiability conditions, the independent components correspond to the sources up to admissible scale and permutation indeterminacies [1].

Two main approaches to ICA have been proposed to date. In the original definition of ICA carried out in early works such as [1] and [2], the independent components are extracted jointly or simultaneously, an approach sometimes called symmetric. On the other hand, the deflation approach estimates the sources one after another [3], and has also been shown to work successfully to separate convolutive mixtures [4]. Due to error accumulation throughout successive deflation stages, it is generally acknowledged that joint algorithms outperform deflationary algorithms without necessarily incurring an excessive computational cost.

The FastICA algorithm [5], [6], [7], originally put forward in deflation mode, appeared when many other ICA methods had already been proposed, such as COM2 [1], JADE [2], COM1 [8], or the deflation methods by Tugnait [4] or Delfosse-Loubaton [3]. A thorough comparative study was carried out in [9], where FastICA is found to fail for weak or highly spatially correlated sources. More recently, its convergence has been shown to slow down or even fail in the presence of saddle points, particularly for short block sizes [10].

The objective of the present contribution is to carry out a brief critical review and experimental assessment of the deflationary kurtosis-based FastICA algorithm. In particular, we aim at evaluating objectively the algorithms' speed and

M.E. Davies et al. (Eds.): ICA 2007, LNCS 4666, pp. 293–300, 2007.

efficiency. For the sake of fairness, FastICA is not compared to joint extraction algorithms [1], [2], [3] but only to a simple modification called RobustICA, possibly the simplest deflation algorithm that can be thought of under the same general conditions.

2 Signal Model

The observed random vector $\mathbf{x} \in \mathbb{C}^L$ is assumed to be generated from the instantaneous linear mixing model:

$$\mathbf{x} = \mathbf{H}\mathbf{s} + \mathbf{n} \tag{1}$$

where the source vector $\mathbf{s} = [s_1, s_2, \ldots, s_K]^{\mathrm{T}} \in \mathbb{C}^K$ is made of $K \leq L$ unknown mutually independent components. The elements of mixing matrix $\mathbf{H} \in \mathbb{C}^{L \times K}$ are also unknown, and so is the noise vector $\mathbf{n}$, which is only assumed to be independent of the sources. Our focus is on block implementations, which, contrary to common belief, are not necessarily more costly than adaptive (recursive, on-line, sample-by-sample) algorithms, and are able to use more effectively the information contained in the observed signal block. Given a sensor-output signal block composed of T samples, ICA aims at estimating the corresponding T-sample realization of the source vector.

3 FastICA Revisited

3.1 Optimality Criteria

In the deflation approach, an extracting vector $\mathbf{w}$ is sought so that the estimate

$$z \stackrel{\text{def}}{=} \mathbf{w}^{\mathrm{H}}\mathbf{x} \tag{2}$$

maximizes some optimality criterion or contrast function, and is hence expected to be a component independent from the others. A widely used contrast is the normalized kurtosis, which can be expressed as:

$$\mathcal{K}(\mathbf{w}) = \frac{\mathrm{E}\{|z|^4\} - 2\mathrm{E}^2\{|z|^2\} - |\mathrm{E}\{z^2\}|^2}{\mathrm{E}^2\{|z|^2\}}. \tag{3}$$

This criterion is easily seen to be insensitive to scale, i.e., $\mathcal{K}(\lambda\mathbf{w}) = \mathcal{K}(\mathbf{w})$, $\forall\lambda \neq 0$. Since this scale indeterminacy is typically unimportant, we can impose, without loss of generality, the normalization $\|\mathbf{w}\| = 1$ for numerical convenience. The kurtosis maximization (KM) criterion started to receive attention with the pioneering work of Donoho [11] and Shalvi-Weinstein [12] on blind equalization, and was later employed for source separation even in the convolutive-mixture case [4]. Contrast (3) is quite general in that it does not require the observations to be prewhitened and can be applied to real- or complex-valued sources without any modification.

To simplify the source extraction, the kurtosis-based FastICA algorithm [5], [6], [7] first applies a prewhitening operation resulting in transformed observations with an identity covariance matrix, $\mathbf{R}_x \stackrel{\text{def}}{=} \mathrm{E}\{\mathbf{x}\mathbf{x}^{\mathrm{H}}\} = \mathbf{I}$. In the real-valued case, contrast (3) then becomes equivalent to the fourth-order moment criterion:

$$\mathcal{M}(\mathbf{w}) = \mathrm{E}\{|z|^4\}, \tag{4}$$

which must be optimized under a constraint, e.g., $\|\mathbf{w}\| = 1$, to avoid arbitrarily large values of z. Under the same constraint, criteria (3) and (4) are also equivalent if the sources are complex-valued but second-order circular, i.e., the non-circular second-moment matrix $\mathbf{C}_x \stackrel{\text{def}}{=} \mathrm{E}\{\mathbf{x}\mathbf{x}^{\mathrm{T}}\}$ is null. Consequently, contrast (4) is less general than criterion (3) in that it requires the observations to be prewhitened and the sources to be real-valued, or complex-valued but circular. Indeed, the extension of the FastICA algorithm to complex signals [13], [14] is only valid for second-order circular sources. In the remainder, we shall restrict our attention to sources fulfilling these requirements.

3.2 Contrast Optimization

Under the constraint $\|\mathbf{w}\| = 1$, the stationary points of $\mathcal{M}(\mathbf{w})$ are obtained as a collinearity condition on $\mathrm{E}\{\mathbf{x}zz^{*2}\}$:

$$\mathrm{E}\{|\mathbf{w}^{\mathrm{H}}\mathbf{x}|^2\mathbf{x}\mathbf{x}^{\mathrm{H}}\}\mathbf{w} = \lambda\mathbf{w} \tag{5}$$

where λ is a Lagrangian multiplier. As opposed to the claims of [5], eqn. (5) is a fixed-point equation only if λ is known, which is not the case here; λ must be determined so as to satisfy the constraint, and thus it depends on $\mathbf{w}_0$, the optimal value of $\mathbf{w}$: $\lambda = \mathcal{M}(|\mathbf{w}_0{}^{\mathrm{H}}\mathbf{x}|^4\}$.

For the sake of simplicity, λ is arbitrarily set to a deterministic fixed value [5], [7], so that FastICA becomes an approximate standard Newton algorithm, as eventually pointed out in [6]. In the real-valued case, the Hessian matrix of $\mathcal{M}(\mathbf{w})$ is approximated as

$$\mathrm{E}\{(\mathbf{w}^{\mathrm{T}}\mathbf{x}\mathbf{x}^{\mathrm{T}}\mathbf{w})\mathbf{x}\mathbf{x}^{\mathrm{T}}\} \approx \mathrm{E}\{\mathbf{w}^{\mathrm{T}}\mathbf{x}\mathbf{x}^{\mathrm{T}}\mathbf{w}\}\mathrm{E}\{\mathbf{x}\mathbf{x}^{\mathrm{T}}\} = \mathbf{w}^{\mathrm{T}}\mathbf{w} = \mathbf{I} \tag{6}$$

As a result, the kurtosis-based FastICA reduces to a gradient-descent algorithm with a judiciously chosen fixed step size leading to cubic convergence:

$$\mathbf{w}^+ = \mathbf{w} - \frac{1}{3}\,\mathrm{E}\{\mathbf{x}(\mathbf{w}^{\mathrm{T}}\mathbf{x})^3\} \tag{7}$$

$$\mathbf{w}^+ \leftarrow \mathbf{w}^+/\|\mathbf{w}^+\|. \tag{8}$$

This is a particular instance of the family of algorithms proposed in [4].

4 RobustICA

A simple quite natural modification of FastICA consists of performing exact line search of the kurtosis contrast (3):

$$\mu_{\mathrm{opt}} = \arg\max_{\mu} \mathcal{K}(\mathbf{w} + \mu\mathbf{g}). \tag{9}$$

The search direction $\mathbf{g}$ is typically (but not necessarily) the gradient: $\mathbf{g} = \nabla_{\mathbf{w}}\mathcal{K}(\mathbf{w})$. Exact line search is in general computationally intensive and presents other limitations [15], which explains why, despite being a well-known optimization method, it is very rarely used. However, for criteria that can be expressed as rational functions of μ, such as the kurtosis, the constant modulus [16], [17] and the constant power [18], [19] contrasts, the optimal step size $\mu_{\rm opt}$ can easily be determined by finding the roots of a low-degree polynomial.

At each iteration, optimal step-size (OS) optimization performs the following steps:

S1) Compute OS polynomial coefficients.

For the kurtosis contrast, the OS polynomial is given by:

$$p(\mu) = \sum_{k=0}^{4} a_k \mu^k. \tag{10}$$

The coefficients $\{a_k\}_{k=0}^4$ can easily can be obtained at each iteration from the observed signal block and the current values of $\mathbf{w}$ and $\mathbf{g}$ (their expressions are not reproduced here due to the lack of space; see [20] for details). Numerical conditioning in the determination of $\mu_{\rm opt}$ can be improved by normalizing the gradient vector.

S2) Extract OS polynomial roots $\{\mu_k\}_{k=1}^4$.

The roots of the 4th-degree polynomial (quartic) can be found at practically no cost using standard algebraic procedures known since the 16th century such as Ferrari's formula [15].

S3) Select the root leading to the absolute maximum:

$$\mu_{\rm opt} = \arg\max_k \mathcal{K}(\mathbf{w} + \mu_k \mathbf{g}).$$

S4) Update $\mathbf{w}^+ = \mathbf{w} + \mu_{\rm opt}\mathbf{g}$.

S5) Normalize as in (8).

For sufficient sample size, the computational cost per iteration is $(5L + 12)T$ flops whereas that of FastICA's iteration (7) is $2(L + 1)T$ flops. A flop is conventionally defined as a real product followed by an addition.

To extract more than one independent component, the Gram-Schmidt-type deflationary orthogonalization procedure proposed for FastICA [5], [6], [7] can also be used in conjunction with RobustICA. After step S4, the updated extracting vector is constrained to lie in the orthogonal subspace of the extracting vectors previously found.

5 Numerical Experiments

The experimental analysis of this section aims at evaluating objectively the speed and efficiency of FastICA and RobustICA in several simulation conditions. The influence of prewhitening on the methods' performance is also assessed.

Performance-complexity trade-off. Noiseless unitary random mixtures of K independent unit-power BPSK sources are observed at the output of an $L = K$ element array in signal blocks of T samples. The search for each extracting vector is initialized with the corresponding canonical basis vector, and is stopped at a fixed number of iterations. The total cost of the extraction can then be computed as the product of the number of iterations, the cost per iteration per source (Sec. 4) and the number of sources. Prewhitening, if used, also adds to the total cost. The complexity per source per sample is given by the total cost divided by KT. As a measure of extraction quality, we employ the signal mean square error (SMSE), a contrast-independent criterion defined as

$$\mathrm{SMSE} = \frac{1}{K}\sum_{k=1}^{K} \mathrm{E}\{|s_k - \hat{s}_k|^2\}. \tag{11}$$

The estimated sources are optimally scaled and permuted before evaluating the SMSE. This performance index is averaged over 1000 independent random realizations of the sources and the mixing matrix. Extraction solutions are computed directly from the observed unitary mixtures (methods labelled as 'FastICA' and 'RobustICA') and after a prewhitening stage based on the SVD of the observed data matrix ('pw+FastICA', 'pw+RobustICA'). The cost of the prewhitening stage is of the order of $2K^2T$ flops.

Fig. 1 summarizes the performance-complexity variation obtained for $T = 150$ samples and different values of the mixture size K. Clearly, the best fastest performance is provided by RobustICA without prewhitening: a given performance

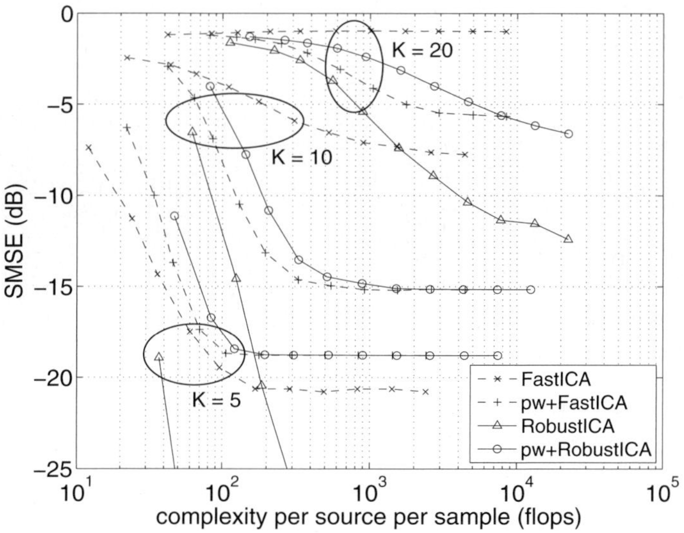

Fig. 1. Average extraction quality against computational cost for different mixture sizes K, with signal blocks of $T = 150$ samples

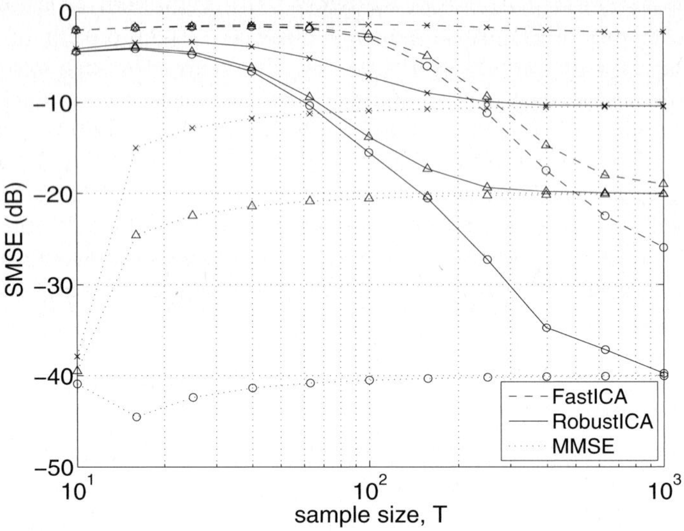

(a)

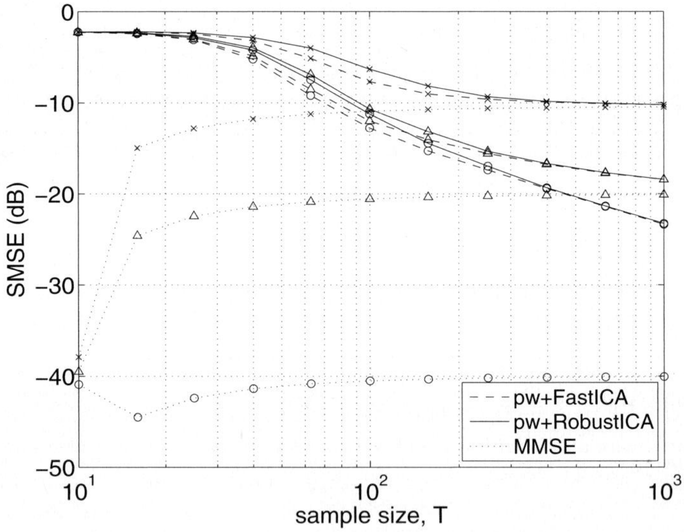

(b)

Fig. 2. Average extraction quality against signal block size for unitary mixtures of $K = 10$ sources and a total complexity of 400 flops/source/sample: (a) without prewhitening, (b) with prewhitening. '$\times$': SNR = 10 dB; '$\triangle$': SNR = 20 dB; '$\circ$': SNR = 40 dB

level is achieved with lower cost or, alternatively, an improved extraction quality is reached with a given complexity. The use of prewhitening worsens RobustICA's performance-complexity trade-off and, due to the finite sample size, imposes the same SMSE bound for two methods. Using prewhitening, FastICA improves considerably and becomes slightly faster than RobustICA.

Efficiency. We now evaluate the methods' performance for a varying block sample size T. Extractions are obtained by limiting the number of iterations per source, as explained above. To make the comparison meaningful, the overall complexity is fixed at 400 flops/source/sample for all tested methods. Accordingly, since RobustICA is more costly per iteration than FastICA, it performs fewer iterations per source. Isotropic additive white real Gaussian noise is present at the sensor output, with signal-to-noise ratio:

$$\mathrm{SNR} = \frac{\mathrm{trace}(\mathbf{H}\mathbf{H}^{\mathrm{T}})}{\sigma_n^2 L}. \tag{12}$$

Results for the minimum mean square error (MMSE) receiver are also obtained by jointly estimating the separating vectors assuming that all transmitted symbols are used for training. The MMSE can be considered as a performance bound for linear extraction.

Fig. 2(a) shows the results without prewhitening for random unitary mixtures of $K = 10$ sources and three different SNR values (10 dB, 20 dB and 40 dB). RobustICA attains the MMSE bound for block sizes of about 1000 samples for the tested SNR levels; the required block size can be shown to decrease for smaller K. FastICA seems to require longer block sizes, particularly for noisier conditions at the given overall complexity. As shown in Fig. 2(b), the use of prewhitening in the same experiment worsens the performance-complexity ratio of RobustICA while improving that of FastICA, making both methods' efficiency comparable.

6 Conclusions

The computational complexity required to reach a given source extraction quality is put forward as a natural objective measure of convergence speed for BSS/ICA algorithms. The kurtosis-based FastICA method can be considered as a gradient-based algorithm with constant step size. Its speed is shown to depend heavily on prewhitening and sometimes on initialization. Without the performance limitations imposed by the second-order preprocessing, a simple algebraic line optimization of the more general kurtosis contrast proves computationally faster and more efficient than FastICA even in scenarios favouring this latter method. Although not demonstrated in this paper, RobustICA is also more robust to initialization [20], and the optimal step-size technique it relies on proves less sensitive to saddle points or local extrema [17], [19].

References

1. Comon, P.: Independent component analysis, a new concept? Signal Processing 36(3), 287–314 (1994)
2. Cardoso, J.F., Souloumiac, A.: Blind beamforming for non-Gaussian signals. IEE Proceedings-F 140(6), 362–370 (1993)

3. Delfosse, N., Loubaton, P.: Adaptive blind separation of independent sources: a deflation approach. Signal Processing 45(1), 59–83 (1995)
4. Tugnait, J.K.: Identification and deconvolution of multichannel non-Gaussian processes using higher order statistics and inverse filter criteria. IEEE Transactions on Signal Processing 45, 658–672 (1997)
5. Hyvärinen, A., Oja, E.: A fast fixed-point algorithm for independent component analysis. Neural Computation 9(7), 1483–1492 (1997)
6. Hyvärinen, A.: Fast and robust fixed-point algorithms for independent component analysis. IEEE Transactions on Neural Networks 10(3), 626–634 (1999)
7. Hyvärinen, A., Karhunen, J., Oja, E.: Independent Component Analysis. John Wiley & Sons, New York (2001)
8. Comon, P., Moreau, E.: Improved contrast dedicated to blind separation in communications. In: Proc. ICASSP-97, 22nd IEEE International Conference on Acoustics, Speech and Signal Processing, Munich, Germany, April 20-24, 1997, pp. 3453–3456. IEEE Computer Society Press, Los Alamitos (1997)
9. Chevalier, P., Albera, L., Comon, P., Ferreol, A.: Comparative performance analysis of eight blind source separation methods on radiocommunications signals. In: Proc. International Joint Conference on Neural Networks, Budapest, Hungary (July 25-29, 2004)
10. Tichavsky, P., Koldovsky, Z., Oja, E.: Performance analysis of the FastICA algorithm and Cramér-Rao bounds for linear independent component analysis. IEEE Transactions on Signal Processing 54(4), 1189–1203 (2006)
11. Donoho, D.: On minimum entropy deconvolution. In: Proc. 2nd Applied Time Series Analysis Symposium, Tulsa, OK, pp. 565–608 (1980)
12. Shalvi, O., Weinstein, E.: New criteria for blind deconvolution of nonminimum phase systems (channels). IEEE Transactions on Information Theory 36(2), 312–321 (1990)
13. Bingham, E., Hyvärinen, A.: A fast fixed-point algorithm for independent component analysis of complex valued signals. International Journal of Neural Systems 10(1), 1–8 (2000)
14. Ristaniemi, T., Joutsensalo, J.: Advanced ICA-based receivers for block fading DS-CDMA channels. Signal Processing 82(3), 417–431 (2002)
15. Press, W.H., Teukolsky, S.A., Vetterling, W.T., Flannery, B.P.: Numerical Recipes in C. The Art of Scientific Computing, 2nd edn. Cambridge University Press, Cambridge (1992)
16. Godard, D.N.: Self-recovering equalization and carrier tracking in two-dimensional data communication systems. IEEE Transactions on Communications 28(11), 1867–1875 (1980)
17. Zarzoso, V., Comon, P.: Optimal step-size constant modulus algorithm. IEEE Transactions on Communications (to appear, 2007)
18. Grellier, O., Comon, P.: Blind separation of discrete sources. IEEE Signal Processing Letters 5(8), 212–214 (1998)
19. Zarzoso, V., Comon, P.: Blind and semi-blind equalization based on the constant power criterion. IEEE Transactions on Signal Processing 53(11), 4363–4375 (2005)
20. Zarzoso, V., Comon, P., Kallel, M.: How fast is FastICA? In: Proc. EUSIPCO-2006, XIV European Signal Processing Conference, Florence, Italy (September 4-8, 2006)

Kernel-Based Nonlinear Independent Component Analysis

Kun Zhang and Laiwan Chan*

Department of Computer Science and Engineering,
The Chinese University of Hong Kong
Shatin, Hong Kong
{kzhang,lwchan}@cse.cuhk.edu.hk

Abstract. We propose the kernel-based nonlinear independent component analysis (ICA) method, which consists of two separate steps. First, we map the data to a high-dimensional feature space and perform dimension reduction to extract the effective subspace, which was achieved by kernel principal component analysis (PCA) and can be considered as a pre-processing step. Second, we need to adjust a linear transformation in this subspace to make the outputs as statistically independent as possible. In this way, nonlinear ICA, a complex nonlinear problem, is decomposed into two relatively standard procedures. Moreover, to overcome the ill-posedness in nonlinear ICA solutions, we utilize the minimal nonlinear distortion (MND) principle for regularization, in addition to the smoothness regularizer. The MND principle states that we would prefer the nonlinear ICA solution with the mixing system of minimal nonlinear distortion, since in practice the nonlinearity in the data generation procedure is usually not very strong.

1 Introduction

Independent component analysis (ICA) aims at recovering independent sources from their mixtures, without knowing the mixing procedure or any specific knowledge of the sources. In particular, in this paper we consider the general nonlinear ICA problem. Assume that the observed data $\mathbf{x} = (x_1, \cdots, x_n)^T$ are generated from an independent random vector $\mathbf{s} = (s_1, \cdots, s_n)^T$ by a nonlinear transformation $\mathbf{x} = \mathcal{H}(\mathbf{s})$, where $\mathcal{H}$ is an unknown real-valued n-component mixing function. (For simplicity, it is usually assumed that the number of observable variables equals that of the original independent variables.) The general nonlinear ICA problem is to find a mapping $\mathcal{G} : \mathbb{R}^n \to \mathbb{R}^n$ such that $\mathbf{y} = \mathcal{G}(\mathbf{x})$ has statistically independent components.

In the general nonlinear ICA problem, in order to model arbitrary nonlinear mappings, one may need to resort to a flexible nonlinear function approximator, such as the multi-layer perceptron (MLP) network or the radius basis function (RBF) network, to represent the nonlinear separation system $\mathcal{G}$ or the mixing system $\mathcal{H}$ (see,

* This work was partially supported by a grant from the Research grants Council of the Hong Kong Special Administration Region, China.

M.E. Davies et al. (Eds.): ICA 2007, LNCS 4666, pp. 301–308, 2007.

e.g. [1]). In such a way, parameters at different locations of the network are adjusted simultaneously. This would probably slow down the learning procedure.

Kernel-based methods has also been considered for solving the nonlinear blind source separation (BSS) problem [5,10].[1] These methods exploit the temporal information of sources for source separation, and do not enforce mutual independence of outputs. In [5], the data are first implicitly mapped to high-dimensional feature space, and the effective subspace in feature space is extracted. TDSEP [13], a BSS algorithm based on temporal decorrelation, is then performed in the extracted subspace. Denote by d the reduced dimension. This method produces d outputs and one needs to select from them n outputs, as an estimate of the original sources. This method produces successful results in many experiments. However, a problem is that its outputs may not contain the estimate of the original sources, due to the effect of spurious outputs. Moreover, this method may fail if some sources lack specific time structures.

In this paper we propose a kernel-based method to solve nonlinear ICA. The separation system $\mathcal{G}$ is constructed using the kernel methods, and unknown parameters are adjusted by minimizing the mutual information between outputs y_i. The first step of this method is similar to that in [5], and kernel principal component analysis (PCA) is adopted to construct the feature subspace of reduced dimension. In the second step we solve a linear problem—we adjust the $n \times d$ linear transformation matrix $\mathbf{W}$ to make the outputs statistically independent. As stated in [5], standard linear ICA algorithms do not work here. We derive the algorithm for learning $\mathbf{W}$, which is in a similar form to the traditional gradient-based ICA algorithm.

We then consider suitable regularization conditions with which the proposed kernel-based nonlinear ICA leads to nonlinear BSS. In the general nonlinear ICA problem, although we do not know the form of the nonlinearity in the data generation procedure, fortunately, the nonlinearity in the generation procedure of natural signals is usually not strong. Hence, provided that the nonlinear ICA outputs are mutually independent, we would prefer the solution with the mixing transformation as close as possible to linear. This information, formulated as the minimal nonlinear distortion (MND) principle [12], can help to reduce the indeterminacies in solutions of nonlinear ICA greatly. MND and smoothness are incorporated for regularization in the kernel-based nonlinear ICA.

2 Kernel-Based Nonlinear ICA

Kernel-based learning has become a popular technique, in that it provides an elegant way to tackle nonlinear algorithms by reducing them to linear ones in some feature space $\mathcal{F}$, which is related to the original input space $\mathbb{R}^n$ by a possibly nonlinear map Φ. Denote by $\mathbf{x}^{(i)}$ the ith sample of $\mathbf{x}$. The dot products of the form $\Phi(\mathbf{x}^{(i)}) \cdot \Phi(\mathbf{x}^{(j)})$ can be computed using kernel representations $k(\mathbf{x}^{(i)}, \mathbf{x}^{(j)}) = \Phi(\mathbf{x}^{(i)}) \cdot \Phi(\mathbf{x}^{(j)})$. Thus, any linear algorithm formulated in terms of dot products can be made nonlinear by making use of the kernel trick, without

[1] Note that kernel ICA [3] actually performs *linear* ICA with the kernel trick.

knowing explicitly the mapping Φ. Unfortunately, ICA could not be kernelized directly, since it can not be carried out using dot products.

However, the kernel trick can still help to perform nonlinear ICA, in an analogous manner to the development of kTDSEP, which is a kernel-based algorithm for nonlinear BSS [5]. Kernel-based nonlinear ICA involves two separate steps. The first step is the same as that in kTDSEP: the data are implicitly mapped to a high-dimensional feature space and its effective subspace is extracted. As a consequence, the nonlinear problem in input space is transformed to a linear one in the reduced feature space. In the second step, a linear transformation in the reduced feature space is constructed such that it produces n statistically independent outputs. In this way nonlinear ICA is performed faithfully, without any assumption on the time structure of sources.

Many techniques can help to find the effective subspace in feature space $\mathcal{F}$. Here we adopt kernel PCA [11], since the subspace it produces gives the smallest reconstruction error in feature space. The effective dimension of feature space, denoted by d, can be determined by inspection on the eigenvalues of the kernel matrix. Let $\mathbf{x}$ be a test point, and let $\tilde{k}(\mathbf{x}^{(i)}, \mathbf{x}) = \tilde{\Phi}(\mathbf{x}^{(i)}) \cdot \tilde{\Phi}(\mathbf{x})$, where $\tilde{\Phi}$ denotes the centered image in feature space. The pth centered nonlinear principal component of $\mathbf{x}$, denoted by z_p, is in the form (for details see [11]):

$$z_p = \sum_{i=1}^{T} \tilde{\alpha}_{pi} \tilde{k}(\mathbf{x}^{(i)}, \mathbf{x}) \tag{1}$$

Let $\mathbf{z} = (z_1, \cdots, z_d)^T$. It contains all principal components of the images $\Phi(\mathbf{x})$ in feature space. Consequently, in the following we just need find a $n \times d$ matrix $\mathbf{W}$ which makes the components of

$$\mathbf{y} = \mathbf{W}\mathbf{z} \tag{2}$$

as statistically independent as possible.

2.1 Can Standard Linear ICA Work in Reduced Feature Space?

As claimed in [5], applying standard linear ICA algorithms, such as JADE [4] and FastICA [6], to the signals $\mathbf{z}$ does not give successful results. In our problem, z_p, $p = 1, \cdots, d$, are nonlinear mixtures of only n independent sources, and we aim at *transforming* z_p *to* n *signals* (generally $n \ll d$) *which are statistically independent*, with a linear transformation. But standard ICA algorithms, such as the natural gradient algorithm [2] and JADE, assume that $\mathbf{W}$ is square and invertible and try to extract d independent signals from z_i. So they can not give successful results in our problem.

Although FastICA, which aims at maximizing the nongaussianity of outputs, can be used in a deflationary manner, its relation to maximum likelihood of the ICA model or minimization of mutual information between outputs is established when the linear ICA model holds and $\mathbf{W}$ is square and invertible [7]. When the linear ICA model does not hold, just like in our problem, nongaussianity of outputs does not necessarily lead to the independence between them. In fact, if we apply FastICA in a deflationary manner to z_i, the outputs y_i will be extremely nongaussian, but they are not necessarily mutually independent. The extreme nongaussianity of y_i is because theoretically, with a properly chosen kernel function, by adjusting the ith row of $\mathbf{W}$ the mapping from $\mathbf{x}$ to y_i covers quite a large class of continuous functions.

2.2 Learning Rule

Now we aim to adjust $\mathbf{W}$ in Eq. 2 to make the outputs y_i as independent as possible. This can be achieved by minimizing the mutual information between y_i, which is defined as $I(\mathbf{y}) = \sum_{i=1}^{n} H(y_i) - H(\mathbf{y})$ where $H(\cdot)$ denotes the differential entropy. Denote by $\mathbf{J}$ the Jacobian matrix of the transformation from $\mathbf{x}$ to $\mathbf{y}$, i.e. $\mathbf{J} = \frac{\partial \mathbf{y}}{\partial \mathbf{x}}$, and by $\mathbf{J}_1$ the Jacobian matrix of the transformation from $\mathbf{x}$ to $\mathbf{z}$, i.e. $\mathbf{J}_1 = \frac{\partial z}{\partial x}$.[2] Due to Eq. 2, one can see $\mathbf{J} = \mathbf{W} \cdot \mathbf{J}_1$. We also have $H(\mathbf{y}) = H(\mathbf{x}) + E \log |\det \mathbf{J}|$. Consequently,

$$I(\mathbf{y}) = \sum_{i=1}^{n} H(y_i) - H(\mathbf{y}) = -\sum_{i=1}^{n} \log p_{y_i}(y_i) - E \log |\det(\mathbf{W} \cdot \mathbf{J}_1)| - H(\mathbf{x})$$

As $H(\mathbf{x})$ does not depend on $\mathbf{W}$, the gradient of $I(\mathbf{y})$ w.r.t. $\mathbf{W}$ is

$$\frac{\partial I(\mathbf{y})}{\partial \mathbf{W}} = E[\boldsymbol{\psi}(\mathbf{y}) \cdot \mathbf{z}^T] - E[\mathbf{J}^{-T} \cdot \mathbf{J}_1^T] \tag{3}$$

where $\boldsymbol{\psi}(\mathbf{y}) = (\psi_1(y_1), \cdots, \psi_n(y_n))^T$ with ψ_i being the score function of p_{y_i}, defined as $\psi_i = -(\log p_{y_i})' = -\frac{p'_{y_i}}{p_{y_i}}$. $\mathbf{W}$ can then be adjusted according to Eq. 3 with the gradient-based method. Note that the gradient in Eq. 3 is in a similar form to that in standard ICA, and the only difference is that the second term becomes $-E[\mathbf{W}^{-T}]$ in standard ICA[3].

In standard ICA, we can obtain correct ICA results even if the estimation of the densities p_{y_i} or the score functions ψ_i is not accurate. But in the nonlinear case, they should be estimated accurately. We use the mixture of 5 Gaussian's to model p_{y_i}. After each iteration of Eq. 3, parameters in the Gaussian mixture model are adjusted by the EM algorithm to adapt the current outputs y_i.

3 With Minimum Nonlinear Distortion

Solutions to nonlinear ICA always exist and are highly non-unique [8]. In fact, in the general nonlinear ICA problem, nonlinear BSS is impossible without additional prior knowledge on the mixing model [9]. Smoothness of the mapping

---

[2] $\mathbf{J}_1$ is involved in the obtained update rule Eq. 3. Since $\tilde{k}(\mathbf{x}^{(i)}, \mathbf{x}) = \tilde{\Phi}(\mathbf{x}^{(i)}) \cdot \tilde{\Phi}(\mathbf{x}) = k(\mathbf{x}^{(i)}, \mathbf{x}) - \frac{1}{T} \sum_{p=1}^{T} k(\mathbf{x}^p, \mathbf{x}) - \frac{1}{T} \sum_{q=1}^{T} k(\mathbf{x}^{(i)}, \mathbf{x}^{(q)}) + \frac{1}{T^2} \sum_{p=1}^{T} \sum_{q=1}^{T} k(\mathbf{x}^{(p)}, \mathbf{x}^{(q)})$. We have $\frac{\partial \tilde{k}(\mathbf{x}^{(i)}, \mathbf{x})}{\partial \mathbf{x}} = \frac{\partial k(\mathbf{x}^{(i)}, \mathbf{x})}{\partial \mathbf{x}} - \frac{1}{T} \sum_{p=1}^{T} \frac{\partial k(\mathbf{x}^{(p)}, \mathbf{x})}{\partial \mathbf{x}}$. According to Eq. 1, the pth row of $\mathbf{J}_1$ is then $\frac{\partial z_p}{\partial \mathbf{x}} = \sum_{i=1}^{T} \tilde{\alpha}_{pi} \frac{\partial \tilde{k}(\mathbf{x}^{(i)}, \mathbf{x})}{\partial \mathbf{x}}$. This can be easily calculated and saved in the first step of our method for later use.

[3] Assuming that $\mathbf{W}$ is square and invertible, the natural gradient ICA algorithm is obtained multiplying the right-hand side of $\frac{\partial I(\mathbf{y})}{\partial \mathbf{W}}$ by $\mathbf{W}^T \mathbf{W}$ [2]. However, as $\mathbf{W}$ in Eq. 2 is $n \times d$, the natural gradient for $\mathbf{W}$ could not be derived in this simple way.

$\mathcal{G}$ provides a useful regularization condition to lead nonlinear ICA to nonlinear BSS [1]. But it seems not sufficient, as shown by the counterexample in [9].

In this paper, in addition to the smoothness regularization, we exploit the "minimal nonlinear distortion" (MND) principle [12] for regularization of nonlinear ICA. MND has exhibited quite good performance for regularizing nonlinear ICA, when the nonlinearity in the data generation procedure is not very strong [12]. The objective function to be minimized thus becomes

$$J(\mathbf{W}) = I(\mathbf{y}) + \lambda_1 R_1(\mathbf{W}) + \lambda_2 R_2(\mathbf{W}) \tag{4}$$

where R_1 denotes the regularization term for achieving MND, R_2 is that for enforcing smoothness, and λ_1 and λ_2 are corresponding regularization parameters.

3.1 Minimum Nonlinear Distortion

MND states that, under the condition that the separation outputs y_i are mutually independent, we prefer the nonlinear mixing mapping $\mathcal{H}$ that is as close as possible to linear. So R_1 is a measure of "closeness to linear" of $\mathcal{H}$. Given a nonlinear mapping $\mathcal{H}$, its deviation from the affine mapping $\mathbf{A}^*$, which fits $\mathcal{H}$ best among all affine mappings $\mathbf{A}$, is an indicator of its "closeness to linear" or the level of its nonlinear distortion. Mean square error (MSE) is adopted to measure the deviation, since it greatly facilitates the following analysis. Let $\mathbf{x}^* = (x_1^*, \cdots, x_n^*)^T = \mathbf{A}^*\mathbf{y}$. R_1 can be defined as the total MSE between x_i and x_i^* (here we assume that both $\mathbf{x}$ and $\mathbf{y}$ are zero-mean):

$$\begin{aligned} R_1 &= E\{(\mathbf{x}-\mathbf{x}^*)^T(\mathbf{x}-\mathbf{x}^*)\}\text{ , where} \\ \mathbf{x}^* &= \mathbf{A}^*\tilde{\mathbf{y}},\text{ and } \mathbf{A}^* = \arg_{\mathbf{A}}\min E\{(\mathbf{x}-\mathbf{A}\mathbf{y})^T(\mathbf{x}-\mathbf{A}\mathbf{y})\} \end{aligned} \tag{5}$$

The derivative of R_1 w.r.t. $\mathbf{A}^*$ is $\frac{\partial R_1}{\partial \mathbf{A}^*} = -2E\{(\mathbf{x}-\mathbf{A}^*\mathbf{y})\mathbf{y}^T\}$. Setting the derivative to $\mathbf{0}$ gives $\mathbf{A}^*$: $E\{(\mathbf{x}-\mathbf{A}^*\tilde{\mathbf{y}})\tilde{\mathbf{y}}^T\} = \mathbf{0} \Leftrightarrow \mathbf{A}^* = E\{\mathbf{x}\mathbf{y}^T\}[E\{\mathbf{y}\mathbf{y}^T\}]^{-1}$. We can see that due to the adoption of MSE, $\mathbf{A}^*$ can be obtained in closed form. This will greatly simplify the derivation of learning rules.

We then have $R_1 = \mathrm{Tr}\big(E[(\mathbf{x}-\mathbf{A}^*\mathbf{y})(\mathbf{x}-\mathbf{A}^*\mathbf{y})^T]\big) = -\mathrm{Tr}\big(E[\mathbf{x}\mathbf{y}^T]\{E[\mathbf{y}\mathbf{y}^T]\}^{-1}\cdot E[\mathbf{y}\mathbf{x}^T]\big) + \text{const}$. Since in the learning process, y_i are approximately independent from each other, they are approximately uncorrelated. We can also normalize the variance of y_i after each iteration. Consequently $E[\mathbf{y}\mathbf{y}^T] = \mathbf{I}$. Let $\mathbf{L} = E[\mathbf{x}\mathbf{z}^T]$. we have $E[\mathbf{x}\mathbf{y}^T] = \mathbf{L}\mathbf{W}^T$. Thus $R_1 = -\mathrm{Tr}(\mathbf{L}\mathbf{W}^T\mathbf{W}\mathbf{L}^T) + \text{const}$. This gives

$$\frac{\partial R_1}{\partial \mathbf{W}} = -2\mathbf{W}\mathbf{L}^T\mathbf{L} \tag{6}$$

It was suggested to initialize λ_1 in Eq. 4 with a large value at the beginning of training and decreasing it to a small constant during the learning process [12]. A large value for λ at the beginning helps to reduce the possibility of getting into unwanted solutions or local optima. As training goes on, the influence of the regularization term is reduced, and $\mathcal{G}$ gains more freedom. In addition, initializing $\mathcal{G}$ to an almost identity mapping would also be useful. This can be achieved by simply initializing $\mathbf{W}$ with $\mathbf{W} = E[\mathbf{x}\mathbf{z}^T]\{E[\mathbf{z}\mathbf{z}^T]\}^{-1}$.

The MND principle can be incorporated in many nonlinear ICA/BSS methods to avoid unwanted solutions, under the condition that the nonlinearity in the mixing procedure is not too strong. As an example, for kTDSEP [5], MND provides a way to select a subset of output components corresponding to the original sources [12].

3.2 Smoothness: Local Minimum Nonlinear Distortion

Both MND and smoothness are used for regularization in our nonlinear ICA method. In fact, the smoothness regularizer exploiting second-order derivatives is related to MND. Particularly, enforcing *local* closeness to linear of the transformation at every point will lead to such a smoothness regularizer [12].

For a one-dimensional sufficiently smooth function $g(\mathbf{x})$, its second-order Taylor expansion in the vicinity of $\mathbf{x}$ is $g(\mathbf{x}+\varepsilon) \approx g(\mathbf{x}) + \left(\frac{\partial g}{\partial \mathbf{x}}\right)^T \cdot \varepsilon + \frac{1}{2}\varepsilon^T \mathbf{H}_\mathbf{x} \varepsilon$. Here ε is a small variation of $\mathbf{x}$ and $\mathbf{H}_\mathbf{x}$ denotes the Hessian matrix of g. Let $\triangledown_{ij} = \frac{\partial^2 g}{\partial x_i \partial x_j}$. It can be shown [12] that if we use the first-order Taylor expansion of g at $\mathbf{x}$ to approximate $g(\mathbf{x}+\varepsilon)$, the square error is

$$\left|\left|g(\mathbf{x}+\varepsilon) - g(\mathbf{x}) - \left(\frac{\partial g}{\partial \mathbf{x}}\right)^T \cdot \varepsilon\right|\right|^2 \approx \frac{1}{4}\left|\left|\varepsilon^T \mathbf{H}_\mathbf{x} \varepsilon\right|\right|^2 = \frac{1}{4}\Big(\sum_{i,j=1}^{n} \triangledown_{ij}\varepsilon_i\varepsilon_j\Big)^2$$

$$\leq \frac{1}{4}\Big(\sum_{i=1}^{n}\triangledown_{ii}^2 + 2\sum_{i,j=1,i\neq j}^{n}\triangledown_{ij}^2\Big)\Big(\sum_{i=1}^{n}\varepsilon_i^4 + 2\sum_{i,j=1,i\neq j}^{n}\varepsilon_i^2\varepsilon_j^2\Big) = \frac{1}{4}||\varepsilon||^4 \cdot \sum_{i,j=1}^{n}\triangledown_{ij}^2$$

The above inequality holds due to the Cauchy's inequality. We can see that in order to make g locally close to linear at every point in the domain of $\mathbf{x}$, we just need minimize $\int_{D_\mathbf{x}} \sum_{i,j=1}^n \triangledown_{ij}^2 d\mathbf{x}$. When the mapping is vector-valued, we need apply this regularizer to each output of the mapping. R_2 in Eq. 4 can then be constructed as $R_2 = \int_{D_\mathbf{x}} \sum_{i,j=1}^n P_{ij} d\mathbf{x}$, where $P_{ij} \triangleq \sum_{l=1}^n \left(\frac{\partial^2 y_l}{\partial x_i \partial x_j}\right)^2$. The derivation of $\frac{\partial R_2}{\partial \mathbf{W}}$ is straightforward. In the result, $\frac{\partial^2 z_p}{\partial x_i \partial x_j}$ is involved. It can be computed and saved in the first step of kernel-based nonlinear ICA.

4 Experiments

According to the experimental results in [1] and our experience, mixtures of subgaussian sources are more difficult to be separated well, than those of supergaussian sources. So for saving space, here we just present some experimental results on separating two subgaussian sources. The sources are a sawtooth signal (s_1) and an amplitude-modulated waveform (s_2), with 1000 samples. x_i are generated in the same form as the example in Sec. 4 of [5], i.e. $\mathbf{x} = \mathbf{B}\mathbf{s} + \mathbf{c}s_1s_2$, but here $\mathbf{c} = (-0.15, 0.5)^T$. The waveforms and scatterplots of s_i and x_i are shown in Fig. 1, from which we can see that the nonlinear effect is significant.

The regularization parameter for enforcing smoothness is $\lambda_2 = 0.2$, and that for enforcing MND, λ_1, decays from 0.3 to 0.01 during the learning process.

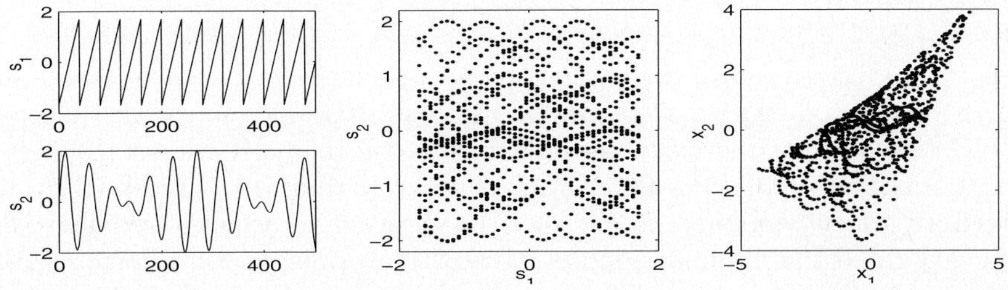

Fig. 1. Source and their nonlinear mixtures. Left: waveforms of sources. Middle: scatterplot of sources. Right: scatterplot of mixtures.

We chose the polynomial kernel of degree 4, i.e. $k(\mathbf{a}, \mathbf{b}) = (\mathbf{a}^T\mathbf{b} + 1)^4$, and found $d = 14$. Here we compare the separation results of four methods/schemes, which are linear ICA (FastICA is adopted), kernel-based nonlinear (kNICA) without explicit regularization, kNICA with only the smoothness regularization, and kNICA with both smoothness and MND regularization. Table 1 shows the SNR of the recovered signals. Numbers in parentheses are the SNR values after trivial indeterminacies are removed.[4] Fig. 2 shows the scatterplots of y_i obtained by various schemes. In this experiment, clearly kNICA with the smoothness and MND regularization gives the best separation result.

Table 1. SNR of the separation results on various methods (schemes)

Channel	FastICA	kNICA (no regu.)	kNICA (smooth)	kNICA (smooth & MND)
No. 1	3.72 (4.59)	9.25(9.69)	11.1(14.4)	**12.1 (16.5)**
No. 2	5.76 (6.04)	6.07(8.19)	8.9(12.7)	**15.4 (25.1)**

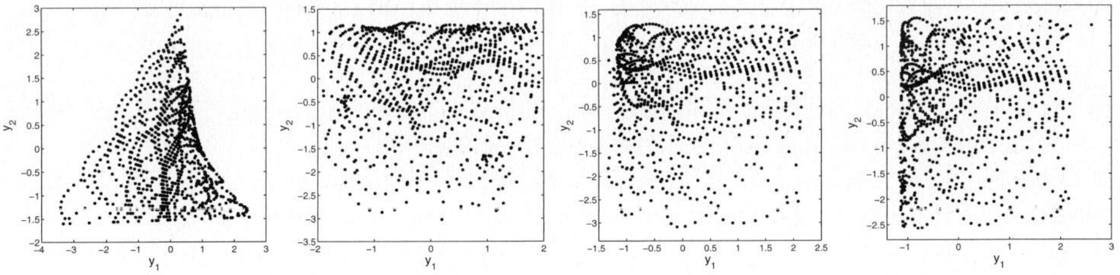

(a) (b) (c) (d)

Fig. 2. Scatterplot of y_i obtained by various methods/schemes. (a) FastICA. (b) kNICA without explicit regularization. (c) kNICA with the smoothness regularizer. (d) kNICA with the smoothness and MND regularization.

[4] We applied a 1-8-1 MLP, denoted by $\mathcal{T}$, to y_i to minimize the square error between s_i and $\mathcal{T}(y_i)$. In this way trivial indeterminacies are removed.

5 Conclusion

We have proposed to solve the nonlinear ICA problem using kernels. In the first step of the method, the data are mapped to high-dimensional feature space and the effective subspace is extracted. Thanks to the kernel trick, in the second step, we need to solve a linear problem. The algorithm in the second step was derived, in a form similar to standard ICA. In order to achieve nonlinear BSS, we incorporated the minimal nonlinear distortion principle and the smoothness regularizer for regularization of the proposed nonlinear ICA method. MND helps to overcome the ill-posedness of nonlinear ICA, under the condition that the nonlinearity in the mixing procedure is not very strong. This condition usually holds for practical problems.

References

1. Almeida, L.B.: MISEP - linear and nonlinear ICA based on mutual information. Journal of Machine Learning Research 4, 1297–1318 (2003)
2. Amari, S., Cichocki, A., Yang, H.H.: A new learning algorithm for blind signal separation. In: Advances in Neural Information Processing Systems (1996)
3. Bach, F.R., Jordan, M.I.: Beyond independent components: trees and clusters. Journal of Machine Learning Research 4, 1205–1233 (2003)
4. Cardoso, J.F., Souloumiac, A.: Blind beamforming for non-Gaussian signals. IEE Proceeding-F 140(6), 362–370 (1993)
5. Harmeling, S., Ziehe, A., Kawanabe, M., Müller, K.R.: Kernel-based nonlinear blind source separation. Neural Computation 15, 1089–1124 (2003)
6. Hyvärinen, A.: Fast and robust fixed-point algorithms for independent component analysis. IEEE Transactions on Neural Networks 10(3), 626–634 (1999)
7. Hyvärinen, A.: The fixed-point algorithm and maximum likelihood estimation for independent component analysis. Neural Processing Letters 10(1), 1–5 (1999)
8. Hyvärinen, A., Pajunen, P.: Nonlinear independent component analysis: Existence and uniqueness results. Neural Networks 12(3), 429–439 (1999)
9. Jutten, C., Karhunen, J.: Advances in nonlinear blind source separation. In: Proc. ICA2003, pp. 245–256 (2003) Invited paper in the special session on nonlinear ICA and BSS
10. Martinez, D., Bray, A.: Nonlinear blind source separation using kernels. IEEE Transaction on Neural Network 14(1), 228–235 (2003)
11. Schölkopf, B., Smola, A., Muller, K.: Nonlinear component analysis as a kernel eigenvalue problem. Neural Computation 10, 1299–1319 (1998)
12. Zhang, K., Chan, L.: Nonlinear independent component analysis with minimum nonlinear distortion. In: ICML 2007, Corvallis, OR, US, pp. 1127–1134 (2007)
13. Ziehe, A., Müller, K.R.: TDSEP – an efficient algorithm for blind separation using time structure. In: Proc. ICANN98, Skövde, Sweden, pp. 675–680 (1998)

Linear Prediction Based Blind Source Extraction Algorithms in Practical Applications

Zhi-Lin Zhang[1,2] and Liqing Zhang[1]

[1] Department of Computer Science and Engineering, Shanghai Jiao Tong University, Shanghai 200240, China
[2] School of Computer Science and Engineering, University of Electronic Science and Technology of China, Chengdu 610054, China
zlzhang@uestc.edu.cn, zhang-lq@cs.sjtu.edu.cn

Abstract. Blind source extraction (BSE) is of advantages over blind source separation (BSS) when obtaining some underlying source signals from high dimensional observed signals. Among a variety of BSE algorithms, a large number of algorithms are based on linear prediction (LP-BSE). In this paper we analyze them from practical point of view. We reveal that they are, in nature, minor component analysis (MCA) algorithms, and thus they have some problems that are inherent in MCA algorithms. We also find a switch phenomenon of online LP-BSE algorithms, showing that different parts of a single extracted signal are the counterparts of different source signals. The two issues should be noticed when one applies these algorithms to practical applications. Computer simulations are given to confirm these observations.

1 Introduction

Blind source extraction (BSE) [1] is a powerful technique that is closely related to blind source separation (BSS). The basic task of BSE is to estimate some of underlying source signals that are linearly combined in observations. Compared with BSS, BSE has some advantages [1]. An attractive one is its ability to extract a small subset of source signals from high-dimensional observed signals. Hence it is often recommended to be used in EEG/MEG fields and alike [1,2,3].

There are many BSE algorithms for extracting source signals based on their temporal structures [1,4]. Among them there is a class of algorithms based on linear prediction. For example, Cichocki, Mandic, and Liu et al. [2,6,7,8] proposed several BSE algorithms based on short-term linear prediction. Barros et al. [5] proposed a BSE algorithm based on long-term linear prediction. Later Smith et al. [3] proposed a BSE algorithm combining short-term prediction and long-term prediction. And recently Liu et al.[9] extended a basic linear prediction based algorithm to the one suitable for noisy environment.

In this paper we consider some possible problems when applying the linear prediction based BSE (LP-BSE) algorithms to practical applications, especially EEG/MEG fields.

M.E. Davies et al. (Eds.): ICA 2007, LNCS 4666, pp. 309–316, 2007.

2 The Linear Prediction Based Algorithms

Suppose that unknown source signals $\mathbf{s}(k) = [s_1(k), \cdots, s_n(k)]^T$ are zero-mean and spatially uncorrelated, and suppose that $\mathbf{x}(k) = [x_1(k), \cdots, x_n(k)]^T$ is a vector of observed signals, which is a linear instantaneous mixture of source signals by $\mathbf{x}(k) = \mathbf{As}(k)$, where k is time index and $\mathbf{A} \in \mathbf{R}^{n\times n}$ is an unknown mixing matrix of full rank. The goal of BSE is to find a demixing vector $\mathbf{w}$ such that $y(k) = \mathbf{w}^T\mathbf{x}(k) = \mathbf{w}^T\mathbf{As}(k)$ is an estimate of a source signal. To cope with ill-conditioned cases and to make algorithms simpler and faster, before extraction whitening [1] is often used to transform the observed signals $\mathbf{x}(k)$ to $\mathbf{z}(k) = \mathbf{Vx}(k)$ such that $E\{\mathbf{z}(k)\mathbf{z}(k)^T\} = \mathbf{I}$, where $\mathbf{V} \in \mathbf{R}^{n\times n}$ is a prewhitening matrix and $\mathbf{VA}$ is an orthogonal matrix.

Assuming that the underlying source signals have temporal structures, the class of LP-BSE algorithms is derived by minimizing the normalized mean square prediction error given by [6,8]

$$J_1 = \frac{E\{e(k)^2\}}{E\{y(k)^2\}} = \frac{E\{(y(k) - \mathbf{b}^T\mathbf{y}(k))^2\}}{E\{y(k)^2\}} \tag{1}$$

where $y(k) = \mathbf{w}^T\mathbf{x}(k)$, $\mathbf{b} = [b_1, b_2, \cdots, b_P]^T$, $\mathbf{y}(k) = [y(k-1), y(k-2), \cdots, y(k-P)]^T$ and P is AR order that is set before running algorithms. If one performs the whitening and normalizes the demixing vector $\mathbf{w}$, the objective function (1) reduces to [2,5]:

$$J_2 = E\{e(k)^2\} = E\{(y(k) - \mathbf{b}^T\mathbf{y}(k))^2\} \tag{2}$$

where $y(k) = \mathbf{w}^T\mathbf{z}(k) = \mathbf{w}^T\mathbf{Vx}(k)$ and $\|\mathbf{w}\| = 1$.

Without loss of generality, we only consider the objective function (2) in the following. After some algebraic calculations, from (2) we obtain

$$J_2 = E\{e(k)^2\} = \mathbf{w}^T\widehat{\mathbf{R}}_\mathbf{Z}\mathbf{w} = \mathbf{w}^T\mathbf{VA}\widehat{\mathbf{R}}_\mathbf{S}\mathbf{A}^T\mathbf{V}^T\mathbf{w} = \mathbf{q}^T\widehat{\mathbf{R}}_\mathbf{S}\mathbf{q}, \tag{3}$$

in which $\mathbf{q} = \mathbf{A}^T\mathbf{V}^T\mathbf{w}$, and

$$\widehat{\mathbf{R}}_\mathbf{Z} = \mathbf{R}_\mathbf{Z}(0) - \sum_{p=1}^{P} b_p\mathbf{R}_\mathbf{Z}(p) - \sum_{q=1}^{P} b_q\mathbf{R}_\mathbf{Z}(-q) + \sum_{p=1}^{P}\sum_{q=1}^{P} b_p b_q\mathbf{R}_\mathbf{Z}(q-p) \tag{4}$$

$$\widehat{\mathbf{R}}_\mathbf{S} = \mathbf{R}_\mathbf{S}(0) - 2\sum_{p=1}^{P} b_p\mathbf{R}_\mathbf{S}(p) + \sum_{p=1}^{P}\sum_{q=1}^{P} b_p b_q\mathbf{R}_\mathbf{S}(q-p) \tag{5}$$

where $\mathbf{R}_\mathbf{Z}(p) = E\{\mathbf{z}(k)\mathbf{z}(k-p)^T\}$, and $\mathbf{R}_\mathbf{S}(p) = E\{\mathbf{s}(k)\mathbf{s}(k-p)^T\}$ is a diagonal matrix due to the assumptions. Also, $\widehat{\mathbf{R}}_\mathbf{S}$ is a diagonal matrix, whose diagonal elements are given by

$$\rho_i = r_i(0) - 2\sum_{p=1}^{P} b_p r_i(p) + \sum_{p=1}^{P}\sum_{q=1}^{P} b_p b_q r_i(q-p),\ i = 1, \cdots, n \tag{6}$$

where r_i is the autocorrelation function of s_i.

Now we calculate the concrete value of ρ_i. Suppose when J_2 achieves its minimum, $\mathbf{b}$ achieves $\mathbf{b}^* = [b_1^*, b_2^*, \cdots, b_p^*]^T$. We express all the source signals as

$$s_i(k) = \sum_{p=1}^{P} b_p^* s_i(k-p) + e_i(k), \; i = 1, \cdots, n \tag{7}$$

where $e_i(k)$ is called residual processes. Then we have

$$\begin{aligned} r_i(0) &= E\Big\{\big(\sum_{p=1}^{P} b_p^* s_i(k-p) + e_i(k)\big)\big(\sum_{q=1}^{P} b_q^* s_i(k-q) + e_i(k)\big)\Big\} \\ &= \sum_{p=1}^{P}\sum_{q=1}^{P} b_p^* b_q^* r_i(q-p) + 2E\{e_i(k)s_i(k)\} - E\{e_i(k)^2\} \end{aligned} \tag{8}$$

where we use the relationship (7). On the other hand, we also have

$$r_i(0) = E\Big\{\big(\sum_{p=1}^{P} b_p^* s_i(k-p) + e_i(k)\big)s_i(k)\Big\} = \sum_{p=1}^{P} b_p^* r_i(p) + E\{e_i(k)s_i(k)\}. \tag{9}$$

Substitute (8) and (9) into (6), we obtain

$$\rho_i = E\{e_i(k)^2\}, \tag{10}$$

implying that $\rho_i (i = 1, \cdots, n)$ are just the powers of residual processes of linear prediction to the source signals given the coefficients $b_p^* (p = 1, \cdots, P)$. Obviously, calculating the minimum of J_2 is equivalently finding the minimum among all $\rho_i (i = 1, \cdots, n)$, which are the eigenvalues of $\widehat{\mathbf{R}}_{\mathbf{S}}$ and are also the ones of $\widehat{\mathbf{R}}_{\mathbf{Z}}$. And the demixing vector $\mathbf{w}$ is the associated eigenvector. Thus the LP-BSE algorithms are in nature the MCA algorithms [10,11,15].

3 Analysis of the LP-BSE Algorithms

It is recognized that MCA algorithms have some flaws in practical applications [10,11]. First, in practice the small eigenvalues of the covariance matrix $\widehat{\mathbf{R}}_{\mathbf{Z}}$ are often close to each other, which reduces the estimate accuracy of associated eigenvectors [14] and brings difficulties to global convergence [10,11]. Moreover the performance of MCA algorithms often suffers from outliers and noise [12].

Naturally, the LP-BSE algorithms inherit some of these flaws when dealing with high dimensional observed signals. Take the extraction of event-related potentials as an example. The number of sensor signals are often larger than 64, and some underlying source signals have similar time structures [13]. According to (10) and (7) the small eigenvalues of $\widehat{\mathbf{R}}_{\mathbf{Z}}$ are close to each other, which makes the estimation of the minor eigenvector sensitive to sensor noise [12].

Now consider online versions of LP-BSE algorithms. Suppose the current extracted source is $s_1(k)$, whose current residual process's power level is $e_1^2(k)$. This implies that given the prediction AR order P in algorithms, $e_1^2(k)$ is the smallest among all $e_j^2(k), j = 1, \cdots, n$. If at time $k+1$, $s_1(k+1)$'s true AR order starts to change but the given prediction AR order does not change, $e_1^2(k+1)$ may become larger[1]. Then there may be another source signal, say $s_2(k+1)$, whose $e_2^2(k+1)$ with the given prediction order is smaller than that of $s_1(k+1)$. Consequently, the algorithms switch to extract $s_2(k+1)$. Therefore the extracted signal is still mixed by the two source signals in the sense that the first part of the extracted signal is the counterpart of s_1 and the second part is the counterpart of s_2. We call this the switch phenomenon. The essential reason to the existence of the switch phenomenon is the use of the fixed prediction order that is set before performing LP-BSE algorithms. Similarly, if the true AR coefficients of source signals vary fast and $b_i (i = 1, \cdots, P)$ cannot be adjusted in the same pace, the switch phenomenon may also occur. Remind that in the EEG data processing, especially in the even-related brain potential extraction, the underlying source signals' AR order and coefficients may quickly vary. Thus the phenomenon may occur in these cases.

4 Simulations

In the first simulation we illustrated unsatisfying performance of LP-BSE algorithms due to their MCA nature. We used the data set ABio7, a benchmark in ICALAB [17]. Three typical LP-BSE algorithms, i.e. the ones in [2,7,8], were used to extract these signals. To make comparison, we intuitively gave a PCA-like BSE algorithm, a variation of our algorithm [4], as follows[2]:

$$\mathbf{w} = PCA_i\Big(\sum_{i=1}^{P} \mathbf{R_Z}(\tau_i)\Big) = PCA_i(\widetilde{\mathbf{R_Z}}), \tag{11}$$

where $\mathbf{R_Z}(\tau_i) = E\{\mathbf{z}(k)\mathbf{z}(k-\tau_i)^T\}$, τ_i was time delay, and $PCA_i(\widetilde{\mathbf{R_Z}})$ was the operator that calculated the i-th principal eigenvector of $\widetilde{\mathbf{R_Z}}$. Using *a priori* knowledge one can choose a specific set of time delays to achieve better performance [4]. Actually, (11) is only a framework, and can be implemented offline or online by using many efficient and robust methods [14,16]. Note that the PCA-like BSE algorithm obtains principal eigenvectors, while the LP-BSE algorithms obtain minor ones. All the algorithms were implemented offline.

The source signals were randomly mixed and whitened. Then each algorithm was performed on these signals. The step-size of the algorithm in [8] was 0.1.

[1] It also may become smaller. So in this case the switch phenomenon does not occur.

[2] Note that we present the PCA-like algorithm in purpose to show that the class of LP-BSE algorithms may not achieve satisfying results when applied to practical applications. Admittedly, better algorithms than the algorithm may be developed, which is not the topic in this paper.

The learning rate parameter μ_0 of the algorithm in [7] (see Equ.(16) in [7]) was 0.5. The extraction performance was measured by

$$PI = \frac{1}{n-1}\Big(\sum_{i=1}^{n} \frac{q_i^2}{\max_i q_i^2} - 1\Big) \tag{12}$$

where $\mathbf{q} = [q_1, \cdots, q_n] = \mathbf{w}^T\mathbf{VA}$ was a global vector, $\mathbf{V}$ was the whitening matrix, $\mathbf{A}$ was the mixing matrix and $\mathbf{w}$ was the demixing vector obtained by algorithms. PI's value lay in [0,1] for any vector $\mathbf{q} = [q_1, \cdots, q_n]$. The smaller it was, the better the extraction performance was. Simulations were independently carried out 50 trials. The results are shown in Table 1, from which we can see that the LP-BSE algorithms generally performed poorly.

Table 1. The averaged performance indexes of the algorithms in the first simulation. For the three LP-BSE algorithms the parameter P was the prediction order, while for the PCA-like algorithm P meant that the time delay set was $\{1, \cdots, P\}$.

P	1	2	3	4	5	6	7	8	9	10	12	20	30	40	50	400
Alg. (11)	0.00	0.00	0.01	0.09	0.02	0.07	0.01	0.01	0.06	0.02	0.01	0.05	0.04	0.07	0.03	0.01
Alg. [8]	0.19	0.17	0.17	0.18	0.18	0.18	0.19	0.19	0.19	0.19	0.19	0.19	0.19	0.19	0.19	0.19
Alg. [2]	0.02	0.00	0.07	0.07	0.06	0.03	0.03	0.02	0.04	0.08	0.10	0.05	0.06	0.07	0.08	0.14
Alg. [7]	0.11	0.14	0.11	0.09	0.12	0.09	0.10	0.11	0.13	0.12	0.13	0.06	0.13	0.08	0.16	0.08

In the second simulation we used the 122-dimension MEG data set (Fig.1 (a)) in [18] to show performance of a typical LP-BSE algorithm in extracting horizontal eye movements, which occurred at about the 4000-th sampling point and the 15500-th sampling point. Since the movements resulted from the same group of muscles, we safely believed that artifacts associated with the movements occurring at different time should appear in the same extracted signal.

After performing the same preprocessing as that in [18], we used the offline LP-BSE algorithm in [8] to extract the artifacts with different levels of data dimension reduction. Its step-size was 0.5 and prediction order was 10. The results are shown in Fig.1 (b), where y_1, y_2, y_3 and y_4 were extracted by the LP-BSE algorithm with data dimension reduced to 120, 80, 60, and 40, respectively. y_5 was extracted by the PCA-like algorithm (11) without data dimension reduction ($\tau_i = \{1, \cdots, 5\}$). y_6 was extracted by FastICA, which was also used in [18]. Since Vigário et al. have shown that FastICA can perfectly extract the horizontal eye movent artifacts, we regarded y_6 as a benchmark. From $y_1 - y_3$ we see that the artifacts were not perfectly extracted, since the horizontal eye movement artifact at about the 15500-th sampling point was not extracted. Although in y_4 all the artifacts were extracted, it was mixed by artifacts resulting from eye blinks [18]. Besides, we see that the extraction performance of the LP-BSE algorithm was affected by the dimension reduction. When the dimension was reduced to a certain degree, the extraction performance became relatively better. In contrast, in y_5 all the horizontal eye movement artifacts were extracted without mixed

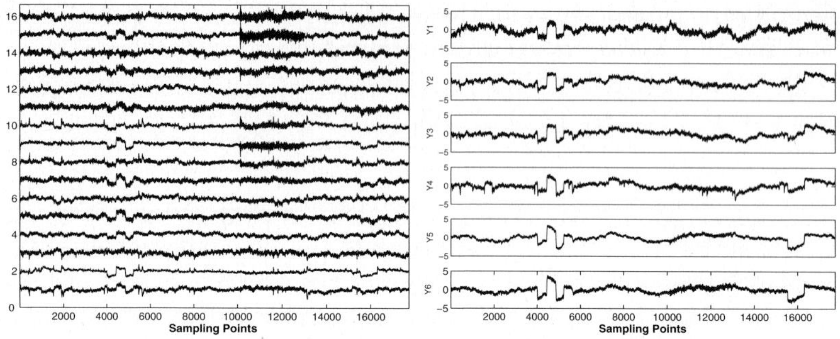

(a) (b)

Fig. 1. A subset of the MEG data set [18] (a) and extracted artifacts (b)

by other artifacts, and we found the extraction quality was not affected by the data dimension (the extraction results with dimension reduction are not shown here due to limited space). We also ran other LP-BSE algorithms and almost obtained the same results. Due to space limit we omit the report.

In the last simulation we showed the switch phenomenon of online LP-BSE algorithms. We generated three AR(6) Gaussian signals of 5-second duration time (Fig.2). Each source signal had zero mean and unit variance. The sampling frequency was 1000 Hz. The AR coefficients of each signal were unchanged during the first 2.5 second, given by:

$$source1 : \mathbf{b} = [-1.6000, 0.9000, -0.2000, 0.0089, 0.0022, -0.0002]$$
$$source2 : \mathbf{b} = [-0.1000, -0.4300, 0.0970, 0.0378, -0.0130, 0.0009]$$
$$source3 : \mathbf{b} = [-2.3000, 2.0400, -0.8860, 0.1985, -0.0216, 0.0009]$$

And hereafter the AR coefficients changed to:

$$source1 : \overline{\mathbf{b}} = [-1.6000, 0.9000, -0.2000, 0.0089, 0.0022, -0.0002]$$
$$source2 : \overline{\mathbf{b}} = [-2.3000, 2.0400, -0.8860, 0.1985, -0.0216, 0.0009]$$
$$source3 : \overline{\mathbf{b}} = [-0.1000, -0.4300, 0.0970, 0.0378, -0.0130, 0.0009]$$

We used the online version of the LP-BSE algorithm in [8] to extract a signal. Its step-size was 0.01 and prediction order was 10. The result is shown in Fig.2 (see y_1), from which we see that the first part of y_1 (before 2.5 second) was the counterpart of source signal s_3, but from 3.6 second or so the signal was clearly the counterpart of source signal s_1. To further confirm this, we measured the similarity between the extracted signal and the source signals, using the performance index $PI_2 = -10\lg(E\{(s(k) - \tilde{s}(k))^2\})(dB)$, where $s(k)$ was the desired source signal, and $\tilde{s}(k)$ was the extracted signal (both of them were normalized to be zero-mean and unit-variance). The higher PI_2 is, the better the performance. Denote by *Part1* the extracted signal's segment from 2.0 s to 2.5 s, and denote by *Part2* the extracted signal's segment from 4.0 s to 5.0 s.

The PI_2 of *Part1* measuring the similarity between *Part1* and the counterpart of s_3 was 18.5 dB, showing *Part1* was very similar to the counterpart of s_3. The PI_2 of *Part2* measuring the similarity between *Part2* and the counterpart of s_1 was 19.7 dB, showing *Part2* was very similar to the counterpart of s_1.

Next we used an online version of the PCA-like algorithm (11), implemented by the OJAN PCA algorithm [16], to extract a source signal. The extracted signal is shown in Fig.2 (see y_2), from which we can see that the extracted signal was just s_3 and the switch phenomenon did not occur. We also calculated the algorithm's PI_2 at *Part1* and *Part2*. The PI_2 of *Part1* measuring the similarity between *Part1* and the counterpart of s_3 was 22.3 dB, showing *Part1* was very similar to the counterpart of s_3. The PI_2 of *Part2* measuring the similarity between *Part2* and the counterpart of s_3 was 19.9 dB, showing *Part2* was very similar to the counterpart of s_3 as well. The results show that the online version has well extracted the whole source signal s_3.

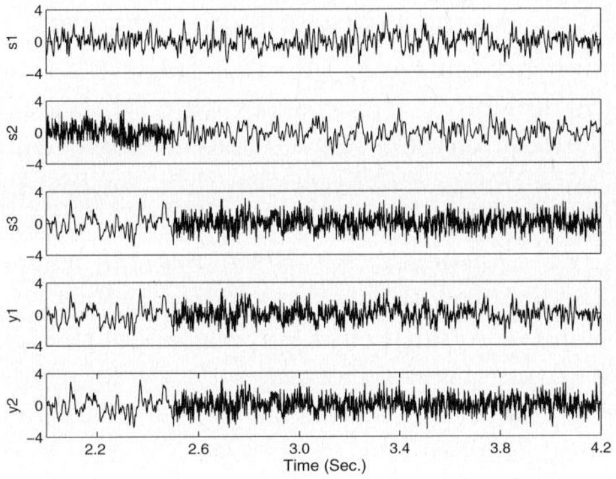

Fig. 2. Segments of the AR source signals (s_1, s_2, s_3) and the extracted signals. y_1 was extracted by the online LP-BSE algorithm in [8], while y_2 was extracted by the online version of the algorithm (11).

5 Conclusion

In this paper we analyze a class of linear prediction based BSE algorithms, revealing that they are in nature the MCA algorithms and showing a switch phenomenon of their online versions. Based on these results, careful attentions should be paid when one applies these algorithms to practical applications such as EEG and MEG fields.

Acknowledgments

The work was supported by the National Basic Research Program of China (Grant No. 2005CB724301) and National Natural Science Foundation of China (Grant No.60375015).

References

1. Cichocki, A., Amari, S.: Adaptive Blind Signal and Image Processing: Learning Algorithms and Applications. John Wiley & Sons, New York (2002)
2. Cichocki, A., et al.: A Blind Extraction of Temporally Correlated but Statistically Dependent Acoustic Signals. In: Proc. of the 2000 IEEE Signal Processing Society Workshop on Neural Networks for Signal Processing X, pp. 455–464. IEEE Computer Society Press, Los Alamitos (2000)
3. Smith, D., Lukasiak, J., Burnett, I.: Blind Speech Separation Using a Joint Model of Speech Production. IEEE Signal Processing Lett. 12(11), 784–787 (2005)
4. Zhang, Z.-L., Yi, Z.: Robust Extraction of Specific Signals with Temporal Structure. Neurocomputing 69(7-9), 888–893 (2006)
5. Barros, A.K., Cichocki, A.: Extraction of Specific Signals with Temporal Structure. Neural Computation 13(9), 1995–2003 (2001)
6. Cichocki, A., Thawonmas, R.: On-line Algorithm for Blind Signal Extraction of Arbitrarily Distributed, but Temporally Correlated Sources Using Second Order Statistics. Neural Processing Letters 12, 91–98 (2000)
7. Mandic, D.P., Cichocki, A.: An Online Algorithm for Blind Extraction of Sources with Different Dynamical Structures. In: Proc. of the 4th Int. Conf. on Independent Component Analysis and Blind Signal Separation (ICA 2003), pp. 645–650 (2003)
8. Liu, W., Mandic, D.P., Cichocki, A.: A Class of Novel Blind Source Extraction Algorithms Based on a Linear Predictor. In: Proc. of ISCAS 2005, pp. 3599–3602 (2005)
9. Liu, W., Mandic, D.P., Cichocki, A.: Blind Second-order Source Extraction of Instantaneous Noisy Mixtures. IEEE Trans. Circuits Syst. II 53(9), 931–935 (2006)
10. Taleb, A., Cirrincione, G.: Against the Convergence of the Minor Component Analysis Neurons. IEEE Trans. Neural Networks 10(1), 207–210 (1999)
11. Feng, D.-Z., Zheng, W.-X., Jia, Y.: Neural Network Learning Algorithms for Tracking Minor Subspace in High-dimensional Data Stream. IEEE Trans. Neural Networks 16(3), 513–521 (2005)
12. Wilkinson, J.H.: The Algebraic Eigenvalue Problem. Oxford Univ. Press, Oxford (1965)
13. Makeig, S., Westerfield, M., Jung, T.-P., et al.: Dynamic Brain Sources of Visual Evoked Responses. Science 295, 690–694 (2002)
14. Golub, G.H., Loan, C.F.V.: Matrix Computation, 3rd edn. The John Hopkins University Press, Baltimore (1996)
15. Zhang, Q., Leung, Y.-W.: A Class of Learning Algorithms for Principal Component Analysis and Minor Component Analysis. IEEE Trans. Neural Networks 11(1), 200–204 (2000)
16. Chatterjee, C.: Adaptive Algorithms for First Principal Eigenvector Computation. Neural Networks 18, 145–159 (2005)
17. ICALAB Toolboxes, Available: http://www.bsp.brain.riken.jp/ICALAB
18. Vigário, R., et al.: Independent Component Analysis for Identification of Artifacts in Magnetoencephalographic Recordings. In: Proc. of NIPS 1997, pp. 229–235 (1997)

Blind Audio Source Separation Using Sparsity Based Criterion for Convolutive Mixture Case

A. Aïssa-El-Bey, K. Abed-Meraim, and Y. Grenier

ENST-Paris, TSI Department, 46 rue Barrault 75634, Paris Cedex 13, France
{elbey,abed,grenier}@tsi.enst.fr

Abstract. In this paper, we are interested in the separation of audio sources from their instantaneous or convolutive mixtures. We propose a new separation method that exploits the sparsity of the audio signals via an ℓ_p-norm based contrast function. A simple and efficient natural gradient technique is used for the optimization of the contrast function in an instantaneous mixture case. We extend this method to the convolutive mixture case, by exploiting the property of the Fourier transform. The resulting algorithm is shown to outperform existing techniques in terms of separation quality and computational cost.

1 Introduction

Blind Source Separation (BSS) is an approach to estimate and recover independent source signals using only the information within the mixtures observed at each channel. Many algorithms have been proposed to solve the standard blind source separation problem in which the mixtures are assumed to be instantaneous. A fundamental and necessary assumption of BSS is that the sources are statistically independent and thus are often separated using higher-order statistical information [1]. If extra information about the sources is available at hand, such as temporal coherency [2], source nonstationarity [3], or source cyclostationarity [4], then one can remain in the second-order statistical scenario, to achieve the BSS.

In the case of non-stationary signals (including audio signals), certain solutions using time-frequency analysis of the observations exist [5]. Other solutions use the statistical independence of the sources assuming a local stationarity to solve the BSS problem [6]. This is a strong assumption that is not always verified [7]. To avoid this problem, we propose a new approach that handles the general linear instantaneous model (possibly noisy) by using the *sparsity* assumption of the sources in the time domain. Then, we extend this algorithm to the convolutive mixture case, by transforming the convolutive problem into instantaneous problem in the frequency domain, and separating the instantaneous mixtures in every frequency bin. The use of sparsity to handle this model, has arisen in several papers in the area of source separation [8,9]. We first present a sparsity contrast function for BSS. Then, in order to achieve BSS, we optimize the considered contrast function using an iterative algorithm based on the relative gradient technique.

M.E. Davies et al. (Eds.): ICA 2007, LNCS 4666, pp. 317–324, 2007.

In the following section, we discuss the data model that formulates our problem. Next, we detail the different steps of the proposed algorithm. In Section 4, some simulations are undertaken to validate our algorithm and to compare its performance to other existing BSS techniques.

2 Instantaneous Mixture Case

2.1 Data Model

Assume that N audio signals impinge on an array of $M \geq N$ sensors. The measured array output is a weighted superposition of the signals, corrupted by additive noise, i.e.

$$\boldsymbol{x}(t) = \boldsymbol{A}\boldsymbol{s}(t) + \boldsymbol{w}(t) \qquad t = 0, \ldots, T-1 \tag{1}$$

where $\boldsymbol{s}(t) = [s_1(t), \cdots, s_N(t)]^T$ is the $N \times 1$ sparse source vector, $\boldsymbol{w}(t) = [w_1(t), \cdots, w_M(t)]^T$ is the $M \times 1$ complex noise vector, $\boldsymbol{A}$ is the $M \times N$ full column rank mixing matrix (i.e., $M \geq N$), and the superscript T denotes the transpose operator. The purpose of blind source separation is to find a separating matrix, i.e. a $N \times M$ matrix $\boldsymbol{B}$ such that $\widehat{\boldsymbol{s}}(t) = \boldsymbol{B}\boldsymbol{x}(t)$ is an estimate of the source signals.

Before proceeding, note that complete blind identification of separating matrix $\boldsymbol{B}$ (or equivalently, the mixing matrix $\boldsymbol{A}$) is impossible in this context, because the exchange of a fixed scalar between the source signal and the corresponding column of $\boldsymbol{A}$ leaves the observations unaffected. Also note that the *numbering* of the signals is immaterial. It follows that the best that can be done is to determine $\boldsymbol{B}$ up to a permutation and scalar shifts of its columns, i.e., $\boldsymbol{B}$ is a separating matrix iff:

$$\boldsymbol{B}\boldsymbol{x}(t) = \boldsymbol{P}\boldsymbol{\Lambda}\boldsymbol{s}(t) \tag{2}$$

where $\boldsymbol{P}$ is a permutation matrix and $\boldsymbol{\Lambda}$ a non-singular diagonal matrix.

2.2 Sparsity-Based BSS Algorithm

Before starting, we propose to use 'an optional' whitening step which set the mixtures to the same energy level and reduces the number of parameters to be estimated. More precisely, the whitening step is applied to the signal mixtures before using our separation algorithm. The whitening is achieved by applying a $N \times M$ matrix $\boldsymbol{W}$ to the signal mixtures in such a way $\mathrm{Cov}(\boldsymbol{W}\boldsymbol{x}) = \boldsymbol{I}$ in the noiseless case, where $\mathrm{Cov}(\cdot)$ stands for the covariance operator. As shown in [2], $\boldsymbol{W}$ can be computed as the inverse square root of the noiseless covariance matrix of the signal mixtures (see [2] for more details). In the following, we apply our separation algorithm on the whitened data:

$$\boldsymbol{x}_w(t) = \boldsymbol{W}\boldsymbol{x}(t).$$

We propose an iterative algorithm for the separation of sparse audio signals, namely the ISBS for Iterative Sparse Blind Separation. It is well known that audio signals are characterized by their sparsity property in the time domain [8,9] which is measured by their ℓ_p norm where $0 \leq p < 2$. More specifically, one can define the following sparsity based contrast function

$$G_p(\boldsymbol{s}) = \sum_{i=1}^{N} \left[\mathcal{J}_p(s_i)\right]^{\frac{1}{p}}, \tag{3}$$

where

$$\mathcal{J}_p(s_i) = \frac{1}{T}\sum_{t=0}^{T-1} |s_i(t)|^p. \tag{4}$$

The algorithm finds a separating matrix $\boldsymbol{B}$ such as,

$$\boldsymbol{B} = \arg\min_{\boldsymbol{B}} \{\mathcal{G}_p(\boldsymbol{B})\}, \tag{5}$$

where

$$\mathcal{G}_p(\boldsymbol{B}) \triangleq G_p(\boldsymbol{z}), \tag{6}$$

and $\boldsymbol{z}(t) \triangleq \boldsymbol{B}\boldsymbol{x}_w(t)$ represents the estimated sources. The approach we choose to solve (5) is inspired from [10]. It is a block technique based on the processing of T received samples and consists in searching iteratively the minimum of (5) in the form:

$$\boldsymbol{B}^{(k+1)} = (\boldsymbol{I} + \boldsymbol{\epsilon}^{(k)})\boldsymbol{B}^{(k)} \tag{7}$$

$$\boldsymbol{z}^{(k+1)}(t) = (\boldsymbol{I} + \boldsymbol{\epsilon}^{(k)})\boldsymbol{z}^{(k)}(t) \tag{8}$$

where $\boldsymbol{I}$ denotes the identity matrix. At iteration k, a matrix $\boldsymbol{\epsilon}^{(k)}$ is determined from a local linearization of $G_p(\boldsymbol{B}^{(k+1)}\boldsymbol{x}_w)$. It is an approximate Newton technique with the benefit that $\boldsymbol{\epsilon}^{(k)}$ can be very simply computed (no Hessian inversion) under the additional assumption that $\boldsymbol{B}^{(k)}$ is close to a separating matrix. This procedure is illustrated in the following:
At the $(k+1)^{th}$ iteration, the proposed criterion (4) can be developed as follows:

$$\begin{aligned}\mathcal{J}_p(z_i^{(k+1)}) &= \frac{1}{T}\sum_{t=0}^{T-1}\left|z_i^{(k)}(t) + \sum_{j=1}^{N}\epsilon_{ij}^{(k)} z_j^{(k)}(t)\right|^p \\ &= \frac{1}{T}\sum_{t=0}^{T-1}|z_i^{(k)}(t)|^p\left|1 + \sum_{j=1}^{N}\epsilon_{ij}^{(k)}\frac{z_j^{(k)}(t)}{z_i^{(k)}(t)}\right|^p.\end{aligned}$$

Under the assumption that $\boldsymbol{B}^{(k)}$ is close to a separating matrix, we have

$$|\epsilon_{ij}^{(k)}| \ll 1$$

and thus, a first order approximation of $\mathcal{J}_p(z_i^{(k+1)})$ is given by:

$$\mathcal{J}_p(z_i^{(k+1)}) \approx \frac{1}{T}\sum_{t=0}^{T-1}|z_i^{(k)}(t)|^p + p\sum_{j=1}^{N}\Re e(\epsilon_{ij}^{(k)})\Re e\left(|z_i^{(k)}(t)|^{p-1}e^{-\jmath\phi_i^{(k)}(t)}z_j^{(k)}(t)\right)$$
$$- \Im m(\epsilon_{ij}^{(k)})\Im m\left(|z_i^{(k)}(t)|^{p-1}e^{-\jmath\phi_i^{(k)}(t)}z_j^{(k)}(t)\right) \tag{9}$$

where $\Re e(x)$ and $\Im m(x)$ denote the real and imaginary parts of x and $\phi_i^{(k)}(t)$ is the argument of the complex number $z_i^{(k)}(t)$.

Using equation (9), equation (3) can be rewritten in more compact form as:

$$\mathcal{G}_p\left(\boldsymbol{B}^{(k+1)}\right) = \mathcal{G}_p\left(\boldsymbol{B}^{(k)}\right) + \Re e\left\{Tr\left(\overline{\boldsymbol{\epsilon}}^{(k)}\boldsymbol{\mathcal{R}}^{(k)H}\boldsymbol{D}^{(k)H}\right)\right\} \tag{10}$$

where $\overline{(\cdot)}$ denotes the conjugate of $(\cdot)$, $Tr(\cdot)$ is the matrix trace operator and the ij^{th} entry of matrix $\boldsymbol{\mathcal{R}}^{(k)}$ is given by:

$$\mathcal{R}_{ij}^{(k)} = \frac{1}{T}\sum_{t=0}^{T-1}|z_i^{(k)}(t)|^{p-1}e^{-\jmath\phi_i^{(k)}(t)}z_j^{(k)}(t) \tag{11}$$

$$\boldsymbol{D}^{(k)} = \left[\mathrm{diag}\left(\mathcal{R}_{11}^{(k)},\ldots,\mathcal{R}_{NN}^{(k)}\right)\right]^{\frac{1}{p}-1}. \tag{12}$$

Using a gradient technique, $\boldsymbol{\epsilon}^{(k)}$ can be chosen as:

$$\boldsymbol{\epsilon}^{(k)} = -\mu\boldsymbol{D}^{(k)}\overline{\boldsymbol{\mathcal{R}}}^{(k)} \tag{13}$$

where $\mu > 0$ is the gradient step. Replacing (13) into (10) leads to,

$$\mathcal{G}_p\left(\boldsymbol{B}^{(k+1)}\right) = \mathcal{G}_p\left(\boldsymbol{B}^{(k)}\right) - \mu\|\boldsymbol{D}^{(k)}\boldsymbol{\mathcal{R}}^{(k)}\|^2. \tag{14}$$

So μ controls the decrement of the criterion. Now, to avoid the algorithm's convergence to the trivial solution $\boldsymbol{B} = \boldsymbol{0}$, one normalizes the outputs of the separating matrix to unit-power, i.e. $\rho_{z_i}^{(k+1)} \triangleq \frac{1}{T}\sum_{t=0}^{T-1}|z_i^{(k+1)}(t)|^2 = 1, \quad \forall\ i$. Using first order approximation, this normalization leads to:

$$\epsilon_{ii}^{(k)} = \frac{1-\rho_{z_i}^{(k)}}{2\rho_{z_i}^{(k)}}. \tag{15}$$

After convergence of the algorithm, the separation matrix $\boldsymbol{B} = \boldsymbol{B}^{(\mathcal{K})}$ is applied to the whitened signal mixtures $\boldsymbol{x}_w$ to obtain an estimation of the original source signals. $\mathcal{K}$ denotes here the number of iterations that can be either chosen a priori or given by a stopping criterion of the form $\|\boldsymbol{B}^{(k+1)} - \boldsymbol{B}^{(k)}\| < \delta$ where δ is a small threshold value.

3 Convolutive Mixture Case

Unfortunately, instantaneous mixing is very rarely encountered in real-world situations, where multipath propagation with large channel delay spread occurs, in which case convolutive mixtures are considered. In this case, the signal can be modeled by the following equation:

$$\boldsymbol{x}(t) = \sum_{l=0}^{L} \boldsymbol{H}(l)\boldsymbol{s}(t-l) + \boldsymbol{w}(t) \tag{16}$$

where $\boldsymbol{H}(l)$ are $M \times N$ matrices for $l \in [0, L]$ representing the impulse response coefficients of the channel and the polynomial matrix $\boldsymbol{H}(z) = \sum_{l=0}^{L} \boldsymbol{H}(l)z^{-l}$ is assumed to be irreducible (i.e. $\boldsymbol{H}(z)$ is of full column rank for all z).

If we apply a short time Fourier transform (STFT) to the observed data $\boldsymbol{x}(t)$, the model in (16) (in the noiseless case) becomes approximately

$$\mathcal{S}_{\boldsymbol{x}}(t, f) \approx \boldsymbol{H}(f)\mathcal{S}_{\boldsymbol{s}}(t, f) \tag{17}$$

where $\mathcal{S}_{\boldsymbol{x}}(t, f)$ is the mixture STFT vector, $\mathcal{S}_{\boldsymbol{s}}(t, f)$ is the source STFT vector and $\boldsymbol{H}(f)$ is the channel Fourier Transform matrix. It shows that, for each frequency bin, the convolutive mixtures reduce to simple instantaneous mixtures. Therefore we can apply our ISBS algorithm for each frequency and separate the signals. As a result, in each frequency bin, we obtain the STFT source estimate

$$\mathcal{S}_{\widehat{\boldsymbol{s}}}(t, f) = \boldsymbol{B}(f)\mathcal{S}_{\boldsymbol{x}}(t, f). \tag{18}$$

It seems natural to reconstruct the separated signals by aligning these $\mathcal{S}_{\widehat{\boldsymbol{s}}}(t, f)$ obtained for each frequency bin and applying the inverse short time Fourier transform. For that we need first to solve the permutation and scaling ambiguities as shown next.

3.1 Removing the Scaling End Permutation Ambiguities

In this stage, the output of the separation filter is processed with the permutation matrix $\boldsymbol{\Pi}(f)$ and the scaling matrix $\boldsymbol{\mathcal{C}}(f)$.

$$\boldsymbol{G}(f) = \boldsymbol{\Pi}(f)\boldsymbol{\mathcal{C}}(f)\boldsymbol{B}(f)\ . \tag{19}$$

The scaling matrix $\boldsymbol{\mathcal{C}}(f)$ is a $N \times N$ diagonal matrix found as in [11] by $\boldsymbol{\mathcal{C}}(f) = \text{diag}[\boldsymbol{B}(f)^{\#}]$. For the permutation matrix $\boldsymbol{\Pi}(f)$, we exploit the continuity property of the acoustic filter in the frequency domain [12]. To align the estimated sources at two successive frequency bins, we test of the closeness of $\boldsymbol{G}(f_n)\boldsymbol{G}(f_{n-1})^{\#}$ to a diagonal matrix. Indeed, by using the representation (19), one can find the permutation matrix by minimizing:

$$\boldsymbol{\Pi}(f_n) = \arg\min_{\widetilde{\boldsymbol{\Pi}}} \left\{ \sum_{i \neq j} \left(\widetilde{\boldsymbol{\Pi}}\boldsymbol{\mathcal{C}}(f_n)\boldsymbol{B}(f_n)\boldsymbol{G}(f_{n-1})^{\#} \right)_{ij}^{2} \right\}. \tag{20}$$

In our simulations, we have used an exhaustive search to solve (20). However, when the number of sources is large, the exhaustive search becomes prohibitive. In that case, one can estimate $\boldsymbol{\Pi}(f_n)$ as the matrix with ones at the ij^{th} entry satisfying $|\boldsymbol{\mathcal{M}}(f_n)|_{ij} = \max_k |\boldsymbol{\mathcal{M}}(f_n)|_{ik}$ and zeros elsewhere with $\boldsymbol{\mathcal{M}}(f_n) = \boldsymbol{\mathcal{C}}(f_n)\boldsymbol{B}(f_n)\boldsymbol{G}(f_{n-1})^{\#}$. This solution has the advantage of simplicity but may lead to erroneous solution in difficult context. An alternative solution would be to decompose $\boldsymbol{\Pi}(f_n)$ as product of elementary permutations[1] $\boldsymbol{\Pi}_{(pq)}$. The latter is considered at a given iteration, only if it decrease criterion (20), if

$$|\boldsymbol{\mathcal{M}}(f_n)|_{pq}^2 + |\boldsymbol{\mathcal{M}}(f_n)|_{qp}^2 > |\boldsymbol{\mathcal{M}}(f_n)|_{pp}^2 + |\boldsymbol{\mathcal{M}}(f_n)|_{qq}^2$$

Finally, we obtain:

$$\boldsymbol{\Pi}(f_n) = \prod_{\text{nb of iterations}} \prod_{1 \le p < q \le N} \widetilde{\boldsymbol{\Pi}}_{(pq)}, \tag{21}$$

$\widetilde{\boldsymbol{\Pi}}_{(pq)}$ being either the identity matrix or the above permutation matrix $\boldsymbol{\Pi}_{(pq)}$ depending on the binary decision rule define above. We stop the iterative process, when all matrices $\widetilde{\boldsymbol{\Pi}}_{(pq)}$ are equal to the identity. We have observed that one or, at most, two iterations are sufficient to get the desired permutation. Finally, we apply the updated separation matrix $\boldsymbol{G}(f)$ to the frequency domain mixture:

$$\mathcal{S}_{\hat{\boldsymbol{s}}}(t,f) = \boldsymbol{G}(f)\mathcal{S}_{\boldsymbol{x}}(t,f). \tag{22}$$

4 Simulation Results

We present here some numerical simulations to evaluate the performance of our algorithm. We consider an array of $M = 2$ sensors receiving two audio signals in the presence of stationary temporally white noise of covariance $\sigma^2\mathbf{I}$ (σ^2 being the noise power). 10000 samples are used with a sampling frequency of 8Khz (this represents 1.25sec recording). In order to evaluate the performance in the instantaneous mixture case, the separation quality is measured using the *Interference to Signal Ratio* (ISR) criterion [2] defined as:

$$ISR \overset{\text{def}}{=} \sum_{p \neq q} \frac{E\left(|(\boldsymbol{BA})_{pq}|^2\right)\rho_q}{E\left(|(\boldsymbol{BA})_{pp}|^2\right)\rho_p} \tag{23}$$

where $\rho_i = E(|s_i(t)|^2)$ is the ith source power evaluated here as $\frac{1}{T}\sum_{t=0}^{T-1}|s_i(t)|^2$. Fig. 1-(a) represents the two original sources and their mixtures in the noiseless case. In Fig. 1-(b), we compare the performance of the proposed algorithm in instantaneous mixture case, to the Relative Newton algorithm developed by Zibulevsky et al. in [9] where the case of sparse sources is considered and to SOBI algorithm developed by Belouchrani et al. in [2]. We plot the residual interference between separated sources (ISR) versus the SNR. It is clearly shown that our algorithm (ISBS) performs better in terms of ISR especially for low SNRs as compared to the two other methods. In Fig. 2-(a), we represent the evolution of

[1] $\boldsymbol{\Pi}_{(pq)}$ is defined such as way that for a given vector $\boldsymbol{y}$, $\widetilde{\boldsymbol{y}} = \boldsymbol{\Pi}_{(pq)}\boldsymbol{y}$ iff $\widetilde{y}(k) = y(k)$, for $k \notin \{p,q\}$, $\widetilde{y}(p) = y(q)$ and $\widetilde{y}(q) = y(p)$.

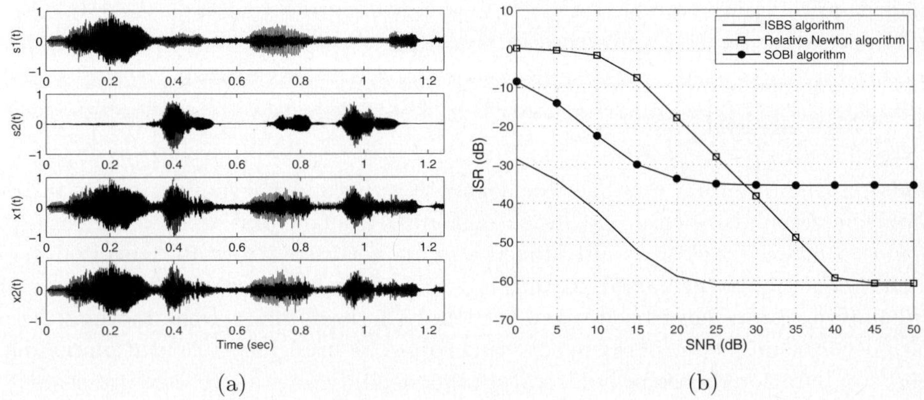

Fig. 1. (a) Up the two original source signals and bottom the two signal mixtures. (b) Interference to Signal Ratio (ISR) versus SNR for 2 audio sources and 2 sensors in instantaneous mixture case.

the ISR as a function of the iteration number. A fast convergence rate is observed. In Fig. 2-(b), we compare, in the 2×2 convolutive mixture case the separation performance of our algorithm, Deville's algorithm in [13], Parra's algorithm in [14] and extended version of Zibulevsky's algorithm to the convolutive mixture case. The filter coefficients are chosen randomly and the channel order is $L = 128$. We use in this experiment the ISR criterion defined for the convolutive case in [14] that takes into account the fact the source estimates are obtained up to a scalar filter. We observe a significant performance gain in favor of the proposed method especially at low SNR values. Moreover, the complexity of the proposed

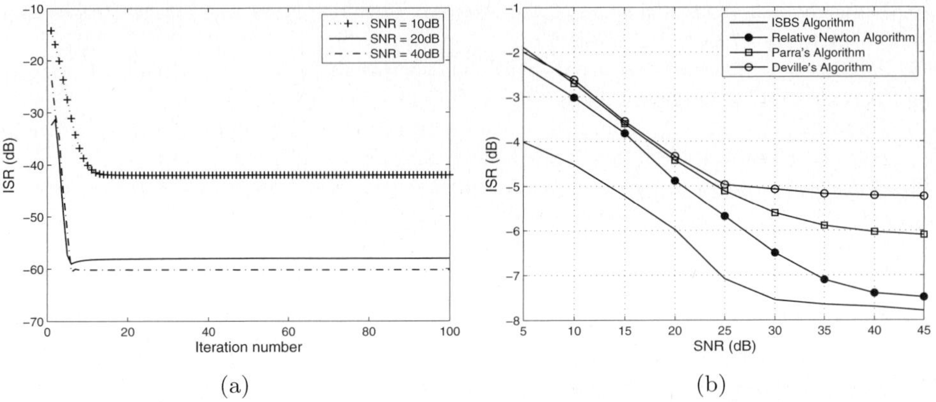

Fig. 2. (a) ISR as a function of the iteration number for 2 audio sources and 2 sensors in instantaneous mixture case. (b) ISR versus SNR for 2×2 convolutive mixture case.

algorithm is equal to $2N^2T + \mathcal{O}(N^2)$ flops per iteration whereas the complexity of the Relative Newton algorithm in [9] is $2N^4 + N^3T + N^6/6$.

5 Conclusion

This paper presents a blind source separation method for sparse sources in instantaneous mixture case and its extension to the convolutive mixture case. A sparse contrast function is introduced and an iterative algorithm based on gradient technique is proposed to minimize it and perform the BSS. Numerical simulation results have been given evidence the usefulness of the method. The proposed technique outperforms existing solutions in terms of separation quality and computational cost in both instantaneous and convolutive mixture cases.

References

1. Cardoso, J.F.: Blind signal separation: statistical principles. In: Proceedings of the IEEE, October 1998, vol. 86(10), pp. 2009–2025. IEEE Computer Society Press, Los Alamitos (1998)
2. Belouchrani, A., Abed-Meraim, K., Cardoso, J.-F., Moulines, E.: A blind source separation technique using second-order statistics. IEEE T-SP 45(2) (1997)
3. Belouchrani, A., Amin, M.G.: Blind source separation based on time-frequency signal representations. IEEE Transactions on Signal Processing 46(11) (1998)
4. Abed-Meraim, K., Xiang, Y., Manton, J.H., Hua, Y.: Blind source separation using second order cyclostationary statistics. IEEE T-SP 49(4), 694–701 (2001)
5. Yilmaz, O., Rickard, S.: Blind separation of speech mixtures via time-frequency masking. IEEE Transactions on Signal Processing 52(7), 1830–1847 (2004)
6. Pham, D.T., Cardoso, J.F.: Blind separation of instantaneous mixtures of non stationary sources. IEEE Transactions on Signal Processing 49, 1837–1848 (2001)
7. Smith, D., Lukasiak, J., Burnett, I.S.: An analysis of the limitations of blind signal separation application with speech. Signal Processing 86(2), 353–359 (2006)
8. Cichocki, A., Amari, S.: Ch. 2. In: Adaptive Blind Signal and Image Processing, Wiley & Sons, Ltd., UK (2003)
9. Zibulevsky, M.: Sparse source separation with relative Newton method. In: Proc. ICA, pp. 897–902 (April 2003)
10. Pham, D.T., Garat, P.: Blind separation of mixture of independent sources through a quasi-maximum likelihood approach. IEEE T-SP 45(7), 1712–1725 (1997)
11. Murata, N., Ikeda, S., Ziehe, A.: An approach to blind source separation based on temporal structure of speech signals. Neurocomputing 41(1-4), 1–24 (2001)
12. Pham, D.T., Serviére, C., Boumaraf, H.: Blind separation of convolutive audio mixtures using nonstationarity. In: Proc. ICA, Nara, Japan, pp. 981–986 (April 2003)
13. Albouy, B., Deville, Y.: Alternative structures and power spectrum criteria for blind segmentation and separation of convolutive speech mixtures. In: Proc. ICA, Nara, Japan, pp. 361–366 (April 2003)
14. Parra, L., Spence, C.: Convolutive blind separation of non-stationary sources. IEEE Transactions on Speech and Audio Processing 8(3), 320–327 (2000)

Maximization of Component Disjointness: A Criterion for Blind Source Separation

Jörn Anemüller

Medical Physics Section
Dept. of Physics
Carl von Ossietzky University Oldenburg
26111 Oldenburg, Germany

Abstract. Blind source separation is commonly based on maximizing measures related to independence of estimated sources such as mutual statistical independence assuming non-Gaussian distributions, decorrelation at different time-lags assuming spectral differences or decorrelation assuming source non-stationarity.

Here, the use of an alternative model for source separation is explored which is based on the assumption that sources emit signal energy at mutually different times. In the limiting case, this corresponds to only a single source being "active" at each point in time, resulting in mutual disjointness of source signal supports and *negative* mutual correlations of source signal envelopes. This assumption will not be fulfilled perfectly for real signals, however, by maximizing disjointness of estimated sources (under a linear mixing/demixing model) we demonstrate that source separation is nevertheless achieved when this assumptions is only partially fulfilled.

The conceptual benefits of the disjointness assumption are that (1) in certain applications it may be desirable to explain observed data in terms of mutually disjoint "parts" and (2) the method presented here preserves the special physical information assigned to amplitude zero of a signal which corresponds to the absence of energy (rather than subtracting the signal mean prior to analysis which for non zero-mean sources destroys this information).

The method of *disjoint component analysis* (DCA) is derived and it is shown that its update equations bear remarkable similarities with maximum likelihood independent component analysis (ICA). Sources with systematically varied degrees of disjointness are constructed and processed by DCA and Infomax and Jade ICA. Results illustrate the behaviour of DCA and ICA under these regimes with two main results: (1) DCA leads to a higher degree of separation than ICA, (2) DCA performs particularly well on positive-valued sources as long as they are at least moderately disjoint, and (3) The performance peak of ICA for zero-mean sources is achieved when sources are disjoint (but not independent)[1].

1 Introduction

Representation of measured data in terms of a number of generating causes or underlying "sources" is an important problem that has gained widespread attention in recent

[1] This research was supported by the EC under the DIRAC integrated project IST-027787.

M.E. Davies et al. (Eds.): ICA 2007, LNCS 4666, pp. 325–332, 2007.

years, either with the goal of extracting known-to-exist sources from measurements (blind source separation), or in order to find an efficient—possibly lower-dimensional—description of given data (exploratory data analysis).

We propose and investigate a novel technique, "disjoint component analysis" (DCA) that is based on the goal of extracting components with maximally disjoint support from given data, i.e., it is sought to describe the data in terms of components of which as few as possible should be activated at any single time (or sample) point. Ideally, only a single source process would account for a single sample of measured data. Since this goal is too strong for real-world data, we demonstrate that it can be significantly relaxed while still retaining the beneficial characteristics of the method.

Disjoint support between generating source processes may constitute a relevant general principle in domains where other assumptions, e.g., statistical independence and the implied effective physical separation of generating source processes, have to be postulated or justified post-hoc rather than deduced a-priori. In some cases such as communicating speakers or densely interconnected nervous cells in the brain, theoretical considerations argue in favor of dependencies between source processes. Even though such dependencies might turn out to be largely negligible in some domains, it does appear to be worthwhile to consider the implications of incorporating such dependencies into the models.

In the opposite direction (and with a different intention than ours), some authors have argued that sources that are often regarded as independent can effectively be modeled as being "w-disjoint orthogonal" [10]. We are demonstrating a close formal link between algorithms derived from striving for independent and disjoint representations, respectively, which may be seen as an indication that both notions may contain similarities that we have not yet fully appreciated.

In relation to existing techniques, DCA differs from ICA [2,4] since the disjointness assumption corresponds to a source model with *dependent* sources. Sparse-coding approaches [9], unlike DCA, impose a sparse prior on each source but do not incorporate a mutual disjointness of sources. Non-negative matrix factorisation approaches [6] and l1-norm minimization methods [5] aim to obtain a parts-based description of the data with fundamentally different algorithms than DCA.

2 Disjoint Component Analysis

2.1 Derivation of Algorithm

Consider N observed signals $\mathbf{x}(t) = [x_1(t), \ldots, x_N(t)]^T$ which are assumed to be generated from N underlying sources $\mathbf{s}(t) = [s_1(t), \ldots, s_N(t)]^T$ by multiplication with a mixing system $\mathbf{A}$ as

$$\mathbf{x}(t) = \mathbf{A}\,\mathbf{s}(t) \tag{1}$$

It is sought to linearly transform the observations by a matrix $\mathbf{W}$ to obtain output signals

$$\mathbf{y}(t) = \mathbf{W}\,\mathbf{x}(t) \tag{2}$$

with components $\mathbf{y}(t) = [y_1(t), \ldots, y_N(t)]^T$. When source reconstruction is desired, these should resemble the sources up to arbitrary rescaling and permutation. When an

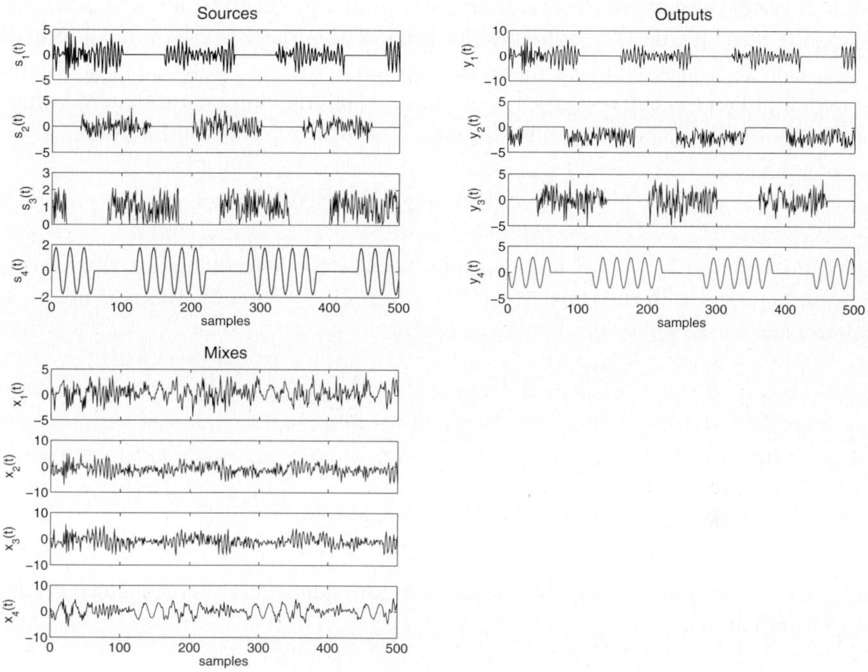

Fig. 1. Disjoint component analysis of four sources (top left) which are not strictly disjoint but exhibit significant overlap. Sources were mixed with a randomly chosen 4×4 mixing matrix to yield observation signals (bottom left) which were successfully separated into the original sources up to arbitrary permutation, rescaling and sign flip (top right) using DCA.

exploratory data analysis view is adopted, the output signals should convey a signal representation that is meaningful in some to-be-specified sense.

A central notion in our approach is the overlap between two output signals y_i and y_j which we define as[2]

$$o_{ij} = E(|y_i|\,|y_j|), \tag{3}$$

where $E(\cdot)$ denotes expectation and sample index t is omitted where convenient. With $o_{ij} \geq 0$ and $o_{ij} = 0$ if and only if $y_i(t)\,y_j(t) = 0$ for all t and $i \neq j$, two signals y_i and y_j have *disjoint support* if $o_{ij} = 0$. In this case, y_i and y_j are called *disjoint*, i.e., at most one of the signals is non-zero at any time.

For strictly disjoint source signals $\mathbf{s}(t)$ and a non-singular matrix $\mathbf{A}$, strictly disjoint outputs can be obtained that resemble the sources up to arbitrary permutation and rescaling. Note that in this case sources are not mutually *independent* but exhibit statistical *dependencies* through the negative correlations of their signal envelopes or signal power time-courses.

[2] Different definitions of the overlap, involving other non-linear functions of the output signals, are possible but beyond the scope of the present paper.

While it is not possible in general to linearly transform an arbitrary signal $\mathbf{x}(t)$ into a signal $\mathbf{y}(t)$ with only disjoint components, finding minimally overlapping outputs is a natural goal as it corresponds to a signal description in terms of processes out of which only a small number is active at any given time. A natural choice to obtain maximally disjoint, minimally overlapping output signals is minimization of the function

$$H = \frac{1}{2} \sum_{i \neq j} o_{ij} = \frac{1}{2} \sum_{i \neq j} E(|y_i|\,|y_j|) \tag{4}$$

The global minimum $H = 0$ is attained only for strictly disjoint signals where for all t any signal $y_i(t) \neq 0$ if and only if $y_j(t) = 0$ for all $j \neq i$. Substituting 2 into 4, the partial derivatives are given by

$$\frac{\partial H}{\partial w_{ij}} = E\Big(\mathrm{sign}(y_i)\, x_j \sum_{k \neq i} |y_k|\Big) \tag{5}$$

which in matrix notation is easily rewritten as

$$\nabla H = E\left(-\mathbf{y}\mathbf{x}^H + ||\mathbf{y}||_1 \mathrm{sign}(\mathbf{y})\mathbf{x}^H\right) \tag{6}$$

where $||\mathbf{y}||_1 = \sum_i |y_i|$ denotes the 1-norm of $\mathbf{y}$.

Right-multiplication with $\mathbf{W}^T\mathbf{W}$ yields an expression similar to the natural gradient ICA algorithm of [1],

$$\tilde{\nabla} H = E\left(-\mathbf{y}\mathbf{y}^H + ||\mathbf{y}||_1 \mathrm{sign}(\mathbf{y})\mathbf{y}^H\right)\mathbf{W}. \tag{7}$$

Gradients (6) and (7) are similar to the corresponding gradients derived from infomax or maximum-likelihood ICA with a sparse prior, however, we emphasize that the mean has not been removed from the output signals (i.e., source estimates) $\mathbf{y}(t)$.

Without constraints the gradients converge to the trivial solution $\mathbf{W} = \mathbf{0}$. To remove the scaling ambiguity each row $\mathbf{w}_i$ of matrix $\mathbf{W}$ is fixed to unit-norm $||\mathbf{w}_i||_2 = 1$. Hence, each row $\mathbf{\Delta}_i$ of ∇H is projected according to

$$\mathbf{\Delta}_i^\perp = \mathbf{\Delta}_i - (\mathbf{\Delta}_i^H \mathbf{w}_i)\,\mathbf{w}_i \tag{8}$$

resulting in the projected gradient matrix $\mathbf{\Delta}^\perp$ that is then used for gradient descent. The final update rule for matrix $\mathbf{W}$ with a step size of η is

$$\mathbf{W} \leftarrow \mathbf{W} - \eta\,\mathbf{\Delta}^\perp \tag{9}$$

for the ordinary gradient (6) and similarly for (7). Periodic row re-normalization of $\mathbf{W}$ is applied to keep it on the constraint manifold for non-infinitesimal η.

3 Evaluation

3.1 Synthetic Data Generation

Disjoint sources $s_i(t)$ are generated from mutually independent signals $\zeta_i(t)$ by multiplying them with disjoint masking functions $\mu_i(t) \in \{0, 1\}$ for all i, t and

$$s_i(t) = \mu_i(t)\,\zeta_i(t) \tag{10}$$

$$E(\mu_i\,\mu_j) = 0 \quad \text{if} \quad i \neq j \tag{11}$$

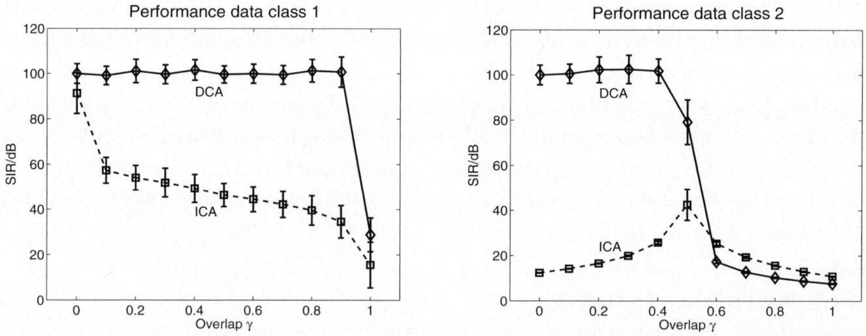

Fig. 2. Separation performance of DCA and ICA in terms of signal-to-interference ratio (SIR) in dB after separation. Performance is given for data class 1 (left panel, sources with positive and negative observation values) and data class 2 (right panel, sources with positive only observation values) as a function of overlap γ. A value of $\gamma = 0$ corresponds to strictly disjoint sources (statistical dependencies between sources through negative correlation of signal envelopes); $\gamma = 0.5$ corresponds to statistically independent sources; and $\gamma = 1.0$ corresponds to fully overlapping, not disjoint sources (statistical dependencies through positive correlation of signal envelopes). Mean and variance of performance for 100 separation runs, each with independently generated data, are given for each condition.

These sources may then be used to generate observations by multiplication with a matrix $\mathbf{A}$ according to Eq. 1.

Strictly disjoint sources with zero overlap are not expected to be an appropriate model for real data. Hence, sources with variable masker overlap γ_{ij}, which may depend on the source pair (i, j),

$$\gamma_{ij} = E(\mu_i \mu_j) \,/\, E(\mu_i^2) \tag{12}$$

with $E(\mu_i^2) = \text{const}$ for all i are also generated. In the experiments reported below masker overlap γ_{ij} is chosen such that a value of $\gamma_{ij} = 1$ corresponds to a source pair (s_i, s_j) exhibiting mutual statistical dependence through maskers with *positive* correlation. The value $\gamma_{ij} = 0$ corresponds to strictly disjoint sources that exhibit mutual statistical dependence through maskers with *negative* correlation. Finally, a value of $\gamma_{ij} = 0.5$ coincides with statistically *independent* sources (s_i, s_j) because of uncorrelated maskers (and statistically independent $\zeta_i(t)$).

The signal generation scheme was inspired by a functional magnetic resonance imaging (fMRI) experiment design [3].

3.2 Separation of Synthetic Sources

Four sources were generated according to the scheme described above, mixed with a randomly chosen mixing matrix and processed with the natural gradient disjoint component analysis algorithm (Eq. 7) with regularization (Eq. 8). The underlying mutually independent signals $\zeta_i(t)$ were chosen as a speech signal (ζ_1), i.i.d. noise from a normal distribution with zero-mean and unit-variance (ζ_2), i.i.d. noise from a uniform distribution on the interval $[0, 1]$ (ζ_3), and a sine wave (ζ_4). The maskers $\mu_i(t)$ were chosen such that $\gamma_{ij} = 0.6$ for source pairs $(1, 2)$, $(2, 3)$, $(3, 4)$, $(1, 4)$, and $\gamma_{ij} = 0.4$ for source pairs $(1, 3)$, $(2, 4)$. Source signals, observed (mixed) signals and output signals are displayed in Fig. 1, demonstrating that the algorithm performs successful separation even though sources are not strictly disjoint but show significant overlap. Similarly, the algorithm successfully separates mixtures of four strictly disjoint sources with $\gamma_{ij} = 0$ for all $i \neq j$ (data not shown here).

3.3 Variable Degree of Overlap

The goal of this experiment was to systematically study the influence of the degree of overlap on the performance of the disjoint component analysis algorithm. Results are reported for the gradient version of the algorithm (Eq. 6) with regularization (Eq. 8). Results for the natural gradient version are virtually identical and not reported separately.

Sources were generated based on two different underlying signal classes. In the first part of the experiment ("data class 1"), two sources s_1 and s_2 were generated from ζ_1 and ζ_2 that were drawn as i.i.d. signals from a zero-mean and unit-variance normal distribution, hence containing positive and negative values. In the second part of the experiment ("data class 2"), ζ_1 and ζ_2 were chosen to be i.i.d. signals from a uniform distribution on the interval $[0, 1]$, hence containing only positive values.

For both data sets the single overlap parameter γ was varied from 0 (no overlap, source dependence through negative masker correlation) via 0.5 (50% overlap, statistically independent sources) to 1.0 (full overlap, source dependence through positive masker correlation) in steps of 0.1.

Hence, 11 data set conditions were generated for each of the two data classes. For each condition, disjoint component analysis was performed on 100 individual datasets drawn independently according to the description above. This resulted in a total of 2200 datasets each with 10000 samples for each of the two sources.

Fig. 2 shows the results with mean and variance of signal separation in dB signal-to-interference ratio (SIR) after separation separately for data class 1 (left panel) and data class 2 (right panel). For data class 1 with sources that adopt positive and negative values, DCA separation performance shows no significant dependence on the overlap parameter γ except (as expected) for complete overlap at $\gamma = 1$ where the algorithm essentially attempts to separate two i.i.d. normally distributed sources which is ill-posed. In all other cases of data class 1, DCA separation is excellent with about 100 dB SIR.

The results look different for data class 2 with positive only source values. Separation remains excellent for data sets with a small overlap ($0.0 \leq \gamma \leq 0.4$), with again about 100 dB SIR. In the case of independent sources at $\gamma = 0.5$, separation is still very good at 80 dB. Performance breaks down for large overlaps ($1.0 \geq \gamma \geq 0.6$), an effect which we attribute to the positivity of the sources.

3.4 Comparison with Independent Component Analysis

The same data generated for section 3.3 was re-analyzed with natural gradient infomax ICA [1,2] using the ICA toolbox [7,8] with logistic function non-linearity. For comparison, a simple gradient approach with fixed step size and sign function non-linearity was also used and gave virtually identical results for data class 1. On data class 2, the fixed step gradient approach gave qualitatively similar results but was outperformed by the referenced ICA toolbox in terms of SIR separation performance. All source signals have been checked to have positive kurtosis. Processing of the same signal with the jade algorithm [4] gave virtually identical results.

Results in Fig. 2 show that in most cases ICA results in a poorer SIR than DCA. For data class 1, ICA shows excellent signal separation for strictly disjoint sources ($\gamma = 0.0$). Performance is significantly lower, though still good, for independent sources, which seems to stand in contradiction to the independence assumption. As expected, performance decreases towards sources with strong overlap ($\gamma = 1.0$).

For data class 2, ICA performs best when sources are independent ($\gamma = 0.5$) with a drop off in performance towards both lower and higher source overlaps, which is plausible due to ICA's independence assumption.

4 Conclusion

Disjoint component analysis (DCA) has been shown to yield good performance for strictly disjoint and moderately disjoint data sets. For data with high overlap between sources (weakly disjoint), performance depends on the specific type of data, with good performance for data sets with sources that take positive and negative observation values, and a break-down of performance in case of purely positive source data.

We have shown that under certain approximations DCA is closely related to independent component analysis (ICA), albeit both start from significantly different assumptions. The empirical algorithm evaluation showed a better separation performance for DCA than for ICA under most conditions. Interestingly, ICA produced the best performance not for statistically independent sources but for strictly disjoint ones (cf. also ICA 2006 oral presentation of I.C. Daubechies).

Results presented here appear to warrant a closer investigation of the differences and similarities of both algorithm classes. It would be desirable to gain experience with a wider range of synthetic and natural data than could be presented here. We are tempted to speculate that DCA might be appropriate in particular for analyzing data where the independence assumption is not strictly fulfilled, where a data representation in terms of disjoint components is preferable to independent components, and where signals are comprised of positive only measurement values. This could be the case, e.g., for brain signals such fMRI, for data from dialog speech signals, and for comparably short signal sequences where independence cannot be fully attained due to finite sample effects.

References

1. Amari, S.-I.: Natural gradient works efficiently in learning. Neural Computation 10, 251–276 (1998)
2. Bell, A.J., Sejnowski, T.J.: An information-maximization approach to blind separation and blind deconvolution. Neural Computation 7, 1129–1159 (1995)
3. Benharrosh, M.S., Takerkart, S., Cohen, J.D., Daubechies, I.C., Richter, W.: Using ICA on fMRI: Does independence matter?. In: Human Brain Mapping, abstract no. 784 (2003)

4. Cardoso, J.-F., Souloumiac, A.: Blind beamforming for non Gaussian signals. IEE Proceedings–F 140, 362–370 (1993)
5. Donoho, D., Elad, M.: Optimally sparse representation in general (nonorthogonal) dictionaries via l^1 minimization. Proc. Nat. Acad. Sci. 100, 2197–2202 (2003)
6. Lee, D.D., Seung, H.S.: Learning the parts of objects by non-negative Matrix Factorization. Nature 40, 788–791 (1999)
7. Makeig, S., et al.: EEGLAB: ICA toolbox for psychophysical research, Swartz Center for Computational Neuroscience, Institute for Neural Computation, University of California, San Diego (2000), `http://www.sccn.ucsd.edu:/eeglab`
8. Makeig, S., Bell, A.J., Jung, T.-P., Sejnowski, T.J.: Independent component analysis of electroencephalographic data. Advances in neural information processing system 8, 145–151 (1996)
9. Olshausen, B.A., Field, D.J.: Sparse coding with an overcomplete basis set: a strategy employed by V1? Vision Research 37, 3311–3325 (1997)
10. Rickard, S., Yilmaz, Z.: On the approximate w-disjoint orthogonality of speech. In: ICASSP '02, pp. I–529–I–532 (2002)

Estimator for Number of Sources Using Minimum Description Length Criterion for Blind Sparse Source Mixtures

Radu Balan

Siemens Corporate Research
755 College Road East
Princeton, NJ 08540
radu.balan@siemens.com

Abstract. In this paper I present a Minimum Description Length Estimator for number of sources in an anechoic mixture of sparse signals. The criterion is roughly equal to the sum of negative normalized maximum log-likelihood and the logarithm of number of sources. Numerical evidence supports this approach and compares favorabily to both the Akaike (AIC) and Bayesian (BIC) Information Criteria.

1 Signal and Mixing Models

Consider the following model in time domain:

$$x_d(t) = \sum_{l=1}^{L} s_l(t - (d-1)\tau_l) + n_d(t) \quad , \quad 1 \leq d \leq D \tag{1}$$

This model corresponds to an anechoic Uniform Linear Array (ULA) with L souces and D sensors. In frequency domain, (1) becomes

$$X_d(k,\omega) = \sum_{l=1}^{L} e^{-i\omega(d-1)\tau_l} S_l(k,\omega) + N_d(k,\omega) \tag{2}$$

We use the following notations: $\mathbf{X}(k,\omega)$ for the D-complex vector of components $(X_d(k,\omega))_d$, $\mathbf{S}(k,\omega)$ for the L-complex valued vector of components $(S_l(k,\omega))_l$, and $A(\omega)$ the $D \times L$ complex matrix whose (d,l) entry is $A_{d,l}(\omega) = e^{-i\omega(d-1)\tau_l}$.

In this paper I make the following statistics assumptions:

1. (H1) Noise signals $(n_d)_{1 \leq d \leq D}$ are Gaussian i.i.d. with zero mean and unknown variance σ^2;
2. (H2) Source Signals are unknown, but for every time-frequency point (k,ω), at most one signal $S_l(k,\omega)$ is nonzero, among the total of L signals;
3. (H3) The number of source signals L is a random variable.

The probem is to design a statistically principled estimator for L, the number of source signals. In this paper I study the Minimum Description Length approach for this problem.

M.E. Davies et al. (Eds.): ICA 2007, LNCS 4666, pp. 333–340, 2007.

For this model, the measured data is $\Xi = \{(X_d(k,\omega))_{1\leq d\leq D}\ ,\ 1\leq k\leq T, 1\leq\omega\leq F\}$. Furthermore the number of sensors D is also known. The rest of parameters are unknown. I denote $\theta = (\theta', L)$, where:

$$\theta' = (\{(S_l(k,\omega))_{1\leq l\leq L}\ ;\ 1\leq k\leq T, 1\leq\omega\leq F\}\ ,\ (\tau_l)_{1\leq l\leq L}\ ,\ \sigma^2) \tag{3}$$

Notice that hypothesis (H2) above imposes a constraint on set $(S_l(k,\omega))_{1\leq l\leq L}$, for every (k,ω). More specifically, the L complex vector $(S_l(k,\omega))_{1\leq l\leq L}$ has to lay in one of the L 1-dimensional coordinate axes (that is, all but one component has to vanish). This fact has a profound implication on estimating the complexity penalty associated to the parameters set. Some real world signals may satisfy (H2) only approximately. For instance [1] studies this assumption for speech signals.

1.1 Prior Works

The signal and mixing model described before has been analyzed by many works before.

In the past series of papers [2,3,4,5,6,7] the authors studied (1), and several generalizations of this model in the following respects. Mixing model: each channel may have an attenuation factor (equivalently, τ_l may be complex); Noise statistics: noise signals may have inter-sensor correlations; Signals: more signals may non-vanish at each time-frequency point (maximum number allowed is $D-1$); more recently we have considered temporal, and time-frequency, dependencies on signal statistics.

A similar model, and a similar sparseness assumption, has been used by the DUET algorithm [1], or by [8], [9].

Similar assumptions to [5] have been made by [10] for an instantaneous mixing model. As the authors mentioned there, as well in [11,12], and several others, a new signal separation class is defined by sparsness assumption, called Sparse Component Analysis (SCA). In this vein, this present paper proposes a look at the Minimum Description Length paradigm in the context of Sparse Component Analysis.

Before discussing the new results of this paper, I would like to comment on other approaches to the BSS problem. Many other works dealt with the mixing model (1), or its generalizations to a fully echoic model. A completely different class of algorithms is furnished by the observation that, in frequency domain, the echoic model simply becomes an instantaneous mixing model. Therefore standard ICA techniques can be applied, as in [13,14] to name a few. Next, one has to connect frequency domain components together for the same source. The permutation ambiguity is the main stumbling block. Several approaches have been proposed, some based on ad-hoc arguments, [15,9]. A more statistically principled approach has been proposed and used by Zibulevsky [16] and in more recent papers, as well as by other authors, by assuming a stochastic prior model for source signals. The Maximum A Posteriori (MAP), or Minimum Mean Square Error (MMSE) estimators can be derived. While principly they are superior to Maximum Likelihood type estimators derived in [4,5], or mixed estimators such as [1,8,9], they require a good prior stochastic model. This makes difficult the comparison between classes of BSS solutions.

In the absence of noise, the number of sources can be estimated straightforwardly by building a histogram of the instantaneous delay (τ), or for a more general model see [10].

As I mention later, the MDL paradigm here may be well applied in conjunction with other signal estimators, in particular with the MAP estimators described before.

2 Estimators

Assume the mixing model (1) and hypotheses (H1),(H2),(H3). Then its associated likelihood is given by

$$\mathcal{L}(\theta) := P(\Xi|\theta) = \prod_{(k,\omega)} \frac{1}{\pi^D \sigma^{2D}} exp\left(-\frac{1}{\sigma^2}\|\mathbf{X}(k,\omega) - A(\omega)\mathbf{S}(k,\omega)\|^2\right) \quad (4)$$

In the next subsection the maximum likelihood estimator for θ', and the maximum likelihood value are going to be derived.

Following a long tradition of statistics papers, consider the following framework. Let $P(X)$ denote the unknown true probability of data (measurements), $P(X|\theta)$ denote the data likelihood given the model (1) and (H1-H3). Then the estimation objective is to minimize the misfit between these two distributions measured by a distance between the two distribution functions. One can choose the Kullback-Leibler divergence, and obtain the following optimization criterion:

$$J(\theta) = D(P_X\|P_{X|\theta}) := \int log\frac{P(X)}{P(X|\theta)} dP(X) = \int \log P(X)\, dP(X) - \int \log P(X|\theta)\, dP(X) \quad (5)$$

Since the first term does not depend on θ, the objective becomes maximization of the second term:

$$\hat{\theta} = argmax_\theta \mathbf{E}[\log P_{X|\theta}(X|\theta)] \quad (6)$$

where the expectation is computed over the true data distribution P_X. However the true distribution is unknown. A first approximation is to replace the expectation $\mathbf{E}$ by average over data points. Thus one obtains the maximum likelihood estimator (MLE):

$$\hat{\theta}_{ML} = argmax_\theta \frac{1}{N} \sum_{t=1}^{N} \log P_{X|\theta}(X_t|\theta) \quad (7)$$

where N is the number of sample points $(X_t)_{1 \le t \le N}$.

As is well known in statistical estimation (see [17,18]), the MLE is usually biased. For discrete parameters, such as number of source signals, this bias has a bootstraping effect that monotonically increases the likelihood and makes the number of parameter estimation impossible through naive MLE. Several approaches proposed to estimate and make correction for this bias. In general, the optimization problem is restated as:

$$\hat{\theta} = argmin_{\theta} \left[-\frac{1}{N} \sum_{t=1}^{N} \log P(X_t|\theta) + \Phi(\theta, N) \right] \tag{8}$$

Following e.g. [18] we call Φ the *regret*. Akaike [17] proposes the following regret:

$$\Phi_{AIC}(\theta, N) = \frac{|\theta|_0}{N} \tag{9}$$

where $|\theta|_0$ represents the total number of parameters. Schwarz [19] proposes a different regret, namely

$$\Phi_{BIC}(\theta, N) = \frac{|\theta|_0 \log N}{2N} \tag{10}$$

In a statistically plausible interpretation of the world, Rissanen [20] obtains for regret the shortest possible description of the model using the universal distribution function of Kolmogorov, hence the name *Minimum Description Length*,

$$\Phi_{MDL}(\theta, N) = Coding\ Length_{Kolmogorov\ p.d.f.}(Model(\theta, N)) \tag{11}$$

Based on this interpretation, $\Phi(\theta, N)$ represents a measure of the model complexity.

My approach here is the following. I propose the following regret function

$$\Phi_{MDL-BSS}(\theta, N) = log_2(L) + \frac{L\, log_2(M)}{N} \tag{12}$$

where M represents precision in optimization estimation of delay parameters τ (for instance the number of grid points of an 1-D exhaustive search). Thus the optimization in (8) is carried out in two steps. First, for fixed L, the log likelihood is optimized over θ':

$$\hat{\theta}'_{MLE}(L) = argmax_{\theta'} P(X|\theta', L) \quad , \quad MLV(L) = P(X|\hat{\theta}'_{MLE}, L) \tag{13}$$

Here MLV denotes the Maximum Likelihood Value. Then L is estimated via:

$$\hat{L}_{MDL-BSS} = armin_L \left[-\log(MLV(L)) \; + \; \log_2(L) \; + \; \frac{L\, log_2(M)}{N} \right] \tag{14}$$

In the next subsection I present the computation of the Maximum Likelihood Value (MLV). Then, in the following subsection I argue the particular form (12) for $\Phi(\theta, N)$ inspired by the MDL interpretation. In same subsection I also present difficulties in a straightforward application of AIC or BIC criteria.

2.1 The Maximum Likelihood Value

The material from this subsection is presented in more detail in [4]. Results are summarized here for the benefit of the reader.

The constraint (H2) assumed in section 1 can be recast by introducing the selection variable $V(k, \omega)$: $V(k, \omega) = l$ iff $S_l(k, \omega) \neq 0$, and the complex amplitudes $G(k, \omega)$. Thus a slightly different parametrization of the model is obtained. The new set of parameters is now $\psi = (\psi', L)$ where

$$\psi' = (\{(G(k, \omega), V(k, \omega)) \; ; \; 1 \leq k \leq T, 1 \leq \omega \leq F\} \; , \; (\tau_d)_{1 \leq d \leq D} \; , \; \sigma^2) \tag{15}$$

The signals in θ' are simply obtained through: $S_{V(k,\omega)}(k, \omega) = G(k, \omega)$, and $S_l(k, \omega) = 0$ for $l \neq V(k, \omega)$.

The likelihood (4) becomes:

$$\mathcal{L}(\psi) = \frac{1}{\pi^{DN}\sigma^{2DN}} exp\left(-\frac{1}{\sigma^2} \sum_{(k,\omega)} \|\mathbf{X}(k,\omega) - G(k,\omega)A_{V(k,\omega)}(\omega)\|^2 \right) \quad (16)$$

where N is the number of time-frequency data points, and $A_l(\omega)$ denotes the lth column of matrix $A(\omega)$. The optimization over G is performed immediately, as a least square problem. The optimum value is replaced in $\mathcal{L}(\psi)$:

$$log\mathcal{L}((V)_{k,\omega},(\tau_l)_l,L) = -DN\,log(\pi) - DN\,log(\sigma^2) - \frac{1}{\sigma^2}\sum_{k,\omega}\left[\|\mathbf{X}(k,\omega)\|^2 - \frac{1}{D}|\langle \mathbf{X}(k,\omega), A_{V(k,\omega)}(\omega)\rangle|^2\right]$$

The optimization over $(V)_{k,\omega}$ and $(\tau_l)_{1\leq l\leq L}$ is performed iteratively as in the K-means algorithm:

- For a fixed set of delays $(\tau_l)_l$, the optimal selection variables are
$$V(k,\omega) = argmax_m |\langle \mathbf{X}(k,\omega), A_m(\omega)\rangle| \quad (17)$$
- For a fixed selection map $(V(k,\omega))_{k,\omega}$, consider the induced partition $\Pi_m = \{(k,\omega)\ ;\ V(k,\omega) = m\}$. Then τ_m is obtained by solving L 1-dimensional optimization problems
$$\tau_m = argmax_\tau \sum_{(k,\omega)\in\Pi_m} |\langle \mathbf{X}(k,\omega), A_m(\omega;\tau)\rangle|^2 \quad (18)$$

This steps are iterated until convergence is reached (usually is a relatively small number of steps, e.g. 10). Denote $\hat{V}_{MLE}(k,\omega)$ and $\hat{\tau}_{lMLE}$ the final values, and replace these values into $\mathcal{L}$. The noise variance parameter is estimated by maximizing $\mathcal{L}$ over σ^2,

$$\hat{\sigma^2}_{MLE} = \frac{1}{N}\sum_{(k,\omega)}\left[\|\mathbf{X}(k,\omega)\|^2 - \frac{1}{D}|\langle \mathbf{X}(k,\omega), A_{\hat{V}_{MLE}(k,\omega)}(\omega;\hat{\tau}_{MLE}\rangle|^2\right] \quad (19)$$

Finally, the log maximum likelihood value becomes:

$$log(MLV(L)) = \frac{1}{N} log(\mathcal{L}(\hat{\psi'}_{MLE};L)) = -D\,log(\pi) - 1 - D\,log(\hat{\sigma^2}_{MLE}) \quad (20)$$

where $\hat{\psi'}_{MLE}$ denoted the optimal parameter set ψ' containing the combined optimal values $(\hat{V}_{MLE}(k,\omega))_{(k,\omega)}$, $(\hat{G}_{MLE}(k,\omega))_{(k,\omega)}$, $(\hat{\tau}_l)_{1\leq l\leq L}$, $\hat{\sigma^2}_{MLE}$.

2.2 Number of Sources Estimation

The next step is to establish the regret function. As mentioned earlier the approach here is to use an estimate of the Minimum Description Length of the model (1) together with hypotheses (H1-H3). In general this is an impossible task since the Kolmogorov's universal distribution function is unkown. However the L-dependent part of the model description is embodied in the mixing parameters $(\tau_l)_{1\leq l\leq L}$, and the selection map $(V(k,\omega))_{(k,\omega)}$. Approximating by a uniform distribution in the space of delays with a finite discretization of, say,

M levels, and no prior preferential treatment of one source signal versus the others, an upper bound on the description length is obtained as the code length of an entropic encoder for this data added to the description length of the entire sequence of models with respect to the Kolmogorov universal distribution:

$$l^*(Model; N) \leq Llog_2(M) + Nlog_2(L) + C(Model) \tag{21}$$

This represents an upper bound since $l^*(Model; N)$ is supposed to represent the optimal description (minimal description) length, whereas the description splits into two parts: the sequence of models parametrized by ψ and N, and then, for a given (L, N) the entropic length of ψ. This clearly represents only one possible way of encoding the pair $(Model(\psi), N)$.

This discussion justifies the following choice for the regret function $\Phi_{MDL-BSS}$

$$\Phi_{MDL-BSS}(L, N) = \frac{L\,log_2(M) + N\,log_2(L)}{N} = log_2(L) + \frac{L\,log_2(M)}{N} \tag{22}$$

as mentioned earlier in (12).

Before presenting experimental evidence supporting this approach, I would like to comment on AIC and BIC criteria. The main difficulty comes from the estimation of the number of parameters. Notice that, using θ description, the number of parameters becomes $LN+L+2$, whereas in ψ description, this number is only $2N + L + 2$. The difference is due to that fact that the set of realizable signal vectors $(S_l)_{1\leq l\leq L}$ lays in a collection of L 1-dimensional spaces. Thus this can be either modeled as a collection of L variables, or by 2 variables: complex amplitude, and a selection map V. Consequently, the regret function for AIC can be either $L + \frac{L+2}{N}$, or $2 + \frac{L+2}{N}$. Similarly, for BIC the regret function can be $L\,log(N)/2 + \frac{(L+2)log(N)}{2N}$, or $log(N) + \frac{(L+2)log(N)}{2N}$. The criterion I propose in (22) interpolates between these two extrema, and, in my opinion, it captures better the actual size of model parametrization.

3 Experimental Evaluation

Consider the following setup. A Uniform Linear Array (ULA) with a variable number of sensors runging from 2 to 5, and distance between adjacent sensors of 5 cm, that records anechoic mixtures of signals coming from $L \leq 6$ sources. The sources are spread uniformly with a minimum of 30 degrees separation. Additive Gaussian noise of average SNR ranging from 10dB to 100dB has been added to recordings. The signals were TIMIT voices sampled at 16 KHz, and each of length 38000 samples (roughly 3 male and female voices saying "She had a dark suit in a greasy wash water all year").

For this setup, the noise was varied in 10dB steps, and number of sources ranged from 1 to 6. The delay optimization (18) was performed through a grid search with step 0.05 samples. Since $\tau_{max} = 2.4$, there were $M = 96$ possible values of τ. Thus $\frac{log_2(M)}{N} = 1.7\,10^{-4}$ and the correction term $L\frac{log_2(M)}{N}$ in $\Phi_{MDL-BSS}$ had no influence. Similarly, the $\frac{L}{N}$ term in AIC and $\frac{L\,log(N)}{N} = 3\,10^{-4}\,L$ in BIC are too small. Therefore the only meaningful AIC and BIC were given by the former regret functions. To summarize, the source number estimator is given by:

$$\hat{L}_{MDL-BSS} = argmin_L \left[-log\, MLV(L) + log_2(L) \right] \tag{23}$$
$$\hat{L}_{AIC} = argmin_L \left[-log\, MLV(L) + L \right] \tag{24}$$
$$\hat{L}_{BIC} = argmin_L \left[-log\, MLV(L) + L\, log(N) \right] \tag{25}$$

where the optimization is done by exhaustive search for L over the range 1 to 10. For a total of 1680 experiments (10 levels of noise x 4 number of sensors x 6 number of sources x 7 realizations), the histogram of estimation error has been obtained. For each of the three estimators, the histogram is rendered in Figure 1. Statistical performance of these estimators is presented in Table at right.

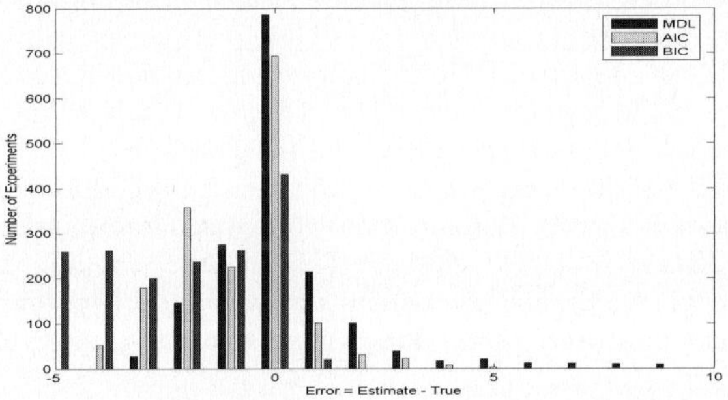

Algorithm	Bias	Variance	Probability of Error
MDL-BSS	0.224	3.17	53 %
AIC	-0.85	2.82	58 %
BIC	-2.189	8.19	74 %

Fig. 1. The histograms of estimation errors for MDL-BSS criterion (left bar), AIC criterion (middle bar), BIC criterion (right bar). Table with statistical performance of the three estimators.

4 Conclusions

The MDL-BSS estimator clearly performed best among the three estimators, since the error distribution is the most concentrated to zero, in every sense: the number of errors is the smallest, the average error is the smallest, the variance is the smallest, the bias is the smallest. Estimation error is explained by a combination of two factors: 1) source signals (voices) do not satisfy the hypothesis (H2), instead there is always an overlap between time-frequency signal supports; and 2) the estimates for location, noise variance, and separated signals were biased; this bias compounded and inverted the minimum position. The other two estimators (AIC, and BIC) were biased towards underestimating the number of sources.

This paper provides a solid theoretical footing for a statistical criterion to estimate number of source signals in an anechoic BSS scenario with sparse signals. Extension to other mixing models (such as instantaneous) is obvious. The regret function stays the same, only the MLV is modified. The same approach can be used to other Sparse Component Analysis, and this analysis will be done elsewhere.

The numerical simulations confirmed the estimation performance.

References

1. Yilmaz, O., Rickard, S.: Blind separation of speech mixtures via time-frequency masking. IEEE Trans. on Sig. Proc. 52(7), 1830–1847 (2004)
2. Rickard, S., Balan, R., Rosca, J.: Real-time time-frequency based blind source separation. In: Proc. ICA, pp. 651–656 (2001)
3. Balan, R., Rosca, J., Rickard, S.: Non-square blind source separation under coherent noise by beamforming and time-frequency masking. In: Proc. ICA (2003)
4. Balan, R., Rosca, J., Rickard, S.: Scalable non-square blind source separation in the presence of noise. In: ICASSP2003, Hong-Kong, China (April 2003)
5. Rosca, J., Borss, C., Balan, R.: Generalized sparse signal mixing model and application to noisy blind source separation. In: Proc. ICASSP (2004)
6. Balan, R., Rosca, J.: Convolutive demixing with sparse discrete prior models for markov sources. In: Proc. BSS-ICA (2006)
7. Balan, R., Rosca, J.: Map source separation using belief propagation networks. In: Proc. ASILOMAR (2006)
8. Aoki, M., Okamoto, M., Aoki, S., Matsui, H.: Sound source segregation based on estimating incident angle of each frequency component of input signals acquired by multiple microphones. Acoust. Sci. & Tech. 22(2), 149–157 (2001)
9. Sawada, H., Mukai, R., Araki, S., Makino, S.: A robust and precise method for solving the permutation problem of frequency-domain blind source separation. IEEE Trans. SAP 12(5), 530–538 (2004)
10. Georgiev, P., Theis, F., Cichocki, A.: Sparse component analysis and blind source separation of underdetermined mixtures. IEEE Tran. Neur.Net. 16(4), 992–996 (2005)
11. Cichocki, A., Li, Y., Georgiev, P., Amari, S.-I.: Beyond ica: Robust sparse signal representations. In: IEEE ISCAS Proc. pp. 684–687 (2004)
12. Cichocki, A., Amari, S.: Adaptive Blind Signal and Image Processing: Learning Algorithms and Applications. Wiley, Chichester (April 2002)
13. Comon, P.: Independent component analysis, a new concept? Signal Processing 36(3), 287–314 (1994)
14. Bell, A.J., Sejnowski, T.J.: An information-maximization approach to blind separation and blind deconvolution. Neural Computation 7, 1129–1159 (1995)
15. Annemuller, J., Kollmeier, B.: Amplitude modulation decorrelation for convolutive blind source separation. In: ICA, pp. 215–220 (2000)
16. Bofill, P., Zibulevsky, M.: Blind separation of more sources than mixtures using sparsity of their short-time Fourier transform. In: Proc. ICA, Helsinki, Finland, pp. 87–92 (June 19-22, 2000)
17. Akaike, H.: A new look at the statistical model identification. IEEE Trans. Aut. Cont. 19(6), 716–723 (1974)
18. Barron, A., Rissanen, J., Yu, B.: The minimum description length principle in coding and modeling. IEEE Trans. Inf. Th. 44(6), 2743–2760 (1998)
19. Schwarz, G.: Estimating the dimension of a model. Ann. Statist. 6(2), 461–464 (1978)
20. Rissanen, J.: Modeling by shortest data description. Automatica 14, 465–471 (1978)

Compressed Sensing and Source Separation

Thomas Blumensath and Mike Davies*

IDCOM & Joint Research Institute for Signal and Image Processing
The University of Edinburgh, The King's Buildings, Edinburgh, EH9 3JL, UK
Tel.: +44(0)131 6505659; Fax.: +44(0)131 6506554
thomas.blumensath@ed.ac.uk, mike.davies@ed.ac.uk

Abstract. Separation of underdetermined mixtures is an important problem in signal processing that has attracted a great deal of attention over the years. Prior knowledge is required to solve such problems and one of the most common forms of structure exploited is sparsity.

Another central problem in signal processing is sampling. Recently, it has been shown that it is possible to sample well below the Nyquist limit whenever the signal has additional structure. This theory is known as compressed sensing or compressive sampling and a wealth of theoretical insight has been gained for signals that permit a sparse representation.

In this paper we point out several similarities between compressed sensing and source separation. We here mainly assume that the mixing system is known, i.e. we do not study *blind* source separation. With a particular view towards source separation, we extend some of the results in compressed sensing to more general overcomplete sparse representations and study the sensitivity of the solution to errors in the mixing system.

1 Compressed Sensing

Compressed sensing or compressive sampling is a new emerging technique in signal processing, coding and information theory. For a good place of departure see for example [1] and [2]. Assume that a signal $\mathbf{y}$ is to be measured. In general $\mathbf{y}$ is assumed to be a function defined on a continuous domain, however, for the discussion here it can be assumed to be a finite vector, i.e. $\mathbf{y} \in \mathbb{R}^{N_y}$ say. In a standard DSP textbook we learn that one has to sample a function on a continuous domain at least at its Nyquist rate. However, assume that we know that $\mathbf{y}$ has a certain structure, for example we assume that $\mathbf{y}$ can be expressed as

$$\mathbf{y} = \mathbf{\Phi s}, \tag{1}$$

where $\mathbf{\Phi} \in \mathbb{R}^{\mathbf{N}_y \times \mathbf{N}_s}$ and where we allow $\mathbf{N}_s \geq \mathbf{N}_y$, i.e. we allow $\mathbf{\Phi s}$ to be an *overcomplete* representation of $\mathbf{y}$. Crucially, we assume $\mathbf{s}$ to be sparse, i.e. we assume that only a small number of elements in $\mathbf{s}$ are non-zero or, more generally,

* This research was supported by EPSRC grant D000246/1. MED acknowledges support of his position from the Scottish Funding Council and their support of the Joint Research Institute with the Heriot-Watt University as a component part of the Edinburgh Research Partnership.

M.E. Davies et al. (Eds.): ICA 2007, LNCS 4666, pp. 341–348, 2007.

that most of the energy in $\mathbf{s}$ is concentrated in a few coefficients. It has recently been shown that, if a signal has such a sparse representation, then it is possible to take less samples (or measurements) from the signal than would be suggested by the Nyquist limit. Furthermore, one is then often still able to reconstruct the original signal using convex optimisation techniques [1] and [2].

The simplest scenario are measurements taken as follows:

$$\mathbf{x} = \mathbf{M}\mathbf{y}, \tag{2}$$

where $\mathbf{x} \in \mathbb{R}^{N_x}$ with $N_x < N_y$. Extensions to noisy measurements can be made [2], i.e. one can consider the problem of approximating $\mathbf{y}$ given a noisy measurement:

$$\tilde{\mathbf{x}} = \mathbf{x} + \mathbf{e} = \mathbf{M}\mathbf{y} + \mathbf{e}. \tag{3}$$

The ability to reconstruct the original signal relies heavily on the structure of $\mathbf{y}$ and different conditions have been derived under which one can exactly or approximately recover $\mathbf{y}$. For example, if $\mathbf{y} = \mathbf{\Phi}\mathbf{s}$ and $\mathbf{s}$ has only a small number of non-zero elements, then linear programming can exactly recover $\mathbf{y}$ if enough measurements have been taken. Similar results have been derived in the case where $\mathbf{s}$ is not exactly sparse, but where the ordered coefficients in $\mathbf{s}$ decay with a power law. In this case $\mathbf{y}$ can be recovered up to some small error. We give examples of these theorems below. More details can be found in for example [2] and the references therein.

2 Relationship to Source Separations

Sparsity has also often been exploited for source separation. In particular, the problem of underdetermined blind source separation has been solved using the fact that an orthogonal transform can often be found in which the data is sparse [3] [4] [5] [6]. More general, possibly over-complete dictionaries have been used for source separation in [7].

Let us assume a quite general source separation scenario. A set of sources, say $\mathbf{g}_1, \mathbf{g}_2, \ldots$, each represented in a column vector, are collected into a matrix

$$\mathbf{G} = [\mathbf{g}_1\ \mathbf{g}_2\ \ldots]^T \tag{4}$$

and similarly, the set of observations $\mathbf{f}_1, \mathbf{f}_2, \ldots$ are gathered in a matrix $\mathbf{F}$. The relationship between the sources $\mathbf{G}$ and the observations F is then modelled by a general linear operator $\mathbf{A}$:

$$\mathbf{F} = \mathbf{A}(\mathbf{G}). \tag{5}$$

Note that this operator does not have to be a matrix and can also represent for example convolutions, so that the above model incorporates a wide range of source separation problems.

The connection to compressive sampling becomes evident if instead of collecting the sources and observations in a matrix, we interleave them into vectors as:

$$\mathbf{x} = [\mathbf{f}_1[1]\ \mathbf{f}_2[1]\ \ldots\ \mathbf{f}_1[2]\ \mathbf{f}_2[2]\ \ldots]^T \tag{6}$$

and

$$\mathbf{y} = [\mathbf{g}_1[1]\ \mathbf{g}_2[1]\ \ldots\ \mathbf{g}_1[2]\ \mathbf{g}_2[2]\ \ldots]^T. \tag{7}$$

We further assume that the operator $\mathbf{A}$ can be expressed in matrix form $\mathbf{M}$ so that the mixing system becomes:

$$\mathbf{x} = \mathbf{M}\mathbf{y}, \tag{8}$$

which is exactly the compressive sampling measurement equation[1].

If we have more sources $\mathbf{g}$ than observations $\mathbf{f}$, where the length of each observation and each source is assumed to be equal, then we have less measurements than samples in $\mathbf{y}$. We therefore require knowledge of additional structure if we want to be able to (approximately) reconstruct $\mathbf{y}$. We can, for example, assume the existence of a sparse representation of $\mathbf{y}$ of the form $\mathbf{y} = \mathbf{\Phi}\mathbf{s}$, where $\mathbf{s}$ is sparse. Note that we do not assume that $\mathbf{\Phi}$ is an orthogonal transform and explicitly allow $\mathbf{\Phi}$ to be overcomplete, i.e. to have more columns than rows.

Let us look at a simple example of an instantaneous mixture. In this case, $\mathbf{A}$ is the $N_f \times N_g$ mixing matrix and the matrix $\mathbf{M}$ becomes matrix diagonal:

$$\mathbf{M} = \begin{bmatrix} \mathbf{A} & \mathbf{0} & \cdots & \mathbf{0} \\ \mathbf{0} & \mathbf{A} & \cdots & \mathbf{0} \\ \vdots & & \ddots & \\ \mathbf{0} & \mathbf{0} & \cdots & \mathbf{A} \end{bmatrix}, \tag{9}$$

Where $\mathbf{0}$ is a $N_f \times N_g$ matrix of zeros. Similarly, assume a convolutive model, in which the impulse responses, say $\mathbf{h}_{1,1}, \mathbf{h}_{2,1}, \mathbf{h}_{3,1}$ and $\mathbf{h}_{1,2}, \mathbf{h}_{2,2}, \mathbf{h}_{3,2}$ and so on are interleaved into the matrix $\mathbf{H}$ as follows:

$$\mathbf{H} = \begin{bmatrix} \mathbf{h}_{1,1}[n] & \mathbf{h}_{2,1}[n] & \mathbf{h}_{3,1}[n] & \mathbf{h}_{1,1}[n-1] & \cdots \\ \mathbf{h}_{1,2}[n] & \mathbf{h}_{2,2}[n] & \mathbf{h}_{3,2}[n] & \mathbf{h}_{1,2}[n-1] & \cdots \\ \vdots & & & & \end{bmatrix}. \tag{10}$$

The measuring matrix then becomes:

$$\mathbf{M} = \begin{bmatrix} \mathbf{H} & & \cdots & \mathbf{0} \\ \mathbf{0} & \mathbf{H} & \cdots & \mathbf{0} \\ \vdots & & \ddots & \\ \mathbf{0} & \mathbf{0} & \cdots & \mathbf{H} \end{bmatrix}, \tag{11}$$

where again $\mathbf{0}$ is a $N_f \times N_g$ matrix of zeros. Also, depending on boundary assumptions, the first and last rows of $\mathbf{M}$ might only contain part of the matrix $\mathbf{H}$.

3 Theoretic Results

The source separation problem is equivalent to the decoding problem faced in compressive sampling and theoretical results from the compressive sampling literature therefore also apply to the source separation problem. However, in

[1] We could have alternatively stacked the vectors $\mathbf{f}_1, \mathbf{f}_2, \ldots$ (and/or $\mathbf{g}_1, \mathbf{g}_2, \ldots$) on top of each other to produce a permutation of the above model.

compressive sampling, the measurement matrix $\mathbf{M}$ can often be 'designed' to fulfil certain conditions[2]. Furthermore, in the current compressive sampling literature, $\mathbf{\Phi}$ is normally assumed to be the identity matrix or an orthogonal transform. In source separation, the measuring system is not normally at our control. Furthermore, orthogonal transform are often not available to sufficiently 'sparsify' many signals of interest.

In this paper we address these problems and extend several results from the compressed sensing literature to more general sparse representations. We start by reviewing some of the important results on compressed sensing [8], which we here write in terms of the m-restricted isometry condition of the matrix $\mathbf{P} = \mathbf{M\Phi}$.

3.1 m-Restricted Isometry

For any matrix $\mathbf{P}$ and integer m, define the m-restricted isometry $\delta_m(\mathbf{P})$ as the smallest quantity such that:

$$(1 - \delta_m(\mathbf{P})) \leq \frac{\|\mathbf{P}_\Gamma \mathbf{y}\|_2^2}{\|\mathbf{y}\|_2^2} \leq (1 + \delta_m(\mathbf{P})), \tag{12}$$

for all $\Gamma : |\Gamma| \leq m$ and all $\mathbf{y}$. Here $|\Gamma|$ is a set of indices and $\mathbf{P}_\Gamma$ the associated submatrix of $\mathbf{P}$ with all columns removed apart from those with indices in Γ. δ_m is then a measure of how much any sub-matrices of $\mathbf{P}$ with size m can change the norm of a vector, hence the name. The quantities $(1 - \delta_m(\mathbf{P}))$ and $(1 + \delta_m(\mathbf{P}))$ can be understood as lower and upper bounds on the squared singular values of all possible sub-matrices of $\mathbf{P}$ with m or less columns.

3.2 Estimation Error Bounds

As examples of the types of theorems available in the compressive sampling literature, we here state two of the fundamental results (these can be found in [2] and references therein), which rely on $\delta_m(\mathbf{M\Phi})$ to be small[3].

Theorem 1. *(Exact Recovery) Assume that* $\mathbf{s}$ *has a maximum of* m *non-zero coefficients and that* $\mathbf{x} = \mathbf{M\Phi s}$ *and that* $\delta_{2m}(\mathbf{M\Phi}) + \delta_{3m}(\mathbf{M\Phi}) < 1$, *then the solution to the linear program:*

$$\min \|\hat{\mathbf{s}}\|, \text{ such that } \mathbf{x} = \mathbf{M\Phi}\hat{\mathbf{s}} \tag{13}$$

recovers the exact representation $\mathbf{y} = \mathbf{\Phi s}$.

---

[2] 'Design' here often means taking a random matrix drawn from certain distributions.

[3] Note that Georgiev et al. [9] have also studied a similar problem in relationship to blind source separation. However, the results in [9] are concerned with identifiability of both $\mathbf{y}$ and $\mathbf{A}$. The theorems given here assume knowledge of $\mathbf{A}$ but are stronger than those in [9] in that they also states that we can use convex optimisation methods to identify the sources. Furthermore, theorem 2 is valid for more general sources and does not require the existence of an exact m-term representation.

A similar result can be derived for noisy observations and a further generalisation was derived in [8] for signals for which the original signal is not m-sparse, but has a power law decay, i.e. the magnitude of the ordered coefficients decays as $|s_{i_k}| \leq Ck^{-\frac{1}{p}}$, where $p \leq 1$. In particular, for an i.i.d. Gaussian observation error with variance σ^2, we have:

Theorem 2. *(Dantzig selector) Assume that* $\mathbf{s}$ *can be reordered so that* $|s_{i_k}| \leq ck^{-\frac{1}{p}}$ *for* $p \leq 1$*. For some* m *assume that* $\delta_{2m}(\mathbf{M\Phi}) + 3\delta_{3m}(\mathbf{M\Phi}) < 1$*. For* $\lambda = \sqrt{2\log N_x}$ *the solution to:*

$$\min \|\hat{\mathbf{y}}\|_1 \ : \ \|\mathbf{M}^T(\mathbf{x} - \mathbf{M}\hat{\mathbf{y}})\|_\infty \leq \lambda\sigma \tag{14}$$

obeys the bound:

$$\|\mathbf{s} - \tilde{\mathbf{s}}\|_2^2 \leq C2(\log N_x)\sigma^p c^{1-\frac{p}{2}}. \tag{15}$$

This can be extend trivially to a bound on the error in the signal space:

$$\|\mathbf{y} - \tilde{\mathbf{y}}\|_2^2 = \|\mathbf{\Phi s} - \mathbf{\Phi}\tilde{\mathbf{s}}\|_2^2 \leq C2(\log N_x)\sigma^p c^{1-\frac{p}{2}}\|\mathbf{\Phi}\|_2^2. \tag{16}$$

3.3 Random Mixing Conditions

In source separation, the mixing system should be considered independently from the dictionary $\mathbf{\Phi}$ in which the signal has a sparse representation. The theorems above are based on $\delta_m(\mathbf{M\Phi})$, which is required to be small. In this and the next section we derive new results that give insight into this quantity by considering $\mathbf{M}$ and $\mathbf{\Phi}$ separately.

The first theorem is a slight modification from [10][4,5]:

Theorem 3. *Assume that* $\mathbf{M} \in \mathbb{R}^{N_x \times N_y}$ *is a random matrix with columns drawn uniformly from the unit sphere and let* $\mathbf{\Phi} \in \mathbb{R}^{N_y \times N_s}$ *have restricted isometry* $\delta_m(\mathbf{\Phi}) < 1$*, then there exists a constant* c*, such that for* $m \leq cN_x \log(N_s/m)$*:*

$$(1 - \delta_{P_m}(\mathbf{M})) \leq \frac{\|\mathbf{M\Phi}_\Gamma \mathbf{s}\|_2^2}{\|\mathbf{\Phi s}\|_2^2} \leq (1 + \delta_{P_m}(\mathbf{M})), \tag{17}$$

and

$$(1 - \delta_m(\mathbf{\Phi}))(1 - \delta_{P_m}(\mathbf{M})) \leq \frac{\|\mathbf{M\Phi}_\Gamma \mathbf{s}\|_2^2}{\|\mathbf{s}\|_2^2} \leq (1 + \delta_{P_m}(\mathbf{M}))(1 + \delta_m(\mathbf{\Phi})), \tag{18}$$

holds with probability

$$\geq 1 - 2(eN_s/m)^m(12/\delta_{P_m})^m e^{-\frac{N_x}{2}(\delta_{P_m}^2/8 - \delta_m^3/24)}. \tag{19}$$

[4] Note, that there are a range of other distributions for which this theorem would hold, see [10] for details.

[5] Since the first submission of this manuscript we became aware of the paper [11], which contains very similar results.

Proof (Outline). The proof that equation (17) holds is similar to the proof given in [10] with the only difference that Theorem 5.1 in [10] can be shown to hold for any m dimensional subspace, and where N in theorem 5.2 in [10] can be replaced by N_s. The restricted isometry in equation (18) then follows by bounding $||\mathbf{\Phi s}||_2^2$ from above and below using the restricted isometry $(1-\delta_m(\mathbf{\Phi})) \leq ||\mathbf{\Phi s}||_2^2 \leq (1+\delta_m(\mathbf{\Phi}))$.

Therefore, for any dictionary $\mathbf{\Phi}$ with $\delta_m(\mathbf{\Phi}) < 1$ and for $\mathbf{M}$ sampled uniformly from the unit sphere, $\delta_m(\mathbf{M\Phi}) \leq \delta_m(\mathbf{\Phi}) + \delta_{P_m}(M) + \delta_m(\mathbf{\Phi})\delta_{P_m}(M)$ with high probability, whenever $m \leq CN_x/\log(N_s/N_x)$.

3.4 Non-random Mixing Matrix Conditions

Unfortunately, theorem 3 assumes randomly generated mixing systems, which is rather restrictive. We therefore derive conditions that relate the measurement matrix M, the dictionary $\mathbf{\Phi}$ and $\delta_m(\mathbf{M\Phi})$.

To bound $\delta_m(\mathbf{M\Phi})$ we define the (to our knowledge novel) concept of M-coherence:

$$\mu_{\mathbf{M}}(\mathbf{\Phi}) = \max_{i,j:i\neq j} |\phi_i^T \mathbf{M}^T \mathbf{M} \phi_j|. \tag{20}$$

This quantity measures the coherence in the dictionary as 'seen through' the measuring matrix. We also need the quantities:

$$a_{min} = \min_i ||\mathbf{M}\phi_i||_2^2 \text{ and } a_{max} = \max_i ||\mathbf{M}\phi_i||_2^2, \tag{21}$$

which measure how much the measuring matrix can deform elements of the dictionary. We assume that $a_{min} \geq m\mu_{\mathbf{M}}(\mathbf{\Phi})$, then by the Gersgorin disk theorem for the eigenvalues of $\mathbf{\Phi}_\Gamma \mathbf{M}^T \mathbf{M} \mathbf{\Phi}_\Gamma$, we find that all squared singular values σ^2 of the matrix $\mathbf{M\Phi}_\Gamma$ with $|\Gamma| \leq m$ are bounded by:

$$a_{min} - m\mu_{\mathbf{M}}(\mathbf{\Phi}) \leq \sigma^2 \leq a_{max} + m\mu_{\mathbf{M}}(\mathbf{\Phi}). \tag{22}$$

We therefore have the bound:

$$a_{min} - m\mu_{\mathbf{M}}(\mathbf{\Phi}) \leq \frac{||\mathbf{M\Phi}_\Gamma \mathbf{s}||_2^2}{||\mathbf{s}||_2^2} \leq a_{max} + m\mu_{\mathbf{M}}(\mathbf{\Phi}). \tag{23}$$

Using $a = \max\{a_{\max} - 1, 1 - a_{min}\}$ we have[6] the bound on $\delta_m(\mathbf{M\Phi})$ of

$$\delta_m(\mathbf{M\Phi}) \leq a + m\mu_{\mathbf{M}}(\mathbf{\Phi}), \tag{24}$$

which is in terms of quantities that are easy to determine for a given dictionary $\mathbf{\Phi}$ and measurement matrix $\mathbf{M}$.

[6] Note that we have the bound $||\mathbf{M}||_2 \geq a_{max} \geq a_{min} \geq 0$.

3.5 Sensitivity to Errors in M

In most source separation applications the mixing system is not given a priori and has to be estimated. This leads to the question of robustness of the method to errors in the estimation of the measuring matrix $\mathbf{M}$.

Assume we have an estimated mixing system $\tilde{\mathbf{M}} = \mathbf{M} + \mathbf{N}$ and estimated sources $\tilde{\mathbf{s}}$ such that $\tilde{\mathbf{x}} = \tilde{\mathbf{M}}\mathbf{\Phi}\tilde{\mathbf{s}}$ and such that $\tilde{\mathbf{s}}$ is supported on $\tilde{m}$ elements. Also assume that $\mathbf{x} = \mathbf{M}\mathbf{\Phi}\mathbf{s}$ is the true generating system with $\mathbf{s}$ supported on m elements. Further assume that $\|\tilde{\mathbf{x}} - \mathbf{x}\|_2 \leq 2\epsilon$.

If $\delta_{m+\tilde{m}}(\tilde{\mathbf{M}}\mathbf{\Phi}) < 1$, then we have the bound:

$$\|\mathbf{y} - \tilde{\mathbf{y}}\|_2 \leq \frac{\|\tilde{\mathbf{M}}\mathbf{\Phi}\mathbf{s} - \tilde{\mathbf{M}}\mathbf{\Phi}\tilde{\mathbf{s}}\|_2 \|\mathbf{\Phi}\|_2}{\sqrt{1 - \delta_{m+\tilde{m}}(\tilde{\mathbf{M}}\mathbf{\Phi})}}. \tag{25}$$

Replacing $\tilde{\mathbf{M}}\mathbf{\Phi}\mathbf{s}$ with $\mathbf{M}\mathbf{\Phi}\mathbf{s} + \mathbf{N}\mathbf{\Phi}\mathbf{s}$ and using $\|\mathbf{M}\mathbf{\Phi}\mathbf{s} - \tilde{\mathbf{M}}\mathbf{\Phi}\tilde{\mathbf{s}}\| \leq 2\epsilon$ together with the triangle inequality, we get the bound:

$$\|\mathbf{y} - \tilde{\mathbf{y}}\|_2 \leq \frac{2\epsilon + \|\mathbf{N}\mathbf{y}\|_2 \|\mathbf{\Phi}\|_2}{\sqrt{1 - \delta_{m+\tilde{m}}(\tilde{\mathbf{M}}\mathbf{\Phi})}} \leq \frac{2\epsilon + \|\mathbf{N}\|_2 \|\mathbf{\Phi}\|_2 \|\mathbf{y}\|_2}{\sqrt{1 - \delta_{m+\tilde{m}}(\tilde{\mathbf{M}}\mathbf{\Phi})}}. \tag{26}$$

4 Discussion and Conclusion

Underdetermined mixtures are a form of compressive sampling. Source separation is therefore equivalent to the decoding problem faced in compressive sampling. This equivalence opens up many new lines of enquiry, both in compressive sampling and in source separation.

On the one hand, as done in this paper, results form compressive sampling shed new insight into the source separation problem. For example, theorems 1 and 2 state that for signals with a sparse underlying representation, whether exact, or with decaying coefficients, convex optimisation techniques can be used to recover or approximate the original signal from a lower dimensional observation. Furthermore, results from compressive sampling give bounds on the estimation error for sources that have a sparse representation with decaying coefficients. The error is a function of this coefficient decay and properties of the matrix mapping the sparse representation into the mixed domain. In the source separation literature, linear programming techniques have been a common approach [3] [4] [7] and the new theory gives additional justification for the application of these techniques and, what is more, provides estimation bounds for certain problems.

The main novel contribution of this paper was an extension of recent results from compressive sampling to allow for more general, possibly over-complete dictionaries for the sparse representation. In source separation, the mixing matrix is in general unrelated to the dictionary and the main contribution of this paper was to derive conditions on the dictionary, the mixing system and their interaction that allow the application of standard compressive sampling results to the more general source separation problem. In particular we have disentangled the dictionary and the measurement matrix and could shown that for randomly generated mixing systems, the required conditions hold with high probability. For more general mixing systems, we have presented bounds on this condition, which are functions of simple to establish properties of the mixing system, the overcomplete dictionary and their interaction. If the mixing system has to be estimated as in many source separation settings, errors in this estimate will influence the estimates of the sources. The theory in subsection 3.5 gives bounds on this error.

Not only does source separation benefit from progress made in compressive sampling, compressive sampling has also much to learn from the extensive work done on source separation. For example, in source separation, the mixing system is not known in general and has to be estimated together with the sources. Many different estimation techniques have therefore been developed in the source separation community able to estimate the mixing system. This suggests an extension of compressive sampling to *blind compressive sampling* (BCS). Different scenarios seem possible depending on the application and, in our notation, either $\mathbf{\Phi}$ or $\mathbf{M}$ (or both) might be unknown or known only approximately. Preliminary work in this direction has shown encouraging first results and more formal studies are currently undertaken.

References

1. Donoho, D.: Compressed sensing. IEEE Trans. on Information Theory 52(4), 1289–1306 (2006)
2. Candès, E.: Compressive sampling. In: Proceedings of the International Congress of Mathematics, Madrid, Spain (2006)
3. Lewicki, M.S., Sejnowski, T.J.: Learning overcomplete representations. Neural Computation 12, 337–365 (2000)
4. Zibulevsky, M., Bofill, P.: Underdetermined blind source separation using sparse representations. Signal Processing 81(11), 2353–2362 (2001)
5. Davies, M., Mitianoudis, N.: A simple mixture model for sparse overcomplete ICA. IEE Proc.-Vision, Image and Signal Processing 151(1), 35–43 (2004)
6. Fevotte, C., Godsill, S.: A bayesian approach for blind separation of sparse sources. IEEE Transactions on Speech and Audio Processing PP(99), 1–15 (2005)
7. Zibulevsky, M., Pearlmutter, B.A.: Blind source separation by sparse decomposition in a signal dictionary. Neural Computation 13(4), 863–882 (2001)
8. Candes, E., Tao, T.: The dantzig selector: statistical estimation when p is larger than n. manuscript (2005)
9. Georgiev, P., Theis, F., Cichocki, A.: Sparse component analysis and blind source separation of underdetermined mixtures. IEEE Transactions on Neural Networks 16(4), 992–996 (2005)
10. Baraniuk, R., Davenport, M., De Vore, R., Wakin, M.: The Johnson-Lindenstrauss lemma meets compressed sensing (2006)
11. Rauhut, H., Schnass, K., Vandergheynst, P.: Compressed sensing and redundant dictionaries. IEEE Transactions on Information Theory (submitted, 2007)

Morphological Diversity and Sparsity in Blind Source Separation

J. Bobin[1], Y. Moudden[1], J. Fadili[2], and J.-L. Starck[1]

[1] CEA-DAPNIA/SEDI, Service d'Astrophysique,
CEA/Saclay, 91191 Gif sur Yvette, France
jerome.bobin@cea.fr, ymoudden@cea.fr, jstarck@cea.fr
[2] GREYC CNRS UMR 6072, Image Processing Group,
ENSICAEN 14050, Caen Cedex, France
jalal.fadili@greyc.ensicaen.fr

Abstract. This paper describes a new blind source separation method for instantaneous linear mixtures. This new method coined GMCA (Generalized Morphological Component Analysis) relies on morphological diversity. It provides new insights on the use of sparsity for blind source separation in a noisy environment. GMCA takes advantage of the sparse representation of structured data in large overcomplete signal dictionaries to separate sources based on their morphology. In this paper, we define morphological diversity and focus on its ability to be a helpful source of diversity between the signals we wish to separate. We introduce the blind GMCA algorithm and we show that it leads to good results in the overdetermined blind source separation problem from noisy mixtures. Both theoretical and algorithmic comparisons between morphological diversity and independence-based separation techniques are given. The effectiveness of the proposed scheme is confirmed in several numerical experiments.

Introduction

Hereafter, we address the classical blind source separation problem. The $m \times t$ data matrix $\mathbf{X}$ is the concatenation of m mixtures $\{x_i\}_{i=1,\cdots,m}$ each of which being the instantaneous linear combination of n sources $\{s_i\}_{i=1,\cdots,n}$ stored in the $n \times t$ matrix $\mathbf{S}$:

$$\mathbf{X} = \mathbf{AS} + \mathbf{N} \tag{1}$$

where $\mathbf{A}$ is the mixing matrix and $\mathbf{N}$ models noise or model imperfections. In this setting, the aim of blind source separation (BSS) techniques is to estimate both the sources $\mathbf{S}$ and the mixing matrix $\mathbf{A}$. BSS is clearly an ill-posed inverse problem which requires additional prior information in order to be solved. Previous work addressing BSS issues clearly emphasized on the need for diversity between the sources to be separated. From a statistical point of view, ICA-like source separation methods use statistical independence (more precisely mutual information) as a kind of "diversity measure" to distinguish between the sources. In [1], the authors proved that maximizing any measure of independence

M.E. Davies et al. (Eds.): ICA 2007, LNCS 4666, pp. 349–356, 2007.

is equivalent to minimizing mutual information. ICA algorithms are then devised according to particular approximations of mutual information.

Recently, sparsity has raised interest in a wide range of applications. Briefly, a signal is said to be sparse in representation $\mathbf{\Phi}$ if most of the entries of α such that $x = \alpha\mathbf{\Phi}$ are almost zero and only a few have significant amplitudes. Sparsity-based BSS methods have recently been devised. In [2], a BSS algorithm is described in which it is taken advantage of sparsity to enhance the diversity between independent sources. Several studies (see [3] and references therein) have explored the extreme sparse case as they considered sources with strictly disjoint (and thus orthogonal) supports. In Section 1, we define a particular sparsity-based diversity measure coined morphological diversity. We propose a new effective BSS algorithm coined GMCA which separates the mixed sources based on their morphological diversity. In Section 2, numerical experiments are given showing how GMCA performs well to separate sources from noisy mixtures.

1 The GMCA Framework

Notations and Definitions

Let x be a $1 \times t$ signal and $\mathbf{\Phi}$ a signal dictionary. For the sake of simplicity, we will first assume that $\mathbf{\Phi}$ is orthonormal. In this case, x has a unique representation α in $\mathbf{\Phi}$ such that $x = \alpha\mathbf{\Phi}$ readily obtained as $\alpha = x\mathbf{\Phi}^T$. The support $\mathcal{S}_{0,\mathbf{\Phi}}(x)$ of x in $\mathbf{\Phi}$ is defined as $\mathcal{S}_{0,\mathbf{\Phi}}(x) = \left\{t \middle| |\alpha[t]| > 0\right\}$ where $\alpha[t]$ is the t-th entry of α. Let us also define the δ-support of x in $\mathbf{\Phi}$ as : $\mathcal{S}_{\delta,\mathbf{\Phi}}(x) = \left\{t \middle| |\alpha[t]| > \delta\|x\|_\infty\right\}$. We then say that two sources s_1 and s_2 are δ-disjoint in $\mathbf{\Phi}$ if $\mathcal{S}_{\delta,\mathbf{\Phi}}(s_1) \cap \mathcal{S}_{\delta,\mathbf{\Phi}}(s_2) = \emptyset$. Sources with strictly disjoint supports in $\mathbf{\Phi}$ are obviously δ-disjoint with $\delta = 0$.

1.1 Generalized Morphological Component Analysis

Sparse coding: Let us first assume that the mixing matrix $\mathbf{A}$ is known. In the GMCA framework, the data are modelled as a linear combination of several sources as in Equation 1. Furthermore, the sources $\{s_i\}_{i=1,\cdots,n}$ are assumed to be the linear combination of so-called morphological components (see [4]) : $s_i = \sum_{k=1}^{D} \varphi_{ik}$. By definition, those morphological components are assumed to be sparse in different orthonormal bases $\{\mathbf{\Phi}_k\}_{k=1,\cdots,D}$. Based on these assumptions, the GMCA algorithm endeavors to estimate the sources *via* the estimation of those morphological components :

$$\{\varphi_{ik}\} = \text{Arg}\min_{\{\varphi_{ik}\}} \|\mathbf{X} - \mathbf{AS}\|_2^2 + 2\lambda \sum_{i=1}^{n} \sum_{k=1}^{D} \|\varphi_{ik}\mathbf{\Phi}_k^T\|_{\ell_1} \tag{2}$$

In [5], we proposed solving this optimization problem by estimating iteratively and alternately each multichannel morphological component $\{\varphi_{ik}\}$ *via* a "block-coordinate"-like algorithm (see [6]). The product $\mathbf{AS}$ is then split into $n \times D$

terms. Introducing the data residual $\mathbf{X}_{ik} = \mathbf{X} - \sum_{\{j,l\} \neq \{i,k\}} a^i \varphi_{jl}$, where a^i is the i-th column of $\mathbf{A}$, the morphological components are estimated one at a time according to : $\varphi_{ik} = \operatorname{Arg\,min}_{\varphi_{ik}} \|\mathbf{X}_{ik} - a^i \varphi_{ik}\|_2^2 + 2\lambda \|\varphi_{ik} \mathbf{\Phi}_k^T\|_{\ell_1}$.

This equation has an exact solution known as soft-thresholding (see [7]). This sparse decomposition is closely linked to a *sparse coding* stage as already exposed in [8].

Dictionary learning: In the previous paragraph, we assumed that the mixing matrix was known and we showed that estimating the morphological components (and thus the sources) boils down to a *sparse coding* step. We consider now that the morphological components are fixed and we want to learn the mixing mixing matrix $\mathbf{A}$. This *dictionary* learning issue has already been addressed by extensive work for a wide range of applications. Refer to [8] and references therein for more on that question. Following the same estimation scheme we introduced previously, we propose to estimate each column of $\mathbf{A}$ assuming the morphological components are fixed as follows: $a^i = \operatorname{Arg\,min}_{a^i} \|\mathbf{X} - \sum_{j \neq i} a^j s_j - a^i s_i\|_2^2$. This update clearly leads to a least-squares estimate of the columns of $\mathbf{A}$: $a^i = \left(\mathbf{X} - \sum_{j \neq i} a^j s_j\right) s_i^T / \|s_i\|_2^2$.

1.2 The GMCA Algorithm for Blind Source Separation

Owing to the "block-coordinate"-like structure of our optimization scheme, for a fixed threshold λ, the *blind* GMCA algorithm estimates alternately the different parameters in the model *i.e.* the columns of $\mathbf{A}$ and the morphological components. The *blind* GMCA algorithm is as follows.

1. Set the number of iterations $I_{\max}$ and threshold $\lambda^{(0)}$
2. While $\lambda^{(h)}$ is higher than a given lower bound $\lambda_{\min}$ (e.g. can depend on the noise variance),

For $i = 1, \cdots, n$

For $k = 1, \cdots, D$

• Compute the residual term $r_{ik}^{(h)}$ assuming the current estimates of $\varphi_{\{pq\} \neq \{ik\}}$, $\tilde{\varphi}_{\{pq\} \neq \{ik\}}^{(h-1)}$ are fixed:

$$r_{ik}^{(h)} = \tilde{a}^{i^{(h-1)T}} \left(\mathbf{X} - \sum_{\{p,q\} \neq \{i,k\}} \tilde{a}^{p^{(h-1)}} \tilde{\varphi}_{\{pq\}}^{(h-1)}\right)$$

• Estimate the current coefficients of $\tilde{\varphi}_{ik}^{(h)}$ by Thresholding with threshold $\lambda^{(h)}$:

$$\tilde{\alpha}_{ik}^{(h)} = \Delta_{\lambda^{(h)}}\left(r_{ik}^{(h)} \Phi_k^T\right)$$

• Get the new estimate of φ_{ik} by reconstructing from the selected coefficients $\tilde{\alpha}_{ik}^{(h)}$:

$$\tilde{\varphi}_{ik}^{(h)} = \tilde{\alpha}_{ik}^{(h)} \Phi_k$$

Update a^i assuming $a^{p \neq k^{(h)}}$ and the morphological components $\tilde{\varphi}_{pq}^{(h)}$ are fixed :

$$\tilde{a}^{i^{(h)}} = \frac{1}{\|\tilde{s}_i^{(h)}\|_2^2}\left(\mathbf{X} - \sum_{p \neq i}^{n} \tilde{a}^{p^{(h-1)}} \tilde{s}_p^{(h)}\right) \tilde{s}_i^{(h)^T} \quad \text{where } \tilde{s}_i^{(h)} = \sum_{k=1}^{D} \tilde{\varphi}_{ik}^{(h)}$$

– Decrease the thresholds $\lambda^{(h)}$ following a given strategy

Note that the value of λ fixes a certain *sparsity level* in the *sparse coding* stage. When λ is "high", the *sparse coding* step will select the most "significant" features in the data which are very likely to belong to the true morphological components. As already introduced in [5] and [7], the threshold λ decreases towards $\lambda_{(min)}$ progressively incorporating new features. The purpose of such a thresholding scheme is twofold: i) it provides numerical stability to the algorithm, ii) it gives robusteness to noise as the morphological components $\{\varphi_{ik}\}$ are first estimated from their most significant coefficients in $\{\mathbf{\Phi}_k\}$. The *sparse coding* step is quite similar to a thresholding-based "denoising". Handling noisy mixtures then boils down to fixing the final threshold λ_{min}. Typically, in the white Gaussian noise case, $\lambda_{\text{min}} = 3\sigma$ where σ is the noise standard deviation.

1.3 A Fast GMCA Algorithm

When $\mathbf{\Phi}$ *is orthogonal:* Note that the above GMCA algorithm is a multichannel extension of MCA (Morphological Component Analysis - see [9,4]) which has been devised in the single channel case. In [4], we showed that MCA is likely to solve the ℓ_0 decomposition (and thus the ℓ_1 decomposition when the two problems are equivalent - see [10] and references therein) of sparse signals in the overcomplete dictionary $\mathbf{\Phi} = [\mathbf{\Phi}_1, \cdots, \mathbf{\Phi}_D]$. Nevertheless, in the multichannel case, this *sparse coding* step requires the use of D transforms for each of the n sources leading to a prohibitive computational cost. Interestingly, if we restrict ourselves to the case where $\mathbf{\Phi}$ is orthonormal, the problem in Equation 2 becomes simpler:

$$\mathbf{\Theta_S} = \operatorname{Arg}\min_{\mathbf{\Theta_S}} \|\mathbf{\Theta_X} - \mathbf{A}\mathbf{\Theta_S}\|_2^2 + 2\lambda \sum_{i=1}^{n} \|\mathbf{\Theta_S}\|_{\ell_1} \tag{3}$$

where $\mathbf{\Theta_S} = \mathbf{S}\mathbf{\Phi}^T$. The sources and the mixing matrix can be estimated in the sparse domain $\mathbf{\Phi}$ which drastically reduces the computational burden. Assuming that $\mathbf{A}$ is nearly orthonormal[1] leads to a very simple *sparse coding* step:

$$\mathbf{\Theta_S} = \Delta_\lambda\left(\mathbf{A}^\dagger \mathbf{\Theta_X}\right) \tag{4}$$

where $\Delta_\lambda(.)$ is the soft-thresholding operator with threshold λ and $\mathbf{A}^\dagger$ is the pseudo-inverse of $\mathbf{A}$. In that setting, estimating $\mathbf{A}$ leads straightforwardly to a simple least-squares estimate :

$$\mathbf{A} = \mathbf{\Theta_X}{\mathbf{\Theta_S}}^T \left(\mathbf{\Theta_S}{\mathbf{\Theta_S}}^T\right)^{-1} \tag{5}$$

Interestingly, we show in [5] that alternating the updates in Equation 4 and Equation 5 provides a fixed-point algorithm the convergence condition of which is the following: $\mathbf{\Theta_S}\Delta_\lambda\left(\mathbf{\Theta_S}\right)^T = \left(\Delta_\lambda\left(\mathbf{\Theta_S}\right)\Delta_\lambda\left(\mathbf{\Theta_S}\right)^T\right)$

In [5], we give heuristics supporting the good convergence of our algorithm. In the same paper, we also show that the same *fast blind* GMCA algorithm can be used with non-orthonormal $\mathbf{\Phi}$.

[1] In practice, even if this assumption is very stringent and seldom true, the algorithm performs well.

1.4 Morphological Diversity

At the beginning of this section we introduced a particular sparsity-based diversity measure to distinguish the sources. A classical sparsity-based diversity measure (see [3]) leads to separate sources with strictly disjoint supports in a sparsifying representation $\mathbf{\Phi}$. Nevertheless, most *natural* signals seldom have strictly disjoint supports in most practical dictionaries $\mathbf{\Phi}$ (*e.g.* discrete cosine, wavelets, bandlets ... etc.). In [7], we slightly relaxed this assumption by considering sources with disjoint supports in an overcomplete signal dictionary made of a union of orthonormal bases $\mathbf{\Phi} = [\mathbf{\Phi_1}^T, \mathbf{\Phi_2}^T]^T$. In this paper, we introduce a new sparsity-based diversity measure which relaxes the strict disjoint support assumption.

The genesis - a deterministic diversity measure: In the next section, we switch from the deterministic point of view we adopted in the above, and examine the concept of morphological diversity from the statistical side. In the former viewpoint, separable sources are such that there exists a sparse representation $\mathbf{\Phi}$ in which these signals have δ-disjoint supports. Heuristically, given sources *e.g.* images which are "visually" and in that sense morphologically different, there exists a dictionary $\mathbf{\Phi}$ and a value of δ for which these sources are sparse and have δ-disjoint supports. In that setting, the way $\mathbf{\Phi}$ is chosen specifies which signals are distinguishable. In practice, in image processing, taking $\mathbf{\Phi}$ to be the union of the curvelet frame [11] and the local discrete cosine representation leads to good separation results for a wide range of images.

From a probabilistic viewpoint: The minimization problem in Equation 2 can also be interpreted as Maximum A Posteriori estimator of the morphological components assuming: 1) the coefficients of the morphological components are generated independently from the same Laplacian law with zero mean and precision λ, 2) the entries of the mixing matrix are uniformly distributed, 3) the additive noise follows a Gaussian distribution with zero mean and identity covariance matrix. Then what is the meaning of morphological diversity in a statistical framework? The point is that sources generated independently from the same *iid* sparse stochastic process are very likely to have δ-disjoint supports for some value of δ. For instance, let us assume that the sources s_1 and s_2 are independently generated from the same Laplacian probability density in the sparse $\mathbf{\Phi}$-domain. Indeed, each entry of the coefficient vector $\alpha_{i=1,2}$ is drawn according to: $P_\alpha(\alpha_i[k]) = \frac{\mu}{2}\exp\left(-\mu|\alpha_i[k]|\right)$. We would like to assess the probability for such sources to have δ-disjoint supports. Define the proposition $H_\tau =$ "$|\alpha_1[t^\star]| = \|\alpha_1\|_\infty > \tau, |\alpha_2[t^\star]| = \|\alpha_2\|_\infty > \tau$ and $\forall t \neq t^\star, \quad |\alpha_{i=1,2}[t]| \leq \tau$". H_τ states that s_1 and s_2 have strictly τ-joint supports; otherwise, if H_τ is false then s_1 and s_2 have at least τ-disjoint supports. We define $P_{|\alpha|>\tau} = P_\alpha(|\alpha| > \tau)$ and $P_{|\alpha|\leq\tau} = P_\alpha(|\alpha| \leq \tau)$. As the entries of each vector $\alpha_{i=1,2}$ are independently generated from the same probability density function P_α, then $P(H_\tau)$ si such that: $P(H_\tau) = T\exp\left(-2\mu\tau\right)\left(1 - \exp\left(-\mu\tau\right)\right)^{2(T-1)}$. As, in practice, $T = dN \gg 1$, sources generated independently from the same sparse probability density function are τ-disjoint. In other words, from a statistical viewpoint, sparse independent sources are morphologically diverse with very high probability. In that sense a separation technique based on morphological diversity is closely related to ICA in a statistical framework.

ICA and GMCA from an algorithmic viewpoint: The *fast blind* GMCA method introduced in section 1.3 can be expressed as a fixed-point algorithm the convergence condition of which asks that the matrix $\mathbf{\Theta_S}\Delta_\lambda\left(\mathbf{\Theta_S}\right)^T$ be symmetric, for all values of λ. Interestingly, as summarized in [12], this condition can be related to the convergence condition of some ICA algorithms which require the symmetry of matrix $\mathbb{E}\left\{f(\mathbf{BX})\mathbf{BX}\right\}$ where $\mathbf{B}$ is a demixing matrix and $f(.)$ is the so-called *score* function. The thresholding operator $\Delta_\lambda\left(.\right)$ in *blind* GMCA is similar in its role to the *score* function of ICA algorithms. A specific and important feature of the thresholding $\Delta_\lambda\left(.\right)$ is that it evolves as the threshold λ decreases from one iteration to the next of the *blind* GMCA algorithm. In [5], we give heuristics showing that this "evolving" *score* function is likely to avoid local *false* mimima of the objective thus providing some numerical stability to the algorithm. A clear difference lies in the estimation of $\mathbf{A}$ instead of a demixing matrix $\mathbf{B}$; this distinction is important as GMCA is also designed to handle data in a noisy environment.

2 Results

The last paragraph emphasized on sparsity as the key for very efficient source separation methods. In this section, we will compare several BSS techniques with GMCA in an image separation context. We choose 3 different reference BSS methods: i) JADE: the well-known ICA (Independent Component Analysis) based on fourth-order statistics (see [13]), ii) Relative Newton Algorithm: the separation technique we already mentioned. This seminal work (see [14]) paved the way for sparsity in Blind Source Separation. In the next experiments, we used the Relative Newton Algorithm (RNA) on the data transformed by a basic orthogonal bidimensional wavelet transform (2D-DWT), iii) EFICA: this separation method improves the FastICA algorithm for sources following generalized Gaussian distributions (which can be well-suited for some sparse signals). EFICA was also applied after a 2D-DWT of the data where the assumptions on the source distributions are appropriate. Figure 1 shows the original sources (top pictures) and the 2 mixtures (bottom pictures). The original sources s_1 and s_2 have unit variance. The matrix $\mathbf{A}$ that mixes the sources is such that $x_1 = 0.25 s_1 + 0.5 s_2 + n_1$ and $x_2 = -0.75 s_1 + 0.5 s_2 + n_2$ where n_1 and n_2 are Gaussian noise vectors (with decorrelated samples) such that the SNR equals 10dB. The noise covariance matrix $\Gamma_\mathbf{N}$ is diagonal. Figure 2 depicts the behavior of the mixing matrix criterion $\Omega_\mathbf{A} = \|\mathbf{I}_n - \mathbf{P}\tilde{\mathbf{A}}^\dagger\mathbf{A}\|_1$ ($\tilde{\mathbf{A}}$ is the estimate of $\mathbf{A}$) as the signal-to-noise ratio (SNR in dB) increases. When the mixing matrix is perfectly estimated, $\Omega_\mathbf{A} = 0$, otherwise $\Omega_\mathbf{A} > 0$. First, JADE does not perform well; it points out the importance of choosing an appropriate diversity measure to separate the sources. Thus, fourth-order statistics are not well suited to the source images in these experiments. Secondly, RNA and EFICA behave

Fig. 1. Left pictures: the 256 × 256 source images. **Right pictures:** two different mixtures. Gaussian noise is added such that the SNR is equal to 10dB.

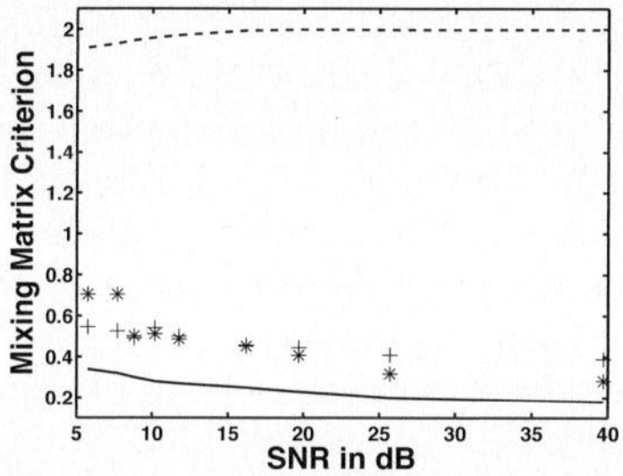

Fig. 2. Evolution of the mixing matrix criterion Δ_A as the noise variance varies: *solid* line: GMCA, *dashed* line: JADE, *(⋆):* EFICA, *(+):* RNA. **Abscissa:** SNR in dB. **Ordinate:** mixing matrix criterion value.

rather similarly. Finally, GMCA provides the best results giving mixing matrix criterion values that are approximately 2 times lower than RNA and EFICA. These results clearly show that i) sparsity enhances the distinguishability of the sources, ii) morphological diversity is a well-performing diversity measure and GMCA is well suited to account for that measure.

3 Conclusion

In this paper we introduced a new *diversity measure* to distinguish between sources: the morphological diversity. It states that morphologically diverse signals should be separated in so-called sparse representations. The recent advances in harmonic analysis and overcomplete representation theory make morphological diversity a practical way to disentangle source processes from mixtures. We proposed a new algorithm coined *blind* GMCA (Generalized Morphological Component Analysis) to address the blind source separation problem based on morphological diversity. Numerical experiments show that GMCA performs notably well. Furthermore, GMCA is an effective algorithm designed to handle noisy mixtures. Further work will focus on extending the algorithm to the underdetermined blind source separation issue.

References

1. Lee, T.W., Girolami, M., Bell, A.J., Sejnowski, T.J.: A unifying information-theoretic framework for independent component analysis (1998)
2. Zibulevsky, M., Pearlmutter, B.B.: Blind source separation by sparse decomposition. Neural Computations 13/4 (2001)
3. Li, Y., Amari, S., Cichocki, A., Ho, D.W.C., Xie, S.: Underdetermined blind source separation based on sparse representation. IEEE Transactions on signal processing 54, 423–437 (2006)
4. Bobin, J., Starck, J-L., Fadili, J., Moudden, Y., Donoho, D.L.: Morphological component analysis: new results. IEEE Transactions on Image Processing - revised (2006), available at `http://perso.orange.fr/jbobin/pubs2.html`
5. Bobin, J., Starck, J.-L., Fadili, J., Moudden, Y.: Sparsity and morphological diversity in blind source separation. IEEE Transactions on Image Processing - submitted (2007), available at `http://perso.orange.fr/jbobin/pubs2.html`
6. Tseng, P.: Convergence of a block coordinate descent method for nondifferentiable minimizations. J. of Optim. Theory and Appl. 109(3), 457–494 (2001)
7. Bobin, J., Moudden, Y., Starck, J.L., Elad, M.: Morphological diversity and source separation. IEEE Signal Processing Letters 13(7), 409–412 (2006)
8. Aharon, M., Elad, M., Bruckstein, A.: k-svd: An algorithm for designing overcomplete dictionaries for sparse representation. IEEE Transactions on Signal Processing 54(11), 4311–4322 (2006)
9. Elad, M., Starck, J.L., Donoho, D., Querre, P.: Simultaneous cartoon and texture image inpainting using morphological component analysis (MCA). ACHA 19(3), 340–358 (2005)
10. Tropp, T.: Greed is good: algorithmic results for sparse approximation. IEEE Transactions on Information Theory 50(10), 2231–2242 (2004)
11. Candes, E., Demanet, L., Donoho, D., Ying, L.: Fast discrete curvelet transforms. SIAM Multiscale Model. Simul. 5/3, 861–899 (2006)
12. Cichocki, A.: New tools for extraction of source signals and denoising. In: Proc. SPIE 2005, Bellingham. vol. 5818, pp. 11–24 (2005)
13. Cardoso, J.F.: Blind signal separation: statistical principles. Proceedings of the IEEE. Special issue on blind identification and estimation 9(10), 2009–2025 (1998)
14. Zibulevski, M.: Blind source separation with relative newton method. In: Procceedings ICA2003, pp. 897–902 (2003)

Identifiability Conditions and Subspace Clustering in Sparse BSS

Pando Georgiev[1], Fabian Theis[2], and Anca Ralescu[1]

[1] Computer Science Department
University of Cincinnati
Cincinnati, OH 45221, USA
{pgeorgie,aralescu}@ececs.uc.edu
[2] Institute of Biophysics, University of Regensburg
D-93040 Regensburg, Germany
fabian@theis.name

Abstract. We give general identifiability conditions on the source matrix in Blind Signal Separation problem. They refine some previously known ones. We develop a subspace clustering algorithm, which is a generalization of the k-plane clustering algorithm, and is suitable for separation of sparse mixtures with bigger sparsity (i.e. when the number of the sensors is bigger at least by 2 than the number of non-zero elements in most of the columns of the source matrix). We demonstrate our algorithm by examples in the square and underdetermined cases. The latter confirms the new identifiability conditions which require less hyperplanes in the data for full recovery of the sources and the mixing matrix.

1 Introduction

A goal of the Blind Signal Separation (BSS) is the recovering of underlying source signals of some given set of observations obtained by an unknown linear mixture of the sources. In order to decompose the data set, different assumptions on the sources have to be made. The most common assumption nowadays is statistical independence of the sources, which leads to the field of *Independent Component Analysis* (ICA), see for instance [2], [6] and references therein. ICA is very successful in the linear *complete* case, when as many signals as underlying sources are observed, and the mixing matrix is non-singular. In [3] it is shown that the mixing matrix and the sources are identifiable except for permutation and scaling. In the *overcomplete* or *underdetermined* case, less observations than sources are given. It can be seen that still the mixing matrix can be recovered [4], but source identifiability does not hold. In order to approximatively detect the sources, additional requirements have to be made, usually sparsity of the sources. We refer to [7,9,10,11] and reference therein for some recent papers on sparsity and underdetermined ICA ($m < n$).

2 Blind Source Separation Using Sparseness

Definition 1. *A vector* $\mathbf{v} \in \mathbf{R}^m$ *is said to be* k*-sparse if* $\mathbf{v}$ *has at least* k *zero entries. A matrix* $\mathbf{S} \in \mathbf{R}^{m \times n}$ *is said to be* k*-sparse if each column of it is* k*-sparse.*

M.E. Davies et al. (Eds.): ICA 2007, LNCS 4666, pp. 357–364, 2007.

The goal of *Blind Signal Separation* of level k (k-BSS) is to decompose a given m-dimensional random vector $\mathbf{X}$ into

$$\mathbf{X} = \mathbf{AS} \tag{1}$$

with a real $m \times n$-matrix $\mathbf{A}$ and an $n \times N$-dimensional k-sparse matrix $\mathbf{S}$. $\mathbf{S}$ is called the *source matrix*, $\mathbf{X}$ the *mixtures* and $\mathbf{A}$ the *mixing matrix*. We speak of *complete*, *overcomplete* or *undercomplete* k-BSS if $m = n$, $m < n$ or $m > n$ respectively.

More generally, when a solution to the above problem doesn't exist, we formulate *least square BSS* problem as follows:

find a best approximation of $\mathbf{X}$ by $\mathbf{AS}$, such that $\mathbf{S}$ is k-sparse, i.e.

$$\text{minimize}|\mathbf{X} - \mathbf{AS}|| \tag{2}$$

$$\text{under constraints} \quad \mathbf{A} \in \mathbb{R}^{m\times n}, \mathbf{S} \in \mathbb{R}^{n\times N} \text{ and } \mathbf{S} \text{ is } k\text{-sparse} \tag{3}$$

In the following without loss of generality we will assume $m \leqslant n$: the undercomplete case can be reduced to the complete case by projection of $\mathbf{X}$.

3 Identifiability Conditions

The following identifiability conditions are extension and refinement of those in [5]. The extension is for the case of bigger sparsity i.e. when the source matrix is at least $(n - m + 2)$-sparse. The refinement can be seen, for instance, when $n \geqslant m + 2$ and the source matrix is $(n - m + 1)$-sparse. The provided examples illustrate both cases. Assume that:

(H1) the indeces $\{1, ..., N\}$ are divided in two groups $\mathcal{N}_1$ and $\mathcal{N}_2$ such that

(a) for any index $i \in \{1, ..., n\}$ there exist $k_i \geqslant m$ different sets of indeces $I_{i,j} \subset \{1, ..., n\}, j = 1, ..., k_i$ each of which contains i, such that $|I_{i,j}| \leqslant m - 1$,

(b) for any set $I_{i,j}$ from (a) there exist $m_{i,j} > |I_{i,j}|$ vectors $\mathbf{s}_1, ..., \mathbf{s}_{m_{i,j}}$ from the set $\mathbf{S}_1 := \{\mathbf{S}(:, j), j \in \mathcal{N}_1\}$, such that each of them has zero elements in places with indeces not in $I_{i,j}$, the rank of the set $\{\mathbf{s}_1, ..., \mathbf{s}_{m_{i,j}}\}$ is full (i.e. equal to $|I_{i,j}|$), and

(c) $\bigcap_{j=1}^{k_i} I_{i,j} = \{i\}$.

(H2) Every m vectors from the group $\{\mathbf{S}(:, j), j \in \mathcal{N}_2\}$ are linearly independent.

Theorem 1 (Matrix identifiability). *Let the assumptions (H1) and (H2) be satisfied. Then, in the set of all matrices with dimension $m \times n$ there is a subset of a full measure $\mathcal{A}$ such that, if $\mathbf{X} = \mathbf{AS}$ and $\mathbf{A} \in \mathcal{A}$, then $\mathbf{A}$ is uniquely determined by $\mathbf{X}$ up to permutation and scaling of the columns.*

Sketch of the proof. The condition (H1) (c) implies that for any column $\mathbf{a}_i$ of the mixing matrix there exists less that $|I_{i,j}|$ detectable subspaces in the data which intersection is this column. One more subspace is reserved for "confirmation" that their intersection is indeed a column of the mixing matrix. Their intersection is stable under small perturbation of the columns, so "most" of the matrices will exclude false intersection (not corresponding to a column). "Detectable subspace" here means that it contains at least $|I_{i,j}| + 1$ data vectors (columns of $\mathbf{X}$) different from zero and any $|I_{i,j}|$ from them are linearly independent (follows from (H1) (b)), so such a subspace can be detected by an algorithm. Condition (H2) ensures that there is no false intersection of subspaces (not corresponding to a column of $\mathbf{A}$). ■

4 Affine Hyperplane Skeletons of a Finite Set of Points

The solution $\{(\mathbf{n}_i^0, b_i^0)\}_{i=1}^k$ of the minimization problem:

$$\text{minimize } f\left(\{(\mathbf{n}_i^T, b_i)\}_{i=1}^k\right) = \sum_{j=1}^N \min_{1\leqslant i\leqslant k} |\mathbf{n}_i^T \mathbf{x}_j - b_i|^l \tag{4}$$

$$\text{subject to } \|\mathbf{n}_i\| = 1, b_i \in \mathbb{R}, i = 1, ..., k, \tag{5}$$

defines *affine hyperplane* $k^{(l)}$*-skeleton* of $\mathbf{X} \in \mathbb{R}^{m\times N}$, introduced for $l = 1$ in [8] and for $l = 2$, in [1]. It consists of a union of k *affine* hyper-planes

$$H_i = \{\mathbf{x} \in \mathbb{R}^m : {\mathbf{n}_i^0}^T \mathbf{x} = b_i^0\}, i = 1, ..., k,$$

such that the sum of minimum distances raised to power l, from every point $\mathbf{x}_j$ to them is minimal. Consider the following minimization problem:

$$\text{minimize } \sum_{j=1}^N \min_{1\leqslant i\leqslant k} |\tilde{\mathbf{n}}_i^T \tilde{\mathbf{x}}_j|^l \tag{6}$$

$$\text{subject to } \|\tilde{\mathbf{n}}_i\| = 1, i = 1, ..., k, \tag{7}$$

where $\tilde{\mathbf{n}}_i = (\mathbf{n}_i, b_i), \tilde{\mathbf{x}}_j = (\mathbf{x}_j, -1)$. Its solution $\tilde{\mathbf{n}}_i$ defines *hyperpane skeleton* in $\mathbb{R}^{m+1}$, consisting of a union of k hyperplanes

$$\tilde{H}_i = \{\tilde{\mathbf{x}} \in \mathbb{R}^{m+1} : \tilde{\mathbf{n}}_i^T \tilde{\mathbf{x}} = 0\}, i = 1, ..., k,$$

such that the sum of minimum distances raised to power l, from every point $\tilde{\mathbf{x}}_j$ to them is minimal. The restriction of $\tilde{H}_i$ to $\mathbb{R}^m$ gives H_i. The usefulness of working in $\mathbb{R}^{m+1}$ can be seen for $l = 2$, if we denote by $\tilde{\mathbf{X}}_i$ the i-th cluster

$$\tilde{\mathbf{X}}_i = \{\tilde{\mathbf{x}}_j : \min_{1\leqslant p\leqslant k} |\tilde{\mathbf{n}}_p^T \tilde{\mathbf{x}}_j|^2 = |\tilde{\mathbf{n}}_i^T \tilde{\mathbf{x}}_j|^2\} \tag{8}$$

for some given $\{\tilde{\mathbf{n}}_i^T\}_{i=1}^k$. Then the following minimization problem

$$\text{minimize } \mathbf{v}_i^T \tilde{\mathbf{X}}_i \tilde{\mathbf{X}}_i^T \mathbf{v}_i \tag{9}$$

$$\text{subject to } \mathbf{v} \in \mathbb{R}^{m+1}, \|\mathbf{v}_i\| = 1, \tag{10}$$

will produce a solution $\{\mathbf{v}_i^0\}_{i=1}^k$ (consisting of eigenvectors corresponding to the minimal eigenvalues) for which f in (4) will not increase, i.e. $f(\{\mathbf{v}_i^0\}_{i=1}^k) \leqslant f(\{\tilde{\mathbf{n}}_i\}_{i=1}^k)$.

So we arrived to the following analogue of the classical k-means clustering algorithm.

Hyperplane clustering algorithm. (see the pseudocode of the subspace clustering algorithm below in the particular case when $r_i = 1$). Apply iteratively the following two steps until convergence:

1) Cluster assignment - forming $\tilde{\mathbf{X}}_i$, $i = 1, ..., k$ by (8),
2) Cluster update - solving the eigenvalue problem (9), (10).

Such kind of analogue of the classical k-means clustering algorithm, (working in $\mathbb{R}^m$ instead in $\mathbb{R}^{m+1}$) was introduced by Bradley and Mangasarian in [1] and called k-plane clustering algorithm.

Since our hyperplane clustering algorithm is equivalent to the k-plane clustering algorithm, it has the same properties, i.e. termination after finite number of steps at a cluster assignment which is locally optimal, i.e. f in (4) cannot be decreased by either reassigning of a point to a different cluster plane, or by defining a new cluster plane for any of the clusters (see Theorem 3.7 in [1]).

Affine Subspace Skeleton
The solution of the following minimization problem

$$\text{minimize} \quad F(\{\mathbf{U}_i\}_{i=1}^k) = \sum_{j=1}^{N} \min_{1 \leqslant i \leqslant k} \sum_{s=1}^{r_i} |\mathbf{u}_{i,s}^T \mathbf{x}_j - b_{i,s}|^l \tag{11}$$

$$\text{subject to} \quad \|\mathbf{u}_{i,s}\| = 1, b_{i,s} \in \mathbb{R}, i = 1, ..., k, \tag{12}$$

$$s = 1, ..., r_i, \mathbf{u}_{i,p}^T \mathbf{u}_{i,q} = 0, p \neq q, \tag{13}$$

where where $\mathbf{U}_i = \{\mathbf{u}_{i,s}\}_{s=1}^{r_i}$, will be called *affine subspace skeleton*. It consists of a union of k affine subspaces $S_i = \{x \in \mathbb{R}^m : \mathbf{u}_{i,s}^T \mathbf{x} = b_{i,s}, s = 1, ..., r_i\}$ each with codimension r_i, $i = 1, ..., k$, such that the sum of minimum distances raised to power l, from every point $\mathbf{x}_j$ to them is minimal. Consider the following minimization problem:

$$\text{minimize} \quad \tilde{F}(\{\tilde{\mathbf{U}}_i\}_{i=1}^k) = \sum_{j=1}^{N} \min_{1 \leqslant i \leqslant k} \sum_{s=1}^{r_i} |\tilde{\mathbf{u}}_{i,s}^T \tilde{\mathbf{x}}_j|^l \tag{14}$$

$$\text{subject to} \quad \|\tilde{\mathbf{u}}_{i,s}\| = 1, \in \mathbb{R}, i = 1, ..., k, \tag{15}$$

$$s = 1, ..., r_i, \tilde{\mathbf{u}}_{i,p}^T \tilde{\mathbf{u}}_{i,q} = 0, p \neq q, \tag{16}$$

where $\tilde{\mathbf{U}}_i = \{\tilde{\mathbf{u}}_{i,s}\}_{s=1}^{r_i}$, $\tilde{\mathbf{u}}_{i,s} = (\mathbf{u}_{i,s}, b_{i,s})$, $\tilde{\mathbf{x}}_j = (\mathbf{x}_j, -1)$. Its solution $\tilde{\mathbf{n}}_{i,s}$ defines a *subspace skeleton* in $\mathbb{R}^{m+1}$, consisting of a union of k subspaces

$$\tilde{S}_i = \{\tilde{\mathbf{x}} \in \mathbb{R}^{m+1} : \tilde{\mathbf{n}}_{i,s}^T \tilde{\mathbf{x}} = 0, s = 1, ..., r_i\}, i = 1, ..., k,$$

such that the sum of minimum distances raised to power l, from every point $\tilde{\mathbf{x}}_j$ to them is minimal. The restriction of $\tilde{S}_i$ to $\mathbb{R}^m$ gives S_i (a non-trivial fact).

Similarly to the hyperspace clustering algorithm, we can write the following analogue of the classical k-means clustering algorithm for finding the subspace skeleton, when $l = 2$.

Subspace clustering algorithm. (see the pseudocode below)
It consists of two steps applied alternatively until convergence:

1) Cluster assignment - forming $\tilde{\mathbf{X}}_i$, $i = 1, ..., k$ by: for given k orthonormal families of vectors $\{\tilde{\mathbf{n}}_{i,s}\}_{s=1}^{r_i}$, $i = 1, ..., k$, put

$$\tilde{\mathbf{X}}_i = \Big\{\tilde{\mathbf{x}}_j : \sum_{s=1}^{r_i} |\tilde{\mathbf{u}}_{i,s}^T \tilde{\mathbf{x}}_j|^2 = \min_{1 \leqslant i \leqslant k} \sum_{s=1}^{r_i} |\tilde{\mathbf{u}}_{i,s}^T \tilde{\mathbf{x}}_j|^2\Big\}. \tag{17}$$

2) Cluster update - for every $i = 1, ..., k$, solving the minimization problem

$$\text{minimize} \quad \sum_{s=1}^{r_i} trace(\mathbf{V}_i^T \tilde{\mathbf{X}}_i \tilde{\mathbf{X}}_i^T \mathbf{V}_i) \tag{18}$$

$$\text{under orthogonality constraints} \quad \mathbf{V}_i^T \mathbf{V}_i = \mathbf{I}_{r_i}, \tag{19}$$

where $\mathbf{V}_i$ has dimensionality $(n + 1 \times r_i)$.

For any i, it will produce a solution $\mathbf{V}_i^0$ - a matrix with columns equal to the eigenvectors corresponding to the r_i smallest eigenvalues of the matrix $\tilde{\mathbf{X}}_i \tilde{\mathbf{X}}_i^T$ (a non-trivial fact) and $\tilde{F}$ in (14) will not increase, i.e. $\tilde{F}(\{\mathbf{V}_i^0\}_{i=1}^k) \leqslant \tilde{F}(\{\tilde{\mathbf{U}}_i\}_{i=1}^k)$.

This algorithm terminates in a finite number of steps (similarly to the Bradley-Mangasarian algorithm) to a solution which is locally optimal, i.e. $\tilde{F}$ in (14) cannot be decreased by either reassigning of a point to a different cluster subspace, or by defining a new cluster subspace for any of the clusters.

Algorithm 1. Subspace clustering algorithm

Data: samples $\mathbf{x}(1), \ldots, \mathbf{x}(T)$
Result: estimated k subspaces $\tilde{S}_i \subset \mathbb{R}^{n+1}, i = 1, ..., k$ given by the orthonormal bases $\{\tilde{\mathbf{u}}_{i,s}\}_{s=1}^{r_i}$ of their orthogonal complements $\tilde{S}_i^{\perp}$.

Initialize $\varepsilon > 0$ and randomly $\tilde{\mathbf{u}}_{i,s} = (\mathbf{u}_{i,s}, b_{i,s})$ with $|\tilde{\mathbf{u}}_{i,s}| = 1, i = 1, \ldots, k$, $s = 1, \ldots, r_i$.
do *(Until the difference of $\tilde{F}$ calculated in two subsequent estimates of $\{\tilde{\mathbf{U}}_i\}_{i=1}^k$ is less than ε)*
  Cluster assignment.
  for $t \leftarrow 1, \ldots, T$ **do**
    Add $\tilde{\mathbf{x}}(t) = (\mathbf{x}(t), -1)$ to cluster $\tilde{\mathbf{X}}_i$ (a matrix), where i is chosen to minimize $\sum_{s=1}^{r_i} |\mathbf{u}_{i,s}^T \mathbf{x}_j - b_{i,s}|^2$ (distance to the subspace S_i).
  end
  Cluster update.
  for $i \leftarrow 1, \ldots, k$ **do**
    Calculate the i-th cluster correlation $\mathbf{C} := \tilde{\mathbf{X}}_i \tilde{\mathbf{X}}_i^T$.
    Choose eigenvectors $\mathbf{v}_s, s = 1, ..., r_i$ of $\mathbf{C}$ corresponding to the r_i minimal eigenvalues.
    Set $\tilde{\mathbf{u}}_{i,s} \leftarrow \mathbf{v}_s / |\mathbf{v}_s|, s = 1, ..., r_i, \tilde{\mathbf{U}}_i = \cup_{s=1}^{r_i} \tilde{\mathbf{u}}_{i,s}$.
  end
end

Algorithm for Identification of the Mixing Matrix

1) Cluster the columns $\{\mathbf{X}(:, j) : j \in \mathcal{N}_1\}$ in k groups $\mathcal{H}_i, i = 1, ..., k$ such that the span of the elements of each group $\mathcal{H}_i$ produces r_i-codimensional subspace and these r_i-codimensional subspaces are different.

2) Calculate any basis of the orthogonal complement of each of these r_i-codimensional subspaces.

3) Find all possible groups such that each of them is composed of the elements of at least m bases in 2), and the vectors in each group form a hyperplane. The number of these hyperplanes gives the number of sources n. The normal vectors to these hyperplanes are estimations of the columns of the mixing matrix $\mathbf{A}$ (up to permutation and scaling).

In practical realization all operations in the above algorithm are performed up to some precision $\varepsilon > 0$.

Source Recovery Algorithm

1. Repeat for $j = 1$ to N:

2.1. Identify the subspace $\mathcal{H}_i$ containing $\mathbf{x}_j := \mathbf{X}(:, j)$, or, in practical situation with presence of noise, identify $\mathcal{H}_i$ to which the distance from $\mathbf{x}_j$ is minimal and project $\mathbf{x}_j$ onto $\mathcal{H}_i$ to $\tilde{\mathbf{x}}_j$;

2.2. if $\mathcal{H}_i$ is produced by the linear hull of column vectors $\mathbf{a}_{p_1}, ..., \mathbf{a}_{p_{m-r_i}}$, then find coefficients $\mathbf{L}_{j,l}$ such that $\tilde{\mathbf{x}}_j = \sum_{l=1}^{m-r_i} \mathbf{L}_{j,l}\mathbf{a}_{p_l}$. These coefficients are uniquely determined generically (i.e. up to measure zero) for $\tilde{\mathbf{x}}_j$ (see [5], Theorem 3).

2.3. Construct the solution $\mathbf{s}_j = \mathbf{S}(:, j)$: it contains $\mathbf{L}_{j,l}$ in the place p_l for $l = 1, ..., m - r_i$, the other its components are zero.

It is clear that such found $(\mathbf{A}, \mathbf{S})$ is a solution to the least square BSS problem defined by (2).

5 Computer Simulation Examples

First example. We created artificially four source signals, sparse of level 2, i.e. each column of the source matrix contains at least 2 zeros (shown in Figure 1). They are mixed with a square nonsingular randomly generated matrix $\mathbf{A}$:

$$\mathbf{A} = \begin{pmatrix} 0.4621 & 0.6285 & 0.5606 & 0.4399 \\ 0.2002 & 0.4921 & 0.5829 & 0.3058 \\ 0.8138 & 0.4558 & 0.2195 & 0.2762 \\ 0.2899 & 0.3938 & 0.5457 & 0.7979 \end{pmatrix}.$$

The mixed sources are shown in Fig.2, the recovered sources by our algorithm are shown in Fig. 3. The created signals are superposition of sinusoidal signals, so it is clear that they are dependent and ICA methods could not separate them (lack of space prevent us to show the results of applying ICA algorithms). The estimated mixing matrix with our subspace clustering algorithm is

$$\mathbf{M} = \begin{pmatrix} 0.6285 & 0.4399 & 0.4621 & 0.5606 \\ 0.4921 & 0.3059 & 0.2002 & 0.5829 \\ 0.4558 & 0.2762 & 0.8138 & 0.2195 \\ 0.3938 & 0.7979 & 0.2899 & 0.5457 \end{pmatrix}.$$

Second Example. Now we create six signals with level of sparsity 4, i.e. each column of the source matrix has 4 zeros. We mixed them with a randomly generated matrix with dimension (3×6):

$$\mathbf{H} = \begin{pmatrix} 0.8256 & 0.3828 & 0.4857 & 0.4053 & 0.7720 & 0.2959 \\ 0.2008 & 0.7021 & 0.0197 & 0.5610 & 0.6182 & 0.6822 \\ 0.5273 & 0.6004 & 0.8739 & 0.7218 & 0.1476 & 0.6686 \end{pmatrix}.$$

The estimated mixing matrix with our subspace hyperplane clustering algorithm is

$$\mathbf{M} = \begin{pmatrix} 0.2959 & 0.4857 & 0.8256 & 0.4053 & 0.3828 & 0.7720 \\ 0.6822 & 0.0197 & 0.2008 & 0.5610 & 0.7021 & 0.6182 \\ 0.6686 & 0.8739 & 0.5273 & 0.7218 & 0.6004 & 0.1476 \end{pmatrix}.$$

Here the number of hyperplanes presented in the data are 9, while the number of all possible hyperplanes (according to the identifiability conditions in [5]) when all combination of 4 zeros in the columns of the source matrix are present, are 15.

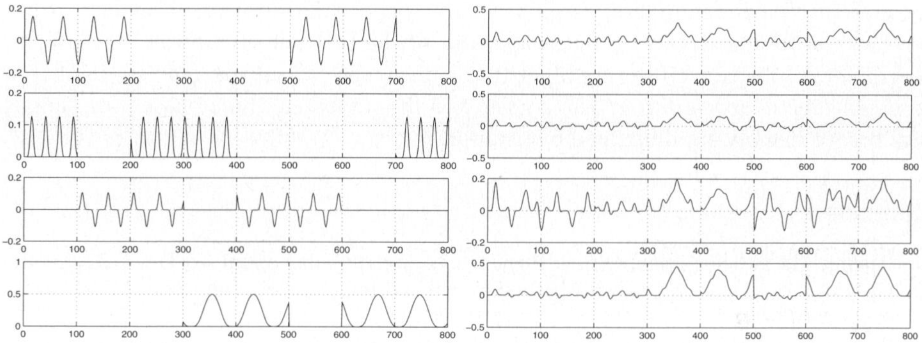

Fig. 1. Example 1: original source signals (left) and mixed ones (right)

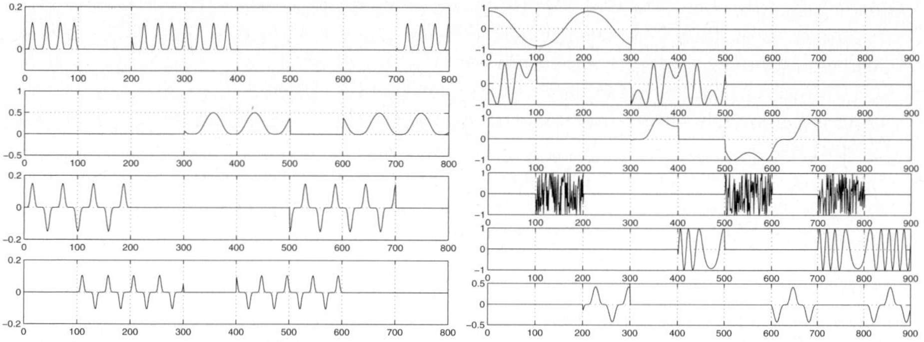

Fig. 2. Left: Example 1 - separated signals. Right: Example 2 - original source signals

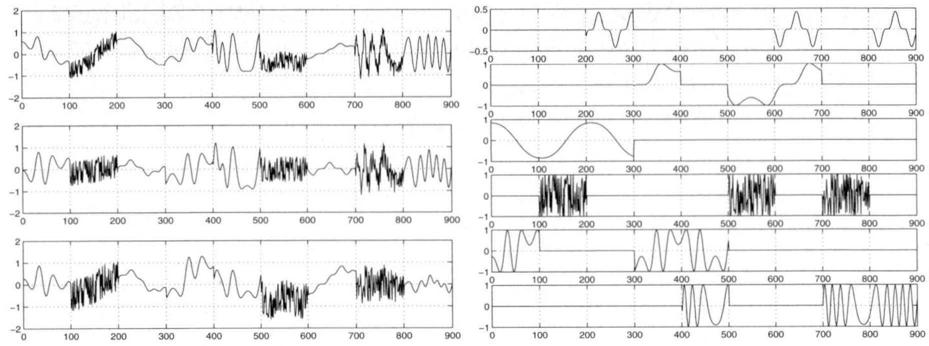

Fig. 3. Example 2: mixed signals (left) and separated signals (right)

6 Conclusion

We presented new identifiability conditions for sparse BSS problem, allowing less hyperplanes in the data points for full recovery of the original sources and the mixing matrix. New subspace clustering algorithm is designed for subspace detection, based on eigenvalue decomposition. The square and the underdetermined cases for signals with bigger sparsity are illustrated by examples.

References

1. Bradley, P.S., Mangasarian, O.L.: k-Plane Clustering. J. Global optim. 16(1), 23–32 (2000)
2. Cichocki, A., Amari, S.: Adaptive Blind Signal and Image Processing. John Wiley, Chichester (2002)
3. Comon, P.: Independent component analysis - a new concept? Signal Processing 36, 287–314 (1994)
4. Eriksson, J., Koivunen, V.: Identifiability and separability of linear ica models revisited. In: Proc. of ICA 2003, pp. 23–27 (2003)
5. Georgiev, P., Theis, F., Cichocki, A.: Sparse Component Analysis and Blind Source Separation of Underdetermined Mixtures. IEEE Trans. of Neural Networks 16(4), 992–996 (2005)
6. Hyvärinen, A., Karhunen, J., Oja, E.: Independent Component Analysis. John Wiley & Sons, Chichester (2001)
7. Lee, T.-W., Lewicki, M.S., Girolami, M., Sejnowski, T.J.: Blind sourse separation of more sources than mixtures using overcomplete representaitons. IEEE Signal Process. Lett. 6(4), 87–90 (1999)
8. Rubinov, A.M., Soukhoroukova, N., Ugon, J.: Minimization of the sum of minia of convex functions and its application to slustering. In: Jeyakumar, V., Rubinov, A. (eds.) Continuous Optimization, Current trends and modern applications, pp. 409–434. Springer, Heidelberg (2005)
9. Theis, F.J., Lang, E.W., Puntonet, C.G.: A geometric algorithm for overcomplete linear ICA. Neurocomputing 56, 381–398 (2004)
10. Waheed, K., Salem, F.: Algebraic Overcomplete Independent Component Analysis. In: Proc. Int. Conf. ICA2003, Nara, Japan, pp. 1077–1082 (2003)
11. Zibulevsky, M., Pearlmutter, B.A.: Blind source separation by sparse decomposition in a signal dictionary. Neural Comput. 13(4), 863–882 (2001)

Two Improved Sparse Decomposition Methods for Blind Source Separation

B. Vikrham Gowreesunker and Ahmed H. Tewfik

Dept. of Electrical and Computer Engineering, University of Minnesota,
Minneapolis, MN 55455
{gowr0001,tewfik}@umn.edu

Abstract. In underdetermined blind source separation problems, it is common practice to exploit the underlying sparsity of the sources for demixing. In this work, we propose two sparse decomposition algorithms for the separation of linear instantaneous speech mixtures. We also show how a properly chosen dictionary can improve the performance of such algorithms by improving the sparsity of the underlying sources. The first algorithm proposes the use of a single channel Bounded Error Subset Selection (BESS) method for robustly estimating the mixing matrix. The second algorithm is a decomposition method that performs a constrained decomposition of the mixtures over a stereo dictionary.

1 Introduction

In the blind source separation(BSS) problem, we have mixtures of several source signals and the goal is to separate them with as little prior information as possible. In this work, we study the instantaneous underdetermined BSS case, where we have more sources than mixtures. We are concerned with separating mixtures of speech signals when the mixing matrix and number of underlying sources are unknown. This problem is intrinsically ill-defined and its solution requires some additional assumptions compared to its overdetermined counterpart. The difficulty of the underdetermined setup can be somewhat alleviated if there exists a representation where all the sources are rarely simultaneously active, which entails finding a representation where the sources are sparse. Some authors have shown that speech signals are sparser in the time-frequency than in the time domain, and that there exists several other representations such as wavelets packets, where different degrees of sparsity can be obtained [12]. It has been shown that better separation can indeed be achieved by exploiting such sparsity [7],[6],[11].

In this paper, we investigate methods for performing BSS using overcomplete dictionaries in the underdetermined case. The fundamental success of the separation depends on two factors, namely the type of dictionary used and the type of decomposition method employed. We study both areas. For the first case, we demonstrate how the underlying sparsity of the sources can be improved using a KSVD-based [3] trained dictionary. For the second case, we propose the use of the Bounded Error Subset Selection(BESS) [5] method as a robust method

M.E. Davies et al. (Eds.): ICA 2007, LNCS 4666, pp. 365–372, 2007.

for estimating the mixing matrix, even in the presence of additive noise in the mixture. Furthermore, we propose a multichannel sparse decomposition method for extracting the underlying sources.

The remainder of this paper is organized as follows. In section 2 we give a mathematical description of the problem and an explanation of how sparsity is used in source separation. In section 3, we show how the right dictionary can improve the underlying sparsity of the sources. In section 4, we illustrate how single channel sparse decomposition algorithms fail to distribute some coefficients to the proper source, setting the stage for multichannel methods. Finally, we detail a decomposition algorithm with directionality constraints and discuss its benefits.

2 Mixture Model for an Arbitrary Dictionary

For a problem where we have M mixtures of N sound sources and $M \leq N$, our goal is to separate the sources into individual tracks. We are concerned with the underdetermined linear instantaneous mixture model, which can be formulated mathematically as follows,

$$x(t) = As(t) + q(t), \tag{1}$$

where $s(t)$ is an unknown $N \times 1$ vector containing the source data, $x(t)$ is a known $M \times 1$ observation vector, $q(t)$ is the $M \times 1$ additive noise vector, t is the sample index and A is an unknown $M \times N$ mixing matrix. Over T time samples, we have the following expression,

$$X = AS + Q, \tag{2}$$

where $X = [x(1])x(2]) \ldots x(T)]$, $S = [s(1)s(2) \ldots s(T])]$, $Q = [q(1)q(2) \ldots q(T)]$.

One approach to solving this problem is to assume that the sources are sufficiently sparse in a given representation. To solve the underdetermined BSS problem using sparsity, one can decompose the signal into a dictionary where the source signals are known to be sparse or use a Sparse Decomposition (SD) algorithm to find a sparse representation for an overcomplete dictionary. In the rest of this section, we illustrate how sparsity can lead to separation, then show how sparsity in alternative representations can be exploited. To simplify the discussion, let us assume without loss of generality that M = 2 and N = 3. Assuming there is no additive noise for illustrative purposes, we can expand Eq. 2 as follows

$$\begin{bmatrix} x_1 \\ x_2 \end{bmatrix} = \begin{bmatrix} a_{11} \; a_{12} \; a_{13} \\ a_{21} \; a_{22} \; a_{23} \end{bmatrix} \times \begin{bmatrix} s_1 \\ s_2 \\ s_3 \end{bmatrix} \tag{3}$$

By looking at the ratio $\frac{x_1}{x_2}$ for the case when only the jth source is active, we get

$$\frac{x_1}{x_2} = \frac{a_{1j}}{a_{2j}} \tag{4}$$

Hence a scatter plot of x_1 v/s x_2 for the case where the sources never overlapped would reveal 3 distinct lines such that the jth source corresponds to the line with gradient $\frac{a_{1j}}{a_{2j}}$. Separation can be easily achieved for Eq. 4. The general thinking is that the sparser the sources are, the less likely they are to be active at the same time, resulting in better separation. The same argument can be applied when the data is represented in a different dictionary. The dictionary can be a basis matrix or an overcomplete matrix and each column is referred to as a dictionary atom. We can express signal,$\{s_j\}_{j=1}^N$, in terms of the dictionary, D and coefficient matrix, C such that

$$S^T = DC. \tag{5}$$

Substituting in Eq. 2, We get

$$X = AC^T D^T, \tag{6}$$

where C is a $K \times M$ coefficient matrix and D is a $T \times K$ matrix. Thus, it is clear that the sparsity of the source is inherently limited by the dictionary used. In the following section we illustrate how the quest for a sparse representation entails finding a good dictionary. In the rest of the paper, we consider the case where we have 2 mixtures and 3 sources.

3 Effect of Dictionary on Signal Sparsity

In this section, we illustrate the importance of dictionary selection for BSS algorithms that rely on sparsity. Although, it has already been well established that speech signal exhibits different degrees of sparsity in different representations [11],[12], we do not know what is the optimally sparse representation for speech signals or whether optimal sparsity offers sufficient improvement in separation quality over standard overcomplete dictionaries such as the Cosine Packet(CP) dictionary[8]. In the absence of an optimal representation, we resort to trained dictionaries to demonstrate that a properly chosen dictionary can offer significant performance improvement. In this section, we present some numerical results to illustrate that indeed the trained dictionary offers better sparsity than the CP dictionary. In section 4, we compare their impact on separation.

3.1 Dictionary Design

The CP dictionary was designed using the Atomizer and Wavelab Matlab toolboxes[1], with dimensions 128×896. The matched dictionary, of the same size as the CP dictionary, was designed using a KSVD method [3]. This method takes after the Vector Quantization technique used in codebook design and tends to promote a sparse structure. The speech data used for training the dictionary had to be first formatted into frames of length T, with a standard windowing technique.

To compare the sparsity improvement of the matched and unmatched dictionaries for speech signals, we use a standard sparse decomposition method, the Orthogonal Matching Pursuit(OMP)[8]. Sparsity for a dictionary was evaluated as the minimum number of coefficients required to represent the signal in that dictionary using OMP. Using three different speakers, we trained three dictionaries. The first, dictionary1, was trained using speaker 1 only, the second, dictionary12, was trained with speaker 1 and 2, and the third was trained using all three speakers. The signal from each speaker was then independently decomposed in the CP dictionary and the three trained dictionaries.

3.2 Sparsity Comparison

We summarize the sparsity of the original sources for each dictionary in Fig. 1. In all three experiments, source1 is clearly much sparser with the KSVD dictionary than the CP. The most important observation is that all speakers exhibit a very high sparsity improvement with dictionary123. A study of the number of dictionary atoms concurrently used by simultaneous sources reveals that indeed the sparsest dictionary, dictionary123, results in the lowest number of overlapping atoms. For space consideration, details are omitted here. Hence, for our purposes, dictionary123 is sparsest available dictionary for the sources under study.

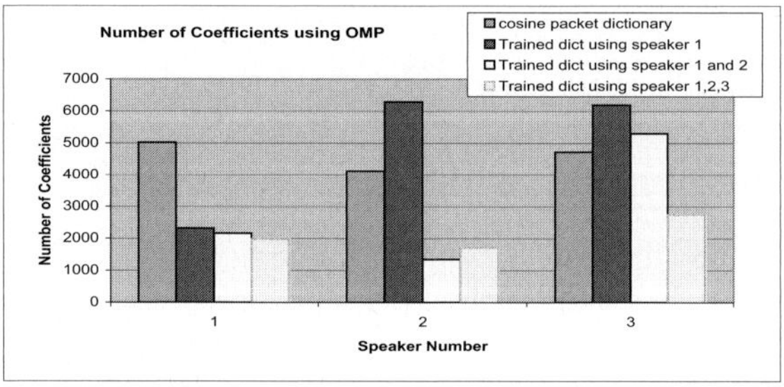

Fig. 1. Number of nonzero coefficients when using OMP. Dictionary1 is trained using source1 only. Dictionary12 is trained using source1 and source2. Dictionary123 is trained using source1, source2 and source3.

4 Sparse Decomposition Methods for Separation

Our goal in this section is to introduce two algorithms for underdetermined BSS. The first algorithm is a single channel BESS, which is shown to be a robust technique for estimating the mixing matrix. The second algorithm performs a constrained decomposition of the two mixture signals over a stereo dictionary.

4.1 Single Channel BESS as a Robust Estimate of the Mixing Matrix

In Fig.2(a), we show the scatter plot for coefficients from performing single channel BESS independently on the each mixture signal. There are two interesting observations about this plot. First, there is a set of coefficients that appear on the axes, and second, there is a different set of coefficients clustered along the mixing matrix columns. The coefficients on the axes are dictionary elements that appear in one mixture but not in the other, and cannot be separated using this method. The coefficients along the matrix orientation share the same dictionary atoms for both mixtures and belong to a single source. We propose using only the dictionary atoms common to both mixtures in estimating the mixing matrix.

A well known method for estimating the mixing matrix is performing a Hough transform on the coefficients of the sparse decomposition. This basically entails finding the histogram of angles, $\arctan \frac{c_1}{c_2}$, for all the coefficients pairs (c_1, c_2) of the decomposed mixtures. The mixing matrix columns show up as peaks on the histograms. The quality of the estimate depends on the how well the peaks can be estimated. We combine this method with the BESS decomposition discussed above, and compare the results with the Hough transform of the coefficients of the Modified Discrete Cosine Transform(MDCT) of the mixtures.

We find that both methods provide equally good estimates, but that the BESS technique is more robust in the presence of additive noise. Fig. 2(b) shows the histogram for MDCT coefficients and fig. 2(c) shows the results for the BESS method, in the presence of noise. We can clearly see that the peaks position gets blurred in the presence of noise with the MDCT transform, while the new method provides more robust estimate of the peaks under noisy conditions.

4.2 A Decomposition Method with Directionality Constraints

Inspired by the stereo dictionary method of Gribonval [7], we propose an algorithm that performs a constrained decomposition of the two mixture signals. This require prior knowledge or estimation of the mixing matrix. We detail the algorithm below and contrast it with stereo dictionary decomposition method.

Given a $T \times K$ dictionary, D, consisting of $\{d_i\}_{i=1}^{K}$ and two mixtures vectors, x_1, and x_2, the decomposition is given as follows,

1. At iteration $M = 1$, the nth residual, R_n^M is initialized to n^{th} mixture signal, x_n. Each channel has a set of coefficients, $c_{n,i} = 0$, where i is the index of the dictionary atom. A list of the available dictionary atoms is kept tracked in list, Li. All indices are included at initialization. At each iteration, a dictionary atom is picked and the coefficients associated with that atom are computed.
2. For picking the dictionary atom, we use the same criteria as [7], i.e., we pick the atom, k, that is maximally correlated to both mixtures channels, $k = argmax_k| < R_1^M, d_k >^2 + < R_2^M, d_k >^2 |$
3. Next we find the projection of the residuals of each channel onto the kth dictionary element. Denoting this projection pairs as, $c_t = [< R_1^M, d_k >, < R_2^M, d_k >]$, we proceed to constrain it to lie along one the mixing matrix columns.
4. By finding the inner product of this projection pair with the mixing matrix columns, we finding the closest column as, $J = argmax_J| < c_t, a_J > |$, where J is the column index.

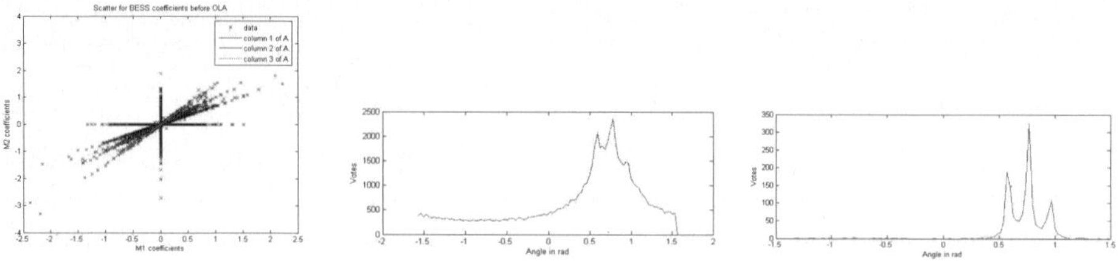

(a) Scatter plot of 2 independent BESS decomposition

(b) Histogram of all MDCT coefficients

(c) Histogram of BESS coefficients, common to both mixtures

Fig. 2. Coefficients of 2 mixtures in the presence of additive noise at sensors

5. We then set the coefficient for the kth atom to be equal to $< c_t, a_J >$, and remove the index of this dictionary element from index list, Li.
6. We check the reconstruction error of both mixtures signal and increment the iteration if necessary.

Stages 1 and 2 are essentially the same as the StereoMP. However, stage 2 can be replaced by some other criteria for picking the dictionary element. In stage 3 through 5, we constrain the coefficients of a particular dictionary elements to lie along one of the columns of the mixing matrix. The difference between doing the projection during the decomposition as opposed to after is that error between $c_t \times d_k$ and $< c_t, a_J > \times c_t \times d_k$ can be reassigned to another dictionary element during the decomposition.

4.3 Results

We computed commonly used performance indices[4] for the above mentioned algorithm for 2 cases, a CP dictionary and a KSVD-trained dictionary. For the purpose of this experiment, we used three 10 seconds long male speech tracks and artificially mixed them using a predefined mixing matrix. The experiments were run on frames of 512 points with a 50% overlap and the dictionary used was of size 512×4608. In Fig. 3, we compared the results for the CP dictionary and the trained dictionary. We find that the trained dictionary offers a Signal-to-Interference Ratio (SIR) improvement of 4 to 9 dB over the CP dictionary and Signal-to-Artifact Ratio (SAR) improvement of of 3 to 4 dB. Also, not shown here for space consideration is an improvement in Signal-to-Distortion Ratio (SDR) similar to the SAR. This confirms the claim in section 3 that dictionary selection is a very important part of the separation process.

We also found that our method to have comparable SIR to Gribonval's method, but with a noticeable improvement numerically and subjectively in distortion and artifacts. This improvement in distortion metrics confirms our previous claim that redistributing the coefficient projection error indeed improves the quality of the separation. The downside, however, is that a larger number of iterations is required.

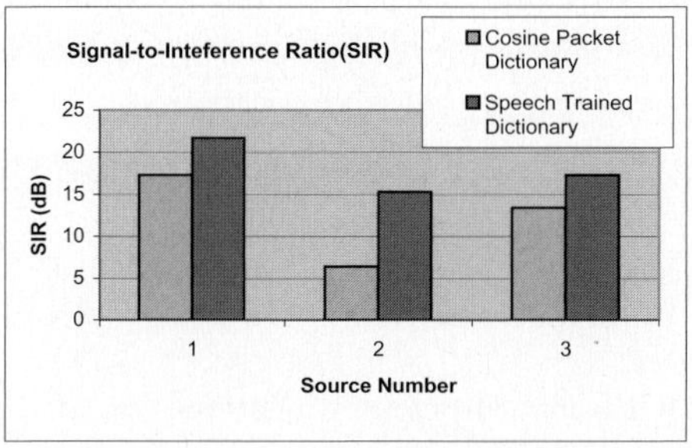

(a) Signal-to-Interference Ratio (dB)

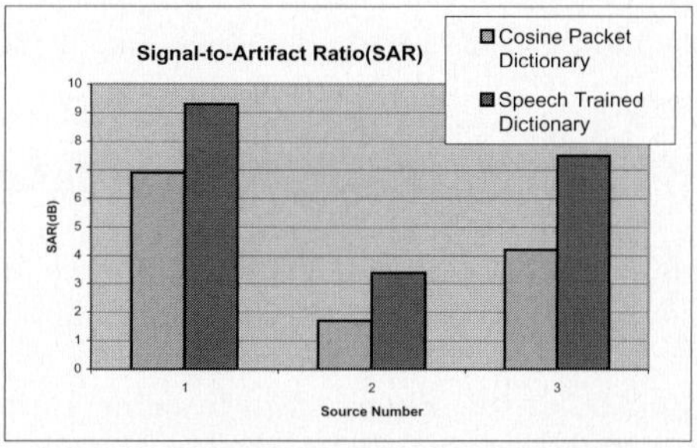

(b) Signal-to-Artifact Ratio

Fig. 3. Performance Comparison between CP and Trained dictionaries for algorithm of section 4.2

5 Conclusion

We discussed the implications of sparsity on source separation and explored different avenues to maximize separation for given signals. We looked at both the dictionary selection problem and selection of decomposition algorithms, and used the BESS method as a robust estimate of the mixing matrix. In this work, we clearly illustrated how the choice of the proper dictionary makes a difference in the sparsity of the coefficients and the resulting separation. Furthermore, we proposed a decomposition method which constrains the coefficients at each iteration, and demonstrated its performance. We currently in process of developing an extension of the BESS method that could be extended to multichannel source separation.

References

1. `http://www-stat.stanford.edu/~wavelab/`
2. `http://sassec.gforge.inria.fr/`
3. Aharon, M., Elad, M., Bruckstein, A.: The K-SVD: An Algorithm for Designing of Overcomplete Dictionaries for Sparse Representation. IEEE Trans. on Signal Processing 54, 4311–4322 (2006)
4. Vincent, E., Gribonval, R., Fevotte, C., et al.: BASS-dB: the blind audio source separation evaluation database, Available on-line `http://www.irisa.fr/metiss/BASS-dB/`
5. Alghoniemy, M., Tewfik, A.H.: Reduced Complexity Bounded Error Subset Selection In: IEEE Int. Conf. Acoustics, Speech and Signal Processing (ICASSP), pp. 725–728 (March 2005)
6. Fevotte, C., Godsill, S.: A Bayesian approach for blind separation of sparse sources. IEEE Trans. on Speech and Audio Processing 14, 2174–2188 (2006)

7. Gribonval, R.: Sparse decomposition of stereo signals with Matching Pursuit and application to blind separation of more than two sources from a stereo mixture. In: IEEE Int. Conf. Acoustics, Speech and Signal Processing (ICASSP), vol. 3, pp. 3057–3060 (May 2002)
8. Mallat, S.: A Wavelet Tour of Signal Processing, 2nd edn. Academic Press, San Diego (1999)
9. Shindo, H., Hirai, Y.: Blind Source Separation by a Geometrical Method. In: Proceedings of the 2002 International Joint Conference on Neural Networks (IJNN), pp. 1109–1114 (May 2002)
10. Mallat, S., Zhang, Z.: Matching Pursuit with Time-frequency Dictionaries. IEEE Trans. on Signal Processing 41, 3397–3415 (1993)
11. Tan, V.Y., Fevotte, C.: A study of the effect of source sparsity for various transforms on blind audio source separation performance. In: Workshop on Signal Processing with Adaptative Sparse Structured Representations (SPARS'05), Rennes, France (November 2005)
12. Zibulevsky, M., Pearlmutter, B.A., Bofill, P., Kisilev, P.: Blind Source Separation by Sparse Decomposition. In: chapter in the book: Roberts, S.J., Everson, R.M. (eds.) Independent Component Analysis: Principles and Practice, Cambridge (2001)

Probabilistic Geometric Approach to Blind Separation of Time-Varying Mixtures

Ran Kaftory and Yehoshua Y. Zeevi

Technion - Israel Institute of Technology, Department of Electrical Engineering, Haifa 32000, Israel
kaftoryr@tx.technion.ac.il, zeevi@ee.technion.ac.il

Abstract. We consider the problem of blindly separating time-varying instantaneous mixtures. It is assumed that the arbitrary time dependency of the mixing coefficient, is known up to a finite number of parameters. Using sparse (or sparsified) sources, we geometrically identify samples of the curves representing the parametric model. The parameters are found using a probabilistic approach of estimating the maximum likelihood of a curve, given the data. After identifying the model parameters, the mixing system is inverted to estimate the sources. The new approach to blind separation of time-varying mixtures is demonstrated using both synthetic and real data.

1 Introduction

Most of the research in the area of blind source separation, has been focused on the instantaneous model. In recent years, more attention is being directed towards convolutive and time-varying problems which represent the majority of real-world mixtures. In blind separation of time-varying instantaneous mixtures, the sources are mixed with time-varying coefficients. It is assumed that the sources are not filtered nor are delayed prior to being mixed.

Many studies have addressed this problem by using an adaptive form of stationary blind source separation algorithms [4,6,7]. Some have even utilized particle filtering in order to track the mixing coefficients [2]. These approaches assume that the variation of the mixing system is small, in order to enable the extraction of the statistical nature of the mixtures. A different idea was employed in [9,10], where a parametric model of the mixing coefficients was used in the case of a linear or periodic dependency of the mixing system on time. This approach can be more efficient for larger variations in comparison to the adaptive algorithms.

The case where the mixing system is an arbitrary function of time, with large variations, is still overlooked. In this paper, the model of the mixing coefficients is arbitrary. We assume that it is known up to a finite number of parameters, and that the sources are sparse (or can be sparsely represented using an appropriate transform). For such sources, the value of the mixing coefficients can be identified over many time instances. We interpret these instances geometrically, as samples of curves representing the parametric model. Estimating the parameters, requires the grouping of the time instances and assigning them to a curve. This is done

M.E. Davies et al. (Eds.): ICA 2007, LNCS 4666, pp. 373–380, 2007.

by using a probabilistic approach of estimating the maximum likelihood of a curve given the data. After identifying the model parameters, the mixing system is inverted to estimate the sources.

2 Time-Varying Instantaneous Mixtures

In the problem of blind separation of time-varying (or position-varying in the case of images) instantaneous mixtures, sensors $x_i(t)$ receive sources, $s_j(t)$, which are linearly mixed by a matrix, $A(t)$, with time-dependent elements $a_{ij}(t)$[1]:

$$x(t) = A(t)s(t) + n(t), \tag{1}$$

where $n(t)$ is an additive noise with known or unknown parameters. We can usually assume the form of $a_{ij}(t)$ by using some knowledge about the physics of the problem[2]:

$$a_{ij}(t) = g_{ij}(\alpha_{ij_k}; t), \tag{2}$$

where the function g_{ij} is known up to some finite number of parameters α_{ij_k}. The index $k = 1, .., N$, where N is the number of unknown parameters.

To illustrate a system of 2 sensors receiving data from position-varying instantaneous mixtures of 2 sources, consider Fig. 1. The semi-reflective glass mixes an image and a reflection. The position-varying mixing is obtained by non-uniform illumination.

The objective of the blind source separation problem is to estimate the sources, s_j, and the unknown mixing parameters, α_{ij_k}, from the observed mixtures, x_i. One of the ways for doing so, is to identify the matrix $A(t)$ and apply its pseudo-inverse to the observations $x(t)$[3]:

$$\hat{s}(t) = A(t)^{\dagger}x(t) = s(t) + A(t)^{\dagger}n(t), \tag{3}$$

where $A(t)^{\dagger}$ stands for the pseudo-inverse of $A(t)$. Similar to the case of instantaneous mixtures, where the separation is up to scaling and permutation of the original sources, we assume that the blind separation of time-varying instantaneous mixtures is up to a time-varying scaling and permutation of the sources. Therefore, we can define our separation problem to account for this fact and rewrite it as follows:

$$x(t) = \tilde{A}(t)\tilde{s}(t) + n(t), \tag{4}$$

where $\tilde{a}_{ij}(t) = a_{ij}(t)/a_{1j}(t)$ and $\tilde{s}_j(t) = a_{1j}s_j(t)$. Similar to Eq. (2), we define $\tilde{g}_{ij}(\alpha_{ij_k}; t)$, such that

$$\tilde{a}_{ij}(t) = \tilde{g}_{ij}(\tilde{\alpha}_{ij_k}; t) = \frac{g_{ij}(\alpha_{ij_k}; t)}{g_{1j}(\alpha_{1j_k}; t)}. \tag{5}$$

[1] In the case of images, the spatial coordinate y replaces t in the equations.
[2] If the function is unknown, we can still assume that it is an analytic function and therefore, can be approximated by a polynomial (representing its Taylor expansion).
[3] Presuming the noise is low.

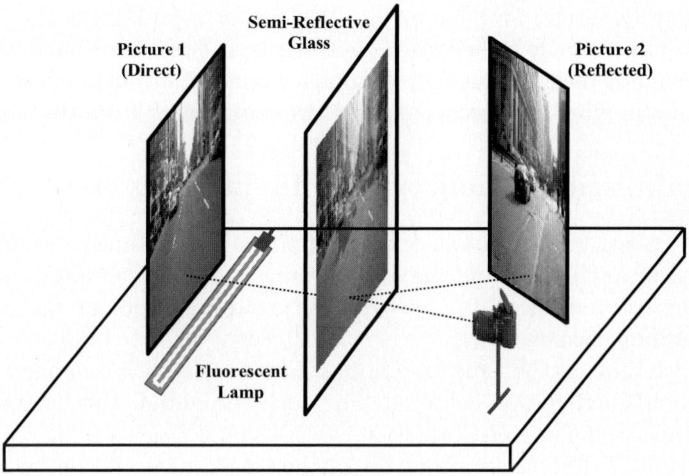

Fig. 1. A setup for acquiring spatial varying instantaneous mixtures: two pictures are positioned opposite to each other while a semi-reflective glass is mounted on the optical axis of one of them. First mixture is acquired using uniform lighting. A second mixture is acquired using a non-uniform illumination by a fluorescent lamp. This setup mimics a physical situation, where the position-varying mixing parameter cannot be avoided.

The time-varying scaled estimated sources, can be found by using

$$\hat{s}(t) = \tilde{A}(t)^{\dagger} x(t). \tag{6}$$

The problem is how to estimate $\tilde{A}(t)$ using only the observed mixtures $x(t)$.

3 Geometric Approach to Separation of Sparse Signals

Sparse signal is characterized by a small percentage of samples with nonzero value. It can be instrumental in estimation of the mixing matrix [5]. If a signal is not sparse, it can be sparsified by using a linear transformation to a domain in which it is sparsely represented [11][4].

Sparse signals can be geometrically interpreted to identify $\tilde{A}(t)$. If we observe during some time instance t_0, the existence of a nonzero signal in the sensors, the most probable assumption would be that the signal was originated from a single source. We can see from Eq. (4) that if only the jth source is active during t_0, $\tilde{a}_{ij}(t_0)$ can be found by:

$$\tilde{a}_{ij}(t_0) = \frac{x_i(t_0) - n_i(t_0)}{x_1(t_0) - n_1(t_0)} \approx \frac{x_i(t_0)}{x_1(t_0)}. \tag{7}$$

[4] By knowing the parametric form of $A(t)$, a transformation can be chosen such that $T[A(t)s(t)] \approx A(t)T[s(t)]$ and the blind source separation can be solved in the transformed sparse domain.

Since s_j is rarely active, we have several non-uniform samples of the continuous curve $\tilde{a}_{ij}(t)$. The obstacle is that when one observes a nonzero signal in the sensors, it is not known to which source it corresponds. Thus, the problem requires to group time instances in which the received signal was originated from the same source.

4 Probabilistic Framework for Identifying $\tilde{A}(t)$

We want to construct a framework for grouping time instances originated from the same sensor and estimating the $\tilde{\alpha}_{ij_k}$ parameters. This framework is necessary since correct estimation of $\tilde{\alpha}_{ij_k}$ solves the grouping problem and vice versa, correct grouping estimates $\tilde{\alpha}_{ij_k}$.

Suppose we take all the time instances $t_0...t_{M-1}$ in which a nonzero (or above some threshold) signal was detected. For each individual time instance t_l, we define the ratio $r_i(t_l) \equiv x_i(t_l)/x_1(t_l)$.

A maximum likelihood approach for estimating $\tilde{\alpha}_{ij_k}$ would be to maximize:

$$\tilde{\alpha}_{ij_k} = \arg\max_{\tilde{\alpha}_{ij_k}} J(\tilde{\alpha}_{ij_k}) \equiv \arg\max_{\tilde{\alpha}_{ij_k}} P[r_i(t_0...t_{M-1}) \mid \tilde{\alpha}_{ij_k}]. \tag{8}$$

Using Bayes' rule, we can calculate the conditional probability of the right hand side of Eq. (8) by:

$$J(\tilde{\alpha}_{ij_k}) \equiv P[r_i(t_0...t_{M-1}) \mid \tilde{\alpha}_{ij_k}] = \frac{P[\tilde{\alpha}_{ij_k} \mid r_i(t_0...t_{M-1})]P[r_i(t_0...t_{M-1})]}{P[\tilde{\alpha}_{ij_k}]}. \tag{9}$$

Since prior information regarding the distribution of $\tilde{\alpha}_{ij_k}$ is not available, a uniform distribution is assumed. Omitting $P[\tilde{\alpha}_{ij_k}]$, which is assumed to be constant, and omitting $P[r_i(t_0...t_{M-1})]$, which does not depend on $\tilde{\alpha}_{ij_k}$, does not affect the maximization of Eq. (9) with respect to $\tilde{\alpha}_{ij_k}$. Therefore, estimating $\tilde{\alpha}_{ij_k}$ using the maximum likelihood approach would be to maximize:

$$\tilde{\alpha}_{ij_k} = \arg\max_{\tilde{\alpha}_{ij_k}} \tilde{J}(\tilde{\alpha}_{ij_k}) \equiv \arg\max_{\tilde{\alpha}_{ij_k}} P[\tilde{\alpha}_{ij_k} \mid r_i(t_0...t_{M-1})]. \tag{10}$$

In order to evaluate the conditional probability $P[\tilde{\alpha}_{ij_k} \mid r_i(t_0...t_{M-1})]$, we first construct a density estimation to get a certain ratio on a specific time, given the measurements $x_i(t_l)$ and $x_1(t_l)$. It can be done using a kernel density estimation:

$$\hat{f}(r_i, t \mid x_i(t_l), x_1(t_l)) = \frac{1}{Mh^r h^t} K\left(\frac{r_i - r_i(t_l)}{h^r}, \frac{t - t_l}{h^t}\right), \tag{11}$$

where M is the number of time instances a signal was detected, K is a multivariate kernel density estimator, and h^r, h^t are the kernel support in the r and t axes respectively.

Using the law of total probability, $\hat{f}(r_i, t)$ can be found by calculating:

$$\hat{f}(r_i, t) = \sum_{l=0}^{M-1} \hat{f}(r_i, t \mid x_i(t_l), x_1(t_l)) P[x_i(t_l), x_1(t_l)]. \tag{12}$$

We interpret the probability $P[x_i(t_l), x_1(t_l)]$ as a measure of the correctness of calculating $\tilde{a}_{ij}(t_l)$ using Eq. (7). If the noise is at least one order of magnitude smaller than the observed signals, the approximation of Eq. (7) holds. If the noise parameters can be estimated, for example in the case of normal distributed noise with a known variance σ^2, the probability of the noise being an order of magnitude smaller than the measurement $x_i(t_l)$ or $x_1(t_l)$, is[5]:

$$P[x_i(t_l), x_1(t_l)] \equiv \min\{ \int_{-|x_i(t_l)|}^{|x_i(t_l)|} \frac{1}{\sigma\sqrt{20\pi}} e^{-\frac{v^2}{20\sigma^2}} dv, \int_{-|x_1(t_l)|}^{|x_1(t_l)|} \frac{1}{\sigma\sqrt{20\pi}} e^{-\frac{v^2}{20\sigma^2}} dv \}. \tag{13}$$

We want to evaluate the conditional probability of $\tilde{a}_{ij}(t)$ represented by $\tilde{\alpha}_{ij_k}$ parameters given the ratio of the measurements $r_i(t_0...t_{M-1})$, i.e. Eq. (10). Since we know from Eq. (5) that $\tilde{a}_{ij}(t) = \tilde{g}_{ij}(\tilde{\alpha}_{ij_k}; t)$, we can calculate this probability using a line integral over a scalar field, being the density $\hat{f}(r_i, t)$:

$$\tilde{J}(\tilde{\alpha}_{ij_k}) \equiv P[\tilde{\alpha}_{ij_k} \mid r_i(t_0...t_{M-1})] = \int_{t_1}^{t^2} \hat{f}(g_{ij}(\tilde{\alpha}_{ij_k}; t), t)\sqrt{1 + [\tilde{g}'_{ij}(\tilde{\alpha}_{ij_k}; t)]^2} dt, \tag{14}$$

where $\tilde{g}'_{ij}$ stands for the derivation with respect to t, and t_1, t_2 are the observation start and stop time respectively.

The optimization can be done using the Newton method, starting from several initial points as follows:

1. Take as an initial guess a vector of $\tilde{\alpha}^{(0)}_{ij_k}$ parameters.
2. Use the vector $\tilde{\alpha}^{(m)}_{ij_k}$ obtained from the previous step and construct the gradient vector $\nabla\tilde{J}(\tilde{\alpha}^{(m)}_{ij_k})$ and the Hessian matrix $H_{\tilde{J}}(\tilde{\alpha}^{(m)}_{ij_k})$ using:

$$\frac{\partial \tilde{J}(\tilde{\alpha}^{(m)}_{ij_k})}{\partial \tilde{\alpha}^{(m)}_{ij_k}} = \frac{\partial P[\tilde{\alpha}^{(m)}_{ij_k} \mid r_i(t_0...t_{M-1})]}{\partial \tilde{\alpha}^{(m)}_{ij_k}} \tag{15}$$

$$\frac{\partial^2 \tilde{J}(\tilde{\alpha}^{(m)}_{ij_k})}{\partial \tilde{\alpha}^{(m)}_{ij_k} \partial \tilde{\alpha}^{(m)}_{ij_p}} = \frac{\partial^2 P[\tilde{\alpha}^{(m)}_{ij_k} \mid r_i(t_0...t_{M-1})]}{\partial \tilde{\alpha}^{(m)}_{ij_k} \partial \tilde{\alpha}^{(m)}_{ij_p}}. \tag{16}$$

3. Update the estimated parameters:

$$\tilde{\alpha}^{(m+1)}_{ij_k} = \tilde{\alpha}^{(m)}_{ij_k} - H_{\tilde{J}}(\tilde{\alpha}^{(m)}_{ij_k})\nabla\tilde{J}(\tilde{\alpha}^{(m)}_{ij_k}). \tag{17}$$

4. Repeat steps 2 and 3 until convergence.

The initial points can be chosen by using an approach similar to that of the Hough transform in image processing [1]. The grouping problem resolved by issuing each time instance to the maximum which maximized Eq. (10) with sufficient probability.

---

[5] In the case where the noise parameters are unknown, we assume that $P[x_i(t_l), x_1(t_l)]$ is constant.

A few implementation remarks are in order:

– The integral of Eq. (14) should be found indefinitely. Therefore, we suggest using the Epanechnikov radially symmetric kernel [8] which is defined as:

$$K(r,t) = \begin{cases} \frac{2}{\pi}(1 - r^2 - t^2) & (r^2 + t^2) \leq 1 \\ 0 & othewise \end{cases} \tag{18}$$

– If $\tilde{g}_{ij}(\tilde{\alpha}_{ij_k})$ makes the indefinite integral of Eq. (14) unsolvable, an approximation for $\tilde{g}_{ij}(\tilde{\alpha}_{ij_k})$ can be used.
– It is usually preferable to define $\tilde{b}_{ij}(t) \equiv arctan(\tilde{a}_{ij}(t))$ in order to eliminate the noise amplification accompanying the calculation of $x_i(t)/x_1(t)$. The definition of $r_i(t_l)$ to be inserted in Eq. (11) should be changed to $r_i(t_l) \equiv arctan(x_i(t_l)/x_1(t_l))$. After optimizing for the $\tilde{\alpha}_{ij_k}$ parameters, $\tilde{a}_{ij}(t)$ can be found using $\tilde{a}_{ij}(t) = tan(\tilde{b}_{ij})$.

5 Results

We tested our approach on both synthetic and real mixtures. For synthetic mixtures, 2 random signals were drawn from an exponential distribution and mixed with the diagonal matrix $A(t)$ with diagonal coefficients: $a_{jj}(t) = \alpha_{jj_1}t^2 + \alpha_{jj_2}t + \alpha_{jj_3}$. Random noise with normal distribution was added to the mixtures yielding a mixture-to-noise ratio of 26 dB. The setup depicted in Fig. 1 was build. Real mixtures were acquired with a Canon D100 digital camera connected to a computer. The PSNR, which was estimated by taking consecutive identical pictures, was 40 dB. We defined $\tilde{b}_{2j}(t) \equiv arctan(\tilde{a}_{2j}(t))$ and assumed that a second

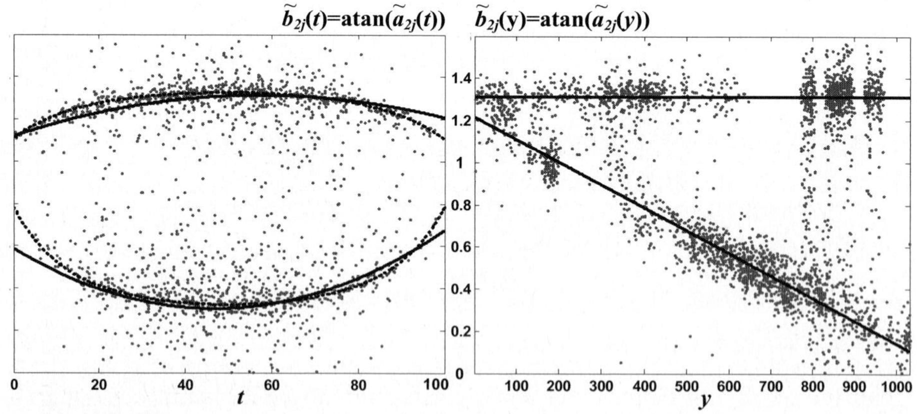

Fig. 2. Estimated $\tilde{b}_{2j}(t)$ for the synthetic (left) mixture and $\tilde{b}_{2j}(y)$ for the real (right) mixture. The estimated $\tilde{b}_{2j}(t)$ or $\tilde{b}_{2j}(y)$ using our algorithm and quadratic approximation is depicted by the solid line. The true $\tilde{b}_{2j}(t)$, for the synthetic mixture, is depicted by the dashed line. Scattered dots represents $r_2(t_n)$ or $r_2(y_n)$, where t_n and y_n are all the time instances or pixels where the signal (or its derivative) was above a threshold.

Table 1. Signal-to-Noise Ratio of the Synthetic Example

Mixture/Estimated	s_1 [dB]	s_2 [dB]
x_1	-15	8
x_2	10	-10
Estimated	16	21

order polynomial is the estimated Taylor expansion of $\tilde{b}_{2j}(t)$. Since images are not naturally sparse, a derivation with respect to the y axis (along the rows) was applied to the image mixtures. We initiated our algorithm with several starting parameters and the algorithm converged to two local maxima. Fig. 2 shows the results of identifying $\tilde{b}_{2j}(t)$ for the synthetic and the real mixtures. It can be seen that the algorithm produces a close approximation. Table 1 summarizes the results of inverting the system of the synthetic example. The signal-to-noise ratio has improved dramatically after the separation.

Fig 3. shows the results of inverting the system of the real mixture to recover the separated images. As it can be seen, the estimated images are separated properly.

Fig. 3. Results of separating the real mixtures: [Top] The mixtures. [Bottom] Estimated separated sources.

6 Conclusion

The proposed approach is general in that it can be applied to any type of position-varying or time-varying mixing model. As such, it is useful in various optical cases where the mixing lens cannot be assumed to acquire distortionless images. Likewise in various time domain problems, the mixing characteristic vary with time. Whereas in simple linear or quadratic forms of varying mixing parameters a simple solution can be obtained according to our approach, in other cases a solution is possible by solving integral equations numerically. Further extension of this study is now in progress, separating time-varying convolutive mixtures by using a proper transformation to a time frequency domain in which the problem is of time-varying instantaneous form (see [3] for details).

Acknowledgement

Research supported in part by the Ollendorff Minerva center, by the HASSIP Research Network Program HPRN-CT-2002-00285, sponsored by the European Commission, and by the Fund for Promotion of Research at the Technion. R. K. gratefully acknowledges the special doctoral fellowship awarded by HP.

References

1. Ballard, D.H.: Generalizing the Hough Transform to Detect Arbitrary Shapes. Pattern Recognition 13(2), 111–122 (1981)
2. Everson, R.M., Roberts, S.J.: Particle Filters for Non-stationary ICA. In: Advances in Independent Component Analysis, pp. 23–41. Springer, Heidelberg (2000)
3. Kaftory, R., Zeevi, Y.Y.: Blind Separation of Time-Varying Signal Mixtures Using Zadeh's Transform. In: Proceedings of EUSIPCO (2006)
4. Kenneth, H., Deniz, E., Jose, P.: Blind Source Separation of Time-Varying Instantaneous Mixtures Using an On-line Algorithm. In: Proc. ICASSP' 02, pp. 993–996 (2001)
5. Kisilev, P., Zibulevsky, M., Zeevi, Y.Y.: Blind Source Separation Using Multinode Sparse Representation. In: Proc. ICIP 2001, pp. 202–205 (2001)
6. Kuivunen, V., Enescu, M., Oja, E.: Adaptive Algorithm for Blind Separation from Noisy Time-Varying Mixtures. Neural Computation. 13, 2339–2358 (2001)
7. Mukai, R., Sawada, H., Araki, S., Makino, S.: Robust Real-Time Blind Source Separation for Moving Speakers in a Room. In: Proc. ICASSP' 03, pp. 469–472 (2003)
8. Scott, D.W.: Multivariate Density Estimation, p. 139. Wiley, Chichester (1992)
9. Weisman, T., Yeredor, A.: Separation of Periodically Time-Varying Mixtures Using Second-Order Statistics. In: Rosca, J., Erdogmus, D., Príncipe, J.C., Haykin, S. (eds.) ICA 2006. LNCS, vol. 3889, Springer, Heidelberg (2006)
10. Yeredor, A.: TV-SOBI: An Expansion of SOBI for Linearly Time-Varying Mixtures. In: Proceedings of the ICA 2003 (2003)
11. Zibulevsky, M., Pearlmutter, B.A., Bofill, P., Kisilev, P.: Blind Source Separation by Sparse Decomposition, Independent Component Analysis: Principles and Practice, Cambridge (2001)

Infinite Sparse Factor Analysis and Infinite Independent Components Analysis

David Knowles and Zoubin Ghahramani

Department of Engineering
University of Cambridge
CB2 1PZ, UK
d.a.knowles.07@cantab.net, zoubin@eng.cam.ac.uk

Abstract. A nonparametric Bayesian extension of Independent Components Analysis (ICA) is proposed where observed data $\mathbf{Y}$ is modelled as a linear superposition, $\mathbf{G}$, of a potentially infinite number of hidden sources, $\mathbf{X}$. Whether a given source is active for a specific data point is specified by an infinite binary matrix, $\mathbf{Z}$. The resulting sparse representation allows increased data reduction compared to standard ICA. We define a prior on $\mathbf{Z}$ using the Indian Buffet Process (IBP). We describe four variants of the model, with Gaussian or Laplacian priors on $\mathbf{X}$ and the one or two-parameter IBPs. We demonstrate Bayesian inference under these models using a Markov Chain Monte Carlo (MCMC) algorithm on synthetic and gene expression data and compare to standard ICA algorithms.

1 Introduction

Independent Components Analysis (ICA) is a model which explains observed data, $\mathbf{y}_t$ (dimension D) in terms of a linear superposition of independent hidden sources, $\mathbf{x}_t$ (dimension K), so $\mathbf{y}_t = \mathbf{G}\mathbf{x}_t + \boldsymbol{\epsilon}_t$, where $\mathbf{G}$ is the mixing matrix and $\boldsymbol{\epsilon}_t$ is Gaussian noise. In the standard ICA model we assume $K = D$ and that there exists $\mathbf{W} = \mathbf{G}^{-1}$. We use FastICA [1], a widely used implementation, as a benchmark. The assumption $K = D$ may be invalid, so Reversible Jump MCMC [2] could be used to infer K. In this paper we propose a sparse implementation which allows a potentially infinite number of components and the choice of whether a hidden source is active for a data point, allowing increased data reduction for systems where sources are intermittently active. Although ICA is not a time-series model it has been used successfully on time-series data such as electroencephalograms [3]. It has also been applied to gene expression data [4], the application we choose for a demonstration.

2 The Model

We define a binary vector $\mathbf{z}_t$ which acts as a mask on $\mathbf{x}_t$. Element z_{kt} specifies whether hidden source k is active for data point t. Thus

$$\mathbf{Y} = \mathbf{G}(\mathbf{Z} \odot \mathbf{X}) + \mathbf{E} \tag{1}$$

M.E. Davies et al. (Eds.): ICA 2007, LNCS 4666, pp. 381–388, 2007.

where $\odot$ denotes element-wise multiplication and $\mathbf{X}$, $\mathbf{Y}$, $\mathbf{Z}$ and $\mathbf{E}$ are concatenated matrices of $\mathbf{x}_t$, $\mathbf{y}_t$, $\mathbf{z}_t$ and $\boldsymbol{\epsilon}_t$ respectively. We allow a potentially infinite number of hidden sources, so that $\mathbf{Z}$ has infinitely many rows, although only a finite number will have non-zero entries. We assume Gaussian noise with variance σ_ϵ^2, which is given an inverse Gamma prior $\mathcal{IG}\left(\sigma_\epsilon^2; a, b\right)$.

We define two variants based on the prior for x_{kt}: *infinite sparse Factor Analysis* (isFA) has a unit Gaussian prior; *infinite Independent Components Analysis* (iICA) has a Laplacian(1) prior. Other heavy tailed distributions are possible but are not explored here. Varying the variance is redundant because we infer the variance of the mixture weights. The prior on the elements of $\mathbf{G}$ is Gaussian with variance σ_G^2, which is given an inverse Gamma prior. We define the prior on $\mathbf{Z}$ using the Indian Buffet Process with parameter α (and later β) as described in Section 2.1 and in more detail in [5]. We place Gamma priors on α and β.

All four variants share

$$\boldsymbol{\epsilon}_t \sim \mathcal{N}\left(0, \sigma_\epsilon^2 \mathbf{I}\right) \qquad \sigma_\epsilon^2 \sim \mathcal{IG}\left(a, b\right) \tag{2}$$

$$\mathbf{g}_k \sim \mathcal{N}\left(0, \sigma_G^2\right) \qquad \sigma_G^2 \sim \mathcal{IG}\left(c, d\right) \tag{3}$$

$$\mathbf{Z} \sim \mathcal{IBP}(\alpha, \beta) \qquad \alpha \sim \mathcal{G}\left(e, f\right) \tag{4}$$

The differences between the variants are summarised here.

	$x_{kt} \sim \mathcal{N}\left(0,1\right)$	$x_{kt} \sim \mathcal{L}(1)$
$\beta = 1$	*isFA*$_1$	*iICA*$_1$
$\beta \sim \mathcal{G}\left(1,2\right)$	*isFA*$_2$	*iICA*$_2$

2.1 Defining a Distribution on an Infinite Binary Matrix

Start with a finite model. We derive our distribution on $\mathbf{Z}$ by defining a finite K model and taking the limit as $K \to \infty$. We then show how the infinite case corresponds to a simple stochastic process.

We assume that the probability of a source k being active is π_k, and that the sources are generated independently. We find

$$P(\mathbf{Z}|\boldsymbol{\pi}) = \prod_{k=1}^{K}\prod_{t=1}^{N} P(z_{kt}|\pi_k) = \prod_{k=1}^{K} \pi_k^{m_k}(1-\pi_k)^{N-m_k} \tag{5}$$

where N is the total number of data points and $m_k = \sum_{t=1}^{N} z_{kt}$ is the number of data points for which source k is active. We put a Beta$(\frac{\alpha}{K}, 1)$ prior on π_k, where α is the strength parameter. Due to the conjugacy between the binomial and beta distributions we are able to integrate out π to find

$$P(\mathbf{Z}) = \prod_{k=1}^{K} \frac{\frac{\alpha}{K}\Gamma(m_k + \frac{\alpha}{K})\Gamma(N - m_k + 1)}{\Gamma(N + 1 + \frac{\alpha}{K})} \tag{6}$$

Take the infinite limit. By defining a scheme to order the non-zero rows of $\mathbf{Z}$ (see [5]) we can take $K \to \infty$ and find

$$P(\mathbf{Z}) = \frac{\alpha^{K_+}}{\prod_{h>0} K_h!} \exp\{-\alpha H_N\} \prod_{k=1}^{K_+} \frac{(N-m_k)!(m_k-1)!}{N!} \tag{7}$$

where K_+ is the number of active features, $H_N = \sum_{j=1}^{N} \frac{1}{j}$ is the N-th harmonic number, and K_h is the number of rows whose entries correspond to the binary number h.

Go to an Indian Buffet. This distribution corresponds to a simple stochastic process, the Indian Buffet Process. Consider a buffet with a seemingly infinite number of dishes (hidden sources) arranged in a line. The first customer (data point) starts at the left and samples Poisson(α) dishes. The ith customer moves from left to right sampling dishes with probability $\frac{m_k}{i}$ where m_k is the number of customers to have previously sampled that dish. Having reached the end of the previously sampled dishes, he tries Poisson($\frac{\alpha}{i}$) new dishes.

If we apply the same ordering scheme to the matrix generated by this process as for the finite model, we recover the correct exchangeable distribution. Since the distribution is exchangeable with respect to the customers we find by considering the last customer that $P(z_{kt} = 1|\mathbf{z}_{-kt}) = \frac{m_{k,-t}}{N}$ where $m_{k,-t} = \sum_{s \neq t} z_{ks}$, which is used in sampling $\mathbf{Z}$. By exchangeability and considering the first customer, the number of active sources for a data point follows a Poisson(α) distribution, and the expected number of entries in $\mathbf{Z}$ is $N\alpha$. We also see that the number of active features, $K_+ = \sum_{t=1}^{N}$ Poisson($\frac{\alpha}{t}$) = Poisson(αH_N) which grows as $\alpha \log N$ for large N.

Two parameter generalisation. A problem with the one parameter IBP is that the number of features per object, α, and the total number of features, $N\alpha$, are both controlled by α and cannot vary independently. Under this model, we cannot tune how likely it is for features to be shared across objects. To overcome this restriction we follow [6], introducing β, a measure of the feature *repulsion*. The ith customer now samples dish k with probability $\frac{m_k}{\beta+i-1}$ and samples Poisson($\frac{\alpha\beta}{\beta+i-1}$) new dishes.

We find $P(z_{kt} = 1|\mathbf{z}_{-kt}, \beta) = \frac{m_{k,-t}}{\beta+N-1}$. The marginal probability of $\mathbf{Z}$ becomes

$$P(\mathbf{Z}|\alpha,\beta) = \frac{(\alpha\beta)^{K_+}}{\prod_{h>0} K_h!} \exp\{-\alpha H_N(\beta)\} \prod_{k=1}^{K_+} B(m_k, N-m_k+\beta) \tag{8}$$

where $H_N(\beta) = \sum_{j=1}^{N} \frac{\beta}{\beta+j-1}$.

3 Inference

Given the observed data $\mathbf{Y}$, we wish to infer the hidden sources $\mathbf{X}$, which sources are active $\mathbf{Z}$, the mixing matrix $\mathbf{G}$, and all hyperparameters. We use Gibbs sampling, but with Metropolis-Hastings (MH) steps for β and sampling new features.

We draw samples from the marginal distribution of the model parameters given the data by successively sampling the conditional distributions of each parameter in turn, given all other parameters.

Hidden sources. We sample each element of $\mathbf{X}$ for which $z_{kt} = 1$. We denote the k-th column of $\mathbf{G}$ by $\mathbf{g}_k$ and $\boldsymbol{\epsilon}_t|_{z_{kt}=0}$ by $\boldsymbol{\epsilon}_{-kt}$. For isFA we find the conditional distribution is a Gaussian:

$$P(x_{kt}|\mathbf{G}, \mathbf{x}_{-kt}, \mathbf{y}_t, \mathbf{z}_t) = \mathcal{N}\left(x_{kt}; \frac{\mathbf{g}_k^T \boldsymbol{\epsilon}_{-kt}}{\sigma_\epsilon^2 + \mathbf{g}_k^T \mathbf{g}_k}, \frac{\sigma_\epsilon^2}{\sigma_\epsilon^2 + \mathbf{g}_k^T \mathbf{g}_k}\right) \tag{9}$$

For iICA we find a piecewise Gaussian distribution, which it is possible to sample from analytically given the Gaussian c.d.f. function F

$$P(x_{kt}|\mathbf{G}, \mathbf{x}_{-kt}, \mathbf{y}_t, \mathbf{z}_t) = \begin{cases} \mathcal{N}\left(x_{kt}; \mu_-, \sigma^2\right) & x_{kt} > 0 \\ \mathcal{N}\left(x_{kt}; \mu_+, \sigma^2\right) & x_{kt} < 0 \end{cases} \tag{10}$$

$$\text{where } \mu_\pm = \frac{\mathbf{g}_k^T \boldsymbol{\epsilon}_{-kt} \pm \sigma_\epsilon^2}{\mathbf{g}_k^T \mathbf{g}_k} \text{ and } \sigma^2 = \frac{\sigma_\epsilon^2}{\mathbf{g}_k^T \mathbf{g}_k} \tag{11}$$

Active sources. To sample $\mathbf{Z}$ we first define the ratio of conditionals, r

$$r = \underbrace{\frac{P(\mathbf{y}_t|\mathbf{G}, \mathbf{x}_{-kt}, \mathbf{z}_{-kt}, z_{kt} = 1, \sigma_\epsilon^2)}{P(\mathbf{y}_t|\mathbf{G}, \mathbf{x}_{-kt}, \mathbf{z}_{-kt}, z_{kt} = 0, \sigma_\epsilon^2)}}_{r_l} \underbrace{\frac{P(z_{kt} = 1|\mathbf{z}_{-kt})}{P(z_{kt} = 0|\mathbf{z}_{-kt})}}_{r_p} \tag{12}$$

so that $P(z_{kt} = 1|\mathbf{G}, \mathbf{X}_{-kt}, \mathbf{Y}, \mathbf{Z}_{-kt}) = \frac{r}{r+1}$. From Section 2.1 the ratio of priors is $r_p = \frac{m_{k,-t}}{\beta+N-1-m_{k,-t}}$. To find $P(\mathbf{y}_t|\mathbf{G}, \mathbf{x}_{-kt}, \mathbf{z}_{-kt}, z_{kt} = 1)$ we must marginalise over all possible values of x_{kt}.

$$P(\mathbf{y}_t|\mathbf{G}, \mathbf{x}_{-kt}, \mathbf{z}_{-kt}, z_{kt} = 1) = \int P(\mathbf{y}_t|\mathbf{G}, \mathbf{x}_t, \mathbf{z}_{-kt}, z_{kt} = 1) P(x_{kt}) \mathrm{d}x_{kt} \tag{13}$$

For isFA, using Equation (9) and integrating we find $r_l = \sigma \exp\left\{\frac{\mu^2}{2\sigma^2}\right\}$. For iICA we use Equation (11) and integrate above and below 0 to find

$$r_l = \sigma\sqrt{\frac{\pi}{2}}\left[F(0; \mu_+, \sigma)\exp\left\{\frac{\mu_+^2}{2\sigma^2}\right\} + (1 - F(0; \mu_-, \sigma))\exp\left\{\frac{\mu_-^2}{2\sigma^2}\right\}\right] \tag{14}$$

Creating new features. $\mathbf{Z}$ is a matrix with infinitely many rows, but only the non-zero rows can be held in memory. However, the zero rows still need to be taken into account. Let κ_t be the number of rows of $\mathbf{Z}$ which contain 1 only in column t, i.e. the number of features which are active only at time t. New features are proposed by sampling κ_t with a MH step. We propose a move $\xi \to \xi^*$ with probability $J(\xi^*|\xi)$, following [7], we set to be equal to the prior on ξ^*. This move is accepted with probability $\min(1, r_{\xi \to \xi^*})$ where

$$r_{\xi \to \xi^*} = \frac{P(\xi^*|\text{rest}) J(\xi|\xi^*)}{P(\xi^*|\text{rest}) J(\xi^*|\xi)} = \frac{P(\text{rest}|\xi^*) P(\xi^*) P(\xi)}{P(\text{rest}|\xi) P(\xi) P(\xi^*)} = \frac{P(\text{rest}|\xi^*)}{P(\text{rest}|\xi)} \tag{15}$$

where *rest* denotes all other variables. By this choice $r_{\xi\rightarrow\xi^*}$ becomes the ratio of likelihoods. From the IBP the prior for κ_t is $P(\kappa_t|\alpha) = \text{Poisson}(\frac{\alpha\beta}{\beta+N-1})$.

For isFA we can integrate out $\mathbf{x}_t'$, the new elements of $\mathbf{x}_t$, but not $\mathbf{G}'$, the new columns of $\mathbf{G}$, so our proposal is $\xi = \{\mathbf{G}', \kappa_t\}$. We find $r_{\xi\rightarrow\xi^*} = |\boldsymbol{\Lambda}|^{-\frac{1}{2}}$ $\exp\left(\frac{1}{2}\boldsymbol{\mu}^T\boldsymbol{\Lambda}\boldsymbol{\mu}\right)$ where $\boldsymbol{\Lambda} = \mathbf{I} + \frac{\mathbf{G}^{*T}\mathbf{G}^*}{\sigma_\epsilon^2}$ and $\boldsymbol{\Lambda}\boldsymbol{\mu} = \frac{1}{\sigma_\epsilon^2}\mathbf{G}^{*T}\boldsymbol{\epsilon}_t$.

For iICA marginalisation is not possible so $\xi = \{\mathbf{G}', \mathbf{x}_t', \kappa_t\}$. From Equation (15) we find

$$r_{\xi\rightarrow\xi^*} = \exp\left\{-\frac{1}{2\sigma_\epsilon^2}\mathbf{x}_t'^T\mathbf{G}^{*T}(\mathbf{G}^*\mathbf{x}_t' - 2\boldsymbol{\epsilon}_t)\right\} \tag{16}$$

Mixture weights. We sample the columns $\mathbf{g}_k$ of $\mathbf{G}$. We denote the kth row of $(\mathbf{Z}\odot\mathbf{X})$ by $\mathbf{x}_k'^T$. We have $P(\mathbf{g}_k|\mathbf{G}_{-k}, \mathbf{X}, \mathbf{Y}, \mathbf{Z}, \sigma_\epsilon^2, \sigma_G^2) \propto P(\mathbf{Y}|\mathbf{G}, \mathbf{X}, \mathbf{Z}, \sigma_\epsilon^2)$ $P(\mathbf{g}_k|\sigma_G^2)$. The total likelihood function has exponent

$$-\frac{1}{2\sigma_\epsilon^2}\,\text{tr}(\mathbf{E}^T\mathbf{E}) = -\frac{1}{2\sigma_\epsilon^2}((\mathbf{x}_k'^T\mathbf{x}_k')(\mathbf{g}_k^T\mathbf{g}_k) - 2\mathbf{g}_k^T\mathbf{E}|_{\mathbf{g}_k=0}) + \text{const} \tag{17}$$

where $\mathbf{E} = \mathbf{Y} - \mathbf{G}(\mathbf{Z}\odot\mathbf{X})$. We thus find the conditional of $\mathbf{g}_k$ is $\mathcal{N}(\boldsymbol{\mu}, \boldsymbol{\Lambda})$ where $\boldsymbol{\mu} = \frac{\sigma_G^2}{\mathbf{x}_k'^T\mathbf{x}_k'\sigma_G^2 + \sigma_\epsilon^2}\mathbf{E}|_{\mathbf{g}_k=0}\mathbf{x}_k'$ and $\boldsymbol{\Lambda} = \left(\frac{\mathbf{x}_k'^T\mathbf{x}_k'}{\sigma_\epsilon^2} + \frac{1}{\sigma_G^2}\right)\mathbf{I}_{D\times D}$.

Learning the noise level. We allow the model to learn the noise level σ_ϵ^2. Applying Bayes' rule we find

$$P(\sigma_\epsilon^2|\mathbf{E}, a, b) \propto P(\mathbf{E}|\sigma_\epsilon^2)P(\sigma_\epsilon^2|a, b) = \mathcal{IG}\left(\sigma_\epsilon^2; a + \frac{ND}{2}, \frac{b}{1 + \frac{b}{2}\,\text{tr}\,(\mathbf{E}^T\mathbf{E})}\right) \tag{18}$$

Inferring the scale of the data. For sampling σ_G^2 the conditional prior on $\mathbf{G}$ acts as the likelihood term

$$P(\sigma_G^2|\mathbf{G}, c, d) \propto P(\mathbf{G}|\sigma_G^2)P(\sigma_G^2|c, d) = \mathcal{IG}\left(\sigma_G^2; c + \frac{DK}{2}, \frac{d}{1 + \frac{d}{2}\,\text{tr}\,(\mathbf{G}^T\mathbf{G})}\right) \tag{19}$$

IBP parameters. We infer the IBP strength parameter α. The conditional prior on $\mathbf{Z}$, given by Equation (8), acts as the likelihood term

$$P(\alpha|\mathbf{Z}, \beta) \propto P(\mathbf{Z}|\alpha, \beta)P(\alpha) = \mathcal{G}\left(\alpha; K_+ + e, \frac{f}{1 + fH_N(\beta)}\right) \tag{20}$$

We sample β by a MH step with acceptance probability $\min(1, r_{\beta\rightarrow\beta^*})$. By Equation (15) setting $J(\beta^*|\beta) = P(\beta^*) = \mathcal{G}(1, 1)$, results in $r_{\beta\rightarrow\beta^*} = \frac{P(Z|\alpha,\beta^*)}{P(Z|\alpha,\beta)}$.

4 Results

4.1 Synthetic Data

We ran all four variants and three FastICA variants (using the *pow3*, *tanh* and *gauss* non-linearities) on 30 sets of randomly generated data with $D = 7, K = 6, N = 200$, the $\mathbf{Z}$ matrix shown in Figure 1(a), and Gaussian or Laplacian source distributions. Figure 1 shows the average inferred $\mathbf{Z}$ matrix and algorithm convergence for a typical 1000 iteration ICA_1 run. $\mathbf{Z}$ is successfully recovered within an arbitrary ordering. The gaps in the inferred $\mathbf{Z}$ are a result of inferring $z_{kt} = 0$ where $x_{kt} = 0$.

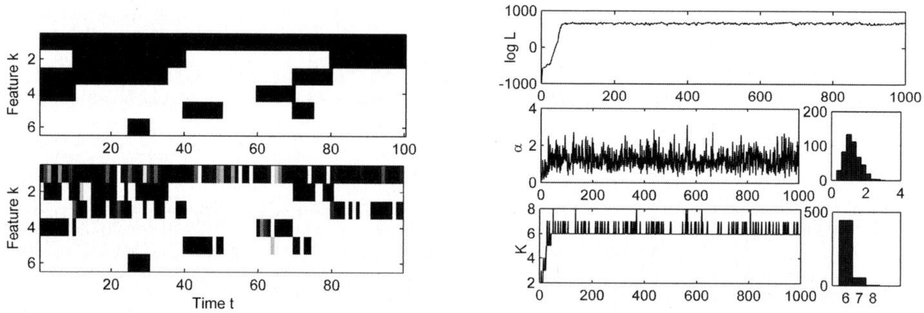

(a) *Top:* True $\mathbf{Z}$. *Bottom:* Inferred $\mathbf{Z}$

(b) Log likelihood, α and K^+ for duration of 1000 iteration run

Fig. 1. True and inferred $\mathbf{Z}$ and algorithm convergence for typical iICA_1 run

Boxplots of the Amari error [8] for each algorithm are shown in Figure 2. Figure 2(a) shows the results when the synthetic data has Gaussian source distributions. All four variants perform significantly better on the sparse synthetic data than any of the FastICA variants, but then we do not expect FastICA to recover Gaussian sources. Figure 2(b) shows the results when the synthetic data has Laplacian source distributions. As expected the FastICA performance is much improved because the sources are heavy tailed. However, isFA_1, iICA_1 and iICA_2 still perform better on average because they correctly model the sparse nature of the data. The performance of isFA_2 is severely effected by having the incorrect source model, suggesting the iICA variants may be more robust to deviations from the assumed source distribution. The two parameter IBP variants of both algorithms actually perform no better than the one parameter versions: $\beta = 1$ happens to be almost optimal for the synthetic $\mathbf{Z}$ used.

4.2 Gene Expression Data

We now apply our model to the microarray data from an ovarian cancer study [4], which represents the expression level of $N = 172$ genes (data points) across

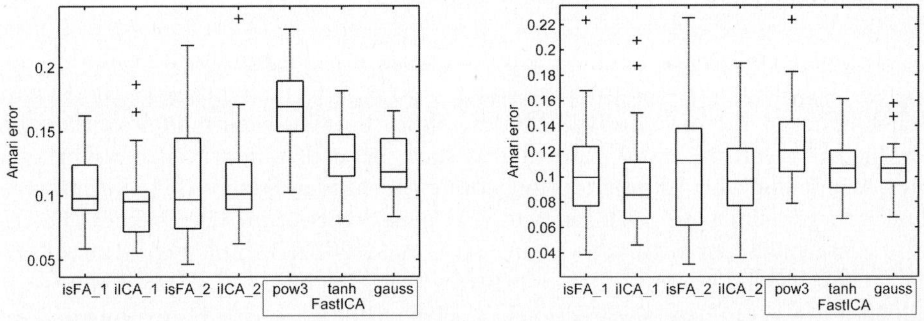

(a) Gaussian sources (b) Laplacian sources

Fig. 2. Boxplots of Amari errors for 30 synthetic data sets with $D = 7, N = 6, N = 100$ analysed using each algorithm variant and FastICA

$D = 17$ tissue samples (observed variables). The tissue samples are grouped into five tissue types: one healthy and four diseased. ICA was applied to this dataset in [4], where the term *gene signature* is used to describe the infered hidden sources. Some of the processes which regulate gene expression, such as DNA methylation, completely *silence* the gene, while others, such as transcription regulation, affect the level at which the gene is expressed. Thus our sparse model is highly valid for this system: **Z** represents which genes are silenced, and **X** represents the expression level of active genes.

Figure 4.2 shows the mean **G** matrix infered by iICA$_1$. Gene signature (hidden source) 1 is expressed across all the tissue samples, accounted for genes shared by all the samples. Signature 7 is specific to the pd-spa tissue type. This is consistent with that found in [4], with the same top 3 genes. Such tissue type dependent signatures could be used for observer independent classification. Signatures such as 5 which is differentially expressed across the pd-spa samples could help subclassify tissue types.

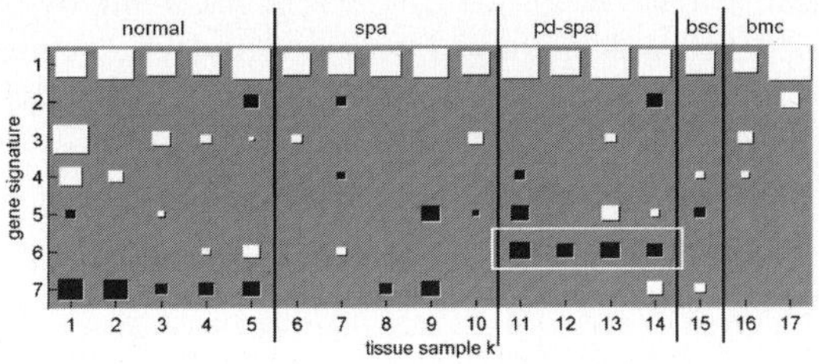

Fig. 3. Hinton diagram of $\mathbf{G}^T$: the expression level of each gene signature within each tissue sample

5 Conclusions and Future Work

In this paper we have defined the Infinite Sparse FA and Infinite ICA models using a distribution over the infinite binary matrix **Z** corresponding to the Indian Buffet Process. We have derived MCMC algorithms for each model to infer the parameters given observed data. These have been demonstrated on synthetic data, where the correct assumption about the hidden source distribution was shown to give optimal performance, and gene expression data, where the results were consistent with those using ICA. A MATLAB implementation of the algorithms will be made available at http://learning.eng.cam.ac.uk/zoubin/.

There are a number of directions in which this work can be extended. The recently developed stick breaking constructions for the IBP will allow a slice sampler to be derived for **Z** which should allow faster mixing than the MH step currently used in sampling new features. Faster partially deterministic algorithms would be useful for online learning in applications such as audio processing. The sparse nature of the model could have useful applications in data compression for storage or data reduction for further analysis.

References

1. Hyvärinen, A.: Fast and robust fixed-point algorithms for independent component analysis. IEEE Transactions on Neural Networks 10(3), 626–634 (1999)
2. Richardson, S., Green, P.J.: On bayesian analysis of mixtures with an unknown number of components. Journal of the Royal Statistical Society 59, 731–792 (1997)
3. Makeig, S., Bell, A.J., Jung, T.P., Sejnowski, T.J.: Independent component analysis of electroencephalographic data. Advances in Neural Information Processing Systems 8, 145–151 (1996)
4. Martoglio, A.M., Miskin, J.W., Smith, S.K., MacKay, D.J.C.: A decomposition model to track gene expression signatures: preview on observer-independent classification of ovarian cancer. Bioinformatics 18(12), 1617–1624 (2002)
5. Griffiths, T., Ghahramani, Z.: Infinite latent feature models and the indian buffet process. Technical Report 1, Gatsby Computational Neuroscience Unit (2005)
6. Ghahramani, Z., Griffiths, T., Sollich, P.: Bayesian nonparametric latent feature models. In: Bayesian Statistics 8, Oxford University Press, Oxford (2007)
7. Meeds, E., Ghahramani, Z., Neal, R., Roweis, S.: Modeling dyadic data with binary latent factors. In: Neural Information Processing Systems. vol. 19 (2006)
8. Amari, S., Cichocki, A., Yang, H.H.: A new learning algorithm for blind signal separation. In: Advances in Neural Information Processing Systems, vol. 8, pp. 757–763. The MIT Press, Cambridge (1996)

Fast Sparse Representation Based on Smoothed ℓ^0 Norm

G. Hosein Mohimani[1], Massoud Babaie-Zadeh[1,*], and Christian Jutten[2]

[1] Electrical Engineering Department, Advanced Communications Research Institute (ACRI), Sharif University of Technology, Tehran, Iran

[2] GIPSA-lab, Department of Images and Signals, National Polytechnic Institute of Grenoble (INPG), France

gh1985im@yahoo.com, mbzadeh@yahoo.com, Christian.Jutten@inpg.fr

Abstract. In this paper, a new algorithm for Sparse Component Analysis (SCA) or atomic decomposition on over-complete dictionaries is presented. The algorithm is essentially a method for obtaining sufficiently sparse solutions of underdetermined systems of linear equations. The solution obtained by the proposed algorithm is compared with the minimum ℓ^1-norm solution achieved by Linear Programming (LP). It is experimentally shown that the proposed algorithm is about two orders of magnitude faster than the state-of-the-art ℓ^1-magic, while providing the same (or better) accuracy.

Keywords: sparse component analysis, over-complete atomic decomposition.

1 Introduction

Obtaining sparse solutions of under-determined systems of linear equations is of significant importance in signal processing and statistics. Despite recent theoretical developments [1,2,3], the computational cost of the methods has remained as the main restriction, especially for large systems (large number of unknowns/equations). In this article, a new approach is proposed which provides a considerable reduction in complexity. To introduce the problem in more details, we will use the context of Sparse Component Analysis (SCA). The discussions, however, may be easily followed in other contexts of application, for example, in finding 'sparse decomposition' of a signal in an over-complete dictionary, which is the goal of the so-called over-complete 'atomic decomposition' [4].

SCA can be viewed as a method to achieve separation of sparse sources. The general Blind Source Separation (BSS) problem is to recover n unknown (statistically independent) sources from m observed mixtures of them, where little or no information is available about the sources (except their statistical independence) and the mixing system. In linear instantaneous (noiseless) model, it is assumed that $\mathbf{x}(t) = \mathbf{A}\mathbf{s}(t)$ in which $\mathbf{x}(t)$ and $\mathbf{s}(t)$ are the $m \times 1$ and $n \times 1$

* This work has been partially funded by Sharif University of Technology, by Center for International Research and Collaboration (ISMO) and by Iran NSF (INSF).

M.E. Davies et al. (Eds.): ICA 2007, LNCS 4666, pp. 389–396, 2007.

vectors of sources and mixtures and $\mathbf{A}$ is the $m \times n$ mixing matrix. The goal of BSS is then to find $\mathbf{s}(t)$ only from $\mathbf{x}(t)$. The general BSS problem is not easy for the case $n > m$. However, if the sources are sparse (i.e., not a totally blind situation), then the problem can be solved in two steps [3,2]: first estimating the mixing matrix [3,2,5,6], and then estimating the sources assuming $\mathbf{A}$ to be known [3,2,7,8]. In this paper we only consider the second step.

To obtain the sparsest solution of $\mathbf{As} = \mathbf{x}$, we may search for a solution of it having minimal ℓ^0 norm, i.e., minimum number of nonzero components. It is usually stated in the literature [4,9,10,3] that searching the minimum ℓ^0 norm is an intractable problem as the dimension increases (because it requires a combinatorial search), and it is too sensitive to noise (because any small amount of noise completely changes the ℓ^0 norm of a vector). Consequently, the researchers look for other approaches to find sparse solution of $\mathbf{As} = \mathbf{x}$ which are tractable. One of the most successful approaches is Basis Pursuit (BP) [11,1,10,3] which finds the minimum ℓ^1 norm (that is, the solution of $\mathbf{As} = \mathbf{x}$ for which $\sum_i |s_i|$ is minimized). Such a solution can be easily found by Linear Programming (LP) methods. The idea of Basis Pursuit is based on the property that for large systems of equations, the minimum ℓ^1 norm solution is also the minimum ℓ^0 norm solution [1,11,12,13]. By utilizing fast LP algorithms, specifically interior-point LP solvers or ℓ^1-magic [14] (which is about one order of magnitude faster), large-scale problems with thousands of sources and mixtures become tractable. However, although this approach is tractable, it is still very time-consuming. Another approach is Matching Pursuit (MP) [15,16,3] which is very fast, but is somewhat heuristic and does not provide good estimation of the sources.

In this article, we present a fast method for finding the sparse solution of an under-determined system of linear equations, which is based on minimization of ℓ^0 norm. The paper is organized as follows. The next section introduces a family of Gaussian sparsity norms and discusses their optimization. The algorithm is then stated in Section 3. Finally, Section 4 provides some experimental results of our algorithm and its comparison with BP.

2 The Main Idea

The main idea of this article is to approximate the ℓ^0 norm by a smooth (continuous) function, which lets us to use gradient based methods for its minimization and solves also the sensitivity of ℓ^0 norm to noise. In this section we introduce a family of smooth approximators of ℓ^0 norm, whose optimization results in a fast algorithm for finding the sparse solution while preserving noise robustness.

The ℓ^0 norm of $\mathbf{s} = [s_1 \dots s_n]^T$ is defined as the number of non-zero components of $\mathbf{s}$. In other words if we define

$$\nu(s) = \begin{cases} 1 & s \neq 0 \\ 0 & s = 0 \end{cases} \tag{1}$$

then

$$\|\mathbf{s}\|_0 = \sum_{i=1}^{n} \nu(s_i). \tag{2}$$

It is clear that the discontinuities of ℓ^0 norm are caused by discontinuities of the function ν. If we replace ν by a smooth estimation of it in (2), we obtain a smooth estimation of ℓ^0 norm. This may also provide some robustness to noise.

Different functions may be utilized for this aim. We use zero-mean Gaussian family of functions which seem to be very useful for this application, because of their differentiability. By defining:

$$f_\sigma(s) = \exp(-s^2/2\sigma^2), \tag{3}$$

we have:

$$\lim_{\sigma \to 0} f_\sigma(s) = \begin{cases} 1 & s = 0 \\ 0 & s \neq 0 \end{cases}. \tag{4}$$

Consequently, $\lim_{\sigma \to 0} f_\sigma(s) = 1 - \nu(s)$, and therefore if we define:

$$F_\sigma(\mathbf{s}) = \sum_{i=1}^{n} f_\sigma(s_i), \tag{5}$$

we have:

$$\lim_{\sigma \to 0} F_\sigma(\mathbf{s}) = \sum_{i=1}^{n} (1 - \nu(s_i)) = n - \|\mathbf{s}\|_0. \tag{6}$$

We take then $n - F_\sigma(\mathbf{s})$ as an approximation to $\|\mathbf{s}\|_0$:

$$\|\mathbf{s}\|_0 \approx n - F_\sigma(\mathbf{s}). \tag{7}$$

The value of σ specifies a trade-off between accuracy and smoothness of the approximation: the smaller σ, the better approximation, and the larger σ, the smoother approximation.

From (6), minimization of ℓ^0 norm is equivalent to maximization of F_σ for sufficiently small σ. This maximization should be done on the affine set $\mathcal{S} = \{\mathbf{s} \mid \mathbf{A}\mathbf{s} = \mathbf{x}\}$.

For small values of σ, F_σ contains a lot of local maxima. Consequently, it is very difficult to directly maximize this function for very small values of σ. However, as value of σ grows, the function becomes smoother and smoother, and for sufficiently large values of σ, as we will show, there is no local maxima (see Theorem 1 of the next section).

Our idea for escaping local maxima is then to decrease the value of σ gradually[1]: for each value of σ we use a steepest ascent algorithm for maximizing F_σ, and *the initial value of this steepest ascent algorithm is the maximizer of* F_σ *obtained for the previous (larger) value of* σ. Since the value of σ changes slowly, the steepest ascent algorithm is initialized not far from the actual maximum. Consequently, we hope that it would not be trapped in the local maxima.

Remark 1. Equation (6) proposes that $F_\sigma(\cdot)$ can be seen as a measure of 'sparsity' of a vector (especially for small values of σ): the sparser $\mathbf{s}$, the larger $F_\sigma(\mathbf{s})$. In this viewpoint, maximizing $F_\sigma(\mathbf{s})$ on a set is equivalent to finding the 'sparsest' element of that set.

---

[1] This idea is similar to simulated annealing, but, here, the sequence of decreasing values is short and easy to define so that the solution is achieved in a few steps, usually less than 10.

– Initialization:
  1. Choose an arbitrary solution from the feasible set $\mathcal{S}$, $\mathbf{v}_0$, e.g., the minimum ℓ^2 norm solution of $\mathbf{As} = \mathbf{x}$ obtained by pseudo-inverse (see the text).
  2. Choose a suitable decreasing sequence for σ, $[\sigma_1 \dots \sigma_K]$.
– for $k = 1, \dots, K$:
  1. Let $\sigma = \sigma_k$.
  2. Maximize (approximately) the function F_σ on the feasible set $\mathcal{S}$ using L iterations of the steepest ascent algorithm (followed by projection onto the feasible set):
     - Initialization: $\mathbf{s} = \mathbf{v}_{k-1}$.
     - for $j = 1 \dots L$ (loop L times):
       (a) Let: $\Delta\mathbf{s} = [s_1 \exp(-s_1^2/2\sigma_k^2), \dots, s_n \exp(-s_n^2/2\sigma_k^2)]^T$.
       (b) Let $\mathbf{s} \leftarrow \mathbf{s} - \mu\Delta\mathbf{s}$ (where μ is a small positive constant).
       (c) Project $\mathbf{s}$ back onto the feasible set $\mathcal{S}$:
$$\mathbf{s} \leftarrow \mathbf{s} - \mathbf{A}^T(\mathbf{A}\mathbf{A}^T)^{-1}(\mathbf{As} - \mathbf{x})$$
  3. Set $\mathbf{v}_k = \mathbf{s}$.
– Final answer is $\mathbf{s} = \mathbf{v}_l$.

Fig. 1. The final algorithm (SL0 algorithm)

3 The Algorithm

Based on the main idea of the previous section, the final algorithm (smoothed ℓ^0 or SL0) is given in Fig. 1. As indicated in the algorithm, the final value of previous estimation is used for the initialization of the next steepest ascent. By choosing a slowly decreasing sequence of σ, we may escape from getting trapped in the local maxima, and obtain the sparsest solution.

Remark 2. The internal loop (steepest ascent for a fixed σ) is repeated a fixed and small number of times (L). In other words, for increasing the speed, we do not wait for the (internal loop of the) steepest ascent algorithm to converge. This may be justified by gradual decrease in value of σ, and the fact that for each value, we do not need the exact maximizer of F_σ. All we need, is to enter a region near the (absolute) maximizer of F_σ for escaping from its local maximizers.

Remark 3. Steepest ascent consists of iterations of the form $\mathbf{s} \leftarrow \mathbf{s} + \mu_k \nabla F_\sigma(\mathbf{s})$. Here, the step-size parameters μ_k should be decreasing, i.e., for smaller values of σ, smaller values of μ_k should be applied. This is because for smaller values of σ, the function F_σ is more 'fluctuating', and hence smaller step-sizes should be used for its maximization. If we set[2] $\mu_k = \mu\sigma_k^2$, we obtain $\mathbf{s} \leftarrow \mathbf{s} - \mu\Delta\mathbf{s}$ as stated in the algorithm of Fig. 1 (note that $\Delta\mathbf{s} \triangleq -\nabla F_\sigma(\mathbf{s})/\sigma^2$).

Remark 4. The algorithm may work by initializing $\mathbf{v}_0$ (initial estimation of the sparse solution) to an arbitrary solution of $\mathbf{As} = \mathbf{x}$. However, the best initial

[2] In fact, we may think about changing the σ in (3) and (5) as looking at the same curve (or surface) at different 'scales', where the scale is proportional to σ^2. For having the same step-sizes of the steepest ascent algorithm in these different scales, μ_k should be proportional to σ^2.

value of $\mathbf{v}_0$ is the minimum ℓ^2 norm solution of $\mathbf{As} = \mathbf{x}$, which is given by the pseudo-inverse of $\mathbf{A}$. It is because this solution is the (unique) maximizer of $F_\sigma(\mathbf{s})$ on the feasible set $\mathcal{S}$, where σ tends to infinity. This is formally stated in the following theorem (refer to appendix for the proof).

Theorem 1. The solution of the problem:

$$\text{Maximize } F_\sigma(\mathbf{s}) \text{ subject to } \mathbf{As} = \mathbf{x},$$

where $\sigma \to \infty$, is the minimum ℓ^2 norm solution of $\mathbf{As} = \mathbf{x}$, that is, $\mathbf{s} = \mathbf{A}^T(\mathbf{AA}^T)^{-1}\mathbf{x}$.

Remark 5. Having initiated the algorithm with the minimum ℓ^2 norm solution (which corresponds to $\sigma = \infty$), the next value for σ (i.e., σ_1) may be chosen about two to four times of the maximum absolute value of the obtained sources ($\max_i |s_i|$). To see the reason, note first that:

$$\exp(-s_i^2/2\sigma^2) = \begin{cases} 1 & , \text{ if } |s_i| \ll \sigma \\ 0 & , \text{ if } |s_i| \gg \sigma \end{cases} . \tag{8}$$

Consequently, if we take, for example, $\sigma > 4\max_i |s_i|$ for all $1 \le i \le n$, then $\exp(-s_i^2/2\sigma^2) > 0.96 \approx 1$, and comparison with (8) shows that this value of σ acts virtually like infinity for all values of s_i, $1 \le i \le n$.

For the next σ_k's ($k \ge 2$), we have used $\sigma_k = \alpha\sigma_{k-1}$, where α is usually between 0.5 and 1.

Remark 6. The final value of σ depends on the noise level. For the noiseless case, it can be decreased arbitrarily to zero (its minimum values is determined by the desired accuracy, and/or machine precision). For the noisy case, it should be terminated about one to two times of energy of the noise. This is because, while σ is in this range, (8) shows that the cost function treats small (noisy) samples as zero (i.e., for which $f_\sigma(s_i) \approx 1$, $1 \le i \le n$). However, below this range, the algorithm tries to 'learn' these noisy values, and moves away from the true answer. Restricting σ to be above energy of the noise, provides the robustness of this approach to noise, which was one of the difficulties of using the exact ℓ^0 norm.

In the simulations of this paper, this noise level was assumed to be known[3].

4 Experimental Results

In this section, we justify performance of the presented approach and compare it with BP. Sparse sources are artificially created using Mixture of Gaussian (MoG) model[4]:

$$s_i \sim p \cdot \mathcal{N}(0, \sigma_{\text{on}}) + (1-p) \cdot \mathcal{N}(0, \sigma_{\text{off}}), \tag{9}$$

[3] Note that its exact value is not necessary, and in practice a rough estimation is sufficient.

[4] The model we have used is also called the Bernoulli-Gaussian model.

Table 1. Progress of SL0, Compared to ℓ^1-magic

itr. #	σ	MSE	SNR (dB)
1	1	$3.75\,e-2$	2.88
2	0.5	$2.19\,e-2$	5.21
3	0.2	$4.28\,e-3$	12.29
4	0.1	$1.67\,e-3$	16.37
5	0.05	$6.18\,e-4$	20.71
6	0.02	$1.91\,e-4$	25.80
7	0.01	$1.87\,e-4$	25.89

algorithm	total time	MSE	SNR (dB)
SL0	0.227 seconds	$2.34\,e-4$	25.67
ℓ^1-magic	20.8 seconds	$4.64\,e-4$	21.95

where p denotes probability of activity of the sources. σ_{on} and σ_{off} are the standard deviations of the sources in active and inactive mode, respectively. In order to have sparse sources, the parameters are required to satisfy the conditions $\sigma_{\text{off}} \ll \sigma_{\text{on}}$ and $p \ll 1$. σ_{off} is to model the noise in sources, and the larger values of σ_{off} produces stronger noise. In the simulation σ_{on} is fixed to 1. Each column of the mixing matrix is randomly generated using the normal distribution which is then normalized to unity.

The mixtures are generated using the noisy model $\mathbf{x} = \mathbf{A}\mathbf{s} + \mathbf{n}$, where $\mathbf{n}$ is an additive white Gaussian noise with variance $\sigma_n \mathbf{I}_m$ ($\mathbf{I}_m$ is the $m \times m$ identity matrix). Note that both σ_{off} and σ_n can be used for modeling the noise and they are both set to 0.01 in the simulation[5].

The values used for the experiment are $n = 1000$, $m = 400$, $p = 0.1$, $\sigma_{\text{off}} = 0.01$, $\sigma_{\text{on}} = 1$, $\sigma_n = 0.01$ and the sequence of σ is fixed to [1, 0.5, 0.2, 0.1, 0.05, 0.02, 0.01]. μ is set equal to 2.5. For each value of σ the gradient-projection loop (the internal loop) is performed three times (i.e., $L = 3$).

We use the CPU time as a measure of complexity. Although, the CPU time is not an exact measure, it can give us a rough estimation of complexity, and lets us roughly compare SL0 (Smoothed ℓ^0 norm) with BP[6].

Table 1 shows the gradual improvement in the output SNR after each iteration, for a typical run of SL0. Moreover, for this run, total time and final SNR have been shown for SL0 and for BP (using ℓ^1-magic). Figure 2 shows the actual source and it's estimations in different iterations for this run of SL0. The experiment is then repeated 100 times (with the same parameters, but for different randomly generated sources and mixing matrices) and the values of SNR (in dB) obtained over these simulations are averaged. For SL0, this averaged SNR was 25.67dB with standard derivation of 1.34dB. For ℓ^1 magic,

[5] Note that, although in the theoretical model only the noiseless case was addressed, because of continuity of the cost functions, the method can work as well in noisy conditions.

[6] Our simulations are performed in MATLAB7 under WinXP using an AMD Athlon sempron 2400+, 1.67GHz processor with 512MB of memory.

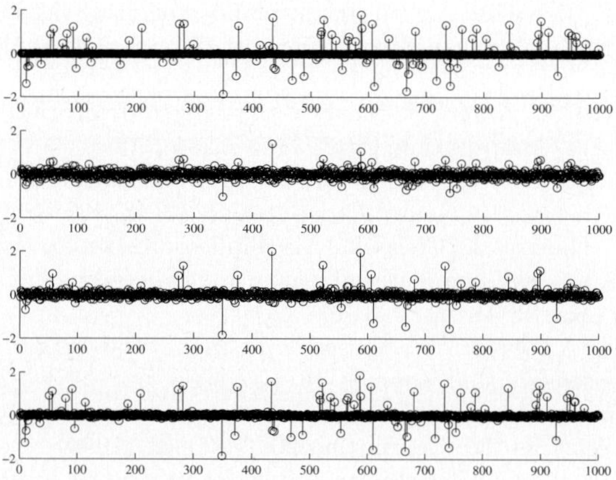

Fig. 2. Evolution of SL0 toward the solution: From top to bottom, first plot corresponds to the actual source, second plot is its estimation at the first level ($\sigma = 1$), third plot is its estimation at the second level ($\sigma = 0.5$), while the last plot is its estimation at third level ($\sigma = 0.2$)

these values were 21.92dB and 1.36dB, respectively. The minimum value of SNR was 20.12dB compared with minimum of 18.51dB for BP.

5 Conclusions

In this article, a fast method for finding sparse solutions of an under-determined system of linear equations was proposed (to be applied in atomic decomposition and SCA). SL0 was based on maximizing a 'smooth' measure of sparsity. SL0 shows to be about two orders of magnitude faster than the ℓ^1-magic, while providing the same (or better) accuracy. The authors conclude that sparse decomposition problem is not computationally as *hard* as suggested by the LP approach.

References

1. Donoho, D.L.: For most large underdetermined systems of linear equations the minimal l^1-norm solution is also the sparsest solution, Tech. Rep. (2004)
2. Bofill, P., Zibulevsky, M.: Underdetermined blind source separation using sparse representations. Signal Processing 81, 2353–2362 (2001)
3. Gribonval, R., Lesage, S.: A survey of sparse component analysis for blind source separation: principles, perspectives, and new challenges. In: Proceedings of ESANN'06, April 2006, pp. 323–330 (2006)
4. Donoho, D.L., Elad, M., Temlyakov, V.: Stable recovery of sparse overcomplete representations in the presence of noise. IEEE Trans. Info. Theory 52(1), 6–18 (2006)
5. Movahedi, F., Mohimani, G.H., Babaie-Zadeh, M., Jutten, C.: Estimating the mixing matrix in sparse component analysis (SCA) based on partial k-dimensional subspace clustering, Neurocomputing (sumitted)

6. Washizawa, Y., Cichocki, A.: on-line k-plane clustering learning algorithm for sparse comopnent analysis. In: Proceedinds of ICASSP'06, Toulouse (France), pp. 681–684 (2006)
7. Li, Y.Q., Cichocki, A., Amari, S.: Analysis of sparse representation and blind source separation. Neural Computation 16(6), 1193–1234 (2004)
8. Zibulevsky, M., Pearlmutter, B.A.: Blind source separation by sparse decomposition in a signal dictionary. Neural Computation 13(4), 863–882 (2001)
9. Georgiev, P.G., Theis, F.J., Cichocki, A.: Blind source separation and sparse component analysis for over-complete mixtures. In: Proceedings of ICASSP'04, Montreal (Canada), May 2004, pp. 493–496 (2004)
10. Li, Y., Cichocki, A., Amari, S.: Sparse component analysis for blind source separation with less sensors than sources. In: ICA2003, pp. 89–94 (2003)
11. Chen, S.S., Donoho, D.L., Saunders, M.A.: Atomic decomposition by basis pursuit. SIAM Journal on Scientific Computing 20(1), 33–61 (1999)
12. Donoho, D.L., Huo, X.: Uncertainty principles and ideal atomic decomposition. IEEE Trans. Inform. Theory 47(7), 2845–2862 (2001)
13. Elad, M., Bruckstein, A.: A generalized uncertainty principle and sparse representations in pairs of bases. IEEE Trans. Inform. Theory 48(9), 2558–2567 (2002)
14. Candes, E., Romberg, J.: ℓ_1-Magic: Recovery of Sparse Signals via Convex Programming, URL: `www.acm.caltech.edu/l1magic/downloads/l1magic.pdf`
15. Mallat, S., Zhang, Z.: Matching pursuits with time-frequency dictionaries. IEEE Trans. on Signal Proc. 41(12), 3397–3415 (1993)
16. Krstulovic, S., Gribonval, R.: MPTK: Matching pursuit made tractable. In: ICASSP'06 (2006)

Appendix: Proof of Theorem 1

Let $\mathbf{g}(\mathbf{s}) \triangleq \mathbf{A}\mathbf{s} - \mathbf{x}$, and consider the method of Lagrange multipliers for maximizing $F_\sigma(\mathbf{s})$ subject to the constraint $\mathbf{g}(\mathbf{s}) = \mathbf{0}$. Setting $\nabla F_\sigma(\mathbf{s}) = \boldsymbol{\lambda}^T \nabla(\mathbf{g}(\mathbf{s}))$, where $\boldsymbol{\lambda} = [\lambda_1, \ldots, \lambda_m]^T$ is the vector collecting the m Lagrange multipliers, along with the constraints $\mathbf{A}\mathbf{s} = \mathbf{x}$, results in the nonlinear system of $m + n$ equations and $m + n$ unknowns:

$$\begin{cases} \mathbf{A}\mathbf{s} = \mathbf{x} \\ \widehat{\boldsymbol{\lambda}}^T \mathbf{A} = [s_1 \exp\left(-s_1^2/2\sigma^2\right) \ldots s_n \exp\left(-s_n^2/2\sigma^2\right)] \end{cases} \tag{10}$$

where $\widehat{\boldsymbol{\lambda}}$ is an $m \times 1$ unknown vector (proportional to $\boldsymbol{\lambda}$). In general, it is not easy to solve this system of nonlinear equations and for small values of σ, the solution is not unique (because of existence of local maxima). However, when σ increases to infinity, the system becomes linear and easy to solve:

$$\begin{cases} \mathbf{A}\mathbf{s} = \mathbf{x} \\ \mathbf{A}^T \widehat{\boldsymbol{\lambda}} = \mathbf{s} \end{cases} \Rightarrow \mathbf{A}\mathbf{A}^T \widehat{\boldsymbol{\lambda}} = \mathbf{x} \Rightarrow \mathbf{s} = \mathbf{A}^T (\mathbf{A}\mathbf{A}^T)^{-1} \mathbf{x} \tag{11}$$

which is the minimum ℓ^2-norm or the pseudo-inverse solution of $\mathbf{A}\mathbf{s} = \mathbf{x}$. □

Estimating the Mixing Matrix in Sparse Component Analysis Based on Converting a Multiple Dominant to a Single Dominant Problem

Nima Noorshams[1], Massoud Babaie-Zadeh[1,⋆], and Christian Jutten[2]

[1] Electrical Engineering Department, Advanced Communications Research Institute (ACRI), Sharif University of Technology, Tehran, Iran
[2] GIPSA-lab, Department of Images and Signals, National Polytechnic Institute of Grenoble (INPG), France
nima_noorshams@yahoo.com, mbzadeh@yahoo.com, Christian.Jutten@inpg.fr

Abstract. We propose a new method for estimating the mixing matrix, $\mathbf{A}$, in the linear model $\mathbf{x}(t) = \mathbf{A}\mathbf{s}(t), t = 1, \ldots, T$, for the problem of underdetermined Sparse Component Analysis (SCA). Contrary to most previous algorithms, there can be more than one dominant source at each instant (we call it a "multiple dominant" problem). The main idea is to convert the multiple dominant problem to a series of single dominant problems, which may be solved by well-known methods. Each of these single dominant problems results in the determination of some columns of $\mathbf{A}$. This results in a huge decrease in computations, which lets us to solve higher dimension problems that were not possible before.

1 Introduction

Sparse Component Analysis (SCA) [1,2,3,4] is a semi-Blind Source Separation problem [5], in which it is a priori known that the source signals are 'sparse'. A sparse signal is a signal whose most samples are nearly zero, and just a few percents take significant values. It has been already shown that such a prior information permits source separation for the case the number of sources exceeds the number of sensors [6,1,2,3,4].

The problem of SCA can be stated as follows. Consider the linear model:

$$\mathbf{x}(t) = \sum_{i=1}^{n} \mathbf{s}_i(t)\mathbf{a}_i = \mathbf{A}\mathbf{s}(t) \quad t = 1, 2, \ldots, T \tag{1}$$

where $\mathbf{A} = [\mathbf{a}_1 \ldots \mathbf{a}_n] \in \mathbb{R}^{m \times n}$ is the mixing matrix, $\mathbf{s}(t)$ and $\mathbf{x}(t)$ are the vectors of all samples of n sources and m observed signals (mixtures) at instant t, T is the number of 'time' samples. The goal of SCA is then to estimate $\mathbf{A}$ and $\mathbf{s}(t)$, only from $\mathbf{x}(t)$, $1 \le t \le T$ and the sparsity assumption. In this paper, we address only the problem of estimation of $\mathbf{A}$ (note that where there are more sources than sensors, it is not equivalent to the estimation of the sources). We call each

⋆ This work has been partially funded by Sharif University of Technology, by Center for International Research and Collaboration (ISMO) and by Iran NSF (INSF).

M.E. Davies et al. (Eds.): ICA 2007, LNCS 4666, pp. 397–405, 2007.

column of the mixing matrix, i.e. each $\mathbf{a}_i$, $1 \le i \le n$, a *mixing vector*. Although the word "time" will be used throughout this paper, the above model may be in another domain, in which the sparsity assumption holds. To see this, let $\mathcal{T}$ be a linear 'sparsifying' transform, and the mixing system is stated as $\mathbf{x} = \mathbf{As}$ in the time domain. Then, we have $\mathcal{T}\{\mathbf{x}\} = \mathbf{A}\,\mathcal{T}\{\mathbf{x}\}$ in the transformed domain, and because of the sparsity of $\mathcal{T}\{\mathbf{s}\}$, it is in the form of (1).

Let k denote the average number of active sources at each instant. In fact, if the probability of inactivity of a source is denoted by p (sparsity implies that $p \approx 1$), then[1] $k = n(1-p)$. Then, two different cases should be distinguished for estimating the mixing matrix: single dominant component and multiple dominant components. In the former, k is equal to one, and the scatter plot of $\mathbf{x}(t)$ ($t = 1, \ldots, T$) geometrically shows the data concentration directions. This can be seen from the fact that at each instant, $\mathbf{x}(t) = \mathbf{As}(t) = s_1(t)\mathbf{a}_1 + \cdots + s_n(t)\mathbf{a}_n$, $t = 1, \ldots, T$; and for most instants, only one of s_i's is dominant and the others are almost zero. Consequently, in most samples, $\mathbf{x}(t)$ is in the direction of one of the mixing vectors. In the latter, k is greater than one and the mixing matrix would not be estimated easily from the scatter plot. Up to now, many papers have been addressed the former case [1,3,4], while only few researchers have considered the latter case [4,7,8]. In this paper, we focus on the case of multiple dominant components.

In the multiple dominant components SCA, the observed data concentrate around k-dimensional subspaces which are spanned by a set of k mixing vectors. We call these subspaces *concentration subspaces* throughout this paper. In a multiple dominant problem finding a k-dimensional concentration subspace is not equivalent to find some of the mixing vectors. All of the existing methods [7,9] need to find most of the concentration subspaces and then estimate the mixing matrix from them (this is not the case for our algorithm).

The main idea of this paper is to show that the multiple dominant problem can be converted to a series of single dominant problems, which may be solved by simple algorithms of the single dominant problem to estimate the mixing matrix. Moreover, by estimating each concentration subspace, some of the mixing vectors are found (contrary to [7,9] in which all or many concentration subspaces were needed to be estimated before starting the estimation of mixing vectors). This results in a low computational cost in comparison to the methods of [7,9] and therefore, problems with higher dimensions can be solved by this algorithm. Up to our best knowledge there is no practical algorithm for solving this problem when $k \ge 3$ but our method can handle dimensions more than this.

Throughout the paper, we suppose that the sources are sparse enough so $k < m/2$ (where m is the number of mixtures), the sources are independent and the probability of activity are the same for all of them. Finally, we assume also that each subset of m columns of $\mathbf{A}$ is linearly independent.

[1] More precisely, in this paper, by the average number of active sources we mean an integer. If $n(1-p)$ is slightly greater than an integer $k = \lfloor n(1-p) \rfloor$ (for example is $n(1-p) = 1.05$, then the k-means algorithm, which has been designed for $k = 1$, still works). In other cases, $k = \lceil n(1-p) \rceil$.

2 The Main Idea

The main idea of converting a multiple dominant problem to a single dominant one comes from the following theorem (the proof is left to the appendix).

Theorem 1. *If $k \leq \frac{m}{2}$ and the sources are statistically independent then the average number of active sources in a k dimensional concentration subspace (denoted by $\widetilde{k}$) is $\widetilde{k} = k(1-p)$.*

The above theorem states that although the average number of active sources $k = n(1-p)$ may be greater than 1, the average number of active sources *within a concentration subspace* $\boldsymbol{B}$ (that is, $\tilde{k} = k(1-p) = n(1-p)^2$) is *one level sparser*. In other words, a multiple dominant problem in the original space may be transformed into a single dominant problem within the subspace $\mathbf{B}$. Consequently in the subset of data points which lie in $\mathbf{B}$, we can use a single dominant algorithm (like that of [2]) for estimating the mixing vectors which are a subset of the mixing vectors of the main problem. If $n(1-p)^2$ does not less than or approximately equal to one, then the single dominant assumption does not hold and the above technique should be used one or several levels.

In summary, our approach for estimating the mixing matrix consists of the following steps:

- **Step 1:** Find a new concentration subspace. A concentration subspace can be found by maximizing a cost function (see Sec. 3). For finding a 'new' concentration subspace, the steepest ascent is initialized by a randomly different starting point (note that there are a lot of concentration subspaces).
- **Step 2:** Determine all data points which lie in this concentration subspace, and run a single dominant algorithm to find the mixing vectors in that subspace, *which are a subset of the mixing vectors of the main problem*. The points whose distances to the desired subspace are less than a specific value are supposed to belong to this subspace.
- **Step 3:** If all of the mixing vectors have been found, the search has been finished. Otherwise, go to step one, and continue. In this paper the number of sources is supposed to be known in advance.

Remark: Assuming that the probability of inactivity (p) is identical for all sources, p^n is the probability of no source being active, and hence p can be estimated as $\hat{p} = (\frac{N}{T})^{1/n}$, where T is the total number of data points, and N is the number of 'active' data points (i.e., $\mathbf{x}$'s whose distances from the origin is greater than a threshold). However, in this paper, p is assumed already known.

3 Finding Concentration Subspaces

Each k-dimensional subspace can be represented by an m by k matrix, whose columns form an orthonormal basis for the subspace. In this paper, we do not distinguish between a subspace and its matrix representation. Let $\mathbf{B} \in \mathbb{R}^{m \times k}$ be the orthonormal matrix representation of an arbitrary k-dimensional subspace.

The following cost function has been presented in [9] to detect whether $\mathbf{B}$ is a concentration subspace or not:

$$f_\sigma(\mathbf{B}) = \sum_{i=1}^{T} \exp\left(\frac{-d^2(\mathbf{x}_i, \mathbf{B})}{2\sigma^2}\right), \tag{2}$$

where $d(\mathbf{x}_i, \mathbf{B})$ is the distance of $\mathbf{x}_i$ from the subspace represented by $\mathbf{B}$ [9].

For small values of $d(\mathbf{x}_i, \mathbf{B})$ compared to σ, $\exp(-d^2(\mathbf{x}_i, \mathbf{B})/2\sigma^2)$ is about 1 and for large values of $d(\mathbf{x}_i, \mathbf{B})$, it is nearly zero. Therefore, for sufficiently small values of σ, the above function *is approximately equal to the number of data points close to* $\mathbf{B}$. Therefore, by maximizing the function f, we actually maximize the number of data points close to $\mathbf{B}$ thus we find a concentration subspace. Moreover, if the set of points are concentrated around several different k-dimensional concentration subspaces, f has a local maximum where $\mathbf{B}$ is close to the basis of each of them.

The idea of [9] for finding a concentration subspace is to maximize the function f_σ for a sufficiently small σ, using steepest ascent method. For very small σ, many local maxima exist which do not correspond to any concentration subspaces. These local maxima correspond to spaces which contains $r < k$ mixing vectors instead of k. On the other hand if σ is large, then the peaks are mixed together. In contrast to [9] which uses an iterative method by considering a sequence of decreasing σ to prevent getting trapped in local maxima, in this paper we use only a medium value for σ. In each step, we find a subset of k mixing vectors which are related to the estimated concentration subspace. As will be discussed in Sec. 7, if an incorrect concentration subspace with r ($r < k$) mixing vectors is estimated, the algorithm detects r mixing vector rather than k and therefore it is robust to these errors.

4 Estimating Mixing Vectors and the Mixing Matrix

Consider a concentration subspace $\mathbf{B}$ and suppose that the points $\mathbf{x}_i$ for $i \in \mathbf{I} \subset \{1...T\}$ belong to this subspace. The fact that $\widetilde{k} < 1$ ensure us that most of these points concentrate along k, 1-dimensional subspaces. Then, we use the same idea of [2] designed for finding the mixing vector in the case $k = 1$: Firstly, data samples are normalized by dividing them by their norms ($\bar{\mathbf{x}}_i = \mathbf{x}_i/\|\mathbf{x}_i\|$), that is, the points are projected onto the unit sphere. Moreover, the sign of the first component is forced to be positive. Then, we have a point distribution on a unit hemisphere. Note that most of these points are concentrated around k points, and hence the mixing vectors (which corresponded to the centroid of these clusters) may be found by a clustering algorithm.

However, there are numerous outliers which do not belong to any clusters. Outlier points make the clustering algorithms inaccurate and increase the probability of error in detecting cluster centers, therefore they have to be removed as more as possible. We say that two points are neighbor if the distance between them is less than a specific value r which is dependent to the energy of the sources. For outlier detection the fact that outliers are alone in the space is used. In other words, they do not have any neighbor, but this is not true for cluster centers because the density around them is high. By this definition, a point is considered as an outlier if it does not have any neighbor.

The method we used in this paper for the clustering is subtractive clustering [10]. In this method each point is considered as a cluster center and its potential for being a cluster center is computed. The point with highest potential is considered as a center and that cluster is removed. This process continues to find all clusters.

5 The Final Algorithm

Putting all together, the final algorithm is summarized as follows.

1. Remove the data samples ($\mathbf{x}(t)$) which are near the origin. In these samples, all of the sources are probably inactive.
2. Estimate k to set the dimension of the concentration subspaces and also p to check that if $\tilde{k}$ is smaller than 1.
3. Assume an appropriate value for the free parameter of the cost function (σ).
4. Maximize $f_\sigma(\mathbf{B})$ with the steepest ascent algorithm in several steps:
   (a) Choose a random starting subspace (an orthonormal m by k matrix $\mathbf{B}_1$).
   (b) Set $\mathbf{B}_{j+1} = \mathbf{B}_j + \mu \partial f_\sigma / \partial \mathbf{B}_j$.[2]
   (c) Orthonormalize $\mathbf{B}_{j+1}$.
   (d) If $\|\mathbf{B}_{j+1} - \mathbf{B}_j\| < 10^{-3}$ go to (5) else $j = j+1$ and go to (b).
5. Consider the points whose distances to $\mathbf{B}$ are less than a specific value (d) and ignore the other points.
6. Normalize the points and force the sign of the first component to be positive.
7. Remove the points with no adjacent (outlier points) by preprocessing.
8. Detect the cluster centers with subtractive clustering algorithm (these vectors are some of the mixing vectors).
9. Compare obtained vectors (in the previous step) with former mixing vectors. If each of these vectors is new[3], then add it up to the list of estimated vectors, else throw it away.
10. If the number of estimated mixing vectors is n, then stop the algorithm, else go back to (4).

6 Experimental Results

In this section, 2 simulations are presented to justify the algorithm. In all of these simulations, sparse sources are generated independently and identically distributed (i.i.d) by the Bernoulli-Gaussian model. In other words, the sources

[2] In all simulations we consider $\mu = .01$.

[3] Two vectors are considered identical if the angle between them is less than a certain amount (5 degree in our simulations).

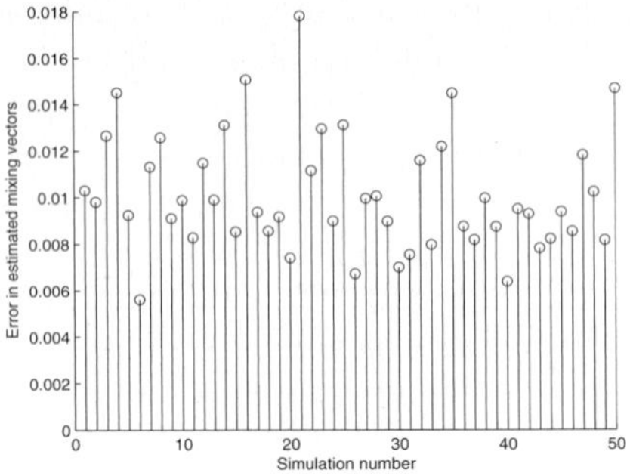

Fig. 1. Error of the overall algorithm for all simulations in the case $n = 10$, $m = 6$, $k = 2$ and $T = 10000$ for 50 different simulations

are inactive with probability p and are active with probability $1 - p$. In the inactive case, their value is a zero mean Gaussian with standard deviation σ_{off}, and in active case it is a zero mean Gaussian with standard deviation σ_{on}. Consequently $s_i \sim \ (1-p)\,\mathcal{N}(0, \sigma_{\text{on}}) + p\,\mathcal{N}(0, \sigma_{\text{off}})$.

In order to have sparse sources, the conditions $\sigma_{\text{on}} \gg \sigma_{\text{off}}$ and $p \approx 1$ should be applied (σ_{off} is to model the noise). In all simulations, the values $\sigma_{\text{on}} = 1$ and $\sigma_{\text{off}} = 0.005$ have been used and each component of the mixing matrix is generated randomly in the $[0, 1]$ interval after that each column of it, is normalized.

All simulations were performed in MATLAB 7 under WindowsXP, using an Intel Pentium IV 2.4 GHz processor with 1 Gigabyte RAM.

Experiment 1: Performance

In this experiment, the performance of our algorithm is demonstrated. 50 simulations for 50 different mixing matrixes are performed for the case $n = 10$, $m = 6$, $k = 2$ ($p = 0.8$) and $T = 10000$. The parameters are chosen as $\sigma = 1/40$, $d = .01$ and $r = .02$.

In all cases, the obtained vectors are compared with the mixing vectors. For comparison the criterion $\mathcal{E} = \min_{\mathbf{P} \in \mathcal{P}} \|\mathbf{A} - \hat{\mathbf{A}}\mathbf{P}\|_2$ is used, where $\mathcal{P}$ is the set of all permutation matrices (this is the same criterion used in [7]). This estimation error is shown in Fig. 1 for all simulations.

The average number of iterations for successfully finding all mixing vectors is around 30, but in 3 simulations this number exceeded 100 iterations and in 1 case more than 500 iterations was required. This may increase the run time of the algorithm. By considering this inefficiency the processes took less than 90 sec in average for estimating a mixing matrix. Moreover the maximum error in the mixing matrix estimation is .018, therefore, the error is negligible.

Experiment 2: Middle and large scale problems

To show that the method is capable of solving medium scale problems, two simulations are performed. In the first simulation, the parameters were $n = 25$,

$m = 15$, $k = 5$ and $T = 100000$, whereas in the second experiment, they were $n = 35$, $m = 20$, $k = 4$ and $T = 50000$. The process took about 1 hour for the first case and 3 hours for the second case. As far as we know there is no algorithm to estimate the mixing vectors in these dimensions ($k = 4$ or 5). In these scales the sources are not so sparse but our algorithm can handle this situation.

To measure the accuracy of the estimation, the angle between each estimated vector and its corresponding actual mixing vector (i.e. inverse cosine of their dot product) were calculated. These n angles were all less than 0.01 radian, showing that all of the mixing matrix have been correctly estimated.

7 Conclusion and Discussion

In this paper, we introduced a method for estimating the mixing matrix in the multiple dominant SCA problem which can handle larger k in comparison to other methods ([7,9]).

At our best knowledge, all existing SCA methods are unable to estimate mixing matrix in large and even medium scales, for the multiple dominant case. However, our method solves the problem at least in the medium scale cases and maybe it can handel larger scales in comparison to other existing methods till now (our algorithm is capable of solving this problem when the averaged number of active sources is up to 5). As observed in the experimental results, all mixing vectors may be detected with good accuracy. However, some mixing vectors might not be found in few iterations, either because of lack of sufficient data, or because some of the actual mixing vectors are close to each other.

As was mentioned in the section 3 a medium value for σ must be considered and for very small σ the chance of error in finding a concentration increases. The subtractive clustering method does not need any prior information about the number of cluster centers, therefore, if the estimated subspace contains r ($r < k$) mixing vectors rather than k, the projected data on the positive normal hemisphere concentrate around r clusters and the clustering method detects r centers instead of k, thus our algorithm is somehow robust to these errors and consequently to σ.

Unfortunately, our algorithm is not efficient to some extent, because some of the mixing vectors are detected several times in order to find all vectors. This may lead to a greater number of iterations and consequently a longer run time. Finding an efficient method for estimating the mixing matrix is a future work.

References

1. Gribonval, R., Lesage, S.: A survey of sparse component analysis for blind source separation: principles, perspectives, and new challenges. In: Proceedings of ESANN'06, April 2006, pp. 323–330 (2006)
2. Zibulevsky, M., Pearlmutter, B.A.: Blind source separation by sparse decomposition in a signal dictionary. Neural Computation 13(4), 863–882 (2001)

3. Bofill, P., Zibulevsky, M.: Underdetermined blind source separation using sparse representations. Signal Processing 81, 2353–2362 (2001)
4. Georgiev, P.G., Theis, F.J., Cichocki, A.: Sparse component analysis and blind source separation of underdetermined mixtures. IEEE Transactions of Neural Networks 16(4), 992–996 (2005)
5. Babaie-Zadeh, M., Jutten, C.: Semi-blind approaches for source separation and independent component analysis. In: Proceedings of ESANN'06 (April 2006)
6. Donoho, D.L.: For most large underdetermined systems of linear equations the minimal l^1-norm solution is also the sparsest solution, Tech. Rep (2004)
7. Washizawa, Y., Cichocki, A.: on-line k-plane clustering learning algorithm for sparse comopnent analysis. In: Proceedinds of ICASSP'06, Toulouse (France), pp. 681–684 (2006)
8. Li, Y., Amari, S., Cichocki, A., Ho, D.W.C., Xie, S.: Underdetermined blind source separation based on sparse representation. IEEE Transactions on Signal Processing 54(2), 423–437 (2006)
9. Movahedi, F., Mohimani, G.H., Babaie-Zadeh, M., Jutten, C.: Estimating the mixing matrix in sparse component analysis (SCA) based on multidimensional subspace clustering. In: ICT'07, Malaysia (May 2007)
10. Chiu, S.: Fuzzy model identification based on cluster estimation. Journal of Intelligent and Fuzzy Systems, 2(3) (September 1994)

Appendix: Proof of Theorem 1

Consider a concentration subspace **B**. Then, by definition, it is formed by a linear combination of k mixing vectors. Let $\mathbf{a}_{l_1} \cdots \mathbf{a}_{l_k}$ be these mixing vectors. Then for every point $\mathbf{x}$ in this subspace, we have $\mathbf{x} = \sum_{i=1}^{k} s_{l_i}\mathbf{a}_{l_i}$ where s_{l_i} $1 \leq i \leq k$ are the sources.

Lemma 1. *If $k \leq \frac{m}{2}$ and the mixing matrix is full rank then, each point in a concentration subspace ($\boldsymbol{B}$), for the sparsest solution, is almost always the linear combination of a set of k fixed mixing vectors. Precisely if $\boldsymbol{x} = \sum_{i=1}^{n} \tilde{s}_i \boldsymbol{a}_i$ then $s_i = 0$ for $i \in \{1, 2, ..., n\} - \{l_1, l_2, ..., l_k\}$.*

To prove this lemma suppose that there is another set of mixing vectors $\{\mathbf{a}_{t_1} \cdots \mathbf{a}_{t_h}\}$ and real valued variables $s'_{t_1}, \cdots, s'_{t_h}$ such that $\mathbf{x} = \sum_{i=1}^{h} s'_{t_i}\mathbf{a}_{t_i}$. Then $\mathbf{x} = \sum_{i=1}^{k} s_{l_i}\mathbf{a}_{l_i} = \sum_{i=1}^{h} s'_{t_i}\mathbf{a}_{t_i}$ shows that the set $\{a_{l_1}, \cdots, a_{l_k}, a_{t_1}, \cdots, a_{t_h}\}$ is not linearly independent. **A** is assumed to be full rank thus each $k \leq m$ different mixing vectors are linearly independent. From this comment we can conclude that $k + h > m$, moreover, $k \leq m/2$ (see section 1) thus $h > m/2$ and we have $h > k$. This is in contrast to our basic assumption that we want to find the sparsest solution to BSS problem. This lemma is in fact similar to the theorem of uniqueness of the sparsest solution [6].

Using the above lemma, the expected value of active sources in **B** is

$$\tilde{k} = \sum_{i=0}^{k} iP\{i \text{ sources from } l_1, \cdots, l_k \text{ active}| \text{ remaining } n-k \text{ sources inactive}\}$$

where $P\{\cdot\}$ denotes the probability and $\widetilde{k}$ is the expected value of the number of active sources in a concentration subspace. Since the sources (and hence their activity status) are assumed to be independent, the above equation is reduced to:

$$\widetilde{k} = \sum_{i=0}^{k} iP\{i \text{ sources of } l_1, \cdots, l_k \text{ active}\} = \sum_{i=0}^{k} i \binom{n}{k} (1-p)^i p^{k-i}$$

This is the expected value of a binomial random variable and hence $\widetilde{k} = k(1-p)$.

Dictionary Learning for L1-Exact Sparse Coding

Mark D. Plumbley

Department of Electronic Engineering, Queen Mary University of London,
Mile End Road, London E1 4NS, United Kingdom
mark.plumbley@elec.qmul.ac.uk

Abstract. We have derived a new algorithm for dictionary learning for sparse coding in the ℓ_1 exact sparse framework. The algorithm does not rely on an approximation residual to operate, but rather uses the special geometry of the ℓ_1 exact sparse solution to give a computationally simple yet conceptually interesting algorithm. A self-normalizing version of the algorithm is also derived, which uses negative feedback to ensure that basis vectors converge to unit norm. The operation of the algorithm is illustrated on a simple numerical example.

1 Introduction

Suppose we have a sequence of observations $\mathbf{X} = [\mathbf{x}^1, \ldots, \mathbf{x}^p]$, $\mathbf{x}^k \in \mathbb{R}^n$. In the sparse coding problem [1] we wish to find a dictionary matrix $\mathbf{A}$ and representation matrix $\mathbf{S}$ such that

$$\mathbf{X} = \mathbf{AS} \tag{1}$$

and where the representations $\mathbf{s}^k \in \mathbb{R}^m$ in the matrix $\mathbf{S} = [\mathbf{s}^1, \ldots, \mathbf{s}^p]$ are *sparse*, i.e. where there are few non-zero entries in each s_j. In the case where we look for solutions $\mathbf{X} = \mathbf{AS}$ with no error, we say that this is an *exact sparse* solution. In the sparse coding problem we typically have $m > n$, so this is closely related to the overcomplete independent component analysis (overcomplete ICA) problem, which has the additional assumption that the components of the representation vectors s_j are statistically independent.

If the dictionary $\mathbf{A}$ is given, then for each sample $\mathbf{x} = \mathbf{x}^k$ we can separately look the sparsest representation

$$\min_{\mathbf{s}} \|\mathbf{s}\|_0 \quad \text{such that} \quad \mathbf{x} = \mathbf{A}\mathbf{s}. \tag{2}$$

However, even this is a hard problem, so one approach is to solve instead the 'relaxed' ℓ_1-norm problem

$$\min_{\mathbf{s}} \|\mathbf{s}\|_1 \quad \text{such that} \quad \mathbf{x} = \mathbf{A}\mathbf{s}. \tag{3}$$

This approach, known in the signal processing literature as Basis Pursuit [2], can be solved efficiently with linear programming methods (see also [3]).

Even if such efficient sparse representation methods exist, learning the dictionary $\mathbf{A}$ is a non-trivial task. Several methods have been proposed in the

M.E. Davies et al. (Eds.): ICA 2007, LNCS 4666, pp. 406–413, 2007.

literature, such as those by Olshausen and Field [4] and Lewicki and Sejnowski [5], and these can be derived within a principled probabilistic framework [1]. A recent alternative is the K-SVD algorithm [6], which is a generalization of the K-means algorithm.

However, many of these algorithms are designed to solve the sparse approximation problem $\mathbf{X} = \mathbf{AS} + \mathbf{R}$ for some nonzero residual term $\mathbf{R}$, rather than the exact sparse problem (1). For example, the Olshausen and Field [4] approximate maximum likelihood algorithm is

$$\Delta \mathbf{A} = \eta E(\mathbf{r}\mathbf{s}^T) \tag{4}$$

where $\mathbf{r} = \mathbf{x} - \mathbf{As}$ is the residual after approximation, and the K-SVD algorithm [1] minimizes the norm $\|\mathbf{R}\|_F = \|\mathbf{X} - \mathbf{AS}\|_F$. If we have a sparse representation algorithm that is successfully able to solve (3) exactly on each data sample $\mathbf{x}^k$, then we have a zero residual $\mathbf{R} = \mathbf{0}$, and there is nothing to 'drive' the dictionary learning algorithm. Some other dictionary learning algorithms have other constraints: for example, the method of Georgiev et al [7] requires at most $m - 1$ nonzero elements in each column of $\mathbf{S}$.

While these algorithms have been successful for practical problems, in this paper we specifically explore the special geometry of the ℓ_1 exact sparse dictionary learning problem. We shall derive a new dictionary learning algorithm for the ℓ_1 exact sparse problem, using the basis vertex $\mathbf{c} = (\overline{\mathbf{A}}^\dagger)^T \mathbf{1}$ associated with a subdictionary (basis set) $\overline{\mathbf{A}}$ identified in the ℓ_1 exact sparse representation problem (3).

2 The Dual Problem and Basis Vertex

The linear program (3) has a corresponding *dual* linear program [2]

$$\max_{\mathbf{c}} \mathbf{x}^T \mathbf{c} \quad \text{such that} \quad \pm \mathbf{a}_j^T \mathbf{c} \leq 1 \quad j = 1, \ldots, m \tag{5}$$

which has an optimum $\mathbf{c}^*$ associated with any optimum $\mathbf{s}^*$ of (3). In a previous paper we explored the polytope geometry of this type of dual problem, and derived an algorithm, Polytope Faces Pursuit (PFP), which searches for the optimal vertex which maximizes $\mathbf{x}^T \mathbf{c}$, and uses that to find the optimal vector $\mathbf{s}$ [8]. Polytope Faces Pursuit is a gradient projection method [9] which iteratively builds a solution basis $\overline{\mathbf{A}}$ consisting of a subset of the signed columns $\sigma_j \mathbf{a}_j$ of $\mathbf{A}$, $\sigma_j \in \{-1, 0, +1\}$, chosen such that $\mathbf{x} = \overline{\mathbf{A}}\bar{\mathbf{s}}$ with $\bar{\mathbf{s}} > \mathbf{0}$ containing the absolute value of the nonzero coefficients of $\mathbf{s}$ at the solution. The algorithm is similar in structure to orthogonal matching pursuit (OMP), but with a modified admission criterion

$$\mathbf{a}' = \arg\max_{\mathbf{a}_i} \frac{\mathbf{a}_i^T \mathbf{r}}{1 - \mathbf{a}_i^T \mathbf{c}} \tag{6}$$

to add a new basis vector $\mathbf{a}'$ to the current basis set, together with an additional rule to switch out basis vectors which are no longer feasible.

The basis vertex $\mathbf{c} = (\overline{\mathbf{A}}^{\dagger})^T \mathbf{1}$ is the solution to the dual problem (5). During the operation of the algorithm $\mathbf{c}$ satisfies $\mathbf{A}^T \mathbf{c} \leq \mathbf{1}$, so it remains *dual-feasible* throughout. For all active atoms $\mathbf{a}_j$ in the current basis set, we have $\mathbf{a}_j^T \mathbf{c} = 1$. Therefore at the minimum ℓ_1 norm solution the following conditions hold:

$$\overline{\mathbf{A}}^T \mathbf{c} = \mathbf{1} \tag{7}$$

$$\mathbf{x} = \overline{\mathbf{A}}\bar{\mathbf{s}} \qquad \bar{\mathbf{s}} > \mathbf{0}. \tag{8}$$

We will use these conditions in our derivation of the dictionary learning algorithm that follows.

3 Dictionary Learning Algorithm

We would like to construct an algorithm to find the matrix $\mathbf{A}$ that minimizes the total ℓ_1 norm

$$J = \sum_{k=1}^{p} \|\mathbf{s}^k\|_1 \tag{9}$$

where $\mathbf{s}^k$ is chosen such that $\mathbf{x}^k = \mathbf{A}\mathbf{s}^k$ for all $k = 1, \ldots, p$, i.e. such that $\mathbf{X} = \mathbf{AS}$, and where the columns of $\mathbf{A}$ are constrained to have unit norm. In particular, we would like to construct an iterative algorithm to adjust $\mathbf{A}$ to reduce the total ℓ_1 norm (9): let us therefore investigate how J depends on $\mathbf{A}$.

For the contribution due to the kth sample we have $J = \sum_k J^k$ where $J^k = \|\mathbf{s}^k\|_1 = \mathbf{1}^T \bar{\mathbf{s}}^k$ since $\bar{\mathbf{s}}^k \geq 0$. Dropping the superscripts k from $\mathbf{x}^k$, $\overline{\mathbf{A}}^k$ and $\bar{\mathbf{s}}^k$ we therefore wish to find how $J^k = \mathbf{1}^T \bar{\mathbf{s}}$ changes with $\overline{\mathbf{A}}$, so taking derivatives of J^k we get

$$dJ^k/dt = \mathbf{1}^T (d\bar{\mathbf{s}}/dt). \tag{10}$$

Now taking the derivative of (8) for fixed $\mathbf{x}$ we get

$$0 = (d\overline{\mathbf{A}}/dt)\bar{\mathbf{s}} + \overline{\mathbf{A}}(d\bar{\mathbf{s}}/dt) \tag{11}$$

and pre-multiplying by $\mathbf{c}^T$ gives us

$$\mathbf{c}^T (d\overline{\mathbf{A}}/dt)\bar{\mathbf{s}} = -\mathbf{c}^T \overline{\mathbf{A}}(d\bar{\mathbf{s}}/dt) \tag{12}$$

$$= -\mathbf{1}^T (d\bar{\mathbf{s}}/dt) \tag{13}$$

$$= -dJ^k/dt \tag{14}$$

where the last two equations follow from (7) and (10). Introducing trace($\cdot$) for the trace of a matrix, we can rearrange this to get

$$\frac{dJ^k}{dt} = -\operatorname{trace}\left(\mathbf{c}^T \frac{d\overline{\mathbf{A}}}{dt}\bar{\mathbf{s}}\right) = -\operatorname{trace}\left((\mathbf{c}\bar{\mathbf{s}}^T)^T \frac{d\overline{\mathbf{A}}}{dt}\right) = -\left\langle \mathbf{c}\bar{\mathbf{s}}^T, \frac{d\overline{\mathbf{A}}}{dt}\right\rangle \tag{15}$$

from which we see that the gradient of J^k with respect to $\overline{\mathbf{A}}$ is given by $\nabla_{\overline{\mathbf{A}}} J^k = -\mathbf{c}\bar{\mathbf{s}}^T$. Summing up over all k and applying to the original matrix $\mathbf{A}$ we get

$$\nabla_{\mathbf{A}} J = -\sum_k \mathbf{c}^k (\mathbf{s}^k)^T = -\mathbf{C}\mathbf{S}^T \tag{16}$$

with $\mathbf{C} = [\mathbf{c}^k]$, a somewhat surprisingly simple result. Therefore the update

$$\Delta\mathbf{A} = \eta \sum_k \mathbf{c}^k (\mathbf{s}^k)^T \tag{17}$$

will perform a steepest descent search for the minimum total ℓ_1 norm J, and any path $d\mathbf{A}/dt$ for which $\langle (d\mathbf{A}/dt), \nabla_{\mathbf{A}} J \rangle < 0$ will cause J to decrease.

3.1 Unit Norm Atom Constraint

Now without any constraint, algorithm (17) will tend to reduce the ℓ_1 norm by causing $\mathbf{A}$ to increase without bound, so we need to impose a constraint on $\mathbf{A}$. A common constraint is to require the columns $\mathbf{a}_j$ of $\mathbf{A}$ to be unit vectors, $\|\mathbf{a}_j\|_2^2 = 1$, i.e. $\mathbf{a}_j^T \mathbf{a}_j = 1$. We therefore require our update to be restricted to paths $d\mathbf{a}_j/dt$ for which $\mathbf{a}_j^T (d\mathbf{a}_j/dt) = 0$.

To find the projection of (16) in this direction, consider the gradient component

$$\mathbf{g}_j = \frac{dJ}{d\mathbf{a}_j} = -\sum_k \mathbf{c}^k s_j^k. \tag{18}$$

The orthogonal projection of $\mathbf{g}_j$ onto the required tangent space is given by

$$\tilde{\mathbf{g}}_j = \mathbf{P}_{\bar{\mathbf{a}}_j} \mathbf{g}_j = \left(\mathbf{I} - \frac{\mathbf{a}_j \mathbf{a}_j^T}{\|\mathbf{a}_j\|_2^2} \right) \mathbf{g}_j = \mathbf{g}_j - \frac{1}{\|\mathbf{a}_j\|_2^2} \mathbf{a}_j (\mathbf{a}_j^T \mathbf{g}_j). \tag{19}$$

Now considering the rightmost factor $\mathbf{a}_j^T \mathbf{g}_j$, from (18) we get

$$\mathbf{a}_j^T \mathbf{g}_j = -\sum_k \mathbf{a}_j^T \mathbf{c}^k s_j^k. \tag{20}$$

Considering just the kth term $\mathbf{a}_j^T \mathbf{c}^k s_j^k$, if $\mathbf{a}_j$ is one of the basis vectors in $\overline{\mathbf{A}}^k$ (with possible change of sign $\sigma_j^k = \text{sign}(s_j^k)$) which forms part of the solution $\mathbf{x}^k = \overline{\mathbf{A}}^k \bar{\mathbf{s}}^k$ found by a minimum ℓ_1 norm solution, then we must have $\sigma_j^k \mathbf{a}_j^T \mathbf{c}^k = 1$ where $\sigma_j^k = \text{sign}(s_j^k)$, because $\overline{\mathbf{A}}^{kT} \mathbf{c} = \mathbf{1}$, so $\mathbf{a}_j^T \mathbf{c}^k s_j^k = \sigma_j^k s_j^k = |s_j^k|$. On the other hand, if $\mathbf{a}_j$ does not form part of $\overline{\mathbf{A}}^k$, then $s_j^k = 0$ so $\mathbf{a}_j^T \mathbf{c}^k s_j^k = 0 = |s_j^k|$. Thus regardless of the involvement of $\mathbf{a}_j$ in $\overline{\mathbf{A}}^k$, we have $\mathbf{a}_j^T \mathbf{c}^k s_j^k = |s_j^k|$, so

$$\mathbf{a}_j^T \mathbf{g}_j = -\sum_k |s_j^k| \tag{21}$$

and therefore

$$\tilde{\mathbf{g}}_j = -\sum_k \left(\mathbf{c}^k s_j^k - \frac{1}{\|\mathbf{a}_j\|_2^2} \mathbf{a}_j |s_j^k| \right). \tag{22}$$

Therefore we have the following 'tangent' update rule:

$$\mathbf{a}_j(T+1) = \mathbf{a}_j(T) + \eta \sum_k \left(\mathbf{c}^k s_j^k - \frac{1}{\|\mathbf{a}_j(T)\|_2^2} \mathbf{a}_j(T) |s_j^k| \right) \tag{23}$$

which will perform a tangent-constrained steepest descent update to find the minimum total ℓ_1 norm J. We should note that the tangent update is not entirely sufficient to constrain $\mathbf{a}_j$ to remain of unit norm, so an occasional renormalization step $\mathbf{a}_j \leftarrow \mathbf{a}_j / \|\mathbf{a}_j\|_2$ will be required after a number of applications of (23).

4 Self-normalizing Algorithm

Based on the well-known negative feedback structure used in PCA algorithms such as the Oja [10] PCA neuron, we can modify algorithm (23) to produce the following self-normalizing algorithm that does not require the explicit renormalization step:

$$\mathbf{a}_j(T+1) = \mathbf{a}_j(T) + \eta \sum_k \left(\mathbf{c}^k s_j^k - \mathbf{a}_j(T) |s_j^k| \right) \tag{24}$$

where we have simply removed the factor $1/\|\mathbf{a}_j(T)\|_2^2$ from the second term in (23). This algorithm is computationally very simple, and suggests an online version $\mathbf{a}_j(k) = \mathbf{a}_j(k-1) + \eta \left(\mathbf{c}^k s_j^k - \mathbf{a}_j(k-1) |s_j^k| \right)$ with the dictionary updated as each data point is presented.

For unit norm basis vectors $\|\mathbf{a}_j(T)\|_2 = 1$, the update produced by algorithm (24) is identical to that produced by the tangent algorithm (23). Therefore, for unit norm basis vectors, algorithm (24) produces a step in a direction which reduces J. (Note that algorithm (24) will not necessarily reduce J when $\mathbf{a}_j$ is not unit norm.)

To show that the norm of the basis vectors $\mathbf{a}_j$ in algorithm (24) converge to unit length, we require that each $\mathbf{a}_j$ must be involved in the representation of at least one pattern $\mathbf{x}^k$, i.e. for some k we have $s_j^k \neq 0$. (If this were not true, that basis vector would have been ignored completely so would not be updated by the algorithm.) Consider the ordinary differential equation (ode) version of (24):

$$\frac{d\mathbf{a}_j}{dt} = \sum_k \left(\mathbf{c}^k s_j^k - \mathbf{a}_j |s_j^k| \right) \tag{25}$$

$$= -\tilde{\mathbf{g}}_j + \left(\frac{1}{\|\mathbf{a}_j\|_2^2} - 1 \right) \mathbf{a}_j \sum_k |s_j^k| \tag{26}$$

$$= -\tilde{\mathbf{g}}_j + \frac{1}{\|\mathbf{a}_j\|_2^2} \left(1 - \|\mathbf{a}_j\|_2^2 \right) \mathbf{a}_j \sum_k |s_j^k| \tag{27}$$

which, noting that $\mathbf{a}_j^T \tilde{\mathbf{g}}_j = \mathbf{a}_j^T \mathbf{P}_{\bar{\mathbf{a}}_j} \mathbf{g}_j = 0$, gives us

$$\mathbf{a}_j^T \frac{d\mathbf{a}_j}{dt} = \frac{1}{\|\mathbf{a}_j\|_2^2} \left(1 - \|\mathbf{a}_j\|_2^2 \right) \mathbf{a}_j^T \mathbf{a}_j \sum_k |s_j^k| = (1 - \|\mathbf{a}_j\|_2^2) \sum_k |s_j^k|. \tag{28}$$

Constructing the Lyapunov function [11] $Q = (1/4)(1 - \|\mathbf{a}_j\|_2^2)^2 \geq 0$, which is zero if and only if $\mathbf{a}_j$ has unit length, we get

$$\begin{aligned} dQ/dt &= -(1 - \|\mathbf{a}_j\|_2^2)\mathbf{a}_j^T(d\mathbf{a}_j/dt) &(29)\\ &= -(1 - \|\mathbf{a}_j\|_2^2)^2 \sum_k |s_j^k| &(30)\\ &\leq 0 &(31)\end{aligned}$$

where $\sum_k |s_j^k| > 0$ since at least one of the s_j^k is nonzero, so equality holds in (31) if and only if $\|\mathbf{a}_j\|_2^2 = 0$. Therefore the ode (25) will cause $\|\mathbf{a}_j\|_2^2$ to converge to 1 for all basis vectors $\mathbf{a}_j$.

While algorithm (24) does not strictly require renormalization, we found experimentally that an explicit unit norm renormalization step did produce slightly more consistent behaviour in reduction of the total ℓ_1 norm J.

Finally we note that at convergence of algorithm (24), the basis vectors must satisfy

$$\mathbf{a}_j = \frac{\sum_k \mathbf{c}^k s_j^k}{\sum_k |s_j^k|} \tag{32}$$

so that $\mathbf{a}_j$ must be a (signed) weighted sum of the basis vertices $\mathbf{c}^k$ in which it is involved. While equation (32) is suggestive of a fixed point algorithm, we have observed that it yields unstable behaviour if used directly. Nevertheless we believe that it would be interesting to explore this in future for the final stages of an algorithm, as it nears convergence.

5 Augmenting Polytope Faces Pursuit

After an update to the dictionary $\mathbf{A}$, it is not necessary to restart the search for the minimum ℓ_1 norm solutions $\mathbf{s}^k$ to $\mathbf{x}^k = \mathbf{A}\mathbf{s}^k$ from $\mathbf{s}^k = \mathbf{0}$. In many cases the dictionary vector will have changed only slightly, so the signs σ_j^k and selected subdictionary $\overline{\mathbf{A}}^k$ may be very similar to the previous solution, before the dictionary update. At the Tth dictionary learning step we can therefore restart the search for $\mathbf{s}^k(T)$ from the basis set selected by the last solution $\mathbf{s}^k(T-1)$.

However, if we start from the same subdictionary selection pattern after a change to the dictionary, we can no longer guarantee that the solution will be *dual-feasible*, i.e. that (5) is always satisfied, which is required for the Polytope Faces Pursuit algorithm [8]. While we will still have $\mathbf{a}_j^T\mathbf{c} = 1$ for all vectors $\mathbf{a}_j$ in the solution subdictionary, we may have increased some other $\mathbf{a}_j$, not in the original solution set, so that now $\mathbf{a}_j^T\mathbf{c} > 1$.

To overcome this problem, if $\mathbf{a}_j^T\mathbf{c}^k > 1$ such that dual feasibility fails for a particular sample k and basis vector $\mathbf{a}_j$, we simply restart the sparse Polytope Faces Pursuit algorithm from $\mathbf{s}^k = 0$ for this particular sample, to guarantee that dual-feasibility is restored. We believe that it may be possible to construct a more efficient method to restore dual-feasibility, based on selectively swapping vectors to bring the solution into feasibility, but it appears to be non-trivial to guarantee that loops will not result.

6 Numerical Illustration

To illustrate the operation of the algorithm, Figure 1 shows a small graphical example. Here four source variables s_j are generated with identical mixture-of-

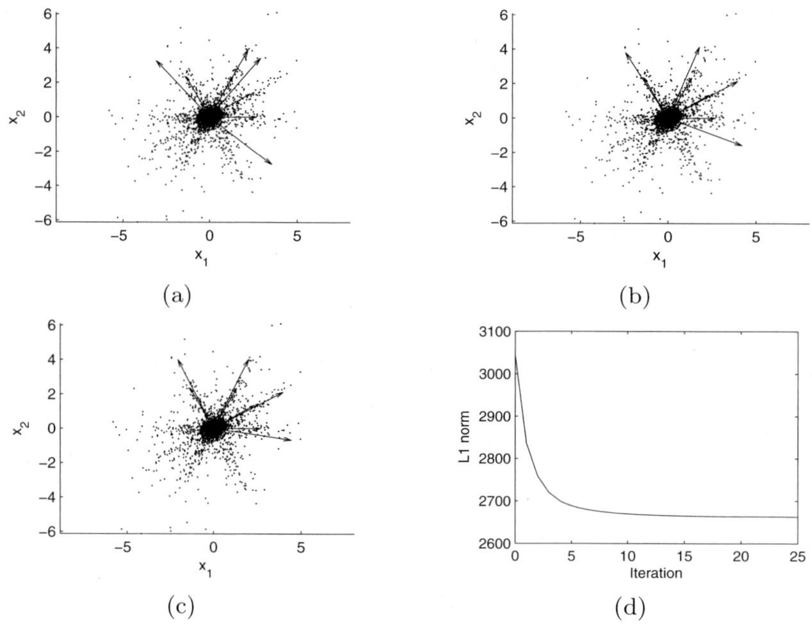

Fig. 1. Illustration of dictionary learning of a $n = 2$ dimensional mixture of $m = 4$ MoG-distributed sources, for $p = 3000$ samples. The plots show (a) the initial condition, and updates after (b) 5 dictionary updates and (c) 25 dictionary updates. The learning curve (d) confirms that the ℓ_1 norm decreases as the algorithm progresses. On (a)-(c), the longer arrows are scaled versions of the learned dictionary vectors $\mathbf{a}_j$, with the shorter arrows showing the directions of the generating dictionary vectors for comparison.

gaussian (MoG) densities in an $n = 2$ dimensional space and added with angles $\theta \in \{0, \pi/6, \pi/3, 4\pi/6\}$. It is important to note that, even in the initial condition Figure 1(a), the basis set spans the input space and optimization of (3) has an exact solution $\mathbf{x}^k = \mathbf{A}\mathbf{s}^k$ for all data samples $\mathbf{x}^k$, at least to within numerical precision of the algorithm. Therefore this situation would not be suitable for any dictionary learning algorithm which relies on a residual $\mathbf{r} = \mathbf{x} - \mathbf{A}\mathbf{s}$.

7 Conclusions

We have derived a new algorithm for dictionary learning for sparse coding in the ℓ_1 exact sparse framework. The algorithm does not rely on an approximation residual to operate, but rather uses the special geometry of the ℓ_1 exact sparse solution to give a computationally simple yet conceptually interesting algorithm. A self-normalizing version of the algorithm is also derived, which uses negative feedback to ensure that basis vectors converge to unit norm.

The operation of the algorithm was illustrated on a simple numerical example. While we emphasize the derivation and geometry of the algorithm in the present paper, we are currently working on applying this new algorithm to practical sparse approximation problems, and will present these results in future work.

Acknowledgements

This work was supported by EPSRC grants GR/S75802/01, GR/S82213/01, GR/S85900/01, EP/C005554/1 and EP/D000246/1.

References

1. Kreutz-Delgado, K., Murray, J.F., Rao, B.D., Engan, K., Lee, T.W., Sejnowski, T.J.: Dictionary learning algorithms for sparse representation. Neural Computation 15, 349–396 (2003)
2. Chen, S.S., Donoho, D.L., Saunders, M.A.: Atomic decomposition by basis pursuit. SIAM Journal on Scientific Computing 20, 33–61 (1998)
3. Bofill, P., Zibulevsky, M.: Underdetermined blind source separation using sparse representations. Signal Processing 81, 2353–2362 (2001)
4. Olshausen, B.A., Field, D.J.: Emergence of simple-cell receptive-field properties by learning a sparse code for natural images. Nature 381, 607–609 (1996)
5. Lewicki, M.S., Sejnowski, T.J.: Learning overcomplete representations. Neural Computation 12, 337–365 (2000)
6. Aharon, M., Elad, M., Bruckstein, A.: K-SVD: Design of dictionaries for sparse representation. In: Proceedings of SPARS'05, Rennes, France, pp. 9–12 (2005)
7. Georgiev, P., Theis, F., Cichocki, A.: Sparse component analysis and blind source separation of underdetermined mixtures. IEEE Transactions on Neural Networks 16, 992–996 (2005)
8. Plumbley, M.D.: Recovery of sparse representations by polytope faces pursuit. In: Rosca, J., Erdogmus, D., Príncipe, J.C., Haykin, S. (eds.) ICA 2006. LNCS, vol. 3889, pp. 206–213. Springer, Heidelberg (2006)
9. Rosen, J.B.: The gradient projection method for nonlinear programming. Part I. Linear constraints. J. Soc. Indust. Appl. Math. 8, 181–217 (1960)
10. Oja, E.: A simplified neuron model as a principal component analyzer. Journal of Mathematical Biology 15, 267–273 (1982)
11. Cook, P.A.: Nonlinear Dynamical Systems. Englewood Cliffs, Englewood Cliffs, NJ (1986)

Supervised and Semi-supervised Separation of Sounds from Single-Channel Mixtures

Paris Smaragdis[1], Bhiksha Raj[1], and Madhusudana Shashanka[2,⋆]

[1] Mitsubishi Electric Research Laboratories
Cambridge MA, USA
[2] Department of Cognitive and Neural Systems
Boston University, Boston MA, USA

Abstract. In this paper we describe a methodology for model-based single channel separation of sounds. We present a sparse latent variable model that can learn sounds based on their distribution of time/frequency energy. This model can then be used to extract known types of sounds from mixtures in two scenarios. One being the case where all sound types in the mixture are known, and the other being being the case where only the target or the interference models are known. The model we propose has close ties to non-negative decompositions and latent variable models commonly used for semantic analysis.

1 Introduction

Separation of sounds from single-channel mixtures can be a daunting task. There is no exact solution nor a process that guarantees good separation behavior. Most approaches in this scenario are model-based and perform separation by splitting the spectrogram of the mixture in parts that correspond to a single source. This approach has been taken in [1,2,3,4] among many others and has been one of the easiest ways to obtain reasonable results. In this paper we employ a similar approach using a new decomposition algorithm which is best suited for spectrogram analysis. We show how this approach can be used both supervised and semi-supervised settings for separation from monophonic mixtures, and draw connections to various types of known analysis methods.

1.1 Probabilistic Latent Component Analysis

In this section we describe the statistical model we will use for acoustic modeling. Probabilistic Latent Component Analysis (PLCA) is a straightforward extension of Probabilistic Latent Semantic Indexing (PLSI) [5] which deals with an arbitrary number of dimensions and can easily extended to exhibit various features such as sparsity or shift-invariance. The basic model is defined as:

$$P(\mathbf{x}) = \sum_z P(z) \prod_{j=1}^{N} P(x^{(j)}|z) \tag{1}$$

⋆ Work performed while at Mitsubishi Electric Research Laboratories. M. Shashanka was partially supported by an AFOSR grant to Prof. Barbara Shinn-Cunningham.

M.E. Davies et al. (Eds.): ICA 2007, LNCS 4666, pp. 414–421, 2007.

where $P(\mathbf{x})$ is a distribution over the N-dimensional random variable $\mathbf{x}$ and $x^{(j)}$ denotes j'th dimension. The random variable $z$is a latent variable, and the $P(x^{(j)}|z)$ are one-dimensional distributions. Effectively this model represents a mixture of marginal distribution products to approximate an N-dimensional distribution. Our objective is to discover the most appropriate marginal distributions. The estimation of the marginals $P(x^{(j)}|z)$ is performed using the EM algorithm. In the expectation step we estimate the posterior probability of the latent variable z:

$$P(z|\mathbf{x}) = \frac{P(z)\prod_{j=1}^{N} P(x^{(j)}|z)}{\sum_{z'} P(z')\prod_{j=1}^{N} P(x^{(j)}|z')} \tag{2}$$

and in a maximization step we re-estimate the marginals using the above weighting to obtain a new and more accurate estimate:

$$P(z) = \int P(\mathbf{x})P(z|\mathbf{x})d\mathbf{x} \tag{3}$$

$$P^*(x^{(j)}|z) = \int \cdots \int P(\mathbf{x})P(z|\mathbf{x})dx^{(k)}, \forall k \neq j \tag{4}$$

$$P(x^{(j)}|z) = \frac{P^*(x^{(j)}|z)}{P(z)} \tag{5}$$

Repeating the above steps in an alternating manner multiple times produces a converging solution for the marginals $P(x^{(j)}|z)$ along each dimension j, and the latent variable priors $P(z)$. In the case where $P(\mathbf{x})$ is discrete we only have to substitute the integrations with summations. Likewise the latent variable z can be continuous valued in which case the summations over z become integrals. In practical applications $P(\mathbf{x})$ and z will both be discrete and we assume that to be the case in the remainder of this paper.

1.2 Sparsity Constraints

In this section we will introduce a modification to the PLCA algorithm which enables us to produce sparse (or maximally non-sparse) estimates of $P(x^{(j)}|z)$. Since the estimated quantities of PLCA are probability distributions, we can directly obtain sparsity by imposing an *entropic prior* instead of obtaining the effect by more traditional means such as L1-norm minimization. This prior can can impose a bias towards estimating a low (or high) entropy $P(x^{(j)}|z)$. We can thus obtain a sparse estimate by requesting low entropy results, a flatter estimate by requesting high entropy results, or any combination of the two cases for different values of the latent variable z.

Let us assume that we wish to manipulate the entropy of the distribution $P(x^{(j)}|z)$. The form of the entropic prior for this distribution is defined as $e^{-\beta\mathcal{H}(P(x^{(j)}|z)} = e^{\beta\sum_i P(x_i^{(j)}|z)logP(x_i^{(j)}|z)}$, where $P(x_i^{(j)}|z)$ denotes the i'th element of the distribution $P(x^{(j)}|z)$. Incorporating the entropic prior in the PLCA model and adding the constraint that $\sum_i P(x_i^{(j)}|z) = 1$ results into optimizing the following function:

$$\frac{P^*(x^{(j)}|z)}{P(x_i^{(j)}|z)} + \beta + \beta log P(x_i^{(j)}|z) + \lambda = 0, \tag{6}$$

where $P^*(x^{(j)}|z)$ is defined in equation 4 and λ is the Langrange multiplier enforcing the unity summation constraint. As shown in [6] this equation can be solved using Lambert's $\mathcal{W}$ function resulting in:

$$P(x^{(j)}|z) = \frac{P^*(x^{(j)}|z)/\beta}{\mathcal{W}(-P^*(x^{(j)}|z)e^{1+\lambda/\beta}/\beta)}. \tag{7}$$

Alternating between the last two equations for a couple of iterations we can obtain a refined estimate of $P(x^{(j)}|z)$ which accommodates the entropy constraint. This process is described in more detail in [7].

2 Applications of PLCA for Source Separation

The two separation scenarios we will introduce in the next sections are both making use of PLCA models of sounds. We will now briefly introduce how we can model a class of sounds using PLCA. One major feature that we can use to describe a sound is that of its frequency distribution. For example we know that speech tends to have a harmonic distribution with most energy towards the low end of the spectrum, whereas, say, a siren would have a more simple timbral profile mostly present at higher frequencies. We can use the PLCA model to obtain a dictionary of spectral profiles that best describe a class of sounds. To do so we consider the 2-d formulation of PLCA when applied on time-frequency distributions of sounds. The model will be:

$$P(f,t) = \sum_z P(z)P(f|z)P(t|z) \tag{8}$$

where $P(f,t)$ is a magnitude spectrogram. The decomposition will result into two sets of marginals, one for the frequency axis and one for the time axis. The time axis marginals are not particularly informative, the frequency axis marginals however will contain a dictionary of spectra which best describe the sound represented by the input spectrogram. To illustrate this operation consider the spectrograms in figure 1 and their corresponding frequency marginals. One can easily see that the extracted marginals are latching on to the specific spectral structure of each sound. These frequency marginals can be used as a model of a class of sounds such as human voice, speech of a specific speaker, a specific type of background noise, etc.

The the next sections we describe how this model can be used for a supervised and semi-supervised source separation.

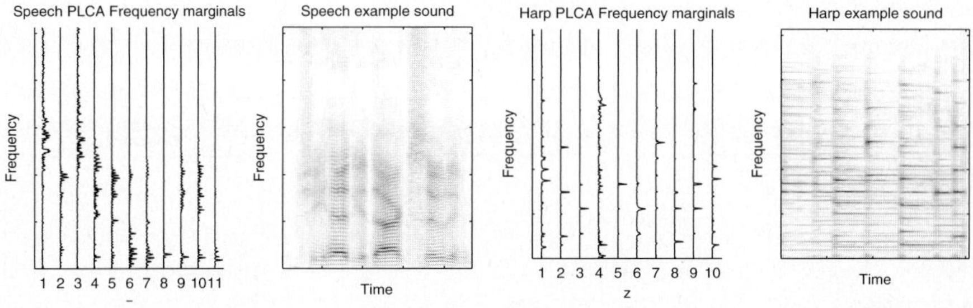

Fig. 1. Example of PLCA models of two different sounds. The two left plots display a spectrogram of speech and a set of speech-derived frequency marginals. Likewise the two right plots display the same information for a harp sound. Note how the derived marginals in both cases extract representative spectra for each sound.

2.1 Supervised Separation

In the case of supervised separation we assume that the mixture we are operating on contains classes of sounds for which we have already trained PLCA models (in the form of frequency marginals, as described above). If the kind of time-frequency distribution that we use is (at least approximately) linearly additive in nature, we can assume that the marginal distributions of our trained models can be used to approximate the mixture's distribution. For the experiments presented in this paper we employ the magnitude short time Fourier transform. Although the linearity assumption does not exactly apply for this transform, it is sufficiently approximately correct in the context of sound mixtures.

In order to perform the separation let us consider a mixture composed out of samples from the two sound classes analyzed in figure 1. Let us denote the already known frequency marginals from these two sounds as $P_1(f|z)$ and $P_2(f|z)$. The spectrogram of the mixture, which we denote by $P(f,t)$, is shown in figure 2. One can easily see elements of both sounds present in it. Once we obtain the spectrogram of the mixture we need to find how to use the already known marginals from prior analysis to approximate it. Doing so it a very simple operation which involves partial use of the training procedure shown above. First we consolidate the marginals of the known sounds into one set $P(f|z) = \{P_1(f|z) \bigcup P_2(f|z)\}$. Since all the of the marginals in $P(f|z)$ should explain the mixture spectrogram $P(f,t)$ we only need to estimate a set of time marginals $P(t|z)$ which will facilitate the approximation. We therefore perform the training outlined in the previous sections, only this time we only estimate $P(t|z)$ and keep $P(f|z)$ fixed to the already known values. After we obtain a satisfactory estimate of $P(t|z)$ we appropriately split it to two sets which correspond to each $P_i(f|t)$. We can then reconstruct the elements of the input spectrogram that correspond to only one sound class by using only the time and frequency marginals that correspond to that sound class. The results in this particular case are shown in figure 2. As is evident the contribution to the mixture from each of the two sources is cleanly separated into two spectrograms. Once the spectrograms of each known sound have been recovered we can easily transform them back to the time domain by using the corresponding phase values from the original mixture.

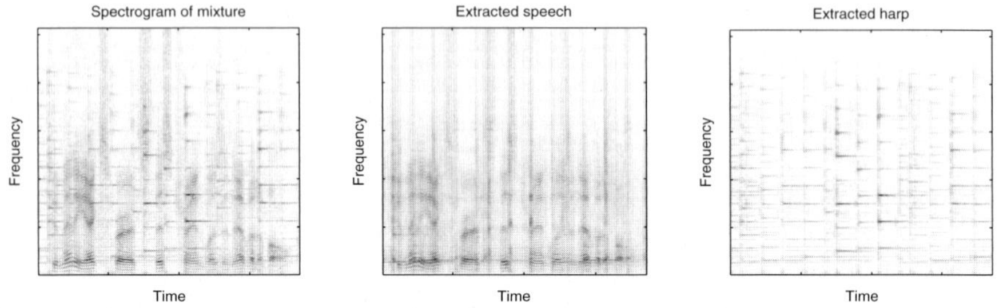

Fig. 2. Example of supervised separation using PLCA. The leftmost plot displays the inputs spectrogram. We can easily see features of the speech and harp sounds. The two remaining plots show the mixture spectrogram as approximated by the speech marginals (center plot), and the harp marginals (right plot).

As one might suspect this approach does not allow the separation of spectrally similar sounds since there will be significant similarity between the marginals of each sound class. The more dissimilar the sounds in the mixture are the better the quality of the separation will be. In this particular example separation was almost flawless since the two sounds had a very different spectral profile. For experiments using $0dB$ speech mixtures the target source improvement ranged from $3dB$ to $10dB$ depending on the similarity between the speakers. Using examples such as various types of ambient noise and speech we often achieved separation of more than $12dB$.

2.2 Semi-supervised Separation

In the case of semi-supervised separation we assume that we only have a PLCA model for one of the sounds in a mixture. In this case we cannot directly use the aforementioned procedure to perform separation. In this section we introduce a methodology which deals with this problem.

Assume that we have a mixture of multiple sounds and we only have a PLCA model for one of them. We can perform PLCA on the mixture using the known frequency marginals for one of the sounds and in the process estimate additional marginals to explain the elements in the mixture we can't already. Doing so with the training procedure we have shown in the previous sections is very easy. We train as we usually do when learning both the frequency and the time marginals, but we make sure that a portion of the frequency marginals are kept fixed as we update only the remaining ones using the same training procedure as before. The fixed marginals are the ones we already know as a model for one of the sounds. Conclusion of training will result into a set of new frequency marginals which are best suited to explain the sources in the mixture other than the one we already know. Since there will most likely be some spectral similarity between the known sound and the rest of the sources we also encourage sparsity on the time marginals to ensure that there is minimal co-occurence of frequency marginals at any time.

Once the marginals of the additional sources have been identified we can revert back to the supervised separation methodology to obtain the results we seek. The additional complication in this scenario is that by having a model of only one of the sources results into the ability to extract either that source by itself or all the other sources as one. This means that we can use this method for applications akin to denoising where we either know the target characteristics, of the background noise characteristics. In our experiments we have used this approach to separate speech from music, where the results often are very impressive[1]. A separation example is shown in figure 3, where a soprano is separated from a piano. We only had a model for the piano and learned the soprano model using the aforementioned methodology. The suppression of the piano was audibly flawless and the only artifact of this approach was a slight coloring of the extracted soprano voice (attributed mostly to the usage of phase of the original mixture).

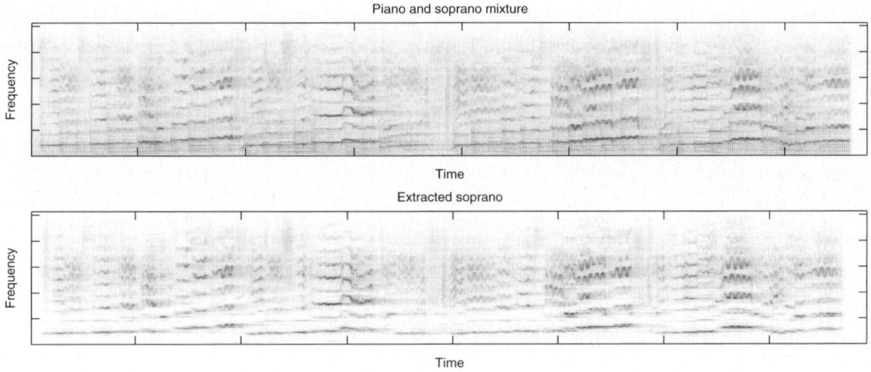

Fig. 3. Example of semi-supervised separation using PLCA. The left plot displays the mixture of a piano and a soprano, the right plot displays the extracted soprano voice. One can easily see that the harmonic series corresponding to the piano notes are strongly suppressed.

3 Discussion

In this section we discuss the selection of parameters and their effect in separation performance and point to some of the relationships of the PLCA model to other known decompositions.

[1] Demonstration sound samples of this approach can be found in `http://www.merl.com/people/paris/sep.html` under the section "PLCA for spectral factoring".

3.1 Parameter Selection

In order to obtain reasonable results we have to make sure that the right parameters are used in the process. First we need to ensure that the time/frequency decomposition we employ is adequate to perform separation. In our experience a $1/10sec$ analysis window is usually a good choice for separation. As this window becomes smaller it results in inadequate frequency resolution, and as it grows larger it results in time smearing. The hop size of the transform also needs to be small enough to ensure a clean reconstruction during the transformation from time/frequency to time (a fourth of the transform size is a good choice). Applying a Hanning window for the frequency analysis is also advised since it minimizes high frequency artifacts which are not part of the sound we model and can result into a skewed representation.

The selection of the PLCA parameters is very important in order to achieve good results. In most of our simulations, sounds were modeled using around 100 marginals (i.e. $z = \{1, 2, 3, ..., 100\}$). Using a small number of marginals results into a poor representation which attains spectrally quantized results, whereas a large number of marginals results into large sets of simple marginals which can also describe elements of the interfering sounds. The tradeoff in this case is between accuracy of model versus separability of models. The sparsity parameter is something we only use in the case of the semi-supervised learning on the time marginals. It ensures that the new marginals that we learn will not overlap as much with the already known ones. Common usage values in our experiments were $\beta = \{0, 0.01, 0.05, 0.1\}$, where larger values were used in harder to separate problems where more spectral overlap between sources was present. The audible effect of using sparsity is a degradation of reconstruction of the sound quality of the source to be learned. Therefore using the sparsity parameter is best when we have a model of the target source and we wish to remove the remaining sources.

3.2 Relation to Similar Decompositions

The PLCA model which we introduced is closely related to a variety of known decompositions. The non-sparse 2-d manifestation is identical to the Probabilistic Latent Semantic Indexing (PLSI) algorithm [5], which itself is a probabilistic generalization of the Singular Value Decomposition. The functional difference is that PLSI/PLCA operate on distributions instead of raw data which means that they can effectively only analyze non-negative inputs. If we rewrite the 2-d PLCA model in terms of matrix operations, this relationship is more evident:

$$P(f,t) = \sum_z P(z)P(f|z)P(t|z) \equiv \mathbf{V} = \mathbf{W} \cdot \mathbf{S} \cdot \mathbf{H} \tag{9}$$

where $\mathbf{V}$ is a matrix containing the distribution $P(f,t)$, $\mathbf{W}$ is a matrix containing in its columns $P(f|z)$ for every z, $\mathbf{S}$ is a diagonal matrix containing in its diagonal the values of $P(z)$, and $\mathbf{H}$ is a matrix containing in its rows $P(t|z)$ for every z.

Additionally if we absorb the values of $\mathbf{S}$ into the two matrices $\mathbf{W}$ and $\mathbf{H}$ so that: $\mathbf{V} = \mathbf{W} \cdot \mathbf{S} \cdot \mathbf{H} = \bar{\mathbf{W}} \cdot \bar{\mathbf{H}}$, we can make a connection to the Non-negative

Matrix Factorization (NMF) decomposition [8]. NMF can employ the Kullback-Leibler divergence to measure how well the factorization $\bar{\mathbf{W}} \cdot \bar{\mathbf{H}}$ approximates the input $\mathbf{V}$. The EM training which we perform also indirectly optimizes the same cost function as it improves the model's log-likelihood. In fact the two training procedures for 2-d PLCA and NMF can be shown to be numerically identical.

Finally we can make a loose connection to non-negative ICA by noting that by using the entropic prior to manipulate the joint entropy of $\mathbf{H}$ we can obtain the equivalent of an ICA mixing matrix in $\mathbf{W}$. Although this is only a conjecture on our part, preliminary results from simulations are encouraging.

4 Conclusions

In this document we introduced a sparse latent variable model which can be employed for the decomposition of time/frequency distributions to perform separation of sources from monophonic recordings. We demonstrated the use of this model for both supervised and semi-supervised source separation, and discussed its relationship with other known decompositions. Our results are very encouraging and amenable to various modifications, such as the use of convolutive bases and transformation invariance, which can help to successfully apply this work to even more challenging source separation problems.

References

1. Casey, M., Westner, A.: Separation of Mixed Audio Sources by Independent Subspace Analysis. In: proceedings ICMC (2000)
2. Roweis, S.T.: One Microphone Source Separation. In: NIPS (2000)
3. Benaroya, L., McDonagh, L., Bimbot, F., Gribonval, R.: Non negative sparse representation for Wiener based source separation with a single sensor. In: proceedings of the ICASSP (2003)
4. Vincent, E., Rodet, X.: Music transcription with ISA and HMM. In: Puntonet, C.G., Prieto, A.G. (eds.) ICA 2004. LNCS, vol. 3195, Springer, Heidelberg (2004)
5. Hofmann, T.: Probabilistic Latent Semantic Indexing. In: proceedings SIGIR'99 (1999)
6. Brand, M.E.: Structure Learning in Conditional Probability Models via an Eutropic Prior and Parameter Extinction. Neural Computation Journal 11(5), 1155–1182 (1999)
7. Shashanka, M.V.S.: A Unified Probabilistic Approach to Modeling and Separating Single-Channel Acoustic Sources, Ph.D. Thesis, Department of Cognitive and Neural Systems. Boston University, Boston (2007)
8. Lee, D.D, Seung, H.S.: Algorithms for Non-negative Matrix Factorization. In: NIPS (2001)

Image Compression by Redundancy Reduction

Carlos Magno Sousa, André Borges Cavalcante, Denner Guilhon,
and Allan Kardec Barros

magno@dee.ufma.br, andre@dee.ufma.br, dennerguilhon@hotmail.com,
allan@ufma.br
http://pib.dee.ufma.br

Abstract. Image compression is achieved by reducing redundancy between neighboring pixels but preserving features such as edges and contours of the original image. Deterministic and statistical models are usually employed to reduce redundancy. Compression methods that use statistics have heavily been influenced by neuroscience research. In this work, we propose an image compression system based on the *efficient coding* concept derived from neural information processing models. The system performance is compared with principal component analysis (PCA) and the discrete cosine transform (DCT) at several compression ratios (CR). Evaluation through both visual inspection and objective measurements showed that the proposed system is more robust to distortions such as ringing and block artifacts than PCA and DCT.

1 Introduction

The goal of data compression is to reduce space or bandwidth required to store or transmit some information [1]. Specifically, in case of images, the data contains a high degree of correlation or redundancy between neighboring samples. This way, compression is achieved by reducing the redundancy of data preserving quality and features such as edges and contours of the original image.

The redundancy reduction principle can be analyzed in both deterministic and statistical fashions [1,2]. In the first, redundancy is understood as data samples that can be inferred without use of statistical information of data. On the second one, redundancy reduction is performed transforming the data into an efficient representation according to statistical independence criterion [2].

There is a large number of deterministic methods used for image compression. For instance, the well-known JPEG scheme employs the discrete cosine transform (DCT) [4,3] to encode images. The DCT is a Fourier-related transform which converts data into frequency components. Contrarily to Fourier's, these components are real coefficients defined as the inner product between the image to be encoded and the DCT basis functions. Although the DCT is easy to implement and fast to compute, it still undergo as many drawbacks as any Fourier-related transform. Gibbs phenomenon [5] is an example of such deficiencies that in case of images, are smoothed edges. Another DCT problem is blocking artifacts [6]. These artifacts are due to the block processing of images and can be understood as luminance discontinuities between block boundaries.

M.E. Davies et al. (Eds.): ICA 2007, LNCS 4666, pp. 422–429, 2007.

On the other hand, compression methods that use statistics have heavily been influenced by neural information processing models [2]. Neuroscience studies suggested that neuron populations process stimuli information according to the concept of "efficient coding" [7]. Under this concept, neuron responses are mutually statistically independent which means that there is no "redundant information" processed throughout the population.

Statistical independence criteria may be explored by two approaches: principal component analysis (PCA) and independent component analysis (ICA). PCA utilizes second-order statistics while ICA uses high-order statistics to obtain an efficient code. For instance, PCA is employed in several image compression systems in order to reduce data dimension [8]. The shortcoming of PCA based systems is that second-order statistics can only provide efficient representations for Gaussian data and images are normally non-Gaussian [7]. To circumvent this problem, new compression systems have used ICA to encode images such as [9]. However, this method does not take into account the non-orthogonality propriety of ICA basis functions in the code estimation.

Therefore the purpose of this work is to propose an image compression system based on the "efficient coding" concept using ICA. In this model, data compression is carried out projecting images onto subspaces learned by ICA where the efficient code is given by the respective projection coefficients. This model is also applied to electrocardiogram data compression in [10].

2 Methods

Let us divide an image into a vector of blocks $\mathbf{y} = [y_1, y_2, \ldots, y_n]^{\mathrm{T}}$ with size of m x m. Now, assume that each block y_i can be reconstructed as a linear combination of vectors from a stochastically learned subspace $\Phi = [\phi_1, \phi_2, \ldots, \phi_n]^{\mathrm{T}}$, where ϕ_i is also called basis function. This process might be written as

$$\hat{y_i} = w_1\phi_1 + w_2\phi_2 + \ldots + w_n\phi_n, \tag{1}$$

where $\hat{y_i}$ is the reconstructed version of block y_i and each component w_i is the projection coefficient on the ith base function.

To learn the subspace Φ and estimate the projection coefficients w_i, we propose the image compression system, shown in Figure 1, which is based on the concept of efficient coding. The following subsections provide a full explanation of the structure of our model.

Efficient Coding. Let us assume that an image block is encoded by the projection coefficients $\mathbf{w} = [w_1, w_2, \ldots, w_n]^{\mathrm{T}}$ in Eq. (1). The goal of efficient coding is to estimate a subspace Φ that reduces the mutual statistical dependence between the coefficients w_i.

An estimation of the subspace Φ may be accomplished by either PCA or ICA. The first assumes that the components are uncorrelated while the second assumes that components are mutually statistically independent. In this work, we use the last approach.

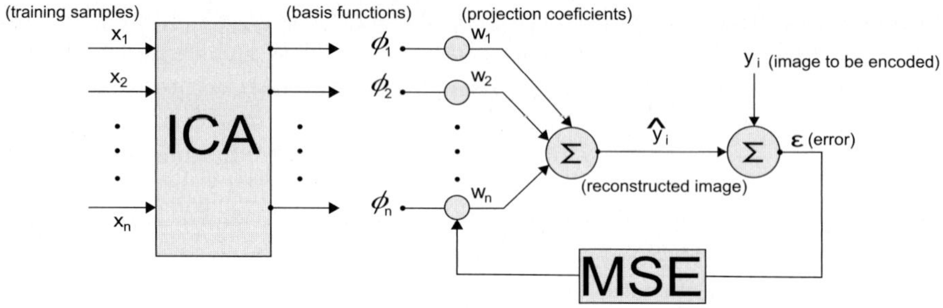

Fig. 1. Proposed image compression system. The system consist of two phases: *learning* and *projection*. In the *learning* phase, we use independent component analysis to *learn* a subspace. In the *projection* phase, we estimate the projections coefficients through a minimum mean square error (MSE) estimation.

Learning a Subspace through ICA. Let $\mathbf{x} = [x_1, x_2, \ldots, x_n]^{\mathrm{T}}$ be a set of observations taken from the same data class and written as in Eq. (2). Using $\mathbf{x}$ as a training input, ICA learns basis functions ϕ_{i} for the data class so that the set of variables which composes vector $\mathbf{a} = [a_1, a_2, \ldots, a_n]^{\mathrm{T}}$ are mutually statistically independent.

$$\mathbf{x} = \mathbf{a}^{\mathrm{T}}\Phi. \tag{2}$$

To achieve statistically independence, ICA algorithms work with higher-order statistics which point out directions where data is maximally independent. Here, we used the FastICA algorithm [11].

Projecting an Image onto Subspaces. The projection phase consists of finding out the image representation for the subspace Φ. This representation is given by the projection vector $\mathbf{w}$ where each coefficient w_i shows how the $i-th$ base function is activated inside the image. In our model, the projection vector is found out through minimum mean square error (MSE) estimation.

Hence, for a given image block y_i, the projection vector is estimated such that it minimizes the MSE between the reconstructed block $\hat{y_i}$ and the original one. For this estimation method, the solution vector is given by

$$\mathbf{w} = \mathrm{E}[\Phi\Phi^{\mathrm{T}}]^{-1}\mathrm{E}[\mathrm{y_i}\Phi]. \tag{3}$$

The term $\mathrm{E}[\Phi\Phi^{\mathrm{T}}]^{-1}$ in Eq. (3) holds information about the angles between every two basis functions of Φ and is necessary to achieve the minimum error once the ICA basis are non-orthogonal. Eq. (3) is also used to select a smaller subspace Ψ from larger subspace Φ with minimum MSE. This process consists of a block-level deflationary basis pursuit which allow us to change the compression rate (CR) using as many coefficients as desired to represent the image. This basis pursuit is summarized in the following procedure:

Step 1: Define an empty subspace Ψ;
Step 2: Repeat next step for $k = 1, 2, ..., n$, where n is the dimension of Φ;

Step 3: Using Eq. (3), find the reconstructed image $\hat{y}_{ik}$ projecting y_i onto the subspace composed of $[\Psi, \phi_k]$, where ϕ_k is the kth base function of Φ;
Step 4: Select the base functions according to the following criteria:

$$\phi_k = \arg\min MSE(\hat{y}_{ik} - y_{ik}); \quad (4)$$

Step 5: Move ϕ_k from Φ to Ψ so that $n = n - 1$;
Step 6: Return to step 2 until Ψ get the desired dimension;

It is important to notice that this basis pursuit is non-orthogonal in contrary to the standard Matching Pursuit [12]. Hence, when Ψ has obtained a dimension higher than one, the term $\mathrm{E}[\Phi\Phi^{\mathrm{T}}]^{-1}$ is used to estimate the correct $\mathbf{w}$.

3 Results

In the learning phase, we have used 200 male and 200 female face images from AR Face Database [13]. This database contains 4000 faces from 126 persons for many expressions and cloths. To form the training set for ICA, the 400 images were divided into 8 x 8 blocks. For each block, two training samples of 64 pixels were obtained. The first sample was obtained performing a line-wise reading of the pixels inside the block and the second one, through column-wise reading. Figure 2 shows an example of ICA subspace learned from face images. Also, for the same training set, the PCA subspace along with the standard DCT's.

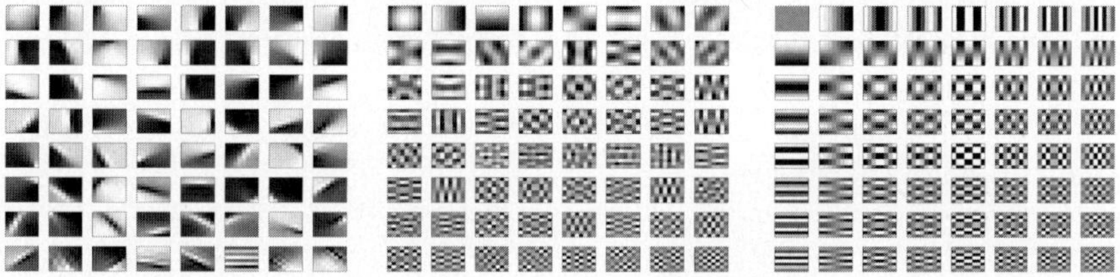

(a) (b) (c)

Fig. 2. Example of subspaces: (a) ICA and (b) PCA subspaces learned from face images, (c) standard DCT subspace

In projection phase, the coefficients were quantized using Lloyds's algorithm [14]. And since we are concerned with losses introduced in the coding process, we used no further lossless methods such as Huffman coding [15].

To evaluate the proposed system performance, we have compressed several images (not used in the learning phase) for ICA, PCA and DCT. Then, the respective reconstructed images were compared using visual inspection and objective measurements. For subjective insight, Figure 3 shows several reconstructed images along with the respective values of CR and *picture quality scale* (PQS) [16].

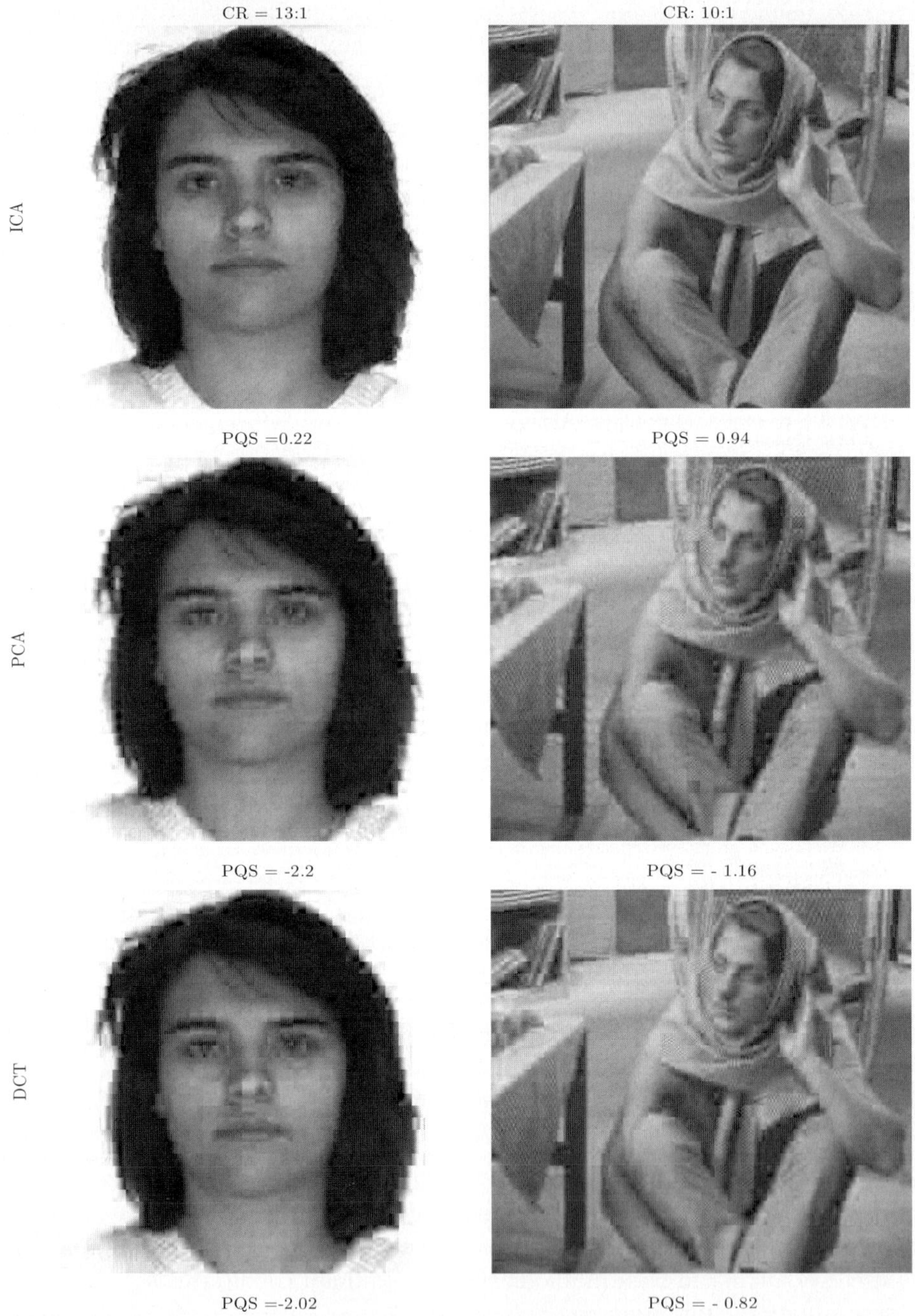

Fig. 3. Reconstructed images for ICA, PCA and DCT. The face images in the first column were all reconstructed using three basis functions and its projections coefficients were quantized using four bits so that the CR = 13:1. For the image in the second column, five basis functions and six bits for quantization, CR = 10:1.

It is important to notice that the images are compressed with ICA, PCA, and DCT at same CR.

The PQS value varies from zero to five and matches the subjective scale MOS, but can assume negative values for poor reconstructions. Further, we used a weighted version of the standard percent root-mean square difference (WPRD) which is defined as

$$\mathrm{WPRD} = \sqrt{\frac{\mathrm{E}[\mathrm{CF}(\hat{y}_i - y_i)^2]}{\mathrm{E}[y_i^2]}}. \tag{5}$$

The WPRD is found out by filtering the spatial frequency of the error image $(\hat{y}_i - y_i)$ with the human contrast sensitivity function (CF) [17]. The weighting aims to match the error influence according to human visual system sensibility for certain frequencies.

Figure 4 shows the average WPRD obtained by increasing the basis functions number used for reconstruction of five face images.

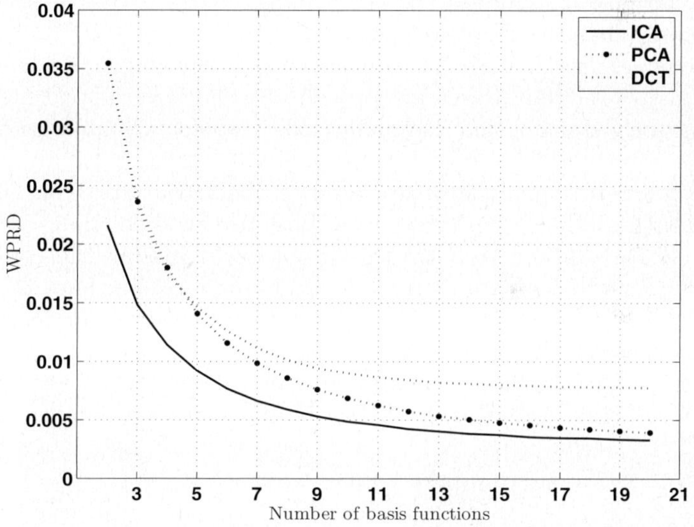

Fig. 4. Average WPRD performance for five reconstructed images which were not used in the learning phase. In the projection phase, the coefficients were quantized using five bits.

4 Discussion

The proposed model has a reasonable physiological foundation: efficient coding. Hence, this model should introduce less perceptible distortions for humans than methods non-based on neural modeling such as DCT or even PCA.

Actually, there are many interesting points to highlight. Firstly, let us analyze the subspaces presented in Figure 2. The statistical independence criteria used in ICA span *edge detectors* as basis functions for face images. These "edges" corresponds to significant variations of the grey level for the training set and are results of the redundancy reduction process [18]. PCA and DCT are both orthogonal transforms so that they similarly distribute data spectrum information between their basis. This fact can be confirmed observing that the shape of their basis functions are alike. Further, we can also see that reconstructed images for PCA and DCT in Figure 3 are very similar.

Secondly, let us analyze an important point in the performance of compression techniques: the presence of ringing and blocking artifacts in reconstructed images. Ringing artifacts are produced by Gibbs phenomenon [5]. This phenomenon can be understood as large oscillations in image areas that have discontinuities such as edges and contours. These oscillations produce a smooth effect on the figure, what can be unacceptable for applications such as medical image compression [19]. From the reconstructed images in the first and second columns of Figure 3, we can see that edges and contours defining image details, such as *pupils*, *eyes*, *nose*, *lips* and *hair wires* of face in the first column; and the *table*, *arms*, *legs* and *face* contours in the second image were better preserved in the proposed model than for PCA and DCT.

The problem of "ringing" might be even worse if blocking artifacts are considered. And in this model, images are processed in non-overlapped blocks which are independently transformed and quantized. Therefore, luminance discontinuities may be introduced between block boundaries, what is highly perceivable by humans when few coefficients are used in the reconstruction. Thus, either the compression method is intrinsically robust to this distortion, or additional methods may be required to solve this problem, increasing the complexity of the compression system. Once more, a neural based modeling should fit. In fact, comparing the images in Figure 3, we can clearly see that our model is more robust to blocking artifacts.

Indeed, since PQS and WPRD takes human sensitivity into account, we can see from the negative values of PQS for the reconstructed images in Figure 3 and the WPRD performance in Figure 4 that our model introduces less errors than PCA and DCT regarding the human perception.

5 Conclusions

We have proposed an image compression system based on the concept of *efficient coding*. Our system consists of two phases: learning and projection. In the learning phase, we used independent component analysis (ICA) to learn a subspace that maximizes the code efficiency. In the projection phase, we found out the code projecting the image onto the subspace. The projection was carried out through a mean square error estimation. The system was compared with principal component Analysis (PCA) and the discrete cosine transform (DCT) through both visual inspection and objective measurements. The results analysis showed that our model is more robust to distortions which are highly perceptible by humans. Several reconstructed images from our model can be found at http://pib.dee.ufma.br

References

1. Kortamnl, C.M.: Redundancy Reduction-A Practical Method of Data Compression. Proc. of The IEEE 55(3), 223–226 (1967)
2. Barros, A.K., Chichocki, A.: Neural Coding by Redundancy Reduction and Correlation. In: Proc. of the VII Brazilian Symposium on Neural Networks (SBRN) (IEEE) (2002)
3. Ahmed, N., Natarajan, T., Rao, K.R.: Discrete Cosine Transform. IEEE Trans. Computers, 90–93 (1974)
4. Wallace, G.K.: The JPEG still-picture compression standard. Commun. ACM 34, 30–44 (1991)
5. Gibbs, J.W.: Fourier Series. Nature, 59 (1898)
6. Coudoux, F.X., Gazalet, M.G., Corlay, P., Rouvaen, J.M.: A Perceptual Approach to the Reduction of Blocking Effect in DCT-Coded Images. Journal of Visual Communication and Image Representation 8(4), 327–337 (1997)
7. Simoncelli, E.P., Olshausen, B.A.: Natural Image statistics and Neural Representation. Annu. Rev. Neurosci. 1193–216 (2001)
8. Dony, R.D., Haykin, S.: Proc. of IEEE Neural Network Approaches to Image Compression, vol. 83(2), pp. 288–303 (1995)
9. Ferreira, A.J, Figueiredo, M.A.T.: On the use of independent component analysis for image compression. Signal Processing: Image Communication 21, 378–389 (2006)
10. Guilhon, D., Barros, A.K.: ECG Compression by Efficient Coding, submitted to ICA (2007)
11. Hyvärinen, A., Karhunen, J., Oja, E.: Independent Component Analysis. John Wiley and Sons, New York (2001)
12. Mallat, S., Zhang, Z.: Matching pursuit with time-frequency dictionaries. IEEE Transactions on Signal Processing 41(12), 3397–3415 (1993)
13. Martinez, A.M., Benavente, R.: The AR Face Database. CVC Technical Report, vol. 24 (1998)
14. Lloyd, S.P.: Least Squares Quantization in PCM. IEEE Transactions on Information Theory IT-28, 129–137 (1982)
15. Huffman, D.A.: A method for the construction of minimum-redundancy codes. In: Proceedings of the I.R.E. pp. 1098–1102 (1952)
16. Miyahara, M., Kotani, K., Algazi, V.: Objective picture quality scale (pqs) for image coding. IEEE Trans. Commun 46, 1215–1226 (1998)
17. Mannos, J.L., Sakrison, D.J.: The Effects of a Visual Fidelity Criterion on the Encoding of Images. IEEE Trans. on Information Theory it-20, 525–536 (1974)
18. Ziou, D., Tabbone, S.: Edge Detection Techniques - An Overview. International Journal of Pattern Recognition and Image Analysis 8, 537–559 (1998)
19. Strijm, J., Cosmanb, P.C.: Medical image compression with lossless regions of interest. Signal Processing 59, 155–171 (1991)

Complex Nonconvex l_p Norm Minimization for Underdetermined Source Separation

Emmanuel Vincent

METISS Group, IRISA-INRIA
Campus de Beaulieu, 35042 Rennes Cedex, France
emmanuel.vincent@irisa.fr

Abstract. Underdetermined source separation methods often rely on the assumption that the time-frequency source coefficients are independent and Laplacian distributed. In this article, we extend these methods by assuming that these coefficients follow a generalized Gaussian prior with shape parameter p. We study mathematical and experimental properties of the resulting complex nonconvex l_p norm optimization problem in a particular case and derive an efficient global optimization algorithm. We show that the best separation performance for three-source stereo convolutive speech mixtures is achieved for small p.

1 Introduction

Underdetermined source separation is the problem of recovering the single-channel source signals $s_j(t)$, $1 \leq j \leq J$, underlying a multichannel mixture signal $x_i(t)$, $1 \leq i \leq I$, with $I < J$. The mixing process can be modeled in the time-frequency domain via the Short-Term Fourier Transform (STFT) as [1]

$$\mathbf{X}(n, f) = \mathbf{A}(f)\mathbf{S}(n, f) \tag{1}$$

where $\mathbf{S}(n, f)$ is the vector of source STFT coefficients in time-frequency bin (n, f), $\mathbf{X}(n, f)$ is the vector of mixture STFT coefficients in the same bin, and $\mathbf{A}(f)$ is a complex mixing matrix. This problem can be tackled by first estimating the mixing matrices and then deriving the Maximum *A Posteriori* (MAP) source coefficients under the constraint (1), based on some prior distribution.

Existing separation methods rely on the assumption that the source coefficients are independent and sparsely distributed, *i.e.* a large proportion of coefficients are close to zero. Examples of sparse priors include mixtures of Dirac impulses and Gaussians [2], mixtures of Gaussians [3], Student t distributions [4] and the Laplacian distribution [5,6,1,7]. The latter is popular since it leads to a convex optimization problem that can be solved efficiently. In this paper, we extend this approach by assuming that the source coefficients follow a generalized Gaussian prior, of which the Laplacian is a special case. This extension is not straightforward, since the resulting criterion can be nonconvex.

The structure of the rest of this paper is as follows. In Section 2, we argue that generalized Gaussian priors are well suited to the modeling of speech signals. We

M.E. Davies et al. (Eds.): ICA 2007, LNCS 4666, pp. 430–437, 2007.

study the properties of the resulting optimization problem in Section 3 in the case where $J = I + 1$ and derive an efficient global optimization algorithm. We evaluate its performance on instantaneous and convolutive speech mixtures in Section 4 and conclude in Section 5.

2 Generalized Gaussian Priors

Generalized Gaussian priors were first introduced in the context of source separation via Independent Component Analysis (ICA) [8,9,10]. The phases of the source STFT coefficients $S_j(n,f)$ are assumed to be uniformly distributed, while their magnitudes are modeled by

$$P(|S_j(n,f)|) = p\frac{\beta^{1/p}}{\Gamma(1/p)}e^{-\beta|S_j(n,f)|^p} \tag{2}$$

where the parameters $p > 0$ and $\beta > 0$ govern respectively the shape and the variance of the prior. This prior includes the Laplacian ($p = 1$) and the Gaussian ($p = 2$) as special cases and its sparsity increases with decreasing p.

In order to assess the benefit of using this prior, we computed the best shape parameters p for 30 speech signals, considering all frequency bins either separately or together. The signals were sampled at 8 kHz and had a duration of 12 s. The STFT was computed using half-overlapping sine windows of various lengths L and each frequency bin was scaled to unit variance. The Maximum Likelihood (ML) parameters were estimated using Matlab `fminunc` optimizer[1].

The observed parameter range is depicted in Figure 1. On average, p varies between 0.4 and 0.9 depending on the window length L and stays almost constant across frequency, except at very low frequencies where background noise dominates. This shows that generalized Gaussian priors with $p < 1$ better fit speech sources than Laplacian priors. Interestingly, the observed value of p reaches a minimum for $L = 64$ ms, which was also previously determined to be the optimal window length for source separation via binary masking [11,12].

3 Properties of the Complex l_p Norm Criterion

Given these results, we now assume that the mixing matrices $\mathbf{A}(f)$ are known and that the source STFT coefficients follow a generalized Gaussian prior with fixed parameters p and β. The MAP source coefficients are given by

$$\widehat{\mathbf{S}}(n,f) = \arg\min_{\mathbf{S}\in\mathbb{C}^J} \|\mathbf{S}\|_p^p \text{ subject to } \mathbf{A}(f)\mathbf{S} = \mathbf{X}(n,f) \tag{3}$$

where $\|\mathbf{S}\|_p$ is the l_p norm of the vector $\mathbf{S}$ defined by $\|\mathbf{S}\|_p^p = \sum_{j=1}^J |S_j|^p$. When $p < 1$, this criterion is nonconvex hence difficult to minimize.

[1] This algorithm is based on a subspace trust region method. For more details, see http://www.mathworks.com/access/helpdesk_r13/help/toolbox/optim/fminunc.html

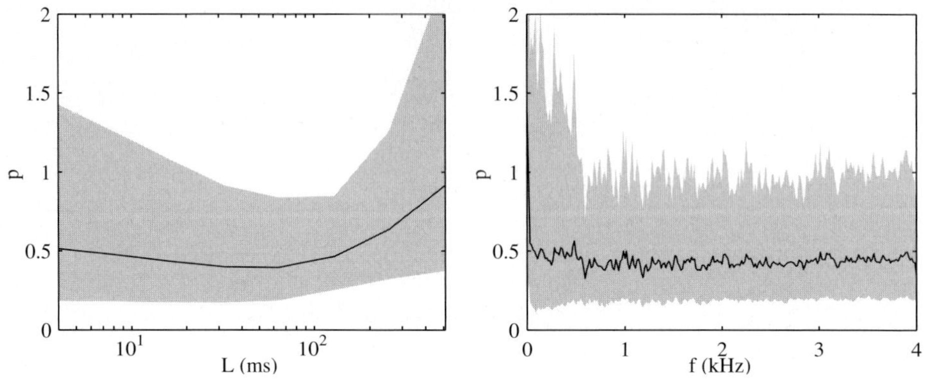

Fig. 1. Variation of the shape parameter p measured on speech STFT coefficients as a function of the window length L (left) and as a function of frequency f with $L = 64$ ms (right). The black curve and the gray area represent respectively the geometric mean and the geometric standard deviation of the measured values. This illustration is motivated by the fact that the measured values are approximately log-Gaussian.

3.1 Unconstrained Expression

We focus in the rest of this article on the simple case where $J = I + 1$ and $\mathbf{A}(f)$ has full row rank. Under this assumption, the constrained optimization problem (3) is equivalent to a one-dimensional unconstrained complex optimization problem [1]. The MAP source coefficients are then expressed as

$$\widehat{\mathbf{S}}(n,f) = \mathbf{O} + \widehat{u}\mathbf{V} \tag{4}$$

where $\mathbf{O}$ is any vector satisfying the constraint, *e.g.* $\mathbf{O} = \mathbf{A}(f)^\dagger \mathbf{X}(n,f)$ with † denoting pseudo-inversion, $\mathbf{V}$ is any vector spanning the null space of $\mathbf{A}(f)$ and

$$\widehat{u} = \arg\min_{u \in \mathbb{C}} \|\mathbf{O} + u\mathbf{V}\|_p^p. \tag{5}$$

This optimization problem has to be solved for each STFT bin (n, f) individually. Using the complex derivative notation [13], the first and second order derivatives of the criterion $\mathcal{L}(u) = \|\mathbf{O} + u\mathbf{V}\|_p^p$ are given by

$$\frac{\partial \mathcal{L}}{\partial \overline{u}} = \overline{\frac{\partial \mathcal{L}}{\partial u}} = \frac{p}{2} \sum_{j=1}^{J} |O_j + uV_j|^{p-2} \overline{V_j} (O_j + uV_j) \tag{6}$$

$$\frac{\partial^2 \mathcal{L}}{\partial u \partial \overline{u}} = \frac{\partial^2 \mathcal{L}}{\partial \overline{u} \partial u} = \frac{p^2}{4} \sum_{j=1}^{J} |O_j + uV_j|^{p-2} |V_j|^2 \tag{7}$$

$$\frac{\partial^2 \mathcal{L}}{\partial \overline{u}^2} = \overline{\frac{\partial^2 \mathcal{L}}{\partial u^2}} = \frac{p(p-2)}{4} \sum_{j=1}^{J} |O_j + uV_j|^{p-4} \overline{V_j}^2 (O_j + uV_j)^2 \tag{8}$$

3.2 Singular and Non-singular Solutions

It is well known that for real variables the global minimum of $\mathcal{L}$ with $p \leq 1$ results in at least one source coefficient being zero and can be found by combinatorial optimization [6,1]. However, this is not true anymore with complex variables, as shown in [1,7] in the particular case $p = 1$. Nevertheless, the local minima of $\mathcal{L}$ can still be characterized using the two lemmas below.

Lemma 1. *Let $\mathcal{J} = \{j : V_j \neq 0\}$. When $p < 1$, the points $z_j = -\frac{O_j}{V_j}$, $j \in \mathcal{J}$, are singular (*i.e.* non-differentiable) local minima of $\mathcal{L}$.*

Proof. Let $\mathcal{Z}_j = \{k : z_k = z_j\}$. The point z_j is characterized by the fact that $S_k(n, f) = 0$ for all $k \in \mathcal{Z}_j$ and $S_k(n, f) \neq 0$ for all $k \notin \mathcal{Z}_j$. By developing the expression of $\mathcal{L}$ around this point when $p < 1$, we get

$$\mathcal{L}(z_j + u) = \mathcal{L}(z_j) + \left(\sum_{k \in \mathcal{Z}_j} |V_k|^p \right) |u|^p + O(u). \tag{9}$$

Thus $\mathcal{L}$ is non-differentiable at z_j and $\mathcal{L}(z_j + u) > \mathcal{L}(z_j)$ for small $u \neq 0$. □

Lemma 2. *The other local minima of $\mathcal{L}$ are non-singular and within the convex hull of z_j, $j \in \mathcal{J}$.*

Proof. If $u \neq z_j$ for all j and u is a local minimum of $\mathcal{L}$, then $\mathcal{L}$ is differentiable at u according to (6) and $\frac{\partial \mathcal{L}}{\partial \bar{u}} = 0$. After rearranging this equality, we get

$$u = \frac{\sum_{j \in \mathcal{J}} |O_j + uV_j|^{p-2} |V_j|^2 z_j}{\sum_{j \in \mathcal{J}} |O_j + uV_j|^{p-2} |V_j|^2}. \tag{10}$$

Thus u can be expressed as a weighted sum of z_j, $j \in \mathcal{J}$, with positive weights summing to one. □

In the following, we use the term "singular" to characterize by extension the local minima of $\mathcal{L}$ where at least one source coefficient is zero, although $\mathcal{L}$ is differentiable at these minima when $p > 1$.

3.3 Critical Value of p for the Existence of Non-singular Solutions

The above distinction between singular and non-singular local minima raises the question whether non-singular minima can exist for all values of p and whether the global minimum can be non-singular. We studied this question experimentally with $I = 2$ and $J = 3$.

We draw 100 independent source coefficient vectors following the generalized Gaussian distribution with shape parameter $p = 0.4$ using the Metropolis-Hastings algorithm [14]. We also draw 100 instantaneous (real) mixing matrices of the form $A_{1j} = \cos(\theta_j)$ and $A_{2j} = \sin(\theta_j)$ with θ_j uniformly distributed in $[-\frac{\pi}{2}, \frac{\pi}{2}]$ and 100 convolutive (complex) mixing matrices of the form $A_{1j} = 1$ and

$A_{2j} = e^{2i\pi\theta_j}$ with θ_j uniformly distributed in $(-\pi, \pi]$. The multiplication of each source coefficient vector by each matrix resulted in a total of 10000 instantaneous mixtures and 10000 convolutive mixtures.

For each mixture, we tested whether non-singular minima of $\mathcal{L}$ existed and whether the global minimum was non-singular as follows. Given Lemma 2, we sampled $\mathcal{L}$ on a discrete grid spanning the convex hull of z_j, $1 \leq j \leq J$, containing points of the form $u = \frac{k_1}{3K} z_1 + \frac{k_2}{3K} z_2 + \frac{3K-k_1-k_2}{3K} z_3$, $1 \leq k_j \leq K$, with $K = 50$, and we selected the global minimum $\widetilde{u}$ on this grid. If $\widetilde{u}$ was non-singular, then the true global minimum was necessarily non-singular. Otherwise, we decided that the global minimum was singular. In the latter case, we also sampled the gradient and the Hessian of $\mathcal{L}$ on the same grid, selected as a potential local minimum the point with the smallest gradient among all points with positive definite Hessian and refined it using the `fminunc` optimizer. We then observed whether the optimizer converged to a non-singular local minimum or not.

The results were very similar for instantaneous and convolutive mixtures. The average percentage of mixture draws resulting in a non-singular local minimum or a non-singular global minimum is depicted in Figure 2 as a function of p. Both quantities decrease with decreasing p, with a large drop around $p = 1$. For $p \lesssim 0.75$, there remains a few non-singular local minima, but no global minima. This can be illustrated in a more general case using the example below.

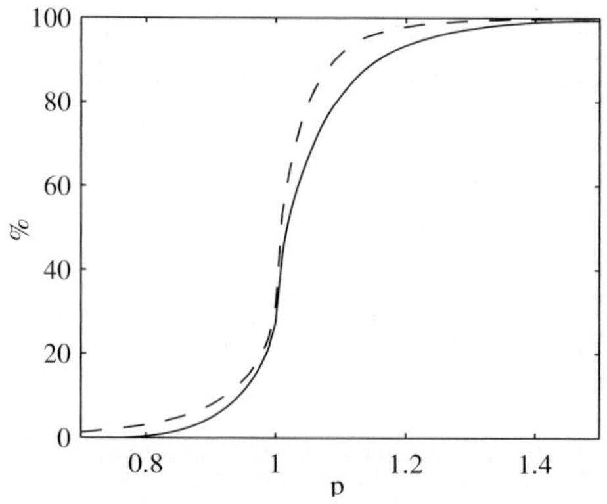

Fig. 2. Percentage of mixture draws resulting in a non-singular local minimum (dashed curve) or a non-singular global minimum (plain curve) of the l_p norm criterion with three-source two-channel mixtures

Example 1. Let $O_j = e^{2i\pi \frac{j}{J}}$ and $V_j = 1$, $1 \leq j \leq J$. Then $u = 0$ is a non-singular local minimum of $\mathcal{L}$ for all $p > 0$. Furthermore, the value of $\mathcal{L}$ at this minimum is smaller than at all singular local minima for $p > p_{\text{crit}}$ where p_{crit} is defined implicitly by $\sum_{j=1}^{J-1} |1 - O_j|^{p_{\text{crit}}} = J$ and equals respectively 0.738, 0.612, 0.534, 0.481 for $J = 3, 4, 5, 6$, and decreases with increasing J.

Proof. Using the fact that $\sum_{j=1}^{J} O_j = \sum_{j=1}^{J} O_j^2 = 0$, the coefficients of the complex gradient, the diagonal coefficients of the complex Hessian and the off-diagonal coefficients of the complex Hessian of $\mathcal{L}$, defined in [13], are given respectively at $u = 0$ by $\frac{\partial \mathcal{L}}{\partial \overline{u}} = \frac{\partial \mathcal{L}}{\partial u} = 0$, $\frac{\partial^2 \mathcal{L}}{\partial u \partial \overline{u}} = \frac{\partial^2 \mathcal{L}}{\partial \overline{u} \partial u} = \frac{Jp^2}{4}$ and $\frac{\partial^2 \mathcal{L}}{\partial \overline{u}^2} = \frac{\partial^2 \mathcal{L}}{\partial u^2} = 0$. Thus the complex gradient is zero and the complex Hessian is positive-definite, which proves that $u = 0$ is a non-singular local minimum of $\mathcal{L}$ [13].

The values of the criterion at this non-singular minimum and at the singular local minima $z_j = -O_j$ are given by $\mathcal{L}(0) = J$ and $\mathcal{L}(z_j) = \sum_{j=1}^{J-1} |1 - O_j|^p$ for all j. The latter is a strictly increasing function of p. Indeed, it can be checked that $\frac{d\mathcal{L}(z_j)}{dp} = \log J > 0$ at $p = 0$ and $\frac{d^2\mathcal{L}(z_j)}{dp^2} > 0$ for all $p > 0$. Thus $\mathcal{L}(0) > \mathcal{L}(z_j)$ if and only if $p > p_{\text{crit}}$ where p_{crit} is the value of p such that $\mathcal{L}(0) = \mathcal{L}(z_j)$. □

This shows that the global minimum of $\mathcal{L}$ can be non-singular when $p > p_{\text{crit}}$. We conjecture that p_{crit} is the lowest value of p for which this can happen.

Conjecture 1. The global minimum of $\mathcal{L}$ with $p < p_{\text{crit}}$ is always singular.

We have not yet managed to prove this conjecture mathematically. However we verified it experimentally with $J = 3$ (see Figure 2) and with $4 \leq J \leq 6$ using the same number of mixture draws and a similar discrete grid for optimization.

3.4 Efficient Optimization Algorithm

Lemmas 1 and 2 and Conjecture 1 suggest the following efficient algorithm for the estimation of the MAP source coefficients.

- If $p \geq 1$, run any gradient-based optimizer initialized randomly using (6)–(8).
- If $p \leq p_{\text{crit}}$, sample the criterion at the singular points z_j, $1 \leq j \leq J$, and select the minimum of the criterion among these points.
- If $p_{\text{crit}} < p < 1$, sample the criterion on a discrete grid spanning the convex hull of the singular points z_j, $1 \leq j \leq J$, and containing these points. Select the minimum of the criterion on this grid. If it is non-singular, refine it via any gradient-based optimizer using (6)–(8).

Provided that Conjecture 1 is true, this algorithm is guaranteed to find the global minimum of the criterion when $p \geq 1$ or $p \leq p_{\text{crit}}$, but also when $p_{\text{crit}} < p < 1$ if the discrete grid is tight enough. Moreover, this algorithm is quite fast, particularly for small p. Using Matlab on a 1.2 GHz CPU with the `fminunc` optimizer and the discrete grid defined in Section 3.3, the computation time equals on average 0.15 s, 0.0065 s and 0.00025 s for $p = 1$, $p = 0.9$ and $p = 0.5$ respectively with $I = 2$ and $J = 3$. By contrast, the optimization via Second Order Cone Programming (SOCP) for $p = 1$ takes about 0.36 s, using the same Matlab toolbox as in [1,7].

4 Source Separation Results

We evaluated the proposed algorithm for the separation of 10 instantaneous and 10 convolutive speech mixtures with $I = 2$ and $J = 3$. The mixture signals were obtained by mixing the source signals of Section 2 either with a matrix of positive coefficients or with a set of simulated room impulse responses corresponding to a reverberation time of 250 ms, as described in [12]. Following [11,12], the STFT window length L was set to 512 (64 ms) for instantaneous mixtures and 2048 (256 ms) for convolutive mixtures. The frequency-domain complex mixing matrices $\mathbf{A}(f)$ were computed by Fourier transform of the mixing filters in the convolutive case. The performance was measured in decibels (dB) for each estimated source by $\mathrm{SDR}_j = 20\log_{10}(\|s_j\|/\|\widehat{s}_j - s_j\|)$ and subsequently averaged. For comparison, we also evaluated the performance when the criterion was optimized over the singular points only, as suggested in [1].

The results are shown in Figure 3. In the instantaneous case, the best SDR is achieved for $p = 1$ and the SDR for smaller values of p is 0.6 dB smaller. In the convolutive case, the best SDR is achieved for $p \to 0$ and it is 1.2 dB larger than for $p = 1$. Note that the ML value of p determined in Section 2, namely $p = 0.4$, does not result in the best SDR. This suggests that algorithms estimating the value of p from the data might not improve performance. Note also that the SDR remains much smaller than the theoretical upper bound computed in [12].

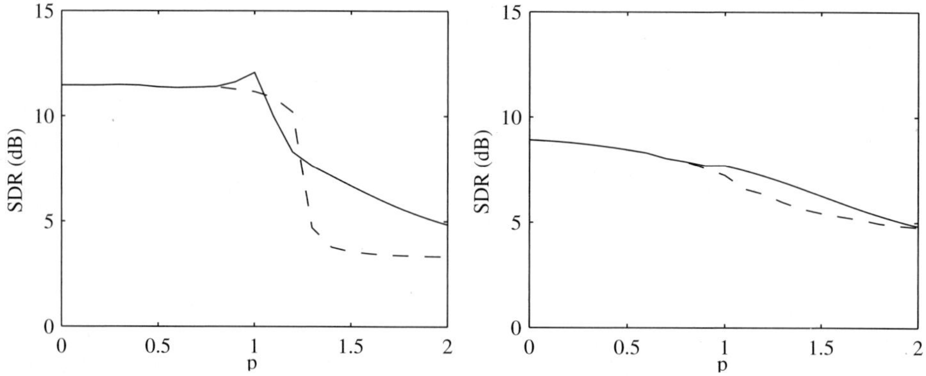

Fig. 3. Average SDR as a function of p for the separation of instantaneous (left) and convolutive (right) three-source two-channel speech mixtures (plain curve: optimization over the full space, dashed curve: optimization over the singular points only)

5 Conclusion

In this article, we investigated the benefit of modeling the sources via generalized Gaussian priors instead of Laplacian priors for underdetermined source separation. This generalization is not straightforward, since the resulting l_p norm criterion is nonconvex for $p < 1$. In the simple case where $J = I + 1$, we characterized mathematically the local minima of this criterion, conjectured that the global maximum is always singular below a critical value of p and derived an efficient global optimization algorithm. We evaluated this algorithm on speech mixtures and showed that small values of p resulted in the best separation performance in the convolutive case, but also in the fastest optimization. This work raises further research issues, including the proof of the above conjecture and the extension of the proposed algorithm when $J > I + 1$.

References

1. Winter, S., Sawada, H., Makino, S.: On real and complex valued l_1-norm minimization for overcomplete blind source separation. In: Proc. IEEE Workshop on Applications of Signal Processing to Audio and Acoustics (WASPAA), pp. 86–89. IEEE Computer Society Press, Los Alamitos (2005)
2. Vielva, L., Erdoğmuş, D., Príncipe, J.C.: Underdetermined blind source separation using a probabilistic source sparsity model. In: Proc. Int. Conf. on Independent Component Analysis and Blind Source Separation (ICA), pp. 675–679 (2001)
3. Davies, M.E., Mitianoudis, N.: Simple mixture model for sparse overcomplete ICA. IEE Proceedings on Vision, Image and Signal Processing 151, 35–43 (2004)
4. Févotte, C., Godsill, S.J.: A Bayesian approach for blind separation of sparse sources. IEEE Trans. on Audio, Speech and Language Processing 14, 2174–2188 (2006)
5. Lee, T.W., Lewicki, M.S., Girolami, M.A., Sejnowski, T.J.: Blind source separation of more sources than mixtures using overcomplete representations. IEEE Signal Processing Letters 6, 87–90 (1999)
6. Zibulevsky, M., Pearlmutter, B.A., Bofill, P., Kisilev, P.: Blind source separation by sparse decomposition in a signal dictionary. In: Independent Component Analysis: Principles and Practice, pp. 181–208. Cambridge Press, Cambridge (2001)
7. Bofill, P., Monte, E.: Underdetermined convoluted source reconstruction using LP and SOCP, and a neural approximator of the optimizer. In: Rosca, J., Erdogmus, D., Príncipe, J.C., Haykin, S. (eds.) ICA 2006. LNCS, vol. 3889, pp. 569–576. Springer, Heidelberg (2006)
8. Choi, S., Cichocki, A., Amari, S.: Flexible independent component analysis. In: Neural Networks for Signal Processing (NNSP 8), pp. 83–92 (1998)
9. Everson, R., Roberts, S.: Independent component analysis: a flexible nonlinearity and decorrelating manifold approach. Neural Computation 11, 1957–1983 (1999)
10. Wu, H.C., Príncipe, J.C.: Generalized anti-Hebbian learning for source separation. In: Proc. IEEE Int. Conf. on Acoustics, Speech and Signal Processing (ICASSP), II–1073–1076 (1999)
11. Yılmaz, O., Rickard, S.T.: Blind separation of speech mixtures via time-frequency masking. IEEE Trans. on Signal Processing 52, 1830–1847 (2004)
12. Vincent, E., Gribonval, R., Plumbley, M.D.: Oracle estimators for the benchmarking of source separation algorithms. Signal Processing 87, 1933–1950 (2007)
13. van den Bos, A.: Complex gradient and Hessian. IEE Proceedings on Vision, Image and Signal Processing 141, 380–382 (1994)
14. Casella, G., Robert, C.P.: Monte Carlo Statistical Methods, 2nd edn. Springer, New York (2005)

Sparse Component Analysis in Presence of Noise Using an Iterative EM-MAP Algorithm

Hadi Zayyani[1], Massoud Babaie-Zadeh[1,⋆], G. Hosein Mohimani[1], and Christian Jutten[2]

[1] Electrical Engineering Department, Advanced Communications Research Institute (ACRI), Sharif University of Technology, Tehran, Iran
[2] GIPSA-lab, Department of Images and Signals, National Polytechnic Institute of Grenoble (INPG), France
zayyani2000@yahoo.com, mbzadeh@yahoo.com, gh1985im@yahoo.com, Christian.Jutten@inpg.fr

Abstract. In this paper, a new algorithm for source recovery in under-determined Sparse Component Analysis (SCA) or atomic decomposition on over-complete dictionaries is presented in the noisy case. The algorithm is essentially a method for obtaining sufficiently sparse solutions of under-determined systems of linear equations with additive Gaussian noise. The method is based on iterative Expectation-Maximization of a Maximum A Posteriori estimation of sources (EM-MAP) and a new steepest-descent method is introduced for the optimization in the M-step. The solution obtained by the proposed algorithm is compared to the minimum ℓ^1-norm solution achieved by Linear Programming (LP). It is experimentally shown that the proposed algorithm is about one order of magnitude faster than the interior-point LP method, while providing better accuracy.

Keywords: sparse component analysis, sparse decomposition, blind source separation, independent component analysis.

1 Introduction

Finding (sufficiently) sparse solutions of under-determined systems of linear equations (possibly in the noisy case) has been studied extensively in recent years [1,2,3,4,5,6,7,8,9,10]. The problem has a growing range of applications in signal processing. One of these applications is the noisy under-determined sparse source separation which is also called Sparse Component Analysis (SCA) [1,2,3,4,5,6]. Another application is the so-called 'atomic decomposition' problem which aims at finding a sparse representation for a signal in an overcomplete dictionary [7,8,9,10]. In this paper, we will mainly use the context of SCA stating our approach. The discussions, however, may be easily followed in other contexts of application such as atomic decomposition.

SCA can be viewed as a method to achieve separation of sparse sources. The Blind Source Separation (BSS) problem is to recover m unknown sources from n

⋆ This work has been partially funded by Sharif University of Technology, by Center for International Research and Collaboration (ISMO) and by Iran NSF (INSF).

M.E. Davies et al. (Eds.): ICA 2007, LNCS 4666, pp. 438–445, 2007.

observed mixtures of them, where little or no information is available about the sources (except their statistical independence) and the mixing system. In this paper we consider the noisy linear instantaneous model:

$$\mathbf{x}(t) = \mathbf{A}\mathbf{s}(t) + \mathbf{n}(t). \tag{1}$$

where $\mathbf{x}(t)$, $\mathbf{s}(t)$ and $\mathbf{n}(t)$ are $n \times 1$, $m \times 1$ and $n \times 1$ vectors of sources, mixtures and white Gaussian noises, respectively, and $\mathbf{A}$ is the $n \times m$ mixing matrix. In the under-determined case ($m > n$), estimating the mixing matrix is not sufficient to recover the sources, since the mixing matrix is not invertible. Then, the estimation of sources requires some prior information on the sources and passes from a blind problem to a semi-blind problem. One such prior information is the sparsity of sources. It means that only a few samples of the sources are nonzero (say they are active) and most of them are almost zero (say they are inactive).

Then SCA can be solved in two steps: first estimating the mixing matrix, and then estimating the sources. The first step may be accomplished by means of clustering [1] or other methods [4]. The second step requires finding the sparse solution of (1) assuming $\mathbf{A}$ to be known [9]. In this paper, we focus on the source estimation assuming $\mathbf{A}$ is known.

In the atomic decomposition viewpoint [10], we have one signal whose samples are collected in the $m \times 1$ signal vector $\mathbf{s}$ and the objective is to express it as a linear combination of a set of predetermined signals where their samples are collected in vector $\{\varphi_i\}_{i=1}^{m}$. After [11], the φ_i's are called atoms and they collectively form a dictionary over which the signal is to be decomposed. In this paper, we also consider a noise term for the decomposition. So we can write $\mathbf{s} = \sum_{i=1}^{m} \alpha_i \varphi_i = \mathbf{\Phi}\alpha + \mathbf{n}$, where $\mathbf{\Phi}$ is the $n \times m$ dictionary (matrix) where the columns are the atoms and α is the $m \times 1$ vector of coefficients. The vector $\mathbf{n}$ can be interpreted as either the noisy term of the original signal that we intend to decompose or the allowed error for the decomposition process.

To obtain the sparse solution of (1), an approach is to search solutions having minimal ℓ^0 norm, *i.e.*, minimum number of nonzero components. This method is intractable when the dimension increases (due to combinatorial search), and it is too sensitive to noise (due to discontinuity of ℓ^0 norm). One of the most successful approaches is Basis Pursuit (BP) [10] which finds the minimum ℓ^1 norm of (1) which can be easily implemented by Linear Programming (LP) methods (especially fast interior-point LP solvers). Another approach is Matching Pursuit (MP) [11] which is very fast, but is somewhat heuristic and does not provide good estimation of sources.

In [5], we proposed a three step (sub-)optimum (in MAP sense) method for SCA in the noisy under-determined case (briefly called MAP) which has the drawback of great complexity and is not tractable for sparse decomposition application (which requires large values of m and n). This problem exists also in [6]. In this article, we propose an iterative method to tackle the great complexity of our MAP method. In the maximization step of our algorithm, we propose here an optimization method based on steepest-descent. Our method results in a fast sparse decomposition (faster than interior point LP) while improving the quality of source recovery because of its optimality in the MAP sense and dealing with noise.

2 System Model

The noise vector in the model (1) is assumed zero-mean Gaussian with covariance matrix $\sigma_n^2\mathbf{I}$. For modeling the sparse sources the following model is used: the sources are inactive with probability p, and are active with probability $1-p$ (sparsity of sources implies that p should be near 1). In the inactive case the sample of sources is zero and in the active case the sample has a Gaussian distribution. We call this model the 'spiky model' which is a special case of the Bernoulli-Gaussian model used in [5]. The probability density function (PDF) of the sources is then:

$$p(s_i) = p\delta(s_i) + (1-p)N(0, \sigma_r^2). \tag{2}$$

In this model, any sample of the sources can be written as $s_i = q_i r_i$ where q_i is a binary variable (with binomial distribution) and r_i is the amplitude of i'th source with Gaussian distribution. So the source vector can be written as:

$$\mathbf{s} = \mathbf{Q}\mathbf{r} \qquad \mathbf{Q} = \mathrm{diag}(\mathbf{q}). \tag{3}$$

We refer the vector $\mathbf{q} \triangleq [q_1, \ldots, q_m]'$ as the 'source activity vector', where $'$ denotes vector/matrix transposition. Each element of this vector shows the activity of the corresponding source. That is:

$$q_i = \begin{cases} 1 \text{ if } s_i \text{ is active with probability } p \\ 0 \text{ if } s_i \text{ is inactive with probability } 1-p \end{cases} \tag{4}$$

The probability of source activity vector $p(\mathbf{q})$ is equal to:

$$p(\mathbf{q}) = (1-p)^{n_a}(p)^{m-n_a}. \tag{5}$$

where n_a is the number of active sources or the number of 1's in $\mathbf{q}$.

3 Review of Our Previous MAP Algorithm

In [5] we proposed a three step MAP algorithm for the noisy sparse component analysis. The parameter estimation step is done by a novel method based on second and fourth order moments of one mixture and an EM algorithm. The source activity estimation step is done with a MAP method that maximizes the posterior probability. This step is the maximization of:

$$p(\mathbf{q})p(\mathbf{x}|\mathbf{q}) = \frac{p(\mathbf{q})}{\sqrt{det(2\pi\mathbf{Q}_q)}} \exp(\frac{-1}{2}\mathbf{x}'\mathbf{Q}_q^{-1}\mathbf{x}). \tag{6}$$

where $\mathbf{Q}_q = \mathbf{A}\mathbf{V}_q\mathbf{A}' + \sigma_n^2\mathbf{I}$, in which, $\mathbf{Q}_q \triangleq E\{\mathbf{x}\mathbf{x}' \mid \mathbf{q}\}$ and $\mathbf{V}_q \triangleq E\{\mathbf{s}\mathbf{s}' \mid \mathbf{q}\} = \sigma_r^2\mathbf{Q}$ are the conditional covariances of observations and sources given $\mathbf{q}$. After finding the optimum source activity vector, the source amplitudes are estimated as:

$$\widehat{\mathbf{r}} = \sigma_r^2\mathbf{Q}\mathbf{A}'(\sigma_r^2\mathbf{A}\mathbf{Q}\mathbf{A}' + \sigma_n^2\mathbf{I})^{-1}\mathbf{x}. \quad (7)$$

Maximization of (6) is done over discrete space of vector $\mathbf{q}$ with 2^m discrete elements. In [5] this maximization had been done through an exhaustive search on all these 2^m cases.

In this paper, this maximization is done by first converting it to a continuous maximization and then to use a steepest descent algorithm (this is similar to the idea used in [9]). To convert our discrete problem to a continuous one, we use a Mixture of two Gaussians model centered around 0 and 1 with sufficient small variances. By this method our discrete binomial variable q_i is converted to a continuous variable. To avoid falling into local maxima of (6) a gradually decreasing variance can be used in the different iterations (similar to simulated annealing methods). But (6) is still very complex to derive for providing an efficient optimization method such as steepest-descent.

4 The Iterative EM-MAP Algorithm

The main idea of our algorithm is that the source estimation is equal to estimation of vectors $\mathbf{q}$ and $\mathbf{r}$, as observed from (3). Estimation of $\mathbf{q}$ and $\mathbf{r}$ can be done iteratively. First, an estimated vector $\widehat{\mathbf{q}}$ is assumed and then the MAP estimate of vector $\mathbf{r}$ based on the known estimated vector $\widehat{\mathbf{q}}$ and the observation vector $\mathbf{x}$ is obtained (we refer to it as $\widehat{\mathbf{r}}$). Secondly, the MAP estimate of vector $\mathbf{q}$ is obtained based on the estimated vector $\widehat{\mathbf{r}}$ and the observation vector $\mathbf{x}$ (we refer to it as vector $\widehat{\mathbf{q}}$). Therefore, the MAP estimation of sources is done in two other MAP estimation steps.

In the first step a source activity vector $\widehat{\mathbf{q}}$ is assumed and the estimation of $\mathbf{r}$ will be computed. Because the vector $\mathbf{r}$ is Gaussian, its MAP estimation is equal to the Linear Least Square (LLS) estimation [13] and can be computed as follows:

$$\widehat{\mathbf{r}}_{\mathrm{MAP}} = \widehat{\mathbf{r}}_{\mathrm{LLS}} = E(\mathbf{r}|\mathbf{x},\widehat{\mathbf{q}}) = E(\mathbf{r}\mathbf{x}'|\widehat{\mathbf{q}})E(\mathbf{x}\mathbf{x}'|\widehat{\mathbf{q}})^{-1}\mathbf{x}. \quad (8)$$

This step can be nominated as Expectation step or Estimation step (E-step). Computation and simplification of (8) (like what done in [5]) leads to the following equation which is similar to (7).

$$\widehat{\mathbf{r}} = \sigma_r^2\widehat{\mathbf{Q}}\mathbf{A}'(\sigma_r^2\mathbf{A}\widehat{\mathbf{Q}}\mathbf{A}' + \sigma_n^2\mathbf{I})^{-1}\mathbf{x}. \quad (9)$$

In the second step we estimate $\mathbf{q}$ based on the known $\widehat{\mathbf{r}}$ and the observed $\mathbf{x}$. The MAP estimation is:

$$\widehat{\mathbf{q}}_{\mathrm{MAP}} = \underset{\mathbf{q}}{\mathrm{argmax}}\, p(\mathbf{q}|\mathbf{x},\widehat{\mathbf{r}}) \equiv p(\mathbf{q}|\widehat{\mathbf{r}})p(\mathbf{x}|\mathbf{q},\widehat{\mathbf{r}}) \equiv p(\mathbf{q})p(\mathbf{x}|\mathbf{q},\widehat{\mathbf{r}}). \quad (10)$$

In (10), $p(\mathbf{q})$ can be computed as a continuous variable:

$$p(\mathbf{q}) = \prod_{i=1}^{m} p(q_i) = \prod_{i=1}^{m} [p \exp(\frac{-q_i^2}{2\sigma_0^2}) + (1-p) \exp(\frac{-(q_i-1)^2}{2\sigma_0^2})]. \tag{11}$$

Also the term $p(\mathbf{x}|q, \widehat{\mathbf{r}})$ in (10) can be computed as:

$$p(\mathbf{x}|\mathbf{q}, \widehat{\mathbf{r}}) = p_n(\mathbf{x} - \mathbf{AQ}\widehat{\mathbf{r}}) = (2\pi\sigma_n^2)^{\frac{-m}{2}} \exp(\frac{-1}{2\sigma_n^2}(\mathbf{x} - \mathbf{AQ}\widehat{\mathbf{r}})'(\mathbf{x} - \mathbf{AQ}\widehat{\mathbf{r}})) . \tag{12}$$

The second step can be called Maximization step (M-step). The maximization can be done over the logarithm of (10). So this step can be simplified as:

$$M - step: \quad \widehat{\mathbf{q}} = \max_{\mathbf{q}} L(\mathbf{q}). \tag{13}$$

where

$$L(\mathbf{q}) = \sum_{i=1}^{m} \log(p(q_i)) + \frac{-1}{2\sigma_n^2}(\mathbf{x} - \mathbf{AQ}\widehat{\mathbf{r}})'(\mathbf{x} - \mathbf{AQ}\widehat{\mathbf{r}}) . \tag{14}$$

Maximization of $L(\mathbf{q})$ in the M-step can be done with the steepest descent method. The main steepest descent iteration is:

$$\mathbf{q}_{k+1} = \mathbf{q}_k - \mu \frac{\partial L(\mathbf{q})}{\partial \mathbf{q}}. \tag{15}$$

In the appendix, we show that the steepest descent algorithm for the M-step is:

$$\mathbf{q}_{k+1} = \mathbf{q}_k + \frac{\mu}{\sigma_0^2}\mathbf{g}(\mathbf{q}) + \frac{\mu}{\sigma_n^2} Diag(\mathbf{A}'\mathbf{AQ}\widehat{\mathbf{r}} - \mathbf{A}'\mathbf{x}).\widehat{\mathbf{r}}. \tag{16}$$

where $\mathbf{g}(\mathbf{q})$ is defined in the appendix. In the successive iterations, we gradually decrease the variance σ_0 in the form $\sigma_0^{(i)} = \alpha\sigma_0^{(i-1)}$ where α is selected between 0.6 and 1. Also, the step-size μ should be decreasing, *i.e.*, for smaller σ's, smaller μ's should be applied. This is because for smaller variances, our function under maximization is more fluctuating. So the step size can be decreased in the similar form as $\mu^{(i)} = \alpha\mu^{(i-1)}$. Our simulations show that for $\alpha = .8$ only about 4 or 5 iterations are sufficient to maximize the expression $L(\mathbf{q})$ in the M-step. Also the EM-step converges at the third or fourth iteration. The first initialization of the EM-MAP method is done with the minimum ℓ^2 norm solution.

As we see from (16) the second summand is responsible for increasing the prior probability $p(\mathbf{q})$ while the third summand is responsible for decreasing the noise power $||\mathbf{x} - \mathbf{As}||$. When σ_0 is much larger than σ_n, the second term is more effective than the third term and as a result exactness of $\mathbf{x} = \mathbf{As}$ is more important than sparsity of $\mathbf{s}$. When σ_0 is decreased to be comparable to σ_n, both terms are effective to yield the equilibrium point between sparsity and noise.

In summary, the overall algorithm is an iterative two step (E-step and M-step in (9) and (13) respectively) algorithm in which the M-step is done iteratively with the steepest descent method in (16).

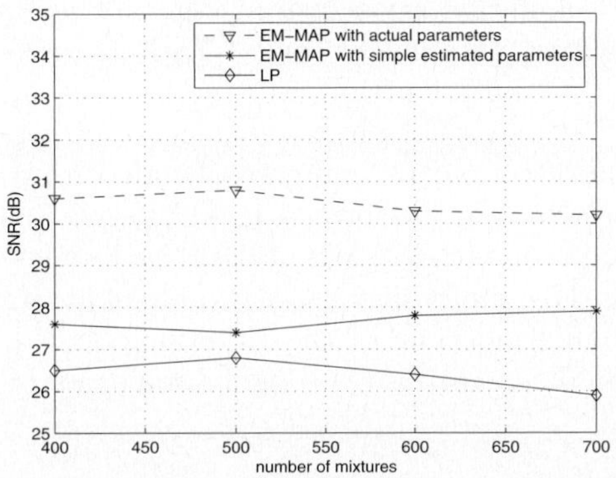

Fig. 1. Comparison of the results of our algorithm in two cases and the (interior-point) LP method. The parameters of simulation are $m = 1000$, $p = .9$, $\sigma_r = 1$, $\sigma_n = .01$, $\alpha = .8$, $\sigma^{(0)} = \widehat{\sigma_r}$ and $\mu^{(0)} = 10^{-6}$. Four iterations are used for EM-step and five iterations for the M-step (steepest descent).

5 Simulation Results

In this section, we examine the performance of our algorithm in two cases (the case of using actual parameters and the case of using a simple method for estimating the parameters which is explained below) and then compare it to the interior-point LP method. Our performance criterion is Signal to Noise Ration (SNR) in dB defined by $\mathrm{SNR} = 10\log_{10}\{\|\mathbf{s}\|^2/\|\hat{\mathbf{s}} - \mathbf{s}\|^2\}$. The simulations have been repeated 50 times (with the same parameters, but for different randomly generated sources and mixing matrix) and the resulting SNR's (in dB) have been averaged.

The values used for the experiment are $m = 1000$, $n = 400, ..., 700$, $p = .9$, $\sigma_r = 1$ and $\sigma_n = .01$. The elements of the mixing matrix are randomly chosen between -1 and 1 and each column normalized to have unit length. In the M-step the value of α is between 0.6 and 1. This parameter effects on the speed of convergence. We use an average value of $\alpha = .8$ in our simulations. The initial value of σ_0 is selected equal to estimated σ_r. The initial value of μ can be selected between 10^{-3} and 10^{-8}. But for small values and large values in this range, the performance is somewhat deteriorated. So we select the value of $\mu = 10^{-6}$. Four iterations are used for the EM-step and five iterations are used for the M-step (steepest descent).

In one of our simulation we use a very simple estimation of the parameters. In this case the parameter p underestimated as $\widehat{p} = .8$. With this assumption and by considering the ergodicity of sources (*i.e.* the mixtures are the ensembles of a random variable $x_j = \sum_{i=1}^{m} a_{ji}s_i + e_j$ where $a_{ji} = b_{ji}/\sqrt{b_{1i}^2 + b_{2i}^2 + ... + b_{ni}^2}$ and b_{ji} is a random variable with uniform distribution on [-1,1] and s_i and e_j are random variables), and by neglecting the noise power, we have $E(x_j^2) = mE(a_{ji}^2)E(s_i^2)$. We know that $\sum_{j=1}^{n} a_{ji}^2 = 1$ (which come from $\sum_{j=1}^{n} a_{ji}^2 = 1$ and the independence of a_{ji}'s), and $E(s_i^2) = (1-p)\sigma_r^2$. With the assumption of $\widehat{p} = .8$, we will have $\widehat{\sigma_r} = \sqrt{\frac{5nE(x_j^2)}{m}}$. For the noise variance, we choose $\widehat{\sigma_n} = \widehat{\sigma_r}/10$.

The results of our simulation are shown in Fig. 5. These results show 3-4 dB improvement (with actual parameters) and 1-2 dB improvement (with simply estimated parameters) of our algorithm over the LP method.

Although, the CPU time is not an exact measure of complexity, it can give us a rough estimation of it, and we compare our algorithm with LP using this measure. Our simulations were performed in MATLAB 7.0 environment using an Intel 2.40 GHz processor with 512 MB of RAM and under Microsoft Windows XP operating system. For one typical simulation, our algorithm takes about 34 seconds while the simulation time of the (interior-point) LP method requires about 204 seconds. So our algorithm is roughly one order of magnitude faster.

6 Conclusions

In this paper, a relatively fast method for finding sparse solution of an underdetermined system of linear equations was proposed. The method was based on the iterative MAP estimation of the sources. This algorithm is approximately one order of magnitude faster than (interior-point) LP, while providing 1-2 dB improvement (with simply estimated parameters). The better performance is obtained due to the optimality of our algorithm which is based on optimum MAP estimation of sources. The simplicity of our algorithm (and its high speed) is obtained due to iterative estimation of source activities and amplitudes and also utilizing an efficient steepest descent for the M-step.

References

1. Zibulevsky, M., Pearlmutter, B.A.: Blind source separation by sparse decomposition in a signal dictionary. Neural Computation 13(4), 863–882 (2001)
2. Gribonval, R., Lesage, S.: A survey of sparse component analysis for blind source separation: principles, perspectives, and new challanges. In: Proceeding of ESANN'06, pp. 323–330 (2006)
3. Davies, M., Mitianoudis, N.: Simple mixture model for sparse overcomplete ICA. In: Puntonet, C.G., Prieto, A.G. (eds.) ICA 2004. LNCS, vol. 3195, pp. 35–43. Springer, Heidelberg (2004)
4. Li, Y.Q., Amari, S., Cichocki, A., Ho, D.W.C, Xie, S.: Underdetermined blind source separation based on sparse representation. IEEE Transaction on Signal Processing 54(2), 423–437 (2006)
5. Zayyani, H., Babaie-Zadeh, M., Jutten, C.: Source estimation in noisy sparse component analysis. Accepted in DSP'2007 (2007)
6. Balan, R., Rosca, J.: Source separation using sparse discrete prior models. In: Proceeding of ICASSP'06 (2006)
7. Donoho, D.L.: For most large underdetermined systems of linear equations the minimal ℓ^1 norm is also the sparsest solution. Technical Report (2004)
8. Donoho, D.L., Elad, M., Temlyakov, V.: Stable recovery of sparse overcomplete representations in the presence of noise. IEEE Transaction on Information theory 52(1), 6–18 (2006)
9. Mohimani, G.H, Babaie-Zadeh, M., Jutten, C.: Fast sparse representation based on smoothed ℓ^0 norm. Accepted in ICA'2007 (2007)

10. Chen, S.S., Donoho, D.L., Saunders, M.A.: Atomic decomposition by basis pursuit. SIAM Journal on Scientific Computing 20(1), 31–61 (1999)
11. Mallat, S., Zhang, Z.: Matching pursuit with time-frequency dictionaries. IEEE Transaction on Signal Processing 41(12), 3397–3415 (1993)
12. Djafari, A.M.: Bayesian source separation: beyond PCA and ICA. In: Proceeding of ESANN'06 (2006)
13. Anderson, B.D., Moor, J.B.: Optimal filtering, 2nd edn. Prentice Hall, Englewood Cliffs (1979)

Appendix: Steepest Descent Algorithm

From (13), we have:

$$\frac{\partial L(\mathbf{q})}{\partial \mathbf{q}} = \frac{\partial}{\partial \mathbf{q}} \sum_{i=1}^{m} \log(p(q_i)) - \frac{1}{2\sigma_n^2} \frac{\partial}{\partial \mathbf{q}} (\mathbf{x} - \mathbf{AQ}\widehat{\mathbf{r}})'(\mathbf{x} - \mathbf{AQ}\widehat{\mathbf{r}}) \; . \tag{17}$$

we define $\mathbf{g}(\mathbf{q}) \triangleq -\sigma_0^2 \frac{\partial}{\partial \mathbf{q}} \sum_{i=1}^{m} \log(p(q_i))$ and $\mathbf{n}(\mathbf{q}) \triangleq (\mathbf{x} - \mathbf{AQ}\widehat{\mathbf{r}})'(\mathbf{x} - \mathbf{AQ}\widehat{\mathbf{r}})$. With these definitions the scalar function $g(q_i)$ and the $\mathbf{n}(\mathbf{q})$ (with omitting the constant terms) can be computed as:

$$g(q_i) = \frac{p q_i \exp(\frac{-q_i^2}{2\sigma_0^2}) + (1-p)(q_i - 1)\exp(\frac{-(q_i-1)^2}{2\sigma_0^2})}{p \exp(\frac{-q_i^2}{2\sigma_0^2}) + (1-p)\exp(\frac{-(q_i-1)^2}{2\sigma_0^2})} \; . \tag{18}$$

$$\mathbf{n}(\mathbf{q}) = -2\mathbf{x}'\mathbf{AQ}\widehat{\mathbf{r}} + \widehat{\mathbf{r}}'\mathbf{QA}'\mathbf{AQ}\widehat{\mathbf{r}} \; . \tag{19}$$

with the definitions $\mathbf{C} \triangleq \mathbf{A}'\mathbf{A}$ and $\mathbf{n_1}(\mathbf{q}) \triangleq -2\mathbf{x}'\mathbf{AQ}\widehat{\mathbf{r}}$ and $\mathbf{n_2}(\mathbf{q}) \triangleq \widehat{\mathbf{r}}'\mathbf{QCQ}\widehat{\mathbf{r}}$ we can write:

$$\frac{\partial n_1(\mathbf{q})}{\partial \mathbf{q}} = \mathrm{diag}(-2\mathbf{x}'\mathbf{A}).\widehat{\mathbf{r}} \; . \tag{20}$$

If we define $\mathbf{W} \triangleq \mathbf{Q}\widehat{\mathbf{r}}$ ($m \times 1$ *vector*) then $\mathbf{n}_2(\mathbf{q}) = \mathbf{W}'\mathbf{CW}$ and so we have:

$$\frac{\partial \mathbf{n}_2(\mathbf{q})}{\partial q_i} = \sum_{j=1}^{m} \frac{\partial \mathbf{n}_2(\mathbf{q})}{\partial W_j} \frac{\partial W_j}{\partial q_i} \; . \tag{21}$$

From the vector derivatives, we have $\frac{\partial \mathbf{n}_2(\mathbf{q})}{\partial \mathbf{W}} = 2\mathbf{CW} \triangleq \mathbf{d}$. Also from the definition of $\mathbf{W}$ we have $\frac{\partial W_j}{\partial q_i} = \widehat{r}_i \delta_{ij}$. So (21) is converted to $\frac{\partial \mathbf{n}_2(\mathbf{q})}{\partial q_i} = \sum_{j=1}^{m} d_j \widehat{r}_i \delta_{ij} = \widehat{r}_i d_i$. So the vector form of (21) is equal to:

$$\frac{\partial \mathbf{n}_2(\mathbf{q})}{\partial \mathbf{q}} = \mathrm{diag}(\mathbf{d}).\widehat{\mathbf{r}} \; . \tag{22}$$

From (20) and (22) and $\mathbf{n}(\mathbf{q}) = \mathbf{n}_1(\mathbf{q}) + \mathbf{n}_2(\mathbf{q})$ and definitions of vectors $\mathbf{d}$ and $\mathbf{C}$, we can write:

$$\frac{\partial \mathbf{n}(\mathbf{q})}{\partial \mathbf{q}} = 2\mathrm{diag}(\mathbf{A}'\mathbf{AQ}\widehat{\mathbf{r}} - \mathbf{A}'\mathbf{x}).\widehat{\mathbf{r}} \; . \tag{23}$$

Finally, (23) and (17) and (15) with the definitions of $\mathbf{n}(\mathbf{q})$ and $\mathbf{g}(\mathbf{q})$ yields the steepest descent iteration in (16).

Mutual Interdependence Analysis (MIA)

Heiko Claussen[1,2], Justinian Rosca[1], and Robert Damper[2]

[1] Siemens Corporate Research Inc.,
755 College Road East, Princeton, New Jersey 08540
{Heiko.Claussen,Justinian.Rosca}@siemens.com
[2] University of Southampton,
School of Electronics and Computer Science,
Southampton SO17 1BJ, UK
{hc05r,rid}@ecs.soton.ac.uk

Abstract. Functional Data Analysis (FDA) is used for datasets that are more meaningfully represented in the functional form. Functional principal component analysis, for instance, is used to extract a set of functions of maximum variance that can represent the data. In this paper, a method of Mutual Interdependence Analysis (MIA) is proposed that can extract an equally correlated function with a set of inputs. Formally, the MIA criterion defines the function whose mean variance of correlations with all inputs is minimized. The meaningfulness of the MIA extraction is proven on real data. In a simple text independent speaker verification example, MIA is used to extract a signature function per each speaker, and results in an equal error rate of 2.9 % in the set of 168 speakers.

1 Introduction

Principal Component Analysis (PCA), discussed in [1] for nonfunctional and [2] for functional data, is an essential feature extraction method. Principal component functions are such that they can approximate new data with minimum mean square error even if only a subset of all components is used. The principal component functions are given by the directions of maximum variance in the projections of the data. The directions of minimum variance, or minor components, have received much less attention in the literature. Minor components correspond to directions that are either orthogonal to the inputs or minimize the variance of their projections. Hence, they represent an invariant or "common" direction of the inputs. Further information about PCA and MCA, including their probabilistic models, can be found in [3], [4], [5] and [6].

[7] acknowledged that minor components are important in some signal processing applications. Examples are spectral estimation, curve and hyper-surface fitting, cognitive perception and computer vision. This work coined the name Minor Component Analysis (MCA) to refer to neural network fitting methods that compute minor components. The authors also suggested that MCA solves the Total Least Square (TLS) problem [8].

Recently, a method called Extreme Component Analysis (XCA) was introduced that combines minor and principal components in order to represent a

M.E. Davies et al. (Eds.): ICA 2007, LNCS 4666, pp. 446–453, 2007.

dataset "optimally". As discussed in [9], the "optimal" representation, if by minor or principal components, is dependent on the dataset. By freely choosing between minor, principal components and their combinations, datasets can be represented with possibly a lower number of components.

As previously mentioned, MCA can be used to extract minor components that represent invariants of the data. However, when it is desired to extract the mutual interdependencies of a set of input functions, MCA suffers because of the necessary preprocessing step of input data centering. In the usual case, where the input functions are linearly independent, centering reduces the span of the data. Therefore, these centered functions can no longer fully represent the inputs. Furthermore, as discussed in [10] the TLS and therefore the MCA solution is non trivial and can usually not be found in closed form.

In this paper, we propose Mutual Interdependence Analysis (MIA) to solve a TLS-like optimization problem in the span of the original inputs in order to extract a mutually interdependent function from the input function set. We prove that in the case of linearly independent input functions, a closed form solution can be found that minimizes the MIA criterion.

In section 2 we define the MIA problem, derive its solution and illustrate its properties. In the experimental section 3, MIA is used for simple text independent speaker recognition. We end the paper with conclusions and directions for further work.

2 Mutual Interdependence Analysis (MIA)

Throughout the paper, we use $\mathbf{x}^T$ to denote the transpose of column vector $\mathbf{x}$. Also, we denote $\mathbf{X}$ to be the matrix whose columns are $\mathbf{x}_i$ with $i = 1, \ldots, D$. We use $\underline{1}$ to represent a vector of ones, $\underline{\underline{1}}$ to represent a matrix of ones and I to be the identity matrix. The dimension will be clear from the context.

Consider D real inputs, $x_i(t_j)$ with $i = 1, \ldots, D$ and $j = 1, \ldots, N$, where each input $\mathbf{x}_i = [x_i(t_1), \ldots, x_i(t_N)]^T$ is viewed as a single entity (i.e. has the intrinsic structure of a function) rather than a series of individual observations. Functional Data Analysis (FDA) normally treats data this way, therefore we refer to each $\mathbf{x}_i$ as an input. In our case, N is typically larger or much larger than D. TLS solves a linear equation $\mathbf{X}^T \cdot \mathbf{s} = \mathbf{b}$ by finding a direction $\mathbf{s}$ that minimizes the squared, orthogonal distances to the data points $\mathbf{x}_i$: $\min_{s} \sum_{i=1}^{D} \frac{|\mathbf{x}_i^T \cdot \mathbf{s} - b_i|^2}{\mathbf{s}^T \cdot \mathbf{s} + 1}$. On the other hand, the ordinary Least Square (LS) problem finds a direction $\mathbf{s}$ that minimizes $\min_{s} \sum_{i=1}^{D} \left|\mathbf{x}_i^T \cdot \mathbf{s} - b_i\right|^2$. While LS assumes that only vector $\mathbf{b}$ to contain errors, the TLS approach models uncertainties in both $\mathbf{b}$ and in the data $\mathbf{X}$. Our goal is different: Extract a new vector (function) $\mathbf{s}$ that "optimally" represents the D input functions. An optimal solution is a function that is maximally correlated with all inputs with the constraint that it is a linear combination of the inputs. We call this problem Mutual Interdependence Analysis (MIA). Formally, the optimality criterion $J(\mathbf{X}|\mathbf{s})$, given a functional data series $\mathbf{s}$, is as follows:

$$J(\mathbf{X}|\mathbf{s}) = \sum_{i=1}^{D} \left(\mathbf{s}^T \cdot \mathbf{x}_i - \frac{1}{D} \sum_{k=1}^{D} \mathbf{s}^T \cdot \mathbf{x}_k \right)^2 \tag{1}$$

Our problem is to find the maximum likelihood vector $\hat{\mathbf{s}}$ of norm one, in the span of the inputs, that minimizes $J(\mathbf{X}|\mathbf{s})$:

$$\hat{\mathbf{s}} = \underset{\mathbf{s}, \|\mathbf{s}\|=1, \mathbf{s}=\sum_{k=1}^{D} c_k \cdot \mathbf{x}_k}{\arg\min} J(\mathbf{X}|\mathbf{s}) \tag{2}$$

2.1 Solution to MIA

Let us find an equivalent formulation for the MIA problem in (2). Consider the mean function $\mathbf{x}^{(m)} = \frac{1}{D} \sum_{i=1}^{D} \mathbf{x}_i$ and the centered functions $\mathbf{x}_i^{(c)} = \mathbf{x}_i - \mathbf{x}^{(m)}$. It can be easily shown that $\mathbf{s}^T \cdot \mathbf{x}_i - \frac{1}{D} \sum_{k=1}^{D} \mathbf{s}^T \cdot \mathbf{x}_k = \mathbf{s}^T \cdot \mathbf{x}_i^{(c)}$. Hence, (2) becomes:

$$\hat{\mathbf{s}} = \underset{\mathbf{s}, \|\mathbf{s}\|=1, \mathbf{s}=\sum_{k=1}^{D} c_k \cdot \mathbf{x}_k}{\arg\min} \left\| \mathbf{s}^T \cdot \mathbf{X}^{(c)} \right\|^2 , \tag{3}$$

where $\mathbf{X}^{(c)}$ has as columns $\mathbf{x}_i^{(c)}$ with $i = 1, \ldots, D$. Then, $\mathbf{x}^{(m)} = \frac{1}{D} \mathbf{X} \cdot \underline{\mathbf{1}}$ and $\mathbf{X}^{(c)} = [\mathbf{x}_1 - \mathbf{x}^{(m)}| \ldots |\mathbf{x}_D - \mathbf{x}^{(m)}] = \mathbf{X} - \mathbf{x}^{(m)} \cdot \underline{\mathbf{1}}^T$. It follows that

$$\mathbf{X}^{(c)} = \mathbf{X} \cdot \mathbf{P} \quad \text{with} \quad \mathbf{P} = \mathbf{I} - \frac{1}{D} \underline{\underline{\mathbf{1}}} \; . \tag{4}$$

Obviously, $\sum_{i=1}^{D} \mathbf{x}_i^{(c)} = \underline{\mathbf{0}}$. Hence, the nullspace $\mathcal{NULL}(\mathbf{x}_1^{(c)}, \mathbf{x}_2^{(c)}, \ldots, \mathbf{x}_D^{(c)})$ is non trivial. All vectors $\mathbf{s} \in \mathcal{NULL}(\mathbf{x}_1^{(c)}, \mathbf{x}_2^{(c)}, \ldots, \mathbf{x}_D^{(c)})$ will minimize $J(\mathbf{X}|\mathbf{s})$. In the next theorem, we show that the problem given by (3) has at least one solution.

Theorem 1. *Assume* $\mathbf{x}_1, \mathbf{x}_2, \ldots, \mathbf{x}_D$ *are linearly independent. Then, there exists* $\mathbf{s} \neq \underline{\mathbf{0}}$ *in* $\mathcal{NULL}(\mathbf{x}_1^{(c)}, \mathbf{x}_2^{(c)}, \ldots, \mathbf{x}_D^{(c)})$ *such that* $\mathbf{s}$ *is in the span of the inputs* $\mathbf{x}_i$*,* $i = 1, \ldots, D$*.*

Proof. A solution $\mathbf{s} \neq \underline{\mathbf{0}}$ and $\mathbf{c} = [c_1, c_2, \ldots, c_D]^T \in \mathbb{R}^D$ of the system of equations:

$$\mathbf{s}^T \cdot \mathbf{X}^{(c)} = \underline{\mathbf{0}} \tag{5}$$

$$\mathbf{s} = \mathbf{X} \cdot \mathbf{c} \tag{6}$$

will also satisfy the theorem and solve the optimization criterion of problem (3). Indeed, (5) is equivalent to the existence of $\mathbf{s}$ such that $\mathbf{s} \in \mathcal{NULL}(\mathbf{x}_1^{(c)}, \mathbf{x}_2^{(c)}, \ldots, \mathbf{x}_D^{(c)})$ and (6) specifies that $\mathbf{s}$ is in the span of the inputs $\mathbf{x}_i$ and $\mathbf{s} \neq \underline{\mathbf{0}}$. Let us substitute $\mathbf{s}$ from (6) and $\mathbf{X}^{(c)}$ from (4) in (5):

$$\Rightarrow \mathbf{c}^T \cdot \mathbf{X}^T \cdot \mathbf{X} \cdot \mathbf{P} = \underline{\mathbf{0}} \tag{7}$$

Given that $\mathbf{G} = (\mathbf{X}^T \cdot \mathbf{X})$ is a Gram matrix formed by linearly independent vectors $\mathbf{x}_1, \mathbf{x}_2, \ldots, \mathbf{x}_D$, $\mathbf{G}$ is invertible (see theorem 7.2.10 in [11]). Let

$$\mathbf{c}^T = \mathbf{d}^T \cdot (\mathbf{X}^T \cdot \mathbf{X})^{-1}. \tag{8}$$

Therefore, (7) becomes: $\mathbf{d}^T \cdot \mathbf{P} = \underline{\mathbf{0}}$ with

$$\mathbf{d} = \zeta\, \underline{\mathbf{1}} \tag{9}$$

and $\zeta \in \mathbb{R}$. When substituting (9) into (8): $\mathbf{c} = \zeta\, (\mathbf{X}^T \cdot \mathbf{X})^{-1} \cdot \underline{\mathbf{1}}$. Hence:

$$\hat{\mathbf{s}} = \zeta\, \mathbf{X} \cdot (\mathbf{X}^T \cdot \mathbf{X})^{-1} \cdot \underline{\mathbf{1}}. \tag{10}$$

Then, $\frac{\hat{\mathbf{s}}}{\|\hat{\mathbf{s}}\|}$ is a solution to (6) and (5) for all $\zeta \in \mathbb{R}$. Therefore, any MIA problem with linearly independent inputs has a solution given by (10). An alternative interpretation of the MIA solution $\frac{\hat{\mathbf{s}}}{\|\hat{\mathbf{s}}\|}$ is that of a direction in the N-dimensional space which minimizes the variance of the projections of all points $\mathbf{x}_i, \quad i = 1, \ldots, D$. □

Should the inputs be translated by a constant $\lambda \in \mathbb{R}$, i.e.

$$\mathbf{x}_i^{'} \leftarrow \mathbf{x}_i - \lambda\, \underline{\mathbf{1}} \quad \forall i \tag{11}$$

then the solution of $J(\mathbf{X}^{'}|s)$ changes. Indeed, it can be easily proven that the criterion (1) itself is invariant to a translation (11), however the constraint of (2) will require that the solution $\hat{\mathbf{s}}$ is in the span of $\mathbf{x}_i^{'}$.

2.2 Example with Synthetic Data

A synthetic example is given below to compare MIA, MCA, PCA and Independent Component Analysis (ICA) (see [12]). Assume three inputs given by:

$$\begin{aligned}
\mathbf{x}_1 &= \mathbf{f}_1 + \mathbf{f}_2 \\
\mathbf{x}_2 &= \mathbf{f}_1 - 0.5\,\mathbf{f}_2 \\
\mathbf{x}_3 &= 2\,\mathbf{f}_1 + 2\,\mathbf{f}_3 + 10\,\underline{\mathbf{1}}
\end{aligned}$$

where $\mathbf{f}_1 = \sin(\frac{2\Pi i}{N}) \quad i = 1, \ldots, N$, $\mathbf{f}_2$ is Gaussian noise $\mathcal{N}(0,1)$ and $\mathbf{f}_3$ is Laplacian noise. The inputs and results of the methods are illustrated in Fig. 1. The MIA solution closely approximates $\mathbf{f}_1$, in contrast to the other methods.

3 Application: Text Independent Speaker Verification

In this section, we apply MIA to the problem of extracting signatures from speech data for the purpose of text independent speaker recognition. This problem is challenging when we need to verify the identity of a person but can not control the way data is acquired (i.e. recording equipment, environment, etc.). For this study, we have used the TIMIT database [13]. Data from 168 speakers was partitioned 50-50 for training and testing. The data was preprocessed by silence removal and normalization of each recording. Data for a given speaker was used as input to MIA in order to generate a speaker signature as described below. We compare the equal error rate (EER) results for speaker verification obtained with MIA versus the PCA and ICA-based methods described in [14].

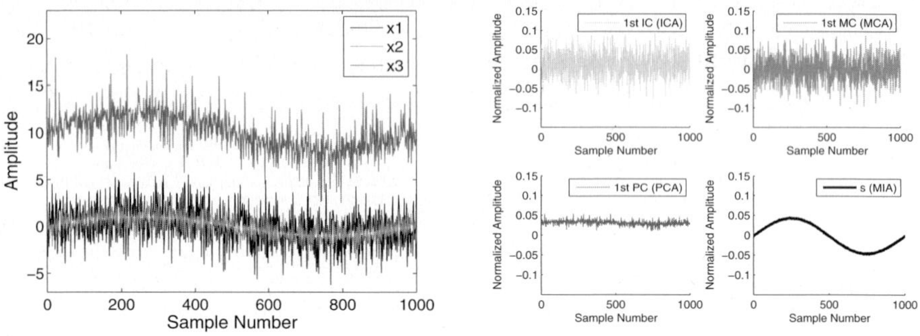

(a) Inputs $\mathbf{x}_i$ (b) Results of Different Methods

Fig. 1. (a) Inputs $\mathbf{x}_i$ are linear combinations of three basis functions. (b) Signals extracted using ICA, MCA, PCA and MIA. The MIA result, with $\lambda = 0$, is meaningful.

3.1 Data Model

A speech signal can be modeled as an excitation that is convolved with a linear dynamic filter which represents the vocal tract. The excitation signal can be modeled for voiced speech as a periodic signal and for unvoiced speech as random noise. It is common to analyze the voiced and unvoiced speech separately [15]. In this example, only the voiced speech is used for speaker recognition. Let $\mathbf{E}^{(p)}$, $\mathbf{H}^{(p)}$ and $\mathbf{V}^{(p)}$ be the spectral representations of the excitation, vocal tract filter and the voiced signal parts of a person p respectively. Moreover, let $\mathbf{M}$ represent speaker independent signal parts in the spectral domain (i.e. recording equipment, environment, etc.). Therefore, the data can be modeled as:

$$\mathbf{V}^{(p)} = \mathbf{E}^{(p)} \cdot \mathbf{H}^{(p)} \cdot \mathbf{M}. \tag{12}$$

By cepstral deconvolution, the model can be represented as a linear combination of its basis functions:

$$\mathbf{x}_i^{(p)} = \log \mathbf{V}^{(p)} = \log \mathbf{E}^{(p)} + \log \mathbf{H}^{(p)} + \log \mathbf{M}. \tag{13}$$

This model suggests that we could use MIA to extract a function that represents the speaker's signature. In practice, we take speech segments of about one second as MIA inputs $\mathbf{x}_i^{(p)}$ in order to achieve spectral accuracy. An example of inputs $\mathbf{x}_i^{(p)}$ is shown in Fig. 2(a). Therefore, MIA will extract signatures that capture typical speaker dependent correlations in the logarithmic spectral domain. Speaker independent signal parts $\mathbf{M}$ will be minimized if they are not equally present in all MIA inputs.

3.2 MIA-Based Text Independent Speaker Verification

We partition the training and testing data for each person p into $D = 8$ segments $\{x_i^{(p)}\}_{i=1,\ldots,D}$ of one second. For each person, we extract a voice signature $s^{(p)}$

using MIA. The cosine distance between the testing data signatures and training data signatures is used as a measure of similarity. A matrix that represents the cosine distances between all signatures in the database is illustrated in Fig. 2(b).

Let N_{FA} and N_{CA} represent the number of false and correct acceptances respectively. N_{U} is the number of registered users. The speaker recognition results are evaluated by a comparison of the false acceptance rate FA with the false rejection rate FR, calculated as follows:

$$\mathrm{FA} = \frac{N_{\mathrm{FA}}}{N_{\mathrm{U}}^2 - N_{\mathrm{U}}} \quad \text{and} \quad \mathrm{FR} = \frac{N_{\mathrm{U}} - N_{\mathrm{CA}}}{N_{\mathrm{U}}}. \tag{14}$$

N_{FA} and N_{CA} result from the entire database by testing each learned speaker signature against all test speaker signatures. This means that a number of N_{U}^2 tests is performed, including N_{U} correct combinations and $N_{\mathrm{U}}^2 - N_{\mathrm{U}}$ impostors. By changing a threshold value, more people can be accepted which results in a trade off between FA and FR. The FA versus FR plot of this example is illustrated in Fig. 3(a). The equal error rate (EER), where FA equals FR, is used to compare to previous results in [14].

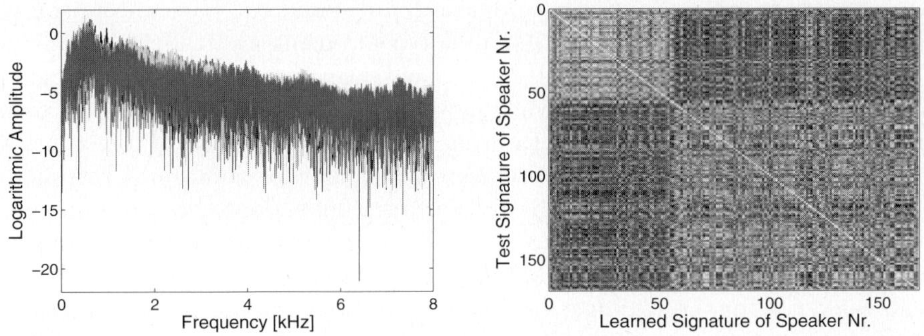

(a) Inputs $\mathbf{x}_i$ in the Fourier Domain (b) Similarity Scores

Fig. 2. MIA applied to text independent speaker recognition. (a) Representation of input data $\{\mathbf{x}_i^{(p)}\}_{i=1,\ldots,8}$, given by speech segments of a single speaker p in the logarithmic Fourier domain. (b) Matrix of similarity scores between different signatures. Bright gray stands for high and dark gray for low similarity between signatures.

3.3 Results

We used a set of 168 speakers from the TIMIT database [13]. For each signature extraction from the training and testing data, we used 8 seconds of voiced speech. The data was partitioned into 8 windows with non overlapping, nearly rectangular windowing functions of one second lengths and Gaussian tails of $\frac{1}{20}$ second. The input functions had their mean subtracted. Thereafter, each extracted signature was down-sampled to 256 points. The mean signature was subtracted

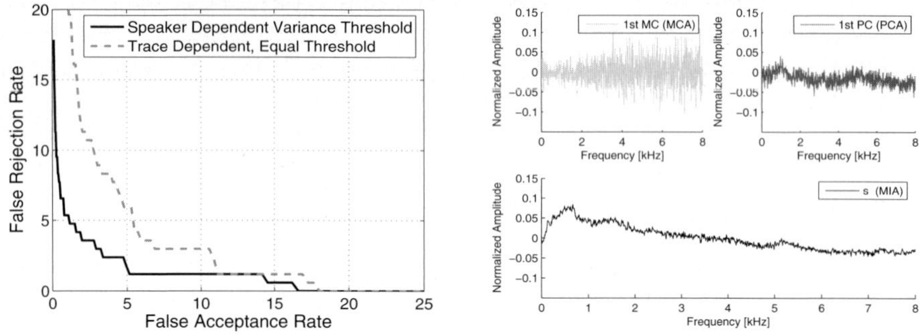

(a) FA vs. FR percentage (b) Results of Different Methods

Fig. 3. Results of MIA-based text independent speaker verification. (a) False rejection (FR) versus false acceptance (FA) rate. (b) Speaker signature extracted by MIA from $\{\mathbf{x}_i^{(p)}\}_{i=1,\ldots,8}$. Also plotted are the first minor component (MCA) and the first principal component (PCA) of the data.

from all signatures to focus on the evaluation of differences. The comparison of the MIA-based signature with the 1st minor component and the 1st principal component is illustrated in Fig. 3(b). Note visually that only MIA extracts signal amplitudes in accordance with the well known result that low frequencies contain most information about a speaker. In order to alter the proportion between FA and FR, two different thresholds were used. First, a trace dependent threshold treats every user the same. Secondly, the variance dependent threshold uses information about the similarity between test cases to learn a speaker dependent weighting. The EER of this MIA-based text independent speaker recognition system was 2.9 % for the variance dependent threshold and 5.4 % for a trace dependent threshold. For a similar experiment, [14] reports EER's between 4.3 % and 6.1 % using ICA and PCA features. Here it has to be noted that this test was done with 462 speakers. However, one person was represented by 16 or 32 features of 128 or 256 samples length. On the other hand, MIA only uses a single signature of 256 samples length per speaker.

4 Conclusion

We proposed a novel feature extraction method, Mutual Interdependence Analysis (MIA), which finds an invariant function of the input function set, representing the direction of minimum variance of input projections. Intuitively, this function is mutually interdependent with all inputs. The proof of the minimization problem exploits the unconstrained span of the original input data to infer a closed form solution. Furthermore, we showed the effect of input data translation by a constant value λ. In this way, one can control the degree of correlatedness with outlier functions in the input set. Indeed, one can choose a value of λ to discriminate between inputs and bias towards a result which correlates only with a subset of them. Further work will analyze the robustness of MIA to noise. Moreover, the effect of changes in the span constraint, or the choice of a basis function set, will be explored.

Acknowledgments

We would like to thank Radu Balan for his invaluable input to this paper.

References

1. Jolliffe, I.T.: Principal Component Analysis, 2nd edn. Springer, Heidelberg (2002)
2. Ramsay, J.O., Silverman, B.W.: Functional Data Analysis, 2nd edn. Springer, Heidelberg (2006)
3. Oja, E.: Principal components, minor components, and linear neural networks. Neural Networks 5, 927–935 (1992)
4. Tipping, M.E., Bishop, C.M.: Probabilistic principal component analysis. Journal of the Royal Statistical Society 61(3), 611–622 (1999) (Series B)
5. Williams, C., Agakov, F.: Products of gaussians and probabilistic minor components analysis. Neural Computation 14(5), 1169–1182 (2002)
6. Thompson, P.A.: An adaptive spectral analysis technique for unbiased frequency estimation in the presence of white noise. In: Systems and Computers, pp. 529–533 (1980)
7. Xu, L., Oja, E., Suen, C.Y.: Modified hebbian learning for curve and surface fitting. Neural Networks 5, 441–457 (1992)
8. Moon, T.K., Stirling, W.C.: Mathematical methods and algorithms for signal processing. Prentice-Hall, Upper Saddle River, NJ (2000)
9. Welling, M., Agakov, F., Williams, C.K.I.: Extreme components analysis. In: Thrun, S., Saul, L., Schölkopf, B. (eds.) Advances in Neural Information Processing Systems 16, MIT Press, Cambridge, MA (2004)
10. Rao, Y.N., Principe, J.: Efficient total least squares method for system modeling using minor component analysis. In: Proceedings of international workshop on neural networks for signal processing, pp. 259–268 (2002)
11. Horn, R.A., Johnson, C.R.: Matrix Analysis. Cambridge University Press, Cambridge (1999)
12. Hyvärinen, A., Karhunen, J., Oja, E.: Independent component analysis. John Wiley and Sons, Chichester (2001)
13. Garofolo, J.S., Lamel, L.F., Fisher, W.M., Fiscus, J.G., Pallett, D.S., Dahlgren, N.L., Zue, V.: Timit acoustic-phonetic continuous speech corpus. CDROM (1993)
14. Rosca, J., Kofmehl, A.: Cepstrum-like ica representations for text independent speaker recognition. In: ICA, pp. 999–1004 (2003)
15. Deng, L., O'Shaughnessy, D.: Speech Processing: A Dynamic and Optimization-Oriented Approach. Signal Processing and Communications. Marcel Dekker, Inc. (2003)

Modeling Perceptual Similarity of Audio Signals for Blind Source Separation Evaluation

Brendan Fox, Andrew Sabin, Bryan Pardo, and Alec Zopf

Northwestern University, Evanston, IL, USA 60201, USA
pardo@northwestern.edu
http://music.cs.northwestern.edu

Abstract. Existing perceptual models of audio quality, such as PEAQ, were designed to measure audio codec performance and are not well suited to evaluation of audio source separation algorithms. The relationship of many other signal quality measures to human perception is not well established. We collected subjective human assessments of distortions encountered when separating audio sources from mixtures of two to four harmonic sources. We then correlated these assessments to 18 machine-measurable parameters. Results show a strong correlation (r=0.96) between a linear combination of a subset of four of these parameters and mean human assessments. This correlation is stronger than that between human assessments and several measures currently in use.

Keywords: Source Separation, Perceptual Model, Music, Audio.

1 Introduction

Blind Source Separation (BSS) is the process of isolating individual source signals, from mixtures of source signals, when the characteristics of the individual sources are not known before-hand. BSS is an active area of research [1,2,3,4,5] and new techniques are continually developing.

The effectiveness of a BSS algorithm is typically measured by comparing the quality of a signal extracted from a mixture (the signal estimate) to the original source signal. Given this methodology, it becomes important to choose an error measure that captures the salient differences between the original and the estimate. Our research [6] focuses on source separation of acoustic sound sources from audio mixtures. Because our ultimate goal is the creation of audio for a human listener, human perception determines what we consider "good" results. Unfortunately, it is not practical to conduct a human listening study each time one varies a parameter of a BSS algorithm. Thus, researchers typically use machine-measurable signal quality measures.

Most BSS researchers for audio applications use existing measures of signal quality such as Signal to Distortion Ratio (SDR) [6] or quality measures specifically for audio source separation, such as Signal to Interference Ratio (SIR) [10]. The relationship between human perception of signal quality and such commonly used machine-measureable statistics remains unstudied studied over the range

M.E. Davies et al. (Eds.): ICA 2007, LNCS 4666, pp. 454–461, 2007.

of distortions introduced by audio source separation algorithms. This makes it difficult to estimate the perceptual effect of a change in the value of such statistics.

One approach to measuring BSS effectiveness for audio applications has been to use the PEAQ (Perceptual Evaluation of Audio Quality) [7] perceptual model for audio codecs. PEAQ calculates a set of statistics about the audio that are fed into a three layer feed-forward perceptron that maps the statistics onto a single quality rating called an Objective Difference Grade (ODG). Vanam and Creusere implemented a version of PEAQ that improves its correlation with subjective human data for intermediate quality codecs [8]. Unfortunately, their improvement depends on the kinds of distortions introduced by particular audio codecs, making it unsuitable for BSS evaluation. Although PEAQ works well for evaluating the small degradations of audio signals introduced by audio compression codecs, the measure has shortcomings when evaluating signals with the larger distortions resulting from source separation. For these signals, PEAQ does not correlate well with subjective human quality assessments and often saturates at the maximum possible rating.

Thus, the relationship between the currently used measures of BSS effectiveness to human perception of audio quality is not well established. In this paper we measure the correlation of 18 existing machine-measurable statistics to human perception of signal quality for sounds extracted from audio mixtures with BSS. We then create a combined model from those statistics that correlate best with human perception.

2 Study of Perceived Sound Similarity

We performed a study to collect human similarity assessments between reference recordings and distorted versions of the references extracted from audio mixtures using BSS algorithms. For this study, each participant was seated at a computer terminal. A series of audio recordings clips, in matched pairs, was played to the participant over headphones. Each pair consisted of a reference audio recording followed by a distorted version of the recording, called the test. The participant had only one chance to hear each pair. For each pair, the participant was asked to rate the similarity of the reference sound to the test sound on a scale from 0 to 10 where the values correspond to the following ratings:

10 – Signals are indistinguishable
8 – Signals are just barely distinguishable
6 – Signals strongly resemble each other but are easily distinguishable
4 – Signals resemble each other
2 – Signals just barely resemble each other
0 – Signals are completely dissimilar

The task began with a short training session of ten pairs to familiarize the participant with the task. Participants then listened to 130 audio pairs and rated the similarity of each pair. The task, complete with instructions, typically took less than one hour per participant. We collected responses from 31 participants drawn from the Northwestern University student, faculty and staff. Median participant age was 22 and the age range was from 18 to 35. Just under half (15) of the participants were male and 16 were female. Participants were screened to ensure they had never been diagnosed with a hearing disorder or language disorder.

2.1 Audio Corpus

The reference audio recordings used in the study are individual long-tones, ranging from 2-4 seconds, played on the alto saxophone, linearly encoded as 16-bit, 44.1 kHz audio. Mixtures of these recordings were created to simulate the stereo microphone pickup of spaced source sounds in an anechoic environment. We assume omni-directional microphones, spaced according to the highest frequency we expect to process. Instruments were placed in a semi-circle around the microphone pair at a distance of one meter. In the two-instrument mixtures, the difference in azimuth angle from the sources to the microphones was 180 degrees. The BSS algorithms we currently study depend on having significant differences in azimuth between sources. A difference of 180 degrees between sources will produces the best results. As the difference tends to 0 degrees, source separation degrades. To generate a range of BSS output from good to bad, instruments were placed at random angles around the microphones in the three and four instrument mixtures.

For each mixture, each source signal was assigned a randomly selected pitch from the 13 pitches on the equal tempered chromatic scale from C4 through C5. We created nine two-instrument mixtures, six three-instrument mixtures, and five four-instrument mixtures in this manner, which is a total of 56 individual instrument comparisons, once extracted. This provides an approximately equal number of single note samples for each type of mixture. Mixtures were separated using the Active Source Estimation (ASE) [6] and DUET [4] source separation algorithms, resulting in 112 extracted sounds.

The corpus also included a set of calibration sounds. For these calibration sounds, the proportion of altered time-frequency frames varied from 0.2 to 1.0 (where 1 means all frames were altered). In altered frames, phase was randomly varied in the full range and amplitude was randomly varied between 8 dB and 20 dB. For eight of the example pairs the test sound was a repeat of the reference sound. The 112 extracted examples, 10 manually distorted examples, and 8 repeat examples, give a total of 130 example pairs for the test corpus.

2.2 Human Study Results

In our listening data we included eight reference-test pairs where the reference and test sounds were identical. We excluded data from three participants who proved unreliable at labeling identical pairs as highly similar (a 9 or a 10). These three participants gave an average similarity score below 8 for the set of identical pairs. This mean fell over two standard deviations below the mean similarity score given to identical pairs by the group as a whole. The group, excluding these three outliers gave a mean similarity rating of 9.6 to the identical pairs.

There was a strong correlation between the remaining 28 participants in the subjective similarity ratings assigned to example pairs of audio. We compared the individual response of each participant to the mean response reported by the group, excluding that participant. The correlation coefficient between each individual and the remainder of the group ranged from 0.8458 to 0.9737 with a median correlation of r = 0.9155. The left panel of Figure 1 illustrates the correlation between the mean group ratings and those of a randomly selected individual. Given the strength of correlation across participants, we based our perceptual model on the mean similarity ratinings for each of the 130 reference-test pairs, averaged across the 28 remaining participants.

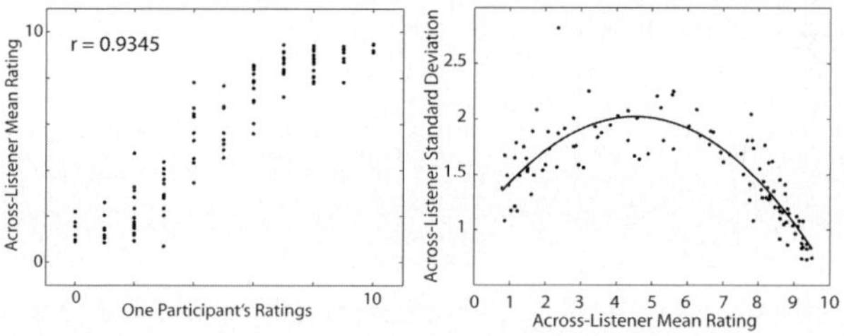

Fig. 1. (Left) Correlation between group mean ratings and those of a randomly selected participant (r = 0.935). Each point is one reference-test pair. (Right) The standard deviation of the range of participant responses, indexed by the mean value of these responses. Each point is one reference-test pair.

The right panel of Figure 1 shows the standard deviation of participant similarity ratings for example pairs, indexed by the mean response value. The line shown is a quadratic polynomial fit to the data with r = 0.83. The standard deviation is quite low for ratings toward the maximum (10) and minimum (0) similarity. There is an increase in across-participant variability at middle values, indicating more agreement on the extremes.

3 Modeling Human Responses

To build a model that effectively predicts human judgments of the similarity between two sounds, one must map machine-quantifiable measures onto human similarity assessments. In our study we consider the measures listed in Table 1. These measures were selected due to their use in the blind source separation community or as inputs to perceptual models used for audio codec evaluation. We refer the reader to the original paper citations for detailed definitions of these measures.

Table 1. Linear correlation of machine-measurable statistics to mean human subject ratings

Machine Measurable Statistic	r value
ISR - Ratio of signal energy to error due to spatial distortion [10]	0.87563
SIR - Ratio of the signal energy to the error due to interference [10]	0.82131
SAR - Ratio of signal energy to the error due to artifacts [10]	0.75001
SDR - Signal to Distortion Ratio [6]	0.72313
ODG - Output of the PEAQ model [7]	0.67735
DIX - A measure of perceived audio quality[7]	0.67074
BWT - Bandwidth of Test signal [7]	0.48946
HSE - Harmonic structure of the error over time [7]	0.13531
NLS - Noise Loudness in Sones [11]	-0.09409
AMD2 - Alternate calculation of average modulation difference [12]	-0.12796
NMR - Noise to mask ratio [7]	-0.34614
MPD - Maximum probability of detection after lowpass filter [12]	-0.36465
THD - Total Harmonic Distortion [7]	-0.48789
WMD - Windowed modulation difference [12]	-0.58947
BWR - Bandwidth of Reference signal (Hz) [7]	-0.67536
AMD - Average modulation difference [12]	-0.75003
RDF - Relative number of Distorted Frames [12]	-0.78455
ADB - Average Distorted Block [12]	-0.81710

As the table shows, the ODG values reported by the PEAQ perceptual model are only loosely correlated to human similarity assessments in our dataset. One might argue that this is because the ODG values may be correlated to a more complex function than a simple linear fit. This hypothesis is not supported when ODG is plotted against mean human assessments. This is shown in Figure 2. The figure also shows a ceiling effect for SDR and poor correlation between THD and mean human assessment. The measures ISR, SIR and ADB all have stronger negative or positive correlation to the mean human similarity assessments than do ODG, SDR or THD.

3.1 Results and Data Analysis

We followed the lead of the PEAQ researchers by mapping objective signal measures to human assessments using a variety of feed-forward, multilayer perceptrons. Every network architecture used all measures except ODG (the output of the PEAQ perceptual model) from Table 1 as input, with one input node for each measure. We varied the number of hidden layers from 0 to 2 and the number of nodes in each hidden layer from 8 to 13. All networks used 11 output nodes network, representing the ratings from 0 through 10 reported by human listeners. In training our neural networks, we applied 6-fold cross validation by dividing our full dataset (130 responses from 31 participants, making 4030 examples) into 6 bins.

Figure 3 show correlation results of the best performing network for each number of hidden layers. For the neural network plots (all except the far right

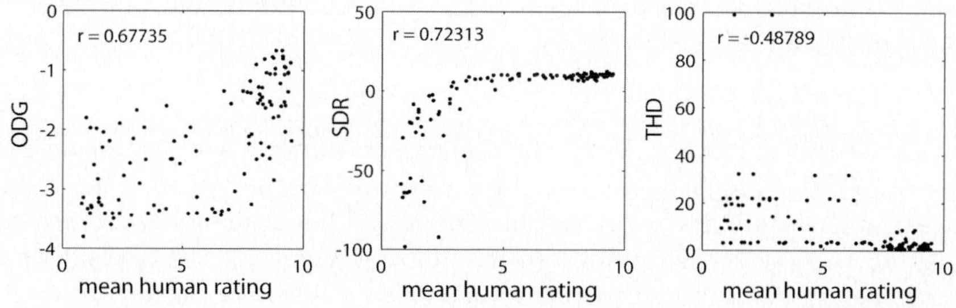

Fig. 2. Correlation of ODG (left), SDR (center) and THD (right) to human perceptions of audio similarity. Each data point indicates one example pair. The vertical axis shows value of the measure and the horizontal axis the mean human similarity assessment.

panel) the vertical coordinate of each point is determined by a weighted average of the output node activations to a given comparison pair. The horizontal coordinate is the mean human response for that pair. Perfect performance for a model would result in a straight line from (0,0) to (10,10), an r value of 1, and root mean squared error (rmse) of 0. Here, the error for a single data point is the difference between the model output and the mean human response. As can be seen from the figure, the performance difference between networks architectures is minor.

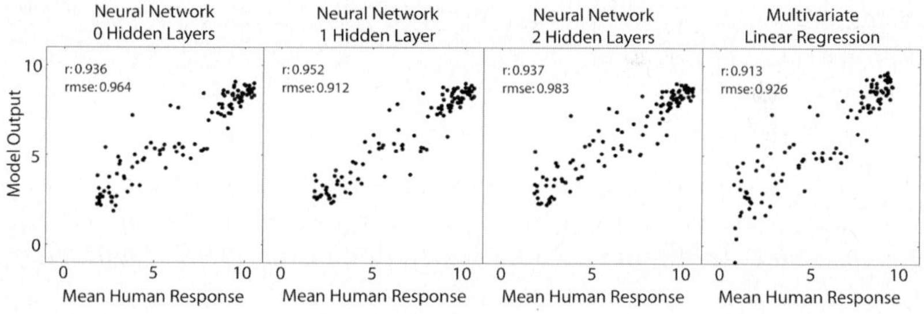

Fig. 3. Correlation of model responses to human responses. In every panel, each data point indicates one example pair. The vertical axis shows the output of the given model. The horizontal axis shows the mean similarity value over the 31 human participants. Both the r value and the root-mean-squared error (rmse) are shown for each network.

Neural networks with no hidden layer can only successfully discriminate linearly seperable classes. Networks with hidden layers can discriminate between classes that are not linearly separable. Since the performance difference between our networks was negligible, we infer that a linear combination of the statistics from Table 1 could be used to map machine measurable statistics of signal similarity onto human estimates of signal similarity. Thus, we fitted human responses to a multivariate linear regression model. As expected, its correlation to human similarity assessments is nearly identical to those of the neural networks. This is shown in Figure 3.

To make a more parsimonious model of human similarity assessments, we performed a stepwise multivariate linear regression on the machine-measurable statistics used in our study. Here, the dependent variable was the mean human response and the independent variables were all the measures from Table 1. Stepwise multivariate linear regression generates a linear model using only those inputs that independently account for the most variance in the dependent variable. After performing this process, we achieved a linear fit to the data with R = 0.96 using only four of the measures from Table 1. The resulting linear correlation is shown in Table 2. The order in which parameters were added to the model is shown by "Entrance Order." Any measure from Table 1 not shown in the Table 2 did not significantly increase correlation between the linear model, given the previous measures already added to the model.

Table 2. Results of stepwise multivariate linear regression using mean human similarity responses as the dependent variable and the measures from Table 1 as the independent variables

Entrance Order	Statistic	Coefficient (b)	Cumulative Correlation (r)
n/a	constant offset	14.968	n/a
1	ISR	0.194	0.876
2	SIR	0.064	0.938
3	SAR	0.103	0.952
4	MPD	-12.787	0.960

The r-value corresponding to the multivariate linear regression model in Figure 3 is 0.913 while the corresponding value in Table 2 is 0.960. This is because results shown in Figure 3 were generated using a 6-fold round robin validation technique where there was no overlap between the the training and testing sets. The correlation in Table 2 was done over the full data set, rather than a subset.

4 Conclusions

We have shown that a linear combination of four machine-measurable statistics can successfully model human similarity assessments for pairs of sounds with a correlation of r=0.96. Three of these statistics (ISR, SIR and SAR) are used in a recent comparison of multichannel audio source separation [10]. Correlation of a linear combination is not improved upon by the nonlinear modeling possible with a multilayer perceptron. For the range of signals under consideration (woodwinds distorted by audio source separation algorithms), the linear model performed much better than the PEAQ (ODG) perceptual model. In future work, we plan to expand the range of test signals over which we study human evaluations of similarity, with a focus on speech, as well as music. If the results of future studies correlate with the current paper, this will provide further evidence that the measures with high correlation in Table 2 are the most useful statistics upon which to measure the effectiveness of source separation for audio applications.

Acknowledgements. This work was funded in part by National Science Foundation Grant number IIS-0643752. We thank John Woodruff and Emmanuel Vincent for their help.

References

1. Anemuller, J., Kollmeier, B.: Amplitude Modulation Decorrelation for Convolutive Blind Source Separation. In: International Symposium on Independent Component Analysis and Blind Source Separation. Helsinki, Finland (1999)
2. O'Grady, P.D., Pearlmutter, B.A., Rickard, S.: Survey of Sparse and Non-Sparse Methods in Source Separation. International Journal of Imaging Systems and Technology 1(15), 18–33 (2005)
3. Virtanen, T., Klapuri, A.: Separation of Harmonic Sounds Using Multipitch Analysis and Iterative Parameter Estimation. In: IEEE Workshop on Applications of Signal Processing to Audio and Acoustics. New Paltz, NY (2001)
4. Yilmaz, O., Rickard, S.: Blind Separation of Speech Mixtures via Time-Frequency Masking. IEEE Transactions on Signal Processing 52(7), 1830–1847 (2004)
5. Master, A.: Stereo Music Source Separation via Bayesian Modeling. Doctoral Thesis. Stanford University, 1–199 (2006)
6. Woodruff, J., Pardo, B.: Using Pitch, Amplitude Modulation and Spatial Cues for Separation of Harmonic Instruments from Stereo Music Recordings. EURASIP Journal on Advances in Signal Processing (Article ID 86369) (2007)
7. Thiede, T., Treurniet, W., Bitto, R., Schmidmer, C., Sporer, T., Beerends, J., Colomes, C., Keyhl, M., Stoll, G., Brandenburg, K., Feiten, B.: PEAQ–The ITU Standard for Objective Measurement of Perceived Audio Quality. Journal of the Audio Engineering Society 48(1/2), 3–29 (2000)
8. Creusere, C.: Evaluating low bitrate scalable audio quality using advanced version of PEAQ and energy equalization approach. Acoustics, Speech, and Signal Processing 3, 189–192 (2005)
9. Vincent, E.: Musical Source Separation Using Time-Frequency Source Priors. IEEE Transactions on Audio, Speech and Language Processing 14(1), 91–98 (2006)
10. Vincent, E., Sawada, H., Bofill, P., Makino, S., Rosca, J.: First Stereo Audio Source Separation Evaluation Campaign: Data, Algorithms and Results. In: International Conference on Independent Component Analysis and Blind Source Separation (ICA) (2007)
11. Schroeder, M., Atal, B., Hall, J.: Optimizing digital speech coders by exploiting masking properties of the human ear. Journal of the Acoustical Society of America 66(6), 1647–1652 (1979)
12. Kabal, P.: An Examination and Interpretation of ITU-R BS.1387: Perceptual Evaluation of Audio Quality. McGill University Technical Report, p. 1–96 (2003)

Beamforming Initialization and Data Prewhitening in Natural Gradient Convolutive Blind Source Separation of Speech Mixtures

Malay Gupta and Scott C. Douglas

Department of Electrical Engineering
Southern Methodist University
Dallas, Texas 75275 USA

Abstract. Successful speech enhancement by convolutive blind source separation (BSS) techniques requires careful design of all aspects of the chosen separation method. The conventional strategy for system initialization in both time- and frequency-domain BSS involves a *diagonal center-spike* FIR filter matrix and no data preprocessing; however, this strategy may not be the best for any chosen separation algorithm. In this paper, we experimentally evaluate two different approaches for potentially-improving the performance of time-domain and frequency-domain natural gradient speech separation algorithms – *prewhitening* of the signal mixtures, and *delay-and-sum* beamforming initialization for the separation system – to determine which of the two classes of algorithms benefit most from them. Our results indicate that frequency-domain-based natural gradient BSS methods generally need geometric information about the system to obtain any reasonable separation quality. For time-domain natural gradient separation algorithms, either beamforming initialization or prewhitening improves separation performance, particularly for larger-scale problems involving three or more sources and sensors.

1 Introduction

Convolutive blind source separation (CBSS) refers to the separation of signals that have been mixed through a dispersive environment using signal processing procedures that do not have specific knowledge of the source properties or the mixing conditions. Due to the dispersive nature of the channel, CBSS algorithms must attempt to undo both spatial and temporal mixing effects. As a result, CBSS algorithms tend to be more complicated than their spatial-only BSS counterparts.

Frequency-domain approaches to CBSS transform the measured mixtures into the discrete frequency-domain via the short-time Fourier transform (STFT) and apply spatial-only (instantaneous) BSS algorithms in each frequency component of the mixtures individually [1]. After separation in the frequency-domain, these signals must be carefully reconstructed before being inverse-Fourier-transformed to recover the time-domain signals. This reconstruction process requires estimating the permutation and scaling ambiguities for all the frequency

M.E. Davies et al. (Eds.): ICA 2007, LNCS 4666, pp. 462–470, 2007.

components of the separated sources. Prior information about the array geometry and directions-of-arrival (DOAs) of the sources at the sensor array is often assumed. Several researchers have offered ways to use this information in the reconstruction process [2]–[6]. Post-processing permutation resolution can be computationally-demanding if more than two sources are being separated. In many cases, a closed form solution is not possible [4].

In contrast, time-domain CBSS algorithms adapt the impulse response of a multichannel linear filter using only as many output signals as the number of sources that are being extracted [7]–[10]. Because they use time-domain convolutions instead of frequency-domain multiplications, these methods tend to be more difficult to code. Note that their computational complexities can be made to be similar to those of frequency-domain approaches through block processing [8]. Since the algorithms employ a separation criterion whose number of outputs equals the number of sources being estimated, time-domain CBSS approaches do not appear to have severe source permutation problems over different extracted frequencies. These time-domain methods tend to converge more slowly, however, if careful strategies for algorithm implementation are not considered. In [11] one simple way to improve convergence performance for the time-domain method in [8] has been described.

In this paper, we compare the use of two well-known strategies for improving the performance of CBSS algorithms: (1) beamforming initialization [5,12], and (2) multichannel prewhitening [8]. Both time-domain and frequency-domain versions of the well-known natural gradient CBSS method are evaluated and their performances compared to other competing approaches using data collected from a controlled laboratory measurement setup. These numerical experiments show that (a) beamforming initialization is required for frequency-domain natural gradient CBSS methods if no other technique is used to resolve permutation ambiguities, (b) prewhitening alone does not improve the performance of frequency-domain natural gradient CBSS methods, and (c) the performance of time-domain natural gradient algorithms improves with either signal prewhitening or beamforming coefficient initialization, and this improvement is significant when dealing with mixtures of more than two sources.

2 Time- and Frequency-Domain Signal Models

For multichannel acoustic recordings, the n-dimensional signal mixtures at time k, $\mathbf{x}(k) = [x_1(k) \cdots x_n(k)]^T$ can be modeled as

$$\mathbf{x}(k) = \sum_{l=-\infty}^{\infty} \mathbf{A}_l \mathbf{s}(k-l), \tag{1}$$

where $\{\mathbf{A}_l\}$ denotes a sequence of $n \times m$ mixing matrices, $\mathbf{A}(z) = \sum_{l=-\infty}^{\infty} \mathbf{A}_l z^{-l}$ is the multichannel system transfer function, and $\mathbf{s}(k) = [s_1(k) \cdots s_m(k)]^T$ is the m-dimensional signal vector at time k. All CBSS algorithms attempt to find a time-varying separating or demixing system $\mathbf{B}(k, z)$ to process the signal mixtures $\mathbf{x}(k) = \mathbf{A}\{\mathbf{s}(k)\}$ such that $\mathbf{y}(k) = \mathbf{B}\{\mathbf{x}(k)\}$ contains the estimates of each of the sources in $\mathbf{s}(k)$ without repetition. Mathematically, this can be represented as

$$\mathbf{y}(k) = \sum_{l=-\infty}^{\infty} \mathbf{B}_l(k)\mathbf{x}(k-l). \tag{2}$$

In practice, a truncated causal approximation to (2) is often employed, where L is a positive integer and

$$\mathbf{y}(k) = \sum_{l=0}^{L} \mathbf{B}_l(k)\mathbf{x}(k-l). \tag{3}$$

Frequency-domain CBSS algorithms use the STFT to transform the time-domain data into the frequency-domain, whereby a separate complex-valued instantaneous demixing system is found for each of the frequency components of the mixed signals. The input data in the lth frequency bin ω_l is given by

$$\mathbf{x}(\omega_l, k) = \mathbf{A}(\omega_l)\mathbf{s}(\omega_l, k), \tag{4}$$

where k denotes the time dependence of the STFT, $\mathbf{s}(\omega_l, k)$ is the transformed source signal vector, and $\mathbf{A}(\omega_l)$ denotes the mixing matrix for the lth frequency bin. As such, the demixing process in each frequency bin is formulated as

$$\mathbf{y}(\omega_l, k) = \mathbf{B}(\omega_l, k)\mathbf{x}(\omega_l, k), \tag{5}$$

where $\mathbf{y}(\omega_l, k)$ and $\mathbf{B}(\omega_l, k)$ are the estimated source signal vector and the demixing matrix, respectively, in the lth frequency bin at time k.

3 Beamforming vs. Prewhitening in Convolutive Blind Source Separation

Both beamforming and prewhitening attempt to solve part of the goal achieved by successful application of CBSS methods in certain contexts.

Beamforming and CBSS attempt to suppress interferences caused by spatially-distinct sources to extract individual source signals when operating on data collected from a uniform linear array. Beamforming methods work by providing maximum gain in the direction of the desired user. CBSS based methods have been observed to place spatio-temporal nulls in the directions of interfering users in some environments [12].

Beamforming methods typically assume a working knowledge of the sensor array manifold and the directions-of-arrival (DOAs). CBSS methods, on the other hand, typically assume no known signal or measurement structure other than a linear dispersive channel for the mixing process. Researchers have suggested the merger of beamforming with CBSS to include prior information about the array manifold and DOAs within CBSS algorithms [3]. These techniques are primarily designed to remove permutation difficulties that lead to lower separation performance. In situations where DOA information is not available *a priori*, researchers have suggested procedures for estimating DOAs as part of the CBSS algorithm being developed [3,6]. In narrowband beamforming, the directional vector associated with a frequency ω for a source impinging on the array from a direction θ is given as

$$\mathbf{d}(\omega,\theta) = [\exp(j\omega\tau_0(\theta)) \ \cdots \ \exp(j\omega\tau_{m-1}(\theta))]^T, \tag{6}$$

where $\tau_l(\theta) = ld\sin(\theta)/c$ is the time delay associated with the lth sensor with respect to the reference sensor, m is the number of sensors in the array, d is the array element separation, and c is the speed of sound. Signals with significant frequency content (*e.g.* audio signals) received at a sensor array will typically have directional vectors associated with each of their frequency components. Thus, clustering of the directional vectors may be needed to obtain a consistent estimate of the source DOAs.

Perhaps the simplest way to employ DOA knowledge to improve CBSS convergence performance is to initialize the separation system coefficients $\{\mathbf{B}_l(0)\}$ or their frequency counterparts $\{\mathbf{B}(\omega_l, 0)\}$ to a series of fixed beamformers in which the mainlobe of each of the beampatterns in each frequency bin for the ith separation system points toward a talker. In this case, we would choose

$$\mathbf{B}(\omega_l, 0) = [\mathbf{d}(\omega_l, \theta_1) \ \mathbf{d}(\omega_l, \theta_2) \ \ldots \ \mathbf{d}(\omega_l, \theta_m)]^H. \tag{7}$$

For time-domain algorithms, we can compute the appropriate initial coefficients by taking the inverse FFTs of the frequency-domain responses in (7) about their points of symmetry. Initializing CBSS algorithms in this way does not modify the algorithm's operation other than choosing its initial state. The main alternative to this coefficient initialization is center-spike initialization, in which

$$\mathbf{B}_l(0) = \begin{cases} \mathbf{I}, & l = \frac{L}{2} \\ \mathbf{0}, & \text{otherwise.} \end{cases} \tag{8}$$

For frequency-domain algorithms, we can compute the appropriate initial coefficients by taking the FFT of the time-domain responses in (8) about their points of symmetry. Center-spike initialization makes no assumption about the source-sensor array geometry.

Prewhitening is a preprocessing strategy whereby the measured signals $\{x_i(k)\}$, $1 \leq i \leq n$ are linearly filtered such that the filtered signals $\{v_i(k)\}$ approximately satisfy

$$E\{v_i(k)v_j(k-l)\} \approx E\{|v_i(k)|^2\}\delta_{i-j}\delta_l, \tag{9}$$

where δ_l is the Kronecker delta function. These prewhitened signals are used in place of $\{x_i(k)\}$ in the separation system. Examples of prewhitening algorithms include the linear phase adaptive procedure in [13] and the least-squares multichannel linear predictor described in [8]. When block processing is used, one can use successive filtering operations to process $\{\mathbf{x}(k)\}$ to produce $\mathbf{v}(k)$, which is likely the most computationally efficient method.

Prewhitening solves part of the CBSS task, as decorrelation is a necessary but not sufficient condition for source separation of mixtures of statistically-independent signals. Hence, it is reasonable to use prewhitening as a preprocessing step to remove any signal correlations contained in the data prior to performing separation with any CBSS method. In this case, the input signals $\{x_i(k-l)\}$ used in the separation system are replaced by the prewhitened signals $\{v_i(k-l)\}$ obtained at the outputs of the prewhitening system.

One can view beamforming initialization and prewhitening as two simple but competing approaches for improving the performances of CBSS methods that do not require significant alteration of the separation algorithm. Note that prewhitening effectively alters the DOAs seen by the separation system within the prewhitened data, so using both prewhitening and beamforming initialization does not make sense unless special constraints are placed on the prewhitening task. It is unclear without performing experiments which procedure is to be preferred, and whether both frequency-domain-based and time-domain-based CBSS algorithms benefit from such procedures. The goal of this paper is to explore these issues through experimental evaluation on real-world speech signal mixtures to see what classes of algorithms benefit most from them.

4 Numerical Experiments

We now present numerical evaluations to illustrate the separate effects that beamforming initialization and data prewhitening have on the behaviors of one class of CBSS algorithms. In order to minimize any performance effects due to choice of separation criterion, we focus on the natural gradient algorithms presented in [4] and [8]. The algorithms attempt to minimize the mutual information of the extracted signals using frequency-domain and time-domain system structures, respectively. For comparison, we show the performance of two other algorithms on this data: one employing decorrelation with geometric beamforming constraints [3], and one using contrast-based optimization with prewhitening [9,10]. These latter algorithms incorporate either beamforming or prewhitening within their structures and are not claimed to work without such pre-processing.

Data for these evaluations was generated in an acoustically-isolated laboratory environment with three loudspeakers playing recordings of talkers (one female and two male) as the sources. The sources were located 127 cm away from the three omnidirectional microphones and were spaced at angles of -30^0, 0^0, and 27.5^0, respectively, from the array normal. The inter-sensor spacing of the microphone array was 4 cm. Acoustic foam was placed on the walls of the room to obtain a reverberation time of 300 ms for the environment. All recordings were made using 7 seconds of data per channel and a 48 kHz sampling rate and were downsampled to an 8 kHz sampling rate for processing. Fig. 1 shows the impulse responses of the loudspeaker/microphone paths for these mixing conditions.

The various algorithms were applied to this measured microphone data for two- and three-source mixtures, whereby the 0^0 source was omitted for the two-source mixture. After separation, least-squares methods were used to estimate

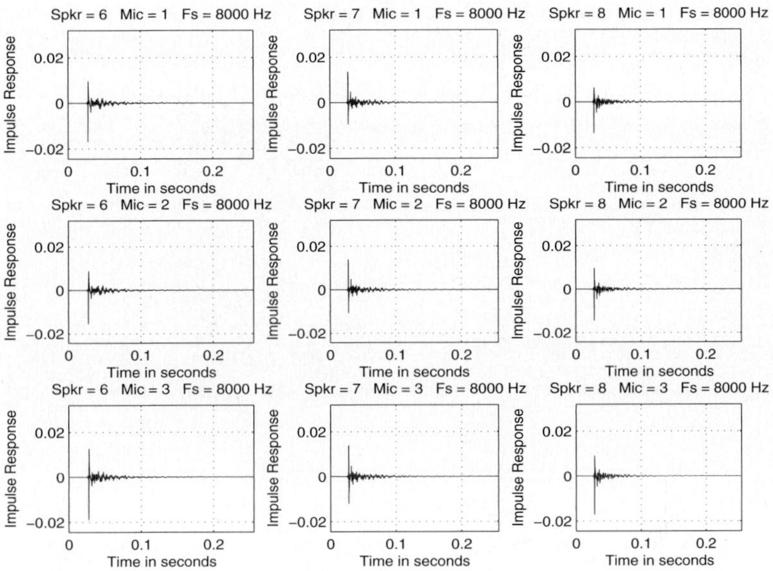

Fig. 1. Acoustic chamber impulse response in a three source, three microphone setup. Room conditions correspond to a reverberation time (RT) of 300 msec.

the contributions of the source recordings to each of the recorded mixtures as well as the output signals from each algorithm. By calculating power ratios from these least-squares estimates, we can compute the average improvement in signal-to-interference-plus-noise ratio (SINR) for each algorithm in each case.

For the normalized natural gradient algorithm in the frequency domain [11], the parameters chosen were $L = 512$ and $\mu = 0.35$, and 200 passes of the algorithm through the data have been used to adapt the filter. For the natural gradient time-domain algorithm [8], we used $L = 512$ and a step size schedule of $\mu = .0009$ for 150 data passes followed by $\mu = 0.0001$ for a single data pass followed by $\mu = 0.00001$ for a second single data pass. The data nonlinearity used in each algorithm was $f(y) = y/|y|$, where y in this case corresponds to the ith frequency bin output or the ith time-domain filter output, respectively.

Table 1 shows the SINR improvements obtained by the various algorithms for the various processing strategies on the two-source mixture data. As can be seen, the frequency-domain natural gradient method does not perform well either with center-spike initialization or with data prewhitening. With beamforming initialization, the algorithm achieves good performance on this data that closely matches the time-domain natural gradient algorithm. The latter algorithm's performance is quite good for center-spike initialization on this data, but improvements of 1.0dB and 2.7dB are obtained with beamforming initialization and data prewhitening, respectively. Shown for comparison are the behaviors of the decorrelation-based method in [3] as well as the contrast-based method with prewhitening in [11]. As can be seen, the time-domain natural gradient

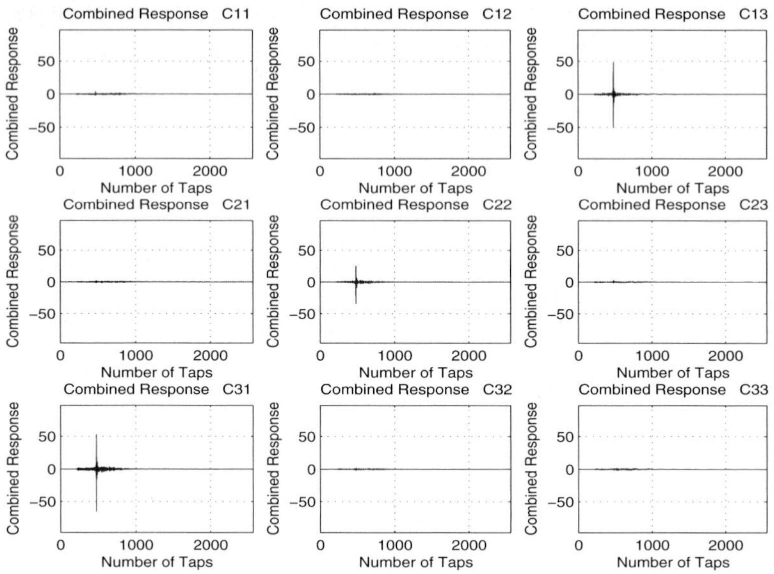

Fig. 2. Combined impulse response of the time-domain truncated natural gradient with delay-and-sum beamforming initialization

Table 1. Improvement in average SINR [dB]; RT=300 ms

Algorithm	TWO SOURCE CASE			THREE SOURCE CASE		
	Center Spike	w/Beam-forming	w/Prewhi-tening	Center Spike	w/Beam-forming	w/Prewhi-tening
SNGFD[11]	0.25	13.56	1.52	3.33	12.55	4.55
NGTD[8]	12.63	13.60	15.34	10.89	17.07	16.80
Parra-GBSSII[3]	–	7.95	–	–	5.42	–
STFICA-Symm[9,10]	–	–	11.23	–	–	12.66

method outperforms both of these competing methods when using the same spatial knowledge of the environment or data pre-processing.

Also shown in Table 1 are the SINR improvements obtained by the various algorithms for the various processing strategies on the three-source mixture data. Similar performance relationships as in the two-source data case are observed in this case. The frequency-domain natural gradient algorithm obtains adequate separation only with beamforming initialization, whereas the time-domain natural gradient algorithm can separate the source mixtures with any of the three strategies employed. The best performance is obtained with beamforming initialization, although separation using data prewhitening is nearly as good. Fig. 2 shows the combined impulse responses at convergence for the natural gradient time-domain algorithm with beamforming initialization when applied to this data, indicating that separation has occurred. It should be noted that prewhitening-based processing strategies can still be used if knowledge of the source-sensor array geometry is not available.

5 Conclusions

In convolutive blind source separation of speech signal mixtures, beamforming initialization and prewhitening are two simple strategies for improving the performance of any separation algorithm not already leveraging this structural knowledge. This paper evaluates the behaviors of two versions of the well-known natural gradient algorithm as implemented in the time- and frequency-domains, respectively, when using each of these strategies. Experiments indicate that the frequency-domain natural gradient algorithms rely on the spatial structure of the source-microphone mixing conditions, and they cannot adequately separate sources without using knowledge of the directions-of-arrival within the algorithm. Prewhitening alone does not help the performance of frequency-domain algorithms. Time-domain natural gradient algorithms can separate without directions-of-arrival knowledge; however, their performances are improved when either beamforming initialization or data prewhitening is employed.

References

1. Smaragdis, P.: Blind separation of convolved mixtures in the frequency domain. Neurocomputing 22(1-3), 21–34 (1998)
2. Parra, L., Spence, C.: Convolutive blind separation of non-stationary sources. IEEE Trans. Speech Audio Processing 8, 320–327 (2000)
3. Parra, L., Alvino, C.: Geometric source separation: Merging convolutve source separation with geometric beamforming. IEEE Trans. Speech Audio Processing 10(6), 352–362 (2002)
4. Mitianoudis, N., Davies, M.E.: Audio source separation of convolutive mixtures. IEEE Trans. Speech Audio Processing 11, 489–497 (2003)
5. Sawada, H., Mukai, R., Araki, S., Makino, S.: A robust and precise method for solving the permutation problem of frequency-domain blind source separation. IEEE Trans. Speech Audio Processing 12, 530–538 (2004)
6. Saruwatari, H., Kawamura, T., Nishikawa, T., Lee, A., Shikano, K.: Blind source separation based on a fast-convergence algorithm combining ICA and beamforming. IEEE Trans. Audio Speech Language Processing 14, 666–678 (2006)
7. Amari, S., Douglas, S.C., Chichocki, A., Yang, H.H.: Multichannel blind deconvolution and equalization using the natural gradient. In: Proc. IEEE Workshop Signal Proc. Adv. Wireless Comm. Paris, France, April 1997, pp. 101–104. IEEE Computer Society Press, Los Alamitos (1997)
8. Douglas, S.C., Sawada, H., Makino, S.: Natural gradient multichannel blind deconvolution and speech separation using causal FIR filters. IEEE Trans. Speech Audio Processing 13, 92–104 (2005)
9. Douglas, S.C., Sawada, H., Makino, S.: A spatio-temporal FastICA algorithm for separating convolutive mixtures. In: Proc. IEEE Int. Conf. Acoust. Speech, Signal Processing, Philadelphia, PA, vol. 5, pp. 165-168 (March 2005)

10. Douglas, S.C., Gupta, M., Sawada, H., Makino, S.: Spatio-temporal FastICA algorithms for the blind separation of convolutive mixtures. IEEE Trans. Speech Audio Language Processing, 15(5) (July 2007)
11. Douglas, S.C., Gupta, M.: Scaled natural gradient algorithms for instantaneous and convolutive blind source separation. In: IEEE Int. Conf. Acoust. Speech, Signal Processing, Honolulu, HI (April 2007) (to appear)
12. Araki, S., Makino, S., Hinamoto, Y., Mukai, R., Nishikawa, T., Saruwatari, H.: Equivalence between frequency-domain blind source separation and frequency-domain adaptive beamforming for convolutive mixtures. EURASIP J. Applied Signal Processing 2003(11), 1157–1166 (2003)
13. Douglas, S.C., Cichocki, A.: Neural networks for blind decorrelation of signals. IEEE Trans. Signal Processing 45, 2829–2842 (1997)

Blind Vector Deconvolution: Convolutive Mixture Models in Short-Time Fourier Transform Domain

Atsuo Hiroe

Intelligent Systems Research Laboratory, Information Technologies Laboratories,
Sony Corporation, 5-1-12 Kitashinagawa, Shinagawa-ku, Tokyo, 141-0001 Japan
Atsuo.Hiroe@jp.sony.com

Abstract. For short-time Fourier Transform (STFT) domain ICA, dealing with reverberant sounds is a significant issue. It often invites a dilemma on STFT frame length: frames shorter than reverberation time (short frames) generate incomplete instantaneous mixtures, while too long frames may disturb the separation.

To improve the separation of such reverberant sounds, the authors propose a new framework which accounts for STFT with short frames. In this framework, time domain convolutive mixtures are transformed to STFT domain convolutive mixtures. For separating the mixtures, an approach of applying another STFT is presented so as to treat them as instantaneous mixtures.

The authors experimentally confirmed that this framework outperforms the conventional STFT domain ICA.

1 Introduction

To separate mixtures of speeches or sounds, Independent Component Analysis (ICA) in short-time Fourier transform (STFT) domain (or frequency domain) has often been used [1]. Compared with time-domain ICA, STFT domain ICA has the advantages of faster convergence and less computation since time domain convolutive mixtures are reduced to instantaneous mixtures in each frequency bin. Also it can generate 'permutation-free' unmixed results by using a measure of independence that is computed from the whole spectrograms [5].

For STFT domain ICA, however, another issue still remains. It is on the length of the STFT analysis frame [2]; STFT frames shorter than real reverberation (short frames) make the conversion to instantaneous mixtures incomplete, while long frames decrease the number of substantive samples since they worsen time resolution in STFT domain. Both features can disturb the separation. As their trade-offs, STFT domain ICA often reaches the peak of the separation performance although the frame is much shorter than the reverberation time.

If a model accounts for STFT domain mixing process more accurately, it can improve the performance of the short frames. Such an approach has been proposed in [3]. Their separation algorithm, however, is simplified to the two-input-two-output case, and the permutation correction is out of their framework.

M.E. Davies et al. (Eds.): ICA 2007, LNCS 4666, pp. 471–479, 2007.

In this paper we present a new framework which includes the mixing process with short frames and the separation without permutation inconsistencies.

2 STFT Domain Convolutive Mixture Models

In this section, we examine what occurs when convolutive mixtures are transformed with short frames.

Let $x_{ki}(t)$ be the contribution from $s_i(t)$, source i at time t, to x_k, observation in sensor k. It is represented as convolution between source $s_i(t)$ and the filter coefficients $a_{ki}(\tau)$:

$$x_{ki}(t) = \sum_{\tau=0}^{p-1} a_{ki}(\tau) s_i(t-\tau). \quad (p: \text{filter length}) \tag{1}$$

Then define $X_{ki}(\omega, r)$ as the STFT of $x_{ki}(t)$ in frequency bin ω and frame r:

$$X_{ki}(\omega, r) \overset{\text{def}}{=} \sum_{t=0}^{L-1} w(t) x_{ki}(rN+t) \exp\left(-2\pi j \frac{\omega-1}{L} t\right), \quad (j\text{: imaginary unit}) \tag{2}$$

where L and N are frame length (number of taps) and shift width respectively.

Now, consider the case $p > L$. According to the discussion in Appendix, $X_{ki}(\omega, r)$ is approximately represented as convolution also in STFT domain:

$$X_{ki}(\omega, r) \approx \sum_{\tau=0}^{P-1} A_{ki}(\omega, \tau) S_i(\omega, r-\tau), \tag{3}$$

where $A_{ki}(\omega, \tau)$ and $S_i(\omega, r-\tau)$ are the STFTs of $a_{ki}(\tau)$ and $s_i(t)$ respectively; P denotes number of frames (frame taps) in STFT domain convolution. Therefore $X_k(\omega, r)$, the STFT domain observations in sensor k, is approximately represented as convolutive mixtures also in STFT domain:

$$X_k(\omega, r) = \sum_{i=1}^{m} X_{ki}(\omega, r) \approx \sum_{i=1}^{m} \sum_{\tau=0}^{P-1} A_{ki}(\omega, \tau) S_i(\omega, r-\tau). \tag{4}$$

Expanding (4) over k and ω, we write them as a single formula:

$$\underbrace{\begin{bmatrix} \boldsymbol{X}_1(r) \\ \vdots \\ \boldsymbol{X}_n(r) \end{bmatrix}}_{\boldsymbol{X}(r)} \approx \sum_{\tau=0}^{P-1} \underbrace{\begin{bmatrix} \boldsymbol{A}_{11}^{[\tau]} & \cdots & \boldsymbol{A}_{1m}^{[\tau]} \\ \vdots & \ddots & \vdots \\ \boldsymbol{A}_{n1}^{[\tau]} & \cdots & \boldsymbol{A}_{nm}^{[\tau]} \end{bmatrix}}_{\boldsymbol{A}^{[\tau]}} \underbrace{\begin{bmatrix} \boldsymbol{S}_1(r-\tau) \\ \vdots \\ \boldsymbol{S}_m(r-\tau) \end{bmatrix}}_{\boldsymbol{S}(r-\tau)}, \tag{5}$$

where $\boldsymbol{X}_k(r) = [X_k(1, r), \ldots, X_k(M, r)]^T$ and $\boldsymbol{S}_k(r) = [S_k(1, r), \ldots, S_k(M, r)]^T$ are spectra of observations and sources respectively; n, m and M are number of sensors, sources and frequency bins respectively; $\boldsymbol{A}_{ki}^{[\tau]} = \text{diag}\{A_{ki}(1, \tau), \ldots, A_{ki}(M, \tau)\}$ is an $M \times M$ diagonal matrix.

Equation (5) gives us another interpretation in STFT domain: each source spectrum $\boldsymbol{S}_i(r)$, a vector of M elements, occurs independently in frame r to arrive at the sensors with at most P frames' delay. It means that $\boldsymbol{S}(r)$ affects P frames' observations $\boldsymbol{X}(r),\ldots,\boldsymbol{X}(r+P-1)$, and equivalently that $\boldsymbol{X}(r)$ is a convolution between the previous P frames' sources $\boldsymbol{S}(r),\ldots,\boldsymbol{S}(r-P+1)$ and mixing matrices $\boldsymbol{A}^{[0]},\ldots,\boldsymbol{A}^{[P-1]}$.

The conventional STFT domain ICA corresponds to the particular case $P=1$ in (5). It is satisfied only when the filter length p is relatively small.

3 Proposed Framework: Blind Vector Deconvolution

As well as the case of time domain [8], we can consider two approaches to estimate sources from STFT domain convolved observation vectors:

1. Convert observations to instantaneous mixtures through another STFT. (Modulation spectrogram domain ICA)
2. Unmix observations directly in STFT domain. (STFT domain deconvolution)

We call them 'Blind Vector Deconvolution' (BVD). In this paper, we address the first approach, Modulation spectrogram domain ICA. This is an application of our previous work [5] to the modulation spectrogram domain.

3.1 Transform to Modulation Spectrograms

Applying another STFT to spectrograms generates modulation spectrograms (MS) [9]. Use of proper length of a frame (L') and shift (N') transforms STFT domain convolutive mixtures (4) to MS domain instantaneous mixtures:

$$X'_k(\omega,\omega_2,r') \approx \sum_{k=1}^{m} A'_{ki}(\omega,\omega_2)S'_i(\omega,\omega_2,r') \tag{6}$$

$$\Leftrightarrow X'_k(\omega',r') \approx \sum_{k=1}^{m} A'_{ki}(\omega',r')S'_i(\omega',r'), \tag{7}$$

where $X'_k()$, $A'_{ki}()$ and $S'_i()$ are MS domain data transformed from $X_k(\omega,r)$, $A_{ki}(\omega)$ and $S_i(\omega,r)$ respectively; ω_2 and r' correspond to newly generated bin and frame respectively. Using a serial index ω' instead of (ω,ω_2), we can write (6) also as (7).

Fig. 1 shows the outline of the conversion to MS: (a) is a set of STFT domain observations; each bin's STFT (2nd STFT) converts them to the MS domain cube structures (b).

Using the same notation as (5), we can write the MS domain mixing process simply as:

$$\boldsymbol{X}'(r') = \boldsymbol{A}'\boldsymbol{S}'(r'). \tag{8}$$

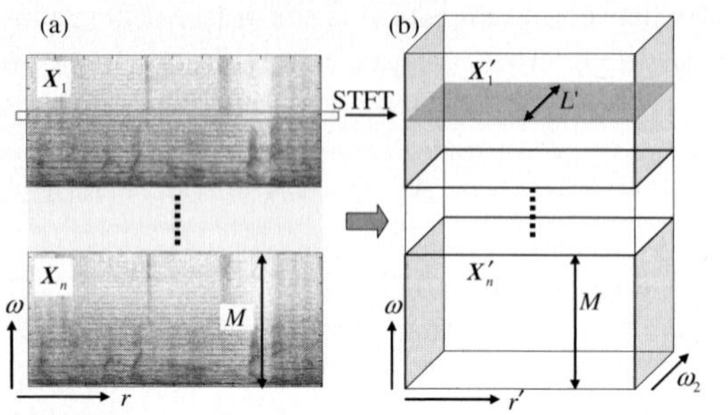

Fig. 1. (a) Spectrograms, (b) Modulation spectrograms

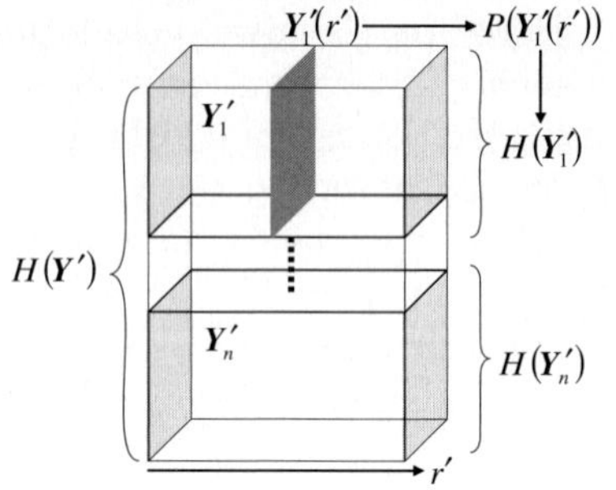

Fig. 2. KL divergence is computed as $\sum_k H(\boldsymbol{Y}'_k) - H(\boldsymbol{Y}')$

3.2 Independence Measure and Learning Rule for MS Domain ICA

Since the MS domain observations are instantaneous mixtures, techniques for the instantaneous ICA apply. In particular, by using the independence measure computed from the whole MSs [5], unmixed results without permutation inconsistencies are generated, as described below.

The unmixing process in channel k and bin ω' is written as (9), where $W'_{ki}(\omega')$ denotes an unmixing coefficient. Expanding (9) over k and ω', as in (5), the whole unmixing process is written as (10).

$$Y'_{ki}(\omega', r') = \sum_{i=1}^{n} W'_{ki}(\omega') X'_i(\omega', r') \tag{9}$$

$$\underbrace{\begin{bmatrix} \boldsymbol{Y}'_1(r') \\ \vdots \\ \boldsymbol{Y}'_n(r') \end{bmatrix}}_{\boldsymbol{Y}'(r')} = \underbrace{\begin{bmatrix} \boldsymbol{W}'_{11} & \cdots & \boldsymbol{W}'_{1n} \\ \vdots & \ddots & \vdots \\ \boldsymbol{W}'_{n1} & \cdots & \boldsymbol{W}'_{nn} \end{bmatrix}}_{\boldsymbol{W}'} \underbrace{\begin{bmatrix} \boldsymbol{X}'_1(r') \\ \vdots \\ \boldsymbol{X}'_n(r') \end{bmatrix}}_{\boldsymbol{X}'(r')} \tag{10}$$

As the measure of independence, we use Kullback-Leibler divergence (KLD) computed from the whole MSs $\boldsymbol{Y}'$. The KLD of $\boldsymbol{Y}'$ is equivalent to the difference between summation of each channel's entropy $H(\boldsymbol{Y}'_k)$ and the joint entropy $H(\boldsymbol{Y}')$:

$$\mathrm{KLD}(\boldsymbol{Y}') = \sum_{k=1}^{n} H(\boldsymbol{Y}'_k) - H(\boldsymbol{Y}') \tag{11}$$

$$= \sum_{k=1}^{n} E_{r'} \left[-\log P\left(\boldsymbol{Y}'_k(r')\right) \right] - \log \left| \det \boldsymbol{W}' \right| - H(\boldsymbol{X}'), \tag{12}$$

where $P\left(\boldsymbol{Y}'_k(r')\right)$ denotes a multivariate probability density function of $\boldsymbol{Y}'_k(r')$, and $E_{r'}[]$ means expectation over r' (Fig. 2).

Applying the natural gradient rule [4] to (12), we obtain a set of learning rules to seek the unmixing matrix $\boldsymbol{W}'$ that makes $\boldsymbol{Y}'_1,\ldots,\boldsymbol{Y}'_n$ mutually independent in the MS domain:

$$\boldsymbol{W}'(\omega') \leftarrow \boldsymbol{W}'(\omega') + \eta \Delta \boldsymbol{W}'(\omega') \quad (\eta : \text{ learning rate}) \tag{13}$$

$$\Delta \boldsymbol{W}'(\omega') = E_r \left[\boldsymbol{I} + \begin{bmatrix} \varphi_{\omega'}\left(\boldsymbol{Y}'_1(r')\right) \\ \vdots \\ \varphi_{\omega'}\left(\boldsymbol{Y}'_n(r')\right) \end{bmatrix} \begin{bmatrix} Y'_1(\omega', r') \\ \vdots \\ Y'_n(\omega', r') \end{bmatrix}^H \right] \boldsymbol{W}'(\omega'), \tag{14}$$

$$\text{where } \varphi_{\omega'}\left(\boldsymbol{Y}'_k(r')\right) = \frac{\partial \log P\left(\boldsymbol{Y}'_k(r')\right)}{\partial Y'_k(\omega', r')}, \; \boldsymbol{W}'(\omega') = \begin{bmatrix} W'_{11}(\omega') & \cdots & W'_{1n}(\omega') \\ \vdots & \ddots & \vdots \\ W'_{n1}(\omega') & \cdots & W'_{nn}(\omega') \end{bmatrix} \tag{15}$$

3.3 Multivariate Probability Density Functions

To compute $H(\boldsymbol{Y}'_k)$, entropy of each channel's MS, we employ $P(\boldsymbol{Y}'_k(r')) \propto \exp(-K||\boldsymbol{Y}'_k(r')||_2)$, as proposed in [5]. We then obtain an activation function:

$$\varphi_{\omega'}\left(\boldsymbol{Y}'_k(r')\right) = -K \frac{Y'_k(\omega', r')}{||\boldsymbol{Y}'_k(r')||_2}. \quad \left(||\boldsymbol{Y}'_k(r')||_2 = \left\{ \sum_{\omega'} |Y'_k(\omega', r')|^2 \right\}^{1/2} \right) \tag{16}$$

3.4 Postprocesses

After the learning, the minimal distortion principle based rescaling [6] is applied, such as $\boldsymbol{W}' \leftarrow \text{diag}(\boldsymbol{W}'^{-1})\boldsymbol{W}'$. Then through the overlap-add based invert STFT [10], the MS domain unmixed results $\boldsymbol{Y}'$ are converted to STFT domain data $\boldsymbol{Y}$. To convert $\boldsymbol{Y}$ to the time domain signals, another invert STFT is applied.

4 Experimental Results

To confirm the separation performance, we performed some experiments using a set of data recorded separately and mixed on a computer. Recording was done in our office room (0.25~0.3s reverberation time) by playing each source from different loudspeakers (Fig. 3). Four or eight seconds of recorded data were used in the experiments. The sampling rate (F_s) was 16k[Hz].

As observations, we generated eight different mixtures as in Fig. 4, where S, F and M denote street noise [7], female speech and male speech, respectively.

To unmix these mixtures, we used the following two methods with STFT parameters shown in Table 1:

Benchmark (BM): STFT domain ICA described in [5].
Proposed: Modulation spectrogram domain ICA described in Section 3.

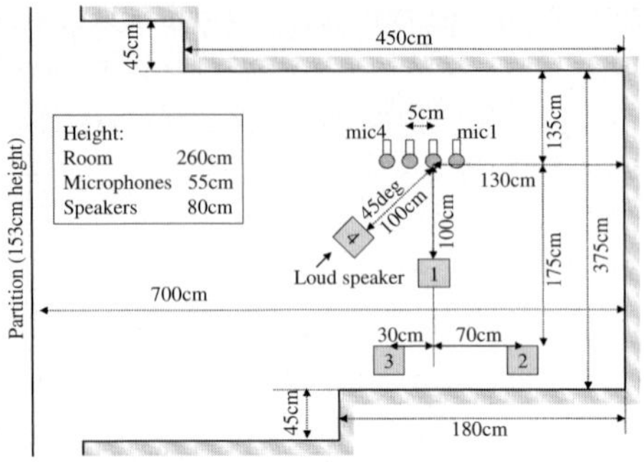

Fig. 3. Recording environment

Test No.	Loudspeaker			
	1	2	3	4
1	S		F	M
2	S		M	F
3	F	S		M
4	M	S		M
5			F	M
6			M	F
7	F			M
8	M			M

Fig. 4. Sources in each test (S: street noise, F/M: female/male speech)

Table 1. Experimental parameters (L, L': Frame length; N, N': Shift width)

	1st STFT		2nd STFT	
	L	N	L'	N'
BM	256 to 8192 (hanning)	$L/4$	—	—
Proposed	512 or 1024 (hanning)	$L/4$	4 to 64 (hamming)	$\lceil L'/8 \rceil$
Common: η=0.3, 400 iteration, K= square root of number of bins				

In the second STFT, we used a hamming window to utilize the both ends of a frame effectively. (In case of $L' = 4$, we used $N' = 1$.)

To evaluate the separation performance in STFT domain, we used Signal-to-Interference Ratio (SIR) in STFT domain instead of time domain. It is computed as follows:

1. Approximate each bin's unmixed result $Y_k(\omega, r)$ as a linear sum of all sources $S_1(\omega, r), \ldots, S_n(\omega, r)$, namely $Y_k(\omega, r) \approx \sum_{i=1}^{m} \alpha_{ki}(\omega) S_i(\omega, r)$.
2. Calculate $\mathrm{SIR}(Y_k, S_i)$, i.e. each output channel's SIR, as $E_\omega[10 \log_{10} E_r\{|\alpha_{ki}(\omega) S_i(\omega, r)|^2\} / E_r\{|\sum_{l \neq i} \alpha_{kl}(\omega) S_l(\omega, r)|^2\}]$.
3. Define SIR_Y, i.e. SIR of the unmixed results, as $E_i\left[\max_k \mathrm{SIR}(Y_k, S_i)\right]$.
4. Define $\mathrm{SIR}_{\mathrm{imp}}$, i.e. improved SIR, as $E_i\left[\max_k \left[\mathrm{SIR}(Y_k, S_i) - \mathrm{SIR}(X_k, S_i)\right]\right]$.

To compare various STFT parameters, we serialized them along with 'time-span', length of observations in second used to estimate single frame's $\boldsymbol{Y}(r)$. It is calculated as:

$$\text{Timespan} = \{L + (L' - 1)N\} / F_s \qquad (\text{or } L/F_s \text{ for the benchmark}) \qquad (17)$$

The average $\mathrm{SIR}_{\mathrm{imp}}$ over eight tests are plotted in Fig. 5. In these plots, dashed vertical lines, dash-dot lines and solid lines show the reverberation time (assuming 0.27[s]), the benchmark and proposed methods respectively. Table 2 shows the best SIRs and corresponding parameters in each method.

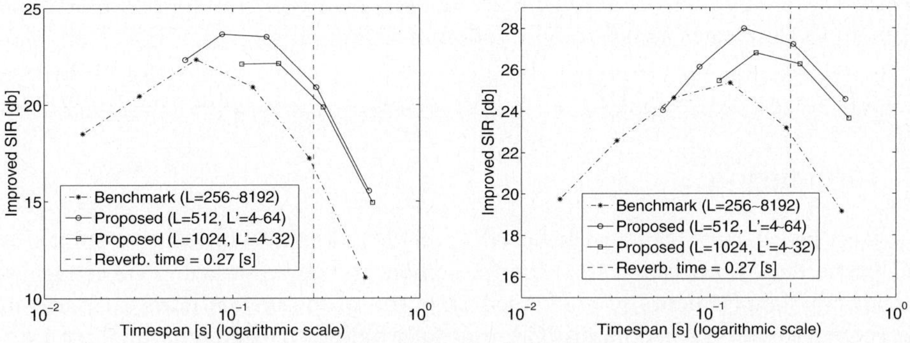

Fig. 5. Plots of SIRs (Left: 4 seconds, Right: 8 seconds)

Table 2. Best SIR in each method

	Length [s]	SIR_Y [dB]	SIR_{imp} [dB]	1st STFT		2nd STFT		Time span [s]
				L	N	L'	N'	
BM	4.0	20.36	22.33	1024	256	—	—	0.064
Proposed	4.0	21.65	23.66	512	128	16	2	0.088
BM	8.0	23.31	25.35	2048	512	—	—	0.128
Proposed	8.0	25.94	27.99	512	128	16	2	0.152

From the above experiments, we have confirmed the followings:

1. For the benchmark, the best SIR is made with $L = 1024$ or $L = 2048$, much shorter than the reverberation time. Longer frames rather worsen SIR, as mentioned in Section 1.
2. For the proposed methods, the best SIR is achieved in longer time span.

5 Discussion on Experimental Results

In the framework of Blind Vector Deconvolution, single frame's unmixed results $\boldsymbol{Y}(r)$ is estimated from observations over L' consecutive frames around frame r, while in the conventional STFT domain ICA, it is estimated from single frame's observations $\boldsymbol{X}(r)$. This property can remove cross-frame interference which are due to using short frames. Moreover, the fact that adjacent frames in STFT domain are overlapped to each other in time domain, prevents the decrease of number of substantive samples in the second STFT. These are reasons why the proposed methods outperform the benchmark.

In longer time-span, however, the proposed methods have also shown the decline in SIR, which is more apparent in the 4 second case. We guess that it is due to the following reasons: 1) In the second STFT, the dilemma of frame length still remains; 2) Longer frames in the second STFT add more cross-frame artifacts to the target signal in STFT domain.

We expect that we can avoid the dilemma by carefully selecting parameters in the first and second STFT, or by introducing STFT domain deconvolution.

6 Conclusion

We proposed a new framework based on STFT domain convolutive mixtures and presented an approach that unmixes them in the modulation spectrogram domain. We experimentally confirmed that the proposed methods outperform the conventional STFT domain ICA due to avoiding the dilemma of the STFT frame length although it still remains in long time-span.

In future works, we plan to evaluate our methods in various environments and to examine STFT domain deconvolution.

References

1. Smaragdis, P.: Blind separation of convolved mixtures in the frequency domain. Neurocomputating 10(2), 251–276 (1998)
2. Araki, S., Makino, S., Mukai, R., Nishikawa, T., Saruwatari, H.: Fundamental limitation of frequency domain blind source separation for convolved mixture of speech. In: Proc. ICA2001, pp. 132–137 (December 2001)
3. Servière, C.: Separation of speech signals under reverberant conditions. In: Proc. EUSIPCO04, pp. 1693–1696 (2004)
4. Amari, S., Cichocki, A., Yang, H.H.: A new learning algorithm for blind signal separation. In: Advances in Neural Information Processing Systems, vol. 8, MIT Press, Cambridge (1996)
5. Hiroe, A.: Solution of Permutation Problem in Frequency Domain ICA, Using Multivariate Probability Density Functions. In: Rosca, J., Erdogmus, D., Príncipe, J.C., Haykin, S. (eds.) ICA 2006. LNCS, vol. 3889, pp. 601–608. Springer, Heidelberg (2006)
6. Matsuoka, K., Nakashima, S.: Minimal distortion principle for blind source separation. In: Proc. ICA 2001, pp.722–727 (December 2001)
7. `http://sound.media.mit.edu/ica-bench/sources/`
8. Hyvarinen, A., Karhunen, J., Oja, E.: Independent Component Analysis (2001)
9. Greenberg, S., Kingsbury, B.E.D.: The Modulation Spectrogram. In: Pursuit Of An Invariant Representation Of Speech. In: Proc. ICASSP-97, pp. 1647–1650 (1997)
10. Rabiner, L., Schafer, R.: Short-Time Fourier Analysis. In: Chapter in Digital Processing of Speech Signals, Prentice-Hall, London (1978)
11. Kim, T., Eltoft, T., Lee, T.-W.: Independent Vector Analysis: an extension of ICA to multivariate components. In: Rosca, J., Erdogmus, D., Príncipe, J.C., Haykin, S. (eds.) ICA 2006. LNCS, vol. 3889, pp. 165–172. Springer, Heidelberg (2006)

Appendix: STFT of Convolved Signals

First we examine STFT with single shift ($N = 1$) [10]. Let $\tilde{S}_i(\omega, R)$ be the STFT of $s_i(t)$ in frame R (also in time R). It is defined in (18). From (1) and (18),

$\tilde{X}_{ki}(\omega, R)$, the STFT of $x_{ki}(t)$, is written as (19). It represents the convolution between $\tilde{S}_i(\omega, R-\tau)$ and $a_{ki}(\tau)$.

$$\tilde{S}_i(\omega, R) \stackrel{\text{def}}{=} \sum_{t=0}^{L-1} w(t) s_i(R+t) \exp\left(-2\pi j \frac{\omega-1}{L} t\right). \tag{18}$$

$$\tilde{X}_{ki}(\omega, R) = \sum_{\tau=0}^{p-1} a_{ki}(\tau) \tilde{S}_i(\omega, R-\tau) \tag{19}$$

Then we apply (19) to STFT with general shift width N. Using integers P, N and L such that $(P-1)N < p \leq (P-1)N + L$, (19) is rewritten as an overlap-add form (20). It can be approximated as in (21), since τ' close to 0 should satisfy (22). Equation (21) shows the convolution between $\tilde{S}_i(\omega, R-\tau N)$ and $\tilde{A}_{ki}(\omega, \tau N)$. It means that as long as the approximation in (22) is sound, STFT with shift N converts time domain convolution to STFT domain convolution.

$$(19) = \sum_{\tau=0}^{P-1} \frac{N}{L} \sum_{\tau'=0}^{L-1} a_{ki}(\tau N + \tau') \tilde{S}_i(\omega, R - \tau N - \tau') \tag{20}$$

$$\approx \sum_{\tau=0}^{P-1} \underbrace{\left\{ \frac{N}{L} \sum_{\tau'=0}^{L-1} a_{ki}(\tau N + \tau') \exp\left(-2\pi j \frac{\omega-1}{L} \tau'\right) \right\}}_{\tilde{A}_{ki}(\omega, \tau N)} \tilde{S}_i(\omega, R - \tau N) \tag{21}$$

$$\tilde{S}_i(\omega, R - \tau N - \tau') \approx \exp\left(-2\pi j \frac{\omega-1}{L} \tau'\right) \tilde{S}_i(\omega, R - \tau N) \quad (0 \leq \tau' < L) \tag{22}$$

Finally, we rewrite (21) as (3) through following replacements:

$$X_{ki}(\omega, r) = \tilde{X}_{ki}(\omega, rN), \quad A_{ki}(\omega, \tau) = \tilde{A}_{ki}(\omega, \tau N), \quad S_i(\omega, r-\tau) = \tilde{S}_i(\omega, R-\tau N)$$

A Batch Algorithm for Blind Source Separation of Acoustic Signals Using ICA and Time-Frequency Masking

Eugen Hoffmann, Dorothea Kolossa, and Reinhold Orglmeister

Berlin University of Technology, Electronics and Medical Signalporcessing Group,
Einsteinufer 17, 10587 Berlin, Germany
{Eugen.Hoffmann.1, Reinhold.Orglmeister}@tu-berlin.de,
d.kolossa@ee.tu-berlin.de

Abstract. The problem of *Blind Source Separation* (BSS) of convolved acoustic signals is of great interest for many classes of applications such as in-car speech recognition, hands-free telephony or hearing devices. Due to the convolutive mixing process, the source separation is performed in the frequency domain, using *Independent Component Analysis* (ICA). However the quality of solution of the ICA-algorithms can be improved by applying *time-frequency masking*. In this paper we present a batch-algorithm for time-frequency masking using the time-frequency structure of separated signals.

1 Introduction

Blind Source Separation (BSS) deals with the problem of recovering the source signals from their mixtures when the mixing process is unknown. Recently, the problem has been widely studied and many methods have been proposed [7].

In this paper we concentrate on the case of BSS for acoustic speech signals observed in a real environment, i.e. convolutive mixtures. Most existing demixing methods are based on Independent Component Analysis (ICA) in the frequency-domain, where the convolutions of the source signals with the room impulse response are reduced to multiplications with the corresponding transfer functions. So for each frequency bin, an individual instantaneous ICA problem arises [2],[7].

The quality of the recovered source signals can be improved by applying time-frequency masking on the ICA outputs. There exist a number of algorithms, calculating the time-frequency mask from the estimated direction of arrival of the separated signals [4],[5], algorithms calculating the power ratio between inputs and outputs of a spatial filter [3] and algorithms based on the cosine distance between a sample vector and the basis vector corresponding to the target [6]. The proposed algorithm estimates the time-frequency mask by comparing the time-frequency structure of the separated signals.

On this basis, we propose a batch-algorithm for separation of two sources by applying the JADE-Algorithm [1] for computation of unmixing filters and

M.E. Davies et al. (Eds.): ICA 2007, LNCS 4666, pp. 480–487, 2007.

time-frequency masking to improve the separation quality. The effectiveness of this approach is shown by the separation results of the algorithm.

2 The Proposed Method

The block diagram of the algorithm is shown in Fig. 1.

First the time-domain signals $\mathbf{x}(t)$ are converted into frequency-domain time-series signals $\mathbf{X}(\Omega, \tau)$ using the *Short-Time Fourier Transform* (STFT), on which the JADE-Algorithm is applied, to compute the unmixing filter matrices $\mathbf{W}(\Omega)$. At the next step the permutation problem is treated, so the unmixing filter matrices can be corrected by multiplication with the permutation matrix $\mathbf{P}(\Omega)$. At the post-processing stage of the algorithm, the time-frequency mask $\mathbf{M}(\Omega)$ is estimated, to minimize the crosstalk components, which could not be eliminated by the ICA-algorithm, and the output vector $\mathbf{y}(t)$ is obtained by transforming the unmixed signals $\mathbf{Y}(\Omega, \tau)$ back into the time-domain.

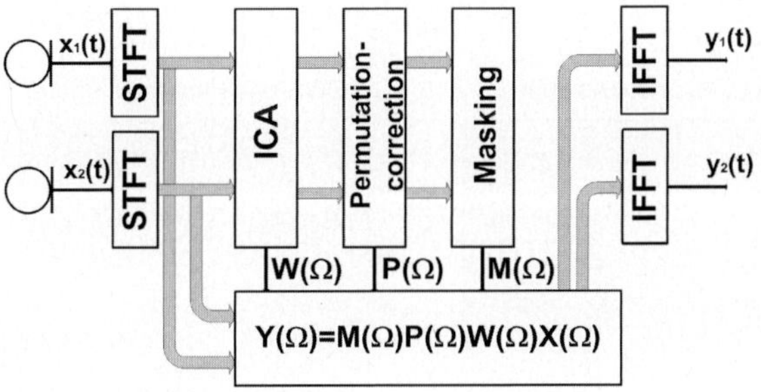

Fig. 1. Overview of the algorithm

2.1 ICA

Acoustic signal mixtures in reverberant environments can be described by

$$\mathbf{x}(t) = \mathbf{A} * \mathbf{s}(t), \tag{1}$$

where $\mathbf{s}(t)$, $\mathbf{x}(t)$ and $\mathbf{A}$ denote the the vector of source signals, the vector of mixed signals and a matrix containing the impulse responses between the sources and the sensors and $*$ denotes the convolution operator. Transforming (1) into the frequency domain reduces the convolutions to multiplications:

$$\mathbf{X}(\Omega, \tau) \approx \mathbf{A}(\Omega)\mathbf{S}(\Omega, \tau), \tag{2}$$

where Ω is the angular frequency, τ represents the frame index, $\mathbf{A}(\Omega)$ is the mixing system in the frequency domain, $\mathbf{S}(\Omega, \tau) = [S_1(\Omega, \tau), \ldots, S_N(\Omega, \tau)]$ represents the source signal, and $\mathbf{X}(\Omega, \tau) = [X_1(\Omega, \tau), \ldots, X_N(\Omega, \tau)]$ denotes the observed signals. So for each frequency bin an instantaneous ICA-Problem has to be solved. For this purpose we use the JADE-algorithm, which is based on joint diagonalization of most significant cumulant matrices of higher order [1].

2.2 Permutation Correction

The filter matrices calculated by ICA can be randomly permutated. To solve the permutation problem, the phase differences in the estimated mixing filter matrices are used [2],[5]. For this purpose we normalize the estimated mixing matrix $\mathbf{A}(\Omega)$ on the first row, so the normalized mixing matrix can be written as:

$$\hat{\mathbf{A}}(\Omega) = \begin{bmatrix} 1 & \cdots & 1 \\ \hat{a}_{21}e^{-j\Omega\delta_{21}} & \cdots & \hat{a}_{2n}e^{-j\Omega\delta_{2n}} \\ \cdots & \cdots & \cdots \end{bmatrix} \tag{3}$$

To correct the permutations, we compare the phase differencies δ_i and sort the columns of the mixing filter matrix, so $\delta_i < \delta_k$ for all $i < k$. Figure 2 shows the results of the permutation correction.

Remark 1. It should be noted, that this approach alone works only in case of signals coming from different directions.

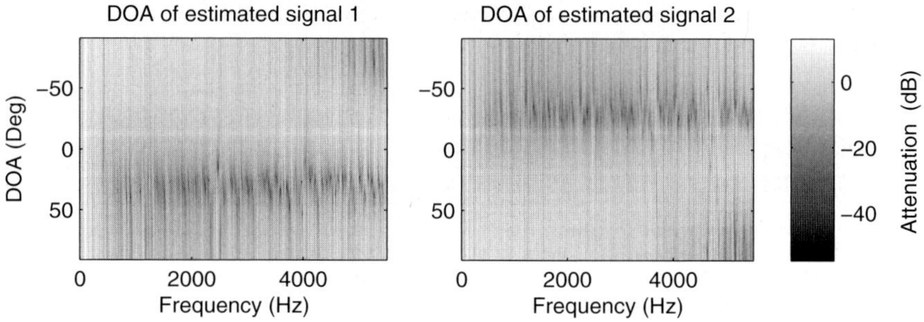

Fig. 2. DOA estimation for two signals using ICA. The gray scale values show the grade of attenuation.

2.3 Time-Frequency Masking

Despite of good performance of the ICA-algorithm some interference remains in the estimated source signals. To minimize the remaining interference, we use time-frequency masking, which is performed as

$$\tilde{\mathbf{Y}}(\Omega,\tau) = \mathbf{M}(\Omega,\tau)\mathbf{Y}(\Omega,\tau), \tag{4}$$

where $0 \leq \mathbf{M}(\Omega,\tau) \leq 1$ is a mask specified for each time index τ in each frequency bin. To calculate the mask, we estimate the interferences

$$\mathbf{U}_k(\Omega,\tau,R_\Omega,R_\tau) = \frac{\left\| \Phi(\Omega,\tau,R_\Omega,R_\tau)(Y_k(\Omega,\tau) - \sum_{m\neq k} Y_m(\Omega,\tau)) \right\|}{\left\| \Phi(\Omega,\tau,R_\Omega,R_\tau)\sum_{m\neq k} Y_m(\Omega,\tau) \right\|} \tag{5}$$

and

$$\mathbf{N}_k(\Omega,\tau,R_\Omega,R_\tau) = \frac{\left\|\Phi(\Omega,\tau,R_\Omega,R_\tau)(Y_k(\Omega,\tau) - \sum_{m\neq k} Y_m(\Omega,\tau))\right\|}{\|\Phi(\Omega,\tau,R_\Omega,R_\tau)Y_k(\Omega,\tau)\|}. \quad (6)$$

between the spectrogram $Y_k(\Omega,\tau)$ and $\sum_{m\neq k} Y_m(\Omega,\tau)$). $\|\cdot\|$ denotes the Euclidean norm operator and

$$\Phi(\Omega,\tau,R_\Omega,R_\tau) = \begin{cases} \mathcal{W}(\Omega-\Omega_0,\tau-\tau_0,R_\Omega,R_\tau), & |\Omega-\Omega_0| \leq R_\Omega, \\ & |\tau-\tau_0| \leq R_\tau \\ 0, & \text{otherwise} \end{cases} \quad (7)$$

uses a two dimensional window function $\mathcal{W}(\Omega-\Omega_0,\tau-\tau_0,R_\Omega,R_\tau)$ of the size $R_\Omega \times R_\tau$ (e.g. a two dimensional Hanning window).

The estimate of the interference gives us different possibilities for mask calculation:

- The interference can be used to estimate the time frequency bins of the signal k, where the signal of interest is active:

$$\mathbf{M}_k(\Omega,\tau) = \begin{cases} 1, & \mathbf{U}_k(\Omega,\tau,R_\Omega,R_\tau) > \lambda_u \\ 0, & \text{otherwise} \end{cases} \quad (8)$$

  where λ_u is the threshold for the signal of interest and $\mathbf{U}_k(\Omega,\tau,R_\Omega,R_\tau)$ is the interference from (5).
- The interference can be used to estimate the time frequency bins of the signal k, where the jammer signal is active, or

$$\mathbf{M}_k(\Omega,\tau) = \begin{cases} 0, & \mathbf{N}_k(\Omega,\tau,R_\Omega,R_\tau) > \lambda_n \\ 1, & \text{otherwise} \end{cases} \quad (9)$$

  where λ_n is the threshold for the jammer signal.
- It is possible to combine both methods, using both estimated interferences $\mathbf{S}_k(\Omega,\tau,R_\Omega,R_\tau)$ and $\mathbf{N}_k(\Omega,\tau,R_\Omega,R_\tau)$:

$$\mathbf{M}_k(\Omega,\tau) = \begin{cases} 1, & \mathbf{N}_k(\Omega,\tau,R_\Omega,R_\tau) < \lambda_n \wedge \mathbf{U}_k(\Omega,\tau,R_\Omega,R_\tau) > \lambda_u \\ 0, & \text{otherwise} \end{cases} \quad (10)$$

Figure 3 shows the results of the interference estimation and masking with (10).

3 Experiments and Results

3.1 Conditions

For the evaluation of the results of the proposed algorithm, the TIDigits [8] database with two different male speakers was used, which was played back and recorded once with two speaker signals simultaneousely and once separately in three different setups of loudspeakers. The recordings were made in a real room with a reverberation time $T_R = 300$ ms (Fig. 4). The loudspeakers were placed with the angles of incidence relative to broadside as shown in the Table 1. The algorithm was tested at a resolution of $NFFT = 512$, the recordings sample rate was 11kHz.

Table 1. Experimental configurations

config	θ_1	θ_2	recordings
A	45°	45°	speaker 1, speaker 2, both speakers
B	45°	25°	speaker 1, speaker 2, both speakers
C	10°	25°	speaker 1, speaker 2, both speakers

3.2 Performance Measures

For calculation of the effectiveness of the proposed algorithm the signal to interference ratio (SIR) was used as a measure of the separation performance and the signal to distortion ratio (SDR) as a measure of the signal quality:

$$\mathrm{SIR}_i = 10 \log \frac{\sum_n y_{is_i}^2(n)}{\sum_{j \neq i} \sum_n y_{is_j}^2(n)} \tag{11}$$

$$\mathrm{SDR}_i = 10 \log \frac{\sum_n x_{ks_i}^2(n)}{\sum_n (x_{ks_i}(n) - \alpha y_{is_i}^2(n - D))^2} \tag{12}$$

where y_{i,s_j} is the i-th separated signal with only the s_j source active, and x_{k,s_j} is the observation obtained by microphone k when only s_j is active. α and D are parameters for phase and amplitude chosen to optimally compensate the difference between y_{i,s_j} and x_{k,s_j}.

3.3 Experimental Results

In this section we compare the results of ICA with and without the proposed time frequency masking algorithm and show the results of time frequency masking for different parameter values and different masks as described in Sect. 2.3.

Table 2 shows the average SIR improvement resulting form different algorithms applied to the configurations from Table 1.

Table 2. Result comparison. Average SIR improvement in dB.

Algorithm	Config. A	Config. B	Config. C
ICA only	11.9 dB	23.1 dB	12.7 dB
ICA and TF-Mask form [3]	31.9 dB	35.4 dB	16.1 dB
ICA and TF-Mask form [6]	30.8 dB	34.6 dB	**18.1 dB**
DUET [5]	31.4 dB	23.6 dB	17.7 dB
ICA and proposed algorithm	**35.3 dB**	**39.4 dB**	**18.1 dB**

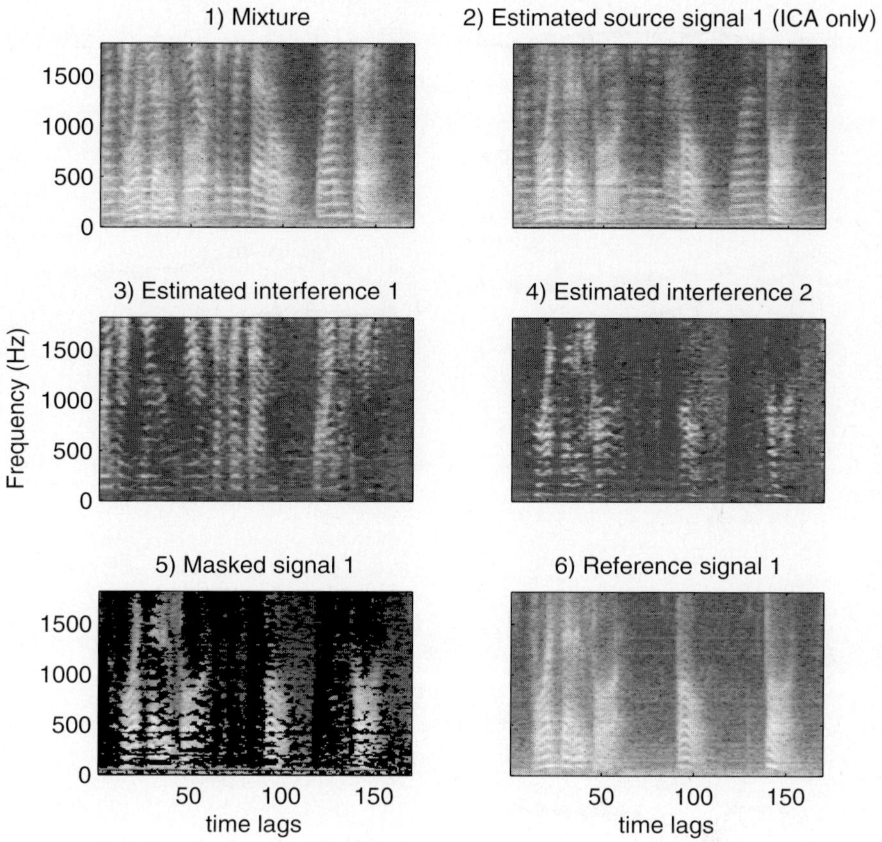

Fig. 3. The lower 200 frequency bins of spectrograms of 1) the mixture $X_1(\Omega, \tau)$, 2) the output signal $Y_1(\Omega, \tau)$ obtained only with ICA, 3) the estimated interference $\mathbf{N}_1(\Omega, \tau, R_\Omega, R_\tau)$, 4) the estimated interference $\mathbf{U}_1(\Omega, \tau, R_\Omega, R_\tau)$, 5) the output signal $Y_1(\Omega, \tau)$ obtained by a combination of ICA and T-F masking calculated with (10), and 6) the clean source signal $S_1(\Omega, \tau)$

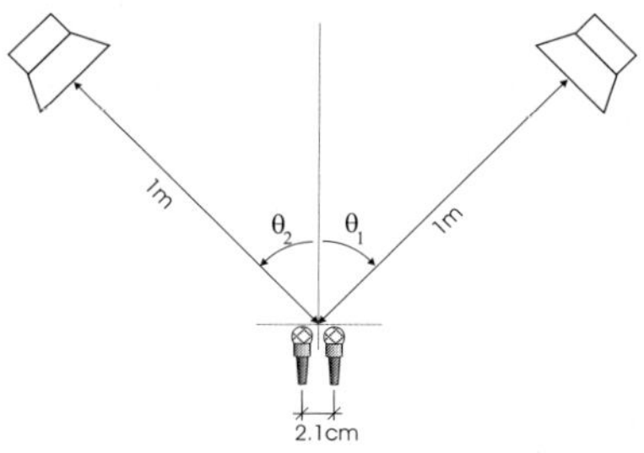

Fig. 4. Experimental setup

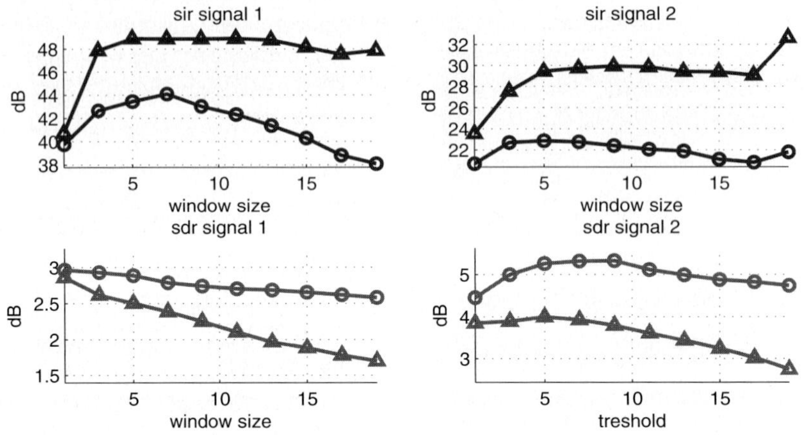

Fig. 5. Comparison of the results of time frequency masking for different window lengths. $\Delta : \lambda_u = \lambda_n = 1.3$, $\circ : \lambda_u = \lambda_n = 2.5$.

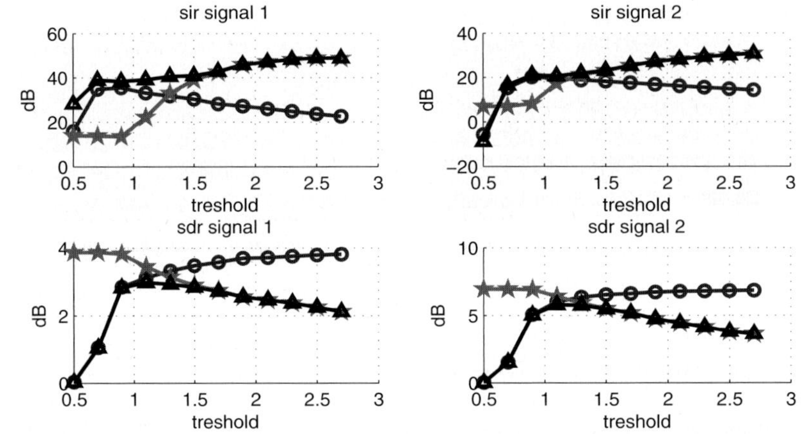

Fig. 6. Comparison of the results of time frequency masking for different window lengths. $*$: the mask calculated with equation (8), $\circ$: the mask calculated with equation (9), Δ: the mask calculated with equation (10) and : $\lambda_u = \lambda_n$.

Figure 5 shows the SIR and SDR for masking over window size. In both cases the mask was calculated with (10) for configuration A in Table 1. Figure 6 shows the SIR and SDR for the masking with different masks and different thresholds.

4 Conclusion

An approach for time frequency masking has been presented, that uses the estimate of the signal interference for mask calculation. The interference is estimated using the normalized correlation of the separated signals.

The proposed algorithm has been tested on real room recordings with a reverberation time of 300 ms, where an SIR-improvement of up to 16dB has been obtained, which was 19dB above ICA performance for the same dataset. The results show, that an significant improvement of separation performance is possible by evaluating residual correlations between separated signals.

References

1. Cardoso, J.-F.: High order contrasts for independent component analysis. Neural computation 11, 157–192 (1999)
2. Baumann, W., Kolossa, D., Orglmeister, R.: Maximum Likelihood Permutation Correction for Convolutive Source Separation. In: ICA 2003, pp. 373–378 (2003)
3. Kolossa, D., Orglmeister, R.: Nonlinear postprocessing for blind speech separation. In: Puntonet, C.G., Prieto, A.G. (eds.) ICA 2004. LNCS, vol. 3195, pp. 832–839. Springer, Heidelberg (2004)
4. Rickard, S., Balan, R., Rosca, J.: Real-time time-frequency based blind source separation. In: Proc. ICA, pp. 651–656 (December 2001)
5. Yilmaz, O., Rickard, S.: Blind separation of speech mixtures via time-frequency masking. IEEE Trans. Signal Process 52(7), 1830–1847 (2004)
6. Sawada, H., Araki, S., Mukai, R., Makino, S.: Blind extraction of a dominant source from mixtures of many sources using ICA and time-frequency masking. In: IEEE International Symposium on Circuits and Systems (ISCAS 2005), pp. 5882–5885 (May 2005)
7. Mansour, A., Kawamoto, M.: ICA Papers Classified According to their Applications and Performances. IEICA Trans. Fundamentals E86-A(3), 620–633 (2003)
8. Leonard, R.G.: A Database for Speaker-Independent Digit Recognition. In: Proc. ICASSP 84, vol. 3, pp. 11–42 (1984)

The Role of High Frequencies in Convolutive Blind Source Separation of Speech Signals

Maria G. Jafari and Mark D. Plumbley*

Centre for Digital Music,
Queen Mary University of London, UK
maria.jafari@elec.qmul.ac.uk
http://www.elec.qmul.ac.uk

Abstract. In this paper, we investigate the importance of the high frequencies in the problem of convolutive blind source separation (BSS) of speech signals. In particular, we focus on frequency domain blind source separation (FD-BSS), and show that when separation is performed in the low frequency bins only, the recovered signals are similar in quality to those extracted when all frequencies are taken into account. The methods are compared through informal listening tests, as well as using an objective measure.

1 Introduction

Convolutive blind source separation is often addressed in the frequency domain, through the short-time fourier transform (STFT), and source separation is performed separately at each frequency bin, thus reducing the problem to that of several instantaneous BSS problems. Although the approximation of convolutions by multiplications result in reduced computational complexity, frequency domain BSS (FD-BSS) remains computationally expensive because source separation has to be carried out on a large number of bins (a typical STFT length is 2048 point), each containing sufficient data samples for the independence assumption to hold. In addition, transforming the problem to several independent instantaneous problems, has the unwelcome side effect of introducing the problem of frequency permutations, whose solution is often quite computationally expensive [1], as it involves the clustering the frequency components of the recovered sources, using methods such as beamforming approaches, e.g. [3,4]. These methods exploit phase information contained in the de-mixing filters identified by the source separation algorithm.

Generally, the characteristics of speech signals are such that little information is contained in the frequencies above 4kHz [9], suggesting a possible approach to BSS for speech mixtures that focuses on the lower frequencies. Motivated by this, and in order to reduce the computational load of FD-BSS algorithms, we consider here the role of high frequencies in source separation of speech signals. We show that high frequencies are not as important as low frequencies,

* This work was funded by EPSRC grant GR/S85900/01.

M.E. Davies et al. (Eds.): ICA 2007, LNCS 4666, pp. 488–494, 2007.

and that intelligibility is preserved even when the high frequency subbands are left umixed, and simply added back onto the separated signal. Other possible approaches would exploit existing methods that assume that high frequencies are not available, such as bandwidth extension. The structure of this paper is as follows: the basic convolutive BSS problem is described in section 2; an overview of FD-ICA is given in section 3, while the role of high frequencies is discussed in section 4. Simulation results are presented in section 5, and conclusions are drawn in section 6.

2 Problem Formulation

The simplest convolutive BSS problem arises when 2 microphones record mixtures $\mathbf{x}(n)$ of 2 sampled real-valued signals, $\mathbf{s}(n)$, which in this paper are considered to be speech signals. The aim of blind source separation is then to recover the sources, from only the 2 convolutive mixtures available. Formally, the signal recorded at the q-th microphone, $x_q(n)$, is

$$x_q(n) = \sum_{p=1}^{2} \sum_{l=1}^{L} a_{qp}(l) s_p(n-l), \quad q = 1, 2 \tag{1}$$

where $s_p(n)$ is the p-th source signal, $a_{qp}(l)$ denotes the impulse response from source p to sensor q, and L is the maximum length of all impulse responses [1]. The source signals are then reconstructed according to

$$y_p(n) = \sum_{q=1}^{2} \sum_{l=1}^{L} w_{qp}(l) x_q(n-l), \quad p = 1, 2 \tag{2}$$

where $y_p(n)$ is the p-th recovered source, and $w_{qp}(l)$, are the unmixing filters which must be estimated.

3 Frequency Domain Blind Source Separation

The convolutive audio source separation is often addressed in the frequency domain. It entails the evaluation of the N-point short-time fourier transform of the observed signals, followed by the use of instantaneous BSS, independently on each of the resulting N subbands. Thus, the mixing and separating models in (1) and (2) become, respectively

$$\mathbf{X}(f,t) = \mathbf{A}(f)\mathbf{S}(f,t) \tag{3}$$

$$\mathbf{Y}(f,t) = \mathbf{W}(f)\mathbf{X}(f,t) \tag{4}$$

where $\mathbf{S}(f,t)$, and $\mathbf{X}(f,t)$ are the STFT representations of the source and mixture vectors respectively, $\mathbf{A}(f)$ and $\mathbf{W}(f)$ are the mixing and separating matrices at frequency bin f, $\mathbf{Y}(f,t)$ is the frequency domain representation of the recovered sources, and t denotes the STFT block index.

FD-BSS has the drawback of introducing the problem of frequency permutations, which is typically solved by clustering the frequency components of the recovered sources, often using beamforming techniques, such as in [1,3,4,5], where the direction of arrival (DOA) of the sources are evaluated from the beamformer directivity patterns

$$F_p(f,\theta) = \sum_{q=1}^{2} W_{qp}^{\text{ICA}}(f) e^{j2\pi f d \sin\theta_p / c}, \quad p = 1,2 \tag{5}$$

where W_{qp}^{ICA} is the ICA de-mixing filter from the q-th sensor to the p-th output, d is the spacing between two sensors, θ_p is the angle of arrival of the p-th source signal, and $c \approx 340$m/s is the speed of sound in air. The frequency permutations are then determined by ensuring that the directivity pattern for each beamformer is approximately aligned along the frequency axis.

The BSS algorithm considered in this paper is given in [6]. It updates the unmixing filters according to

$$\begin{aligned} \varDelta\mathbf{W}(f) &= D\left[diag(-\alpha_i) + E\left\{\phi(\mathbf{y}(f,t))\mathbf{y}^H(f,t)\right\}\right]\mathbf{W}(f) \\ \mathbf{W}(f) &\leftarrow \mathbf{W}(f)(\mathbf{W}(f)^H\mathbf{W}(f))^{-0.5} \end{aligned} \tag{6}$$

where $\mathbf{y}^H$ is the conjugate transpose of $\mathbf{y}$, $\alpha_i = E\{y_i(f,t)\phi(y_i(f,t))\}$, $D = diag(1/(\alpha_i - E\{\phi'(y_i(f,t)\}))$, and the activation function $\phi(\mathbf{y}(f,t))$ is given by

$$\phi(\mathbf{y}(f,t)) = \frac{\mathbf{y}(f,t)}{|\mathbf{y}(f,t)|}, \quad \forall|\mathbf{y}(f,t)| \neq 0 \tag{7}$$

with its derivative approximated by $\phi'(\mathbf{y}(f,t)) \approx |\mathbf{y}(f,t)|^{-1} - \mathbf{y}(f,t)^2|\mathbf{y}(f,t)|^{-3}$ [6]. Moreover, the algorithm (6) requires that the mixtures $\mathbf{x}(f,t)$ be pre-whitened; we refer to it as MD2003.

4 The Role of High Frequencies

In this paper, we aim to investigate the role of the high frequencies in convolutive blind source separation of speech signals, whose characteristics are such that little information is contained in the frequencies above a certain cut-off frequency [9], which we define in this paper as f_c. Here, we consider the following decomposition of the observed signal

$$\mathbf{X}(f,t) = \mathbf{X}_{LFs}(f,t) + \mathbf{X}(f,t)_{HFs} \tag{8}$$

where $\mathbf{X}_{LFs}(f,t)$ is the STFT representation of the mixtures with the subbands corresponding to the high frequencies ($f > f_c$) set to zero, and similarly $\mathbf{X}(f,t)_{HFs}$ has the low frequencies subbands ($f \leq f_c$) set to zero. Defining the recovered signal as $\mathbf{Y}(f,t) = \mathbf{Y}_{LFs}(f,t) + \mathbf{Y}(f,t)_{HFs}$, the following four scenarios are considered, in which source separation is performed using MD2003:

1. on **all** frequency bins (MD2003): $\mathbf{Y}(f,t) = \mathbf{Y}_{LFs}(f,t) + \mathbf{Y}(f,t)_{HFs}$
2. on the **low** frequency bins only; the high frequencies are set to **zero** (LF): $\mathbf{Y}(f,t) = \mathbf{Y}_{LFs}(f,t)$

3. on the **low** frequency bins; the high frequency components are extracted using a beamformer $\mathbf{W}_B F(f)$ based on the DOAs estimated from the low frequency components (LF-BF): $\mathbf{Y}(f,t) = \mathbf{Y}_{LFs}(f,t) + \mathbf{W}_B F(f)$ $\mathbf{X}(f,t)_{HFs}$
4. on the **low** frequency bins; the high frequency components are left mixed, and they are added back to the separated low frequencies prior to applying the inverse STFT (LF-BF): $\mathbf{Y}(f,t) = \mathbf{Y}_{LFs}(f,t) + \mathbf{X}(f,t)_{HFs}$

Figure 1 illustrates the four methods described above.

5 Simulation Results

In this section, we consider the separation of two speech signals, from two male speakers, sampled at 16kHz. The sources were mixed using simulated room impulse responses, determined by the image method [2] using MGovern's RIR Matlab function[1], with a room reverberation time of 160 ms. The STFT frame length used was set to 2048 in all cases. The performance of the FD-BSS method in [6] (MD2003) was compared for the four methods described in section 4, and permutations were aligned as in [3]. We set $f_c = 4.7$kHz, so that the low frequency bands are between 0 to 4.7kHz, while the high frequencies are above 4.7kHz. This value was obtained empirically by inspecting the frequency content of the mixtures, and with the aim of ensuring that as much information as possible is preserved in the low frequencies.

Table 1. Signal-to-distortion (SDR), signal-to-interference (SIR), and signal-to-artifact ratios (SAR), for the four methods separating the sources signals: At all frequencies - MD2003; At low frequencies only - LF; At low frequencies; BF applied at high frequencies - LF-BF; At low frequencies; high frequencies added still mixed - LF-HF, for a cut off of 4.7kHz

Method	SDR (dB)	SIR (dB)	SAR (dB)	Listening Tests
MD2003 [6]	5.37	19.17	6.08	+++
LF	5.37	19.66	5.59	+
LF-BF	5.15	17.33	5.52	++
LF-HF	5.04	13.16	6.14	++++

The performance of each method was evaluated using the objective criteria of Signal-to-Distortion Ratio (SDR), Signal-to-Interference Ratio (SIR) and Signal-to-Artefacts Ratio (SAR), as defined in [7]. SDR, SIR and SAR measure, respectively, the level of the total distortion in the estimated source, with respect to the target source, the distortion due to interfering sources, and other

[1] Available from:http://2pi.us/code/rir.m

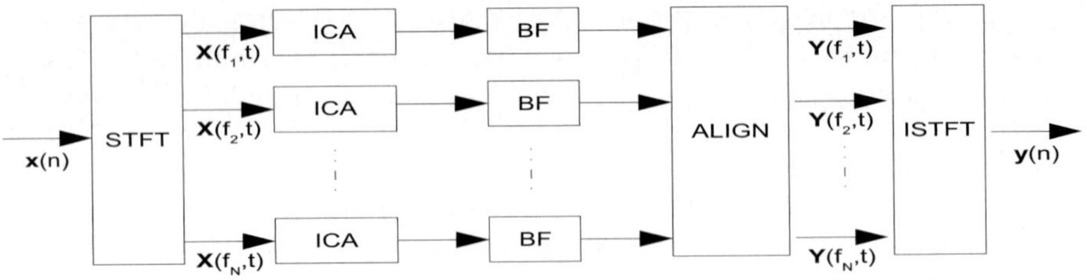

(a) Separation of all frequency bins (MD2003): $\mathbf{Y}(f,t) = \mathbf{Y}_{LFs}(f,t) + \mathbf{Y}(f,t)_{HFs}$

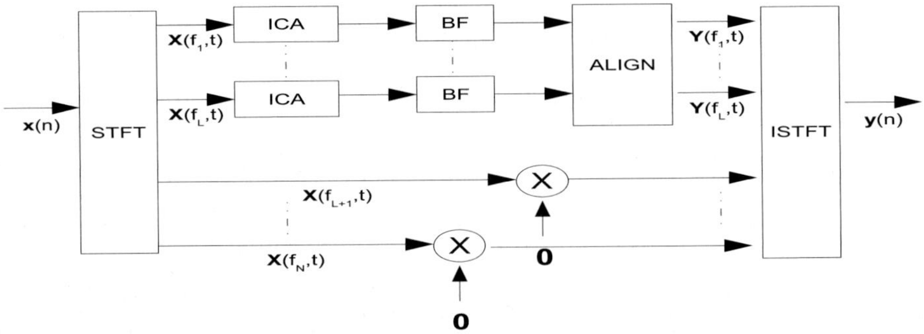

(b) Separation of low frequency bins only (LF): $\mathbf{Y}(f,t) = \mathbf{Y}_{LFs}(f,t)$

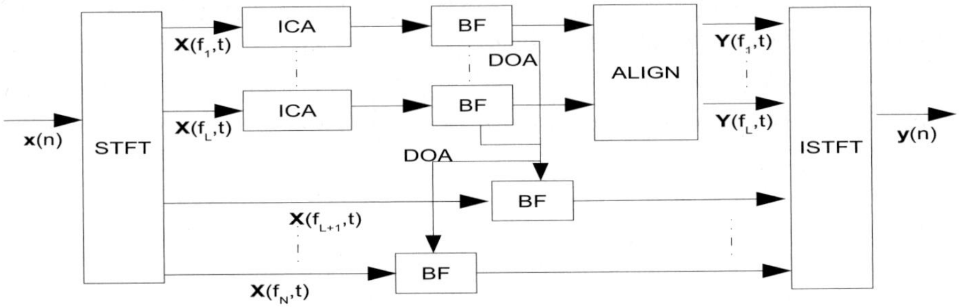

(c) Separation of low frequency bins, with beamforming in the high frequencies (LF-BF): $\mathbf{Y}(f,t) = \mathbf{Y}_{LFs}(f,t) + \mathbf{W}_BF(f)\mathbf{Y}(f,t)_{HFs}$

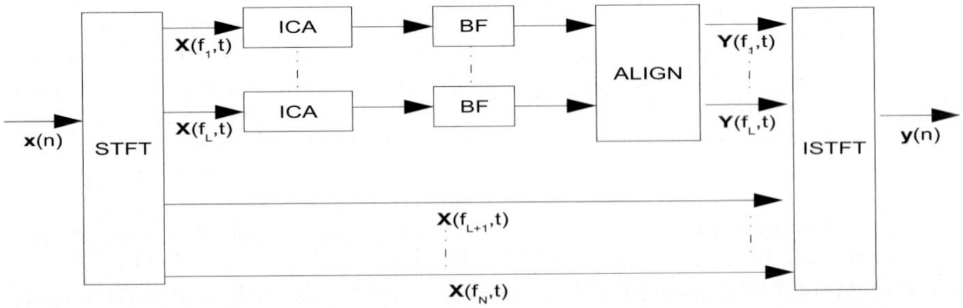

(d) Separation of low frequency bins. High frequency are added back without separation (LF-HF): $\mathbf{Y}(f,t) = \mathbf{Y}_{LFs}(f,t) + \mathbf{X}(f,t)_{HFs}$

Fig. 1. Illustration of the four methods compared

remaining artefacts. The evaluation criteria allows for the recovered sources to be modified by a permitted distortion, and we considered a time-invariant filter of length 512 samples, when calculating the performance measures. This length was chosen so that the filter would cover the reverberation time. We obtained SDR, SIR and SAR figures for the four methods, and for all sources and microphones. The results are shown in Table 1, where the single figure was produced by averaging the criteria across all microphones and all sources.

The SDRs in Table 1 show that the total distortion for all methods is essentially the same. Distortion increases for LF-HF, due to the high frequencies not being separated, and therefore re-introducing some level of distortion. This is supported by the corresponding SIR figure for the same method, which shows that a higher level of interference from the other source is present. The values for SAR indicate that most artefacts are introduced when separation is performed on the low frequency (LF) components only, and when the high frequency components are extracted using beamforming (LF-BF). This is hardly surprising, since both methods can have quite severe effects on the data. The most interesting result is observed from the SIR figures. They show that separating only the low frequency components, and truncating the high frequency ones, has the effect of removing more interference from the undesired source signal than when working with all frequencies, while not introducing any additional distortion (SDR is unchanged), although the level of artefacts present increases. This result is rather counterintuitive, as it suggests that there is little to be gained from performing separation in the high frequencies. This might be explained by the fact that source separation methods perform worse on high frequency components, which are generally lower in amplitude; using beamforming methods to deal with the permutation problem also yields poor results due to phase ambiguity in the high frequencies [8].

Informal listening tests were performed, to corroborate the outcome of the objective criteria. They indicated that the ratios are a good guide to the audible performance. The outputs of LF were found to sound the least natural among all the recovered signals, due to the high frequencies not being present, while the sources separated with LF-HF were found to sound somehow better than the outputs of MD2003. However, the crucial point is that the outputs of all methods sounded similar in quality, suggesting that they all have similar performance. The last column in Table 1 shows a classification of the recovered sources, with the number of + indicating how good the quality of the separated signal is. In general, LF-HF gave the best results, and LF is the worst only because it it not as natural as the others. Nonetheless, the output of LF is equally as intelligible as the others.

We can conclude from these results that performing separation in all subbands is not always the best approach. Especially for speech signals, it might be more advantageous to apply BSS only in the low frequencies, hence reducing, or even halving, the computational burden of some frequency domain algorithms.

6 Conclusions

In this paper, we discussed the role of the high frequencies in frequency domain blind source separation of speech signals. We found that when the high frequencies are ignored, the separated sources remain quite clear, albeit they do not always sound very natural. Our findings were supported by objective criteria, and informal listening tests, which have suggested that it might be a good strategy to separate the mixtures in the low frequencies only, and then add on the high frequency components, without performing any processing on them. This approach may bring significant advantages in terms of reduced computational complexity.

References

1. Sawada, H., Mukai, R., Araki, S., Makino, S.: A robust and precise method for solving the permutation problem of frequency-domain blind source separation. IEEE Trans. on Speech and Audio Processing 12, 530–538 (2004)
2. McGovern, S.: A model for room acoustics (2003), Available at `http//2pi.us/rir.html`
3. Mitianoudis, N., Davies, M.: Permutation alignment for frequency domain ICA using subspace beamforming methods. In: Puntonet, C.G., Prieto, A.G. (eds.) ICA 2004. LNCS, vol. 3195, pp. 669–676. Springer, Heidelberg (2004)
4. Saruwatari, H., Kurita, S., Takeda, K.: Blind source separation combining frequency-domain ICA and beamformning. In: Proc. ICASSP, vol. 5, pp. 2733–2736 (2001)
5. Ikram, M., Morgan, D.: A beamforming approach to permutation alignment for multichannel frequency-domain blind speech separation. In: Proc. ICASSP, vol. 1, pp. 881–884 (2002)
6. Mitianoudis, N., Davies, M.: Audio source separation of convolutive mixtures. IEEE Trans. on Audio and Speech Processing 11, 489–497 (2003)
7. Févotte, C., Gribonval, R., Vincent, E.: BSS_EVAL Toolbox User Guide, IRISA Technical Report 1706 (April 2005), `http://www.irisa.fr/metiss/bss_eval/`
8. Jafari, M.G., Adballah, S.A., Plumbley, M.D., Davies, M.E.: Sparse coding for convolutive blind audio source separation. In: Rosca, J., Erdogmus, D., Príncipe, J.C., Haykin, S. (eds.) ICA 2006. LNCS, vol. 3889, pp. 132–139. Springer, Heidelberg (2006)
9. Balcan, D., Rosca, J.: Independent component analysis for speech enhancement with missing TF content. In: Rosca, J., Erdogmus, D., Príncipe, J.C., Haykin, S. (eds.) ICA 2006. LNCS, vol. 3889, pp. 552–560. Springer, Heidelberg (2006)

Signal Separation by Integrating Adaptive Beamforming with Blind Deconvolution

Kostas Kokkinakis* and Philipos C. Loizou

Center for Robust Speech Systems,
Department of Electrical Engineering,
University of Texas at Dallas,
Richardson, TX 75083, USA
{kokkinak,loizou}@utdallas.edu

Abstract. In this paper, we present a broadband two-microphone blind spatial separation technique by efficiently combining adaptive beamforming (ABF) with multichannel blind deconvolution (MBD). First, the inaccessible source signal streams are partially identified by simple time-delay steering and then are spatially separated through an MBD structure. The proposed spatio-temporal ABF-MBD algorithm exhibits fast convergence properties and high computational efficiency. Numerical experiments illustrate the practical appeal of the proposed method in separating convolutive mixtures of speech within nearly anechoic and also highly reverberant enclosures.

1 Introduction

Multi-microphone speech enhancement methods have a very strong potential for use in a variety of applications such as automatic speech recognition, hearing aid devices and hands-free telephony. In such applications, speech from various sources is often collected simultaneously over two spatially distributed microphones or through a closely-spaced multi-sensor array. The key challenge here is to develop algorithms that can maximize speech intelligibility in both anechoic and modest-to-severe reverberant scenarios by recovering and perceptually enhancing the waveform of the desired (or target) source signal, while relying only on an observed set of composite (or mixed) signals. One such prominent technique, is *blind source separation* (BSS). Practically, BSS can blindly recover a set of unknown signals, the so-called sources from their observed mixtures, based on very little to almost no prior knowledge about the source characteristics or the mixing structure itself. Recent work on BSS has been encouraging even in long reverberation cases. However, most BSS techniques suffer from (1) ambiguities (e.g., scaling and permutation) in the independence criterion when the problem is treated exclusively in the frequency-domain or (2) are inherently slow and computationally inefficient when operating in the time-domain, especially if in this latter case long filters are needed to reach an adequate level of separation.

* Work supported in part by Grant R01–DC07527 awarded from the NIDCD/NIH.

M.E. Davies et al. (Eds.): ICA 2007, LNCS 4666, pp. 495–503, 2007.

To overcome these problems, various authors have proposed spatial filtering techniques that combine BSS with *adaptive beamforming* (ABF). Beamforming has long been used in many areas, such as radar, medical imaging and hearing aids [5]. By definition, the purpose of beamforming is to pick up and amplify sounds from one direction, while suppressing undesired interferences and reverberation arriving from all other directions. Exploiting the similarity between ABF and BSS, Parra and Alvino [14], resorted to beamforming and incorporated geometric constraints to solve permutations between adjacent frequency bands. Their method, called geometric source separation (GSS), was shown to work well in reverberant conditions, albeit at the expense of adding multiple sensors to estimate the directions of arrival (DOAs) of the sources. The potential of using the directivity (or gain) pattern formed by two parallel beamformers to yield maximally independent outputs, was also later explored in [1] (time-domain) and [3] (frequency-domain). Unlike most frequency-domain methods, the latter technique did not suffer from different permutations along the frequency bands. More recently, similar techniques have also achieved a correct permutation alignment, based on information acquired from a beamformer about the direction of the nulls (sidelobes) in different frequency bins (e.g., see [13], [16]).

In this paper, we use *multichannel blind deconvolution* (MBD) for convolutive BSS. In stark contrast to frequency-domain BSS, our MBD approach operates in the z-domain partially and thus remains immune to *permutation* disparities [12]. Also, *scaling* indeterminacies that cause *whitening* are fully alleviated to retain intelligible source contributions at the output [9], [10]. Nonetheless, MBD approaches are computationally demanding when long filters are used. To ameliorate this and speed up convergence, we use ABF to place spatial nulls at the location of the interfering sources before passing the signals through MBD.

2 Signal Model

To isolate the original or "true" sources in a *multipath* propagation scenario, one needs to rely solely on information extracted from the *convolutive* mixtures of the original signal streams $\mathbf{x}(t) = [x_1(t), \ldots, x_m(t)]^T \in \mathbb{R}^m$ given by

$$\mathbf{x}(t) = \sum_{\ell=0}^{\infty} \mathbf{H}(\ell)\, \mathbf{s}(t-\ell), \quad t = 1, 2, \ldots \tag{1}$$

where $\mathbf{H}(\ell)$ is the unknown linear-time invariant (LTI) multiple-input multiple-output (MIMO) mixing system that models the acoustic channel. MBD can blindly achieve the recovery of the sources $\mathbf{s}(t)$, by processing measurements at the sensors, such that the system outputs $\mathbf{u}(t) = [u_1(t), \ldots, u_n(t)]^T \in \mathbb{R}^n$ read

$$\mathbf{u}(t) = \sum_{\ell=0}^{L-1} \mathbf{W}(\ell)\, \mathbf{x}(t-\ell), \quad t = 1, 2, \ldots \tag{2}$$

where $\mathbf{W}(\ell)$ is the unmixing matrix linking the jth source estimate $u_j(t)$ with the ith sensor observation $x_i(t)$, composed of sufficiently long finite impulse response

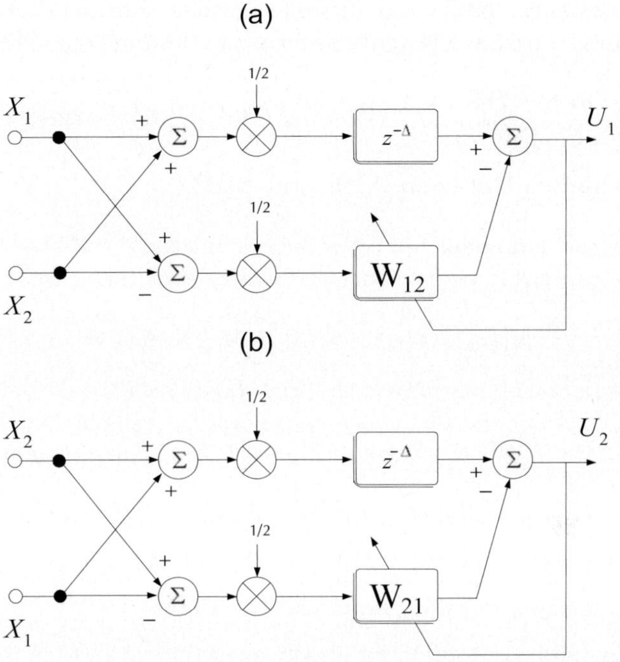

Fig. 1. Setup of two-microphone Griffiths-Jim beamformers. The microphone signals are added and subtracted to form the sum signal and the difference signal. (a) A null pattern is formed towards source S_2. (b) A null pattern is formed towards source S_1.

(FIR) filters with each element given by vector $[w_{ji}(0), w_{ji}(1), \ldots, w_{ji}(L-1)]$ for all coefficients $0 \leq \ell \leq L-1$ with $j = 1, 2, \ldots, n$ and $i = 1, 2, \ldots, m$.

3 Beamforming with Multichannel Blind Deconvolution

3.1 Stage 1. ABF

The basis of our multi-microphone algorithm is the Griffiths-Jim beamformer [6]. Although, any multiple-input single-output (MISO) algorithm can be used to adapt the filter coefficients, here we choose the least-mean-squares (LMS) algorithm to continuously adjust the filter weights. Since our purpose is to segregate two signals S_1 and S_2 with two microphones, we use two such MISO beamformers (see Figure 1). The ABF shown in Figure 1(a) can identify S_1 by forming a null directivity pattern towards S_2 using W_{12}. Similarly, the ABF in Figure 1(b) can focus on S_2 by using spatial filter W_{21} to attenuate source S_1.

The upper ABF structure forms a sum and difference signal by adding and subtracting the microphone signals X_1 and X_2. The sum signal is passed through the delay element $z^{-\Delta}$. Typically, if nothing is known *a priori* about the setup of the sources and the microphone locations, $\Delta = L/2$. The difference (or interference) signal is filtered through the $(L+1)$-point adaptive filter W_{12} to form an interference cancellation signal, which is then subtracted from the delayed sum signal in the primary channel to form the desired source estimate U_1. The weights of the filter taps are adapted using LMS to minimize the error, and ultimately yield U_1. The same process is repeated while recovering source U_2.

3.2 Equivalence Between ABF and MBD

Focusing on the z-domain for convenience, in the 2×2 scenario depicted in Figure 2, we can easily deduce that the mixtures at the sensor inputs are

$$\begin{aligned} X_1(z) &= H_{11}(z)\, S_1(z) + H_{12}(z)\, S_2(z) \\ X_2(z) &= H_{21}(z)\, S_1(z) + H_{22}(z)\, S_2(z) \end{aligned} \tag{3}$$

By observing the ABF structures in Figure 1, the source estimates are equal to

$$\begin{aligned} U_1(z) &= \big[X_1(z) + X_2(z)\big] - W_{12}(z)\,\big[X_1(z) - X_2(z)\big] \\ U_2(z) &= \big[X_1(z) + X_2(z)\big] - W_{21}(z)\,\big[X_2(z) - X_1(z)\big] \end{aligned} \tag{4}$$

If we assume that the transfer functions from the sources to the microphones are similar (a valid assumption for closely-spaced arrays), then upon subtracting one from the other, the two microphone signals will (ideally) cancel the contribution from the undesired source out. Hence, the difference signal in the upper ABF structure can be approximated as $X_1(z) - X_2(z) \approx U_2(z)$ and accordingly in the lower ABF structure as $X_2(z) - X_1(z) \approx U_1(z)$. In addition, summing the microphone inputs will result in obtaining filtered versions of $X_1(z)$ (upper ABF) and $X_2(z)$ (lower ABF). Based on such simplifications, (4) reduces to

$$\begin{aligned} U_1(z) &\approx X_1(z) - W_{12}(z)\, U_2(z) \\ U_2(z) &\approx X_2(z) - W_{21}(z)\, U_1(z) \end{aligned} \tag{5}$$

which resembles a feedback network of FIR filters in the MBD configuration. A drawback of ABF is that due to signal subtraction in the auxiliary channel, the output or target signals will almost always exhibit a typical high-pass characteristic resulting in loss of frequencies below 1 kHz (see also [8]). In order to compensate for this effect, the outputs from the beamformer can be low-pass filtered before any further processing [15]. Another drawback is that the effectiveness of ABF in realistic conditions is only limited to zero-to-moderate reverberation settings. Several authors have suggested that the presence of reverberant energy severely degrades the performance of beamforming [5]. As a general rule, the more reverberation present in the environment, the more difficulty this algorithm has in placing nulls at the locations of the undesired signals. Still, after performing spatial filtering with ABF, the tasks of multichannel separation and deconvolution are expected to become easier, leading to a significant increase in separation performance and faster convergence to the optimal filter coefficients.

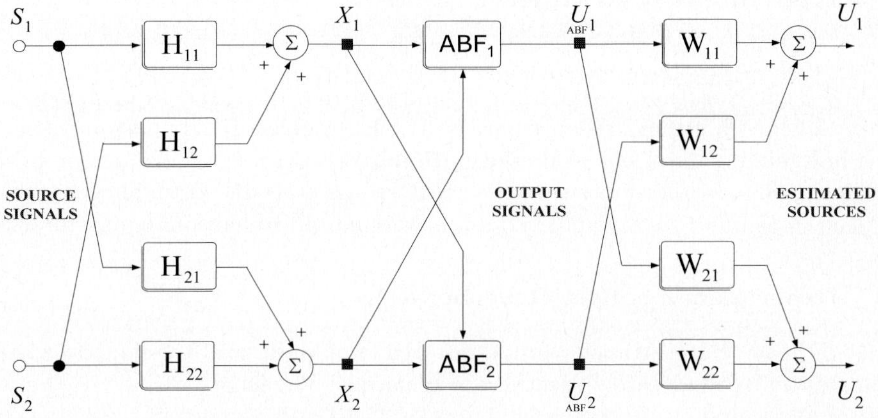

Fig. 2. Cascaded mixing and unmixing system configuration in the two-source and two-sensor scenario when integrating two ABF structures (see Figure 1) with MBD

3.3 Stage 2. MBD

Based on the *isomorphic* mapping between scalar and FIR polynomial matrices (e.g., see [12]), several adaptation rules derived from the entropy maximization principle [4], have been extended to MBD. Such an efficient update is the linear prediction-based NGA (LP-NGA) algorithm (e.g., see [9], [10]), which stems from the natural gradient algorithm (NGA) introduced by Amari *et al.* [2]. As shown in Figure 2, the partly identified signal estimates $U_1(z)$ and $U_2(z)$ obtained from the ABF stage are then fed as inputs to the LP-NGA algorithm, which reads

$$\underline{\mathbf{W}}_{k+1}(z) = \underline{\mathbf{W}}_k(z) + \mu\, \Delta\underline{\mathbf{W}}_k(z) \tag{6}$$

where $\underline{\mathbf{W}}(\cdot)$ is the unmixing FIR polynomial matrix, μ denotes the step-size and

$$\Delta\underline{\mathbf{W}}_k(z) = \left[\begin{pmatrix} \bar{1} & \bar{0} \\ \bar{0} & \bar{1} \end{pmatrix} - \mathrm{FFT}[\boldsymbol{\varphi}(\boldsymbol{u})]\, \boldsymbol{u}^H\right] \begin{bmatrix} W_{11}(z) & W_{12}(z) \\ W_{21}(z) & W_{22}(z) \end{bmatrix} \tag{7}$$

$W_{ji}(\cdot)$ are the unmixing FIR filters in the z-domain, $(\cdot)^H$ is the Hermitian operator, the matrix composed of a sequence of all ones $(\bar{1})$ in the main diagonal and all zeros $(\bar{0})$ elsewhere is the identity (unit) FIR matrix, whereas the term FFT $[\boldsymbol{\varphi}(\boldsymbol{u})]$ denotes the score function vector $\boldsymbol{\varphi}(\boldsymbol{u})$, operating in the time domain

$$\varphi_i(u_i) = -\frac{d}{du_i} \log p_{u_i}(u_i), \quad i = 1, 2. \tag{8}$$

with the ABF-MBD spatially separated source outputs written as

$$\underset{\text{MBD}}{\boldsymbol{u}}(z) = [U_1(z), U_2(z)]^T = \underline{\mathbf{W}}(z)\underset{\text{ABF}}{\boldsymbol{u}}(z) \tag{9}$$

4 Experimental Results

The source signals were sentences of one male and one female speaker, approximately 3 s in duration, recorded at a sampling rate of 8 kHz. In total, we processed 20 speech stimuli taken from the IEEE database [7]. The sound level of each individual source was also adjusted relative to the fixed level of the other, yielding a signal-to-interference ratio (SIR) equal to 0 dB. Both speech signals had the same onset, and where necessary were edited to have an equal duration.

4.1 Experiment 1. Short Reverberation

A set of head-related transfer functions (HRTFs) were used to simulate speech mixtures under moderately reverberant scenarios. The length of the HRTFs was 256 samples, amounting to a small delay of 32 ms and very short reverberation.

4.2 Experiment 2. Long Reverberation

The speech signals were convolved with a set of binaural room impulse responses (BRIRs) (e.g., see [17]). In contrast to the relatively smooth and nearly free-field HRTFs used in Experiment 1, these exhibit rapid variations both in phase and magnitude and are, in general, difficult to invert with FIR filters [11]. The BRIRs were measured in a $5 \times 9 \times 3.5$ m classroom using a KEMAR positioned at 1.5 m above the floor at ear level [17]. By convolving the speech signals with the pre-measured impulse responses, one source is placed directly to the front and the other at an angle of 30° to the right, while both are 1 m away from the KEMAR. In this case, the broadband reverberation time of the room is $T_{60} = 300$ ms.

4.3 Results and Discussion

The convolutive speech mixtures were processed with four different spatial processing schemes, ABF only ('ABF'), MBD only with a single pass (on-line mode) ('MBD [1]'), MBD only with 10 passes ('MBD [10]') corresponding to 30 s of total training time (off-line mode), and the new processing scheme merging ABF with MBD ('ABF-MBD'). Note that in the latter case, the algorithm was allowed only one pass through the data. The algorithms were executed with 128- and 512-sample point adaptive FIR filters for Experiments 1 and 2, respectively, and a fixed step-size maximized up to the stability margin. The overlap between successive frames (or blocks) of data was set to 50%. To assess the separation ability of the algorithms in different reverberation scenarios, we used the signal-to-interference-ratio improvement (SIRI) and measured the overall amount of crosstalk reduction in dB, before (SIR_i) and after (SIR_o) separation, as in [11].

The SIRI values averaged for both sources and across all sentences are plotted in Figure 3 (Experiment 1) and Figure 4 (Experiment 2). According to Figure 3, the MBD algorithm yields a substantial improvement in SIR, when allowed multiple passes through the mixed speech data. Separation performance for ABF-MBD was slightly lower (around 25%) compared to MBD with 10 passes. The

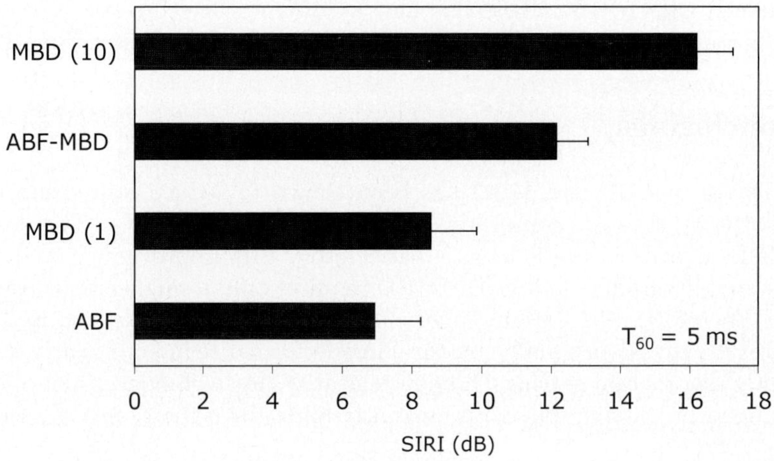

Fig. 3. Mean SIRI values (dB) for 10 IEEE sentences ($T_{60} = 5\,\text{ms}$). Error bars indicate standard deviations.

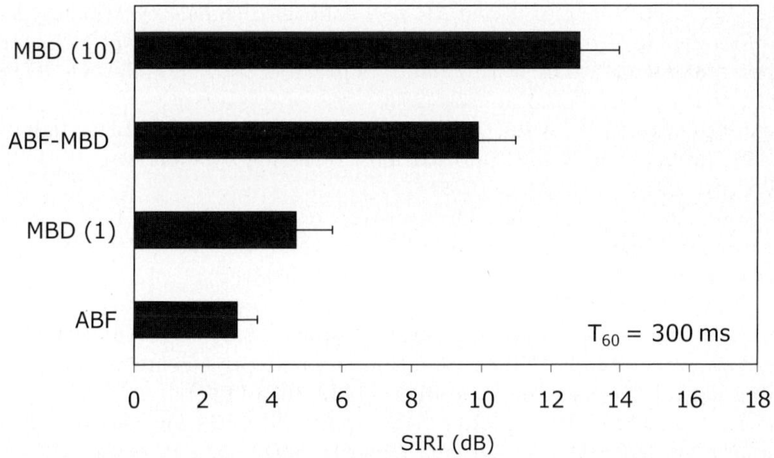

Fig. 4. Mean SIRI values (dB) for 10 IEEE sentences ($T_{60} = 300\,\text{ms}$). Error bars indicate standard deviations.

performance of MBD with one pass, was significantly lower than ABF-MBD for the same amount of training. Also, ABF alone only partially managed to recover the source estimates and provided a marginal improvement. By observing the results shown in Figure 4, we notice that the overall separation performance decreases when reverberation energy increases for all spatial separation schemes tested here. MBD performs fairly well, but only with an adequate amount of training. In contrast, the benefit of ABF was found to be negligible. The separation performance obtained with ABF-MBD is still lower by about 20% than the one obtained with MBD, albeit this new adaptive processing strategy requires much less training and uses relatively short FIR filters to equalize long BRIRs.

5 Conclusions

The joint use of ABF and MBD has been shown to reduce computational demands without severely compromising separation performance. The proposed ABF-MBD algorithm can achieve a satisfactory SIR improvement and can be implemented in on-line mode. ABF-MBD requires only a single pass through the data and therefore is potentially amenable to real-time implementation. Experimental results reveal an equally encouraging performance in both nearly anechoic and highly reverberant settings. The potential of this technique when operating in a subband processing scheme is currently under investigation (e.g., see [11]).

References

1. Aichner, R., Araki, S., Makino, S., Nishikawa, T., Saruwatari, H.: Time domain blind source separation of non-stationary convolved signals by utilizing geometric beamforming. In: Proc. 12th IEEE Int. Workshop on Neural Networks for Signal Process, Martigny, Valais, Switzerland, September 4-6, pp. 445–454. IEEE Computer Society Press, Los Alamitos (2002)
2. Amari, S., Cichocki, A., Yang, H.: A new learning algorithm for blind signal separation. In: Advances in Neural Information Processing Systems, vol. 8, pp. 757–763. MIT Press, Cambridge, MA (1996)
3. Baumann, W., Kolossa, D., Orglmeister, R.: Beamforming-based convolutive source separation. In: Proc. 28th IEEE Int. Conf. on Acoust. Speech and Signal Process, Hong Kong, April 6-10, pp. 357–360 (2003)
4. Bell, A.J., Sejnowski, T.J.: An information maximization approach to blind separation and blind deconvolution. Neural Computat. 7(6), 1129–1159 (1995)
5. Greenberg, Z.E., Zurek, P.M.: Evaluation of an adaptive beamforming method for hearing aids. J. Acoust. Soc. Am. 91(3), 1662–1676 (1992)
6. Griffiths, L.J., Jim, C.W.: An alternative approach to linearly constrained adaptive beamforming. IEEE Trans. Antennas Propag. 30(1), 27–34 (1982)
7. IEEE Subcommittee, IEEE recommended practice speech quality measurements. IEEE Trans. Audio Electroacoust. vol. 17(3), pp. 225–246 (1969)
8. Joho, M., Moschytz, G.S.: On the design of the target-signal filter in adaptive beamforming. IEEE Trans. Circuits Syst. 46(7), 963–966 (1999)
9. Kokkinakis, K., Nandi, A.K.: Optimal blind separation of convolutive audio mixtures without temporal constraints. In: Proc. 29th IEEE Int. Conf. on Acoust. Speech and Signal Process. Montréal, Canada, May 17-21, pp. 217–220 (2004)
10. Kokkinakis, K., Nandi, A.K.: Multichannel blind deconvolution for source separation in convolutive mixtures of speech. IEEE Trans. Audio, Speech, Lang. Process 14(1), 200–212 (2006)
11. Kokkinakis, K., Loizou, P.C.: Subband-based blind signal processing for source separation in convolutive mixtures of speech. In: Proc. 32nd IEEE Int. Conf. on Acoust. Speech and Signal Process, Honolulu, Hawaii, April 15-20, pp. 917–920 (2007)

12. Lambert, R.H.: Multichannel Blind Deconvolution: FIR Matrix Algebra and Separation of Multipath Mixtures. Ph.D. Thesis, University of Southern California (May 1996)
13. Mitianoudis, N., Davies, M.: Permutation alignment for frequency domain ICA using subspace beamforming methods. In: Proc. 5th Int. Conf. on ICA and BSS, Granada, Spain, September 22-24, pp. 669–676 (2004)
14. Parra, L., Alvino, C.V.: Geometric source separation: Merging convolutive source separation with geometric beamforming. IEEE Trans. Speech Audio Process 10(6), 352–362 (2002)
15. Puder, H.: Adaptive signal processing for interference cancellation in hearing aids. Signal Process 86(6), 1239–1253 (2006)
16. Saruwatari, H., Kawamura, T., Nishikawa, T., Lee, A., Shikano, K.: Blind source separation based on a fast-convergence algorithm combining ICA and beamforming. IEEE Trans. Speech Audio Process 14(2), 666–678 (2006)
17. Shinn-Cunningham, B., Kopco, N., Martin, T.: Localizing nearby sound sources in a classroom: Binaural room impulse responses. J. Acoust. Soc. Am. 117(5), 3100–3115 (2005)

Blind Signal Deconvolution as an Instantaneous Blind Separation of Statistically Dependent Sources

Ivica Kopriva

Rudjer Bošković Institute, Bijenička cesta 54, P.O. Box 180, 10002 Zagreb, Croatia
ikopriva@irb.hr

Abstract. We propose a novel approach to blind signal deconvolution. It is based on the approximation of the source signal by Taylor series expansion and use of a filter bank-like transform to obtain multichannel representation of the observed signal. Currently, as an *ad hoc* choice a wavelet packets filter bank has been used for that purpose. This leads to multi-channel instantaneous linear mixture model (LMM) of the observed signal and its temporal derivatives converting single channel blind deconvolution (BD) problem into instantaneous blind source separation (BSS) problem with statistically dependent sources. The source signal is recovered provided it is a non-Gaussian, non-stationary and non- independent identically distributed (i.i.d.) process. The important property of the proposed approach is that order of the channel filter does not have to be known or estimated. We demonstrate viability of the proposed concept by blind deconvolution of the speech and music signals passed through a linear low-pass channel.

Keywords: Blind deconvolution, Blind source separation, Independent component analysis, Instantaneous mixture model, Statistically dependent sources.

1 Introduction

The problem of single channel BD is to reconstruct the original signal from its filtered version also termed observed signal, where only observed signal is available. Neglecting the noise term the process is modeled as a convolution of the unknown causal channel impulse response $h(t)$ with an original source signal $s(t)$ as:

$$x(t) = \sum_{\tau=0}^{T} h(\tau)s(t-\tau) \tag{1}$$

where T denotes the order of the channel filter. Standard algorithms for blind deconvolution are capable of recovering source signal $s(t)$ based on the observed signal $x(t)$ only, provided that $s(t)$ is a non-Gaussian i.i.d. process, [1]. We shall demonstrate here that proposed concept is capable of blind deconvolution of signals with colored statistics such as speech. The original signal $s(t-\tau)$ can be approximated by Taylor series expansion around $s(t)$ giving:

$$s(t-\tau) = \sum_{n=0}^{N}\left((-\tau)^n/n!\right)s^{(n)}(t) + \text{H.O.T.} \tag{2}$$

M.E. Davies et al. (Eds.): ICA 2007, LNCS 4666, pp. 504–511, 2007.

where $s^{(n)}(t)$ denotes *n-th* order temporal derivative of *s*(*t*) and H.O.T. denotes higher-order-terms. It is assumed that $s^{(0)}(t) = s(t)$. Inserting (2) into (1) yields:

$$x(t) \cong \sum_{n=0}^{N} a_{1(n+1)} s^{(n)}(t) \tag{3}$$

where $a_{11} = \sum_{\tau=0}^{T} h(\tau)$, $a_{12} = -\sum_{\tau=0}^{T} \tau h(\tau)$, $a_{13} = \sum_{\tau=0}^{T} (\tau^2/2) h(\tau)$, etc. Evidently, quality of the approximations (2) and (3) depends on the number of terms in the Taylor series expansion of the source signal *s*(*t*). However, *x*(*t*) in (3) can be also obtained as an inverse Fourier transform of the expression $H(j\omega)S(j\omega)$ where $H(j\omega)$ and $S(j\omega)$ respectively represent Fourier transforms of the channel impulse response and source signal. Owing to the fact that *h*(*t*) is an aperiodic sequence $H(j\omega)$ is obtained as

$$H(j\omega) = \sum_{\tau=0}^{T} h(\tau) e^{-j\omega\tau} \cong \sum_{\tau=0}^{T} h(\tau) - j\omega \sum_{\tau=0}^{T} \tau h(\tau) + \frac{(j\omega)^2}{2} \sum_{\tau=0}^{T} \tau^2 h(\tau) \tag{4}$$

that yields

$$X(j\omega) \cong \left(\sum_{\tau=0}^{T} h(\tau) \right) S(j\omega) - \left(\sum_{\tau=0}^{T} \tau h(\tau) \right) j\omega S(j\omega) + \left(\sum_{\tau=0}^{T} \tau^2 h(\tau) \right) \frac{(j\omega)^2}{2} S(j\omega) \tag{5}$$

Evidently, number of terms in the expansions (4) and (5) depends on the property of the channel: size of the support *T* of the impulse response *h*(*t*), but also on the property of the signal: size of its support Ω in the frequency domain i.e. $|S(j\omega)| \cong 0$ for $\omega > \Omega$. For example, for either *T*=0 or Ω =0 relation (3) and inverse Fourier transform of (5) yield the same result. Thus, channels with the maximal delay that is small relative to the coherence time of the signal, i.e. $T \ll (2\pi/\Omega)$, will demand small number of terms, *N*, in the approximation (3) and vice versa.

Taylor series expansion has already been used in [2]-[7] to convert multichannel convolutive BSS problem into instantaneous BSS problem. Two cases can be distinguished. In [2]-[5] authors assumed sensor array that is smallest than the shortest wavelength of the sources. This allows to keep only the first order derivative in the Taylor series expansion in Eq.(2). This is due to the fact that delay is defined relative to the center of the array and is therefore always smaller than the coherence time of the sources. Under this assumption another array that calculates spatial gradients of the observed signal converts the convolutive BSS problem into instantaneous BSS problem with the first order temporal derivatives of the source signals acting as sources. Once they are recovered by instantaneous ICA, the true sources are obtained by their temporal integration. In [6] and [7] Taylor series expansion is also used to convert multichannel convolutive BSS problem into instantaneous BSS problem. In [6] it is assumed that delay *T* is smaller than the coherence time of the source signals which allows to use only first order temporal derivative in the Taylor series expansion Eq.(2). Assuming that signal and its first order derivative are statistically independent, that is actually proven for stationary signals only [8][9], the instantaneous BSS problem is solved by some of the standard ICA methods. However, assumption that delay is smaller than the coherence time of the source signals is too restrictive for realistic reverberant environments. That was realized in [7]. In that case higher order temporal derivatives exist in the Taylor series expansion Eq.(2), and they are statistically dependent. An algorithm is derived in [7] for grouping dependent sources and extracting source signals from each group.

The algorithm proposed here solves single channel BD problem by converting it into instantaneous BSS problem with statistically dependent sources. No special assumption is made on the amount of delay. Thus, higher order derivatives in the Taylor series expansion are allowed. The problem of their statistical dependence is solved by means of independence enhancement technique, which is based on innovations of the multichannel version of the observed signal. However, another transform such as high-pass filtering, [10], may be used for independence enhancement purpose as well.

A BD capable of recovering temporally dependent signals is derived in [11]. It is based on the measure of temporal predictability and argumentation that an output of the low-pass channel is smoother and therefore more predictable than the input to the channel. Thus, the BD problem is formulated as temporal predictability minimization problem and numerically solved as general eigenvalue problem. Equivalent solution of the instantaneous BSS problem by looking for maximum of the temporal predictability is defined in [12]. In relation to the proposed Taylor series expansion BD method, the temporal predictability approach suffers from the fact that order of the deconvolution filter has to be defined based on some *a priori* knowledge. Because the order of the generalized eigenvalue problem equals the order of the deconvolution filter the temporal predictability based algorithm can become numerically very demanding. Temporal predictability itself is defined for deconvolved signal $y(t)=\sum_{k=0}^{M} w(k)x(t-k)$ as:

$$F\left(\{y(t)\}\right)=\log\frac{V\left(\{y(t)\}\right)}{U\left(\{y(t)\}\right)}=\log\frac{\sum_{t}^{t_{\max}}\left(\bar{y}(t)-y(t)\right)^{2}}{\sum_{t}^{t_{\max}}\left(\tilde{y}(t)-y(t)\right)^{2}} \tag{6}$$

where V reflects the extent to which y(t) is predicted by long term moving average $\bar{y}(t)$ and U reflects the extent to which y(t) is predicted by short term moving average $\tilde{y}(t)$, [12][11].

2 Formulation of the Instantaneous Linear Mixture Model

We now apply a filter bank-like transform on (3) in order to obtain a multichannel representation, **x**, of the observed signal $x(t)$. It is the matter of further analysis to find out which type of the transform is optimal. Here, in order to illustrate the concept, as an *ad hoc* choice we have used a non-decimated wavelet packets filter bank with two decomposition levels that results in L=6 filters. In order to have clear notation let us introduce $x_1(t)=x(t)$. When filters are applied on observed signal $x(t)$ we obtain:

$$x_{l+1}(t)\cong a_{l+1,1}s(t)+a_{l+1,2}s^{(1)}(t)+a_{l+1,3}s^{(2)}(t)\quad l=1,..,L \tag{7}$$

where $a_{l+1,1}=\sum_{\tau=0}^{\bar{T}}\bar{h}_l(\tau)$, $a_{l+1,2}=-\sum_{\tau=0}^{\bar{T}}\tau\bar{h}_l(\tau)$, $a_{l3}=\sum_{\tau=0}^{\bar{T}}\left(\tau^2/2\right)\bar{h}_l(\tau)$, where $\bar{h}_l(t)$ represents convolution of the appropriate l-th filter with $h(t)$, $\bar{T}=T+M+2$ and M is an order of the filter. Observed signal and its filtered versions can be represented in a form of the following instantaneous LMM:

$$\mathbf{x}(t)=\begin{bmatrix}x_1(t)\\x_2(t)\\\cdots\\x_{L+1}(t)\end{bmatrix}\cong\begin{bmatrix}a_{11} & a_{12} & a_{13}\cdots\cdots\cdots a_{1,N+1}\\a_{21} & a_{22} & a_{23}\cdots\cdots\cdots a_{2,N+1}\\\cdots\cdots\cdots\cdots\cdots\cdots\cdots\cdots\\a_{L+1,1} & a_{L+1,2} & a_{L+1,3}\cdots\cdots a_{L+1,N+1}\end{bmatrix}\begin{bmatrix}s(t)\\s^{(1)}(t)\\s^{(2)}(t)\\\cdots\\s^{(N)}(t)\end{bmatrix}=\mathbf{A}\mathbf{s}(t) \tag{8}$$

where $\mathbf{x}\in\mathbf{R}^{(L+1)\times K}$, $\mathbf{A}\in\mathbf{R}^{(L+1)\times(N+1)}$, $\mathbf{s}\in\mathbf{R}^{(N+1)\times K}$, K represents number of samples and N represents an unknown number of temporal derivatives of the source signal. We have used inspection of the singular values of the sample data covariance matrix $\hat{\mathbf{R}}_{\mathbf{xx}}=(1/K)\mathbf{x}\mathbf{x}^{\mathrm{T}}$ to estimate overall number of sources, N+1. ICA algorithms can be applied to the LMM given by Eq. (8) in order to extract the source signal $s(t)$, with the benefits that the order T of the channel impulse response $h(t)$ is absorbed in the mixing matrix **A** and does not have to be known or estimated. The *source* signals have to be non-Gaussain and statistically independent but not i.i.d. This has important practical consequence because BD of signals with colored statistics is possible. This is demonstrated in the section 4 where simulation results are presented.

3 Statistical Properties of the Source Signal: Implications to Deconvolution Results

We reproduce here results and conditions from [8][14] necessary for the stochastic differentiability of the random source signal $s(t)$. We emphasize that conclusions drawn from this analysis can in principle be generalized to blind image deconvolution problem due to the existence of the space filling curves (Peano-Hillbert curves) that enable 2D to 1D mapping and vice versa by preserving local or neighborhood statistics [13]. First we present two important results that relate (non-)stationarity and linear signal representation. If the signal $s(t)$ is stationary it can be represented by the linear time invariant generative model:

$$s(t)=\sum_{\nu=0}^{\infty}b(\nu)\varepsilon(t-\nu) \tag{9}$$

where $\varepsilon(t)$ is an i.i.d. driving signal. If the signal $s(t)$ is non-stationary the linear signal model becomes time variant:

$$s(t)=\sum_{\nu=0}^{\infty}b(t,\nu)\varepsilon(t-\nu) \tag{10}$$

First order derivative $s^{(1)}(t)$ of the stationary signal $s(t)$ is defined if the first order derivative of the autocorrelation function at the time lag zero is zero i.e. $\rho_s^{(1)}(0)=0$, [8]. $\rho_s^{(1)}(0)$ is always zero for non-i.i.d. process due to symmetry of $\rho_s(\tau)$. According to [8] the stronger condition for the existence of $s^{(1)}(t)$ is $\rho_s^{(2)}(0)\neq 0$. If this is true then from [9] it is also true $\rho_s^{(2)}(\tau)\neq 0,\ \forall\tau$. Analogously, condition for existence of $s^{(2)}(t)$

assumes $\rho_s^{(3)}(0) \neq 0$. If the first order derivative of the stationary signal $s(t)$ exists then [8]:

$$E\left[s(t)s^{(1)}(t)\right] = 0 \tag{11}$$

where E represents mathematical expectation. We now interpret these results for the three types of the source signal $s(t)$.

Source signal is a stationary i.i.d. process. In this case a condition $\rho_s^{(1)}(0) = 0$ is not fulfilled. The reason is that autocorrelation function of the i.i.d. process is delta function i.e. $\rho_s(\tau) = \sigma_s^2\delta_\tau$. Therefore, Taylor series expansion (2) for such a signal does not exist. Consequently, the LMM model (8) also does not exist. Thus, i.i.d. signals can not be blindly deconvolved by the proposed algorithm. However, this is not a drawback since a number of blind deconvolution methods solve this problem, [1][15].

Source signal is a stationary non-i.i.d. process. As it has been said such signal has first order derivative. Under previously defined conditions second order derivative also exists. However, we have to emphasize that stationary signals, that are represented by linear time invariant generative signal model (9), can also not be blindly deconvoloved by the proposed algorithm. Assuming that $b(t)$ represents impulse response of the linear time invariant signal generative model, it is impossible to distinguish the channel filter $h(t)$ from the linear convolution of the channel filter and modeling filter $h(t)*b(t)$. Thus, proposed algorithm will deconvolve the i.i.d. driving sequence $\varepsilon(t)$, i.e. the algorithm will have the whitening effect on the stationary non-i.i.d. signal.

Source signal is a non-stationary and non-i.i.d. process. Although, conditions required for stochastic differentiability are derived for stationary signals only we can use the linear generative model of the non-stationary signal (10) and derive derivatives of the non-stationary signal $s(t)$ provided that time varying filter $b(t, \nu)$ is stationary with respect to the independent variable t. In such a case we define:

$$s^{(m)}(t) = \sum_{\nu=0}^{\infty} b^{(m)}(t,\nu)\varepsilon(t-\nu) \tag{12}$$

where $b^{(m)}(t,\nu) = \left(d^m b(t,\nu)/dt^m\right)$. Thus, Taylor series expansion (2) and the LMM (8) do exist. However, we can not make conclusion regarding statistical independence between $s(t)$, $s^{(1)}(t)$, $s^{(2)}(t)$, etc, as it was the case with a stationary signal, (11). Thus, it is justified to use some of the methods derived to enhance statistical independence between the hidden variables in the LMM (8). One of them that is computationally efficient is based on innovations, [16]. It is known that innovations are more non-Gaussian and more statistically independent than original processes. These conditions are of essential importance for the success of the ICA algorithms. Innovation process of the hidden components of **s** is

$$\tilde{s}_n(t) = s_n(t) - E\left[s_n(t)\middle|t, s_n(t-1), s_n(t-2), \ldots\right] \quad s_n \in \left\{s, s^{(1)}, s^{(2)}, \ldots\right\} \tag{13}$$

where the second term in Eq.(13) represents conditional expectation. If both sides of (13) are multiplied by the unknown basis matrix **A** we obtain

$$\tilde{\mathbf{x}}(t) = \mathbf{A}\tilde{\mathbf{s}}(t) \tag{14}$$

Eq.(14) implies that innovations preserve the basis matrix **A**. The innovations based multichannel model (14) enables more accurate estimation of the mixing matrix **A** by means of ICA algorithms, than when ICA algorithms are applied directly on the LMM (8). The expectation is in practice replaced by the autoregressive (AR) model of the finite order yielding:

$$\tilde{x}_l(t) = \sum\nolimits_{j=0}^{J} g_j x_l(t-j) \tag{15}$$

where J represents order of the AR model and g_0=1. The coefficients of the prediction-error filter g_j are efficiently estimated by means of Levinson's algorithm, [17]. We identify for the LMM model (8) L+1 filters and obtain the prediction-error filter in (15) as an average of all identified filters. Hidden variables are then recovered by applying the Moor-Penrose pseudoinverse $\mathbf{A}^{\dagger}$ on the originally observed process **x**. The temporal predictability measure Eq.(6) could be used as a criteria for the selection of the recovered source signal $\hat{s}(t)$ after solution of BSS problem (8).

4 Simulation Results

We have conducted the following experiments: BD of the female speech signal and BD of the choir singing passed through a lowpass channels. 2nd order Butterworth lowpass filter has been used to model the channel response. Figure 1 shows one hundred time points of the true female speech source signal, signal recovered by temporal predictability based algorithm, [12] and signal recovered by the proposed algorithm. For temporal predictability based algorithm we have shown the best result obtained after experimenting with several values for the order of the deconvolution filter. Normalized correlation coefficients between the source and mixed signal, source signal and signal recovered by the proposed algorithm, and source signal and

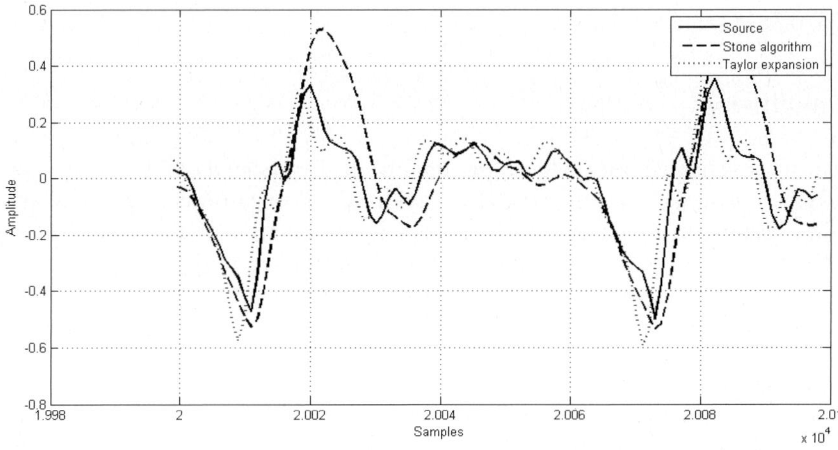

Fig. 1. One hundred time samples of the source signal (solid), signal recovered by temporal predictability based algorithm (dashed), [12], and proposed algorithm based on the Taylor series expansion (dotted)

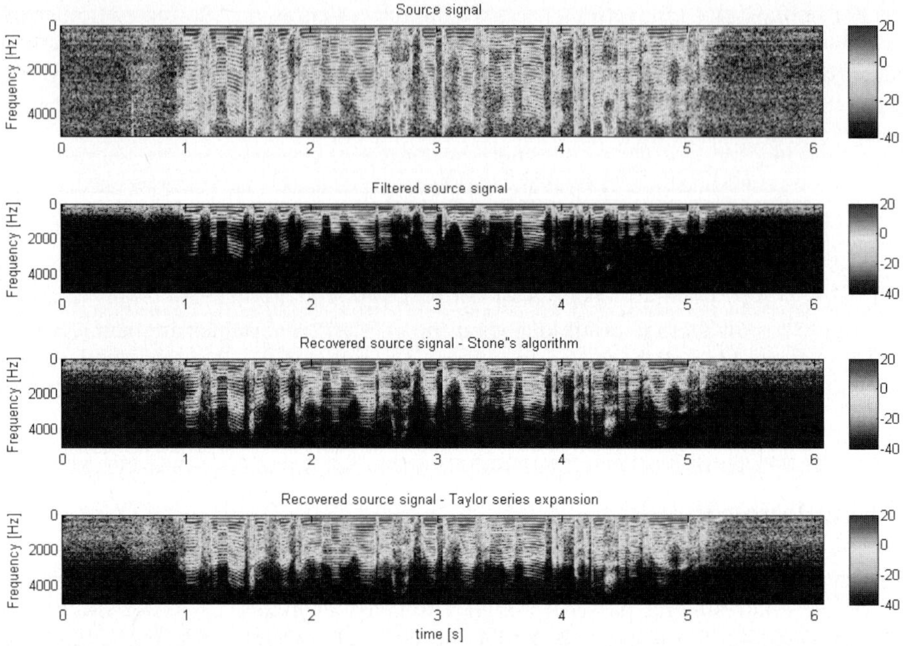

Fig. 2. From top to bottom: spectrograms of the source signal, observed signal, signal recovered by temporal predictability based algorithm, [12], and signal recovered by the proposed algorithm

signal recovered by algorithm [12] are respectively: 0.71774, 0.88658 and 0.75476. Spectrograms of these signals are shown in Figure 2. Regarding the choir-singing signal the normalized correlation coefficients in the same order as before were 0.5276, 0.86152 and 0.84015.

5 Conclusion

Novel single channel BD algorithm has been formulated. It is based on the approximation of the source signal by Taylor series expansion and use of a filter bank-like transform to yield a multichannel representation of the observed single-sensor signal. This yields instantaneous LMM and converts the single channel BD problem into instantaneous BSS problem with statistically dependent sources with the important property that channel order does not have to be known. It has been shown that signal amenable for BD by proposed method must be non-stationary and non-i.i.d. non-Gaussian process. As yet unresolved issues remain: optimality of the linear transforms used to yield a multivariate representation of the observed signal and efficiency of the linear transforms used to enhance statistical independence among hidden variables of the LMM. The later issue might affect performance of the proposed algorithm when degradations are strong as it can be expected for real acoustic channels.

Acknowledgment. Part of this work was supported through grant 098-0982903-2558 funded by the Ministry of Science, Education and Sport, Republic of Croatia. Initial suggestion by Professor Andrzej Cichocki is also gratefully acknowledged.

References

1. Haykin, S. (ed.): Unsupervised Adaptive Filtering – Blind Deconvolution, vol. II. John Wiley, Chichester (2000)
2. Cauwenberghs, C., Stanacevic, M., Zweig, G.: Blind Broadband Localization and Separation in Miniature Sensor Arrays. In: Proc. International Symposium Circuits and Systems (ISCAS'2001), Sydney, Australia, pp. 193–196 (2001)
3. Stanacevic, M., Cauwenberghs, C., Zweig, G.: Gradient flow broadband beamforming and source separation. In: Proc. ICA 2001, San Diego, pp. 49–52 (2001)
4. Stanacevic, M., Cauwenberghs, C., Zweig, G.: Gradient flow adaptive beamforming and signal separation in a miniature microphone array. In: Proc. ICASSP, pp. 4016–4019 (2002)
5. Stanacevic, M., Cauwenberghs, C.: Gradient Flow Bearing Estimation with Blind Identification of Multiple Signals and Interference. In: Proc. International Symposium Circuits and Systems, vol. 5, pp. 5–8 (2004)
6. Barrére, J., Chabriel, G.: A Compact Sensor Array for Blind Separation of Sources. IEEE Trans. Circuits and Systems-I: Fundamental Theory and Applications 49, 565–574 (2002)
7. Chabriel, G., Barrére, J.: An Instantaneous Formulation of Mixtures for Blind Separation of Propagating Waves. IEEE Trans. on Signal Processing 54, 49–58 (2006)
8. Priestley, M.B.: Spectral Analysis and Time Series. Academic Press, London (1981)
9. Yaglom, A.M.: Introduction to the Theory of Stationary Random Functions. Prentice-Hall, Englewood Cliffs (1962)
10. Cichocki, A., Georgiev, P.: Blind source separation algorithms with matrix constraints. IEICE Transactions on Fundamentals of Electronics Communications and Computer Sciences E86-A, 522–531 (2003)
11. Stone, J.V.: Blind Source Separation Using Temporal Predictability. Neural Computation 13, 1559–1574 (2001)
12. Stone, J.V.: Blind deconvolution using temporal predictability. Neurocomputing 49, 79–86 (2002)
13. Lam, W.M., Shapiro, J.M.: A Class of Fast Algorithms for the Peano-Hillbert Space Filling Curve. In: Proceedings of the IEEE International Conference Image Processing (ICIP-94) vol. 1, pp. 638–641 (1994)
14. Kopriva, I.: Approach to Blind Image Deconvolution by Multiscale Subband Decomposition and Independent Component Analysis. Journal Optical Society of America A 24, 973–983 (2007)
15. Cichocki, A., Amari, S.: Adaptive Blind Signal and Image Processing. John Wiley, Chichester (2002)
16. Hyvärinen, A.: Independent component analysis for time-dependent stochastic processes. In: Proceedings of the International Conference on Artificial Neural Networks (ICANN'98) Skovde, Sweden, pp. 541–546 (1998)
17. Orfanidis, S.J.: Optimum Signal Processing – An Introduction, 2nd edn. MacMillan Publishing Comp. New York (1988)

Solving the Permutation Problem in Convolutive Blind Source Separation*

Radoslaw Mazur and Alfred Mertins

Institute for Signal Processing, University of Lübeck, 23538 Lübeck, Germany
{mazur,mertins}@isip.uni-luebeck.de

Abstract. This paper presents a new algorithm for solving the permutation ambiguity in convolutive blind source separation. When transformed to the frequency domain, the source separation problem reduces to independent instantaneous separation in each frequency bin, which can be efficiently solved by existing algorithms. But this independency leads to the problem of correct alignment of these single bins which is still not entirely solved. The algorithm proposed in this paper models the frequency-domain separated signals using the generalized Gaussian distribution and utilizes the small deviation of the exponent between neighboring bins for the detection of correct permutations.

1 Introduction

Blind Source Separation (BSS) is used to recover signals from observed mixtures without prior knowledge of the sources nor the mixing system. For the case of linear instantaneous mixtures, a number of different efficient approaches has been proposed [1,2]. When aiming at real-world mixtures of audio signals like speech, the situation becomes much more difficult. In this case, the mixing process is convolutive and can be modeled using FIR filters, where, for realistic scenarios, the length of these filters can be up to several thousands taps. The unmixing then has to be done using FIR filters of similar length. It is possible to calculate such filters directly in the time domain [3,4], but this approach suffers from high computational cost and difficulties of convergence. The most successful approach is to transform the signals to the frequency domain, where the convolution becomes multiplication [5]. Then the separation can be done independently in each frequency bin, which is a much simpler task. The major drawback of this approach is that the separated bins usually have different scaling and are arbitrarily permuted. Therefore they have to be correctly equalized and aligned, because otherwise the entire process of separation will fail.

While it is possible to obtain a proper scaling for the frequency components [6], there is still no algorithm that can tackle the permutation problem in all cases. One idea for solving the permutation problem is based on the assumption that neighboring bins have alike time structure [7]. Correlation coefficients for neighboring

* This work has been supported by the German Research Foundation under Grant No. ME 1170/1.

M.E. Davies et al. (Eds.): ICA 2007, LNCS 4666, pp. 512–519, 2007.

bins then yield a criterion for correct permutation. Another approach uses the unmixing matrices as beamformer. After computation of the directions of arrival for all bins, most of them can be aligned properly [8]. Unfortunately, if there are more than two sensors in a nonuniform array, the computation becomes very difficult.

In this paper we present a new approach for solving the permutation problem based solely on the statistics of the signals. The new algorithm models the single frequency bins using the generalized Gaussian Distribution (GGD) and utilizes the small changes of the shape parameter of the GGD between neighboring bins.

2 Model and Methods

2.1 BSS for Instantaneous Mixtures

In the instantaneous case the mixing process of N sources into N observations can be modeled by an $N \times N$ matrix $\boldsymbol{A}$. Given the source vector $\boldsymbol{s}(n) = [s_1(n), \ldots, s_N(n)]^T$ and assuming negligible measurement noise, the vector of observation signals $\boldsymbol{x}(n) = [x_1(n), \ldots, x_N(n)]^T$ can be described as

$$\boldsymbol{x}(n) = \boldsymbol{A} \cdot \boldsymbol{s}(n). \tag{1}$$

The separation can be written as a multiplication with a $N \times N$ matrix $\boldsymbol{B}$:

$$\boldsymbol{y}(n) = [y_1(n), \ldots, y_N(n)]^T = \boldsymbol{B} \cdot \boldsymbol{x}(n) \tag{2}$$

The aim of BSS is to find $\mathbf{B}$ from the observed process $\boldsymbol{x}(n)$ so that $\mathbf{BA} = \mathbf{D\Pi}$ where $\mathbf{\Pi}$ is a permutation matrix and $\mathbf{D}$ an arbitrary diagonal matrix. These matrices represent the two ambiguities of BSS: (a) the separated signals appear in arbitrary order and (b) they are scaled versions of the sources.

We here consider the well known gradient-based update rule [1]

$$\Delta \boldsymbol{B} \propto (\boldsymbol{I} + E\{\boldsymbol{g}(\boldsymbol{y})\boldsymbol{y}^T\})\boldsymbol{B} \tag{3}$$

with $\boldsymbol{g}(\boldsymbol{y}) = (g_i(y_i), \ldots, g_n(y_n))$ being a component-wise vector function of nonlinear score functions g_i of the assumed source probability densities $p_i(s_i)$:

$$g_i = \frac{p_i'(s_i)}{p_i(s_i)} \tag{4}$$

In order to achieve good separation performance, the probability density function of the sources has to be known or at least well approximated [9].

2.2 Statistical Source Models and Estimators

Speech signals usually follow a Laplacian distribution. Therefore, for instantaneous mixtures, the nonlinear function $g_i(\cdot)$ reduces to

$$g_i(y) = \frac{\mathrm{sgn}(y)}{\sigma}. \tag{5}$$

Unfortunately, this assumption does not hold for the time-frequency representation $\boldsymbol{X}(\omega_k, n)$. The probability density functions of the components in the bins ω_k can vary in a large range from being sub- to super-Gaussian. A sufficient approximation can be achieved by the generalized Gaussian distribution (GGD) [10]:

$$p_y(y) = \frac{\beta}{2\alpha\Gamma(1/\beta)} e^{-(|y|/\alpha)^\beta} \tag{6}$$

with $\alpha, \beta > 0$ and the Gamma function given by $\Gamma(y) = \int_0^\infty x^{y-1}e^{-x}dx$. The β-parameter of the GGD describes the overall structure of the distribution. With $\beta = 2$ the GGD reduces to standard Gaussian distribution, with $\beta = 1$ to a Laplacian distribution and with $\beta = 0.5$ to a Gamma distribution. Generally, a large value of β indicates a flat distribution, whereas a small value yields a spiky distribution. α is the generalized measure of the standard deviation.

Usually, nonlinearities for super-Gaussian distributions utilize sigmoidal functions like sgn() or tanh(). Using the GGD, this model can be more generalized. The nonlinear function $g_i(\cdot)$ becomes $g_i(x) = |x|^{\beta-1}\text{sgn}(x)$, and using $\text{sgn}(x) = x/|x|$, we obtain

$$g_i(x) = \frac{x}{|x|^{2-\beta}}. \tag{7}$$

As shown in [9], based on this nonlinear function, even mixtures of sub- and super-Gaussian signals can be separated. Although the authors used fixed values for β they could achieve good results.

The above approach has been extended in [11], where an adaptive algorithm has been proposed. Because, in the blind scenario, the sources are not available and therefore an accurate estimation of β is not possible, the authors proposed to calculate β based on the statistics of the separated signals. They used the method of moments [12] to estimate β after each iteration of (3) and used this new value for the next step. It was shown that the approach leads to improved overall performance in terms of better separation and faster convergence.

2.3 Convolutive Mixtures

In real-world acoustic scenarios, the mixing channels can be modeled by FIR filters of length L, where L can be 2000 or more, depending on the reverberation time and sampling rate. The convolutive mixing model reads

$$\boldsymbol{x}(n) = \boldsymbol{H}(n) * \boldsymbol{s}(n) = \sum_{l=0}^{L-1} \boldsymbol{H}(l)\boldsymbol{s}(n-l) \tag{8}$$

where $\boldsymbol{H}(n)$ is a sequence of $N \times N$ matrices containing the impulse responses of the mixing channels. For the separation we use FIR filters of length $M \geq L-1$ and obtain

$$\boldsymbol{y}(n) = \boldsymbol{W}(n) * \boldsymbol{x}(n) = \sum_{l=0}^{M-1} \boldsymbol{W}(l)\boldsymbol{x}(n-l) \tag{9}$$

with $\boldsymbol{W}(n)$ containing the unmixing coefficients.

Estimating $\boldsymbol{W}(n)$ in the time domain is a very difficult task, because the number of unknowns, MN^2, can reach several tens of thousands. Although there exist approaches to this problem [3,4] the results are not satisfying because of distortions introduced by the unmixing system.

Due to this problem another approach is widely used. After transforming the signals to the frequency domain, for example using the blockwise Short-Time-Fourier-Transform (STFT), the convolution becomes a multiplication [5]:

$$\boldsymbol{Y}(\omega_k, n) = \boldsymbol{W}(\omega_k)\boldsymbol{X}(\omega_k, n) \tag{10}$$

Instead of estimating all coefficients at once, in the frequency domain it is possible to separate every bin independently. However, since there is the scaling and permutation ambiguity in every bin, we obtain

$$\boldsymbol{Y}(\omega_k, n) = \boldsymbol{W}(\omega_k)\boldsymbol{X}(\omega_k, n) = \boldsymbol{D}(\omega_k)\boldsymbol{\Pi}(\omega_k)\boldsymbol{S}(\omega_k, n) \tag{11}$$

with $\mathbf{\Pi}(\omega_k)$ being a permutation matrix and $\mathbf{D}(\omega_k)$ a diagonal scaling matrix for frequency ω_k. Therefore, it is necessary to correct the amplitudes and solve the permutation before transforming the signals back to the time domain.

The scaling ambiguity can be resolved to an acceptable degree using the method proposed by Ikeda and Murata [6]. The central idea is to recover the signals as they have been recorded by the sensors. Matusuoka and Nakashima [13] showed that this is the optimal approach, as it minimizes $E\{|y(t) - x(t)|^2\}$. Their Minimal Distortion Principle uses the following unmixing matrix:

$$\boldsymbol{W}'(\omega_k) = \mathrm{diag}(\boldsymbol{W}^{-1}(\omega_k)) \cdot \boldsymbol{W}(\omega_k) \tag{12}$$

with diag(·) returning the argument with all off-diagonal elements set to zero.

The correction of the permutation ambiguity is even more important. Even if every bin is perfectly separated, different permutations at different frequencies make both signals appear in every output channel.

3 Resolving the Permutation Ambiguity

One of the first ideas used for the permutation problem is based on the statistics of the separated signals [6,7]. The key assumption is that the envelopes of all bins of one source are highly correlated. With $\boldsymbol{V}(\omega_k, n) = |\boldsymbol{Y}(\omega_k, n)|$ the correlation between two bins k, l is defined as

$$\rho_{qp}(\omega_k, \omega_l) = \frac{\sum_{n=0}^{N-1} \boldsymbol{V}(\omega_k, n)\boldsymbol{V}(\omega_l, n)}{\sqrt{\sum_{n=0}^{N-1} \boldsymbol{V}^2(\omega_k, n)}\sqrt{\sum_{n=0}^{N-1} \boldsymbol{V}^2(\omega_l, n)}} \tag{13}$$

with p, q being the indices of the separated signals. To decide if two bins are permutated equally, the value of

$$r = \frac{\rho_{pp}(\omega_k, \omega_l) + \rho_{qq}(\omega_k, \omega_l)}{\rho_{pq}(\omega_k, \omega_l) + \rho_{qp}(\omega_k, \omega_l)} \tag{14}$$

can be used. If $r > 1$, then the bins are sorted correctly. Otherwise, with $r < 1$, a permutation has occurred. With more than two sources the value of r has to be estimated for all pairs, which means that $N!$ calculations have to be performed.

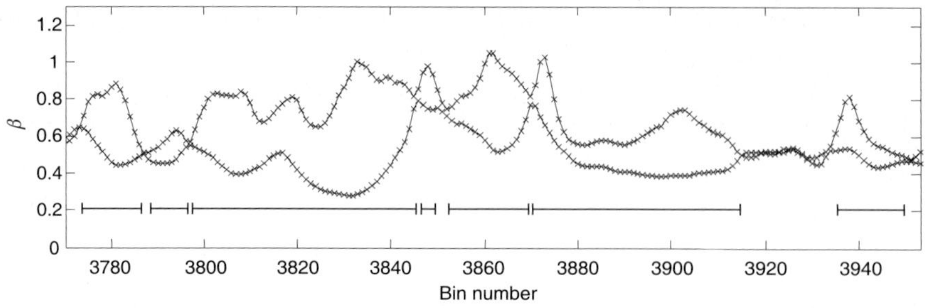

Fig. 1. Beta values of two signals over the frequency index. The detected clusters are indicated with bars ⊢⊣.

Although there are algorithms with less complexity, the practical use is restricted to only few sources [7].

Trying to sort all bins with respect to r for all p and q usually does not work for speech signals. The reason for this is that the key assumption of highly correlated envelopes often does not hold for frequencies which are not close together. Restricting the test to only neighboring frequencies is also not a solution, because at some frequencies, the envelopes of the individual signals do not differ enough to allow for correct sorting. A compromise is the dyadic sorting [7], which starts with pairwise correlation of two neighboring bins and then successively builds groups of bins in a dyadic fashion. This algorithm utilizes the fact that, in a sorted group, a few outliers do not preponderate, and the groups can be aligned properly. But like other proposals that rely only on the correlation of the separated signals, this algorithm suffers if there are too many poorly separated bins close to each other. Because of this, some of the first small groups are often not sorted properly, which then propagates while building the larger groups. The results are block permutations and the separation of the whole signal fails.

4 The Proposed Method

In this paper we propose to use the smoothness of the exponent β of the GGD. For this, we approximate the statistics of every bin by (6). Although the values of β vary in a significant way, the values in neighboring bins do not differ much. Furthermore, two different signals usually have distinct values in most bins, as can be seen in Fig. 1 for a typical situation. However, it can also be seen in Fig. 1 that there are some bins with almost the same value of β, like the bins around 3920. In this range, no differentiation of the two signals is possible on the basis of the value of β. But huge ranges like the bins 3800-3840, can be clustered with certainty. These clusters can be used to correctly de-permute wide frequency ranges. Afterwards, the remaining bins can be de-permuted using alternative methods.

The proposed method consists of three parts: (1) estimation of the boundaries of the clusters, (2) calculation of the permutation between the clusters, and (3) aligning the remaining bins. The algorithm is at first derived for two signals and then extended for multiple signals.

4.1 Calculation of the Cluster Boundaries

The first step is to estimate the β values for all bins of both estimated sources. One possibility is to estimate this parameter in every iteration of the BSS algorithm mentioned in Section 2.1. Alternatively, any other known BSS algorithms can be used, because β can be also estimated after separation.

The second step is to make a simple grouping. The bins are compared pairwise: the ones with higher values of β are assigned to one and the ones with lower values are assigned to the other source.

The third step is to determine the actual clusters. The idea for a simple and fast method is the following: Take an existing cluster and find out if the neighboring bin can be added to it. The decision is based on the assumption of the values of β being distinct and smooth.

The actual implementation is as follows:

1. Start at bin $l = 1$.
2. Test by comparing the β-values if the next bin $l + 1$ can be added.
3. If yes, then add this bin to the cluster, increase l and go to Step 2.
4. If not, then the end of the cluster has been found. If the cluster is large enough, mark it as being correctly permutated. Increase l, mark l as the beginning of a new cluster and go to Step 2.

The result of this algorithm is shown in Fig. 1.

If there are more than two signals, the algorithm can be extended. For this, the β values are sorted, and the two largest ones are assigned to $\beta_H(\omega_k)$ and $\beta_L(\omega_k)$, respectively. After clustering and removing $\beta_H(\omega_k)$, the same procedure can be applied to the remaining bins. An analogous procedure can be applied to the bottommost values for increased performance.

4.2 Calculation of Cluster Correlations and Aligning the Remaining Bins

The next step after the identification of the clusters is to determine the permutation between them. As the gaps between clusters are usually much smaller than the clusters themselves, the assumption of highly correlated envelopes can be used. Here we follow the idea of dyadic sorting and calculate the value of r, as defined in (14), for all combinations of all bins of two clusters. As the bins within the clusters are de-permuted with high confidence, the correct permutation between clusters can be determined by the highest or lowest value of r, as for the dyadic sorting in [7].

After calculating the correct permutation for the clusters, the remaining bins also have to be aligned. Again, a comparison of the correlation coefficients r for these bins with all coefficients for the bins in the neighboring clusters can be used.

5 Simulations

In a first simulation, the algorithm has been used on unmixed audio signals, which have been arbitrarily permuted in the frequency domain. This should simulate the behavior of the algorithm in ideal conditions, as if the blind separation stage in each frequency bin would be able to work perfectly. In this case, the algorithm was able to correctly de-permute all bins.

When using real-world data, the separation in the single bins is not always perfect. Therefore, the estimation of correct permutations is harder. In the experiments we used a data set where the individual contributions from the sources to the microphones were available [14], and the separation performance could be estimated using the signal-to-interference ratio

$$SIR_{y_i} = 10log_{10}\frac{E[(g_{ii}(n) * s_i(n))^2]}{E[(\sum_{j=1,j\neq i}^{N} g_{ij}(n) * s_j(n))^2]} \tag{15}$$

with $g_{ij}(n) = w_i(n) * h_j(n)$. In Fig. 2, the separation performance for the single bins is given.

As the individual sources are known, the best possible unmixing can be estimated. In Fig. 3, the difference between this best approach and the result of the proposed algorithm is shown. As we can see, above 300 Hz the proposed algorithm produces exactly the same output as the ideal de-permutation. Below this frequency there occur permutations, but this is a frequency range where the separation has failed in several bins. Further inspection of the data showed that the estimation of clusters worked, but the cluster correlations were incorrect. This is a typical behavior for correlation-based approaches, when the separation is not perfect. The overall performance with swapped bins is an SIR of 13.16 dB. When leaving the low frequencies out and recovering only the signal components above 300 Hz, the overall performance increases to 20.03 dB.

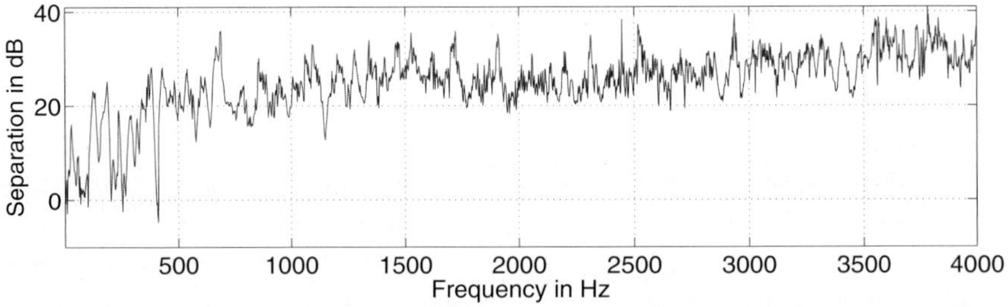

Fig. 2. Separation Performance

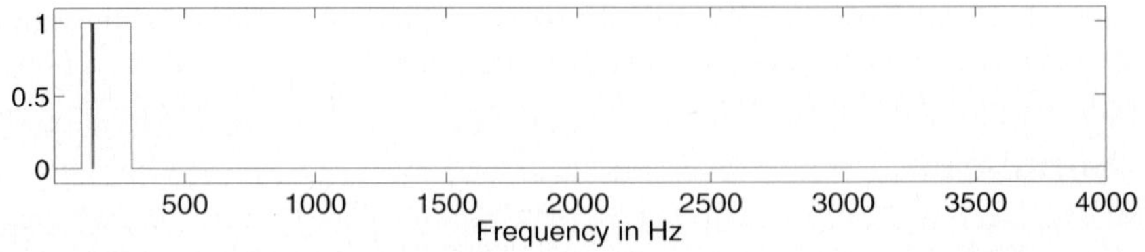

Fig. 3. Swap errors

6 Summary

In this paper, we presented a new approach for resolving the permutation problem, which occurs in convolutive blind source separation. For this we modeled every bin using the generalized Gaussian Distribution and used the exponent β for estimating the correct permutation. The performance of the algorithm has been studied on artificial and real word data.

References

1. Amari, S., Cichocki, A., Yang, H.H.: A new learning algorithm for blind signal separation. In: Touretzky, D.S., Mozer, M.C., Hasselmo, M.E. (eds.) Advances in Neural Information Processing Systems, vol. 8, pp. 757–763. The MIT Press, Cambridge (1996)
2. Hyvärinen, A., Oja, E.: A fast fixed-point algorithm for independent component analysis. Neural Computation 9, 1483–1492 (1997)
3. Douglas, S.C., Sawada, H., Makino, S.: Natural gradient multichannel blind deconvolution and speech separation using causal fir filters. IEEE Trans. Speech and Audio Processing 13(1), 92–104 (2005)
4. Aichner, R., Buchner, H., Araki, S., Makino, S.: On-line time-domain blind source separation of nonstationary convolved signals. In: Proc. 4th Int. Symp. on Independent Component Analysis and Blind Signal Separation (ICA2003), Nara, Japan (April 2003), pp. 987–992 (2003)
5. Smaragdis, P.: Blind separation of convolved mixtures in the frequency domain. Neurocomputing 22(1-3), 21–34 (1998)
6. Ikeda, S., Murata, N.: A method of blind separation based on temporal structure of signals. In: Proc. Int. Conf. on Neural Information Processing, pp. 737–742 (1998)
7. Rahbar, K., Reilly, J.: A frequency domain method for blind source separation of convolutive audio mixtures. IEEE Trans. Speech and Audio Processing 13(5), 832–844 (2005)
8. Sawada, H., Mukai, R., Araki, S., Makino, S.: A robust and precise method for solving the permutation problem of frequency-domain blind source separation. IEEE Trans. Speech and Audio Processing 12(5), 530–538 (2004)
9. Choi, S., Cichocki, A., Amari, S.: Flexible independent component analysis. In: Constantinides, T., Kung, S.Y., Niranjan, M., Wilson, E. (eds.) Neural Networks for Signal Processing VIII, pp. 83–92 (1998)
10. Gazor, S., Zhang, W.: Speech probability distribution. IEEE Signal Processing Letters 10(7), 204–207 (2003)
11. Kokkinakis, K., Nandi, A.K.: Multichannel Speech Separation Using Adaptive Parameterization of Source PDFs. In: ICA 2004. LNCS, vol. 3195, pp. 486–493. Springer, Heidelberg (2004)
12. Varanasi, M.K., Aazhang, B.: Parametric generalized gaussian density estimation. Acoustical Society of America Journal 86(4), 1404–1415 (1989)
13. Matsuoka, K.: Minimal distortion principle for blind source separation. In: Proceedings of the 41st SICE Annual Conference. vol. 4, pp. 2138–2143 (August 5-7, 2002)
14. `http://www.kecl.ntt.co.jp/icl/signal/sawada/demo/bss2to4/index.html`

Discovering Convolutive Speech Phones Using Sparseness and Non-negativity

Paul D. O'Grady[1] and Barak A. Pearlmutter[2]

[1] Complex & Adaptive Systems Laboratory, University College Dublin, Belfield, Dublin 4, Ireland
[2] Hamilton Institute, National University of Ireland Maynooth, Co. Kildare, Ireland
paul.d.ogrady@ucd.ie, barak@cs.nuim.ie

Abstract. Discovering a representation that allows auditory data to be parsimoniously represented is useful for many machine learning and signal processing tasks. Such a representation can be constructed by Non-negative Matrix Factorisation (NMF), which is a method for finding parts-based representations of non-negative data. Here, we present a convolutive NMF algorithm that includes a sparseness constraint on the activations and has multiplicative updates. In combination with a spectral magnitude transform of speech, this method extracts speech phones that exhibit sparse activation patterns, which we use in a supervised separation scheme for monophonic mixtures.

1 Introduction

A preliminary step in many data analysis tasks is to find a suitable representation of the data. Typically, methods exploit the latent structure in the data. For example, ICA [1] reduces the redundancy of the data by projecting the data onto its independent components, which can be discovered by maximising a statistical measure such as independence or non-Gaussianity.

Non-Negative Matrix Factorisation (NMF) approximately decomposes a non-negative matrix $\mathbf{V}$ into a product of two non-negative matrices $\mathbf{W}$ and $\mathbf{H}$ [2, 3]. NMF is a parts-based approach that does not make a statistical assumption about the data. Instead, it assumes that for the domain at hand, negative numbers are physically meaningless. Data that contains negative components, for example audio, must be transformed into a non-negative form before NMF can be applied. Here, we use a magnitude spectrogram. Spectrograms have been used in audio analysis for many years and in combination with NMF have been applied to a variety of problems such as sound separation [4] and automatic transcription of music [5].

In this paper, we combine a previous convolutive extension of NMF [4] with a sparseness constraint on $\mathbf{H}$, and present an algorithm that has multiplicative updates. This paper is structured as follows: We overview convolutive NMF in Section 2 and present sparse convolutive NMF in Section 3. In Section 4 we apply sparse convolutive NMF to speech spectrograms, and extract phones that

M.E. Davies et al. (Eds.): ICA 2007, LNCS 4666, pp. 520–527, 2007.

have sparse activation patterns. We use these phones in a supervised separation scheme for monophonic mixtures, and demonstrate the superior separation performance achieved over those extracted by convolutive NMF in Section 5.

2 Convolutive NMF

NMF [3] is a linear non-negative approximate factorisation, and is formulated as follows. Given a non-negative matrix $\mathbf{V} \in \mathbb{R}^{\geq 0, M \times T}$ the goal is to approximate $\mathbf{V}$ as a product of two non-negative matrices $\mathbf{W} \in \mathbb{R}^{\geq 0, M \times R}$ (basis) and $\mathbf{H} \in \mathbb{R}^{\geq 0, R \times T}$ (activations), $\mathbf{V} \approx \mathbf{WH}$, where $R \leq M$, such that the reconstruction error is minimised. For our purposes we require a convolutive basis, such a model has previously been used to extend NMF [4], which we review in this section.

For conventional NMF each object is described by its spectrum and corresponding activation in time, while for convolutive NMF each object has a sequence of successive spectra and corresponding activation pattern across time. The conventional NMF model is extended to the convolutive case:

$$\mathbf{V} \approx \sum_{t=0}^{T_o - 1} \mathbf{W}_t \overset{t\rightarrow}{\mathbf{H}} \qquad v_{ik} \approx \sum_{t=0}^{T_o-1} \sum_{j=1}^{R} w_{ijt} (\overset{t\rightarrow}{h_{jk}}) \tag{1}$$

where T_o is the length of each spectrum sequence and the j-th column of $\mathbf{W}_t$ describes the spectrum of the j-th object t time steps after the object has begun. The function $\overset{i\rightarrow}{(\cdot)}$ denotes a column shift operator that moves its argument i places to the right; as each column is shifted off to the right the leftmost columns are zero filled. Conversely, the $\overset{\leftarrow i}{(\cdot)}$ operator shifts columns off to the left, with zero filling on the right. We use the beta divergence, which is a parameterisable divergence, as the reconstruction objective,

$$D_{\text{BD}}(\mathbf{V} \| \mathbf{\Lambda}, \beta) = \sum_{ik} \Big(v_{ik} \frac{v_{ik}^{\beta-1} - [\mathbf{\Lambda}]_{ik}{}^{\beta-1}}{\beta(\beta-1)} + [\mathbf{\Lambda}]_{ik}{}^{\beta-1} \frac{[\mathbf{\Lambda}]_{ik} - v_{ik}}{\beta} \Big), \tag{2}$$

where β controls reconstruction penalty and $\mathbf{\Lambda}$ is the current estimate of $\mathbf{V}$, $\mathbf{\Lambda} = \sum_{t=0}^{T_o-1} \mathbf{W}_t \overset{t\rightarrow}{\mathbf{H}}$. The choice of the β parameter depends on the statistical distribution of the data, and requires prior knowledge, see [6, Chapter 3]. For $\beta = 2$, Squared Euclidean Distance is obtained; for $\beta \rightarrow 1$, the divergence tends to the Kullback-Leibler Divergence; and for $\beta \rightarrow 0$, it tends to Itakura-Saito Divergence. It is evident that Eq. 1 can be viewed as a summation of T_o conventional NMF operations. Consequently, as opposed to updating two matrices ($\mathbf{W}$ and $\mathbf{H}$) as in conventional NMF, $T_o + 1$ matrices require an update ($\mathbf{W}_0, \ldots, \mathbf{W}_{T_o-1}$ and $\mathbf{H}$). The resultant convolutive NMF update equations are

$$w_{ijt} \leftarrow w_{ijt} \frac{\sum_{k=1}^{T} (v_{ik} / [\mathbf{\Lambda}]_{ik}^{2-\beta}) \overset{t\rightarrow}{h_{jk}}}{\sum_{k=1}^{T} [\mathbf{\Lambda}]_{ik}^{\beta-1} \overset{t\rightarrow}{h_{jk}}}, \quad h_{jk} \leftarrow h_{jk} \frac{\sum_{i=1}^{M} w_{ijt} \overset{\longleftarrow t}{(v_{ik}/[\mathbf{\Lambda}]_{ik}^{2-\beta})}}{\sum_{i=1}^{M} w_{ijt} \overset{\leftarrow t}{[\mathbf{\Lambda}]}_{ik}^{\beta-1}}, \tag{3}$$

where $\mathbf{H}$ is updated to the average result of its updates for all t. When $T_o = 1$ this reduces to conventional NMF.

3 Sparse Convolutive NMF

Combining our reconstruction objective (Eq. 2) with a sparseness constraint on $\mathbf{H}$ results in the following objective function:

$$G(\mathbf{V}\|\mathbf{\Lambda}, \mathbf{H}, \beta) = D_{\mathrm{BD}}(\mathbf{V}\|\mathbf{\Lambda}, \beta) + \lambda \sum_{jk} h_{jk}, \tag{4}$$

where the left term of the objective function corresponds to convolutive NMF, and the right term is an additional constraint on $\mathbf{H}$ that enforces sparsity by minimising the L_1-norm of its elements. The parameter λ controls the trade off between sparseness and accurate reconstruction.

3.1 Basis Normalisation

The objective of Eq. 4 creates a new problem: The right term is a strictly increasing function of the absolute value of its argument, so it is possible that the objective can be decreased by scaling w_{ijt} up and $\mathbf{H}$ down ($w_{ijt} \mapsto \alpha w_{ijt}$ and $\mathbf{H} \mapsto (1/\alpha)\mathbf{H}$, with $\alpha > 1$). This situation does not alter the left term in the objective function, but will cause the right term to decrease, resulting in the elements of w_{ijt} growing without bound and $\mathbf{H}$ tending toward zero. Consequently, the solution arrived at by the optimisation algorithm is not influenced by the sparseness constraint.

To avoid the scaling misbehaviour of Eq. 4 another constraint is needed; by normalising the convolutive bases we can control the scale of the elements in w_{ijt} and $\mathbf{H}$. Here, normalisation is performed for each object matrix, $\mathbf{W}_j$, by rescaling it to the unit L_2-norm, $\bar{\mathbf{W}}_j = \frac{\mathbf{W}_j}{\|\mathbf{W}_j\|}, j = 1, \ldots, R$, where the matrix $\mathbf{W}_j$ is constructed from the j-th column of w_{ijt} at each time step, $t = 0, 1, \ldots, T_o - 1$.

3.2 Multiplicative Updates

Multiplicative updates can be obtained by including the normalisation requirement in the objective. Previously, this has been achieved for conventional NMF using the Squared Euclidean Distance reconstruction objective [7]. Here, we present the multiplicative updates for a convolutive NMF algorithm utilising beta divergence. Our new reconstruction objective is a modification of Eq. 2 where each object, $\mathbf{W}_j$, is normalised, $\bar{\mathbf{W}}_j$, resulting in the following generative model: $\mathbf{\Delta} = \sum_{t=0}^{T_o-1} \sum_{j=1}^{R} \bar{\mathbf{w}}_{jt}(\overset{t\rightarrow}{\mathbf{h}_j})$. By substituting $\mathbf{\Lambda}$ for $\mathbf{\Delta}$ in Eq. 4 we obtain [6, Chapter 4] the following multiplicative update rules for $\mathbf{H}$ and $\mathbf{W}$:

$$h_{jk} \leftarrow h_{jk} \frac{\sum_{i=1}^{M} \bar{w}_{ijt} (v_{ik} / \overset{\longleftarrow t}{[\mathbf{\Delta}]}_{ik}^{2-\beta})}{\sum_{i=1}^{M} \bar{w}_{ijt} \overset{\leftarrow t}{[\mathbf{\Delta}]}_{ik}^{\beta-1} + \lambda}, \tag{5}$$

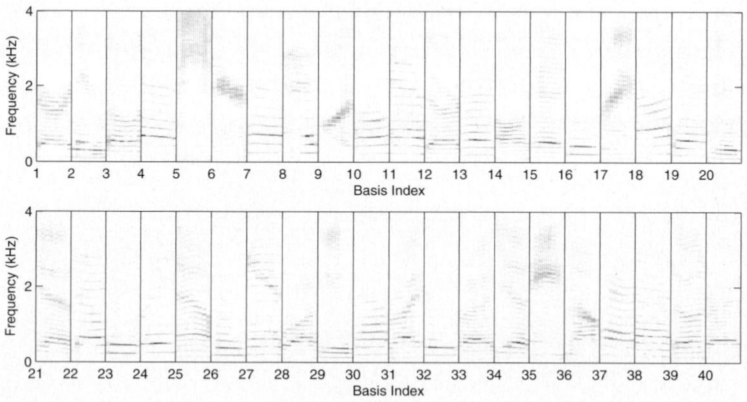

Fig. 1. A collection of 40 phone-like basis functions for a mixture of a male (`DMT0`) and female speaker (`SMA0`) taken from the TIMIT speech database

$$w_{ijt} \leftarrow w_{ijt} \frac{\sum_{k=1}^{T} \overset{t\rightarrow}{h_{jk}} \left[(v_{ik}/[\boldsymbol{\Delta}]_{ik}^{2-\beta}) + \bar{w}_{ijt} (\bar{w}_{ijt} [\boldsymbol{\Delta}]_{ik}^{\beta-1}) \right]}{\sum_{k=1}^{T} \overset{t\rightarrow}{h_{jk}} \left[[\boldsymbol{\Delta}]_{ik}^{\beta-1} + \bar{w}_{ijt} (\bar{w}_{ijt} (v_{ik}/[\boldsymbol{\Delta}]_{ik}^{2-\beta})) \right]} . \quad (6)$$

4 Sparse Convolutive NMF Applied to Speech Spectra

We apply sparse convolutive NMF to speech, and present a learned basis for the sparse representation of speech using the TIMIT database. Recently, such work has been presented for convolutive NMF [8].

4.1 Discovering a Phone-Like Basis

To illustrate the differences between the phones extracted by convolutive NMF and sparse convolutive NMF we perform the following experiment for both algorithms: We take around 15 seconds of speech from a male (`DMT0`) and female (`SMA0`) speaker to create a contiguous mixture. The data is normalised to unit variance, down-sampled from 16 kHz to 8 kHz and a magnitude spectrogram of the data is constructed. We use a FFT frame size of 512, a frame overlap of 384 and a Hamming window to reduce the presence of sidelobes. We extract 40 bases, $R = 40$, with a temporal extent of 0.176 seconds, $T_o = 8$, and run convolutive NMF (with $\beta = 1$) for 200 iterations. The extracted bases are presented in Figure 1. The experiment is repeated for sparse convolutive NMF with $\lambda = 15$, and the corresponding bases are presented in Figure 2.

For convolutive NMF, it is evident that the extracted bases correspond to speech phones. The verification of which, can be achieved by listening to an audible reconstruction. Most of the phones represent harmonic series with differing pitch inflections, while a smaller subset of phones contain wideband components that correspond to consonant sounds. It is evident for the harmonic phones that

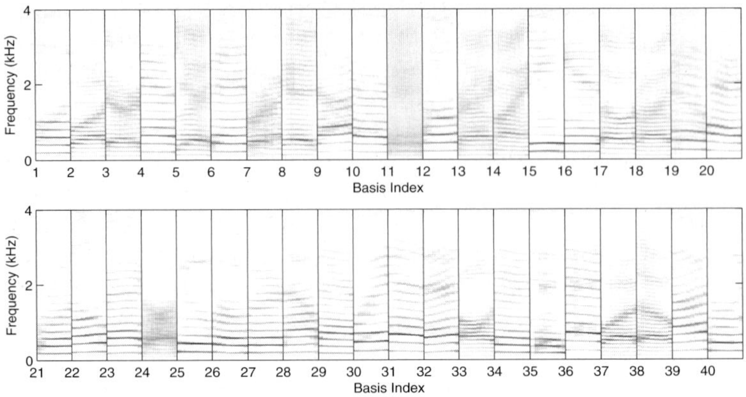

Fig. 2. A collection of 40 phone-like basis functions for a a mixture of a male (`DMTO`) and female speaker (`SMAO`) taken from the TIMIT speech database. The basis is extracted using Spare Convolutive NMF with $\lambda = 15$.

some bases have harmonics that are spaced much closer together, which is indicative of a lower pitched male voice, while others are farther apart, indicating a higher pitched female voice. Therefore, it is evident that the extracted phones correspond to either the male or female speaker, which indicates that the timbral characteristics of the male and female speaker are sufficiently different, such that phones that are representative of both cannot be extracted.

By placing a sparseness constraint on the activations of the basis functions, we specify that the expressive power of each basis be extended such that it is capable of representing phones parsimoniously, much like an over-complete dictionary. The result is that the extracted phones exhibit a structure that is rich in phonetic content, where harmonics at higher frequencies have a much greater intensity than seen in the phones extracted by convolutive NMF. Analysis of the male and female sparse phone set reveals another important difference between the two speakers. In addition to difference in harmonic spacing, it is evident that the structure of the male phones are of a more complex nature, where changes over time are much more varied than for the female phone set.

5 Supervised Method for the Separation of Speakers

As illustrated in our previous experiments, the structure of the bases that are extracted from the speech spectrogram are uniquely dependent on the speaker (given the same algorithm parameters). In the context of speech separation, it is not unreasonable to expect that the bases extracted for a specific speaker adequately characterise the speaker, such that they can be used to discriminate them from other speakers. For a monophonic mixture where a number of speakers are summed together, it is possible to separate the speakers in the mixture by constructing an individual magnitude spectrogram from each speaker, using the phones specific to that speaker. More formally, we use the following procedure for the separation of a known male and female speaker from a monophonic mixture:

1. Obtain training data for the male, $s_m(t)$, and female, $s_f(t)$, speaker, create a magnitude spectrogram for both, and extract corresponding phone sets, $\mathbf{W}_t^m$ and $\mathbf{W}_t^f$, using sparse convolutive NMF.
2. Construct a combined basis set $\mathbf{W}_t^{mf} = [\mathbf{W}_t^m|\mathbf{W}_t^f]$, which results in a basis that is twice as big as R.
3. Take a mixture that is composed of two unknown sentences voiced by our selected speakers, and create a magnitude spectrogram of the mixture. Fit the mixture to $\mathbf{W}_t^{mf}$ by performing sparse convolutive NMF with $\mathbf{W}_t$ fixed to $\mathbf{W}_t^{mf}$, and learn only the associated activations $\mathbf{H}$.
4. Partition $\mathbf{H}$ such that the activations are split into male, $\mathbf{H}^m$, and female, $\mathbf{H}^f$, parts that correspond to their associated bases, $\mathbf{H} = [\mathbf{H}^m|\mathbf{H}^f]^\mathsf{T}$.
5. Construct a magnitude spectrogram for both speakers, using their respective bases and activations: $\mathbf{S}^m = \sum_{t=0}^{T_o-1} \mathbf{W}_t^m \mathbf{H}^m$ and $\mathbf{S}^f = \sum_{t=0}^{T_o-1} \mathbf{W}_t^f \mathbf{H}^f$.
6. Use the phase information from the mixture to create an audible reconstruction for both speakers.

This procedure may also be used for convolutive NMF, and can be generalised for more than two speakers, and speakers of the same gender.

5.1 Separation Experiments

Here, we compare the separation performance of convolutive NMF and sparse convolutive NMF. For an extensive study of the relationship between parameter selection and separation performance for convolutive NMF, see [8].

We select three male (`ABCO`, `BJVO`, `DWMO`) and three female (`EXMO`, `KLHO`, `REHO`) speakers from the TIMIT database, and create a training set for each that includes all but one sentence voiced by that speaker. We artificially generate a monophonic mixture by summing the remaining sentences for a selected male-female pair, for a total of nine mixtures. Each sentence pair is normalised to unit variance, down-sampled from 16 kHz to 8 kHz, and summed together. A magnitude spectrogram of each mixture is constructed using an FFT frame size of 512, a frame overlap of 256 and a Hamming window.

The separation performance for both algorithms is evaluated for each mixture over a selection of values for R ($R = \{40\ 80\ 140\ 220\}$). For both algorithms the temporal extent of each phone is set to 0.224 seconds ($T_o = 6$), the number of iterations is 150, β is set to 1 and each experiment is repeated for 10 Monte Carlo runs. For convolutive NMF, a total of 24 speaker phone sets are extracted and used in 360 ($9 \times 4 \times 10$) separation experiments. For sparse convolutive NMF separation performance is tested for $\lambda = \{0.01\ 0.1\ 0.3\ 1.0\ 2.0\}$; resulting in 120 ($6 \times 4 \times 5$) speaker phone sets and 1800 ($9 \times 4 \times 5 \times 10$) separation experiments.

For the purposes of ease of comparison with existing separation methods, we evaluate the separation performance of both algorithms using the *Source-to-Distortion Ratio* (SDR) measure provided by the `BSS_EVAL` toolbox [9]; SDR

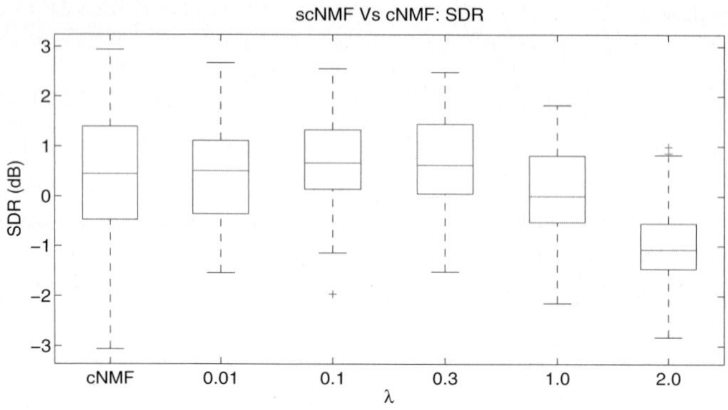

Fig. 3. A comparison of the SDR results obtained by convolutive and sparse convolutive NMF: Box plots are used to illustrate the performance results, where each box represents the median and the interquartile range of the results. It is evident that for $\lambda = 0.1$, a better spread of results is obtained, indicating that sparse convolutive NMF achieves superior overall performance.

indicates overall separation performance and is expressed in dB, with higher performance values indicating better quality estimates. An extensive investigation utilising all measures provided by the toolbox is presented in [6, Chapter 4].

5.2 Separation Performance

We statistically analyse the performance of convolutive NMF and sparse convolutive NMF by collating the results from all experiments and presenting the results using box plots: Each box presents information about the median and the statistical dispersion of the results. The top and bottom of each box represents the upper and lower quartiles, while the length between them is the interquartile range; the whiskers represent the extent of the rest of the data, and outliers are represented by +. Box plots for SDR are presented in Figure 3.

The SDR results indicate that for $\lambda = \{0.1, 0.3\}$, the median performance obtained (0.66 dB, 0.62 dB) exceeds convolutive NMF (0.44 dB), for our given algorithm parameters. It is also evident that a better spread of results is produced for sparse convolutive NMF; demonstrating that when λ is chosen appropriately, sparse convolutive NMF achieves superior overall performance. However, audible reconstructions reveal that convolutive NMF is more resilient to artifacts; this may reflect the fact that each sparse phone set exhibits phones that are rich in features, which may manifest as artifacts in the resultant source estimates. It is also evident that the performance of the sparse convolutive algorithm degrades significantly for large λ, so much so, that it renders the results useless, for our data this is especially evident for $\lambda > 1$.

6 Conclusion

In this paper, we presented a sparse convolutive NMF algorithm with multiplicative updates, which effectively discovers a sparse parts-based convolutive representation for non-negative data. This method extends the convolutive NMF objective by including a sparseness constraint on the activation patterns, enabling the discovery of over-complete representations. Furthermore, we demonstrated the superiority of sparse convolutive NMF over convolutive NMF, when applied to a supervised monophonic speech separation task.

Acknowledgements

Supported by Higher Education Authority of Ireland (An tÚdarás Um Ard-Oideachas), and Science Foundation Ireland grant 00/PI.1/C067.

References

[1] Comon, P.: Independent component analysis: A new concept. Signal Processing 36, 287–314 (1994)
[2] Paatero, P., Tapper, U.: Positive matrix factorization: A nonnegative factor model with optimal utilization of error estimates of data values. Environmetrics 5, 111–126 (1994)
[3] Lee, D.D., Seung, H.S.: Algorithms for non-negative matrix factorization. In: Adv. in Neu. Info. Proc. Sys. 13, pp. 556–562. MIT Press, Cambridge (2001), URL `citeseer.ist.psu.edu/lee00algorithms.html`
[4] Smaragdis, P.: Non-negative matrix factor deconvolution; extraction of multiple sound sources from monophonic inputs. In: Puntonet, C.G., Prieto, A.G. (eds.) ICA 2004. LNCS, vol. 3195, pp. 494–499. Springer, Heidelberg (2004)
[5] Abdallah, S.A., Plumbley, M.D.: Polyphonic transcription by non-negative sparse coding of power spectra. In: Proceedings of the 5th International Conference on Music Information Retrieval (ISMIR 2004), pp. 318–325 (2004)
[6] O'Grady, P.D.: Sparse Separation of Under-Determined Speech Mixtures. PhD thesis, National University of Ireland Maynooth (2007), URL `http://ee.ucd.ie/~pogrady/ogrady2007_phd.pdf`
[7] Eggert, J., Körner, E.: Sparse coding and NMF. In: IEEE International Joint Conference on Neural Networks, Proceedings, July 2004, vol. 4, pp. 2529–2533. IEEE, Los Alamitos (2004)
[8] Smaragdis, P.: Convolutive speech bases and their application to supervised speech separation. IEEE Transaction on Audio, Speech and Language Processing (2007)
[9] Févotte, C., Gribonval, R., Vincent, E.: BSS_EVAL toolbox user guide. Technical Report 1706, IRISA (2005)

Frequency-Domain Implementation of a Time-Domain Blind Separation Algorithm for Convolutive Mixtures of Sources

Masashi Ohata and Kiyotoshi Matsuoka

Department of Brain Science and Engineering, Kyushu Institute of Technology, 2-4 Hibikino, Wakamatsu-ku, Kitakyushu city, 808-0196, Japan

Abstract. This paper proposes a way to implement a time-domain blind separation algorithm for convolutive mixtures of source signals. The approach provides another form of the algorithm by discrete Fourier transform and has the possibility of designing a separating filter in the frequency domain, without bothering about the permutation problem inherent in frequency-domain blind separation approach. This paper also shows a technique to improve separation performance in the frequency domain. The validity of our approach was demonstrated by performing an experiment on separation for convolutive mixtures of two speeches.

Keywords: source separation, convolutive models, time-frequency representations, normalization.

1 Introduction

Blind source separation (BSS) for convolutive mixtures of sources is formulated as follows. Let us consider a case where M sensors receive convolutive mixtures of M statistically independent signals referred to as source signals. The relationship between the source signals and their mixtures are expressed as

$$\mathbf{x}(n)=\sum_{l=0}^{K-1}\mathbf{A}_l\mathbf{s}(n-l), \tag{1}$$

where $\mathbf{s}(n)$ and $\mathbf{x}(n)$ are M-dimensional real-valued vectors, representing collections of the source signals and the observations at discrete time n, respectively. Matrices $\mathbf{A}_l$ represents the impulse response of the channel from the sources to the sensors. Although the channel is not known beforehand, it is assumed to satisfy, 1) the channel is invertible; $\mathbf{A}(z)=\sum_{l=0}^{K-1}\mathbf{A}_l z^{-l}$ is nonsingular for every complex variable z on the unit circle $|z|=1$, 2) each source signal is a stationary non-Gaussian process with zero mean and contains every frequency component.

Approaches to the separation problem can be classified into two types. One is referred as frequency-domain BSS (FD-BSS) and the other as time-domain BSS (TD-BSS); various separation methods are collected in [1]. The former cuts out

M.E. Davies et al. (Eds.): ICA 2007, LNCS 4666, pp. 528–535, 2007.

a series of frames with appropriate length L_s from sequence $\{\mathbf{x}(n)\}$ and converts the frames to the series of frequency data by discrete Fourier transform (DFT):

$$\mathbf{x}[n,k]=\sum_{m=0}^{L_s-1}\mathbf{x}(n+m)e^{-j\omega_k m}\approx\mathbf{A}[k]\mathbf{s}[n,k]\ \ (k=0,1,\ldots,N-1). \tag{2}$$

Here, ω_k denotes a discrete angular frequency: $\omega_k=2\pi k/N$. N is a number by which the interval $[0,2\pi)$ is divided, and is 2 to the power of positive integer large enough ($L_s\le N$). Symbol j is the unit imaginary number. $\{\mathbf{A}[k]\}$ is the DFT of $\{\mathbf{A}_l\}$ and is a collection of $M\times M$ nonsingular matrices. $\mathbf{x}[n,k]$ and $\mathbf{s}[n,k]$ are the DFTs of the segmented sequences of $\mathbf{x}(n)$ and $\mathbf{s}(n)$, respectively. By applying an ICA (independent component analysis) method to the converted sequence individually, independent components can be obtained at each frequency:

$$\mathbf{y}[n,k]=\mathbf{B}[k]\mathbf{x}[n,k]\ \ (k=0,1,\ldots,N-1). \tag{3}$$

Here, $\mathbf{B}[k]$ are $M\times M$ nonsingular matrices. We refer to $\{\mathbf{B}[k]\}$ as a separating filter or shortly a separator hereafter. This approach evaluates the statistical independence at each frequency. The ICA solutions are given in the form of $\tilde{\mathbf{B}}[k]=\mathbf{P}[k]\mathbf{D}[k]\mathbf{A}^{-1}[k]$, where $\mathbf{P}[k]$ and $\mathbf{D}[k]$ represent a permutation matrix and an invertible diagonal matrix, respectively. These two matrices cannot be determined from the statistical independence. When converting back the frequency components to the time domain expression by inverse discrete transform (IDFT), the independent components at multiple frequencies have to be aligned in such a way that the IDFTs of the frequency components correspond to the source signals; it is necessary to set a constant matrix $\mathbf{P}$ to $\mathbf{P}[k]$ over every frequency. To solve the alignment problem, various methods have been proposed; for example, using a time structure of speech signal in [2] and a beamforming technique of microphone array in [3],[4]. Advantages of this approach are that the algorithm is simple and can be parallelized on a computer, and that the separation performance can be improved in the frequency domain. On the other hand, the TD-BSS approach does not require such an alignment step. However, separation performance in the frequency domain is not taken into account.

To make the most of the advantages of these two approaches, this paper shows a way to implement a TD-BSS algorithm in the frequency domain. More specifically, our method evaluates the statistical independence among signals in the time domain, but designs a separating filter in the frequency domain.

2 Time-Domain Blind Source Separation

2.1 Basic Algorithm

The output sequence of a separator, i.e., the IDFT of Eq.(3) is given as

$$\mathbf{y}(n+m)=\frac{1}{N}\sum_{k=0}^{N-1}\mathbf{y}[n,k]e^{j\omega_k m}\ \ (m=0,1,\ldots,N-1), \tag{4}$$

where $\mathbf{y}(n) = [y_1(n), \ldots, y_M(n)]^T$ (superscript 'T' denotes the transpose of a vector or a matrix). Since $\mathbf{x}(n)$ is real-valued, its DFT satisfies $\mathbf{x}[n,k] = \mathbf{x}^*[n, N-k]$, where superscript '$*$' denotes the conjugate of a complex value. Equation (4) is real-valued if and only if the DFT sequence $\{\mathbf{B}[k]\}$ satisfies $\mathbf{B}[k] = \mathbf{B}^*[N-k]$ $(k=0,1,\ldots,N-1)$.

Several contrast functions have been proposed to solve blind source separation problem. In this paper, an information-theoretical approach is employed:

$$C(n, \{\mathbf{B}[k]\}) = -\sum_{m=0}^{N-1}\sum_{p=1}^{M} \log r_p(y_p(n+m)) - \frac{1}{2}\sum_{k=0}^{N-1} \log\det \mathbf{B}[k]\mathbf{B}^H[k]. \quad (5)$$

Here, superscript 'H' denotes the conjugate transpose of a vector or a matrix, and $r_p(y)$ is a model for the probability density function (pdf) of source p. This is the DFT expression of the contrast function mentioned in [5]. Since DFT and IDFT can be expressed using nonsingular matrices, the optimization of the contrast function in the discrete time domain is equivalent to that in the discrete frequency domain (when the length of an optimized filter is N). The contrast function can evaluate the independence among the filter outputs to input $\{\mathbf{x}(n), \ldots, \mathbf{x}(n+L_s-1)\}$.

To search for a desired separator, the function is minimized with respect to matrices $\mathbf{B}[k]$. It is necessary that sequence $\{\mathbf{B}[k]\}$ should satisfy $\mathbf{B}[k] = \mathbf{B}^*[N-k]$ during the minimization. Let $\mathbf{B}[n,k]$ be an updated filter of $\mathbf{B}[k]$ at the n-th iteration and $\Delta\mathbf{B}[n,k]$ be an update value for the filter: $\mathbf{B}[n+1,k] = \mathbf{B}[n,k] + \Delta\mathbf{B}[n,k]$. The derivative of $(-\log r_p(y))$ is denoted by $\varphi_p(y)$. Let $\varphi(\mathbf{y}(n))$ be the M-dimensional column vector whose elements are $\varphi_p(y_p(n))$. Define $\mathbf{f}[n,k] = [\varphi_1[n,k], \ldots, \varphi_M[n,k]]^T = \sum_{m=0}^{N-1} \varphi(\mathbf{y}(n+m))e^{-j\omega_k m}$. Using the natural gradient method proposed by Amari [6], the following rule is obtained:

$$\begin{aligned} \Delta\mathbf{B}[n,k] &= \alpha\left\{N\mathbf{I} - \mathbf{f}[n,k]\mathbf{y}^H[n,k]\right\}\mathbf{B}[n,k], \\ \Delta\mathbf{B}[n,N-k] &= \Delta\mathbf{B}^*[n,k] \ (k=0,1,\ldots,N/2). \end{aligned} \quad (6)$$

Here, α is a small positive constant and $\mathbf{I}$ is the $M \times M$ identity matrix. This rule is available to the case of independent, identically distributed (i.i.d.) source signals. But, the rule is undesirable for color sources.

It is possible to obtain a set of unwhitened independent signals by employing the nonholonomic constraint method proposed by Amari et al.[7]. Specifically, by applying diag $\Delta\mathbf{B}[n,k]\mathbf{B}^{-1}[n,k] = \mathbf{O}$ to Eq.(6), the following rule is obtained:

$$\Delta\mathbf{B}[n,k] = -\alpha \text{ off-diag } \mathbf{f}[n,k]\mathbf{y}^H[n,k] \cdot \mathbf{B}[n,k]. \quad (7)$$

Here, diag $\mathbf{Z}$ (off-diag $\mathbf{Z}$) represents the matrix whose diagonal (off-diagonal) elements are identical to those of square matrix $\mathbf{Z}$ and other elements are zeros.

2.2 Proposed Algorithm

Algorithm (7) is affected by the magnitudes of $y_p[n,k]$ and $\varphi_p[n,k]$, because they are distributed in a wide range. The learning rate requires being carefully set to obtain sufficient separation results. This may induce good separation in some frequency ranges, but poor separation in other ranges. To improve the separation performance, we modify the algorithm as follows. Let $\gamma_p^2[n,k]$ and $\sigma_p^2[n,k]$ be estimates of the variances of $\varphi_p[n,k]$ and $y_p[n,k]$, respectively. The estimates are updated in accordance with

$$\gamma_p^2[n,k]=(1-\beta_1)\gamma_p^2[n-1,k]+\beta_1|\varphi_p[n,k]|^2, \quad (8)$$

$$\sigma_p^2[n,k]=(1-\beta_2)\sigma_p^2[n-1,k]+\beta_2|y_p[n,k]|^2, \quad (9)$$

where β_1 and β_2 are positive constants smaller than one. Define $M \times M$ diagonal matrices $\mathbf{\Gamma}[n,k] = \mathrm{diag}\{1/\sqrt{\gamma_1^2[n,k]}, \ldots, 1/\sqrt{\gamma_M^2[n,k]}\}$ and $\mathbf{\Xi}[n,k] = \mathrm{diag}\{1/\sqrt{\sigma_1^2[n,k]}, \ldots, 1/\sqrt{\sigma_M^2[n,k]}\}$. By using these positive-definite matrices and a method proposed in [8], the separation algorithm can be extended as

$$\Delta\mathbf{B}[n,k] = -\alpha \text{ off-diag } \mathbf{\Gamma}[n,k]\mathbf{f}[n,k]\mathbf{y}^H[n,k]\mathbf{\Xi}[n,k] \cdot \mathbf{B}[n,k]. \quad (10)$$

The elements of the time average of $\mathbf{\Gamma}[n,k]\mathbf{f}[n,k]\mathbf{y}^H[n,k]\mathbf{\Xi}[n,k]$ are the correlation coefficients corresponding to those of $\mathbf{f}[n,k]\mathbf{y}^H[n,k]$.

Learning rules (7) and (10) cannot obtain a separator with unique $\mathbf{D}[k]$. It is possible to remove the ambiguity on $\mathbf{D}[k]$ by using the minimal distortion principle (MDP), which is originally proposed by Matsuoka and Nakashima [9]. The MDP separator can be obtained as

$$\hat{\mathbf{B}}[n+1,k]=\mathrm{diag}\,\mathbf{B}^{-1}[n+1,k] \cdot \mathbf{B}[n+1,k]e^{-j\omega_k\tau}, \quad (11)$$

where τ is a positive integer representing a delay (the similar equation is mentioned in [4]). The MDP solution can be also iteratively obtained without calculating the inverse of a matrix; the procedure is omitted in this paper.

A given sequence is segmented into a series of frames with length L_s by shifting a frame by interval R and then the shifted segments are transformed to frequency-sequences by DFT. Let l be an integer variable representing the frame position. Setting $n = lR$ in Eq.(10), the following iterative procedure is obtained:

P1: Initialization: $l=0, \mathbf{B}[0,k], \gamma_p^2[0,k], \sigma_p^2[0,k]$ $(k=0,1,\ldots,N/2)$,
P2: Obtain $\mathbf{x}[lR,k]=\sum_{m=0}^{L_s-1}\mathbf{x}(lR+m)e^{-j\omega_k m}$,
P3: Calculate $\mathbf{y}[lR,k]=\mathbf{B}[lR,k]\mathbf{x}[lR,k](\mathbf{y}[lR,N-k]=\mathbf{y}^*[lR,k])$,
P4: Obtain the IDFT $\{\mathbf{y}(lR),\ldots,\mathbf{y}(lR+N-1)\}$ of $\{\mathbf{y}[lR,k]\}$.
P5: Calculate $\{\mathbf{f}[lR,k]\}$: the DFT of $\{\varphi(\mathbf{y}(lR)),\ldots,\varphi(\mathbf{y}(lR+N-1))\}$,
P6: Update $\gamma_p^2[lR,k]$ and $\sigma_p^2[lR,k]$ by Eqs.(8) and (9), respectively,
P7: Update separator in accordance with Eq.(10),
P8: Increment l: $l \leftarrow l+1$,
P9: Go to P2.

The convolution using DFT is basically not linear convolution, but circular convolution as mentioned in [10]. The output $\mathbf{y}(n+m)(m = 0,\ldots,N-1)$ obtained by Eqs.(3) and (4) is a linear convolution between $\{\mathbf{x}(n),\ldots,\mathbf{x}(n+$

$L_s - 1)\}$ and impulse response $\{\mathbf{B}_l\}$ corresponding to the filter $\{\mathbf{B}[k]\}$ if the length L_f of the impulse response and L_s satisfy $L_f + L_s - 1 \leq N$. That may affect the search of a desired separator because the length of the IDFT of $\{\mathbf{B}[k]\}$ does not necessarily satisfy the inequality; the length is N in general. During the learning of a separator, it is necessary that the inequality holds. Thus, the impulse response $\{\mathbf{B}_l\}$ should be truncated at fixed intervals during the learning.

3 Experiment

An experiment on speech separation was performed to demonstrate the validity of our algorithm.

Two different speeches were produced with two loudspeakers and their mixtures were recorded with two omnidirectional microphones ($M = 2$) in a room. The loudspeakers and the microphones were arranged as shown in Fig.1 (the reverberation time was approximately 550 ms). They were set 1.2 m high from the floor. While the position of Loudspeaker 2 was fixed, that of Loudspeaker 1 was changed as $\theta = 0, 15, 30, 45, 60, 75, 90$ [deg].

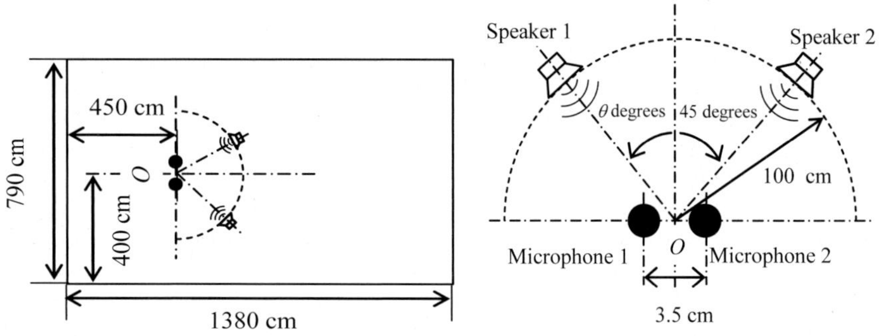

(a) Positions of the microphones and the speakers in a room

(b) Arrangement of the microphones and the speakers

Fig. 1. Experimental setup

We used four kinds of speech, consisting of two men voices and two women voices. We selected every combination of two different speeches from them. The number of the combinations is ${}_4C_2 \times 2 = 12$, which includes permutations of speeches. To easily evaluate separation performance of an obtained separator, we individually recorded two different speeches produced with the loudspeakers by the microphones at each position of Loudspeaker 1. The speeches were recorded at a sampling rate of 8 kHz for ten seconds. The recorded signals in which case a speech is produced with Loudspeaker q is represented as $\mathbf{x}_q(n) = [x_{1q}(n)\ x_{2q}(n)]^T$. We summed these recorded signals on a computer and used the summations as convolutive mixtures of two different speeches:

$\mathbf{x}(n)=\mathbf{x}_1(n)+\mathbf{x}_2(n)$. Let $\mathbf{y}_q(n)=[y_{1q}(n)\ y_{2q}(n)]^T$ be the outputs of separator $\mathbf{B}'(z)$ to inputs $\mathbf{x}_q(n)(q=1,2)$: $\mathbf{y}(n)=\mathbf{y}_1(n)+\mathbf{y}_2(n)$.

The separation performance of the obtained separators can be evaluated in terms of the signal-to-interference ratio (SIR): $\text{SIR}[\mathbf{B}'(z)]=\frac{1}{2}\sum_{p=1}^{2}\text{SIR}_p[\mathbf{B}'(z)]=\frac{1}{2}\sum_{p=1}^{2}10\log_{10}(\max_q\langle y_{pq}^2(n)\rangle/\min_q\langle y_{pq}^2(n)\rangle)$, where $\langle\cdot\rangle$ represents the time average. If a desired separator were obtained, then the ratio would become infinity. A separator with a larger SIR is closer to a desired separator. Let $P_{pq}^{xx}(f)$, $P_r^{yy}(f)$ and $P_{r,pq}^{yx}(f)$ be the power and cross spectra of $x_{pq}(n)$ and $y_r(n)(p,q,r=1,2)$. The coherence between $x_{pq}(n)$ and $y_r(n)$ is defined as

$$Coh_{r,pq}^{yx}(f)=\frac{|P_{r,pq}^{yx}(f)|^2}{P_r^{yy}(f)P_{pq}^{xx}(f)}.$$

This takes a value in the range $[0,1]$. To investigate the separation performance in the frequency domain, we used the function defined as $Coh_{rq}^{yx}(f)=\frac{1}{2}\sum_{p=1}^{2}Coh_{r,pq}^{yx}(f)$. If this function took one in the whole frequency range, then the output signal $y_r(n)$ would be identical to signal $s_q(n)$.

For both the algorithms, the initial value of separator was set to

$$\mathbf{B}[0,k]=\begin{bmatrix}1 & -0.9e^{-j\omega_k}\\ -0.9e^{-j\omega_k} & 1\end{bmatrix}e^{-j\omega_k\tau}\quad(k=0,1,\ldots,N-1).$$

This setting came from a consideration on the basis of a microphone arrey technique (the detail is omitted due to the limited space). For algorithm (10), the initial values of $\gamma_p^2[n,k]$ and $\sigma_p^2[n,k]$ were set to 10.0. The values of the learning rates and other parameters are given in the Table 1. The repetition number in the table represents the frequency at which algorithm is applied to mixtures of speeches. Algorithms (7) and (10) were applied to the speech data and SIRs were calculated for the obtained separators and the MDP separators obtained by Eq.(11). The SIRs were averaged over all the combinations of two different speeches for each position of Loudspeaker 1.

Table 1. Parameters in experiments

Parameters	Algorithm (7)	Algorithm (10)
Sample numbers of obserbations L_s	4096	
Filter length L_f	4000	
Length of DFT N	8192	
Shifting samples R	2750	
Delay τ	2000	
Repetition number	5	
Nonliner function $\varphi_p(y)$	$\tanh(75y)$	
Learning rate for separator α	1.5×10^{-5}	8.5×10^{-3}
Learning rate in Eq.(8) β_1	–	0.15
Learning rate in Eq.(9) β_2	–	0.005

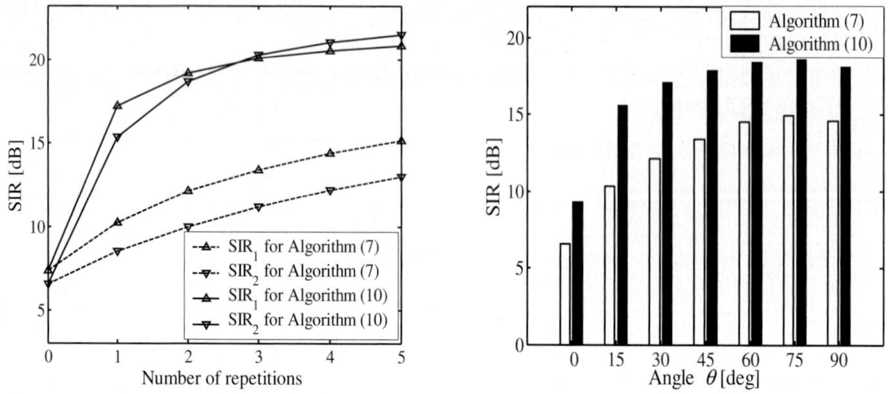

(a) Variations of the averaged SIRs for algorithms (7) and (10) at $\theta = 30$(deg.)

(b) Averaged SIRs for the MDP separators (at the fifth repetition)

Fig. 2. Separation performance in SIR

Figure 2(a) illustrates the variations of the averaged SIRs for the separators obtained by the algorithms with $\theta = 30$ [deg]; the separator is not minimal distortion solutions. This suggests that algorithm (10) can obtain a better separating filter faster than algorithm (7). Figure 2(b) depictes the averaged SIRs for the MDP separator obtained at each location of Loudspeaker 1. This shows that our setting of an initial separator is effective in other cases than $\theta = 0, 15$ [deg]. Figure 3 depictes $Coh_{rq}^{yx}(f)$ for the MDP separators. This result states that, Eq.(10) can design better separator in the frequency domain than Eq.(7).

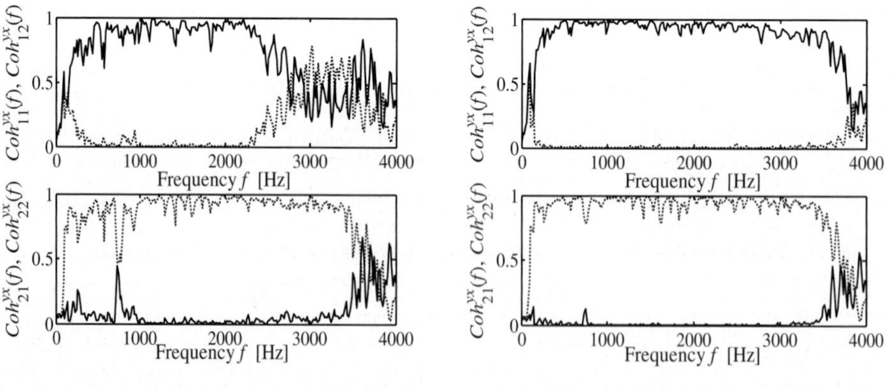

(a) MDP filter $\hat{\mathbf{B}}'(z)$ obtained by (7)

(b) MDP filter $\hat{\mathbf{B}}'(z)$ obtained by (10)

Fig. 3. Coherences for the MDP filters ($\theta = 30$); the solid lines represent $Coh_{11}^{yx}(f)$ and $Coh_{21}^{yx}(f)$, and the dashed lines represent $Coh_{12}^{yx}(f)$ and $Coh_{22}^{yx}(f)$

We programed the above algorithms in MATLAB code. It took approximatly 13.4 second to perform the program of Eq.(10) for a repetition (when the algorithm was applied to ten second data once) on a computer with Pentium 4 CPU-3.0 GHz (Intel Co.).

4 Conclusion

This paper proposed a frequency domain implementation for a time-domain blind separation algorithm. Our method provides the possibility of improving the separation peformance in the frequency domain. This paper shows a modification of the performance by normalizing the terms evaluating the statistical independence in the algorithm. The result of the experiment on speech separation shows that our algorithm can separate source signals from thier mixtures without solving the permutation problem inhelent in FD-BSS.

References

1. Cichocki, A., Amari, S.: Adaptive blind signal and imagine processing. John Wiley & Sons, Lid. Chichester (2002)
2. Murata, N., Ikeda, S., Ziehe, A.: An approach to blind source separation based on temporal structure of speech signal. Neurocomputing 41(4), 1–24 (2001)
3. Saruwatari, H., Kurita, S., Takeda, K., Itakura, F., Nishikawa, T., Shikano, K.: Blind source separation combining independent analysis and beamforming. EURASIP Journal on Applied Signal Processing, No.11, 1135–1146 (2003)
4. Sawada, H., Mukai, R., Araki, S., Makino, S.: Robust and precise method for solving the permutation problem of frequency-domain blind source separation. IEEE Trans. on Speech and Audio Processing 12(5), 530–538 (2004)
5. Amari, S., Douglas, S.C., Cichocki, A., Yang, H.H.: Multichannel blind deconvolution and equalization using the natural gradient. In: Proc. of IEEE International Workshop on Signal processing advances in Wireless Communications, pp. 101–104. IEEE Computer Society Press, Los Alamitos (1997)
6. Amari, S.: Natural gradient learning works efficiently in learning. Neural Computation 10(2), 251–276 (1998)
7. Amari, S., Chen, T.-P., Cichocki, A.: Nonholonomic orthogonal learning algorithms for blind source separation. Neural Computation 12(6), 1463–1484 (2000)
8. Georgiev, P., Cichocki, A., Amati, S.: On some extensions of the natural gradient algorithm. In: Proc. of the Third international conference on Independent Component Analysis and Blind Signal Separation, December 9-12, San Diego, CA, pp. 581–585 (2001)
9. Matsuoka, K., Nakashima, S.: Minimal distortion principle for blind source separation. In: Proc. of the Third international conference on Independent Component Analysis and Blind Signal Separation, December 9-12, San Diego, CA, pp.722–727 (2001)
10. Oppenheim, A.V., Schafer, R.W., Buck, J.R.: Discrete-time signal processing the second edition. Prentice Hall International, Inc, Englewood Cliffs (1998)

Phase-Aware Non-negative Spectrogram Factorization

R. Mitchell Parry and Irfan Essa

Georgia Institute of Technology
College of Computing / GVU Center
85 Fifth Street NW, Atlanta, GA USA
{parry,irfan}@cc.gatech.edu
http://www.cc.gatech.edu/

Abstract. Non-negative spectrogram factorization has been proposed for single-channel source separation tasks. These methods operate on the magnitude or power spectrogram of the input mixture and estimate the magnitude or power spectrogram of source components. The usual assumption is that the mixture spectrogram is well approximated by the sum of source components. However, this relationship additionally depends on the unknown phase of the sources. Using a probabilistic representation of phase, we derive a cost function that incorporates this uncertainty. We compare this cost function against four standard approaches for a variety of spectrogram sizes, numbers of components, and component distributions. This phase-aware cost function reduces the estimation error but is more affected by detection errors.

Keywords: audio processing, source separation, sparse representations, time-frequency representations, unsupervised learning.

1 Introduction

Non-negative spectrogram factorization (NSF) has been proposed for single-channel source separation [1, 2, 3], music transcription [4, 5], and speech recognition [6]. The input mixture is first transformed into a time-frequency representation such as the short-time Fourier transform (STFT). Because of phase-invariant aspects of human hearing the phase information in the STFT is removed yielding the absolute value or absolute square of the STFT (*i.e.*, magnitude or power spectrogram) [2]. The resulting spectrogram matrix is then factored into the sum of rank-one component spectrograms using independent component analysis (ICA) or non-negative matrix factorization (NMF). Each component comprises a static spectral shape and time-varying amplitude envelope. Ideally, each component contains information unique to a particular source for separation or a particular event for transcription. We focus on this basic approach although various other algorithms incorporate sparseness, convolution, or multiple channels [4, 7, 8].

NSF methods commonly assume that the mixture magnitude or power spectrogram is well approximated by the sum of source components. ICA forces this

M.E. Davies et al. (Eds.): ICA 2007, LNCS 4666, pp. 536–543, 2007.

relationship while maximizing the independence of the spectral components [1], whereas NMF minimizes a cost function between the mixture spectrogram and the sum of spectral components [9]. However, because of the nonlinearity of the absolute value function a mixture spectrogram is not the sum of the component spectrograms. Instead, the mixture spectrogram depends on the component spectrograms and their phases. We derive a cost function suitable for NSF by treating the phase as a uniform random variable and maximizing the likelihood of the mixture spectrogram. In previous work, we derived the explicit likelihood function for the case of two components [10]. In this paper, we extend this result to the case of more than two components and show that it is analogous to the multiplicative noise model employed by Abdallah and Plumbley [4]. Even though this cost function is specifically tailored to non-negative spectrograms, the Euclidean distance or generalized Kullback-Leibler divergence is more commonly used for NSF. We compare each cost function based on its ability to estimate the component spectrograms for a variety of spectrogram sizes, numbers of components, and component distributions.

2 Non-negative Matrix Factorization

Non-negative matrix factorization (NMF) was first proposed for the decomposition of images [11]. Image data is inherently non-negative and a single image can be regarded as a linear combination of underlying image parts. NMF estimates these components by minimizing the distance between a set of mixture images contained in the columns of a matrix, A, and the sum of the component matrices, B. The two common distance functions are the Euclidean distance:

$$\|A - B\|^2 = \sum_{ij} (A_{ij} - B_{ij})^2 \tag{1}$$

and a generalized version of the Kullback-Leibler divergence:

$$D(A\|B) = \sum_{ij} \left(A_{ij} \log \frac{A_{ij}}{B_{ij}} - A_{ij} + B_{ij} \right). \tag{2}$$

When applied to non-negative spectrograms, A represents the mixture spectrogram and B represents the sum of component spectrograms. Instead of decomposing multiple images, spectrogram factorization decomposes multiple spectral frames contained in the columns of A. Although magnitude or power spectrograms are non-negative they are *not* a linear combination of underlying component spectrograms because of the nonlinearity of the absolute value function used to generate them.

3 Non-negative Spectrograms

A popular way to transform an audio signal into a series of image-like representations is to extract its frequency spectrum at multiple time-points. We consider the case of one mixture signal and model it as the sum of R source component signals:

$$x(t) = \sum_{r=1}^{R} s_r(t). \tag{3}$$

The short-time Fourier transform (STFT) is a linear transformation into the frequency domain that preserves this relationship:

$$\mathcal{F}_x(k,t) = \sum_{r=1}^{R} \mathcal{F}_{s_r}(k,t). \tag{4}$$

The magnitude spectrogram is the absolute value of the complex-valued STFT:

$$X_{kt} = |\mathcal{F}_x(k,t)| \qquad [S_r]_{kt} = |\mathcal{F}_{s_r}(k,t)|. \tag{5}$$

The original STFT contains additional phase information:

$$\mathcal{F}_x(k,t) = X_{kt}(\cos\Theta_{kt} + i\sin\Theta_{kt}) = \sum_r [S_r]_{kt}(\cos[\Theta_r]_{kt} + i\sin[\Theta_r]_{kt}). \tag{6}$$

When applied to non-negative spectrograms, ICA and NMF estimate rank-one component spectrograms. The columns of a $K \times R$ matrix W specify the spectral shapes and the rows of an $R \times T$ matrix H represent the amplitude envelopes of all the component spectrograms:

$$[S_r]_{kt} = W_{kr}H_{rt}. \tag{7}$$

The various algorithms for NSF vary in the way that they estimate W and H.

4 Non-negative Spectrogram Factorization

The vast majority of NSF methods treat each column of a magnitude or power spectrogram matrix as though it were an image and use ICA or NMF to estimate the components. To our knowledge, there has been only one cost function specifically designed for non-negative spectrograms, namely that of Abdallah and Plumbley [4]. They derive a divergence function based on a multiplicative noise model for estimating the variance (*i.e.*, power) at each time-frequency bin. In this paper, we define the mixture magnitude spectrogram in terms of the component magnitude spectrograms and their phases. Using a probabilistic representation of the phase, we derive an analogous divergence function.

Both ICA- and NMF-based techniques implicitly assume that the mixture non-negative spectrogram, X, is well approximated by the sum of the spectral components, S_r. However, by incorporating the phase of the components, Θ_r, we make this relationship precise:

$$X_{kt} = \sqrt{\sum_{qr} [S_q]_{kt}[S_r]_{kt}\cos\left([\Theta_q]_{kt} - [\Theta_r]_{kt}\right)}. \tag{8}$$

The mixture magnitude spectrogram does not equal the sum of component magnitude spectrograms unless at most *one* component is active at a time or all active components have the *same* phase.

5 Probabilistic Representation of Phase

Given the mixture spectrogram's dependence on the phase in Equation 8, we represent the phase as a uniform random variable. We also make the simplifying assumption that the phase is independent at different time-frequency points. To some degree, this is true. However, the unwrapped phase of a steady state signal can be approximated from the previous two time-steps [12]. Although this violates the independence assumption, we have found that the resulting approach works well in practice.

We wish to maximize the likelihood of the mixture magnitude spectrogram as a function of the source component magnitude spectrograms. For the case of two components, Equation 8 is a function of one random variable (*i.e.*, $\Theta_d = \Theta_1 - \Theta_2$) and it is relatively straightforward to derive $p(X|S_1, S_2)$ directly [10]. However, for more components it becomes increasingly difficult to derive the precise likelihood function. Instead, we estimate the likelihood using the central limit theorem to capture the shape of the distribution for a large number of components.

The probability density function for a complex random variable with magnitude S_r and uniform random phase has a mean of zero and a variance of S_r^2. According to the Lindeberg-Feller central limit theorem [13], the sum of many such variables tends toward a complex Gaussian with zero mean and a variance of $\sum_r S_r^2$. This theorem is valid under the Lindeberg condition, which states that the component variances, S_r^2, are small relative to their sum [13]. Applied to magnitude spectrograms we have the following:

$$p(\mathcal{F}_x|S_1,\ldots,S_R) = \prod_{kt} \frac{1}{\pi \Lambda_{kt}} \exp\left(-\frac{X_{kt}^2}{\Lambda_{kt}}\right), \tag{9}$$

where $\Lambda_{kt} = \sum_r [S_r^2]_{kt}$. We find the likelihood of X by integrating with respect to phase, resulting in a Rayleigh distribution:

$$p(X|S_1,\ldots,S_R) = \prod_{kt} \frac{2X_{kt}}{\Lambda_{kt}} \exp\left(-\frac{X_{kt}^2}{\Lambda_{kt}}\right). \tag{10}$$

6 Maximum Likelihood

In order to estimate S_r, we propose minimizing the negative log likelihood of X:

$$-\log p(X|S_1,\ldots,S_R) = -\sum_{kt}\left[\log\left(\frac{2X_{kt}}{\Lambda_{kt}}\right) - \frac{X_{kt}^2}{\Lambda_{kt}}\right]. \tag{11}$$

For comparison, we frame our maximum likelihood approach in terms of a divergence function. The minimum of Equation 11 is $1 - \log(2/X_{kt})$ at $\Lambda_{kt} = X_{kt}^2$. By subtracting this value we find a divergence function that is non-negative reaching zero only when all $\Lambda_{kt} = X_{kt}^2$:

$$D_s = D(1\|X^2/\Lambda) = \sum_{kt} \frac{X_{kt}^2}{\Lambda_{kt}} - 1 + \log\left(\frac{\Lambda_{kt}}{X_{kt}^2}\right), \tag{12}$$

which is equivalent to Equation 8 in Abdallah and Plumbley [4]. We derive the gradient for D_s with respect to W_{kr}^2 and H_{rt}^2:

$$\frac{\partial D_s}{\partial (W_{kr}^2)} = \sum_t H_{rt}^2 \left(\frac{\Lambda_{kt} - X_{kt}^2}{\Lambda_{kt}^2} \right) \qquad \frac{\partial D_s}{\partial (H_{rt}^2)} = \sum_k W_{kr}^2 \left(\frac{\Lambda_{kt} - X_{kt}^2}{\Lambda_{kt}^2} \right), \tag{13}$$

where $\Lambda_{kt} = \sum_r W_{kr}^2 H_{rt}^2$. Although D_s is not convex with respect to W_{kr}^2 or H_{rt}^2, we find local minima using the following multiplicative update rules:

$$W_{kr}^2 \leftarrow W_{kr}^2 \frac{\sum_t H_{rt}^2 X_{kt}^2 / \Lambda_{kt}^2}{\sum_t H_{rt}^2 / \Lambda_{kt}} \qquad H_{rt}^2 \leftarrow H_{rt}^2 \frac{\sum_k W_{kr}^2 X_{kt}^2 / \Lambda_{kt}^2}{\sum_k W_{kr}^2 / \Lambda_{kt}} . \tag{14}$$

7 Results

We compare the phase-aware cost function, D_s, to four other cost functions based on Euclidean or Kullback-Leibler divergence for magnitude or power spectrograms. Figure 1 plots the shape of the likelihood functions for each of the cost functions with $X = 1$. Magnitude spectrogram methods (E_m and D_m) reach a maximum on the line $S_1 + S_2 = X$. Power spectrogram methods (E_p, D_p, and D_s) reach a maximum on the circle $S_1^2 + S_2^2 = X^2$. When $X = 1$, the sum of S_1 and S_2 must be greater than one. D_s encourages this result by penalizing solutions near the origin more than the other cost functions.

In our experiment, we evaluate the performance of the cost functions for a variety of spectrogram sizes, numbers of components, and component distributions. Specifically, we construct square spectrograms and vary their size with

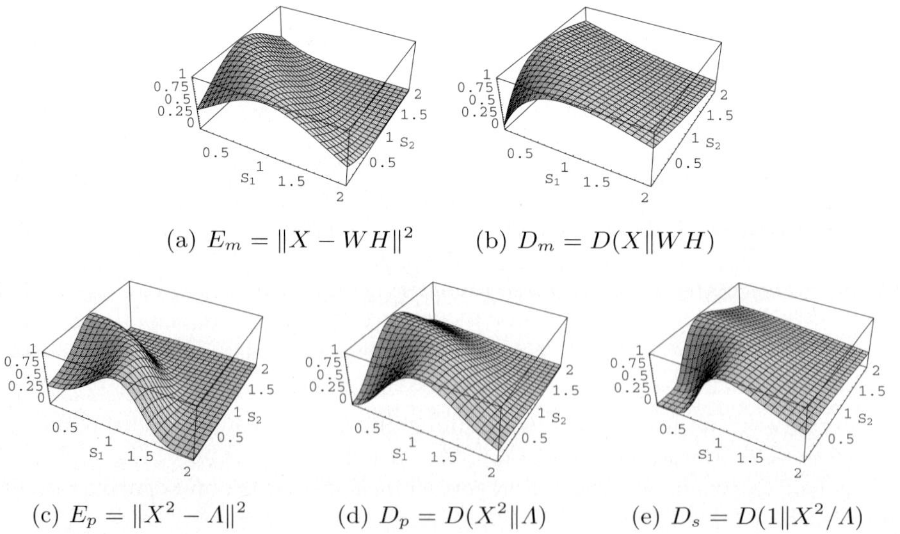

Fig. 1. The shape of the likelihood functions derived from the 5 labeled cost functions for the case of two components and $X = 1$

$K = T \in [32, 64, 128, 256, 512, 1024]$, $R \in [1, \ldots, 30]$, and W and H drawn from the uniform, positive normal, or exponential distribution. After drawing W and H from the specified distribution, we construct X using Equations 5–7 with uniformly distributed random phase, Θ_r. We then estimate W and H using each cost function with multiplicative update rules derived in Section 6 or by Lee and Seung [9]. Because scaling W by α and H by $1/\alpha$ produces the same cost, we normalize the rows of H to unit L_2 norm after every update.

We evaluate each cost function according to the mean square error between the original and estimated $\{S_r\}$. Because the factorization technique is permutation invariant, we must determine the mapping between each estimated and original S_r. For this purpose, we use a greedy algorithm that matches the two most similar components (one original and one estimated) and then removes them from consideration. The process repeats until the mapping is complete.

Figure 2 plots the average performance over ten trials for each configuration of parameters. For space considerations, we only show $R \in [1, \ldots, 10]$ and W and H drawn from the uniform distribution. Each of the 60 $[R, K]$ pairs are sorted along the x-axis in order of increasing minimum error among the five cost functions. Clearly, the problem becomes more difficult as R increases or as K decreases.

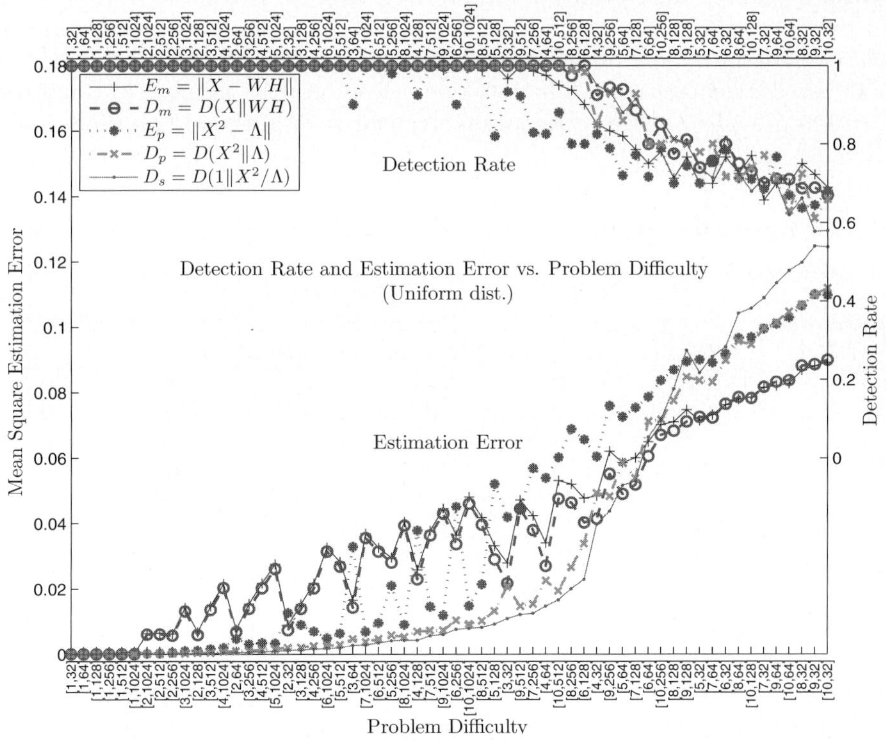

Fig. 2. Estimation error and detection rate for the five cost functions

The bottom of Figure 2 plots the mean square estimation error. For simpler versions of the problem, D_s outperforms the rest. However, toward the right of the plot the performance becomes markedly worse and E_m and D_m perform better. This inversion of performance is linked to the detection rate.

The top of Figure 2 plots the detection rate. When each estimated component uniquely matches a real component, the detection rate is 100%. However, when none of the estimated components match one of the real components, that component is not detected. We compute the detection rate as the fraction of real components that are the closest match (in the mean square sense) for at least one estimated component. At $[R, K] = [4, 32]$, the detection rate for D_s drops below 100% for the first time and this corresponds to the first large increase in estimation error. After that, the estimation rate for D_s accelerates until it is the worst of the group. We speculate that if 100% detection could be maintained, D_s would continue to outperform the others.

The underlying distribution of W and H also affects estimation and detection. As presented, the cost functions implicity assume a uniform prior distribution on W and H in the maximum likelihood framework. Therefore, as the component distributions diverge from the uniform distribution (*e.g.*, become more sparse) the maximum likelihood approach becomes less realistic. The aggregated mean square error for the uniform, positive normal (more sparse), and exponential (most sparse) distribution is 0.036, 0.19, and 0.44, respectively. However, sparseness has the opposite effect on detection. All of the cost functions attain 100% detection for more problems as sparseness increases. Table 1 lists the number of problems that resulted in 100% detection and the number of times each algorithm provides the best estimation error for each of the distributions and R between 2 and 10.

Table 1. Summary of detection rate and lowest estimation error for $R = [2, 10]$

Distribution:	Uniform		Positive Normal		Exponential	
Cost func.	100% det.	Best est.	100% det.	Best est.	100% det.	Best est.
E_m	27	9	37	3	44	0
D_m	34	8	43	6	47	6
E_p	23	0	29	0	30	0
D_p	33	0	38	4	41	3
D_s	35	37	40	41	42	45
Total	152	54	187	54	204	54

8 Conclusion

We present a new derivation of a divergence function, D_s, specifically tuned to non-negative spectrogram factorization. We compare its performance against four standard approaches for a variety of spectrogram sizes, numbers of components, and sparseness. We show that this divergence improves the estimation of the source components. However, it is more affected by detection error. Algorithms aimed at improving detection rates (*e.g.*, a prior distribution on W and H) are likely to improve D_s.

References

1. Casey, M., Westner, W.: Separation of mixed audio sources by independent subspace analysis. In: Proc. of the Int'l. Computer Music Conf. (2000)
2. Smaragdis, P.: Redundancy Reduction for Computational Audition, a Unifying Approach. PhD thesis, MAS Dept. Massachusetts Institute of Technology (2001)
3. Wang, B., Plumbley, M.D.: Investigating single-channel audio source separation methods based on non-negative matrix factorization. In: Rosca, J., Erdogmus, D., Príncipe, J.C., Haykin, S. (eds.) ICA 2006. LNCS, vol. 3889, pp. 17–20. Springer, Heidelberg (2006)
4. Abdallah, S.A., Plumbley, M.D.: Polyphonic transcription by non-negative sparse coding of power spectra. In: Proc. of the Int'l. Conf. on Music Information Retrieval, pp. 318–325 (2004)
5. FitzGerald, D., Coyle, E., Laylor, B.: Sub-band independent subspace analysis for drum transcription. In: Proc. of Int'l. Conf. on Digital Audio Effects, pp. 65–69 (2002)
6. Raj, B., Singh, R., Smaragdis, P.: Recognizing speech from simultaneous speakers. In: Eurospeech (2005)
7. Virtanen, T.: Separation of sound sources by convolutive sparse coding. In: ISCA Tutorial & Research Wkshp on Statistical & Perceptual Audio Processing (2004)
8. FitzGerald, D., Cranitch, M., Coyle, E.: Sound source separation using shifted non-negative tensor factorisation. In: Proc. of the IEEE ICASSP (2006)
9. Lee, D.D., Seung, H.S.: Algorithms for non-negative matrix factorization. In: Advances in NIPS 13, pp. 556–562. MIT Press, Cambridge (2001)
10. Parry, R.M., Essa, I.: Incorporating phase information for source separation via spectrogram factorization. In: Proc. of IEEE ICASSP (2007)
11. Lee, D.D., Seung, H.S.: Learning the parts of objects by non-negative matrix factorization. Nature 401, 788–791 (1999)
12. Bello, J.P., Sandler, M.B.: Phase-based note onset detection for music signals. In: Proc. of the IEEE ICASSP. vol. 5, pp. 441–444 (2003)
13. Feller, W.: An Introduction to Probability Theory and Its Applications. Wiley, New York (1971)

Probabilistic Amplitude Demodulation

Richard E. Turner and Maneesh Sahani

Gatsby Computational Neuroscience Unit, UCL,
Alexandra House, 17 Queen Square, London, U.K.
{turner,maneesh}@gatsby.ucl.ac.uk
http://www.gatsby.ucl.ac.uk

Abstract. Auditory scene analysis is extremely challenging. One approach, perhaps that adopted by the brain, is to shape useful representations of sounds on prior knowledge about their statistical structure. For example, sounds with harmonic sections are common and so time-frequency representations are efficient. Most current representations concentrate on the shorter components. Here, we propose representations for structures on longer time-scales, like the phonemes and sentences of speech. We decompose a sound into a product of processes, each with its own characteristic time-scale. This demodulation cascade relates to classical amplitude demodulation, but traditional algorithms fail to realise the representation fully. A new approach, probabilistic amplitude demodulation, is shown to out-perform the established methods, and to easily extend to representation of a full demodulation cascade.

Keywords: audio processing, dynamic and temporal models, hierarchical models, sparse representations, unsupervised learning.

1 Introduction

Natural sounds are structured on many time-scales. A typical segment of speech, for example, contains features that span four orders of magnitude: Sentences (~ 1 s); phonemes ($\sim 10^{-1}$ s); glottal pulses ($\sim 10^{-2}$ s); and formants ($\sim 10^{-3}$ s or less). This temporal diversity results directly from the diversity of physical processes that support and control sound production. If the impact of these many processes could be expressed in a single efficient representation, then difficult problems like source separation and auditory scene analysis, that are routinely solved by the brain, might become more accessible to machine audition. However, the diversity of structures and time-scales makes this hard, and most work has concentrated on shorter-time features (e.g. time-frequency representations). Here, we introduce representations that capture longer-range temporal structure in natural sounds. For speech, which will be the running example throughout, this means the sentence and phoneme structure. The basic idea is to represent a sound as a product of processes drawn from a hierarchy, or cascade, of progressively longer time-scale modulators. For speech this might involve three processes: representing sentences on top, phonemes in the middle, and pitch and formants at the bottom (e.g. fig. 2A). To construct such a representation, one

M.E. Davies et al. (Eds.): ICA 2007, LNCS 4666, pp. 544–551, 2007.

might start with a traditional amplitude demodulation algorithm, which decomposes a signal into a quickly-varying carrier and more slowly-varying envelope. The cascade could then be built by applying the same algorithm to the (possibly transformed) envelope, and then to the envelope that results from this, and so on. This procedure is only stable, however, if *both* the carrier *and* the envelope found by the demodulation algorithm are well-behaved. In section 2 we show that traditional methods return a suitable carrier *or* envelope, but not both. A new approach to amplitude demodulation is thus called for.

Fundamentally, amplitude demodulation is ill-posed: there are infinitely many decompositions of a signal into a slow positive modulator and quickly varying carrier. Ill-posed problems cannot be solved without assumptions, and a deficiency of traditional methods is to make these assumptions implicit. The approach developed here (section 3) is quite different: Demodulation is viewed as a task of probabilistic inference, to which prior information is integral. Our Bayesian approach thus serves to make the unavoidable assumptions, that determine the solution, explicit [2]. This tack yields many benefits. One is that we can tap into the extensive collection of methods developed for probabilistic inference. These are used to construct a family of new algorithms that out-perform traditional amplitude demodulation methods. A second is that the approach generalises easily, for instance to hierarchies and to multidimensional time-series.

2 Traditional Amplitude Demodulation

We begin by briefly reviewing two traditional methods for amplitude demodulation. The first is to obtain the envelope by low-pass filtering a non-linear transformation of the stimulus (for example, square and low pass (SLP)). Roughly, this works because the non-linearity moves energy associated with the modulation from high to low frequencies (via a self convolution in the case of SLP), where it can then be extracted by a low-pass filter. By tuning the filter cut-off one can recover a good estimate for the modulator (see fig. 1A). This type of algorithm derives from applications in radio engineering, where the carrier is a pure sinusoid of known frequency. However, for more general carriers, the demodulated carrier (obtained by point-wise division) is often badly behaved, with large spikes and a non-stationary variance. While the method drives the envelope to be smooth, it places no useful constraint on the carrier waveform.

The second method is based on a quantity called the analytic signal. The goal is to express the original signal $y(t)$ in terms of a time varying amplitude and phase such that, $y(t) = \Re\left[a(t)\exp\left[i\theta(t)\right]\right]$. In general, however, this problem is ill posed and the solution is shaped by assumption. One choice might be to constrain the time-scale of $a(t)$ (see section 3). More commonly, however, the imaginary part is chosen to make the signal analytic, by setting it to the Hilbert transform of $y(t)$. This does mean that, in certain circumstances, the amplitude of the analytic signal (AAS), $a(t)$, is restricted to lower frequencies than the phase component $\exp\left[i\theta(t)\right]$ [1], and this property might well be desirable. In general, however, the particular signals that result may not correspond to intuition. Indeed, by contrast to SLP, the modulators often seem poor (see fig. 1B), but the demodulated carriers are good, at least in that they have a stationary variance.

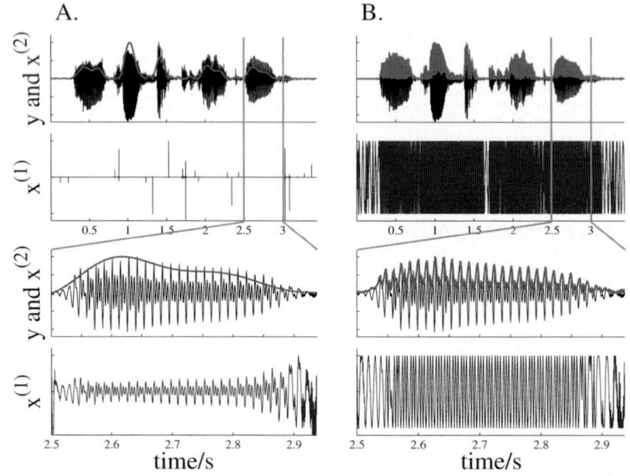

Fig. 1. The result of applying two traditional demodulation schemes to a spoken sentence (black), shown at two different scales (top and bottom). Both the envelopes (red) and carriers (blue) are shown. A) SLP: The cut-off of the filter was chosen to have a time-scale of 0.1s corresponding to the phoneme structures. The extracted envelope is good, but the carrier contains large spikes in regions where the envelope is zero, but the signal is non-zero. B) AAS: This demodulates the pitch period and not the phoneme structure. The carrier variance is stationary.

To summarise, SLP was derived from the perspective of feed-forward processing and, broadly speaking, extracts good estimates for the envelopes, but poor carriers. By contrast, the AAS, a demodulation method by accident rather than design, extracts good carriers, but poor envelopes. SLP has a tunable parameter (the cut-off of the low pass filter), which is useful to select one of several different envelopes present in a signal. However, it might be advantageous to automate the setting of such a slowness parameter. The AAS has no free parameters which is favourable when we want a method to work quickly.

3 Probabilistic Amplitude Demodulation (PAD)

One conclusion from the previous section is that a good demodulation algorithm should recover not only a smooth and slow envelope, but also a carrier with stationary variance. The new approach we propose explicitly utilises these two types of prior knowledge in order to find an optimal envelope and carrier.

A natural framework in which to leverage prior information to solve an ill-posed problem is provided by Bayesian methods [2]. These are based on a probabilistic model for the observed signal, which in our case includes a model for the two latent variables, the carrier ($\mathrm{X}^{(1)}$) and the modulator ($\mathrm{X}^{(2)}$), and for the dependence of the observed data (Y) on these. Our prior beliefs about the variables ($p(\mathrm{X}^{(1)})$, $p(\mathrm{X}^{(2)})$) are expressed through probability distributions; so for example, the distribution over envelopes may assign higher probability to slow processes than to fast ones. Having specified this model for $p(\mathrm{Y}, \mathrm{X}^{(1)}, \mathrm{X}^{(2)}|\theta)$, the calculus of probabilities leads naturally to algorithms to infer the latent variables, and to learn parameters. Whilst the integrals required to form such quantities may be analytically intractable, there are a variety of well known approximation schemes that can be exploited.

3.1 The Generative Model

Perhaps the simplest generative model for amplitude modulation is as follows,

$$p\left(\mathrm{z}_0^{(2)}\right) = \mathrm{Norm}\left(0, 1\right), \quad p\left(\mathrm{z}_\mathrm{t}^{(2)}|\mathrm{z}_{\mathrm{t}-1}^{(2)}\right) = \mathrm{Norm}\left(\lambda \mathrm{z}_{\mathrm{t}-1}^{(2)}, \sigma\right) \quad \forall t > 0, \tag{1}$$

$$\mathrm{x}_\mathrm{t}^{(2)} = f_{\mathrm{a}^{(2)}}\left(\mathrm{z}_\mathrm{t}^{(2)}\right), \quad \mathrm{x}_\mathrm{t}^{(1)} = \mathrm{Norm}\left(0, 1\right), \quad y_t = \mathrm{x}_\mathrm{t}^{(2)}\mathrm{x}_\mathrm{t}^{(1)}. \tag{2}$$

This expresses the generation of the envelope in two steps. First a slowly varying, but symmetric, process is produced ($\mathrm{Z}^{(2)}$); the Gaussian random-walk gives this an effective length-scale determined by λ, $\mathrm{l}_{\mathrm{eff}} = -\log(\lambda)$, which is inherited by $\mathrm{X}^{(2)}$. This length-scale is learnt from data, and is typically long (i.e. λ is close to one and $\mathrm{l}_{\mathrm{eff}} = -\log(1-\delta) \approx \frac{1}{\delta}$). The positive envelope is obtained using point-wise non-linearity, here given by

$$f_{\mathrm{a}^{(2)}}\left(\mathrm{z}_\mathrm{t}^{(2)}\right) = \log\left(\exp\left(\mathrm{z}_\mathrm{t}^{(2)} + \mathrm{a}^{(2)}\right) + 1\right), \tag{3}$$

which is logarithmic for large negative values of $\mathrm{z}_\mathrm{t}^{(2)}$, and linear for large positive values. This transforms the Gaussian distribution over $\mathrm{Z}^{(2)}$ into a sparse distribution, which is a good match to the marginal distributions of natural envelopes. The parameter $\mathrm{a}^{(2)}$ controls exactly where the transition from log to linear occurs, and consequently alters the degree of sparsity.

Having generated the envelope, the carrier is simply Gaussian white noise. The observations Y are generated by a point-wise product of the envelope and carrier. A typical draw from this generative model can be seen in Fig. 2B. This model is a fairly crude one for speech. For example, the speech carrier process will be structured (containing formant and pitch information) and yet it is modelled as Gaussian white noise. Whilst more complex models can certainly be developed, surprisingly even this very simple model is excellent at demodulation.

3.2 Learning and Results

The joint probability of both latent and observed signals is:

$$p(\mathrm{Y}, \mathrm{X}^{(2)}, \mathrm{X}^{(1)}|\lambda, \sigma) = p(\mathrm{x}_0^{(2)}) \prod_{t=1}^{T} p(\mathrm{y}_\mathrm{t}|\mathrm{x}_\mathrm{t}^{(1)}, \mathrm{x}_\mathrm{t}^{(2)}) p(\mathrm{x}_\mathrm{t}^{(2)}|\mathrm{x}_{\mathrm{t}-1}^{(2)}) p(\mathrm{x}_\mathrm{t}^{(1)}), \tag{4}$$

$$\text{with} \quad p(\mathrm{x}_\mathrm{t}^{(2)}|\mathrm{x}_{\mathrm{t}-1}^{(2)}) = p(\mathrm{z}_\mathrm{t}^{(2)}|\mathrm{z}_{\mathrm{t}-1}^{(2)}) \left|\frac{\mathrm{dz}_\mathrm{t}^{(2)}}{\mathrm{dx}_\mathrm{t}^{(2)}}\right|. \tag{5}$$

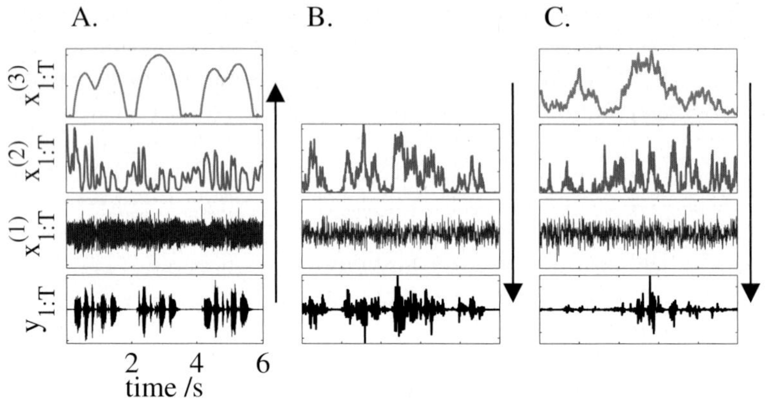

Fig. 2. An example of a modulation-cascade representation of speech (A) and typical samples from generative models used to derive that representation (B and C). A) The spoken-speech waveform (black) is represented as the product of a carrier (blue), a phoneme modulator (red) and a sentence modulator (magenta). (Derived using the method described in section 4.) B) The standard generative model with an envelope (red), a carrier (blue), and the waveform (black). C) The extended model ($M = 3$) with an additional slowly varying envelope (magenta). For sounds drawn from generative model and processed using PAD see [5].

As $p\big(\mathrm{y_t}|\mathrm{x}_\mathrm{t}^{(1)}, \mathrm{x}_\mathrm{t}^{(2)}\big) = \delta\big(\mathrm{y_t} - \mathrm{x}_\mathrm{t}^{(1)}\mathrm{x}_\mathrm{t}^{(2)}\big)$, we can integrate out the carrier from this expression which yields,

$$p(\mathrm{Y}, \mathrm{X}^{(2)}|\lambda, \sigma) = p(\mathrm{z}_0^{(2)}) \prod_{t=0}^{T} \left| \frac{\mathrm{dz}_\mathrm{t}^{(2)}}{\mathrm{dx}_\mathrm{t}^{(2)}} \right| \prod_{t=1}^{T} \frac{1}{\mathrm{x}_\mathrm{t}^{(2)}} p\big(\mathrm{z}_\mathrm{t}^{(2)}|\mathrm{z}_{\mathrm{t}-1}^{(2)}\big) p\left(\mathrm{x}_\mathrm{t}^{(1)} = \frac{\mathrm{y_t}}{\mathrm{x}_\mathrm{t}^{(2)}}\right). \quad (6)$$

Unfortunately, this expression cannot be analytically marginalised with respect to the envelope $\mathrm{X}^{(2)}$, and so an approximation is needed. One approach is to assume the distribution over $\mathrm{X}^{(2)}$ is highly peaked and to approximate the integral by its value at the peak: the *maximum a posteriori* (MAP) value, $p\left(\mathrm{Y}|\lambda, \sigma\right) \approx p\left(\mathrm{Y}, \mathrm{X}^{(2)}{}_{\mathrm{MAP}}|\lambda, \sigma\right)$. This is a coarse approximation, but it is well established and resembles a zero-temperature form of expectation maximisation. An alternative, to which we will return, is to approximate the integral by sampling.

Before discussing these approximations in more detail, we describe a final improvement to the model. In the Bayesian methodology parameters have the same status as latent variables and accordingly, may also be integrated out. In the present case, this is possible for either of the parameters controlling $\mathrm{Z}^{(2)}$; σ^2 or λ. More general models might have multidimensional λ, and so we choose to integrate over this, but for the simple model both methods work equally well. There are several specific advantages to this approach in the present application. Firstly, while the old model had one smoothness, the new model is more flexible, being essentially a weighted sum of the old models with different smoothnesses (see eq. 7). Secondly, it is not possible to learn λ, σ^2 *and* $\mathrm{X}^{(2)}$ using the old approach: a trivial solution ($\mathrm{x}_{\mathrm{t}}^{(2)} = 1$, $\lambda = 1$, $\sigma^2 = 0$) causes the likelihood to diverge. However, this maximum has infinitesimal width and therefore essentially no mass, so integration over λ removes this deficiency.

Practically the integration proceeds as follows: A conjugate Gaussian prior is placed on the smoothnesses $p(\lambda) = \mathrm{Norm}(\mu_\lambda^{\mathrm{pri}}, \sigma_\lambda^{\mathrm{pri}})$, and the integral that results is Gaussian,

$$p(\mathrm{Y}, \mathrm{X}^{(2)} | \mu_\lambda^{\mathrm{pri}}, \sigma_\lambda^{\mathrm{pri}}, \sigma) = \int \mathrm{d}\lambda \, p(\mathrm{Y}, \mathrm{X}^{(2)} | \lambda, \sigma) p(\lambda | \mu_\lambda^{\mathrm{pri}}, \sigma_\lambda^{\mathrm{pri}}). \tag{7}$$

Completing this integral yields the following objective function,

$$\log p(\mathrm{Y}, \mathrm{X}^{(2)} | \mu_\lambda^{\mathrm{pri}}, \sigma_\lambda^{\mathrm{pri}}, \sigma) = -\frac{1}{2} \sum_{t=1}^{T} \left[2 \log \mathrm{x}_{\mathrm{t}}^{(2)} + \frac{1}{\sigma^2} \left(\mathrm{z}_{\mathrm{t}}^{(2)} \right)^2 + \frac{\mathrm{y}_{\mathrm{t}}^2}{\left(\mathrm{x}_{\mathrm{t}}^{(2)} \right)^2} \right]$$
$$+ \sum_{t=0}^{T} \log \left| \frac{\mathrm{dz}_{\mathrm{t}}^{(2)}}{\mathrm{dx}_{\mathrm{t}}^{(2)}} \right| - \frac{1}{2} \left(\mathrm{z}_0^{(2)} \right)^2 - \frac{T}{2} \log \sigma^2 - \frac{1}{2} \left(\frac{\mu_\lambda^{\mathrm{pri}}}{\sigma_\lambda^{\mathrm{pri}}} \right)^2 + \frac{1}{2} \left(\frac{\mu_\lambda^{\mathrm{post}}}{\sigma_\lambda^{\mathrm{post}}} \right)^2$$
$$+ \log \frac{\sigma_\lambda^{\mathrm{post}}}{\sigma_\lambda^{\mathrm{pri}}} + \left(\frac{3}{2} T + 1 \right) \log 2\pi, \tag{8}$$

where $\mu_\lambda^{\mathrm{post}}$ and $\left(\sigma_\lambda^{\mathrm{post}}\right)^2$ are the posterior mean and variance over the smoothness parameter, which are given by,

$$\left(\sigma_\lambda^{\mathrm{post}}\right)^2 = \frac{\left(\sigma_\lambda^{\mathrm{pri}}\right)^2 \sigma^2}{\sigma^2 + \left(\sigma_\lambda^{\mathrm{pri}}\right)^2 \sum_{t=1}^{T} \left(\mathrm{z}_{\mathrm{t-1}}^{(2)}\right)^2}, \quad \mu_\lambda^{\mathrm{post}} = \frac{\left(\sigma_\lambda^{\mathrm{pri}}\right)^2 \sum_{t=1}^{T} \mathrm{z}_{\mathrm{t-1}}^{(2)} \mathrm{z}_{\mathrm{t}}^{(2)} + \mu_\lambda^{\mathrm{pri}} \sigma^2}{\sigma^2 + \left(\sigma_\lambda^{\mathrm{pri}}\right)^2 \sum_{t=1}^{T} \left(\mathrm{z}_{\mathrm{t-1}}^{(2)}\right)^2}. \tag{9}$$

The MAP estimate of the envelope can be found by gradient-based optimisation of this cost function. We used a conjugate-gradient algorithm on the log of the envelope (to ensure positivity). Depending on the application we can optimise the remaining parameter and hyperparameters too, or set them by hand. Results are shown for both approaches. In fig. 3A all parameters and hyper-parameters have been optimised and the envelope picks off the sentence structure. In fig. 3B the priors and the variances have been fixed in order that the algorithm picks off a rather faster envelope (this occurs for a wide range of prior/parameters settings and does not require fine tuning). In this case the phonemes are discovered. Qualitatively the performance appears far superior to that of traditional algorithms: a smooth envelope and a demodulated carrier of approximately constant variance are recovered reliably.

4 Extensions to Probabilistic Amplitude Demodulation

Our original motivation was to develop new representations for the long-temporal structures in sounds, particularly those based on a product of processes. A

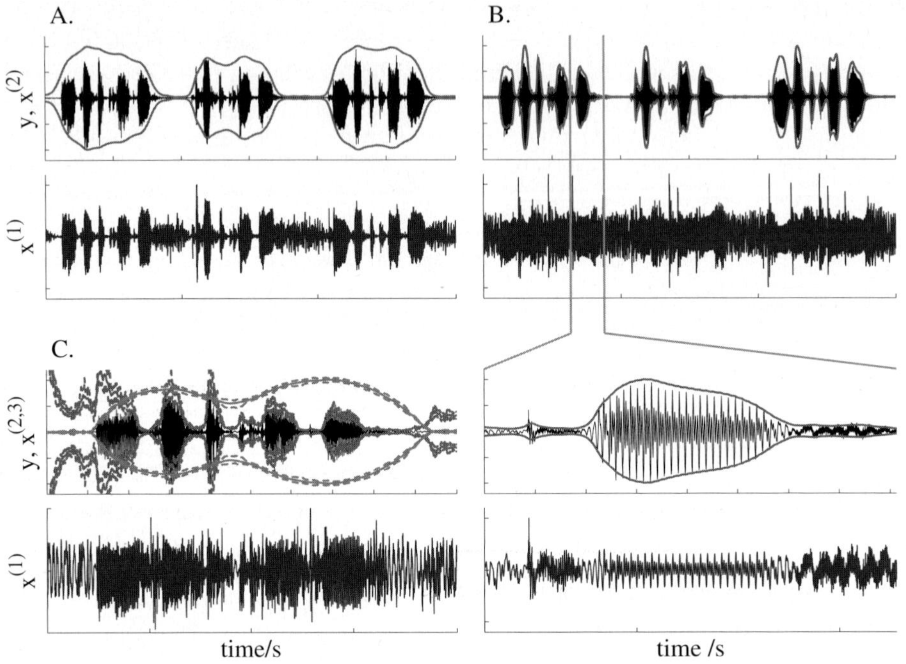

Fig. 3. Carriers (blue) and modulators (red and magenta) extracted by probabilistic amplitude demodulation (PAD) from a spoken sentence (black). A) Vanilla PAD selects a slow sentence envelope, but the carrier is still significantly modulated. B) Fixing the priors and variances leads to a faster, phoneme envelope, and results in a carrier that is more demodulated. C) The cascaded version of PAD, using sampling to generate error-bars on the extracted processes, provides an elegant representation of the sound.

necessary stepping stone along this path was the development of new methods for amplitude demodulation. We have already outlined a recursive procedure, which can use these new algorithms, for deriving such a representation (see section 1). The approach was to successively remove the fastest remaining process. However, ideally we would like to estimate the processes concurrently. Fortunately this can be done by extending the probabilistic method to a cascade of M processes:

$$p\left(\mathrm{z}_0^{(\mathrm{m})}\right) = \mathrm{Norm}\left(0, 1\right), \quad p\left(\mathrm{z}_\mathrm{t}^{(\mathrm{m})}|\mathrm{z}_{\mathrm{t}-1}^{(\mathrm{m})}\right) = \mathrm{Norm}\left(\lambda_m \mathrm{z}_{\mathrm{t}-1}^{(\mathrm{m})}, \sigma_m^2\right) \; \forall t > 0, \quad (10)$$

$$\mathrm{x}_\mathrm{t}^{(\mathrm{m})} = f_{\mathrm{a}^{(\mathrm{m})}}\left(\mathrm{z}_\mathrm{t}^{(\mathrm{m})}\right) \; \forall m > 1, \qquad \mathrm{x}_\mathrm{t}^{(1)} = \mathrm{Norm}\left(0, 1\right), \qquad y_t = \prod_{m=1}^{M} \mathrm{x}_\mathrm{t}^{(\mathrm{m})}. \quad (11)$$

A suitable model for speech might have $M{=}3$ with a "sentence" modulator ($\mathrm{X}^{(3)}$) and a "phoneme" modulator ($\mathrm{X}^{(2)}$). *A priori* we would expect $\lambda_{m+1} > \lambda_m$.

Learning and inference in this model can be completed in an analogous manner to that described in the previous section. That is; integrate out the carrier $\mathrm{X}^{(1)}$

and the dynamics $\lambda_{1:M}$ and optimise over the modulators $\mathrm{X}^{(2:\mathrm{M})}$ simultaneously (this was how fig. 2A was generated). However, an alternative method is to use sampling to integrate out the modulators, approximately. The most amenable method is Hamiltonian Monte Carlo as it requires the exact same evaluations as required by a gradient-based optimiser, namely evaluations of the PAD objective function and its derivatives. There are several potential advantages of the sampling approach, which gives back samples from the posterior distribution over the modulators $p(\mathrm{X}^{(2:\mathrm{M})}|\mathrm{Y})$, rather than just the maximum value of this distribution. Firstly, using information about the whole distribution might help learn better parameters. Secondly, we can now put error-bars on our inferences for the envelope. Thirdly we can check whether the mode of the posterior is typical of the distribution, and therefore assess the merits of the previous approach.

This sampling procedure was used to learn the cascade model with $M = 3$. The results are shown in fig. 3C. The algorithm extracts a sentence process and a phoneme process, and provides an elegant representation of the speech sound. Empirically, the mode and the mean of the distribution over envelopes is found to be in a similar location, and the parameter values discovered by both methods similar. This indicates that the MAP approximation might not be too severe.

5 Conclusion

The contributions of this paper are two fold. Firstly we provide a family of algorithms for probabilistic amplitude demodulation that out perform traditional methods. Secondly, and more generally, we propose an elegant new representation for the long time-scale temporal structure in sounds based on a cascade of modulatory processes. The goal of future research will be to wed this model for phonemes and sentences, to one that models pitch and formant information (e.g. [4]), to solve hard machine-audition tasks like blind-source separation and auditory scene analysis.

Acknowledgements. We would like to thank Pietro Berkes for helpful discussions. This work was supported by the Gatsby Charitable Foundation.

References

1. Cohen, L.: Time-Frequency Analysis. Prentice Hall Signal Processing Series (1995)
2. Mackay, D.J.C.: Information Theory, Inference, and Learning Algorithms. Cambridge University Press, Cambridge (2003)
3. Lawrence, N.: Probabilistic Non-linear Principal Component Analysis with Gaussian Process Latent Variable Models. J. Mach. Learn. Res. 6, 1783–1816 (2005)
4. Turner, R.E., Sahani, M.: Modeling Natural Sounds with Gaussian Modulation Cascade Processes. In: Advances in Models for Acoustic Processing Workshop (2006)
5. Turner, R.E., Sahani, M.: Probabilitic Amplidude Demodulation Technical Report. GCNU TR 2007-002 (2007), `http://www.gatsby.ucl.ac.uk/publications/tr/`

First Stereo Audio Source Separation Evaluation Campaign: Data, Algorithms and Results

Emmanuel Vincent[1], Hiroshi Sawada[2], Pau Bofill[3], Shoji Makino[2], and Justinian P. Rosca[4]

[1] METISS Group, IRISA-INRIA
Campus de Beaulieu, 35042 Rennes Cedex, France
emmanuel.vincent@irisa.fr
[2] Signal Processing Research Group, NTT Communication Science Labs
2-4, Hikaridai, Seika-cho, Soraku-gun, Kyoto 619-0237, Japan
[3] Departament d'Arquitectura de Computadors, Universitat Politècnica de Catalunya
Campus Nord Mòdul D6, Jordi Girona 1-3, 08034 Barcelona, Spain
[4] Siemens Corporate Research
755 College Road East, Princeton NJ 08540, USA

Abstract. This article provides an overview of the first stereo audio source separation evaluation campaign, organized by the authors. Fifteen underdetermined stereo source separation algorithms have been applied to various audio data, including instantaneous, convolutive and real mixtures of speech or music sources. The data and the algorithms are presented and the estimated source signals are compared to reference signals using several objective performance criteria.

1 Introduction

Large-scale evaluations facilitate progress in a field by revealing the effects of different choices in algorithm design, promoting common test data and evaluation criteria and attracting the interest of funding bodies. Several evaluations of audio source separation algorithms have been conducted recently, focusing on single-channel speech mixtures[1] or multichannel over-determined speech mixtures[2,3,4]. This article provides an overview of the complementary evaluation campaign for stereo underdetermined audio mixtures organized by the authors. Detailed results of the campaign are available at http://sassec.gforge.inria.fr/.

We define the source separation task and describe test data and evaluation criteria in Section 2. Then we present the algorithms submitted by the participants in Section 3 and summarize their results in Section 4. We conclude in Section 5.

[1] http://www.dcs.shef.ac.uk/~martin/SpeechSeparationChallenge.htm
[2] http://bme.engr.ccny.cuny.edu/faculty/parra/bss/
[3] http://homepages.inf.ed.ac.uk/mlincol1/SSC2/
[4] http://mlsp2007.conwiz.dk/index.php?id=43

M.E. Davies et al. (Eds.): ICA 2007, LNCS 4666, pp. 552–559, 2007.

2 Data and Evaluation Criteria

2.1 The Stereo Underdetermined Source Separation Task

Common audio signals, *e.g.* radio, television, music CDs and MP3s, are typically available in *stereo* (two-channel) format and consist of a mixture of more than two sound sources. Denoting by $J > 2$ the number of sources, each channel $x_i(t)$ $(1 \leq i \leq 2)$ of the mixture signal can be expressed as [1]

$$x_i(t) = \sum_{j=1}^{J} s_{ij}^{\mathrm{img}}(t) \tag{1}$$

where $s_{ij}^{\mathrm{img}}(t)$ is the *spatial image* of source j $(1 \leq j \leq J)$ on channel i, that is the contribution of this source to the observed mixture in this channel.

Different types of mixtures can be distinguished. *Instantaneous* mixtures are generated via (1) using a mixing desk or dedicated software by constraining the spatial images of each source j to $s_{ij}^{\mathrm{img}}(t) = a_{ij}s_j(t)$, where $s_j(t)$ is a single-channel source signal and a_{ij} are positive mixing gains. Synthetic *convolutive* mixtures are obtained similarly via $s_{ij}^{\mathrm{img}}(t) = \sum_\tau a_{ij}(\tau)s_j(t-\tau)$, where $a_{ij}(\tau)$ are mixing filters. *Live recordings* are acquired by recording all the sources simultaneously in a room using a pair of microphones. These recordings may also be obtained by recording the sources one at a time in the same room and adding the resulting source images together within each channel [2].

We define the source separation task as that of estimating the spatial images $s_{ij}^{\mathrm{img}}(t)$ of all sources j on all channels i from the two channels $x_i(t)$ of a mixture. This definition has two advantages: it is valid for all types of mixtures, even with spatially extended sources that cannot be represented as single-channel signals, and potential gain or filtering indeterminacies about the estimated single-channel source signals $s_j(t)$ disappear when considering their spatial images instead [1].

2.2 Development and Test Data

The development and test data used for the evaluation campaign involved four classes of signals: male speech, female speech, non-percussive music and music including drums. Music mixtures involved three sources taken from synchronized multitrack recordings, while speech mixtures involved four independent sources. All the source signals were sampled at 16 kHz and had a duration of 10 s.

The development data consisted of one instantaneous mixture, two synthetic convolutive mixtures and two live recordings per class. Instantaneous mixtures were generated by scaling the source signals by positive gains. Live recordings were acquired by playing the source signals through loudspeakers in a room at NTT with $\mathrm{RT}_{60} = 250\,\mathrm{ms}$ reverberation time and recording them using two pairs of omnidirectional microphones with spacings of 5 cm and 1 m. Figure 1 depicts the arrangement of loudspeakers and microphones. Synthetic convolutive mixtures were obtained by filtering the sources with simulated room impulse responses computed for the same arrangement using Roomsim[5]. Ground truth

[5] http://media.paisley.ac.uk/~campbell/Roomsim/

data, *i.e.* the source signals, their spatial images and the mixing filters or gains, were distributed with the mixture signals at `http://sassec.gforge.inria.fr/`.

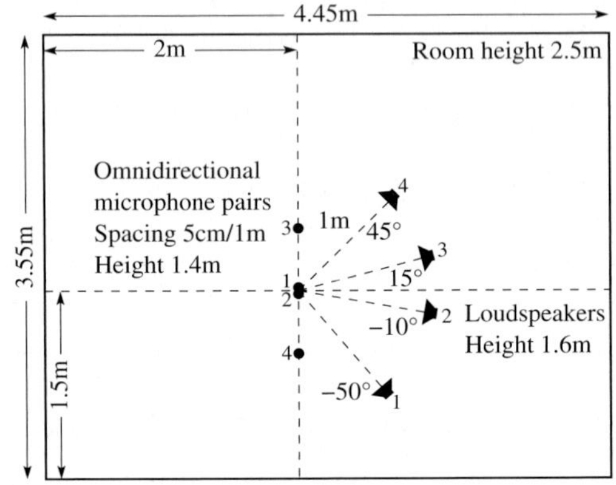

Fig. 1. Recording arrangement used for development data. Only three of the four loudspeakers were used for music mixtures.

The same number of test data was obtained similarly to the development data, using different source signals and positions for each mixture. The distances of the sources from the center of the microphone pairs were drawn randomly between 80 cm and 1.2 m and their angles of arrival between $-60°$ and $+60°$ with a minimal spacing of 15°. The mixture signals were made available, but ground truth data, including the exact source positions, was kept hidden[6].

2.3 Objective Performance Criteria

The participants were asked to provide estimates $\hat{s}_{ij}^{\text{img}}(t)$ of the spatial images of all sources j for some test mixtures. The quality of these estimates was then evaluated by comparison with the true source images $s_{ij}^{\text{img}}(t)$ using four objective performance criteria, inspired from criteria previously designed for single-channel source estimates [3]. By contrast with other existing measures [4,5], the proposed criteria can be computed for all types of separation algorithms and do not necessitate knowledge of the separating filters or masks.

The criteria derive from the decomposition of an estimated source image as

$$\hat{s}_{ij}^{\text{img}}(t) = s_{ij}^{\text{img}}(t) + e_{ij}^{\text{spat}}(t) + e_{ij}^{\text{interf}}(t) + e_{ij}^{\text{artif}}(t) \quad (2)$$

where $s_{ij}^{\text{img}}(t)$ is the true source image and $e_{ij}^{\text{spat}}(t)$, $e_{ij}^{\text{interf}}(t)$ and $e_{ij}^{\text{artif}}(t)$ are distinct error components representing spatial (or filtering) distortion, interference and artifacts. This decomposition is motivated by the auditory distinction between sounds from the target source, sounds from other sources and "gurgling"

[6] Only the first two authors of this article had potentially access to these data.

noise, corresponding to the signals $s_{ij}^{\text{img}}(t) + e_{ij}^{\text{spat}}(t)$, $e_{ij}^{\text{interf}}(t)$ and $e_{ij}^{\text{artif}}(t)$ respectively. The computational modeling of this auditory segregation process is an open issue so far. For simplicity, we chose to express spatial distortion and interference components as filtered versions of the true source images, computed by least-squares projection of the estimated source image onto the corresponding signal subspaces [3]

$$e_{ij}^{\text{spat}}(t) = P_j^L(\hat{s}_{ij}^{\text{img}})(t) - s_{ij}^{\text{img}}(t) \tag{3}$$

$$e_{ij}^{\text{interf}}(t) = P_{\text{all}}^L(\hat{s}_{ij}^{\text{img}})(t) - P_j^L(\hat{s}_{ij}^{\text{img}})(t) \tag{4}$$

$$e_{ij}^{\text{artif}}(t) = \hat{s}_{ij}^{\text{img}}(t) - P_{\text{all}}^L(\hat{s}_{ij}^{\text{img}})(t) \tag{5}$$

where P_j^L is the least-squares projector onto the subspace spanned by $s_{kj}^{\text{img}}(t-\tau)$, $1 \le k \le I, 0 \le \tau \le L-1$, and P_{all}^L is the least-squares projector onto the subspace spanned by $s_{kl}^{\text{img}}(t-\tau)$, $1 \le k \le I$, $1 \le l \le J$, $0 \le \tau \le L-1$. The filter length L was set to 512 (32 ms), which was the maximal tractable length.

The relative amounts of spatial distortion, interference and artifacts were then measured using three energy ratio criteria expressed in decibels (dB): the source Image to Spatial distortion Ratio (ISR), the Source to Interference Ratio (SIR) and the Sources to Artifacts Ratio (SAR), defined by

$$\text{ISR}_j = 10 \log_{10} \frac{\sum_{i=1}^I \sum_t s_{ij}^{\text{img}}(t)^2}{\sum_{i=1}^I \sum_t e_{ij}^{\text{spat}}(t)^2} \tag{6}$$

$$\text{SIR}_j = 10 \log_{10} \frac{\sum_{i=1}^I \sum_t (s_{ij}^{\text{img}}(t) + e_{ij}^{\text{spat}}(t))^2}{\sum_{i=1}^I \sum_t e_{ij}^{\text{interf}}(t)^2} \tag{7}$$

$$\text{SAR}_j = 10 \log_{10} \frac{\sum_{i=1}^I \sum_t (s_{ij}^{\text{img}}(t) + e_{ij}^{\text{spat}}(t) + e_{ij}^{\text{interf}}(t))^2}{\sum_{i=1}^I \sum_t e_{ij}^{\text{artif}}(t)^2}. \tag{8}$$

The total error was also measured by the Signal to Distortion Ratio (SDR)

$$\text{SDR}_j = 10 \log_{10} \frac{\sum_{i=1}^I \sum_t s_{ij}^{\text{img}}(t)^2}{\sum_{i=1}^I \sum_t (e_{ij}^{\text{spat}}(t) + e_{ij}^{\text{interf}}(t) + e_{ij}^{\text{artif}}(t))^2} \tag{9}$$

We emphasize that this measure is arbitrary, in the sense that it weights the three error components equally. In practice, each component should be given a different weight depending on the application. For instance, spatial distortion is of little importance for most applications, except for karaoke where it can result in imperfect source cancellation, even in the absence of interference or artifacts. Similarly, artifacts are crucial for hearing aid applications, for which "gurgling" noise should be avoided at the cost of increased interference. These criteria were implemented in Matlab and distributed at `http://sassec.gforge.inria.fr/`.

3 Algorithms

The campaign involved thirteen participants, who submitted the results of fifteen source separation algorithms. The underlying approaches are summarized in

Table 1. Submitted source separation algorithms

N°	Submitter Name	Source localization	Source signal estimation
		Algorithms for instantaneous mixtures only	
1	D. Barry ADRess	Manual IID clustering from a magnitude-weighted histogram with auditory feedback [6]	Source magnitude estimation in the STFT bins associated with each IID cluster [6]
2	P. Bofill	Peak picking on a smoothed IID histogram [7] with STFT bins selected as in [8]	Minimization of the l_1 norm of the real and imaginary parts of the source STFTs [9]
3	A. Ehmann	Manual peak picking on an IID histogram	Binary STFT masking with different resolutions at high/low frequencies
4	V. Gowreesunker	Peak picking on a thresholded IID histogram [10]	Binwise MDCT projection onto the nearest IID subspace [10]
5	M. Kleffner	Peak picking on a thresholded IID histogram [11] with STFT bins selected as in [12]	Online FFT-domain minimum-variance beamforming [13]
6	N. Mitianoudis	Soft IID clustering given the number of sources [14]	Binwise MDCT projection onto the nearest IID subspace [14]
7	H. Sawada	Hard IID clustering given the number of sources	Binary STFT masking
8	E. Vincent	Manual peak picking on an IID histogram weighted as in [12]	Minimization of the l_0 norm of the source STFTs [15]
9	M. Xiao SABM+SSDP	Hard fixed-width IID clustering on selected STFT bins [8]	Mixing inversion with 2 sources per time frame estimated from the mixture covariance [16]
10	M. Xiao SABM+SNSDP	Hard fixed-width IID clustering on selected STFT bins [8]	Extension of [16] with more active sources in some time frames
		Algorithms for instantaneous and/or convolutive mixtures	
11	S. Araki	Soft (IID,ITD) clustering given the number of sources [17]	Maximum SNR beamforming [18] and soft STFT masking [19]
12	Y. Izumi	Soft clustering of the mixture STFT bins based on (IID,IPD) given the number of sources [20]	Soft STFT masking by cluster probabilities [20]
13	T. Kim		FFT-domain independent component analysis [21] and soft masking (two sources only)
14	R. Weiss & M. Mandel	Soft (IID,IPD) clustering given the number of sources [22]	Soft STFT masking by cluster probabilities [22]
15	H. Sawada	Frequency-wise (IID,IPD) clustering given the number of sources as in [17] and sorting [23]	Binary STFT masking

Table 1. All algorithms except n°13 could be broken into (possibly iterated) source localization and source signal estimation steps. These two steps were conducted in the time-frequency domain via a Short-Time Fourier Transform (STFT) or a

Table 2. Results for instantaneous mixtures

Algorithm	1	2	3	4	5[7]	6	7	8	9	10	14
SDR (dB)	4.0	4.2	6.8	3.5	-23.4	-16.0	7.2	10.3	5.8	2.7	-2.4
ISR (dB)	7.5	8.2	13.9	6.2	-21.8	-12.8	14.6	19.2	15.9	20.0	4.1
SIR (dB)	13.2	12.9	15.5	14.4	12.8	13.2	15.9	16.0	10.7	6.8	-3.0
SAR (dB)	5.3	10.8	7.8	5.5	5.9	5.3	8.1	12.2	5.8	8.7	4.2
Time (s)	1	300	5	10	600	200	9	5	2	2	1000

Table 3. Results for synthetic convolutive mixtures and live recordings with two different microphone spacings

Mixtures	Synth 5 cm			Synth 1 m		Live 5 cm					Live 1 m		
Algorithm	11[7]	14	15	14	15	11[7]	12[7]	13[8]	14	15	13[8]	14	15
SDR (dB)	2.5	0.9	0.2	0.7	0.6	2.6	-23.2	-20.3	1.2	1.8	-19.0	2.1	3.6
ISR (dB)	6.0	2.8	4.6	2.8	4.4	5.9	-19.2	-17.0	4.0	7.0	-15.5	4.9	8.4
SIR (dB)	5.8	-2.7	4.4	-0.4	4.2	4.6	1.3	2.9	-1.9	4.2	2.9	0.8	6.9
SAR (dB)	4.9	14.1	7.5	10.7	7.5	5.4	6.2	6.2	13.0	6.8	5.8	8.0	6.8
Time (min)	1	20	0.6	20	0.6	1	1	4	20	0.6	4	20	0.6

Modified Discrete Cosine Transform (MDCT), except for algorithms n°9 and 10 where source estimation was directly performed in the time domain. The directions of the sources were modeled by the Interchannel Intensity Difference (IID) or variants thereof in the instantaneous case. The Interchannel Time Difference (ITD) or the Interchannel Phase Difference (IPD) were additionally used in the convolutive case. Algorithms n°2, 4, 5, 9 and 10 were fully blind, while others required manual input of the number of sources or the source directions.

4 Results

The performance of each algorithm was assessed by sorting the estimated source image signals so as to maximize the average SIR and successively averaging the measured SDR, ISR, SIR and SAR over the sources and over the mixtures. The resulting figures are given in Tables 2 and 3 for instantaneous and convolutive mixtures respectively, along with platform-specific computation times. The large negative SDR and ISR figures for algorithms n°5, 6, 12 and 13 are due to incorrect scaling of the submitted source images. Detailed results and sound files are available at `http://sassec.gforge.inria.fr/`.

In the instantaneous case, most algorithms provided similar SIR and SAR values clustered around 13 dB and 6 dB respectively, denoting high interference rejection but clear artifacts. Algorithms n°2 and 8 resulted in fewer artifacts, while algorithms n°10 and 14 provided more interference. Note that blind

[7] Average performance for speech mixtures only.

[8] Average performance over the two estimated sources for speech mixtures only.

algorithms n°9 and 10 achieved similar source localization accuracy as non-blind algorithms n°3, 7 and 8, as shown by large ISR values.

In the convolutive case, most algorithms provided again similar SIR and SAR values but around 4 dB and 6 dB respectively, indicating both strong interference and artifacts. Algorithms n°11 and 15 resulted in slightly less interference, while algorithm n°14 provided much more interference but less artifacts. Interestingly, performance did not vary much between synthetic convolutive mixtures and live recordings or with different microphone spacings.

5 Conclusion

In this article, we described the test data and objective performance criteria used in the context of the first stereo audio source separation evaluation campaign and summarized the approaches behind the fifteen submitted algorithms and their results. We are currently planning to complement objective performance figures by listening tests and present detailed results on the campaign website. We hope that this campaign fosters interest for evaluation in the source separation community and that larger-scale regular campaigns will take place in the future. The creation of a collaborative organization framework appears crucial to this aim, since it would allow sharing between the participants of time-consuming tasks such as the collection of test data under appropriate licenses and the recording of live mixtures.

References

1. Cardoso, J.F.: Multidimensional independent component analysis. In: Proc. IEEE Int. Conf. on Acoustics, Speech and Signal Processing (ICASSP), IV–1941–1944 (1998)
2. Schobben, D., Torkkola, K., Smaragdis, P.: Evaluation of blind signal separation methods. In: Proc. Int. Conf. on Independent Component Analysis and Blind Source Separation (ICA), pp. 261–266 (1999)
3. Vincent, E., Gribonval, R., Févotte, C.: Performance measurement in blind audio source separation. IEEE Trans. on Audio, Speech and Language Processing 14, 1462–1469 (2006)
4. Mansour, A., Kawamoto, M., Ohnishi, N.: A survey of the performance indexes of ICA algorithms. In: Proc. IASTED Int. Conf. on Modelling, Identification and Control (MIC), pp. 660–666 (2002)
5. Yılmaz, O., Rickard, S.T.: Blind separation of speech mixtures via time-frequency masking. IEEE Trans. on Signal Processing 52, 1830–1847 (2004)
6. Barry, D., Coyle, E., Lawlor, B.: Real-time sound source separation using azimuth discrimination and resynthesis. In: Proc. 117th AES Convention. (2004) (preprint 6258)
7. Bofill, P., Zibulevsky, M.: Underdetermined blind source separation using sparse representations. Signal Processing 81, 2353–2362 (2001)
8. Xiao, M., Xie, S., Fu, Y.: A novel approach for underdetermined blind source separation in the frequency domain. In: Wang, J., Liao, X.-F., Yi, Z. (eds.) ISNN 2005. LNCS, vol. 3498, pp. 484–489. Springer, Heidelberg (2005)

9. Bofill, P., Monte, E.: Underdetermined convoluted source reconstruction using LP and SOCP, and a neural approximator of the optimizer. In: Rosca, J., Erdogmus, D., Príncipe, J.C., Haykin, S. (eds.) ICA 2006. LNCS, vol. 3889, pp. 569–576. Springer, Heidelberg (2006)
10. Gowreesunker, B.V., Tewfik, A.H.: Two improved sparse decomposition methods for blind source separation. In: Proc. Int. Conf. on Independent Component Analysis and Blind Source Separation (ICA) (2007)
11. Mohan, S., Kramer, M.L., Wheeler, B.C., Jones, D.L.: Localization of nonstationary sources using a coherence test. In: Proc. IEEE Workshop on Statistical Signal Processing (SSP), pp. 470–473 (2003)
12. Arberet, S., Gribonval, R., Bimbot, F.: A robust method to count and locate audio sources in a stereophonic linear instantaneous mixture. In: Rosca, J., Erdogmus, D., Príncipe, J.C., Haykin, S. (eds.) ICA 2006. LNCS, vol. 3889, pp. 536–543. Springer, Heidelberg (2006)
13. Lockwood, M.E., Jones, D.L., Bilger, R.C., Lansing, C.R., O'Brien Jr., W.D., Wheeler, B.C., Feng, A.S.: Performance of time- and frequency-domain binaural beamformers based on recorded signals from real rooms. Journal of the Acoustical Society of America 115, 379–391 (2004)
14. Mitianoudis, N., Stathaki, T.: Underdetermined source separation using mixtures of warped Laplacians. In: Proc. Int. Conf. on Independent Component Analysis and Blind Source Separation (ICA) (2007)
15. Vincent, E.: Complex nonconvex l_p norm minimization for underdetermined source separation. In: Proc. Int. Conf. on Independent Component Analysis and Blind Source Separation (ICA) (2007)
16. Xiao, M., Xie, S., Fu, Y.: A statistically sparse decomposition principle for underdetermined blind source separation. In: Proc. Int. Symp. on Intelligent Signal Processing and Communication Systems (ISPACS), pp. 165–168 (2005)
17. O'Grady, P.D., Pearlmutter, B.A.: Soft-LOST: EM on a mixture of oriented lines. In: Puntonet, C.G., Prieto, A.G. (eds.) ICA 2004. LNCS, vol. 3195, pp. 428–435. Springer, Heidelberg (2004)
18. Araki, S., Sawada, H., Makino, S.: Blind speech separation in a meeting situation with maximum SNR beamformers. In: Proc. IEEE Int. Conf. on Acoustics, Speech and Signal Processing (ICASSP), vol. I, pp. 41–44 (2007)
19. Cermak, J., Araki, S., Sawada, H., Makino, S.: Blind source separation based on beamformer array and time-frequency binary masking. In: Proc. IEEE Int. Conf. on Acoustics, Speech and Signal Processing (ICASSP), vol. I, pp. 145–148 (2007)
20. Izumi, Y., Ono, N., Sagayama, S.: Sparseness-based 2ch BSS using the EM algorithm in reverberant environment. In: Submitted to IEEE Workshop on Applications of Signal Processing to Audio and Acoustics (WASPAA) (2007)
21. Kim, T., Attias, H.T., Lee, S.Y., Lee, T.W.: Blind source separation exploiting higher-order frequency dependencies. IEEE Trans. on Audio, Speech and Language Processing 15, 70–79 (2007)
22. Mandel, M.I., Ellis, D.P.W., Jebara, T.: An EM algorithm for localizing multiple sound sources in reverberant environments. In: Advances in Neural Information Processing Systems (NIPS 19) (2007)
23. Sawada, H., Araki, S., Makino, S.: Measuring dependence of bin-wise separated signals for permutation alignment in frequency-domain BSS. In: Proc. IEEE Int. Symp. on Circuits and Systems (ISCAS), pp. 3247–3250 (2007)

'Shadow BSS' for Blind Source Separation in Rapidly Time-Varying Acoustic Scenes

S. Wehr[1], A. Lombard[1], H. Buchner[2], and W. Kellermann[1]

[1] University Erlangen-Nuremberg
Multimedia Communications and Signal Processing
Cauerstraße 7, 91058 Erlangen, Germany
{Wehr,Lombard,WK}@LNT.de
[2] Deutsche Telekom Laboratories
Technical University Berlin
Ernst-Reuter-Platz 7, 10587 Berlin, Germany
hb@buchner-net.com

Abstract. This paper addresses the tracking capability of blind source separation algorithms for rapidly time-varying sensor or source positions. Based on a known algorithm for blind source separation, which also allows for simultaneous localization of multiple active sources in reverberant environments, the source separation performance will be investigated for abrupt microphone array rotations representing the *worst case.* After illustrating the deficiencies in source-tracking with the given efficient implementation of the BSS algorithm, a method to ensure robust source separation even with abrupt microphone array rotations is proposed. Experimental results illustrate the efficiency of the proposed concept.

1 Introduction

This paper is motivated by the so-called *cocktail-party problem* which arises when convolutive mixtures of multiple simultaneously active speakers are recorded by multiple microphones. In many applications (e.g. hands-free human-machine interfaces, [1]), we need to focus on one single source and try to suppress interfering sources. We address this problem here by *blind source separation* (BSS) algorithms which can deal well with unknown microphone and source positions [2]. Furthermore, BSS provides us with several separated source signals which may be individually selected for further processing.

We briefly review the generic ICA-based BSS framework for convolutive mixtures called TRINICON [3,4], which is also capable of simultaneously localizing multiple active sources [5,6]. The motivation for considering it here is that most of the known state-of-the-art BSS algorithms may be seen as certain *approximations* of this concept. As a fairly recent and advanced approximate practical algorithm, we investigate here [7]. This algorithm serves thus as a good representative for many of the major ICA algorithms. It is based on a special choice of Sylvester constraint, the correlation method, the natural gradient, and on an

M.E. Davies et al. (Eds.): ICA 2007, LNCS 4666, pp. 560–568, 2007.

approximated normalization. This allows for an efficient implementation, but may be responsible for the observed deficiencies in certain situations. Although the investigated algorithm is designed for Q active sources, we consider only two active sources in this paper. In Section 2, we demonstrate by simulations that both the separation performance of the considered BSS algorithm as well as the performance of the BSS-based source localization may significantly degrade for rapidly time-varying sensor positions. Analysis of this scenario leads us to proposing the so-called *shadow-BSS* system, which runs in parallel to the main BSS algorithm. Simulation results confirm the efficiency of the proposed extension.

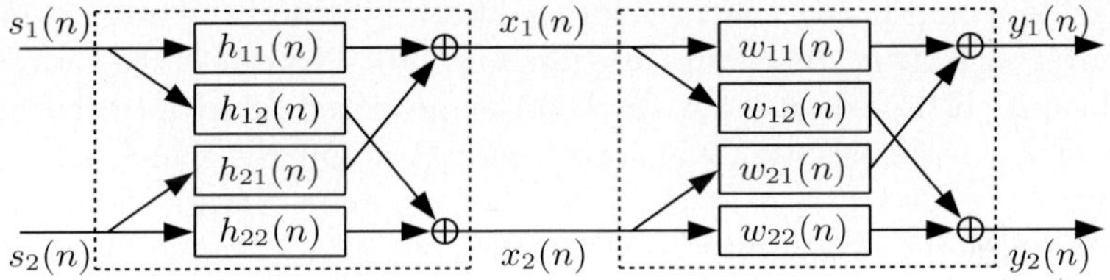

Fig. 1. 2-Channel Mixing and Demixing System

Figure 1 illustrates the BSS scheme for two sources with the source signals $s_i(n)$, the sensor signals $x_i(n)$, and the BSS output signals $y_i(n)$, respectively $(i = 1, 2)$. The unknown mixing system is modeled by M-tap room impulse responses $h_{ij}(n)$ and the demixing system determined by BSS is modeled by L-tap FIR filters $w_{ij}(n)$ $(j = 1, 2)$. The microphone signals and BSS output signals can then be written as:

$$x_i(n) = \sum_{j=1}^{2} \sum_{\kappa=0}^{M-1} h_{ji}(\kappa) s_j(n - \kappa) \tag{1}$$

$$y_i(n) = \sum_{j=1}^{2} \sum_{\kappa=0}^{L-1} w_{ji}(\kappa) x_j(n - \kappa). \tag{2}$$

The source separation problem is then solved by appropriately determined demixing filters. Further details on the adaptation of the demixing filters are given in, e.g., [7].

As shown in [6], TRINICON-based BSS inherently identifies the (unknown) mixing system up to a scaling for $Q = 2$:

$$w_{ji}(n) = -\alpha_{ji} \cdot h_{ji}(n) \qquad \text{and} \qquad w_{jj}(n) = \alpha_{ii} \cdot h_{ii}(n), \qquad i \neq j. \tag{3}$$

Based on this system identification, the TDOA (*Time Difference of Arrival*) can be derived from the demixing filters simultaneously for both active sources from the main peaks of $w_{ij}(n)$ [6].

2 BSS and Source Localization in Rapidly Time-Varying Scenarios

We now investigate the separation and localization performance of the chosen BSS algorithm for rapidly time-varying scenarios. The experimental setup is as follows: We assume abrupt rotations of a microphone array consisting of two sensors, which represents a *worst case* for time-variant scenarios. Dealing with rapidly rotating microphone arrays is important in many applications of BSS, e.g. when the microphone array is held and moved by a person. Figure 2 illustrates the DOAs (*Direction of Arrival*) for two typical scenarios: In the first scenario, the broadside of the microphone array points between the two sources and the array is rotated by $\pm 30°$. In the second scenario, one source is located broadside and the other source is on the side after each turn, where the rotations are $\pm 80°$. Note that the array orientation is abruptly changed and that the DOA is measured relative to the broadside direction of the array. We use clean speech signals as sources, which are convolved by measured impulse responses of a low-echoic chamber ($T_{60} \approx 50$ms) to represent the microphone array inputs. All simulations are performed with sampling rate $f_s = 16$kHz and demixing filter length $L = 1024$.

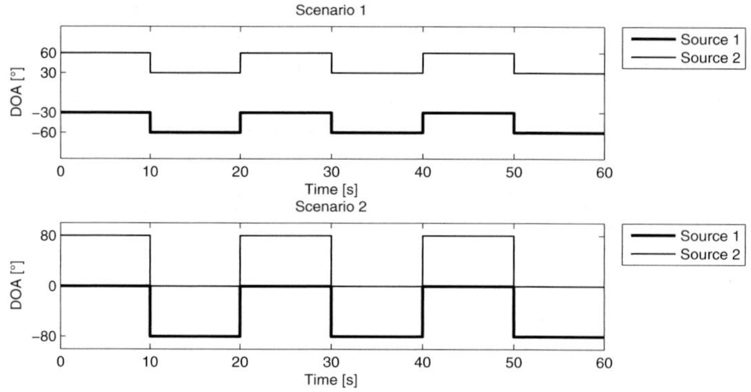

Fig. 2. DOAs for two scenarios with abrupt microphone array rotations

Figures 3 and 4 depict both the BSS gain and the results of the source localization obtained with the chosen BSS algorithm for both scenarios, where the BSS gain in dB represents the suppression of the interfering source in each BSS channel. The vertical dashed lines indicate the time instants, when the array orientation is changed. The channel-averaged BSS gain for each array orientation is displayed in the framed boxes. In scenario 1, the chosen BSS algorithm exhibits the expected good source separation and localization performance: After each array rotation, the demixing filters allow for good source separation and the estimated TDOAs follow the DOAs given by scenario 1. In scenario 2, we observe that the investigated BSS algorithm is only able to track the first array rotation (see the time period 10s-20s), but already with a degradation in separation and

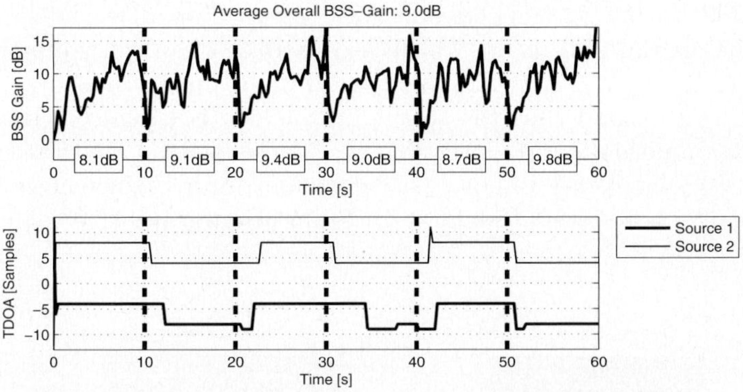

Fig. 3. BSS gain (top) and Estimated TDOAs (bottom) obtained with the chosen BSS algorithm, Scenario 1

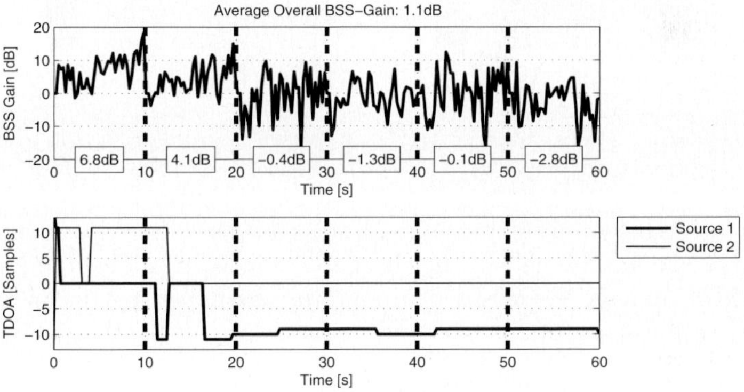

Fig. 4. BSS gain (top) and Estimated TDOAs (bottom) obtained with the chosen BSS algorithm, Scenario 2

localization performance. However, the chosen BSS algorithm fails to separate and locate the two sources after the subsequent array turns: The BSS gain is even negative and the estimated TDOAs seem to be "frozen". The results of the TDOA estimation suggest that the chosen BSS algorithm is not able to adapt the demixing filters after the second array rotation. Therefore, we investigate the demixing filters. The first 32 coefficients of the demixing filters $w_{12}(n)$ and $w_{21}(n)$ are depicted in Figure 5. The filter coefficients of $w_{11}(n)$ and $w_{22}(n)$ are not depicted, because they mainly consist of distinct positive peaks at 16 samples, which leads to a delayed but mainly unfiltered contribution of the microphone signals to the two BSS outputs. The vertical dashed lines indicate again array rotations. The horizontal dashed lines mark the filter coefficients which are important for suppressing sources from -80°, 0°, and 80°, respectively. By appropriately placing negative peaks in the demixing filters $w_{12}(n)$ and $w_{21}(n)$, spatial nulls are formed and source 1 and source 2 are suppressed in BSS output

2 and in BSS output 1, respectively. We observe that after the second array rotation, the spatial null in 0°, which is caused by the negative peak at filter coefficient 16 in $w_{21}(n)$, is fixed. This spatial null cancels the source located in broadside direction and thus the source on the side is enhanced. However, the BSS adaptation does not form a significant negative peak in $w_{12}(n)$ to cancel the source at ±80°. Instead, a minor positive peak at filter coefficient 16 is formed, which basically corresponds to a filter-and-sum beamformer at 0° enhancing the broadside source.

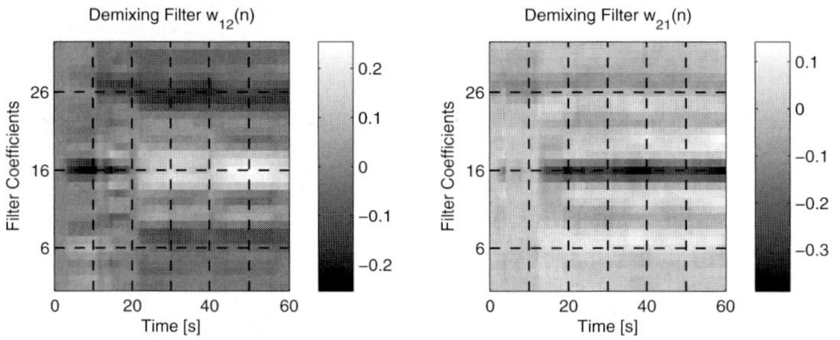

Fig. 5. Demixing filters obtained with the chosen BSS algorithm in scenario 2

In order to further analyze and understand the encountered problem, we take a close look at the update rule of the chosen BSS algorithm given in [4]. For each online block m, matrix $\mathbf{W}(m)$, which contains the demixing filters $w_{ij}(n)$ in a so-called *Sylvester structure*, is updated as follows:

$$\mathbf{W}(m) = \mathbf{W}(m-1) - \mu \triangle \mathbf{W}(m). \tag{4}$$

Incorporating the *natural gradient*, the update $\triangle \mathbf{W}$ becomes

$$\triangle \mathbf{W} = 2 \sum_{i=0}^{\infty} \beta(i,m) \mathbf{W} \left\{ \mathbf{R_{yy}} - \text{bdiag}\left\{ \mathbf{R_{yy}} \right\} \right\} \cdot \left(\text{bdiag}\left\{ \mathbf{R_{yy}} \right\} \right)^{-1}, \tag{5}$$

where the weighting function $\beta(i,m)$ allows offline and online implementations. The operator bdiag $\{\mathbf{R_{yy}}\}$ returns the block-diagonal elements of $\mathbf{R_{yy}}$, which is the cross-correlation matrix of the BSS outputs. After abrupt array rotations, the update $\triangle \mathbf{W}$ is then based on badly adapted demixing filters and on BSS output signals, which will not exhibit any source separation. Due to this improper data, the chosen BSS algorithm is not able to adapt the demixing filters and thus fails to track the sources after abrupt array rotations in scenario 2. This deficiency might be caused by the approximations described in [7].

3 Shadow-BSS

In Section 2, we studied a scenario, where the chosen BSS algorithm was not able to track abrupt array rotations. However, we could always observe that

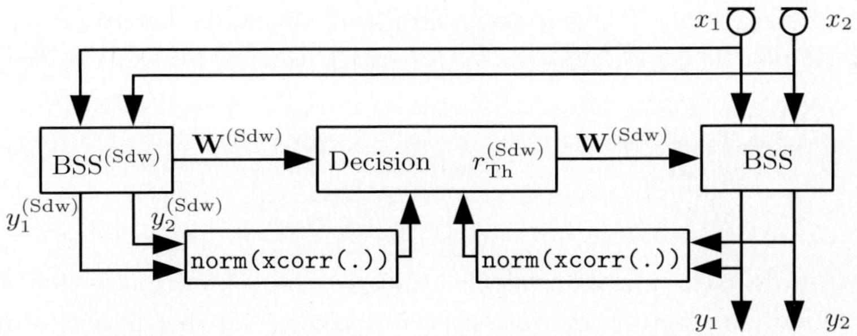

Fig. 6. Block diagram of the Shadow-BSS algorithm

the chosen BSS algorithm is capable to converge to well-separating demixing filters after blind initialization. Therefore, we propose the usage of a so-called *shadow-BSS* system, which is periodically blindly initialized and – in the case of outperforming the signal separation of the main BSS system – the demixing filters of the shadow-BSS are transferred for use in the main BSS system. The shadow-BSS system is motivated by the successful usage of shadow systems in, e.g., adaptive echo cancellation [8]. The effectiveness of the shadow-BSS scheme is demonstrated by simulations.

3.1 Algorithm

Figure 6 shows the block diagram of the proposed shadow-BSS system, where the dependency on time has been omitted for notational convenience. BSS and $\mathrm{BSS}^{(\mathrm{Sdw})}$ denote the main BSS system and the shadow-BSS system. Note that the demixing filter length in the shadow-BSS system $L^{(\mathrm{Sdw})}$ can be chosen independently from L. Both systems use the two microphone signals x_1 and x_2 and perform source separation, which leads to the output signals $y_{1,2}$ and $y_{1,2}^{(\mathrm{Sdw})}$. The shadow-BSS system is periodically blindly reinitialized at multiples of period $T^{(\mathrm{Sdw})}$. We now investigate the method for replacing the demixing filters of BSS by the demixing filters of $\mathrm{BSS}^{(\mathrm{Sdw})}$, if $\mathrm{BSS}^{(\mathrm{Sdw})}$ performs better source separation.

Based on the two pairs of output signals, the blocks `norm(xcorr(.))` compute the norms of the cross-correlations as follows:

$$\mathrm{norm}\left\{R_{yy}(m)\right\} = \sqrt{\sum_{\tau=-D}^{D} \left|R_{yy}(m,\tau)\right|^2} \tag{6}$$

$$\mathrm{norm}\left\{R_{yy}^{(\mathrm{Sdw})}(m)\right\} = \sqrt{\sum_{\tau=-D}^{D} \left|R_{yy}^{(\mathrm{Sdw})}(m,\tau)\right|^2}. \tag{7}$$

The parameter τ denotes the time lags of the cross-correlation. The cross-correlation norms may now be considered as a quantity measuring the separation performance of BSS and $\mathrm{BSS}^{(\mathrm{Sdw})}$. In the case of good source separation, the two output signals of a separation system are sufficiently uncorrelated and the according norm in Equations (6), (7) is small. Hence the ratio of both cross-correlation norms $r(m)$,

$$r(m) = \frac{\text{norm}\left\{R_{yy}^{(\text{Sdw})}(m)\right\}}{\text{norm}\left\{R_{yy}(m)\right\}} \tag{8}$$

is used as a decision variable, which indicates the separation system (BSS or BSS$^{(\text{Sdw})}$) with the better separation performance. To avoid unnecessary transfers of the demixing filters from BSS$^{(\text{Sdw})}$ to BSS, we can average $r(m)$ with an exponentially decaying forgetting factor $\lambda^{(\text{Sdw})}$:

$$r(m) = \lambda^{(\text{Sdw})} r(m-1) + \left(1 - \lambda^{(\text{Sdw})}\right) \frac{\text{norm}\left\{R_{yy}^{(\text{Sdw})}(m)\right\}}{\text{norm}\left\{R_{yy}(m)\right\}}. \tag{9}$$

Comparing $r(m)$ to the threshold $r_{\text{Th}}^{(\text{Sdw})}$ allows for a decision:

$r(m) < r_{\text{Th}}^{(\text{Sdw})} \Rightarrow$ Transfer demixing filters from shadow-BSS to BSS
$r(m) \geq r_{\text{Th}}^{(\text{Sdw})} \Rightarrow$ Keep demixing filters of BSS

The sensitivity of the overall system may be adjusted by the threshold $r_{\text{Th}}^{(\text{Sdw})}$. Note that the latter $L - L^{(\text{Sdw})}$ filter coefficients of BSS are set to zero, if the demixing filters are transferred from shadow-BSS to BSS.

3.2 Simulations

We now present both the separation performance and the results of the TDOA estimation obtained with the proposed algorithm based on a shadow-BSS system for scenario 2 described in Section 2. The demixing filter lengths are $L = 1024$

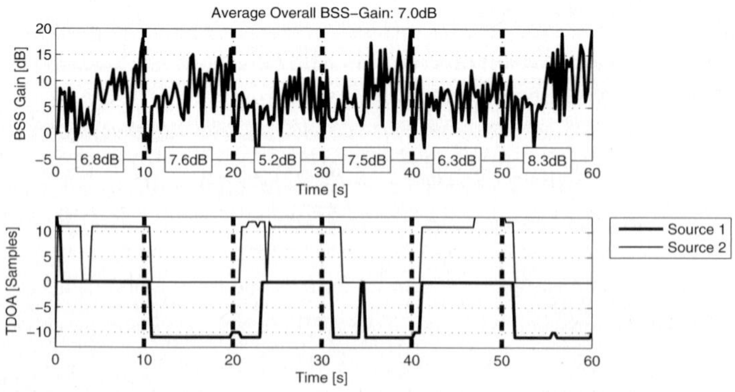

Fig. 7. BSS gain (top) and Estimated TDOAs (bottom) obtained with the shadow-BSS system

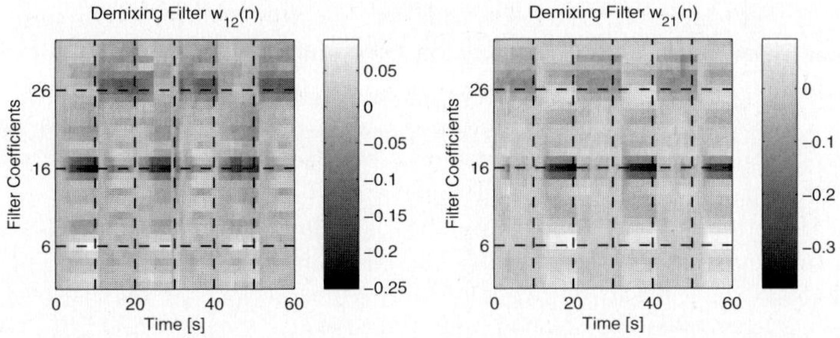

Fig. 8. Demixing filters of BSS influenced by shadow-BSS

and $L^{(\mathrm{Sdw})} = 30$. Moreover, the configuration of the shadow-BSS system is $T^{(\mathrm{Sdw})} = 2\mathrm{s}$, $\lambda^{(\mathrm{Sdw})} = 0.6$, and $r_{\mathrm{Th}}^{(\mathrm{Sdw})} = 0.8$.

Both the separation performance of BSS and the estimated TDOAs depicted in Figure 7 illustrate that the proposed shadow-BSS system is capable to track even the worst case of abrupt microphone array rotations. After a few seconds, the estimated TDOAs represent the true DOAs illustrated by Figure 2. Again, we investigated the demixing filters of BSS influenced by shadow-BSS, where we especially focus on the two cross-filters $w_{12}(n)$ and $w_{21}(n)$. As we see from Figure 8, the spatial nulls formed by the demixing filters now follow the alternating DOAs of the sources caused by the abrupt array rotations. Finally, it should be mentioned that no audible artefacts could be observed when the demixing filters are transferred from shadow-BSS to BSS.

4 Conclusions

In this paper, we first investigated the source separation and localization performance of the chosen BSS algorithm in the case of abrupt array rotations. We found that for certain relevant cases the chosen BSS algorithm was not able to maintain the usually high separation performance and thus fails to localize the sources correctly. We then proposed an extension to BSS, which incorporates a periodically blindly initialized shadow-BSS system. If the separation performance of the shadow-BSS system outperforms the main BSS system, the demixing filters were transferred from the shadow-BSS system to the main BSS system. An appropriate method for comparing the separation performance of both systems was introduced. Finally, we would like to mention that the proposed shadow-BSS system may also be applied in the case of rapidly moving sources.

References

1. Brandstein, M.S., Ward, D.B. (eds.): Microphone Arrays: Signal Processing Techniques and Application. Springer, Berlin (2001)
2. Hyvarinen, A., Karhunen, J., Oja, E.: Independent Component Analysis. John Wiley & Sons, New York (2001)

3. Buchner, H., Aichner, R., Kellermann, W.: Audio Signal Processing for Next-Generation Multimedia Communication Systems. Kluwer Academic Publishers, Boston (2004)
4. Buchner, H., Aichner, R., Kellermann, W.: A generalization of blind source separation algorithms for convolutive mixtures based on second order statistics. IEEE Transactions on Speech and Audio Processing 13(1), 120–134 (2005)
5. Buchner, H., Aichner, R., Kellermann, W.: Relation between blind system identification and convolutive blind source separation. In: Conf. Rec. Joint Workshop for Hands-Free Speech Communication and Microphone Arrays (HSCMA) (March 2005)
6. Buchner, H., Aichner, R., Stenglein, J., Teutsch, H., Kellermann, W.: Simultaneous localization of multiple sound sources using blind adaptive MIMO filtering. In: IEEE Intl. Conf. on Acoustics, Speech, and Signal Processing (ICASSP) (March 2005)
7. Aichner, R., Buchner, H., Kellermann, W.: Exploiting narrowband efficiency for broadband convolutive blind source separation. EURASIP Journal on Applied Signal Processing 2007, 1–9 (2006)
8. Ochiai, K., Araseki, T., Ogihara, T.: Echo canceler with two echo path models. IEEE Transactions On. Communications COM-25(6), 589–595 (1977)

Evaluation of Propofol Effects in Atrial Fibrillation Using Principal and Independent Component Analysis

Raquel Cervigón, Conor Heneghan, Javier Moreno, Francisco Castells, and José Millet

Innovation in Bioengineering Research Group, Universiy of Castilla-La Mancha, Camino del Pozuelo sn. 16071 Cuenca, Spain
School of Electrical, Electronic and Mechanical Engineering, University College Dublin, Belfield, Dublin 4, Ireland
{raquel.cervigon}@uclm.es,
{conor.heneghan}@ucd.ie,
{javier.moreno}@secardiologia.es,
{fcastells,jmillet}@eln.upv.es

Abstract. The mechanisms responsible for the initiation, maintenance and spontaneous termination of atrial fibrillation (AF) are not yet completely understood. Though much of the underlying physiology has been well determined, it has been demonstrated in numerous clinical investigations that the autonomic nervous system plays an important role in AF genesis and maintenance. In this work the effects of a widely used anaesthetic (propofol) in AF therapies has been studied. ECG recording and 12 intracardiac bipolar leads were recorded from 17 patients diagnosed with AF at both baseline and during anaesthetic infusion, in order to study its effects on AF behavior. By considering all intracardiac leads, the dominant atrial cycle length found at baseline was higher than during propofol infusion, but this difference was not statistically significant. However, the process of averaging results over all 12 leads may obscure clinically significant changes. In order to try to emphasize any differences which may exist, Principal Component Analysis (PCA) and Independent Component Analysis (ICA) were applied. This statistical analysis did show a significant difference between both groups. The shorter cycle lengths found in this study at baseline are consistent with parasympathetic and/or other physiological modulation during anaesthetic infusion and suggest that propofol may have antiarrhythmic properties.

Keywords: Atrial fibrillation, anaesthetic, Independent and Principal Component Analysis (ICA and PCA).

1 Introduction

Atrial fibrillation (AF) is the most commonly encountered arrhythmia in clinical practice, with prevalence rising to near 10% in the elderly. AF is originated at the atria (the upper heart chambers), and is considered to be due to the coexistence

M.E. Davies et al. (Eds.): ICA 2007, LNCS 4666, pp. 569–576, 2007.

of multiple re-entrant atrial wavelets which are often initiated by arrhythmogenic foci located within the pulmonary veins [1,2]. Among the factors contributing to the genesis or maintenance of circulating wavelets, the Autonomic Nervous System (ANS) may play a significant pro-arrhythmic role [3].

Since AF is associated with an elevated heart rate, a common treatment strategy is heart rate control, although there is still clinical controversy over whether rhythm control (i.e., conversion to normal sinus rhythm) is more desirable. Cardioversion is the process of restoring the AF rhythm to normal sinus rhythm [4]. However, pharmaceutically-induced cardioversion is only marginally effective in treating this arrhythmia, and may have the potential for serious side effects, including life-threatening pro-arrhythmic effects. Therefore, in recent years catheter-based ablation of atrial tissue by application of energy through intracardiac catheters has become a widely used therapeutic method in patients with both persistent and paroxysmal AF [5].

The radiofrequency (RF) ablation procedure consists of generating electrical barriers in various sites within the atria by altering the tissue properties in the vicinity of the ablating catheter tip. The extent of the altered tissue depends on the power and duration of the application, as well as on the characteristics of the tissue itself.

During the RF catheter ablation procedure (and other cardiac electrophysiological studies), patients are typically under the influence of anaesthetic agents. One of the most useful agents is propofol (2,6-diisopropylphenol) which is a new, rapidly acting intravenous anaesthetic. The rapid redistribution and metabolism of propofol results in a short elimination half-life of approximately one hour, and suggests that the drug could be suitable for use in short procedures. This is the reason for the recent interest in whether (and how) propofol may affect the electrophysiological properties of cardiac tissue, and hence alter the electrical activity within the atria.

The purpose of this study was, therefore, to explore the short-term influence of propofol on the electrical activity within the atria in patients with AF, who were undergoing catheter ablation.

A working hypothesis is that the cumulative effect of the numerous wavelets circulating within the atria affect the spatiotemporal organization of AF, and hence determine the overall fibrillatory wavefront observed within the atria. The time course of these circulating electrical wavelets is determined by the local refractoriness of the tissue (i.e., its inability to generate action potentials at an arbitrarily high rate). Propofol may act by altering atrial refractoriness, but its exact effects are unknown. A good index of refractoriness is the overall atrial cycle length (which is the inverse of the dominant frequency of the fibrillatory waveform). This has previously been shown to be a local index of atrial refractoriness during fibrillation [6,7,4]. Accordingly, in this paper we attempt to measure the dominant fibrillatory frequency (and hence atrial refractoriness) both before and during the administration of propofol.

Since the localized intracardiac electrograms (IEGMs) recorded during the procedure are a mixture of both local and global cardiac activity, it can be hard to distinguish overall trends. Therefore, we have used Principal Component Analysis and and Independent Component Analysis to emphasize the most significant trends in the measured set of IEGMs.

2 Data Acquisition

IEGMs were recorded in 17 patients (13 paroxysmal AF and 4 persistent AF) before and during anesthesia with propofol (bolus of 100-180 mg/kg intravenously with incremental doses depending on the weight and time to hypnosis). These patients were undergoing radio-frequency catheter ablation. Informed consent was provided by all patients, and the study was approved by the Hospital Ethics Review Board. A bipolar catheter was positioned in the high right atrium, and a 24-pole catheter (Orbiter, Bard Electrophysiology, 2-7-2 mm electrode spacing) was positioned with electrodes at the level of the atrial septum and the left and the right atriums. Since these are bi-polar electrograms, the results are in a set of 12 signals, which we refer to as lead or dipole 1-2, 3-4, etc. While the exact positioning of the catheter will vary from subject to subject, in general the leads 1-2, 3-4, and 5-6 are in the left atrium, leads 7-8, 9-10, 11-12 are in the septum, and leads from 15-16, 17-18, ... to 23-24 are in the right atrium.

Contact catheter data and surface three-lead ECGs (I, aVF, V1) (1 kHz sampling rate, 16 bit A/D conversion, equipment supplied by Siemens-Elema AB, Solna, Sweden), were recorded simultaneously on the Duo Laboratory system (Bard) over 30 to 60 seconds before and during the anaesthetic effect.

Note that bipolar electrograms, collected from a pair of closely spaced electrodes, measure the differential voltage between the two electrodes. This differential measurement is equivalent to a spatially high-pass filtered version of the underlying activation traversing the two electrodes. Thus it is sensitive to the direction of nearby depolarization and repolarization wavefronts.

3 Methods

3.1 Frequency Domain Analysis

A fast Fourier transform (FFT) was calculated on the wave over a 4-s Hamming window of 4096 points just prior to energy delivery, overlapping 50% between adjacent windowed sections. The maximum peak of the resulting magnitude spectrum was identified and the positions of the harmonic peaks were determined based on the position of the maximum peak [8].

3.2 Principal Component Analysis

PCA is a popular data processing and dimension reduction technique [9]. PCA seeks the linear combinations of the original variables such that the derived variables capture maximal variance. The objective is to find a (linear) transformation of the original variables, ordered by high proportion of the variation of the old variables, in a set of new uncorrelated variables, the principal components (PCs). PCA can be done via the singular value decomposition (SVD) of the data matrix. Biological expression data are currently rather noisy, and SVD can detect and extract small signals from noisy data.

In detail, let the data $\mathbf{X}$ be a $n \times p$ matrix, where n and p are the number of observations or samples and the number of variables, respectively. Without loss of generality, assume the column means of $\mathbf{X}$ are all 0. Suppose we have the SVD of $\mathbf{X}$ as

$$\mathbf{X} = \mathbf{UDV}^T, \tag{1}$$

where $\mathbf{U}$ are the PCs of unit length, and the columns of $\mathbf{V}$ are the corresponding loadings of the PCs. The variance of the ith PC, is d_i^2, with d_i being the ith element in the diagonal of $\mathbf{D}$. Usually the first q ($q << p$) PCs are chosen to represent the data, thus a great dimensionality reduction can be achieved.

3.3 Independent Component Analysis

ICA is a technique recently developed for the analysis of multidimensional signals [10,11]. It has been employed in numerous biomedical applications, such as electroencephalography (EEG), magnetoencephalography (MEG), functional magnetic resonance imaging (fMRI) electrocardiography (ECG) and magnetocardiography (MCG), among others [11,12].

PCA is a common pre-processing technique for signal processing before using ICA [10]. The first step is to center $\mathbf{X}$, subtract its mean vector to make $\mathbf{X}$ a zero mean variable. The second is using PCA to find a smaller set of variables with less redundancy. The application of ICA to the study assumes that the sources are non-Gaussian and mutually independent [13,14]. In the ICA model, the observed data $\mathbf{X}$ has been generated from source data through a linear process $\mathbf{X}$=$\mathbf{AS}$, where both the mixing matrix $\mathbf{A}$ and the source $\mathbf{S}$ are unknown. It is assumed that $\mathbf{X}$ is the observed data vector with a dimension n, number of the atrial signals recorded at different electrodes, $\mathbf{S}$ is the m-dimensional source vector, therefore $\mathbf{A}$ requires to be an $n \times m$ matrix of full rank with $n \geq m$. FastICA has been the algorithm employed. It is a computationally highly efficient method for performing the estimation of ICA which uses an approximation of negentropy and a fixed-point iteration scheme to carry out the optimization of the contrast function.

3.4 Statistical Analysis

The parameters are expressed as $mean \pm SD$. Paired and unpaired t-tests were used for comparison between the 2 groups of results. Comparison of serial measures was obtained by repeated measures ANOVA coupled with the Student-Newman-Keuls test. Results were considered to be statistically significant at $p < 0.05$. All statistical analyses were performed with the SPSS program.

4 Result

4.1 Results from Complete Set of 12 Lead Electrograms

At first, by applying frequency domain analysis to the signals (imposing a range between 4-9Hz for the atrial frequency), the average dominant frequency from each recording was calculated, before propofol administration and during its effect. The results showed a trend in which the frequency was reduced 6.14 ± 0.93 versus 5.99 ± 0.79 Hz, but it was not statistically significant ($p = 0.14$).

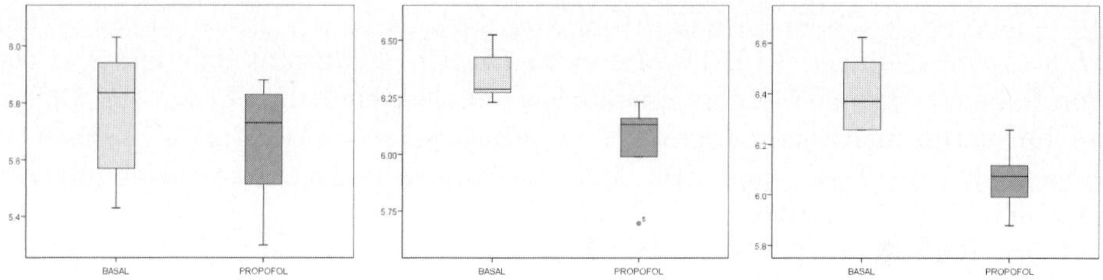

(a) Fp from original signal (b) Fp extracted from PCA (c) Fp extracted from ICA

Fig. 1. Average Main Frequency (Fp) from original leads (left), 5PCs (middle), and 5ICs (right) in non-anaesthetic (basal) and anaesthetic (propofol) states

In order to improve the differences between both groups, PCA and ICA were applied. The chosen parameter to discriminate between both groups was the main atrial fibrillatory frequency. In order to discard a noise/signal separation, the dimensionality of the data was reduced using the first eigenvectors, because most of the signal is contained in the first few PCs. There were statistically significant differences between the frequencies pre and post-propofol when the analysis was carried out on the reduced set of PCs. With 7 PCs the frequency varied from (6.26 ± 0.72 Hz at baseline to 5.98 ± 0.64 Hz,($p = 0.009$) at peak propofol effect. These 7 PCs captured 95% of the signal energy. With 5 components the change was from 6.35 ± 0.72 to 6.04 ± 0.57 Hz, $p = 0.012$), and these 5 PCs typically captured 85% of the energy (Fig. 1). Therefore PCA is a useful tool in more clearly indicating statistically significant trends.

The results of applying ICA after PCA, to the index of the last eigenvalue retained with PCA, were similar in significance, when using dominant fibrillatory frequency as a classification parameter, obtaining at basal 6.26 ± 0.67 vs. 6.03 ± 0.80 Hz during propofol infusion ($p = 0.012$) for 7 components and 6.41 ± 0.63 vs. 6.06 ± 0.73 Hz for 5 components ($p = 0.001$).

4.2 Results from the Different Atrial Regions

At each different region inside the atrium, the average frequency of activation was calculated. For simplicity, we will consider three different regions: right atrium

(RA), left atrium (LA) and the area between them called the septum area (SA). In each atrial region the main peak frequency was calculated.

The results from original signals showed a more distinctive difference between both states in the RA, 6.32 ± 0.96 Hz at basal state vs. 6.08 ± 0.79 Hz during propofol infusion, ($p = 0.069$), however, there was any substantial change of dominant frequency in the LA.

After the application of PCA on the LA (dipoles 1-2, 3-4 and 5-6), statistically significant differences between both groups were found in the main frequency of the first two components (6.14 ± 0.54 basal state vs. 5.83 ± 0.79 Hz during propofol infusion, $p = 0.012$), and after ICA processing the significance decreased to $p = 0.05$ (6.08 ± 0.50 vs. 5.75 ± 0.80 Hz).

On the SA (dipoles 9-10, 11-12 and 13-14), the significant differences of the main frequency from PCA components are a little higher than in the LA. Before and during the anaesthetic infusion these values were 6.04 ± 0.64 vs. 5.71 ± 0.65 Hz respectively ($p = 0.038$), and after ICA, the changes had a statistical significance of $p = 0.042$ (6.05 ± 0.66 vs. 5.76 ± 0.51 Hz).

On the RA (dipoles 15-16, 17-18, 19-20 and 21-22), the most important differences in main frequency were obtained by extracting 2 components using PCA $p = 0.005$ (6.49 ± 0.95 vs. 6.03 ± 0.76 Hz) and with ICA $p = 0.014$ (6.48 ± 0.91 vs. 6.08 ± 0.78 Hz). If the SA and RA groups are joined (leads 9-10 to 21-22), the significance of the differences in frequency using the first 4 components extracted from PCA increases to $p = 0.001$ (6.44 ± 0.68 vs. 6.00 ± 0.60 Hz), and with ICA $p = 0.002$ (6.44 ± 0.67 vs. 6.05 ± 0.64 Hz).

It is possible that this processing leads to an error, because of a different covariance matrix for the groups before and during anaesthetic infusion, so the transformation applied to both groups could be different. In order to counteract this possibility, the signals before and after the anesthetic treatment were concatenated prior to PCA and ICA processing, so that a unique transformation matrix was obtained, which is applicable to both groups. After computing the fundamental frequency of the PCs and ICs for the corresponding segments, it was concluded that the differences were also significant, as indicated in Table 1.

Table 1. Statistical Significance of the Change in Main Peak Frequency(Fp) for PCs and ICs with a Transformation Matrix Based on Basal and Propofol States

	PCA Fp Significance	ICA Fp Significance
12 leads PCA 7 comp	0.05	0.03
LA 3 leads (1,2-5,6) 2comp	> 0.05	> 0.05
SA 2 leads (9,10-11,12) 2comp	0.05	0.04
RA 4 leads (15,16-21,22) 2comp	0.04	0.02
SRA 7 leads (9,10-21,22) 4comp	0.02	0.01

In addition, it has been possible to find differences between the patients that have paroxysmal AF and those that have persistent AF. During persistent AF the differences between both states are low, and the highest changes are seen in the left atrial region (Table 2).

Table 2. Statistical Significance of Main Peak Frequency(Fp) with different combinations of PCA and ICA components in Paroxysmal and Persistent AF

	Paroxysmal AF Fp Signification	Persistent AF Fp Signification
Original leads LA (1,2-5,6)	0.89	0.89
Original leads SA (9,10-11,12)	0.25	0.36
Original leads RA (15,16-21,22)	0.23	0.38
LA 3 leads (1,2-5,6) PCA 2comp	0.07	0.02
LA 3 leads (1,2-5,6) ICA 2comp	0.94	0.78
SA 2 leads (9,10-11,12) PCA 2comp	0.04	0.36
SA 2 leads (9,10-11,12) ICA 2comp	0.08	0.28
RA 4 leads (15,16-21,22) PCA 2comp	0.01	0.29
RA 4 leads (15,16-21,22) ICA 2comp	0.03	0.18

5 Discussion and Conclusion

Atrial refractoriness may be affected by the administration of anaesthetic agents. The influence of propofol on the electrophysiological properties of the myocardium is sparsely reported in the literature. Results from *in vitro* experiments on isolated rabbit sinoatrial node preparation showed that propofol had only small effects on atrial conduction at $10\mu g \cdot ml^{-1}$, but that it reduced conduction drastically at $33\mu g \cdot ml^{-1}$ and caused complete block at $100\mu g \cdot ml^{-1}$ [15].

The conclusion from this study is that propofol does affect the electrical properties of the heart in patients with AF. Its effects in patients with AF are such that the atrial rate is consistently decreased as a result of the anaesthesia. The observation of these differences was optimized after the application of PCA and ICA algorithms, with rates of the first components in the range of the atrial activity in AF, discarding noise and improving the statistical significance between both groups.

These results correlate with those extracted from a study that analyzes the variability of AF during circadian cycles, where the AF cycle length during chronic AF exhibits diurnal variability, with longer cycle lengths occurring at night [16].

Although the exact mechanisms of the alteration of the atrial rate remain to be fully understood, the results in the present study have important implications. The atrial cycle length is markedly modulated during anaesthesia compared to the resting conscious state immediately before. This modulation may affect the activation and the repolarization sequence in the heart. More information about the effect of individual intra-operative factors would help to evaluate these phenomena, but it is strongly suggested that the differences between electrophysiological properties before and during anaesthetic infusion are due to true changes in the physiological conditions.

Acknowledgments. This work was supported by Castilla-La Mancha research scheme (Ref: PAC-05-008-1) and by the Minister of Education and Science of Spain (Ref: TEC2005-08401-C02-01).

References

1. Moe, G.K.: On multiple wavelet hypothesis of atrial fibrillation. Archives Internationales de Pharmacodynamie et de Therapie 140, 183–188 (1962)
2. Konings, K., Allessie, M., Wijels, M.: Atrial arrhythmias: State of the art. In: DiMarco, J.P., Prytowsky, E.N. (eds.) chapter Electrophysiological mechanisms of atrial fibrillation, pp. 155–161. Futura Publishing Company, Armonk (1995)
3. Akselrod, S., Gordon, D., Ubel, F.A., Shannon, D.C., Berger, A.C., Cohen, R.J.: Power spectrum analysis of heart rate fluctuation: a quantitative probe of beat-to-beat cardiovascular control. Science 213, 220–222 (1981)
4. Coumel, P.: Paroxysmal atrial fibrillation: a disorder of autonomic tone? Eur Heart J 15(Suppl A), 9–16 (1994)
5. Pappone, C., Rosanio, S., Oreto, G.: A new anatomic approach for curing atrial fibrillation. Circulation 102, 2619–2628 (2000)
6. Kim, K.-B., Rodefeld, M.D., Schuessler, R.B., Cox, J.L., Boineau, J.P.: Relationship between local atrial fibrillation interval and refractory period in the isolated canine atrium. Circulation 94, 2961–2967 (1996)
7. Capucci, A., Biffi, M., Boriani, G., Ravelli, F., Nollo, G., Sabbatani, P., Orsi, C., Magnani, B.: Dinamic electrophysiological behaviour of humann atria during paroxysmal atrial fibrillation. Circulation 92, 1193–1202 (1995)
8. Welch, P.D.: Use of fast fourier transform for estimation of power spectra: A method based on time averaging over short modified periodograms. IEEE Trans. Audio Electroacoust AE-15, 70–73 (1967)
9. Joliffe, I.T: Principal component analysis. Springer, Heidelberg (2002)
10. Hyvrinen, A., Oja, E.: Independent component analysis: A tutorial. helsinki university of technology. Laboratory of Computer and Information Science (1999)
11. Cichocki, A., Amari, S.I.: Adaptive blind signal and image processing. Sons Inc. (2002)
12. Cardoso, J.F., Souloumiac, A.: Blind beamforming for non gaussian signals. In: IEEE Proc F, vol. 140, pp. 362–370 (1993)
13. Papoulis, A.: Probability, random variables and stochastic processes. McGraw-Hill, New York (1991)
14. Comon, P.: Independent component analysis a new concept? Signal Process 36, 287–314 (1994)
15. Briggs, I., Heapy, C.G., Pickering, L.: Electrophysiological effects of propofol on isolated sinoatrial node preparations and isolated atrial conduction in vitro. Br J Pharmacol 97, 504 (1989)
16. Meurling, C., Sornmo, L., Stridh, M., Olsson, B.: Non invesive assessment of atrial fibrillation (af) cycle length in man: potential application for studyimg af. Ist Super Sanita 37(3), 341–349 (2001)

Space-Time ICA and EM Brain Signals

Mike Davies*, Christopher James, and Suogang Wang

IDCOM & Joint Research Institute for Signal and Image Processing, University of Edinburgh, Scotland, UK
mike.davies@ed.ac.uk
Signal Processing and Control Group, ISVR, University of Southampton, UK
{c.james,sgw}@soton.ac.uk

Abstract. Recently Single Channel ICA has been proposed where it can be shown that the algorithms learn temporal filters for separating the different components. Here we consider the natural extension to learning a set of space-time separating filters. We argue that these are capable of separation above and beyond that possible using only spatial or temporal methods alone. We then consider the potential of these ideas when applied to Ictal Electroencephalographic (EEG) data and Brain Computer Interaction (BCI).

1 Introduction

Independent Component Analysis (ICA) was originally proposed for the blind separation of vector-valued observations into independent sources. However it has also been used to learn 'features' or codebooks for single channel data that provide a more efficient representation of a signal than a fixed (non-adapted) representation [1,2]. Empirical evidence suggested that it was possible to also use this for signal separation (e.g. [3,4,5]) and it was recently shown [6] that under certain restricted conditions this is indeed the case.

Here we examine the natural extension of the model in [6] to consider a combination of space time vectors: Space-Time ICA (ST-ICA). Previously this model has be studied empirically for the convolutional source separation problem [7,8]. Here we consider the circumstances under which signals are separable within this model. As with single channel ICA, separation of more sources than sensors is possible. It therefore provides an appealing alternative to the usual computationally demanding solutions to the more sources than sensors problem.

Before we embark we note that these ideas are distinct from the Spatiotemporal ICA proposed in [9] and it is not equivalent to using temporal information to learn a spatial unmixing matrix, e.g. as in [10]. In particular ST-ICA takes advantage of *full* spatial-temporal filtering.

* MD acknowledges support of his position from the Scottish Funding Council and their support of the Joint Research Institute with the Heriot-Watt University as a component part of the Edinburgh Research Partnership.

M.E. Davies et al. (Eds.): ICA 2007, LNCS 4666, pp. 577–584, 2007.

2 Mathematical Framework

Traditional ICA assumes that we observe a sequence of vectors $\mathbf{x}(t) \in \mathbb{R}^Q$ which are a linear mixture of independent components: $\mathbf{x}(t) = \mathbf{A}\mathbf{s}(t)$. Using nonGaussianity, nonstationarity or temporal structure we can then generally estimate A blindly up to the usual ambiguities and hence estimate the individual sources: $\hat{\mathbf{s}}(t) = \hat{\mathbf{A}}^{-1}\mathbf{x}(t)$.

An alternative way to view the problem, first proposed by Cardoso [11], is to treat it as an additive model:

$$\mathbf{x}(t) = \sum_{p}^{C} \mathbf{x}_p(t) \tag{1}$$

which can be related back to the original ICA model with $\mathbf{x}_p(t) = \mathbf{a}_p^T s_p(t)$ where $\mathbf{A} = [\mathbf{a}_1, \ldots, \mathbf{a}_Q]^T$. As this alternative approach only defines the independent sources within the observation domain it generalises ICA and allows $\mathbf{x}_p(t)$ to be intrinsically multi-dimensional. Let E_p define an n_p-dimensional subspace containing the ith component: $\mathbf{x}_i(t) \in E_i$. Then, if the sources are non-Gaussian (or non-stationary etc.) and as long as all the subspaces, E_i, are linearly independent the sources can still be separated. Furthermore separation can be performed using a standard ICA algorithm followed by a component grouping step. See Cardoso [11] for details.

2.1 Single Channel and Space-Time ICA

In [6] we showed that when the input data is formed from a delay vector of samples, $\mathbf{x}(t) = [x(t), x(t-1), \ldots, x(t-N+1)]^T$ taken from a single channel, source separation is still possible and the resulting single channel ICA can be seen as a special instance of MICA. This means that it is possible to separate multiple sources from a single channel. However this model carries a rather restrictive separability requirement. For the MICA subspaces E_p to be independent, the sources (assuming stationarity) must have disjoint spectral support. This assumption is generally over-optimistic, although under certain circumstances it may hold approximately. We can, however relax this requirement in ST-ICA as we descibe now.

Suppose that we observe a vector valued sequence $\mathbf{x}(t) \in \mathbb{R}^Q$ that we believe is composed of multiple independent stationary sources, as in the MICA model (1). In the same manner as single channel ICA, we can augment the dimension of the observation space by including delayed copies of observations. Let us define the $Q \times N$-dimensional space-time vector $\tilde{\mathbf{x}}(t)$ as:

$$\tilde{\mathbf{x}}(t) = [\mathbf{x}(t), \mathbf{x}(t-1), \ldots, \mathbf{x}(t-N+1)]^T \tag{2}$$

where N is the number of 'taps' in our delay vector. We can now treat this as a $Q \times N$ dimensional MICA problem as opposed to a Q dimensional one. As with MICA and single channel ICA source separation can be performed by using a standard ICA algorithm followed by component grouping, [11]. We call this ST-ICA. Note that spatial ICA, spatial MICA and convolutional ICA are all restrictions of the ST-ICA model. The link between ST-ICA and convolutional ICA modelling is further explored in [7,8].

3 Separability of Sources

Like single channel ICA, ST-ICA can provide an advantage over spatial ICA (and MICA) when the independent sources have a finite spectral support. Many signals exhibit this property, at least approximately, as is evidenced by the popular use of Singluar Systems Analysis [12]. In this context we can define ST-ICA separability requirements in terms of MICA-type requirements at each frequency.

If we assume that the subspaces, E_p, have already been correctly identified it only remains to determine conditions for which they are linearly independent. This means that the space-time correlation matrix for each source, $R_{x_p} = \mathcal{E}\{\mathbf{x}_p\mathbf{x}_p^T\}$, must be rank deficient and the sources characterisable as multichannel singular systems.

If we further assume that the sources are stationary then we know that the correlation matrices, R_{x_p}, are block Toeplitz which, in the large window limit ($N \to \infty$) are block diagonalizable via the Fourier transform giving the matrix valued power spectra, $S_{x_p}(\omega)$.

Transforming the MICA model into the Fourier domain gives:

$$X(\omega) = \sum_p X_p(\omega) \tag{3}$$

and due to the block diagonal structure of the correlation matrices we can now consider each frequency, ω, separately. Let us assume that each $X_p(\omega)$ is restricted to a subspace $E_p(\omega) \subset \mathbb{C}^Q$. Without any further restriction on the model we therefore have the following separability requirement:

> *Separability*: A Q-dimensional random process, $\mathbf{x}(t) = \sum_p \mathbf{x}_p(t)$, composed of independent stationary random processes, $\mathbf{x}_p(t)$, is linearly separable if and only if the subspaces $E_p(\omega)$ such that $X_p(\omega) \subset E_p(\omega)$ are linearly independent.

This allows there to be more sources than sensors with the restriction that there should be no more than Q sources present at any given frequency.

4 Applications to EM Brain Signals

The technique will be trialled on two different examples of EEG data. In the first set, clinical data in the form of epileptic (ictal) EEG will be analysed with the goal of extracting epileptic seizure components from cortical recording channels placed either over the epileptic focus (focal) or further away (extra focal).

In BCI brain signals are interpreted to provide a means of communication. One of the more popular applications of BCI is called the P300 word speller introduced by [13]. The P300 evoked potential is a late positive wave that occurs over the parietal cortex at about 300 ms after the onset of a meaningful stimulus. The P300 word speller presents a matrix of letters, numbers, and other symbols (generally 6x6), whereby over short intervals one of the rows or columns of the matrix is randomly flashed (the stimulus). The user selects a character by focusing attention on individual characters in the matrix. The premise is that the P300 recording in the EEG is prominent only in those responses elicited by focusing attention on the desired character, these are, however, buried in the ongoing brain activity, and artifacts (such as movement artifacts, eyeblinks, etc.). Stimulus locked coherent averaging of the P300 responses is the usual method of enhancing the SNR, however this requires trails to be repeated several times. Not only is this costly in terms of time but it is also widely believed that due to habituation this may affect the signal as well as the noise components.

4.1 Ictal EEG

The data were recorded during pre-surgical evaluation at the Epilepsy Center of the University Hospital of Freiburg, Germany. Intracranial grid-, strip-, and depth-electrodes were used. The EEG data were acquired using a Neurofile NT digital video EEG system with 128 channels, 256 Hz sampling rate, and a 16 bit analogue-to-digital converter. 23 EEG recordings with simple partial, complex partial and generalized tonic-clonic seizures from patients with focal epilepsy originated in the temporal region were recorded and made available for dissemination [14].[1] We depict the results when applied to a seizure recording of one specific patient with temporal lobe epilepsy with 6 cortical recording channels. Figure 1a depicts the data, channels 1,2 & 3 are focal to the seizure and channels 4,5 & 6 are extra-focal. The vertical lines in each figure represent the start and stop of the seizure as indicated by an epileptologist. Channel 3 is the channel with the strongest visible seizure involvement, the extra-focal electrodes show no visible seizure recording.

Figure 1b depicts the outputs of ensemble ICA (Fast ICA) on all 6 recording channels. It can be seen that the unmixing process is not sucessful in completely isolating seizure activity, most probably because the recordings are already fairly independent of each other this is apparent in the mixing matrix as shown in Figure 1c where the focal (seizure) components map almost 1:1 onto their corresponding recording channels.

Next, the pair of electrodes 3 & 6 are analysed with ST-ICA. (The space-time vector had dimension $190 = 2\times95$, and a data reduction was performed to reduce the data-matrix to rank 30 through SVD). Figure 1d (upper) depicts the mixing filters learned by the ST-ICA process; two per independent component (IC). These can be manually clustered into similar groups of shifted filters. Figure 1d

[1] We are grateful to the Epilepsy Center of the University Hospital of Freiburg, Germany for their permission to use the invasive EEG recordings in this study.

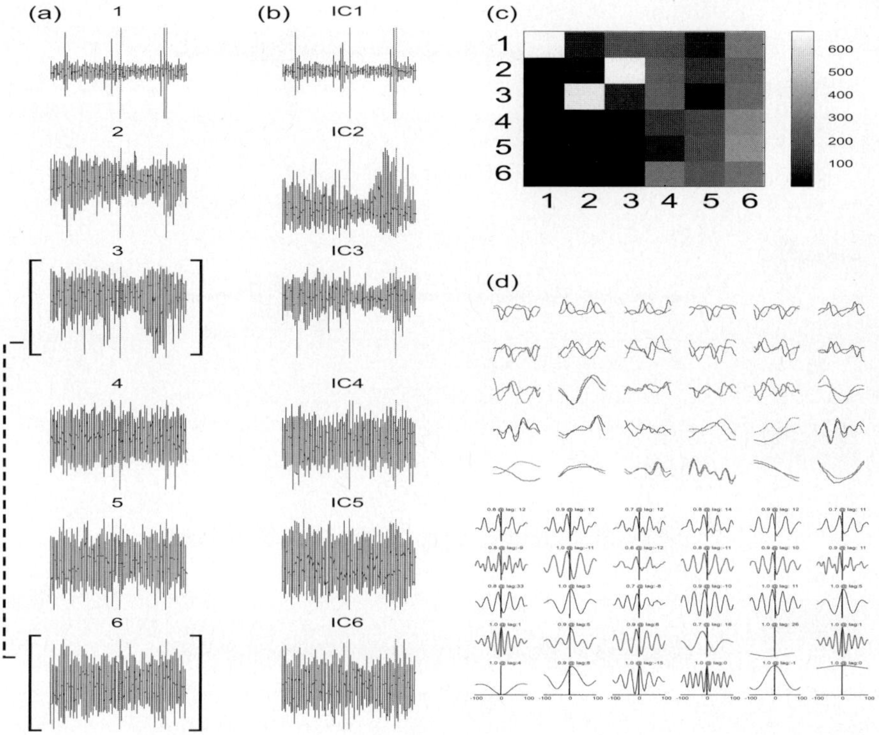

Fig. 1. (a) 6 channels of ictal EEG recorded from the cortex; (b) following ensemble ICA and (c) mixing matrix. (d) The mixing filters and their cross-correlations following ST-ICA on channels 3 and 6.

(lower) shows the cross-correlation between the two mixing filters for each IC for 100 lags - in each case greatest (absolute) value of cross-correlation is indicated along with the lag at which it occurs. It can be seen that a number of filters peak at a lag of around 10-12 samples (39.1-46.9 ms) with the filters of channel 3 leading those of channel 6, whilst others are maximal at around 0-5 samples (0-19.5 ms) with channel 6 leading channel 3.

Figure 2a shows resulting waveforms following manual clustering of the filters into 3 groups, S1-S4. S1 & S2 depict seizure components, and S3 & S4 show no evidence of seizure activity. Figures 2b, 2c and 2d depict an examplar waveform extracted for each of the main clusters Seizure 1, Seizure 2 and non-seizure. In each case the relative amplitude scales have been fixed and the cross-correlation between both depicted. It can be seen that for the seizure component, the signal over the focal area is strongest and leads a weaker signal in the extra-focal electrode which lags by about 50 ms. The non-seizure components are equally present in both channels with no discernable lags.

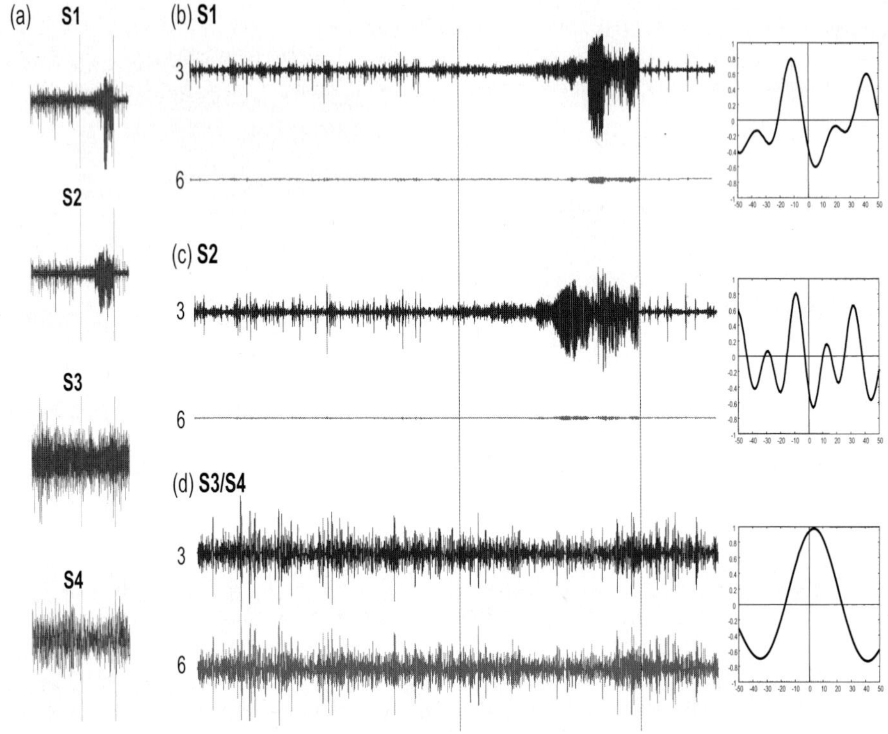

Fig. 2. (a) Reconstructed sources following ST-ICA; (b)-(d) depict seizure and noise components projected to both recording channels 3 and 6 along with their cross-correlations

4.2 BCI P300 Speller

For this demonstration we used data from data set IIb (P300 speller paradigm) obtained from the BCI competition 2003 data bank. The data was collected from one subject with 64 scalp electrodes (10/20 System) and sampled at 240 Hz. We demonstrate the proposed methods on the data by randomly selecting only *five* 1.5 s-epochs with possible P300 patterns and concatenate them to form a 7.5s trial. We apply ST-ICA to channels C3 and C4 (channels C3 & C4 cover the P300 focus). Finally we test using a similar setup whilst replacing alternating epochs with P300, non-P300, P300 etc. Figure 3 depicts the two channels C3 and C4 with the 5 epochs highlighted, within each epoch the first vertical line represents the presentation of the visual stimulus and the second line - 300 ms afterwards - represents the location where the P300 response (if present) should be maximal. The second part of Figure 3 depicts the two resulting P300 components after applying ST-ICA - 16 independent components were manually identified as contributing towards P300 responses and they were projected back to the measurement space. Although still relatively noisy, clear peaks can be seen around the 300 ms mark following stimulus presentation. In order to test

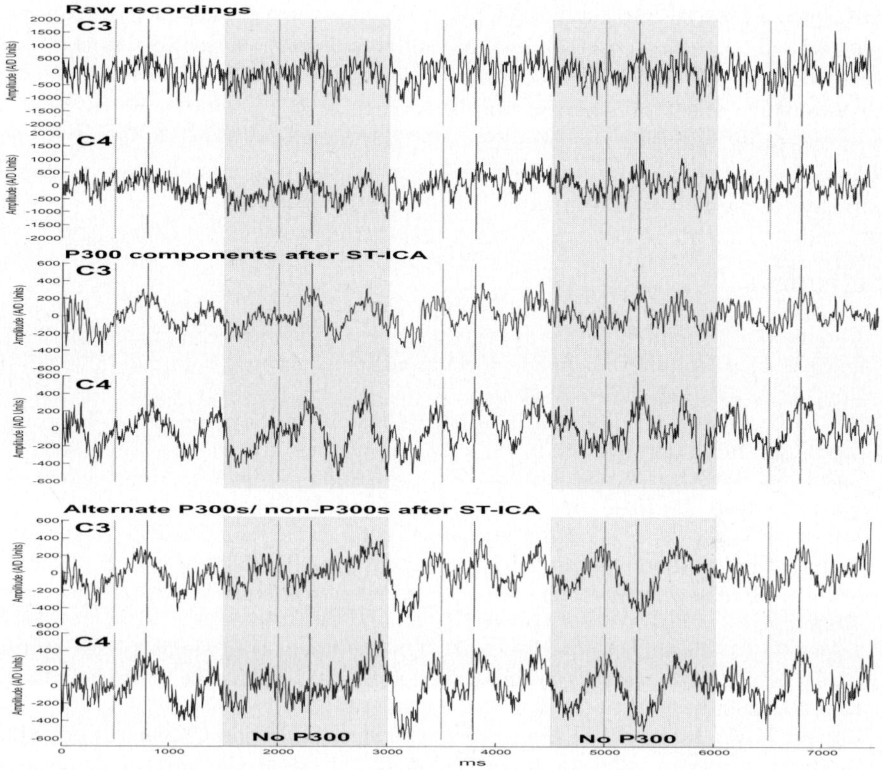

Fig. 3. 2 channels of EEG (C3 and C4) made up of concatenated P300 epochs (5 epochs). The extracted P300s are depicted following ST-ICA. Alternate P300/ non-P300 epochs are analyzed in the same way.

the reliablity of the process in the absence of P300 responses the final section of Figure 3 depicts the same analysis as before, however the dataset is chosen such that a P300 response appears in an alternating pattern. i.e. P300 - non-P300 - P300 - non-P300 - P300. It can be seen that whilst P300 responses are present where expected, the two epochs where there was no visual stimulus present, show no P300 response at all.

5 Discussion and Conclusion

ST-ICA provides a method of exploiting both spatial and temporal means of discrimination between independent sources, thereby allowing more sources to be extracted from fewer observation channels. We have shown that applied to quite different types of brain signals the ST-ICA technique yields very insightful results. This technique will be extremely useful in areas of brain signal analysis where multiple brain sources underly few channel recordings - either by design or through necessity. For the ictal EEG we have shown how an extra-focal electrode can still detect the presence of an epileptic seizure elsewhere in the brain, indicating that few channel recordings could be successfully used for systems such as seizure onset predictors. Similarly, in the BCI scenario we see that we can successfully enhance the presence of P300 evoked potentials using as little as 5 epochs - this is paramount for the design of BCI systems whose overall aims are always to be fast and accurate.

References

1. Abdallah, S., Plumbley, M.D.: If edges are the independent components of natural images, what are the independent components of natural sounds? In: Proc. Int Conf. ICA 2001, pp. 534–539 (2001)
2. Bell, A.J., Sejnowski, T.J.: An information maximization approach to blind separation and blind deconvolution. Neural Computation 7(6), 1129–1159 (1995)
3. Casey, M., Westner, A.: Separation of Mixed Audio Sources by Independent Subspace Analysis. In: Proc. Int. Comp. Music Conf. Berlin (2000)
4. James, C.J., Lowe, D.: Extracting information from multichannel versus single channel EEG data in epilepsy analysis. In: Hyder, A.K., Shahbazian, E., Waltz, E. (eds.) Multisensor Fusion, pp. 889–895. Kluwer Academic Publishers, Dordrecht (2002)
5. James, C.J., Gibson, O., Davies, M.E.: On the analysis of single versus multiple channels of electromagnetic brain signals. Artificial Intelligence in Medicine 37(2), 131–143 (2006)
6. Davies, M.E., James, C.J.: Source Separation using Single Channel ICA., Signal Processing, special issue on Advances on Independent Component Analysis (in press, 2007)
7. Abdallah, S., Plumbley, M.D.: Application of geometric dependency analysis to the separation of convolved mixtures. In: Puntonet, C.G., Prieto, A.G. (eds.) ICA 2004. LNCS, vol. 3195, pp. 540–547. Springer, Heidelberg (2004)
8. Davies, M.E., Jafari, M., Abdallah, S., Vincent, E., Plumbley, M.D.: Blind Source Separation using Space-Time Independent Component Analysis. In: Makino, S., Lee, T-W., Sawada, H. (eds.) Blind Speech Separation, Springer, Heidelberg (to appear, 2007)
9. Stone, J.V., Porrill, J., Buchel, C., Friston, K.: Spatial, Temporal, and Spatiotemporal Independent Component Analysis of fMRI Data. In: 18th Leeds Statistical Research Workshop on Spatial-temporal modelling and its applications (July 1999)
10. Belouchrani, A., Abed-Meraim, K., Cardoso, J.F., Moulines, E.: A blind source separation technique using second-order statistics. IEEE Trans. SP 45(2), 434–444 (1997)
11. Cardoso, J.-F.: Multidimensional independent component analysis. In: Proc. ICASSP'98, Seattle, WA, pp. 1941–1944 (1998)
12. Golyandina, N., Nekrutkin, V., Zhigljavsky, A.: Analysis of Time Series Structure SSA and Related Techniques. Chapman and Hall, Sydney (2001)
13. Farwell, L.A., Donchin, E.: Talking off the top of your head: Toward a mental prosthesis utilizing event-related brain potentials. Electroencephalogr. Clin. Neurophysiol 70, 510–523 (1988)
14. `epilepsy.uni-freiburg.de/freiburg-seizure-prediction-project/eeg-database`

Extraction of Gastric Electrical Response Activity from Magnetogastrographic Recordings by DCA

C.A. Estombelo-Montesco[1], D.B. De Araujo[1], A.C. Roque[1], E.R. Moraes[1], A.K. Barros[2], R.T. Wakai[3], and O. Baffa[1]

[1] Department of Physics and Mathematics, FFCLRP, University of Sao Paulo, Ribeirao Preto, SP, Brazil, 14040-901
estombelo@pg.ffclrp.usp.br, draulio@usp.br, antonior@.ffclrp.usp.br, eder@ffclrp.usp.br, baffa@usp.br
[2] Department of Electrical Engineering, Federal University of Maranhao, Sao Luis, Maranhao, Brazil
allan@ufma.br
[3] Department of Medical Physics, Medical School, University of Wisconsin, Madison-WI, USA
rtwakai@wisc.edu

Abstract. The detection of the basic electric rhythm (BER), composed of 3 cycles/minute oscillation, can be performed using SQUID sensors. However the electric response activity (ERA), which is generated when the stomach is performing a mechanical activity, was detected mainly by invasive electrical measurements and only recently one report was published dealing with its detection by magnetic measurements. This study was performed with the aim to detect and extract the ERA and ECA noninvasively before and after a meal. After acquire MGG recordings the signals were processed to extract both source components and remove cardiac interference and others interferences by an algorithm based on Dependent Component Analysis (DCA) then autoregressive and wavelet analysis was performed. Therefore, first, we can compare their relative amplitudes in the time or frequency domain, and get evidences of ERA signal. Second, we can get the spatial contribution from each channel to the source signal extracted. Finally, results have shown that there is an increase in the signal power at higher frequencies around (0.6-1.3 Hz) from ERA source component usually associated with the basic electric rhythm (ECA source component). We show that the method is effective in removing interference signals of MGG recordings, and is computationally efficient.

1 Introduction

The first non-invasive measurement of the stomach's electrical activity was made by Alvarez in 1922. Since then this field has grown considerably due to the extensive exploration of information from electrogastrography (EGG) and more recently from magnetogastrography (MGG) [1]. While the literature shows many contributions leading to a better understanding of gastric electrical activity (GEA), the EGG is difficult to measure, mostly because it is superimposed with other electrical signals that

M.E. Davies et al. (Eds.): ICA 2007, LNCS 4666, pp. 585–592, 2007.

are difficult to discriminate [2]. The MGG records the magnetic field produced by the GEA, and has been measured using SQUIDs (Superconducting Quantum Interference Device). The magnetic signals are less affected by tissue conductivity than the electric signals and show a stronger dependence of source-to-sensor distance [3]. MGG, thus, can provide higher spatial resolution than EGG [3].

Some evidence suggests that most gastrointestinal diseases are related to gastric motility impairment (or mechanical activity) [4]. Previous invasive studies have shown that the electrical extracellular signal detected with serosal electrodes has two distinct components. One, often referred to as 'electrical control activity' (ECA) or 'slow wave', is an omnipresent periodic activity not necessarily related to contractile motion. The second component, called electrical response activity (ERA) or 'spike activity', is time-locked to the ECA, but only occurs in conjunction with phasic contractile activity[5;6]. A better characterization of this ERA is considered a subject of major importance that had not been investigated satisfactorily in noninvasive human studies. Unfortunately, MGG signals are highly contaminated.

Recently, major advances in signal processing have been achieved with blind source separation (BSS). In most cases, extraction of all the source signals is unnecessary; instead, a priori information can be applied to extract only the signal of interest. Here we propose a strategy based on the algorithm of Barros and Cichocki [7;8] to separate the ECA and ERA signal from the other interferences, even in cases of low signal-to-noise-ratio, which we call dependent component analysis (DCA). For a quasi-periodic signal, DCA identifies the signal component based on the time delay determined from the temporal characteristics of the ECA in MGG measurement.

2 Material and Methods

The recordings were made with a 74-channel first-order gradiometer system (Magnes, Biomagnetic Technologies, Inc) housed within a magnetic shielded room. The system consists of two sensor units, A and B, each containing 37 channels uniformly distributed over circular areas of diameter 13.7 and 14.4 cm respectively. Seven asymptomatic subjects volunteered for the study.

The subjects lay with the stomach over the B sensor through a special bed with an opening such that the stomach could lie directly on the B sensor. The A sensor was positioned over the back of the subject. With this experimental arrangement it was possible to acquire signals from the stomach at the closest possible distance. Three epochs of duration10 minutes were acquired. The first was acquired before the ingestion of the test meal (pre-prandial). After that, a standard test meal comprised of a cheese sandwich with 250 kcal (110 kcal bread + 140 kcal cheese) was given to the subjects immediately before the second measurement (first post-prandial). Then, the last 10 minute acquisition was made (second post-prandial). The dc-coupled MGG signals were sampled at 73.1Hz and stored for subsequent analysis.

2.1 Blind Source Separation Using Temporal Structure

The proposed strategy to extract ECA and ERA components is based on Dependent Component analysis (DCA) method. DCA is a method based on multivariate analysis that uses a priori the delay based on the temporal characteristics of the ECA and ERA signal to be extracted (see Figure 3). The method for artifact removal is fully described in [8] and a short description follows. Consider n sources $\mathbf{S} = [s_1, s_2, ..., s_n]^T$ that are mixed into vector x through the following linear combination: $\boldsymbol{X} = \boldsymbol{AS}$, where **A** is an $n \times n$ invertible matrix. Our goal here is to find the source of interest, s_i. In general, the number of independent components can be as large as the dimension of **X**. The method proposed by Barros and Cichocki[7] aims at extracting only the desired component with a given characteristic, instead of all sources, and has been successfully applied to other applications [8]. The method can be described briefly as follows. As we wish to extract only a single source, we can write the signal as $y(k) = \boldsymbol{w}^T \boldsymbol{x}(k)$, where $\boldsymbol{w}$ is a weight vector for that single source. Defining the error as $\varepsilon_a(k) = y(k) - y(k-p)$ and minimizing the mean squared error $\xi(\boldsymbol{w}) = E[\varepsilon_a^2]$, we find: $\boldsymbol{w} = E[\boldsymbol{x} y_p]$, where $y_p = y(k-p)$. We will use sequential signal extraction along with a priori information about the autocorrelation function. One practical problem is how to estimate the optimal time delay. A simple solution is to calculate the autocorrelation function of the sensor signals and find the feature, in our case a peak with appropriate time-lag, corresponding to the signal of interest. In order to accomplish this, we model the system using autoregression, as described next.

2.2 The Adaptive Autoregressive Model for Spectrum Estimation

To characterize the power spectrum of the signal we use the well known autoregressive (AR) model [1;8]. This power spectrum will be used to estimate the optimal time delay (before DCA processing) for each desired component and at the end (after DCA processing) to compare energies of the estimated source signals in pre-prandial versus post-prandial measurements.

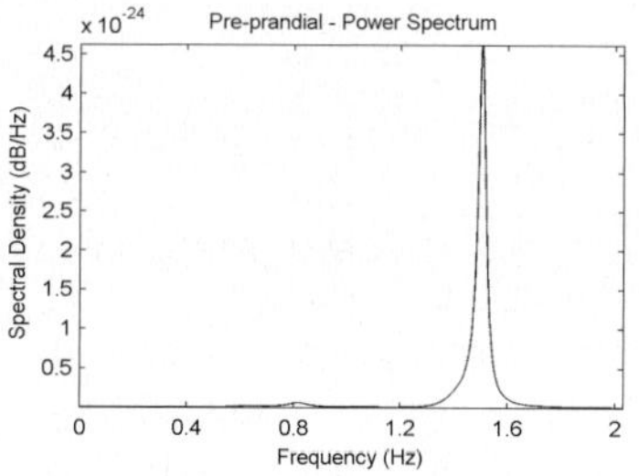

Fig. 1. ERA power spectrum of a pre-prandial single recording from raw data. ERA signal has weak energy compared to heart component at higher frequency.

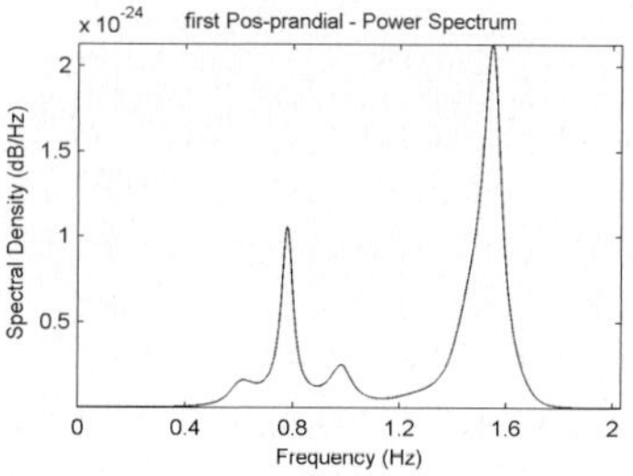

Fig. 2. Power spectrum of a first post-prandial single recording from raw data. Two peaks corresponding to the ERA and heart components.

In Figure 1 and Figure 2 we show representative power spectra of segments of the raw signal from a pre-prandial and post-prandial recording respectively. These representative power spectra show peaks near 0.8 Hz (ERA), indicating a different state compared to pre-prandial measurement in Figure 1. For ECA signal the process is similar where we found that the power is at 0.05Hz (or 3cpm). This *a priori* information used to estimate the time-delay for DCA filtering.

2.3 Avoiding Scaling Factor Problem by Projection Approach

The estimation of the signal extracted by DCA might be scaled at the output. Subsequently this estimation can lead to an erroneous comparison between pre-prandial and post-prandial energies if we don't determine the proper scale factor. To overcome this problem we need an estimation method for the scale factor for each signal component extracted by DCA. Consider an output vector, $y = wx = wAs$. It has an indeterminacy that can be expressed as $y = \alpha s$, where α is a scaling factor that needs to be estimated for the signal extracted from each epoch (see Figure 3). After extracting the desired source component, y , one can project the source signal of interest back onto sensor array signals, calculating the scale factor as follows. First let us define the following error: $\varepsilon_b = x_i - \alpha_i y$, where x_i is the desired signal and y is the output of the DCA filter. Next we estimate a scale factor α_i that minimizes the mean squared error $\xi(\alpha_i) = E[\varepsilon_b^2]$. Then we have $\xi(\alpha_i) = x_i^2 - 2x_i\alpha_i y + (\alpha_i y)^2$. The minimum will be reached yielding the following scale factor: $\alpha_i = \mathrm{E}[y^2]^{-1}\,\mathrm{E}[xy]$. The scaling factor provides two valuable pieces of information. First if we take the high absolute value of α , then we have the scaling factor for the output of DCA (estimated source signal). Therefore we can compare their relative amplitudes in the time or frequency domain, and get evidences of ERA signal. Second, we can get the contribution from each channel to the source signal extracted. It allows spatial localization over 37 channels of the estimated source signal.

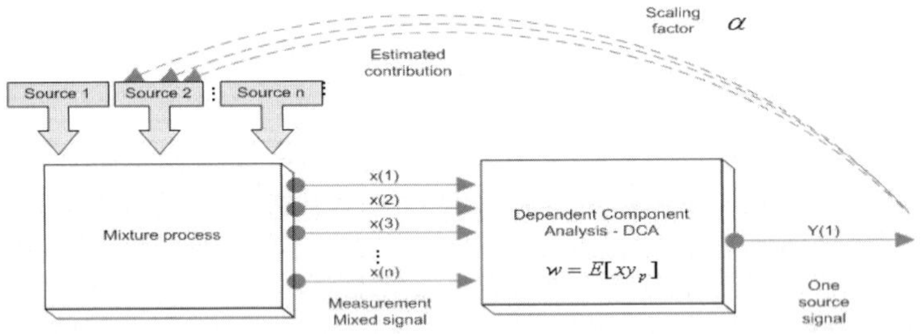

Fig. 3. Schematics for scaling factor determination. a) First each source (stomach, heart, tissue, artifacts, etc) produces a magnetic signal, actually not seen directly. b) Then each source produced is mixed with other sources, in our case we consider a linear mixture and it is represented by the block mixture process. c) When the signals are acquired actually we are measuring the signals from the mixture process obtaining one time series for each sensor. d) Then separation/extraction of the source of interest can be done and further evaluated through DCA process. e) At the output there is a single time series of interest where a scale factor needs to be calculated to estimate the relative amplitude. These steps were applied for every epoch: pre and post-prandial measurements.

3 Results: ECA and ERA Components by DCA

The upper diagram of Figure 4 and Figure 5 shows the ECA(dotted line) and ERA(solid line) components extracted at each epoch by DCA. The upper diagram of Figure 4 shows the extracted components for the pre-prandial epoch and The upper diagram of Figure 5 the extracted components for the first post-prandial epochs.

Figure 4 and Figure 5 show that ECA component are always present in the stomach. Furthermore, it can be noted that ERA components have higher amplitude than the pre-prandial component, especially during plateau phase of the ECA near 50 seconds, 70 seconds, 90 seconds and 110 seconds of Figure 5.

The right side shows the ECA scale and left side shows the ERA scale. The ECA signal component consists of an upstroke followed by a plateau and then by a slow depolarization phase with approximate frequency 3 cpm. Note the difference of amplitude between the ERA signals in pre-prandial epoch and the ERA signals of the post-prandial epochs.

The upper diagram of Figure 4 shows the ERA and ECA components during a pre-prandial epoch. The lower diagram of Figure 4 shows a white solid line, obtained by summing the ERA and ECA time series, superimposed on the time-frequency representation (TFR) of the ERA component, obtained by the Morlet wavelet transformation [9;10]. The y-scale of the TFR is from 0.5 Hz to 1.3Hz. A few localized high energy regions can be observed for the ERA component, but they are not necessarily time-locked with the ECA component.

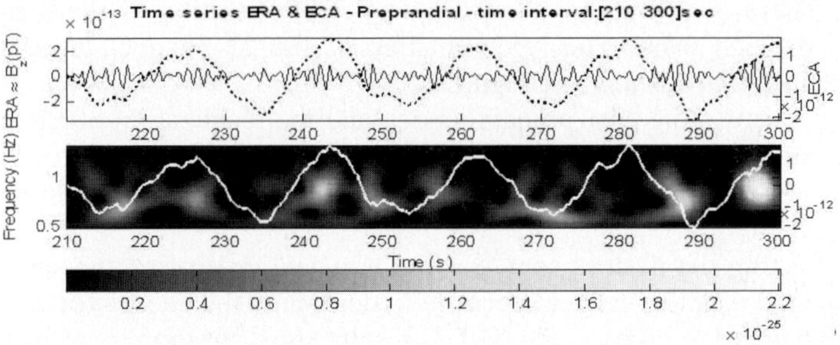

Fig. 4. Time interval of pre-prandial epoch with the ECA and ERA components in the upper panel and the TFR of the ERA component in the lower panel. Superimposed on the TFR with a solid line is the sum of the ECA component plus the ERA component, which are shown in the upper panel. The scale of the left side is for the TFR of the ERA and the scale on the right side is for the summed time series. The x-scale is in seconds.

In Figure 5, upper panel, shows the ERA and ECA components during the first post-prandial epoch. Here, in contrast, the amplitude of the ERA component is greater than during the pre-prandial epoch. The lower panel of Figure 5 shows high energy spots of the ERA component that are time-locked with the ECA component. This characteristic is very important to verify the existence of the ERA component. Another characteristic to note is the fundamental frequency of the ERA component; although it concentrates at 0.8 Hz it can vary with time increasing up to 1.30 Hz, but it preserves the time-locking with the ECA component at the plateau.

During the second post-prandial epoch we can observed that the ERA amplitude remains high, as in the first post-prandial epoch. Differences are related to a more diffuse energy distribution than in the previous epoch. However, it still preserves the time-locking with the ECA component, despite the energy decrease.

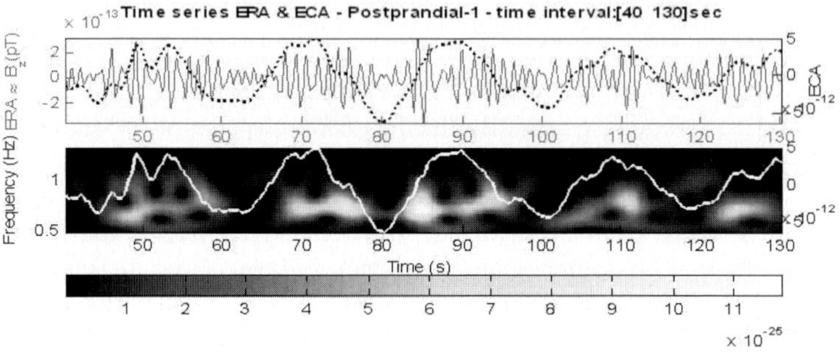

Fig. 5. Time interval during the first post-prandial epoch with ECA and ERA components. In the upper panel the spikes of the ERA component can be observed, which are reflected in the TFR in the lower panel and are time-locked with the ECA component. The frequency of the ERA component in the TFR varies from 0.6 Hz to 1.0 Hz. The scales of this figure are similar to those of Figure 4. The x-scale is in seconds.

The energy contribution from each channel can be used to construct isocontribution maps. These contributions show a spatial representation of the area where the source signal came from. The representation of the 37 channel layout from ERA (post-prandial epoch) is shown in Figure 6.

In Figure 6 shows the post-prandial epoch and the localization of ERA after DCA extraction. An increase in the contribution energy in a number of channels on right side is observed, whereas the contribution of these channels was low in the pre-prandial epoch.

After extracting the desired source with DCA and estimating the scale factor, we can calculate the adaptive power spectrum to determine the energy for each epoch using the autoregressive (AR) method. The results show an amplitude increase of the signal around (0.6-1.0 Hz) (Figure 7) with a dominant frequency at 0.8Hz, usually correlated with the higher intensity of the ECA rhythm.

The integrated power spectrum in the frequency band of (0.5-1.33) Hz was used to generate an index of ERA. Signal acquired from all volunteers shown an increment of ERA index, when the pre and post-prandial were compared. These results can be seen in Table 1.

4 Discussion

In the present study, the amplitudes of the ECA and ERA signals were different, especially in comparison with the cardiac signal. Moreover, due to the overlap of the cardiac and ERA signals in the frequency domain, it is not possible to use a classical filter to remove the cardiac component. Finally, analysis of MGG signals is complicated by biological interferences such as respiration, small and large intestine, and duodenum magnetic signals.

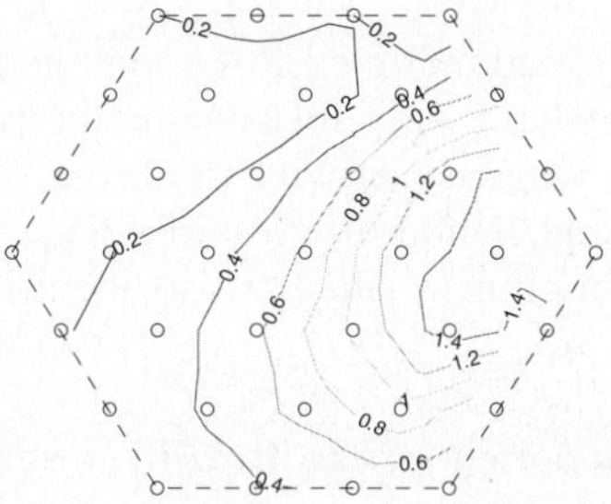

Fig. 6. Isocontribution map of the ERA component from the first post-prandial epoch

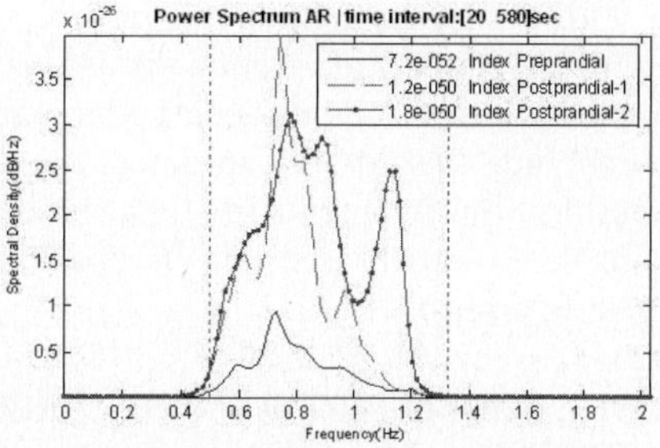

Fig. 7. Autoregressive power spectrum of pre-prandial and post-prandial epochs after signal extraction from one subject. The inset shows the index of each epochs.

Table 1. Index for each epoch (between 0.5Hz and 1.33Hz) x 10^{51}

	Pre-prandial	Post-prandial (1)	Post-prandial (2)
Volunteer – 1	8.0	1100	930
Volunteer – 2	0.72	12	18
Volunteer – 3	9.0	91	120
Volunteer – 4	27	29	44
Volunteer – 5	16	300	83
Volunteer – 6	4.2	2800	5100
Volunteer – 7	8.6	99	180

These difficulties were overcome by using DCA to extract only the desired component with a specified periodicity, rather than extracting all sources. DCA can be applied even to a low signal-to-noise ratio recording. One problem, however, is that the scale of the extracted component can be altered, leading to an inaccurate energy. To avoid the scaling factor problem, a projection approach can be applied. Generally scaling factors have not been a problem in many applications of BSS that involve a single measurement or experimental condition. But this is not the case in our study, which includes three epochs, one pre-prandial and two post-prandial. It is therefore necessary to compute the scaling factor to obtain coherent relative amplitudes for each epoch, which was accomplished here by the projection approach.

A previous work [2] described and stated confidently the detection and analysis of a type of human ERA signal; however, based on qualitative analysis the authors proposed future work to validate their results with simultaneous measurement involving serosal or mucosal electrode recordings. In our study, the extracted signals satisfy the properties of the ERA signals reported in the literature in which invasive recordings were made in animals, apparently requiring no further experiments.

It is important to notice that through the DCA process the ECA and ERA components can be extracted in time domain. The energy increase can be seen in all volunteers, as shown in Figure 7 and Table 1. This method and result have not been reported previously.

Recordings using invasive methods [**6**] show that the ERA component is time-locked with the ECA component. In this work, using MGG, a non-invasive method, along with DCA and time-frequency analysis, we found that the ERA component was time-locked with ECA component, which agrees with previous invasive methods.

We conclude that MGG can detect the electric response activity in normal volunteers. Further improvements in signal processing and standardization of signal acquisition are necessary to ascertain its possible use in clinical situations to identify and study gastric diseases.

Acknowledgements. This work was partially supported by the Brazillian agencies CNPq, FAPESP and CAPES.

References

[1] Moraes, E.R., Troncon, L.E.A., Baffa, O., Oba-Kunyioshi, A.S., Wakai, R., Leuthold, A.: Adaptive, autoregressive spectral estimation for analysis of electrical signals of gastric origin. Physiological Measurement 24(1), 91–106 (2003)

[2] Irimia, A., Richards, W.O., Bradshaw, L.: Magnetogastrographic detection of gastric electrical response activity in humans. Physics in Medicine and Biology 51(5), 1347–1360 (2006)

[3] Andrä, W., Nowak, H.: Magnetism in Medicine: A Handbook. Wiley-vch, Chichester (2006)

[4] Chen, J.D., McCallum, R.W.: Clinical-Applications of Electrogastrography. American Journal of Gastroenterology 88(9), 1324–1336 (1993)

[5] Smout, A.J.P.M., Vanderschee, E.J., Grashuis, J.L.: What Is Measured in Electrogastrography. Digestive Diseases and Sciences 25(3), 179–187 (1980)

[6] Akin, A., Sun, H.H.: Time-frequency methods for detecting spike activity of stomach. Medical & Biological Engineering & Computing 37(3), 381–390 (1999)

[7] Barros, A.K., Cichocki, A.: Extraction of specific signals with temporal structure. Neural Computation 13(9), 1995–2003 (2001)

[8] de Araujo, D.B., Barros, A.K., Estombelo-Montesco, C., Zhao, H., da Silva, A.C.R., Baffa, O., et al.: Fetal source extraction from magnetocardiographic recordings by dependent component analysis. Physics in Medicine and Biology 50(19), 4457–4464 (2005)

[9] TallonBaudry, C., Bertrand, O., Delpuech, C., Pernier, J.: Oscillatory gamma-band (30-70 Hz) activity induced by a visual search task in humans. Journal of Neuroscience 17(2), 722–734 (1997)

[10] Jensen, O., Gelfand, J., Kounios, J., Lisman, J.E: Oscillations in the alpha band (9-12 Hz) increase with memory load during retention in a short-term memory task. Cerebral Cortex 12(8), 877–882 (2002)

ECG Compression by Efficient Coding

Denner Guilhon[2], Allan K. Barros[3,⋆], and Silvia Comani[1,2]

[1] Department of Clinical Sciences and Bio-imaging, Chieti University, Italy
[2] ITAB - Institute of Advanced Biomedical Technologies, University Foundation 'G.D'Annunzio', Chieti University, Italy
comani@itab.unich.it, dennerguilhon@gmail.com
http://www.unich.it/itab
[3] Federal University of Maranhão - UFMA, São Luís – MA, Brazil
akbarros@ieee.org
http://pib.dee.ufma.br

Abstract. The continuous demand for high performance and low cost electrocardiogram (ECG) processing systems have required the elaboration of more and more efficient and reliable ECG compression techniques. Such techniques face a tradeoff between compression ratio and retrieved quality, where the decrease of the last can compromise the subsequent use of the signal for clinical purposes. The objective of this work is to evaluate the validity and performance of an independent component analysis (ICA) based scheme used to efficiently compress ECG signals while introducing tests for a different type of record of the electrical activity of the heart, such as fetal magnetocardiogram (fMCG). As a result, the reconstructed signals underwent negligible visual deterioration, while achieving promising compression ratios.

Keywords: independent component analysis, efficient coding, electrocardiogram, fetal magnetocardiogram.

1 Introduction

As the need for constantly larger quantities of ECG data increases, more efficient compression methods are required. Whether to optimize storage or to make online transmission possible over the public phone network, many efforts have been made in order to enhance ECG compression techniques. Consequently, throughout this process many other methods have emerged [1]. Recent works on ECG compression are related mainly to transform methods, as Karhunen-Loeve (KL) transform [2] and wavelet transforms [3][4][5].

Similarly to other compression techniques, electrocardiogram (ECG) compression aims to reduce data volume while preserving the morphological features of the signal after reconstruction. It implies that signal redundancy must be minimized without loss of the information contained therein.

When discussing about the primary visual cortex, neuroscientists [6] argued that a primary function of visual neurons is to re-code the input in a way that

⋆ Corresponding author.

M.E. Davies et al. (Eds.): ICA 2007, LNCS 4666, pp. 593–600, 2007.

reduces redundancy and maximizes the information transmitted to the output. It requires a redundancy reduction process in which the activation of each visual feature is supposed to be as statistically independent from the others as possible [7]. As for natural scenes, an efficient ECG compression method must take into account the high-order statistical dependencies in the data and safeguards its information content, in order to seek a minimum-entropy code [8].

In this work, we discuss a compression method that, for a given ECG signal, finds its basis functions (features) and then the coefficients of the projection of this signal onto a vector subspace spanned by the basis functions. This technique aims to obtain a less redundant and, therefore, more efficient code representation of the ECG source. To achieve this, independent component analysis (ICA) is used to find the vector subspace. Then the signal is projected on that subspace, estimating the projection coefficients in such a manner that they minimize a mean-square-error (MSE) cost function. The same reasoning has led to promising results on image compression [9].

Additionally, initial tests for the compression of fetal magnetocardiograms (fMCG) signals are introduced. Magnetocardiography is a noninvasive and risk-free technique that allows recording the magnetic fields associated with the spontaneous electrophysiological activity of the fetal heart during the second half of gestation [10]. ICA is particularly efficient to process the fMCG and to retrieve fetal cardiac signals that are undistorted by the tissues interposed between the fetal heart and the sensors. The retrieved fetal signals can complement diagnostic methods for pregnancies at risk, such as in the case of intra-uterine fetal growth retardation, or in the presence of fetal arrhythmias [11][12][13].

The use of ICA to extract the fetal signal from fMCG, in conjunction with the application of an ICA based compression algorithm, might permit the online reconstruction of the fetal cardiac signal, which would reveal extremely useful in those clinical cases for which the online monitoring of the fetal cardiac activity is required, such as life threatening fetal arrhythmias.

2 Methods

2.1 Proposed Solution

Let $e(t)$ be an ECG signal and assume that it can be divided in m windows of fixed length n. Let us also assume that we can find through ICA a vectorial subspace $\mathbf{\Phi} = [\phi_1(t), \ldots, \phi_n(t)]$, where the columns $\phi_i(t)$ are defined as the basis functions of $e(t)$ (see Figure 1). Given that the projection of the ith window of $e(t)$ into $\mathbf{\Phi}$ is expressed by [14]

$$\hat{e_i}(t) = w_1\phi_1(t) + w_2\phi_2(t) + \ldots + w_n\phi_n(t) \qquad i = 1, \ldots, m, \tag{1}$$

where $\hat{e_i(t)}$ is the estimated version of $e_i(t)$ and each component w_i is the projection coefficient for the ith base function of the subspace $\mathbf{\Phi}$. We will drop time index t, for convenience.

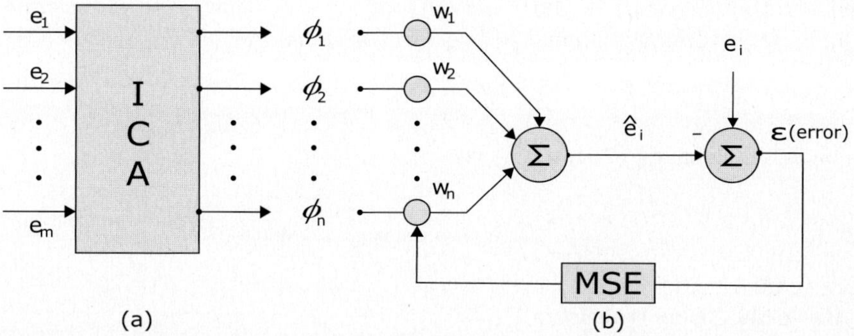

Fig. 1. System block diagram. (a) The learning phase where the system learns the basis functions ϕ_i through the ICA algorithm. (b) The test phase, where the coefficients of the projection are calculated, through a simple mean-square error estimator.

Those coefficients can be calculated through mean-square estimation, where the signal e_i is the desired one and input to the estimator. The vector which finds the minimum of the mean-square error, $E[\varepsilon^2]$, is given by

$$w_i = E[\mathbf{\Phi}\mathbf{\Phi}^T]^{-1}E[e_i\mathbf{\Phi}] \tag{2}$$

Here we assume that the desired signal spans the same subspace as the training one, otherwise the output would be null, and that the length of the training input has to be small enough so that in a specific time window, the signal is stationary [15]. The chosen ICA algorithm was *FastICA* [11][16].

2.2 Efficient Coding

Let the mutual information of the random variables $\mathbf{e}_1, \ldots, \mathbf{e}_m$ be defined as

$$I(\mathbf{e}_1, \ldots, \mathbf{e}_m) = \sum_{i=1}^{m} H(\mathbf{e}_i) - H(\mathbf{e}_1, \ldots, \mathbf{e}_m), \tag{3}$$

where $H(\mathbf{e}_1, \ldots, \mathbf{e}_m)$ is the joint entropy of $\mathbf{e}_1, \ldots, \mathbf{e}_m$. The mutual information gives a measure of the dependency among variables. Since information cannot be lost, we recall that

$$I(\mathbf{e}_1, \ldots, \mathbf{e}_m) \geq 0. \tag{4}$$

Substituting (4) into (3) yields

$$\sum_{i=1}^{m} H(\mathbf{e}_i) \geq H(\mathbf{e}_1, \ldots, \mathbf{e}_m). \tag{5}$$

Likewise, let the mutual information of the random variables $\mathbf{w}_1, \ldots, \mathbf{w}_m$ be defined as

$$I(\mathbf{w}_1, \ldots, \mathbf{w}_m) = \sum_{i=1}^{m} H(\mathbf{w}_i) - H(\mathbf{w}_1, \ldots, \mathbf{w}_m), \tag{6}$$

where $H(\mathbf{w}_1, \ldots, \mathbf{w}_m)$ is the joint entropy of $\mathbf{w}_1, \ldots, \mathbf{w}_m$. If we assume that $\mathbf{w}_1, \ldots, \mathbf{w}_m$ are independents [8], we can state that [17]

$$I(\mathbf{w}_1, \ldots, \mathbf{w}_m) = 0. \tag{7}$$

Then substituting (7) into (6) we get

$$\sum_{i=1}^{m} H(\mathbf{w}_i) = H(\mathbf{w}_1, \ldots, \mathbf{w}_m). \tag{8}$$

Given the linear transform

$$\mathbf{e}_i = \mathbf{\Phi} \mathbf{w}_i, \tag{9}$$

we have [18]

$$H(\mathbf{e}_1, \ldots, \mathbf{e}_m) = H(\mathbf{w}_1, \ldots, \mathbf{w}_m)^1. \tag{10}$$

Hence, from (5), (8) and (10) we obtain

$$\sum_{i=1}^{m} H(\mathbf{e}_i) \geq H(\mathbf{w}_1, \ldots, \mathbf{w}_m) \tag{11a}$$

$$\Longrightarrow \sum_{i=1}^{m} H(\mathbf{e}_i) \geq \sum_{i=1}^{m} H(\mathbf{w}_i). \tag{11b}$$

Since $\overline{L}_{min} = H(\vartheta)$, i.e., the average code length is minimum when it equals the entropy of the set, we can conclude that

$$\sum_{i=1}^{m} \overline{L}_{min}(\mathbf{e}_i) \geq \sum_{i=1}^{m} \overline{L}_{min}(\mathbf{w}_i), \tag{12}$$

Eq.(12) establishes a relationship between the total code length required to represent $\mathbf{e}$, by means of either $\mathbf{e}_i$ or $\mathbf{w}_i$. We observe that both representations have the same length only when the total code length of $\mathbf{e}_i$ is the minimum possible, for the representation of $\mathbf{w}_i$ is already efficient [7]. Otherwise, the total code length of $\mathbf{e}_i$ would be increased due to the presence of redundancies, that were not present in $\mathbf{w}_i$, according to Eq. (7).

3 Results

Our approach was first tested for the MIT Normal Sinus Rhythm Database and the MIT Supraventricular Arrhythmia Database, 15 records each. The first 30 minutes of each record were used in the learning phase, while the test phase used two minutes [14].

[1] The equantion is valid since the random variable $\mathbf{w}_i$ is of discrete type and the transformation $\mathbf{e}_i = g(\mathbf{w}_i)$ has a unique inverse.

Then, similar tests were performed for 15 fetal signals reconstructed from fMCG data recorded for a pregnancy at 35 weeks. The fetal cardiac signals were extracted according to [19]. The first 100 seconds of each record were used in the learning phase, while the test phase used 10 seconds.

The coefficients found through Eq. (2) were quantized using as many levels as needed to properly reconstruct the ECG and the fMCG signals. Figure 2 shows that the method does not introduce errors that result in significant visual differences, indicating that the morphological characteristics of the fetal magnetocardiogram are preserved. The reconstruction errors, also shown in figure 2, confirm those results. Figures 3 and 4 show the mean values of 50 repetitions of the percent root-mean-square-difference (PRD) upon reconstruction of each record, defined as

$$\mathrm{PRD} = \sqrt{\frac{\sum_{i=1}^{n} \left[sig_{orig}(i) - sig_{rec}(i)\right]^2}{\sum_{i=1}^{n} sig_{orig}^2(i)}} * 100\% \qquad (13)$$

where $sig_{orig}(i)$ is the original signal, and sig_{rec} is the reconstructed one.

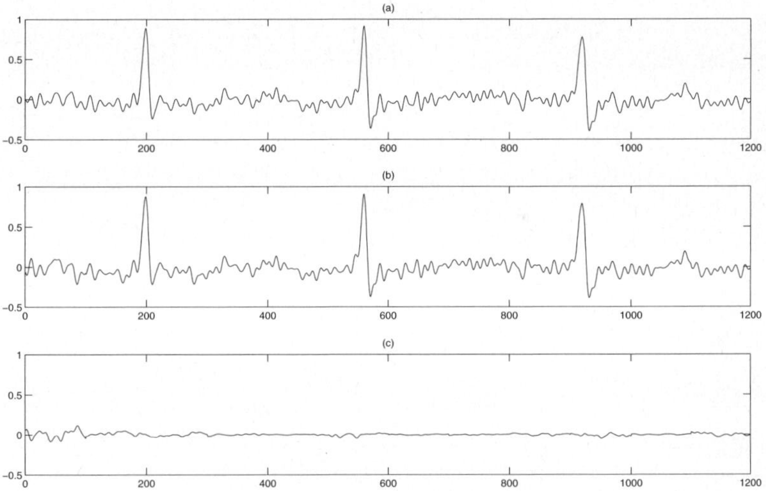

Fig. 2. Result of 1200 samples of a fMCG record. (a) Original signal. (b) Reconstructed signal, with CR = 2.66:1 and PRD = 3.43%. (c) Reconstruction error.

4 Discussion

ICA can be used to find a vectorial subspace where the component projections of the ECG signal are mutually independents. Therefore, the coefficients of the signal projected onto this subspace are independent as well. Coding that projection, as a whole or by its parts, results in the same code length [17].

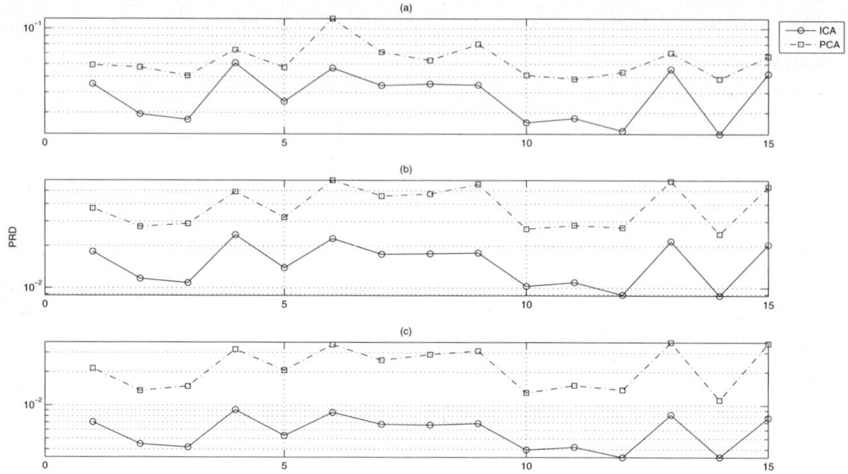

Fig. 3. Percent root-mean-square-difference upon reconstruction of 15 records of the MIT Normal Sinus Rhythm Database. Full lines with circle marker stand for ICA results, whereas dashed-dot lines with square markers stand for PCA results. (a), (b) and (c) show results for compression ratios fixed at 3, 2.4 and 2, respectively. Notice the logarithm scale.

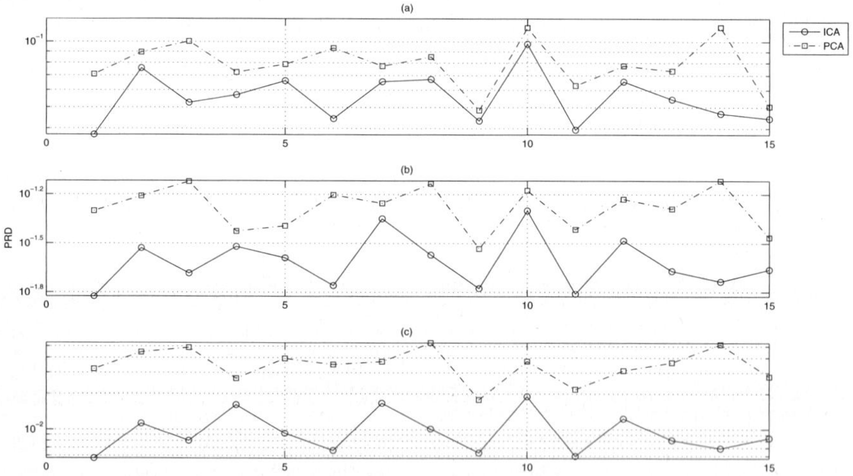

Fig. 4. Percent root-mean-square-difference upon reconstruction of 15 records of the MIT Supraventricular Arrhythmia Database. Full lines with circle marker stand for ICA results, whereas dashed-dot lines with square markers stand for PCA results. (a), (b) and (c) show results for compression ratios fixed at 2.5, 2 and 1.667, respectively. Notice the logarithm scale.

Figures 3 an 4 refer to the reconstruction errors obtained when using either ICA or PCA to find a vectorial subspace associated with each one of the 15 ECG records of both databases. These results reflect the performance simulation of our method when compared to that of [2], excluding the contributions of classic compression algorithms. By using ICA, it is observed that the ECG reconstruction errors, even after the quantization process, are smaller than those calculated using PCA. Again, one can see that the method achieves data compression without adding relevant distortion to the signal, what can be confirmed by figure 2.

Moreover, from (12) we observe that the average code length required to represent the original signal is larger than that required for represent its coefficients, unless the first is already efficiently coded. In that case, the left side of the equation equals the right one, which means that our method do not alter the representation code length, because it is minimum. Otherwise, due to the presence of redundancies, the code length of the original data representation would be larger than that achieved using our method.

5 Conclusion

We evaluated the validity and performance of an ICA based scheme used to compress ECG and fMCG data. It was verified that such a tool efficiently compresses those signals by means of non-redundant representations of them, ensuring both the reduction of the total data volume and the preservation of the morphological characteristics of the signals.

From the perspective of clinical applications, the shown effectiveness of the described compression algorithm would be beneficial not only for adult ECG, but also for prenatal online monitoring of the fetal cardiac activity.

A further step in the development of this tool might be its use as a preprocessing step for a classic compression algorithm; furthermore, a selection criterion might be included to allow also the reduction of the numbers of coefficients that should be stored.

Acknowledgements

This work was supported by FAPEMA.

References

[1] Jalaleddine, S.M., Hutchens, C.G., Strattan, R.D., Coberly, W.A.: ECG data compression techniques-A unified approach. IEEE Trans. Biomed. Eng. 37(4), 329–343 (1990)

[2] Olmos, S., Millan, M., Garcia, J., Laguna, P.: ECG data compression with the Karhunen-Loeve transform. In: Computers in Cardiology, Menlo Park, CA, pp. 253–256. IEEE Comput. Soc. Press, Los Alamitos (1996)

[3] Miaou, S.G., Chao, S.N.: Wavelet-based lossy-to-lossless ECG compression in a unified vector quantization framework. IEEE Trans. Biomed. Eng. 52(3), 539–543 (2005)
[4] Blanco-Velasco, M., Cruz-Roldan, F., Godino-Llorente, J.I., Barner, K.E.: ECG compression with retrieved quality guaranteed. Electronics Letters 40(23), 1466–1467 (2004)
[5] Rajoub, B.A.: An efficient coding algorithm for the compression of ECG signals using the wavelet transform. IEEE Trans. Biomed. Eng. 49(4), 355–362 (2002)
[6] Sekuler, A.B., Bennet, P.J.: Visual neuroscience: Resonating to natural images. Curr. Biol. 11, R733–R736 (2001)
[7] Bell, A.J., Sejnowski, T.J.: Edges are the independent components of natural scenes. In: Advances in neural information processing systems, vol. 9, pp. 831–837. MIT Press, Cambridge (1997)
[8] Olshausen, B.A., Field, D.J.: Emergence of simple-cell receptive field properties by learning a sparse code for natural images. Nature 381, 607–609 (1996)
[9] Souza, C.M., Cavalcante, A.B., Guilhon, D., Barros, A.K.: Image compression by Redundancy Reduction, submited to ICA (2007)
[10] Tavarozzi, I., et al.: Magnetocardiography: current status and perspectives: Part I. Physical principles and instrumentation, Ital. Heart J. 3, 75–85 (2002)
[11] Hild II, K.E., Alleva, G., Nagarajan, S.S., Comani, S.: Performance comparison of six Independent Components Analysis algorithms for fetal signal extraction from real fMCG data. Phys. Med. Biol. 52, 449–462 (2007)
[12] Comani, S., et al.: Time course reconstruction of fetal cardiac signals from fMCG: Independent Component Analysis vs. Adaptive Maternal Beat Subtraction, Physiol. Meas 25, 1305–1321 (2004)
[13] Comani, S., et al.: Characterization of fetal arrhythmias by means of fetal magnetocardiography in three cases of difficult ultrasonographic imaging. Pacing. Clin. Electrophysiol. 27, 1647–1655 (2004)
[14] Guilhon, D., Medeiros, E., Barros, A.K.: ECG Data Compression by Independent Component Analysis. In: IEEE Workshop on Machine Learning for Signal Processing, September 28-30, 2005, pp. 189–193 (2005)
[15] Barros, A.K., Ohnishi, N.: Single channel speech enhancement by efficient coding. Signal Processing 85(9), 1805–1812 (2005)
[16] Hyvarinen, A., Oja, E.: A fast fixed-point algorithm for independent component analysis. Neural Computation 9, 1483–1492 (1997)
[17] Hyvarinen, A., Karhunen, J., Oja, E.: Independent Component Analysis. John Wiley & Sons, New York (2001)
[18] Papoulis, A., Pillai, S.U.: Probability, Random Variables and Stochastic Processes, 4th edn. McGraw-Hill, New York (2002)
[19] Comani, S., et al.: Independent component analysis: fetal signal reconstruction from magnetocardiographic recordings. Comput. Meth. Prog. Biomed. 75, 163–177 (2004)

Detection of Paroxysmal EEG Discharges Using Multitaper Blind Signal Source Separation

Jonathan J. Halford

Medical University of South Carolina, Department of Neuroscience,
Adult Neurology Division, 96 Jonathan Lucas St., Suite 307 Clinical Science Building
Charleston, SC 29425
halfordj@musc.edu

Abstract. The routine electroencephalogram (rEEG) is a useful diagnostic test for neurologists. But this test is frequently misinterpreted by neurologists due to a lack of systematic understanding of paroxysmal electroencephalographic discharges (PEDs), one of the most important features of EEG. A heuristic algorithm is described which uses conventional blind signal source separation (BSSS) algorithms to detect PEDs in a routine EEG recording. This algorithm treats BSSS as a 'black box' and applies it in a computationally-intensive multitaper algorithm in order to detect PEDs without a pre-specification of signal morphology or scalp distribution. The algorithm also attempts to overcome some of the limitations of conventional BSSS as applied to the study of neurophysiology datasets, specifically the 'over-completeness problem' and the 'non-stationarity problem'.

1 Introduction

The routine electroencephalogram (rEEG) recorded from the scalp is an important diagnostic test used by neurologists for the medical management of patients with undetermined spells, epileptic seizures, altered mental status, and coma [1]. Unfortunately, the misinterpretation of rEEG is common, since most rEEG studies are interpreted by neurologists who lack subspecialty fellowship training in clinical neurophysiology. These misinterpretations sometimes cause medical mismanagement and harm to patients [2]. Clinically-significant misinterpretations of rEEG studies usually involve the misinterpretation of paroxysmal EEG discharges (PEDs) – short bursts of electrical activity usually lasting between 0.1 and 2.0 seconds which have higher signal amplitude than the surrounding background EEG activity. The manifestation of PEDs in rEEG are varied and complex. The morphological features which describe the boundaries of normality for PEDs are subtle (particularly in children), not rigorously defined, and vary between experts and clinical neurophysiology training programs [3]. The subspecialty training of clinical neurophysiologists in the interpretation of PEDs is much like the training of a baseball umpire who gradually develops a 'strike-zone' through observation, attention to oral history, and supervision by instructors.

M.E. Davies et al. (Eds.): ICA 2007, LNCS 4666, pp. 601–608, 2007.

In order to promote the rigorous scientific study of PEDs, it would be useful to be able to collect all PED activity in a rEEG recording objectively using a computerized detection algorithm. A database of the PEDs present in a rEEG could then be characterized in signal morphology and scalp distribution and submitted, along with clinical data, to clustering algorithms. This could provide a more objective definition of what constitutes a normal or abnormal constellation of PEDs in a given rEEG, for both clinical and research purposes. To the authors knowledge, a computer algorithm which detects all PED signal activity in rEEG in a non-specific way (with little regard to signal morphology or scalp distribution) does not exist. Previous scientific study of PEDs in rEEG has focused on other algorithmic approaches. In many studies, the signal morphology and scalp distribution of normal or abnormal types of PEDs are defined *a priori* and algorithms for detecting PEDs are created based on these morphologic and topographic characteristics [4,5]. This causes a significant selection bias in the detection of PEDs since 'you find only what you are looking for'. Other studies use a neural network approach to categorize a sample of rEEG data as normal or abnormal based on a training signal from human experts [6]. This does not require an *a priori* specification of the characteristics of the PEDs to be analyzed, but it does not provide an objective method of collecting and studying individual PEDs either. Algorithms which enable the objective detection and capture of all PED activity in an rEEG record are needed to provide an improved first-stage to advanced pattern-recognition algorithms which could be useful for rEEG research and clinical interpretation. Some clinical neurophysiology researchers have begun using BSSS as a first-stage in their pattern recognition algorithms [7].

Routine EEG recordings consist of a mixture of electrical signals generated by a myriad of both intracranial and extracranial biological and artifactual sources. Blind signal source separation (BSSS) algorithms are a method for separating signals of interest from a mixture of signals [8]. BSSS is an ideal foundation for algorithms to detect PEDs in a non-specific way, since BSSS enables some separation of the PED signal from other concurrent EEG signals. But there are limitations to conventional BSSS algorithms when applied to neurophysiologic recordings. The classic description of BSSS is the 'cocktail party problem' in which BSSS analysis separates the voices of n number of speakers in a cocktail party using n number of microphones placed throughout the room [9]. Conventional BSSS algorithms assume that the number of guests at the cocktail party is limited to the number of microphones and that the location of the guests do not change over time. But the 'cocktail party problem' facing clinical interpreters of EEG is very different from this classic one. In the 'cocktail parties' of rEEG recordings, there are an unknown number of guests at the party, the number of guests is likely larger than the number of microphones, and the number of guests probably changes over time. This leads to a significant 'over-completeness problem'. Also, many of the guests are probably walking about the room while they are talking (causing their location to be 'non-stationary' throughout data acquisition), leading to a significant 'non-stationarity problem' [10]. This paper describes a computationally-demanding heuristic algorithm created to use conventional BSSS algorithms to:

1. Detect PEDs in neurophysiologic datasets in a non-specific way (with minimal regard to signal morphology or scalp distribution)
2. Circumvent the over-completeness problem of BSSS
3. Partially address the non-stationarity problem of BSSS.

2 Algorithm

2.1 Step 1. Multitaper Blind Signal Source Separation

Because rEEG recordings are often lengthy and obtained with a limited number of recording electrodes (approximately 20), they are assumed to be overcomplete. This overcompleteness could cause a BSSS analysis of an entire rEEG dataset to fail to resolve an individual PED into one or more components of the output signal. Also, although the source locations for brain activity are anatomically fixed, the sequential activations of interconnected neurons which occur during brain activations frequently cause non-stationarity in measured rEEG signals [10]. BSSS analysis of an entire rEEG dataset will represent an individual PED, and often a group of similar PEDs, in a limited number of components. If the source generators for a PED are non-stationary and vary slightly from PED-to-PED, each individual PED could be better represented by a unique set of scalp distribution vectors (SDVs) associated with different ranges of time-points during the time-of-occurence of the individual PED. A simple (but computationally-intensive) algorithm is implemented to do this. In the multitaper BSSS algorithm (mBSSS), a large number of BSSS operations are performed throughtout the rEEG dataset using a range of window lengths at incremental temporal locations in a multitaper method. Because the author has implemented the mBSSS algorithm with Infomax independent component analysis (ICA), the output of the algorithm will be referred to as independent components (ICs) and SDVs [11]. By performing many BSSS operations in an evenly-spread pattern of window lengths and temporal locations, mBSSS attempts to detect each PED in one or more ICs, with one or more associated unique SDVs. Any BSSS algorithm may be 'plugged-in' to the mBSSS algorithm, as long as the number of output ICs (and SVDs) match the number of input sources and as long as it converges on a solution very reliably.

A number of mBSSS parameters must be chosen empirically. These parameters describe how many BSSS operations are performed, where they are performed in the dataset, and how much the windows for BSSS overlap. This is a list of parameters used for the algorithm:

w_{min}	shortest window length for BSSS, given in datapoints
w_{max}	longest window length for BSSS, given in datapoints
w_{ef}	window length expansion factor
γ	BSSS window overlap parameter
τ	parameter defining number of 'zones' within each BSSS window

These are the sets used for the algorithm:

$X(n,t)$	time versus amplitude rEEG dataset of n channels of length t
W	BSSS 'window' lengths to be used for mBSSS
R	number of BSSS operations ('repetitions') performed for each window length
B	start time positions in X ('begin points') for all BSSS windows
S	ICs produced by mBSSS
A	SDVs produced by mBSSS

The set of window lengths W to be used for BSSS range exponentially from

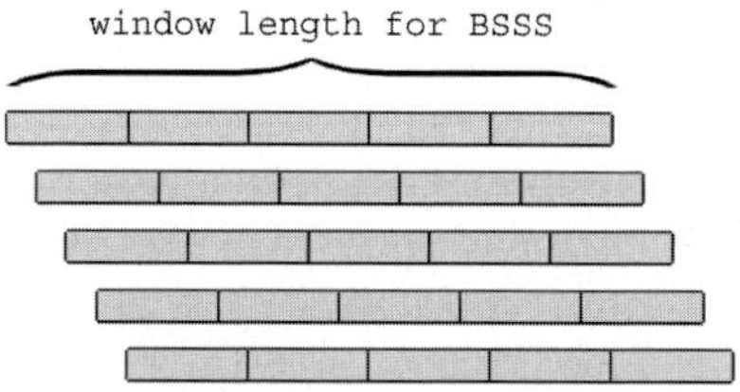

Fig. 1. Parameter settings of $\tau = 2$ and $\gamma = 4$. At a setting of $\tau = 2$, the IC is divided into five zones. The window for BSSS is moved forward by a length equal to a zone length divided by γ.

w_{min} to w_{max} with an expansion factor of w_{ef}. The γ parameter, which can be any non-zero integer, defines the extent to which consecutive windows for BSSS overlap. The start-points for each of the consecutive windows for BSSS are distributed equally throughout the dataset. Each window for BSSS is partitioned into $2\tau + 1$ 'zones' each of equal length. For a given window length, the mBSSS algorithm performs consecutive BSSS operations progressively through the dataset, translating the location of the window for BSSS a $\frac{1}{\gamma(2\tau+1)}$ fraction of the window length at each increment. The output sets for the algorithm are S, which is a large set of ICs, and A, which is a large set of the respective SDVs.

2.2 Step 2. Paroxysmal Event Detection

Since the magnitude of the signals in both sets A and S are related indirectly and unpredictably to the amplitude of the source signals X, the A and S sets are normalized to create A_n and S_n, respectively, using a normalization parameter. ICs in set S_n with a paroxysmal appearance are retained and all of the other ICs in set S_n are discarded along with their respective SDVs in A_n. Each window for BSSS is partitioned into $2\tau + 1$ 'zones' each of equal length (i.e., a central zone with tau zones on either side, as shown in Figure 2). An IC is considered paroxysmal (and therefore may represent a PED) if it has a higher root-mean-square

(RMS) power in the central zone of the IC ('zone B') than in the surroundings zones ('zone A' and 'zone C'). This is illustrated in Figure 2. The RMS power

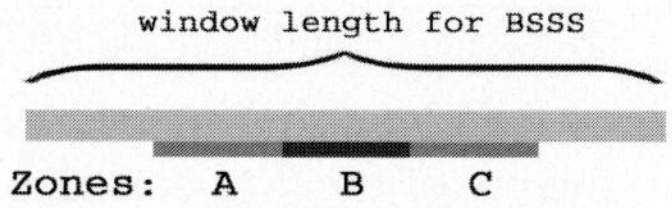

Fig. 2. At a setting of $\tau = 2$, the IC is divided into five zones. Zones A, B, and C are always assigned as the central three zones within the IC, regardless of the value of tau (τ).

of each IC in set S_n in zones A, B, and C is defined as P_A, P_B, and P_C. The 'paroxysmal event index' (PEI) termed Υ for each IC in S_n is calculated:

$$\Upsilon = \frac{\sqrt{2P_B^2 - P_A^2 - P_C^2}}{W} \tag{1}$$

The PEI (Υ) threshold parameter υ_p is selected empirically. Only those independent components in S_n (along with their respective SDVs in A_n) which have an $\Upsilon > \upsilon_p$ are retained and placed in set S_{np} (and their respective SDVs and placed in set A_{np}). All other ICs in S_n (and respective SDVs in A_n) are discarded.

2.3 Step 3. IC Redundancy Reduction and Representation

The database of independent components (ICs) in set S_{np} often contains multiple ICs representing each single PED. ICs which represent the same PED can be identified by simple cross-multiplication of the ICs which overlap temporally by 50 percent of the length of at least one of the two ICs. The simplest approach to removing this redundancy is to pick out and retain the IC (and its respective SDV) which has the highest Υ (or some other characteristic) within the group of ICs with the same 'event number'. Another more complicated method is to develop a hierarchical representation of each group of ICs with the same 'event number' which contains a longer IC and possibly several shorter 'subcomponent' ICs. This provides a description of how the SDV of the PED changes over time, helping to address the 'non-stationarity' problem of applying BSSS to neurophysiology datasets.

2.4 Step 4. Visualization of PEDs Detected Using mBSSS

Depending on the values of the parameters selected, when the mBSSS algorithm is applied to a rEEG dataset, a large database of ICs and their respective SDVs (which represent PEDs) can be created. In the author's experience, is it possible to capture hundreds of PEDs in a typical 30-minute routine EEG recording. Visualization of this many ICs is a challenge. The individual IC signals and

and SDVs can be visualized in groups based on their 'event number'. Also, single-value secondary characteristics of the zone B of each IC in set S_{np} can be calculated. These secondary characteristics may include typical quantitative measures such as peak spectral frequency and 'spikiness' morphology but may also include more complex measures such as peak wavelet power using various mother wavelets. Single-value secondary characteristics of the SDV such as location of peak IC activity and degree of focality can also be calculated. Secondary characteristics of each IC (and the respective SDV) captured by mBSSS can be plotted in a graph with time on the x-axis, IC length on the y-axis, and a single-value IC or SDV secondary characteristic on the z-axis (color).

3 Medical Application

One year of routine clinical EEGs performed at the MUSC Neurophysiology Laboratory were reviewed by the author (over 1000 EEGs). These digital EEGs were acquired at 256 Hz with 19 channels using the standard 10-20 electrode placement. A database of 101 PEDs was collected by the author from these clinical EEG recordings. This database of PEDs consists of 61 PEDs from the EEGs of 50 patients without of history of epilepsy which were judged by the author to be normal but difficult to interpret. This database also included 40 PEDs from the EEGs of 29 patients with known epilepsy judged by the author to be abnormal PEDs which were subtle due to their relatively low amplitude. Thirty-second EEG epochs containing these 101 PEDs were de-identified and assimilated into a single digital EEG file termed the 'source EEG dataset'. Each 30-second segment contained the PED-of-interest in approximately the temporal midpoint of the segment. Another parallel database of 101 tag signals was created using segments of EEG from just one EEG channel which the author thought best represented each PED. (The length of these 'tag signals' varied based on the length of each PED.) The mBSSS analysis was performed on the source EEG dataset in the sequence described below using Matlab code composed by the author and using various EEGLab Matlab scripts including Infomax ICA [12]:

1. mBSSS analysis was performed over four-second windows of EEG data centered temporally over all of the 101 PEDs. The parameters for mBSSS included $\tau = 5$, $\gamma = 6$, and 14 BSSS window lengths varying exponentially from approximately 0.5 to 10 seconds (each window length 10% greater than the next). This mBSSS analysis produced 991 PED ICs and their respective SDVs.
2. The 991 PED ICs were compared with each other and it was determined that they represented 379 unique PEDs in the source EEG dataset.
3. One of 991 PED ICs was selected out of each of the 379 groups of PED ICs to represent each of the 379 detected PEDs (producing 379 'representative PED ICs'), based on its PEI (Υ) value and other secondary characteristics.
4. Each of the 379 representative PED ICs were compared to the 101 EEG tag signals using the methods described in #2 above. The representative PED ICs which matched one of the 101 PED tag signals were labelled as such.

The mICA analysis detected 98 of the 101 target PEDs in the source EEG dataset. All of the PEDs which were not detected by mBSSS were normal PEDs. Secondary characteristics of the detected PEDs and the sensitivity and specificity for categorizing the target PEDs as normal and abnormal based on these characteristics have recently been presented elsewhere as a poster presentation and can be viewed at http://www.drivehq.com/web/halfordjj/HalfordAESposter2006.ppt [13].

4 Conclusion

Based on the results of the preliminary testing presented above, the mBSSS algorithm is able to detect many PEDs in routine EEG datasets without a prespecification of signal morphology or scalp distribution. But the mBSSS algorithm has many limitations. First, it has been developed heuristically and is not based on the fundamental mathematical principles of BSSS. Secondly, the definition of PED is subjective and needs to be better defined both mathematically and clinically. This would require the creation of standardized EEG datasets to verify the accuracy of PED detection with mBSSS. Third, the mBSSS algorithm is very computationally intensive and therefore not practical for routine clinical or research use at the present time.

If the algorithm is to be useful for research or clinical practice, many improvements need to be made. First, the mBSSS algorithm needs to be restructured to minimize the number of empirically-set algorithm parameters. Secondly, since the current implementation uses un-compiled Matlab scripts, the algorithm needs to be C-coded and implemented with an efficient ICA algorithm which possibly not only detects supra-Gaussian sources but also sub-Gaussian sources as well. Third, experiments with a range of parameter settings need to be performed for parameter optimization. These experiments will require standardized clinical rEEG datasets. In order to test if the mBSSS algorithm can detect PED activity visible to the human eye, rEEG datasets need to be developed in which all visible PEDs have been marked by several expert rEEG interpreters. Information retrieval statistics could be used to test the precision, recall, and accuracy of the mBSSS algorithm in detecting PEDs. Also, in order to determine if the algorithm can capture PED activity which is too subtle for the human eye to detect, datasets with intracranial EEG or magnetoencephalogram (MEG) data acquired concurrently with rEEG data need to be studied in a similar fashion. The high computational requirement of the algorithm may become less of a problem over time. The cost of computational power continues to decrease at a somewhat predictable rate. If the trend toward the development of multi-core microprocessors continues, substantial computational power for parallel processing could be available in personal computers within a few decades. The code for algorithms which use multiple BSSS computations can be easily multithreaded. These algorithms could be implemented on small local clusters or used via a telemedicine approach in which clinical EEG datasets are moved to and from remote centralized data processing centers using the Web. BSSS algorithms which detect PEDs could help provide an objective structure to the computer analysis of PED activity in routine clinical EEG. This could lead to an improved understanding of clinical EEG and an improvement in neurology patient care.

Acknowledgement. Special thanks to Martin J. McKeown MD for inspiration and to Jose C. Principe PhD and Brian C. Dean PhD for advice.

References

1. Zifkin, B.G., Cracco, R.Q.: An Orderly Approach to the Abnormal Electroencephalogram. In: Current Practice of Clinical Electroencephalography, pp. 288–302. Lippincott Williams & Wilkins, Philadelphia, PA (2003)
2. Benbadis, S.R.: The eeg in nonepileptic seizures. Journal of Clinical Neurophysiology 23, 340–352 (2006)
3. Westmoreland, B.F.: Benign Electroencephalographic Variants and Patterns of Uncertain Clinical Significance. In: Current Practice of Clinical Electroencephalography, pp. 235–245. Lippincott Williams & Wilkins, Philadelphia (2003)
4. Wilson, S.B., Emerson, R.: Spike detection: a review and comparison of algorithms. Clinical Neurophysiology 113, 1873–1881 (2002)
5. da Lopes Silva, F.: Computer-Assisted EEG Diagnosis: Pattern Recognition and Brain Mapping. In: Electroencephalography: Basic Principles, Clinical Applications, and Related Fields, vol. 2, pp. 1233–1264. Lippincott Williams & Wilkins, Philadelphia, PA (2005)
6. Acir, N., Oztura, I., Kuntalp, M., Baklan, B., Guzeli, C.: Automatic detection of epileptiform events in eeg by a three-stage procedure based on artificial neural networks. IEEE Transactions on Biomedical Engineering 52(1), 30–40 (2005)
7. LeVan, P., Urrestarazu, E., Gotman, J.: A system for automatic artifact removal in ictal scalp eeg based on independent component analysis and bayesian classification. Clinical Neurophysiology 117(117), 912–927 (2006)
8. Haykin, S., Chen, Z.: The cocktail party problem. Neural Computation 17, 1875–1902 (2005)
9. Brown, G.D., Yamada, S., Sejnowski, T.J.: Independent component analysis at the neural cocktail party. Trends in Neurosciences 24, 54–63 (2001)
10. Unsworth, C., Spowart, J., Lawson, G., Brown, J., Mulgrew, B., Minns, R., Clark, M.: Redundancy of independent component analysis in four common types of childhood epileptic seizure. Journal of Clinical Neurophysiology 23(3), 245–253 (2006)
11. Bell, A.J., Sejnowski, T.J.: An information-maximization approach to blind separation and blind deconvolution. Neural Computing 7, 1129–1159 (1995)
12. Delorme, A., Makeig, S.: Eeglab: an open source toolbox for analysis of single-trial eeg dynamics. Journal of Neuroscience Methods 134, 9–21 (2004)
13. Halford, J.J.: Analysis of temporal lobe paroxysmal events using multitaper independent component analysis. Abstract Presentation at the American Epilepsy Society Annual Meeting. San Diago, CA (December 2006)

Constrained ICA for the Analysis of High Stimulus Rate Auditory Evoked Potentials

James M. Harte

Centre for Applied Hearing Research, Acoustic Technology, Ørsted•DTU, Technical University of Denmark, DK-2800, Lyngby, Denmark
jha@oersted.dtu.dk

Abstract. A temporally-constrained blind-source-separation algorithm was used to analyse auditory evoked potentials, evoked from impulse trains with inter-stimulus rates of 95 and 198 Hz. A nonstationarity of variance contrast function was used, and a simulation run showing its ability to extract sources based on a simple convolved model of auditory brainstem and middle latency responses. For a stimulus rate of 95 Hz, where no neural adaptation occurs, this approach was partially successful for experimental data. For the higher rate of 198 Hz particularly poor results were observed for brainstem responses. It is hypothesised that this may be due to the neural adaptation process and/or an inappropriate choice of source model.

1 Introduction

Auditory evoked potentials are the summed response from many remotely located neurons recorded via scalp electrodes. They can be recorded from all levels of the auditory pathway, from the auditory nerve, the brainstem up to the cortex. They are typically grouped in terms of time of occurrence after stimulus offset and thus are known as; auditory brainstem responses (ABRs) recorded between 1 and 7 ms after stimulus offset; middle latency responses (MLRs) recorded in the interval 15-50 ms after acoustic stimulus; and auditory late response (ALR) recorded in the interval 75-200 ms after stimulus. To date no studies, to the author's knowledge, exist on successfully using independent component analysis (ICA) for examining ABR responses. Though examples exist for using ICA on MLR and ALR responses [4,9].

It is the goal of this paper to develop a tool for characterising the phenomenon known as neural adaptation, where a reduced neural output is observed due to prolonged or repeated stimulation, in each stage of the auditory pathway. This is by no means a trivial task and the work presented here represents a first step toward this goal. Adaptation has an important function in auditory perception models [1], and is therefore of interest in audiological science. It is often investigated as a function of stimulus rate of repeated impulse or chirp stimuli [6]. Here, ICA with reference [7,8] will be used to subtract two components attributed to ABR and MLR based sources, using a correlation based distance

M.E. Davies et al. (Eds.): ICA 2007, LNCS 4666, pp. 609–616, 2007.

metric [5]. A nonstationarity of variance [2] based approach for finding maximally independent components will be used. It is assumed that any approach using temporal structure in the data is more appropriate for evoked potentials, with well known temporal features. If it is possible to separate signals based on a principled semi-blind source separation algorithm such as the one proposed here, then it may prove a useful tool for non-invasively investigating the human auditory pathways.

2 Constrained Blind Source Separation

The classic blind source separation problem assumes that we have a series of mixtures obtained via an unknown mixing matrix on a set of underlying sources, $\mathbf{x} = \mathbf{As}$. The goal is to find an unmixing matrix $\mathbf{W}$ such that the estimate of the sources is given by $\hat{\mathbf{s}} = \mathbf{Wx}$. It is common to pre-process and reduce the dimension of the linear mixtures via principle component analysis (PCA) $\mathbf{z} = \mathbf{Ex}$. All discussion from this point of the mixtures assumes PCA and dimension reduction has been performed.

Hyvärinen [2] showed that maximization of the nonstationarity of variance, measured by the absolute value of the 4th cross-cumulant, of a linear combination of the observed mixtures allows for the estimation of one source signal. The 4th-order cross-cumulant corresponds to the autocorrelation of energy in a given signal, and is given by

$$\mathrm{cum}(y, y, y_\tau, y_\tau) = E\{y^2 y_\tau^2\} - E\{y^2\}E\{y_\tau^2\} - 2\left(E\{y y_\tau\}\right)^2 \tag{1}$$

where the time dependence on $y(t)$ is omitted for brevity and y_τ represents the delayed signal $y(t - \tau)$, this convention will be used for the rest of this paper. Lu and Rajapakse [7,8] developed a framework for constrained independent component analysis (cICA), to incorporate additional requirements and prior information in the form of constraints on the ICA contrast function. The goal is to extract the component closest to some user specified reference signal $r(t)$, via some distance metric. The closeness constraint to be used here will be correlation at zero lag:

$$g(\mathbf{w}) = \zeta - E\left\{r(\mathbf{w}^T\mathbf{z})\right\} \leq 0 \tag{2}$$

where $\mathbf{w}$ is a single (one-unit) demixing weight vector, such that $y(t) = \mathbf{w}^T\mathbf{z}$, and ζ is the threshold that defines the lower bound of the optimum correlation [5]. The one-unit constrained optimization problem for a nonstationarity of variance contrast function $J(y)$ is given by:

$$\begin{aligned} &\text{maximise} \quad J(y) = \left|\mathrm{cum}\left(\mathbf{w}^{\mathrm{T}}\mathbf{z}, \mathbf{w}^{\mathrm{T}}\mathbf{z}, \mathbf{w}^{\mathrm{T}}\mathbf{z}_\tau, \mathbf{w}^{\mathrm{T}}\mathbf{z}_\tau\right)\right| \\ &\text{subject to} \quad g(\mathbf{w}) \leq 0,\ h(\mathbf{w}) = E\{(\mathbf{w}^{\mathrm{T}}\mathbf{z})^2\} - 1 = 0 \end{aligned} \tag{3}$$

The equality constraint $h(\mathbf{w})$ bounds the scale of the output y and the weight vector $\mathbf{w}$, and is needed as cumulants are not scale invariant. The specific algorithm used in this paper is shown in appendix A.

3 Experimental Methods and AEP Data

In this preliminary study, only a single test subject was used. The stimulus was generated in MATLAB and A/D conversion made through an ADI-8 Pro 24-bit converter, the levels were set via a DT PA5 programmable attenuator, and the stimuli presented to the left ear of the test subject via an ER-2 insert earphone. EEG activity was recorded differentially between the vertex and 28 electrodes distributed over a 61-channel triangulated equidistant arrangement headcap, with the ground electrode placed on the forehead. Silver/silver chloride electrodes were used, and an inter-electrode impedance was maintained below 5kΩ. The EEG activity was recorded on a SynAmps 5803 amplifier, providing 74 dB of gain before a low-pass filter stage (cut-off of 2 kHz), with a sampling rate of 10 kHz. After recording the EEG-data was epoched and filtered again from 10 to 1500 Hz using a 200 tap FIR filter. The epochs were averaged using an iterative weighted-averaging algorithm [11].

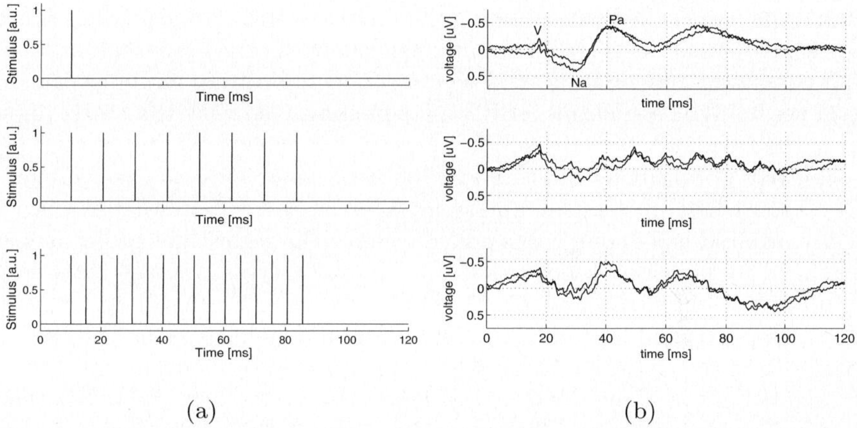

Fig. 1. (a) Experiment stimuli waveforms and (b) event-related potentials for a single exemplary channel located on the ipsilateral mastoid, showing combined ABR and MLR responses to (top) single impulse; (middle) 95 Hz impulse train; and (bottom) 198 Hz impulse train. The two curves represent two repeat recordings with 4000 averages and show good repeatability.

The basic stimuli used in this experiment was an 83 μs duration impulse. Three sets of stimuli conditions, see Fig. 1(a), were presented at a constant inter-epoch rate of $\approx$ 8.33 Hz (i.e. a duration of 120 ms). The first stimuli set was a single impulse to act as a control where both ABR and MLR responses could be seen. Stimuli set 2 presented a train of impulses with a within train rate of 95 Hz, shown in [6] to produce an unadapted ABR response. The MLR response will be convolved over the epoch, as the inter-epoch rate was chosen with no jitter so circular convolution would occur. Stimuli set 3 presents the impulses at a rate of 198 Hz, ensuring the ABRs would be adapted over the impulse train. A total of 4000 averages were made per stimulus type and repeated twice to ensure repeatability of results. The stimuli were all presented at a level of 60 dB pe SPL, to ensure good SNR and test subject comfort.

An illustrative event-related potential for a single channel located at the ipsilateral mastoid for the three stimulus conditions is shown in Fig. 1(b). The results for the single impulse stimuli (top) show typical features expected in the ABR and MLR auditory evoked potentials, namely the ABR wave V is clearly located around 6 ms after stimulus offset. Typical MLRs are observed with the first negativity Na around 25 ms after stimulus offset and the first positivity Pa around 35 ms. The two high stimulus rate ERPs are shown in the middle and bottom panes of Fig. 1(b). It can be seen that the slow MLR response are highly convolved within the event related potential.

4 Simulation

A simulation was carried out assuming three sources and square mixing. Source 1 used the ABR within the single impulse experiment evoked potential, re-filtered with a pass-band 100-1500 Hz, and windowed temporally to minimise the influence of the MLR. This 'clean' ABR was then convolved with the 95 Hz impulse train in the frequency domain to obtain circular convolution. The second source was similarly obtained by filtering (10-100 Hz), windowing and convolving the MLR response from the single impulse experiment evoked potential. A third biologically inspired noise source was added to limit the algorithms performance. It was defined to have a pink power spectra (i.e. $1/f$), typical of raw EEG recordings, limited to the pass-band $10 - 1500$ Hz of interest in this study. The noise had a Gaussian distribution as one might expect after synchronous averaging and filtering due to the central limit theorem. The sources are shown in Fig. 2(a). The cICA algorithm was given the exact MLR source as reference, and a deflationary orthogonalization procedure [3] was used to extract the remaining components in the data. Fig. 2(b) shows the output from the cICA algorithm for a delay of $\tau = 1$. The first component, obtained with the reference and constraints, was extracted with a correlation of $\zeta_{rs} = 1.00$, i.e. perfectly. The ABR component, obtained from the deflationary orthogonalization, appears to be adequately extracted. However, some corruption with the noise source is apparent. Similar results were obtained for the 198 Hz inter-impulse rate simulation.

Blind-source-separation via nonstationarity of variance assumes the 4th cross-cumulant for a source is non-zero, i.e. some structure is required in the energy of the signal. Hyvärinen [2] gave examples with smoothly changing variances where the envelopes were random and low-pass filtered, and thus have sharply decaying cross-cumulants for increasing delay τ. In this experiment matters are complicated by the introduction of a set of convolved or repeated energy profiles, thus the cumulant for the source becomes periodic as a function of delay τ. We also assume two (or more) sources with the same periodicity, i.e. phase locked to the periodic stimuli. The maxima in the 4th cross-cumulant for such a mixing model

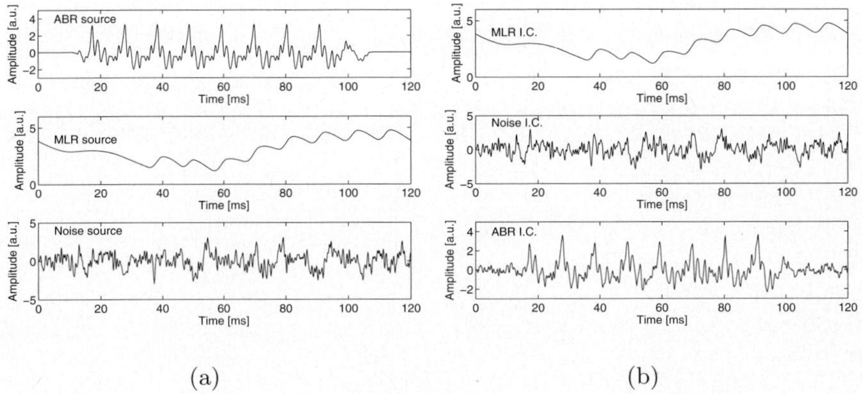

Fig. 2. Simulation results for an 95Hz inter-impulse rate: (a) Original sources used; (b) extracted maximally independent components

become functions of delay τ. As indicated by Hyvärinen [2] the choice of delay τ becomes difficult, yet theoretically crucial. The success of the algorithm can be judged through the use of the convergence index suggested in [2]. Defined as the sums of the absolute value of the matrix $\mathbf{WA} - 3$, this will be zero if the mixing matrix $\mathbf{A}$ was estimated perfectly. The simulation was carried out 1000 times for delays $\tau = 1 \rightarrow 50$ to obtain a mean convergence index as a function of delay. It was hoped that this might suggest a meaningful delay to use when analysing the experimental results. There was significant variance for each τ equal to the variance across τ so no useful optimal delay could be found from this simple analysis. Due to this a delay of $\tau = 1$ was chosen for the experimental results.

5 Constrained ICA Results

As indicated in section 2 the data was whitened prior to applying the cICA algorithm. A dimension reduction was carried out from 28 mixtures to only 9 principle components, accounting for 99% of the data variance. The two main benefits for dimension reduction are in reducing noise and preventing overlearning.

The previous section discussed a simulation where it was assumed that the ABR and MLR source signals could be modeled as the response to a single impulse convolved with the stimulus impulse train. This further assumes that the ABR and MLRs can themselves be represented by a single source. The adaptation process known to exist for high stimulus rates was not included in the simulation or source model. As a first approximation this simple model may be useful in analysing the experimental data. The MLR and ABR based sources from the simulation were used as two reference signals for the cICA algorithm. A deflationary orthogonalization procedure was used, where the first independent component was extracted using the MLR reference, then the second using the ABR reference, and finally the remaining unconstrained components. Fig. 3(a)

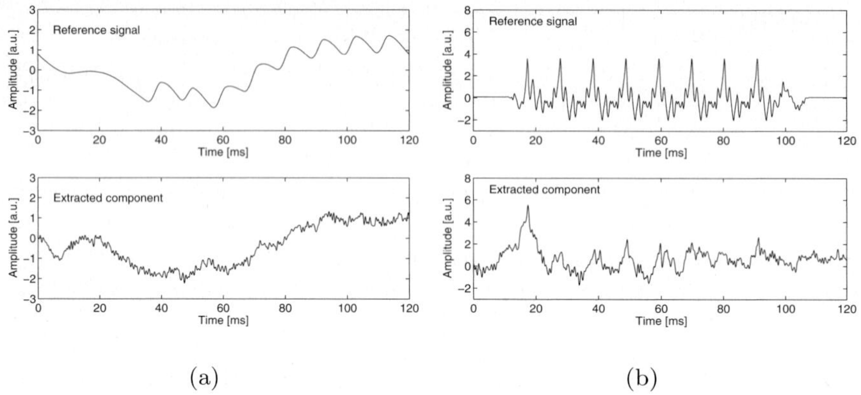

Fig. 3. Extracted independent component for (a) convolved MLR based reference (correlation of $\zeta_{ry} = 0.94$); and (b)convolved ABR based reference (correlation of $\zeta_{ry} = 0.38$)

shows the reference and extracted component for the convolved MLR reference for the stimulus set 2, with an inter-impulse rate of 95 Hz. A correlation of $\zeta_{ry} = 0.94$ was obtained between the reference and the extracted component. The second reference and extracted component is shown in Fig. 3(b). Here a correlation of only $\zeta_{ry} = 0.38$ was maximally obtained after a number of runs of the algorithm. The results for the 198 Hz inter-impulse rate data were similar for the MLR based source($\zeta_{ry} = 0.92$), however the ABR source never obtained a correlation greater than $\zeta_{ry} = 0.11$.

6 Discussion

The poor results obtained for the ABR extracted components potentially indicates one or more problems: (1) The nonstationarity of variance contrast function is not appropriate here, due to the periodic 4th cross-cumulant. The results from the simulation would tend not to agree with this. (2) The correlation constraint metric used might be inappropriate for the data. However, the constraint only guides the algorithm toward an independent component, as defined by the contrast function. (3) The underlying source model assumed here might be inappropriate. For the 198 Hz stimulus rates then neural adaptation effects might explain poor performance. However, each ABR response is known to have a contribution from multiple auditory brainstem structures [10]. The predominant recorded surface electrode potentials for ABRs are believed to come from a propagating action potential along the auditory pathway, rather than more cortically located post-synaptic potentials. Thus ABRs are generated by a time-varying source propagating through the neural structures from the brainstem up to the mid-brain. This nonstationarity manifests as small changes in the morphology of the ABR seen across the channels in the averaged event related potential, i.e. the peak of the wave V may occur at a latency of 6 ms, say in an ipsilaterally located electrode, and a latency of 6.2 ms in a contralateral electrode site. This fact may limit performance of the source separation algorithm used here. This nonstationarity of variance based extraction method may be sensitive to the small nonstationarity associated with ABRs over time across the electrodes. The author also applied the ICA with reference algorithm from [7,8,5], with their negentropy contrast function. This approach might be expected to yield better results due to its insensitivity to the temporal structure of the data. Negentropy is a measure of nongaussianity, and thus time shifting the source will yield identical results to the original. However, almost identical results were obtained to the nonstationarity of variance contrast function. Thus the nonstationarity of the ABR generating mechanisms may not play a roll in its identifiability. Also both algorithms may 'incorrectly' separate the ABR source in the remaining unconstrained components, though this did not appear to be the case through observation. More work needs to be done to verify or challenge these observations.

The choice of delay or lag for the cumulant estimation is very important. The convolved model used here is too simple and did not aid the extraction of the real sources from the mixtures. Ironically the *de facto* choice of delay $\tau = 1$ used here produced the best results. It is not clearly understood why this is and therefore more work is needed.

References

1. Dau, T., Püschel, D., Kohlrausch, A.: A quantitative model of the effective signal processing in the auditory system. I. Model structure. J. Acoust. Soc. Am. 99(6), 3615–3622 (1996)
2. Hynvärinen, A.: Blind Source Separation by Nonstationarity of Variance: A Cumulant-Based Approach. IEEE transactions on neural networks 12(6), 1471–1474 (2001)
3. Hynvärinen, A., Karhunen, J., Oja, E.: Independent component analysis. Wiley-Interscience, Chichester (2001)
4. Iyer, D., Boutros, N.N., Zouridakis, G.: Independent Component Analysis of Multichannel Auditory Evoked Potentials. In: Proc. of the 2nd Joint EMBS/BMES Conference, pp. 204–205 (2002)
5. James, C., Gibson, O.J.: Temporally Constrained ICA: An application to artifact rejection in electromagnetic brain signal analysis. IEEE Transactions on Biomedical Engineering 50, 1108–1116 (2003)
6. Junius, D., Dau, T.: Influence of cochlear traveling wave and neural adaptation on auditory brainstem responses. Hearing Research 205, 53–67 (2005)
7. Lu, W., Rajapakse, J.C.: ICA with reference. In: Proc. 3rd Int. Conf. Independent Component Analysis and Blind Signal Separation: ICA2001, pp. 120–125 (2001)
8. Lu, W., Rajapakse, J.C.: Approach and Applications of Constrained ICA. IEEE Transactions on Neural Networks 16(1), 203–212 (2005)
9. Makeig, S., Jung, T.P., Bell, A., Ghahremani, D., Sejnowski, T.J.: Blind Separation of Auditory Event-Related Brain Responses into Independent Components. PNAS 94(20), 10980–10984 (1997)
10. Melcher, J.R., Kiang, N.Y.S.: Generators of the brainstem auditory evoked potential in cat III: identified cell populations. Hearing Research 93, 52–71 (1996)

11. Riedel, H., Granzow, M., Kollmeier, B.: Single-sweep-based methods to improve the quality of auditory brainstem responses. Part II: Averaging methods. Z. Audiol 40(2), 62–85 (2001)

A Constrained ICA Algorithm

The constrained optimization problem can be solved using the method of Lagrangian multipliers, and transforming the inequality constraint $g(\mathbf{w}) \leq 0$ to an equality constraint by introducing a slack variable. Following [7,8], the augmented Lagrangian function $\mathcal{L}_1(\mathbf{w}, \alpha, \beta)$ is given by:

$$\mathcal{L}_1(\mathbf{w}, \alpha, \beta) = J(y) - \frac{1}{2\gamma}[\max^2\{\alpha + \gamma\left(\zeta - E\left\{r(\mathbf{w}^T\mathbf{z})\right\}\right)] \\ - \beta\left(E\{(\mathbf{w}^{\mathrm{T}}\mathbf{z})^2\} - 1\right) - \frac{1}{2}\gamma\left\|E\{(\mathbf{w}^{\mathrm{T}}\mathbf{z})^2\} - 1\right\|^2 \quad (4)$$

where α and β are the Lagrange multipliers, γ is the scalar penalty parameter, $\|.\|$ denotes the Euclidean norm, and $\frac{1}{2}\gamma\|h(\mathbf{w})\|^2$ is a penalty term to help the convergence of the algorithm [7]. The maximum of the augmented Lagrangian is found via a Newton learning algorithm:

$$\mathbf{w}_{k+1} = \mathbf{w}_k - \eta\left(\mathcal{L}_{1\mathbf{w}_k^2}^{''}\right)^{-1}\mathcal{L}_{1\mathbf{w}_k}^{'} \quad (5)$$

where k is the iteration index, η is a step size parameter that can change with the iteration count until some satisfactory convergence occurs. $\mathcal{L}_{1\mathbf{w}_k}^{'}$ and $\mathcal{L}_{1\mathbf{w}_k^2}^{''}$ are respectively the Jacobian and Hessian for the augmented Lagrangian function $\mathcal{L}_1(\mathbf{w}, \alpha, \beta)$. It can be shown that the Jacobian, $\mathcal{L}_{1\mathbf{w}_k}^{'}$, is given by:

$$\mathcal{L}_{1\mathbf{w}_k}^{'} = \chi\Big(2E\{\mathbf{z}(\mathbf{w}^{\mathrm{T}}\mathbf{z})(\mathbf{w}^{\mathrm{T}}\mathbf{z}_\tau)^2\} + 2E\{\mathbf{z}_\tau(\mathbf{w}^{\mathrm{T}}\mathbf{z}_\tau)(\mathbf{w}^{\mathrm{T}}\mathbf{z})^2\} \\ - 4\mathbf{M}\mathbf{w}E\{(\mathbf{w}^{\mathrm{T}}\mathbf{z})(\mathbf{w}^{\mathrm{T}}\mathbf{z}_\tau)\} - 4\mathbf{w}\Big) - \frac{\alpha}{2}E\{r\mathbf{z}\} - \beta\mathbf{w} \quad (6)$$

where $\chi = \mathrm{sgn}\left(\mathrm{cum}\left(\mathbf{w}^{\mathrm{T}}\mathbf{z}, \mathbf{w}^{\mathrm{T}}\mathbf{z}, \mathbf{w}^{\mathrm{T}}\mathbf{z}_\tau, \mathbf{w}^{\mathrm{T}}\mathbf{z}_\tau\right)\right)$, and sgn(.) is the sign function, $\mathbf{M}$ is symmetric and is the sum of the covariance matrices, $\mathbf{C}_{\mathbf{z}\mathbf{z}_\tau} = E\{\mathbf{z}\mathbf{z}_\tau\}$ and $\mathbf{C}_{\mathbf{z}_\tau\mathbf{z}} = E\{\mathbf{z}_\tau\mathbf{z}\}$. Similarly through a little therapeutic algebra it is possible to approximate the Hessian by:

$$\mathcal{L}_{1\mathbf{w}_k^2}^{''} \approx -\left(4\chi\mathbf{M}\mathbf{w}\mathbf{w}^{\mathbf{T}}\mathbf{M} + \beta\mathbf{I}\right) \quad (7)$$

where $\mathbf{I}$ is the identity matrix. The optimum Lagrangian multipliers can be found by using an iterative gradient-ascent method[7,8]:

$$\alpha_{k+1} = \max\left\{0, \alpha_k + \gamma\left(\zeta - E\left\{r(\mathbf{w}^T\mathbf{z})\right\}\right)\right\}, \\ \beta_{k+1} = \beta_k + \gamma\left(E\{(\mathbf{w}^{\mathrm{T}}\mathbf{z})^2\} - 1\right). \quad (8)$$

Thus the learning algorithm to find the maximum of the augmented Lagrangian can be formulated from Eqns. 5 – 8.

Gradient Convolution Kernel Compensation Applied to Surface Electromyograms

Aleš Holobar[1] and Damjan Zazula[2]

[1] LISiN, Politecnico di Torino, Torino, Italy
ales.holobar@delen.polito.it
http://storm.uni-mb.si
[2] Faculty of Electrical Engineering and Computer Science,University of Maribor, Maribor, Slovenia

Abstract. This paper introduces gradient based method for robust assessment of the sparse pulse sources, such as motor unit innervation pulse trains in the filed of electromyography. The method employs multichannel recordings and is based on Convolution Kernel Compensation (CKC). In the first step, the unknown mixing channels (convolution kernels) are compensated, while in the second step the natural gradient algorithm is used to blindly optimize the estimated source pulse trains. The method was tested on the simulated mixtures with random mixing matrices, on synthetic surface electromyograms and on real surface electromyograms, recorded from the external anal sphincter muscle. The results prove the method is highly robust to noise and enables complete reconstruction of up to 10 concurrently active motor units.

1 Introduction

Biomedical signals are important, but very complex source of information. They typically comprise contributions of many concurrently active sources, such as neurons and muscle fibers. The sources are commonly considered statistically independent (or at most weakly correlated), but their mixing process is virtually unknown. Therefore, the acquired signals must be decomposed blindly.

When it comes to neurophysiology, electromyograms (EMG) are one of the most active research areas. They measure electrical activity of skeletal muscles and provide an insight into peripheral properties of skeletal muscles and control strategies of human motor cortex [1]. Their field of application ranges from clinical assessments of neuromuscular disorders and objective evaluations of medical treatments to basic investigations of different physiological phenomena (e.g. cramps, muscle reinnervation, etc.). Comprising action potentials (AP) from several tens of concurrently active motor units (MU), EMG signal is commonly considered a random interference pattern which is very difficult to interpret [2]. This is especially true in the case of surface electromyography [1], which deals with measuring the electrical activity of human muscles on the surface of the skin. Bad electrode-to-skin contact and presence of noise hinder the decomposition of surface EMG and make the extraction of clinically relevant information difficult.

M.E. Davies et al. (Eds.): ICA 2007, LNCS 4666, pp. 617–624, 2007.

Recent development of high-density surface electrode arrays enabled acquisition of several tens or even hundreds of EMG channels. Different pattern recognition techniques, capable of dealing with such amount of data were also proposed. Kleine et al. [3] studied the importance of two-dimensional spatial filters, Gazzoni et al. [4] introduced the template matching segmentation and classification technique, while Wood et al. [5] employed the finite element analysis. Blind source separation decomposition techniques have also been proposed. Garcia et al. [6] modelled the EMG signal as an instantaneous mixture of motor unit action potential (MUAP) trains, while Holobar and Zazula [7] proposed Convolution Kernel Compensation (CKC) to deal with the convolutive mixtures of motor unit innervation pulse trains (IPT). The latter proved to be highly efficient as it enables the complete reconstruction of up to 30 concurrently active MUs from a good quality multichannel surface EMG.

In this paper the gradient-based extension of the CKC method [7] applied to the low-quality noisy signals is addressed. This extension is of paramount importance for clinical practice, where recoding environment cannot be strictly controlled. This paper is organized in five sections. In Section 2, the assumed data model is presented and the classic CKC approach is briefly summarized. Section 3 introduces its gradient-based extension, while in Section 4 the results of tests on simulated and real EMG signals are presented. Section 5 discusses the results and concludes the paper.

2 Data Model and Convolution Kernel Compensation

Suppose M convolutive measurements are simultaneously observed and denote their sampled vector by $\mathbf{x}(n) = [x_1(n), ..., x_M(n)]^T$, where $x_i(n)$ stands for the n-th sample of the i-th measurement. In the case of linear time-invariant (LTI) multiple-input multiple-output (MIMO) system, $\mathbf{x}(n)$ can be written as:

$$\mathbf{x}(n) = \mathbf{H}\bar{\mathbf{t}}(n) + \omega(n) \tag{1}$$

where $\omega(n) = [\omega_1(n), ..., \omega_M(n)]^T$ is a zero-mean spatially and temporally white additive noise vector, $\bar{\mathbf{t}}(n) = [t_1(n), t_1(n-1), ..., t_1(n-L+1), ..., t_N(n), ...$ $t_N(n-L+1)]^T$ is the extended version of vector of input signals from N sources $\mathbf{t}(n) = [t_1(n), ..., t_N(n)]^T$, and the mixing matrix $\mathbf{H}$ comprises all the channel responses (convolution kernels) $\mathbf{h}_{ij} = [h_{ij}(0), ..., h_{ij}(L-1)]$ of length of L samples:

$$\mathbf{H} = \begin{bmatrix} h_{11}(0) & \dots & h_{11}(L-1) & h_{12}(0) & \dots & h_{12}(L-1) & \dots \\ h_{21}(0) & \dots & h_{21}(L-1) & h_{22}(0) & \dots & h_{22}(L-1) & \dots \\ \vdots & \dots & \vdots & \vdots & \dots & \vdots & \dots \\ h_{M1}(0) & \dots & h_{M1}(L-1) & h_{M2}(0) & \dots & h_{M2}(L-1) & \dots \end{bmatrix} \tag{2}$$

In surface electromyography, the channel response $\mathbf{h}_{ij}$ corresponds to the j-th MUAP, as detected by the i-th measurement, while each model input $t_j(n)$ is a sample of an IPT, modelled as a binary pulse sequence carrying the information about the MUAP triggering times only:

$$t_j(n) = \sum_{k=-\infty}^{\infty} \delta\left[n - T_j(k)\right] \tag{3}$$

where $\delta(.)$ denotes the Dirac impulse and $T_j(k)$ stands for the time instant in which the k-th MUAP of the j-th MU appears.

2.1 Convolution Kernel Compensation

CKC method compensates the unknown mixing matrix $\mathbf{H}$ in model (1) and directly estimates the innervation pulse trains $\hat{t}_j(n)$ [7]:

$$\hat{t}_j(n) = \mathbf{c}_{t_j\mathbf{x}}^T \mathbf{C}_{\mathbf{xx}}^{-1} \mathbf{x}(n) \tag{4}$$

where $\mathbf{C}_{\mathbf{xx}} = E\left(\mathbf{x}(n)\mathbf{x}^T(n)\right)$ is the correlation matrix of measurements, $\mathbf{c}_{t_j\mathbf{x}} = E\left(\mathbf{x}(n)t_j^T(n)\right)$ is cross-correlation vector, and $E(.)$ denotes mathematical expectation. Estimator (4) is linear minimum mean square error (LMMSE) estimator of the j-th IPT and requires the cross-correlation vector $\mathbf{c}_{t_j\mathbf{x}}$ to be known in advance. This is rarely the case and Holobar and Zazula [7] proposed probabilistic iterative procedure for its blind estimation. In the first iteration step, the unknown cross-correlation vector $\mathbf{c}_{t_j\mathbf{x}}$ is approximated by vector of measurements $\hat{\mathbf{c}}_{t_j\mathbf{x}} = \mathbf{x}(n_1)$ where, without loss of generality, we assumed the j-th MU discharged at time instant n_1. Then, the first estimation of the j-th IPT yields

$$\hat{t}_j(n) = \hat{\mathbf{c}}_{t_j\mathbf{x}}^T \mathbf{C}_{\mathbf{xx}}^{-1} \mathbf{x}(n) \tag{5}$$

In the second step, the largest peak in $\hat{t}_j(n)$ is selected as the most probable candidate for the second discharge of the j-th source, $n_2 = \max\limits_{\arg}(t_j(n))\,|_{n_2 \neq n_1}$, and the vector $\hat{\mathbf{c}}_{t_j\mathbf{x}}$ is updated as:

$$\hat{\mathbf{c}}_{t_j\mathbf{x}} = \frac{\hat{\mathbf{c}}_{t_j\mathbf{x}} + \mathbf{x}(n_2)}{2} \tag{6}$$

This procedure is then repeated, with a special attention paid to the separation of superimposed pulse sources (note that more than one source may be active at instant n_1). Interested reader is referred to [7] for further description of classic CKC approach.

3 Gradient Descent Optimization of Estimated Pulse Trains

Let us use shorthand notation $\mathbf{w}_{j,k} = \mathbf{C}_{\mathbf{xx}}^{-1}\hat{\mathbf{c}}_{t_j\mathbf{x}}$ to denote the estimation of the j-th separation vector in the k-th iteration step and let $F(\hat{t}_j) = \sum\limits_m f(\hat{t}_j(m))$ denote sample cost function in the manifold of pulse trains, with arbitrary differentiable scalar function $f(t)$ applied to each sample of pulse train $\hat{t}_j(n)$. General iteration step for natural gradient descent algorithm is defined as [8]:

$$\mathbf{w}_{j,k+1} = \mathbf{w}_{j,k} - \eta(k)\mathbf{G}\Delta\mathbf{w}_{j,k} \tag{7}$$

where $\eta(k)$ is the learning rate and $\mathbf{G}$ is Riemannian metric tensor embedding the manifold of pulse trains $\mathbf{t}\,(n)$ into the manifold of measurements $\mathbf{x}\,(n)$. For smooth manifolds, the induced metric tensor $\mathbf{G}$ can be computed as:

$$\mathbf{G} = \mathbf{H}^{-1}\mathbf{H}^{-T} \tag{8}$$

where, in our case, $\mathbf{H}^{-1}$ stands for the Jacobian of the system $\mathbf{t}\,(n) = \mathbf{H}^{-1}\mathbf{x}\,(n)$. Taking the common convention on the amplitude ambiguity of trains $\mathbf{t}\,(n)$ into account, we assume $\mathbf{C}_{\overline{\mathbf{t}}\overline{\mathbf{t}}} = \mathbf{I}$. Hence, $\mathbf{G}$ can be written as $\mathbf{G} = \mathbf{H}^{-1}\mathbf{C}_{\overline{\mathbf{t}}\overline{\mathbf{t}}}^{-1}\mathbf{H}^{-T} = \mathbf{C}_{\mathbf{xx}}^{-1}$. Finally, by expressing the gradient $\Delta\mathbf{w}_{j,k}$ in terms of cost function $F(\hat{t}_j)$ we derive to the following gradient update rule:

$$\mathbf{w}_{j,k+1} = \mathbf{w}_{j,k} - \eta(k)\sum_m \frac{\partial f(\hat{t}_j(m))}{\partial \hat{t}_j(m)}\mathbf{C}_{\mathbf{xx}}^{-1}\mathbf{x}(m) \tag{9}$$

There is yet another possible interpretation of update rule (9). By rewriting (9) in terms of (5) we get the update rule for the cross-correlation vector $\hat{\mathbf{c}}_{t_j\mathbf{x}}$:

$$\hat{\mathbf{c}}_{t_j\mathbf{x}} = \hat{\mathbf{c}}_{t_j\mathbf{x}} - \eta(k)\sum_m \frac{\partial f(\hat{t}_j(m))}{\partial \hat{t}_j(m)}\mathbf{x}(m) \tag{10}$$

which provides insight into the convergence properties of the algorithm (9). Namely, by selecting the $\frac{\partial f(t)}{\partial t}$ to be concave even function, e.g. $\frac{\partial f(t)}{\partial t} = t^2$, the peaks in $\hat{t}_j(n)$ get reinforced (i.e. the corresponding measurement vectors $\mathbf{x}(m)$ in (10) get multiplied by large weights), while the base-line noise (i.e. values close to zero) is suppressed. The steeper the function $\frac{\partial f(t)}{\partial t}$, the larger the weights multiplying the peaks in $\hat{t}_j(n)$, and the faster the convergence of (10). However, by setting the peak weights too high, we jeopardize the stability of convergence as it may happen that the highest peak in $\hat{t}_j(n)$ outweights all the others. In such a case, $\hat{\mathbf{c}}_{t_j\mathbf{x}}$ converges to $\mathbf{x}(m_p)$, where m_p denotes the time instant of the largest peak in $\hat{t}_j(n)$. Typically, $\frac{\partial f(t)}{\partial t} = |t|$ or $\frac{\partial f(t)}{\partial t} = t^2$ prove to be a good compromise between the speed and stability of convergence, yielding the cost functions $F(\hat{t}_j) = \frac{1}{2}\sum_m \hat{t}_j(m)\sqrt{\hat{t}_j^2(m)}$ and $F(\hat{t}_j) = \frac{1}{3}\sum_m \hat{t}_j^3(m)$, respectively.

Gradient descent algorithm (9) still requires a good initial approximation of $t_j(n)$ in order to converge to the genuine solution (i.e. the peaks in $\hat{t}_j(n)$ must represent the true train pulses). As demonstrated in the next section, the CKC approximation (5) with $\hat{\mathbf{c}}_{t_j\mathbf{x}} = \mathbf{x}(n_1)$ proves to be a good initialization point. Required initial time instants n_1 can be selected from the activity index $I_A(n) = \mathbf{x}^T(n)\mathbf{C}_{\mathbf{xx}}^{-1}\mathbf{x}(n)$, as suggested in [7].

4 Simulation and Experimental Results

Gradient CKC method was applied to three different sets of test signals. The first two experiments evaluated the influence of noise in the case of multichannel synthetic measurements with well-conditioned random mixing matrix $\mathbf{H}$ and in the case of badly-conditioned synthetic surface EMG measurements, while in the third experiment, the method was applied to recordings of external anal sphincter muscle. In all three experiments the scalar function $f(t) = \frac{1}{3}t^3$ was used in (9), while in each iteration step the adaptive learning rate $\eta(k)$ was adjusted according to bisection. The sensitivity of gradient CKC algorithm, false alarm rate and the number of reconstructed pulse trains were observed and compared to the results of classic CKC approach.

Experiment 1: Ten simulation runs were performed, with the number of sources N set equal to 10. In each run, random input pulse trains $t_j(n) = \sum_{k=1}^{200} \delta(n - k \cdot 100 + T_j(k))$ were generated with the mean inter-pulse interval (IPI) set equal to 100 samples and the values $T_j(k); k = 1, 2, \ldots, 200$ uniformly distributed on the interval $[-10, 10]$. The length of simulated pulse trains $\mathbf{t}$ was 20,000 samples. Random zero-mean mixing matrix $\mathbf{H}$ was generated, with $L = 10$ samples long convolution kernels. The number of observations M was set equal to 25. Seven realizations of Gaussian zero-mean noise per each generated signal (1) were simulated, with SNR ranging from 20 dB to -10 dB. In order to increase the number of measurements, 9 delayed repetitions of each original measurement were used as additional measurements [7]. As a result, the number of extended pulse trains increased to 190, while the number of observations was fixed at 250. Each mixture was decomposed two times - by classic and gradient CKC. The results are summarized in Fig. 1. The CKC gradient method converged after 15 iterations, on average.

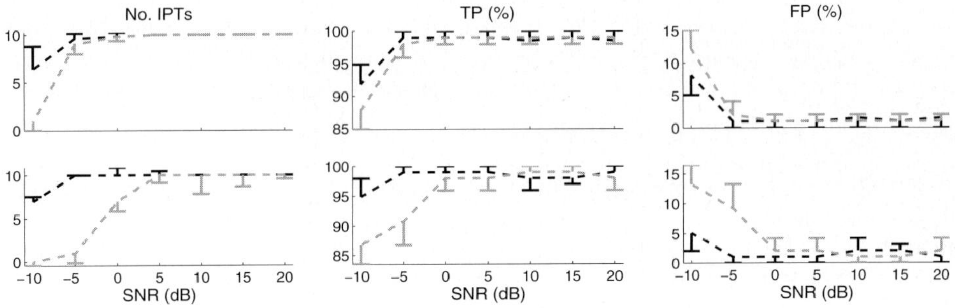

Fig. 1. Number of reconstructed IPTs (*left column*), True Positive rate (TP) (*central column*) and False Positive rate (FP) (*right column*) for gradient CKC (*black line*) and classic CKC (*gray line*). The results are averaged over 10 simulation runs (*error bars* indicate std. deviations). In each run, random mixing matrix $\mathbf{H}$ was generated, with condition number set equal to 240 ± 10 (*upper row*) and 1200 ± 50 (*bottom row*).

Experiment 2: Synthetic surface EMG signals were generated by cylindrical volume conductor model consisting of bone, muscle, subcutaneous, and skin tissues [2]. Biceps Brachii muscle with 200 MUs and 200 mm^2 cross-section was simulated. The distribution of the MU locations was random and the fibers of a MU were randomly scattered in circular MU territory, with a density of 20 fibers/mm^2. Exponentially distributed innervation numbers ranged from 25 to 2500. The surface-recorded MUAP comprised the sum of the action potentials of the muscle fibers belonging to the MU. The MUs had muscle fiber conduction velocities of 4 ± 0.3 m/s. The recording system was a grid of 13×5 electrodes of circular shape (1-mm radius) with 5-mm interelectrode distance in both directions. The EMG signals were additionally corrupted by additive zero-mean Gaussian noise between 20 and 0 dB SNR. Ten seconds long contraction at 10% excitation level (constant over time) was simulated what resulted in 105 active MUs. MU IPTs were based on a motor unit population recruitment model [3] with the recruitment and the peak discharge rate set to 8 and 35 pulses per second. In likefashion to the *Experiment 1*,9 delayed repetitions of each original measurement were added to the original set of measurements. The results of CKC decomposition, averaged over 25 Monte Carlo runs are depicted in Fig. 2.

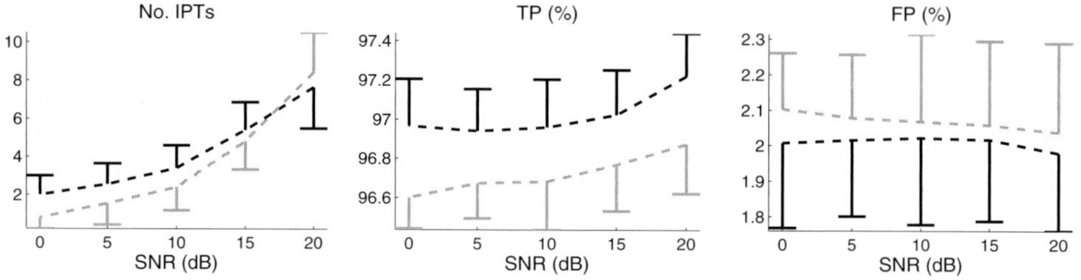

Fig. 2. Number of reconstructed IPTs (*left panel*), True Positive rate (TP) (*central panel*) and False Positive rate (FP) (*right panel*) for gradient CKC (*black line*) and classic CKC (*gray line*) when decomposing synthetic surface EMG (*Experiment 2*). The results are averaged over 25 simulation runs (*error bars* indicate std. deviations).

Experiment 3: The last experiment was conducted on real surface EMG signals, recorded by a 48-channel cylindrical anal probe from the external anal sphincter muscle. The electrodes (1×10 mm) were arranged in 3 circumferential arrays of 16 electrodes each. Interelectrode and inter-array distance was 2.7 mm and 5 mm, respectively. The experiment was conducted in Gynecological Clinic at University of Tübingen, Germany, and was approved by the local ethics committee. Six subjects participated to the experiment. The signals were acquired during three 10 s long maximum voluntary contractions. The EMG signals were amplified, band-pass filtered (3 dB bandwidth, 10 Hz-500 Hz), sampled at 2 kHz, and converted to digital form by a 12-bit A/D converter. The acquired set of measurements was additionally extended by 9 delayed repetitions of each

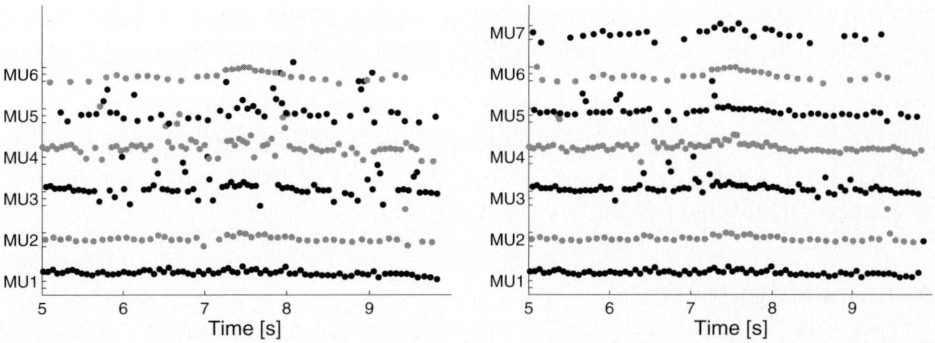

Fig. 3. IPTs reconstructed from real surface EMG by classic CKC (*left panel*) and by gradient CKC (*right panel*). Each plotted dot corresponds to single MU discharge. EMG signals were recorded from the external anal sphincter muscle.

Table 1. Number of MUs (mean ± std. dev.) reconstructed from real EMG signals

Subject	A	B	C	D	E	F
classic CKC	2.7 ± 0.6	3.7 ± 1.2	3.3 ± 0.6	3.3 ± 0.6	3.3 ± 0.6	2.3 ± 1.2
gradient CKC	6.3 ± 1.2	8.3 ± 1.2	4.7 ± 0.6	5.7 ± 1.2	8.0 ± 2.0	3.3 ± 0.6

measurements. The results of decomposition are summarized in Table 1 and exemplified by Fig. 3.

5 Discussion and Conclusions

The gradient CKC method proved to be highly efficient. In low noise environments, it is equivalent to the classic CKC approach. In the presence of severe noise, however, it provides superior accuracy. The results on synthetic measurements with random mixing matrix $\mathbf{H}$ proved that almost complete reconstruction of pulse trains at the SNR of -5 dB is possible. In the case of synthetic surface EMG, the robustness to noise was reduced. This was mainly due to the badly conditioned mixing process (in the case of surface EMG, typical condition number of the mixing matrix $\mathbf{H}$ is about 10^7). Nevertheless, up to 5 pulse trains were reconstructed down to the SNR of 10 dB, with the pulse accuracy exceeding the 97%. When compared to the classic CKC approach at low SNR, the gradient CKC yielded 30% increase in the number of reconstructed IPTs and 0.5% increase in the accuracy of the reconstructed pulse trains. Similar results were observed in the case of real surface EMG signals, acquired from the external anal sphincter. The general quality of acquired signals was low, mainly due to the bad electrode-mucosa contact and movement of the anal probe with respect to the muscle fibers. The gradient CKC technique reconstructed 6.2 ± 2.3 MUs per contraction (compared to 3.1 ± 0.8 MUs reconstructed by the classic CKC).

MU discharge patterns reconstructed by gradient CKC also exhibited higher regularity than those of the classic CKC method, as verified by careful visual inspection.

It is concluded that gradient CKC is highly robust to noise. In low SNR environment, it yields the performance superior to the classic CKC approach and has the potential to be used in regular clinical practice, where the quality of acquired signals cannot be strictly controlled.

Acknowledgment

The authors are sincerely grateful to Prof. R. Merletti from Politecnico di Torino, Italy, and to Prof. P. Enck, Dr. H Hinninghofen and Dr. T. Kreutz from University of Tübingen, Germany, for their measurements of the external anal sphincter. This research was supported by a Marie Curie Intra-European Fellowships within the 6th European Community Framework Programme (Contract No. MEIF-CT-2006-023537), and by German-Italian research project TASI.

References

1. Merletti, R., Parker, P.A.: Electromyography: physiology, engineering, and non-invasive applications, IEEE Press and John Wiley & Sons (2004)
2. Farina, D., Merletti, R.: A novel approach for precise simulation of the EMG signals detected by surface electrodes. IEEE trans. Biomed. Eng. 48, 637–646 (2001)
3. Kleine, B.U., van Dijk, J.P., Lapatki, B.G., Zwarts, M.J., Stegman, D.F.: Using two-dimensional spatial information in decomposition of surface EMG signals, J. of Electromy and Kinesiol (2006)
4. Gazzoni, M., Farina, D., Merletti, R.: A new method for the extraction and classification of single motor unit action potentials from surface EMG signals. J. of Neurosc. Methods 136, 165–177 (2004)
5. Wood, S.M., Jarratt, J.A., Barker, A.T., Brown, B.H.: Surface electromyography using electrode arrays: a study of motor neuron diseases. Muscle Nerve 24, 223–230 (2001)
6. Gracia, G.A., Okuno, R., Akazawa, K.: A Decomposition Algorithm for Surface Electrode-Array Electromyogram: A Noninvasive, Three-Steep Approach to Analyze Surface EMG Signals. IEEE Eng. in Med. Biol. Magazine 5, 63–71 (2005)
7. Holobar, A., Zazula, D.: Multichannel Blind Source Separation Using Convolution Kernel Compensation, IEEE Trans.Sig. Process., 56 (2007)
8. Amari, S.I.: Natural Gradient Works Efficiently in Learning. Neural Computation 10, 251–276 (1998)
9. Fuglevand, A.J., Winter, D.A., Patla, A.E.: Models of recruitment and rate coding organization in motor unit pools. J. Neurophysiol 70, 2470–2488 (1993)

Independent Component Analysis of Functional Magnetic Resonance Imaging Data Using Wavelet Dictionaries

Robert Johnson[1,2,*], Jonathan Marchini[1], Stephen Smith[2], and Christian Beckmann[2]

[1] Department of Statistics, University of Oxford, 1 South Parks Road, Oxford OX1 3TG, UK
johnson@stats.ox.ac.uk
[2] FMRIB (Oxford Centre for Functional Magnetic Resonance Imaging of the Brain), John Radcliffe Hospital, Headington, Oxford OX3 9DU, UK

Abstract. Functional Magnetic Resonance Imaging (FMRI) allows indirect observation of brain activity through changes in blood oxygenation, which are driven by neural activity. ICA has become a popular exploratory analysis approach due its advantages over regression methods in accounting for structured noise as well as signals of interest. However, standard ICA in FMRI ignores some of the spatial and temporal structure contained in such data. Using prior knowledge that the Blood Oxygenation Level Dependent (BOLD) response is spatially smooth and manifests itself on certain spatial scales, we estimate the unmixing matrix using only the coarse coefficients of a 3D Discrete Wavelet Transform (DWT). We utilise prior biophysical knowledge that the BOLD response manifests itself mainly at the spatial scales we use for unmixing. Tests on realistic synthetic FMRI data show improved accuracy, greater robustness to misspecification of underlying dimensionality, and an approximate fourfold speed increase; in addition the algorithm becomes parallelizable.

Keywords: Functional Magnetic Resonance Imaging, Independent Component Analysis, Biophysical Prior, Sparse Dictionaries.

1 Introduction

ICA offers several advantages over regression in FMRI: it can account for the large amounts of structured noise found in FMRI data such as MR acquisition artefacts, physiological noise, and often stimulus correlated head motion occur. Furthermore, ICA has been used to study resting state networks[3], investigate the mechanisms of memory and decision-making, and brain activity before an epileptic seizure[9]. In all cases the timings of task-related activity are difficult to specify accurately which makes it difficult - if not impossible - to use simple linear regression.

* Corresponding author.

M.E. Davies et al. (Eds.): ICA 2007, LNCS 4666, pp. 625–632, 2007.

Given an FMRI experiment with n voxels[1] measured at p different timepoints, after the pre-processing steps of motion correction, removal of low-frequency drift, voxelwise de-meaning and voxelwise variance normalisation, the spatial structure is discarded to allow the construction of a $p \times n$ matrix $\mathbf{X}$. We follow the methodology outlined in [2]: the generative model is that of linear mixing with additive Gaussian noise $\mathbf{N}$

$$\mathbf{X} = \mathbf{AS} + \mathbf{N} \tag{1}$$

where $\mathbf{S}$ includes not only BOLD components but also structured noise of physical and physiological origin. $\mathbf{A}$ is the linear mixing matrix. Typically, a Singular Value Decomposition is performed to obtain a factorisation into orthogonal regression timecourses and their corresponding spatial maps. The data is partitioned into major and minor PCA subspaces, using the eigenspectrum of the data covariance matrix to estimate the underlying dimensionality q (number of components). Within the standard PICA framework[2] ICA unmixing based on maximisation of non-Gaussianity is performed in the major subspace within the spatial domain, with the minor subspace used as the noise component for subsequent hypothesis testing (see [2] for details). In the resulting decomposition, $\mathbf{A}$ is a $p \times q$ linear mixing matrix and the ICs span the major subspace, with each voxel having factor loadings of the ICs associated with it.

Such an approach partially ignores knowledge about the spatiotemporal characteristics of FMRI data. In particular, standard linear decomposition techniques ignore neighbourhood relationships in the spatial and temporal domains once the data is represented as a 2D matrix $\mathbf{X}$. We have physiological knowledge that BOLD activation is spatially sparse due to the way the brain is organised and smooth due to the fluid characteristics of blood and the diffusion of oxygen from the bloodstream. We also know that functional brain anatomy is on a reasonably coarse scale compared to FMRI voxel sizes. In the temporal domain we know that the BOLD response is smooth at the timescales we observe, again because oxygen is supplied by diffusion from the bloodstream.

In previous work[10] we have incorporated assumptions about the temporal smoothness of the signals via a smoothness constraint in the temporal domain, substituting Regularised Principal Component Analysis (RPCA)[16] for PCA[2]. Such an approach is computationally efficient and has been shown to provide an increase in accuracy for block paradigms. However, a disadvantage is that the temporal variance is represented less effectively and that non-smooth components are estimated less efficiently. The non-smooth components may correspond to MR-physics related effects and can co-exist with smooth components which relate to physiological processes. Other researchers have applied a temporal smoothness constraint within the ICA algorithm[5,19] on only some components of interest, although they have assumed that they are analysing

[1] A voxel is a cube of tissue, typically of size 3mm^3 in FMRI, compared to 1mm^3 in structural MRI.

[2] RPCA constrains the factors to be represented by a B-spline basis set and applies a roughness penalty in addition to the existing orthogonality constraint.

experiments with a clearly defined experimental design to enable the identification of components of interest. This introduces some of the disadvantages of a regression analysis into the modified ICA algorithm.

A way of incorporating prior biophysical knowledge while not being dependent on identifiable design timecourse(s) is to use a prior in the *spatial* domain. In section 2 we describe an algorithm which uses Discrete Wavelet Transforms (DWTs) to incorporate our prior biophysical knowledge that the effect we attempt to observe is spatially smooth, sparse, and occurs at certain scales. We evaluate our proposed algorithm with realistic synthetic data in section 3. Finally, in section 4 we summarise our findings and discuss the demonstrated and potential advantages to constraining the ICA solution in the spatial domain rather than the temporal domain.

2 Algorithm

It has previously been demonstrated that ICA on natural signals operates more effectively when the signals are represented using a sparse dictionary[20]. We use a 3 level 3D separable DWT using 'Farras' wavelets[1,18] as our (complete) spatial dictionary $\mathbf{\Phi}$, since the multiresolution property proves useful. We zero-pad the FMRI volumes so they have dimensions which are multiples of 2^3 to enable a 3 level decomposition, and perform the DWT on each FMRI volume. The filterbank analysis and synthesis algorithms mean that we never have to explicitly calculate the $n \times n$ matrix $\mathbf{\Phi}$. If $\mathbf{X}$, $\mathbf{S}$ and $\mathbf{N}$ have $p \times n$ coefficient matrices in the dictionary $\mathbf{\Phi}$, denoted by $\mathbf{C}_{mixures}$, $\mathbf{C}_{sources}$ and $\mathbf{C}_{noise}$, we have

$$\mathbf{X} = \mathbf{C}_{mixtures}\mathbf{\Phi} \tag{2}$$

with similar equations for $\mathbf{C}_{sources}$ and $\mathbf{C}_{noise}$. Having postmultiplied by $\mathbf{\Phi}^{-1}$ we can restate Eqn. 1 as

$$\mathbf{C}_{mixtures} = \mathbf{A}\mathbf{C}_{sources} + \mathbf{C}_{noise} \tag{3}$$

We can incorporate our spatial prior within the ICA unmixing by estimating the unknown mixing matrix $\mathbf{A}$ from only the father wavelet coefficients. Conveniently, Gaussian noise will present itself mainly in the more detailed wavelet levels, a property used for wavelet denoising. This representation differs from [20]'s notion of sparseness because instead of the majority of the wavelet coefficients being close to zero, which can only be assumed for the mother wavelets, we have summarised the important features of the data in the father coefficients. The father coefficients themselves are not sparse, but do form a sparse representation of the data since we are ignoring the mother coefficients. Denoting the number of father coefficients as r, we find that we will be estimating $\mathbf{A}$ using only $\frac{1}{512}$ of the wavelet coefficients — since it is a dyadic transform, we raise 2 to the power of 3 for the scale level and then to the power of 3 again for the number of dimensions. Denoting the truncated $p \times r$ coefficient matrices by $\mathbf{C}'_{mixtures}$, $\mathbf{C}'_{sources}$ and $\mathbf{C}'_{noise}$ and the $r \times n$ truncated dictionary matrix by

$\mathbf{\Phi}'$, we can restate Eqn. 1 with $\mathbf{A}$ constrained to lie in the subspace of the father coefficients as

$$\mathbf{C}'_{mixtures} = \mathbf{A}\mathbf{C}'_{sources} + \mathbf{C}'_{noise}. \tag{4}$$

We previously used PCA[2,10] in the temporal domain to constrain the number of ICs to be equal to the estimated dimensionality of the data. Noiseless ICA is performed in the major PCA subspace, while the minor PCA subspace is treated as a Gaussian noise term for later hypothesis testing. If we perform PCA in the temporal domain on Eqn. 4 and retain only the major subspace we now have $q \times r$ coefficient matrices denoted by $\mathbf{C}^*_{mixtures}$, etc. The q rows contain factor loadings for regressed timecourses and r columns contain the spatial coefficients. We now have the generative model:

$$\mathbf{C}^*_{mixtures} = \mathbf{A}\mathbf{C}^*_{sources}. \tag{5}$$

We make a number of assumptions in order for ICA to be applicable:

- The underlying sources of our FMRI data are statistically independent.
- The independent components have non-Gaussian distributions (the minor PCA subspace we discarded is assumed to contain the Gaussian noise).
- We correctly identified the dimensionality q of the data, so our unknown mixing matrix is square.

$\mathbf{A}$ is estimated by performing noiseless ICA in the spatial domain on Eqn. 5 using the natural gradient Infomax algorithm[15] implemented in [14]. Next, we project $\mathbf{C}_{mixtures}$ onto the regressed PCA timecourses and use $\hat{\mathbf{A}}^{-1}$ to unmix the result, and finally apply the inverse DWT to the source estimates. Our BOLD ICs should now be estimated more accurately since unmixing has been performed in the scale subspace where the BOLD signals represent themselves most strongly.

3 Evaluation

Two FMRI datasets were obtained – from a subject at rest, and from the same subject hearing a 30 second on, 30 second off boxcar auditory stimulus. The second dataset was analysed within the general linear model framework as implemented in FEAT[17] and the activation maps found were thresholded at a fixed Z-statistic level. The intensity of the activation map remaining was linearly mapped to the range [0.1,1] and combined in the temporal domain with a simulated boxcar stimulus convolved with a gamma based haemodynamic response function. The resulting 4D activation was embedded in the resting data to produce a synthetic FMRI dataset with known spatial activation maps and timecourse. This was done at Contrast to Noise Ratios (CNRs) from 0% CNR to 200% CNR in 10% increments by scaling the timecourse amplitude accordingly.

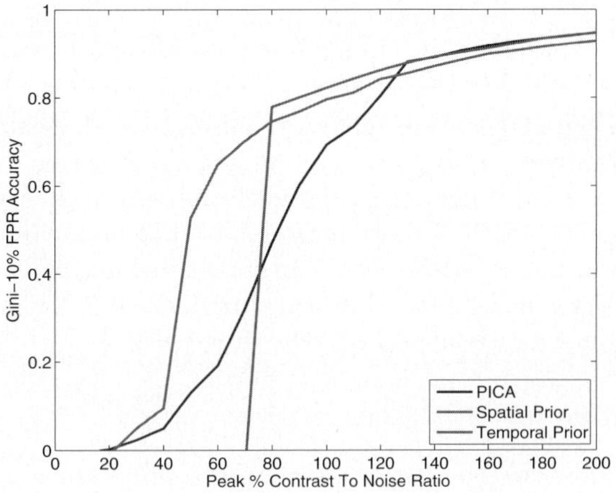

Fig. 1. Comparison of performance of PICA with spatial prior proposed in this paper and PICA with temporal prior implemented by using RPCA instead of PCA against standard PICA. For the spatial prior a 'step' is visible: nothing is resolved below 80% CNR, much higher accuracy is achieved above 80%. The temporal prior outperforms standard PICA in the CNR range 30%-130% and from 140% does slightly worse, the smoothness constraint outweighing the benefits when the BOLD signal is very strong.

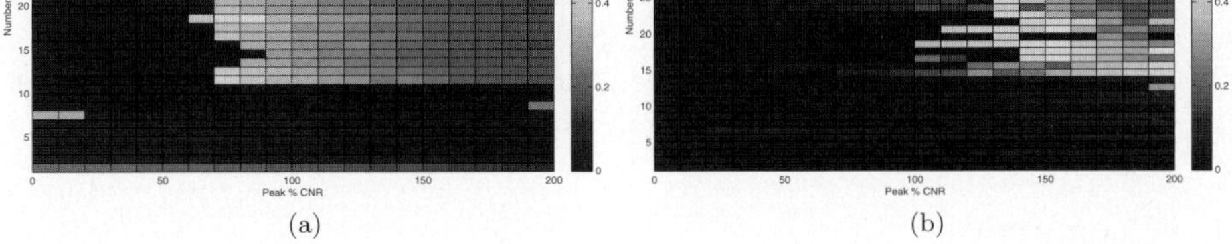

Fig. 2. Accuracy across number of components retained and peak % CNR. Results for a simple GLM are given in the bars at the top. A trade-off is apparent between greater sensitivity at lower CNRs for lower dimensionality estimates in the range 20–30 and greater accuracy at higher CNRs for higher estimates of dimensionality of 35 upwards. (a) Accuracy of PICA with spatial prior proposed in this paper. (b) Accuracy of standard PICA.

3.1 Test Metrics

Receiver operating characteristic (ROC) plots are frequently used to illustrate the accuracy of FMRI analysis techniques. Here, we calculated the Gini coefficient[8] as our measure of signal detection accuracy. The Gini coefficient provides a scalar measure, but typically integrates over the entire False Positive Rate (FPR) range. For the evaluation of the detection of signals from FMRI experiments however, a range of 0% up to 10% FPR is more reasonable since no meaningful conclusions can be drawn if a large percentage of the brain is known to be falsely classified as active. We therefore calculate the Gini measure in the 0–10% False Positive Rate (FPR) range and normalised it to lie in the range [0,1].

3.2 Test Results

Figures 1 and 2 illustrate the performance of the proposed methodology on the simulated data. Figure 1 shows a line plot of the accuracy over CNRs obtained by standard PICA, the new algorithm we have described, and the results of our previous work using a temporal smoothness prior. We speculate that the jump in accuracy using the spatial scale prior occurs because below a certain level of CNR too little information is available to enable effective unmixing. A similar effect can be observed on the plots in Fig. 2 when too few PCs are retained.

Figure 2 shows plots for accuracy (colour-coded) across CNR and number of components retained. With 11 or more PCs retained our proposed methodology gives values of above 0.6 accuracy at CNRs of 80% upwards, with only 2 exceptions. This indicates that our proposed methodology would be more robust to errors in specifying the underlying dimensionality of the data. Figure 2(b) shows that the standard methodology can vary considerably in its performance with small changes in dimensionality estimation.

The algorithm proposed here is faster than the standard PICA technique – the time taken to perform the DWTs is more than offset by a considerable reduction in the time taken to perform the ICA step, since the computationally intensive unmixing is performed on hundreds rather than tens of thousands of spatial coefficients, leading to an overall speed increase of more than four times, and the DWTs are easily parallelizable while the ICA unmixing is not.

4 Conclusion

We have described a method of incorporating a spatial biophysical prior into ICA for FMRI and tested it on realistic synthetic data. Incorporating a prior in the spatial domain offers the advantages over the temporal domain of increased accuracy, an increase in speed and the fact that no experimental design needs to be specified, unlike some methods of incorporating a temporal prior.

Note that the alternative approach of using all the wavelet coefficients or any other wavelet detail level yielded no advantage over the standard PICA method, as we would expect if the benefit comes from the scale prior rather than merely the sparse representation in the wavelet domain increasing the efficiency of ICA.

Elsewhere[11] we have evaluated using the Dual-Tree Complex DWT to denoise FMRI statistical maps obtained using ICA. Initial results suggest an increase in estimation accuracy to approximately 0.85 from a CNR level of 80% upwards. We are encouraged by this because 80% CNR is typical of the FMRI BOLD signals we observe, and at this level on our test data the standard PICA methodology in current use only achieved an accuracy measure of approximately 0.20 in our tests.

An adaptive technique is probably required to incorporate some of the more detailed coefficients in the estimation step and overcome the poor accuracy achieved by our algorithm at 70% CNR and below in Fig. 1. One disadvantage of our method is that while BOLD signals of interest are estimated more accurately, there may be a decrease in the accuracy of estimating non-BOLD structured noise which presents itself mainly at the scale levels we discard. A solution to this may be to estimate different classes of components using different subsets of wavelet coefficients.

Neurological conclusions are typically drawn from only about a hundred functionally defined regions of the brain and areas on the cortical surface. Performing inference using anatomically informed basis functions[12] to inform the unmixing might lead to further improvements. A spatial prior could also be incorporated into the analysis of multi-subject FMRI data. Particularly when using simultaneous group analysis techniques such as tensorial extensions to ICA[4], the unmixing of coefficients from a DWT would decrease the computational and memory requirements.

Acknowledgments. We are grateful for EPSRC and MRC funding for our research and for the helpful comments from our anonymous reviewers.

References

1. Farras Abdelnour, A., Selesnick, I.: Symmetric Nearly Shift-Invariant Tight Frame Wavelets. IEEE Trans. on Signal Processing 53(1), 231–239 (2005)
2. Beckmann, C., Smith, S.: Probabilistic Independent Component Analysis for Functional Magnetic Resonance Imaging. IEEE Trans. on Medical Imaging 23(2), 137–152 (2005)
3. Beckmann, C., DeLuca, M., Devlin, J., Smith, S.: Investigations into resting-state connectivity using independent component analysis. Philos. Trans. R. Soc. Lond. B. Biol. Sci. 360(1457), 1001–1013 (2005)
4. Beckmann, C., Smith, S.: Tensorial Extensions of Independent Component Analysis for Group FMRI Data Analysis. NeuroImage 25(1), 294–311 (2005)
5. Calhoun, V., Adali, T., Stevens, M., Kiehl, K., Pekar, J.: Semi-blind ICA of fMRI: a Method for Utilizing Hypothesis-Derived Time Courses in a Spatial ICA Analysis. NeuroImage 25, 527–538 (2005)
6. Damoiseaux, J., Rombouts, S., Barkhof, F., Scheltens, P., Stam, C., Smith, S., Beckmann, C.: Consistent Resting-State Networks across Healthy Subjects. PNAS 103(37), 13848–13853 (2006)
7. Donoho, D., Johnstone, I.: Adapting to Unknown Smoothness via Wavelet Shrinkage. Journal of the American Statistical Association 90, 1200–1224 (1995)

8. Fawcett, T.: ROC Graphs: Notes and Practical Considerations for Researchers. Technical report, HP Laboratories, Palo Alto, CA (2004)
9. Federico, P., Abbott, D., Briellmann, R., Harvey, A., Jackson, G.: Functional MRI of the pre-ictal state. Brain 128(8), 1811–1817 (2005)
10. Johnson, R., Marchini, J., Smith, S., Beckmann, C.: Temporal Regularisation for PCA Applied to FMRI. In: Twelfth Annual Meeting of the Organization for Human Brain Mapping (2006)
11. Johnson, R., Marchini, J., Smith, S., Beckmann, C.: Wavelet denoising as a post-processing step for ICA applied to FMRI. In: Thirteenth Annual Meeting of the Organization for Human Brain Mapping (2007)
12. Kiebel, S., Goebel, R., Friston, K.: Anatomically Informed Basis Functions. Neuroimage 11(6), 656–667 (2005)
13. Kingsbury, N.: Image processing with complex wavelets. Phil. Trans. Royal Society London A 357(1760), 2543–2560 (1999)
14. Makeig, S., et al.: EEGLAB: ICA Toolbox for Psychophysiological Research WWW Site, Swartz Center for Computational Neuroscience, Institute of Neural Computation, University of San Diego California, `http://www.sccn.ucsd.edu/eeglab/`
15. Makeig, S., Bell, A., Jung, T.-P., Sejnowski, T.: Independent component analysis of electroencephalographic data. In: Touretzky, D., Mozer, M., Hasselmo, M. (eds.) Advances in Neural Information Processing Systems, vol. 8, pp. 145–151. MIT Press, Cambridge, MA (1996)
16. Ramsay, J., Silverman, B.: Functional Data Analysis. Springer, New York (1997)
17. Smith, S.M., et al.: FEAT, part of FSL suite, Accessed (March 2007), `http://www.fmrib.ox.ac.uk/fsl/feat5`
18. Cai, S., Li, K., Selesnick, I.: Matlab implementation of wavelet transforms (accessed May 2007), `http://taco.poly.edu/WaveletSoftware/`
19. Valente, G., De Martino, F., Balsi, M., Formisano, E.: Optimizing ICA Using Generic Knowledge of the Sources. Technical report, Universita degli Studi di Roma La Sapienza, Rome, Italy (2005)
20. Zibulevsky, M., Pearlmutter, B., Bofill, P., Kisilev, P.: Independent Component Analysis. In: Principles and Practice, chapter Blind Source Separation by Sparse Decomposition in a Signal Dictionary, pp. 181–208. Cambridge University Press, Cambridge (2001)

Multivariate Analysis of fMRI Group Data Using Independent Vector Analysis

Jong-Hwan Lee[1], Te-Won Lee[2], Ferenc A. Jolesz[1], and Seung-Schik Yoo[1,3]

[1] Dept. of Radiology, Brigham and Women's Hospital, Harvard Medical School, MA, USA
jhlee@bwh.harvard.edu
[2] Institute of Neural Computation, University of California at San Diego, La Jolla, CA, USA
[3] Dept. of BioSystems, Korea Advanced Insitute of Science and Techonology, Daejeon, Korea

Abstract. A multivariate non-parametric approach for the processing of fMRI group data is important to address variability of hemodynamic responses across subjects, sessions, and brain regions. Independent component analysis (ICA) has a limitation during the inference of group effects due to a permutation problem of independent components. In order to address this limitation, we present an independent vector analysis (IVA) for the processing of fMRI group data. Compared to the ICA, the IVA offers an extra dimension for the dependent parameters, which can be assigned for the automated grouping of dependent activation patterns across subjects. The IVA was applied to the fMRI data obtained from 12 subjects performing a left-hand motor task. In comparison with conventional univariate methods, IVA successfully characterized the group-representative activation time courses (as component vectors) without extra data processing schemes to circumvent the permutation problem, while effectively detecting the areas with hemodynamic responses deviating from canonical, model-driven ones.

Keywords: Independent Vector Analysis, fMRI, Neuroimaging, Group Study, Multivariate Analysis.

1 Introduction

Functional MRI (fMRI) measures the blood-oxygenation-level-dependent (BOLD) signal changes associated with neural activity. Thus, temporal dynamics of the BOLD signal (time series; TS), called as hemodynamic response function (HRF), is the key element in analyzing fMRI data. Typically, univariate approaches such as the generalized linear model (GLM) or regression analysis, are performed to estimate the conformity of a measured voxel-wise BOLD TS to the fixed, canonical HRF [1]. However, the BOLD TS may not be fully appreciated from the univariate methods due to the variations across subjects, scans, and brain regions [2].

ICA [3], as one of the multivariate approaches, has provided flexibility in data processing compared to the hypothesis-driven, univariate methods. Such flexibility applies especially when observed hemodynamic responses deviated from expected (hypothesized) HRF [4]. Task-related activation components, often similar in their

M.E. Davies et al. (Eds.): ICA 2007, LNCS 4666, pp. 633–640, 2007.

spatial/temporal patterns across subjects/sessions, may not be inherently generalized from individual to group level analysis since the ICA algorithm permutes the order of output components. Therefore, the task-related components across subjects/sessions were manually inspected and grouped in the previous study [5]. This method may require careful selection of the component-of-interest from the large number of subjects/sessions.

In the present study, we propose a novel fMRI analysis method for group processing based on independent vector analysis (IVA) [6]. IVA was originally proposed to address the limitation of the conventional ICA approach during the blind source/signal separation (BSS) in the frequency domain (*i.e.* permutation of extracted independent components across frequency bins). IVA correctly indexed the independent components (ICs) that were identified during the BSS in the frequency domain by utilizing the mutual dependency among the extracted ICs across frequency bins. Intuitively, the IVA model offers an extra dimension for processing dependent components, compared to the ICA model. In the fMRI study, this extra dimension can be assigned for automated grouping of similar IC maps across subjects.

Fundamentally, IVA is an extension of ICA, whereby the component of an input and an output stage forms a vector (instead of a scalar value as in the case of ICA). IVA assumes and, therefore, attempts to increase independency across output vector components, while maintaining dependency among scalar elements within each vector. The 'dependency' in the fMRI study is analogous to mutual activation patterns across the subjects, comparable to the group trend in similar spatial activation patterns. Using the IVA algorithm, the spatially-similar trend in activation maps across subjects (dependent, thus representing group-trend) can be derived as the output vector components. As a result, one can avoid the complications of manual selection of task-related IC maps/time-courses (TCs) across subjects, thus rendering the whole process user-independent. In order to show the utility of the proposed method, we implemented and applied the IVA algorithm to analyze fMRI data of a left-hand motor clenching task using a short-time trial-based paradigm design. The obtained result was compared with the result from the generalized linear model (GLM) in SPM2 (Wellcome Department of Imaging Neuroscience, University College London, London, UK; www.fil.ion.ucl.ac.uk/spm) .

2 Methods

2.1 IVA Model and Learning Algorithm

Figure 1 shows a schematic diagram of concurrent synthesis (generative) and analysis models for group data processing using IVA. In the synthesis model, the weighting values at the vth voxels associated with IC maps (assuming mutual dependence across subjects) were grouped into a single vector array assuming the spatial similarity across the subjects. Through the mixing matrix $\mathbf{A}$ (TCs represented by each column), the arrays of these vectors are linearly combined to form sets of other vector arrays (equal to the number of temporal points of the BOLD TS), each containing the measured BOLD signal from a specific (v^{th}) voxel location across all of the subjects. In the analysis model, the unknown weighting values (also associated with IC map) can be estimated via a corresponding unmixing matrix $\mathbf{W}$. The matrix $\mathbf{W}$ can be obtained by applying a learning rule to increase independence among output vector arrays (thus, sorting out the activation patterns). This is accompanied by maintaining dependence of weighting values within each output vector array, thus deriving mutual activations across the subjects for group inference.

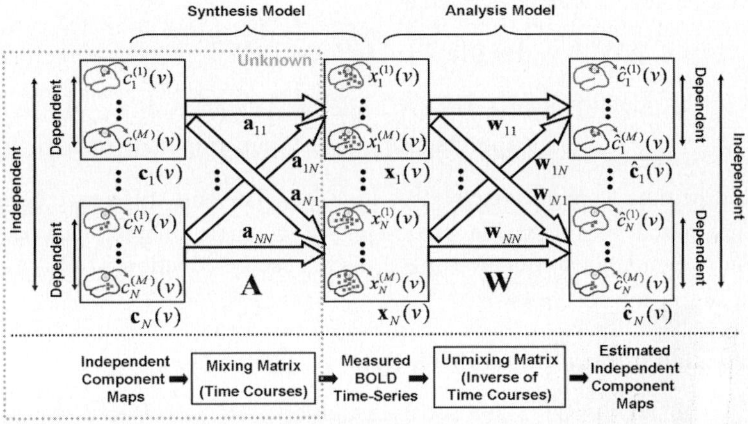

Fig. 1. The schematic diagrams of synthesis and analysis models for fMRI group data processing using independent vector analysis. $\mathbf{a}_{ji} = \left[a_{ji}^{(1)}, \cdots, a_{ji}^{(M)}\right]$ and $\mathbf{w}_{kj} = \left[w_{kj}^{(1)}, \cdots, w_{kj}^{(M)}\right]$.

From the synthesis model in Fig. 1, the measured BOLD TS at the vth voxel of subject m can be represented as

$$\mathbf{x}^{(m)}(v) = \mathbf{A}^{(m)} \mathbf{c}^{(m)}(v). \tag{1}$$

Each subject has its own mixing matrix (*i.e.* TCs) with independent activations (*i.e.* weighting values associated with IC maps) for the measured BOLD TS and does not share a mixing matrix with other subjects. In addition, the dependence among weighting values across subjects, within each unknown vector component, is assumed by the multivariate probability density function (p.d.f.) $p(\mathbf{c}_i(v))\left(= p(c_i^{(1)}(v), \cdots, c_i^{(M)}(v))\right)$. The boldface and lightface represent a vector/matrix and scalar value, respectively. The superscript and subscript denote the indices of the subjects and the IC map/fMRI volumes, respectively. M is the number of subjects ($m \in \{1, \cdots, M\}$), N is the number of fMRI volume acquisitions corresponding to the unknown IC maps ($i, j, k \in \{1, \cdots, N\}$), and V is the number of voxels within a brain region ($v \in \{1, \cdots, V\}$). The number of unknown IC maps was assumed to be the same as the number of volume acquisitions in Fig. 1. The assumed number of IC maps can be further reduced using a dimension reduction scheme such as principle component analysis (PCA) [3,5].

Then, by applying an unmixing matrix in the analysis model (*i.e.* inverse of TCs), the weighting values at the vth voxel (across IC maps) of subject m can be estimated as

$$\hat{\mathbf{c}}^{(m)}(v) = \mathbf{W}^{(m)} \mathbf{x}^{(m)}(v). \tag{2}$$

In order to derive learning algorithm, as presented in [6], the Kullback-Leibler (KL) divergence was adopted as a measure of independence of output vector components. Also, variance-dependent multivariate p.d.f. was utilized as a measure of dependence among elements within output vector component. By following the procedure in [6], the algorithm for calculating an update term, $\Delta\mathbf{W}^{(m)}$ corresponding to the unmixing matrix of the subject *m,* can be derived as

$$\Delta\mathbf{W}^{(m)} \propto \left[\mathbf{I} - \varphi\left(\hat{\mathbf{c}}^{(m)}(v)\right)\left(\hat{\mathbf{c}}^{(m)}(v)\right)^{\mathrm{T}}\right]\mathbf{W}^{(m)}, \tag{3}$$

where $\mathbf{I}$ is an identity matrix ($N \times N$), $\varphi\left(\hat{\mathbf{c}}^{(m)}(v)\right) = \left[\varphi\left(\hat{c}_1^{(m)}(v)\right) \ \cdots \ \varphi\left(\hat{c}_N^{(m)}(v)\right)\right]^{\mathrm{T}}$, and $\varphi\left(\hat{c}_k^{(m)}(v)\right) = \hat{c}_k^{(m)}(v) \Big/ \sqrt{\sum_{l=1}^{M} \left|\hat{c}_k^{(l)}(v)\right|^2}$. Applying Eq. (3), the unmixing matrix can be iteratively updated for the data from all voxels ($v = 1, \cdots, V$). The only difference compared to the Infomax-based ICA (for the processing of a single subject data [3]) is the application of a nonlinear function $\varphi\left(\hat{\mathbf{c}}^{(m)}(v)\right)$ (*c.f.* score function in ICA), which is dependent across subjects in IVA.

2.2 Application to Trial-Based fMRI Data

This study was approved by the local Institutional Review Board. Twelve right-handed subjects (aged 24.7±4.5, 5 females) performed one session of a left hand (LH) clenching (2 times/sec) task based on a short-time trial-based paradigm design (65-sec duration excluding 10-sec of dummy scans; task onset occurred at 15-sec followed by a 3-sec task-period). For the start/end of the task, a pre-recorded sound cue was played to the subject in the MRI system via an auditory headset (Avotec, Jensen Beach, FL). The fMRI data was obtained in a 3-Tesla clinical scanner (Signa VH, GE Medical Systems) using a single channel, standard birdcage, head coil.

To obtain functional data, an EPI sequence was applied to image most of the brain volume (13 axial slices, flip angle=80°, TE/TR=40/1000msec, 64 frequency and phase encoding: 64×64 in-plane voxels, 5mm thickness with a 1mm gap, 240mm square field-of-view) for detection of the BOLD TS associated with neural activity. Prior to group processing, individual EPI data was standardized to the MNI (Montreal Neurological Institute) space by following preprocessing steps in SPM2 (*i.e.* in order: slice timing correction, realignment, normalization, and spatial smoothing using an 8mm full-width-at-half-maximum 3-D Gaussian kernel). Before processing using IVA algorithm, a PCA-based dimension reduction scheme [3,5] was applied to reduce the number of IC maps/TCs to 50. The sum of corresponding 50 eigenvalues was more than 99% of a sum of total 65 eigenvalues for each subject.

Using the IVA algorithm in Eq. (3) to a normalized set of dimension-reduced fMRI group data, a semi-batch learning scheme [3] was applied to update the unmixing matrix of each subject. A batch of a 10×10×10mm^3 isotropic cluster (5×5×5=125 voxels due to the 2×2×2mm^3 isotropic voxel) was used, assuming dependencies of neural activations within this cluster across subjects. The learning rate (η) was set to 10^{-3} throughout iterations. The iteration was stopped when a ratio of weight change ($\eta\Delta\mathbf{W}^{(m)}/\mathbf{W}^{(m)}$) was stabilized ($3.6\times10^{-5} \sim 7.5\times10^{-4}$). After the algorithm converged, the resulting IC map (*i.e.* weighting values of TC) was transformed into a z-scored map by subtracting a mean value and dividing by the standard deviation [3].

Sign ambiguity of IC maps/TCs that typically arise from ICA-based methods [3, 4, 5, 7] also applied to the case of IVA. In IVA, the voxel-wise correlation coefficients between (1) the IC z-map within the activated regions (|z|>threshold) and (2) the original fMRI volumes were utilized. If the averaged (across time points) value of the correlation coefficients was negative, the sign of the IC z-map (& corresponding TC) was inverted.

In order to find task-related components, a spatial sorting scheme was employed, whereby a voxel-wise correlation value between the sign-corrected IC z-map and cross-correlation (CC) map (obtained from the temporal correlation coefficients using the task-related canonical HRF) was regarded as the degree of task relevance of each IC z-map for each subject. This similarity measure (*i.e.* the voxel-wise correlation value) was then averaged across all subjects for each output vector array of IVA and subsequently sorted in descending order. The resulting five most task-related IC z-maps across subjects were processed using one-sample t-test implemented in SPM2 by considering a random effect (RFX) model [8]. These resulting task-related group activation maps of IVA were compared to the group activation map of GLM obtained from SPM2. After obtaining a group inference, the group activation maps were qualitatively compared in terms of location of activation. The areas with an activation volume greater than $5 \times 5 \times 5 mm^3$ ($p<10^{-3}$) were identified and labeled from the Brodmann's Area (BA) and Automated Anatomical Labeling (AAL) templates (provided by MRIcro; www.mricro.com).

3 Results

Figure 2 shows the task-related group activation maps obtained from GLM and IVA. In the results of GLM (Fig.2A), which is a univariate approach, a single task-related group activation map was acquired. In the results of IVA, the 5 most task-related components were selected after the reordering of output components based on the spatial sorting strategy explained in Section 2.2. The group activations ($p<10^{-2}$ ~ $p<10^{-7}$) were coded with a color gradient. The labeled anatomical areas of the group activations were listed in Table 1. From the analysis by GLM, the directly task-related areas (*e.g.* right primary motor area: M1, supplementary motor area: SMA, primary sensory area: S1, and cingulate gyrus) and paradigm-related areas (auditory; superior temporal) showed more significant activations compared to basal ganglia (caudate/putamen/pellidum) and thalamus (see Fig. 2A). In the results of IVA, the task-related activations (*e.g.* the right M1, SMA, S1, cingulate gyrus, and sup./mid. temporal gyrus) were extracted as the 1st task-related group activation map $\hat{\mathbf{c}}_1$. The group activations in the remaining components were dominant in the primary auditory area ($\hat{\mathbf{c}}_2$), basal ganglia & thalamus ($\hat{\mathbf{c}}_3$), inf. frontal & insular cortex ($\hat{\mathbf{c}}_4$), and mid. frontal & cingulate gyrus ($\hat{\mathbf{c}}_5$).

From the comparison of group activation maps between GLM and IVA (Fig. 2), some of the activations revealed by IVA were underestimated by GLM. For example, the size of activated regions obtained in the auditory area by GLM (examples were shown with green circles in Fig. 2A) was reduced compared to the area detected by IVA (marked as green in Fig. 2B). Also, the activations in the basal ganglia/thalamus identified from IVA (blue area in Fig. 2B) were not detected when processed with GLM (examples were shown with blue circles in Fig. 2A).

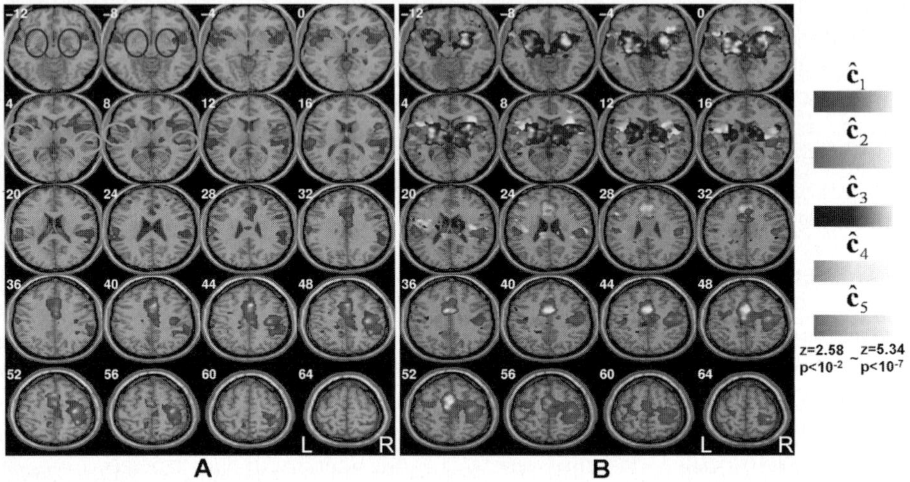

Fig. 2. Group activation maps obtained by (A) GLM and (B) IVA methods. For IVA, five most task-related component maps ($\hat{\mathbf{c}}_1 \sim \hat{\mathbf{c}}_5$) were color-coded.

Table 1. The strongly activated cortical areas inside group activation map. If any activated cluster ($p<10^{-3}$) was bigger than $5\times5\times5mm^3$, the center of this cluster was registered using BA and AAL indices (provided by MRIcro; www.mricro.com).

	Left-Hemisphere	**Right-Hemisphere**
GLM	Inf. frontal / SMA / Insula / Mid. Cingulate / S1 / SupraMarginal / Sup. Temp. **BA: 6, 23, 24, 32, 42, 47, 48*	M1 / Inf. Frontal / SMA / Insula / Mid. Cingulate / S1 / Parietal (Sup., Inf) / SupraMarginal / Sup. Temp. **BA: 2, 3, 4, 6, 8, 23, 24, 32, 38, 40, 42, 44, 47, 48*
IVA	$\hat{\mathbf{c}}_1$: SMA / Mid. Cingulate / S1 / SupraMarginal / Temp. (Sup., Mid.) **BA: 3, 6, 8, 22, 23, 24, 32, 41, 42, 48* $\hat{\mathbf{c}}_2$: Insula / SupraMarginal / Heschl / Temporal (Sup., Mid.) **BA: 22, 42, 48* $\hat{\mathbf{c}}_3$: Frontal (Sup., Inf.) / Insula / Hippocampus / Amygdala / Putamen / Pallidum / Thalamus / Heschl / Sup.Temp. **BA: 11, 20, 25, 34, 38, 47, 48* $\hat{\mathbf{c}}_4$: Inf. Frontal / Insula / Putamen **BA: 6, 48* $\hat{\mathbf{c}}_5$: Mid. Frontal / Cingulate (Ant., Mid.) / Sup. Temp. **BA: 24, 32, 38, 45, 46*	$\hat{\mathbf{c}}_1$: M1 / Frontal (Sup., Mid.) / SMA / Mid. Cingulate / S1 / Parietal (Sup., Inf) / SupraMarginal / Sup. Temp. **BA: 2, 3, 4, 6, 23, 24, 32, 40, 48* $\hat{\mathbf{c}}_2$: Heschl / Sup. Temp. **BA: 22, 48* $\hat{\mathbf{c}}_3$: Insula / Hippocampus / Amygdala / Caudate / Putamen / Pallidum / Thalamus **BA: 11, 20, 27, 34, 37, 38, 48* $\hat{\mathbf{c}}_4$: Inf. Frontal / Insula / SupraMarginal / Putamen / Heschl **BA: 45, 47, 48* $\hat{\mathbf{c}}_5$: Insula / Cingulate (Ant., Mid.) / Sup. Temp. **BA: 24, 32, 48*

We conjectured that the differences in activation maps between GLM and IVA were caused by the individual differences in temporal patterns of hemodynamic responses from the areas. Therefore, individual differences in the obtained TCs were further compared across subjects (Note that the TC represents a dominant feature of the BOLD TS corresponding to the activated voxels in the IC map). Figure 3 shows the individual TCs corresponding to three highly task-related group activation maps ($\hat{\mathbf{c}}_1 \sim \hat{\mathbf{c}}_3$ in Fig. 2B) by IVA. Here, we adopted the convention introduced by Duann *et al.*, [8], whereby the normalized TC (0~1) was coded in gray scale (black: 0 & white: 1) so that relative amplitudes can be readily discriminated within/across subjects. First of all, the TCs (Fig. 3A) corresponding to the 1st (highly) task-related group activation map $\hat{\mathbf{c}}_1$ was in good agreement with the hypothesized HRF in GLM (yellow box: task-related period; correlation coefficient between the averaged TC across subjects and the hypothesized HRF: 0.88). On the other hand, the TCs (Fig. 3B&C) corresponding to $\hat{\mathbf{c}}_2 \& \hat{\mathbf{c}}_3$ showed some degree of variations in peak position within the task-related period. It is also notable that additional peaks during rest-periods were observed (examples are shown with arrows) with reduced correlation coefficients of 0.56 & 0.63, compared to the $\hat{\mathbf{c}}_1$. Because of these large variations between the actual hemodynamic responses (analogous to TCs) and the hypothesized HRF across subjects, the activations in the corresponding areas may not be detected by GLM.

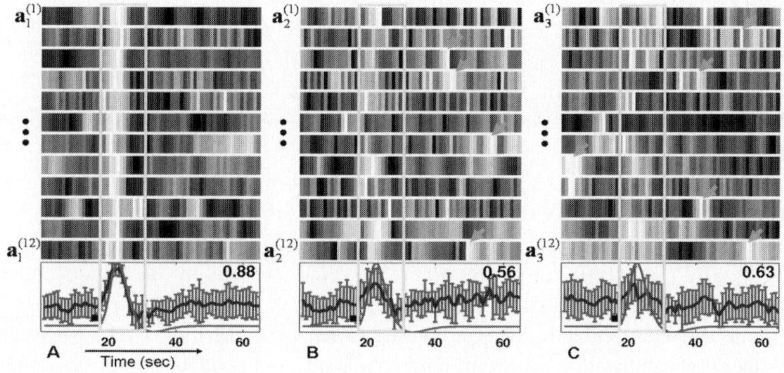

Fig. 3. Image plots of TCs across subjects corresponding to 3 highly task-related group activation maps ($\hat{\mathbf{c}}_1 \sim \hat{\mathbf{c}}_3$). Each TC was normalized between 0 (black) and 1 (white). A yellow box indicates the period of task-related response. Green arrows indicate examples of the peaks during the rest-period. The plots in the bottom are the hypothesized HRF (green line), along with the averaged TC (blue line), and standard deviations (red bars) across all subjects. A correlation coefficient between the averaged TC and the hypothesized HRF is shown in the top-right corner of each averaged time plot. A task-period (3-sec) is marked with thick black bar.

4 Conclusion

In this study, we have proposed the use of IVA to infer the group-activation pattern from fMRI data. The IVA algorithm was applied for multiple subjects' BOLD signals and the spatially-similar trend in activation maps across subjects (dependent, thus representing group-trend) were derived as the output vector components. From the application of the proposed method to fMRI data, the resulting IC maps/TCs of individual subjects provided reliable task-related information for a further group level inference. In addition, IVA provided more robust activation patterns than GLM (based on the hypothesized univariate HRF), especially when the HRF deviated from the hypothesized HRF (*e.g.* from the basal ganglia/thalamus). These results show the feasibility of IVA to be used in fMRI group studies as potential alternative to conventional univariate approach. The proposed model may also be adopted to find out multivariate common activation patterns across multiple trials/sessions from a single subject data by substituting the index of a subject for the index of a trial/session.

The IVA approach demands computational load since an individual unmixing matrix is iteratively trained using the results from other subjects (represented as nonlinear function, $\varphi\left(\hat{\mathbf{c}}^{(m)}(v)\right)$), and thus, all of the unmixing matrices should be parallely updated. This increased computational demand can be alleviated by the increasing hardware memory. In order to achieve the fully-automated group processing using IVA, elaborate sets of optimization in terms of learning parameters and sorting schemes (for the selection of task-related features) are needed.

Acknowledgments. This work was partially supported in part by grants from NIH (R01-NS048242 to Yoo, SS and NIH U41RR019703 to Jolesz FA).

References

1. Worsley, K.J., Friston, K.J.: Analysis of fMRI time-series revisited—again. Neuroimage. 2, 173–181 (1995)
2. Aguirre, G.K., Zarahn, E., D'Esposito, M.: The variability of human, BOLD hemodynamic responses. NeuroImage 8, 360–369 (1998)
3. McKeown, M.J., Makeig, S., Brown, G.G., Jung, T.P., Kindermann, S.S., Bell, A.J., Sejnowski, T.J.: Analysis of fMRI data by blind separation into independent spatial components. Hum. Brain Mapp. 6, 160–188 (1998)
4. Quigley, M.A., Haughton, V.M., Carew, J., Cordes, D., Moritz, C.H., Meyerand, M.E.: Comparison of independent component analysis and conventional hypothesis-driven analysis for clinical functional MR image processing. AJNR Am. J. Neuroradiol. 23, 49–58 (2002)
5. Calhoun, V.D., Adali, T., McGinty, V.B., Pekar, J.J., Watson, T.D., Pearlson, G.D.: fMRI activation in a visual-perception task: network of areas detected using the generalized linear model and independent component analysis. Neuroimage 14, 1080–1088 (2001)
6. Kim, T.S., Attias, H.T., Lee, S.Y., Lee, T.W.: Blind source separation exploiting higher-order frequency dependencies. IEEE Trans. Audio, Speech and Language Process. 15, 70–79 (2007)
7. Duann, J.R., Jung, T.P., Kuo, W.J., Yeh, T.C., Makeig, S., Hsieh, J.C., Sejnowski, T.J.: Single-trial variability in event-related BOLD signals. Neuroimage 15, 823–835 (2002)
8. Friston, K.J., Holmes, A.P., Worsley, K.J.: How many subjects constitute a study? Neuroimage 10, 1–5 (1999)

Extraction of Atrial Activity from the ECG by Spectrally Constrained ICA Based on Kurtosis Sign

Ronald Phlypo[1,2,*], Vicente Zarzoso[1], Pierre Comon[1], Yves D'Asseler[2], and Ignace Lemahieu[2]

[1] Laboratoire I3S, CNRS/UNSA Les Algorithmes - Euclide-B, BP 121, 2000 Route des Lucioles, 06903 Sophia Antipolis Cedex, France
{phlypo,zarzoso,pcomon}@i3s.unice.fr

[2] MEDISIP-IBBT, ELIS/UGent, IBiTech - Campus Heymans, De Pintelaan 185, B-9000 Ghent, Belgium
{ronald.phlypo,yves.dasseler,ignace.lemahieu}@ugent.be

Abstract. This paper deals with the problem of estimating atrial activity during atrial fibrillation periods in the electrocardiogram (ECG). Since the signal of interest differs in kurtosis sign from the dominant sources in the ECG, we propose an independent component analysis method for source extraction based on the different kurtosis sign and extend it with a constraint of spectral concentration in the 3-12Hz frequency band. Results show that we are able to estimate the atrial fibrillation with a single algorithm having low computational complexity ($\mathcal{O}$(7n-7)T).

1 Introduction

This paper describes a method to recover narrow band independent signals from a linear mixture model where high impulsive (high kurtosis) signals are the main source of interference. This set-up is a commonly encountered problem in biomedical signal analysis, e.g. when considering spectral bands of activity in electroencephalographic recordings or atrial fibrillation (AF) signals in the electro cardiogram (ECG) where the main sources of interference are respectively the ocular activity and the QRS(-T) complex.

The focus here is on the recovery of atrial activity during AF from an ECG recording, whatever the conditions of noise or interference from other physiological signals (e.g. QRS complex). Since we consider narrow band spectra in a volume conductor, a linear approximation of the electromagnetic Maxwell equations is valid and hence we may suppose that a general linear mixing model holds. This mixture model translates the measured potentials at the chest or body surface into bio-electrical source signals (generally specified by their currents) and *vice versa*. If we consider the mixing-demixing model, all relations exhibit the

* R. Phlypo would like to thank H. Rix and V. Zarzoso for their kind hospitality at the laboratories of I3S.

M.E. Davies et al. (Eds.): ICA 2007, LNCS 4666, pp. 641–648, 2007.

characteristics of a linear model and thus we can rewrite our system of measurements $\boldsymbol{y}$ into an equivalent set of potentials $\boldsymbol{x}$, where each x_i is associated with a column of $\boldsymbol{A}$, $\boldsymbol{a}_i$. The latter represent the mappings of the sources on the measurement surface (known as source topographies). Or, in matrix notation:

$$\boldsymbol{y}(t) = \boldsymbol{A}\boldsymbol{x}(t) + \eta(t), \tag{1}$$

which explains the relations between the measurements $\boldsymbol{y} \in \mathbb{R}^{m \times 1}$, the mixing matrix $\boldsymbol{A} \in \mathbb{R}^{m \times n}$, the sources $\boldsymbol{x} \in \mathbb{R}^{n \times 1}$ and the noise $\eta \in \mathbb{R}^{m \times 1}$.

The measurements and the sources can be seen as realisations of random variables. Therefore we will drop the time index in the subsequent work to improve readability. For the biomedical case we might assume that these sources are quasi statistically independent. In the case of atrial fibrillation, we can restrict the AF source characteristics even further by imposing the extra constraint that the AF signal should have a narrow band spectrum, thus having platokurtic statistics. This is in contrast to the QRS(-T) complex - the main masking source - which is highly leptokurtic (see e.g. [1]). In the rest of this paper we will develop this idea further, sketching a framework in which we can extract the independent AF source based on the difference in kurtosis sign and under the constraint of narrow band source spectra. The solution is given as the ensemble of subsequent algebraic solutions to the pairwise separation problem, subjected to a conditional update. This guarantees a robust algorithm, with only few parameters to estimate and omitting the need for exhaustive search algorithms.

2 Methods

2.1 The Kurtic Difference as an Object Function to ICA

ICA. The solution to the ICA problem has been proposed by different authors, using different contrast functions. Despite the diversity at the basis of the algorithms, the solution space is almost always given by components whose higher order cross cumulants vanish [2,3], which in its turn is equivalent to a reduction of the mutual information between the components [4,5]. Solutions have been proposed to solve the problem by deflation approaches [6] - estimating source by source - or to interact on the whole subset at once. The deflation approach offers the ability to solve for independence in a component-by-component way, sorted according to the value they take in the cost function, see e.g. RobustICA [7], FastICA [8]. However, when considering the *a priori* constraint of a narrow spectrum, most algorithms lack the possibility to include this without going to excessive computational complexity, see e.g. the number of tensor slices or correlation matrices needed in JADE [4]-like, respectively SOBI [9]-like, algorithms especially when applied to high data dimensionalities n.

Givens Rotations. The method proposed here is an extraction (or deflation) approach with pairwise optimisation. The advantage is that there exists an algebraic expression able to update the source estimates, avoiding computationally unattractive search methods. Moreover, since the signals are prewhitened (i.e. mutually decorrelated), it suffices to find an orthogonal matrix to find maximally independent source estimates. We can thus constrain our parameter space to only one parameter per signal pair if we do not take into account permutation and scaling, which are irrelevant parameters when considering the independence criterion. Our search space can thus be limited to the optimal rotation angle for each pair to process [4,3]. This amounts to the following algorithm for prewhitened signals $\hat{\boldsymbol{z}} \in \mathbb{R}^{n\times 1}$:

$$\hat{\boldsymbol{x}}_{ij} = \boldsymbol{Q}\left(\theta_\star\right)\boldsymbol{x}_{ij}\ , \quad \text{where } \boldsymbol{Q}\left(\theta_\star\right) = \begin{pmatrix} \cos\theta_\star & \sin\theta_\star \\ -\sin\theta_\star & \cos\theta_\star \end{pmatrix}, \tag{2}$$

where $\theta_\star$ is the optimal rotation angle that is to be specified, and the matrix $\boldsymbol{Q}\left(\theta_\star\right)$ represents a plane rotation, also known as Givens rotation. $\boldsymbol{x}_{ij}$ and $\hat{\boldsymbol{x}}_{ij}$ are the ith and jth component of $\boldsymbol{x}$, respectively $\hat{\boldsymbol{x}}$. The result of the left multiplication of the data $\hat{\boldsymbol{x}}_{ij}$ by $\boldsymbol{Q}^{-1} = \boldsymbol{Q}^T$ would thus results in the standarized sources $\hat{\boldsymbol{x}}$, with additional constraints imposed by the objective function to which $\theta_\star$ is a solution. If the objective function is chosen well, these sources are maximally independent, an assumption that is believed to hold true for many, if not all, bioelectrical source signals. When the objective function meets the requirements of being maximal if and only if the components are independent, while being blind to possible permutations and scaling, it becomes a contrast function for ICA [3].

Kurtic Difference as a Contrast. We have shown in [10] that the objective function

$$\Psi\left(\boldsymbol{Q}\right) = \sum_{i=1}^{n} \epsilon_i \kappa_{iiii}^{\hat{\boldsymbol{x}}} \tag{3}$$

fulfils all requirements to be a contrast function for ICA, where ϵ_i is the sign of the fourth order auto cumulant of the ith source and $\kappa_{iiii}^{\hat{\boldsymbol{x}}}$ the fourth order cumulant of the ith output. Based on this fact, together with the assumption that the atrial activity caused by AF is a (the sole) platokurtic source in the ECG, the contrast would translate into:

$$\Psi_{AF}\left(\boldsymbol{Q}\right) = \left(\sum_{i=2}^{n} \kappa_{iiii}^{\hat{\boldsymbol{x}}}\right) - \kappa_{1111}^{\hat{\boldsymbol{x}}}, \tag{4}$$

which can be solved using subsequent Givens rotations as defined above. The next paragraph gives the algebraic solution for θ when a pair of signals is considered.

The Optimal $\boldsymbol{\theta}$-value: $\boldsymbol{\theta_\star}$. We can now obtain the optimal value for θ, $\theta_\star$, by calculating the stationary point of our contrast function Ψ_{AF} (4) by setting its derivative to zero. It is sufficient to consider the pairs with opposite kurtosis signs, the other cases being known. As a function of the observed whitened signals $\hat{\boldsymbol{x}} = \boldsymbol{Q}\boldsymbol{x}$ we obtain for Ψ_{AF}:

$$\Psi_{AF}\left(\theta\right) = \lambda_2 - \lambda_1 = \alpha\cos 2\theta + 2\beta\sin 2\theta, \tag{5}$$

where α and β are given by $\left(\kappa^{\hat{\boldsymbol{x}}}_{1111} - \kappa^{\hat{\boldsymbol{x}}}_{2222}\right)$ and $\left(\kappa^{\hat{\boldsymbol{x}}}_{1112} + \kappa^{\hat{\boldsymbol{x}}}_{1222}\right)$, respectively and λ_1, λ_2 are the kurtosis values of $\hat{\boldsymbol{x}}_{ij}$, which can be written as a multi linear function of the source kurtosis values [3] using Eq. (2). Equation 5 has its stationary points at

$$2\theta_\star = \arctan \frac{2\beta}{\alpha}, \tag{6}$$

where $\theta_\star$ is the rotation angle to be found.

Equation 6 is also the equation obtained in [11] based on centroid estimators.

2.2 Inclusion of the Spectral Concentration Constraint

AF is typically characterised by a sinusoidal to triangular waveform (depending on the relative power in the harmonics) with a frequency and amplitude modulation. This spectrally rather narrow banded signal has its main frequency in the 3 to 12 Hz band. This enables us to create additional constraints regarding the spectra, forcing us to redefine the update sequence of the pairwise processing for source extraction as given in [10]. The method extends the natural sweeping procedure for source extraction to a criterion based sweeping procedure. The update criterion is given as

Criterion 1. *Replace the source estimates $\hat{x}_i$ and $\hat{x}_j$ with their updates $\hat{x}_i^\star$ and $\hat{x}_j^\star$ by using the relation $\hat{\boldsymbol{x}}_{ij}^\star = \boldsymbol{Q}^T \hat{\boldsymbol{x}}_{ij}$ iff the spectral concentration in the 3-12Hz band of one of the new estimates exceeds the spectral concentration of the reference source estimate.*

The spectral concentration in criterion 1 is taken as the ratio of spectral density in a $\pm 10\%$ band around the center frequency f_c to the total energy in the signal's spectrum, i.e. $SC = \int_{.9f_c}^{1.1f_c} P(\tau)\mathrm{e}^{-2\pi\tau f}\mathrm{d}f \Big/ \int_0^{Fs/2} P(\tau)\mathrm{e}^{-2\pi\tau f}\mathrm{d}f$, if $f_c \in [3\mathrm{Hz}, 12\mathrm{Hz}]$, otherwise $SC = 0$. With this information we can define the sweep procedure as described in table 1, where the stopping criterion is defined to be positive when a sweep occurs without update of the reference source estimate.

Table 1. The pseudo-code for the sweep algorithm

```
Initialise reference source with x̂1
While false(stopping criterion)
   StartSweep: For j from 2 to m
      Compute θ⋆ for the reference source estimate x̂1 and estimate x̂j
      Compute spectral concentration for x̂1⋆ and x̂j⋆
      If criterion 1: replace x̂1 (x̂j) with x̂⋆ having highest (lowest) SC
   EndSweep
```

3 Results

3.1 Data

Patient Data. The data upon which the algorithm was run consists of 51 patient registrations with known AF. All ECG sets are standard 12 lead ECG measurements consisting of the leads I-III, aVR, aVL, aLL and the potentials at the electrodes V1-V6. The dataset is by definition overdetermined for the bio-potentials since I-III, aVR, aVL and aLL can all be expressed in terms of the left arm (LA), right arm (RA) and left leg (LL) electrode potentials. This means that there is a redundancy of factor 2 in the leads. If we take the LL electrode as the reference electrode for all measurements (which is generally the physical measurement setup), then we are left with 8 independent variables. We can thus reduce our set to 8 recording sites or derivations only without compromising the information in the data. Taking V1-V6 and extracting the potentials at LA and RA from the leads would eliminate this data redundancy.

One has to be careful though in highly noisy environments where the noise at the electrodes is not stationary and the noise term would thus take a sufficiently high amount of the total data subspace. In that case it might not suffice to take only the 8 electrode potentials and the extra derivations might add extra information to solve the ill-conditioned problem.

Simulated Data. To have an idea of the quality of separation we introduce a simulated dataset. This dataset contains an AF signal constructed following the method in [12], whereas the QRS-T simulation has been done using high kurtosis components using the model:

$$\text{QRS-T}_i\,(t) = \sum_j \tan\left(a_j\,(t)\sin\left(j\omega\,(t)\,t\right)\right). \tag{7}$$

The model allows for amplitude modulation in a_j, where $\max_t \sum_j |a_j\,(t)\,| \leq 1$, and modulation in $\omega\,(t)$. By changing the number of harmonics and the parameters in the modulations we can change the statistics of the total time series. Additionally we added two sources that are of no physiological meaning but have a positive kurtosis value, so we do not violate the model assumptions. The randomly drawn square mixing matrix is orthonormal, avoiding the need of the prior whitening step as described in the introduction without restricting the generality.

3.2 Estimating the Central Frequency

To evaluate our method, we focus on the value of the main frequency estimated by our method. The main frequency is defined as the frequency at which the power spectral density is the highest in the 3 - 12Hz band. For the 51 patient registrations we compare the resulting frequencies with those found from a combined FastICA and SOBI approach as it was applied in [13]. We compare the results as well for AEML, EML and the combEML methods [11] in the same constrained updating framework. In table 2 the results of the central frequency difference among the methods with their respective standard deviation are displayed. To exclude biasing toward short time artefactual instances in the data, the data has been resampled at each iteration before calculation of the optimal rotation angle $\theta_\star$ based on a bootstrap sampling, allowing for overlap. We choose a bootstrap sample size of $2 \cdot 10^3$. To exclude the bootstrap based differences between the methods, we consider 100 Monte Carlo runs per method over the whole dataset of which the mean of the results so obtained are used as the frequencies to construct table 2.

Table 2. Differences in main frequency estimation. The upper right triangle displays the results when 12 leads are considered, in the lower left triangle displays the results for the reduced set of 8 electrode potentials are given.

	fastICA+SOBI	AEMLa	AEMLc	combEML
fastICA+SOBI	0	-0.026 (0.444)	0.083 (0.294)	0.425 (0.429)
AEMLa	0.467 (1.025)	0	0.142 (0.384)	0.032 (0.291)
AEMLc	0.291 (0.526)	-0.085 (0.755)	0	-0.110 (0.338)
combEML	0.425 (0.761)	0.063 (0.767)	0.148 (0.390)	0

Visualisation of the Results. To evaluate the performance on simulated and real datasets, we present the artificial mixture, respectively the observations of the electrode potentials and the source extraction results in Figs. (1) & (2).

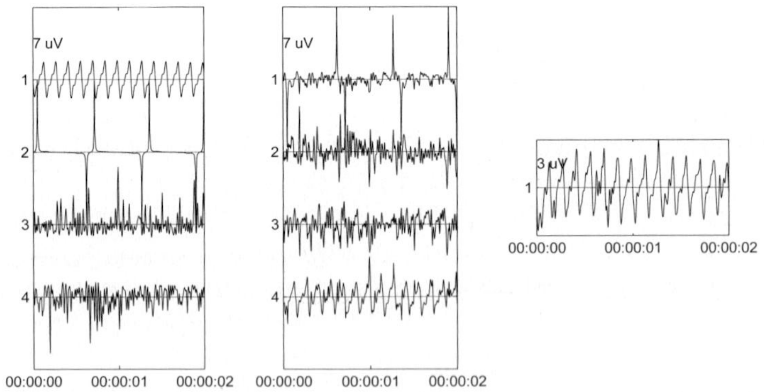

Fig. 1. Extraction of an AF like signal from an artificially generated mixture. left: the original sources; center: the mixture; right: the extracted source.

Fig. (3) gives the PSD for both source estimates in Figs. (1) & (2).

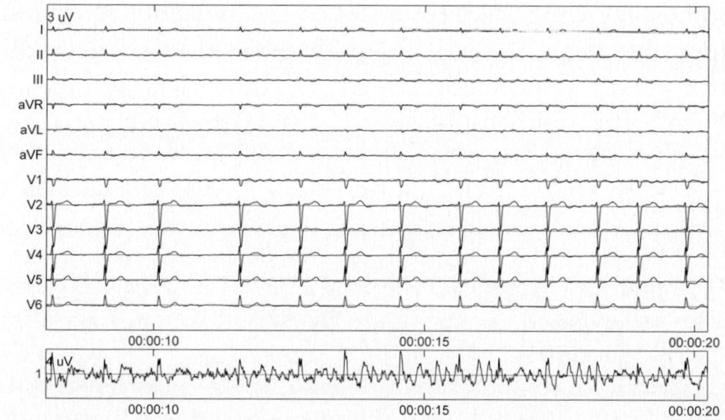

Fig. 2. Extraction of an AF signal from the ECG. upper: 12 channel ECG signal; lower: extracted AF signal.

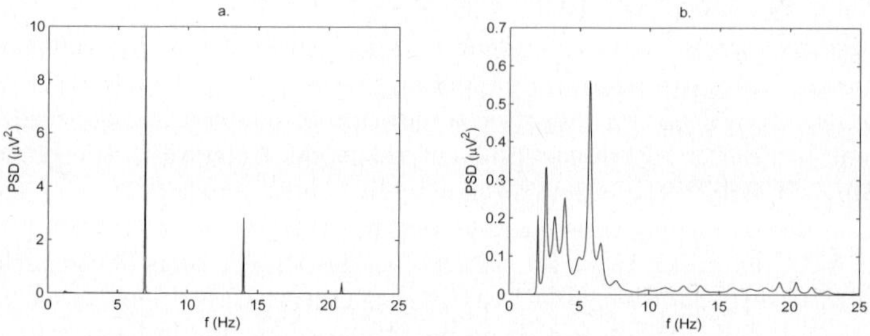

Fig. 3. PSD for the extracted sources from (a) the artificially generated mixture in Fig. (1) and (b) the ECG signal in Fig. (2)

4 Discussion

From Table 2 it is clear that From the figures we can see that the restriction of the spectral concentration does not prevent to extract signals with multiple harmonics, a result that is supported by the fact that the main harmonic is still the central frequency. Moreover, the sources are maximally independent, being a solution to the contrast in [10] subjected to the constraint of spectral concentration in the 3-12Hz band. Moreover, since our technique is based on source extraction, there is no need to do a full decomposition with *a posteriori* source selection, which is computationally attractive, since the overall complexity is of order $\mathcal{O}$ (7n-7)T.

5 Conclusion

The results of the method based on constrained extended AEML are promising toward the extraction of AF from the ECG. Although the presented values and figures are already showing the strengths of the method, it remains to explore how to obtain a quantitative and objective measure for the evaluation of the proposed source extraction technique against the widely accepted techniques of unmasking the AF through suppression of the QRS-T complex.

References

1. Castells, F., Igual, J., Millet, J., Rieta, J.: Atrial activity extraction from atrial fibrillation episodes based on maximum likelihood source separation. Signal Processing 85, 523–535 (2005)
2. Comon, P.: Analyse en composantes indépendantes et identification aveugle. Traitement du signal 7(3), 435–450 (1990) (Numero special non lineaire et non gaussien)
3. Comon, P.: Independent component analysis, a new concept? Signal Processing 36, 287–314 (1994)
4. Cardoso, J.F.: High-order contrasts for independent component analysis. Neural Computation 11, 157–192 (1999)
5. Lee, T.W., Girolami, M., Bell, A.J., Sejnowski, T.J.: A unifying information-theoretic framework for independent component analysis. International Journal on Mathematical and Computer Modeling (1998)
6. Delfosse, N., Loubaton, P.: Adaptive separation of independent sources: a deflation approach. In: IEEE International Conference on Acoustics, Speech, and Signal Processing (ICASSP'94), Adelaide, Australia, 19-22 April 1994, vol. 4, pp. 41–44. IEEE Computer Society Press, Los Alamitos (1994)
7. Zarzoso, V., Comon, P.: How fast is FastICA? In: Proceedings of the 14th European Signal Processing Conference (EUSIPCO), Firenze, Italy (September 2006)
8. Hyvärinen, A., Oja, E.: A fast fixed-point algorithm for independent component analysis. Neur. Comp. 9, 1483–1492 (1997)
9. Belouchrani, A., Abed-meraim, K., Cardoso, J., Moulines, E.: A blind source separation technique using second order statistics. IEEE Trans on Sign. Proc. 45(2), 434–444 (1997)
10. Phlypo, R., Zarzoso, V., Comon, P., D'Asseler, Y., Lemahieu, I.: ISRN I3S/RR-2007-13-FR: A contrast for ICA based on the knowledge of source kurtosis signs. Technical report, I3S, Sophia Antipolis, France (2007), `http://www.i3s.unice.fr/~mh/RR/2007/liste-2007.html`
11. Zarzoso, V., Nandi, A.K., Hermann, F., Millet-Roig, J.: Combined estimation scheme for blind source separation with arbitrary source PDFs. IEE Electronics Letters 37(2), 132–133 (2001)
12. Stridh, M., Sörnmo, L.: Spatiotemporal qrst cancellation techniques for analysis of atrial fibrillation. IEEE Trans. Biomed. Eng. 48(1), 105–111 (2001)
13. Castells, F., Rieta, J., Millet, J., Zarzoso, V.: Spatiotemporal blind source separation approach to atrial activity estimation in atrial tachyarrhythmias. Biomedical Engineering, IEEE Transactions on 52(2), 258–267 (2005)

Blind Matrix Decomposition Techniques to Identify Marker Genes from Microarrays

R. Schachtner[1], D. Lutter[1], F.J. Theis[1], E.W. Lang[1], A.M. Tomé[2], J.M. Gorriz Saez[3], and C.G. Puntonet[3]

[1] Institute of Biophysics, University of Regensburg, 93040 Regensburg, Germany
elmar.lang@biologie.uni-regensburg.de
[2] DETI / IEETA, Universidade de Aveiro, 3810-Aveiro, Portugal
[3] DATC / ETSI, Universidad de Granada, 18371 Granada, Spain

Abstract. Exploratory matrix factorization methods like PCA, ICA and sparseNMF are applied to identify marker genes and classify gene expression data sets into different categories for diagnostic purposes or group genes into functional categories for further investigation of related regulatory pathways. Gene expression levels of either human breast cancer (HBC) cell lines [6] or the famous leucemia data set [10] are considered.

1 Introduction

The *transcriptom* comprises all cellular units and molecules needed to read out the genetic information encoded in the DNA. Among others, the level of messenger RNA (mRNA), specific to each gene, depends on environmental stimuli or the internal state of the cell and represents the gene expression profile (GEP) of the cell. High-throughput genome-wide measurements of gene transcript levels have become available with the recent development of microarray technology [1]. Microarray data sets are characterized by many variables (the GEPs) on only few observations (environmental conditions). Traditionally two strategies exist to analyze such data sets: a) *Supervised approaches* can identify gene expression patterns, called features, specific to each class but also classify new samples. b) *Unsupervised approaches* like PCA [3], ICA or NMF [2] represent exploratory matrix decomposition techniques for microarray analysis. Both approaches can be joined to build classifiers which allow to classify GEPs into different classes. We apply PCA, ICA and NMF to two well-characterized microarray data sets to identify marker genes and classify the data sets according to the diagnostic classes they represent.

2 The Data Sets

2.1 Breast Cancer Cell Lines - Bone Metastasis

The data set was taken from the supplemental data to [6]. The study investigated the ability of human breast cancer (BC) cells (MDA-MB-231 cell line) to form

M.E. Davies et al. (Eds.): ICA 2007, LNCS 4666, pp. 649–656, 2007.

bone metastasis. Data set 1 comprised 14 samples; experiments 1-8 showed weak (7 and 8 mild) metastasis ability, while experiments 9-14 were highly active. Data set 2 consists of 11 experiments, 5 among them of high and 6 showing weak metastasis ability. Both data sets carry measured expression levels of 22283 genes using the Affymetrix U133a chip. For each measurement, the flags A(absent) or P(present) are provided. All genes showing more than 40% absent calls in one of the two data sets were removed. The remaining data sets contained the same 10529 genes. The authors published a list of 16 potential marker genes, 14 of which were still contained in the reduced data set.

2.2 Leukemia

Leukemia (LK) is a form of cancer in which white blood cells (leukocytes) show abnormal behavior. Pathologically, leukemia is split into acute and chronic forms. Acute leukemia types can be divided into *Acute Myelogenous Leukemia* (AML) and *Acute Lymphoblastic Leukemia* (ALL). Furthermore lymphocytes can be classified by their cell surface into B-cells and T-cells. ALL leukemia cells can thus be divided into the subtypes ALL-B- and ALL-T-leukemia. The data set was taken from the well-known supplemental data to [10] which comprises a training set of 38 experiments and a test set of 34 samples. The training set contains 27 ALL samples from childhood ALL patients, and 11 adult AML samples. The test set in addition contains peripheral blood samples and cases of childhood AML. The original training and test data sets contain measurements of 7129 genes. All data were hybridized on an affy-HU6800 Affymetrix-chip. The training set contains negative expression levels. In a later study [5], a non-negative version of the training data set was provided. It contains expression profiles of 5000 genes. In this study, this latter data set was used.

3 Data Analysis

The gene expression profiles are commonly represented by an $(N \times M)$ data matrix $\mathbf{X} = [\mathbf{x}_1 \cdots \mathbf{x}_M]$ with each column $\mathbf{x}_m$ representing the expression levels of all genes in one of the M experiments conducted. Note that the data matrix is non-square with $N \approx 10^3 \cdot M$ typically. This renders a transposition of the data matrix necessary when techniques like PCA and ICA are applied. Hence ICA follows the data model $\mathbf{X}^T = \mathbf{AS}$. Then each row represents the expression profile of all genes within one experiment. The rows of $\mathbf{S}$ contain the nearly independent component expression profiles, called *expression modes*, and the columns of $\mathbf{A}$ the corresponding basis vectors containing the mixing coefficients. In this study the JADE-algorithm [4] was used throughout, though with the natural gradient and the fastICA algorithm equivalent results were obtained. With NMF, a decomposition is sought according to $\mathbf{X} = \mathbf{WH}$ which is not unique, of course, and needs further specification. The columns of $\mathbf{W}$ are usually called *metagenes* and the rows of $\mathbf{H}$ are called *meta-experiments*. The local NMF (LNMF) algorithm [8] was applied in this study.

3.1 ICA - Analysis

We propose a new method based on basic properties of the matrix decomposition model as well as on available diagnostic information to build a diagnostic classifier. ICA essentially seeks a decomposition $\mathbf{X}^T = \mathbf{AS}$ of the data matrix. Column $\mathbf{a}_m$ of $\mathbf{A}$ can be associated with *expression mode* $\mathbf{s}_m$, representing the m-th row of $\mathbf{S}$. The m-th row of the matrix $\mathbf{A}$ contains the weights with which the $k \leq M$ expression levels of each of the N genes, forming the columns of $\mathbf{S}$, contribute to the m-th observed expression profile. Hence a concise analysis of matrix $\mathbf{A}$ hopefully provides insight into the structure of the data set.

Each microarray data set investigated here represents at least two different diagnostic classes. If the M expression profiles of $\mathbf{X}^T$ are ordered according to their class labels, this assignment is also valid for the rows of $\mathbf{A}$. Suppose one of the independent *expression modes* $\mathbf{s}_m$ is characteristic of a putative cellular gene regulation process, which is related to the difference between the classes. Then in all experiments, this characteristic profile should only contribute substantially to experiments of one class and less so to the experiments of the other class (or vice versa). Since the m-th column of $\mathbf{A}$ contains the weights with which $\mathbf{s}_m$ contributes to all observations, this column should show large/small entries according to their class labels. In contrast to the method used by [7], the clinical diagnosis of the experiments is taken into account. The strategy concentrates on the identification of a column of $\mathbf{A}$, which shows a class specific signature. Informative columns were identified using the correlation of each column vector of $\mathbf{A}$ with a design vector $\mathbf{d}$ whose i-th entry is $d_i = \pm 1$, according to the class label of experiment $\mathbf{x}_i$.

3.2 Local NMF - Analysis

With NMF, each column of $\mathbf{X}$ comprises the expression profile resulting from one experiment. After applying the LNMF- algorithm [8], at least one column of $\mathbf{W}$, called a *metagene* is expected to be characteristic of a regulatory process, which is related to the class specific signature of the experiments. Its contribution to the observed expression profiles is contained in a corresponding row of matrix $\mathbf{H}$, called a *meta-experiment*. The correlation coefficients $c(\mathbf{h}^j, \mathbf{d})$ between every *meta-experiment* $\mathbf{h}^j$ and $\mathbf{d}$ are then computed. Empirically, $c > 0.9$ signifies a satisfactory similarity between a *meta-experiment* and the design vector. The number of extracted basis components k, i.e. the *metagenes*, controls the structure of $\mathbf{W}$ and $\mathbf{H}$. For several decompositions $\mathbf{X} = \mathbf{WH}$ using different numbers k of *metagenes*, the rows of $\mathbf{H}$ are studied with respect to their correlation with the design vector. A *metagene* is considered **informative** only if **all** entries of the corresponding *meta-experiment* which belong to class 1 are smaller than **all** other entries of that *meta-experiment* (or vice versa). After 5000 iterations, the cost function of the LNMF algorithm did not show noticeable changes with any of the data sets investigated. For $k = 2, \ldots, 49$, ten separate simulations were carried out and only the simulation showing the smallest reconstruction error was stored. Further matrix decompositions with $k = 50, \ldots, 400$ *metagenes* were examined. In the latter case, three simulations were performed only for each k.

4 Results

4.1 Breast Cancer Data Set

In order to test the idea of a diagnostic classifier, we first selected the set of expression profiles from bone metastasis mediating breast cancer cell lines provided by [6].

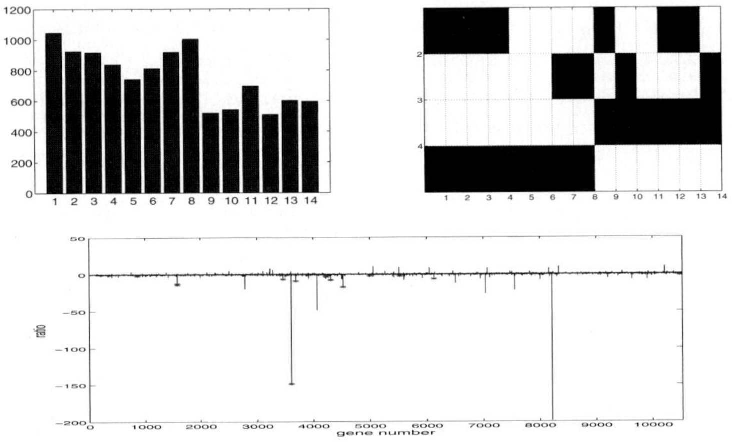

Fig. 1. *Top Left*: Entries of the 9-th column of mixing matrix **A** as obtained with JADE, *Top Right*: Matrix **H** resulting from a decomposition into $k = 4$ *metagenes* using LNMF. Row 3 and 4 show a clear separation between columns $1, \ldots, 8$ and columns $9, \ldots, 14$ *Bottom*:Componentwise ratio of row $\mathbf{x}_9$ with $a_{n9}\mathbf{s}_9$. The genes of [6] are marked.

ICA Analysis. The analysis of the 14×14 matrix **A** identified one column with $|c((a)_9, \mathbf{d})| = 0.89$ (see Fig. 1). Hence $\mathbf{s}_9$ should contain genes which provide diagnostic markers for the metastasis forming ability of the cell lines considered. In [6], a list of 16 putatively informative genes is provided. As shown in Tab. 1 the expression levels (taken from **S**) across all M experiments of many of these genes exhibit a high correlation with the design vector **d** indicating a rather high single discriminative power. Many of these genes show large negative expression levels in *expression mode* 9. An even more revealing picture appears if one divides componentwise the rows of the data matrix by the weighted row of the informative *expression mode*. The resulting diagram marks genes which contribute most to the observation. Many of the genes listed by [6] stick out as informative here.

NMF Analysis. The same data set was also analyzed using the LNMF algorithm. The decomposition is very robust and highly accurate. Considering the correlation between any row of matrix **H** and the design vector **d**, a decomposition into $k = 4$ *metagenes* yielded two rows of the 4×14 matrix **H** which

Table 1. The correlation coefficient c of the gene vector $\mathbf{s}_n$ with the design vector $\mathbf{d}$ for the 16 genes suggested by [6]. *number* denotes the column index in the data set $\mathbf{X}$, *gene name* denotes the affymetrix-ids, — genes missing in the reduced data set.

number	affymetrix-id	gene name	c-value	number	affymetrix-id	gene name	c-value
3611	204749-at	NAP1L3	-0.96	3694	204948-s-at	FST	-0.89
1586	201859-at	PRG1	-0.95	10480	222162-s-at	ADAMTS1	-0.86
5007	209101-at	CTGF	-0.94	6133	211919-s-at	CXCR4	-0.81
4311	207345-at	FST	-0.93	4233	206926-s-at	IL11	-0.57
1585	201858-s-at	PRG1	-0.92	3469	204475-at	MMP1	-0.47
4529	208378-x-at	FGF5	-0.92	4232	206924-at	IL11	-0.43
5532	209949-at	NCF2	-0.92	—	210310-s-at	FGF5	—
860	201041-s-at	DUSP1	-0.89	—	209201-x-at	CXCR4	—

show an excellent correlation to the design vector $\mathbf{d}$, see Fig. 1, with coefficients $c(\text{row } 3, \mathbf{d}) = -0.91$ and $c(\text{row } 4, \mathbf{d}) = 0.91$, respectively. Thus, a decomposition in a comparatively small set of *metagenes* perfectly displays the diagnostic structure of the breast cancer data set. For the sake of comparison, a decomposition into $k = 20$ metagenes revealed four informative *meta-experiments* and their related *metagenes*. A comparison of the ten most expressed genes in each of the four identified metagenes shows, that 5 genes were also identified in case of $k = 4$, while 7 genes were also identified with ICA and 9 genes were identified with a SVM [9] approach as well. These genes are spread over all four *metagenes*.

4.2 Leukemia Data Set

Two different types of diagnostic classes are present in this data set: a) The leukemia-types ALL (experiments $1, \ldots, 27$) and AML (experiments $28, \ldots, 38$) of the training set. A putative reference list of informative genes is available here from [10], b) ALL-leukemia can further be split into subtypes ALL-B (exp. $1, \ldots, 19$) and ALL-T (exp. $20, \ldots, 27$).

ICA Analysis. During the whitening step of the JADE-algorithm, a reduction, i.e. ($k < M$), in the number of extracted *expression modes* can be performed. For $k = 2, \ldots, 38$, the maximal correlation coefficient $\|c\|$ between any column vector of $\mathbf{A}_k$ and the design vectors $\mathbf{d}^i$, $i = 1, 2$ for cases (1) and (2) was computed. These correlation coefficients peak at $c = 0.86$ and $k = 17$ in case 1 (AML vs. ALL), and at $c = 0.97$ and $k = 12$ in case 2 (ALL-T vs. ALL-B/AML). The first 17 principal components cover 93.5% of the total variance of the data set, while the first 12 principal components still represent 89.8% of the variance. In case 1 the 9-th column of matrix $\mathbf{A}$ shows a very pronounced discrimination between AML and ALL cells, and in case 2 the 8-th column of matrix $\mathbf{A}$ shows a very pronounced discrimination between the ALL-T cells and all others. In [10], a list of 24 significantly up-regulated and 24 significantly down-regulated genes is provided, considering the discrimination of the ALL-AML classes. Most of the down-regulated genes appear in the clusters derived from *expression mode* 9*neg*, while the up-regulated genes are contained in 9*pos* mostly. Six of them even belong to the 10 most strongly expressed genes in *expression modes* 9*neg* and 9*pos*. Concerning a discrimination of the subtypes

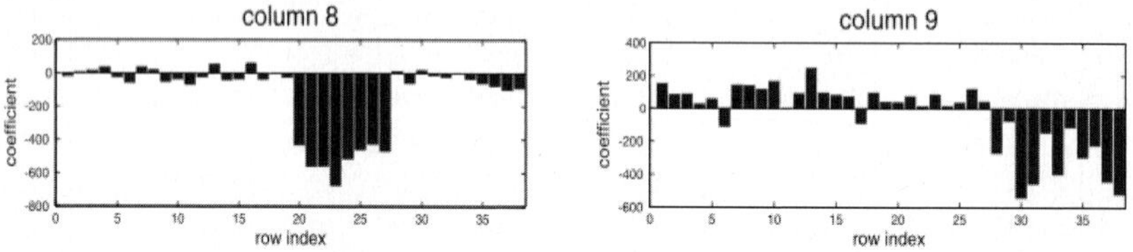

Fig. 2. *Left*: Mixing coefficients $a_{i8}, i = 1, \ldots, 38$ resulting in a signature of column $\mathbf{a}_8$ of matrix $\mathbf{A}$ with a high correlation to the design vector $\mathbf{d}_2$. *Right*: Corresponding signature of column $\mathbf{a}_9$ for the discrimination of AML vs ALL.

ALL-B/ALL-T, a decomposition in $k = 12$ *expression modes* revealed a strong correlation of column 8 of matrix $\mathbf{A}_{12}$ with the design vector $\mathbf{d}_2$, see Fig. 2. The most expressed genes in mode 8*pos* and 8*neg*, respectively, thus form diagnostic marker genes to differentiate between subtypes of the ALL-type Leukemia.

LNMF Analysis. In case AML vs ALL, class 1 was chosen to represent AML-leukemia, while in case ALL-T vs ALL-B/AML, class 1 was considered to be ALL-T, and class 2 consequently comprised all ALL-B and AML probes.

Case AML vs ALL: First, the structure of *meta-experiments* related to the case AML vs. ALL was investigated. Correlations between the *meta-experiments* and the design vector $\mathbf{d}$ were determined for all $k \leq 400$. Strong correlations ($c > 0.85$) could be obtained for several decompositions. The strongest correlation was found for $k = 100$. Fig. 3 shows the most informative *meta-experiment* and its related *metagene*. From the latter the 10 most strongly expressed genes are listed in Table 2. Six of them could be identified also with a SVM-classifier [9] or by applying ICA, or belonged to the list published in [10].

Case ALL-T vs ALL-B/AML: In case $ALL - B < ALL - T$, at least one *meta-experiment* was found with $c > 0.9$ for any $k > 20$. In the reverse situation $ALL - T < ALL - B$ highly correlated ($|c| > 0.85$) *meta-experiments* were found for all $k > 70$. Fig. 3 shows the result of a factorization into $k = 300$ *metagenes*. Only one informative *metagene* could be identified. The two most strongly expressed genes of this *metagene* also have been identified as potential marker genes using ICA.

In contrast to the BC data set, only few informative *meta-experiments* could be identified with the LK data set. However, the related *metagenes* obtained for different decompositions were very consistent as measured by their respective correlation coefficient. Obviously, the LNMF decomposition detects rather similar *metagenes* related to the diversity of ALL-T and -B-subtypes when the number k of extracted *metagenes* is varied.

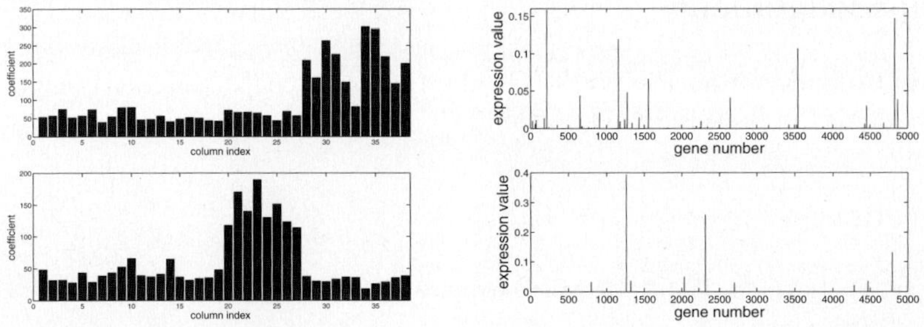

Fig. 3. *Top*: case AML vs ALL, $k = 100$: Signature of the most informative *meta-experiment* and the related *metagene*, *Bottom*: case ALL-T vs ALL-B, $k = 300$: Signature of the most informative *meta-experiment* and corresponding *metagene*

Table 2. List of the 10 most strongly expressed genes of *metagene* 100. (* mentioned in the original paper [10], ** identified with an SVM-classifier [9], *** identified by ICA).

	gene nr.	affy-id		gene nr.	affy-id
1	* 215	M96326-rna1-at	6	1274	D88422-at
2	***,*4828	M27891-at	7	651	M27783-s-at
3	1157	J04990-at	8	**,*2646	X95735-at
4	**3543	M19507-at	9	*4870	M57710-at
5	**1555	M84526-at	10	4454	M20203-s-at

5 Conclusion

The application of matrix decomposition techniques like ICA and NMF to microarray data explores the possibility to extract features like statistically independent *expression modes* or strictly positive and sparsely encoded *metagenes*. Combined with a design function reflecting the experimental protocol, biomedical knowledge is incorporated into the data analysis task which allows to construct a classifier for diagnostic purposes based on a global analysis of the whole data set rather than a statistical analysis based on single gene expression levels. This global analysis is based on the columns (ICA) or rows (NMF) of a matrix which contains the weights with which the underlying *expression modes* or *metagenes* contribute to any given observation in response to an applied environmental stimulus. It was shown on two benchmark data sets that if the signature of these column or row vectors matches the experimental design vector, the related *expression mode* or *metagene* contains genes with a high discriminative power. These genes represent biomarkers for diagnostic purposes. Knowledge of such marker genes allows to construct a simple and cheap chip for diagnostic purposes.

Acknowledgement

Support by grants from the DFG (Graduate College 638) and the DAAD (PPP Luso - Alemã) is gratefully acknowledged.

References

1. Baldi, P., Hatfield, W.: DNA Microarrays and Gene Epression. Cambridge University Press, Cambridge (2002)
2. Cichocki, A., Amari, S.-I.: Adaptive Blind Signal and Image Processing. Wiley, Chichester (2002)
3. Diamantaras, K.I., Kung, S.Y.: Principal Component Neural Networks, Theory and Applications. Wiley, Chichester (1996)
4. Souloumiac, A., Cardoso, J.-F.: Blind beamforming for non-gaussian signals. IEEE Proc. 140, 362–370 (1993)
5. Golub, T., Mesirov, J.P., Brunet, J.-P., Tamayo, P.: Metagenes and molecular pattern discovery using matrix factorization. PNAS 101, 4164–4169 (2004)
6. Kang, Y., Siegel, P.M., Shu, A., Drobnjak, M., Kakonen, S.M., Cordón, C., Guise, Th.A., Massagué, J.: A multigenic program mdeiating breast cancer metastasis to bone. Cancer Cell 3, 537–549 (2003)
7. Lee, S.-I., Batzoglou, S.: Application of independent component analysis to microarrays. Genome Biology, 4, R76.1–R76.21 (2003)
8. Zhang, H.J., Cheng, Q., Li, S.Z., Hou, X.W.: Learning spatially localized, parts-based representation. In: IEEE (2001)
9. Schachtner, R.: Machine Learning Approaches to the Analysis of Microarray Data. Diploma Thesis, University of Regensburg (2006)
10. Tamayo, P., Huard, C., Gaasenbeek, M., Mesirov, J.P., Coller, H., Loh, M.L., Downing, J.R., Caligiuri, M.A., Bloomeld, C.D., Lander, E.S., Golub, T.R., Slonim, D.K.: Molecular classification of cancer: Class discovery and class prediction by gene expression monitoring. Science, 286 (1999)

Perception of Transformation-Invariance in the Visual Pathway

Wenlu Yang[1,2], Liqing Zhang[1,⋆], and Libo Ma[1]

[1] Department of Computer Science and Engineering,
Shanghai Jiao Tong University, Shanghai 200240, China
wenluyang@online.sh.cn, lqzhang@sjtu.edu.cn
[2] Department of Electronic Engineering,
Shanghai Maritime University, Shanghai 200135, China

Abstract. Visual perception of transformation invariance, such as translation, rotation and scaling, is one of the important functions of processing visual information in the Brain. To simulate this perception property, we propose a computational model for perception of transformation. First, we briefly introduce the transformation-invariant basis functions learned from natural scenes using Independent Component Analysis (ICA). Then we use these basis functions to construct the perceptual model. By using the correlation coefficients of two neural responses as the measure of transformation-invariance, the model is able to perform the task of perception of transformation. Comparisons with Bilinear Sparse Coding presented by Grimes and Rao and Topo-ICA by Hayvarinen show that the proposed perceptual model has some advantages such as simple to implement and more robust to transformation invariance. Computer simulation results demonstrate that the model successfully simulates the mechanism for visual perception of transformation invariance.

1 Introduction

We can recognize an object regardless of its distance, position or rotation. In the mathematical term, object recognition is not influenced by its transformation, such as translation, rotation or scaling. Many recent researches in the fields of neuroscience, neurophysiology and psychology show that such a transformation-invariant preprocessing could be a necessary step to achieve transformation invariant classification or detection in a hierarchical computational system. In this paper, we will focus on the computational mechanism for transformation invariance. We will propose a hierarchical model that simulates the mechanism in the visual pathway. On the other hand, due to biological evolution from nature in the long term, this mechanism has an important correlation with statistical properties of natural scenes. Following this way, Barlow[1,2] found that the role of early sensory neurons in the visual pathway is to remove statistical redundancy in the sensory inputs, suggesting that Redundancy Reduction is an important processing principle in the neural system. Based on this principle, Gabor-like features

⋆ To whom correspondence should be addressed.

M.E. Davies et al. (Eds.): ICA 2007, LNCS 4666, pp. 657–664, 2007.

resembling the receptive fields of simple cells in the primary visual cortex(V1) have been derived either by imposing sparse over-complete representations[6] or statistical independence as in Independent Component Analysis(ICA)[8].

However, these studies have not taken transformation invariance into account, and the question is how well this line of research predicts the full spatiotemporal receptive fields of simple cells. For example, when an image rotates within receptive fields of simple cells, how do the simple cells and complex cells response? Some researchers have begun to bring this question into consideration. Hyvarinen and Hoyer[9,10] modelled receptive fields of complex cells and Van Hateren [12]obtained spatiotemporal receptive fields of complex cells. Grimes and Rao[14] proposed a bilinear generative model to study the translation-invariance. Berkes[7] investigated temporal slowness as a learning principle for receptive fields using slow feature analysis. However, there are few models in the literatures perceiving transformations of objects or images. To investigate the problem, we apply ICA to learning from natural scenes the transformation-invariant features, and then use these features to construct a model for transformation-invariant perception. The goal of the model is to perceive transformation of patches from natural images.

The rest of the paper is organized as follows. Section 2 introduces a method for learning transformation-invariant basis functions and then propose a model for perception of transformation invariance. In section 3, we will demonstrate these basis functions and perceptual simulation results. The final section gives the comparison with other related works and models.

2 The Invariance Perception Model

In this section, we first introduce the method for learning the transformation-invariant basis functions. Then we propose a perceptual model for perception of transformation invariance.

2.1 Method for Learning Invariant Basis Functions

To obtain transformation-invariant basis functions, the training data sets should have the possession of transformational properties. The method for generating the training data will be introduced in section 3.1. Applying ICA on the training data, sequences of patches with the parameter $\alpha_i (i = 1, ..., M)$, yields transformation-invariant basis functions with parameters same as patches. For simple explanation of the method, shown in Fig.1, we use the rotation transformation and the resulting rotation basis functions.

We briefly introduce the learning algorithm of ICA for training sparse basis functions. For the standard ICA model $x = \mathbf{W}u$, Cichocki et al.[5] used the Kullback-Leibler divergence between the distribution $p(\mathbf{x}; \mathbf{W})$ of obtained by the actual value $\mathbf{W}$ and the reference distribution $q(\mathbf{x})$ to give the cost function as

$$R(\mathbf{x},\mathbf{W}) = -\frac{1}{2} \log |det(\mathbf{W}\mathbf{W}^T)| - \sum_{i=1}^{n} E \log q_i(x_i). \tag{1}$$

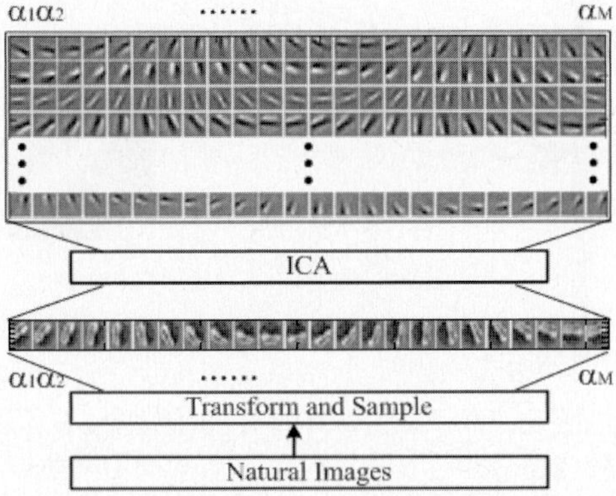

Fig. 1. Method for learning transformation-invariant basis functions. The input data is a set of natural images. Patches selected from the images are transformed and feed to the ICA algorithm.

Applying the Natural Gradient rule to the cost function, the learning algorithm of $\mathbf{W}$(the corresponding basis functions $\mathbf{A} = \mathbf{W}^{-1}$) can be described[3,4] as

$$\triangle \mathbf{W} = -\eta(t)\frac{\partial R}{\partial \mathbf{W}}\mathbf{W}^T\mathbf{W} = \eta(t)[\mathbf{I} - \langle \varphi[\mathbf{x}(k)]\mathbf{x}^T(k)\rangle]\mathbf{W}, \tag{2}$$

where, $\varphi_i(x_i) = -\frac{q_i'(x_i)}{q_i(x_i)}$. $q(x_i)$ is a supergaussian probability distribution, for instance, the Laplace pdf.

2.2 Model for Perception of Transformation Invariance

In this section, we will propose a model for transformation-invariant perception, shown in Fig. 2. The invariance perception model consists of three layers. The first layer is to receive the input patterns which are two patches with parameters of α_i and α_j, respectively. Here, α_i and α_j belong to a same group of parameters. For rotation, α is in the range of zero and three hundred sixty degree by an interval of fifteen. For scaling, α in the range of from one to two times by ten percent. And, for translation, α in the range of size of input images. For simplicity, we only discuss in detail rotational samples and basis functions in the model. The middle layer of the model is to sparsely represent input patterns with a group of basis functions which is one of three groups respectively including translational, rotational, and scaling bases, shown in Figs.{3,4,5}.

After the neurons respond to the stimuli u_{α_i} at time t_1 and u_{α_j} at time t_2 , the final layer of the model calculates the correlation coefficients between any two responses $\mathbf{X}_{\alpha_i}^{t_1}(i = 1, ..., M)$ and $\mathbf{X}_{\alpha_j}^{t_2}(j = 1, ..., M)$, and of which the maximum is selected to determine the relative dispersion. The index (i, j) of the maximum in coefficient matrix will tell us the relative dispersion such as counter-clockwise rotation angle $\Delta\theta$, translational distance Δd, and scaling ratio

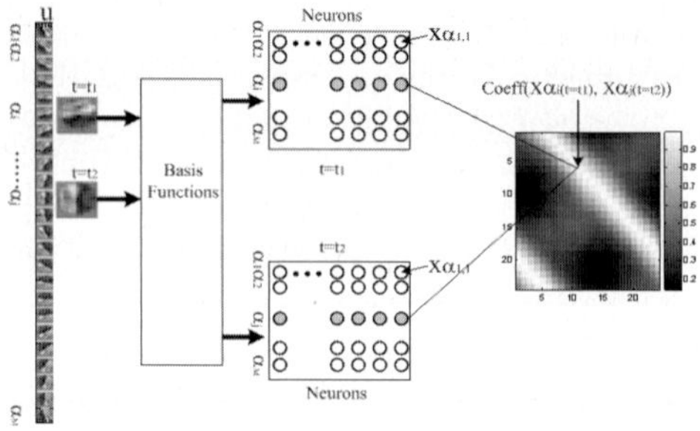

Fig. 2. Model for transformation-invariant perception. For example, the input patterns are the rotational data. $x^{t_1}_{\alpha_{i,k}}(k = 1, 2, \cdots, N)$ denotes the response of the k-th neuron in the row α_i responding to stimulus u_{α_i} at time t_1 through the basis function $\alpha_{i,k}$. And so does response $x^{t_2}_{\alpha_{j,l}}(l = 1, 2, \cdots, N)$ at time t_2. $\mathbf{X}^{t_1}_{\alpha_i}$ denotes the vector of responses that the neurons in the row α_i respond to stimulus u_{α_i} at time t_1 through the subsets α_i of basis functions. Namely, $\mathbf{X}^{t_1}_{\alpha_i} = [x^{t_1}_{\alpha_{i,1}}, x^{t_1}_{\alpha_{i,2}}, \cdots, x^{t_1}_{\alpha_{i,N}}]^T$.

Δr. It is necessary to note that we only need the relative transformation, not the absolute value of parameters of the stimuli. For rotation, if $j \geq i$, $\Delta\theta = (j - i) \times 360/M$; otherwise, $\Delta\theta = (M + j - i) \times 360/M$. For translation, i and j have their corresponding coordinates (x_i, y_i) and (x_j, y_j), respectively. We can calculate the relative translation distance $\Delta d = \sqrt{(x_i - x_j)^2 + (y_i - y_j)^2}$ and the moving direction according the relative position of coordinates (x_i, y_i) and (x_j, y_j). For scaling, if $j \geq i$, $\Delta r = r_j - r_i$; otherwise, $\Delta r = r_i - r_j$. Here, r_j and r_i denote the scaling ratio of the *j-th* and *i-th* subsets of basis functions, respectively.

3 Simulations and Results

We present experimental results to verify the performance of our proposed model and the learning algorithm. First we present the basis functions of transformation invariance including translation, rotation and scaling. Then, as an example, the rotation-invariant perception is discussed.

3.1 Training Data

To learn basis functions from natural scenes, we sample a sequence of small patches of size 10×10 from a set of big natural images by three methods of transformations such as translating, rotating, and scaling. This three data sets are used to learn transformation-invariant basis functions. For example, the sampling method of rotational data set is described in detail as follows.

A sampling window is randomly located on a big natural scene and a patch is selected. Then fix the same center, clockwise rotate the sampling window by an interval of 15 degree, another patch is sampled. Again, rotate the window and sample next one, till twenty-four times. Similarly, the total twenty-four of patches are sampled and then reshaped to one column vector as a sample, size of 2400-by-1.

We select patches from a set of big natural images by the above sampling methods and generate three data sets which are composed of 20000 samples respectively. All data sets are then low-pass filtered by reducing the dimension of the data vector by principle component analysis (PCA), retaining the 100 principal components with the largest variances, after which the data is whitened by normalizing the variances of the principal components. These preprocessing steps are essentially similar to those used in [6,9].

3.2 Transformation-Invariant Basis Functions

Respectively using the translational, scaling, and rotational training data to learn transformation-invariant basis functions, the translation-, scaling-, and rotation-invariant basis functions are yielded, shown in Figs.{3,4,5}. From these figures, we note that the Gabor-like basis functions, which are localized, oriented, and bandpass, resemble receptive fields of simple cells found in V1[13]. Meanwhile, there also are different characteristics, as follows, among three types of basis functions.

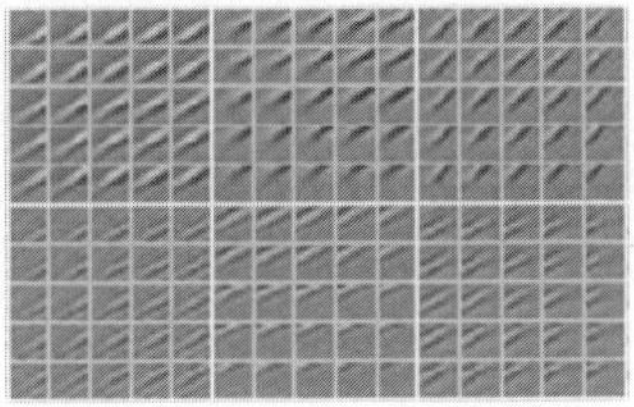

Fig. 3. Subsets of translation-invariant basis functions

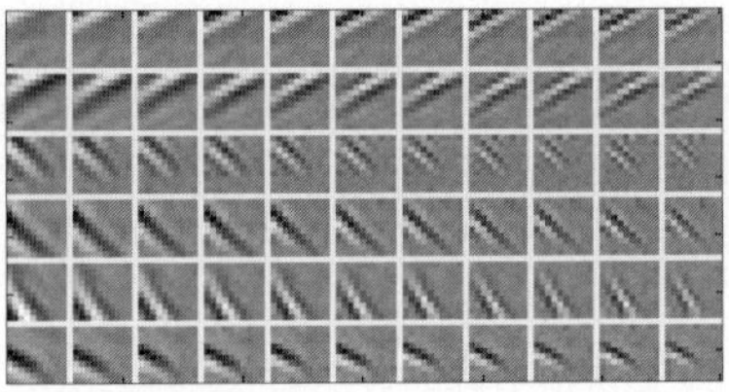

Fig. 4. Subsets of scaling-invariant basis functions

Translation-invariant basis functions. In Fig. 3, bigger rectangles composed of 5×5 basis functions with the same orientation are similar to the receptive fields of complex cells which activate while the same orientational contents are moving within their receptive fields.

Scaling-invariant basis functions. Fig. 4 shows that basis functions in one row represent the receptive field of a complex cell which performs the perception of scaling invariance. Those in one row are subsets with the same scaling. The scaling interval is 10%.

Rotation-invariant basis functions. In Fig. 5(*Left*), a group of basis functions in one row is similar to the receptive field of a complex cell which performs perception of rotation invariance. The neighboring basis functions in a row have an interval of fifteen degree of counter-clockwise rotation. Basis functions in a

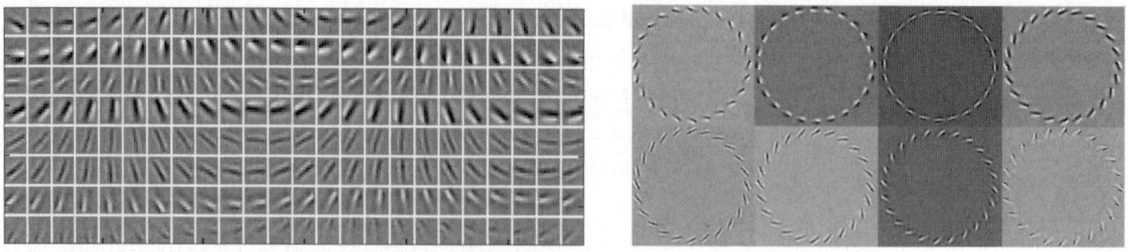

Fig. 5. Subsets of rotation-invariant basis functions. *Right*: basis functions (in the *left*) are rearranged in counter-clockwise along the circumference with an interval of fifteen degree. Every circle includes basis functions in one row (*left*).

column are a group of which elements are used to reconstruct the input patterns while given corresponding activities of simple cells. For the convenience of viewing the regularity, these basis functions are arranged in counter-clockwise along the circumference with an interval of fifteen degree, shown in Fig. 5(*Right*). Every circle resembles the receptive field of a complex cell.

3.3 Perception Experiments

For anyone of three transformations: translation, rotation, and scaling, the same experimental method is used and the invariant results are easy to obtain. Here, the invariance means that complex cells maintain their existing states, while a patch is moving, rotating and scaling within their receptive fields. In other words, we can recognize the same object however it moves, rotates, and scales within our field of vision.

An example of rotation perception is introduced and its goal is to calculate the relative rotation angle. According the perception model in section 2.2, two input patterns with different rotational angles are needed.

Randomly select two image patches u_{α_i} and u_{α_j}(i.e. i=6, j=11) from a sample data which is composed of twenty-four patches, shown in Fig. 6. For example, the sixth and eleventh of stimuli represents, respectively, rotational angles of ninety and one hundred and sixty-five in degree. The sixth patch is first input to the perception model at time $t_1$and the eleventh is second at time t_2.

Computing the responses of neurons at time t_1 and t_2 and the matrix of the correlation coefficients $\text{Coeff}(X_{\alpha_k}^{t_1}, X_{\alpha_l}^{t_2})(k, l{=}1, 2, \cdots, M)$, here M=24. Find the max value from any row in the matrix and obtain its corresponding index of its row and column, i.e. at the first row, the index of max value is (1,6). So, we know the relative rotation angle is (6-1)×15=75 degree. It is necessary to note that we only need the relative transformation, not the absolute value of angles of the stimuli.

In more detail, at time t_1 and t_2, the responses $X_{\alpha_6}^{t_1}$ and $X_{\alpha_{11}}^{t_2}$ are plotted in the last column of Fig.6. It is easy to know $X_{\alpha_6}^{t_1}$ and $X_{\alpha_{11}}^{t_2}$ are very similar to each other. The difference between $X_{\alpha_6}^{t_1}$ and $X_{\alpha_{11}}^{t_2}$, plotted in Fig. 6, shows

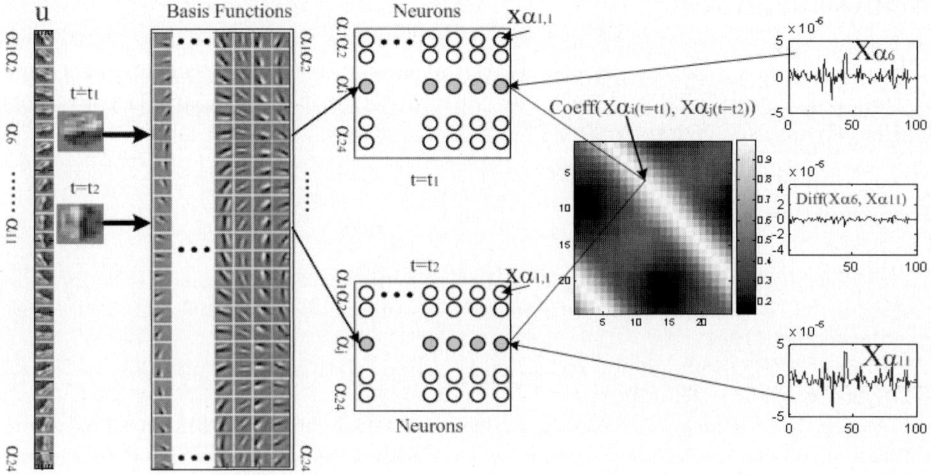

Fig. 6. Rotation-invariant perception

the rotation invariance of neuronal responses while the input pattern is rotating from time t_1 to t_2.

4 Discussions and Conclusions

We have proposed a method for learning transformation-invariant basis functions and a model for perception of transformation invariance. Computer simulation results show that our proposed model do work well in simulating the perceptual function of transformation invariance in the brain. Our proposed model has some different properties compared with others such as bilinear generative models [14] and Topo-ICA[11].

First, bilinear generative models[14] proposed by Grimes and Rao only study the translation invariance and however, ours is able to provide more transformation invariant basis functions such as translational, rotational and scaling basis functions. Our model also performs perception of the three types of transformations. On the other hand, our algorithm is much simpler whereas that of the bilinear model is more complex. Second, the Topo-ICA model[11] provided by Hyvarinen et al. considered the second-order correlation of responses of simple cells, but the Topo-ICA model cannot produce overcomplete basis functions because of constrains of orthogonality.

Our future work will focus on learning other transformational basis functions such as three dimensional geometry transformations and on transformational perception of more complex stimuli. We are also going to extend the model to a framework for learning other transformation-invariant basis functions and perception of other transformation such as view changes.

Acknowledgment

The work was supported by the National Basic Research Program of China (Grant No. 2005CB724301) and the National High-Tech Research Program of China (Grant No. 252006AA01Z125).

References

1. Barlow, H.B.: Possible principles underlying the transformations of sensory messages. In: Rosenblith, W.A. (ed.) Sensory Communication, pp. 217–234. MIT Press, Cambridge (1961)
2. Barlow, H.B.: Redundancy reduction revisited. Network: Computation in Neural Systems 12, 241–253 (2001)
3. Zhang, L., Cichocki, A., Amari, S.: Natural Gradient Algorithm to Blind Separation of Over-determined Mixture with Additive Noises. IEEE Signal Processing Letters 6(11), 293–295 (1999)
4. Zhang, L., Cichocki, A., Amari, S.: Self-Adaptive Blind Source Separation Based on Activation function Adaptation. IEEE Transactions on Neural Networks 15(2), 233–244 (2004)
5. Cichocki, A., Zhang, L.: Two-stage bink deconvolution using state-space models. In: proceedings of the Fifth International Conference on Neural Information Processing, Kitakyushu, Japan, pp. 729–732 (1998)
6. Olshausen, B.A., Field, D.J: Sparse coding with an overcomplete basis set: A strategy employed by V1? Vision Research 37, 3311–3325 (1997)
7. Berkes, P., Wiskott, L.: Slow feature analysis yields a rich repertoire of complex cell properties. Journal of Vision 5, 579–602 (2005)
8. Bell, A.J., Sejnowski, T.J.: The independent component of natural scenes are dege filters. Vision Research 37, 3327–3338 (1997)
9. Hyvarinen, A., Hoyer, P.O.: Emergence of phase and shift invariant features by decomposition of natural images into independent feature subspaces. Neur. Comp. 12(7), 1705–1720 (2000)
10. Hoyer, P.O., Hyvarinen, A.: A multi-layer sparse coding network learns contour coding from natural images. Vision Research 42(12), 1593–1605 (2002)
11. Hyvarinen, A., Hoyer, P.O.: Emergence of Topography and Complex Cell Properties from Natural Images using Extensions of ICA. Advances in Neural Information Processing Systems (NIPS*99) 12, 827–833 (2000)
12. Van Hateren, J.H., Ruderman, D.L.: Independent component analysis of natural image sequences yields spatio-temporal filters similar to simple cells in primary visual cortex. Proc. R. Soc. London B 265, 2315–2320 (1998)
13. Hubel, D.H., Wiesel, T.N.: Receptive fields and functional architecture of monkey striate cortex. Journal of Physiology (London) 195, 215–243 (1968)
14. Grimes David, B., RAO, R.P.N.: Bilinear sparse coding for invariant vision. In: Neural computation, vol. 17(11), pp. 47–73. MIT Press, Cambridge (2005)

Subspaces of Spatially Varying Independent Components in fMRI

Jarkko Ylipaavalniemi[1,2] and Ricardo Vigário[2,3]

[1] jarkko.ylipaavalniemi@tkk.fi
[2] Adaptive Informatics Research Centre,
Laboratory of Computer and Information Science,
Helsinki University of Technology, P.O. Box 5400, FI-02015 TKK, Finland
[3] Advanced Magnetic Imaging Centre,
Helsinki University of Technology, P.O. Box 3000, FI-02015 TKK, Finland

Abstract. In contrast to the traditional hypothesis-driven methods, independent component analysis (ICA) is commonly used in functional magnetic resonance imaging (fMRI) studies to identify, in a blind manner, spatially independent elements of functional brain activity. ICA is particularly useful in studies with multi-modal stimuli or natural environments, where the brain responses are poorly predictable, and their individual elements may not be directly relatable to the given stimuli. This paper extends earlier work on analyzing the consistency of ICA estimates, by focusing on the spatial variability of the components, and presents a novel method for reliably identifying subspaces of functionally related independent components. Furthermore, two approaches are considered for refining the decomposition within the subspaces. Blind refinement is based on clustering all estimates in the subspace to reveal its internal structure. Guided refinement, incorporating the temporal dynamics of the stimulation, finds particular projections that maximally correlate with the stimuli.

1 Introduction

Functional magnetic resonance imaging (fMRI) is one of the most successful methods for studying the living human brain. Traditionally, fMRI analysis relies on artificially generated stimuli, coupled with hypothesis-driven statistical signal processing (*cf.*, [1]).

Independent component analysis (ICA) (see, *e.g.*, [2]) of fMRI data, as first proposed in [3], has recently gained considerable attention for its ability to blindly decompose the measured brain activity into spatially independent functional elements. The corresponding mixing vectors reveal the temporal dynamics of each element. However, the individual elements are often not directly relatable to a given stimulus. This is particularly true in studies using multi-modal stimuli, such as in natural environments, where the brain responses are poorly predictable. Furthermore, it has been proposed that such functional elements can participate in varying networks, to perform complex tasks [4].

M.E. Davies et al. (Eds.): ICA 2007, LNCS 4666, pp. 665–672, 2007.

The optimization landscape of ICA is defined by structure of the data, noise, as well as the objective function used. The landscape can form elongated or branched valleys, containing many strong points, instead of singular local optima. Previous studies [5,6] have analyzed the consistency of independent components, and suggested that some components can have a characteristic variability. The goal was to provide additional insight into the components, that is not possible to attain with single run approaches. Complex valleys can also be considered as separate subspaces, where statistical independence is not necessarily the best objective for decomposition.

In this paper, we present a novel method to reliably identify subspaces formed by independent components, and illustrate two approaches to further refine the decomposition into functionally meaningful components. The subspace detection is based on analyzing the spatial variability under a similar consistent ICA as in the previous studies. The subspaces reveal connections between the individual functional elements. One refinement method uses clustering to distinguish the internal structure of the subspace. Another method is based on finding the coordinate system inside the subspace that maximally correlates with the temporal dynamics of the stimulation. The directions are found with canonical correlation analysis (CCA) [7].

Related canonical correlation approaches have been recently suggested for fMRI (see, *e.g.*, [8,9,10]). However, the goals have been to utilize several stimulation time-courses to simply rank the individual components found by ICA, or to extend the purely hypothesis-driven methods into multivariate analyses.

2 Materials and Methods

The analysis uses data from a recent fMRI study carried out by Malinen et al., at the Advanced Magnetic Imaging Centre [11]. The study combined auditory, visual, and tactile stimuli, in a continuous manner. The stimuli were presented in 6–33 s blocks, with no resting periods in between. Fig. 1 illustrates the block design of the sequence, which has a duration of 8 min 15 s.

2.1 Measured and Preprocessed fMRI Data

The recordings, thoroughly described in [11], were made with a Signa VH/i 3.0 T MRI scanner (General Electric, Milwaukee, WI, USA). Functional images were acquired using gradient echo-planar-imaging sequence (TR 3 s, TE 32 ms, matrix 64×64, 44 oblique axial slices, voxel size $3 \times 3 \times 3$ mm^3, FOV 20 cm, flip angle 90°) producing 165 volumes including 4 dummy scans, which were excluded from further analysis. Structural images were scanned with 3-D T1 spoiled gradient imaging (TR 9 ms, TE 1.9 ms, matrix 256×256, slice thickness 1.4 mm, FOV 26 cm, flip angle 15°, preparation time 300 ms, number of excitations 2).

Preprocessing of the data using SPM2 [12] included realignment, normalization and smoothing with a 6 mm (full-width half maximum) Gaussian filter. Skull stripping was also performed. For further details, see [11].

2.2 Consistent Spatial ICA

Independent component analysis is one of the most popular methods for solving the blind source separation (BSS) problem. It consists of finding solutions to the mixture $\mathbf{X} = \mathbf{AS}$, where only the observed data $\mathbf{X}$ is known. ICA assumes only statistical independence of the sources $\mathbf{S}$, and full rank of the mixing $\mathbf{A}$. In the context of fMRI, independence is considered in the spatial domain, and the mixing reveals the temporal activation patterns of the corresponding sources. The reliable ICA approach, proposed in [5], is based on multiple runs of FastICA [13] in a bootstrapping framework, *i.e.*, with resampled data and randomized initializations.

In this study, FastICA was run 100 times with *tanh* nonlinearity in *symmetric* mode. On each run, the data was whitened to 80 dimensions and 40 independent components were extracted. The estimated mixing vectors from all runs were normalized to have zero mean and unit variance, and grouped using correlation. The correlation matrix was thresholded by 0.85 and raised to a power of 4 (see [5] for further details). The parameter values were selected heuristically. Starting with a few dimensions, the dimensionality was increased until the new components were all overfits, appearing only once. Similarly, starting with a high value, the correlation was lowered as long as the most consistent components, appearing 100 times, did not split into many groups.

2.3 Subspace Canonical Correlation Analysis

The emergence of a subspace in ICA means that the coordinate system within the subspace can not be identified, based solely on statistical independence. Even if there is a strong relation between the subspace as a whole and the stimulation, this relation may not be readily visible as a high correlation between any given component and the stimuli.

Canonical correlation analysis seeks for covariations between two spaces. In the current work, they are the independent subspace and the stimulation design. Such relation is found through maximally correlated linear transformations of both spaces. Let $\mathbf{Y}$ be a set of columns of the mixing matrix $\mathbf{A}$, corresponding to an independent subspace, and $\mathbf{Z}$ the set of stimulation time-courses. The goal of CCA is to maximize $corr(\mathbf{W_y}^T\mathbf{Y}, \mathbf{W_z}^T\mathbf{Z})$ with respect to $\mathbf{W_y}$ and $\mathbf{W_z}$, which are the transformation projections. As a result, the coordinate system within the subspace is fixed according to maximal correlation to the stimuli, rather than independence.

3 Results

Fig. 2 shows a set of independent components (ICs), strongly related to auditory stimulation. Each IC is consistent, appearing in all or most of the 100 runs. The mixing variability is also minimal. However, the spatial variance reveals a coincident location of variability, shared by all ICs. The variability links the ICs

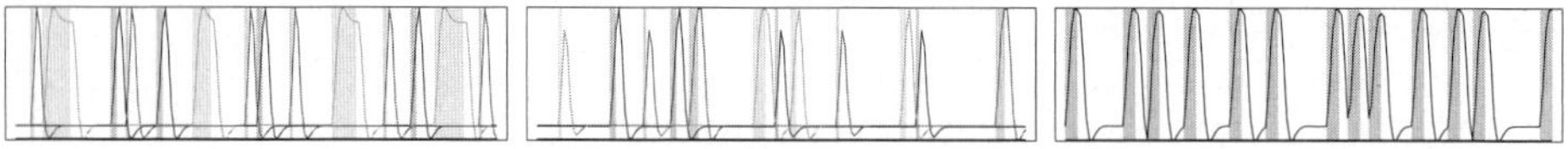

(a) Auditory stimulation (b) Visual stimulation (c) Tactile stimulation

Fig. 1. Stimulation block design with hemodynamically convolved time-courses. (a) Auditory stimulation with tone pips, spoken history text and spoken instruction text (represented by red, green and blue in the color version). (b) Visual stimulation with scenes dominated by buildings, faces and hands (represented by red, green and blue in the color version). (c) Tactile stimulation.

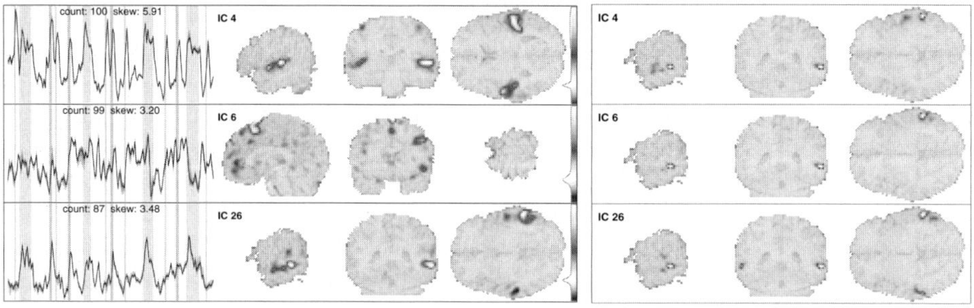

(a) Temporal and spatial mean (b) Spatial variance

Fig. 2. A set of independent components identified as a subspace through the shared variance, with strongly auditory stimulus-related time-courses. (a) The mean spatial maps and time-courses of each component. (b) The spatial variance maps of the corresponding components. A sagittal, coronal and axial slice of each volume is shown with the histogram of the mean volume. Consistency counts and skewness of the histograms are shown as text, and the reference blocks for the time-courses are from Fig. 1(a).

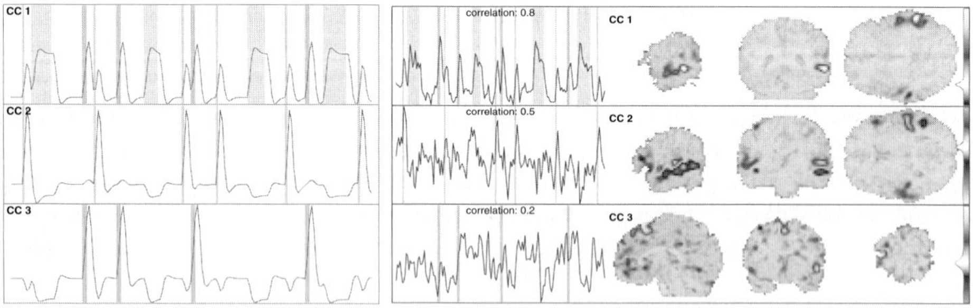

(a) Stimulus recombination (b) Temporal and spatial recombination

Fig. 3. A set of linear combinations that maximize the correlation between the mean time-courses of the subspace components shown in Fig. 2, and the stimulation time-courses shown in Fig. 1(a). (a) The time-courses combined from the stimulation design. (b) The spatial maps and time-courses of the corresponding, maximally correlated, combinations of the independent components. Other details as in Fig. 2.

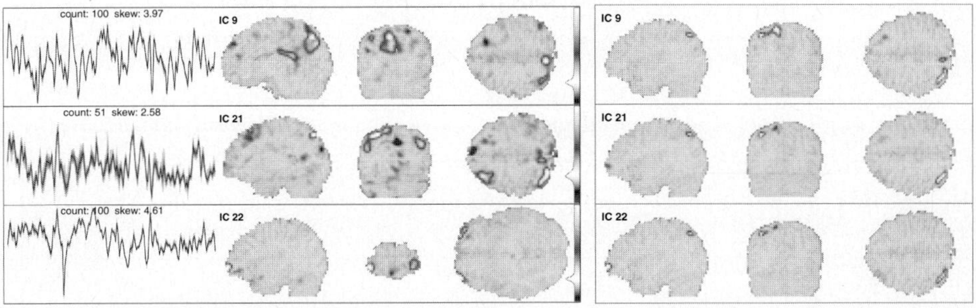

(a) Temporal and spatial mean (b) Spatial variance

Fig. 4. A set of independent components identified as a subspace through the shared variance, with weakly stimulus-related time-courses. Other details as in Fig. 2, except no reference blocks are shown.

into a three dimensional subspace, even though ICA has consistently identified directions within the subspace.

The subspace in Fig. 2 was further analyzed with CCA using all auditory references, shown in Fig. 1(a). Fig. 3 shows the canonical components (CCs) identified within the subspace. Compared to the ICs, the CCs reveal the best stimulation-matching decomposition within the subspace. A thorough physiological interpretation of the results is out of the scope of this paper, but the

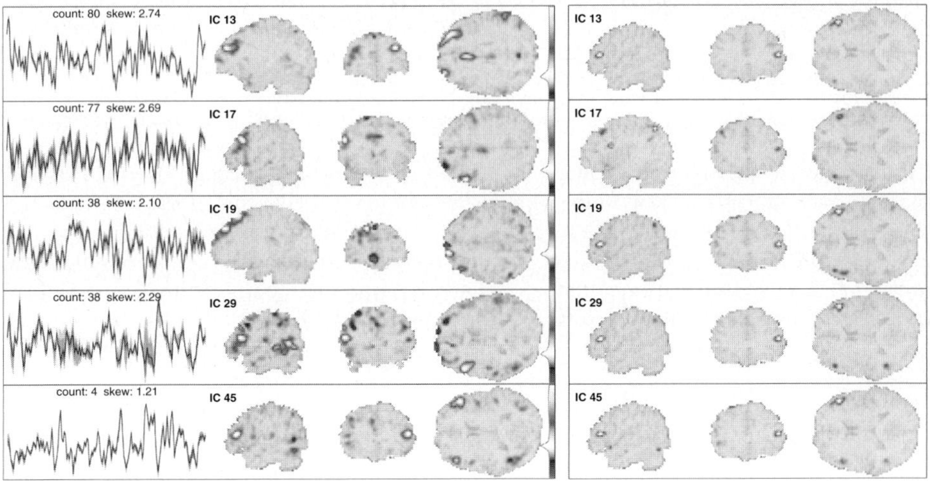

(a) Temporal and spatial mean (b) Spatial variance

Fig. 5. A set of independent components identified as a subspace through the shared variance, with transiently stimulus-related time-courses. Other details as in Fig. 2, except no reference blocks are shown.

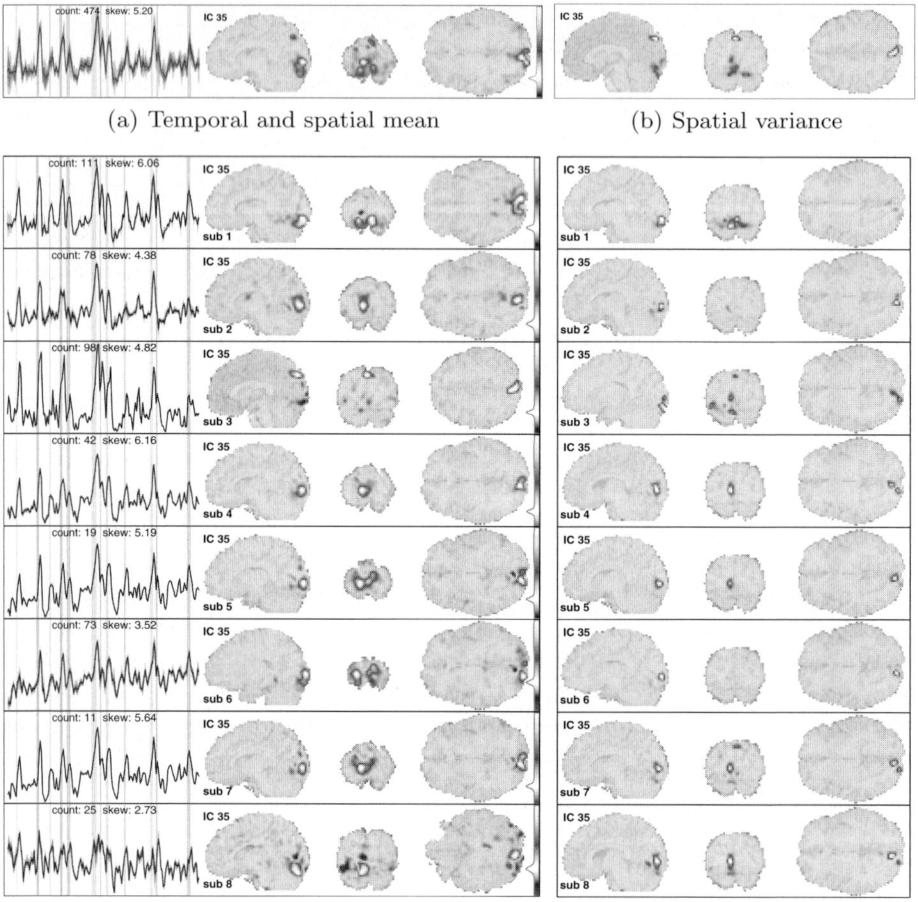

(a) Temporal and spatial mean (b) Spatial variance

(c) Reclustered temporal and spatial mean (d) Reclustered spatial variance

Fig. 6. One independent component, identified as a subspace through overall variability, with strongly visual stimulus-related time-course. (a) The mean spatial map and time-course of the component. (b) The spatial variance map of the component. (c) The mean spatial maps and time-courses of components from reclustering within the subspace. (d) The spatial variance maps of the corresponding components. Other details as in Fig. 2, except the reference blocks are from Fig. 1(b).

decomposition appears refined. The first, and highest correlating, CC depicts a baseline of activity related to all types of auditory stimulation. The second CC reveals a clear deviation from the baseline, occurring during the tone pip stimuli. It includes two brain regions, associated with auditory processing, having opposite signs in the spatial map. The last CC appears quite scattered, containing most of the activity within the subspace that is not explained by the other two CCs, as indicated by the low correlation.

Another example of a subspace linked through spatial variance is shown in Fig. 4, which appears weakly stimulus-related. The last IC in the subspace presents a potential artifact, with a sharp peak at a single time instance.

Fig. 5 shows a more complex set of activity, also identified as a subspace by the shared spatial variance. In this case, the ICs themselves are less consistent, and have considerable mixing variability. As no single component appears in all 100 runs, ICA can not identify consistent directions within the subspace. Some of the ICs are weakly stimulus-related, so a meaningful coordinate system inside the subspace could be fixed with CCA. However, the given stimulus design is not rich enough to decompose the 5 dimensional subspace.

The last example, shown in Fig. 6, is identified as a subspace already by the consistent ICA method. The strong mixing variability, together with the count of 474 estimates suggest that ICA can separate the subspace from the other components, but roughly 5 arbitrary directions from the subspace appear on each run. Additionally, the spatial variance coincides with the component itself, rather than being shared with other ICs. To further analyze the consistency of the strongly stimulus-related subspace, the 474 estimates within the subspace were clustered again, now using a higher threshold of 0.95. Fig. 6 also shows the set of 8 most consistent directions within the subspace. The directions are not strictly independent, since the clustering does not take into account from which run the estimates are taken. The subspace directions appear functionally meaningful, representing separate brain regions of the visual processing stream, including the primary visual cortex and other areas along the occipital lobes. Again, with a richer set of stimulus references, CCA could offer further refinement.

In addition to the illustrated subspaces, several other were identified, either through the overall variability of the components or by their shared spatial variance. The complete set of 46 consistent ICs also included several that were not part of a subspace.

4 Conclusions

Analyzing the variability of independent components, under a consistent ICA framework, can reveal characteristic information related to the underlying phenomena that is otherwise not visible. As shown by the results, components can be roughly divided into 3 classes based on spatial variance: individual and consistent components, with distributed variance due to noise; consistent members of a subspace, with focal variance coincident with the variance of the other members (see Fig. 2); and unconsistent subspaces, with variances coincident with their own mean (see Fig. 6). Such subspaces can provide information on networks of related activity in a purely data-driven manner.

Directions within each subspace can be further refined either blindly by clustering them into semi-independent constituents, or by using CCA with additional data. More than just refining the subspace decomposition, CCA provides a direct link to the set of related stimuli. However, the use of CCA is limited by the richness of the stimulation design. A more supervised approach was recently presented, with the goal of relating networks of brain activity with given complex stimulus features [4].

Acknowledgments. We thank Sanna Malinen and Riitta Hari for discussions on the subject of the paper, and both, together with Yevhen Hlushchuk for the data. All are from the Advanced Magnetic Imaging Centre (AMI Centre) and the Brain Research Unit in the Low Temperature Laboratory of Helsinki University of Technology.

References

1. Haacke, E.M., Brown, R.W., Thompson, M.R., Venkatesan, R.: Magnetic Resonance Imaging: Physical Principles and Sequence Design, 1st edn. Wiley-Interscience, New York (1999)
2. Hyvärinen, A., Karhunen, J., Oja, E.: Independent Component Analysis, 1st edn. Wiley-Interscience, New York (2001)
3. McKeown, M.J., Makeig, S., Brown, G.G., Jung, T.P., Kindermann, S.S., Bell, A.J., Sejnowski, T.J.: Analysis of fMRI Data by Blind Separation Into Independent Spatial Components. Human Brain Mapping 6(3), 160–188 (1998)
4. Ylipaavalniemi, J., Savia, E., Vigário, R., Kaski, S.: Functional Elements and Networks in fMRI. In: Proceedings of the 15th European Symposium on Artificial Neural Networks (ESANN 2007), Bruges, Belgium (April 2007), pp. 561–566 (2007)
5. Ylipaavalniemi, J., Vigário, R.: Analysis of Auditory fMRI Recordings via ICA: A Study on Consistency. In: Proceedings of the, International Joint Conference on Neural Networks (IJCNN 2004). Budapest, Hungary (July 2004), vol. 1, pp. 249–254 (2004)
6. Ylipaavalniemi, J., Mattila, S., Tarkiainen, A., Vigário, R.: Brains and Phantoms: An ICA Study of fMRI. In: Rosca, J., Erdogmus, D., Príncipe, J.C., Haykin, S. (eds.) ICA 2006. LNCS, vol. 3889, pp. 503–510. Springer, Heidelberg (2006)
7. Timm, N.H.: Applied Multivariate Analysis, 1st edn. Springer, New York (2002)
8. Friman, O., Carlsson, J., Lundberg, P., Borga, M., Knutsson, H.: Detection of neural activity in functional MRI using canonical correlation analysis. Magnetic Resonance in Medicine 45(2), 323–330 (2001)
9. Youssef, T., Youssef, A.B.M., LaConte, S.M., Hu, X.P., Kadah, Y.M.: Robust ordering of independent components in functional magnetic resonance imaging time series data using Canonical correlation analysis. In: Proceedings of the SPIE Medical Imaging,: Physiology and Function: Methods, Systems, and Applications. San Diego, CA (February 2003), vol. 5031, pp. 332–340 (2003)
10. Hardoon, D.R., Mourão-Miranda, J., Brammer, M., Shawe-Taylor, J.: Unsupervised fMRI Analysis. In: NIPS Workshop on New Directions on Decoding Mental States from fMRI Data, Whistler, Canada (December 2006)
11. Malinen, S., Hlushchuk, Y., Hari, R.: Towards natural stimulation in fMRI – Issues of data analysis. NeuroImage 35(1), 131–139 (2007)
12. SPM2: MATLAB^{TM} Package (2002), `http://www.fil.ion.ucl.ac.uk/spm`
13. FastICA: MATLAB^{TM} Package (1998), `http://www.cis.hut.fi/research/ica/fastica`

Multi-modal ICA Exemplified on Simultaneously Measured MEG and EEG Data*

Heriberto Zavala-Fernandez[1], Tilmann H. Sander[2], Martin Burghoff[2], Reinhold Orglmeister[1], and Lutz Trahms[2]

[1] Technische Universität Berlin, Fachgebiet Elektronik und medizinische Signalverarbeitung, Einsteinufer 17, 10587 Berlin, Germany
hzavimed@mailbox.tu-berlin.de

[2] Physikalisch-Technische Bundesanstalt, Abbestr. 2-10, 10587 Berlin, Germany

Abstract. A multi-modal linear mixing model is suggested for simultaneously measured MEG and EEG data. On the basis of this model an ICA decomposition is calculated for a combined MEG and EEG signal vector using the TDSEP algorithm. A single modality demixing procedure is developed to classify ICA components to be multi-modality sources detected by EEG and MEG simultaneously or to be single mode sources. Under this premise, data from 10 subjects are analysed and four exemplary types of sources are selected. We found that these sources represent physically meaningful multi- and single-mode signals: Alpha oscillations, heart activity, eye blinks, and slow signal drifts.

1 Introduction

Magnetoencephalography (MEG) and Electroencephalography (EEG) are non-invasive methods to measure the electrical activity of groups of neurons in the human brain with a ms temporal resolution. This electrical activity leads to potential differences at the scalp (EEG) and a magnetic field outside of the head (MEG). It is well known that the sensitivty profiles of EEG and MEG are quite different with respect to the orientation and the location of neural currents. This is utilized in studies such as [1] [2] [3] [4], where the aim was an improved source localization by a combined MEG and EEG data analysis.

The value of applying ICA to single modality data, i.e., either MEG or EEG, has been demonstrated frequently (see [5] for an overview). Here we want to apply ICA specifically to combined EEG and MEG datasets and evaluate a different mixing model compared to the subspcace channel ICA approach in [6]. Given the results of an ICA applied to multi-modal data sets it is important to determine, whether a component represents a source recorded by both modalities, MEG and EEG, or a source recorded exclusively by one of the modalities.

* This work was supported by the DAAD (German Academic Exchange Service) scholarship for H.Z.F (A0421558) and the Berlin Neuroimaging Centre (BMBF 01GO 0503 BNIC).

M.E. Davies et al. (Eds.): ICA 2007, LNCS 4666, pp. 673–680, 2007.

To answer this question post-hoc for a mixing matrix $\mathbf{A}$ estimated from experimental data is not easy due to the permutation problem of ICA and the lack of a common scale for multi-modal data sets such as combined MEG and EEG data. For this context we propose a way of checking the origin of the estimated sources without modifying an ICA algorithm itself.

2 Multi-modal ICA

2.1 Mixing Model for Multi-modal Data Sets

The basic model of ICA is well known, where signals $\boldsymbol{x}(t)$ are described as the product of an unknown mixing matrix $\mathbf{A}$ and an unknown source vector $\boldsymbol{s}(t)$. ICA aims to estimate $\mathbf{A}$ from statistical properties of $\boldsymbol{x}(t)$. With an estimate of $\mathbf{A}$ the source signals can be computed from

$$\boldsymbol{s}(t) = \mathbf{A}^{-1} \cdot \boldsymbol{x}(t) = \mathbf{W} \cdot \boldsymbol{x}(t). \tag{1}$$

Extending the basic model to the (multi-modal) case of two simultaneously measured quantities, EEG and MEG with related content, the data vector "$\boldsymbol{x}(t)$" now reads (superscript M or E always denotes MEG respectively EEG, it does not denote power):

$$\boldsymbol{x}^{M,E}(t) = \begin{bmatrix} \boldsymbol{x}^M(t) \\ \boldsymbol{x}^E(t) \end{bmatrix}. \tag{2}$$

For the mixing process at least two cases have to be considered, firstly the number of sources is known and less than the sum of the m MEG and n EEG channels. In this case the mixing matrix $\mathbf{A}$ contains $m + n$ rows and as many columns as there are sources. This case is discussed generally in [6]. It is well known that single brain sources can contribute to MEG and EEG signals at the same time. Nevertheless the number of brain sources is generally not known and considering the strong background and noise signals in MEG and EEG data the more general case of a square matrix of dimension $(m + n,\ m + n)$ is considered here. In case that the MEG and EEG signals are without common sources the combined mixing matrix has to have the following block form

$$\mathbf{A} = \begin{pmatrix} \mathbf{A^M} & 0 \\ 0 & \mathbf{A^E} \end{pmatrix}, \tag{3}$$

where $\mathbf{A^M} \in \mathbb{R}^{m \times m}$ for the magnetic and $\mathbf{A^E} \in \mathbb{R}^{n \times n}$ for the electric sources.

Due to an arbitrary permutation of the columns of $\mathbf{A}$ estimated by ICA and the lack of a common physical scale for MEG and EEG data the simple form (3) cannot be compared to a matrix $\mathbf{A}$ resulting from experimental data. Naturally the more general case of common sources in MEG and EEG data is of interest here and the mixing matrix has the general form given by

$$\mathbf{A} = \begin{pmatrix} \mathbf{C^M} & \mathbf{D^M} & 0 \\ \mathbf{C^E} & 0 & \mathbf{D^E} \end{pmatrix}, \tag{4}$$

where inner matrices $\mathbf{C}^{\mathbf{M}}$ and $\mathbf{C}^{\mathbf{E}}$ with columns $\boldsymbol{c}_j$ represent sources recorded by both modalities and therefore a gain in information in comparison to a separate analysis (3). Similarly, columns $\boldsymbol{d}_k$ and $\boldsymbol{d}_l$ of inner matrices $\mathbf{D}^{\mathbf{M}}$ and $\mathbf{D}^{\mathbf{E}}$, respectively, are then sources recorded exclusively by one of the modalities. Consequently, $j+k+l = m+n$. The columns $\boldsymbol{a}_i$ of $\mathbf{A}$ now represent in the upper part, rows $1 \ldots m$, a so called MEG map, i.e. sensor weights for the MEG channels, and in the lower part, rows $m+1 \ldots m+n$, an EEG map.

2.2 Single Modality Demixing

To characterize whether a multi-modal ICA component corresponds in fact to a single modality signal we devised a procedure termed "single modality demixing". The aim is to avoid a rescaling of the experimental data while removing the inherent ICA scaling ambiguity between mixing matrix $\mathbf{A}$ and sources $\boldsymbol{s}(t)$. Therefore a first step is to move all signal energy into the mixing matrix. This is done by creating sources $\boldsymbol{s}'(t)$ that have a root-mean-square (rms) value of unity: $\boldsymbol{s}_i(t) = b_i \cdot \boldsymbol{s}_i{}'(t)$, where $rms(\boldsymbol{s}_i{}') = 1$ (equivalent to a scaling of the distribution of the observed variable to unit variance). Combining the b_i into a diagonal matrix $\mathbf{B}$ the sources $\boldsymbol{s}(t)$ are replaced by the $\boldsymbol{s}'(t)$ in (1) and $\boldsymbol{x}(t) = \mathbf{B} \cdot \mathbf{A} \cdot \boldsymbol{s}'(t) = \mathbf{A}' \cdot \boldsymbol{s}'(t)$. Multiplying the mixing matrix $\mathbf{A}$ by $\mathbf{B}$, a new mixing matrix results $\mathbf{A}'$, which is meant in the text from now on. This rescaling of the mixing matrix should not be confused with the whitening procedure used as a first step in most ICA algorithms.

The single modality demixing is now defined as setting either $\boldsymbol{x}^M$ or $\boldsymbol{x}^E$ in (2) to zero and then applying (1) and to obtain single modality demixed time series $\boldsymbol{s}^{M,E=0}(t)$ respectively $\boldsymbol{s}^{M=0,E}(t)$:

$$\boldsymbol{s}^{M,E=0} = \mathbf{W}' \cdot \begin{bmatrix} \boldsymbol{x}^M \\ 0 \end{bmatrix}, \qquad \boldsymbol{s}^{M=0,E} = \mathbf{W}' \cdot \begin{bmatrix} 0 \\ \boldsymbol{x}^E \end{bmatrix}. \tag{5}$$

This single modality demixing is motivated by the block mixing form in (3). In this case only the MEG sources are non-zero for $\boldsymbol{s}^{M,E=0}$ and vice versa for $\boldsymbol{s}^{M=0,E}$. These single modality demixed time series can then be compared with each other or with the time series from the full demixing, $\boldsymbol{s}^{M,E}$. The ICA component vector $\boldsymbol{a}_i$ is not affected by the single modality demixing. The single modality mixing should not be confused with completely separate ICA calculations for MEG and EEG.

The rms values of $\boldsymbol{s}^{M,E=0}(t)$ and $\boldsymbol{s}^{M=0,E}(t)$ can be computed and compared with each other and the $rms = 1$ of $\boldsymbol{s}'(t)$, the original source. The rms of the single modality demixed time series indicates the relative contribution of this source to the MEG and EEG data. Another measure to be used is the correlation between single modality demixed time series and the time series resulting from the full demixing.

The single modality demixing approach can be exemplified assuming two MEG channels and one EEG channel and then depicting the demixing matrix

$\mathbf{W}'$. Assuming $w_{33} = 0$ in (6) it follows $s_3^{M,E} = w_{31}x_1^M + w_{32}x_2^M$ irrespective of x_3^E, i.e., s_3 represents a pure MEG source.

$$\begin{pmatrix} s_1^{M,E}(t) \\ s_2^{M,E}(t) \\ s_3^{M,E}(t) \end{pmatrix} = \begin{pmatrix} w_{11} \; w_{12} \; w_{13} \\ w_{21} \; w_{22} \; w_{23} \\ w_{31} \; w_{32} \; w_{33} \end{pmatrix} \cdot \begin{pmatrix} x_1^M(t) \\ x_2^M(t) \\ x_3^E(t) \end{pmatrix} \tag{6}$$

3 Experimental Procedure

3.1 MEG and EEG Data Acquisition

Using a typical recording configuration, e.g. $m > n$, the MEG was measured using $m = 93$ channel SQUID system (ET160 Eagle Systems) installed in a shielded room. Within the MEG helmet the EEG was recorded from $n = 28$ sites at positions given by the international 10-20 system for electrode placement. MEG and EEG channels were connected to a 128 channel data acquisition system to ensure a simultaneous recording. Ten healthy subjects had to listen to 30 tones of 2 Khz and 30 tones of 1 Khz presented in a random order and with a random interstimulus interval in a session lasting 305 s. This task was part of a larger set of measurements investigating the auditory N100 in relation to attention. The results with respect to the N100 are not discussed here. The data were recorded with a sampling rate of 2 kHz and then off-line down sampled to 250 Hz. The data were off-line filtered with a zero-phase delay band pass from 0.5 to 40 Hz preserving typical brain signals.

3.2 Data Processing Using TDSEP-ICA

The Second Order Blind Identification algorithm (SOBI[7], TDSEP[8]) is suitable for data with a rich temporal structure and consequently colored spectra. The basis of the TDSEP algorithm is a set of time-lagged covariance matrices $R_x(\tau) = \langle x(t+\tau) \cdot x^T(t) \rangle$ with $\tau \neq 0$. For independent sources these matrices have to be diagonal. To estimate the sources a joint diagonalization of the time-lagged covariance matrices is performed. To use a set of τ values is an empirically established procedure to avoid an inferior source separation as there is no theoretically proven choice of τ values.

The TDSEP ICA algorithm was applied to the combined MEG and EEG data vector given in (2). As time delays τ the set $\tau = 1, 5, 10, 15, \ldots, 500$ was chosen in the TDSEP calculation and consequently a set of 101 time-lagged covariance matrices had to be simultaneously diagonalized. The calculation was performed for each of the subjects separately.

Given that TDSEP probes the temporal structure of signals by correlating EEG and MEG signals with each other, any phase difference due to the technical recording equipment has to be avoided. This was tested and no delay was found between any of the channels of both modalities.

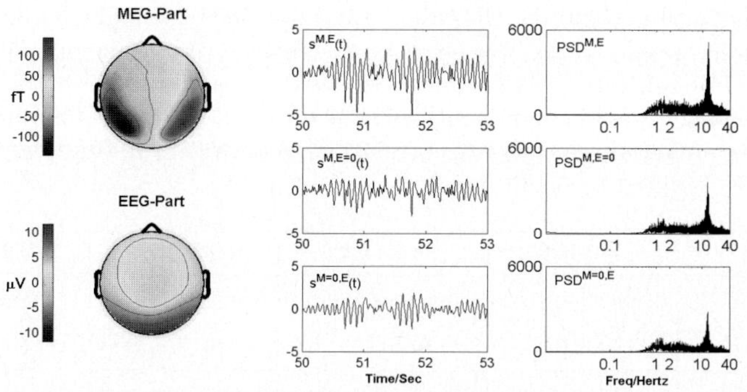

Fig. 1. Occipital alpha component of the multi-modal ICA applied to simultaneously measured MEG and EEG data. The left column shows the MEG and EEG part of the component vector, the top time trace of the middle column is a short section of the component time series $s^{M,E}$ and the spectrum is shown to the right. The other two time series and spectra are the single modality demixed signals ($s^{M,E=0}$ respectively $s^{M=0,E}$, see text).

4 Results

The time series and power spectra of the single modality demixed signals, $s^{M,E=0}$ respectively $s^{M=0,E}$, are shown below the data of $s^{M,E}$ in Figs. 1-3. Note that the multi-modal approach excludes the use of terminology like EOG or MOG. In the multi-modal case different physical quantities are combined in the mixing matrix and consequently the ICA time series is dimensionless. A physical unit can be given for the scaling of the MEG and EEG maps as indicated in Figs. 1-3. Figure 1 presents a component, in which both single modality demixed signals have spectra with 10 Hz peaks, and the time series are similar to the fully demixed signal. The 3 s section of the associated time series $s^{M,E}$ (top of middle column) shows the corresponding amplitude modulation of about 10 Hz.

A quantitative comparison is given in Table 1, where the *rms* values of the single modality demixed signals and the correlations ρ between the signals are mean values for the group of subjects. The *rms* values are about 0.6 and 0.7 indicating an equal contribution of MEG and EEG for the component of Fig. 1. The correlations between $s^{M,E}$ and the single modality demixed signals are fairly high in agreement with the visual impression from Fig. 1. Interestingly the correlation between $s^{M,E=0}$ and $s^{M=0,E}$ is fairly small. Considering that one row of (6) reads $s_1^{M,E} = w_{11}x_1^M + w_{12}x_2^M + w_{13}x_3^E$ the *rms* values of $s^{M,E=0}$ and $s^{M=0,E}$ and the correlation between them cannot be close to one at the same time as this contradicts the normalization $s^{M,E} = 1.0$. Thc ICA component shown in Fig. 1 has MEG and EEG part maps (left side of Fig. 1) with peaks at the back of the head.

The most common spontaneous oscillatory signal detected in normal subjects is an occipital alpha wave, which has a spectral range between 8 and 13 Hz and

Table 1. Quantitative results for the time signals obtained by single modality demixing (ALPHA = alpha waves, HEART = heart beat, EYE = eye movements, DRIFT = slow signal drifts). The top two rows are mean *rms* values of the single modality demixed sources for the four types of components and the group of subjects (standard deviations indicated), the bottom three rows are correlations ρ between the time signals including the time series obtained by a full demixing.

Type	ALPHA	HEART	EYE	DRIFT
$\langle rms(s^{M,0}) \rangle$	0.5983 ± 0.1697	0.9881 ± 0.0610	0.4029 ± 0.1188	1.0006 ± 0.0022
$\langle rms(s^{0,E}) \rangle$	0.7195 ± 0.1197	0.4716 ± 0.1908	0.8006 ± 0.1155	0.0806 ± 0.0150
$\langle \rho(s^{M,0}, s^{0,E}) \rangle$	0.1867 ± 0.2736	-0.2403 ± 0.0750	0.3400 ± 0.3172	-0.0453 ± 0.0299
$\langle \rho(s^{M,E}, s^{M,0}) \rangle$	0.7131 ± 0.1254	0.8665 ± 0.1228	0.6530 ± 0.1926	0.9966 ± 0.0012
$\langle \rho(s^{M,E}, s^{0,E}) \rangle$	0.7999 ± 0.1464	0.2346 ± 0.1715	0.9232 ± 0.0529	0.0352 ± 0.0206

can be seen in relaxed awake subjects. The occipital lobes are believed to be the source of alpha activity [9]. Comparing these properties with the features of the data of Fig. 1, this ICA component can be attributed to alpha oscillations. The fact that the MEG and EEG part maps are orthogonal to each other indicates that the magnetic and electric field are due to the same current configuration.

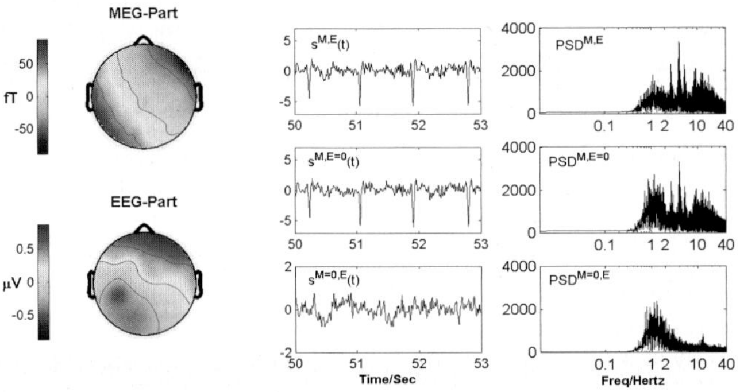

Fig. 2. Heart beat component of the multi-modal ICA. The figure is structured as Fig. 1. The map of the MEG part is typical for heart activity and the $s^{M,E}$ and $s^{M,E=0}$ time series show the heart beat. It is notably absent in $s^{M=0,E}$.

Figure 2 shows a component with a time series that obviously reflects the QRS activity of the heart. Heart activity is an inherent artifact known to be present in MEG [10] [12]. In contrast to that, the QRS complex is not visible in most raw EEG data and of little relevance for EEG studies [11]. In our multimodal ICA, the heart artifact is only found in the MEG part as can be seen in Fig. 2, where $s^{M,E}$ and $s^{M,E=0}$ show the heart beat, but $s^{M=0,E}$ does not. As given in Table 1 the single modality demixing *rms* values are 0.99 and 0.47 for

MEG and EEG, respectively, indicating a dominance of the MEG contribution. A correlation value of 0.87 between $s^{M,E}$ and $s^{M,E=0}$ and 0.23 between $s^{M,E}$ and $s^{M=0,E}$ confirms this behaviour.

Figure 3 shows a component that is obviously generated by a source in the frontal region of the head. The corresponding tabulated values for the standard deviations of rms and correlations in Table 1 indicate that the multi-modal ICA result is quite homogeneous for the group of subjects investigated and electric field are due to the same current configuration. We attribute this component to eye blinks considering the fact that the eyes and related muscles are themselves magnetical and electrical sources MEG and EEG data can be contaminated heavily by eye blink signals especially in frontal regions [12].

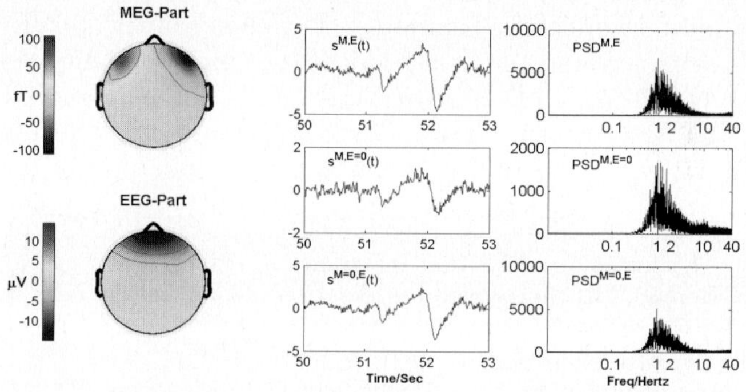

Fig. 3. Eye movement component of the multi-modal ICA. The figure is structured as Fig. 1. The map of the MEG and EEG part have their peaks at the front and can be attributed to the magnetical and electrical signature of eye blinks (muscle and eye dipole).

In addition to these components, there is another component that exhibits little correlation between MEG and EEG channels. The rms values are 1 and 0.04 for MEG and EEG, respectively, indicating a dominance of the MEG contribution as for the heart beat component. We refer this component to magnetic field drifts. Slow signal drifts are often present in MEG measurements due to the reduced shielding capabilities of mu-metallic rooms at frequencies below 1 Hz. Typical EEG artifact signals such as loose electrodes were easily identified in the same manner as the MEG signal drifts.

5 Conclusion

By our analysis, a component associated with the auditory evoked M100/N100 was not identified. This is probably due to the random interstimulus interval by which the auditory stimulus were presented. However, we found four characteristic sources in almost all of the ten subjects studied using multi-modal ICA on common

MEG and EEG data. These four types of signals were easily found by manual inspection in almost all subjects of the group. The common source for occipital alpha oscillations was found without relying on any physical modeling of currents in the brain. A similar result was obtained for signals due to eye blinks. One heart beat component was found, which was detected mainly in the MEG as deduced from the correlation and *rms* values associated with the single modality demixing.

This behaviour agrees with the properties of the electric potential and the magnetic field of the heart muscle. We have demonstrated that multi-modal ICA and single modality demixing are powerful tools to identify common sources in MEG and EEG data.

The next step will be a comparison between multi-modal ICA results and single modality ICA decompositions and the usage of a physical model to estimate the source location from the combined ICA EEG and MEG maps. Future works will assess the influnce of channels number. The choice of the TDSEP algorithm was somewhat arbitrary and the usefulness of the single modality demixing concept should be tested using other algorithms and other multi-modal data sets.

References

1. Baillet, S., Garnero, L., Marin, G., Hugonin, J.P.: Combined MEG and EEG Source Imaging by Minimization of Mutual Information. Biomedical Eng. 46(5) (1999)
2. Dale, A.M., Sereno, M.I.: Improved Localization of Cortical Activity by Combining EEG and MEG with MRI Cortical Surface Reconstruction: A Linear Approach. Journal of Cognitive Neuroscience 5(2), 162–176 (1993)
3. Huizenga, H.M., van Zuijen, T.L., Heslenfeld, D.J., Molenaar, P.C.M.: Simultaneous MEG and EEG source analysis. Phys. Med. Biol. 46, 1737–1751 (2001)
4. Yoshinaga, H., Nakahori, T., Ohtsuka, Y., Oka, E., Kitamura, Y., Kiriyama, H., Kinugasa, K., Miyamoto, K., Hoshida, T.: Benefit of Simultaneous Recording of EEG and MEG in Dipole Localization. Epilepsia 43(8), 924–928 (2002)
5. James, C., Hesse, C.: ICA for Biomedical Signals. Physiol. Meas. 26, 15–39 (2005)
6. Anemüller, J.: Second-Order Separation of Multidimensional Sources with Constrained Mixing System. In: Rosca, J., Erdogmus, D., Príncipe, J.C., Haykin, S. (eds.) ICA 2006. LNCS, vol. 3889, pp. 16–23. Springer, Heidelberg (2006)
7. Belouchrani, A., Abed-Meraim, K., Cardoso, J.-F., Moulines, E.: A BSS Technique Based on Second-Order Statistics. IEEE Trans. on Sig. Proc. 45, 434–444 (1997)
8. Ziehe, A., Müller, K.-R.: TDSEP–An Efficient Algorithm for Blind Separation Using Time Structure. In: Proc. of the 8th ICANN, pp. 675–680. Springer, Heidelberg (1998)
9. Hari, R., Salmelin, R.: Human cortical oscillations: a neuromagnetic view through the skull. Trends Neurosci. 20(1), 44–49 (1997)
10. Sander, T.H., Wuebbeler, G., Lueschow, A., Curio, G., Trahms, L.: Cardiac Artifact Subspace Identification and Elimination in Cognitive MEG-Data Using Time-Delayed Decorrelation. IEEE Trans. Biomed. Eng. 49, 345–354 (2002)
11. Dirlich, G., Vogl, L., Plaschke, M., Strian, F.: Cardiac field effects on the EEG. Electroencephalography and clinical Neurophysiology 102, 307–315 (1997)
12. Zavala-Fernandez, H., Sander, T., Burghoff, M., Orglmeister, R., Trahms, L.: Comparison of ICA Algorithms for the Isolation of Biological Artifacts in Magnetoencephalography. In: Rosca, J., Erdogmus, D., Príncipe, J.C., Haykin, S. (eds.) ICA 2006. LNCS, vol. 3889, pp. 511–518. Springer, Heidelberg (2006)

Blind Signal Separation Methods for the Identification of Interstellar Carbonaceous Nanoparticles*

O. Berné[1,2], Y. Deville[2], and C. Joblin[1]

[1] Centre d'Etude Spatiale des Rayonnements, CNRS et Université Paul Sabatier Toulouse 3, Observatoire Midi-Pyrénées, 9 Av. du Colonel Roche, 31028 Toulouse cedex 04, France
olivier.berne@cesr.fr, christine.joblin@cesr.fr

[2] Laboratoire d'Astrophysique de Toulouse-Tarbes, CNRS et Université Paul Sabatier Toulouse 3, Observatoire Midi-Pyrénées, 14 Av. Edouard Belin, 31400 Toulouse, France
ydeville@ast.obs-mip.fr

Abstract. The use of Blind Signal Separation methods (ICA and other approaches) for the analysis of astrophysical data remains quite unexplored. In this paper, we present a new approach for analyzing the infrared emission spectra of interstellar dust, obtained with NASA's Spitzer Space Telescope, using *FastICA* and Non-negative Matrix Factorization (NMF). Using these two methods, we were able to unveil the *source* spectra of three different types of carbonaceous nanoparticles present in interstellar space. These spectra can then constitute a basis for the interpretation of the mid-infrared emission spectra of interstellar dust in the Milky Way and nearby galaxies. We also show how to use these extracted spectra to derive the spatial distribution of these nanoparticles.

1 Introduction

The Spitzer Space Telescope (*Spitzer*) comprises one of today's best instruments to probe the mid-infrared (mid-IR) emission of interstellar dust in the Milky Way and nearby galaxies. This emission is mainly carried by very small (nanometric) interstellar dust particles. One of the goals of infrared astronomy is to identify the physical/chemical nature of these species, as they play a fundamental role in the evolution of galaxies. Unfortunately, the observed spectra are mixtures of the emission from various dust populations. The strategy presented in this paper is to apply Blind Signal Separation (BSS) methods i.e. *FastICA* and NMF to a set of *Spitzer* mid-IR (5-30 μm) spectra obtained with the InfraRed Spectrograph (IRS), in order to extract the genuine spectrum of each population of nanoparticles. We first present these observations in Sect. 2, then we apply the

* This work is based on observations made with the Spitzer Space Telescope, which is operated by the Jet Propulsion Laboratory, California Institute of Technology under a contract with NASA.

M.E. Davies et al. (Eds.): ICA 2007, LNCS 4666, pp. 681–688, 2007.

two BSS methods to these observations and finally give an example of how the extracted spectra can be used to trace the evolution of dust, in the Milky Way and external galaxies.

2 Observations

We have observed with *Spitzer* nearby photo-dissociation regions (PDRs), which consist of a star illuminating the border of dense clouds of gas and dust. The physical conditions (UV field intensity and hardness, cloud density) strongly vary from a PDR to another as well as inside each PDR depending on the considered position. These variations are extremely useful to probe the nature of dust particles which are altered by the local physical conditions [1]. Therefore, we have observed 11 PDRs as part of the SPECPDR program using the IRS in "spectral mapping" mode. This mode enabled us to construct one dataset for each PDR. This dataset is a spectral cube, with two spatial dimensions and one spectral dimension (see Fig. 1). Each spectral cube is thus a 3-dimensional matrix $C(p_x, p_y, \lambda)$, which defines the variations of the recorded data with respect to the wavelength λ, for each considered position with coordinates (p_x, p_y) in the cube. The dimensions of these cubes are generally about 30×30 positions and 250 points in wavelength ranging between 5 and 30 μm.

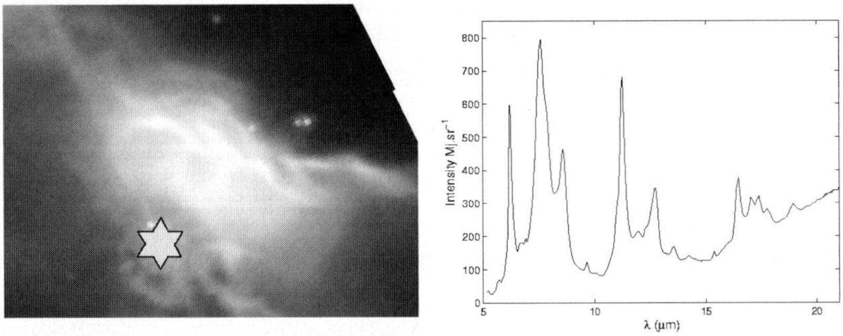

Fig. 1. *Left*: Infrared (8 μm) view of the NGC 7023 North PDR. The star is illuminating the cloud situated in the upper part of the image. *Right*: Mid-IR spectrum for a given position in the spectral cube of NGC 7023.

3 Blind Separation of Interstellar Dust Spectra

BSS is commonly used to restore a set of unknown "source" signals from a set of observed signals which are mixtures of these source signals, with unknown mixture parameters [2]. BSS is most often achieved using ICA methods such as *FastICA* [3]. An alternative class of methods for achieving BSS is NMF, which was introduced in [4] and then extended by a few authors. In the astrophysical community, ICA has been successfully used for spectra discrimination in infrared spectro-imagery of Mars ices [5], elimination of artifacts in astronomical images [6] or extraction of cosmic microwave background signal in *Planck* simulated data [7]. To our knowledge, NMF has not yet been applied to astrophysical problems. However, it has been used to separate spectra in other application fields, e.g. for magnetic resonance chemical shift imaging of the human brain [8] or for analyzing wheat grain spectra [9].

The simplest version of the BSS problem concerns so-called "linear instantaneous" mixtures. It is modeled as follows:

$$X = AS \tag{1}$$

where X is an $m \times n$ matrix containing n samples of m observed signals, A is an $m \times r$ mixing matrix and S is an $r \times n$ matrix containing n samples of r source signals. The observed signal samples are considered to be linear combinations of the source signal samples (with the same sample index). It is assumed that $r \leq m$ in most investigations, including this paper. The objective of BSS algorithms is then to recover the source matrix S and/or the mixing matrix A from X, up to BSS indeterminacies.

The correspondence between the generic BSS data model (1) and the 3-dimensional spectral cube $C(p_x, p_y, \lambda)$ to be analyzed in the present paper may be defined as follows. In this paper, the sample index is associated to the wavelength λ, and each observed signal consists of the overall spectrum recorded for a cube pixel (p_x, p_y). Each one of these signals defines a row of the matrix X in Eq. (1). Moreover, each observed spectrum is a linear combination of "source spectra" (see Sect. 3.1), which are respectively associated to each of the (unknown) types of nanoparticles that contribute to the recorded spectral cube. Therefore, the recorded spectra may here be expressed according to (1), with unknown combination coefficients in A, unknown source spectra in S and an unknown number r of source spectra.

3.1 Suitability of BSS Methods for the Analysis of *Spitzer*-IRS Cubes

In order to apply the NMF or *FastICA* to the IRS data cubes, it is necessary to make sure that the "linear instantaneous" mixture condition is fulfilled. Here we consider that each observed spectrum is a linear combination of "source spectra", which are due to the emission of different populations of dust nanoparticles. The main effect that can disturb the linearity of the model is radiative transfer as shown by [10], because of the non-linearity of the equations. In our case however, this effect is completely negligible because the emission spectra we observe come from the surface of clouds and are therefore not altered by radiative transfer.

3.2 Considered BSS Methods

In this section, we detail which particular BSS methods we have applied to the observed data.

NMF. We used NMF as presented in [11]. The matrix of observed spectra X is approximated using

$$WH, \tag{2}$$

where W and H are non-negative matrices, with the same dimensions as in (1). This approximation is optimized by adapting the matrices W and H using the algorithm of [11] in order to minimize the divergence between X and WH. We implemented the algorithm with Matlab. Convergence is reached after about 1000 iterations (which takes less than one minute with a 3.2 GHz processor). The value of r (number of "source" spectra) is not imposed by the NMF method. Our strategy for setting it so as to extract the sources was the following:

- Apply the algorithm to a given dataset, with the minimum number of assumed sources, i.e. $\hat{r} = 2$, providing 2 sources.
  → If the found solutions are physically coherent and linearly independent, we consider that at least $\hat{r} = 2$ sources can be extracted.
  → Else, we consider that the algorithm is not suited for analysis (this never occurred in our tests).
- Try the algorithm on the same dataset but with $\hat{r} = 3$ sources.
  → If the found solutions are physically coherent and linearly independent, we consider that at least $\hat{r} = 3$ sources can be extracted.
  → Else, we consider that only two sources can be extracted, extraction was over with $\hat{r} = 2$ and thus $r = 2$.
- Same as previous step but with $\hat{r} = 4$ sources.
  → If the found solutions are physically coherent and linearly independent, we consider that at least $\hat{r} = 4$ sources can be extracted.
  → Else we consider that only three sources can be extracted, extraction was over with $\hat{r} = 3$ and thus $r = 3$.

...

Physically incoherent spectra exhibit sparse peaks (spikes) which cannot be PDR gas lines. We found $r = 3$ for NGC 7023-NW and $r = 2$ for the other PDRs, implying that we could respectively extract 3 and 2 spectra from these data cubes.

FastICA. We used *FastICA* in the deflation version [3] in which each source is extracted one after the other and subtracted from the observations until all sources are extracted. The advantage of this *FastICA* method is that it is not necessary to fix, before running the algorithm, the number r of sources that we want to extract, as it is for NMF. The extraction of the sources takes less than one minute using *FastICA* coded with Matlab, and with a 3.2 GHz processor.

3.3 Results

Using the BSS methods presented in this paper, we were able to extract up to three source spectra from the *Spitzer* observations. The number r of sources found in a given PDR is always the same with NMF and *FastICA*. The three extracted spectra in NGC 7023 North are presented in Fig. 2. Two of them exhibit the series of aromatic bands which have previously been attributed to Polycyclic Aromatic Hydrocarbons (PAHs, [12] and [13]). These two spectra show different band intensity ratios. One is the spectrum of neutral PAHs (PAH^0) while the other is due to ionized PAHs (PAH^+). The last spectrum exhibits a continuum and aromatic bands, which can be attributed to very small carbonaceous grains (VSGs), possibly PAH clusters [14].

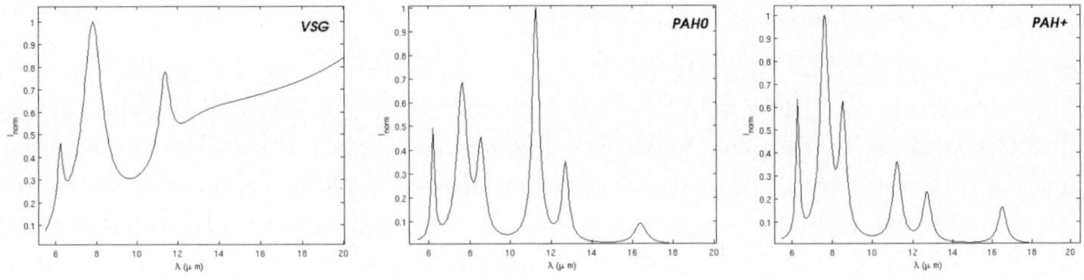

Fig. 2. The three BSS-extracted spectra from our study on PDRs

3.4 *FastICA* vs. NMF for Our Application

As mentioned in Sect. 3.3, we were able to extract the source spectra from our data using both *FastICA* and NMF. However, the extracted spectra are not exactly the same for both methods. We conducted several tests in order to be able to evaluate which one of the two methods is more appropriate for our application. We created a set of 2/3 artificial carbonaceous nanoparticle spectra, to which we added a variable level of white, spatially homogeneous noise. We mixed these spectra with a random matrix to create a set of 100 artificial observed spectra. We then applied the two BSS methods considered in this paper. With a noise level at zero, both methods recover the original signals with high efficiency (correlation coefficients between original and extracted signals above 0.995). When adding noise, this efficiency decreases but remains acceptable down to a noise level corresponding to a SNR of 3dB (which is much lower that the average SNR of the *Spitzer spectra*). We note however that the efficiency of FastICA drops slightly faster than the one of NMF under the effect of an increasing noise, and drops dramatically below a SNR of 3dB, while NMF can still partly recover the original signals. Finally, with both methods we observe that the power of the residuals (i.e. observed signal minus signal reconstructed from the estimated sources and mixing coefficients) has the same level as Spitzer noise.

We have shown in Sect. 3.3 that there are two main populations: one with a continuum (VSGs) and one with bands only (PAHs). Using *FastICA*, we sometimes find a residual continuum in the BSS-extracted PAH spectrum, which we interpret as an incomplete separation. It is possible that the criterion of NMF is more appropriate in our case because less restrictive. Indeed, NMF only requires non-negativity of the sources and mixing coefficients, which is in essence the case for emission spectra, while *FastICA* is based on the statistical independence and non-gaussianity of the sources, which is more difficult to prove. As a conclusion, we would like to stress the fact that both methods are very efficient for the first task presented in this paper. We however note that NMF seems slightly better for this particular application.

4 Deriving the Spatial Distribution of Carbonaceous Nanoparticles

The next step of our analysis consists in using our extracted source spectra (Fig. 2) in order to determine the spatial distribution of the three populations in galactic clouds or in external galaxies. The *Spitzer* observations on-line archive contains hundreds of mid-IR spectral cubes of such regions which can be interpreted in this way. Our strategy consists in calculating the correlation parameter $c_p = E[Obs(p_x, p_y, \lambda)y_p(\lambda)]$ between an observed spectrum $Obs(p_x, p_y, \lambda)$ at a position (p_x, p_y) in a spectral cube and one of our extracted source spectra $y_p(\lambda)$, where $E[.]$ stands for expectation. With the considered (i.e. linear instantaneous) mixture model, each observed spectrum reads

$$Obs(p_x, p_y, \lambda) = \sum_n w(p_x, p_y)_n S_n(\lambda) \tag{3}$$

where $S_n(\lambda)$ is the nth source spectrum and $w(p_x, p_y)_n$ are the mixing coefficients associated to that source. Moreover, BSS methods extract the sources up to arbirary scale factors, i.e. they provide $y_p(\lambda) = \eta_p S_p(\lambda)$, where η_p is an unknown scale factor and $S_p(\lambda)$ is the pth source. By centering the observations and thus the extracted spectra, and assuming that the sources are not correlated, the above-defined correlation parameter becomes

$$c_p = \eta_p w(p_x, p_y)_p E[S_p(\lambda)^2]. \tag{4}$$

This coefficient c_p is calculated for all the positions (p_x, p_y), therefore yielding a 2D correlation map. Eq (4) shows that this map is proportional to $w(p_x, p_y)_p$ and thus defines the spatial distribution of the considered extracted source $y_p(\lambda) = \eta_p S_p(\lambda)$. We applied this approach to the spectral cube of NGC 7023 North (Fig. 1) and obtained the correlation maps presented in Fig. 3. We find that the three nanoparticle populations emit in very different regions. It appears from the maps of Fig 3 that there is an evolution from a population of VSGs to PAH^0 and then PAH^+ while approaching the star. This reveals the processing of the nanoparticles by the UV stellar radiation. The same strategy was tested using the cubes of external galaxies from the *SINGS* program which provides a database of mid-IR spectral cubes for tens of nearby galaxies. Fig. 4 presents a map of the ratio of the two correlation parameters, resp. of PAH^0 and PAH^+, obtained for the Evil Eye galaxy. This method provides a unique way to spatially trace the ionization fraction of PAHs which, combined with other tracers, is fundamental to understand the evolution of galaxies.

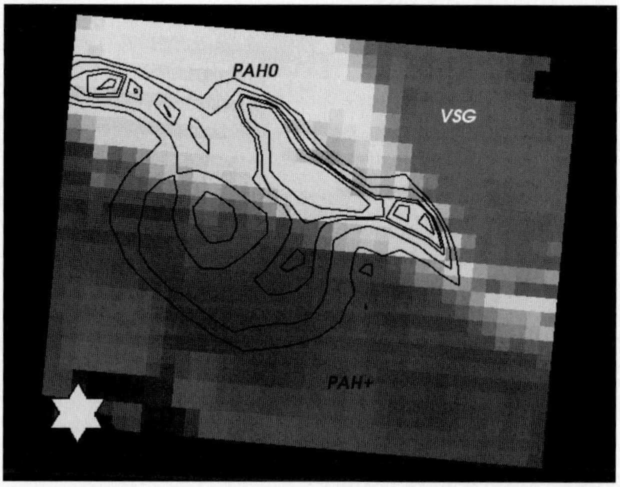

Fig. 3. Correlation maps of the three populations of nanoparticles in NGC 7023 North: VSGs in red, PAH^0 in green and PAH^+ in blue. The contours in black show the emission at 8 μm from Fig. 1. The slight correlation of VSGs with observations seen near the star is an artifact.

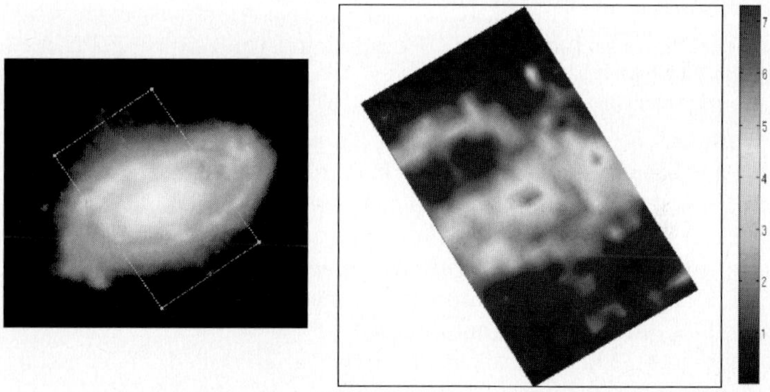

Fig. 4. *Left*: Infrared (8 μm) view of the NGC 4826 (Evil Eye) Galaxy. The rectangle indicates the region observed in spectral mapping with IRS. *Right*: Map of the ratio of PAH^0 over PAH^+ in NGC 4826 achieved using the BSS-extracted spectra (Fig.2).

5 Conclusion

Using two BSS methods, we were able to identify the genuine mid-IR spectra of three propulations of carbonaceous nanoparticles in the interstellar medium. We have shown that both *FastICA* and NMF are efficient for this task, although NMF is found to be sligthly more appropriate. The extracted spectra enable us to study the evolution of carbonaceous nanoparticles in the interstellar medium with unprecedented precision, including in external galaxies. These results stress the fact that BSS methods have much to reveal in the field of observational astrophysics. We are currently analyzing more spectral cubes observations from the *Spitzer* database using the strategy presented in this paper.

References

1. Rapacioli, M., Joblin, C., Boissel, P.: Spectroscopy of polycyclic aromatic hydrocarbons and very small grains in photodissociation regions. Astronomy and Astrophysics 429, 193–204 (2005)
2. Hyvarinen, A., Karhunen, J., Oja, E. In: Wiley (ed.): Independent Component Analysis (2001)
3. Hyvarinen, A.: A Fast Robust Fixed Point Algorithm for Independent Component Analysis. IEEE Transactions on Neural Networks 10, 626–934 (1999)
4. Lee, D.D., Seung, H.S.: Learning the parts of objects by non-negative matrix factorization. Nature 401, 788–791 (1999)
5. Forni, O., Poulet, F., Bibring, J.P.: The Omega Science Team: Component Separation of OMEGA Spectra with ICA. In: 36th Annual Lunar and Planetary Science Conference (March 2005), p. 1623 (2005)
6. Funaro, M., Erkki, O., Valpola, H.: Independent component analysis for artifact separation in astrophysical images. Neural Networks 16, 469–478 (2003)
7. Maino, D., Farusi, A., Baccigalupi, C., Perrotta, F., Banday, A.J., Bedini, L., Burigana, C., De Zotti, G., Górski, K.M., Salerno, E.: All-sky astrophysical component separation with Fast Independent Component Analysis (FASTICA) (July 2002) 334, 53–68 (2002)
8. Sajda, P., Du, S., Brown, T.R., Stoyanova, R., Shungu, D.C., Mao, X., Parra, L.C.: IEEE Transactions on Medical Imaging 23, 1453–1465 (December 2004)
9. Gobinet, A., Elhafid, A., Vrabie, V., Huez, R., Nuzillard, D.: In: Proceedings of the 13th European Signal processing Conference (September 2005)
10. Nuzillard, D., Bijaoui, A.: Blind source separation and analysis of multispectral astronomical images. Astronomy and Astrophysics, Supplement 147, 129–138 (2000)
11. Lee, D.D., Seung, H.S.: Algorithms for non-negative matrix factorization. In: NIPS, vol. 13, p. 556. MIT Press, Cambridge (2001)
12. Léger, A., Puget, J.L.: Identification of the 'unidentified' IR emission features of interstellar dust? Astronomy and Astrophysics 137, L5–L8 (1984)
13. Allamandola, L.J., Tielens, A.G.G.M., Barker, J.R.: Polycyclic aromatic hydrocarbons and the unidentified infrared emission bands - Auto exhaust along the Milky Way. The Astrophysical Journal, Letters 290, L25–L28 (1985)
14. Rapacioli, M., Calvo, F., Joblin, C., Parneix, P., Toublanc, D., Spiegelman, F.: Formation and destruction of polycyclic aromatic hydrocarbon clusters in the interstellar medium. Astronomy and Astrophysics 460, 519–531 (2006)

Specific Circumstances on the Ability of Linguistic Feature Extraction Based on Context Preprocessing by ICA

Markus Borschbach and Martin Pyka

Dept. of Mathematics and Computer Science, Institute for Computer Science, University of Münster, Einsteinstr. 62, D-48149, Germany
{markus.borschbach,pyka}@uni-muenster.de

Abstract. Blind Signal Separation (BSS) based on Independent Component Analysis (ICA) is an emerging approach which application is not limited to the signal processing research, where its application principle is rather straight forward. For an increasing amount of information processing fields, ICA has meaningful application which are still undiscovered. The aim of this paper is to investigate the ability of linguistic feature extraction based on word context preprocessing by ICA. The work refers to a first brief analysis in which ICA was applied to an English corpus. We continue this analysis depending on the number of components and the amount of syntactical information that we take into account. Furthermore we discuss to which extent the results deliver general linguistic features, or linguistic features giving us information about the text.

1 Introduction

In various information processing fields, Independent Component Analysis (ICA) [1] has emerged as a key notion of the underlying principle, often leading to an alternative preprocessing step or filter of the observed data streams. The different cases of assumed underlying models, - from a scalar and linear case, to a convolutive, a nonlinear or the case of single channel ICA, have enabled a broad spectrum of suitable different application areas. In many signal processing fields, the task of Blind Signal Separation based on ICA can be directly applied to a given number of observations and the assumed underlying sources can be extracted. In all but a few information processing areas, the meaning of the independent components is rather undiscovered so far. All assumed underlying ICA models are comparable in the sense of identification of similarities expressed by the condition of independence. At first, this paper contributes by summarizing approaches for linguistic text analysis based on ICA and expressing the role of the independent components. More specific, ICA is used as a preprocessing step to identify similarities in the context of all words of different types of text. The paper is organized as follows. Section 2 briefly reviews the objective function of ICA in an existing approach for linguistic feature extraction. The operations step of ICA-preprocessing and their strengths and weaknesses are outlined in section 3.

M.E. Davies et al. (Eds.): ICA 2007, LNCS 4666, pp. 689–696, 2007.

The simulation results based on this approach with two kinds of preprocessed texts are presented in section 4 and discussed in section 5.

2 Related Work

2.1 Word-Based Linguistic Feature Extraction

The approach which is examined in this paper was introduced primarily in [2] and applied on a corpus of 4,921,934 tokens (words in the running text) including 117,283 types (different unique words). At the preprocessing all uppercase letters were replaced by the corresponding lowercase letters and punctuation was removed. For the creation of the context matrix, one hundred common words were manually selected. The context of these words was analyzed using the 2000 most common words of the text in the following way: If one of the hundred words appears in the text followed by a word of the 2000 most common words, the corresponding value c_{ij}, where i is the index of the manually selected word and j the index of the context word, is increased by 1. Therefore the context matrix displays the number of occurrences of each context word in the immediate context of the manually selected words. If BSS is assumed as the classical ICA signal processing application task, the dimension of the mixture vectors is 100 and the samples size is 2000. In the last preprocessing step the logarithm was taken in order to reduce the effect of the very common words in the text.

The FastICA-Algorithm [4] was used to reduce and extract the 10 strongest independent components based on the greatest eigenvalues (using Principle Component Analysis). Now each of the 2000 selected context words is not represented by its number of occurrences in the context of different index words but by a ten dimensional vector. Depending on the amplitude of the values in each dimension the syntactical function of the words can be categorized. In the given experiment all plural nouns had a significant greater value in component five than in other components for instance. Adjectives, verbs and even verb forms of “to be” and “to have” were represented by other categories.

The results show that the ICA approach was able to extract emergent linguistic features although the analysis reveals that the assignments of the ICs to the linguistic features are still quite loose. Therefore we consider it necessary to investigate, how the results can be improved by changing the context distance, the size of the corpora and the number of independent components.

2.2 Morpheme-Based Linguistic Feature Extraction

The same approach was applied on a Finnish corpus taking into account morphemes as the smallest unit [3]. During preprocessing a Finnish newspaper text with 30 million words and 1.3 million unique words was segmented into morphemes by applying the segmentation tool Hutmegs [10]. For a selection of 3759 data morphemes a context matrix was generated representing the distribution of the 506 most common morphemes in the immediate context of the data morphemes. The first 50 independent components were obtained via the FastICA algorithm. The results showed that morphemes with a common grammatical feature had a similar 50-dimensional vector with high values (+ or -) in the components which indicate their grammatical functions. For instance morphemes of country names had three components with high values. A further analysis showed that this group of morphemes can be extended to words which refer to languages (ranska/ranskaksa: France/frensh), to inhabitants (ranska/ranskalainen: France/ frensh-man) or to words with the same kind of endings. Ambiguous words had a higher number of active components representing the different meanings.

3 Simulation Onset

In this paper, we tried to reproduce and to improve the results of the mentioned approach in order to figure out possible application fields. Therefore we increased the number of independent components up to 30 and - in a second step - we ran the same test with the full stop as sentence separator which should increase the amount of linguistic information that was used to create the context matrix.

3.1 Settings and Selections

In the first run we tried to reproduce the same or at least similar results with a corpus of articles of the newspaper Times from 1992, including 4,261,330 words and 97,936 unique tokens. Therefore we used the 2000 most common word of the text as contextwords and 100 manually selected so-called indexwords in order to create a context matrix in which each element of the matrix contains the number of occurrences of a contextword after an indexword. Figure 1 illustrates this approach. After that the number of dimensions was reduced to 10 by principle component analysis.

The analysis of all components reveals, that there are five clear distinguishable syntactical categories and four categories that have overlapping syntactical features. For one component a meaningful linguistic binding could not be identified.

While the first five components correspond very well to syntactical categories, the last four categories shown in the table (and the category that did not match to any recognizable linguistic feature binding) seem to have overlapping syntactical features which can be explained by the limited number of independent components that we chose. Nevertheless, even for those components plausible syntactic categories could be assigned. For example, component six contains verbs like 'should', 'says' or 'took'. Their common feature are the group of words, mainly pronouns, which are prefixed. All verbs in this category are used in expressions like 'he should...', 'he says...' or 'it took...'. In category eight a similar phenomenon appeared. It includes words like 'november', 'edingburgh' or '1986'. They have in common that they are mainly used with 'on', 'in' and 'of'. Although time- and location-related words dominate the results, we found words like 'corporation' or 'principle' as well. Therefore, the dependency of the kind of syntactical categories depend on the principles for establishing the context

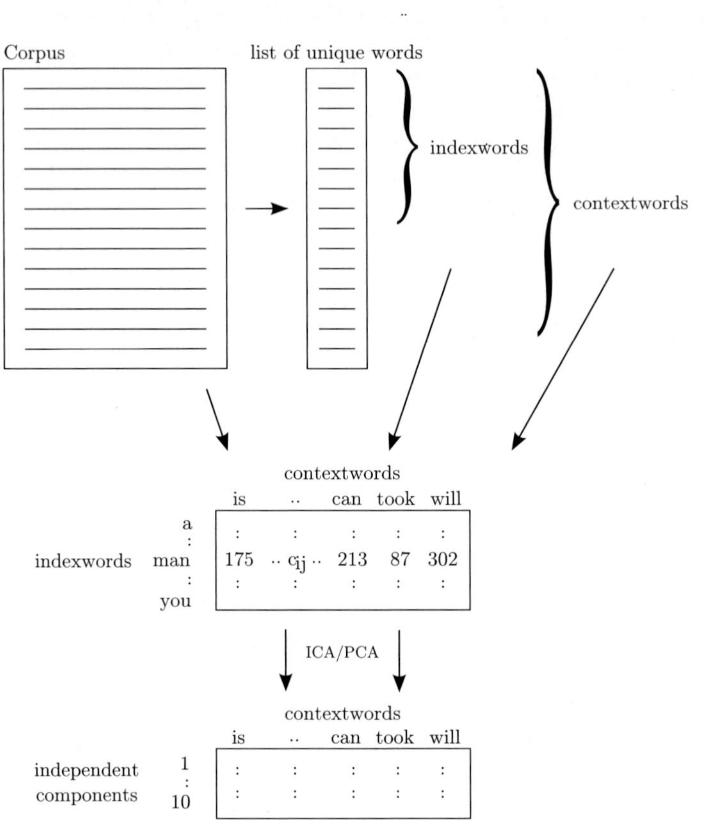

Fig. 1. All unique words are extracted out of a corpus. After a manual selection of the index- and contextwords a contextmatrix is computed. Through ICA with PCA it is reduced to a given number of independent components.

1	Verbs in Past Tense
2	Verbs in Infinitive
3	Numbers
4	Nouns in plural
5	Proper names
6	Verbs in various forms
7	Prepositions, conjunctions, articles
8	Nouns, that follow 'in' and 'on', mainly names of cities and months
9	Words which appear frequently with 'the' and 'an'

Fig. 2. Clear distinctive categories

matrix and the number of independent components. The chances and weaknesses of this behavior are discussed in the following section.

4 Results

4.1 Settings

Based on the same context matrix the number of components was increased in order to see how good further linguistic categories can be extracted and to which extend the overlapping components can be minimized. First 20 categories were calculated by principle component analysis and their results investigated. Then an analysis with 30 categories was compared with the results of the one before. In general it could be stated: The more independent components one computes, the more specialized are the categories that can be found in the data. They do not necessarily correspond to typical linguistic categories. Their dependency on the collocations in the text leads to various groups in which one can only vaguely discern any linguistic or even semantical categories. First the results of this experiment are examined in more detail and then the possible application fields will be discussed.

4.2 Implications

The results are summarized in table 1. The categories 1, 2, 3, 4, 5 and 16 can be contemplated as the most clear syntactical categories that could be extracted with ICA. They relate to a clear syntactical environment in which they occur and they are discernible for humans as a part of speech as well. Although it seems self-evident that singular nouns are a strong syntactical group as well, they do not occur as a whole group. In the categories of the 20 and 30 components we find them divided into several sub groups, like category 10 in which there are all nouns that follow after "on", "in" or "last", like "on sunday", "in december" or "last week". Category number 11 contains nouns that follow prevalently after words like "former", "national", "party" and some others. Typical collocations are "former chairman", "national institute" and "party leader".

These words seem to share a semantical feature, which is vaguely related to the domain 'corporate structures, business, economy'. This kind of category and the words in it, reflect topics of the articles in the Times newspaper. So the categories that we receive are on one hand the result of an emergent unsupervised learning process of the English language. We get distinctive features that have clear syntactical properties. On the other hand they give us a notion of the topics that are discussed in the articles. We have to admit that they do not qualify for completeness, but at least it seems that a particular selection of context- and indexwords can lead to various components in which we can find key words of a text. This has to be analyzed in more detail with various indexwords and different kinds of texts.

Due to the fact that we build the context matrix as described in [5], the emergent feature categories do not always match completely to classical categories.

Table 1. Linguistic categories in 10, 20 and 30 components

No.	10 components	20 components	30 components
1	Verbs in Past Tense	Verbs in Past Tense	Verbs in Past Tense
2	Verbs in Infinitive	Verbs in Infinitive	Verbs in Infinitive
3	Numbers	Numbers	Numbers
4	Nouns in plural	Nouns in plural	Nouns in plural
5	Proper names	Proper names	Proper names
6	Verbs in various forms	Verbs in various forms	Verbs in various forms
7	Prep., conj., articles	Prep., conj., articles	Prep., conj., articles
8	Nouns, that follow after "in" and "on", mainly names of cities and months	Nouns, that follow after "in" and "on", mainly names of cities and months	Nouns, that follow after "in" and "to", mainly years and months (some cities)
9	Words which appear frequently with "the" and "an"	Verbs like "going", "based" or "able" that follow after "was" or "is"	Verbs like "going", "based" or "able" that follow after "was" or "is"
10		Nouns, that follow after "on" and "last": "january", "february", "monday'	Nouns, that follow after "on" and "last": "january", "february", "monday"
11		Words that follow after "former" or "national": "chief", "director", "institute'	Words that follow after "former" or "national": "chief", "director", "institute"
12		Words that follow after numbers or forenames: "per", "million", "clarke"	Words that follow after numbers: "weeks", "years", "goals"
13		Numbers that begin with zero: 01, 02, 05 (part of the structure of the Times corpus)	Numbers that begin with zero: 01, 02, 05 (part of the structure of the Times corpus)
14		Words that follow after mainly possessive pronouns: "father", "mother", "greatest"	Words that follow after mainly possessive pronouns: "father", "mother", "greatest"
15		Verbs that mainly occur with personal pronouns: "could", "might", "hope"	Words that can occur with possessive pronouns or "the": "biggest", "largest", "eyes"
16		Adjectives	Adjectives
17			Words that follow after "last", "first", "next": "week", "game", "century"
18			Words that follow after auxiliaries or conjunctions, mainly adverbs: "hardly", "usually", "not"
19			Words that follow after "national", "international, "british": "market", "group", "football"
20			Words that follow after "new": "zealand", "york", "ideas"
21			Verbs that follow after "he", "she", "it", mainly verbs in third person singular: "seems", "plans", "did"
22			A second category of proper names, mainly forenames: "richard", "george", "professor"
23			Words that follow after "for": "himself", "lunch", "example"

An interesting example can be found in category 14, which consists of words that follow after possessive pronouns like “my”, “his”, “her” etc.. It includes words like “mother”, “father”, “brother”, “colleagues” and body parts like “eyes” and “hands”. In general these are nouns in this category. As there must be a big amount of collocations like “my greatest...” or “my entire...”, the words “greatest” and “entire” and several others are in category 14 as well. Here we expect to reduce the effect of overlapping syntactical categories by using a different context matrix which takes the previous and the following word of an index word into account.

4.3 Using Punctuation

In the previous chapter the punctuation marks were not taken into account. Instead we followed the guideline described in [5]. Hence the punctuation marks might be an important syntactic information for the analysis and were therefore left for the second test in the text. So only question marks and brackets were removed, everything else was treated like a word. The extraction of the 30 strongest components showed that basically the same results as in the analysis of the text without punctuation marks can be extracted. But instead of 23 categories we were able to find syntactical similarities in 26 of 30 components. 18 of the 23 categories were in both tests identical[1]. The punctuation marks as separator in the text prevented the connection between two successive words that are not located in the same sentence. Nevertheless the analysis of the words without punctuation marks was still quite effective.

5 Conclusion

The increase of the number of components led to a more detailed syntactical segmentation of the 2000 selected words. This segmentation gives us information about the context in which the words are generally used. Although some of this categories relate even to lexical categories, there are always a few outstanding words that occur through exceptions in the grammatical use of the language. The more components we extract with ICA, the more specific are the syntactical categories.

In corpora with an emphasis on certain topics we were able to extract collocations that are not typical for the English language in general but for this specific topic. For example, we extracted a category with words that are used in collocations like “former director” or “national institute” (category 11), and a category with words that follow after “international”, 'british' and several others. They are used in collocations like “international group” or “british market”. This segmentation does not only reflect the syntactical features of the English language, but also the mainly used terms and phrasings of a specific text. Due to

[1] Due to the fact that FastICA starts with a random initialization, the results can differ from run to run. But in all tests that were made with 30 components we got at least 14 categories that occur in each test.

their disproportionately high occurrences, they were grouped in a few categories. Further experiments with different context distances and different types of text will show if this could be a way of extracting topic related information from a text as well.

The segmentation-ability of this approach could be interesting in the following way as well. Each context word that is used in this experiment is encoded in a 10- up to 30-dimensional vector which represents its functionality in the text. It is an abstracted description of the word. These codes could be used for finding certain patterns in English sentences in an unsupervised and emergent manner as well. For instance, it should be possible to extract grammatical rules from the text through the encoded words because their vectors identify them as words with particular syntactical features. There are many possibilities to increase the amount of syntactical information that might be important for ICA to extract linguistic features. One is the handling of punctuation marks as single words which allowed us to extract a few clear distinguishable linguistic features more. Further experiments with different context matrices, greater corpora, more context words and more independent components should show to which extend we could extract linguistic features that are maybe usable in other applications.

References

1. Hyvärinen, A., Karhunen, J., Oja, E.: Independent Component Analysis. Wiley & Sons, Chichester (2001)
2. Honkela, T., Hyvärinen, A., Väyrynen J.: Emergence of Linguistic Features: Independent Component Analysis of Contexts. In: Proc. of the 9th Neural Computation and Psychology Workshop (NCPW9), pp. 129–138 (2005)
3. Lagus, K., Creutz, M., Virpioja, S.: Latent Linguistic Codes for Morphemes using Independent Component Analysis. In: Proc. of the 9th Neural Computation and Psychology Workshop (NCPW9), pp. 129–138 (2005)
4. Hyvärinen, A.: Fast and Robust Fixed-Point Algorithms for Independent Component Analysis. IEEE Transactions on Neural Networks 10(3), 626–634 (1999)
5. Honkela, T., Hyvärinen, A.: Linguistic Feature Extraction using Independent Component Analysis. In: Proc. of Int. Joint Conf. on Neural Networks (IJCNN), pp. 279–284 (2004)
6. Rapp, R.: Mining Text for Word Senses using Independent Component Analysis. In: Proc. of Int. Conf. on Data Mining (2004)
7. Yarowsky, D.: Unsupervised Word Sense Disambiguation rivaling Supervised Methods. In: Proc. of the 33rd Annual Meeting of the ACL, pp. 189–196 (1995)
8. Steiner, P.: Wortarten und Korpus. Dissertation at the Westfälische Wilhelms-Universität Münster. Shaker Verlag (2004)
9. Rapp, R.: Automatic Identification of Word Translations from unrelated English and German Corpora. In: Proc. of the 37th Annual Meeting of the ACL, pp. 519–526 (1999)
10. Creutz, M., Lindén, K.: Morpheme Segmentation Gold Standards for Finnish and English. Techreport Helsinki University of Technology, Publications in Computer and Information Science (2005)

Conjugate Gamma Markov Random Fields for Modelling Nonstationary Sources

A. Taylan Cemgil[1] and Onur Dikmen[2,⋆]

[1] Engineering Dept., University of Cambridge, CB2 1PZ, Cambridge, UK
atc27@eng.cam.ac.uk
http://www-sigproc.eng.cam.ac.uk/∼atc27
[2] Dept. of Computer Eng., Bogazici University, 80815 Bebek, Istanbul, Turkey

Abstract. In modelling nonstationary sources, one possible strategy is to define a latent process of strictly positive variables to model variations in second order statistics of the underlying process. This can be achieved, for example, by passing a Gaussian process through a positive nonlinearity or defining a discrete state Markov chain where each state encodes a certain regime. However, models with such constructs turn out to be either not very flexible or non-conjugate, making inference somewhat harder. In this paper, we introduce a conjugate (inverse-) gamma Markov Random field model that allows random fluctuations on variances which are useful as priors for nonstationary time-frequency energy distributions. The main idea is to introduce auxiliary variables such that full conditional distributions and sufficient statistics are readily available as closed form expressions. This allows straightforward implementation of a Gibbs sampler or a variational algorithm. We illustrate our approach on denoising and single channel source separation.

1 Introduction

In the Bayesian framework, various signal estimation problems can be cast into posterior inference problems. For example, source separation [6,5,9,11,3,2], can be stated as

$$p(\mathbf{s}|\mathbf{x}) = \frac{1}{Z_x} \int d\Theta_o d\Theta_s p(\mathbf{x}|\mathbf{s}, \Theta_o) p(\mathbf{s}|\Theta_s) p(\Theta_o) p(\Theta_s) \qquad (1)$$

where $\mathbf{s} \equiv s_{1:K,1:N}$ and $\mathbf{x} \equiv x_{1:K,1:M}$. Here, the task is to infer N *source signals* $s_{k,n}$ given M *observed signals* $x_{k,m}$ where $n = 1 \dots N$, $m = 1 \dots M$ at each index k where $k = 1 \dots K$. Here, k typically denotes time or a time-frequency atom in a linear transform domain. In Eq.(1), the (possibly degenerate, deterministic) conditional distribution $p(\mathbf{x}|\mathbf{s}, \Theta_o)$ specifies the *observation model* where Θ_o denotes

⋆ This research is funded by EPSRC (Engineering and Physical Sciences Research Council of UK) under the grant EP/D03261X/1 entitled "Probabilistic Modelling of Musical Audio for Machine Listening" and by TUBITAK (Scientific and Technological Research Council of Turkey) under the grant entitled "Time Series Modelling for Bayesian Source Separation".

M.E. Davies et al. (Eds.): ICA 2007, LNCS 4666, pp. 697–705, 2007.

the collection of mixing parameters such as the mixing matrix, observation noise variance, etc. The prior term $p(\mathbf{s}|\Theta_s)$, the *source model*, describes the statistical properties of the sources via their own prior parameters Θ_s. The normalisation term $Z_x = p(\mathbf{x})$ is the marginal probability (*evidence*) of the data under the model and plays a key role for model order selection (such as determining the number of sources) [8]. The hierarchical model is completed by postulating hyper-priors over the nuisance parameters Θ_s and Θ_o. Estimates of the sources can be obtained from posterior features such as marginal maximum-a-posteriori (MMAP) or minimum-mean-square-error (MMSE) estimate[1]

$$\mathbf{s}^* = \underset{\mathbf{s}}{\mathrm{argmax}}\, p(\mathbf{s}|\mathbf{x}) \qquad \langle \mathbf{s} \rangle_{p(\mathbf{s}|\mathbf{x})} = \int \mathbf{s} p(\mathbf{s}|\mathbf{x}) d\mathbf{s}$$

Unfortunately, exact calculation of these quantities is intractable for almost all relevant observation and source models, even under conditionally Gaussian and independence assumptions. Hence, approximate numerical integration techniques have to be employed.

In applications, the key object is often the source model $p(\mathbf{s}|\Theta_s)$. Indeed, many popular signal estimation algorithms can be obtained by choosing a particular source prior and applying Bayes rule. If the sources have some known structure, one can design more realistic prior models to improve the estimates. In this paper, we explore generic source models that explicitly model nonstationarity.

Perhaps the prototypical example of a nonstationary process is one where the conditional variance is slowly changing in time. In finance literature, such models are known as *stochastic volatility* models and are important to characterise non-stationary behaviour observed in financial markets [12]. In spatial statistics, similar constructions are needed in 2-D where one is interested in changing intensity over a region [15]. In audio processing, the energy content of a signal is typically time-varying hence it is natural to model audio with a process with a time varying power spectral density on a time frequency plane [10,14,4].

In the sequel, we introduce an alternative model that is useful for modelling a random walk on variances. The main idea is to introduce auxiliary variables such that full conditional distributions and sufficient statistics are readily available as closed form expressions. This allows straightforward implementation of a Gibbs sampler or a variational algorithm. Consequently we extend the model to 2-D random fields, which is useful for modelling nonstationary time-frequency energy distributions or intensity functions that need to be strictly positive. We illustrate our approach on various denoising and source separation scenarios.

2 Model

The *inverse Gamma* distribution with shape parameter a and scale parameter z is defined as

$$\mathcal{IG}(v; a, z) \equiv \exp((a+1)\log v^{-1} - z^{-1}v^{-1} + a \log z^{-1} - \log \Gamma(a))$$

[1] Here, and elsewhere the notation $\langle f(x) \rangle_{p(x)}$ will denote the expectation of the function $f(x)$ under the distribution $p(x)$, i.e. $\langle f(x) \rangle_p \equiv \int dx f(x) p(x)$.

Here, Γ is the gamma (generalised factorial) function. The sufficient statistics of the inverse-Gamma distribution are given by $\left\langle v^{-1}\right\rangle_{\mathcal{IG}} = az$ and $\left\langle \log v^{-1}\right\rangle_{\mathcal{IG}} = \Psi(a) - \log z^{-1}$ where Ψ is the digamma function defined as $\Psi(a) \equiv d\log\Gamma(a)/da$. The inverse gamma distribution is the conjugate prior for the variance v of a Gaussian distribution $\mathcal{N}(s;\mu,v) \equiv \exp\left(-(s-\mu)^2 v^{-1}/2 + \log v^{-1}/2 - \log(2\pi)/2\right)$. When the prior $p(v)$ is inverse Gamma, the posterior distribution $p(v|s)$ can be represented as an inverse Gamma distribution since the logarithm of a Gaussian is a polynomial in v^{-1} and $\log v^{-1}$. Similarly, the *Gamma* distribution with shape parameter a and scale parameter z is defined as

$$\mathcal{G}(\lambda; a, z) \equiv \exp((a-1)\log\lambda - z^{-1}\lambda + a\log z^{-1} - \log\Gamma(a))$$

The sufficient statistics of the Gamma distribution are given by $\langle\lambda\rangle_{\mathcal{G}} = az$ and $\langle\log\lambda\rangle_{\mathcal{G}} = \Psi(a) - \log z^{-1}$. Gamma distribution is the conjugate prior for the precision parameter (inverse variance) of a Gaussian distribution as well as for the intensity parameter λ of a Poisson distribution

$$c \sim \mathcal{PO}(c;\lambda) \equiv e^{-\lambda}\lambda^c/c! = \exp\left(c\log\lambda - \lambda - \log\Gamma(c+1)\right)$$

We will exploit this property to estimate intensity functions of non-homogeneous Poisson processes.

2.1 Markov Chain Models

It is possible to define a Markov chain on inverse Gamma random variables in a straightforward way by $v_k|v_{k-1} \sim \mathcal{IG}(v_k; a, v_{k-1}/a)$. The full conditional distribution $p(v_k|v_{k-1}, v_{k+1})$ is conjugate, i.e. it is also inverse Gamma. However, by this construction it is not possible to attain positive correlation between v_k and v_{k-1}. Positive correlations can be obtained by conditioning on the reciprocal of v_{k-1} and defining $p(v_k|v_{k-1}) = \mathcal{IG}(v_k; a, (v_{k-1}a)^{-1})$; however in this case the full conditional distribution $p(v_k|v_{k-1}, v_{k+1})$ becomes non-conjugate since it has v_k, $1/v_k$ and $\log v_k$ terms. The basic idea is to introduce latent auxiliary variables z_k between v_k and v_{k-1} such that when z_k are integrated out we restore positive correlation between v_k and v_{k-1} while retaining conjugacy. We define an *Inverse Gamma-Markov* chain (IGMC) for $k = 1 \ldots K$ as follows

$$z_1 \sim \mathcal{IG}(z_1; a_z, b_z/a_z) \quad v_k|z_k \sim \mathcal{IG}(v_k; a, z_k/a) \quad z_{k+1}|v_k \sim \mathcal{IG}(z_{k+1}; a_z, v_k/a_z)$$

Here, z_k are auxiliary variables that ensure the full conditionals

$$p(v_k|z_k, z_{k+1}) \propto \exp\left((a + a_z + 1)\log v_k^{-1} - (az_k^{-1} + a_z z_{k+1}^{-1})v_k^{-1}\right) \qquad (2)$$

and $p(z_k|v_k, v_{k-1})$ are inverse Gamma. By integrating out over the auxiliary variable z_k we obtain the effective transition kernel of the Markov chain, where it can be easily shown that

$$p(v_k|v_{k-1}) = \int dz_k p(v_k|z_k)p(z_k|v_{k-1}) = \int dz_k \mathcal{IG}(v_k; a, z_k/a)\mathcal{IG}(z_k; a_z, v_{k-1}/a_z)$$

$$= \frac{\Gamma(a + a_z)}{\Gamma(a_z)\Gamma(a)} \frac{(a_z v_{k-1}^{-1})^{a_z}(a v_k^{-1})^a}{(a_z v_{k-1}^{-1} + a v_k^{-1})^{(a_z + a)}} v_k^{-1} \qquad (3)$$

This distribution, which in our knowledge does not have a designated name, is a scale mixture of inverse Gamma distributions where the scaling function is also inverse Gamma. The transition kernel $p(v_k|v_{k-1})$ has positive correlation for various shape parameters a_z and a. The absolute value of a_z and a control the strength of the correlation and the ratio a_z/a controls the skewness. For $a_z/a < 1$ ($a_z/a > 1$), the probability mass is shifted towards the interval $v_k < v_{k-1}$ ($v_k > v_{k-1}$) hence, typical trajectories from a IGMC will exhibit a systematic negative (positive) drift. Using an exactly analogous construction, we define a *Gamma-Markov* chain (GMC) as $z_1 \sim \mathcal{G}(z_1; a_z, (b_z a_z)^{-1})$, $\lambda_k|z_k \sim \mathcal{G}(\lambda_k; a_\lambda, (z_k a_\lambda)^{-1})$, $z_{k+1}|\lambda_k \sim \mathcal{G}(z_{k+1}; a_z, (\lambda_k a_z)^{-1})$. The effective transition kernel has a very similar expression as in Eq.3.

Example 1, Nonstationary Gaussian Process: We define a non-stationary Gaussian process $\{y_k\}_{k=1,2,\ldots}$ by drawing the variances $\{v_k\}_{k=1,2,\ldots}$ from an IGMC and drawing $y_k|v_k \sim \mathcal{N}(y_k; 0, v_k)$ In Figure 1(a)-top, we show a realisation of $v_{1:K}$ from the IGMC, labelled as "true" and generate $y_{1:K}$ conditionally Figure 1(a)-bottom. Given a realisation $y_{1:K}$, we can estimated the posterior variance $\langle v_k|y_{1:K}\rangle$. In this case, inference is carried out with variational Bayes as will be detailed in section 3.

Example 2, Nonhomogeneous Poisson Process: We partition an interval $\mathbb{I}$ on the real line into small disjoint regions R_k of area L such that $\mathbb{I} = \cup_{k=1}^K R_k$. We assume that the unknown intensity function of the process is piecewise constant and has the value λ_k on region R_k. The intensity function $\{\lambda_k\}_{k=1,2,\ldots}$ is drawn according to a GMC. The number points in R_k, given the intensity function, is denoted by the Poisson random variable $c_k|\lambda_k \sim \mathcal{PO}(c_k; \lambda_k L)$ To generate a realisation from the Poisson process, we can uniformly draw c_k points in each region R_k. In Figure 1(b), we show a realisation from the model. Given the number of events in each region R_k, we can estimate the value of the intensity function on R_k by calculating $\langle \lambda_k|c_{1:K}\rangle$.

2.2 (Inverse) Gamma Markov Random Fields – (I)GMRF

We have defined the IGMC and GMC in the previous section using conditional distributions. An alternative but equivalent factorisation, that encodes the same distribution but corresponds to

$$p(\mathbf{z}, \mathbf{v}) \propto \psi(b_z^{-1}, a_z z_1^{-1}) \prod_k \phi(v_k^{-1}; a + a_z)\phi(z_k^{-1}; a + a_z)\psi(a z_k^{-1}, v_k^{-1})\psi(a_z v_k^{-1}, z_{k+1}^{-1})$$

where we specify singleton potentials ϕ and pairwise potentials ψ as

$$\phi(\xi; \alpha) = \exp((\alpha + 1)\log \xi) \qquad\qquad \psi(\xi, \eta) = \exp(-\xi\eta)$$

Generalising this to a general undirected graph with vertex set $\mathcal{V}$ and undirected edge set $\mathcal{E}$, we define an IGMRF on $\boldsymbol{\xi} = \{\xi_i\}_{i\in\mathcal{V}}$ by a set of connection weights $\mathbf{a} = \{a_{i,j}\}_{(i,j)\in\mathcal{E}}$ for $i, j \in \mathcal{V}$ and $i \neq j$

$$p(\boldsymbol{\xi}; \mathbf{a}) = \frac{1}{Z_\mathbf{a}} \prod_{i\in\mathcal{V}} \phi(\xi_i^{-1}; \sum_j a_{i,j}) \prod_{(i,j)\in\mathcal{E}} \psi(\xi_i^{-1}, (a_{i,j}/2)\xi_j^{-1}) \equiv \frac{1}{Z_\mathbf{a}} p_\mathbf{a}^*(\boldsymbol{\xi}) \quad (4)$$

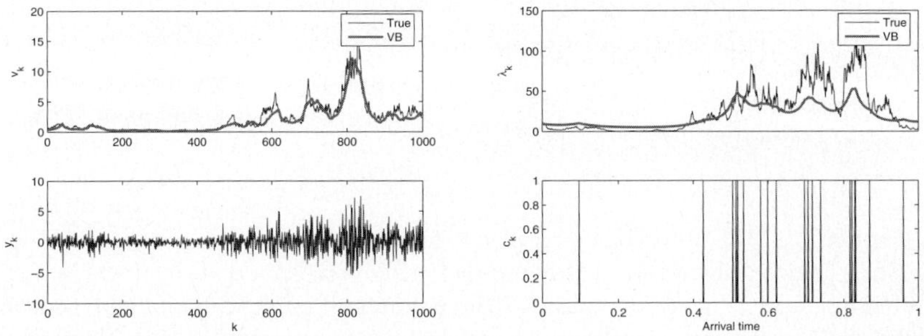

(a) (b)

Fig. 1. Synthetic examples generated from the model. The thick line shows the result of the variational inference(a) Non-stationary Gaussian Process. (a-Top) a typical draw of $v_{1:K}$ from the IGMC $b_z = 1, a = a_z = 100$. (a-Bottom) Draw from the Gaussian process $y_{1:K}$ given $v_{1:K}$. (b) Non-homogeneous Poisson Process. (b-Top) a typical draw from the GMC with $b_z = 10, a = a_z = 100$ and frame length $\mu(R_k) = L = 0.001$. (b-Bottom) Number of events in each R_k.

where ϕ and ψ are defined above. A GMRF is defined similarly by the potentials $\phi(\xi_i; \sum_j a_{i,j})$ and $\psi(\xi_i, (a_{i,j}/2)\xi_j)$ but with $\phi(\xi; \alpha) = \exp((\alpha - 1)\log \xi)$.

3 Inference

Exact inference in random fields is in general intractable and various numerical methods have been developed, based on sampling (Monte Carlo-stochastic) or analytic approximation (Variational-deterministic). Here, we focus on a particularly simple variational algorithm (mean field - variational Bayes [1,13]) – but application of Monte Carlo methods, such as the Gibbs sampler [7] is algorithmically very similar[2]. Variational methods have been applied extensively, notably in machine learning for inference in large models. While lacking theoretical guarantees of Monte Carlo approaches, variational methods have been viable alternatives in several practical situations where only a fixed amount of CPU budget is available.

The main idea in variational Bayes is to approximate a target distribution $\mathcal{P} \equiv p^*(\boldsymbol{\xi})/Z$ (such as the IGMRF defined in Eq.(4)) with a simple distribution $\mathcal{Q}$. The variational distribution $\mathcal{Q}$ is chosen such that its expectations can be obtained easily, preferably in closed form. One such distribution is a factorised one $\mathcal{Q}(\boldsymbol{\xi}) = \prod_{i \in \mathcal{V}} \mathcal{Q}_i(\xi_i)$. An intuitive interpretation of mean field method is minimising the KL divergence with respect to (the parameters of) $\mathcal{Q}$ where $KL(\mathcal{Q}||\mathcal{P}) = \langle \log \mathcal{Q} \rangle_{\mathcal{Q}} - \langle \log p^*/Z_{\mathbf{a}} \rangle_{\mathcal{Q}}$. Using non-negativity of KL, one can obtain a lower bound on log-normalisation constant

$$\log Z_{\mathbf{a}} \geq \langle \log p^* \rangle_{\mathcal{Q}} - \langle \log \mathcal{Q} \rangle_{\mathcal{Q}} \tag{5}$$

The maximisation of this lower bound is equivalent to finding the "nearest" $\mathcal{Q}$ to $\mathcal{P}$ in terms of KL divergence. Whilst the solution is in general not available in

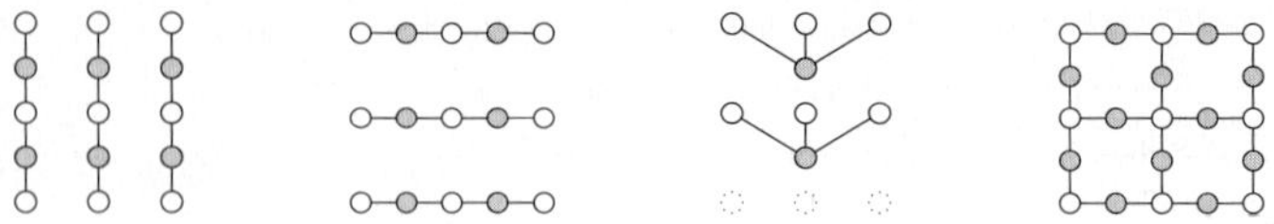

Fig. 2. Various IGMRF topologies as priors for time frequency energy distributions. White nodes (placed always on a rectangular grid) correspond to $v_{\nu,\tau}$ where the vertical and horizontal axis corresponds to the frequency band index ν and time index τ respectively. Gray nodes correspond to the auxiliary variables z. (a)-(d) `vertical, horizontal, band, grid`.

closed form, it can be easily shown, e.g. see [13], that each factor $\mathcal{Q}_i$ of the optimal approximating distribution should satisfy the following fixed point equation

$$\mathcal{Q}_i \propto \exp\left(\langle \log p^* \rangle_{\mathcal{Q}_{-i}}\right) \tag{6}$$

where $\mathcal{Q}_{-i} \equiv \mathcal{Q}/\mathcal{Q}_i$, that is the joint distribution of all factors excluding $\mathcal{Q}_i$. Hence, the mean field approach leads to a set of (deterministic) fixed point equations that need to be iterated until convergence. For a MRF, this fixed point expression is efficient to evaluate since it depends only on the neighbouring variables $j \in \mathcal{N}(i)$. Finally, for conjugate models the factors are available in closed form; for example IGMRF leads to the factors $\mathcal{Q}_i^{(t)}(\xi_i) = \mathcal{IG}(\xi_i; \alpha_i^{(t)}, \beta_i^{(t)})$ with

$$\alpha_i^{(t)} = \theta_{\alpha,i} + \sum_{j \in \mathcal{N}(i)} a_{i,j} \qquad \beta_i^{-1(t)} = \theta_{\beta,i} + \sum_{j \in \mathcal{N}(i)} a_{i,j} \left\langle \xi_j^{-1} \right\rangle_{\mathcal{Q}_j^{(t-1)}}$$

Here, $\theta_{\alpha,i}$ and $\theta_{\beta,i}$ denote the data contributions when a IGMRF is used as a prior where some of the ξ are observed via a conjugate observation model. For example, in the conditionally Gaussian observation model of section 2 we have $\theta_{\alpha,i} = 1/2$ and $\theta_{\beta,i} = y_i^2/2$. Similarly, the Poisson model with a GMRF prior has $\theta_{\alpha,i} = c_i$ and $\theta_{\beta,i} = L$.

3.1 Simulation Experiments

In Figure 1, we show the results of variational inference for two synthetic examples. In the following, we will illustrate the IGMRF model used as a prior for time-frequency energy distributions of nonstationary sources.

Linear time-frequency representations decompose a signal $y(t)$ as a linear decomposition of form $y(t) = \sum_{(\nu,\tau)} s_{(\nu,\tau)} f_{\nu,\tau}(t)$ where $s_{(\nu,\tau)}$ is the expansion coefficient corresponding to the basis function $f_{\nu,\tau}(t)$. Somewhat succinctly we can write $\mathbf{y} = F\mathbf{s}$, where the collection of basis functions is denoted by a matrix F where each column corresponds to a particular $f_{\nu,\tau}$. The well known Gabor representation or modified cosine transform (MDCT) have this form and can be computed using fast transforms where τ corresponds to time and ν corresponds to frequency.In the sequel, we will impose a conditionally Gaussian prior on transform coefficients $\mathcal{N}(s_{(\nu,\tau)}; 0, v_{(\nu,\tau)})$ where the covariance structure will be drawn from a IGMRF.

Denoising: In the first experiment, we illustrate the denoising performance of 4 MRF topologies on a set of 5 audio clips (`speech, piano, guitar, perc1, perc2`) in 3 different noise conditions `low, medium, high`. We transform each clip via MDCT to $s^{true}_{(\nu,\tau)}$ and add independent white Gaussian noise with variance $r \sim \mathcal{IG}(r; a_r, b_r)$ to obtain $x_{(\nu,\tau)}$. Note that since MDCT is an orthonormal linear transform, we could have added noise in time domain and the noise characteristics would have remained unaltered. As inference engine, we use variational Bayes. The task of the inference algorithm is to infer the latent source coefficients $s_{(\nu,\tau)}$ by integrating out the noise variance r and the IGMRF variables $\boldsymbol{\xi}$. The optimisation of MRF parameters $\mathbf{a}$ is carried out by the Newton method where we maximise the lower bound in Eq.5. We assume homogeneous MRF structure where the coupling values are the same throughout the network [2]. The signal to noise ratio of reconstructions and inference results are given in Figure 3-(a). The SNR results do not show big variations across topologies, with the grid model consistently providing good reconstruction, especially in high and medium noise conditions. We note that SNR may not be the best metric to measure perceptual quality and the reader is invited to listen to the audio examples provided online at `http://www.cmpe.boun.edu.tr/~dikmen/ICA07/`. In informal subjective listening tests, we perceive a somewhat better reconstruction quality with the grid topology.

Source Separation: In the second experiment, we illustrate the viability of the approach on a single channel separation task with $j = 1 \dots J$ sources. At each time-frequency location $k \equiv (\nu, \tau)$, the generative model is $\mathbf{v}_j \sim \text{IGMRF}_j$, $s_{k,j} \sim \mathcal{N}(s_k; 0; v_{k,j})$, $x_k = \sum_{j=1}^{J} s_{k,j}$ In this scenario, the reconstruction equations have a particularly simple form. Given the variance estimate $V_{k,j} = \langle 1/v_{k,j} \rangle^{-1}$ at k, we define a positive quantity, which we shall name as *responsibility* also know as filter factors, $\kappa_{k,j} = V_{k,j}/(\sum_{j'} V_{k,j'})$, where by construction $0 < \kappa_j$ for all j and $\sum_j \kappa_j \leq 1$. It turns out that the sufficient statistics can be compactly written as

$$\langle s_{k,j} \rangle = \kappa_{k,j} x_k \qquad \left\langle s^2_{k,j} \right\rangle = V_{k,j}(1 - \kappa_{k,j}) + \kappa^2_{k,j} x^2_k$$

We illustrate this approach to separate a piano sound into its constituent components. We assume that $J = 2$ components are generated independently by two IGMRF models with vertical and horizontal topology. In figure 3-(b), we observe that the model is able to separate transients and harmonic components.

[2] In (a) (`vertical`), each white node is connected to two gray nodes with a_{north} or a_{south} and in (*b*) (`horizontal`) with a_{west} or a_{east}. The `grid` topology (d) has couplings in four directions and in (c) (`band`), we use a single a.

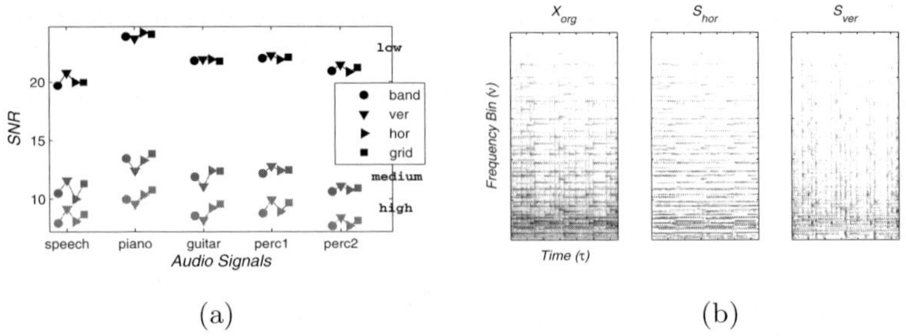

Fig. 3. (a) Signal-to-Noise ratio results for reconstructions obtained from the audio clips in `low, medium,high` noise conditions. (b) Single channel Source Separation example, left to right, log-MDCT coefficients of the original signal and reconstruction with horizontal and vertical IGMRF models.

3.2 Discussion

We have introduced a conjugate gamma Markov random field model for modelling nonstationary sources. The conjugacy makes it possible to design fast inference algorithms in a straightforward way. The simple MRF topologies considered here are quite generic, yet provide good results without any hand tuned parameters. One can envision more structured models with a larger set of parameters to capture physical reality, certainly for acoustical signals. We are currently investigating further applications such as restoration, transcription or tracking time varying intensity functions.

References

1. Attias, H.: Independent factor analysis. Neural Computation 11(4), 803–851 (1999)
2. Cemgil, A.T., Fevotte, C., Godsill, S.J.: Variational and Stochastic Inference for Bayesian Source Separation. Digital Signal Processing (in print, 2007)
3. Cichocki, A., Amari, S.I.: Adaptive Blind Signal and Image Processing: Learning Algorithms and Applications, revised version edition. Wiley, Chichester (2003)
4. Févotte, C., Daudet, L., Godsill, S.J., Torrésani, B.: Sparse regression with structured priors: Application to audio denoising. In: Proc. ICASSP, Toulouse, France (May 2006)
5. Hyvärinen, A., Karhunen, J., Oja, E.: Independent Component Analysis. John Wiley & Sons, New York (2001)
6. Knuth, K.H.: Bayesian source separation and localization. In: SPIE'98: Bayesian Inference for Inverse Problems, San diego, pp. 147–158 (July 1998)
7. Liu, J.S.: Monte Carlo strategies in scientific computing. Springer, Heidelberg (2004)
8. MacKay, D.J.C.: Information Theory, Inference and Learning Algorithms. Cambridge University Press, Cambridge (2003)
9. Miskin, J., Mackay, D.: Ensemble learning for blind source separation. In: Roberts, S.J., Everson, R.M. (eds.) Independent Component Analysis, pp. 209–233. Cambridge University Press, Cambridge (2001)

10. Reyes-Gomez, M., Jojic, N., Ellis, D.: Deformable spectrograms. In: AI and Statistics Conference, Barbados (2005)
11. Rowe, D.B.: A Bayesian approach to blind source separation. Journal of Interdisciplinary Mathematics 5(1), 49–76 (2002)
12. Shepard, N. (ed.): Stochastic Volatility, Selected Readings. Oxford University Press, Oxford (2005)
13. Wainwright, M., Jordan, M.I.: Graphical models, exponential families, and variational inference. Technical Report 649, Department of Statistics, UC Berkeley (September 2003)
14. Wolfe, P.J., Godsill, S.J., Ng, W.J.: Bayesian variable selection and regularisation for time-frequency surface estimation. Journal of the Royal Statistical Society (2004)
15. Wolpert, R.L., Ickstadt, K.: Poisson/gamma random field models for spatial statistics. Biometrica (1998)

Blind Separation of Quantum States: Estimating Two Qubits from an Isotropic Heisenberg Spin Coupling Model

Yannick Deville[1] and Alain Deville[2]

[1] Université Paul Sabatier Toulouse 3 - CNRS, Laboratoire d'Astrophysique de Toulouse-Tarbes (UMR 5572), 14 Av. Edouard Belin, 31400 Toulouse, France
ydeville@ast.obs-mip.fr

[2] Université de Provence, Bâtiment IRPHE, 49 Rue Joliot Curie, BP146, 13384 Marseille Cedex 13, France
alain.deville2@wanadoo.fr

Abstract. Blind source separation (BSS) and Quantum Information Processing (QIP) are two recent and rapidly evolving fields. No connection has ever been made between them to our knowledge. However, future practical QIP systems will probably involve "observed mixtures", in the BSS sense, of quantum states (qubits), e.g. associated to coupled spins. We here investigate how individual qubits may be retrieved from Heisenberg-coupled versions of them, and we show the relationship between this problem and classical BSS. We thus introduce new nonlinear mixture models for qubits, motivated by actual quantum physical devices. We analyze the invertibility and ambiguities of these models. We propose practical data processing methods for performing inversions.

1 Introduction

Various areas in the information processing field developed very rapidly during the last decades. This includes the generic Blind Source Separation (BSS) problem [1], which consists in estimating a set of unknown source signals from a set of observed (i.e. measured) signals which are "mixtures" of these source signals. BSS methods thus apply to a wide range of signal denoising and component extraction problems. This especially concerns communications, e.g. when a set of radio-frequency antennas provide linear combinations, i.e. "mixtures", of several emitted signals and one aims at retrieving each emitted signal only from their available mixtures (see e.g. [2] for an implementation of this approach).

Another growing area is Quantum Information Processing (QIP), which is closely related to Quantum Physics (QP) [3]. QIP uses abstract representations of systems whose behavior is requested to obey the laws of QP. This already made it possible to develop new and powerful information processing methods, to be contrasted with "classical", i.e. non-quantum, methods such as the above-mentioned BSS approaches. Their effective implementation then requires to develop corresponding practical quantum systems, which is only an emerging topic today.

M.E. Davies et al. (Eds.): ICA 2007, LNCS 4666, pp. 706–713, 2007.

To our knowledge, no connection has ever been made between the BSS and QIP/QP areas. One may expect, however, that "coupling" between individual "signals" (i.e. states) will also have to be considered in the QIP/QP area. Such couplings indeed occur even in the basic situation when two spins interact according to the isotropic Heisenberg model. In this paper, we consider this configuration, we investigate how each spin may be retrieved from the coupled version of both of them, and we show the relationship between this problem and classical BSS. The relevance of this approach also stems from the fact that, to a large extent, classical BSS belongs to the more general Statistical Signal Processing (SSP) field. Since QIP and QP are essentially based on a *probabilistic* view of physical phenomena, trying to bridge the gap between SSP/BSS and QIP/QP is a priori a reasonable attempt.

2 Quantum and SSP Points of View for One Qubit

The fundamental concept used in abstract QIP is the quantum bit, or qubit [3]. A qubit has a state $|\psi>$, which may be expressed in the basis defined by two vectors, that we denote $|+>$ and $|->$ hereafter. This state thus reads

$$|\psi>= \alpha|+> +\beta|-> \tag{1}$$

where α and β are two complex-valued coefficients, which are requested to be such that the state $|\psi>$ is normalized, i.e.

$$|\alpha|^2 + |\beta|^2 = 1\ . \tag{2}$$

From a QP point of view, this abstract mathematical model especially concerns the spin of an electron, which is a quantum (i.e. non-classical) quantity. The component of this spin along a given arbitrary axis Oz defines a two-dimensional linear operator s_z. The two eigenvalues of this operator are equal to $+\frac{1}{2}$ and $-\frac{1}{2}$ in normalized units, and the corresponding eigenvectors are therefore denoted $|+>$ and $|->$. The value obtained when measuring this spin component can only be $+\frac{1}{2}$ or $-\frac{1}{2}$. Moreover, assume this spin is in the state $|\psi>$ defined by (1) when performing such a measurement. Then, the probability that the measured value is equal to $+\frac{1}{2}$ (resp. $-\frac{1}{2}$) is equal to $|\alpha|^2$ (resp. $|\beta|^2$), i.e. to the squared modulus of the coefficient in (1) of the associated eigenvector $|+>$ (resp. $|->$).

The above discussion concerns the state of the considered spin at a given time. In addition, this state evolves with time. During the time interval when this spin (or the two coupled spins in the next section) is considered, it is supposed to be isolated or placed in a magnetic field, so that this system has a Hamiltonian. The spin state thus evolves according to Schrödinger's equation. Briefly, if this state $|\psi(t_0)>$ is defined by (1) at time $t = t_0$, its value at any time t is then

$$|\psi(t)>= \alpha e^{-i\omega_p(t-t_0)}|+> +\beta e^{-i\omega_m(t-t_0)}|-> \tag{3}$$

where $i = (-1)^{\frac{1}{2}}$ and the real (angular) frequencies ω_p and ω_m depend on the considered physical setup.

While we summarized above well-known concepts from QIP and QP points of view, we now propose a way to link them with a SSP perspective. We essentially aim at handling the situation when a first user (called "W" hereafter) "writes" the considered spin, i.e. initializes its state at a given time t_w with the value

$$|\psi(t_w)> = \alpha|+> + \beta|-> \tag{4}$$

and a second user (called "R") does not know this state and aims at "reading" it, i.e. at estimating it, at another time t_r. From a SSP point of view, the above description suggests that this may be achieved by introducing a "repeated write/read" (RWR) procedure, which consists in performing K times the same "write/read" step. Each occurence k of this step, with $k = 1 \ldots K$, consists in first letting user W write the spin at a time $t_w(k)$ with the state of interest, i.e.

$$|\psi(t_w(k))> = \alpha|+> + \beta|-> . \tag{5}$$

Due to (3), this spin state becomes at time $t_r(k)$

$$|\psi(t_r(k))> = \alpha e^{-i\omega_p T(k)}|+> + \beta e^{-i\omega_m T(k)}|-> \tag{6}$$

with $T(k) = t_r(k) - t_w(k)$. User R reads this state at time $t_r(k)$. As explained above, this measurement of the spin component along axis Oz can only yield one of the values $+\frac{1}{2}$ and $-\frac{1}{2}$, resp. with probabilities p_1 and p_2 equal to the squared moduli of the coefficients associated to the states $|+>$ and $|->$ in (6), i.e.

$$\left|\alpha e^{-i\omega_p T(k)}\right|^2 = p_1 \tag{7}$$

$$\left|\beta e^{-i\omega_m T(k)}\right|^2 = p_2 . \tag{8}$$

These equations do not depend on their phase factors, i.e. they reduce to

$$|\alpha|^2 = p_1 \tag{9}$$

$$|\beta|^2 = p_2 . \tag{10}$$

Estimates of p_1 and p_2 may be straightforwardly obtained as the relative frequencies of occurence of the values $+\frac{1}{2}$ and $-\frac{1}{2}$ resp. in the measurements. Eq. (9)-(10) then directly provide the squared moduli (up to estimation errors) of the parameters α and β that user R aims at determining. This leaves a phase ambiguity, which may then be reduced or completely avoided as follows. First consider a configuration where we constrain user W to only write the spin with real-valued α and β. Eq. (9)-(10) then make it possible to estimate these parameters up to only a sign ambiguity. Now consider another configuration, where only real *and positive* (including zero) values of α and β are used. The approach that we proposed above then yields these parameters without any ambiguity.

We thus defined a procedure which allows user R to read a spin "blindly", i.e. without any knowledge about it except, possibly, about the domain where α and β are requested to be situated. Let us stress that this procedure may be actually implemented in practice, i.e. the component of a spin along a given axis may be measured, although this requires a complex physical setup. Building upon this solution for a single spin, we now aim at extending it to two coupled spins.

3 Estimating Two Qubits with Heisenberg Spin Coupling

3.1 Considered Qubits: Quantum Point of View

Future QIP systems will simultaneously handle several qubits, which will e.g. be physically implemented as sets of spins. One may expect that undesired coupling (i.e. "mixture", using BSS terminology[1]) between these spins will appear in quantum physical setups, as in current classical signal processing systems, such as the one outlined in Section 1 for communication applications. Indeed, a well-known type of mixture between two spins consists of isotropic Heisenberg coupling, which may be defined as follows [4]. Assume two spins, called spin 1 and spin 2 hereafter, are resp. initialized with states

$$(\alpha_1|+> +\beta_1|->) \quad \text{and} \quad (\alpha_2|+> +\beta_2|->) \tag{11}$$

at a given time t_0 and coupled according to Heisenberg model from then on. Hereafter, we consider the state $|\psi(t)>$ of the overall system composed of these two identifiable spins. At time t_0, this state is equal to the tensor product of the states (11) of both spins. It may be expressed as

$$|\psi(t_0)>= \alpha_1\alpha_2|++> +\alpha_1\beta_2|+-> +\beta_1\alpha_2|-+> +\beta_1\beta_2|--> \tag{12}$$

in the four-dimensional basis $\mathcal{B}_+ = \{|++>, |+->, |-+>, |-->\}$ which corresponds to the operators s_{1z} and s_{2z} resp. associated to the components of the two spins along a given axis Oz. This state may also be expressed in the four-dimensional basis composed of the eigenvectors of Heisenberg coupling's Hamiltonian. We here denote this basis $\mathcal{B}_1 = \{|1,1>, |1,-1>, |1,0>, |0,0>\}$. Using the known expression of $\mathcal{B}_+$ with respect to $\mathcal{B}_1$, (12) yields

$$|\psi(t_0)>= \alpha_1\alpha_2|1,1>+\beta_1\beta_2|1,-1>+\frac{\alpha_1\beta_2+\beta_1\alpha_2}{\sqrt{2}}|1,0>+\frac{\alpha_1\beta_2-\beta_1\alpha_2}{\sqrt{2}}|0,0>. \tag{13}$$

The temporal evolution of this state then corresponds to phase rotations for each eigenvector, as in (3). The state at any time t then reads in basis $\mathcal{B}_1$

$$\begin{aligned}|\psi(t)> = {} & \alpha_1\alpha_2 e^{-i\omega_{1,1}(t-t_0)}|1,1> +\beta_1\beta_2 e^{-i\omega_{1,-1}(t-t_0)}|1,-1> \\ & +\frac{\alpha_1\beta_2+\beta_1\alpha_2}{\sqrt{2}} e^{-i\omega_{1,0}(t-t_0)}|1,0> +\frac{\alpha_1\beta_2-\beta_1\alpha_2}{\sqrt{2}} e^{-i\omega_{0,0}(t-t_0)}|0,0>\end{aligned} \tag{14}$$

where $\omega_{i,j}$ is the real frequency associated to the phase rotation for each eigenvector $|i,j>$. Using the expression of $\mathcal{B}_1$ with respect to $\mathcal{B}_+$ then yields the expression of the system state at any time t in basis $\mathcal{B}_+$

$$\begin{aligned}|\psi(t)> = {} & \alpha_1\alpha_2 e^{-i\omega_{1,1}(t-t_0)}|++> +\beta_1\beta_2 e^{-i\omega_{1,-1}(t-t_0)}|--> \\ & +\frac{1}{2}\left[(\alpha_1\beta_2+\beta_1\alpha_2)e^{-i\omega_{1,0}(t-t_0)} + (\alpha_1\beta_2-\beta_1\alpha_2)e^{-i\omega_{0,0}(t-t_0)}\right]|+-> \\ & +\frac{1}{2}\left[(\alpha_1\beta_2+\beta_1\alpha_2)e^{-i\omega_{1,0}(t-t_0)} - (\alpha_1\beta_2-\beta_1\alpha_2)e^{-i\omega_{0,0}(t-t_0)}\right]|-+>.\end{aligned} \tag{15}$$

[1] This should not be confused with "statistical mixtures" used in QP. From a QP point of view, this paper only concerns pure states, as opposed to statistical mixtures.

Note that this state $|\psi(t)>$ is more easily expressed in basis $\mathcal{B}_1$ than in $\mathcal{B}_+$. We have to consider the latter expression however, because only this basis corresponds to variables which may be measured in practice, i.e. s_{1z} and s_{2z}.

We here started from a concrete (i.e. physical) setup, thus considering a QP point of view. This led us to the state expression (15). From here on, we may therefore move to an abstract QIP point of view, only considering the couple of qubits defined by this state expression (15) and aiming at estimating each of these qubits from their coupled version (15).

3.2 Considered Qubits: SSP Point of View

The first step that we propose towards the estimation of the considered qubits is again based on a SSP approach. It extends to two qubits the RWR procedure that we introduced for one qubit in Section 2. The resulting method operates as follows. In each occurence k of the write/read step, user W first writes both qubits at time $t_w(k)$, resp. with the states defined in (11), and user R then reads at time $t_r(k)$ the state of the system composed of the two coupled qubits, which is defined by (15) except that $(t - t_0)$ is replaced by $T(k) = t_r(k) - t_w(k)$. Reading this state means that user R measures the couple of values associated to s_{1z} and s_{2z}. This couple is then equal to one of the four possible values $(+\frac{1}{2}, +\frac{1}{2})$, $(-\frac{1}{2}, -\frac{1}{2})$, $(+\frac{1}{2}, -\frac{1}{2})$ and $(-\frac{1}{2}, +\frac{1}{2})$, resp. with probabilities p_1, p_2, p_3 and p_4 equal to the squared moduli of the coefficients associated to the states composing $\mathcal{B}_+$ which appear in the considered modified version of (15), i.e.

$$\left|\alpha_1\alpha_2 e^{-i\omega_{1,1}T(k)}\right|^2 = p_1 \tag{16}$$

$$\left|\beta_1\beta_2 e^{-i\omega_{1,-1}T(k)}\right|^2 = p_2 \tag{17}$$

$$\frac{1}{4}\left|(\alpha_1\beta_2 + \beta_1\alpha_2)e^{-i\omega_{1,0}T(k)} + (\alpha_1\beta_2 - \beta_1\alpha_2)e^{-i\omega_{0,0}T(k)}\right|^2 = p_3 \tag{18}$$

$$\frac{1}{4}\left|(\alpha_1\beta_2 + \beta_1\alpha_2)e^{-i\omega_{1,0}T(k)} - (\alpha_1\beta_2 - \beta_1\alpha_2)e^{-i\omega_{0,0}T(k)}\right|^2 = p_4. \tag{19}$$

Again, (16)-(17) do not depend on their phase factors, i.e. they reduce to

$$|\alpha_1\alpha_2|^2 = p_1 \tag{20}$$

$$|\beta_1\beta_2|^2 = p_2. \tag{21}$$

In order to use our SSP approach, (18)-(19) should involve the same parameter values in all occurences k of the write/read step. The write-read time interval $T(k)$ should therefore be the same for all occurences. It is denoted T hereafter.

Estimates of p_1 to p_4 may be straightforwardly obtained as the relative frequencies of occurence of the four values $(+\frac{1}{2}, +\frac{1}{2})$ to $(-\frac{1}{2}, +\frac{1}{2})$ resp. in the measurements. However, unlike in Section 2, these estimated probabilities do not directly yield the parameters α_i and β_i that user R aims at determining, i.e. the two considered qubits are still "mixed" in these measured data. This therefore defines a new nonlinear BSS-like problem (a survey of nonlinear BSS is e.g.

available in [5]), where the observed data consist of the measured probabilities p_1 to p_4, the "source signals" to be extracted from them are the parameters α_i and β_i and the unknown coefficients of the considered set of nonlinear mixing equations are the frequencies $\omega_{i,j}$. This quantum mixture model (16)-(19), or its slightly simplified form involving (20)-(21), may be referred to as "natural complex-valued isotropic Heisenberg spin coupling model" or more briefly "DD1 model" (for Deville & Deville quantum model no. 1 from BSS point of view). Note that the equations in this model are partly redundant:

$$p_1 + p_2 + p_3 + p_4 = 1 \tag{22}$$

because the initial states (11) are normalized, so that the state $|\psi(t_w(k)) >$ defined by (12) is normalized, and this state then evolves according to Schrödinger's equation, which keeps norm unchanged. We now show how to separate these sources, depending on how these qubits are initialized.

3.3 Retrieving Qubits with Real-Valued Initialization

We first consider the case when the parameters α_i and β_i which define the initial states of both qubits are constrained to be real-valued. (20)-(21) then reduce to

$$\alpha_1^2 \alpha_2^2 = p_1 \tag{23}$$
$$\beta_1^2 \beta_2^2 = p_2. \tag{24}$$

These equations are sufficient for extracting the two qubits[2], as shown below. This "real-valued isotropic Heisenberg spin coupling model" (or sub-model) is called "DD2" below. It may be inverted as follows. Each initial qubit state meets the normalization condition (2), so that

$$\beta_i^2 = 1 - \alpha_i^2. \tag{25}$$

Inserting it in (24) yields

$$\alpha_2^2 = 1 - \frac{p_2}{1 - \alpha_1^2}. \tag{26}$$

Inserting (26) in (23) and multiplying by $(1 - \alpha_2^2)$ yields a second-order polynomial equation for α_1^2, whose roots read

$$\alpha_1^2 = \frac{1}{2}\left[(1 + p_1 - p_2) \pm \sqrt{(1 + p_1 - p_2)^2 - 4p_1}\right]. \tag{27}$$

The corresponding values of α_2^2, may be derived from (26), but this is not needed: it may be shown that if α_1^2 is set to one of the roots of (27), then the corresponding value of α_2^2 defined by (26) is equal to the other root of (27). As for α_1^2 and α_2^2, the considered problem then has a single solution, up to a permutation.

[2] At least up to some ambiguities. The possibility to reduce these ambiguities by using (18)-(19) will be presented in a future paper, due to space limitations. The use of (18) is also discussed in Section 3.4, in a slightly different context.

Besides, each value of α_i^2 yields a single value β_i^2, due to (25). This leads to the following results for the overall qubit values: (i) if both qubits are constrained to be initialized with positive values, then the above equations yield both qubit parameters (α_1, β_1) and (α_2, β_2) only up to a permutation ambiguity, which is natural due to the symmetry of (23)-(24) with respect to both qubits, (ii) if the signs of the real parameters of the initial states of both qubits are not constrained, then a sign ambiguity appears in addition, because each above value of α_i^2 or β_i^2 yields two opposite solutions for α_i or β_i. One may note the similarity of these results with (i) those for a single qubit obtained in Section 2, (ii) the ambiguities in classical linear instantaneous BSS, i.e. permutation and scale, with scale ambiguity reducing to sign ambiguity for normalized signals.

3.4 Retrieving Qubits with Complex-Valued Initialization

We now come back to the general DD1 model, i.e. we consider arbitrary complex-valued qubit initializations. We here express each qubit parameter in polar form

$$\alpha_1 = r_1 e^{i\theta_1} \qquad \beta_1 = q_1 e^{i\phi_1} \qquad \alpha_2 = r_2 e^{i\theta_2} \qquad \beta_2 = q_2 e^{i\phi_2}. \tag{28}$$

Eq. (16) and (17) are then easily shown to be equivalent to

$$r_1^2 r_2^2 = p_1 \tag{29}$$

$$q_1^2 q_2^2 = p_2. \tag{30}$$

Longer calculations using phase factorizations show that (18) is equivalent to

$$(r_1 q_2 \cos \Delta_E)^2 + (q_1 r_2 \sin \Delta_E)^2 - 2 r_1 r_2 q_1 q_2 \cos \Delta_E \sin \Delta_E \sin \Delta_I = p_3 \tag{31}$$

$$\text{where} \quad \Delta_I = (\phi_2 - \phi_1) - (\theta_2 - \theta_1) \quad \text{and} \quad \Delta_E = \frac{(\omega_{1,0} - \omega_{0,0})T}{2}. \tag{32}$$

A similar expression may be derived from (19) but, due to (22), this expression is redundant with (29)-(31). The latter equations then form our "polar complex-valued isotropic Heisenberg spin coupling model" (or sub-model), called "DD3".

Eq. (29)-(30) only concern the moduli of the qubits. They have already been solved above, since they turn out to be identical to (23)-(24), here with positive parameters so that they yield a single solution, up to a permutation. The phase part of the qubits is then addressed by (31), which yields the following conclusions. Only the combination Δ_I of the qubit phases, defined in (32), may be retrieved from (31). To avoid ambiguities, one may therefore fix three of the phase parameters θ_1, ϕ_1, θ_2, and ϕ_2 (e.g. to 0) and only use the fourth parameter to store information. Eq. (31) only yields $\sin \Delta_I$, i.e. it provides Δ_I up to the ambiguities of the sine function. To derive $\sin \Delta_I$ from (31), all quantities r_1, r_2, q_1, q_2, $\cos \Delta_E$ and $\sin \Delta_E$ must be non-zero. To derive $\sin \Delta_I$ from (31), all other quantities should be known a priori or estimated. p_1 to p_3 are estimated from measurements and r_1, r_2, q_1, q_2, are derived from them as explained above. In a given standard configuration, Δ_E is fixed but not known a priori, because this requires very detailed knowledge of the system's physical properties. We here aim at estimating it blindly with SSP methods, i.e. from a sequence of different coupled-qubit values, where these values are unknown but some of their statistical properties are assumed to be known. This need for SSP methods should be contrasted with the previous aspects of this paper, where the "mixing" equations could be solved from a *single* couple of qubits. For example, a simple SSP approach consists in using a realization of a sequence (indexed by n) of identically distributed, mutually statistically independent, random variables $r_1(n)$, $r_2(n)$, $q_1(n)$, $q_2(n)$ and $\Delta_I(n)$ and considering the expectation of (31). This yields

$$E\{r_1^2(n)\}E\{q_2^2(n)\}\cos^2\Delta_E + E\{q_1^2(n)\}E\{r_2^2(n)\}\sin^2\Delta_E \qquad (33)$$
$$-2E\{r_1(n)\}E\{r_2(n)\}E\{q_1(n)\}E\{q_2(n)\}\cos\Delta_E \sin\Delta_E E\{\sin\Delta_I(n)\} = E\{p_3(n)\}.$$

The moduli $r_i(n)$ and $q_i(n)$ for each couple of qubits in the sequence may be estimated as explained above, and then used to estimate the statistics of these moduli used in (33). $E\{p_3(n)\}$ is also estimated from this sequence. By constraining the sequence of processed data (i.e. qubit values) to be such that their statistical parameter $E\{\sin\Delta_I(n)\}$ has a known value, (33) makes it possible to estimate Δ_E. Note that the solution is quite simple when $E\{\sin\Delta_I(n)\} = 0$.

4 Conclusion

In this theoretical paper, we bridged the gap between the QIP/QP and SSP/BSS domains. We thus introduced what may become the "Blind Quantum Source Separation" (BQSS) field. From a BSS point of view, we proposed several nonlinear mixture models, motivated by actual quantum physical systems. We analyzed the invertibility and ambiguities of these models and we proposed practical methods for restoring qubits from their coupled versions. The next stages of this work, not detailed here due to space limitations, will consist in extending these BSS methods, testing them with simulations of quantum systems and introducing more general quantum mixture models which result in higher needs for SSP methods.

References

1. Hyvarinen, A., Karhunen, J., Oja, E.: Independent Component Analysis. Wiley, New York (2001)
2. Deville, Y., Damour, J., Charkani, N.: Multi-tag radio-frequency identification systems based on new blind source separation neural networks. Neurocomputing 49, 369–388 (2002)
3. Nielsen, M.A., Chuang, I.L.: Quantum computation and quantum information. Cambridge University Press, Cambridge (2000)
4. Caspers, W.J.: Spin Systems. World Scientific, Singapore (1989)
5. Jutten, C., Karhunen, J.: Advances in Blind Source Separation (BSS) and Independent Component Analysis (ICA) for Nonlinear Mixtures. International Journal of Neural Systems 14(5), 267–292 (2004)

An Application of ICA to BSS in a Container Gantry Crane Cabin's Model

Juan-José González de-la-Rosa[1,2], Carlos G. Puntonet[3], A. Moreno Muñoz[2,4], A. Illana[2], and J.A. Carmona[2]

[1] University of Cádiz, Electronics Area, EPSA, Av. Ramón Puyol S/N. E-11202, Algeciras-Cádiz, Spain
juanjose.delarosa@uca.es
[2] Research Group TIC168-Computational Instrumentation and Industrial Electronics
[3] University of Granada, Dept. of Architecture and Computers Technology, ESII, C/Periodista Daniel Saucedo. 18071, Granada, Spain
carlos@atc.ugr.es
[4] University of Córdoba, Electronics Area
Edf. Albert Einstein. Campus Rabanales, Córdoba, Spain
amoreno@uco.es

Abstract. This paper deals with the simulation of a ship-containers' gantry crane cabin behavior, during an operation of load releasing and the BSS via ICA de-noising and movements separation. The goal consists of obtaining a reliable model of the cabin, with the aim of reducing the non-desired cabin vibrations. We present the *Simulink*-based simulation results and the result of the signal separation algorithms when the load is released by the crane in the containers' ship. We conclude that the mass center position of the cabin affects dramatically to the vibrations of the crane. A set of graphs are presented involving displacements and rotations of the cabin to illustrate the effect of the mass center position's bias and the results of the ICA action.

1 Introduction

The study of the vibrations in a gantry crane used in a containers terminal is an issue related to the security of the crane operator and to the durability of the design. The vibrations take place mostly in the operator cabin. The main problem is that a short amplitude vibration in the trolley may produce high amplitude values in the cabin, which may affect the operator's safety. Numerous achievements have been made in the field of the control for overhead crane systems, which have proven to be an improvement in the position accuracy, safety and stabilization control [1,2,3,4,5].

With the goal of adapting the developed control schemes to portainers (container gantry cranes), the modeling of the system has to be developed. In this paper we present an innovative Simulink model of a real-life gantry crane cabin, like one shown in Fig. 1, and its emulated performance when a container is released into the ship.

M.E. Davies et al. (Eds.): ICA 2007, LNCS 4666, pp. 714–721, 2007.

Fig. 1. Container Gantry Cranes at Algeciras harbor

The crucial role of FastICA consist of extracting any parasitic vibration which may remain in the system, affecting the human operator's comfort, along with the de-noising of the signals in order to perform a further analysis of the vibration modes of the cabin in an ulterior cabin's model. Similar results of ICA have been observed in [6] and in [7], which indicates the suitability of the method for signal extraction. The inedit of this paper is the application in gantry cranes.

The results show a new set of signals that may be used in a future vibration control scheme and the power of ICA in extracting parasitic vibrations and de-noising (SNR=20 dB). The paper is structured as follows. In Section 2 we present the *Simulink* model of the portainer's cabin; Section 3 summarizes ICA foundations and FastICA. Section 4 comprises the set of the simulation results and the ICA analysis which in fact are "the guts" of the paper; finally conclusions are drawn in Section 5.

2 The Simulink Model

2.1 Model Equations

Fig. 2 shows an scheme of the complete crane structure where we can see the cabin, whose dimensions are detailed in Fig. 3.

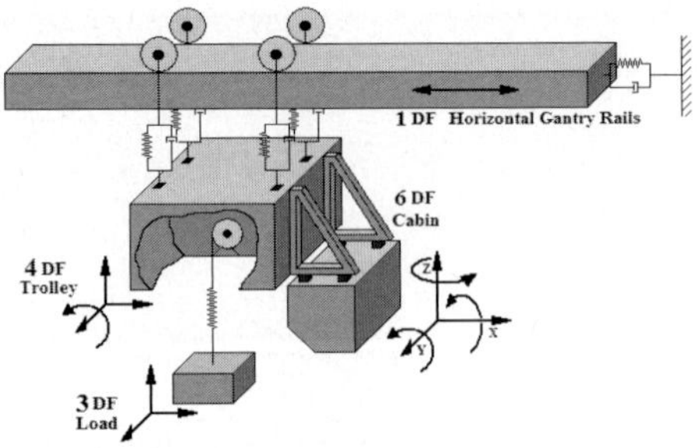

Fig. 2. Gantry crane model scheme

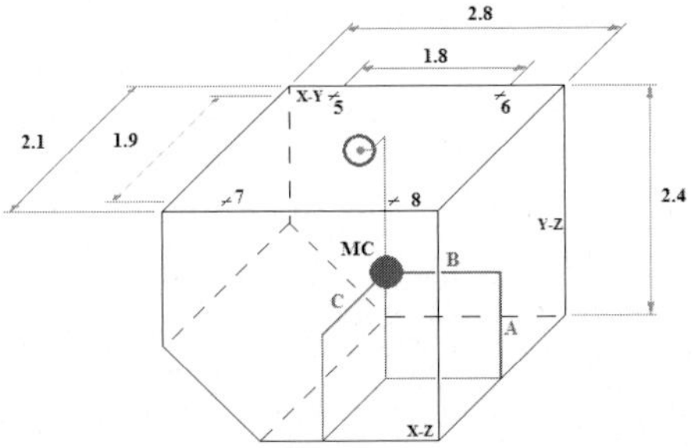

Fig. 3. Gantry crane cabin dimensions. Units in meters. Note where the mass center is and where it should be. Points 5-8 play a special role in the equations that model the dynamics.

The six degrees of freedom of the cabin are solved using the well-known Newton equations, applied to the mass center of the cabin, three of them for forces and other three for torques, from Eq. (1) to Eq. (6); where all the variables and points are referred to Fig. 3.

$$\begin{aligned} \sum_{i\in\{5,6,7,8\}} F_{i,x} &= M\ddot{x}_{mc} \\ F_{i,x} &= C_{i,x}(x_{i,r}-x_{i,b})' \pm C_{i,xy}(y_{i,r}-y_{i,b})' \\ &\quad + K_{i,x}(x_{i,r}-x_{i,b}) \pm K_{i,xy}(y_{i,r}-y_{i,b}) \end{aligned} \tag{1}$$

$$\begin{aligned} \sum_{i\in\{5,6,7,8\}} F_{i,y} &= M\ddot{y}_{mc} \\ F_{i,y} &= C_{i,y}(y_{i,r}-y_{i,b})' \pm C_{i,xy}(x_{i,r}-x_{i,b})' \\ &\quad + K_{i,y}(y_{i,r}-y_{i,b}) \pm K_{i,xy}(x_{i,r}-x_{i,b}) \end{aligned} \tag{2}$$

$$\begin{aligned} \sum_{i\in\{5,6,7,8\}} F_{i,z} &= M\ddot{z}_{mc} \\ F_{i,z} &= C_{i,z}(z_{i,r}-z_{i,b})' + K_{i,z}(z_{i,r}-z_{i,b}) \end{aligned} \tag{3}$$

$$\sum_{i\in\{5,6,7,8\}} M_{i,x} = I_x \ddot{w}_{x,mc} - (I_y - I_z) w_{y,mc} w_{z,mc}$$
$$\sum_{i\in\{5,6,7,8\}} M_{i,x} = \sum (F_{i,z} d_{i,y} + F_{i,y} d_{i,y} d_{i,z}) \tag{4}$$

$$\sum_{i\in\{5,6,7,8\}} M_{i,y} = I_y \ddot{w}_{y,mc} - (I_z - I_x) w_{z,mc} w_{x,mc}$$
$$\sum_{i\in\{5,6,7,8\}} M_{i,y} = \sum (F_{i,z} d_{i,x} + F_{i,x} d_{i,z}) \tag{5}$$

$$\sum_{i\in\{5,6,7,8\}} M_{i,z} = I_z \ddot{w}_{z,mc} - (I_x - I_y) w_{x,mc} w_{y,mc}$$
$$\sum_{i\in\{5,6,7,8\}} M_{i,z} = \sum (F_{i,x} d_{i,y} + F_{i,y} d_{i,x}) \tag{6}$$

Some remarks are to be made in this set of equations. The "±" refers to index i=5,8, respectively, the "-" sign refers to i=6,7. Subindex "mc" refers to the mass center, "r" refers to the trolley and "b" to the cabin. "w" are angles, "F" forces, "M" torques, "I" inertias, "K" are for springs, "C" are for dampers; "d" symbolizes distances.

2.2 Simulink Scheme

The *Simulink* model solves and plots the displacements, velocities and accelerations of each one of the six degrees of freedom of the cabin. To do that the trolley movements and the system's physical constants (mass, inertias, spring and damper values and mass center position) have to be considered. Fig. 4 presents a detail of the model, concretely the forces and torques solver block.

The model is mainly divided into four blocks. The forces and torques solver block (Fig. 4) receives all the constants and positions of the system and solves every force and torque. The second block is the equations' solver. It receives forces, torques, mass, inertias and angles to solve every acceleration of the mass center of the cabin. The third block converts accelerations into velocities and positions of the mass center, which are the outputs of the system. Finally the fourth block calculates positions and velocities of the four cabin-trolley connection points, using cabin and trolley positions and velocities; finally it connects them to the first block, so the new forces and torques may be calculated.

3 The ICA Model and Algorithms

3.1 Outline of ICA

BSS by ICA is receiving attention because of its applications in many fields such as speech recognition, medicine and telecommunications [8]. Statistical

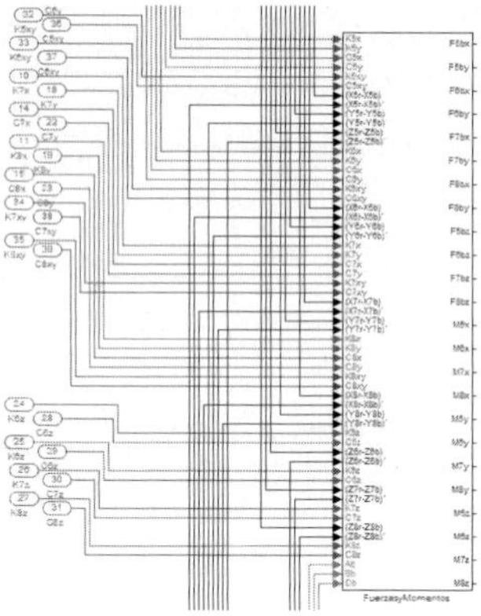

Fig. 4. Coarser representation of the *Simulink* model. Forces and torques solver block. This is valid only to get an approximate idea of the complex model.

methods in BSS are based in the probability distributions and the cumulants of the mixtures. The recovered signals (the source estimators) have to satisfy a condition which is modeled by a contrast function. The underlying assumptions are the mutual independence among sources and the non-singularity of the mixing matrix [6],[9].

Let $\mathbf{s}(t) = [s_1(t), s_2(t), \ldots, s_m(t)]^T$ be the transposed vector of sources (statistically independent). The mixture of the sources is modelled via

$$\mathbf{x}(t) = \mathbf{A} \cdot \mathbf{s}(t) \tag{7}$$

where $\mathbf{x}(t) = [x_1(t), x_2(t), \ldots, x_m(t)]^T$ is the available vector of observations and $\mathbf{A} = [a_{ij}] \in \Re^{m \times n}$ is the unknown mixing matrix, modelling the environment in which signals are mixed, transmitted and measured [10]. We assume that $\mathbf{A}$ is a non-singular n×n square matrix. The goal of ICA is to find a non-singular n×m separating matrix $\mathbf{B}$ such that extracts sources via

$$\hat{\mathbf{s}}(t) = \mathbf{y}(t) = \mathbf{B} \cdot \mathbf{x}(t) = \mathbf{B} \cdot \mathbf{A} \cdot \mathbf{s}(t) \tag{8}$$

where $\mathbf{y}(t) = [y_1(t), y_2(t), \ldots, y_m(t)]^T$ is an estimator of the sources. The separating matrix has a scaling freedom on each row because the relative amplitudes of sources in $\mathbf{s}(t)$ and columns of $\mathbf{A}$ are unknown [9]. The transfer matrix $\mathbf{G} \equiv \mathbf{BA}$ relates the vector of independent (original) signals to its estimators.

3.2 FastICA

One of the independent components is estimated by $y = \mathbf{b}^T\mathbf{x}$. The goal of FastICA is to take the vector $\mathbf{b}$ that maximizes the non-Gaussianity (independence)of y, by finding the maxima of its negentropy [9]. The algorithm scheme is an approximative Newton iteration, resulting from the application of the *Kuhn-Tucker* conditions. This leads to the Eq. (9)

$$E\{\mathbf{x}g(\mathbf{b}^T\mathbf{x}) - \beta\mathbf{b} = 0\} \tag{9}$$

where g is a non-quadratic function and β is an iteration parameter.

Provided with the mathematical foundations the experimental results are outlined.

4 Results

We present the set of results in the form of graphics due to the interest and inedit results. We have introduced, in the simulated model, a real-life bias in the position of the mass center (A = 1 m, B = 1.35 m, C = 1 m), in order to asses the real cabin behavior. A delay of 1 sec is introduced to enhance the visualization of the graphs. The initial conditions are null for all the variables involved in the differential equations. A step-type input (5 cm amplitude) is chosen to assess the outputs of the system. This input emulates the behavior of the sudden bump in the trolley when the load is released in the container ship.

The left column in the matrix of signals in Fig. 5 shows all the signals involved, and the right column shows the ICA outputs. First of all we analyze the measured signals (left column). Regarding the X displacement of the mass center of the cabin, it can be seen than the system is not able to dump it adequately. This movement is produced by the horizontal bias; it has the peculiarity that the vertical bias of the mass center also affects this horizontal movement in a critical way. But we have to point out that the unique presence of a vertical bias is not enough to start this movement.

Another coupling effect is the one produced by the vertical input, this time in the Y axis. The high frequency component of the signal is rapidly attenuated while the low frequency component is not attenuated at all and remains as a parasitic vibration in the system. This fact has also been shown in the X direction.

The displacement of the system in the Z-axis is the only one that behaves like a typical response to a step-like input. We must point out than the amplitude of the movement nearly doubles the input; so, an immediate conclusion is that the system's behavior is far from its original aim of isolate the cabin from the trolley vibrations. in other words, this movement has the peculiarity of not being fully dumped.

Finally, the rotations of the cabin are not attenuated. These rotations affect to the X, Y, Z movements, and will not be extinguished due to the geometric disposition of the dumps.

This scheme fits a BSS scenario. The ICA results show the separation of the remaining oscillation (IC♯6). The noise is extracted in IC♯5 (a SNR=20 dB is simulated). The z-angle movement is extracted in IC♯4. IC♯3 is the pure y-angle movement. IC♯2(1) is the y(x)-displacement without the parasitic vibration.

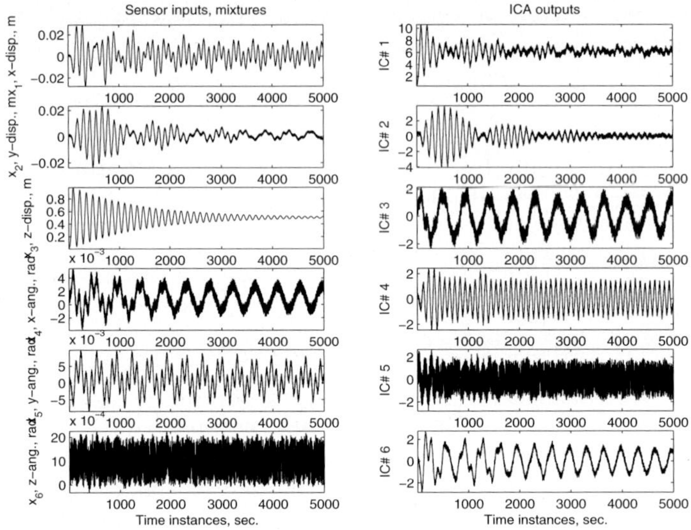

Fig. 5. Measured-simulated signals (left column) and ICA outputs (Independent Components, ICs, right column)

5 Conclusions

We conclude that the system is not able to dump the cabin vibrations, as every real cabin mass center has a bias. Even with very high values of dump constants, the time needed to attenuate the vibrations is high. Our direct real-life experience in those cabins shows that they continually work in a transitory vibration state, often leading the system to resonances.

A real cabin prototype is being built to adequate our model to the reality, and solutions to the vibration matter will be tested in it. The critical influence of the mass center position showed in this paper lead us to think that cabin-trolley connection points must be placed around the real mass center position instead than on the top of the cabin, where the vertical distance to the mass center makes the system to behave like a pendulum.

ICA extracts the parasitic vibration and the uniform noise process added in the simulation, leading us to get the real mechanical vibration modes of the cabin's operator.

Acknowledgement

We would like to acknowledge the Spanish Ministry of Education and Science for funding the projects DPI2003-00878 and PETRI-95-0824-OP, and to the Andalusian Government for funding the project TIC155. We would also like to acknowledge Mr. John Thonsem for the trust shown in the research group

PAI-TIC-168 from the University of Cádiz, which works in the MAERSK containers' terminal in the Algeciras harbor.

References

1. Ju, F., Choo, Y., Cui, F.: Dynamic response of tower crane induced by the pendulum motion of the payload. International Journal of Solids and Structures, 376–389 (2006)
2. Hua, Y.J., Shine, Y.K.: Adaptive coupling control for overhead crane systems. Mechatronics (in Press, 2007)
3. Lee, D.H., Cao, Z., Meng, Q.: Scheduling of two-transtainer systems for loading outbound containers in port container terminals with simulated annealing algorithm. Int. J. Production Economics (in Press, 2007)
4. Benhidjeb, A., Gissinger, G.: Fuzzy control of an overhead crane performance comparison with classic control. In: Proceedings of Control Eng. Practice, pp. 168–796 (1995)
5. Jianqiang, Y., Naoyoshi, Y., Kaoru, H.: Anti-swing and positioning control of overhead traveling crane. Inform. Sci. 19–42 (2003)
6. De la Rosa, J.J.G., Puntonet, C.G., Lloret, I.: An application of the independent component analysis to monitor acoustic emission signals generated by termite activity in wood. In: Measurement, vol. 37, pp. 63–76. Elsevier, Amsterdam (2005) (Available online, October 12, 2004)
7. Antoni, J.: Blind separation of vibration components: Principles and demonstrations. In: Mechanical Systems and Signal Processing, vol. 19, pp. 1166–1180. Elsevier, Amsterdam (2005)
8. Puntonet, C.G., de la Rosa, J.J.G., Galiana, I.L., Sáez, J.M.G.: On the performance of a hos-based ica algorithm in bss of acoustic emission signals. In: Rosca, J., Erdogmus, D., Príncipe, J.C., Haykin, S. (eds.) ICA 2006. LNCS, vol. 3889, pp. 400–405. Springer, Heidelberg (2006)
9. Hyvärinen, A., Oja, E.: Independent Components Analysis: A Tutorial. Helsinki University of Technology, Laboratory of Computer and Information Science (1999)
10. De la Rosa, J.J.G., Puntonet, C.G., Górriz, J.M., Lloret, I.: An application of ICA to identify vibratory low-level signals generated by termites. In: Puntonet, C.G., Prieto, A.G. (eds.) ICA 2004. LNCS, vol. 3195, pp. 1126–1133. Springer, Heidelberg (2004)

Blind Separation of Non-stationary Images Using Markov Models

Rima Guidara, Shahram Hosseini, and Yannick Deville

Laboratoire d'Astrophysique de Toulouse-Tarbes (LATT) - UMR 5572
Université Paul Sabatier Toulouse 3 - CNRS
14 Avenue Edouard Belin, 31400 Toulouse, France
{rguidara,shosseini,ydeville}@ast.obs-mip.fr

Abstract. In recent works, we presented a blind image separation method based on a maximum likelihood approach, where we supposed the sources to be stationary, spatially autocorrelated and following Markov models. To make this method more adapted to real-world images, we here propose to extend it to non-stationary image separation. Two approaches, respectively based on blocking and kernel smoothing, are then used for the estimation of source score functions required for implementing the maximum likelihood approach, in order to allow them to vary within images. The performance of the proposed algorithm, tested on both artificial and real images, is compared to the stationary Markovian approach, and then to some classical blind source separation methods.

1 Introduction

In a previous work [1], we proposed a blind source separation method based on a maximum likelihood approach, where Markov processes are used to model temporal autocorrelations of *stationary* sources. This method exploits both non-Gaussianity and temporal autocorrelation of mutually independent sources and provides an asymptotically efficient estimator.

We then extended this approach in [2] to bi-dimensional sources, where second-order Markov Random Fields (MRFs) were used to describe the spatial autocorrelation within each source image. The likelihood function was maximized using a modified equivariant Newton-Raphson algorithm, and a parametric polynomial estimator was introduced in order to reduce the computational cost of conditional score function estimation, required for implementing the method.

In [2], we supposed the source images to be stationary, so that their statistics did not vary over the image pixels. This hypothesis is, however, not realistic for the majority of real-world images, so that the algorithm sometimes fails to separate highly mixed images.

In this paper, we propose to modify the method presented in [2] so that the Probability Density Functions (PDFs) of the pixels may vary within each source image. As a result, the images may be *non-stationary* and the proposed method can then simultaneously exploit non-Gaussianity, non-stationarity and spatial autocorrelation in a quasi-optimal manner.

M.E. Davies et al. (Eds.): ICA 2007, LNCS 4666, pp. 722–729, 2007.

The non-stationarity of the sources has already been exploited in the literature [3,4,5,6,7,8] for blind source separation. Nevertheless, most of these methods do not exploit the non-Gaussianity and/or the autocorrelation of the sources and are usually based on variance non-stationarity while our method can also exploit higher-order non-stationarities.

2 Markovian Separation Method

In this paper, we consider the blind image separation problem in its simplest form, where the observations are linear instantaneous mixtures of source images. In a noiseless and determined context, this problem can be formulated as follows. Having $N = N_1 \times N_2$ pixels of K linear transforms of K source images, we consider the mixture model $\mathbf{x}(n_1, n_2) = \mathbf{A}\mathbf{s}(n_1, n_2)$, where $\mathbf{x}(n_1, n_2)$ and $\mathbf{s}(n_1, n_2)$ are, respectively, the K-dimensional observation and source vectors, and $\mathbf{A}$ is an unkown $K \times K$ invertible mixing matrix.

The Maximum Likelihood (ML) approach can be used to estimate the separating matrix $\mathbf{B} = \mathbf{A}^{-1}$ up to a diagonal matrix and a permutation matrix. It consists in maximizing, with respect to the matrix $\mathbf{B}$, the joint PDF of all the pixels of all the images in the observation vector $\mathbf{x}$

$$f_{\mathbf{x}}(x_1(1,1), \cdots, x_K(1,1), \cdots, x_1(N_1, N_2), \cdots, x_K(N_1, N_2)) \ . \tag{1}$$

Assuming source images are independent and described by second-order MRFs according to the sweeping scheme defined in [2], this joint PDF can be approximated by

$$\Big(\frac{1}{|\det(\mathbf{B}^{-1})|}\Big)^N \prod_{i=1}^{K} \prod_{n_1=2}^{N_1} \prod_{n_2=2}^{N_2-1} f_{s_i(n_1,n_2)}(s_i(n_1,n_2)|s_i(n_1,n_2-1), \\ s_i(n_1-1,n_2+1), s_i(n_1-1,n_2), s_i(n_1-1,n_2-1)). \tag{2}$$

Defining then the conditional score function $\psi^{k,l}_{s_i(n_1,n_2)}$ of a source s_i, with respect to the pixel $s_i(n_1-k, n_2-l)$, by

$$\psi^{k,l}_{s_i(n_1,n_2)}(n_1,n_2) = \frac{-\partial}{\partial s_i(n_1-k,n_2-l)} \log f_{s_i(n_1,n_2)}(s_i(n_1,n_2)|s_i(n_1,n_2-1), \\ s_i(n_1-1,n_2+1), s_i(n_1-1,n_2), s_i(n_1-1,n_2-1)) \tag{3}$$

the maximization of the logarithm of (2) leads finally to a system of $K(K-1)$ estimating equations defined as

$$E_N[\sum_{(k,l)\in\Upsilon} \psi^{k,l}_{s_i(n_1,n_2)}(n_1,n_2).s_j(n_1-k,n_2-l)] = 0 \qquad i \neq j = 1, \cdots, K \tag{4}$$

where $\Upsilon = \{(0,0),(0,1),(1,-1),(1,0),(1,1)\}$ corresponds to the considered sweeping scheme over the image [2] and $E_N[.]$ is a spatial average operator defined by $E_N[.] = \frac{1}{N}\sum_{n_1=2}^{N_1}\sum_{n_2=2}^{N_2-1}[.]$.

In practice, the sources s_i, actually unknown, are replaced using an iterative algorithm by the estimated sources $\hat{s}_i(n_1, n_2) = \mathbf{e_i}^T\hat{\mathbf{B}}\mathbf{x}(n_1, n_2)$, where $\mathbf{e_i}$ is the ith column of the identity matrix. The separating matrix $\mathbf{B}$ may be estimated via the resolution of the system of equations (4), using for example the modified equivariant version of the Newton-Raphson algorithm presented in [2].

The score functions and their derivatives, required for the computation of the system coefficients, may be estimated using a parametric polynomial estimator.

In [2], we supposed that source images were stationary, so that the score functions $\psi^{k,l}_{s_i(n_1,n_2)}(.)$ did not vary with n_1 and n_2, *i.e.* reduced to $\psi^{k,l}_{s_i}(.)$. However, this condition being unrealistic for most real-world images, we propose, in this paper, to extend this approach to non-stationary sources, allowing the statistics to be dependent on the pixel position. Two methods, based respectively on blocking and kernel smoothing, are introduced in the following section in order to adapt the score function estimation to non-stationary images.

3 Non-stationary Estimation of the Score Functions

For simplicity, we denote the conditional score functions $\psi^{k,l}_{s_i(n_1,n_2)}(n_1, n_2) \triangleq \psi^{k,l}_{s_i(n_1,n_2)}(s_i(n_1, n_2)|s_i(n_1, n_2-1), s_i(n_1-1, n_2+1), s_i(n_1-1, n_2), s_i(n_1-1, n_2-1))$ by $\psi^{k,l}_{s_i(n)}(\xi_0|\xi_1, \ldots, \xi_4)$. This function can be rewritten as follows

$$\psi^{k,l}_{s_i(n)}(\xi_0|\xi_1, \ldots, \xi_4) = \psi^{k,l}_{s_i(n)}(\xi_0, \ldots, \xi_4) - \psi^{k,l}_{s_i(n)}(\xi_1 \ldots, \xi_4) \ . \tag{5}$$

We propose to estimate each of the non-stationary joint score functions in (5) using a parametric third-order polynomial estimator. The polynomial function order is chosen so that the resulting approximation induces low computational cost without decreasing the estimation performance. Thus, assuming $g^{k,l}_{s_i(n)}(\xi_0, \ldots, \xi_4, \mathbf{W})$ is a polynomial estimator of the joint score function $\psi^{k,l}_{s_i(n)}(\xi_0, \ldots, \xi_4)$, we can define this estimator by

$$g^{k,l}_{s_i(n)}(\xi_0, \ldots, \xi_4, \mathbf{W}) = \sum_j w^{k,l}_j(s_i(n))h_j(\xi_0, \ldots, \xi_4) = \mathbf{h}^T\mathbf{W}^{k,l}(s_i(n))$$

where $h_j(\xi_0, \ldots, \xi_4)$ and $w^{k,l}_j(s_i(n))$ are respectively the monomial functions and the coefficients. The polynomial coefficients $w^{k,l}_j(s_i(n))$ should be selected so that $g^{k,l}_{s_i(n)}(\xi_0, \ldots, \xi_4, \mathbf{W})$ is the least mean-square estimator of the joint score function $\psi^{k,l}_{s_i(n)}(\xi_0, \ldots, \xi_4)$. Consequently, $w^{k,l}_j(s_i(n))$ are solutions of the following optimization problem

$$\mathbf{W}^{k,l}(s_i(n)) = \text{argmin} E\{[\psi^{k,l}_{s_i(n)}(\xi_0, \ldots, \xi_4) - g^{k,l}_{s_i(n)}(\xi_0, \ldots, \xi_4, \mathbf{W})]^2\} \ . \tag{6}$$

Using the theorem mentioned in Section 4 of [2], we obtain

$$\mathbf{W}^{k,l}(s_i(n)) = \text{argmin}\left\{E\left[\mathbf{h}^T\mathbf{W}^{k,l}(s_i(n))[\mathbf{W}^{k,l}(s_i(n))]^T\mathbf{h}\right] - 2E\left[\frac{\partial \mathbf{h}^T}{\partial \xi^{k,l}}.\mathbf{W}^{k,l}(s_i(n))\right]\right\}$$

where $\xi^{k,l} \in \{\xi_0, \ldots, \xi_4\}$ and represents $s_i(n_1 - k, n_2 - l)$ according to (3) and (5). The minimum of the above function is finally given by

$$\mathbf{W}(s_i(n))^{k,l} = \left(E\left[\mathbf{h}^T\mathbf{h}\right]\right)^{-1} E\left[\frac{\partial \mathbf{h}}{\partial \xi^{k,l}}\right] . \tag{7}$$

A parametric estimator for the second joint score function $\psi^{l,k}_{s_i(n)}(\xi_1 \ldots, \xi_4)$ may be obtained following the same calculus, or simply deduced from the first estimator by regression.

In the initial version of this estimation procedure [2], we supposed the sources were stationary. As a consequence, the mathematical expectations in Eq. (7) were replaced by a spatial average over all the pixels in the image. To extend this estimation to non-stationary sources, two methods, proposed initially in [8] for non-stationary *temporally uncorrelated one-dimensional* sources, are here adapted to model the statistics of non-stationary *spatially autocorrelated* images.

Blocking method: Each observed image is split into $M_1 \times M_2$ sub-images I_j, supposing that the score function variations, in each block I_j, are small. Each of these sub-images may be then supposed to be stationary, so that the mathematical expectations, required in Eq. (7), can be locally approximated by a spatial average over each sub-image. The coefficients of the parametric estimator of the score functions being unchanged within I_j, we can write

$$\psi^{l,k}_{s_i(n_1,n_2)} = \psi^{l,k}_{s_i}(j), \quad \forall \quad (n_1, n_2) \in I_j \ .$$

Kernel smoothing method: In a kernel smoothing approach, score functions are approximated at each pixel of the image by a local spatial average all around this pixel.

Using a general notation, we denote $E = E(\phi(\xi_0(n_1, n_2), \ldots, \xi_4(n_1, n_2)))$ the mathematical expectations required for the estimation of the score function parameters at pixel $s_i(n_1, n_2)$, where $(\xi_0(n_1, n_2), \ldots, \xi_4(n_1, n_2))$ represents $(s_i(n_1, n_2), s_i(n_1, n_2 - 1), s_i(n_1 - 1, n_2 + 1), s_i(n_1 - 1, n_2), s_i(n_1 - 1, n_2 - 1))$ and $\phi(.)$ each of the functions involved in Eq. (7). Contrary to the blocking method, these expectations are estimated in each pixel using the formula

$$\hat{E} = \frac{\sum_{\mu_1=2}^{N_1-1} \sum_{\mu_2=2}^{N_2-1} \kappa(\frac{\mu_1 - n_1}{\nu}, \frac{\mu_2 - n_2}{\nu}) \phi(\xi_0(\mu_1, \mu_2), \ldots, \xi_4(\mu_1, \mu_2))}{\sum_{\mu_1=2}^{N_1-1} \sum_{\mu_2=2}^{N_2-1} \kappa(\frac{\mu_1 - n_1}{\nu}, \frac{\mu_2 - n_2}{\nu})} \tag{8}$$

where $\kappa(.)$ is a kernel function and ν a window width parameter.

The kernel smoothing method should be advantageous in cases when the score functions vary rapidly within the image. In fact, in such cases, the size of nearly stationary blocks is so small that there are not enough pixels for a reliable estimation of score functions using the blocking method. However, the kernel smoothing method is computationally too expensive, since a new estimation is required at each pixel. To reduce the algorithm complexity, we can approximate the estimator (8) by a sparser one, according to the formula

$$\hat{E} = \frac{\sum_{l_1=l_{11}}^{L_1} \sum_{l_2=l_{22}}^{L_2} \kappa(\frac{\frac{l_1 Q_1}{L_1}-n_1}{\nu}, \frac{\frac{l_2 Q_2}{L_2}-n_2}{\nu})\phi(\xi_0(\frac{l_1 Q_1}{L_1}, \frac{l_2 Q_2}{L_2}), \ldots, \xi_4(\frac{l_1 Q_1}{L_1}, \frac{l_2 Q_2}{L_2}))}{\sum_{l_1=l_{11}}^{L_1} \sum_{l_2=l_{22}}^{L_2} \kappa(\frac{\frac{l_1 Q_1}{L_1}-n_1}{\nu}, \frac{\frac{l_2 Q_2}{L_2}-n_2}{\nu})}$$

where $Q_1 = N_1 - 1$, $Q_2 = N_2 - 1$, L_1 and L_2 are chosen so that $\frac{Q_1}{L_1}$ and $\frac{Q_2}{L_2}$ are integers, and l_{11} and l_{22} are the first integers greater than $\frac{2L_1}{Q_1}$ and $\frac{2L_2}{Q_2}$, respectively. The choice of sparsness parameters L_1 and L_2 should be adapted to the smoothness of the signal.

4 Experimental Results

4.1 Artificial Images

In our first simulations, we want to compare the proposed non-stationary Markovian blocking method to the stationary Markovian image separation approach, presented in [2].

In the first experiment, we apply our blocking method to two artificial mixtures of two non-stationary source images, following exactly second-order MRF models. Two independent white and uniformly distributed noise images with size 200×200, $e_1(n_1, n_2)$ and $e_2(n_1, n_2)$, are therefore generated and filtered by two Infinite Impulse Response (IIR) filters, according to the following scheme

$$\begin{aligned} \varsigma_i(n_1, n_2) = e_i(n_1, n_2) + \rho_{0,1}^i s_i(n_1, n_2 - 1) + \rho_{1,-1}^i s_i(n_1 - 1, n_2 + 1) \\ + \rho_{1,0}^i s_i(n_1 - 1, n_2) + \rho_{1,1}^i s_i(n_1 - 1, n_2 - 1) \; . \end{aligned} \quad (9)$$

The coefficients ρ^i of the first and second filters are fixed to $\{-0.5, 0.3, 0.5, -0.29\}$ and $\{-0.5, 0.4, 0.5, 0.3\}$, respectively, and satisfy the filter stability conditions. The resulting images are then split into $L_1 \times L_2$ sub-images, and each sub-image is multiplied by a different coefficient α_p^i, $p = 1, \ldots, L_1 \times L_2$. The resulting autocorrelated, non-stationary sources are finally mixed using the mixing matrix $\mathbf{A} = \begin{pmatrix} 1 & 0.99 \\ 0.99 & 1 \end{pmatrix}$. The Markovian non-stationary blocking method is then used to separate the sources. After normalizing the separated sources $\hat{s}_i(n_1, n_2)$ so that they have the same variances and signs as the source signals $s_i(n_1, n_2)$, the output Signal to Interference Ratio (in dB) is computed using the formula

$$SIR = \frac{1}{K} \sum_{i=1}^{K} 10 \log_{10} \frac{E[s_i^2(n_1, n_2)]}{E[(\hat{s}_i(n_1, n_2) - s_i(n_1, n_2))^2]}$$

where $K = 2$ in the above experiment. Choosing $L_1 = L_2 = 4$, we computed the mean of SIR over 100 Monte Carlo simulations for different values of the algorithm parameters M_1 and M_2. The results are shown in Fig. 1 as a function of the number of sub-images ($M_1 \times M_2$), with $M_1 = M_2 = M$.

The one-block image case, *i.e.* $M_1 = M_2 = 1$, corresponding to the stationary Markovian method, led only to an average SIR of 22 dB. The separation performance increases rapidly when we use a greater number of blocks, and it reaches its maximum, with an average SIR of 99 dB, when the model takes the same number of blocks as in the source image, *i.e.* $M_1 = L_1$ and $M_2 = L_2$. Moreover, it can be noticed that over-blocking the image does not have a great effect on the separation result, provided that the number of pixels in each sub-image is sufficient for the score function estimation.

In comparison to the stationary Markovian algorithm, the blocking approach significantly reduces both memory and time consumption. In the above experiment, the stationary Markovian algorithm needs 192 Mbytes of memory to estimate score functions while the blocking algorithm with $M_1 = M_2 = 4$ needs less than 12 Mbytes. The time consumption reduction induced by the blocking method is significant with large-size images. For example, using mixtures of 400×400-sized non-stationary images, the running times of the blocking algorithm with $M_1 = M_2 = 4$ and the stationary Markovian algorithm on a 1.53 GHz AMD-Athlon PC are 48 seconds and 1647 seconds, respectively, for each iteration.

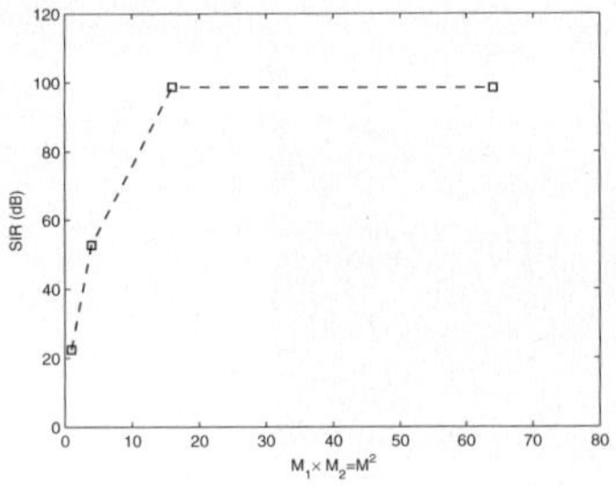

Fig. 1. Mean of SIR vs. the number of sub-images using IIR filtered sources

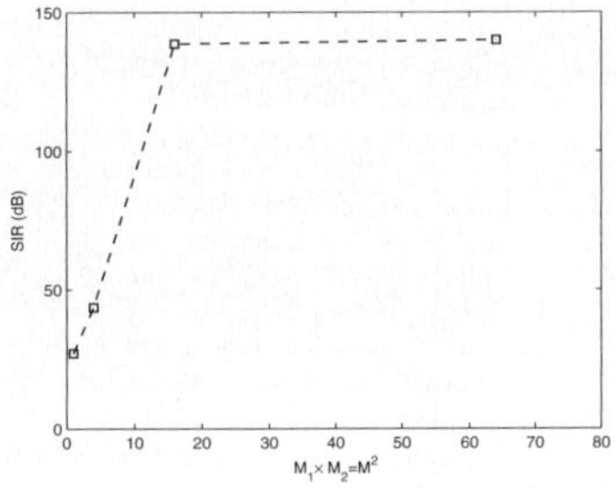

Fig. 2. Mean of SIR vs. the number of sub-images using FIR filtered sources

In the second experiment, we want to test the robustness of our blocking method with respect to the Markov model assumption. As in the first simulation, two noise images with size 200×200 are generated but filtered this time by two symmetrical bidimensional Finite Impulse Response (FIR) filters. Thus, the filtered images can no longer be modeled exactly by second-order MRFs. The resulting images are then split into $L^2 = 16$ square blocks, so that $L_1 = L_2 = 4$, and each sub-image is multiplied by a different coefficient α_p^i. Finally, the same mixing matrix $\mathbf{A}$ as in the first experiment is used and the average SIR over 100 Monte Carlo simulations is computed, and shown in Fig. 2 as a function of the number of blocked sub-images $M_1 \times M_2 = M^2$. Results show clearly the high performance of our algorithm, even when the Markov model is not satisfied. The average SIR achieved by the stationary Markovian algorithm is only 27 dB, whereas the blocking algorithm led to 140 dB for $M^2 = 16$ sub-images.

In the third simulation, our goal is to highlight the advantage of the kernel smoothing method compared to the blocking one for images whose statistics are rapidly varying. The kernel smoothing algorithm being very time consuming, we can only use it for small-sized images. Two images with size 32×32, shown in Fig. 3, are artificially mixed using the same matrix $\mathbf{A}$ as in the first experiment. The kernel smoothing method, using a Gaussian kernel, is then applied to the mixture and the results are compared to those achieved by the blocking method. The kernel smoothing method, with a kernel standard deviation $\sigma = 10$, leads to 57-dB SIR, whereas the blocking method completely fails to separate the sources in this case because they are too non-stationary.

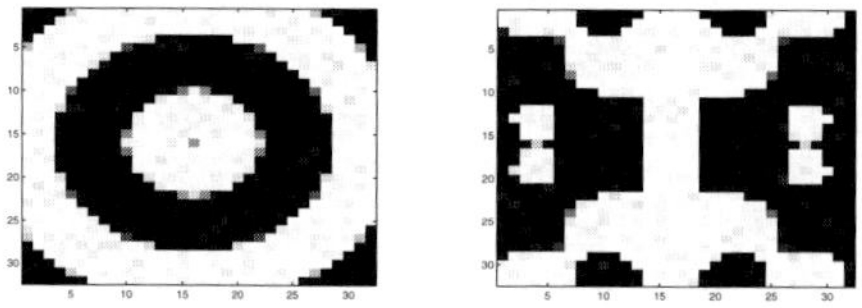

Fig. 3. Two small-sized images with highly non-stationary variations

Fig. 4. Two photographic real-world images

4.2 Real-World Images

In the last experiment, we want to evaluate the performance of our blocking algorithm using real-world images. Two photographic images, provided by [9], are artificially mixed using the matrix $\mathbf{A} = \begin{pmatrix} 1 & 0.99 \\ 0.99 & 1 \end{pmatrix}$. The two images, shown in Fig. 4, are 320×420-sized and clearly non-stationary. The blocking method, using different values for the number of blocks, is first applied and the SIR is compared to the SIR achieved by the stationary Markovian method. Provided we select an adequate number of sub-images, which corresponds to $M_1 = M_2 = 10$ in this case, the blocking method led to nearly 60 dB SIR, whereas the separation completely failed with the stationary Markovian algorithm.

In the second step, we compared our blocking method to the 15 classical algorithms available in the ICALAB Toolbox [10,11]. Our method outperforms all 15 algorithms, which led to 37 dB SIR at best, with the SOBI-RO method.

5 Conclusion

In this paper, we extended to non-stationary source images our Markovian maximum likelihood approach for blind image separation. The proposed algorithms exploit simultaneously non-Gaussianity, spatial autocorrelation and non-stationarity of the sources in a quasi-optimal manner. To handle non-stationarity, we adapt to our problem two approaches, respectively based on blocking and kernel smoothing. Experimental results, using both artificial and real images, show clearly the better performance of our blocking method compared to the stationary Markovian algorithm and the classical algorithms available in the ICALAB Toolbox. Moreover, tests proved the high performance of our kernel smoothing method, even with highly non-stationary images. However, this method can only be applied to small-sized images, since it is very time consuming. Therefore, we are currently working on reducing its computational cost.

References

1. Hosseini, S., Jutten, C., Pham, D.-T.: Markovian source separation. IEEE Trans. on Signal Processing 51, 3009–3019 (2003)
2. Hosseini, S., Guidara, R., Deville, Y., Jutten, C.: Markovian Blind Image Separation. In: Rosca, J., Erdogmus, D., Príncipe, J.C., Haykin, S. (eds.) ICA 2006. LNCS, vol. 3889, pp. 106–114. Springer, Heidelberg (2006)
3. Matsuoka, K., Ohya, M., Kawamoto, M.: A neural net for blind separation of non-stationary signals. Neural Networks 8(3), 411–419 (1995)
4. Souloumiac, A.: Blind source detection and separation using second-order non-stationarity. In: Proc. ICASSP, pp. 1912–1915 (1995)
5. Choi, S., Cichoki, A., Belouchrani, A.: Second order non-stationary source separation. Journal of VLSI Signal Processing 32(1-2), 93–104 (2002)
6. Hyvarinen, A.: Blind source separation by non-stationarity of variance: a cumulant based approach. IEEE Trans. on Neural Networks 12(6), 1471–1474 (2001)
7. Pham, D.-T., Cardoso, J.-F.: Blind separation of independent mixtures of non-stationary sources. IEEE Trans. on Signal Processing, 49(9) (2001)
8. Pham, D.-T.: Blind Separation of non stationary non Gaussian sources. In: Proc. European Signal Processing Conference (EUSIPCO 2002), Toulouse, France (September 2002)
9. Web site of ICACentral: `http://www.tsi.enst.fr/icacentral/base_multi.html`
10. Cichoki, A., Amari, S., Siwek, K., Tanaka, T., et al.: ICALAB Toolboxes, `http://www.bsp.brain.riken.jp/ICALAB`
11. Cichocki, A., Amari, S.: Adaptive Blind Signal and Image Processing: Learning Algorithms and Applications. Wiley, Chichester (2003)

Blind Instantaneous Noisy Mixture Separation with Best Interference-Plus-Noise Rejection*

Zbyněk Koldovský[1,2] and Petr Tichavský[1]

[1] Institute of Information Theory and Automation, Pod vodárenskou věží 4, P.O. Box 18, 182 08 Praha 8, Czech Republic
zbynek.koldovsky@tul.cz
[2] Faculty of Mechatronic and Interdisciplinary Studies
Technical University of Liberec, Hálkova 6, 461 17 Liberec, Czech Republic

Abstract. In this paper, a variant of the well known algorithm FastICA is proposed to be used for blind source separation in off-line (block processing) setup and a noisy environment. The algorithm combines a symmetric FastICA with test of saddle points to achieve fast global convergence and a one-unit refinement to obtain high noise rejection ability. A novel test of saddle points is designed for separation of complex-valued signals. The bias of the proposed algorithm due to additive noise can be shown to be asymptotically proportional to σ^3 for small σ, where σ^2 is the variance of the additive noise. Since the bias of the other methods (namely the bias of all methods using the orthogonality constraint, and even of recently proposed algorithm EFICA) is asymptotically proportional to σ^2, the proposed method has usually a lower bias, and consequently it exhibits a lower symbol-error rate, when applied to blind separation of finite alphabet signals, typical for communication systems.

1 Introduction

The noisy model of Independent Component Analysis (ICA) considered in this paper, is

$$\mathbf{X} = \mathbf{A}\mathbf{S} + \sigma\mathbf{N}, \tag{1}$$

where $\mathbf{S}$ denotes a vector of d independent random variables representing the original signals, $\mathbf{A}$ is an unknown regular $d \times d$ mixing matrix, and $\mathbf{X}$ represents the observed mixed signals. The noise $\mathbf{N}$ denotes a vector of independent variables having the covariance matrix $\mathbf{\Sigma}$. Without loss of generality, we will further assume that $\mathbf{\Sigma}$ equals to the identity matrix $\mathbf{I}$. Consequently, σ^2 is the variance of the added noise to the mixed signals. All signals considered here are i.i.d. sequences, i.e., they are assumed to be white in the analysis.

It is characteristic for most ICA methods that they were derived for the noiseless case, so to solve the task of estimating the mixing matrix $\mathbf{A}$ or its inversion

* This work was supported by Ministry of Education, Youth and Sports of the Czech Republic through the project 1M0572 and through the Grant 102/07/P384 of the Grant Agency of the Czech Republic.

M.E. Davies et al. (Eds.): ICA 2007, LNCS 4666, pp. 730–737, 2007.

$\mathbf{W} = \mathbf{A}^{-1}$. Then, abilities to separate noised data are studied experimentally, and the non-vanishing estimation error as $N \to +\infty$, N being length of data, is taken for a bias caused by the noise. To compensate such bias, several techniques were proposed [6]. Unfortunately, these methods have a drawback that the covariance structure of the noise needs to be known *a priori* [4].

In accord with [7], we suggest to measure the separation quality not through accuracy of estimation of the mixing mechanism but through the achieved interference + noise to signal ratio (INSR) or its inverse SINR. In separating the finite alphabet signals, the ultimate criterion should be the symbol error rate (SER). Computation of the INSR or the SER assumes that the permutation, scale, and sign or phase ambiguities were resolved by minimizing the INSR.

In the case $\mathbf{\Sigma} = \mathbf{I}$, the INSR of a k-th estimated signal can be computed as

$$\mathrm{INSR}_k = \frac{\sum_{i \neq k}^{d} (\mathbf{BA})_{ki}^2 + \sigma^2 \sum_{i=1}^{d} \mathbf{B}_{ki}^2}{(\mathbf{BA})_{kk}^2}, \tag{2}$$

where $\mathbf{B}$ is the separating transformation [7]. The solutions that minimize (2) are known to be given by the MMSE separating matrix, denoted by $\mathbf{W}^{\mathrm{MMSE}}$, that takes the form

$$\mathbf{W}^{\mathrm{MMSE}} = \mathbf{A}^H (\mathbf{A}\mathbf{A}^H + \sigma^2 \mathbf{I})^{-1} \tag{3}$$

where ^{H} denotes the conjugate (Hermitian) transpose. Signals given by $\mathbf{W}^{\mathrm{MMSE}}\mathbf{X}$ will be further called the *MMSE solution.* Note that these signals may not be necessarily normalized to have unit variance, unlike outcome of common blind separation methods, that produce normalized components. For exact comparisons, we introduce a matrix $\mathbf{W}^{\mathrm{NMMSE}}$ such that $\mathbf{W}^{\mathrm{NMMSE}}\mathbf{X}$ are the normalized MMSE signals.

The paper is organized as follows. In section 2, we briefly describe several variants of algorithm FastICA and the proposed method, including a novel test of saddle points for separating complex-valued signals. Section 3 presents analytic expressions for an asymptotic bias of solutions obtained by real domain FastICA variants [5,8] from the MMSE solution. Specifically, we study the biases of estimates of de-mixing matrix $\mathbf{W}$ from $\mathbf{W}^{\mathrm{NMMSE}}$, and the one-unit FastICA and the proposed algorithm is shown to be less biased than the other methods. Simulations in Section 4 demonstrate drawbacks of the unbiased algorithm [6] (further referred to as *unbiased FastICA*) following from required knowledge of $\mathbf{\Sigma}$ and/or σ. Conversely, the proposed algorithm with one-unit FastICA-like performance is shown to be the best blind MMSE estimator when separating noisy finite-alphabet signals.

2 FastICA and Its Variants

Common FastICA algorithms work with the decorrelated data $\mathbf{Z} = \mathbf{C}^{-1/2}\mathbf{X}$, where $\mathbf{C} = \mathrm{E}[\mathbf{X}\mathbf{X}^H]$ is the data covariance matrix. Only the unbiased FastICA [6] that aims at unbiased estimation of $\mathbf{A}^{-1}$ assuming that the noise has a known covariance matrix $\mathbf{\Sigma}$, uses the preprocessing $\mathbf{Z} = (\mathbf{C} - \mathbf{\Sigma})^{-1/2}\mathbf{X}$.

One-unit FastICA in real domain [5] estimates one de-mixing vector $\mathbf{w}_k^{1\mathrm{U}}$ iteratively via the recursion

$$\mathbf{w}_k^+ \leftarrow \mathrm{E}[\mathbf{Z}g(\mathbf{w}_k^{1\mathrm{U}^T}\mathbf{Z})]\} - \mathbf{w}_k^{1\mathrm{U}}\mathrm{E}\{g'(\mathbf{w}_k^{1\mathrm{U}^T}\mathbf{Z}))\}, \qquad \mathbf{w}_k^{1\mathrm{U}} \leftarrow \mathbf{w}_k^+/\|\mathbf{w}_k^+\| \quad (4)$$

until convergence is achieved. Here $g(\cdot)$ is a smooth nonlinear function that approximates/surrogates the score function corresponding to the distribution of the original signals [11]. The theoretical expectation values in (4) are, in practice, replaced by their sample-based counterparts.

Similar recursion was proposed for one-unit FastICA in the complex domain [1]. The symmetric (real or complex) variant performs the one-unit iterations in parallel for all d separating vectors, but the normalization in (4) is replaced by a symmetric orthogonalization.

The algorithm EFICA [8] combines the symmetric approach with the test of saddle points, an adaptive choice of nonlinearity $g_k(\cdot)$ for each signal separately, and it does the refinement step that relaxes the orthogonal constraint introduced by the symmetric approach and is designed towards asymptotic efficiency.

The unbiased FastICA [6] uses the recursion

$$\mathbf{w}_k^+ \leftarrow \mathrm{E}[\mathbf{Z}g(\mathbf{w}_k^{\mathrm{unb}^T}\mathbf{Z})] - (\mathbf{I}+\widetilde{\boldsymbol{\Sigma}})\mathbf{w}_k^{\mathrm{unb}}\,\mathrm{E}[g'(\mathbf{w}_k^{\mathrm{unb}^T}\mathbf{Z})],$$

where $\widetilde{\boldsymbol{\Sigma}} = (\mathbf{C}-\boldsymbol{\Sigma})^{-1/2}\boldsymbol{\Sigma}(\mathbf{C}-\boldsymbol{\Sigma})^{-1/2}$. Both approaches (one-unit and symmetric) can be considered; in simulations, we use the one-unit variant, and the resulting de-mixing matrix will be denoted by $\mathbf{W}^{\mathrm{UNB}}$. In order to compare performance of the unbiased FastICA by means of (2) with the other techniques fairly, it is necessary to consider a MMSE estimate derived from $\mathbf{W}^{\mathrm{UNB}}$, namely

$$\mathbf{W}^{\text{MMSE-UNB}} = \boldsymbol{\Sigma}^{-1}(\mathbf{W}^{\mathrm{UNB}})^{-T} \times [(\mathbf{W}^{\mathrm{UNB}})^{-1}\boldsymbol{\Sigma}^{-1}(\mathbf{W}^{\mathrm{UNB}})^{-T} + \sigma^2\mathbf{I}]^{-1} \quad (5)$$

2.1 Proposed Algorithm

The proposed algorithm is a combination of symmetric FastICA, test of saddle points, and one-unit FastICA as a refinement. Usually, one unit FastICA is used in a deflation way, when the estimated components are subtracted from the mixture one by one. This is computationally effective method, but accuracy of the later separated components might be compromised. Therefore, we propose to initialize the algorithm using symmetric FastICA, that is known for having very good global convergence and allows equal separation precision for all components.

The test of saddle points was first proposed in [11] to improve probability of the symmetric FastICA to converge to the true global maximum of the cost function $[\mathrm{E}\{G(\mathbf{w}^T\mathbf{Z})\} - G_0]^2$ where $G(\cdot)$. is a primitive function of $g(\cdot)$ and $G_0 = \mathrm{E}\{G(\xi)\}$, where ξ is a standard Gaussian random variable.

In short, the test of saddle points consists in checking all pairs of the estimated components $(\mathbf{u}_k, \mathbf{u}_\ell)$, whether or not other pair of signals $(\mathbf{u}'_k, \mathbf{u}'_\ell)$ gives a higher value of the cost function

$$c(\mathbf{u}_k, \mathbf{u}_\ell) = [\mathrm{E}\{G(\mathbf{u}_k)\} - G_0]^2 + [\mathrm{E}\{G(\mathbf{u}_\ell)\} - G_0]^2, \quad (6)$$

where $\mathbf{u}'_k = (\mathbf{u}_k + \mathbf{u}_\ell)/\sqrt{2}$ and $\mathbf{u}'_\ell = (\mathbf{u}_k - \mathbf{u}_\ell)/\sqrt{2}$.

The motivation is that a random initialization of the algorithm may begin at a point of zero gradient of the cost function (a saddle point / an unstable point of the iteration) and terminate there, despite being not the desired stable solution. See [11] for details.

In the complex domain, the situation is a bit more tricky, because if $(\mathbf{u}_k, \mathbf{u}_\ell)$ is the pair of valid independent components in the mixture, not only their weighted sum and a difference represent a false (unstable) point of the iteration. *All* pairs $(\mathbf{u}'_k, \mathbf{u}'_\ell)$ of the form $\mathbf{u}'_k = (\mathbf{u}_k + e^{i\alpha}\mathbf{u}_\ell)/\sqrt{2}$ and $\mathbf{u}'_\ell = (\mathbf{u}_k - e^{i\alpha}\mathbf{u}_\ell)/\sqrt{2}$ are stationary but unstable for any phase factor $e^{i\alpha}$, $\alpha \in \mathcal{R}$.

Therefore we propose to do a phase shift of each separated component so that the real part and the imaginary part of the signal are as much independent each of other as possible before the test of the saddle points. This phase shift can be easily performed using a two-dimensional symmetric FastICA in the real domain applied to the real and imaginary part of the component. After this preprocessing, it is sufficient to perform the test of saddle points exactly as in the real-valued case, i.e. to check all pairs $(\mathbf{u}'_k, \mathbf{u}'_\ell)$ with $\mathbf{u}'_k = (\mathbf{u}_k + \mathbf{u}_\ell)/\sqrt{2}$ and $\mathbf{u}'_\ell = (\mathbf{u}_k - \mathbf{u}_\ell)/\sqrt{2}$, whether they give a higher value of the cost function (6) or not.

Validity of the above described complex domain test of the saddle points can be easily confirmed in simulations by starting the algorithm from the pairs $\mathbf{u}'_k = (\mathbf{u}_k + e^{i\alpha}\mathbf{u}_\ell)/\sqrt{2}$ and $\mathbf{u}'_\ell = (\mathbf{u}_k - e^{i\alpha}\mathbf{u}_\ell)/\sqrt{2}$ with an arbitrary $\alpha \in \mathcal{R}$ where $\mathbf{u}_k$ and $\mathbf{u}_\ell$ are the true independent sources. We have successfully tested this approach on separation of complex-valued finite alphabet sources known in communications (QAM, V27).

The resultant algorithm (symmetric FastICA + test of saddle points + one unit refinements) will be referred to as 1FICA.

3 Bias of the FastICA Variants

In this section, asymptotic expressions for bias of algorithms described in previous section working in the real domain will be presented. (The complex-domain FastICA exhibits a similar behavior in simulations.) For details of analysis, the reader is referred to [9] due to lack of space.

In brief, the theoretical analysis is done for "small" σ and infinite number of samples. Similarly to [11], for theoretical considerations, it is assumed that the analyzed method starts from the MMSE solution and stops after one iteration. This assumption is reasonable due to the following facts: (1) deviation of the global maximizer $\widehat{\mathbf{W}}$ of the FastICA cost function from $\mathbf{W}^{\mathrm{MMSE}}$ is of the order $O(\sigma^2)$, and (2) convergence of the algorithm is at least quadratic [10]. Therefore, after performing the one iteration, the deviation of the estimate from the global maximizer $\widehat{\mathbf{W}}$ is of the order $O(\sigma^4)$ and, hence, is negligible.

The bias of the algorithm will be studied in terms of the deviation of $\widehat{\mathbf{W}}(\mathbf{W}^{\mathrm{MMSE}})^{-1}$ from a diagonal matrix. More precisely, the bias is equal to the difference between $\mathrm{E}[\widehat{\mathbf{W}}](\mathbf{W}^{\mathrm{MMSE}})^{-1}$ and $\mathbf{D} = \mathbf{W}^{\mathrm{NMMSE}}[\mathbf{W}^{\mathrm{MMSE}}]^{-1}$, where $\mathbf{D}$ is the diagonal matrix that normalizes the MMSE signals $\mathbf{S}^{\mathrm{MMSE}} = \mathbf{W}^{\mathrm{MMSE}}\mathbf{X}$.

It holds that

$$\mathbf{D} = \mathbf{I} + \frac{1}{2}\sigma^2 \mathrm{diag}[\mathbf{V}_{11}, \ldots, \mathbf{V}_{dd}] + O(\sigma^3). \tag{7}$$

From here we use the notation $\mathbf{W} = \mathbf{A}^{-1}$ and $\mathbf{V} = \mathbf{W}\mathbf{W}^T$. Finally, for a matrix $\widehat{\mathbf{W}}$ that separates the data $\mathbf{S}^{\mathrm{MMSE}}$, the bias is $\mathrm{E}[\widehat{\mathbf{W}}] - \mathbf{D}$.

3.1 Bias of the One-Unit FastICA and 1FICA

It can be shown that the de-mixing vector $\mathbf{w}_k^{1\mathrm{U}}$ resulting from the one-unit FastICA (applied to the data $\mathbf{S}^{\mathrm{MMSE}}$), for $N \to +\infty$, is proportional to

$$\mathbf{w}_k^{1\mathrm{U}} = \tau_k \mathbf{e}_k + \frac{1}{2}\sigma^2 \mathbf{V}_{kk}(\tau_k + \delta_k)\mathbf{e}_k + O(\sigma^3) \tag{8}$$

where $\tau_k = \mathrm{E}[s_k g(s_k) - g'(s_k)]$, and δ_k is a scalar that depends on the distribution of s_k and on the nonlinear function g and its derivatives to the third order. Since (8) is a scalar multiple of $\mathbf{e}_k$ (the k-th column of the identity matrix), it follows that the asymptotic bias of the one-unit approach is $O(\sigma^3)$. Prospectively, the separating matrix $\mathbf{W}^{1\mathrm{F}}$ given by the proposed 1FICA has the same bias. Simulations confirm this expectation [9].

3.2 Bias of the Inversion Solution

It is interesting to compare the previous result with the solution that is given by exact inversion of the mixing matrix, i.e. $\mathbf{W}\mathbf{X} = \mathbf{S} + \sigma\mathbf{W}\mathbf{N}$; the signals will be called *the inversion solution.* From

$$\mathbf{W}(\mathbf{W}^{\mathrm{MMSE}})^{-1} = \mathbf{W}(\mathbf{A}\mathbf{A}^T + \sigma^2\mathbf{I})\mathbf{W}^T = \mathbf{I} + \sigma^2\mathbf{V}$$

it follows that the "bias" of the inversion solution is proportional to σ^2 and in general it is greater than that of 1FICA. In other words, **the algorithm 1FICA produces components that are asymptotically closer to the MMSE solution than to the inversion solution**.

3.3 Bias of Algorithms Using the Orthogonal Constraint

Large number of ICA algorithms (e.g. JADE [2], symmetric FastICA, etc.) use an orthogonal constraint, i.e., they enforce the separated components to have sample correlations equal to zero. Since the second-order statistics cannot be estimated perfectly, this constraint compromises the separation quality [3,11]. Here we show that the bias of all ICA algorithms that use the constraint has the asymptotic order $O(\sigma^2)$.

The orthogonality constraint can be written as

$$\mathrm{E}[\widehat{\mathbf{W}}\mathbf{X}(\widehat{\mathbf{W}}\mathbf{X})^T] = \widehat{\mathbf{W}}(\mathbf{A}\mathbf{A}^T + \sigma^2\mathbf{I})\widehat{\mathbf{W}}^T = \mathbf{I}\ . \tag{9}$$

It follows that the bias of all constrained algorithms is lower bounded by

$$\min_{\widehat{\mathbf{W}}(\mathbf{A}\mathbf{A}^T + \sigma^2\mathbf{I})\widehat{\mathbf{W}}^T = \mathbf{I}} \|\widehat{\mathbf{W}}(\mathbf{W}^{\mathrm{MMSE}})^{-1} - \mathbf{D}\|_F = O(\sigma^2) \tag{10}$$

where the minimization proceeds for $\widehat{\mathbf{W}}$. The matrix $\mathbf{D}$ in (10) is the same as in (7). For the minimizer $\widehat{\mathbf{W}}$ of (10) it holds that $\widehat{\mathbf{W}}(\mathbf{W}^{\mathrm{MMSE}})^{-1} = \mathbf{I}+\sigma^2\mathbf{\Gamma}+O(\sigma^3)$, where $\mathbf{\Gamma}$ is a nonzero matrix obeying $\mathbf{\Gamma} + \mathbf{\Gamma}^T = \mathbf{V}$; see [9] for details. This result can be interpreted in the way that the algorithms using the orthogonality constraint may prefer some of the separated components to give them a zero bias, but the total average bias for all components has the order $O(\sigma^2)$.

3.4 Bias of the Symmetric FastICA and EFICA

The biases of the algorithms can be expressed as

$$\mathrm{E}[\widehat{\mathbf{W}}](\mathbf{W}^{\mathrm{MMSE}})^{-1} - \mathbf{D} = \frac{1}{2}\sigma^2\mathbf{V} \odot (\mathbf{1}_{d\times d} - \mathbf{I} + \mathbf{H}) + O(\sigma^3), \tag{11}$$

where $\mathbf{H}_{k\ell} = \frac{|\tau_\ell|-|\tau_k|}{|\tau_k|+|\tau_\ell|}$ for the symmetric FastICA, and $\mathbf{H}_{k\ell} = \frac{c_{k\ell}|\tau_\ell|-|\tau_k|}{|\tau_k|+c_{k\ell}|\tau_\ell|}$ for EFICA, where $c_{k\ell} = \frac{|\tau_\ell|\gamma_k}{|\tau_k|(\gamma_\ell+\tau_\ell^2)}$ for $k \neq \ell$ and $c_{kk} = 1$. Here, $\gamma_k = \mathrm{E}[g_k^2(s_k)] - \mathrm{E}^2[s_k g_k(s_k)]$, and g_k is the nonlinear function chosen for the k-th signal.

It can be seen that the bias of both of the algorithms has the order $O(\sigma^2)$.

4 Simulations

In this section, we present results of two experiments to demonstrate and compare the performance of the proposed algorithm 1FICA with competing methods: The symmetric FastICA (marked by SYMM), the unbiased FastICA (unbiased FICA), EFICA, and JADE [2]. Results given by "oracle" MMSE solution and the inversion solution are included as well. Examples with complex signals are not included due to lack of space.

In the first example, we separate 10 randomly mixed [7] BPSK signals with added Gaussian noise, first, for various length of data N (Fig. 1(a)) and, second, for varying input signal-to-noise ratio (SNR) defined as $1/\sigma^2$ (Fig. 1(b)). The experiment encompasses several extremal conditions: In the first scenario, where SNR=5dB ($\sigma \doteq 0.56$), N goes from 100, which is quite low for the dimension $d = 10$. The second situation examines $N = 200$ and SNR going down to 0dB.

Note that the bias may be less important than the estimation variance when the data length N is low. Therefore, in simulations, we have included two slightly changed versions of 1FICA and EFICA algorithm, denoted by "1FICA-biga" and "EFICA-biga", respectively. The modifications consist in that the used nonlinear function g is equal to the score function of marginal pdfs of the signals to-be estimated (i.e., noisy BPSK that have **bi**modal **Ga**ussian distribution, therefore, "biga" in the acronym). Adopted from the noiseless case [11], better performance of the modified algorithms may be expected.

Figure 1 shows superior performance of the proposed algorithm 1FICA and of its modified version. The same performance is achieved by the modified EFICA for $N \leq 200$, but it is lower due to the bias when N is higher. The unbiased FastICA achieves the same accuracy for $N \geq 500$ but is unstable when N is low.

The average performance of an algorithm is often spoiled due to poorer stability, which occurs in high dimensions and low N cases, mainly. In this issue, we highlight positive effect of the test of saddle points that is included in the proposed 1FICA or in EFICA. For instance, the results achieved by the symmetric FastICA would be significantly improved if the test was included in it.

The second example demonstrates conditions when the covariance of the noise is not exactly known or varying. To this end, the noise level was changed randomly from trial to trial. Five BPSK signals of the length $N = 50000$ were mixed with a random matrix and disturbed by Gaussian noise with covariance $\sigma^2\mathbf{I}$, where σ was randomly taken from interval $[0, 1]$, and then blindly separated. The mean value of the noise covariance matrix, i.e. $\mathbf{I}/3$, was used as the input parameter of the unbiased FastICA. Note that INSR and BER of this method were computed for solutions given by $\mathbf{W}^{\text{MMSE-UNB}}$ defined in (5).

The following table shows the average INSR and bit error rate (BER) that were achieved in 1000 trials. The performance of the proposed 1FICA is almost the same like that of "oracle" MMSE separator, because, here, N is very high, and the estimation error is caused by the bias only. The unbiased FastICA significantly suffers from inaccurate information about the noise intensity.

algorithm	average INSR [dB]	BER [%]
1FICA	**-5,98**	**3,19**
Symmetric FastICA	-5,68	3,55
unbiased FastICA	*6,79*	*5,25*
EFICA	-5,79	3,41
MMSE solution	**-5,98**	**3,19**
inversion solution	-4,76	4,71
JADE	-5,68	3,55

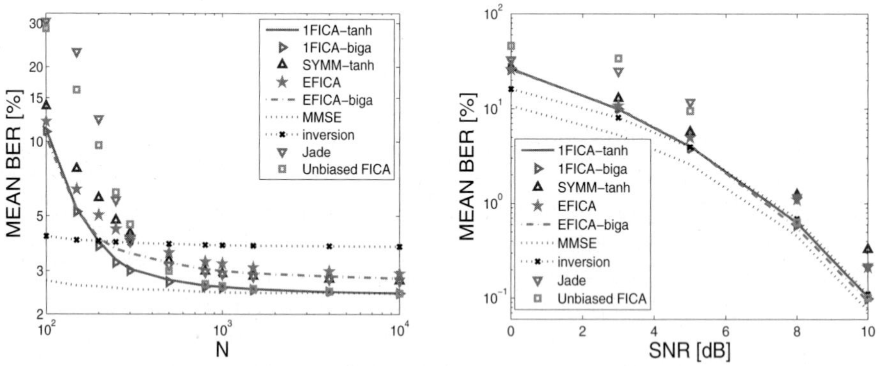

(a) (b)

Fig. 1. Average BER of 10 separated BPSK signals when (a) SNR is fixed to 5dB and (b) a fixed number of data samples is $N = 200$. Averages are taken from 1000 independent trials for each settings.

5 Conclusions

This paper presents novel results from analysis of bias of several FastICA variants, whereby the one-unit FastICA was shown to be minimally biased from the MMSE solution, i.e., it achieves the best interference-plus-noise rejection rate for $N \to +\infty$.

By virtue of the theoretical results, a new variant of FastICA algorithm, called 1FICA, was derived to have the same global convergence as symmetric FastICA with the test of saddle points, and a noise rejection like the one-unit FastICA. Computer simulations show superior performance of the method when separating binary (BPSK) signals. Unlike the unbiased FastICA, it does not require prior knowledge of covariance of the noise to achieve the best MMSE separation.The Matlab codes for 1FICA in real and in complex domains can be downloaded from the first author's homepage, http://itakura.kes.tul.cz/zbynek/downloads.htm.

References

1. Bingham, E., Hyvärinen, A.: A fast fixed-point algorithm for independent component analysis of complex valued signals. Int. J. Neural Systems 10, 1–8 (2000)
2. Cardoso, J.-F., Souloumiac, A.: Blind Beamforming from non-Gaussian Signals. IEE Proc.-F 140, 362–370 (1993)
3. Cardoso, J.-F.: On the performance of orthogonal source separation algorithms. In: Proc. EUSIPCO, Edinburgh, pp. 776–779 (1994)
4. Davies, M.: Identifiability Issues in Noisy ICA. IEEE Signal Processing Letters 11, 470–473 (2005)
5. Hyvärinen, A., Oja, E.: A fast fixed-point algorithm for independent component analysis. Neural Computation 9, 1483–1492 (1997)
6. Hyvärinen, A.: Gaussian Moments for Noisy Independent Component Analysis. IEEE Signal Processing Letters 6, 145–147 (1999)
7. Koldovský, Z., Tichavský, P.: Methods of Fair Comparison of Performance of Linear ICA Techniques in Presence of Additive Noise. In: Proc. of ICASSP 2006, Toulouse, pp. 873–876 (2006)
8. Koldovský, Z., Tichavský, P., Oja, E.: Efficient Variant of Algorithm FastICA for Independent Component Analysis Attaining the Cramér-Rao Lower Bound. IEEE Tr. Neural Networks 17, 1265–1277 (2006)
9. Koldovský, Z., Tichavský, P.: Asymptotic analysis of bias of FastICA-based algorithms in presence of additive noise, Technical report no 2181, ÚTIA, AV ČR (2007), Available at `http://itakura.kes.tul.cz/zbynek/publications.htm`
10. Oja, E., Yuan, Z.: The FastICA algorithm revisited: Convergence analysis. IEEE Trans. on Neural Networks 17, 1370–1381 (2006)
11. Tichavský, P., Koldovský, Z., Oja, E.: Performance Analysis of the FastICA Algorithm and Cramér-Rao Bounds for Linear Independent Component Analysis. IEEE Tr. on Signal Processing 54, 1189–1203 (2006)

Compact Representations of Market Securities Using Smooth Component Extraction

Hariton Korizis, Nikolaos Mitianoudis, and Anthony G. Constantinides

Communications and Signal Processing Group, Imperial College London,
Exhibition Road, London SW7 2AZ, UK
{hariton.korizis99,n.mitianoudis,a.constantinides}@ic.ac.uk

Abstract. Independent Component Analysis (ICA) is a statistical method for expressing an observed set of random vectors as a linear combination of statistically independent components. This paper tackles the task of comparing two ICA algorithms, in terms of their efficiency for compact representation of market securities. A recently developed sequential blind signal extraction algorithm, SmoothICA, is contrasted to a classical implementation of ICA, FastICA. SmoothICA uses an additional 2nd order constraint aiming at identifying temporally smooth components in the data set. This paper demonstrates the superiority of this novel smooth component extraction algorithm in terms of global and local approximation capability, applied to a portfolio of 60 NASDAQ securities, by utilizing common ordering algorithms for financial signals.

Keywords: Independent Component Analysis, SmoothICA, FastICA, market securities, Finance.

1 Introduction

The goal of Independent Component Analysis is to find a linear representation of non-Gaussian variables. Finding such a representation provides an insight to the underlying structure of many signal processing problems. The ICA problem is equivalent to establishing the following generating model for the data:

$$x = As \tag{1}$$

where x and s are n-dimensional random vectors, and components s are assumed mutually independent. A is a constant $n \times n$ full rank matrix, denoting the unknown mixing matrix. Relevant to our investigation is the formulation that x consists of a set of observation vectors generated in the financial markets, which are driven by the hidden underlying sources s. The driving mechanisms s are mixed and contaminated among others by elements, such as news and expectations related to results of companies and sectors, domestic and foreign politics that affect exchange and interest rates, consumer confidence, unexpected events and even the weather that affects the commodities' prices. The transformation:

$$s = Wx \tag{2}$$

can be defined, with W the demixing matrix and $A = W^{-1}$. This method allows at most one Gaussian component, concentrating all the signal innovations which cannot

M.E. Davies et al. (Eds.): ICA 2007, LNCS 4666, pp. 738–745, 2007.

be accounted for by the original problem assumptions. In the case of signals originating from the financial markets, this assumption can be considered valid for a great majority of the cases, as purely gaussian financial signals are rarely generated.

The assumption of statistical independence of the source signals, can be assumed to be valid in the scope of global economy and the hugely diverse micro- and macroeconomic factors that affect financial processes. Unexplained noise, as well as the markets' response to large trades can be also of significance to researchers and traders. However it is logical that the underlying driving sources' independence assumption in the financial markets can be debated, as every source might exert a small influence on all others. A review of various contrast functions can be found in [1], as the sources s can be separated using various interpretations of statistical independence.

In a recent paper, the authors in [2] proposed a sequential blind signal extraction algorithm incorporating a smoothness constraint based on the original FastICA algorithm [3]. Along with the negentropy cost function, the added temporal constraint seeks to find smooth orthogonal projections in the mixtures vector x. In [4], this algorithm is referred to as SmoothICA and contrasted to the performance of FastICA, in the search for temporal structure in the underlying sources that give rise to stock evolutions. A portfolio of 20 NASDAQ securities was analyzed and possible advantages of this novel approach were highlighted over FastICA, as it produced components with smoother temporal structure. A small degree of correlation was present among the components extracted, introduced by the balancing of the 4th and 2nd order temporal constraint.

In Principal Component Analysis (PCA) the component ranking can be achieved according to the eigenvalues of the source vectors. In ICA the same task is not straightforward, as the corresponding projection vector consists of normalized rows to unity [3]. PCA is optimal for dimensionality reduction to a set of uncorrelated components, in terms of variance. However, the idea here is the existence of independent underlying sources, thus examining only algorithms that seek independence. Common ordering algorithms for financial signals are used and their error performance are contrasted in approximating a given portfolio using reduced set of components.

This text is organized in the following way. The next section contains an overview of ICA ordering techniques for financial time series. Section 3 contains an overview of the SmoothICA algorithm. Section 4 presents the ordering algorithms that are utilized in the experimental Section 5. An ordering method for selecting a reduced number of components for reconstruction of a whole portfolio of securities is considered (global approximation), as well as two methods for ordering the components' contributions to each security signal and reconstructing accordingly. Section 6 concludes this study.

2 Overview of Independent Component Ordering in Finance

Several investigations of ICA with application to finance have been performed. The most influential was done in [5], examining the portfolio returns of 28 Japanese stocks. PCA and ICA performances were compared for such signals. From the operation of ICA, the components produced have $var(s_i) = 1$. It is therefore assumed that any information about the contribution of each individual component, to a mixture's variation is engulfed in the mixing matrix A. The authors used the maximum norm L_∞ to sort the rows of the A, and thus to determine which ICs have the maximum contribution to a selected signal's amplitude. Such a measure is applied in this research.

Cheung and Xu [6] presented a criterion for ordering source signals, according to their contribution to the trend reservation of each observed signal. This algorithm uses the MSE criterion and is named Testing-and-Acceptance (TnA), and when applied to foreign exchange rates it produces superior results over the L_∞ norm method. This is the second method which will be used in this paper. The same authors have presented a criterion to select the appropriate dimension for the source signal subset to approximate a portfolio of foreign exchange rates in [8]. An algorithm using the Relative Hamming Distance (RHD) instead, was proposed in [7].

A consequence of the increased interest in this type of component extraction and its demonstrated superiority in terms of source separation over PCA, are applications utilizing its capabilities in econometrics and finance; from prediction approaches [9] and Factor Model estimation [10] to the computation of the risk of a portfolio of securities [11] and the application of ICA in the context of state space models for interbank foreign exchange rates to obtain a better separation of the observation noise and the "true" price [12]. It is worth focusing on [11] where the contribution of an individual independent component to the variance of the whole portfolio of securities is calculated. The ICs are ordered according to that contribution, and this operates as a preprocessing step for dimensionality reduction before switching back to the prices' space. This is the third method examined in the current paper, testing global approximation performance.

3 The SmoothICA Algorithm

After an initial prewhitening step, SmoothICA solves the following inequality constrained optimization problem:

$$\max_{w} \quad J_1(\underline{w}) \tag{3}$$

$$\text{subject to } J_2(\underline{w}) \leq 0 \tag{4}$$

$$J_3(\underline{w}) = 0 \tag{5}$$

where $\underline{w}$ the projection operator of the white data $\underline{z}$, $J_1(\cdot)$ is the approximated negentropy as proposed by Hyvarinen [3], $J_2(\underline{w}) = \mathcal{E}\{(\underline{w}^T\underline{\Delta z})^2\} - \rho\mathcal{E}\{(\underline{w}^T\underline{z})^2\}$ is the second-order smoothness criterion, $J_3(\cdot)$ is the unit-norm constraint, $\rho \in [0,1]$ defines the degree of smoothness [4] and $\underline{\Delta z}$ is the whitened data differences matrix. Modifying the inequality constraint to the equality constraint $\max(J_2(\underline{w}), 0) = 0$, one can find the desired optima using alternating unconstrained maximization of the Lagrangian function $J_1(\underline{w}) + \lambda\max(J_2(\underline{w}), 0) + \kappa J_3(\underline{w})$, where λ, κ are the Lagrange multipliers. The following Newton-step provides an update:

$$\underline{w}^+ \leftarrow \underline{w} - \left[\frac{\partial^2 J}{\partial \underline{w}^2}\right]^{-1} \frac{\partial J}{\partial \underline{w}} \tag{6}$$

where, in this case, the gradient vector and the Hessian matrix are estimated using the following updates :

$$\frac{\partial J}{\partial \underline{w}} = \mu\mathcal{E}\{\underline{z}G'(\underline{w}^T\underline{z})\} + \lambda(\mathcal{E}\{(\underline{w}^T\underline{\Delta z})\underline{\Delta z} - \rho(\underline{w}^T\underline{z})\underline{z}\})(\text{sgn}(J_2)+1)$$

$$\frac{\partial^2 J}{\partial \underline{w}^2} = \mu \mathcal{E}\{G''(\underline{w}^T \underline{z})\}I + \lambda(C_{\Delta z} - \rho I)(\text{sgn}(J_2) + 1)$$

where $\underline{u} = \underline{w}^T \underline{z}$, $G(\underline{u}) = logcosh(\underline{u})$, $\mu = \text{sgn}(\mathcal{E}\{G(\underline{u})\} - \mathcal{E}\{G(v)\})$ and v a zero-mean, unit-norm Gaussian variable. After calculating the estimate for $\underline{w}$, we calculate estimates for λ via alternating optimization. The unit-norm constraint $\underline{w}^+ \leftarrow \underline{w}^+/||\underline{w}^+||$ is then imposed as a projection of the $\underline{w}$ estimate on the unit hypersphere, to ensure that rotation and not scale deformation is performed. The orthogonal deflation procedure is used to extract subsequent smooth components [1].

4 Ordering Methods for Independent Component Analysis

In Finance dimensionality reduction is applied for various purposes. It is performed to remove unwanted information and hence get a clearer picture of an underlying process, allowing better modeling and understanding of its statistical nature. It it also applied to represent a large set of assets by an appropriate subset that best defines it and reduce memory requirements and computational burden. Unlike PCA, ICA is not constructed to have an inherent ordering of the ICs. The methods below follow two notions; approximation of a particular security using a few ICs (selected according to their contribution to that particular security) and approximation of a whole portfolio of securities by selecting an appropriate subset of independent components.

4.1 Global Approximation

In ICA the components produced are scaled to unit variance. This means that the additional information about individual contributions of the ICs to the observed signals lies in the mixing matrix A [11]. The variance of the security i is σ_i^2 and the amount of total variance V_j explained by each component s_j can be derived from:

$$\sigma_i^2 = \sum_{i=1}^{n} a_{ij}^2 \quad \text{and} \quad V_j = \frac{\sum_{j=1}^{n} a_{ij}^2}{\sum_{i,j=1}^{n} a_{ij}^2} \tag{7}$$

Thus by ordering the ICs according to their individual contributions to the whole portfolio, we can approximate efficiently by selecting a reduced number of components.

4.2 Local Approximation

The L_∞ norm: The weighted ICs, given by (9), with the largest amplitudes are defined to be the dominant ICs. This of course presents an ordering criterion, as these ICs have the largest effect on the securities. The reconstruction of the ith security from the estimates of the source signals is:

$$\widehat{x}_i = \sum_{k=1}^{n} a_{ik} s_k \tag{8}$$

where s_k is the kth estimated IC and a_{ik} is the weight in the ith row, kth column of A. The weighted ICs are therefore obtained from:

$$\widehat{s}_{ik} = a_{ik} s_k \quad k = 1..n \tag{9}$$

The L_∞ norm was used in [5] to order the weighted ICs for each particular stock, as this reveals the magnitude contribution of each source signal to a particular stock.

The Testing-and-Acceptance algorithm: The TnA algorithm in [6] aims at creating a list L_i, whose elements are the component subscripts decided according to decreasing contribution to a specified security signal. Initially, the IC which introduces the minimum MSE error of reconstruction of the selected security if omitted, is selected from the m components. The reconstructed security, while the ith component is omitted, is $\{\widehat{y}_j\}_{j=1,j\neq i}^{m}$. The subscript of this IC is put last in the list L. The next step of the iteration starts with a subset of the ICs that do not include the previously selected component. It finds the next component that, while omitted, causes minimum MSE error of approximation, and puts it second to last in L, and so on. It is a suboptimal heuristic method compared with the exhaustive search, however the TnA algorithm involves just $\frac{m(m+1)}{2} - 1$ compared to $(m+1)!$ steps.

The algorithm operates as follows:

1. Let the set of independent component subscripts $Z = \{j \mid 1 \leq j \leq m\}$, $d = 0$, and the order list $L_i = ()$.
2. For each $j \in Z$ and N being the signal's length, let:

$$v_{ij(t)} = \sum_{m\neq j, m\in Z} \widehat{s}_{im(t)} \quad , 1 \leq t \leq N \tag{10}$$

   The β which will be stored as the dth element of L_i and removed from the set Z, is selected according to:

$$\beta = \arg\min_{j\in Z} MSE(x_i, v_{ij}) \tag{11}$$

$$d^{new} = d^{old} + 1$$

$$\text{Then let:} \quad L_i^{new} = L_i^{old} + \beta$$

$$Z^{new} = Z^{old} - \{\beta\}$$

3. If $Z \neq \{\}$, goto Step 2; otherwise stop. In order to make the list ordered according to descending contribution, flip it.

5 Experiments

5.1 Description of the Data

The experiments are performed with daily closing prices of a portfolio of 60 US technology stocks[1], for the period ranging from 01/01/2002 to 05/04/2005. The data is

[1] The portfolio consists of the first 60 stocks (alphabetically) of the NASDAQ US Exchange.

centered and whitened so that uncorrelated, unit variance signals are obtained. The SmoothICA algorithm is performed on the centered and whitened data as outlined in [2] and [4]. Extraction of smoother components than FastICA is achieved, although more computational time is required. Flexibility is added with ρ starting at 0.05 and increasing progressively in the case of non converging components. Coefficient ρ is initialized at this level, as this corresponds to a long-period sinusoidal signal with low levels of noise. Using a large portfolio (a high number of mixtures) the issue of low correlation among the components [4] is avoided. The correlation matrix among the sources is now a proper identity matrix. The FastICA algorithm is also applied, due to its convergence efficiency, which gives the reference results for approximation fitness comparison for the all the ranges of subset orders possible.

5.2 Global Approximation Comparison

After obtaining successful convergence for both algorithms, the percentages of variance contribution of each of their components are calculated, using the expression for V_j in (7). The result is presented on Figure 1. While in the FastICA case there is an almost equal parsing of the variance contributions among the ICs, a significant amount of variance is concentrated in approximately the first 20 ICs that SmoothICA extracts. Indicatively, 35 FastICA ICs contain 70% of the portfolio's variance, while by using the additional smoothness temporal constraint only 16 components are required and 90% of the variance in just 21 contrasted to 49 components. This signifies the great advantage in terms of dimensionality reduction and global approximation using a smaller subset of signals. SmoothICA, as observed, produces components that have an inherent ordering of the source signals and can provide a more efficient representation of a portfolio of securities. It can be used among other tasks, as an alternative to dimensionality reduction for simpler modeling or extraction of seasonal and structural variations (currently done during the pre-whitening step by PCA).

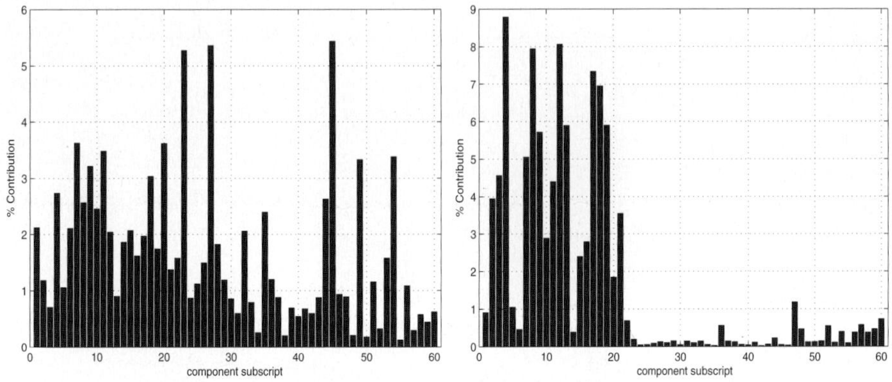

(a) For the FastICA case.

(b) For the SmoothICA case.

Fig. 1. Global approximation performance. Percentage variance contributions of each source signal.

5.3 Local Approximation Comparison

The local approximation performances of both algorithms are compared using both ordering methods presented above. Two error criteria produce different mean approximation errors across all securities on the portfolio. The error criteria calculated are the Mean Squared Error (MSE) and the Mean Absolute Percentage Error (MAPE). The former is a commonly used fitness measure penalizing large deviations from observed security prices in a greater extent, while the latter being an easily understood intuitive measure. On the x-axis lie the numbers of ICs used for approximation of each security signal; from only 1 to all the ICs (60). The lists containing the ordered contributions are calculated for both algorithms, using both L_∞ norm measure and TnA heuristic algorithm. The results on Figure 2 demonstrate a clearly superior local approximation performance of SmoothICA. Equally consistent results are obtained for the Root Mean Squared Error (RMSE) and the Mean Percentage Error (MPE) not presented here for economy of space. The differences in the performances of L_∞ and TnA seem marginal, however the former is significantly less computationally demanding, thus preferred.

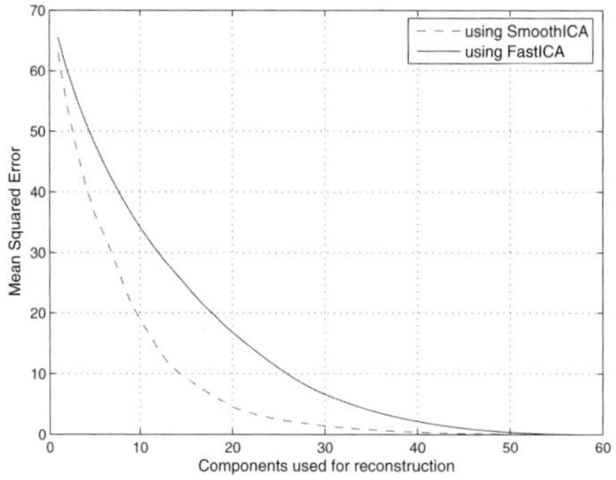

(a) Mean MSE using L_∞ norm ordering.

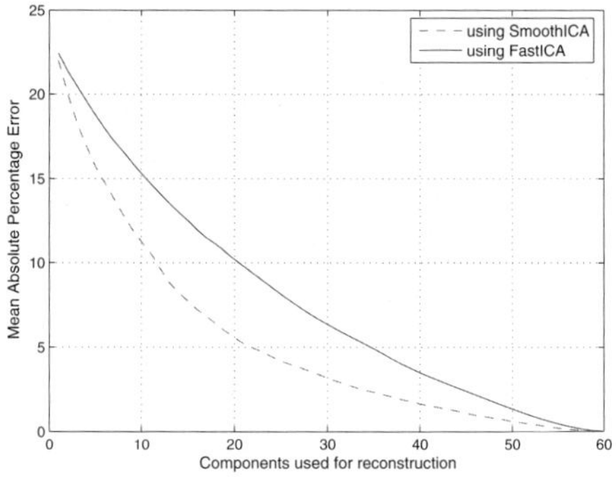

(b) Mean MAPE using L_∞ norm ordering.

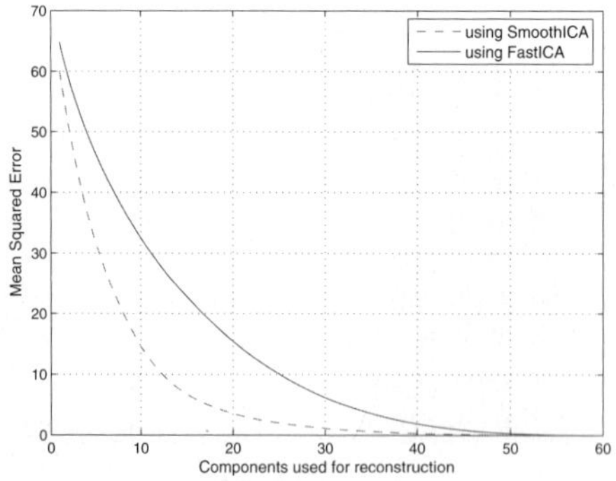

(c) Mean MSE using TnA algorithm.

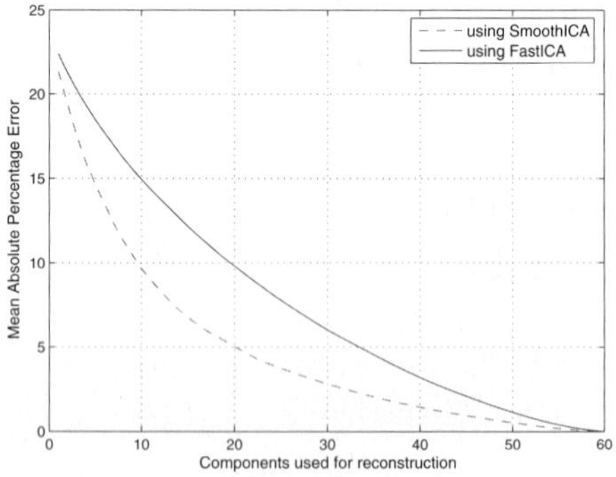

(d) Mean MAPE using TnA algorithm.

Fig. 2. Local approximation performance. SmoothICA is shown to be superior in terms of more efficient representation of each source signal.

6 Conclusions

Through the addition of the 2nd order temporal constraint, which seeks to identify temporally smooth underlying sources, the SmoothICA algorithm is more efficient than the FastICA in approximating a portfolio of securities from an appropriate subset of the estimated sources (section 5.2). This novel algorithm estimates smoother underlying sources that have an inherent ordering, as a high percentage of the portfolio's variance is contained in the first few components, compared to FastICA which has a significantly higher variance spreading among its ICs. To contain 70% of the portfolio's variance in just 16 components, while the classical FastICA requires 35, and 90% of the variance in just 21 contrasted to 49 components, is a significant improvement in terms of global approximation. In the local approximation part of this paper (section 5.3), each security in the portfolio is reconstructed by appropriate subsets of the source signals of dimensions 1 to 60. For each dimension selected the mean MSE and MAPE approximation error across the portfolio is plotted against the subset dimension. For both component ordering methods examined, the errors calculated show consistent superiority of the SmoothICA algorithm for efficient compact representation of a portfolio of securities. Furthermore, the gradients in the plots of Figure 2 support the global case conclusions.

References

1. Hyvarinen, A., Oja, E.: Independent component analysis: algorithms and applications. Neural Networks 13(4-5), 411–430 (2000)
2. Mitianoudis, N., Stathaki, T., Constantinides, A.G.: Smooth signal extraction from instantaneous mixtures. IEEE Signal Processing Letters 14(4) (2007)
3. Hyvarinen, A.: Fast and robust fixed-point algorithms for independent component analysis. IEEE Transactions on Neural Networks 10(3), 626–634 (1999)
4. Korizis, H., Constantinides, A., Christofides, N.: Smooth Component Extraction from a Set of Financial Data Mixtures. In: Proc. Signal Processing, Pattern recognition and Applications, Innsbruck Austria, pp. 136–554 (2007)
5. Back, A.D., Weigend, A.S.: A First Application of Independent Component Analysis to Extracting Structure from Stock Returns. Int. J. Neural Systems 8(4), 473–484 (1997)
6. Cheung, Y.M., Xu, L.: The MSE Reconstruction Criterion for Independent Component Ordering in ICA Time Series Analysis. NSIP 8(4), 793–797 (1999)
7. Lai, Z.B., Cheung, Y.M., Xu, L.: Independent Component Ordering in ICA Analysis of Financial Data. In: Computational Finance, Ch. 14, pp. 201–212. The MIT Press, Cambridge (1999)
8. Cheung, Y.M., Xu, L.: An empirical method to select dominant independent components in ICA for time series analysis. In: Proc. Int. Joint Conf. on Neural Networks '99, Washington DC, pp. 3883–3887 (1999)
9. Malaroiu, S., Kiviluoto, K., Oja, E.: Time series prediction with Independent Component Analysis. In: Proc. '99 Conf. on Advanced Investment Technologies, Gold Coast Australia (2000)
10. Cha, S.M., Chan, L.W.: Applying Independent Component Analysis to Factor Model in Finance. In: Leung, K.-S., Chan, L., Meng, H. (eds.) IDEAL 2000. LNCS, vol. 1983, pp. 538–544. Springer, Heidelberg (2000)
11. Chin, E., Weigend, A., Zimmermann, H.: Computing portfolio risk using gaussian mixtures and independent component analysis. In: Proc. CIFEr '99, New York, pp. 74–117 (1999)
12. Moody, J., Wu, L.: What is the true price? State space models for high frequency FX data. In: Proc. CIFEr '97, New York (1997)

Bayesian Estimation of Overcomplete Independent Feature Subspaces for Natural Images

Libo Ma and Liqing Zhang

Department of Computer Science and Engineering,
Shanghai Jiao Tong University
800 Dong Chuan Road, Shanghai 200240, China
malibo@sjtu.edu.cn, zhang-lq@cs.sjtu.edu.cn

Abstract. In this paper, we propose a Bayesian estimation approach to extend independent subspace analysis (ISA) for an overcomplete representation without imposing the orthogonal constraint. Our method is based on a synthesis of ISA [1] and overcomplete independent component analysis [2] developed by Hyvärinen et al. By introducing the variables of dot products (between basis vectors and whitened observed data vectors), we investigate the energy correlations of dot products in each subspace. Based on the prior probability of quasi-orthogonal basis vectors, the MAP (maximum a posteriori) estimation method is used for learning overcomplete independent feature subspaces. A gradient ascent algorithm is derived to maximize the posterior probability of the mixing matrix. Simulation results on natural images demonstrate that the proposed model can yield overcomplete independent feature subspaces and the emergence of phase- and limited shift-invariant features—the principal properties of visual complex cells.

1 Introduction

Recent linear implementations of efficient coding hypothesis [3,4], such as independent component analysis (ICA) [5] and sparse coding [6], have provided functional explanations for the early visual system, especially neurons in the primary visual cortex (V1). Nevertheless, there are many complex nonlinear statistical structures in the natural signals, which are not able to be extracted by a linear model. For instance, Schwartz et al. have observed that, for natural images, there are significant statistical dependencies among the variances of filter outputs [7]. Several algorithms have been proposed to extend the linear ICA model to capture such residual nonlinear dependencies [1,8,7,9]. Hyvärinen et al. developed the independent subspace analysis (ISA) method, which generalizes the assumption of component independence to subspace independence [1]. However, this method is limited to the complete case. The orthogonality requirement of the mixing matrix restricts the generalization to the overcomplete representation. In the overcomplete representation, the dimension of the feature vector is *larger* than the dimension of the input. Overcomplete representations present several potential advantages. High-dimensional representations are more flexible in capturing inherent structures in signals. Overcomplete representations generally provide more efficient representations than the complete case [10]. Furthermore, studies of human visual cortex have shown interesting implications of overcomplete representations in the visual system [11].

M.E. Davies et al. (Eds.): ICA 2007, LNCS 4666, pp. 746–753, 2007.

In this paper, we combine ISA [1] and overcomplete independent component analysis [2] to extend ISA for overcomplete representations. We apply a Bayesian inference to estimating overcomplete independent feature subspaces of natural images. In order to derive the prior probability of the mixing matrix, the quasi-orthogonality of the dot product between two basis vectors is investigated. Moreover, we assume that the probability density of the dot products (between basis vectors and whitened observed data vectors) in one subspace depends only on the norms of the projections of the data onto the subspace. Then, a learning rule based on gradient ascent algorithm is derived to maximize the posterior probability. Simulation results on natural image data are provided to demonstrate the performance of overcomplete representations for independent subspace analysis. Furthermore, our model can lead to the emergence of phase- and limited shift-invariant features—principal properties of visual complex cells as well.

This paper is organized as follows: In section 2, we propose a Bayesian approach to estimate the overcomplete independent feature subspaces. The learning rule is given as well. In section 3, some experimental results on natural images are presented. Finally, some discussions on representation performance of the proposed method are given in section 4.

2 Model

2.1 Bayesian Inference

In this section, we apply Bayesian MAP (maximum a posteriori) approach to estimating overcomplete independent feature subspaces. The basic ICA model can be expressed as:

$$\mathbf{x} = \mathbf{A}\mathbf{s} = \sum_{i=1}^{N} \mathbf{a}_i s_i, \tag{1}$$

where $\mathbf{x} = (x_1, x_2, ..., x_M)^T$ is a vector of observed data, $\mathbf{s} = (s_1, s_2, ..., s_N)^T$ is a vector of components, and $\mathbf{A}$ is the mixing matrix. $\mathbf{a}_i$ is i^{th} the column of $\mathbf{A}$, and it is often called basis function or basis vector. In our model, the observed data vector $\mathbf{x}$ is whitened to vector data $\mathbf{z}$, just as the preprocessing step in most ICA methods. Furthermore, instead of considering the independent components, as in most ICA, we consider the dot product between the ith basis vector and the whitened data vector. For simplicity, it is assumed that the norms of the basis vectors are set to be unity and that the variances of the sources can differ from unity. Then, the dot product is

$$y_i = \mathbf{a}_i^T \mathbf{z} = \mathbf{a}_i^T \mathbf{A} s = s_i + \sum_{j \neq i} \mathbf{a}_i^T \mathbf{a}_j s_j, \tag{2}$$

where s_i is the ith independent component. Given the overcomplete representations of our model (there is a large number of components in a high-dimensional space), the second term approximately follows Gaussian distribution. Moreover there is no component whose variance is considerably larger than others. Therefore the marginal distributions of dot products should be maximally sparse (super-Gaussian). And maximizing the non-Gaussianities of these dot products is sufficient to provide an approximation of basis vectors. Thus, we we can replace the component s_i by the dot product y_i to estimate independent feature subspaces. Considering the dot product vector $\mathbf{y} = (y_1, ..., y_N)^T = \mathbf{A}^T\mathbf{z}$, the probability for $\mathbf{z}$ given $\mathbf{A}$ can be approximated by

$$p(\mathbf{z}(t)|\mathbf{A}) = p(\mathbf{y}) \approx C\prod_{i=1}^{N} p_{y_i}(y_i) = C\prod_{i=1}^{N} p_{y_i}(\mathbf{a}_i^T z(t)), \tag{3}$$

where C is a constant. Obviously, the accuracy of the prior probability p_{y_i} is important, especially for overcomplete representations [10]. Several choices of prior on the basis coefficients $P(\mathbf{s})$ have been applied in classical linear models respectively. Bell and Sejnowski utilize the prior $P(s_i) \propto \text{sech}(s_i)$, which is corresponding to the hyperbolic tangent nonlinearity [5]. Olshausen and Field use a generalized Cauchy prior [6]. Whereas van Hateren and van der Schaaf simply explore non-Gaussianity [12]. Nevertheless, all these choices of prior is derived under a single-layer network of linear model. Surely, it is desirable to capture nonlinear dependencies by a second or third stage in a hierarchical fashion.

In our model, we apply the prior probability p_{y_i} proposed in the ISA algorithm, in which the basis function coefficients in each subspace have the energy correlations [1]. A diagram of feature subspaces is given in Figure 1.

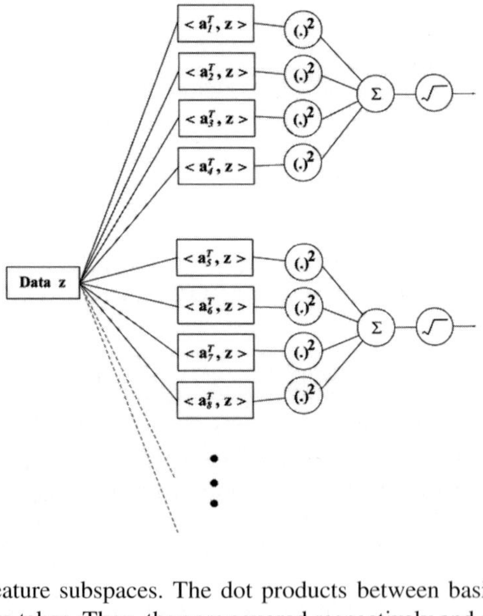

Fig. 1. Illustration of feature subspaces. The dot products between basis vectors and whitened observed data vectors are taken. Then, they are squared respectively and summed inside the same feature subspace. Square roots are taken for normalization.

The dot product (neuronal response) y_i is assumed to be divided into n-tuples, so that y_i inside a given n-tuple may be dependent on each other, but different n-tuples are mutually independent. The subspaces model introduces a certain dependency structure for different components. Let $\Omega_j, j = 1, ..., J$ denote the set of independent feature subspaces, where J is the number of subspaces. The probability distributions for n-tuples of y_i are spherically symmetric. In other words, the probability density $p_{y_j}(.)$ of n-tuple can be expressed as a function of the sum of the squares of y_i, $i \in \Omega_j$ only. And, for simplicity, we assume $p_{y_j}(.)$ are identical for all subspaces. Therefore, the probability density inside the jth n-tuple of y_i can be calculated:

$$p_{y_j}(y_j) = \exp\Big(G\Big(\sum_{i\in\Omega_j} y_i^2\Big)\Big), \tag{4}$$

where the function $G(y)$ should be convex for non-negative y. For example, one could use the form of $G(.)$ as: $G(y) = -\alpha_1\sqrt{y} + \beta_1$, where α_1 is the scaling constant and β_1 is the normalization constant. These constants are unimportant for the learning process.

Overcomplete representations mean that there is a large number of basis vectors. In other words, the basis vectors are randomly distributed in a high-dimensional space. In order to approximate the prior probability of basis vectors, we employ a result presented by Hecht-Nielsen [13]: the number of almost orthogonal directions is much larger than that of orthorgonal directions. This property is called quasi-orthogonality [2]. Therefore, in a high-dimensional space even vectors having random directions might be sufficiently close to orthogonality. Thus, the prior probability of the mixing matrix $\mathbf{A}$ can be obtained in terms of the quasi-orthogonality as follows:

$$p(\mathbf{A}) = c_m \prod_{i<j} \left(1 - (\mathbf{a}_i^T\mathbf{a}_j)^2\right)^{\frac{m-3}{2}}, \tag{5}$$

where c_m is a constant. The detailed derivation of Equation (5) can be obtained in [2].

Bayes' Theorem allows one to describe the probability of the model in terms of the likelihood of the data and the prior probability of the model. Thus, given observation $\mathbf{z}$, the posterior probability $p(\mathbf{A}|\mathbf{z})$ can be derived as follows:

$$p(\mathbf{A}|\mathbf{z}) = \frac{p(\mathbf{z}|\mathbf{A})p(\mathbf{A})}{p(\mathbf{z})}, \tag{6}$$

where $p(\mathbf{z})$ is constant with respect to $\mathbf{A}$.

It is easier to estimate the mixing matrix that maximize the logarithm of posterior probability $p(\mathbf{A}|\mathbf{z})$. Thus, taking the logarithm of Equation (6) and combining Equation (5) with Equation (3) and (4), we obtain the approximation of log-probability of the posterior:

$$\log p(\mathbf{A}|\mathbf{z}(t), t = 1, .., T) \propto \sum_{t=1}^{T}\sum_{j=1}^{J} G\Big(\sum_{i\in\Omega_j} y_i^2\Big) + \alpha T \sum_{i<j} \log(1 - (\mathbf{a}_i^T\mathbf{a}_j)^2) + C \tag{7}$$

where α is a constant related to c_m.

2.2 Learning Rule

Gradient ascent maximization of posterior probability over basis vector $\mathbf{a}_k$ yields the following learning rule:

$$\Delta\mathbf{a}_k \propto \eta\Big(\sum_{t=1}^{T}\mathbf{z}(t)(\mathbf{a}_k^T\mathbf{z}(t))g\Big(\sum_{i\in\Omega_{j(k)}}(\mathbf{a}_i^T\mathbf{z}(t))^2\Big)+\alpha T\sum_{i<j}\frac{-2\mathbf{a}_i^T\mathbf{a}_j}{1-(\mathbf{a}_i^T\mathbf{a}_j)^2}\mathbf{b}_k\Big), \quad (8)$$

where η is the learning rate, and $\Omega_{j(k)}$ is the subspace to which $\mathbf{a}_k$ belongs. $\mathbf{b}_k$ is the kth column vector of matrix $\mathbf{B} = [0, ..., \mathbf{a}_j, ..., \mathbf{a}_i, ...0]$, $\mathbf{a}_j$ is the ith column vector, and $\mathbf{a}_i$ is the jth column vector. The function g is the derivative of G. After each iteration in equation (8), the norm of the basis vector $\mathbf{a}_k$ needs to be set to unity. This is different from ordinary ISA, where the mixing matrix is orthonormalized.

3 Simulations

We tested the algorithm for overcomplete independent subspace analysis on natural image data. The training set of images consists of 50,000 patches of size 16×16 that were randomly extracted from thirteen 256×512 pixel gray images. We use the natural images in [1], which is available on http://www.cis.hut.fi/projects/ica/data/images/. The mean gray-scale value of data (i.e., the DC component) was removed. The dimension of data was reduced by principle component analysis, projecting onto the leading 160 eigenvectors of the data covariance matrix. Then, the data vectors were whitened as in most ICA methods. The log posterior probability was maximized by an ordinary gradient method to estimate **A**, using the averaged version of the learning rule in equation (8). Note that there was no constraint of orthogonality of basis vectors during each

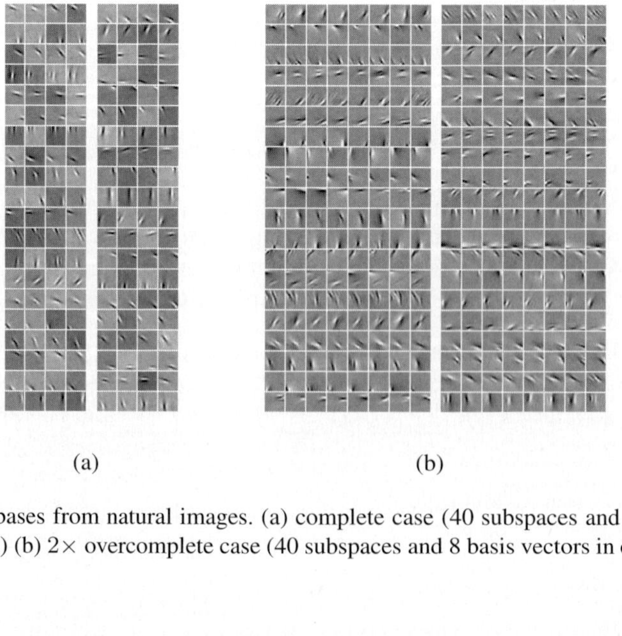

(a) (b)

Fig. 2. Learned bases from natural images. (a) complete case (40 subspaces and 4 basis vectors in each subspace) (b) 2× overcomplete case (40 subspaces and 8 basis vectors in each subspace).

iteration. Only the norms of basis vectors were set to unity. The random initial value was set for mixing matrix.

The effects of varying the level of overcompleteness and the dimension of subspaces were investigated in depth. The basis was set to be complete and 2× overcomplete. The dimension of components is 160 and 320, respectively. Figure 2 shows the estimated basis vectors, which is the complete case of four-dimensional subspaces and 2× overcomplete case of eight-dimensional subspaces.

To analyze the tiling properties of the estimated basis vectors, we fitted each basis vector with a Gabor function by minimizing the squared error between the estimated basis vectors and the model Gabor. Figure 3 shows the distribution of parameters obtained by fitting Gabor functions to complete and 2× overcomplete basis vectors. We can see that, with the increasing of the level of overcompleteness, the scattering points in the plot of location, spatial frequency and orientation become denser and more uniform. And the distribution of phase is much closer to uniform.

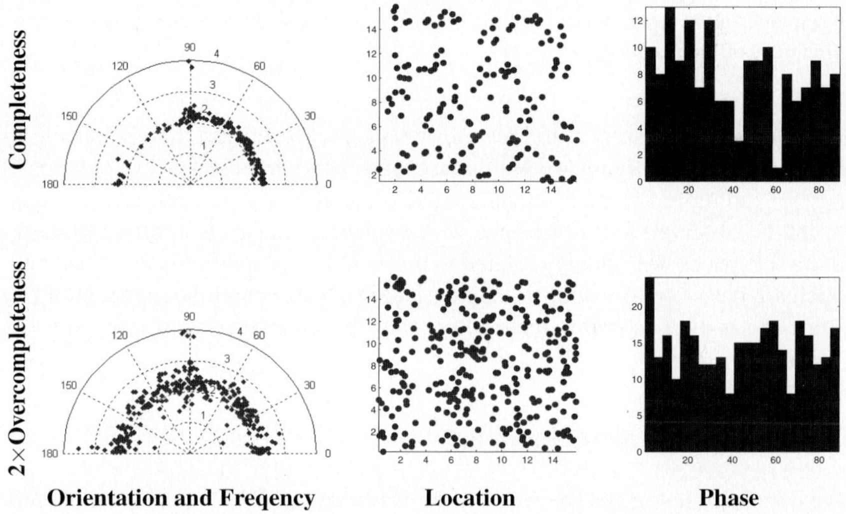

Fig. 3. The distributions of parameters derived by fitting Gabor functions with completeness and 2× overcompleteness. (a) Center location of Gabor fitted within a patch. (b) Joint distribution of orientation and spatial frequency (plotted in the upper-half plane) (c) Histogram of phase of Gabor fitted (mapped to range 0 °~ 90 °).

Furthermore, we compare the responses of all the feature subspace and the corresponding linear filters for different stimulus cases. First, an optimal stimulus for the feature subspace was computed in the set of Gabor filters. The tested stimuli for the subspace was calculated in the set of Gabor functions. In each time, only one stimuli parameter was changed to see how the response changes. The tested parameters were location (shift), orientation, and phase. Figure 4 shows the median responses of the whole population of 40 subspace and 320 linear filters corresponding to 2× overcomplete case. The top row shows the absolute responses of the linear filters, and the

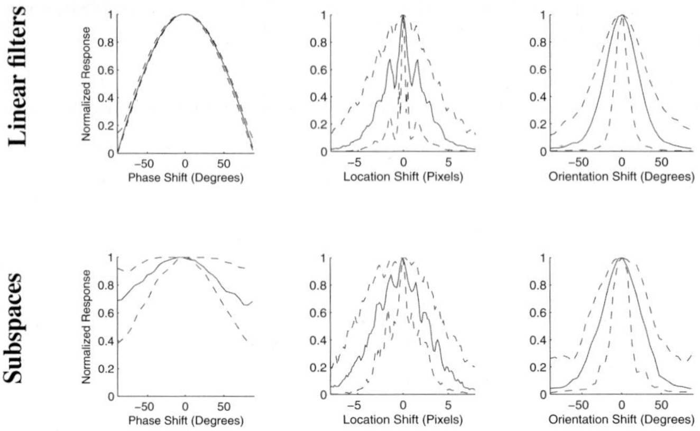

Fig. 4. Statistical curves for whole population and linear filers while shifting different Gabor parameters: orientation, frequency, and phase with 2× overcompleteness. The solid line gives the median response in the population of all filters or subspaces. The dashed lines give the 90% and 10% percentiles of the responses.

bottom row shows the results of the feature subspaces. We can see that the responses of subspaces are considerably invariant to phase, and somewhat invariant to position. The sharpness of tuning to orientation and spatial frequency remains roughly unchanged. Thus it can be observed that invariance with respect to phases is a strong property of the feature subspaces. It is closely related to the response properties of complex cells in V1, which are based on location, frequency, and orientation and independent of phase. In contrast, the responses of the linear filters show no invariance with respect to any of these parameters.

4 Discussions and Conclusions

We have demonstrated in this paper how the Bayesian approach can be employed for learning overcomplete representations by utilizing the quasi-orthogonal property of basis vectors in a high-dimensional space, whereas ordinary ISA can only provide complete representations of basis functions. In addition, we examine the dot products (between basis vectors and whitened observed data vectors) instead of the basis function coefficients. Furthermore, our model need not impose the constraint of orthogonality on basis vectors. Only the norms of basis vectors were set to unity during the learning process. In contrast, basis vectors have to be orthogonal in ordinary ISA. Compared with the methods for estimating overcomplete bases by using maximum likelihood estimation, our method is as computationally effective as basic ICA estimation.

Another issue addressed in this paper is the relevance of the learned codes to neurobiological plausibilities. Both complete and overcomplete basis functions adapted to natural images suggest functional similarities to neurons of V1 receptive fields. Simulation results on natural image data demonstrate that our model can lead to the emergence of phase- and shift-invariant features—principal properties of visual complex cells as well. This method shows promising prospects in extended applications of our method to higher levels of cortical representations.

An important concern in our model is the accuracy of the coefficient prior probability. Our overcomplete ISA algorithm can capture the underlying statistical structure of images, i.e., the energy correlations of coefficients in each subspace. However, a Laplacian prior probability as in overcomplete ICA algorithms can not capture well higher-order statistics, such as dependencies among the variances of filter outputs. This method finds compact descriptions of overcomplete representation and has the potential in a wide varieties of applications, such as image processing and pattern recognition.

Acknowledgments

The work was supported by the National Basic Research Program of China (Grant No. 2005CB724301) and the National High-Tech Research Program of China (Grant No.252006AA01Z125).

References

1. Hyvärinen, A., Hoyer, P.: Emergence of phase-and shift-invariant features by decomposition of natural images into independent feature subspaces. Neural Computation 12(7), 1705–1720 (2000)
2. Hyvärinen, A., Inki, M.: Estimating Overcomplete Independent Component Bases for Image Windows. Journal of Mathematical Imaging and Vision 17(2), 139–152 (2002)
3. Attneave, F.: Some informational aspects of visual perception. Psychol. Rev. 61(3), 183–193 (1954)
4. Barlow, H.B.: Possible principles underlying the transformation of sensory messages. Sensory Communication, 217–234 (1961)
5. Bell, A.J., Sejnowski, T.J.: The independent components of natural scenes are edge filters. Vision Research 37(23), 3327–3338 (1997)
6. Olshausen, B., Field, D.: Emergence of simple-cell receptive field properties by learning a sparse code for natural images. Nature 381(6583), 607–609 (1996)
7. Schwartz, O., Simoncelli, E.: Natural signal statistics and sensory gain control. Nature Neuroscience 4, 819–825 (2001)
8. Hyvärinen, A., Hoyer, P.O., Inki, M.: Topographic Independent Component Analysis. Neural Computation 13(7), 1527–1558 (2001)
9. Karklin, Y., Lewicki, M.S.: A Hierarchical Bayesian Model for Learning Nonlinear Statistical Regularities in Nonstationary Natural Signals (2005)
10. Lewicki, M., Olshausen, B.: Probabilistic framework for the adaptation and comparison of image codes. Journal of the Optical Society of America A 16(7), 1587–1601 (1999)
11. Popovic, Z., Sjostrand, J.: Resolution, separation of retinal ganglion cells, and cortical magni fication in humans. Vision Research 41(10-11), 1313–1319 (2001)
12. van Hateren, J.H.: Independent component filters of natural images compared with simple cells in primary visual cortex. Proceedings: Biological Sciences 265(1394), 359–366 (1998)
13. Hecht-Nielsen, R.: Context vectors: general purpose approximate meaning representations self-organized from raw data. Computational Intelligence: Imitating Life, 43–56 (1994)

ICA-Based Image Analysis for Robot Vision

Naoya Ohnishi[1] and Atsushi Imiya[2]

[1] School of Science and Technology, Chiba University, Japan
Yayoicho 1-33, Inage-ku, Chiba, 263-8522, Japan
ohnishi@graduate.chiba-u.jp

[2] Institute of Media and Information Technology, Chiba University, Japan
Yayoi-cho 1-33, Inage-ku, Chiba, 263-8522, Japan
imiya@faculty.chiba-u.jp

Abstract. In this paper, we develop an ICA-based obstacle detection and 3D-environment understanding for a mobile robot navigation. From a camera mounted on a mobile robot, the robot observes a sequence of images. This sequence of images allows the robot to compute optical flow, which is the apparent motion of each point on the image. We apply ICA to the optical flow field computed from images captured by the camera mounted on the robot. ICA-based separation of optical flow derives a obstacle region and a ground plane region in a space. For these applications, we also introduce an ordering criterion of independent components using its variances.

1 Introduction

Independent Component Analysis(ICA) [6] extracts statistically independent features from signals and still images. In this paper, we apply independent component analysis to a dynamic image sequence for autonomous robot navigation, that is, ICA is applied to optical flow [1] computed from a image sequence. The optical flow [1] is the apparent motion of successive images and is independent of the features in images, unlike edges or corner points in images. Furthermore, optical flow is considered to be fundamental information for navigation and obstacle avoidance in the context of biological data processing [11]. Therefore, the use of optical flow is valid for the robot navigation using the vision system.

In neuroscience, it is known that the medial superior temporal (MST) area performs visual motion processing. For motion cognition at the MST area in the brain[7,12], it is shown that independent components of optical flow are used. Furthermore, since the optical flow field on an image can be represented as a linear combination of independent components of optical flow, we can use ICA for the detection of the dominant plane by separating obstacles and the dominant part in an image. Our application of ICA separates the planes from image sequences.

Statistical analysis of optical flow are addressed in [3,5] for the robust motion understanding for optical flow. Our ICA algorithm separates optical flow on the dominant plane and the obstacle areas in optical flow observed through an uncalibrated camera mounted on a mobile robot. First, an optical flow is computed

M.E. Davies et al. (Eds.): ICA 2007, LNCS 4666, pp. 754–761, 2007.

from a pair of successive images observed through a camera mounted on a mobile robot. Optical flow is used for the estimation of homography on the dominant plane. Homography can be approximately calculated by affine transform. Then, we estimate optical flow on the dominant plane using the obtained homography. The details of the algorithm are described in reference [9]. Next, optical flow computed from images and the estimated optical flow on the dominant plane are used as input signals in ICA. Since two signals are input in ICA, two signals are output.

For the concurrent detection of a local and global motion, we use independent components of optical flow fields on pyramidal layers. It is known that animals, insects, and human beings use the independent components of optical flow fields for visual behavior [7,11]. In human object recognition, the hierarchical model is proposed [4]. Furthermore, for the computation of optical flow, the pyramid transform of an image sequence is used for the analysis of a global motion and local motion [2,8]. The pyramid transform generates multiple-resolution images as layered images. These layered images are used for computation of optical flow in its original images from the image in the lowest layer. This idea based on the assertion that a global motion is described as the collection of a local motion. We introduce the application of hierarchical image expression for motion analysis, that is, we develop an algorithm for the detection layered optical flows from a multi resolution image sequence.

2 ICA of Optical Flow Field

In this section, we introduce an algorithm for applying ICA to optical flow fields.

The optical flow is apparent motion of each points computed from successive two images [1]. Setting $I(x, y, t)$ to be time-varying image, the optical flow is computed by solving the equation

$$I_x \dot{x} + I_y \dot{y} + I_t = 0, \quad (1)$$

where $(\dot{x}, \dot{y})^\top$ is the optical flow vector. To solve this singular equation, we adopt the Lucas and Kanade method with the pyramid transform [1,2,9,10].

Similar to ICA separating mixture signals into independent components, the MST area in the brain separates the motion fields from visual perception into independent components [7,12]. As previously introduced, we accept the assumption that optical flow fields observed by the moving camera are linear combinations of optical flow fields of the dominant plane and the obstacles. That is, setting $\dot{\boldsymbol{u}}_{\mathrm{dominant}}$ and $\dot{\boldsymbol{u}}_{\mathrm{obstacle}}$ to be optical flow fields of the dominant plane and the obstacles, respectively, the observed optical flow field $\dot{\boldsymbol{u}}$ is approximately expressed by a linear combination of $\dot{\boldsymbol{u}}_{\mathrm{dominant}}$ and $\dot{\boldsymbol{u}}_{\mathrm{obstacle}}$ as

$$\dot{\boldsymbol{u}} = a_1 \dot{\boldsymbol{u}}_{\mathrm{dominant}} + a_2 \dot{\boldsymbol{u}}_{\mathrm{obstacle}}, \quad (2)$$

where a_1 and a_2 are the mixture coefficients, as shown in Fig. 1. This assumption is numerically and geometrically acceptable if motion displacement is small

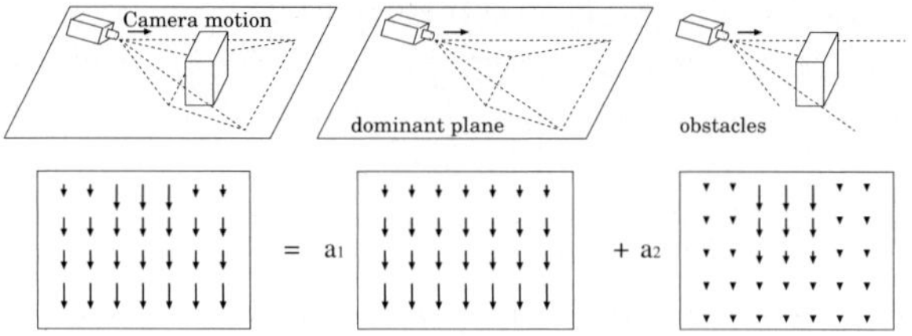

Fig. 1. Linear combination of optical flow field in the scene. Example of camera displacement and these optical flow fields. a_1 and a_2 are mixture coefficients.

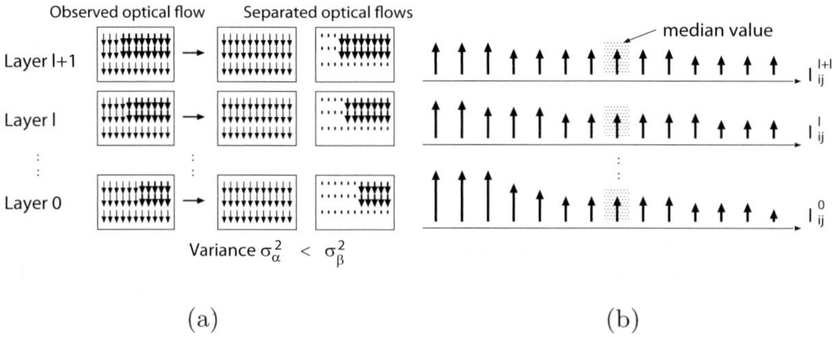

Fig. 2. Ordering of independent components of optical flow fields. (a)Difference in the motions of the dominant plane and obstacles. The order of the components can be determined by using variances σ_α^2 and σ_β^2. (b)Sorting using the norm $\boldsymbol{l}$ for determination of output order.

compared with the size of obstacles, as shown in a numerical experiment. Therefore, ICA is suitable for the separation of optical flow into the independent flow components. For each image in a sequence, we consider that optical flow vectors in the dominant plane correspond to independent components.

For ICA of optical flow fields, we align the matrix of two-dimensional vectors to a one-dimensional array as

$$\dot{\boldsymbol{u}} \rightarrow ((\dot{u},\dot{v})_1 \ \cdots (\dot{u},\dot{v})_k \cdots (\dot{u},\dot{v})_n)^\top \rightarrow (\dot{u}_1 \cdots \dot{u}_n \ \dot{v}_1 \cdots \dot{v}_n)^\top = \mathrm{vec}\dot{\boldsymbol{u}}. \quad (3)$$

Since the relation $\dot{\boldsymbol{w}} = \alpha\dot{\boldsymbol{u}} + \beta\dot{\boldsymbol{v}}$ leads to the relation $\mathrm{vec}\dot{\boldsymbol{w}} = \alpha\mathrm{vec}\dot{\boldsymbol{u}} + \beta\mathrm{vec}\dot{\boldsymbol{v}}$. These steps are invertible. Therefore, it is possible to extract regions corresponding to $\dot{\boldsymbol{u}}$ and $\dot{\boldsymbol{v}}$ if the observation $\dot{\boldsymbol{w}}$ is decomposed into two independent components $\dot{\boldsymbol{u}}$ and $\dot{\boldsymbol{v}}$. We use this vector, derived from a vector-valued image, as input to ICA.

3 ICA-Based Obstacle Detection

In this section, we present algorithms for obstacle detection using ICA and optical flow. The first algorithm is ground-plane detection for the mobile robot navigation, The details of the algorithm is described in [10]. The second algorithm detects multiple planes by extension of the first one. The third algorithm is estimate multi-resolution obstacle using multi-layer optical flow fields.

3.1 Dominant-Plane Detection by ICA

For the detection of the dominant plane, ICA requires at least two input signals for separation into two independent components. Then, we use optical flow field $\dot{\boldsymbol{u}} = \{(\dot{u}, \dot{v})_{ij}^{\top}\}_{i=1,j=1}^{h,w}$ and planar flow field $\hat{\boldsymbol{u}} = \{(\hat{u}, \hat{v})_{ij}^{\top}\}_{i=1,j=1}^{h,w}$ as the input vectors of ICA, where w and h are the width and the height of an image. Since planar flow is the motion of the dominant plane relative to the robot motion, the use of planar flow is suitable for separation into the dominant plane and obstacles.

Setting $\boldsymbol{v}_\alpha$ and $\boldsymbol{v}_\beta$ to be the output vectors, $\boldsymbol{v}_\alpha$ and $\boldsymbol{v}_\beta$ have ambiguities in those order and length of each component. We are required to determine whether components have optical flow of the dominant plane or of obstacle areas. We solve this problem using the difference between the variances of the norms of $\boldsymbol{v}_\alpha$ and $\boldsymbol{v}_\beta$.

Setting $\boldsymbol{l}_{\alpha,\beta} = \{l_{ij}\}_{i=1,j=1}^{h,w}$ to be the norm of $\boldsymbol{v}_{\alpha,\beta} = \{(\dot{u}, \dot{v})_{ij}\}_{i=1,j=1}^{h,w}$, that is, $l_{ij} = |(\dot{u}, \dot{v})_{ij}|$ and the variance σ^2 is computed as

$$\sigma^2 = \frac{1}{hw} \sum_{i=1,j=1}^{h,w} (l_{ij} - \bar{l})^2, \quad \text{where } \bar{l} = \frac{1}{hw} \sum_{i=1,j=1}^{h,w} l_{ij}. \tag{4}$$

The motions of the dominant plane and obstacles in the images are different, and the dominant-plane motion is smooth on the images compared with obstacle motion, as shown in Fig. 2. Consequently, the output signal of obstacle motion has larger variance than the output signal of dominant-plane motion. Therefore, if $\sigma_\alpha^2 > \sigma_\beta^2$, we use the norm $\boldsymbol{l}_\alpha$ of output flow field $\boldsymbol{v}_\alpha$ for dominant-plane detection; else we use the norm $\boldsymbol{l}_\beta$ of output flow field $\boldsymbol{v}_\beta$.

Since the planar flow field is subtracted from the optical flow field including obstacle motion, $\boldsymbol{l}$ is constant on the dominant plane. However, the length of $\boldsymbol{l}$ is ambiguous. Then, we use the median value of $\boldsymbol{l}$ for the detection of the dominant plane. Since the dominant plane occupies the largest domain in the image, we compute the distance between $\boldsymbol{l}$ and the median of $\boldsymbol{l}$, as shown in Fig. 2(b). The area which has the median value of the component is detected as the dominant plane. Setting m to be the median value of the elements in $\boldsymbol{l}$, the distance $\boldsymbol{d} = \{d_{ij}\}_{i=1,j=1}^{h,w}$ is

$$d_{ij} = |l_{ij} - m|. \tag{5}$$

We detect the area in which $d_{ij} \approx 0$ as the dominant plane.

The procedure for dominant-plane detection by ICA is summarized as follows.

1. Input optical flow field $\dot{\boldsymbol{u}}$ and planar flow field $\hat{\boldsymbol{u}}$ to ICA, and output the optical flow fields $\boldsymbol{v}_\alpha$ and $\boldsymbol{v}_\beta$.
2. Compute the norms $\boldsymbol{l}_\alpha$ and $_\beta$ from $\boldsymbol{v}_\alpha$ and $\boldsymbol{v}_\beta$, respectively.

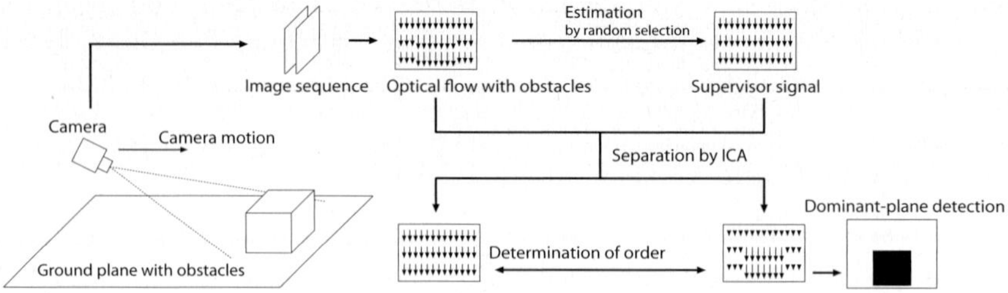

Fig. 3. Procedure for dominant-plane detection

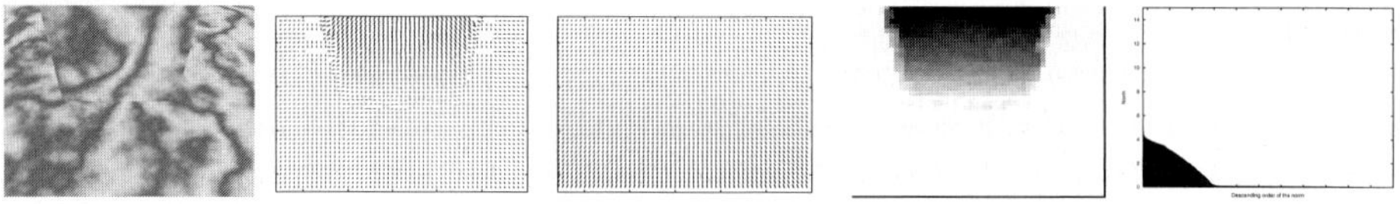

Fig. 4. Input optical flow fields to ICA for translational motion in an environment with one obstacle. Left to right: captured image, optical flow field $\dot{\boldsymbol{u}}$, planar flow field $\hat{\boldsymbol{u}}$, detected dominant plane, and sorted norm $\boldsymbol{l}$ of output $\boldsymbol{v}_\alpha$. Variances of $\boldsymbol{v}_\alpha$ and $\boldsymbol{v}_\beta$ are $\sigma_\alpha^2 = 1.60$ and $\sigma_\beta^2 = 0.51$, respectively. The median value $m = 0.09$. The area where norm $\boldsymbol{l}$ is large corresponds to an obstacle, and the area where $l_{ij} \approx m$ corresponds to the dominant plane.

3. Compute the variances σ_α^2 and $\sigma_b^2 eta$ from $\boldsymbol{l}_\alpha$ and $\boldsymbol{l}_\beta$, respectively.
4. If $\sigma_\alpha^2 > \sigma_\beta^2$, then $\boldsymbol{l} = \boldsymbol{l}_\alpha$, else $\boldsymbol{l} = \boldsymbol{l}_\beta$.
5. Compute the distance $\boldsymbol{d}$ between $\boldsymbol{l}$ and the median of $\boldsymbol{l}$.
6. Detect the area in which $d_{ij} \approx 0$ as the dominant plane.

Figure 3 shows the procedure of dominant-plane detection from the image sequence using ICA. Figures 4 and 5 are experimental results on detecting the dominant plane.

3.2 Iterative Multiple Plane Segmentation

Using the dominant-plane-detection algorithm iteratively, we develop an algorithm for multiple-plane segmentation in an image. After removing the region corresponding to the dominant plane from an image, we can extract the second dominant planar region from the image. Then, it is possible to extract the the third dominant plane by removing the second dominant planar area. This process is expressed as

$$D_k = \begin{cases} \mathbf{A}(R \setminus D_{k-1}), & k \geq 2, \\ \mathbf{A}(R), & k = 1, \end{cases} \tag{6}$$

where $\mathbf{A}$, R, D_k stand for the dominant-plane-extraction algorithm, the region of interest observed by a camera, and the k-th dominant planar area, respectively.

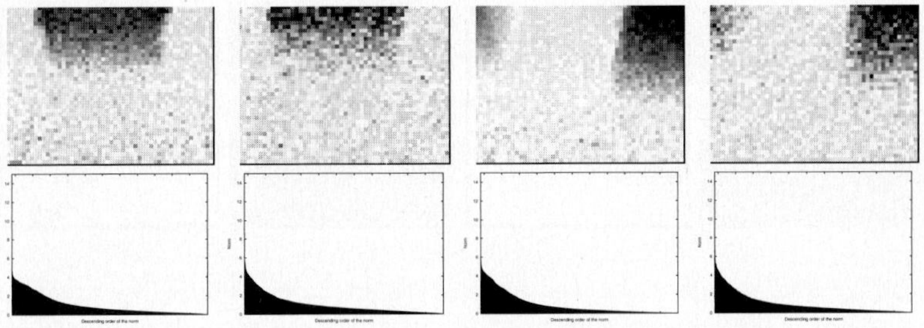

(a)$m = 0.60$ (b)$m = 0.68$ (c)$m = 0.45$ (d)$m = 0.60$

Fig. 5. Results obtained using optical flows with error. (a) Translational motion in an environment with one obstacle. (b) Rotational motion in an environment with one obstacle. (c) Translational motion in an environment with two obstacles. (d) Rotational motion in an environment with two obstacles. Graphs in bottom row are sorted norm $\boldsymbol{l}$ of output $\boldsymbol{v}_\alpha$.

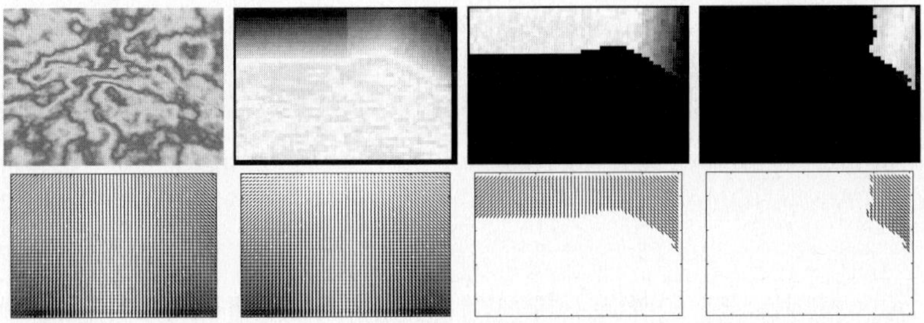

Fig. 6. Top row: captured image, dominant plane d at first, second, and third steps. Bottom row: optical flow, planar flow fields at first, second, and third steps

The algorithm is stopped after iterated to a pre-determined iteration time or the size of k-th dominant plane is smaller th pre-determined size.

Setting R to be the root of the tree, this process derives a binary tree such that

$$R\langle D_1, R \setminus D_1 \langle D_2, R_2 \setminus D_2 \langle \cdots, \rangle \rangle \qquad (7)$$

Assuming that D_1 is the ground plane on which the robot moves, D_k for $k \geq 2$ is the planar areas on the obstacles. Therefore, this tree expresses the hierarchical structure of planar areas on the obstacles. We call this tree the binary tree of planes. Using this tree constructed by the dominant-plane detection algorithm, we obtain geometrical properties of planes in a scene. For example, even if an object exists in a scene and it lies on D_k $k \geq 2$, the robot can navigate ignoring this object, using the tree of planes. Figures 6 and 7 are experimental results on detecting multiple planes.

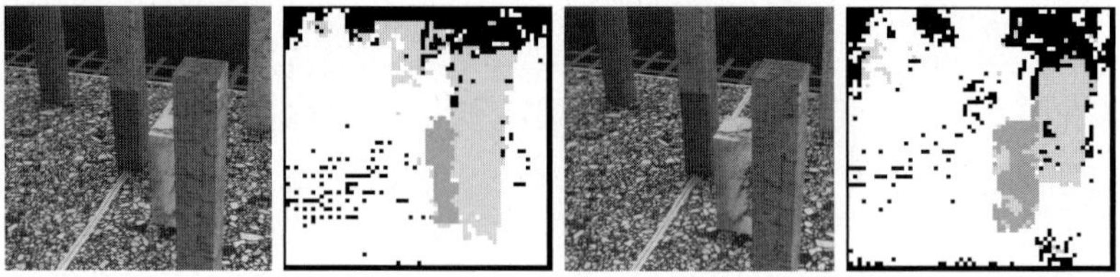

Fig. 7. Image and results estimated from Marbled-Block. The white area is the first dominant plane. The light-gray and dark-gray areas are second and third dominant plane.

3.3 Obstacle Detection Using ICA on Pyramid Layers

Our algorithm is processed at layers $l = 0, \cdots, L$ in the pyramid transform. Using the optical flow field $\boldsymbol{u}^l(x, y, t)$ at layer l, we detect obstacles in a image sequence.

Figure 8 shows that, setting O_l to be the obstacle region on the l-th layer, the hierarchical expression of obstacles satisfies the relations

$$O_0 \subset O_1 \subset \cdots \subset O_L \quad \text{and} \quad D^L \subset D^{L-1} \subset \cdots \subset D^0 \tag{8}$$

for the dominant plane $D^K = R_k^0$. These relations imply that a pair $C_l = (D^l, O_l)$ shows global and local configuration in the work space for a larger and a

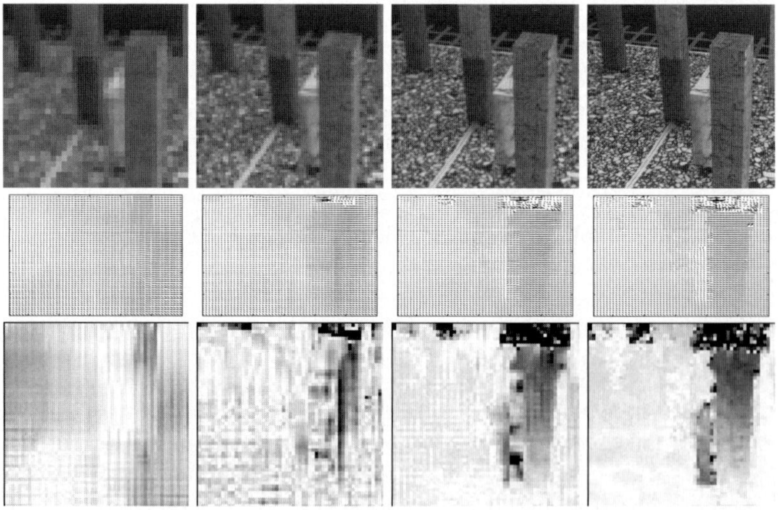

Fig. 8. Pyramidal representation of the Marbled-Block images in a simulated environment. Computed layer optical flow fields from the Marbled-Block images. Detected obstacle at each layer.

smaller l, respectively. This hierarchical relation is automatically detected from a pyramid-based hierarchical expression of images for optical flow computation. The system uses selectively C_l for navigation and spatial perception.

4 Conclusion

Our algorithm is an application of ICA for vector-valued images. This is an extension of ICA-based image analysis to vector-valued images. We proposed an ordering technique for independent components of vector-valued images. The ordering allows us to separate obstacles and ground planes from a sequence of images captured by a camera mounted on an autonomous robot. For each image in a sequence, it is shown that the dominant plane corresponds to an independent component. This relationship provides a statistical definition of the dominant plane. Combination of our ICA based image analysis and multi-resolution image representation provides a method which allows to detect hierarchical configuration of obstacles. This hierarchical processing is achieved concurrently for the simultaneous detection of local and global obstacle-configuration in the work space.

References

1. Barron, J.L., Fleet, D.J., Beauchemin, S.S.: Performance of optical flow techniques. International Journal of Computer Vision 12, 43–77 (1994)
2. Bouguet, J.-Y.: Pyramidal implementation of the Lucas Kanade feature tracker description of the algorithm. Intel Corporation, Microprocessor Research Labs, OpenCV Documents (1999)
3. Calow, D., Krüger, N., Wörgötter, F., Lappe, M.: Statistics of optic flow for self-motion through natural scenes. Dynamic Perception, 133–138 (2004)
4. Domenella, R.G., Plebe, A.: A neural model of human object recognition development. In: De Gregorio, M., Di Maio, V., Frucci, M., Musio, C. (eds.) BVAI 2005. LNCS, vol. 3704, pp. 116–125. Springer, Heidelberg (2005)
5. Fermüller, C., Shulman, D., Aloimonos, Y.: The statistics of optical flow. Computer Vision and Image Understanding 82, 1–32 (2001)
6. Hyvarinen, A., Oja, E.: Independent component analysis: algorithms and application. Neural Networks 13, 411–430 (2000)
7. Jabri, M.A., Park, K.-Y., Lee, S.-Y., Sejnowski, T.J.: Properties of independent components of self-motion optical flow. In: Proc. of IEEE Int. Symp. on Multiple-Valued Logic, pp. 355–362. IEEE Computer Society Press, Los Alamitos (2000)
8. Mahzoun, M.R., Kim, J., Sawazaki, S., Okazaki, K., Tamura, S.: A scaled multigrid optical flow algorithm based on the least RMS error between real and estimated second images. Pattern Recognition 32, 657–670 (1999)
9. Ohnishi, N., Imiya, A.: Featureless robot navigation using optical flow. Connection Science 17, 23–46 (2005)
10. Ohnishi, N., Imiya, A.: Dominant plane detection using optical flow and independent component analysis. In: Perner, P., Imiya, A. (eds.) MLDM 2005. LNCS (LNAI), vol. 3587, pp. 497–506. Springer, Heidelberg (2005)
11. Vaina, L.M., Beardsley, S.A., Rushton, S.K.: Optic flow and beyond. Kluwer Academic Publishers, Dordrecht (2004)
12. Zemel, R.S., Sejnowski, T.J.: A model for encoding multiple object motions and self-motion in area mst of primate visual cortex. Neuroscience 18, 531–547 (1998)

Learning of Translation-Invariant Independent Components: Multivariate Anechoic Mixtures

Lars Omlor[1] and Martin A. Giese[1,2]

[1] ARL, Hertie Institute for Clinical Brain Sciences, Tübingen, Germany
[2] School of Psychology, University of Wales, Bangor, UK

Abstract. For the extraction of sources with unsupervised learning techniques invariance under certain transformations, such as shifts, rotations or scaling, is often a desirable property. A straight-forward approach for accomplishing this goal is to include these transformations and its parameters into the mixing model. For the case of one-dimensional signals in presence of shifts this problem has been termed anechoic demixing, and several algorithms for the analysis of time series have been proposed. Here, we generalize this approach for sources depending on multi-dimensional arguments and apply it for learning of translation-invariant features from higher-dimensional data, such as images. A new algorithm for the solution of such high-dimensional anechoic demixing problems based on the Wigner-Ville distribution is presented. It solves the multi-dimensional problem by projection onto multiple one-dimensional problems. The feasibility of this algorithm is demonstrated by learning independent features from sets of real images.

1 Introduction

Many common approaches in blind source separation (BSS) are based on linear instantaneous mixture models, where the output signals result from the linear superposition of source signals, time-point by time-point. In spite of its great success in feature extraction (see [3] for review) this simple model usually fails when features appear under transformations, such as scaling, rotation or shifts. Invariance against such transformations can be achieved by embedding them into the mixture model, requiring the estimation of additional parameters. This approach has been applied to acoustic data for modeling the transmission delays of different microphones, applying a generative model of the form:

$$x_i(t) = \sum_{j=1}^{d} \alpha_{ij} \cdot s_j(t - \tau_{ij}) \quad i = 1, \ldots, m \tag{1}$$

The time functions $x_i(t)$ signify the original signals, $s_j(t)$ the source signals, and α_{ij} the mixing weights. The constants τ_{ij} are the temporal shifts (delays) that need to be estimated. Since this model allows for delays, but not for reverberations of the sounds, it has been termed *anechoic mixing* model. Most existing algorithms have treated this problem for the *under-determined* case,

M.E. Davies et al. (Eds.): ICA 2007, LNCS 4666, pp. 762–769, 2007.

where sources outnumber the sensors (i.e. $m \leq d$) [6,14,15,16,17]. Much fewer algorithms exist for the *over-determined* case $m \geq d$ (cf. [8]).

The invariance of the estimated model under translation (time shifts) of the source signals is obvious from equation (1). This property remains valid if the time arguments are replaced by vectors, resulting in multivariate functions $x_i(t) = x_i(t_1, \ldots, t_n)$. In this case the scalar delays have to be replaced by displacement vectors $\tau_{ij} = \overrightarrow{\tau_{ij}} = \left(\tau_{ij_1}, \cdots, \tau_{ij_n}\right) \in \mathbb{R}^n$. Such a multivariate model is suitable for feature learning that accounts for shift-invariance in higher-dimensional spaces.

In this paper, we present an algorithm for the solution of such generalized anechoic mixing problems for multivariate signals. The algorithm is derived by applying methods from stochastic time-frequency analysis to the generalized mixture model. These techniques allow the projection of the multi-dimensional mixture equation onto multiple one-dimensional problems. For their solution we introduce a modification of non-negative matrix factorization(NMF) [4,9] that extends this method for convolutive models. The efficiency of the developed algorithm is demonstrated by shift-invariant learning of features from sets of gray-level images.

2 Derivation of the Algorithm

Notation: Throughout the paper the following notations will be used:

- $e_k \in \mathbb{R}^n$ denotes the kth canonical unit vector
- E denotes the expectation value
- $\mathcal{F} = \mathcal{F}^1$ denotes the (multidimensional) Fourier transform, $\mathcal{F}^{-1}$ its inverse
- T_τ, M_f denote the time and frequency shift operators, e.g. $(T_\tau M_f x)(t) := e^{-2\pi \mathrm{i} f(t-\tau)} x(t-\tau)$
- $\mathcal{R}_{\boldsymbol{\xi}}$ denotes the multidimensional Radon transform; with h being a square-integrable multivariate function and $\|\boldsymbol{\xi}\| = 1$ it is defined [5] by:

$$\mathcal{R}_{\boldsymbol{\xi}} h(u) := \int h(\boldsymbol{t}) \delta(\boldsymbol{\xi} \cdot \boldsymbol{t} - u) d\boldsymbol{t}$$

- $\mathcal{F}^a$ denotes the (multidimensional) fractional Fourier transform. $\mathcal{F}^a$ is the ath power of the classical Fourier operator [13]. Thus $\mathcal{F}^a$ is the operator with the same eigenfunctions as the Fourier transform (Hermite-Guassians ψ_n) but with the eigenvalues $(e^{-\mathrm{i}n\pi/2})^a$ instead of $e^{-\mathrm{i}n\pi/2}$.
- The notations $\mathcal{F}$ and $\mathcal{F}^a$ will also be used for the discrete versions of the fractional Fourier operators.

The algorithm for the solution of the multivariate anechoic mixture problem is based on the *Wigner-Ville Spectrum* (WVS), a time-frequency integral transform with particularly suitable properties for the solution of anechoic mixture problems [8]. In the following, we introduce first this transform, apply it to the mixture model, and derive the algorithm for a solution by projection onto one-dimensional problems. An efficient algorithm for the solution of the one-dimensional problems is presented in section 2.3.

2.1 Wigner Ville Spectrum

Stochastic time-frequency distributions provide powerful mathematical tools for the analysis of non-stationary random processes. The most prominent quadratic distribution is the Wigner-Ville spectrum (WVS) defined as [11]:

$$W_x(t,\omega) = \int_\tau E\left\{x\left(t+\frac{\tau}{2}\right)x^*\left(t-\frac{\tau}{2}\right)\right\} e^{-2\pi i\omega\tau} d\tau \qquad (2)$$

where $x(t)$ is a random process and $x^*(t) = \bar{x}(t)$ the conjugated process. The WVS can loosely be interpreted as a time-frequency distribution of the mean energy of $x(t)$. This definition implies many useful properties. The following derivations rely in particular on two properties. The first is the time-frequency shift covariance given by:

$$W_{(T_\tau M_f x)}(t,\omega) = W_x(t-\tau,\omega-f)$$

The second property is a relationship between the Radon transform of the WVS and the fractal Fourier transform of the original signal. In the case of a one-dimensional signal h this relationship can be expressed as (cf. [13]):

$$|\mathcal{F}^a h(t)|^2 = \mathcal{R}_{a\pi/2}(W_x(t,\omega))$$

Both properties can be easily generalized for the multidimensional case [2,12].

2.2 Application of the WVS to the Multivariate Mixture Model

Introducing the vectorial notation $\tau_{ij} = \overrightarrow{\tau_{ij}} = \left(\tau_{ij_1},\ldots,\tau_{ij_n}\right) \in \mathbb{R}^n$ and with $t = \left(t_1,\ldots,t_n\right) \in \mathbb{R}^n$ the multivariate anechoic mixing model can be written compactly:

$$x_i(t) = \sum_{j=1}^{d} \alpha_{ij} \cdot s_j(t-\tau_{ij}) \quad i = 1,\cdots,m \qquad (3)$$

With the the assumption that the multivariate source signals are independent and have zero means, i.e. $E\{s_i\} = 0$ and $E\{s_i s_j\} = 0\ \forall i \neq j$, the mixture model (3) can be mapped into the time-frequency domain by application of the WVS:

$$W_{x_i}(t,\omega) = \int \mathrm{E}\Big\{ \sum_{j,k=1}^{d} \alpha_{ij}\overline{\alpha_{ik}} s_j(t+\frac{\tau}{2}-\tau_{ij}) s_k^*(t-\frac{\tau}{2}-\tau_{ik})\Big\} e^{-2\pi i\omega\tau} d\tau$$

$$= \sum_{j=1}^{d} |\alpha|_{ij}^2 W_{s_j}(t-\tau_{ij},\omega) \quad i = 1,\ldots,m \qquad (4)$$

A direct evaluation of this equation is possible only in the univariate case. Even for two-dimensional time arguments the computational costs (in time and memory) grow prohibitively. However, equation (4) is redundant and can be solved by computing a set of projections onto lower dimensional spaces that specify the same information as the original problem (3). Projections by integrating over unbounded domains with respect to the vectorial time parameter t are particularly useful as they eliminate the dependence of the unknown shifts τ_{ij}.

If a sufficient number of integral projections is computed the inversion theorem for the Radon transform guarantees that the solution of problem (4), and thus of equation (3), can be uniquely recovered. Let $\boldsymbol{\xi} \in \mathbb{R}^n$ be an arbitrary unit vector, then the radon transform of equation (4) can be computed explicitly:

$$\mathcal{R}_{\frac{\pi}{2}\boldsymbol{\xi}}(W_{x_i}(t,\omega))(u) = E\{|\mathcal{F}^{\boldsymbol{\xi}} x_i(u)|^2\} = \sum_{j=1}^{d} |\alpha|_{ij}^2 \mathcal{R}_{\frac{\pi}{2}\boldsymbol{\xi}}(W_{(T_{\tau_{ij}} s_j)}(t,\omega))(u) \tag{5}$$

$$= \sum_{j=1}^{d} |\alpha|_{ij}^2 E\{|\mathcal{F}^{\boldsymbol{\xi}} s_j(u - (\tau_{ij_l} \cos \xi_l)_l)|^2\} \tag{6}$$

In contrast to the power spectrum of the ordinary Fourier transform, the fractional power spectrum is not shift invariant. Instead, all displacement vectors τ_{ij} are scaled with the factors $\cos(\boldsymbol{\xi}) = (\cos(\xi_1), \ldots, \cos(\xi_n))$ [13]. The special choice $\xi_l = \frac{\pi}{2} \forall l$ thus eliminates all delays from (6), since in this case the fractional Fourier transform reduces to the ordinary Fourier transform.

Moreover, the special form of (6) makes it possible to specify conditions for which the projections are reversible. For every dimension at least two fractional power spectra with distinct exponents suffice for inversion [10]. This implies that the special choice $\boldsymbol{\xi} \in \{\varphi_k = (\frac{\pi}{2}, \ldots, \frac{\pi}{2}) - \varepsilon_k e_k \ \forall k\}$ (with the free parameters $\varepsilon_k \neq 0$) defines an invertible family of anechoic mixing problems. Each of these problems depends solely on a single component of the delay vectors τ_{ij}. For example this implies:

$$E\{|\mathcal{F}^{(1,\ldots,1,\varepsilon_k,1,\ldots)} x_i(u)|^2\} = \sum_{j=1}^{d} |\alpha|_{ij}^2 E\{|\mathcal{F}^{(1,\ldots,1,\varepsilon_k,1,\ldots)} s_j(u - \tau_{ij_k} \cos \varepsilon_k e_k)|^2\} \tag{7}$$

By transformation of all multidimensional variables into column vectors the equations (7) can be vectorized. Each of them specifying a one-dimensional anechoic mixture with positivity constraints for all variables.

2.3 Solution of the One-Dimensional Problems

The one-dimensional problems derived from (4) can be written more compactly with the non-negative time series $y_i(t)$, $i = 1, \ldots, m$ and sources $\sigma_j(t)$, $j = 1, \ldots, d$, $t \in \mathbb{R}$. The solution of these problems can be obtained by solving the optimization problem:

$$\min_{a_{ij},\sigma_j,\tau_{ij}} \|y_i(t) - \sum_{j=1}^{d} a_{ij}\sigma_j(t - \tau_{ij})\| \quad \text{subject to } \sigma_j \geq 0 \text{ , } a_{ij} \geq 0 \, \forall i,j$$

This is a special case of a positive deconvolution problem:

$$\min_{\nu_{ij},\sigma_j} \|y_i(t) - \sum_{j=1}^{n} (\nu_{ij} * \sigma_j)(t)\| \quad \text{subject to } \nu_{ij} \geq 0 \text{ , } \sigma_j \geq 0 \, \forall i,j \tag{8}$$

For the implementation time is discretized. In this case it is more convenient to adopt a matrix-vector notation. If $\mathbf{Y}, \mathbf{A}, \boldsymbol{\Sigma}$ signify the vectorized time-discrete variables $y_i(l), \nu_{ij}(l), \sigma_j(l)$, and $\mathcal{A}, \mathcal{X}$ the block circulant matrices that represent the sums of convolutions with ν_{ij} and σ_j, equation (8) can be rewritten:

$$\min_{\mathcal{A},\boldsymbol{\Sigma}} \|\mathbf{Y} - \mathcal{A}\boldsymbol{\Sigma}\| = \min_{\mathcal{X},\mathbf{A}} \|\mathbf{Y} - \mathcal{X}\mathbf{A}\| \text{ subject to } \mathbf{A}, \mathcal{X}, \boldsymbol{\Sigma}, \mathcal{A} \geq 0 \; \forall i, j$$

This shows that the (discretized) deconvolution (8) is a special case of a non-negative matrix factorization problem. Depending on the exact error-function or norm used for the minimization, it is possible to adopt different multiplicative update rules [4]. Since the adjoint of a convolution operator is again a convolution, fast implementations of many update rules can be obtained exploiting Fast Fourier Transform (FFT). A standard rule based on the Euclidian distance [9] can be written:

$$\nu_{ij} \leftarrow \nu_{ij} \frac{\left(\mathcal{F}^{-1}\left(\overline{\mathcal{F}\sigma_j}\mathcal{F}y_i\right)\right)}{\left(\mathcal{F}^{-1}\left(\sum_k \overline{\mathcal{F}\sigma_j}\mathcal{F}\nu_{ik}\mathcal{F}\sigma_k\right)\right)} \; \forall i, j, \quad \sigma_j \leftarrow \sigma_j \frac{\mathcal{F}^{-1}(\sum_k \overline{\mathcal{F}\nu_{kj}} \cdot \mathcal{F}y_k)}{\mathcal{F}^{-1}(\sum_{p,l} \overline{\mathcal{F}\nu_{pj}} \cdot \mathcal{F}\nu_{pl} \cdot \mathcal{F}\sigma_l)} \; \forall j$$

More suitable for the anechoic case are update rules that allow a control of the sparseness of the estimated sources or filters [4]:

$$\nu_{ij} \leftarrow \left[\nu_{ij}\left(\mathcal{F}^{-1}\left(\overline{\mathcal{F}\sigma_j}\mathcal{F}\left[y_i/\mathcal{F}^{-1}(\sum_k \mathcal{F}\nu_{ik}\mathcal{F}\sigma_k)\right]^{\beta}\right)\right)^{\frac{\mu}{\beta}}\right]^{1+\lambda_1} \tag{9}$$

$$\sigma_j \leftarrow \left[\sigma_j\left(\mathcal{F}^{-1}\sum_k \overline{\mathcal{F}\nu_{jk}}\mathcal{F}\left[y_k/\mathcal{F}^{-1}\left(\sum_l \mathcal{F}\nu_{kl}\mathcal{F}\sigma_l\right)\right]^{\beta}\right)^{\frac{\mu}{\beta}}\right]^{1+\lambda_2} \tag{10}$$

These learning rules are quite flexible, and a variety of regularizers can be included if additional information about the sources is available. The specific choice $\mu = 1.9, \beta = 2, \lambda_1 = 0.02, \lambda_2 = 0$ results in sparse features [4].

Summarizing the individual steps defines the complete algorithm for the solution of the one-dimensional positive anechoic mixture problem:

Input: Data y_i for $i = 1, \ldots, m$
Initialize: Choose random values $\nu_{ij}, \sigma_j \geq 0$
  for iter1 = 1 : maxiter1,
    Update σ_j using (10).
    for iter2 = 1 : maxiter2,   % Sparse update for the filters
      Update ν_{ij} using (9).
      Normalize $\nu_{ij} = \frac{\nu_{ij}}{\|\nu_{ij}\|}$.
    end
    $\nu_{ij} = \max(\nu_{ij})\, e_l$ with $l \in \{l | \nu_{ij}(l) = \max(\nu_{ij})\}$
                    % Sparse filter replaced with delta function
  end

2.4 Algorithm for Multivariate Anechoic Mixtures

The algorithm for solving the multivariate anechoic mixing problem can be summarized:

1. **Input** Data $x_i \in \mathbb{R}^n, i = 1, \ldots, m$
   Parameters $\varepsilon_k \neq 0, k = 1, \ldots, n$

2. **Compute** (fractional) Fourier transforms $\mathcal{F}x_i$ and $\mathcal{F}^{(\varepsilon_k e_k)}(\mathcal{F}x_i) =: \mathcal{F}^{\varphi_k}x_i$
3. **Solve** the 1D-anechoic mixture problems:

$$|\mathcal{F}^{\varphi_k}x_i(\omega)|^2 = \sum_{j=1}^{d} |\alpha|_{ij}^2 |\mathcal{F}^{\varphi_k}s_j(\omega - \tau_{ij_k} e_k \cos\varepsilon_k)|^2$$

   These subproblems can be treated with the deconvolutive NMF algorithm presented in section 2.3. This step provides estimates for the linear weights α_{ij}, the delays τ_{ij}, and the fractional power spectra $|\mathcal{F}^{\varphi_k}s_j|$.
4. **Compute** Radon inversion, applying one of the following methods:
   - *Gerchberg-Saxton phase retrieval:*
     Given $n+1$ different fractional Fourier intensity spectra $|\mathcal{F}^{\varphi_k}s_j|$, the signals s_j can be reconstructed by a modification of the original Gerchberg-Saxton phase recovery algorithm [10].
   - *Deconvolution:*
     For known weights and delays (3) specifies a normal deconvolution problem with known mixing filters. Applying standard deconvolution algorithms (e.g. Wiener filter) the sources s_j can be retrieved by least squares estimation.

In the last step of the algorithm several other methods can be applied for the integration of the solutions of the one-dimensional problems. For small parameters ε_k one can also compute the angular derivative of the fractional Fourier transforms. The unknown phases could then be recovered by integration [1].

3 Applications in Image Processing

The described algorithm has a broad application spectrum. In principle, it permits invariant learning of mixture models in an arbitrary number of dimensions. The algorithm described in section 2.4 is adjustable to under-, over- and even-determined mixtures, and can be easily modified by inclusion of sparseness or positivity constraints. Specifically, the last two steps of the algorithm can be implemented using a broad range of established methods for non-negative matrix factorization and inverse Radon transformation. Given the limited available space, we present here only two examples from image processing, where the method is applied for the extraction of image components that reappear at different positions in different images.

One set of images was generated by pasting two objects from an image data basis at different randomly chosen positions of images with a size of 150×150 pixels. The generated images are shown in figure 1. Goal of the application of the algorithm is the extraction of the original objects from the image set. For the last step of the algorithm two implementations were compared (Gerchberg-Saxton algorithm and the deconvolution method). Figure 2 shows the results of the feature extraction. Both implementations retrieve the original objects. However, the deconvolution approach is clearly superior to the phase retrieval. This partially due to the very slow convergence of the Gerchberg-Saxton algorithm, especially for small fractional powers and for inaccurate estimates of the power spectra. In addition, the deconvolution method exploits for a second time the specific structure of the mixture model. A quantitative comparison shows that images predicted from the extracted components predict 95% of the variance of the original images for the deconvolution method, but only 72% for the phase retrieval method.

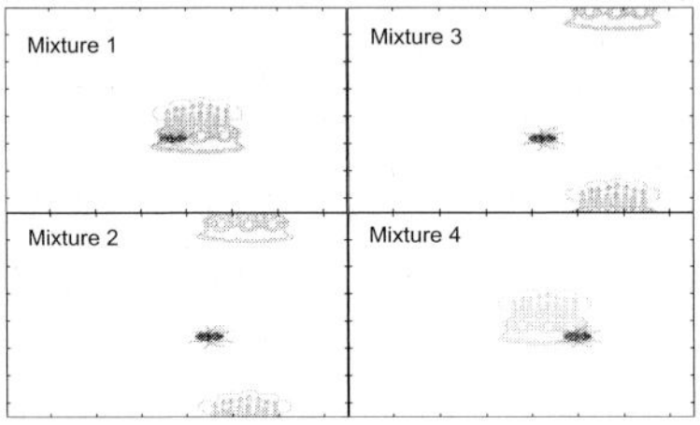

Fig. 1. Example images defining an (over-determined) anechoic mixture in two dimensions

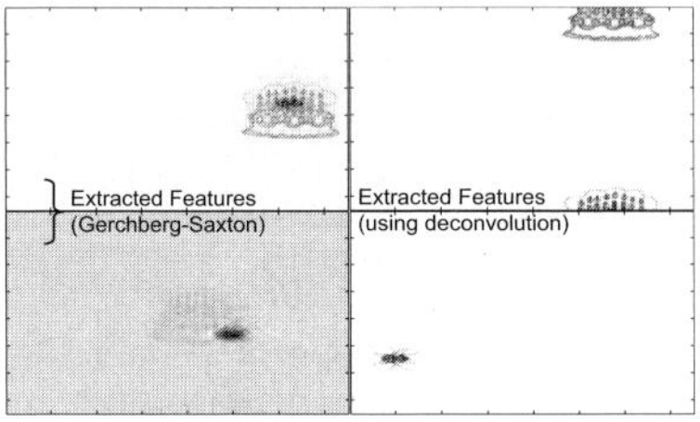

Fig. 2. Extracted features from the image set in figure 1 using two different algorithms for Radon inversion

The second data set consisted of four gray-scale images taken with a digital camera and resampled with a resolution of 200 × 200 pixels (cf. figure 3). The photographs show two objects (scissors and a cup) that were placed at different positions on a wooden surface. Before application of the algorithms the images were whitened [7] to level the correlation statistics of natural images, removing strong correlations between features on small spatial scales. In this case only the deconvolution method was implemented. The reconstruction explains 85% of the of the pre-whitened training images and recovers the original objects with reasonable accuracy.

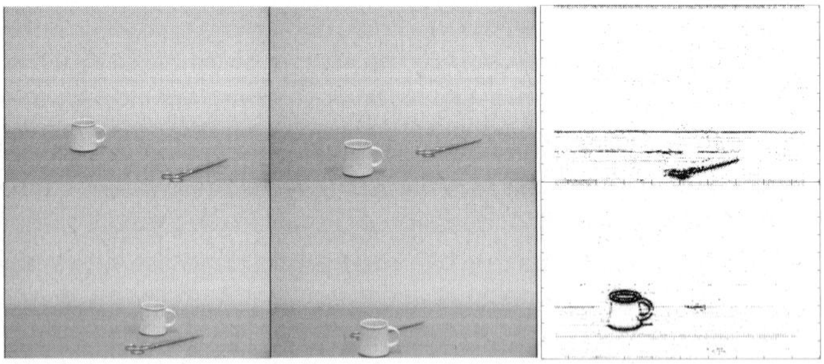

Fig. 3. Left: Real Images. Right: Extracted Features from pre-whitened images.

4 Conclusion

We presented a generalization of the anechoic mixture problem for the multivariate case, which allows learning of independent components with invariance against translation in multi-dimensional spaces. An efficient algorithm was presented that makes the high-dimensional problem tractable by projection onto one-dimensional problems, resulting in a computational complexity that grows linearly in the number of dimensions. We demonstrated the efficiency of this algorithm by learning independent features that reappear at different positions of real images. Invariant unsupervised learning of independent features has a vast amount of other applications, e.g. in computer vision and computer graphics, 3D data analysis, and neural data analysis. Future work will extend the work to new applications and optimize the computational steps. In addition, the robustness of the algorithm against different levels of reverberations will be tested.

Acknowledgments. This work was supported by HFSP, DFG, the Volkswagenstiftung, and the EU FP6 Project 'COBOL'.

References

1. Alieva, T., et al.: Signal reconstruction from two close fractional Fourier power spectra. SPIE Milestone Series 181, 566–577 (2006)
2. Alieva, T., Bastiaans, M.J.: Wigner distribution and fractional Fourier transform for two-dimensional symmetric optical beams. J. Opt. Soc. Am. 17, 2319–2323 (2000)
3. Choi, S., et al.: Blind source separation and independent component analysis: A review. Neural Information Processing - Letters and Review 6, 1–57 (2005)
4. Cichocki, A., et al.: New Algorithms for Non-Negative Matrix Factorization in Applications to Blind Source Separation. In: ICASSP-06, pp. 621–625 (2006)
5. Cuyt, A., et al.: Multidimensional Integral Inversion, with Applications in Shape Reconstruction. SIAM J. Sci. Comput. 27, 1058–1070 (2005)
6. Emile, B., Comon, P.: Estimation of time delays between unknown colored signals. Signal Processing 69, 93–100 (1998)
7. Gluckman, J.: Higher order whitening of natural images. In: IEEE Computer Society Conference vol. 2, pp. 354–360 (2005)
8. Omlor, L., Giese, M.A.: Blind source separation for over-determined delayed mixtures. In: Neural Information Processing Systems, vol. 19, MIT Press, Cambridge (2007)
9. Lee, D.D., Seung, H.S.: Learning the parts of objects by Non-Negative Matrix Factorization. Nature 401, 788–791 (1999)
10. Tian-He, L., et al.: Image recovery from double amplitudes in fractional Fourier domain. Chinese Phys. 15, 347–352 (2006)
11. Martin, W.: Time-frequency analysis of random signals. In: Proc. IEEE Int. Conf. Acoust. Speech, Sig. Proc. vol. 82, pp. 1325–1328 (1982)
12. Matz, G., Hlawatsch, F.: Wigner distributions (nearly) everywhere. Signal Processing 83, 1355–1378 (2003)
13. Ozaktas, H.M., et al.: The Fractional Fourier Transform. John Wiley & Sons, Chichester (2001)
14. Roy, R., Kailath, T.: ESPRIT-Estimation of signal parameters via rotational invariance techniques. IEEE Trans. Acoust. Speech, Sig. Proc. 37, 984–995 (1989)
15. Torkkola, K.: Blind separation of delayed sources based on information maximization. In: Proc. IEEE Int. Conf. Acoust. Speech, Sig. Proc. vol. 96, pp. 3509–3512 (1996)
16. Yeredor, A.: Time-delay estimation in mixtures. Acoustics, Speech, and Signal Processing 5, 237–240 (2003)
17. Ylmaz, Ö., Rickard, S.: Blind Separation of Speech Mixtures via Time-Frequency Masking. IEEE Transactions On Signal Processing 52, 1830–1847 (2004)

Channel Estimation for O-STBC MISO Systems Using Fourth-Order Cross-Cumulants*

Héctor J. Pérez-Iglesias and Adriana Dapena

Departamento de Electrónica y Sistemas, Universidade da Coruña
Campus de Elviña 5, 15071 A Coruña, Spain
Tel.: ++34-981-167000, Fax: ++34-981-167160
hperez@udc.es, adriana@udc.es

Abstract. This paper proposes several algorithms to recover the transmitted signals in systems with multiple antennas that make use of orthogonal space time block code (O-STBC) to attain full transmit diversity. We interpret the scheme proposed by Alamouti in [1] and half-rate systems presented in [2] as classic blind source separation (BSS) problems where the received signals (observations) are instantaneous mixtures of the transmitted signals (sources). In order to recover the sources, we first propose to perform an eigenvalue decomposition of matrices containing fourth-order cross-cumulants of the observations. Subsequently, we show that the performance of this approach can be improved by doing a simultaneous diagonalization of the cumulant matrices. This second approach can be interpreted as a particular case of Joint Approximate Diagonalization of Eigen-matrices (JADE) algorithm for systems where the mixing matrix is orthogonal.

1 Introduction

Wireless communication systems that employ multiple antennas at both transmission and reception are commonly referred to as Multiple Input Multiple Output (MIMO) systems. One of the major advantages of MIMO systems is their ability to provide spatial diversity gains to decrease the Symbol-Error-Rate (SER) in multipath fading channels [3]. Diversity gain results from combining signals that experience independent signal fades.

Achieving the promised performance gains, even in practical operating conditions, requires for specific Space-Time Coding (STC) techniques that spread the transmitted symbols over the space and time dimensions. The popularity of STC is due to the techniques known as Orthogonal Space-Time Block Codes (O-STBC), where different versions of the original data are transmitted by

* This work has been supported by Xunta de Galicia, Ministerio de Educación y Ciencia of Spain and FEDER funds of the European Union under grants number PGIDT05PXIC10502PN and TEC2004-06451-C05-01.

M.E. Davies et al. (Eds.): ICA 2007, LNCS 4666, pp. 770–777, 2007.

several transmitting antennas across several time-slots. In general, O-STBC does not provided coding gain but, however, the decoding method is very simple and it can be performed using linear processing.

In addressing the issue of decoding complexity, Alamouti has proposed in [1] a remarkable O-STBC scheme for transmission with two antennas, that is currently part of both the W-CDMA and CDMA-2000 standards [4]. This code achieves a transmission rate equal to one by transmitting a pair of symbols in a time-slot and the same pair with a different phase in the next time-slot. This scheme supports Maximum-Likelihood (ML) detection only based on linear processing at the receiver. Tarokh et al. [2,5] has developed a theory to design O-STBC which also supports ML detection with linear processing at the receiver. For any number of transmitting antennas, these codes achieve the maximum possible transmission rate when the symbols correspond to any arbitrary real constellation and 1/2 for complex constellations. For the specific case of three or four transmitting antennas, it is possible to achieve 3/4 of the maximum transmission using any complex constellation. The simulation results presented in [5] show that significant gains can be achieved by increasing the number of transmitting antennas with very little decoding complexity.

This paper shows in Section 2 that the system proposed by Alamouti and the half-rate coding schemes presented by Tarokh et al. can be interpreted as classic problems of BSS where a set of unknown signals (sources) must be recovered from instantaneous mixtures of them without resorting to pilot symbols. Using this model, we propose in Section 3 to estimate the channel matrix by performing an eigenvalue decomposition of fourth-order cross-cumulant matrices. Subsequently, we present an approach based on performing a simultaneous diagonalization of the cumulant matrices. Simulations results are presented in Section 4 to compare the performance of the proposed approaches. Finally, Section 5 is devoted to the conclusions.

2 O-STBC Schemes

We consider a Multiple Input Single Output (MISO) system, a particular case of MIMO systems, with P transmitting antennas and only one receiving antenna. Let s_i, $i = 1, ..., N$ be N zero-mean complex-valued statistically independent signals (sources). An O-STBC is defined by a $K \times P$ transmission matrix $\mathbf{G}^{(P)}$ formed by orthogonal columns containing linear combination of these N sources, and their conjugates. Each row represents a time slot and each column represents one transmitting antenna. Since K time-slots are used to transmit N signals, the code rate is defined as $R = N/K$.

In this paper, we will consider the O-STBC code proposed by Alamouti in [1] and the half-rate code presented in [2] for four transmitting antennas. This codes are, respectively, characterised by the matrices

$$\mathbf{G}^{(2)} = \begin{bmatrix} s_1 & s_2 \\ -s_2^* & s_1^* \end{bmatrix} \qquad \mathbf{G}^{(4)} = \begin{bmatrix} s_1 & s_2 & s_3 & s_4 \\ -s_2 & s_1 & -s_4 & s_3 \\ -s_3 & s_4 & s_1 & -s_2 \\ -s_4 & -s_3 & s_2 & s_1 \\ s_1^* & s_2^* & s_3^* & s_4^* \\ -s_2^* & s_1^* & -s_4^* & s_3^* \\ -s_3^* & s_4^* & s_1^* & -s_2^* \\ -s_4^* & -s_3^* & s_2^* & s_1^* \end{bmatrix} \tag{1}$$

Denoting by x_k the received signal at the k-th time slot, we can written $x_k = \sum_{i=1}^{K} h_i \mathbf{G}_{k,i}^{(P)} + n_k$, $k = 1, ..., K$ where $P = 2, 4$ is the number of transmitting antennas, h_i represents the channel path from the i-th transmitting antenna to the receiving antenna and n_k is modelled as additive white Gaussian noise (AWGN). The term $\mathbf{G}_{k,i}^{(P)}$ denotes the element into the k-th row, i-th column of $\mathbf{G}^{(P)}$.

By defining the observation vector $\mathbf{x} = [x_1, ..., x_{K/2}, x_{K/2+1}^* ..., x_K^*]^T$, the source vector $\mathbf{s} = [s_1, ..., s_N]^T$ and the noise vector $\mathbf{n} = [n_1, ..., n_{K/2}, n_{K/2+1}^* ..., n_K^*]^T$, the relationship between the observations and the sources can be written as

$$\mathbf{x} = \mathbf{H}^{(P)}\ \mathbf{s} + \mathbf{n} \tag{2}$$

where $\mathbf{H}^{(P)}$ is the channel matrix depending of the number of transmitting antennas,

$$\mathbf{H}^{(2)} = \begin{bmatrix} h_1 & h_2 \\ h_2^* & -h_1^* \end{bmatrix} \qquad \mathbf{H}^{(4)} = \begin{bmatrix} h_1 & h_2 & h_3 & h_4 \\ h_2 & -h_1 & h_4 & -h_3 \\ h_3 & -h_4 & -h_1 & h_2 \\ h_4 & h_3 & -h_2 & -h_1 \\ h_1^* & h_2^* & h_3^* & h_4^* \\ h_2^* & -h_1^* & h_4^* & -h_3^* \\ h_3^* & -h_4^* & -h_1^* & h_2^* \\ h_4^* & h_3^* & -h_2^* & -h_1^* \end{bmatrix} \tag{3}$$

For simplicity in the notation we will remove the superindex (P) from $\mathbf{H}^{(P)}$. Note that the matrices $\mathbf{H}$ in (3) are orthogonal, i.e.,

$$\mathbf{H}^H\ \mathbf{H} = ||\mathbf{h}||^2\ \mathbf{I}_P \tag{4}$$

where $||\mathbf{h}||^2 = \sum_{k=1}^{P} |h_k|^2$ for the Alamouti's coding scheme and $||\mathbf{h}||^2 = 2\sum_{k=1}^{P} |h_k|^2$ for the half-rate system, $\mathbf{I}_P$ is the $P \times P$ identity matrix and ^{H} is the Hermitian operator. For the Alamouti's coding system it is also verified $\mathbf{H}\ \mathbf{H}^H = ||\mathbf{h}||^2\ \mathbf{I}_P$.

Since the observations can be interpreted as instantaneous mixtures of the sources (equation (2)), the channel matrix can be found by using many existing BSS algorithms (see, for instance, [6] and references therein). Most of these algorithms only take into account the independence of the sources and they do not consider other properties of the channels. Recently, algorithms based on second-order statistics have been developed for blind channel estimations in O-STBC transmissions [7,8]. In practice, when these methods are used for the Alamouti's coding scheme, the communication system must be modified by including a precoder before the O-STBC encoder. In this paper we investigate the performance of approaches based on fourth-order cross-cumulants which do not required to include additional modules in the encoder.

3 Proposed Approaches

In this section we present several methods to estimate the channel matrix by diagonalizing matrices containing fourth-order cross-cumulants of the observations. We will consider that the sources have the same power and the same kurtosis. The fourth-order cross-cumulant matrices is defined by

$$\mathbf{C}[k,l] = c_4(\mathbf{x}, \mathbf{x}^H, x_k, x_l^*), \quad k,l = 1,...,K \tag{5}$$

$$\Rightarrow \mathbf{C}[k,l](i,j) = c_4(x_i, x_j^*, x_k, x_l^*) \tag{6}$$

where $c_4(x_1, x_2, x_3, x_4) = E[x_1x_2x_3x_4] - E[x_1x_2]E[x_3x_4] - E[x_1x_3]E[x_2x_4] - E[x_1x_4]E[x_2x_3]$. This matrices can be decomposed in

$$\mathbf{C}[k,l] = \beta \mathbf{H} \ \mathbf{\Delta}[k,l] \ \mathbf{H}^H \tag{7}$$

where β is a complex valued number and $\mathbf{\Delta}[k,l]$ is a $P \times K$ diagonal matrix with the form

$$\mathbf{\Delta}[k,l] = diag(\mathbf{H}_{k,1}\mathbf{H}_{l,1}^*, \ \mathbf{H}_{k,2}\mathbf{H}_{l,2}^*, \ \ldots, \ \mathbf{H}_{k,N}\mathbf{H}_{l,N}^*) \tag{8}$$

where $\mathbf{H}_{m,n}$ denotes the element in the m-th row, n-th column of matrix $\mathbf{H}$.

3.1 Eigenvalue Decomposition

A way to estimate the channel matrix consists in computing the eigenvectors, $\mathbf{U}$, of the fourth-order cross-cumulants matrices. For the Alamouti's coding scheme, $\mathbf{U}$ is a squared matrix of dimension 2×2. In the case of the half-rate code, $\mathbf{U}$ contains the P eigenvectors of dimension $K \times 1$ corresponding to the largest eigenvalues.

The sources are recovered using $\hat{\mathbf{s}} = \mathbf{U}^H\mathbf{s}$. From equation (8) we deduce that the condition to guarantee that the mixing system be identifiable using an eigenvalue decomposition is that the matrix $\mathbf{\Delta}[k,l]$ has different real values into its diagonal. In particular, Beres and Adve have proposed in [9] to identify the channel matrix for the Alamouti's coding scheme by using of matrices $\mathbf{C}[1,1]$ or $\mathbf{C}[2,2]$. Note, however, that this approach fails when $|h_1| = |h_2|$ because $\mathbf{\Delta}[k,k]$ for both $k = 1$ and $k = 2$ have equal entries into its diagonal and, therefore, $\mathbf{C}[k,k] = \rho_4|h_1|^2\mathbf{I}_2$.

Now, we focus our attention in the matrix $\mathbf{C}[1,2]$ for the Alamouti's coding system. For this case the cross-cumulant matrix can be written as equation (7) where $\beta = \rho_4 h_1 h_2$, $\rho_4 = c_4(s_1, s_1^*, s_1, s_1^*) = c_4(s_2, s_2^*, s_2, s_2^*)$ is the kurtosis and $\mathbf{\Delta}[1,2] = diag(1,-1)$. Therefore, we deduce that the mixing system is always identifiable independently of channel path values.

Note that for the half-rate code, the matrix $\mathbf{C}[k,l],\ k \neq l$ can be also written as equation (7) but, however, the elements into the diagonal matrix $\mathbf{\Delta}[k,l]$ are complex valued and they cannot be identified using an eigenvalue decomposition.

3.2 Joint Diagonalization

The mixing matrix can be also computed by performing a simultaneous diagonalization of some fourth-order cross-cumulant matrices. Towards this aim, we have used the extension of the Jacobi technique described in [10][1]. This idea can be also interpreted as an particularisation of the Joint Approximate Diagonalization of Eigen-matrices (JADE) algorithm [11] obtained by including the orthogonality of the mixing matrix.

For the Alamouti's coding scheme we have only diagonalized the matrices $\mathbf{C}_4[1,1]$ and $\mathbf{C}_4[1,2]$. For the half-rate code, we have 64 matrices with 8×8 fourth-order cross-cumulants. It is apparent the high computational load associated to compute these matrices and to perform their diagonalization. However, exists a similarity between the matrices: $\mathbf{C}[k,l] = \mathbf{C}^*[l,k]$, $\mathbf{C}[k,k] = \mathbf{C}[k-4,k-4]$ with $k = 5,...,8$ and $\mathbf{C}[k,l] = \mathbf{C}^*[k-4,l-4]$ for $k,l = 5,...,8$. As a consequence, the number of matrices can be considerably reduced. In the simulations we have considered the following cases:

- Approach I: $joint-diag(\mathbf{C}[1,1], \mathbf{C}[1,2], ..., \mathbf{C}[1,8])$
- Approach II: $joint–diag(\mathbf{C}[1,1], \mathbf{C}[1,2],..., \mathbf{C}[1,8], \mathbf{C}[2,3], \mathbf{C}[2,4],..., \mathbf{C}[2,8])$
- Approach III: $joint - diag(\mathbf{C}[1,1], \mathbf{C}[1,2], ..., \mathbf{C}[1,8], \mathbf{C}[2,3], \mathbf{C}[2,4], ..., \mathbf{C}[2,8], \mathbf{C}[3,4], \mathbf{C}[3,5], ..., \mathbf{C}[3,8])$
- Approach IV: $joint - diag(\mathbf{C}[1,1], \mathbf{C}[1,2], ..., \mathbf{C}[1,8], \mathbf{C}[2,3], \mathbf{C}[2,4], ..., \mathbf{C}[2,8], \mathbf{C}[3,4], \mathbf{C}[3,5], ..., \mathbf{C}[3,8], \mathbf{C}[4,5], ..., \mathbf{C}[4,8])$.

4 Simulation Results

This section presents the results of several computer simulations carried out to verify the estimation algorithms explained in previous sections. The experiments have been performed by transmitting QPSK over Rayleigh-distributed randomly generated block fading channels. We assume no Intersymbol Interference (ISI), perfect synchronisation and sampling to symbol period. The statistics in (5) have been calculated for each block by sample averaging over the block symbols. The performance has been measured in terms of the Symbol Error Rate (SER). For comparison purposes we also present the SER obtained with Perfect Channel State Information (Perfect CSI). The simulations have been performed using `MATLAB` code running on an `Athlon XP 3200+`.

[1] The matlab code is available in *www.tsi.enst.fr/˜ cardoso/jointdiag.html*

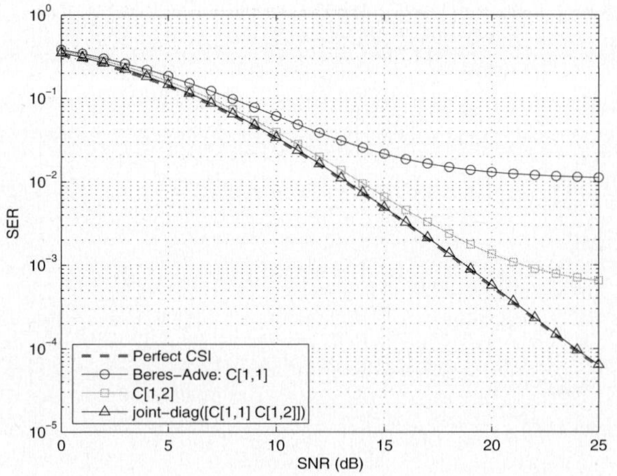

Fig. 1. SER versus SNR obtained for the Alamouti's coding scheme using packets of 500 symbols

In the first set of simulations, the QPSK signal has been coded using the Alamouti' scheme over two transmitting antennas. Figure 1 shows the performance of the proposed approaches in terms of SER versus the SNR. The SER has been obtained by averaging the results for 100, 000 blocks of $L = 500$ symbols. In the figure we can see that the joint diagonalization method achieves the optimum performance while the other approaches present a flooring effect for high SNR. For a SNR of 20 dB, Figure 2 shows the SER versus the packet size

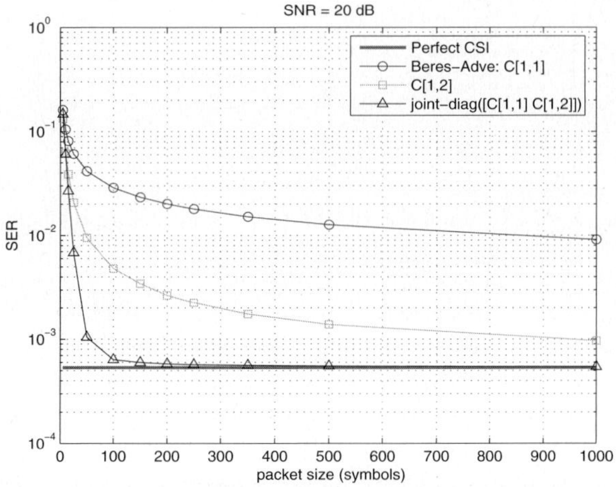

Fig. 2. SER versus the packet size for the Alamouti's coding scheme obtained for a SNR of 20 dB

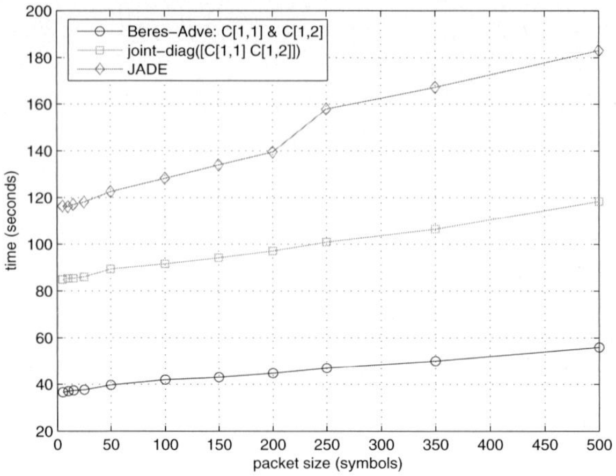

Fig. 3. Time requires to process 10^5 packets for the Alamouti's coding scheme

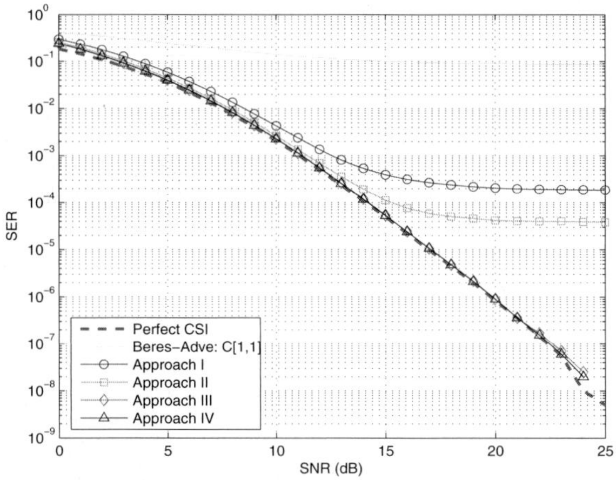

Fig. 4. SER versus SNR obtained for the half-rate code using packets of 500 symbols

and Figure 3 shows the time needed to process 10^5 packets. We can see that the joint diagonalization method requires few symbols (about 200 symbols) to achieve the optimum SER but the time required to process one packet is higher than the needed with the other approaches. For comparison, Figure 3 also shows the time used by JADE to process one block. It is apparent the difference with respect to the joint-diagonalization method proposed in this paper.

In the second set of simulations, the QPSK signal have been coded using half-rate code with four transmitting antennas. Figure 4 plots the SER versus SNR obtained using the method proposed by Beres and Adve and the results obtained with the joint diagonalization method presented in Subsection 3.2. It can be seen that it is possible to achieve the Perfect CSI using the Approach III and the Approach IV. Comparing with the result obtained with the Alamouti's code (Figure 1), we conclude that a significant gain can be achieved by increasing the number of transmitting antennas.

5 Conclusions

This paper proposes several algorithms to recover the transmitted signals without using pilot symbols in systems that makes use of O-STBC without using of training sequences. The basic idea is to estimate the channel parameters by calculating the eigenvectors of a square matrix formed by fourth-order cross-cumulants obtained from the signals received in different time-slots. Simulation results show that the best performance is obtained by performing a simultaneous diagonalization of the fourth-order cross-cumulants matrices.

References

1. Alamouti, S.M.: A simple transmit diversity technique for wireless communications. IEEE Journal Select. Areas Communications 16, 1451–1458 (1998)
2. Tarokh, V., Jafarkhani, H., Calderbank, A.R.: Space-Time Block Codes from Orthogonal Designs. IEEE Transactions on Information Theory 45(5), 1456–1467 (1999)
3. Goldsmith, A.: Wireless Communications. Cambridge University Press, Cambridge (2005)
4. Karim, M.R., Sarraf, M.: W-CDMA and cdma2000 for 3G mobile networks. McGraw-Hill, New York (2002)
5. Tarokh, V., Jafarkhani, H., Calderbank, A.R.: Space-Time Block Coding for Wireless Communications: Performance Results. IEEE Journal on Select Areas in Communications 17(3), 451–460 (1999)
6. Hyvarinen, A., Oja, E.: Independent Component Analysis: Algorithms and Applications. Neural Networks 13(4-5), 411–430 (2000)
7. Shahbazpanah, S., Gershman, A.B., Manto, J.: Closed-form Blind MIMO Channel Estimation for Orthogonal Space-Time Block Codes. IEEE Trans. on Signal Processing 53(12), 4506–4516 (2005)
8. Vía, J., Santamaría, I., Pérez, J., Ramírez, D.: Blind Decoding of MISO-OSTBC Systems based on Principal Component Analysis. In: Proc. of International Conference on Acoustic, Speech and Signal Processing, vol. IV, pp. 545–549 (May 2006)
9. Beres, E., Adve, R.: Blind Channel Estimation for Orthogonal STBC in MISO Systems. In: Proc. of Global Telecommunications Conference, 2004, vol. 4, pp. 2323–2328 (November 2004)
10. Cardoso, J.-F., Souloumiac, A.: Jacobi angles for simultaneous diagonalization. SIAM Journal on Matrics Analysis and Applications 17(1), 161–164 (1996)
11. Cardoso, J.-F., Souloumiac, A.: Blind beamforming for non-Gaussian signals. IEE Proceedings F 140(46), 362–370 (1993)

System Identification in Structural Dynamics Using Blind Source Separation Techniques

F. Poncelet[1], G. Kerschen[1], J.C. Golinval[1], and F. Marin[2]

[1] Aerospace and Mechanical Engineering Department (LTAS), University of Liège,
1 Chemin des Chevreuils, B-4000 Liège, Belgium
{FPoncelet,G.Kerschen and JC.Golinval}@ulg.ac.be
http://www.ltas-vis.ulg.ac.be
[2] V_2i S.A., Liège, Belgium
F.Marin@V2i.be
http://www.V2i.be

Abstract. This paper proposes to explore the potential of Blind Source Separation (BSS) techniques for the estimation of modal parameters, namely the resonant frequencies, vibration modes and damping ratios. The concept of *virtual sources*, which was introduced in recent publications, allows to consider BSS as a simple way of doing output-only modal analysis. This work illustrates the proposed methodology using free and random responses of an experimental truss structure.

Keywords: Blind Source Separation, Second-Order Blind Identification, Structural Dynamics, Experimental Application.

1 Introduction

Blind Source Separation (BSS) techniques were initially developed for signal processing in the early 80's, but during the last decade the number of the application fields never stops increasing. This success certainly comes from two of their intrinsic characteristics. Firstly, the ambition of BSS (which is to recover unobserved source signals from their observed mixtures) is shared with many other research domains. Secondly, the small number of necessary assumptions allows to consider the application of the methodology to various kinds of data sets (resulting from fields as diverse as finance, image or speech processing, astrophysics, and even medicine).

However, if BSS techniques proved useful in numerous application domains, they were quite underused for many years in structural dynamics. Some applications were naturally carried out such as damage detection, condition monitoring and discrimination between pure tones and sharp-pointed resonances, but the modal parameter estimation remained quite marginal in these studies.

Recently, using the concept of virtual sources, a one-to-one relationship between the vibration modes and the BSS modes (i.e.. the mixing matrix) was demonstrated [1], allowing the use of BSS for modal analysis. Since then, two algorithms were tested, and one of them (namely the Second-Order Blind Identification) seemed to perform quite well [2].

M.E. Davies et al. (Eds.): ICA 2007, LNCS 4666, pp. 778–785, 2007.

This paper proposes to explore this possibility from an experimental case submitted to an impulse and a random excitation. The results are compared with those of a well-established modal analysis method, the so-called Stochastic Subspace Identification [3].

2 System Identification in Structural Dynamics: Modal Analysis

2.1 What Is Modal Analysis?

The precise knowledge of the dynamics characteristics is essential for the design and validation of many engineering products. The usual modeling of a dynamical system is defined by modal parameters, namely the natural frequencies f_i, vibration modes $\mathbf{n}_i$ and damping ratios ξ_i. A modal parameter prediction is usually performed using numerical techniques such as the finite element method. But because of possible uncertainties on the material behavior, boundary conditions and joint modeling, experimental validations are often necessary. These ones are performed using Experimental or Operational Modal Analysis (EMA or OMA).

Modal analysis quickly proved to be very popular, and numerous approaches were developed to estimate modal parameter from the structural dynamic response. In 1973, Ibrahim proposed a robust time domain method [4]. The stochastic subspace identification method (and its variant versions) [3] and more recently the polyreference least-squared complex frequency domain method [5] are also efficient modal analysis methods. For further information about modal analysis, the interested reader may consult [6].

2.2 Why Another Method?

The lack of a priori knowledge about the number of existing natural frequencies in the range of interest usually prevents the classical modal analysis methods from a direct identification of modal parameters. A model order (linked to the number of identified frequencies) has to be chosen and progressively increased. A diagram, the so-called stabilization diagram, collecting all these frequencies for increasing order is then plotted. This step allows to separate the physical natural frequencies from the numerical ones, introduced by the algorithm. Unfortunately, the separation between numerical and stabilized frequencies may require a great deal of expertise and can be cumbersome.

Even if many modal analysis methods already exist, a method combining an automatic selection process (with a confidence criteria for each mode) and a physical interpretation of this choice should be interesting. The application of BSS techniques to perform modal analysis partially meets these objectives.

During, the last years, other statistical signal processing techniques have been considered for the analysis of structural dynamic responses. The proper orthogonal decomposition (POD) is one of these, and its modes, the so-called proper orthogonal modes (POMs), have been linked to the structural normal modes [7]. Some additional information can be found in [8,9].

3 From Signal Processing to Modal Analysis

3.1 Second-Order Blind Identification (SOBI)

The basic idea of BSS is to recover the unobservable inputs of a system, called the sources s_i, only from the measured outputs x_i even though very little, if anything, is known about the mixing system. The simplest BSS model assumes the existence of n sources signals $s_1(t), ..., s_n(t)$ and the observation of as many mixtures $x_1(t), ..., x_n(t)$. Note that we focus on systems with linear and static mixtures. Using matrix notations the noisy model can be expressed as

$$\mathbf{x}(t) = \mathbf{A} \cdot \mathbf{s}(t) + \boldsymbol{\sigma}(t) \tag{1}$$

where $\boldsymbol{A}$ is referred to as the mixing matrix, and $\boldsymbol{\sigma}$ is the noise vector corrupting the data.

Most BSS approaches are based on a model in which the sources are independent and identically distributed variables. Independent component analysis (ICA, [10]) does not escape the rule; the sample order has no importance in the method. The objective of SOBI is to take advantage, whenever possible, of the temporal structure of the sources for facilitating their separation. The SOBI algorithm consists in constructing several time-lagged covariance matrices $\mathbf{R}(\tau)$ from the measured data and to find a matrix $\mathbf{U}$ which jointly diagonalizes all the covariance matrices. This matrix corresponds to the mixing matrix $\mathbf{A}$ of (1).

$$\mathbf{R}(\tau) = E[\mathbf{x}(t + \tau) \cdot \mathbf{x}^*(t)] \tag{2}$$

For further detail about the SOBI method, the reader can refer to [11].

3.2 Concept of Virtual Source

The dynamic response of mechanical systems which are considered in this study is described by the equation

$$\mathbf{M} \cdot \ddot{\mathbf{x}}(t) + \mathbf{C} \cdot \dot{\mathbf{x}}(t) + \mathbf{K} \cdot \mathbf{x}(t) = \mathbf{f}(t) \tag{3}$$

where $\mathbf{M}$, $\mathbf{C}$ and $\mathbf{K}$ are the mass, damping and stiffness matrices, respectively. The vector $\mathbf{f}$ represents the real excitation sources applied to the structure. The system response $\mathbf{x}(t)$ may be expressed as a mixture of these **real sources** $\mathbf{f}(t)$. Unfortunately, this mixture is a convolutive product between the impulse response function, denoted $\mathbf{h}(t)$, and the sources $\mathbf{f}(t)$, and the separation of convolutive mixtures of sources is not yet completely solved.

An interesting alternative is to use the modal expansion. Indeed, the m normal modes $\mathbf{n}_{(i)}$ form a complete basis for the expansion of any m-dimensional vector (if m is the number of degrees of freedom). Then the response can be expressed using modal superposition

$$\mathbf{x}(t) = \sum_{i=1}^{m} \mathbf{n}_{(i)} \cdot \eta_i(t) = \mathbf{N} \cdot \boldsymbol{\eta}(t) \tag{4}$$

where the weight coefficients η_i are in fact the modal coordinates and represent the amplitude modulation of the corresponding normal modes $\mathbf{n}_{(i)}$. The similarity between equations (1) and (4) shows that the modal coordinates may act as **virtual sources** (which are statistically independent as proved in [1,2]) regardless of the number and type of physical excitation forces. In addition, the time response can be interpreted as a static mixture of these virtual sources, which renders the application of the BSS techniques possible.

The SOBI algorithm (which requires sources with different spectral contents) is particulary appropriate for the separation of these sources. In the free response case of the system (3), the theoretical expression of the normal coordinates is an exponentially damped harmonic function

$$\eta_i(t) = Y \cdot exp(-\xi_i \cdot \omega_i \cdot t) \cdot cos(\sqrt{1-\xi_i^2} \cdot \omega_i \cdot t + \alpha_i) \tag{5}$$

where ω_i and ξ_i are the natural frequency and damping ratio of the ith mode, respectively. The amplitude Y and the phase α are constants depending on the initial conditions. The modal coordinates are then monochromatic, with different spectral contents.

3.3 Procedure Details

In summary, a simple modal analysis procedure is proposed, using the modal coordinates as virtual sources. The procedure is as follows:

1. Perform experimental measurements of the structure response to obtain time series at different sensing position.
2. Apply SOBI directly to the measured time series to estimate the mixing matrix $\mathbf{A}$ and the sources $\mathbf{s}(t)$.
3. The mode shapes are simply contained in the mixing matrix $\mathbf{A}$.
4. In the case of random excitation, the identified (random) sources are transformed into free decaying responses using NExT (Natural Excitation Technique) algorithm [12].
5. The identification of the other modal parameters (frequencies and damping ratios) is carried out by fitting the time series of the sources $\mathbf{s}(t)$ with the theoretical expression (5).
6. The fitting error between the identified and fitted sources is then computed which allows to reject the non-reliable virtual sources easily.

4 Experimental Demonstration

In this paper the proposed modal analysis technique is applied to the response of the truss structure depicted in Figure 1. The free and random response cases are considered. At each corner, on each storey, two accelerometers measure the horizontal responses. The quality of the identification results is evaluated comparing with the covariance-driven stochastic subspace identification (SSI) method [3].

Fig. 1. Experimental fixture mounted on a 26kN electrodynamic shaker

4.1 Free Response

The free response was obtained using a hammer which provided a short impulse to the system. The sampling frequency was set to 5120 Hz, and the first 6000 samples of the measured time series were taken into account. The SOBI identification requires the definition of delays (for the construction of correlation matrices (2)). 20 delays were chosen uniformly distributed between 0.0025 and 0.1 seconds, which covers the whole frequency range of interest.

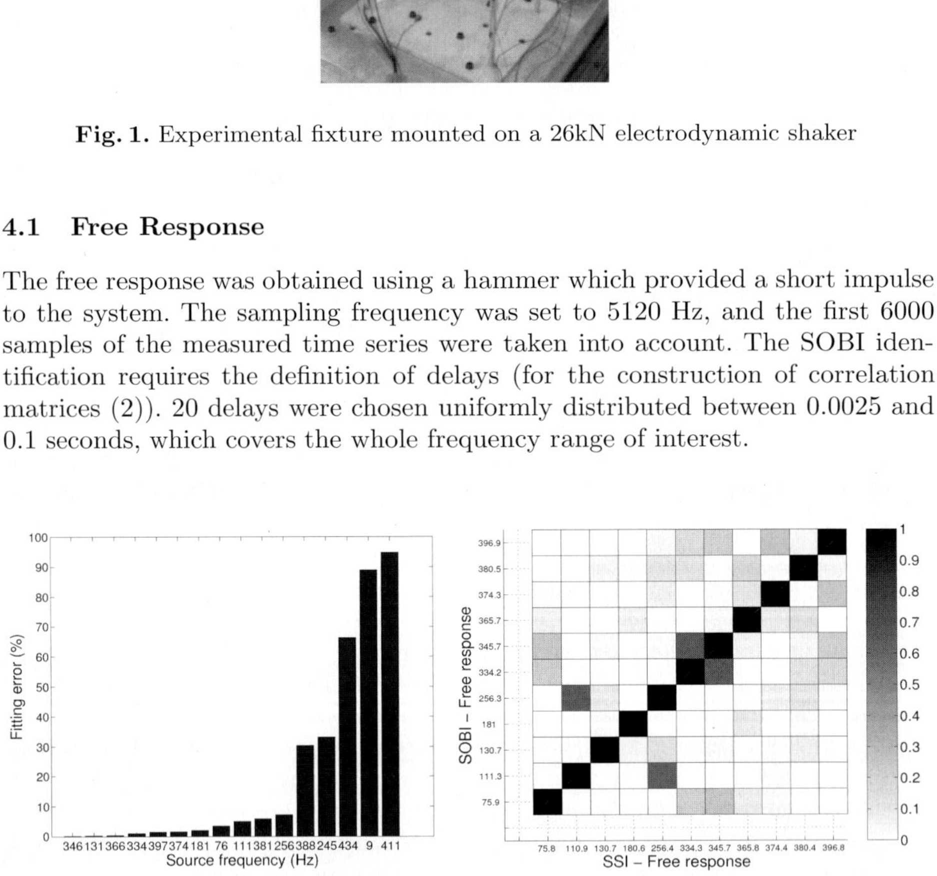

(a) (b)

Fig. 2. Fitting error of the 16 SOBI identified sources for the free response (a) and MAC comparison between SOBI and SSI modes (b)

Because there are 16 measurement locations, a total of 16 virtual sources can be considered. The fitting error of each source is shown, in Figure 2(a). 11 sources have a fitting error below 7% and can be safely retained. The sum of their participation in the system response is above 97.7%. The identification results are listed in Table 1. Concerning the frequency and damping ratio identification, the SOBI results are totally similar to those of the SSI method. Note that the damping ratios of SSI are presented as intervals because the value changes according to the chosen model order. The comparison between the two methods for the mode shapes is performed using the Modal Assurance Criterion in the Figure 2(b). The closer the value to 1, the higher the correspondence. We can see a complete correlation between the modes identified using SSI and SOBI.

Table 1. Identified natural frequencies and damping ratios for the free response

SOBI Freq. [Hz]	SSI Freq. [Hz]	SOBI Damping Ratio [%]	SSI Damping Ratio [%]
75.94	75.82	0.20	[0.05 - 0.12]
111.37	110.99	0.37	[0.40 - 0.60]
130.75	130.76	0.21	[0.20 - 0.28]
181.06	180.69	0.18	[0.20 - 0.28]
256.30	256.48	0.18	[0.10 - 0.15]
334.24	334.32	0.05	[0.02 - 0.05]
345.75	345.76	0.04	[0.04 - 0.05]
365.79	365.81	0.05	[0.05 - 0.06]
374.34	374.45	0.15	[0.10 - 0.30]
380.55	380.45	0.16	[0.20 - 0.40]
396.91	396.81	0.08	[0.07 - 0.10]

4.2 Forced Response

For the random response, the structure was mounted on a 26kN electrodynamic shaker (see Figure 1). The sampling frequency was set to 5120 Hz, and 160000 samples were considered for the measured time series. The same parameters as previously were chosen for the SSI and SOBI methods. The fitting error of each identified source was computed and is presented in Figure 3(a). This time, 14 sources have a fitting error below 8%.

Table 2 lists all the reliable identified results and Figure 3(b) compares the corresponding mode shapes. Once more the correspondence between both methods is remarkable, except for the mode at 75 Hz. If the results obtained using SSI in the free response case are taken as a reference we can note that none of the methods seems able to accurately estimate this mode. The MAC values SOBI random/SSI free and SSI random/SSI free are both lower than 0.65.

Finally, we note that the SSI method was able to identify 4 more modes in the frequency range considered (around 162, 189, 204 and 294 Hz). Nonetheless, because the participation in the system response of the 14 sources identified using SOBI amounts to 93%, these four modes have necessarily a very low participation in the system response.

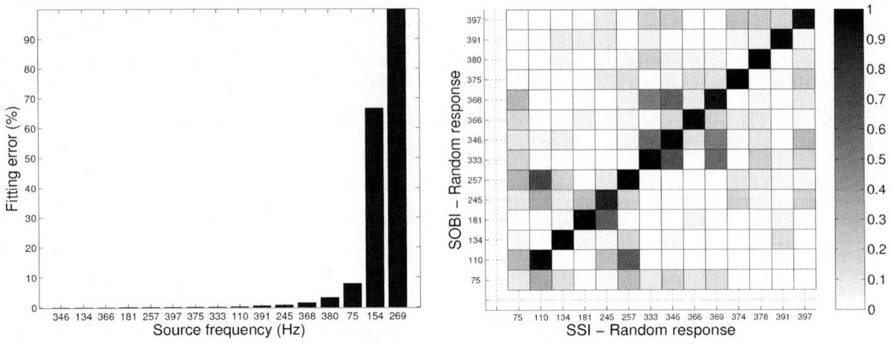

(a) (b)

Fig. 3. Fitting error of the 16 SOBI identified sources for the random response (a) and MAC comparison between SOBI and SSI modes (b)

Table 2. Identified natural frequencies and damping ratios for the random response

SOBI Freq. [Hz]	SSI Freq. [Hz]	SOBI Damping Ratio [%]	SSI Damping Ratio [%]
74.75	74.68	2.15	[1.70 - 2.00]
110.06	110.28	2.03	[1.50 - 2.00]
133.59	133.77	0.85	[0.60 - 0.80]
180.87	180.98	0.23	[0.20 - 0.30]
245.29	245.38	0.16	[0.01 - 0.05]
257.47	257.47	0.11	[0.09 - 0.11]
333.21	333.34	0.12	[0.05 - 0.10]
345.64	345.51	0.09	[0.10 - 0.12]
365.60	365.76	0.12	[0.07 - 0.15]
368.19	369.53	0.33	[0.15 - 0.30]
374.34	374.69	0.16	[0.20 - 0.40]
380.06	378.34	0.71	[0.50 - 0.70]
390.81	390.95	0.33	[0.45 - 0.50]
396.83	397.20	0.17	[0.15 - 0.25]

5 Conclusions

Based on the virtual source concept, a new application is developed for the BSS methods, and particulary for the SOBI algorithm, in the field of structural dynamics. An output-only modal analysis technique is proposed. The experimental application shows that the method holds promise for identification of mechanical system for free as well as for forced response.

- A truly simple identification scheme is proposed for the modal parameters, due to the straightforward application of SOBI to the measured data.
- A seemingly robust criterion has been developed for the selection of reliable sources. The use of stabilization charts, which always require a great deal of expertise, is therefore avoided. In addition, the selection of a model order, a common issue for conventional modal analysis techniques such as SSI, is not necessary.
- Compared to SSI, the computation load is very reduced, which makes the method a potential candidate for online modal analysis.

A possible limitation of the method is that sensors should always be chosen in number greater or equal to the number of active modes. This will be addressed in subsequent studies.

Acknowledgments. The author F. Poncelet is supported by a grant from the Walloon Government (Contract N 516108-VAMOSNL). The author G. Kerschen is supported by a grant form the Belgian National Fund for Scientific Research (FNRS).

References

1. Kerschen, G., Poncelet, F., Golinval, J.C.: Physical interpretation of independent component analysis in structural dynamics. Mechanical System and Signal Processing 21, 1561–1575 (2007)
2. Poncelet, F., Kerschen, G., Golinval, J.C., Verhelst, D.: Output-only modal analysis using blind source separation techniques. Mechanical Systems and Signal Processing. Corrected Proof. Available online (January 5, 2007) (in press)
3. Van Overschee, P., De Moor, B.: Subspace Identification for Linear Systems: Theory, Implementation, Applications. Kluwer Academic Publishers, Dordrecht (1996)
4. Ibrahim, S.R., Mikulcik, E.C.: A time domain modal vibration test technique. Shock and Vibration Bulletin 43, 21–37 (1973)
5. Peeters, B., Van Der Auweraer, H., Guillaume, P.: The PolyMAX frequency domain method: a new standard for modal parameter estimation. Shock and Vibration 11, 395–409 (2004)
6. Ewins, D.J.: Modal Testing: Theory, Practice and Application, 2nd edn. Research Studies Press LTD, Hertfordshire (2000)
7. Feeny, B.F., Kappagantu, R.: On the physical interpretatino of proper orthogonal modes in vibrations. Journal of Sound and Vibration 211, 607–616 (1998)
8. Kerschen, G., Golinval, J.C.: Physical interpretatino of the proper orthogonal modes using the singular value decompostion. Journal of Sound and Vibration 249, 849–865 (2002)
9. Chelidze, D., Zhou, W.: Smooth orthogonal decomposition-based vibration mode identification. Journal of Sound and Vibration 292, 461–473 (2006)
10. Comon, P.: ICA: a new concept? Signal Processing 36, 287–314 (1994)
11. Belouchrani, A., Abed-Meraim, K., Cardoso, J.F., Molines, E.: A BSS technique using 2nd-order statistics. IEEE Trans. on Signal Processing 45, 434–444 (1997)
12. James, G.H., Carne, T.G., Lauffer, J.P.: The natural excitation technique for modal parameter extraction from operating wind turbines. SAND92-1666 UC–261 (1993)

Image Similarity Based on Hierarchies of ICA Mixtures

Arturo Serrano, Addisson Salazar, Jorge Igual, and Luis Vergara

Universidad Politécnica de Valencia, Departamento de Comunicaciones, Camino de Vera s/n, 46022 Valencia, Spain
asalazar@dcom.upv.es, jigual@dcom.upv.es

Abstract. This paper presents a novel algorithm to build hierarchies from independent component analyzer mixtures and its application to image similarity measure. The hierarchy algorithm composes an agglomerative (bottom-up) clustering from the estimated parameters (basis vectors and bias terms) of the ICA mixture. Merging at different levels of the hierarchy is made using the Kullback-Leibler distance between clusters. The procedure is applied to merge similar patches on a natural image, to group different images of an object, and to create hierarchical levels of clustering from images of different objects. Results show suitable image hierarchies obtained by clustering from basis functions to higher-level structures.

1 Introduction

Independent component analyzers (ICA) mixture models were introduced in [1] considering a source model switching between Laplacian and bimodal densities. Recently this model has been relaxed using generalised exponential sources [2], self-similar areas as a mixture of Gaussians sub-features [3], and sources with non-Gaussian structures recovered by a learning algorithm using Beta divergence [4]. Real applications of those works span: separation of eye-movement artefacts from EEG recordings, separating 'back-ground' brain tissue, fluids and tumours in fMRI images, and the separation of voices and background music in conversations.

It is well known that local edge detectors can be extracted from natural scenes by standard ICA algorithms as Infomax [5], or fastICA [6] or new approaches as Linear Multilayer ICA [7]. In addition there is neurophysiological evidence that suggest relation of primary visual cortex activities with the detection of edges, and some theoretical dynamic models of abstraction process from visual cortex to higher-level abstraction has been proposed [8].

The contribution of this paper is to provide a new algorithm to process the parameters of ICA mixtures in order to obtain hierarchical structures from the basis function level (edges) to higher levels of clustering. Particularly the algorithm is applied to image analysis obtaining promising results in discerning object similarity and suitable levels of hierarchies by processing image patches. This kind of feedforward process would suggest some relation with abstraction. The algorithm is agglomerative and uses the symmetric Kullback-Leibler distance [9] to select the grouping of the clusters at each level.

M.E. Davies et al. (Eds.): ICA 2007, LNCS 4666, pp. 786–793, 2007.

2 Hierarchy of ICA Mixtures

2.1 Estimation of the ICA Mixture Parameters

In the ICA mixture model, the observation vectors $\mathbf{x}$ are modelled as the result of applying a linear transformation $\mathbf{A}_k$ ($\mathbf{W}_k = \mathbf{A}_k^{-1}$ is the filter matrix) to a vector $\mathbf{s}_k$ (sources), whose elements are independent random variables, plus a bias vector $\mathbf{b}_k$, for all the classes $C_k, (k = 1 \ldots K$ number of ICAs$)$. The probability of every available observation vector can be separated into the contributions due to every class.

An iterative learning algorithm based on maximum-likelihood estimation (MLE) is used to adapt the parameters of the ICA mixtures, i.e., the basis functions and the bias terms of each class, using gradient ascent [1]. To estimate the probability density function of the sources different priors could be used as Laplacian [1] or non-parametric densities [10].

2.2 Agglomerative Clustering

From the estimated ICA mixture parameters, a procedure that follows a bottom-up agglomerative scheme for merging the mixtures was developed.

The conditional probability density of $\mathbf{x}$ for cluster $C_k^h, k = 1, 2, ..., K - h + 1$ in level $h = 1, 2, ..., K$ is $p(\mathbf{x}/C_k^h)$. At the first level, $h = 1$, it is modelled by K ICA mixtures, i.e., $p(\mathbf{x}/C_k^1)$ is:

$$p(\mathbf{x}/C_k^1) = \left|\det \mathbf{A}_k^{-1}\right| p(\mathbf{s}_k), \ \mathbf{s}_k = \mathbf{A}_k^{-1}(\mathbf{x} - \mathbf{b}_k) \tag{1}$$

At each consecutive level, two clusters are merged according to some minimum distance measure until we reach at level $h = K$ only one cluster.

As distance measure we use the symmetric Kullback-Leibler distance between the ICA mixtures. It is defined for the clusters u, v by:

$$D_{\mathrm{KL}}(C_u^h, C_v^h) = \int p(\mathbf{x}/C_u^h) \log \frac{p(\mathbf{x}/C_u^h)}{p(\mathbf{x}/C_v^h)} d\mathbf{x} + \int p(\mathbf{x}/C_v^h) \log \frac{p(\mathbf{x}/C_v^h)}{p(\mathbf{x}/C_u^h)} d\mathbf{x} \tag{2}$$

For level $h = 1$, from (2), we can obtain (we write $p_{\mathbf{x}_u}(\mathbf{x}) = p(\mathbf{x}/C_u^1)$ and omit the superscript $h = 1$ for brevity):

$$D_{\mathrm{KL}}(C_u, C_v) = D_{\mathrm{KL}}(p_{\mathbf{x}_u}(\mathbf{x}) // p_{\mathbf{x}_v}(\mathbf{x})) = \int p_{\mathbf{x}_u}(\mathbf{x}) \log \frac{p_{\mathbf{x}_u}(\mathbf{x})}{p_{\mathbf{x}_v}(\mathbf{x})} d\mathbf{x} + \int p_{\mathbf{x}_v}(\mathbf{x}) \log \frac{p_{\mathbf{x}_v}(\mathbf{x})}{p_{\mathbf{x}_u}(\mathbf{x})} d\mathbf{x} \tag{3}$$

where, imposing the independence hypothesis and supposing that both clusters have the same number of sources M for simplicity (assuming that sources follow the same model and the data they draw are on the same space):

$$p_{\mathbf{x}_u}(\mathbf{x}) = \frac{\prod_{i=1}^{M} p_{s_{u_i}}(s_{u_i})}{|\det \mathbf{A}_u|}, \quad s_{u_i} = \mathbf{A}_{u_i}^{-1}(\mathbf{x} - \mathbf{b}_{u_i})$$
$$p_{\mathbf{x}_v}(\mathbf{x}) = \frac{\prod_{j=1}^{M} p_{s_{v_j}}(s_{v_j})}{|\det \mathbf{A}_v|}, \quad s_{v_j} = \mathbf{A}_{v_j}^{-1}(\mathbf{x} - \mathbf{b}_{v_j}) \tag{4}$$

The pdf of the sources is approximated by a non-parametric kernel-based density for both clusters:

$$p_{s_{u_i}}(s_{u_i}) = \sum_{n=1}^{N} a e^{-\frac{1}{2}\left(\frac{s_{u_i} - s_{u_i}(n)}{h}\right)^2}, \; p_{s_{v_j}}(s_{v_j}) = \sum_{n=1}^{N} a e^{-\frac{1}{2}\left(\frac{s_{v_j} - s_{v_j}(n)}{h}\right)^2} \tag{5}$$

where again for simplicity we have assumed the same kernel function for all the clusters, with the parameters a, h and number of samples N adapted to each cluster. Note that this corresponds to a Gaussian mixture model where the number of Gaussians is maximum (one for every observation) and the weights are equal. Reducing to standard mixture of Gaussians does not help in order to compute the Kullback-Leibler distance because there is not analytical solution to it. Therefore, we prefer to maintain the non parametric approximation of the pdf in order to model more complex distributions than a mixture of a small finite number of Gaussians.

The symmetric Kullback-Leibler distance between the clusters u, v can be expressed such as:

$$D_{\mathrm{KL}}(p_{\mathbf{x}_u}(\mathbf{x}) // p_{\mathbf{x}_v}(\mathbf{x})) = -H(\mathbf{x}_u) - H(\mathbf{x}_v) - \int p_{\mathbf{x}_u}(\mathbf{x}) \log p_{\mathbf{x}_v}(\mathbf{x}) d\mathbf{x} - \int p_{\mathbf{x}_v}(\mathbf{x}) \log p_{\mathbf{x}_u}(\mathbf{x}) d\mathbf{x} \tag{6}$$

where $H(\mathbf{x})$ is the entropy, defined as $H(\mathbf{x}) = -E[\log p_{\mathbf{x}}(\mathbf{x})]$. To obtain the distance, we have to calculate the entropy for both clusters and the cross-entropy terms $E_{\mathbf{x}_v}\left[\log p_{\mathbf{x}_u}(\mathbf{x})\right], E_{\mathbf{x}_u}\left[\log p_{\mathbf{x}_v}(\mathbf{x})\right]$.

The entropy for the cluster u can be calculated through the entropy of the sources of that cluster considering the linear transformation of the random variables and their independence (4):

$$H(\mathbf{x}_u) = \sum_{i=1}^{M} H(s_{u_i}) + \log|\det \mathbf{A}_u| \tag{7}$$

The entropy of the sources can not be analytically calculated. Instead, we can obtain a sample estimate $\hat{H}(s_{u_i})$ using the training data. Denote the i-th source obtained for the cluster u by $\{s_{u_i}(1), s_{u_i}(2), \ldots, s_{u_i}(Q_i)\}$. The entropy can be approximated as follows:

$$\hat{H}(s_{u_i}) = -\hat{E}\left[\log p_{s_{u_i}}(s_{u_i})\right] = -\frac{1}{Q_i} \sum_{n=1}^{Q_i} \log p_{s_{u_i}}(s_{u_i}(n)),$$
$$p_{s_{u_i}}(s_{u_i}(n)) = \sum_{l=1}^{N} a e^{-\frac{1}{2}\left(\frac{s_{u_i}(n) - s_{u_i}(l)}{h}\right)^2} \tag{8}$$

The entropy of $H(\mathbf{x}_v)$ is obtained analogously:

$$H(\mathbf{x}_v) = \sum_{i=1}^{M} H(s_{v_i}) + \log\left|\det \mathbf{A}_v\right| \approx \sum_{i=1}^{M} \hat{H}(s_{v_i}) + \log\left|\det \mathbf{A}_v\right|$$
$$\hat{H}(s_{v_i}) = -\frac{1}{Q_i}\sum_{n=1}^{Q_i} \log p_{s_{v_i}}(s_{v_i}(n)),\ p_{s_{v_i}}(s_{v_i}(n)) = \sum_{l=1}^{N} a e^{-\frac{1}{2}\left(\frac{s_{v_i}(n)-s_{v_i}(l)}{h}\right)^2} \tag{9}$$

with $\hat{H}(s_{v_i})$ defined analogously to (8). Following the same procedure for *j*-th source we can estimate $\hat{H}(s_{v_j})$.

Once the entropy is computed, we have to obtain the cross-entropy terms. After some operations and considering the relationships $\mathbf{x} = \mathbf{A}_u\mathbf{s}_u + \mathbf{b}_u$, $\mathbf{x} = \mathbf{A}_v\mathbf{s}_v + \mathbf{b}_v$ and thus $\mathbf{s}_v = \mathbf{A}_v^{-1}\left(\mathbf{A}_u\mathbf{s}_u + \mathbf{b}_u - \mathbf{b}_v\right)$, the independence of the sources, and that the samples for clusters u, v follow the corresponding distribution $\left\{s_{u_i}(1), s_{u_i}(2), \ldots, s_{u_i}(Q_i)\right\}$, $i = 1,\ldots,M$, $\left\{s_{v_j}(1), s_{v_j}(2), \ldots, s_{v_j}(Q_j)\right\}$, $j = 1,\ldots,M$; we can estimate,

$$\hat{H}(\mathbf{s}_v, s_{u_i}) = \frac{1}{\prod_{i=1}^{M} Q_i} \cdot \sum_{s_{v_1}=1}^{Q_M} \cdots \sum_{s_{v_M}=1}^{Q_1} \log \sum_{n=1}^{N} a e^{-\frac{1}{2}\left(\frac{\left[\mathbf{A}_u^{-1}\left(\mathbf{A}_v\mathbf{s}_v + \mathbf{b}_v - \mathbf{b}_u\right)\right]_i - s_{u_i}(n)}{h}\right)^2} \tag{10}$$

and $\hat{H}(\mathbf{s}_u, s_{v_j})$ defined analogously to (10), with $\mathbf{s}_u = \mathbf{A}_u^{-1}\left(\mathbf{A}_v\mathbf{s}_v + \mathbf{b}_v - \mathbf{b}_u\right)$.

Using the terms obtained above, we can estimate the symmetric Kullback-Leibler distance between the clusters u, v:

$$D_{\mathrm{KL}}(p_{\mathbf{x}_u}(\mathbf{x}) // p_{\mathbf{x}_v}(\mathbf{x})) = -\sum_{i=1}^{M} \hat{H}(s_{u_i}) - \sum_{j=1}^{M} \hat{H}(s_{v_j}) - \sum_{i=1}^{M} \hat{H}(\mathbf{s}_v, s_{u_i}) - \sum_{j=1}^{M} \hat{H}(\mathbf{s}_u, s_{v_j}) \tag{11}$$

As we can observe, the similarity between clusters depends not only on the similarity between the bias term, but the similarity between the distributions and the mixing matrices.

Once the distances are obtained for all the clusters, the two clusters with minimum distance are merged in level $h = 2$. This is repeated in every step of the hierarchy until we reach one cluster in the level $h = K$. To merge cluster in level h we can calculate the distances from the distances of level $h-1$. Suppose that from level $h-1$ to h the clusters C_u^{h-1}, C_v^{h-1} are merged in cluster C_w^h. Then, the density for the merged cluster at level h is:

$$p_h(\mathbf{x}/C_w^h) = \frac{p_{h-1}(C_u^{h-1})p_{h-1}(\mathbf{x}/C_u^{h-1}) + p_{h-1}(C_v^{h-1})p_{h-1}(\mathbf{x}/C_v^{h-1})}{p_{h-1}(C_u^{h-1}) + p_{h-1}(C_v^{h-1})} \tag{12}$$

where $p_{h-1}(C_u^{h-1})$, $p_{h-1}(C_v^{h-1})$ are the priors or proportions of the clusters u, v at level $h-1$. The rest of terms are the same in the mixture model at level h that at level $h-1$. The only difference from one level to the next one in the hierarchy is that there is one cluster less and the prior for the new cluster is the sum of the priors of its components and the density the weighted average of the densities that are merged to form it. Therefore, the estimation of the distance at level h can be done easily starting from the distances at level $h-1$ and so on until level $h=1$. Consequently, we can calculate the distances at level h from a cluster C_z^h to a merged cluster C_w^h obtained by the agglomeration of clusters C_u^{h-1}, C_v^{h-1} at level $h-1$ as the distance to its components weighted by the mixing proportions:

$$D_h(p_h(\mathbf{x}/C_w^h)//p_h(\mathbf{x}/C_z^h)) = \frac{p_{h-1}(C_u^{h-1})\cdot D_{h-1}(p_{h-1}(\mathbf{x}/C_u^{h-1})//p_{h-1}(\mathbf{x}/C_z^{h-1}))}{p_{h-1}(C_u^{h-1})+p_{h-1}(C_v^{h-1})} + \frac{p_{h-1}(C_v^{h-1})\cdot D_{h-1}(p_{h-1}(\mathbf{x}/C_v^{h-1})//p_{h-1}(\mathbf{x}/C_z^{h-1}))}{p_{h-1}(C_u^{h-1})+p_{h-1}(C_v^{h-1})}. \quad (13)$$

3 Application on Image Data

ICA can be used to analyze image patches as a linear superposition of basis functions. Those vectors have been related with the detection of borders in natural images [5]. Therefore basis functions have a physical relation with objects and they can be used to measure the similarity between objects based on ICA decomposition. In image patches decomposition, the set of independent components is larger than what can be estimated at one time, and what we get at one time is an arbitrarily chosen subset [11]. Nevertheless ICA has been applied successfully in several image applications [1].

3.1 Object Similarity

For the hierarchical classification of images of objects, the COIL-100 database was used [12]. The database consists of different views of objects over a dark background. The method applied to preprocess the images was this. The images were converted to greyscale, and grouped in different views in order to obtain several images to train up to three classes per object. From each image, patches of 8 by 8 pixels were randomly taken to estimate the basis function previous a whitening process, with a reduction to 40 components. A total of 1000 patches per object were extracted [5].

The basis functions of each class were then calculated with the ICA mixtures algorithm, considering supervision, and using the Laplacian prior to estimate the source pdfs. Fig. 1 shows the 40 basis functions of six classes corresponding to different views of two objects. The basis functions of Fig. 1a correspond to a box with a label inscribed whereas Fig. 1b corresponds to an apple. We can observe the similarity between the functions of each object and differences, for instance, the lower frequency in the pattern corresponding to a natural object versus the frequency in the pattern of a more artificial object.

The same data were used to measure the distance between classes estimating the symmetric Kullback-Leibler distance from the mixture matrices calculated previously, as we explain in Section 2. Distances reveal that basis functions allow finding the similarity (short distances) between classes corresponding to the same object (intra-object), whereas distances are much longer between classes of different objects (inter-object), see Table 1.

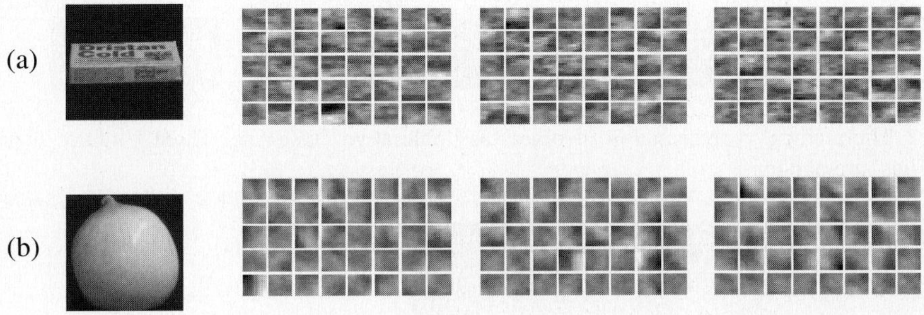

Fig. 1. Two groups of basis functions corresponding to two different objects. Basis functions at top are from a little box and basis functions at bottom are from an apple.

Table 1. Mean distances inter-object and intra-object of Fig. 1

Object	box (a)	apple (b)
box (a)	12.89	114.90
apple (b)	114.90	13.81

Additionally, experiments in order to create a hierarchical classification of objects were performed. Thus, patches were sampled from a large number of objects, some of them very similar among themselves. A hierarchical representation was then created applying the agglomerative clustering algorithm. Fig. 2 shows an example of classification of eight objects, with three main kinds of objects. The tree outlined by the dendrogram positively shows grouping of objects based on similarity content, and suitable similarities between 'families' of objects, e.g., cars were more alike with cans than with apples.

3.2 Natural Images

The proposed algorithm was applied to natural images in order to obtain a bottom-up structure merging several zones of an image. Fig. 3 shows an image with 9 zones, some of then clearly different and others more or less similar each other. Dendrogram of Fig. 3 shows how the zones are merged from the patches. It shows two broad kinds of basis functions that correspond to the part of the image that mainly contains portions of sky, and those zones that correspond to patches where there is a predominant portion of stairs (high frequency).

The dendrogram also shows the distances at which the clusters are merged, it can be used as a similarity measure of the zones of the image. The bottom zones are merged at low distances due to the high similarity in borders.

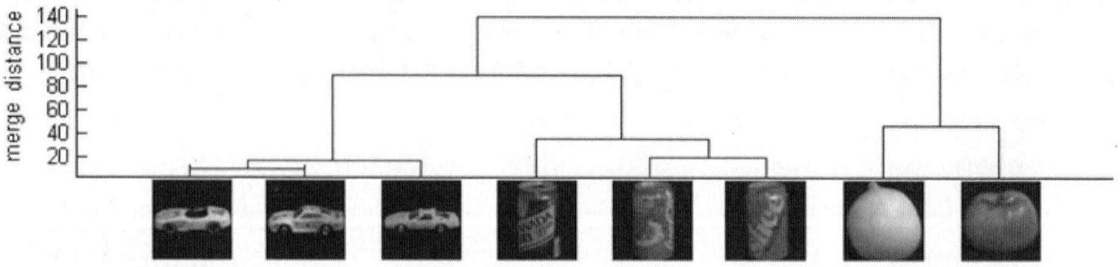

Fig. 2. Hierarchical representation of object agglomerative clustering. Three kinds of object 'families' are obtained.

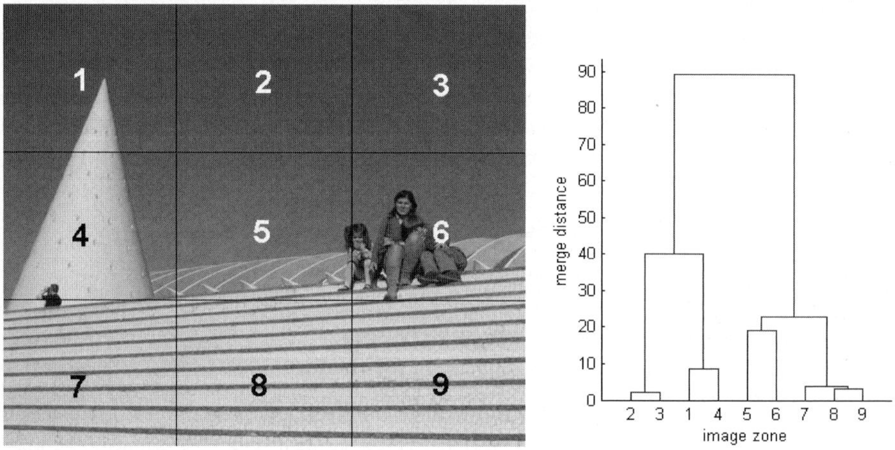

Fig. 3. (Left) Image divided in nine zones. (Right) Hierarchical representation of the zones of the image based on basis functions similarity. It shows two broad groups of zones.

4 Conclusions

The new algorithm for hierarchical ICA mixtures uses the mixture matrices to calculate distances between the distributions of the independent sources based on a symmetric Kullback-Leibler distance. The estimation of the source pdfs is made using a non-parametric kernel-based approach allowing adaptation to several kinds of densities. Clusters are merged using a bottom-up strategy defining hierarchical levels creating higher-level structures.

Results of the hierarchical algorithm application demonstrated its suitability to process image data. Image content similarity between objects based on ICA basis functions allows learning an organization of objects in higher-levels of abstraction where the more separated hierarchical levels more different the objects. Experiments with natural images showed application to image segmentation based on similarity of the different zones. The application of the procedure could be extended to unsupervised or semi-supervised classification of images in order to discover meaningful hierarchical levels.

Many potential applications of the procedure could be approached as defect classification in non-destructive testing. Hierarchical levels would represent concepts as material condition, kind of defect, defect orientation, or defect dimension [13].

Acknowledgements

This work was supported by Spanish Administration under grant TEC 2005-01820.

References

1. Lee, T.W., Lewicki, M.S., Sejnowski, T.J.: ICA mixture models for unsupervised classification of non-gaussian classes and automatic context switching in blind signal separation. IEEE Trans. on Patt. Analysis and Machine Intelligence 22(10), 1078–1089 (2000)
2. Penny, W.D., Roberts, S.: Mixtures of independent component analyzers. In: Dorffner, G., Bischof, H., Hornik, K. (eds.) ICANN 2001. LNCS, vol. 2130, pp. 527–534. Springer, Heidelberg (2001)
3. Choudrey, R., Roberts, S.: Variational Mixture of Bayesian Independent Component Analysers. Neural Computation 15(1), 213–252 (2003)
4. Mollah, N.H., Minami, M., Eguchi, S.: Exploring Latent Structure of Mixture ICA Models by the Minimum ß-Divergence Method. Neural Computation 18(1), 166–190 (2006)
5. Bell, A.J., Sejnowski, T.J.: The Independent Components of natural scenes are edge filters. Vision Research 37(23), 3327–3338 (1997)
6. Van Hateren, J.H., van der Shaaf, A.: Independent component filters of natural images compared with simple cells in primary visual cortex. Proceedings of Royal Society of London: B 265, 359–366 (1998)
7. Matsuda, Y., Yamaguchi, K.: Linear multilayer ICA generating hierarchical edge detectors. Neural Computation 19(1), 218–230 (2007)
8. Lee, T.S., Mumford, D.: Hierarchical Bayesian inference in the visual cortex. Journal of the Optical Society of America A 20(7), 1434–1448 (2003)
9. Mackay, D.J.: Information theory, inference, and learning algorithms. Cambridge University Press, Cambridge (2004)
10. Vergara, L., Salazar, A., Igual, J., Serrano, A.: Data Clustering Methods Based on Mixture of Independent Component Analyzers. In: Proc. of ICA Research Network International Workshop, ICArn, Liverpool, pp. 127–130 (2006)
11. Hyvarinen, A., Hoyer, P.O., Inki, M.: Topographic independent component analysis. Neural Computation 13(7), 1527–1558 (2001)
12. Nene, S.A., Nayar, S.K., Murase, H.: Columbia Object Image Library (COIL-100), Technical Report CUCS-006-96 (February 1996)
13. Salazar, A., Unió, J.M., Serrano, A., Gosalbez, J.: Neural networks for defect detection in non-destructive evaluation by sonic signals. In: IWANN 2007, LNCS, vol. 4507, pp. 631–638. Springer, Heidelberg (2007)

Text Clustering on Latent Thematic Spaces: Variants, Strengths and Weaknesses

Xavier Sevillano, Germán Cobo, Francesc Alías, and Joan Claudi Socoró

GPMM - Grup de Recerca en Processament Multimodal
Enginyeria i Arquitectura La Salle. Universitat Ramon Llull
Quatre Camins, 2. 08022 - Barcelona, Spain
{xavis,gcobo,falias,jclaudi}@salle.url.edu

Abstract. Deriving a thematically meaningful partition of an unlabeled text corpus is a challenging task. In comparison to classic term-based document indexing, the use of document representations based on latent thematic generative models can lead to improved clustering. However, determining *a priori* the optimal indexing technique is not straightforward, as it depends on the clustering problem faced and the partitioning strategy adopted. So as to overcome this indeterminacy, we propose deriving a consensus labeling upon the results of clustering processes executed on several document representations. Experiments conducted on subsets of two standard text corpora evaluate distinct clustering strategies based on latent thematic spaces and highlight the usefulness of consensus clustering to overcome the optimal document indexing indeterminacy.

1 Introduction

The increasingly growing number of unlabeled digital text documents available calls for the development of automatic tools, such as document clustering systems, capable of organizing unlabeled document collections thematically.

However, when facing any text clustering problem, practitioners must blindly make several decisions that largely condition the quality of the clustering results, such as selecting *i)* the document indexing technique used for representing the documents (including its dimensionality), *ii)* the clustering strategy employed for partitioning the data, or *iii)* the number of clusters to be found.

As regards the former aspect, it is a commonplace that the application of feature transformations can improve clustering results significantly [6]. Thus, the influence of distinct indexing techniques on the performance of text clustering systems has been analyzed elsewhere [14, 17].

In this context, latent thematic generative models, such as Independent Component Analysis [8], Latent Semantic Analysis [1] or Non-negative Matrix Factorization [10] constitute an interesting option for finding document projections on low dimensional spaces where clustering can be conducted more efficiently and effectively than in the original, high-dimensional term vector space. For this reason, this work presents an extensive comparison between document clustering strategies based on the aforementioned latent thematic generative models.

M.E. Davies et al. (Eds.): ICA 2007, LNCS 4666, pp. 794–801, 2007.

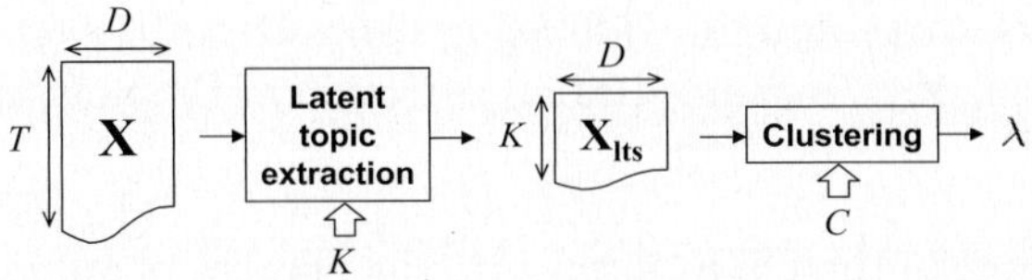

Fig. 1. Extracting K latent thematic sources for clustering D documents into C clusters

Unfortunately, the conclusions drawn from this study make it difficult to generalize, as no indexing technique guarantees a universally superior performance across distinct clustering problems, giving rise to what we call the *data representation dependence effect* [13]. In order to overcome this indeterminacy, we present a strategy based on consensus clustering which allows to set text clustering practitioners free from the obligation of selecting a single document representation blindly, while still obtaining reasonably good clustering results.

This paper is organized as follows: section 2 describes two clustering strategies based on latent thematic models, and section 3 presents several experiments regarding these strategies. Section 4 describes the consensus clustering proposal for overcoming the data representation dependence effect and presents related experiments. Finally, the conclusions of our work are discussed in section 5.

2 Clustering Strategies Based on Latent Thematic Models

The rationale behind latent thematic generative models establishes an analogy between the blind source separation (BSS) problem and the generation of text collections: the mixing of several latent random topics —or *thematic sources*— gives rise to a set of D documents (equivalent to the *observations* in the BSS scenario) [8, 18]. Therefore, as the goal of text clustering systems is to group documents into C thematically homogeneous clusters, recovering those latent thematic sources may lead to improved clustering.

Figure 1 illustrates the application of latent topic extraction methods for document clustering. Let $\mathbf{X}$ denote the $T \times D$ term-by-document matrix representing the document corpus in the original high-dimensional vector space model (where T stands for the vocabulary size) [11]. The extraction of $K << T$ latent thematic sources yields a K-dimensional representation of the D documents in the latent thematic space, $\mathbf{X_{lts}}$. Subsequently, the documents are grouped into C clusters by applying a clustering process on $\mathbf{X_{lts}}$, which yields the D-dimensional *labeling vector* λ, whose i-th component λ_i contains a numeric label identifying the cluster the i-th document is assigned to, i.e. $\lambda_i \in \{1, 2, \ldots, C\}, \forall i = \{1, 2, \ldots D\}$.

As regards the use of the latent thematic space document representation $\mathbf{X_{lts}}$ for clustering, two main approaches can be followed: firstly, in what we call *latent source driven clustering* (LSDC), the latent topics are not used as document representations, but rather as cluster membership indicators (i.e. documents are assigned to a specific cluster depending on their maximally active latent source in

$\mathbf{X_{lts}}$) [7, 9, 12, 18]. And secondly, following what we call *latent thematic feature clustering* (LTFC), $\mathbf{X_{lts}}$ is deemed as a projection of the documents onto a new feature space, where a clustering algorithm is applied [13, 14, 17]. Despite sharing a common conceptual background, both approaches differ significantly from a practical viewpoint. In the LSDC approach, the number of latent topics retrieved must be tuned to match the desired number of clusters, i.e. $K = C$. Moreover, the clustering stage simply boils down to a cluster assignment based on finding the maxima in $\mathbf{X_{lts}}$ [7, 9]. In contrast, in LTFC, the number of clusters to be found is a parameter affecting the clustering stage rather than the latent topic extraction process, i.e. K is not necessarily equal to C. Furthermore, the grouping of the documents is conducted in this case by applying a standard partitioning algorithm on the latent thematic feature space where $\mathbf{X_{lts}}$ lies, which usually increases the computational cost of the clustering stage in comparison with the LSDC approach.

3 Experiments on LSDC and LTFC

The following experiments compare three well-known unsupervised latent thematic generative models applied to the document clustering task –following both the LSDC and LTFC approaches–, namely: Latent Semantic Analysis (LSA), Independent Component Analysis (ICA) and Non-negative Matrix Factorization (NMF)[1].

Two single-class balanced clustering problems have been created upon subsets of the standard miniNewsgroups [4] and OHSUMED [3] document collections. Table 1 summarizes the main aspects of both corpora: the predefined number of categories C –which is assumed to be known throughout all the experiments–, the number of documents D, their vocabulary size T and the average number of terms per document, T_d.

Experimental results are evaluated by comparing the labeling vector λ delivered by the clustering processes with the original labeling of the documents (enclosed in vector κ) in terms of the Normalized Mutual Information ($\phi^{(\mathrm{NMI})}$), a similarity measure ranging in value from 0 to 1 [16]:

$$\phi^{(\mathrm{NMI})}(\kappa,\lambda) = \frac{\sum_{h=1}^{C}\sum_{l=1}^{C} n_{h,l} \log\left(\frac{D \cdot n_{h,l}}{n_h^{(\kappa)}\, n_l^{(\lambda)}}\right)}{\sqrt{\left(\sum_{h=1}^{C} n_h^{(\kappa)} \log\frac{n_h^{(\kappa)}}{D}\right)\left(\sum_{l=1}^{C} n_l^{(\lambda)} \log\frac{n_l^{(\lambda)}}{D}\right)}} \tag{1}$$

where $n_h^{(\kappa)}$ is the number of objects in cluster h according to κ, $n_l^{(\lambda)}$ is the number of objects in cluster l according to λ, $n_{h,l}$ denotes the number of objects in cluster h according to κ as well as in group l according to λ [16].

[1] The ICA document representation is created by applying a version of the FastICA algorithm that maximizes skewness [7], using LSA (implemented by singular value decomposition) for pre-whitening and dimension reduction. The NMF-based document representation is created by applying a mean square reconstruction error minimization algorithm from NMFPACK [5].

Table 1. Document corpora subsets description

Corpus	C	D	T	T_d
miniNewsgroups	6	600	3735	99
OHSUMED	11	1100	4705	120

Table 2. Latent source driven clustering results using LSA, ICA and NMF

Corpus	**LSA**	**ICA**	**NMF**
miniNewsgroups	$.342 \pm .018$	$.472 \pm .001$	$\mathbf{.474 \pm .038}$
OHSUMED	$.188 \pm .004$	$\mathbf{.229 \pm .001}$	$.213 \pm .013$

3.1 Latent Source Driven Clustering

This experiment evaluates the performance of LSA, ICA and NMF in the context of latent source driven clustering (i.e. as many latent thematic sources as desired clusters -C- are extracted and their value is used as a cluster membership indicator). The mean values and standard deviations of the $\phi^{(\mathrm{NMI})}$ scores obtained across 10 independent runs of this experiment are presented in table 2.

For the miniNewsgroups corpus, the best result in average is achieved by NMF-based LSDC (in boldface in table 2). In contrast, ICA is the latent source extraction method that yields the best results in the OHSUMED experiment. Note that, for both document collections, LSA is clearly outperformed by ICA and NMF, which suggests that the latent topics recovered by these two techniques are better aligned with the real thematic contents of the documents.

So as to illustrate this fact, figure 2 compares the six latent thematic sources extracted by means of LSA and NMF with the categories in the miniNewsgroups corpus (100 documents per topic). Notice that the NMF_1, NMF_5 and NMF_6 latent sources clearly identify the `comp.graphics`, `sci.crypt` and `misc.forsale` topics –in sharp contrast with the less defined pattern of the LSA latent sources–, which somehow justifies the superiority of NMF in this experiment.

3.2 Latent Thematic Feature Clustering

This experiment compares the results of applying four state-of-the-art clustering algorithms –group average agglomerative clustering (AC), graph-based clustering (GC), direct clustering (DC) and repeated bisecting clustering (BC)[2]– on *i)* the original T-dimensional term vector space and *ii)* on LSA, ICA and NMF latent thematic feature spaces of dimensionalities ranging from $K = 2$ to $K = 50$.

Figure 3 presents the results of this experiment, averaged across 10 independent runs. An interesting observation is that LTFC delivers better results than LSDC. For the miniNewsgroups corpus (figure 3a), the best clustering results are obtained using the original term-based representation except with agglomerative

[2] Implementations provided by the CluTo clustering package, available online at `http://glaros.dtc.umn.edu/gkhome/views/cluto`

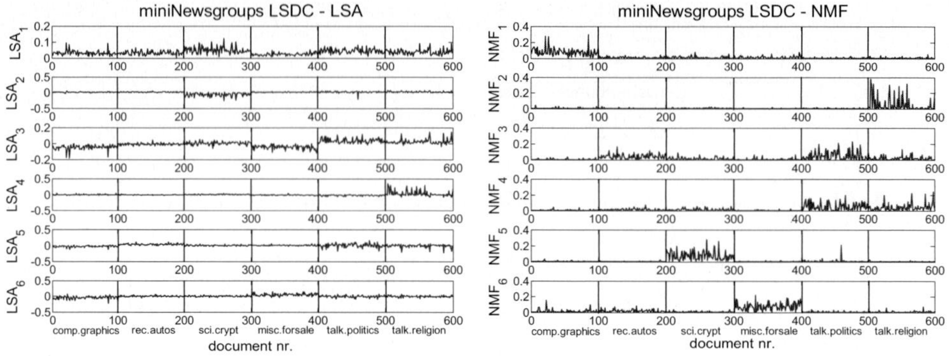

Fig. 2. Latent thematic sources extracted by LSA and NMF on the miniNewsgroups corpus

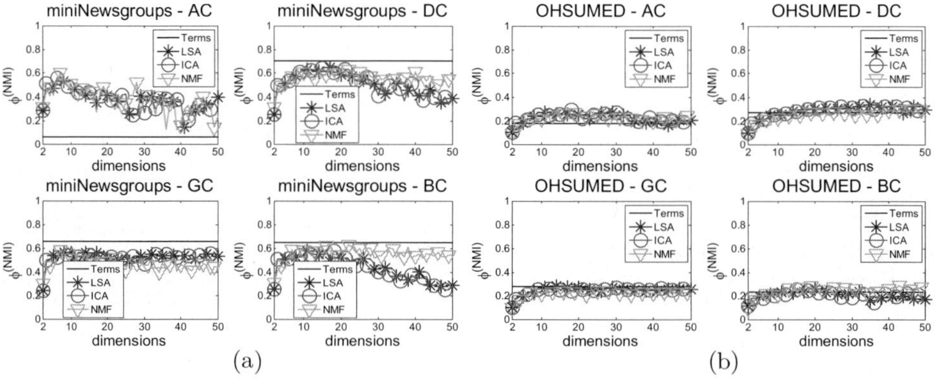

Fig. 3. Latent thematic feature clustering results for both corpora (miniNewsgroups on the left, OHSUMED on the right) applying four clustering strategies on latent feature spaces of varying dimensionality

clustering. In this case, the highest $\phi^{(\mathrm{NMI})}$ is achieved when AC is conducted on a 7-dimensional NMF feature space. Note that fairly diverse results are obtained across the latent thematic feature space dimensionality range. Moreover, notice that nearly identical clustering results are yielded by all the clustering strategies when operating on the LSA and ICA feature spaces, which is in clear contrast with the situation found in LSDC (see table 2). In contrast, for the OHSUMED corpus (figure 3b), none of the clustering algorithms achieves the best results when operating on the term-based representation, and there exists an indeterminacy regarding both the optimal type of feature and the dimensionality of the latent thematic feature space. Most of all, it is to note that these optimality properties depend on the particular partitioning strategy employed and the clustering problem faced. These results suggest that it is not possible to claim for the universal superiority of any latent thematic model. So as to overcome this indeterminacy, we propose applying a consensus clustering strategy.

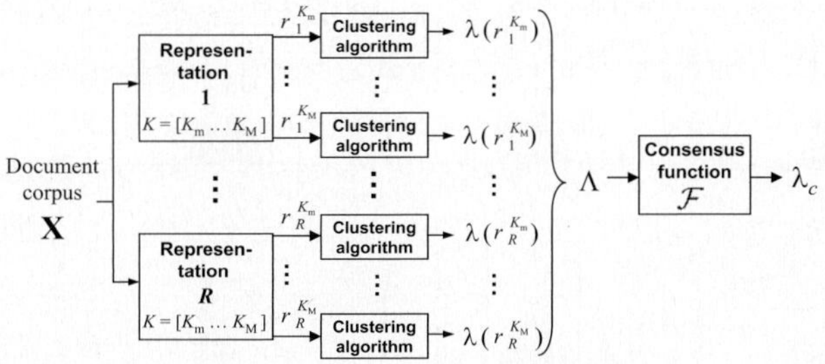

Fig. 4. Construction of a cluster ensemble $\mathbf{\Lambda}$ upon R document representations of dimensionalities $K = \{K_{\rm m}, \ldots, K_{\rm M}\}$, and subsequent creation of a consensus clustering λ_c by means of a consensus function $\mathcal{F}$

4 Consensus Clustering

Being the unsupervised counterpart of classifier committees, consensus clustering is the task of creating a consensus labeling λ_c by applying a consensus function $\mathcal{F}$ on a cluster ensemble $\mathbf{\Lambda}$ that collects the labelings output by several partitioning processes. Typical applications of cluster ensembles include clustering reuse besides distributed and robust clustering [16]—the aim being, in this latter case, that λ_c approximates or even improves the best clustering in the ensemble.

Following this principle, we propose overcoming the data representation dependence effect by building a cluster ensemble upon a set of R document indexing techniques, and subsequently constructing a consensus labeling λ_c (see figure 4).

Several consensus functions have been proposed in the literature [2, 15, 16], and their general rationale is the application of cluster identification plus voting strategies across the labelings in the ensemble. For the sake of space, the reader is referred to [16] for a general introduction to consensus clustering.

4.1 Experiments on Consensus Clustering

This experiment analyzes the ability of consensus clustering for overcoming the uncertainty regarding optimal document indexing. To that effect, the 148 labelings delivered by each clustering algorithm in the LTFC experiment (section 3.2) have been collected into four cluster ensembles $\mathbf{\Lambda}$ (one per clustering strategy). Then, we have applied three state-of-the-art consensus functions on these ensembles: Cluster-Similarity Partitioning Algorithm (CSPA), Hyper-Graph Partitioning Algorithm (HGPA) and Meta-Clustering Algorithm (MCLA) [16].

It is important to note that the robustness to the data representation dependence effect is proportional to the closeness between the $\phi^{(\mathrm{NMI})}$ of the consensus labeling and the *maximum* $\phi^{(\mathrm{NMI})}$ of the labelings in the cluster ensemble. For this reason, figure 5 presents –through a $\phi^{(\mathrm{NMI})}$ histogram (averaged across 10 experiment runs)– a visual comparison between the labelings in the ensemble and the labelings derived by each consensus function. Complementarily, table 3 presents the relative $\phi^{(\mathrm{NMI})}$ differences between the consensus labelings and

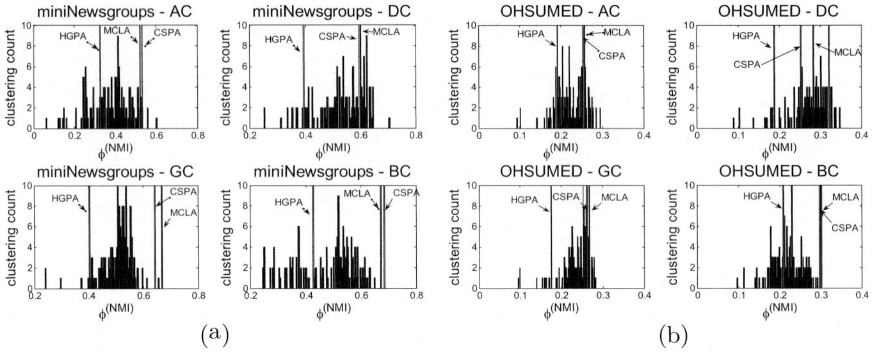

Fig. 5. $\phi^{(\text{NMI})}$ histogram of the cluster ensembles and consensus clustering for both corpora (miniNewsgroups on the left, OHSUMED on the right) applying four clustering strategies and three consensus functions

Table 3. Relative $\phi^{(\text{NMI})}$ differences (mean value $\pm$ standard deviation) between the consensus labelings and the average and best individual labelings in the ensemble

Corpus	**miniNewsgroups**			**OHSUMED**		
$\mathcal{F}$	CSPA	HGPA	MCLA	CSPA	HGPA	MCLA
$\Delta\phi^{(\text{NMI})}$ w.r.t. ALE	+31.4%	-16 ± 15%	$+31.6 \pm 0.1$%	+13.2%	-18.6 ± 6.2%	$+17.2 \pm 0.1$%
$\Delta\phi^{(\text{NMI})}$ w.r.t. BLE	−6%	-40 ± 10%	-5.7 ± 0.1%	−11.6%	-36.5 ± 4.8%	-8.4 ± 0.1%

the average and best labelings in the ensemble (referred to as ALE and BLE, respectively). It can be observed that, in comparison to HGPA, the CSPA and MCLA consensus functions *i)* achieve notable robustness to the data representation dependence effect (i.e. the consensus labelings derived by CSPA and MCLA are closer to the BLE, being even better in some cases –see miniNewsgroups-GC or OHSUMED-BC in figure 5), and *ii)* show a more stable behaviour (i.e. smaller standard deviations) across the 10 experiment runs –in fact, CSPA yields repetitive results when operating on the same data.

5 Conclusions

One of the main difficulties encountered by text clustering practitioners is the uncertainty regarding the selection of a document indexing technique. In this context, document representations based on latent thematic generative models such as LSA, ICA or NMF can be advantageous with respect to the original term-based representation. However, the optimality of a single indexing technique is not a universal property, which gives rise to the so-called data representation dependence effect. Consensus clustering constitutes a reliable strategy to overcome this problem, and it can easily be extended to deal with additional indeterminacies, e.g. the one regarding the optimal clustering algorithm [13].

References

[1] Deerwester, S., Dumais, S.T., Furnas, G.W., Landauer, T.K., Harshman, R.: Indexing by Latent Semantic Analysis. J. American Society Information Science 6(41), 391–407 (1990)
[2] Fred, A., Jain, A.K.: Combining Multiple Clusterings Using Evidence Accumulation. IEEE Trans. on Pattern Analysis and Machine Intelligence 27(6), 835–850 (2005)
[3] Hersh, W., Buckley, C., Leone, T., Hichman, D.: OHSUMED: an interactive retrieval evaluation and new large text collection for research. In: Proc. of the 17th ACM SIGIR Conference, pp. 192–201 (1994)
[4] Hettich, S., Bay, S.D.: The UCI KDD Archive. University of California at Irvine, Dept. of Information and Computer Science (1999), http://kdd.ics.uci.edu
[5] Hoyer, P.O.: Non-Negative Matrix Factorization with Sparseness Constraints. J. Machine Learning Research 5, 1457–1469 (2004)
[6] Jain, A.K., Murty, M.N., Flynn, P.J.: Data Clustering: a Survey. ACM Computing Surveys 31(3), 264–323 (1999)
[7] Kabán, A., Girolami, M.: Unsupervised Topic Separation and Keyword Identification in Document Collections: a Projection Approach. Dept. of Computing and Information Systems, University of Paisley. Technical Report Nr. 10 (2000)
[8] Kolenda, T., Hansen, L.K., Sigurdsson, S.: Independent Components in Text. In: Girolami, M. (ed.) Advances in Independent Component Analysis, pp. 241–262. Springer, Heidelberg (2000)
[9] Kolenda, T.: Clustering text using Independent Component Analysis. Inst. of Informatics and Mathematical Modelling, Tech. University of Denmark. T.R (2002)
[10] Lee, D.D., Seung, H.S.: Learning the Parts of Objects by Non-Negative Matrix Factorization. Nature 401, 788–791 (1999)
[11] Sebastiani, F.: Machine Learning in Automated Text Categorisation. ACM Computing Surveys 34(1), 1–47 (2002)
[12] Sevillano, X., Alías, F., Socoró, J.C.: Reliability in ICA-Based Text Classification. In: Proc. of the 5th Intl. Conference on Independent Component Analysis and Blind Signal Separation, pp. 1210–1217 (2004)
[13] Sevillano, X., Cobo, G., Alías, F., Socoró, J.C.: A Hierarchical Consensus Architecture for Robust Document Clustering. In: Proc. of the 29th ECIR Conference, pp. 741–744 (2007)
[14] Shafiei, M., Wang, S., Zhang, R., Milios, E., Tang, B., Tougas, J., Spiteri, R.: A Systematic Study of Document Representation and Dimension Reduction for Text Clustering. Technical Report CS-2006-05. Dalhousie University (2006)
[15] Siersdorfer, S., Sizov, S.: Restrictive Clustering and Metaclustering for Self-Organizing Document Collections. In: Proc. of the 27th ACM SIGIR Conference, pp. 226–233 (2004)
[16] Strehl, A., Ghosh, J.: Cluster Ensembles – A Knowledge Reuse Framework for Combining Multiple Partitions. J. Machine Learning Research 3, 583–617 (2002)
[17] Tang, B., Shepherd, M., Milios, E., Heywood, M.I.: Comparing and Combining Dimension Reduction Techniques for Efficient Text Clustering. In: Proc. of the Intl. Workshop on Feature Selection for Data Mining, pp. 17–26 (2005)
[18] Xu, W., Liu, X., Gong, Y.: Document Clustering Based on Non-Negative Matrix Factorization. In: Proc. of the 26th ACM SIGIR Conference, vol. 2, pp. 267–273 (2003)

Top-Down Versus Bottom-Up Processing in the Human Brain: Distinct Directional Influences Revealed by Integrating SOBI and Granger Causality

Akaysha C. Tang[1], Matthew T. Sutherland[1], Peng Sun [2], Yan Zhang[3], Masato Nakazawa[1], Amy Korzekwa[1], Zhen Yang[1], and Mingzhou Ding[3]

[1] University of New Mexico, Department of Psychology, Albuquerque, NM, USA
[2] University of New Mexico, Department of Electrical Engineering, Albuquerque, NM, USA
[3] University of Florida, J. Crayton Pruitt Family Department of Biomedical Engineering, Gainesville, FL, USA
akaysha@unm.edu, msuther@unm.edu, pengsun@unm.edu, zhyang @unm.edu, maliszt@unm.edu, akorzekw@unm.edu, mding@bme.flu.edu

Abstract. Top-down and bottom-up processing are two distinct yet highly interactive modes of neuronal activity underlying normal and abnormal human cognition. Here we characterize the dynamic processes that contribute to these two modes of cognitive operation. We used a blind source separation algorithm called second-order blind identification (SOBI [1]) to extract from high-density scalp EEG (128 channels) two components that index neuronal activity in two distinct local networks: one in the occipital lobe and one in the frontal lobe. We then applied Granger causality analysis to the SOBI-recovered neuronal signals from these two local networks to characterize feed-forward and feedback influences between them. With three repeated observations made at least one week apart, we show that feed-forward influence is dominated by alpha while feedback influence is dominated by theta band activity and that this direction-selective dominance pattern is jointly modulated by situational familiarity and demand for visual processing.

Keywords: electroencephalogram, second-order blind identification (SOBI), coherence, Granger causality, top-down, bottom-up, feed-forward, feedback.

1 Introduction

Second-order blind identification (SOBI) [1] is an emerging signal processing technique that can be used to facilitate source analysis from high-density EEG. Similar to other ICA algorithms that have been applied to EEG data [2], [3], SOBI can be used to isolate and remove ocular artifact [4]. In our laboratory, we have conducted extensive investigations to demonstrate the utility of SOBI in aiding source analysis from high-density EEG. Specifically, we have shown that: (1) SOBI can correctly recover known noise sources (noisy sensors and artificially injected noise at

M.E. Davies et al. (Eds.): ICA 2007, LNCS 4666, pp. 802–809, 2007.

known electrodes) and known neuronal sources (SI activation by median nerve stimulation) [5]; (2) SOBI can increase signal to noise ratios leading to improved performance in single-trial ERP classification [6]; (3) SOBI can recover neuronal sources whose activations are correlated [7]; (4) SOBI can recover neuronal sources using EEG collected when the brain is in its default mode (i.e., the "resting" state) [8]; (5) SOBI can recover neuronal sources during free viewing of continuous streams of visual information [9]; and (6) SOBI can recover weak neuronal signals that temporally overlap with much stronger signals (e.g. signals associated with ipsilateral activation of primary somatosensory cortex) [10].

In this paper, we set out to achieve three goals. First, we seek to provide further validation for SOBI recovered neuronal sources by investigating whether the same neuronal sources can be recovered from repeated EEG measures that are obtained days and weeks apart. Second, we combine SOBI with Granger causality analysis to show distinct patterns of theta(θ)/alpha(α) contributions in the feed-forward and feedback influences between the frontal and occipital cortices. Third, we investigate how such asymmetrical influence between the frontal and occipital cortices is modulated by sensory processing and by situational familiarity.

2 Methods

Eight right-handed subjects volunteered to participate in the present study. All subjects were free of any history of neurological or psychological disorders. The experimental procedures were conducted in accordance with the Human Research Review Committee at the University of New Mexico. Each subject was tested in three sessions at Week 0, Week 1, and Week 4 or later. Up to 7 min of continuous 128-channel EEG data were collected at 1000 Hz during: (1) eyes-closed "resting"; (2) eyes-open "resting"; (3) video-viewing (a silently played nature video); (4) listening to only the audio track of the video; and (5) forming mental images of scenes from the video. This paper limits the discussion to conditions 1-3.

SOBI was applied to the continuous EEG data $\mathbf{x}(t)$, across all conditions to extract the continuous time course of activation from two types of neuronal components--- an anterior (A) and a posterior (P) component. For details on SOBI application, see [5]. Briefly, SOBI recovers the underlying sources, $\mathbf{s}(t)$, by minimizing the sum squared cross-correlations between $\mathbf{s}_i(t)$ and $\mathbf{s}_j(t + \tau)$, across all pairs of sources and across multiple time delays, τs. A subset of SOBI-recovered components can be verified as neuronal sources via source localization using a forward model (e.g. BESA 5.0) [3]. Here we focused our analysis on two such neuronal components that correspond to focal regions within the frontal and occipital lobes.

Feed-forward (FF) and feedback (FB) influences were quantified by Granger causality between the two components, reflecting long-distance *directional* influences between the frontal and occipital cortices. Granger causality analysis was carried out on the continuous time courses, $\mathbf{s}_i(t)$, from the selected A and P components according to methods detailed in [11], [12]. As Granger causality can be decomposed into its frequency content, we computed Granger causality spectrum and measured power within the θ (4-7 Hz) and α (8-14 Hz) bands using a moving window of 30-sec with

5-sec increments. Power in the θ and α bands from the A and P components were also computed as indicators of synchronization within the local networks.

3 Results

Reliable Extraction and Identification of Neuronal Components from Repeated Measures made Weeks Apart. In all 8 subjects, across all 3 sessions, we were able to recover SOBI components that corresponded to two distinct neuronal sources, one localized to a rather focal region within the frontal cortex, in or near anterior cingulate cortex (ACC) and the other to focal regions within the occipital lobe (occipital gyrus). Repeated-measure ANOVA revealed no statistically significant differences in the location of the corresponding ECD models across the 3 recording sessions. As no session-to-session difference was found, the averaged locations across the 3 sessions are shown in Fig. 1. ECDs for each of the 8 subjects are superimposed in the figure revealing a tight clustering of ECDs across subjects. This result demonstrates that SOBI can reliably recover components that correspond to anatomically well defined brain regions even when the recording sessions were made weeks apart.

It is important to emphasize that the recovery of these two neuronal sources was achieved without imposing constraints of fixation or use of event-related stimulation paradigms. Instead, subjects were allowed to freely move or blink their eyes as needed during the recording conditions. No segment of the EEG data was excluded prior to SOBI application. These unique features of SOBI processing have non-trivial

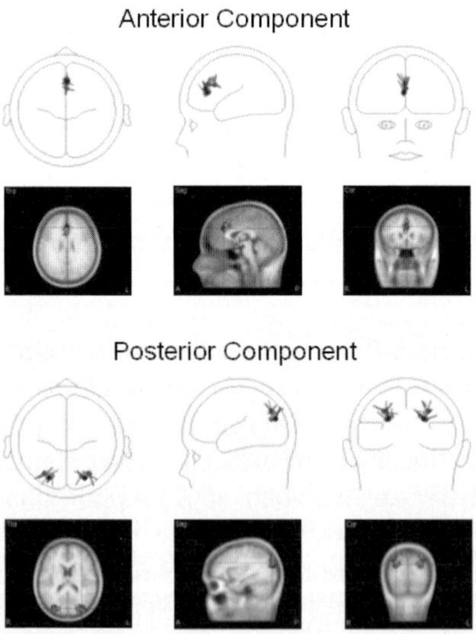

Fig. 1. Equivalent current dipole (ECD) locations for the SOBI recovered A and P components

implications for the study of mental disorders and the study of early development or aging where subjects are often unable to conform to typical experimental constraints.

Theoretically, this result implies that SOBI's ability to recover anatomically well-defined neuronal sources does not depend upon the use of an event-related stimulation paradigm. Thus, fast electrical brain activity in the default mode [13] can be investigated in terms of neuronal signals originating from specific, focal cortical areas. In comparison to default mode activity revealed by fMRI, the default mode activity revealed with SOBI and EEG will offer millisecond temporal resolution, allowing for the characterization of default mode brain dynamics within a new temporal domain.

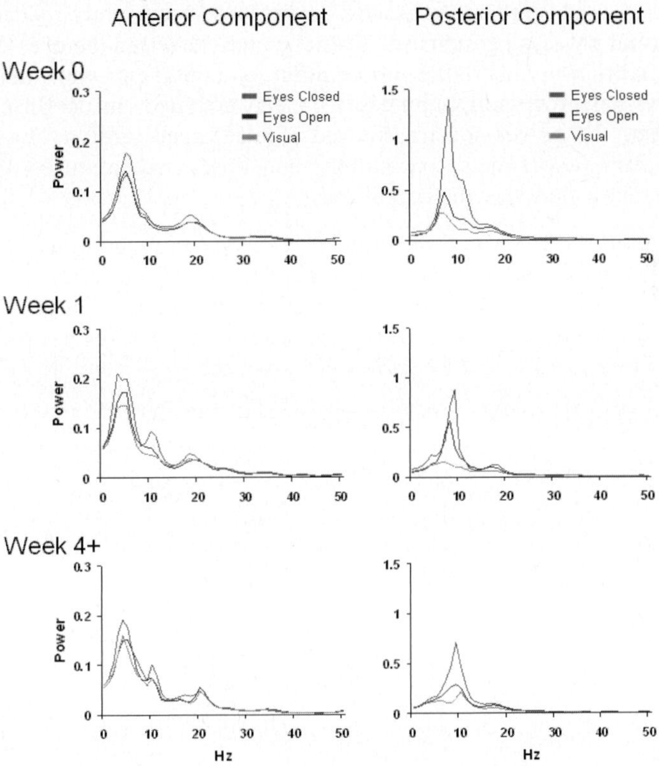

Fig. 2. Median power spectra of two SOBI-neuronal components as a function of repeated exposures to the same experimental situation. Session 1: week 0; Session 2: week 1; Session 3: week 4+.

Local Network Synchrony Shows Distinct Patterns of Change across 3 Repeated Exposures to the same Experimental Situation. For each of the 3 recording sessions, power spectra from the component time courses were computed for ~5-min segments during which the subjects had their eyes-closed (red), eyes-open (blue), or viewed a nature video (green), respectively (Fig. 2).

The anterior component had peak power within the θ band while the posterior component had peak power within the α band, indicated by a significant main effect of Region in the θ-to-α ratio ($F[1,7] = 52.12$, $p < 0.001$, partial $\eta^2 = 0.88$). This is consistent with the well established fact that the posterior and anterior parts of the brain are major sources of α and θ generators, respectively.

Power spectra in these two components were differentially modulated by sessions and experimental conditions [interaction effect: Region x Session (contrast coefficients: 1, -1, 0) x Condition (1, 0, -1), $F(1,7) = 3.52$, $p = 0.05$, 1-tailed, partial $\eta^2=0.33$)]. For the P component, the power spectra revealed a systematic effect of session and experimental condition. Across the 3 repeated exposures to the same experimental conditions, peak α power decreased as the testing situation became increasingly familiar.

Across the 3 experimental conditions, the highest peak α power was associated with the eyes closed condition and the peak α power was successively reduced when the demand for visual processing increased from the eyes-closed to the eyes-open and video-viewing conditions. This latter observation is consistent with the known observation that visual processing suppresses α band activity. In contrast, for the anterior component, the power spectra showed a relative insensitivity to repeated exposures to the same experimental conditions and little modulation by the eyes-closed, eyes-open, and video-viewing conditions.

Differential Modulation of θ/α Contribution to Feed-Forward and Feedback Influences by Situational Familiarity and Visual Processing. FF(posterior-to-anterior) and FB (anterior-to-posterior) influences were measured by Granger causality in the θ and α band activity separately. FF and FB Granger causality measures were plotted as a function of time (Fig. 3). For the FF influence, when the eyes were closed, α band

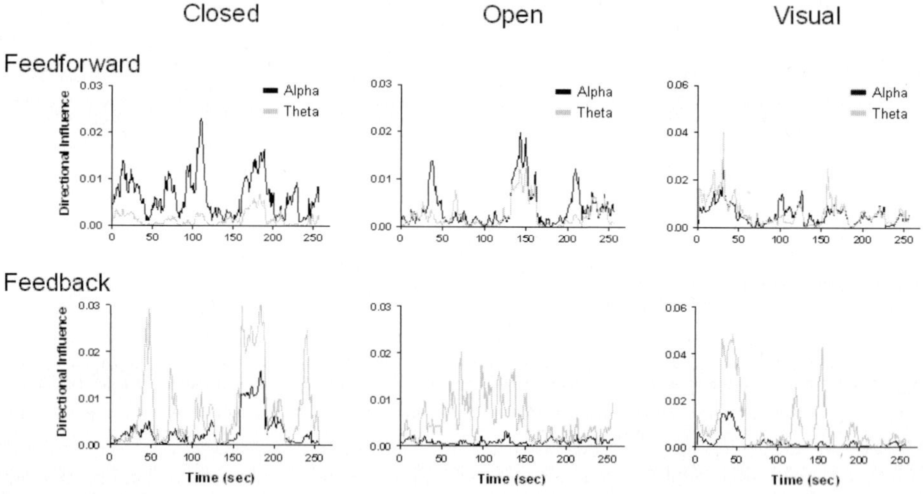

Fig. 3. Theta dominance over alpha in the anterior-to-posterior feedback (lower) influence and its reversal in the posterior-to-anterior feed-forward influence (upper) from a single-subject

activity clearly dominated as indicated by the α waveforms (black) having greater area underneath the curve than the θ waveforms (grey). This α dominance was clearly reduced when the eyes were open and was further reduced to nearly non-existent when the subjects viewed a video. For the FB influence, the pattern of α dominance over θ was reversed showing uniform θ dominance over α across all 3 experimental conditions.

Using the area underneath the curve as a dependent measure, we summarize results from all 8 subjects across all 3 recording sessions in Fig. 4. To determine whether θ and α band activity contribute differentially to the FF and FB influences and how such differential contributions are modulated by situational familiarity and sensory processing, we performed an ANOVA on the θ/α ratio.

The θ/α ratio differed significantly between the FF and FB influences with a greater ratio for FB influence than for the FF influence (main effect of Direction, $F[1,7] = 34.64$, $p < 0.001$, partial $\eta^2 = 0.83$), i.e. a θ dominance in FB influence. This can be seen by the higher measures for the θ band activity than the α band activity for the FB influences in most of the 9 conditions and clear reversal or reduction of this θ dominance in the FF influence (Fig. 4).

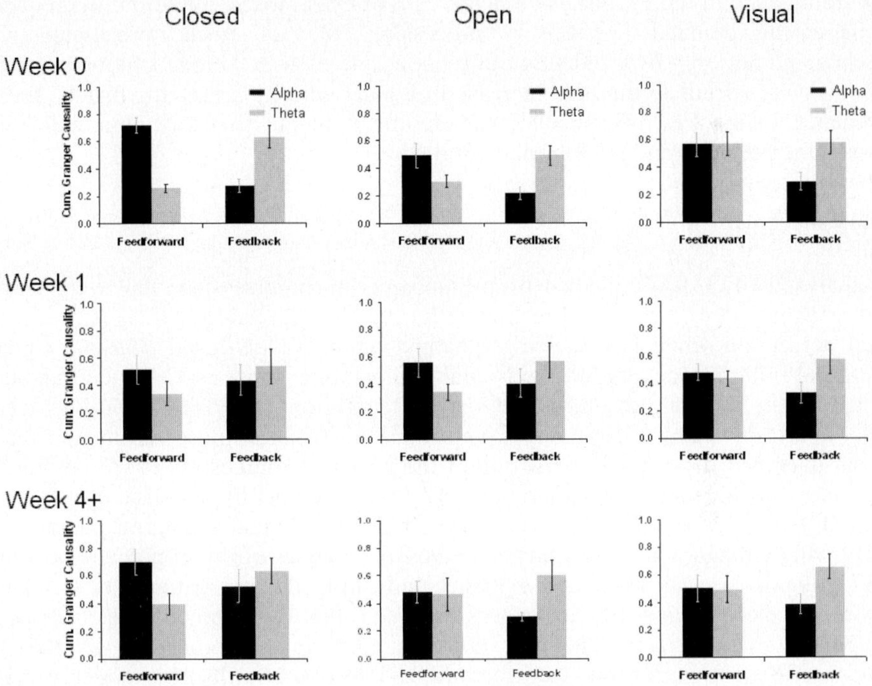

Fig. 4. Cumulative Granger Causality (area underneath the curve in Fig. 3) in the θ and α band as a function of situational familiarity (repeated sessions) and a function of visual processing (eyes-closed, eyes-open, video-viewing).

This reversal of θ dominance from FF and FB influences was significantly modulated by the familiarity of the situation [Direction x Session (contrast coefficients: 1, -1, 0), $F(1,7) = 11.97$, $p=0.005$, 1-tailed, partial $\eta^2=0.63$]. The reversal is more prominent when the situation was novel (Week 0) than when it became more familiar (Week 1 and 4+). This is best seen in the case of eyes-closed condition. The magnitude of reversal is clearly reduced from Week 0 in comparison to Week 4+.

For the eyes-open condition, the θ dominance was reversed in Week 0 and 1 and reduced in Week 4+. For the video-viewing condition, the reversal of θ dominance does not appear to be influenced by the increasing situational familiarity. These patterns indicate that the FF/FB contrast is dependent upon the amount of visual information processing involved. When the subjects were engaged in visual perception during video-viewing, θ dominance in the FB influence and θ-α balance in the feed-forward influence are maintained across recording sessions. This visual processing-dependent effect is supported by a significant 3-way interaction [Direction x Session (1, -1, 0) x Condition (1, 0, -1), $F(1,7) = 7.13$, $p = 0.02$, 1-tailed partial η^2=0.63].

Within Week0 when the recording situation was novel (which is comparable to most studies that do not deal with the issue of task familiarity), θ dominance in the FB influence was maintained despite varying demand for visual processing. In contrast, the α dominance in the FF influence in the case of eyes-closed condition was reduced by increasing demand for sensory processing. In fact, visual processing was accompanied not only by a reduction in α but an increase in θ band activity in the FF influence. We speculate that this increase in θ band activity serves to "match" the θ-dominance in the FB influence to mediate the dynamic two-way communication between the posterior and anterior parts of the brain.

4 Discussion

We analyzed EEG data collected from 8 subjects in three sessions that were weeks apart, each including a period of resting with eyes-closed, resting with eyes-open, and visual perception while free viewing a nature video. We extracted neuronal signals from focal brain regions within the frontal and occipital lobes and showed that such extraction can be achieved under free viewing conditions and from recordings made weeks apart. As many intervening events must have taken place during the inter-session intervals, the reliable extraction of the neuronal sources raises the possibility that such a wide range of variations may be overcome by the use of SOBI in longitudinal experimental designs necessary for developmental and aging studies.

Applying Granger causality analysis to the time courses of the frontal and occipital SOBI components, we presented evidence indicating distinct patterns of θ/α band activity in the FF and FB influences between the two components, with a θ dominance characterizing the FB influence and an α dominance in the FF influence. By comparing the feed-forward and feedback influences under varying degrees of situational familiarity (sessions) and under conditions of varying degrees of visual processing (eyes-closed, eyes-open, and video viewing), we presented evidence that the balance in θ-α band activity between the FF and FB influences is modulated by two factors. First, situational familiarity can reduce the degrees of θ and α dominance in the FB and FF influences, respectively (as in the case of eyes-closed). Second, the amount of sensory processing increases the θ band contribution and decreases α band contribution to FF influence but has little effect on FB influence. Finally, situational familiarity and sensory processing jointly determine the θ-α balance. Increasing familiarity and increasing visual processing both *increases* θ band contribution to *FF* influence. In contrast, for *FB* influences, increasing familiarity *decreases* θ band contribution when there is little demand for visual processing (eyes-closed) and has no effect on θ band contribution when there is high demand for visual processing (visual).

Together, these findings demonstrate a novel non-invasive approach to the assessment of top-down and bottom-up influences in the human brain. These findings may particularly benefit those clinicians and researchers who are interested in how bottom-up and top-down influences interact in both diseased and normal brains. Future work will extend this analysis to networks involving more functionally distinct brain regions.

Acknowledgments. Supported by a grant from Sandia National Labs to A.C.T.

References

1. Belouchrani, A., Abed-Meraim, K., Cardoso, J.-F., Moulines, E.: A blind source separation technique using second-order statistics. IEEE Trans. Signal Process. 5, 434–444 (1997)
2. Bell, A.J., Sejnowski, T.J.: An information-maximization approach to blind separation and blind deconvolution. Neural Computation 1, 1129–1159 (1995)
3. Hyvarinen, A., Oja, E.: A fast fixed-point algorithm for independent component analysis. Neural Computation 9, 1483–1492 (1997)
4. Joyce, C.A., Gorodnitsky, I.F., Kutas, M.: Automatic removal of eye movement and blink artifacts from EEG data using blind component separation. Psychophysiology 41, 313–325 (2004)
5. Tang, A.C., Sutherland, M.T., McKinney, C.J.: Validation of SOBI components from high density EEG. NeuroImage 25, 539–553 (2005)
6. Tang, A.C., Sutherland, M.T., Wang, Y.: Contrasting single-trial ERPs between experimental manipulations: Improving differentiability by blind source separation. NeuroImage 29, 335–346 (2006)
7. Tang, A.C., Liu, J.Y., Sutherland, M.T.: Recovery of correlated neuronal sources from EEG: The good and bad ways of using SOBI. NeuroImage 28, 507–519 (2005)
8. Sutherland, M.T., Tang, A.C.: Blind source separation can recover systematically distributed neuronal sources from resting EEG. In: Eurasip Proceedings of the Second International Symposium on Communications, Control, and Signal Processing (ISCCSP 2006), Marrakech, Morocco (March 13-15, 2006), http://www.eurasip.org/content/Eusipco/isccsp06/defevent/papers/cr1307.pdf
9. Tang, A.C., Sutherland, M.T., McKinney, C.J., Liu, J.Y., Wang, Y., Parra, L.C., Gerson, A.D., Sajda, P.: Classifying single-trial ERPs from visual and frontal cortex during free viewing. In: IEEE Proceedings of the International Joint Conference on Neural Networks (IJCNN 2006), Vancouver, BC, Canada, July 16-21, pp. 1376–1383. IEEE Computer Society Press, Los Alamitos (2006)
10. Sutherland, M.T., Tang, A.C.: Reliable detection of bilateral activation in human primary somatosensory cortex by unilateral median nerve stimulation. NeuroImage 33, 1042–1054 (2006)
11. Ding, M.: Short-window spectral analysis of cortical event-related potentials by adaptive multivariate autoregressive modeling: Data preprocessing, model validation, and variability assessment. Biological Cybernetics 83, 35–45 (2000)
12. Ding, M., Chen, Y., Bressler, S.L.: Granger causality: Basic theory and application to neuroscience. In: Winterhalder, M., Schelter, B., Timmer, J. (eds.) Handbook of Time Series Analysis, Wiley, Chichester (2006)
13. Gusnard, D.A., Raichle, M.E.: Searching for a baseline: Functional imaging and the resting human brain. Nature Review Neuroscience 102, 685–694 (2001)

Noisy Independent Component Analysis as a Method of Rotating the Factor Scores

Steffen Unkel and Nickolay T. Trendafilov

Department of Statistics, Faculty of Mathematics & Computing,
The Open University, Walton Hall, Milton Keynes, MK7 6AA, United Kingdom
{S.Unkel,N.Trendafilov}@open.ac.uk

Abstract. Noisy independent component analysis (ICA) is viewed as a method of factor rotation in exploratory factor analysis (EFA). Starting from an initial EFA solution, rather than rotating the loadings towards simplicity, the factors are rotated orthogonally towards independence. An application to Thurstone's box problem in psychometrics is presented using a new data matrix containing measurement error. Results show that the proposed rotational approach to noisy ICA recovers the components used to generate the mixtures quite accurately and also produces simple loadings.

Keywords: Independent component analysis, Exploratory factor analysis, Factor rotation, Factor scores, Gradient projection algorithm.

1 Introduction

The key difference between EFA and noisy ICA is that in the latter model the common factors are assumed to be both independent and non-normal. Since only second-order statistics are analyzed, the loadings in the EFA model can only be estimated up to an orthogonal rotation. Hence, EFA is not able to separate linear mixtures into their independent components. In contrast, the non-normality of the common factors allows ICA to perform blind source separation. The rotational redundancy of the EFA model is removed, using supplementary information not contained in the sample covariance or correlation matrix.

Several authors use EFA merely for quasi-sphering the data before doing an ICA analysis [8,14]. ICA can be considered as a method for factor rotation seeking a rotation matrix that maximizes the independence between the common factors [6,7]. This connection was first explored in [13], where a varimax-based criterion was proposed to implement noise-free ICA. The current paper implements noisy ICA from an EFA perspective by exploiting this link with factor rotation. Starting from an initial EFA solution, the predicted factor scores are rotated orthogonally towards independence. This is done using an appropriate rotation criterion and an orthogonal rotation algorithm. Recently, the rotational approach to ICA proposed here was introduced for the noise-free case using methods from principal components analysis (PCA) [10]. In the sequel, this rotational approach is applied for studying the less developed noisy version of ICA.

M.E. Davies et al. (Eds.): ICA 2007, LNCS 4666, pp. 810–817, 2007.

2 Independent Non-normal Factor Analysis Model

Consider the following linear latent variable model in which all variables are assumed to be measured at least on an interval scale:

$$\mathbf{x} = \boldsymbol{\mu} + \boldsymbol{\Lambda}\mathbf{f} + \mathbf{u} \ , \tag{1}$$

where $\mathbf{x} \in \mathbb{R}^{p\times 1}$ is a random vector of manifest variables with mean vector $\boldsymbol{\mu}$, $\mathbf{f} \in \mathbb{R}^{k\times 1}$ is a random vector of $k \ll p$ latent variables called common factors, $\boldsymbol{\Lambda} \in \mathbb{R}^{p\times k}$ is a matrix of fixed coefficients referred to as factor loadings, and $\mathbf{u} \in \mathbb{R}^{p\times 1}$ is a random vector of latent variables called unique factors. In EFA, the choice of k is subject to some limitations, which will not be discussed here [5]. The loading matrix $\boldsymbol{\Lambda}$ is required to have full column rank. Assume that $\mathrm{E}(\mathbf{f}) = \mathbf{0}$. Furthermore, let $\mathbf{u} \sim \mathcal{N}_p(\mathbf{0}, \boldsymbol{\Psi})$, where $\boldsymbol{\Psi}$ is assumed a positive definite diagonal matrix. Finally, suppose that $\mathrm{E}(\mathbf{ff}') = \mathbf{I}_k$ and $\mathrm{E}(\mathbf{fu}') = \mathbf{0}_{k\times p}$. Thus, all the factors are uncorrelated with one another and the variances of the common factors equal unity. Using these assumptions, the model (1) represents an EFA model with orthogonal (uncorrelated) common factors [5].

The idea of model (1) is that the common factors account for the covariance structure among the set of manifest variables, while each unique factor corresponds to that portion of a particular manifest variable which cannot be accounted for by the common factors. As such, a unique factor contains the specificity of that variable as well as errors in measurement or noise.

In EFA, it is often covenient to assume that not only $\mathbf{u}$ but also $\mathbf{f}$ and hence $\mathbf{x}$ are multinormally distributed. This assumption is usually made for purposes of statistical inference [15]. The elements of $\mathbf{f}$ being normally distributed and uncorrelated are thus statistically independent random variables. Unlike EFA, ICA assumes that the k common factors are both mutually independent and non-normal or at least all but one non-normal [3]. With this key difference, the model (1) is similar to a noisy ICA model [7].

Given a multivariate sample of n independent observations on $\mathbf{x} = (x_1, \ldots, x_p)'$, the k-factor model (1) can be written as

$$\mathbf{X} = \mathbf{M} + \mathbf{F}\boldsymbol{\Lambda}' + \mathbf{U}, \tag{2}$$

where $\mathbf{X} = (\mathbf{x}_1, \ldots, \mathbf{x}_p) \in \mathbb{R}^{n\times p}$ is the observed data matrix in which $\mathbf{x}_j = (x_{1j}, \ldots, x_{nj})'$ $(j = 1, \ldots, p)$, $\mathbf{M} = \mathbf{1}_n\boldsymbol{\mu}' \in \mathbb{R}^{n\times p}$ is the matrix of location parameters, and $\mathbf{F} = (\mathbf{f}_1, \ldots, \mathbf{f}_k) \in \mathbb{R}^{n\times k}$ and $\mathbf{U} = (\mathbf{u}_1, \ldots, \mathbf{u}_p) \in \mathbb{R}^{n\times p}$ denote the unknown matrices of factor scores of the k common factors and unobserved values for the p unique factors on n observations, respectively.

The aim of noisy ICA based on model (2) is to recover $\mathbf{F}$ from $\mathbf{X}$ alone without knowing $\boldsymbol{\Lambda}$, $\boldsymbol{\Psi}$, and the distributions of the common factors [4]. This problem can be transformed into a specific EFA task.

The matrix $\mathbf{M}$ can easily be estimated by $\hat{\mathbf{M}}$ which consists of n constant rows $\bar{\mathbf{x}}'$, where $\bar{\mathbf{x}} = (\bar{x}_{.1}, \ldots, \bar{x}_{.p})'$ denotes the $(p \times 1)$ sample mean vector with $\bar{x}_{.j} = \frac{1}{n}\sum_{i=1}^{n} x_{ij}$ $(j = 1, \ldots, p)$ being the sample means for each variable.

Assume without changing notation that $\mathbf{X}$ has been mean-corrected and that the column vectors of $\mathbf{X}$ are standardized to unit variance. The model in (1) and the assumptions imply the following model correlation structure, $\mathbf{R}$:

$$\mathbf{R} = \mathbf{\Lambda}\mathbf{\Lambda}' + \mathbf{\Psi} \ . \tag{3}$$

The estimation of the parameters in EFA is a problem of finding the pair $\{\hat{\mathbf{\Lambda}}, \hat{\mathbf{\Psi}}\}$ which gives the best fit for certain k to the sample correlation matrix $\mathbf{C} = \mathbf{X}'\mathbf{X}/(n-1)$ with respect to some goodness-of-fit measure.

To find estimates of $\mathbf{\Lambda}$ and $\mathbf{\Psi}$, several factor extraction methods can be employed [5]. A natural and popular choice is the (unweighted) least squares (LS) approach. It can be formulated as the following optimization problem [11]:

$$\min_{\mathbf{\Lambda},\mathbf{\Psi}} ||(\mathbf{C} - \mathbf{\Lambda}\mathbf{\Lambda}' - \mathbf{\Psi})||^2 \quad \text{s.t.} \quad \mathbf{\Lambda}'\mathbf{\Lambda} \text{ a diagonal matrix,} \tag{4}$$

where $||\mathbf{A}|| = \sqrt{\text{trace}(\mathbf{A}'\mathbf{A})}$ denotes the Frobenius norm of $\mathbf{A}$. In EFA, the constraint in (4) eliminates the indeterminacy in (3). This indeterminacy-elimination feature is not always helpful in EFA, because such solutions are usually difficult to interpret [15]. Instead, the parameter estimation is usually followed by some kind of 'simple structure' rotation [5], which in turn gives solutions violating (4). Recall that in ICA 'simple structure' rotation is not necessary. The constraint in (4) is invoked to facilitate the algorithms for numerical solution of the LS problem. The standard numerical solutions of the optimization problem in (4) are iterative, usually based on a Newton-Raphson procedure [11].

3 Factor Scores and Rotation Towards Independence

After estimates $\hat{\mathbf{\Lambda}}$ and $\hat{\mathbf{\Psi}}$ of the parameters $\mathbf{\Lambda}$ and $\mathbf{\Psi}$ have been found, (initial) factor scores can be predicted in a second step:

$$\hat{\mathbf{F}} = \mathbf{X}\hat{\mathbf{\Psi}}^{-1}\hat{\mathbf{\Lambda}}\left(\hat{\mathbf{\Lambda}}'\hat{\mathbf{\Psi}}^{-1}\mathbf{C}\hat{\mathbf{\Psi}}^{-1}\hat{\mathbf{\Lambda}}\right)^{-\frac{1}{2}} . \tag{5}$$

This set of factor scores was proposed by [1]. Equation (5) produces predicted factor scores which are orthogonal. The factor scores are also valid, which means that the predictions do have high correlations with the factors being measured. However, the factor scores are neither univocal, that is, they do not have the property of not correlating with any of the factors except those they were designed to measure nor are they unbiased estimators [5].

So far, finding $\{\hat{\mathbf{\Lambda}}, \hat{\mathbf{\Psi}}\}$ and $\hat{\mathbf{F}}$ is a standard EFA problem. To solve the corresponding ICA problem one needs to go one step further. The initial factor scores are rotated towards independence, that is,

$$\tilde{\mathbf{F}} = \hat{\mathbf{F}}\mathbf{T}, \tag{6}$$

for some orthogonal matrix $\mathbf{T}$. To find the matrix $\mathbf{T}$ that leads to (approximately) independent factor scores, an appropriate rotation criterion is set up.

Recall that if the common factors are independent their squares are also independent. Thus, the (model) covariance matrix of the squared components is diagonal. Let $\mathbf{V}$ be an arbitrary orthogonal matrix and let

$$\mathbf{G} = \hat{\mathbf{F}}\mathbf{V}. \tag{7}$$

The sample covariance matrix between the elementwise squares of $\mathbf{G}$ is

$$\mathbf{S} = \frac{1}{n-1}(\mathbf{G} \odot \mathbf{G})' (\mathbf{I}_k - n^{-1}\mathbf{1}_k\mathbf{1}_k')(\mathbf{G} \odot \mathbf{G}), \tag{8}$$

where $\odot$ denotes the element-wise (Hadamard) matrix product.

Consider the following rotation criterion to be minimized [10]:

$$\mathcal{F}(\mathbf{V}) = \text{trace}\left(\mathbf{S}'(\mathbf{S} \odot \mathbf{N})\right), \tag{9}$$

where $\mathbf{N}$ is a square matrix with zeros on the diagonal and ones elsewhere. The aim is to minimize the sum of the squared off-diagonal elements of $\mathbf{S}$ over all orthogonal rotations $\mathbf{V}$ of $\hat{\mathbf{F}}$.

The gradient projection algorithm proposed by [9] is used to find $\mathbf{T}$ that minimizes $\mathcal{F}$. Let $\mathcal{M}$ be the manifold of all orthogonal matrices. Given a current value of $\mathbf{V}$, this algorithm computes the gradient of $\mathcal{F}$ at $\mathbf{V}$ and moves α units in the negative gradient direction from $\mathbf{V}$. The result is projected on $\mathcal{M}$. The algorithm proceeds iteratively, it is strictly descending and converges from any starting point to a stationary point. At a stationary point of $\mathcal{F}$ restricted to $\mathcal{M}$, the Frobenius norm of the gradient after projection onto the plane tangent to $\mathcal{M}$ at the current value of $\mathbf{V}$ is zero. The algorithm stops when the norm is less than some prescribed precision, say 10^{-5}. Summarizing, the proposed EFA approach to noisy ICA is as follows:

1. Set up the number of common factors, k, prescribed or estimated.
2. Estimate the parameters $\mathbf{\Lambda}$ and $\mathbf{\Psi}$ by a factor extraction method.
3. Calculate (initial) predicted factor scores, $\hat{\mathbf{F}}$, using (5).
4. Find an orthogonal matrix $\mathbf{T}$ that minimizes $\mathcal{F}$ in (9).
5. Calculate the approximately independent factors as $\tilde{\mathbf{F}} = \hat{\mathbf{F}}\mathbf{T}$.
6. Obtain the ICA mixing matrix by $\tilde{\mathbf{\Lambda}} = \hat{\mathbf{\Lambda}}\mathbf{T}$.

4 Application

Developing analytical methods for factor rotation has a long history in factor analysis [2]. It is motivated by both solving the indeterminacy problem and facilitating the factors' interpretation. Thurstone's 26-variable box problem [16] was notorious for being difficult to solve by any analytic rotation method. In this data set, the boxes constitute the observational units.

Table 1 shows the three dimensions f_1 (length), f_2 (width) and f_3 (height) for each box. As in [10], seven additional boxes, whose dimensions are given in Tab. 2, are added to the 20 boxes to form an independent set of boxes which in turn is well-suited for an ICA analysis.

Table 1. Dimensions f_1, f_2, and f_3 of Thurstone's original 20 box-set

	1	2	3	4	5	6	7	8	9	10	11	12	13	14	15	16	17	18	19	20
f_1	3	3	3	3	3	4	4	4	4	4	4	4	4	5	5	5	5	5	5	5
f_2	2	2	3	3	3	2	2	3	3	3	4	4	4	2	2	3	3	4	4	4
f_3	1	2	1	2	3	1	2	1	2	3	1	2	3	1	2	2	3	1	2	3

Table 2. Dimensions f_1, f_2, and f_3 of the 7 additional boxes

	21	22	23	24	25	26	27
f_1	3	3	3	3	4	5	5
f_2	4	4	4	2	2	3	2
f_3	1	2	3	3	3	1	3

Twenty-six functions of these dimensions represent the variables of the study: f_1, f_2, f_3, f_1f_2, f_1f_3, f_2f_3, $f_1^2f_2$, $f_1f_2^2$, $f_1^2f_3$, $f_1f_3^2$, $f_2^2f_3$, $f_2f_3^2$, f_1/f_2, f_2/f_1, f_1/f_3, f_3/f_1, f_2/f_3, f_3/f_2, $2f_1+2f_2$, $2f_1+2f_3$, $2f_2+2f_3$, $\sqrt{f_1^2+f_2^2}$, $\sqrt{f_1^2+f_3^2}$, $\sqrt{f_2^2+f_3^2}$, $f_1f_2f_3$, and $\sqrt{f_1^2+f_2^2+f_3^2}$.

The analytic rotations' aim is to find loadings with simple structure which identify the dimensions of the boxes. As the three dimensions are independent, the problem seems quite appropriate to be attacked by ICA instead. Then, one can expect to find loadings with such simple structure as a side effect.

The resulting 27×26 data matrix contains no measurement error. A less artificial data set is proposed in [12] in which the remedy is to double the number of boxes and add a 54×26 error matrix made up of random normal deviates to the enlarged data matrix. The boxes are doubled to make the results less-dependent on the pseudo-random numbers added. This procedure injects measurement error to the data, giving the problem a greater degree of realism. A cute side-effect is obtained. Since the new data matrix yields a non-singular correlation matrix, it can be used with factor extraction methods such as maximum likelihood factor analysis, canonical factor analysis, image factoring, et cetera.

The matrix of measurement errors is mean-centered and also sphered to ensure that the columns of the error matrix are uncorrelated. After that the columns are scaled to have variance 1/19 of the variance of the corresponding columns of the original data matrix. This yields a reliability of (approximately) 95% for each variable represented in the new data matrix. The 54×26 data matrix, $\mathbf{X}$, is finally mean-centered and standardized to unit variance.

The first few eigenvalues of $\mathbf{C}$ sorted in decreasing order are 11.8906, 6.7467, 5.4132, 0.4015. As expected, three eigenvalues are considerably greater than one, which is the Kaiser's solution for the number of common factors.

The prodecure described in the last section was applied to get $\tilde{\mathbf{F}} = \hat{\mathbf{F}}\mathbf{T}$. The columns $\tilde{\mathbf{f}}_1$, $\tilde{\mathbf{f}}_2$, and $\tilde{\mathbf{f}}_3$ of $\tilde{\mathbf{F}}$ are the rotated factor scores and estimates of the standardized form of the three dimensions $\mathbf{f}_1$, $\mathbf{f}_2$, and $\mathbf{f}_3$ used to generate the mixtures.

Table 3. Covariances (diagonal and above) and correlations (below diagonal) between the element-wise squares of $\tilde{\mathbf{f}}_1$, $\tilde{\mathbf{f}}_2$, and $\tilde{\mathbf{f}}_3$

0.5847561	0.0000001	-0.0000139
0.0000001	0.6107587	0.0000016
-0.0000238	0.0000027	0.5868985

According to Tab. 3, $\tilde{\mathbf{f}}_1$, $\tilde{\mathbf{f}}_2$, and $\tilde{\mathbf{f}}_3$ are quite independent. The off-diagonal elements of the correlation matrix for the element-wise squares of $\tilde{\mathbf{f}}_1$, $\tilde{\mathbf{f}}_2$, and $\tilde{\mathbf{f}}_3$ are all very small (below 3×10^{-5}).

The gradient projection algorithm converged after 18 iterations to a stationary point and the value of $\mathcal{F}$ at the minimum is 1.97×10^{-10}.

Figure 1 displays that the EFA approach to ICA has quite accurately recovered the dimensions for each of the 54 boxes despite the noise introduced to the model. Quite common for an ICA analysis, permutation ambiguities were revealed. The first factor $\mathbf{f}_1$ corresponds to the third column $\tilde{\mathbf{f}}_3$ of $\tilde{\mathbf{F}}$ and vice versa.

Note that some of the manifest variables are non-linear functions of the dimensions of the boxes. However, as [10] point out, the non-linear functions are nearly linear over the values $\mathbf{f}_1$, $\mathbf{f}_2$, and $\mathbf{f}_3$ used to generate the mixtures.

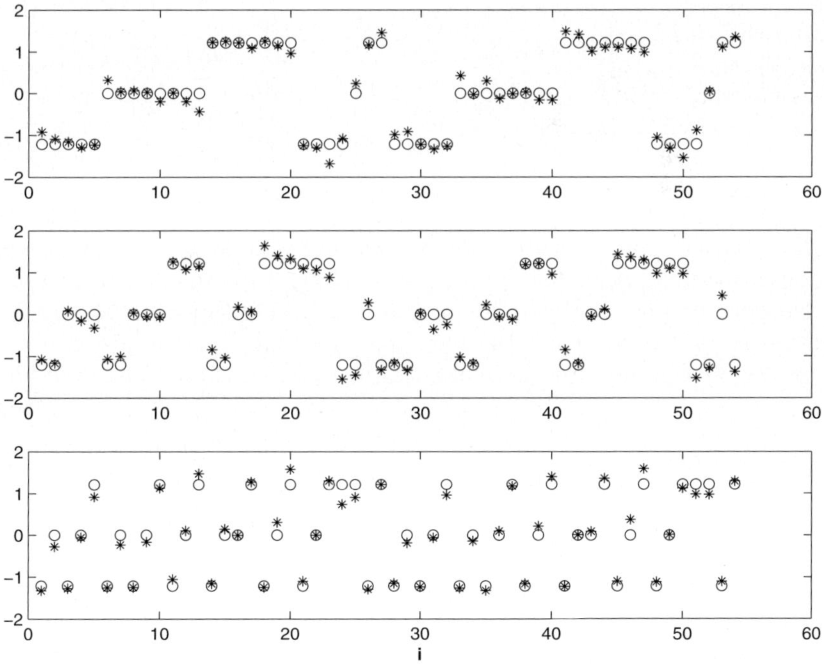

Fig. 1. Standardized box dimensions 'o' and their estimates '*' for each dimension f_1 (upper panel), f_2 (middle panel), and f_3 (lower panel) and each box i ($i = 1, \ldots, 54$)

Table 4. ICA (orthogonal) and Geomin (oblique) rotated loadings for the box data

Function	ICA ($\tilde{\mathbf{\Lambda}}$)			Geomin		
f_1	.97	.09	.09	.97	-.06	.02
f_2	-.13	.96	.10	.02	.98	.02
f_3	-.07	-.10	.96	-.02	-.01	.97
$f_1 f_2$	.45	.86	.12	.58	.78	-.03
$f_1 f_3$	.33	-.06	.89	.37	-.05	.86
$f_2 f_3$	-.19	.45	.82	-.07	.54	.77
$f_1^2 f_2$	.69	.64	.11	.79	.52	-.04
$f_1 f_2^2$	.24	.91	.16	.39	.86	.01
$f_1^2 f_3$	.57	-.01	.77	.61	-.04	.70
$f_1 f_3^2$	.21	-.01	.92	.27	.02	.89
$f_2^2 f_3$	-.15	.65	.68	.00	.72	.60
$f_2 f_3^2$	-.14	.27	.90	-.04	.36	.87
f_1/f_2	.66	-.68	-.05	.54	-.78	-.02
f_2/f_1	-.63	.71	.01	-.51	.81	-.03
f_1/f_3	.43	.14	-.83	.40	.00	-.88
f_3/f_1	-.49	-.15	.77	-.46	.00	.83
f_2/f_3	-.06	.63	-.69	-.01	.57	-.76
f_3/f_2	.06	-.56	.74	.02	-.50	.80
$2f_1 + 2f_2$	.56	.77	.17	.69	.68	.01
$2f_1 + 2f_3$	.68	.03	.70	.72	-.03	.62
$2f_2 + 2f_3$	-.16	.60	.74	-.02	.67	.67
$\sqrt{f_1^2 + f_2^2}$	.70	.67	.10	.80	.55	-.05
$\sqrt{f_1^2 + f_3^2}$	.84	.09	.47	.87	-.01	.37
$\sqrt{f_2^2 + f_3^2}$	-.16	.71	.63	.00	.77	.55
$f_1 f_2 f_3$	.22	.43	.83	.34	.45	.74
$\sqrt{f_1^2 + f_2^2 + f_3^2}$	.57	.58	.51	.69	.51	.37

If one rotates the factor scores, the loadings are rotated as well. Table 4 shows the rotated loadings $\tilde{\mathbf{\Lambda}}$. If one ignores all loadings with magnitude .19 or less, the remaining loadings perfectly identify the subsets of the variables f_1, f_2, and f_3 that were used to generate the mixtures. The simple structure is nearly as good as the one obtained by the more sophisticated method of Geomin [2].

5 Discussion

In this paper, noisy ICA was implemented from an EFA perspective. The approach was applied to the notorious Thurstone's box problem. By rotating the factors towards independence, a simple structure of the loadings was achieved. The criterion for rotating the factor scores towards independence requires minimization of squared fourth-order statistics. Optimization was easily carried out using the gradient projection algorithm. Other methods can also be applied to rotate $\hat{\mathbf{F}}$ towards independence, as for example the varimax-based criterion [13] or the FastICA algorithm [7]. The proposed approach has to be compared to these and/or other ICA methods. One might ask whether the factor analysis' approach is able to recover the common factors and produces simple loadings if the sources are dependent rather than independent and if real correlated data is considered. Tackling these issues will be the subject of future work.

Acknowledgments. The authors would like to thank the anonymous reviewers for their helpful comments and suggestions.

References

1. Anderson, R.D., Rubin, H.: Statistical inference in factor analysis. In: Neyman, J. (ed.) Proceedings of the Berkeley Symposium on Mathematical Statistics and Probability, vol. V, pp. 111–150. University of California Press, Berkeley (1956)
2. Browne, M.W.: An overview of analytic rotation in exploratory factor analysis. Multivariate Behavioral Research 36, 111–150 (2001)
3. Comon, P.: Independent component analysis, a new concept? Signal Processing 36, 287–314 (1994)
4. Davies, M.: Identifiability issues in noisy ICA. In: Puntonet, C.G., Prieto, A.G. (eds.) ICA 2004. LNCS, vol. 3195, pp. 470–473. Springer, Heidelberg (2004)
5. Harman, H.H.: Modern Factor Analysis, 3rd edn. University of Chicago Press, Chicago (1976)
6. Hastie, T., Tibshirani, R., Friedman, J.H.: The Elements of Statistical Learning: Data Mining, Inference, and Prediction, 3rd edn. Springer, New York (2001)
7. Hyvärinen, A., Karhunen, J., Oja, E.: Independent Component Analysis. John Wiley & Sons, New York (2001)
8. Ikeda, S.: ICA on noisy data: a factor analysis approach. In: Girolami, M. (ed.) Advances in Independent Component Analysis, pp. 201–215. Springer, Berlin (2000)
9. Jennrich, R.I.: A simple general procedure for orthogonal rotation. Psychometrika 66, 289–306 (2001)
10. Jennrich, R.I., Trendafilov, N.T.: Independent component analysis as a rotation method: A very different solution to Thurstone's box problem. British Journal of Mathematical and Statistical Psychology 58, 199–208 (2005)
11. Jöreskog, K.G.: Factor analysis by least-squares and maximum likelihood methods. In: Enslein, K., Ralston, A., Wilf, H.S. (eds.) Mathematical methods for digital computers, vol. 3, pp. 125–153. John Wiley & Sons, New York (1977)
12. Kaiser, H.F., Horst, P.: A score matrix for Thurstone's box problem. Multivariate Behavioral Research 10, 17–25 (1975)
13. Kano, Y., Miyamoto, Y., Shimizu, S.: Factor rotation and ICA. In: Proceedings of the 4th International Symposium on Independent Component Analysis and Blind Source Separation (ICA 2003), Nara, Japan, pp. 101–105 (2003)
14. Kawanabe, M., Murata, N.: Independent component analysis in the presence of Gaussian noise based on estimating functions. In: Proceedings of the 2nd International Workshop on Independent Component Analysis and Blind Signal Separation (ICA 2000), Helsinki, Finland, pp. 39–44 (2000)
15. Mardia, K.V., Kent, J.T., Bibby, J.M.: Multivariate Analysis. Academic Press, London (1979)
16. Thurstone, L.L.: Multiple Factor Analysis. The University of Chicago Press, Chicago (1947)

Multilinear (Tensor) ICA and Dimensionality Reduction

M. Alex O. Vasilescu[1,3] and Demetri Terzopoulos[2,3]

[1] Massachusetts Institute of Technology, Cambridge, MA 02139, USA
[2] University of California, Los Angeles, CA 90095, USA
[3] University of Toronto, Toronto, ON M5S 3G4, Canada

Abstract. Multiple factors related to scene structure, illumination, and imaging contribute to image formation. Independent Components Analysis (ICA) maximizes the statistical independence of the representational components of a training image ensemble, but it cannot distinguish between these different factors, or modes. To address this problem, we introduce a nonlinear, multifactor model that generalizes ICA. Our *Multilinear ICA* model of image ensembles learns the statistically independent components of each of the multiple factors. We present an associated dimensionality reduction algorithm for multifactor subspace analysis. As an application, we consider the multilinear analysis of ensembles of facial images that combine several modes, including different facial geometries (people), expressions, head poses, and lighting conditions. For the purposes of face recognition, we introduce a *multilinear projection algorithm* that simultaneously projects an unknown test image into the multiple constituent mode spaces in order to infer its mode labels. We show that multilinear ICA computes a set of factor subspaces that yield improved recognition rates.

1 Introduction

Historically, linear models that capture the statistical properties of image or other signal data have been broadly applied in pattern recognition. For example, the linear, appearance-based face recognition method known as "Eigenfaces" is founded on the principal components analysis (PCA) of facial image ensembles [1]. PCA encodes pairwise relationships between pixels—the second-order, correlational structure of the training image ensemble—but it ignores higher-order pixel statistics. By contrast, independent components analysis (ICA) [2,3] learns a set of statistically independent components by also considering these higher-order dependencies in the training data.

However, ICA cannot distinguish between higher-order statistics associated with different factors, or modes, inherent to image formation—factors pertaining to scene structure, illumination, and imaging. In particular, ICA has been employed in face recognition [4] and, like PCA, it works best when person identity is the only factor that is permitted to vary. If additional factors, such as illumination, viewpoint, and expression can modify facial images, recognition rates deteriorate dramatically.

We propose a multilinear framework that addresses the aforementioned problems. Specifically, we introduce a nonlinear, multifactor model of image ensembles that generalizes conventional ICA.[1] Unlike its conventional, linear counterpart, our *Multilinear*

[1] A preliminary description of this work appeared as an extended abstract in the *Learning 2004 Workshop*, Snowbird, UT, April, 2004.

M.E. Davies et al. (Eds.): ICA 2007, LNCS 4666, pp. 818–826, 2007.

ICA model exploits multilinear (tensor) algebra in order to learn the interactions of multiple factors inherent to image formation and separately encode the higher-order statistics of each of these factors. By contrast, the multilinear generalization of Eigenfaces, dubbed TensorFaces [5], encodes only their second-order statistics. Our multilinear ICA should not be confused with existing tensorial algorithms for computing the conventional, linear ICA models [6,7].

We demonstrate the application of multilinear ICA to the problem of face recognition under varying viewpoint and illumination, obtaining significantly improved recognition rates. In this context, our second contribution is a novel, *multilinear projection algorithm*. It projects an unknown test image into the multiple factor representation spaces to infer the person, viewpoint, illumination, and other mode labels associated with the test image.

After reviewing the mathematical foundations of our tensor approach in Section 2, we motivate our work by discussing PCA and multilinear PCA in Section 3. Next, we generalize ICA (Section 4), developing our multilinear ICA algorithm in Section 5. Section 6 introduces the multilinear projection algorithm for recognition. Section 7 presents our experiments and results and Section 8 concludes the paper.

2 Multilinear (Tensor) Algebraic Fundamentals

Definition 1 (Tensor). *A tensor, or n-way array, is a generalization of a vector (first-order tensor) and a matrix (second-order tensor).*[2] *Tensors are multilinear mappings over a set of vector spaces. The order of tensor $\mathcal{A} \in \mathbb{R}^{I_1 \times \ldots \times I_N}$ is N. An element of $\mathcal{A}$ is denoted as $\mathcal{A}_{i_1 \ldots i_n \ldots i_N}$ or $a_{i_1 \ldots i_n \ldots i_N}$, where $1 \leq i_n \leq I_n$.*

Definition 2 (Mode-n Vectors). *The mode-n vectors (or fibers) of an $N^{th}-$order tensor $\mathcal{A} \in \mathbb{R}^{I_1 \times \ldots \times I_N}$ are the I_n-dimensional vectors obtained from $\mathcal{A}$ by varying index i_n while keeping the other indices fixed. They are the column vectors of matrix $\mathbf{A}_{(n)} \in \mathbb{R}^{I_n \times (I_{n+1} \ldots I_N I_1 \ldots I_{n-1})}$ that results from flattening the tensor $\mathcal{A}$ (Fig. 1).*

Definition 3 (Mode-n Orthonormal Matrices). *Mode matrix $\mathbf{U}_n$ contains the orthonormal vectors spanning the column space of matrix $\mathbf{A}_{(n)}$ resulting from the mode-n flattening of $\mathcal{A}$.*

Definition 4 (Mode-n Rank). *The mode-n rank R_n of $\mathcal{A} \in \mathbb{R}^{I_1 \times \ldots \times I_N}$ is defined as the dimension of the vector space generated by the mode-n vectors: $R_n = \mathrm{rank}_n(\mathcal{A}) = \mathrm{rank}(\mathbf{A}_{(n)})$.*

Definition 5 (Mode-n Product, $\times_n$). *The mode-n product of a tensor $\mathcal{A} \in \mathbb{R}^{I_1 \times \ldots I_n \times \ldots I_N}$ and a matrix $\mathbf{M} \in \mathbb{R}^{J_n \times I_n}$, denoted by $\mathcal{A} \times_n \mathbf{M}$, is a tensor of dimensionality $\mathbb{R}^{I_1 \times \ldots I_{n-1} \times J_n \times I_{n+1} \times \ldots I_N}$ whose entries are computed by $(\mathcal{A} \times_n \mathbf{M})_{i_1 \ldots i_{n-1} j_n i_{n+1} \ldots i_N} = \sum_{i_n} a_{i_1 \ldots i_{n-1} i_n i_{n+1} \ldots i_N} m_{j_n x i_n}$. It can be expressed in terms of flattened matrices as $\mathbf{B}_{(n)} = \mathbf{M}\mathbf{A}_{(n)}$.*

[2] We denote scalars by italic lowercase letters $(a, b, \ldots)$, vectors by bold lowercase letters $(\mathbf{a}, \mathbf{b}, \ldots)$, matrices by bold uppercase letters $(\mathbf{A}, \mathbf{B}, \ldots)$, and higher-order tensors by calligraphic uppercase letters $(\mathcal{A}, \mathcal{B}, \ldots)$.

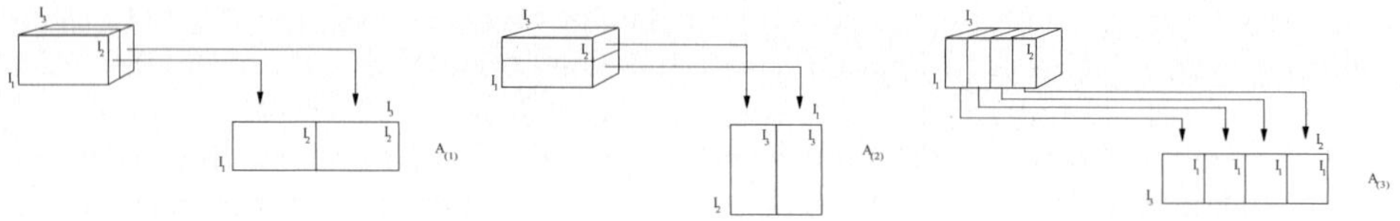

Fig. 1. Flattening a (3rd-order) tensor. The tensor can be flattened in 3 ways to obtain matrices comprising its mode-1, mode-2, and mode-3 vectors.

A matrix representation of the mode-n product of a tensor $\mathcal{A} \in \mathbb{R}^{I_1 \times \ldots \times I_n \times \ldots \times I_N}$ and a set of N matrices, $\mathbf{F}_n \in \mathbb{R}^{J_n \times I_n}$ can be obtained as follows:

$$\mathcal{B} = \mathcal{A} \times_1 \mathbf{F}_1 \ldots \times_n \mathbf{F}_n \ldots \times_N \mathbf{F}_N, \quad \text{or in terms of flattened tensors,}$$
$$\mathbf{B}_{(n)} = \mathbf{F}_n \mathbf{A}_{(n)} (\mathbf{F}_{n-1} \otimes \ldots \mathbf{F}_1 \otimes \mathbf{F}_N \otimes \ldots \mathbf{F}_{n+1})^T,$$

where $\otimes$ denotes the matrix Kronecker product. The *Frobenius norm* of a tensor $\mathcal{A}$ is given by $\|\mathcal{A}\| = \sqrt{\langle \mathcal{A}, \mathcal{A} \rangle}$.

3 PCA and Multilinear PCA

The principal components analysis of an ensemble of I_2 images is computed by performing an SVD on a $I_1 \times I_2$ *data matrix* $\mathbf{D}$ whose columns are the "vectorized" I_1-pixel "centered" images. The matrix $\mathbf{D} \in \mathbb{R}^{I_1 \times I_2}$ is a two-mode mathematical object that has two associated vector spaces, a row space and a column space. In a PCA analysis of $\mathbf{D}$, the SVD orthogonalizes these two spaces and decomposes the matrix as $\mathbf{D} = \mathbf{U}\mathbf{\Sigma}\mathbf{V}^T$. Using mode-$n$ products, the SVD can be rewritten as $\mathbf{D} = \mathbf{\Sigma} \times_1 \mathbf{U} \times_2 \mathbf{V}$. The eigenvectors $\mathbf{U}$ are called the *principal component* directions of $\mathbf{D}$ (Fig. 3(a)).

The analysis of an ensemble of images resulting from the confluence of multiple factors, or modes, related to scene structure, illumination, and viewpoint is a problem in multilinear algebra [5]. Within this mathematical framework, the image ensemble is represented as a higher-order tensor. This image data tensor $\mathcal{D}$, Fig. 2(b), must be decomposed in order to separate and parsimoniously represent the constituent factors. This can be achieved by employing the N*-mode SVD*, a multilinear extension of the aforementioned conventional matrix SVD [8,9].

$\mathcal{D}$ is an N-dimensional matrix comprising N spaces. The N-mode SVD orthogonalizes these N spaces and decomposes the tensor as the mode-n product of N orthogonal spaces:

$$\mathcal{D} = \mathcal{Z} \times_1 \mathbf{U}_1 \times_2 \mathbf{U}_2 \ldots \times_n \mathbf{U}_n \ldots \times_N \mathbf{U}_N. \quad (1)$$

Tensor $\mathcal{Z}$, known as the *core tensor*, is analogous to the diagonal singular value matrix in conventional matrix SVD (although it does not have a simple, diagonal structure). The core tensor governs the interaction between the *mode matrices* $\mathbf{U}_1, \ldots, \mathbf{U}_N$. Mode matrix $\mathbf{U}_n$ contains the orthonormal basis vectors spanning the column space of matrix $\mathbf{D}_{(n)}$ resulting from the *mode-n flattening* of $\mathcal{D}$. The tensor basis associated with this multilinear PCA is displayed in Fig. 3(b).

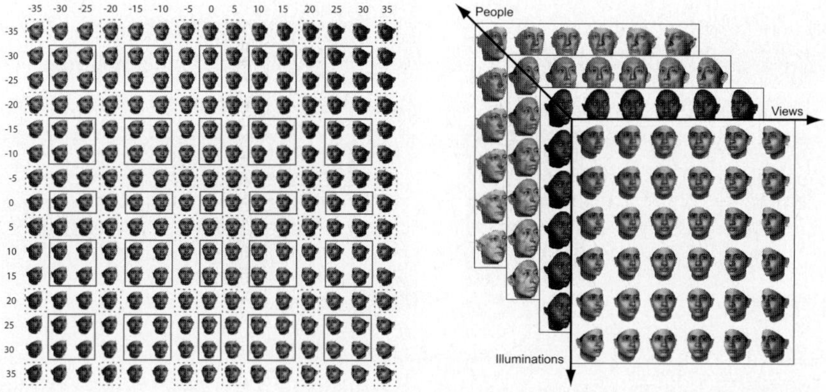

(a) (b)

Fig. 2. A facial image dataset of 2,700 training images out of 16,875 images. 3D scans of 75 subjects, recorded using a Cyberware^{TM} 3030PS laser scanner as part of the University of Freiburg 3D morphable faces database [10]. A portion of the 4th-order data tensor $\mathcal{D}$ for the image ensemble formed from the dash-boxed images, Fig. 2(a), of each person. Only 4 of the 75 people are shown.

The N**-mode SVD algorithm** for decomposing $\mathcal{D}$ according to equation (1) is as follows:

1. For $n = 1, \ldots, N$, compute matrix $\mathbf{U}_n$ in (1) by computing the SVD of the flattened matrix $\mathbf{D}_{(n)}$ and setting $\mathbf{U}_n$ to be the left matrix of the SVD.
2. Solve for the core tensor: $\mathcal{Z} = \mathcal{D} \times_1 \mathbf{U}_1^T \times_2 \mathbf{U}_2^T \ldots \times_n \mathbf{U}_n^T \ldots \times_N \mathbf{U}_N^T$.

---

4 ICA

The independent components analysis of multivariate data can be applied in two ways [4]: 1) to $\mathbf{D}^T$, each of whose rows is a different image, which finds a spatially independent basis set that reflects the local properties of imaged objects; 2) to $\mathbf{D}$, which finds a set of coefficients that are statistically independent while the basis reflects the global properties of imaged objects.

ICA Approach 1: ICA starts essentially from the PCA solution and computes a transformation of the principal components such that they become *independent components*

$$\mathbf{D}^T = \mathbf{V}\mathbf{\Sigma}\mathbf{U}^T = \left(\mathbf{V}\mathbf{\Sigma}\mathbf{W}_{s_1}^{-1}\right)\left(\mathbf{W}_{s_1}\mathbf{U}^T\right) = \mathbf{K}^T\mathbf{C}^T, \tag{2}$$

where every column of $\mathbf{D}$ is a different image, $\mathbf{W}_{s_1}$ is an invertible transformation matrix that is computed by the ICA algorithm, $\mathbf{C} = \mathbf{U}\mathbf{W}_{s_1}^T$ are the independent components (Fig. 3(a)), and $\mathbf{K} = \mathbf{W}_{s_1}^{-T}\mathbf{\Sigma}\mathbf{V}^T$ are the coefficients. Various objective functions, such as those based on mutual information, negentropy, higher-order cumulants, etc., are presented in the literature for computing the independent components along with different optimization methods for extremizing the objective functions [3].

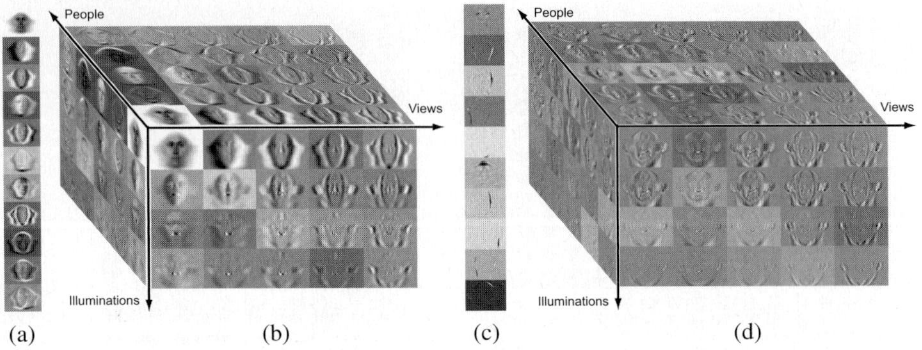

Fig. 3. Eigenfaces and TensorFaces bases for an ensemble of 2,700 facial images spanning 75 people, each imaged under 6 viewing and 6 illumination conditions (see Section 7). (a) PCA eigenvectors (eigenfaces), which are the principal axes of variation across all images. (b) A partial visualization of the $75 \times 6 \times 6 \times 8560$ TensorFaces representation of $\mathcal{D}$, obtained as $\mathcal{T} = \mathcal{Z} \times_4 \mathbf{U}_{\text{pixels}}$ which captures viewpoint varaition, illumination variation and people variation. (c) Independent components $\mathbf{C}_{\text{pixels}}$. (d) A partial visualization of the $75 \times 6 \times 6 \times 8560$ multilinear ICA representation of $\mathcal{D}$, obtained as $\mathcal{B} = \mathcal{S} \times_4 \mathbf{C}_{\text{pixels}}$.

ICA Approach 2: Alternatively, ICA can be applied to $\mathbf{D}$, and it transforms the principal components directions such that the coefficients are statistically independent, as follows:

$$\mathbf{D} = \mathbf{U}\boldsymbol{\Sigma}\mathbf{V}^T = \left(\mathbf{U}\mathbf{W}_{s_2}^{-1}\right)\left(\mathbf{W}_{s_2}\boldsymbol{\Sigma}\mathbf{V}^T\right) = \mathbf{C}\mathbf{K}, \tag{3}$$

where $\mathbf{C} = \mathbf{U}\mathbf{W}_{s_2}^{-1}$ is the basis matrix and $\mathbf{K} = \mathbf{W}_{s_2}\boldsymbol{\Sigma}\mathbf{V}^T$ are the statistically independent coefficients.

Note that $\mathbf{C}$, $\mathbf{K}$ and $\mathbf{W}$ are computed differently in the two approaches. Approach 1 yields statistically independent bases, whereas Approach 2 yields a "factorial code".

5 Multilinear ICA

Like PCA, ICA is a linear analysis method, hence it is not well suited to the representation of multi-factor image ensembles. To address this shortcoming, we next propose a novel multilinear generalization of ICA. Multilinear ICA is obtained by decomposing the data tensor $\mathcal{D}$ as the mode-n product of N mode matrices $\mathbf{C}_n$ and a core tensor $\mathcal{S}$, as follows:

$$\mathcal{D} = \mathcal{S} \times_1 \mathbf{C}_1 \times_2 \mathbf{C}_2 \ldots \times_n \mathbf{C}_n \ldots \times_N \mathbf{C}_N. \tag{4}$$

The N**-mode ICA algorithm** is as follows:

1. For $n = 1, \ldots, N$, compute the mode matrix $\mathbf{C}_n$ in (4) in one of two ways, by (5)–(6) or by (12)–(13).
2. Solve for the core tensor: $\mathcal{S} = \mathcal{D} \times_1 \mathbf{C}_1^{-1} \times_2 \mathbf{C}_2^{-1} \ldots \times_n \mathbf{C}_n^{-1} \ldots \times_N \mathbf{C}_N^{-1}$.

As in ICA, there are two approaches for multilinear ICA.

Multilinear ICA Approach 1: Transposing the flattened data tensor $\mathcal{D}$ in mode n and computing the ICA as in (2), we obtain:

$$\mathbf{D}_{(n)}^T = \mathbf{V}_n \boldsymbol{\Sigma}_n \mathbf{U}_n^T = \left(\mathbf{V}_n \boldsymbol{\Sigma}_n \mathbf{W}_n^{-1}\right)\left(\mathbf{W}_n \mathbf{U}_n^T\right) = \mathbf{K}_n^T \mathbf{C}_n^T, \tag{5}$$

where the mode matrices are given by

$$\mathbf{C}_n = \mathbf{U}_n \mathbf{W}_n^T. \tag{6}$$

The columns associated with each of the mode matrices $\mathbf{C}_n$ are statistically independent; i.e., a factorial code representation is computed for each mode of variation. We can derive the relationship between N-mode ICA and N-mode SVD (1) in the context of this approach as follows:

$$\begin{aligned}
\mathcal{D} &= \mathcal{Z} \times_1 \mathbf{U}_1 \ldots \times_N \mathbf{U}_N & (7)\\
&= \mathcal{Z} \times_1 \mathbf{U}_1 \mathbf{W}_1^T \mathbf{W}_1^{-T} \ldots \times_N \mathbf{U}_N \mathbf{W}_N^T \mathbf{W}_N^{-T} & (8)\\
&= \mathcal{Z} \times_1 \mathbf{C}_1 \mathbf{W}_1^{-T} \ldots \times_N \mathbf{C}_N \mathbf{W}_N^{-T} & (9)\\
&= (\mathcal{Z} \times_1 \mathbf{W}_1^{-T} \ldots \times_N \mathbf{W}_N^{-T}) \times_1 \mathbf{C}_1 \ldots \times_N \mathbf{C}_N & (10)\\
&= \mathcal{S} \times_1 \mathbf{C}_1 \ldots \times_N \mathbf{C}_N, & (11)
\end{aligned}$$

where the core tensor $\mathcal{S} = \mathcal{Z} \times_1 \mathbf{W}_1^{-T} \ldots \times_N \mathbf{W}_N^{-T}$.

Multilinear ICA Approach 2: Alternatively, flattening the data tensor $\mathcal{D}$ in mode n and computing the ICA as in (3), we obtain:

$$\mathbf{D}_{(n)} = \mathbf{U}_n \boldsymbol{\Sigma}_n \mathbf{V}_n^T = \left(\mathbf{U}_n \mathbf{W}_n^{-1}\right)\left(\mathbf{W}_n \boldsymbol{\Sigma}_n \mathbf{V}_n^T\right) = \mathbf{C}_n \mathbf{K}_n, \tag{12}$$

where the mode matrices are given by

$$\mathbf{C}_n = \mathbf{U}_n \mathbf{W}_n^{-1}. \tag{13}$$

This second approach results in a set of basis vectors that are statistically independent across the different modes.

Note that the $\mathbf{W}_n$ in (13) differs from the $\mathbf{W}_n$ in (6); the latter is analogous to $\mathbf{W}_{s_1}$ in (2) while the former is analogous to $\mathbf{W}_{s_2}$ in (3). We can derive the relationship between N-mode ICA and N-mode SVD (1) in the context of the second approach as follows:

$$\begin{aligned}
\mathcal{D} &= \mathcal{Z} \times_1 \mathbf{U}_1 \ldots \times_N \mathbf{U}_N & (14)\\
&= \mathcal{Z} \times_1 \mathbf{U}_1 \mathbf{W}_1^{-1} \mathbf{W}_1 \ldots \times_N \mathbf{U}_N \mathbf{W}_N^{-1} \mathbf{W}_N & (15)\\
&= \mathcal{Z} \times_1 \mathbf{C}_1 \mathbf{W}_1 \ldots \times_N \mathbf{C}_N \mathbf{W}_N & (16)\\
&= (\mathcal{Z} \times_1 \mathbf{W}_1 \ldots \times_N \mathbf{W}_N) \times_1 \mathbf{C}_1 \ldots \times_N \mathbf{C}_N & (17)\\
&= \mathcal{S} \times_1 \mathbf{C}_1 \ldots \times_N \mathbf{C}_N, & (18)
\end{aligned}$$

where the core tensor $\mathcal{S} = \mathcal{Z} \times_1 \mathbf{W}_1 \ldots \times_N \mathbf{W}_N$.

Optimal dimensionality reduction in multilinear ICA, which yields the approximation $\hat{\mathcal{D}} = \hat{\mathcal{S}} \times_1 \hat{\mathbf{C}}_1 \ldots \times_N \hat{\mathbf{C}}_N$, is achieved by optimizing iteratively, mode per mode using alternating least squares, by holding fixed all mode components except one and solving for the remaining mode, with an additional step that computes the transformation matrix $\mathbf{W}$. The error function minimized is:

$$e = \|\mathcal{D} - \hat{\mathcal{D}}\| = \|\mathcal{D} - (\hat{\mathcal{S}} \times_1 \hat{\mathbf{C}}_1 \ldots \times_N \hat{\mathbf{C}}_N)\|. \tag{19}$$

The N**-mode ICA dimensionality reduction algorithm** is as follows:

1. **Initialize:** Apply Step 1 of the N-mode ICA algorithm to $\mathcal{D}$; truncate each mode matrix $\mathbf{U}_n$, for $n = 1, 2, \ldots, N$, to R_n columns, and compute the mode matrix $\mathbf{C}_n$, thus obtaining the initial ($k = 0$) mode matrices $\mathbf{C}_1^0, \mathbf{C}_2^0, \ldots, \mathbf{C}_N^0$.
2. Iterate, for $k = 1, 2, \ldots$, until convergence:
   Alternating Least Squares: Compute matrix $\mathbf{C}_n$, $1 \leq n \leq N$:
   Set $\tilde{\mathcal{C}}_n^k = \mathcal{D} \times_1 \mathbf{C}_1^{k+} \ldots \times_{n-1} \mathbf{C}_{n-1}^{k\;+} \times_{n+1} \mathbf{C}_{n+1}^{k-1+} \ldots \times_N \mathbf{C}_N^{k-1+}$; mode-$n$ flatten $\tilde{\mathcal{C}}_n^k$ to obtain the matrix $\tilde{\mathbf{C}}_n^k$; compute $\mathbf{C}_n^k$ according to (5) or (12) by setting $\mathbf{D}_{(n)} = \tilde{\mathbf{C}}_n^k$.
3. Set the converged mode matrices to $\hat{\mathbf{C}}_1, \hat{\mathbf{C}}_2, \ldots, \hat{\mathbf{C}}_N$. Compute the core tensor $\hat{\mathcal{S}} = \tilde{\mathcal{C}}_N \times_N \hat{\mathbf{C}}_N^+$. The approximation of $\mathcal{D}$ is $\hat{\mathcal{D}} = \hat{\mathcal{S}} \times_1 \hat{\mathbf{C}}_1 \times_2 \hat{\mathbf{C}}_2 \ldots \times_N \hat{\mathbf{C}}_N$.

---

6 Multilinear Projection

We will now develop a multilinear method for simultaneously inferring the identity, illumination, viewpoint, etc., coefficient vectors of an unlabeled, test image. Multilinear ICA represents the unlabeled, test image by a set of unknown coefficient vectors, $\mathbf{d}^T = \mathcal{B} \times_1 \mathbf{c}_p^T \times_2 \mathbf{c}_v^T \times_3 \mathbf{c}_l^T$, where the coefficient vector $\mathbf{c}_p$ encodes the person, the coefficient vector $\mathbf{c}_v$ encodes the viewpoint, and the coefficient vector $\mathbf{c}_l$ encodes the illumination.

The **multilinear projection algorithm** is as follows:

1. Compute the projection transformation $\mathcal{P}$. In matrix form, $\mathbf{P}_{\text{(mode)}} = \mathbf{B}_{\text{(pixels)}}^{T+}$.
2. Compute the response tensor $\mathcal{R} = \mathcal{P} \times_{\text{pixels}} \mathbf{d}^T$:

$$\overbrace{\mathcal{P} \times_{\text{pixels}} \mathbf{d}^T}^{\mathcal{R}} \approx \overbrace{\mathcal{I} \times_{\text{pixels}} (\mathbf{c}_l^T \otimes \mathbf{c}_v^T \otimes \mathbf{c}_p^T)}^{\mathcal{C}}$$
$$= \mathbf{c}_p \circ \mathbf{c}_v \circ \mathbf{c}_l,$$

3. Since $\mathcal{R} = \mathcal{C}$ and has rank-$(1, \ldots, 1)$, the coefficients are extracted by factorizing the response tensor using the N-mode SVD algorithm.

Intuitively, the unknowns $\mathbf{c}_p$, $\mathbf{c}_v$, and $\mathbf{c}_l$ need to be estimated from $\mathbf{d}$ and $\mathcal{B}$. This involves computing a pseudo-inverse tensor. Projecting $\mathbf{d}$ onto the pixel mode of $\mathcal{B}$ yields the image projection tensor $\mathcal{R} = \mathcal{P} \times_4 \mathbf{d}^T \approx \mathcal{C}$, where the "projection transformation" $\mathcal{P}$ is obtained by re-tensorizing matrix $\mathbf{P}_{\text{(pixels)}} = \mathbf{B}_{\text{(pixels)}}^{+T}$ (the matrix $\mathbf{B}_{\text{(pixels)}}$ is the pixel-mode flattening of tensor $\mathcal{B}$). The tensor $\mathcal{R}$ has the structure $(\mathbf{c}_p \circ \mathbf{c}_v \circ \mathbf{c}_l)$, the outer product of the coefficient vectors associated with each factor inherent to the data $\mathbf{d}^T$; hence, it is of rank-$(1,\ldots,1)$. The rank of $\mathcal{R}$ and the fact that the coefficient vectors are unit vectors enables us to compute the three coefficient vectors via a tensor decomposition using the N-mode SVD algorithm. In principle, $\mathcal{R}=\mathcal{C}$, but sometimes in practice $\mathcal{R}\approx\mathcal{C}$. Thus, the optimal dimensionally-reduced rank-$(1,\ldots,1)$ decomposition must be computed by optimizing the objective function: $\|\mathbf{d}^T - \hat{\mathcal{B}} \times_1 \hat{\mathbf{c}}_1^T \times_2 \ldots \times_N \hat{\mathbf{c}}_N^T\|$.

7 Experiments

In our face recognition experiments, each subject is imaged from 15 different viewpoints ($\theta = -35^\circ$ to $+35^\circ$ in 5° steps on the horizontal plane $\phi = 0^\circ$) under 15 different illuminations ($\theta = -35^\circ$ to $+35^\circ$ in 5° steps on an inclined plane $\phi = 45^\circ$). Fig. 2(a) shows the full set of 225 images for one of the subjects with viewpoints arrayed horizontally and illuminations arrayed vertically. The image set was rendered from 3D scans of 75 subjects. Of the 16,875 images, we employed 2,700 as training images. The training data tensor, a 4th-order tensor, $\mathcal{D}$ is shown in Fig. 2.

Multilinear ICA yields better recognition rates (98.14%) than PCA (eigenfaces) (83.9%), conventional ICA (89.5%) and even multilinear PCA (93.4%) in scenarios involving the recognition of people imaged in previously unseen viewpoints and illuminations. Fig. 3(d) illustrates the multilinear ICA basis derived from the training ensemble, while Fig. 3(c) illustrates the conventional ICA basis.

8 Conclusion

We presented a multilinear generalization of ICA. We applied our new multilinear ICA algorithm along with a novel, multilinear projection method to face recognition involving multiple people imaged under different viewpoints and illuminations. Multilinear ICA disentangles the multiple factors inherent to image formation and explicitly represents the higher-order statistics associated with each factor, thus yielding improved recognition rates relative to related prior methods.

References

1. Sirovich, L., Kirby, M.: Low dimensional procedure for the characterization of human faces. Journal of the Optical Society of America A. 4, 519–524 (1987)
2. Comon, P.: Independent component analysis, a new concept? Signal Processing 36, 287–314 (1994)
3. Hyvärinen, A., Karhunen, J., Oja, E.: Independent Component Analysis. Wiley, Chichester (2001)
4. Bartlett, M.: Face Image Analysis by Unsupervised Learning. Kluwer, Boston (2001)
5. Vasilescu, M., Terzopoulos, D.: Multilinear analysis for facial image recognition. In: Proc. Int. Conf. on Pattern Recognition. Quebec City, vol. 2, pp. 511–514 (2002)
6. Cardoso, J.F., Comon, P.: Tensor-based independent component analysis. Signal Processing V: Theories and Applications 5, 673–676 (1990)

7. De Lathauwer, L., De Moor, B., Vandewalle, J.: Independent component analysis and (simultaneous) third-order tensor diagonalization. IEEE Trans. Signal Processing 49(10), 2262–2271 (2001)
8. Tucker, L.R.: Some mathematical notes on three-mode factor analysis. Psychometrika 31, 279–311 (1966)
9. De Lathauwer, L., De Moor, B., Vandewalle, J.: A multilinear singular value decomposition. SIAM J. Matrix Anal. Appl. 21(4), 1253–1278 (2000)
10. Blanz, V., Vetter, T.: A morphable model for the synthesis of 3D faces. In: SIGGRAPH 99 Conference Proceedings, ACM SIGGRAPH, pp. 187–194 (1999)

ICA in Boolean XOR Mixtures

Arie Yeredor

School of Electrical Engineering, Tel-Aviv University, Israel
arie@eng.tau.ac.il

Abstract. We consider Independent Component Analysis (ICA) for the case of binary sources, where addition has the meaning of the boolean "Exclusive Or" (XOR) operation. Thus, each mixture-signal is given by the XOR of one or more of the source-signals. While such mixtures can be considered linear transformations over the finite Galois Field of order 2, they are certainly nonlinear over the field of real-valued numbers, so classical ICA principles may be inapplicable in this framework. Nevertheless, we show that if none of the independent random sources is uniform (i.e., neither one has probability 0.5 for 1/0), then any invertible mixing is identifiable (up to permutation ambiguity). We then propose a practical deflation algorithm for source separation based on entropy minimization, and present empirical performance results by simulation.

1 Introduction and Problem Formulation

The classical Independent Components Analysis (ICA) framework usually assumes linear combinations of independent sources over the field of real-valued numbers $\mathbb{R}$, with some exceptions (e.g., [1]) that assume the field of complex-valued numbers $\mathbb{C}$. It might be interesting, at least from a theoretical point of view, to explore the applicability of ICA principles to other algebraic fields.

Let us consider the field (often denoted Galois Field of order 2, GF(2)) of binary numbers $\{0, 1\}$, where, for $x, y \in \{0, 1\}$, addition is defined as the "Exclusive Or" (XOR) operation, denoted $z = x \oplus y$, where z equals 1 if and only if $x \neq y$ (and equals zero otherwise). Multiplication in this field (either by 0 or by 1) is defined and denoted in the "usual" way, $z = xy$. These values and operations trivially satisfy all the requirements that constitute a field [2], namely: associativity, commutativity, distributivity, existence of an additive and of a multiplicative identity element (0 and 1, resp.) and of additive and multiplicative inverses ($-0 = 0$, $-1 = 1$; and $1^{-1} = 1$, resp.).

Naturally, all random variables (RVs) in this field are binary, and any probability distribution is uniquely defined by a single parameter p, denoting the probability with which the RV takes the value 1. We shall refer to p as the "1-probability" of the RV.

Assume now that there are K statistically independent random sources denoted $\boldsymbol{s}[n] = [s_1[n] \;\; s_2[n] \;\; \cdots \;\; s_K[n]]^T$, with respective fixed, unknown 1-probabilities $\boldsymbol{p} = [p_1 \; p_2 \;\; \cdots \;\; p_K]^T$. For simplicity we shall further assume that the samples of each source are independent, identically distribute (iid) in time.

M.E. Davies et al. (Eds.): ICA 2007, LNCS 4666, pp. 827–835, 2007.

Naturally, just like in "classical" ICA it is also possible to extend this basic model to temporally-correlated or non-stationary sources (e.g., [3] or [4], respectively, for classical ICA), but for now we choose to concentrate on this basic, iid model.

Let these sources be mixed (over GF(2)) by an unknown, square ($K \times K$) mixing matrix $\boldsymbol{A}$ (whose elements also belong to GF(2)),

$$\boldsymbol{x}[n] = \boldsymbol{A} \circ \boldsymbol{s}[n], \tag{1}$$

where "$\circ$" denotes matrix/vector multiplication over the field, such that the k-th element of $\boldsymbol{x}[n]$ is given by

$$x_k[n] = a_{k1}s_1[n] \oplus a_{k2}s_2[n] \oplus \cdots \oplus a_{kK}s_K[n] \quad k = 1, 2, ..., K. \tag{2}$$

We further assume that $\boldsymbol{A}$ is invertible over the field, namely that it has a unique inverse in GF(2), denoted $\boldsymbol{B} \stackrel{\triangle}{=} \boldsymbol{A}^{-1}$, satisfying $\boldsymbol{B} \circ \boldsymbol{A} = \boldsymbol{A} \circ \boldsymbol{B} = \boldsymbol{I}$, where $\boldsymbol{I}$ denotes the $K \times K$ identity matrix. Like in "classical" linear algebra (over $\mathbb{R}$), $\boldsymbol{A}$ is non-singular (invertible) if and only if (iff) its determinant[1] is non-zero (namely 1). Equivalently, $\boldsymbol{A}$ is singular iff there exists (in GF(2)) a nonzero vector $\boldsymbol{u}$, such that $\boldsymbol{A} \circ \boldsymbol{u} = \boldsymbol{0}$ (an all-zeros vector).

We are interested in the possibility to recover the source signals $\boldsymbol{s}[n]$ from the observations (mixtures) $\boldsymbol{x}[n]$ under this "blind" scenario, where the only available knowledge is that the sources are statistically independent. Admittedly, this problem is not directly related to any specific application, but it is possible to think, e.g., of a hypothetical situation in a digital communication system, where cross-talk between channels might have the effect of a XOR combination (e.g., in a binary symmetric channel (BSC, [5]), the output can be considered as a XOR operation between the signal and noise processes).

Note that although the mixing is linear over our GF(2) field, it is certainly not linear over the "standard" ICA fields $\mathbb{R}$ or $\mathbb{C}$. Therefore, clearly not all classical results from the ICA theory and practice are applicable to this problem.

2 Identifiability

Let us first address the issue of identifiability (possibly up to some tolerable ambiguities) of $\boldsymbol{A}$ (or, equivalently, of its inverse $\boldsymbol{B}$) from the set of observations $\boldsymbol{x}[n]$, $n = 1, 2, ...N$, under asymptotic conditions, namely when $N \to \infty$. Due to the assumption of iid samples for each source (implying ergodicity), the joint statistics of the observations can be fully and consistently estimated from the available data. Therefore, the assumption of asymptotic conditions implies full and exact knowledge of the joint probability distribution of the observation vector $\boldsymbol{x}$ (we dropped the time-index n here, due to the stationarity). Before we proceed, let us consider the characterization of statistical properties of an arbitrary random vector in GF(2).

[1] The determinant over GF(2) can be calculated just like over $\mathbb{R}$, but with the ordinary addition / subtraction replaced by the XOR operation.

2.1 Statistical Characterization of Random Vectors

For any $K \times 1$ random vector $\boldsymbol{y}$ with elements in GF(2), the probability function can be fully described in a K-way tensor (K-dimensional array) $\boldsymbol{\mathcal{P}}^{(y)}$, with two elements in each direction, indexed as 0 or 1 for convenience, such that

$$\boldsymbol{\mathcal{P}}^{(y)}_{i_1,i_2,\ldots,i_K} \triangleq \text{Prob}\{y_1 = i_1, y_2 = i_2, ..., y_K = i_K\} \ , \ \ i_1, i_2, \ldots, i_K \in \{0,1\}. \quad (3)$$

For convenience, we may concatenate the K indices into a $K \times 1$ "index-vector" $\boldsymbol{i} \triangleq [i_1 \ i_2 \ \cdots \ i_K]^T$ and use the notation $\boldsymbol{\mathcal{P}}^{(y)}(\boldsymbol{i}) \equiv \boldsymbol{\mathcal{P}}^{(y)}_{i_1,i_2,\ldots,i_K}$, leading to $\boldsymbol{\mathcal{P}}^{(y)}(\boldsymbol{i}) = \text{Prob}\{\boldsymbol{y} = \boldsymbol{i}\}$. Evidently, $\boldsymbol{\mathcal{P}}^{(y)}$ has 2^K elements, the sum of which is always 1. Given N iid realizations $\boldsymbol{y}[n]$ of $\boldsymbol{y}$, a consistent estimate of $\boldsymbol{\mathcal{P}}^{(y)}(\boldsymbol{i})$ can be easily obtained, for all 2^K possible values of $\boldsymbol{i}$, from

$$\widehat{\boldsymbol{\mathcal{P}}}^{(y)}(\boldsymbol{i}) = \frac{1}{N}\sum_{n=1}^{N} \mathcal{I}\{\boldsymbol{y}[n] = \boldsymbol{i}\} \quad (4)$$

where $\mathcal{I}\{\cdot\}$ denotes the Indicator function (being 1 if the condition in its argument is satisfied and 0 otherwise).

An alternative, but generally *incomplete* characterization of the statistics of $\boldsymbol{y}$ can be described by its first and second joint moments, namely by

$$\boldsymbol{\eta}^{(y)} \triangleq E[\boldsymbol{y}] \quad \text{and} \quad \boldsymbol{\Lambda}^{(y)} \triangleq E[\boldsymbol{y}\boldsymbol{y}^T], \quad (5)$$

respectively. Note that due to the 0/1 values in $\boldsymbol{y}$, the elements of $\boldsymbol{\eta}^{(y)}$ and of $\boldsymbol{\Lambda}^{(y)}$ also carry explicit probabilistic interpretations:

$$\eta^{(y)}_k = \text{Prob}\{y_k = 1\}, \quad \text{and} \quad \Lambda^{(y)}_{k,\ell} = \text{Prob}\{y_k = 1, x_\ell = 1\}. \quad (6)$$

Consistent estimates can be similarly obtained from N iid realizations,

$$\widehat{\boldsymbol{\eta}}^{(y)} = \frac{1}{N}\sum_{n=1}^{N} \boldsymbol{y}[n] \quad \text{and} \quad \widehat{\boldsymbol{\Lambda}}^{(y)} = \frac{1}{N}\sum_{n=1}^{N} \boldsymbol{y}[n]\boldsymbol{y}^T[n]. \quad (7)$$

We say that $\boldsymbol{y}$ has *independent components* if and only if the joint probability of any combination of the K elements equals the product of their marginal probabilities. This condition can be expressed as (recall that $i_k \in \{0,1\}$)

$$\boldsymbol{\mathcal{P}}^{(y)}(\boldsymbol{i}) = \prod_{k=1}^{K} (\eta^{(y)}_k)^{i_k}(1 - \eta^{(y)}_k)^{(1-i_k)} \triangleq (\boldsymbol{\eta}^{(y)})^{\boldsymbol{i}}(\boldsymbol{1} - \boldsymbol{\eta}^{(y)})^{(\boldsymbol{1}-\boldsymbol{i})}, \quad (8)$$

where the notation $\boldsymbol{a}^{\boldsymbol{b}}$ is shorthand for $a_1^{b_1} \cdot a_2^{b_2} \cdots a_K^{b_K}$. If $\boldsymbol{y}$ has independent components then they are all uncorrelated, namely the covariance matrix

$$\boldsymbol{C}^{(y)} \triangleq \boldsymbol{\Lambda}^{(y)} - \boldsymbol{\eta}^{(y)}(\boldsymbol{\eta}^{(y)})^T \quad (9)$$

is diagonal. Note, however, that although a diagonal covariance matrix implies *pairwise independence* of the components of $\boldsymbol{y}$, it does not, in general, imply *full independence* of these components.

We now return to the identifiability problem.

2.2 Identifiability Through Decorrelation

A natural approach for exploring the identifiability is to look for possible linear transformations $\hat{\boldsymbol{B}}$ such that the vector $\boldsymbol{y} = \hat{\boldsymbol{B}} \circ \boldsymbol{x}$ has independent components. Due to the invertibility of $\boldsymbol{A}$, it is clear that there exists at least one such matrix, $\hat{\boldsymbol{B}} = \boldsymbol{B} = \boldsymbol{A}^{-1}$. Moreover, it is clear that any $\hat{\boldsymbol{B}} = \boldsymbol{\Pi}\boldsymbol{B}$, where $\boldsymbol{\Pi}$ is any permutation matrix, also produces independent components in $\boldsymbol{y}$ - in accordance with the well-known inherent permutation ambiguity in ICA (fortunately, in our binary framework the classical scaling ambiguity is irrelevant and does not exist). The key question for establishing identifiability is to determine whether (and under what conditions) no other such transformations exist; namely, under what conditions independent components in $\boldsymbol{y}$ imply that the overall mixing-unmixing matrix $\boldsymbol{D} \triangleq \hat{\boldsymbol{B}} \circ \boldsymbol{A}$ is a permutation matrix[2].

The following Theorem establishes our main identifiability result.

Theorem 1. *Let $\boldsymbol{s} = [s_1\ s_2\ \cdots\ s_K]$ denote K statistically independent sources in GF(2), the k-th source having 1-probability p_k. Let $\boldsymbol{y} = \boldsymbol{D} \circ \boldsymbol{s}$ denote a linear transformation of $\boldsymbol{s}$ over GF(2), where $\boldsymbol{D}$ is a $K \times K$ matrix (with elements in GF(2)). Let $\boldsymbol{\eta}^{(y)}$ and $\boldsymbol{C}^{(y)}$ denote the mean and covariance (resp.) of $\boldsymbol{y}$. If:*

1. *All sources are non-degenerate, namely $0 < p_k < 1$, $k = 1, 2, \ldots, K$;*
2. *None of the sources is uniform, namely $p_k \neq 0.5$, $k = 1, 2, \ldots, K$;*
3. *All elements of $\boldsymbol{\eta}^{(y)}$ are nonzero, $\eta_k^{(y)} > 0$, $k = 1, 2, \ldots, K$;*
4. *$\boldsymbol{C}^{(y)}$ is diagonal,*

Then $\boldsymbol{D}$ is a permutation matrix.

Proof. Let us first establish the following lemma.

Lemma 1. *Let u and v be two RVs in GF(2) with 1-probabilities p and q (resp.), and let $w \triangleq u \oplus v$. If u and v are independent, non-degenerate ($0 < p, q < 1$) and non-uniform ($p, q \neq 0.5$), then w is also non-degenerate and non-uniform.*

To show this nearly trivial (and intuitively appealing) property, note that w has 1-probability $r = p(1-q) + q(1-p)$. It can then be easily shown that the only valid values of (p, q) with which $r = 0$ are $(0, 0)$ or $(1, 1)$; with which $r = 1$ are $(1, 0)$ or $(0, 1)$; and that $r = 0.5$ iff either p, q or both equal 0.5.

Under conditions 1 and 2 of Theorem 1, Lemma 1 establishes that no XOR sum of any (two, or more, by induction) of the sources can produce a degenerate or a uniform RV (however, if any of the sources are uniform, then any XOR sum involving one or more of these sources is also uniform).

Let us assume now that $\boldsymbol{D}$ is a general matrix, and consider any pair y_k and y_ℓ ($k \neq \ell$) in $\boldsymbol{y}$. y_k and y_ℓ are linear combinations of respective subgroups of the sources, indexed by the 1-s in $\boldsymbol{D}_{k,:}$ and $\boldsymbol{D}_{\ell,:}$, the k-th and ℓ-th rows (resp.) of $\boldsymbol{D}$. These two subgroups define, in turn, three other subgroups (some of which may be empty):

[2] We include $\boldsymbol{I}$ in the set of permutation matrices.

1. Sub-group 1: Sources common to $\boldsymbol{D}_{k,:}$ and $\boldsymbol{D}_{\ell,:}$. Denote the (XOR) sum of these sources as u;
2. Sub-group 2: Sources included in $\boldsymbol{D}_{k,:}$ but excluded from $\boldsymbol{D}_{\ell,:}$. Denote the (XOR) sum of these sources as v_1;
3. Sub-group 3: Sources included in $\boldsymbol{D}_{\ell,:}$ but excluded from $\boldsymbol{D}_{k,:}$. Denote the (XOR) sum of these sources as v_2.

For example, if (for $K = 6$) $\boldsymbol{D}_{k,:} = \begin{bmatrix} 0 \, 1 \, 1 \, 1 \, 1 \, 1 \end{bmatrix}$ and $\boldsymbol{D}_{\ell,:} = \begin{bmatrix} 1 \, 1 \, 0 \, 0 \, 1 \, 1 \end{bmatrix}$, then $u = s_2 \oplus s_5 \oplus s_6$, $v_1 = s_3 \oplus s_4$ and $v_2 = s_1$.

Note that by construction, the RVs u, v_1 and v_2 are statistically independent, and are also non-uniform. Obviously, $y_k = u + v_1$ and $y_\ell = u + v_2$. Let us denote the 1-probabilities of u, v_1 and v_2 as p, q_1 and q_2, respectively. Then the 1-probabilities of y_k, y_ℓ are given (resp.) by

$$\eta_k^{(y)} = p + q_1 - 2pq_1 \quad \text{and} \quad \eta_\ell^{(y)} = p + q_2 - 2pq_2. \tag{10}$$

We're further interested in $R_{k,\ell}^{(y)}$, the probability that *both* y_k and y_ℓ are 1. This happens if $u = 1$ and $v_1 = v_2 = 0$, or if $u = 0$ and $v_1 = v_2 = 1$, so

$$R_{k,\ell}^{(y)} = p(1 - q_1)(1 - q_2) + (1 - p)q_1q_2 = p(1 - q1 - q2) + q_1q_2. \tag{11}$$

Now, condition 4 of Theorem 1 implies that $C_{k,\ell}^{(y)} = R_{k,\ell}^{(y)} - \eta_k^{(y)}\eta_\ell^{(y)} = 0$, so

$$p(1 - q1 - q2) + q_1q_2 - (p + q_1 - 2pq_1)(p + q_2 - 2pq_2) = 0, \tag{12}$$

which with slight manipulations on the left-hand side can be written as

$$p(1 - p)(1 - 2q_1)(1 - 2q_2) = 0. \tag{13}$$

Since we've established that $q_1 \neq 0.5$ and $q_2 \neq 0.5$, (13) can only be satisfied if $p = 0$ or if $p = 1$. Since no non-trivial linear combination of the sources can be degenerate, $p = 1$ is also ruled out. The only option left is $p = 0$, which can happen if and only if sub-group 1 is empty, namely, iff the two rows $\boldsymbol{D}_{k,:}$ and $\boldsymbol{D}_{\ell,:}$ do not share common sources, or, in other words, iff there is no column m in $\boldsymbol{D}$ such that both $\boldsymbol{D}_{k,m}$ and $\boldsymbol{D}_{\ell,m}$ are 1.

Applying this to all possible pairs of $k \neq \ell$ (for which $C_{k,\ell}^{(y)} = 0$), and recalling that due to condition 3 of the theorem, $\boldsymbol{D}$ cannot have any all-zeros row, we immediately arrive at the conclusion that each row and each column of $\boldsymbol{D}$ must contain exactly one 1, meaning that $\boldsymbol{D}$ is a permutation matrix. □

We therefore conclude that if a transformation matrix $\hat{\boldsymbol{B}}$ is found such that

$$\boldsymbol{y} = \hat{\boldsymbol{B}} \circ \boldsymbol{x} = \hat{\boldsymbol{B}} \circ (\boldsymbol{A} \circ \boldsymbol{s}) = (\hat{\boldsymbol{B}} \circ \boldsymbol{A}) \circ \boldsymbol{s} = \boldsymbol{D} \circ \boldsymbol{s} \tag{14}$$

has uncorrelated, non-degenerate components, then if none of the sources is degenerate or uniform, $\boldsymbol{D}$ must be a permutation matrix. This means that the sources are fully separated, and are given by the elements of $\boldsymbol{y}$ (up to immaterial permutation ambiguity).

Note that as mentioned earlier, the decorrelation condition only implies pairwise independence (in GF(2)), which for a general random vector does not necessarily imply full independence. However, our theorem evidently asserts, that when $\boldsymbol{y}$ is a linear combination of independent, non-degenerate and non-uniform sources, pairwise independence indeed implies full independence.

3 Separation Algorithm

The theoretical identifiability theorem gives rise to at least one theoretically feasible separation strategy. Luckily, unlike classical ICA, in GF(2) there exists a finite number ($2^{(K^2)}$) of possible separation matrices $\hat{\boldsymbol{B}}$. Many of these matrices are singular, and thus cannot be considered as candidates for the inverse of $\boldsymbol{A}$ (naturally, the conditions of Theorem 1 cannot be satisfied with a singular $\hat{\boldsymbol{B}}$). So a possible "theoretical" algorithm would be to run an exhaustive search among all $M < 2^{(K^2)}$ nonsingular $K \times K$ matrices, looking for those which make the transformed observations $\boldsymbol{y}[n] = \hat{\boldsymbol{B}} \circ \boldsymbol{x}[n]$ as "empirically uncorrelated" as possible. However, this approach is practically inapplicable for values of K above, say, 5, due to the huge number ($2^{(K^2)}$) of potential matrices $\hat{\boldsymbol{B}}$ to check.

Instead, we now propose an entirely different approach, based on properties of the *entropies* of the mixtures. Let u be a binary RV with 1-probability p. Its entropy is $H(u) = -p \log_2 p - (1-p)\log_2(1-p)$, and it is easy to show [5] that $H(u)$ takes its maximum value (of 1) when u is uniform ($p = 0.5$). Now, let u and v be two statistically independent binary RVs and let $w = u \oplus v$. It can be easily shown[3] that $H(w)$ is greater or equal to both $H(u)$ and $H(v)$, where equality holds iff at least one of the RVs u and v is uniform.

Consequently, if $\boldsymbol{A}$ is invertible and none of the sources is uniform, then the source with the minimal entropy can be recovered by searching for the (non-trivial) linear combination (over GF(2)) of components of $\boldsymbol{x}$ which has the minimal entropy. If there are several sources with the same (minimal) entropy, there would be just as many entropy-minimizing linear combinations, each recovering its respective source. The number of potential (non-trivial) linear combinations of components in $\boldsymbol{x}$ would be $2^K - 1$, usually far less than the $O(2^{(K^2)})$ trials required in the previous algorithm.

Let $\hat{\boldsymbol{b}}$ denote a non-zero $K \times 1$ vector (with elements in GF(2)), containing prospective linear combination coefficients. To estimate the marginal probability (hence the entropy) of the linear combination $y = \hat{\boldsymbol{b}}^T \circ \boldsymbol{x}$ one may use time-averaging over the series $y[n] = \hat{\boldsymbol{b}}^T \circ \boldsymbol{x}[n]$, but this approach unnecessarily requires computation of the N-long series $y[n]$ for each tested $\hat{\boldsymbol{b}}$. Instead, following initial estimation of the observations' probabilities tensor ($\widehat{\boldsymbol{\mathcal{P}}}^{(x)}$ can be constructed using (4)), the 1-probability p of $y[n]$ can be obtained from

$$\hat{p} = \sum_{\boldsymbol{i}} (\boldsymbol{b}^T \circ \boldsymbol{i}) \cdot \widehat{\boldsymbol{\mathcal{P}}}^{(x)}(\boldsymbol{i}), \tag{15}$$

[3] E.g., a proof involving conditional entropy: $H(w) = H(u \oplus v) \geq H(u \oplus v|v) = H(u)$.

where the summation extends over all 2^K possible values of $\boldsymbol{i}$. Having obtained $\hat{p}$ for each of the $2^K - 1$ possible values of $\hat{\boldsymbol{b}}$, we select the $\hat{\boldsymbol{b}}$ which produced $\hat{p}$ with the minimal entropy[4].

While this approach only enables to extract the component(s) with the smallest entropy, we may proceed by taking a "deflation approach" (e.g., [6]). To this end, we would like to first eliminate the extracted source from the mixtures, by subtracting (or adding, it doesn't matter in GF(2)) that source from the mixture components in which it participates. Thus, given an extracted source, we have to decide, for each component in $\boldsymbol{x}$, say x_k, whether or not the extracted source y is part of the linear combination that created x_k.

To decide, all we have to do is to examine whether the entropy of x_k is smaller or larger than the entropy of $x_k \oplus y$. Again, this can be done by direct empirical estimation of the 1-probabilities from both series ($x_k[n]$ and $x_k[n] + y[n]$), or, preferably, from the estimated probabilities tensor $\widehat{\boldsymbol{\mathcal{P}}}^{(x)}$: Denoting by $\boldsymbol{e}_k$ the k-th column of $\boldsymbol{I}$, we have

$$\hat{p}_k\{0\} = \sum_{\boldsymbol{i}} (\boldsymbol{e}_k^T \circ \boldsymbol{i}) \widehat{\boldsymbol{\mathcal{P}}}^{(x)}(\boldsymbol{i}) \quad \text{and} \quad \hat{p}_k\{1\} = \sum_{\boldsymbol{i}} ((\boldsymbol{e}_k^T \oplus \boldsymbol{b}^T) \circ \boldsymbol{i}) \widehat{\boldsymbol{\mathcal{P}}}^{(x)}(\boldsymbol{i}) \tag{16}$$

where $\boldsymbol{b}$ denotes the coefficients vector that was selected for the extraction of the first source. Here $\hat{p}_k\{0\}$ and $\hat{p}_k\{1\}$ denote the estimated 1-probabilities of the series $x_k[n]$ and $x_k[n] + y[n]$, respectively. We then chose the one farther from 0.5: if this is $\hat{p}_k\{1\}$, we create a new observation $x'_k[n] = x_k[n] \oplus y[n]$, otherwise we simply set $x'_k[n] = x_k[n]$. Repeating the procedure for each k, we obtain a new set of observations $\boldsymbol{x}'[n] = [x'_1[n]\ x'_2[n],\ \ldots\ x'_K[n]]^T$, which is related to the original set by

$$\boldsymbol{x}'[n] = (\boldsymbol{I} \oplus (\boldsymbol{j} \cdot \boldsymbol{b}^T)) \circ \boldsymbol{x}, \tag{17}$$

where $\boldsymbol{j}$ is a $K \times 1$ vector containing 1-s in the indices corresponding to values of k for which $\hat{p}_k\{1\}$ was farther from 0.5 than $\hat{p}_k\{0\}$.

We may now repeat the process by applying the same procedure to the new set of observations $\boldsymbol{x}'$. As a first step, we have to construct an updated $\widehat{\boldsymbol{\mathcal{P}}}^{(x')}$ from $\widehat{\boldsymbol{\mathcal{P}}}^{(x)}$. To do this, we first set $\widehat{\boldsymbol{\mathcal{P}}}^{(x')} = \boldsymbol{0}$ (an all-zeros tensor), and then, running over all 2^K possible values of $\boldsymbol{i}$, we update

$$\widehat{\boldsymbol{\mathcal{P}}}^{(x')}((\boldsymbol{I} \oplus (\boldsymbol{j} \cdot \boldsymbol{b}^T)) \circ \boldsymbol{i}) = \widehat{\boldsymbol{\mathcal{P}}}^{(x')}((\boldsymbol{I} \oplus (\boldsymbol{j} \cdot \boldsymbol{b}^T)) \circ \boldsymbol{i}) + \widehat{\boldsymbol{\mathcal{P}}}^{(x)}(\boldsymbol{i}). \tag{18}$$

We then substitute $\boldsymbol{x} = \boldsymbol{x}'$, $\widehat{\boldsymbol{\mathcal{P}}}^{(x)} = \widehat{\boldsymbol{\mathcal{P}}}^{(x')}$, and return to the initial step, repeating the process $K - 2$ times (after the $(K-1)$-th pass, the last (maximum entropy) source would appear unmixed in one or more of the components of $\boldsymbol{x}'$).

It has to be noted that after each (k-th) pass $\boldsymbol{x}$ is an over-determined set: It still has K components, but they are now mixtures of $K - k$ sources. Consequently, there exist non-trivial linear combination coefficients $\hat{\boldsymbol{b}}$ that produce null

[4] There's no real need to compute the entropy: since $H(\hat{p})$ is a monotonically decreasing function of the distance of $\hat{p}$ from 0.5, it is sufficient to monitor $|\hat{p} - 0.5|$.

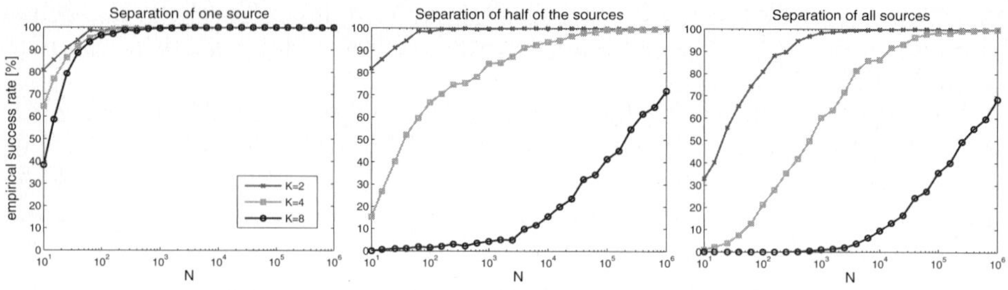

Fig. 1. Empirical success rate (1000 trials) of MEXICO for $K = 2, 4, 8$ vs. N. In each trial source probabilities were independently drawn in $(0, 0.4) \vee (0.6, 1)$, and a nonsingular mixing matrix was drawn as the product of upper- and lower-triangular matrices with random independent uniform binary elements above/below the diagonals.

components $y[n] = \hat{\boldsymbol{b}}^T \circ \boldsymbol{x}'[n] = 0$ (or even one or more of the K observations $\boldsymbol{x}$ themselves might be null). In principle, we may apply some order reduction, but this is not necessary: we can easily detect null combinations (characterized by $\hat{p} = 0$) and exclude the respective $\hat{\boldsymbol{b}}$-s from the search for minimum entropy. At the end of each pass, the respective source can be extracted as $y[n] = \boldsymbol{b}^T \circ \boldsymbol{x}[n]$. Sources would be extracted in order of non-decreasing entropy.

The algorithm is given the acronym MEXICO: Minimizing Entropies of Xored Independent COmponents. Its separation performance naturally depends on the accuracy of $\widehat{\boldsymbol{\mathcal{P}}}^{(x)}$ (when the true $\boldsymbol{\mathcal{P}}^{(x)}$ is used, perfect separation is obtained, as long as none of the sources is degenerate or uniform), which in turn depends on the length N and on the true probabilities $\boldsymbol{p}$. A rigorous performance analysis is quite involved, and falls beyond the scope of this paper. Instead, we present some simulation results: performance is shown in Fig.1 in terms of the empirical probability of success, namely the percentage of trials attaining perfect source reconstruction. We present success-rates in separating the first (minimum-entropy) source (left); half of all sources (middle); and all of the sources (right). See the figure's caption for details of the simulation setup.

To conclude: We derived a necessary and sufficient condition for identifiability of invertible linear mixtures over GF(2). A separation algorithm was proposed, capable of (asymptotically) perfect separation whenever the condition is satisfied (and failing otherwise). Extensions to higher order Galois Fields are also possible.

References

1. Eriksson, J., Koivunen, V.: Complex random vectors and ica models: identifiability, uniqueness, and separability. IEEE Trans. Information Theory 52, 1017–1029 (2006)
2. Ribenboim, P.: Classical theory of algebraic numbers. Springer, Heidelberg (2001)

3. Belouchrani, A., Abed-Meraim, K., Cardoso, J.F., Moulines, E.: A blind source separation technique using second-order statistics. IEEE Trans. Signal Processing 45, 434–444 (1997)
4. Pham, D.T., Cardoso, J.F.: Blind separation of instantaneous mixtures of nonstationary sources. IEEE Trans. Signal Processing 49, 1837–1848 (2001)
5. Cover, T.: Elements of Information Theory. Wiley-Interscience, Chichester (2006)
6. Delfosse, N., Loubaton, P.: Adaptive blind separation of independent sources: a deflation approach. Signal Processing 45, 59–83 (1995)

A Novel ICA-Based Image/Video Processing Method

Qiang Zhang, Jiande Sun, Ju Liu, and Xinghua Sun

School of Information Science and Engineering, Shandong University
Jinan 250100, Shandong, China
jd_sun@sdu.edu.cn

Abstract. Since Independent Component Analysis was developed, it has been a hotspot in the field of signal processing, and has received increasing attention in feature extraction, data compression, and so on. In this paper, a novel ICA-based image/video processing method, called ICA transform (ICAT), is proposed. Instead of the traditional blocking, ICAT derives more than one sub-images/sub-videos from one original image/video by down-sampling, and features are obtained from these sub-images/sub-videos by using ICA. That helps ICAT extracts features with the global characteristics of the original. And the comparison between ICAT and Digital Wavelet Transform (DWT) is performed in image/video processing, which exhibits that the results obtained by using ICAT has something similar to those of DWT, even something superior. And the comparison also demonstrates that ICAT is promising in image/video processing.

Keywords: Independent Component Analysis, Image/Video Processing, Digital Wavelet Transform.

1 Introduction

Independent Component Analysis (ICA) is a useful signal processing and data analysis method developed in the research of blind signals separation. Using ICA, even without any information of the source signals and the coefficients of transmission channels, people can recover or extract the source signals only from the observations according the stochastic property of the input signals. It has been one of the most important methods of blind source separation and received increasing attentions in pattern recognition, data compression, image analyzing and so on, because the ICA process derives features that best present the data via a set of components that are as statistically independent as possible and characterizes the data in a natural way [1-9].

In ICA model, more than one observation signals are needed to achieve the analysis, so when ICA is used to image/video processing, how to generate observations from one image must be firstly considered. At present, blocking is the prevalent manner and the features obtained in such way have been applied in many areas[4-9]. Hateren divided the image into blocks with size of 8×8 or 16×16, respectively, and all the blocks are taken as the observations of ICA model, and then features can be extracted by using ICA, which are proved to be some directional

M.E. Davies et al. (Eds.): ICA 2007, LNCS 4666, pp. 836–842, 2007.

edges and can be used as the bases to reconstruct the image. Furthermore, Hateren proved that these features are similar to what human visual cortex captures, and such experiments on video also result some similar conclusions. Though such kinds of features are applied widely in current researches [4-7], their meaning is still ambiguous, because blocking destroys the global properties of an image.

Down-sampling is a common signal processing method, through which sub-signals, approximate to the original, can be obtained. Down-sampling can help us to derive some similar sub-signals from only one original signal. And the sub-signals and the original have the same global properties.

In this paper, we propose a new image/video processing method, called ICA Transform (ICAT), based on down-sampling instead of blocking. In ICAT, the original image/video is down-sampled into sub-images/sub-videos, which are looked on as the observations in ICA model. Thus features can be derived from the sub-images/sub-videos by using ICA. A comparison between ICAT and Digital Wavelet Transform (DWT) helps us to understand the properties of such kind of features. The comparison shows that ICAT can obtain features similar to those extracted by DWT, furthermore has superiorities in some aspects.

2 Comparison Between ICAT and DWT in Image Processing

2.1 2 Dimensional Digital Wavelet Transform (2D DWT)

Digital Wavelet Transform (DWT) is a time-frequency analysis method. Because of its characteristic of multi-resolution, DWT has been used widely in signal processing, since it came into being. Fig. 1 shows the original image, peppers, and the four sub-bands obtained through 1-level 2D-DWT, which are the approximate one and the details containing vertical, horizontal, and diagonal information respectively [10, 11].

(a) (b)

Fig. 1. (a) is the original image, peppers, and (b) is 1-level DWT of the image, peppers. The left-top, LL sub-band, is the approximate component, and the others, LH, HL, HH sub-bands, are details in the vertical, horizontal, and diagonal directions respectively.

2.2 Independent Component Analysis Transform (ICAT)

As we all know that according to ICA model, there are more than one observation signals, so how to derive them from only one image is what have to be resolved firstly. In this paper, we down-sample the original image into sub-images. Assuming the size of the original image is $n \times m$, after down-sampling with factor 2 as shown in Fig. 2, the four sub-images are:

$$\begin{aligned} \boldsymbol{I}_{sub1}(i,j) &= \boldsymbol{I}(2i-1, 2j-1) \\ \boldsymbol{I}_{sub2}(i,j) &= \boldsymbol{I}(2i-1, 2j) \\ \boldsymbol{I}_{sub3}(i,j) &= \boldsymbol{I}(2i, 2j-1) \\ \boldsymbol{I}_{sub4}(i,j) &= \boldsymbol{I}(2i, 2j) \end{aligned} \tag{1}$$

where $\boldsymbol{I}$ is the original image, $i = 1,2 \cdots n/2$, $j = 1,2 \cdots m/2$. And then we take the sub-images as the observations to perform ICA. That is what is called ICA Transform (ICAT) in this paper. The four Feature Images (FI) obtained by using ICAT, four sub-bands like, are shown in Fig. 3.

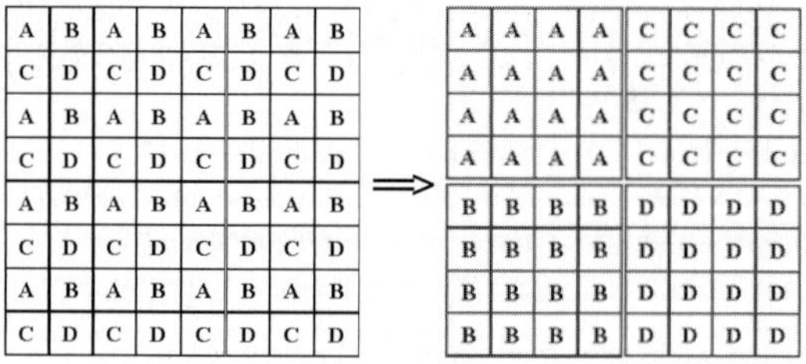

Fig. 2. The diagram of image down-sampling, which results four sub-images

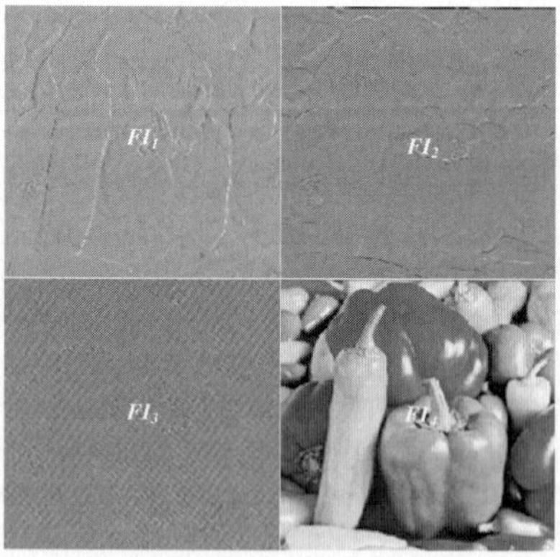

Fig. 3. The four FIs obtained by ICAT. FI_4 is the approximate component of the original image, while the FI_1, FI_2, FI_3 are the details of the original image. That is very similar to what in DWT as shown in Fig. 1(b).

2.3 Comparison Between ICAT and 2D DWT

In the comparison, we use the 512×512 standard image, peppers, as the original image, and choose Daubechies-4 as the mother wavelet. And down-sampling with factor 2 is adopted in ICAT.

Here in order to analyze the property of the FIs extracted by ICAT, the following procedure is performed. The approximate component is used to take the place of the details each at once, and the inverse transform is executed after each replacement. The results of inverse DWT(IDWT) and inverse ICAT(IICAT) are shown in Fig. 4 and Fig. 5 respectively. From Fig. 4, we can see that each result of such kind of procedure has the obvious textures in a certain direction, e.g. Fig. 4(a), which has heavy vertical textures, shows the vertical detail has been replaced by the approximate component.

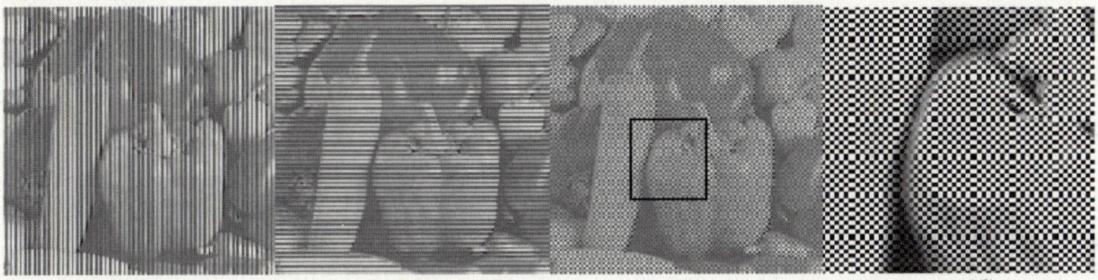

(a) (b) (c) (d)

Fig. 4. (a), (b), and (c) are the results of IDWT after the replacement of LH, HL, and HH respectively. And (d) is the enlarged of the rectangle part in (c).

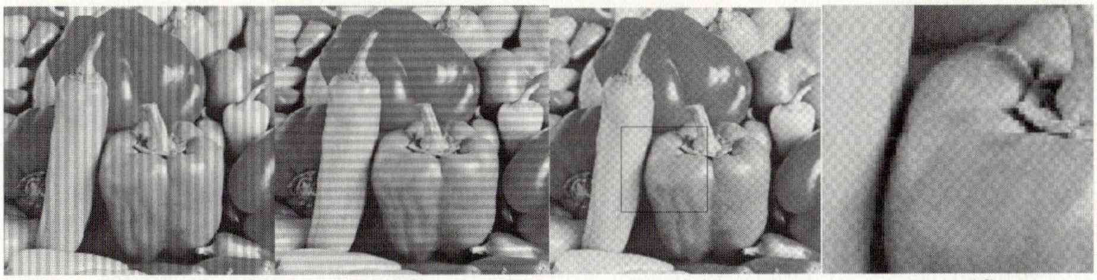

(a) (b) (c) (d)

Fig. 5. (a), (b), and (c) are the results of IICAT after the replacement of FI_1, FI_2, and FI_3 in Fig. 3 respectively. And (d) is the enlarged of the rectangle part in (c).

Given the analysis of Fig. 4, Fig. 5 shows that FI_1, FI_2, and FI_3 in Fig. 3 are also some directional details of the original image. Furthermore, DCT is performed on the approximate component obtained by DWT and ICAT, respectively, to analyze their frequency characteristics. Among the AC coefficients of the DWT approximate component, the first **5627** lowest frequency AC coefficients occupy the 95% in energy, while in the case of ICAT approximate component, the number is only **3451**, which is only about **61%** of that in DWT. That suggests that ICAT must superior to DWT in image compression.

2.4 Discussion

Given the above comparison between ICAT and DWT, we can see that the features derived by ICAT are very similar to those derived by DWT. Furthermore, ICAT has at least two superiorities to DWT in image processing. First, when an image is analyzed by different mother wavelet, the size of sub-bands may be different. But it is not the case in ICAT, so ICAT may be more compatible for analysis. Second, the comparison on the occupation ratio of low frequency shows that ICAT is superior to DWT in image compression. Besides, one of the reasons that DCT is used in compression is that DWT can reduce the redundancy by removing the correlation [11], while ICA is to obtain independent components.

So from the comparison above, we can conclude that ICAT will be a promising method to image and video processing. It must be widely used in the field of video feature analysis, motion object extraction, compression, and so on.

3 Temporal DWT and ICAT on Video Processing

In this comparison, temporal DWT and ICAT are used to analyze the same video in time. And in this video, a hand plays notes on a piano and there is only 5 keys played by each finger respectively. The key pressed by thumb is defined as the No.1 key, and the one pressed by forefinger is the No.2 key, and so on. The keys are pressed in the order of 1-2-3-4-2-3-1-5. The total frames are 50.

3.1 Temporal DWT on Video

Temporal DWT is often used in video analysis in order to extract the motion features from the video, since motion is considered as the temporal details of a video, and the background is the approximate one [12]. Db4 is used here, and two video feature sequences are obtained. The one with motion information is shown in Fig. 6.

3.2 ICAT on Video

We perform ICAT to video by down-sampling in temporally with the factor 2, and two video feature sequences are obtained. The feature sequence with motion information is shown in Fig. 7 corresponding as a counterpart of Fig. 6.

3.3 Discussion

According to Fig. 6 and Fig. 7, Temporal DWT and ICAT with temporal down-sampling can extract the motion of the original video. But given the two features sequences shown above, ICAT with temporal down-sampling has two superiorities to temporal DWT. One is that the frame number of ICAT is invariant and it is the half of original video. But the frame number of temporal DWT will be different with various wavelets. The other one is that the motion features obtained by ICAT have the same time order as that of original video. But this time order is lost in the temporal DWT.

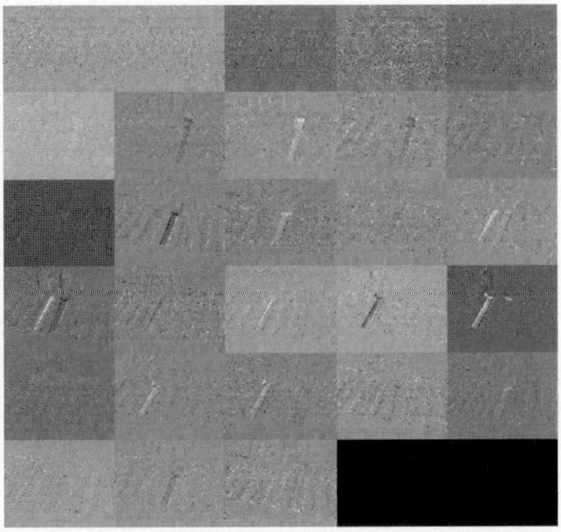

Fig. 6. The DWT features sequence with motion information is shown in the format of frames. The frames are ordered timely from left the top left to the lower right. And the total feature frames are 28, more than half of the total frame number in original video. In this sequence, there are some feature frames which show only the pressed keys in the original video, e.g. frame in row 2 column 2, in row 2 column 3, and so on. That means temporal DWT can filter the motion objects out.

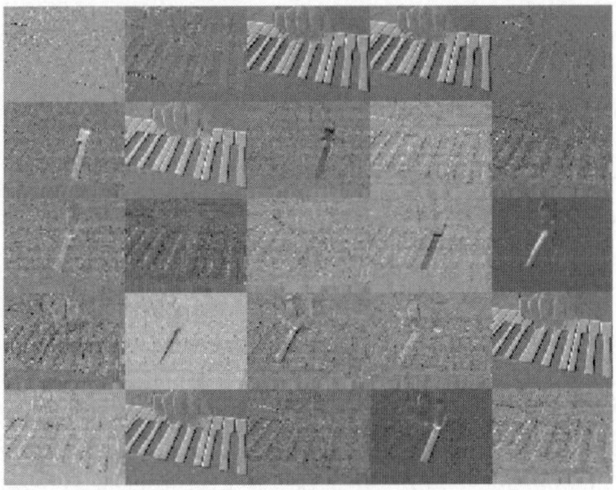

Fig. 7. The ICAT feature sequence with motion information is shown in the format of frames. The frames are ordered timely from left the top left to the lower right. From this sequence, we can find the features only with the motion objects and some only with something like background. That means that ICAT can extract the motion objects, the pressed keys, and the background as well. By the way, the number of frames in feature sequence is exactly the half of that of original video.

4 Conclusion

In this paper, a novel image/video processing method, ICA Transform, is proposed. Given the comparison with DWT in image/video processing, we can see that using

ICAT we can obtain some features like what is extracted by DWT in both image processing and video processing. What is more, ICAT has something superior to DWT to some extent. ICAT will be promising in image/video processing. And how to overcome the indeterminacy of ICA with the help of image/video properties is the next research interest.

Acknowledgment

The work is supported by Program for New Century Excellent Talents in University (NCET-05-0582), Specialized Research Fund for the Doctoral Program of Higher Education (Grant No. 20050422017) and the Project sponsored by SRF for ROCS, SEM ([2005]55). The contact author is Jiande Sun (jd_sun@sdu.edu.cn).

References

1. Hyvarinen, A.: Survey on Independent Component Analysis. Neural Computing Surveys 2, 94–128 (1999)
2. Hyvarinen, A., Oja, E.: Independent Component Analysis: Algorithms and Applications. Neural Networks 13(4-5), 411–430 (2000)
3. Hyvarinen, A., Oja, E.: A Fast Fixed-Point Algorithm for Independent Component Analysis. Neural Computation 9(7), 1483–1492 (1997)
4. Bartlett, M.S., Movellan, J.R., Sejnowski, T.J.: Face Recognition by Independent Component Analysis. IEEE Transactions on Neural Networks 16(6), 1450–1464 (2002)
5. Larsen, J., Hansen, L.K., Kolenda, T., Nielsen, F.A.: Independent Component Analysis in Multimedia Modeling. In: 4th International Symposium on Independent Component Analysis and Blind Signal Separation (ICA2003), Japan, pp. 687–695 (2003)
6. Hurri, J., Hyvarinen, A., Karhunen, J., Oja, E.: Image Feature Extraction Using Independent Component Analysis. In: Proc. IEEE Nordic Signal Processing Symposium, Espoo Finland (1996)
7. Artur, J.F., Mário, A.T.F.: On the Use of Independent Component Analysis for Image Compression. Signal Processing: Image Communication 21, 378–389 (2006)
8. van Hateren, J.H., van der Schaaf, A.: Independent Component Filters Of Natural Images Compared With Simple Cells In Primary Visual Cortex. Proceedings Royal Society of London: Biological Sciences 265(1394), 359–366 (1998)
9. van Hateren, J.H., Ruderman, D.L.: Independent Component Analysis of Natural Image Sequences Yields Spatio-Temporal Filters Similar to Simple Cells in Primary Visual Cortex. Proceedings of the Royal Society of London B 265(1412), 2315–2320 (1998)
10. Mallat, S.: Wavelets for A Vision. Proceedings of the IEEE 84(4), 604–614 (1996)
11. Antonini, M., Barlaud, M., Mathieu, P.: Daubechies: Image Coding Using Wavelet Transform. IEEE Transactions on Image Processing 1(2), 205–220 (1992)
12. Swanson, M.D., Zhu, B., Tewfik, A.H.: Multiresolution Scene-Based Video Watermarking Using Perceptual Models. IEEE Journal on Selected Areas in Communications 16(4), 540–550 (1998)

Blind Audio Source Separation Based on Independent Component Analysis

Shoji Makino, Hiroshi Sawada, and Shoko Araki

NTT Communication Science Laboratories, Kyoto, Japan
maki@cslab.kecl.ntt.co.jp

Abstract. This keynote talk describes a state-of-the-art method for the blind source separation (BSS) of convolutive mixtures of audio signals. Independent component analysis (ICA) is used as a major statistical tool for separating the mixtures. We provide examples to show how ICA criteria change as the number of audio sources increases. We then discuss a frequency-domain approach where simple instantaneous ICA is employed in each frequency bin. A directivity pattern analysis of the ICA solutions provides us with a physical interpretation of the ICA-based separation. It tells us the relationship between ICA-based BSS and adaptive beamforming. In order to obtain properly separated signals with the frequency-domain approach, the permutation and scaling ambiguity of the ICA solutions should be aligned appropriately. We describe two complementary methods for aligning the permutations, i.e., collecting separated frequency components originating from the same source. The first method exploits the signal envelope dependence of the same source across frequencies. The second method relies on the spatial diversity of the sources, and is closely related to source localization techniques. Finally, we describe methods for sparse source separation, which can be applied even to an underdetermined case.

M.E. Davies et al. (Eds.): ICA 2007, LNCS 4666, p. 843, 2007.

Author Index

Printing: Mercedes-Druck, Berlin
Binding: Stein+Lehmann, Berlin

Lecture Notes in Computer Science

For information about Vols. 1–4403

please contact your bookseller or Springer

Vol. 4742: I. Stojmenovic, R.K. Thulasiram, L.T. Yang, W. Jia, M. Guo, R.F. de Mello (Eds.), Parallel and Distributed Processing and Applications. XX, 995 pages. 2007.

Vol. 4708: L. Kučera, A. Kučera (Eds.), Mathematical Foundations of Computer Science 2007. XVIII, 764 pages. 2007.

Vol. 4707: O. Gervasi, M.L. Gavrilova (Eds.), Computational Science and Its Applications – ICCSA 2007, Part III. XXIV, 1205 pages. 2007.

Vol. 4706: O. Gervasi, M.L. Gavrilova (Eds.), Computational Science and Its Applications – ICCSA 2007, Part II. XXIII, 1129 pages. 2007.

Vol. 4705: O. Gervasi, M.L. Gavrilova (Eds.), Computational Science and Its Applications – ICCSA 2007, Part I. XLIV, 1169 pages. 2007.

Vol. 4703: L. Caires, V.T. Vasconcelos (Eds.), CONCUR 2007 – Concurrency Theory. XIII, 507 pages. 2007.

Vol. 4697: L. Choi, Y. Paek, S. Cho (Eds.), Advances in Computer Systems Architecture. XIII, 400 pages. 2007.

Vol. 4684: L. Kang, Y. Liu, S. Zeng (Eds.), Evolvable Systems: From Biology to Hardware. XIV, 446 pages. 2007.

Vol. 4683: L. Kang, Y. Liu, S. Zeng (Eds.), Intelligence Computation and Applications. XVII, 663 pages. 2007.

Vol. 4681: D.-S. Huang, L. Heutte, M. Loog (Eds.), Advanced Intelligent Computing Theories and Applications. XXVI, 1379 pages. 2007.

Vol. 4671: V. Malyshkin (Ed.), Parallel Computing Technologies. XIV, 635 pages. 2007.

Vol. 4666: M.E. Davies, C.J. James, S.A. Abdallah, M.D. Plumbley (Eds.), Independent Component Analysis and Signal Separation. XIX, 847 pages. 2007.

Vol. 4649: V. Diekert, M.V. Volkov, A. Voronkov (Eds.), Computer Science – Theory and Applications. XIII, 420 pages. 2007.

Vol. 4647: R. Martin, M. Sabin, J. Winkler (Eds.), Mathematics of Surfaces XII. IX, 509 pages. 2007.

Vol. 4644: N. Azemard, L. Svensson (Eds.), Integrated Circuit and System Design. XIV, 583 pages. 2007.

Vol. 4641: A.-M. Kermarrec, L. Bougé, T. Priol (Eds.), Euro-Par 2007 Parallel Processing. XXVII, 974 pages. 2007.

Vol. 4639: E. Csuhaj-Varjú, Z. Ésik (Eds.), Fundamentals of Computation Theory. XIV, 508 pages. 2007.

Vol. 4638: T. Stützle, M. Birattari, H.H. Hoos (Eds.), Engineering Stochastic Local Search Algorithms. X, 223 pages. 2007.

Vol. 4628: L.N. de Castro, F.J. Von Zuben, H. Knidel (Eds.), Artificial Immune Systems. XII, 438 pages. 2007.

Vol. 4627: M. Charikar, K. Jansen, O. Reingold, J.D.P. Rolim (Eds.), Approximation, Randomization, and Combinatorial Optimization. XII, 626 pages. 2007.

Vol. 4624: T. Mossakowski, U. Montanari, M. Haveraaen (Eds.), Algebra and Coalgebra in Computer Science. XI, 463 pages. 2007.

Vol. 4619: F. Dehne, J.-R. Sack, N. Zeh (Eds.), Algorithms and Data Structures. XVI, 662 pages. 2007.

Vol. 4618: S.G. Akl, C.S. Calude, M.J. Dinneen, G. Rozenberg, H.T. Wareham (Eds.), Unconventional Computation. X, 243 pages. 2007.

Vol. 4616: A. Dress, Y. Xu, B. Zhu (Eds.), Combinatorial Optimization and Applications. XI, 390 pages. 2007.

Vol. 4613: F.P. Preparata, Q. Fang (Eds.), Frontiers in Algorithmics. XI, 348 pages. 2007.

Vol. 4600: H. Comon-Lundh, C. Kirchner, H. Kirchner (Eds.), Rewriting, Computation and Proof. XVI, 273 pages. 2007.

Vol. 4599: S. Vassiliadis, M. Berekovic, T.D. Hämäläinen (Eds.), Embedded Computer Systems: Architectures, Modeling, and Simulation. XVIII, 466 pages. 2007.

Vol. 4598: G. Lin (Ed.), Computing and Combinatorics. XII, 570 pages. 2007.

Vol. 4596: L. Arge, C. Cachin, T. Jurdziński, A. Tarlecki (Eds.), Automata, Languages and Programming. XVII, 953 pages. 2007.

Vol. 4595: D. Bošnački, S. Edelkamp (Eds.), Model Checking Software. X, 285 pages. 2007.

Vol. 4590: W. Damm, H. Hermanns (Eds.), Computer Aided Verification. XV, 562 pages. 2007.

Vol. 4588: T. Harju, J. Karhumäki, A. Lepistö (Eds.), Developments in Language Theory. XI, 423 pages. 2007.

Vol. 4583: S.R. Della Rocca (Ed.), Typed Lambda Calculi and Applications. X, 397 pages. 2007.

Vol. 4580: B. Ma, K. Zhang (Eds.), Combinatorial Pattern Matching. XII, 366 pages. 2007.

Vol. 4576: D. Leivant, R. de Queiroz (Eds.), Logic, Language, Information and Computation. X, 363 pages. 2007.

Vol. 4547: C. Carlet, B. Sunar (Eds.), Arithmetic of Finite Fields. XI, 355 pages. 2007.

Vol. 4546: J. Kleijn, A. Yakovlev (Eds.), Petri Nets and Other Models of Concurrency – ICATPN 2007. XI, 515 pages. 2007.

Vol. 4545: H. Anai, K. Horimoto, T. Kutsia (Eds.), Algebraic Biology. XIII, 379 pages. 2007.

Vol. 4533: F. Baader (Ed.), Term Rewriting and Applications. XII, 419 pages. 2007.

Vol. 4528: J. Mira, J.R. Álvarez (Eds.), Nature Inspired Problem-Solving Methods in Knowledge Engineering, Part II. XXII, 650 pages. 2007.

Vol. 4527: J. Mira, J.R. Álvarez (Eds.), Bio-inspired Modeling of Cognitive Tasks, Part I. XXII, 630 pages. 2007.

Vol. 4525: C. Demetrescu (Ed.), Experimental Algorithms. XIII, 448 pages. 2007.

Vol. 4514: S.N. Artemov, A. Nerode (Eds.), Logical Foundations of Computer Science. XI, 513 pages. 2007.

Vol. 4513: M. Fischetti, D.P. Williamson (Eds.), Integer Programming and Combinatorial Optimization. IX, 500 pages. 2007.

Vol. 4510: P. Van Hentenryck, L.A. Wolsey (Eds.), Integration of AI and OR Techniques in Constraint Programming for Combinatorial Optimization Problems. X, 391 pages. 2007.

Vol. 4507: F. Sandoval, A. Prieto, J. Cabestany, M. Graña (Eds.), Computational and Ambient Intelligence. XXVI, 1167 pages. 2007.

Vol. 4501: J. Marques-Silva, K.A. Sakallah (Eds.), Theory and Applications of Satisfiability Testing – SAT 2007. XI, 384 pages. 2007.

Vol. 4497: S.B. Cooper, B. Löwe, A. Sorbi (Eds.), Computation and Logic in the Real World. XVIII, 826 pages. 2007.

Vol. 4494: H. Jin, O.F. Rana, Y. Pan, V.K. Prasanna (Eds.), Algorithms and Architectures for Parallel Processing. XIV, 508 pages. 2007.

Vol. 4493: D. Liu, S. Fei, Z. Hou, H. Zhang, C. Sun (Eds.), Advances in Neural Networks – ISNN 2007, Part III. XXVI, 1215 pages. 2007.

Vol. 4492: D. Liu, S. Fei, Z. Hou, H. Zhang, C. Sun (Eds.), Advances in Neural Networks – ISNN 2007, Part II. XXVII, 1321 pages. 2007.

Vol. 4491: D. Liu, S. Fei, Z.-G. Hou, H. Zhang, C. Sun (Eds.), Advances in Neural Networks – ISNN 2007, Part I. LIV, 1365 pages. 2007.

Vol. 4490: Y. Shi, G.D. van Albada, J. Dongarra, P.M.A. Sloot (Eds.), Computational Science – ICCS 2007, Part IV. XXXVII, 1211 pages. 2007.

Vol. 4489: Y. Shi, G.D. van Albada, J. Dongarra, P.M.A. Sloot (Eds.), Computational Science – ICCS 2007, Part III. XXXVII, 1257 pages. 2007.

Vol. 4488: Y. Shi, G.D. van Albada, J. Dongarra, P.M.A. Sloot (Eds.), Computational Science – ICCS 2007, Part II. XXXV, 1251 pages. 2007.

Vol. 4487: Y. Shi, G.D. van Albada, J. Dongarra, P.M.A. Sloot (Eds.), Computational Science – ICCS 2007, Part I. LXXXI, 1275 pages. 2007.

Vol. 4484: J.-Y. Cai, S.B. Cooper, H. Zhu (Eds.), Theory and Applications of Models of Computation. XIII, 772 pages. 2007.

Vol. 4475: P. Crescenzi, G. Prencipe, G. Pucci (Eds.), Fun with Algorithms. X, 273 pages. 2007.

Vol. 4474: G. Prencipe, S. Zaks (Eds.), Structural Information and Communication Complexity. XI, 342 pages. 2007.

Vol. 4459: C. Cérin, K.-C. Li (Eds.), Advances in Grid and Pervasive Computing. XVI, 759 pages. 2007.

Vol. 4449: Z. Horváth, V. Zsók, A. Butterfield (Eds.), Implementation and Application of Functional Languages. X, 271 pages. 2007.

Vol. 4448: M. Giacobini (Ed.), Applications of Evolutionary Computing. XXIII, 755 pages. 2007.

Vol. 4447: E. Marchiori, J.H. Moore, J.C. Rajapakse (Eds.), Evolutionary Computation, Machine Learning and Data Mining in Bioinformatics. XI, 302 pages. 2007.

Vol. 4446: C. Cotta, J. van Hemert (Eds.), Evolutionary Computation in Combinatorial Optimization. XII, 241 pages. 2007.

Vol. 4445: M. Ebner, M. O'Neill, A. Ekárt, L. Vanneschi, A.I. Esparcia-Alcázar (Eds.), Genetic Programming. XI, 382 pages. 2007.

Vol. 4436: C.R. Stephens, M. Toussaint, D. Whitley, P.F. Stadler (Eds.), Foundations of Genetic Algorithms. IX, 213 pages. 2007.

Vol. 4433: E. Şahin, W.M. Spears, A.F.T. Winfield (Eds.), Swarm Robotics. XII, 221 pages. 2007.

Vol. 4432: B. Beliczynski, A. Dzielinski, M. Iwanowski, B. Ribeiro (Eds.), Adaptive and Natural Computing Algorithms, Part II. XXVI, 761 pages. 2007.

Vol. 4431: B. Beliczynski, A. Dzielinski, M. Iwanowski, B. Ribeiro (Eds.), Adaptive and Natural Computing Algorithms, Part I. XXV, 851 pages. 2007.

Vol. 4424: O. Grumberg, M. Huth (Eds.), Tools and Algorithms for the Construction and Analysis of Systems. XX, 738 pages. 2007.

Vol. 4423: H. Seidl (Ed.), Foundations of Software Science and Computational Structures. XVI, 379 pages. 2007.

Vol. 4422: M.B. Dwyer, A. Lopes (Eds.), Fundamental Approaches to Software Engineering. XV, 440 pages. 2007.

Vol. 4421: R. De Nicola (Ed.), Programming Languages and Systems. XVII, 538 pages. 2007.

Vol. 4420: S. Krishnamurthi, M. Odersky (Eds.), Compiler Construction. XIV, 233 pages. 2007.

Vol. 4419: P.C. Diniz, E. Marques, K. Bertels, M.M. Fernandes, J.M.P. Cardoso (Eds.), Reconfigurable Computing: Architectures, Tools and Applications. XIV, 391 pages. 2007.

Vol. 4416: A. Bemporad, A. Bicchi, G. Buttazzo (Eds.), Hybrid Systems: Computation and Control. XVII, 797 pages. 2007.

Vol. 4415: P. Lukowicz, L. Thiele, G. Tröster (Eds.), Architecture of Computing Systems - ARCS 2007. X, 297 pages. 2007.

Vol. 4409: J.L. Fiadeiro, P.-Y. Schobbens (Eds.), Recent Trends in Algebraic Development Techniques. VII, 171 pages. 2007.

Vol. 4407: G. Puebla (Ed.), Logic-Based Program Synthesis and Transformation. VIII, 237 pages. 2007.